ROCK MECHANICS
MODELS AND MEASUREMENTS
CHALLENGES FROM INDUSTRY

PROCEEDINGS OF THE 1ST NORTH AMERICAN ROCK MECHANICS SYMPOSIUM / THE UNIVERSITY OF TEXAS AT AUSTIN / 1-3 JUNE 1994

Rock Mechanics Models and Measurements Challenges from Industry

Edited by

PRISCILLA P. NELSON

Department of Civil Engineering, The University of Texas at Austin

STEPHEN E. LAUBACH

Bureau of Economic Geology, The University of Texas at Austin

A.A. BALKEMA / ROTTERDAM / BROOKFIELD / 1994

Cover photograph of N15 Shaft on the SSC project
Courtesy of John Bird, SSC Laboratory

The texts of the various papers in this volume were set individually by typists under the supervision of each of the authors concerned.

Published by
A.A. Balkema, P.O. Box 1675, 3000 BR Rotterdam, Netherlands
A.A. Balkema Publishers, Old Post Road, Brookfield, VT 05036, USA

ISBN 90 5410 380 8

Printed in the Netherlands

Rock Mechanics, Nelson & Laubach (eds) © 1994 Balkema, Rotterdam, ISBN 90 5410 380 8

Preface

The North American Rock Mechanics Symposium is the principal forum for critical peer review of research in progress and engineering achievements of the North American rock mechanics community. The U.S. National Committee on Rock Mechanics, Sociedad Mexicana de Mecanica de Rocas, and the Canadian Rock Mechanics Association established the symposium as a biennial conference for exploring all aspects of rock mechanics that affect the three countries. This volume documents the first symposium's technical presentations.

The first North American Rock Mechanics symposium focused on the challenges facing the petroleum, civil, and mining industries and the rock mechanics research efforts that can meet these challenges. In this volume we present papers that describe the most recent developments in research and applications. These papers are timely, original, and succinct samples of the best new work in North American rock mechanics. Many papers have direct industrial applications, others will allow the industries to meet challenges in the years ahead. Papers span the spectrum of rock mechanics, including such diverse topics as engineering experience gained from drilling horizontal wells in oil and gas reservoirs, new numerical models for design of large underground spaces, and stress and fluid flow in deep mines.

Yet rock mechanics is essentially and ultimately a discipline that cuts across science and engineering cultures. To those unfamiliar with the study of rock mechanics the diversity of the topics in this volume may obscure the underlying unity of challenges that is imposed on rock engineering by that most inhomogeneous of engineering media—rock. Lessons presented in this volume—models, methods, and approaches—transcend the confining categories of the various engineering disciplines. Symposium participants and readers of this volume will find ideas that have been formulated and tested to meet demands of one branch of rock mechanics that can readily be adapted for use in other branches.

The NARM symposium provides a unique opportunity for dialogue among the practitioners of various disciplines of rock mechanics. To promote this dialogue, special programs designed for audience-expert interaction will be held during an afternoon retreat on the second day of the symposium. During the panel and poster sessions of this conference-within-a-conference, symposium participants will focus on ideas and issues on the leading edge of rock mechanics. Among the subjects of special panel discussions will be the rock mechanics of weak rock; the interrelationship of fracture networks, stress, and fluid flow; borehole stability, stress profiles, and hydraulic fracture; and the future of studies of microseismic and acoustic emissions of deforming rock. Informal discussions can be among the most valuable exchanges that occur at technical conferences, and our goal in designing these special sessions is to bring such discussions before the entire symposium. Accounts of such discussions cannot, of course, be found in proceedings volumes prepared prior to a conference, but we trust that the ideas exchanged in this part of the symposium will lead to publishable results.

The local organizing committee would like to recognize the efforts of individuals and groups that made development of the symposium a pleasure. Sitharam Thallak and Jon Holder were part of the central committee that oversaw all aspects of symposium organization. We thank Peter Smeallie, National Research Council, for his key role in launching this symposium, and the three representatives of the national rock mechanics societies who formed the NARMS coordinating committee. We are grateful for the support of Peter K. Kaiser, ISRM Vice-President for North America. We would like to thank those who so freely gave their time and energy to organize panel discussions and other interactive sessions. We also wish to thank the staff of the Bureau of Economic Geology and the Department of Civil Engineering, The University of Texas at Austin, for their contributions to developing and processing various symposium documents. Finally, we are grateful to the Gas Research Institute, the U.S. Department of Energy, Dowell-Schlumberger, and Exxon Production Research for their substantial support of the symposium.

Priscilla P. Nelson
The University of Texas at Austin
Chair of the First NARM Symposium

Stephen E. Laubach
Bureau of Economic Geology
Co-Chair of the First NARM Symposium

Rock Mechanics, Nelson & Laubach (eds) © 1994 Balkema, Rotterdam, ISBN 90 5410 380 8

Table of contents

2 *Borehole stability and hydraulic fracture*

3 *Rock engineering for underground excavations*

4 *New developments in blasting*

7 *Mechanical properties of intact rock – Laboratory testing*

8 *Static and dynamic properties of intact rock*

9 *Uncertainty and reliability in rock engineering*

10 *Numerical modeling – Discontinuum analysis*

15 *Mechanics of poorly consolidated weak rock*

16 *Deep mine design in South Africa and Canada*

17 *Deep mine design/bursting/pillar design*

20 *Rock mechanics in the design of rock slopes*

Rock Mechanics, Nelson & Laubach (eds) © 1994 Balkema, Rotterdam, ISBN 90 5410 380 8

Acknowledgements

The Local Organizing Committee of the First North American Rock Mechanics Symposium gratefully acknowledges the financial contributions of:

Benefactors

Dowell Schlumberger, Inc.
Exxon Production Research Company
Gas Research Institute
U. S. Department of Energy, Office of Basic Energy Sciences

Sponsors

J. F. T. Agapito & Associates
MTS Systems, Inc.
Woodward-Clyde Consultants
RE/SPEC, Inc.

Supporters

Elsevier Science Inc.
Halliburton Energy Service
RocTest, Inc.
Slope Indicator Company
Structural Behavior Engineering Laboratory, Inc.
TerraTek, Inc.
The University of Texas at Austin
    Bureau of Economic Geology
    Department of Civil Engineering
    Department of Petroleum Engineering

Rock Mechanics, Nelson & Laubach (eds) © 1994 Balkema, Rotterdam, ISBN 90 5410 380 8

Organization of the Symposium

Local Organizing Committee

Priscilla P. Nelson, Chair
Stephen E. Laubach, Co-Chair

Robert Gilbert
Steven Glaser
Jon Holder
Larry Lake
Steve Morriss
Augustus Podio
Gregory Rodin
Mukul Sharma
Kenneth Stokoe
Sitharam Thallak
Dave Triana
Roger Tyler

Invited Keynote Lecturers

D. Nathan Meehan
Union Pacific Resources, Ft. Worth, TX, USA

William Dershowitz and Paul LaPointe
Golder Associates Inc., Seattle, WA, USA

Alfred J. Hendron
University of Illinois, Champaign-Urbana, IL, USA

Arturo A. Bello Maldonado
Consulting Engineer, Mexico City, MEXICO

Malcolm Scoble
McGill University, Montreal, Quebec, CANADA

J. Nielen van der Merwe
Sasol Coal, Johannesburg, SOUTH AFRICA

U.S. National Committee for Rock Mechanics (USNC/RM)

Commission on Engineering and Technical Systems
National Research Council

Barry H. G. Brady, Chair

Bernard Amadei
University of Colorado

Herbert H. Einstein
Massachusetts Institute of Technology

Bezalel C. Haimson
University of Wisconsin-Madison

Jane C. S. Long
Lawrence Berkeley Laboratory

James E. Monsees
The PB/MK Team

Ronald P. Nordgren
Rice University

Miklos G. Salamon
Colorado School of Mines

Lawrence W. Teufel
Sandia National Laboratories

Donald L. Turcotte
Cornell University

Lewis V. Wade
U.S. Bureau of Mines

Peter H. Smeallie
Geotechnical Board

Mahadevan Mani
Director, DIEEE, NAS

Executive of the Canadian Rock Mechanics Association (CARMA)

Will Bawden, Chair Dougal McCreath, Secretary-Treasurer

Michel Aubertin, Canadian Geotechnical Society

Parviz Mottahed, Canadian Institute of Mining and Metallurgy

Sociedad Mexicana de Mecanica de Rocas (SMMR)

Alberto Menache Varela, President

Gustavo Diaz Velarde Seres, Vice President Ismael Sanchez Mora, Vice President

Jose Luis Garrido Uribe, Secretary Jorge Albarran Ascencio, Treasurer

International Society for Rock Mechanics (ISRM)

Peter K. Kaiser, ISRM Vice President for North America

Committee for NARMS Session Development

Ahmed Abou-Sayed
B. P. Exploration

Bernard Amadei
University of Colorado

Michel Aubertin
l'Universite de Montreal

Michael Batzle
ARCO Exploration

Robert A. Bauer
Illinois State Geological Survey

Will F. Bawden
Queen's University

Tom Boone
Imperial Oil Resources Ltd.

Mel Friedman
Texas A&M University

Jane Geller
Lawrence Berkeley Laboratory

Alex Grenoble
Stone and Webster, Inc.

John Hansen
Gas Research Institute

Jon Holder
The University of Texas at Austin

Neil Hurley
Marathon Oil Company

Moses Karakouzian
The University of Nevada Las Vegas

John Kemeny
The University of Arizona

Andreas Kronenberg
Texas A&M University

P. H. S. W. Kulatilake
The University of Arizona

Jane Long
Lawrence Berkeley Laboratory

John Lorenz
Sandia National Laboratories

Derek Martin
AECL Research

Gregory P. Matthews
Malcolm Pirnie, Inc.

Kent Perry
Gas Research Institute

Richard A. Plumb
Schlumberger Cambridge Research

Dale Preece
Sandia National Laboratories

Joe Ratigan
RE/SPEC, Inc.

J.-C. Roegiers
The University of Oklahoma

Birger Schmidt
Parsons Brinckerhof Quade & Douglas

Douglas Schmitt
University of Alberta

Dan Schultz-Ela
Bureau of Economic Geology

Gregg A. Scott
U. S. Bureau of Reclamation

Mukul Sharma
The University of Texas at Austin

David Summers
The University of Missouri - Rolla

Sitharam Thallak
The University of Texas at Austin

Roger Tyler
Bureau of Economic Geology

Luis Vieitez Uteza
Centro de Investigacion Sismica

Bruno Vendeville
Bureau of Economic Geology

Jiannan Xiang
Bureau of Economic Geology

David Yale
Mobil Research and Development

Rock Mechanics, Nelson & Laubach (eds) © 1994 Balkema, Rotterdam, ISBN 90 5410 380 8

The North American Rock Mechanics Symposium Coordinating Committee

Increased mutual dependence of the economies of Canada, the United States and Mexico has now been recognized formally by agreements between the respective national governments. Noting the basic economic role of rock mechanics in the resource recovery and construction industries, it is appropriate that the First North American Rock Mechanics Symposium should confirm mutual interest in rock mechanics research and engineering practice in the neighboring countries. Different government and industrial emphases in the NAFTA countries lead to complementary strengths in their research and engineering programs. The First NARM Symposium is our first opportunity to explore thoroughly, within the scope of a single meeting, rock mechanics research in progress and engineering achievements in the three countries.

In addition to the convergence of economic interests in North America, motivation for organization of the NARM Symposium was provided by the International Society for Rock Mechanics. Sensing a proliferation of rock mechanics symposia, the ISRM has endorsed the coordination of national meetings in regional symposia, to draw on a broader base of prospective contributors and to promote more effective international communication.

In 1992, the Canadian Rock Mechanics Association, the Sociedad Mexicana de Mecanica de Rocas and the United States National Committee for Rock Mechanics agreed on the terms, purposes and outline of organizational arrangements for a series of North American rock mechanics symposia. They posit biennial meetings providing a forum for presentation of papers on rock mechanics research and engineering applications from authors throughout North America. A specific concern is to encourage interaction between researchers and practitioners in civil, mining, petroleum and geological engineering and to provide for review of research work in progress in these fields and the related scientific disciplines. Industrial participation is to be encouraged by the organization of short courses and workshops, and special effort is to be made to engage graduate students in the meetings.

Having accepted an invitation to conduct a rock mechanics symposium at The University of Texas at Austin in 1994, the Local Organizing Committee led by Drs Priscilla Nelson and Stephen Laubach decided to extend the scope of the meeting and assume responsibility for conducting the First NARM Symposium. The ISRM national groups of Canada, the United States and Mexico record their gratitude to the LOC at The University of Texas for the enterprise, effort and commitment they have shown in bringing organization of the symposium to realization. We are confident it will be the first in a series better serving the interests of our rock mechanics constituencies throughout North America. We welcome you to the First NARM Symposium and look forward with keen anticipation to future symposia in Canada and Mexico.

Guillermo Botas Espinoza	Barry H. G. Brady	Dougal McCreath
ICI, v. 10 de Mayo	Manager, Appl. Mechanics	Geomech. Research Centre
No. 96-303	Dowell Schlumberger, Inc.	Laurentian University
Col. San Pedro de los Pines	P.O. Box 2710	Ramsey Lake Road
C. P. 03800 Mexico, D. F.	Tulsa, OK 74101	Sudbury, ON P3E 2C6
MEXICO	USA	CANADA

Rock Mechanics, Nelson & Laubach (eds) © 1994 Balkema, Rotterdam, ISBN 90 5410 380 8

The NARMS paper review process

The organizers of the first North American Rock Mechanics Symposium used a paper selection and review process that was aimed at ensuring consistently high-quality content without compromising timeliness and originality. Papers were selected for inclusion in this proceedings volume based on review of extended paper proposals. Paper proposals were expected to define the scope and status of the work, to clearly summarize important results, to explain technical significance, and to indicate any industrial application. Paper proposals were judged on the basis of originality, technical content, timeliness, and clarity of expression. More than twice as many paper proposals were submitted to the symposium as could be accepted for publication. One of our goals was to achieve broad coverage of the field of rock mechanics. Thus, many excellent paper proposals could not be included because the proposal covered a topic that was oversubscribed in the program.

Paper proposals were received in September 1993. All paper proposals were reviewed by the co-chairs of the symposium and in many cases by another member of the local organizing committee. In addition, at least two members of a panel of national experts—the committee for session development—produced a written review of each proposal. Finally, each paper proposal was reviewed by a technical expert representing each of the three national rock mechanics organizations. Including late submissions that received fewer reviews, all proposals received between four and eight written reviews. The local organizing committee compiled all the reviews and made an initial ranking of paper proposals. At a meeting held in Austin in November 1993, members of the committee for session development and the three national rock mechanics organizations reviewed paper proposals, inspected proposal reviews, and participated in the final ranking of paper proposals prior to paper acceptance. From this second list, the symposium co-chairs selected the papers for acceptance in the proceedings volume.

Review of papers continued after acceptance of paper proposals. Papers were divided into sessions based on topic, and teams of two or three session developers were assigned to each session. These session developers, all eminent scientists with technical expertise in the topic in question, were responsible for monitoring paper progress and holding papers to scientific standards of top research and engineering journals. Session developers also assisted authors trying to complete high-quality papers within a short deadline. Final, camera-ready papers were due in Austin on February 1, 1994, only 2 months after paper acceptance. The symposium co-chairs carried out the final step in the review process by reading the completed papers. As a result of the diligent efforts of the session developers, few problems remained, and the co-chairs were able to accept nearly all of the papers approved by session developers.

Keynote papers from five of the six keynote speakers are included in this volume. Keynote speakers were given wide latitude to survey their field of rock mechanics. Some of these review papers are personal and philosophical; others are technical summaries of important recent developments. This volume does not include records of presentations at short courses, workshops, and field trips, or the symposium panel discussions or poster session. Abstracts of poster presentations were distributed to symposium participants in an abstracts and program volume. We would like to thank the individuals who prepared these elements of the symposium program.

The following list includes all persons who participated in the review process. We thank them all for their contributions to the content of this volume.

Mikko Ahola
Bernard Amadei
J. Archibald
Tim Aston
Michel Aubertin
Michael Batzle
Will Bawden
Robert Bauer
Arturo Bello
Marc Betournay
Tom Boone
Guillermo Botas
Barry Brady
Richard K. Brummer
Neil Chandler
Asadul Chowdhury
Jaak Daeman
E. DeSouza
Emmanuel Detournay
Pat Dickerson
Herbert Einstein
Russell Ewy
Mel Friedman
Amitava Ghosh
Bezalel Haimson
Michael Hardy
F. P. Hassini
Sergio Herrera
Carlos Huerta
Neil Hurley
Jon Holder
Simon Hsiung
Moses Karakouzian
John Kemeny
Kunsoo Kim
Tony Klym
Andy Kronenberg
P. Kulatilake
Joseph Labuz
E. Lajtai
Stephen Laubach
Jane Long
John Lorenz
Mark Mack
Sean Maloney
Derek Martin
Gregory Matthews
Pat McLellan
Dougal McCreath
Alberto Menache V.
Douglas Morrison
Priscilla Nelson
Mike Neumann
Ron Nordgren
Goodluck Ofoegbu
Juan Manuel Orozco
William Pariseau
Wesley Patrick
Richard Plumb
Dale Preece
Joseph Ratigan
J.-C. Roegiers
K. Saleh
Birger Schmidt
Dan Schultz-Ela
Gregg Scott
Mukul Sharma
Douglas Stead
Brian Stimpson
David Summers
Graham Swan
Dwayne Tannant
Larry Teufel
Sitharam Thallak
Kam Tsui
Roger Tyler
Bruno Vendeville
Luis Vieitez
Lewis Wade
Jiannan Xiang
David Yale

Abstracts of USNC/RM Award Winning Papers for 1993

Rock Mechanics, Nelson & Laubach (eds) © 1994 Balkema, Rotterdam, ISBN 90 5410 380 8

PhD Thesis Award in Rock Mechanics

"Micromechanics of Seismic Wave Propagation in Granular Rocks" by Kurt Toshimi Nihei, PhD Thesis, University of California at Berkeley, 1992

This thesis investigates the details of seismic wave propagation in granular rocks by examining the micromechanical processes which take place at the grain level. Grain contacts are identified as the primary sites of attenuation in dry and fluid-saturated rocks. In many sedimentary rocks such as sandstones and limestones, the process of diagenesis leaves the grains only partially cemented together. When viewed at the micron scale, grain contacts are non-welded interfaces similar in nature to large scale joints and faults. Using a lumped properties approximation, the macroscopic properties of partially cemented grain contacts are modeled using a displacement-discontinuity boundary condition. This model is used to estimate the magnitude and the frequency dependence of the grain contact scattering attenuation for an idealized grain packing geometry.

Ultrasonic P- and S-wave group velocity and attenuation measurements on sintered glass beads, alundum, and Berea sandstone were performed to determine the effects of stress, frequency, and pore fluid properties in granular materials with sintered and partially sintered grain contacts. P- and S-wave attenuation displayed the same overall trends for tests with n-decane, water, silicone oil, and glycerol. The magnitudes of the attenuation coefficients were, in general, higher for S-waves. The experimental measurements reveal that viscosity-dependent attenuation dominates in materials with sintered grain contacts. Viscosity-dependent attenuation is also observed in Berea sandstone but only at hydrostatic stresses in excess of 15 MPa where the grain contacts are highly stiffened. Fluid surface chemistry-related attenuation was observed in Berea sandstone loaded uniaxially. These measurements suggest that attenuation in fluid-saturated rocks with partially cemented grain contacts is dependent on both the fluid properties and the state of stress at the grain contacts.

A numerical method for simulating seismic wave propagation in granular materials was developed to examine the influence of the grain packing geometry, grain shape, and grain contact properties on seismic velocities and attenuation. The *multi-grain method* is formulated using a dynamic boundary integral equation method in which individual grains are coupled together using a displacement-discontinuity boundary condition. Viscous fluid-flow losses which take place in the grain contacts are modeled using a complex interfacial stiffness. Preliminary simulations demonstrate the utility of the method for investigating the micromechanical processes of seismic wave propagation in granular rocks.

Rock Mechanics, Nelson & Laubach (eds) © 1994 Balkema, Rotterdam, ISBN 90 5410 380 8

M.S. Thesis Award in Rock Mechanics

"A Numerical Model for the Study of Deformable, Naturally Fractured Reservoirs" by Maria A. Vieira, M.S. Thesis, University of Oklahoma, 1992

Conventional reservoir simulation proves in general inadequate when attempting to analyze the response of naturally fractured reservoirs due to the complexity of the flow systems (with large contrasts in porosity and permeability), and the strong fluid-rock interactions involved. In addition, the fractures represent non-linear discontinuities that are highly sensitive to changes in the state of stresses due to variations in fluid pressures.

An alternate approach to this complex problem is to compute the flow rate inside each fracture using the parallel plate flow model, and to consider the mass transfer from the matrix to the fractures as the result of changes in fluid storage. Two numerical models were developed which differ only in the approach used to calculate changes in fracture volume. One model considers the variations in fracture volume as proportional to the fluid pressure changes, using an average compressibility factor for the fractures. The second model explicitly calculates the changes in fracture aperture based on the stress-strain characteristics of the fracture-fluid system. Both models assume the rock mass to be two-dimensional and dissected by a large number of fractures; they are also limited to the single-phase, pseudo-steady state flow of undersaturated, slightly compressible oil.

The models were used to study the response of a small "sugar block" reservoir subjected to various field conditions and production rates. In general, it can be concluded that the modeling of the fracture system as an equivalent porous medium with finite permeability does not adequately represent the flow conditions. In addition, the fracture deformability can play a fundamental role in the reservoir performance, and the associated compaction can represent an additional drive mechanism. However, significant reductions in fracture porosity and permeability, as a function of pressure depletion, can be expected.

Rock Mechanics, Nelson & Laubach (eds) © 1994 Balkema, Rotterdam, ISBN 90 5410 380 8

Case History Award

"Effect of reservoir depletion and pore pressure drawdown on in situ stress and deformation in the Ekofisk Field, North Sea," by Lawrence W. Teufel, Douglas W. Rhett, and Helen E. Farrell, Proc., 32nd U. S. Symposium on Rock Mechanics, Rock Mechanics as a Multidisciplinary Science, Roegiers (ed.), Ch. 28, pp 63 - 72, 1991

Knowledge of in situ stress and how stress changes with reservoir depletion and pore pressure drawdown is important in a multi-disciplinary approach to reservoir characterization, reservoir management, and enhanced oil recovery projects. Over 20 years of petroleum production from the Ekofisk Field has resulted in a 21 - 24 MPa reduction in reservoir pore pressure. The effect of pore pressure drawdown on the minimum horizontal in situ stress in the Ekofisk Field has been determined from shut-in pressure data of 32 hydraulic fractures. The effective stresses in the reservoir increase linearly with pore pressure drawdown, but at different rates. The ratio of the change in effective minimum horizontal stress to the change in effective vertical (overburden) stress is approximately 0.20. Laboratory experiments, which simulate the stress path followed by reservoir rock during the production history of the Ekofisk Field, clearly indicate that shear failure has occurred during compaction of high porosity chalk as the shear stress increased with pore pressure drawdown. It is suggested that shear failure during primary production has increased fracture density and reduced matrix block dimensions, and has therefore maintained reservoir permeability, which may account for the continued good producibility of the Ekofisk Field, in spite of compaction.

Applied Research Award

"Identifying Probable Failure Modes for Underground Openings Using a Neural Network," by Chengho Lee and Raymond Sterling, Int. J. Rock Mech. Min. Sci & Geomech. Abstr., Vol. 29, No. 1, pp. 49-67, Jan. 1992.

The design of underground openings in complex geological environments is restrained by our limitations in defining the geological environment in which the opening will be created and in modelling the response of the geological environment to the excavation created and support procedures used. This paper described the use of a neural network to identify possible failure modes for underground openings from prior case history information. This step in geotechnical design is a critical step in identifying the dominant geologic parameters to be included in a simplified geologic and engineering model.

The structure of the neural network adopted and the "learning" algorithm by which the neural network obtains its knowledge from case histories are described. The results of "learning" are then used to examine the operational characteristics of the neural network. Four experiments are designed to test its abilities in inferring possible failure modes, retrieving patterns from partial cues (content-addressability) and in resistance to faulty input data. Limitations of and possible improvements on the neural network are also described.

Use of the knowledge obtained by the neural network learning in a geotechnical design context is demonstrated by a tunnel design assistance system. This neural network approach differs from conventional rule-based expert system in the manner of knowledge representation and the problem solving process. Instead of applying rules and facts to end up with conclusions, the approach solves problems by pattern matching and allows input information to be incomplete and vague.

Rock Mechanics, Nelson & Laubach (eds) © 1994 Balkema, Rotterdam, ISBN 90 5410 380 8

Basic Research Award

"Acoustic Emissions Produced by Discrete Fracture in Rock" by Steven D. Glaser and Priscilla P. Nelson, Int. J. Rock Mech. Min. Sci & Geomech. Abstr., Vol. 29, No. 3, pp. 237-265, Mar. 1992.

"Part 1 - Source Location and Orientation Effects"

An experimental programme was undertaken to develop a servo-controlled loading and high-fidelity data acquisition system to investigate acoustic emission (AE) waveforms produced during discrete fracture propagation in rock. Large (0.3 x 0.3 x 1m) chevron-notched specimens were loaded in controlled conditions, and characteristic acoustic emissions associated with Mode I and Mode II fracture were isolated. Two different rocks, granite and dolostone, were tested.

Only the initial segment of each AE waveform was used for analysis, ensuring an analysis free from the effects of sample size and shape, with propagation effectively through an infinite half-space. High-fidelity, NIST-type, piezoelectric displacement transducers were built and used. The entire experimental system was calibrated, and the system inverse filter was applied to recorded signals. The system had little effect on the recorded signals and deconvolution was not necessary for these tests.

The effects of AE source position and orientation relative to the transducer were studied. A reproducible step impulse source was used. Changes in source orientation were found to result in two types of waveform from a common source -- a step-like waveform which changes to a ramp-like waveform when the source was directed with a large angle to the surface being monitored. The effects of source position were small.

"Part 2 - Kinematics of Crack Growth During Controlled Mode I and Mode II Loading of Rock"

A well-characterized experimental programme was undertaken to study acoustic emission (AE) waveforms produced during discrete fracture propagation in rock. Large chevron-notched specimens were loaded in controlled conditions, and characteristic AE associated with Mode I and II fracture were isolated. Tested rock included medium-grained granite and micritic dolostone.

The AE waveforms belonged to five characteristic classes. The variety of AE event classes was the same in Mode I and II loadings for both rock types. By far the most common three classes are related to basic step impulses. The mechanism for crack propagation in rock is indicated to be due to localized tensile failure for both tensile and shear loadings.

Relative toughness of the crack surface corresponded closely with relative distributions of the different characteristic event classes. A ratio of event types was formulated which is a good indicator of the amount of out-of-plane crack growth.

Rates of occurrence of AE events varied during loading modes, and between the two materials. Very few events occurred during load increase to peak or during unloading. Event rates increased as the crack propagated through the notch, and AE rates for subsequent constant-length crack propagation were constant.

Rock Mechanics, Nelson & Laubach (eds) © 1994 Balkema, Rotterdam, ISBN 90 5410 380 8

Vitae

Helen E. Farrell received a BSc degree in Geology from Exeter University, England in 1975. She received a MSc degree in Structural Geology and Rock Mechanics from Imperial College, London in 1976. Her PhD, also from Imperial College, was awarded in 1980 and dealt with fault rock textures and microstructures. She has worked as a structural geologist for Phillips Petroleum in Bartlesville, OK since 1980. During this time she has been involved in reservoir characterization, in particular in fracture analysis. Other areas of interest include borehole imaging, structural analysis of dipmeter data and multi-disciplinary integration aspects of reservoir characterization methods.

Steven D. Glaser is Assistant Professor of Engineering at the Colorado School of Mines. He is expanding on his earlier work on the interpretation of acoustic emission waveforms from discrete rock fracture. Dr. Glaser's involvement with civil engineering began at age seventeen when he began working as a union construction laborer in the Washington D.C. area. While earning his degree in philosophy at Clark University, Worcester, MA, he became an apprentice operation engineer. He worked as a driller for eight years in the D.C. area and Iraq, with time spent in the tunnels of the Metro project. This practical experience with geotechnical engineering led to Dr. Glaser enrolling in the civil engineering program at The University of Texas in 1981. Dr. Glaser was a research civil engineer at the National Institute of Standards and Technology from 1991 to 1993. Extensive work was undertaken on the application of System Identification methods to estimate soil properties at large strain.

Chengho Lee obtained his PhD degree from the University of Minnesota in April 1992. Since completion of his degree, he has been working with Itasca Consulting Group, Inc. in Minneapolis as a research engineer. Dr. Lee has developed and applied numerical models (distinct element and finite difference) to rock mechanics for tunnels and boreholes. He has also developed algorithms and constitutive models for borehole breakout analysis as well as a design approach which couples a neural network and numerical models for analysis of underground excavations.

Priscilla P. Nelson received a BS degree in Geology from the University of Rochester in 1970, and continued study in Geology at Indiana University through 1972, earning a masters degree. She served as a Peace Corps volunteer and followed with a two year stint field engineering on the Trans-Alaska Pipeline. She received her MS degree in Structural Engineering from the University of Oklahoma in 1979, and her PhD in Geotechnical Engineering was completed at Cornell University in 1983. She has been on the faculty at The University of Texas at Austin since 1983, currently holding the Murray Fellowship in Engineering. In addition to the topic addressed in the award paper, her research interests include mechanics of weak rock, design of foundations in heterogeneous rock materials, underground construction and mechanized excavation, and experimental evaluation of fracture material properties of rock.

Kurt T. Nihei received a BA degree in Geophysics in 1986, an MS degree in Applied Geophysics in 1989, and a PhD in Geomechanics in 1992, all from the University of California at Berkeley. His PhD research on the micromechanics of elastic wave propagation in granular rocks was conducted under the guidance of Prof. N. G. W. Cook and Dr. L. R. Myer. His current research interests are in damage mechanics, guided waves, and elastic wave propagation in fractured rock. He is currently a post-doctoral fellow in the Earth Sciences Division at Lawrence Berkeley Laboratory.

Douglas W. Rhett received his BA degree in Geology from Middlebury College in 1965, and his PhD in Geology from Rutgers University in 1974. He has been an employee of Phillips Petroleum Company since 1975. He worked in the areas of uranium and geothermal exploration research and petroleum exploration research prior to 1985, when he transferred to production research. He worked in experimental rock mechanics from 1985 until the present. His current duties include application of experimental and theoretical rock mechanics to problems in petroleum production and reservoir management.

Raymond L. Sterling received his B.Eng. degree in Civil and Structural Engineering from the University of Sheffield (England) in 1970. After several years work in the consulting industry in England, he moved to Minnesota where he completed his MS work in Geological Engineering in 1975, and received his PhD in 1977 from the University of Minnesota. Dr. Sterling has been a faculty member in the Department of Civil and Mineral Engineering at Minnesota, and Director of the Underground Space Center since 1977. He is the retiring Chairman of the U.S. National Committee on Tunneling Technology, and his research interests include rock mechanics, underground construction, underground space utilization, and energy use in buildings.

Lawrence W. Teufel received his BS degree in Geology from Syracuse University in 1973. He transferred to Texas A&M University and completed his MS degree work in 1976, and received his PhD in 1979. He has been at Sandia National Laboratories since 1979, currently as Distinguished Member of the Technical Staff. Dr. Teufel is also on the faculty at the New Mexico Institute of Mining and Technology, holder of the Langdon Taylor Chair in Petroleum and Natural Gas Engineering. His research interests include: the influence of pore pressure, geologic structure, and lithology on in situ stress; poroelastic behavior of matrix and fractured rock; coupled deformation and fluid-flow behavior of naturally fractured reservoirs; and reservoir compaction and subsidence. He is a member of the U.S. National Committee for Rock Mechanics.

Maria A. Vieira holds a BS degree in Mining Engineering from the Instituto Politecnico di Torino, Italy, and she completed her MS work in the Department of Petroleum Engineering at the University of Oklahoma in 1992. She began working as a Reservoir Engineer in the E&P Reservoir Engineering Department of Maraven S. A. (a PDVSA affiliate company) in Caracas, Venezuela in 1985, and returned to her home country of Venezuela and work at Maraven following completion of her graduate studies.

Keynote lectures

Rock Mechanics, Nelson & Laubach (eds) © 1994 Balkema, Rotterdam, ISBN 90 5410 380 8

Rock mechanics issues in petroleum engineering

D. Nathan Meehan
Union Pacific Resources, Ft. Worth, Tex., USA

ABSTRACT: Historically, petroleum engineers have neglected the role of the physical porous medium except for its storage properties (porosity, thickness, areal extent) and its delivery capabilities (permeability, natural fractures, susceptibility to damage). Engineers charged with maximizing hydrocarbon recovery address rock mechanics issues only of dire necessity and then with little enthusiasm. Rock stresses and failure criteria are the most critical aspects of rock engineering for petroleum engineers. Horizontal drilling has presented a host of new issues that are more pervasive than are vertical well issues. This paper addresses several areas, including:

- an overview of rock mechanics issues of importance to petroleum engineers,
- a discussion of typical stress magnitude measurements routinely used,
- a brief review of stress orientation methods, and
- a discussion of the rock mechanics issues critical to horizontal wells.

1 ROCK MECHANICS ISSUES IN PETROLEUM ENGINEERING

Rock properties are important to drilling, production, and reservoir engineers for many reasons. Concerns range from drilling, completion, stimulation, production, and monitoring or reservoir processes. The state-of-the-art in borehole stability is reviewed in a recent National Research Council publication [NRC 1993].

Drilling: Vertical wells have typical diameters ranging from 0.1-0.4 m. with depths from 1,000 to 5,000+ m. Rock layers encountered have a variety of pore pressures, permeabilities, mechanical properties, and fluid contents. Most drilling techniques maintain fluid pressures in the wellbore that equal or exceed those of the pore pressures in the permeable zones penetrated (with the exception of air, mist, and foam rotary drilling and certain methods for very shallow wells). Fluid pressure gradients range from subnormal pressures below 6 Kpa/m (0.27 psi/ft) to extremely abnormal pressures of 20 KPa/m (0.9 psi/ft). Borehole pressures must be high enough to prevent excessive fluid influx into the wellbore and be low enough to prevent "lost circulation", the rapid loss of wellbore fluids either to very high permeability formations or through induced hydraulic fractures.

Drilling rates are generally more rapid with lower mud weights and higher circulation rates. Many of the penetrated strata are chemically sensitive to the mud used (*e.g.,* shales may swell when exposed to aqueous muds) leading to potential wellbore failure. Alternative muds (including oil-based or oil-water emulsion-based) may mitigate some problems at higher costs and with increased environmental risks. Methods of predicting wellbore failure based on linear elastic models neglecting plastic flow and rock creep overestimate mud weight requirements, slowing drilling and increasing the risks of tensile failure.

Production problems involving the formation include sand control, stimulation issues, perforation design, and monitoring of rock movement related to oilfield operations that alter reservoir stresses. In some cases, permeability and porosity are strong functions of the effective stresses and change significantly with variations in fluid pressure.

Sand control is a particularly onerous problem in which friable formations fail. Formation failure ranges from rock matrix collapse to the production of formation material. Formation failure involves rock stresses, material strength, fluid velocities, and chemical interactions. Sand production can exceed 10 kg daily. The problems associated with sand production include decreased efficiency or destruction of artificial lift equipment, erosion of downhole and surface equipment, decreases in well productivity, and complexities in the separation and disposition of the produced material. Erosion impacts are most severe at high pressures and high flow rates where sand production poses safety hazards. At the opposite extreme, certain tar sands and heavy oil producers find the "coproduction" of sand to be positively correlated to well productivity.

A host of remedial techniques to repair or mitigate sand production are widely used. The most common of these involve combinations of wire wrapped screens placed inside the casing opposite the perforated intervals, gravel packing (in which a dense slurry of sized sand particles are forced into the annular region between the screen and into the perforations), and pre-packed screens containing sized sands between two screens. Resin coated sands are occasionally used in either application. The resin coatings vary in detail, but are designed to consolidate the sand grains. Sand consolidation treatments involve injecting liquids designed to permanently connect sand grains when the liquid materials harden. All of these techniques routinely *decrease* well productivity.

One problem associated with sand control is predicting the onset of sand production problems. While highly friable sands are ultimately likely to fail, they may produce "sand-free" for years. For less friable formations, it is not always possible to predict when (or if) wellbore failure will occur. Many wells fail only after water production commences. Extremely overpressured wells often fail when wellbore pressures decline below normal pressure gradients. Gravel packing a well at the time of its initial completion is less expensive than packing in a subsequent workover, whether or not the gravel pack is needed initially. With decreased reservoir pressure at later times, subsequent gravel pack operations will be more difficult to clean up damage caused by the water and polymer materials used to transport the sand. However, unnecessary gravel packing of a well decreases its productivity, often significantly. Pre-emptive gravel packing decreases the productivity when high rates are most needed from a Net Present Value (NPV) economic basis.

Well stimulation is a means of improving the reservoir flow properties to overcome damage or improve the flow capacity of low permeability wells. Horizontal wells compete with *stimulated* vertical wells in economic and recovery terms. The primary stimulation techniques for vertical wells are *acidization* and *hydraulic fracturing.* Acids are used to dissolve carbonate minerals, silicates, clays, and feldspars. One effect of acidization is to remove formation damage caused by [Economides], [Meehan 1989]:

- invasion of drilling mud solids or filtrate, damages from cement slurries, washers, and spacers,
- perforation damage from overbalanced perforations, charge debris, or inadequate perforation length,
- damage from completion and workover fluids,

- damage to gravel packs by improper slurry placement, fines migration, plugging by foreign objects, or improper design of gravel size,

- damage during production by high flow rates and consequent fines movement, excessive drawdown leading to rock failure or permeability reduction, deposition of organic or inorganics materials associated with pressure reductions, or

- damage due to other wellbore treatments including injected water, fracturing fluids, or prior acid treatments.

Additionally, well stimulation is achieved by increasing near-well permeability to levels in excess of the native values. Matrix acidization is a near-wellbore treatment where injection rates and pressures do not cause rock tensile failure. Acid fracturing is a technique similar to hydraulic fracturing in that a hydraulically induced fracture is created with conductivity at fracture closure due to uneven etching of the material along the fracture walls.

Hydraulic fracturing is one of the most common stimulation techniques and one of the most important applications of rock mechanics in petroleum engineering: Hubbert and Willis [Hubbert] explained the nature of hydraulic fracturing mechanics and concluded what is taken axiomatically today: Hydraulically induced fractures propagate in a direction normal to the least principal compressive stress. In most areas at depths greater than a few thousand feet, this least stress (σ_{min}) will be approximately horizontal and fractures will be vertical to subvertical. At shallow depths, the reverse is often true: the least stress will be essentially equal to the overburden pressure and fractures will be horizontal.

The pressure required to *initiate* the fracture is a function of rock tensile strength, hole geometry, rock stresses, and fluid properties. Pressures required to *continue* fracture propagation are largely determined by rock stresses. In a typical field, vertical fractures should have similar azimuths, or compass orientations. Large variations in effective stress directions, large pore pressure gradients, or alteration of effective stresses may alter the fracture azimuths created. Small differences between σ_H and σ_h (the maximum and minimum horizontal compressive stresses) tend to increase the possibilities for altering fracture propagation directions due to interwell interference. In practice, σ_h is measured with reasonable accuracy on many wells, but measurements of σ_H are rare. Knowledge of σ_H is of minimal importance in vertical wells but is critical for drilling and fracturing highly deviated and horizontal wells.

For very low permeability reservoirs, Massive Hydraulic Fractures (MHF) are created by pumping enormous quantities of sand-laden gelled fluids to create hydraulic fractures. Although they may have tip-to-tip lengths in excess of 2000 feet, actual created hydraulic fracture lengths can rarely be measured directly. Model estimates of created and propped hydraulic fracture lengths vary significantly [Warpinski 1993]; however, models that match the net fracturing pressure generally have reasonable agreement as to fracture lengths [Cleary 1994]. Net fracturing pressure is defined as:

$$\Delta p_f = p_f - \sigma_h \qquad (1)$$

where p_f is the actual fracture pressure. Determining the actual value of p_f requires a good estimate of the bottom-hole treating pressure (BHTP), perforation friction [Wright], and near-wellbore tortuosity [Johnson]. Some fracturing issues that remain poorly understood include:

- Fracture tip processes – It is widely agreed that the fracture tip consumes more energy than previously thought due to dilatancy [Cleary 1991], *in situ* fracture toughness orders of magnitudes higher than laboratory measurements [Shlyapobersky], or other reason [Valko].

- Factors controlling fracture height growth are not well understood for shales, coals, and naturally fractured media.

- Near wellbore friction due to tortuosity can be severe and is difficult to predict but is easy to measure.

The dimensionless fracture conductivity ratio, F_{CD} is defined as:

$$F_{CD} = \frac{k_f b_f}{x_f k} = (k_f b_f)_D \qquad (2)$$

where k_f and b_f are the permeability and width of the fracture, x_f is fracture half-length (well axis to fracture tip), and k is formation permeability. This ratio, F_{CD}, must be large to have a substantive, long-term increase in production. For low permeability formations, the denominator becomes small, and efforts to make high conductivity fractures are less important. However, long-term effects of potential proppant crushing, imbedment, fines migration, and other damage to the fracture must be considered. Additionally, long hydraulic fractures have decreased fracture conductivities. Therefore, a maximum effective fracture length may occur due to diminishing fracture conductivity. The advantage of very high values of F_{CD} diminishes at values above about 50, with significant increases in productivity from 2 to 50. Fracturing of very high permeability formations for sand control and increased productivity is increasingly popular. Very short fractures ($2x_f = h$) and large reservoir permeabilities necessitate very large values of $k_f b_f$ to obtain F_{CD} values greater than unity.

Formation mechanical strength and other rock properties, along with *in situ* stress conditions are particularly important. Without sufficient contrast in rock mechanical properties and/or stresses between layers, creating extremely long propped fractures is impossible because of unconstrained vertical growth. Data from "mini-frac" tests and pressure monitoring during fracturing can be used to improve created and propped length, to avoid early screenouts (a premature end to the hydraulic fracturing treatment associated with excessive treating pressures) or uncontrolled vertical growth.

Pad volume (the quantity of proppant-less fluid pumped initially) is important since it both creates the fracture and cools the formation, enabling the fluid containing the proppant to maintain its viscosity. Excessive pad volumes result in higher-than-required fracture costs and inadequate propped fracture lengths.

Perforation issues involving rock mechanics are numerous[Warpinski 1992]. Oriented perforations may mitigate excess near-wellbore friction observed in many vertical and *most* highly deviated wells. Optimum perforation length, density, and cleanup techniques are far from resolved. Extreme overbalanced perforating (EOB) coupled with near-wellbore proppants [Handren] is a recent variation of earlier EOB techniques that is proving succesful in a wide variety of applications.

Monitoring of reservoir processes involves rock mechanics for many specialized applications, including:

- Subsidence monitoring– Reduced reservoir pressures lead to reduced stresses; decreased reservoir thickness can be communicated to the surface, especially for shallow reservoirs. Surface subsidence in excess of 10 m. has occurred, with much smaller subsidence posing large costs and safety hazards. While fluid injection for pressure maintenance can minimize subsequent subsidence, there is a need for continuous monitoring of subsidence prone areas. Surface tilt and observation well logging can be used to measure the impact of field operations on subsidence.

- Monitoring enhanced recovery– Enhanced recovery requires injection of various fluids to displace oil. Knowledge of where the injected fluids actually are is essential to optimal economics; numerous EOR failures are directly attributable to failure to understand reservoir heterogeneity. If the location of injected fluids can be directly detected at early times, sweep and recovery efficiency can be improved. Injection of steam, CO_2, or even water can significantly alter the acoustic impedance of reservoir rocks and fluids, making EOR monitoring by surface seismic possible.

2 DETERMINATION OF *IN SITU* STRESSES

Formation stresses result from an interaction of overburden, tectonic stresses, and fluid pressures. The vertical compressive stress is due to the weight of the overburden and at depth Z can be expressed as:

$$\sigma_v = g\int_o^Z \rho_f(z)dZ \tag{3}$$

where ρ_f is obtained from density logs. Equation 3 can be significantly in error for steeply dipping beds and mountainous areas. The *effective* vertical stress is diminished from this value in porous media because the fluids also support the overburden.

$$\sigma_v' = \sigma_v - \alpha p_{pore} \tag{4}$$

Minimum horizontal stress can be related to the vertical stress through the Poisson equation or by Mohr-Coloumb theory; either method predicts effective horizontal stresses of about one-third of σ_v'. Horizontal stress anisotropy is caused by tectonic forces. The value of σ_H can range from near σ_h (normal faulting regimes) to values significantly in excess of σ_v (strike-slip and thrust faulting regimes).

In practice, σ_h (at depths where $\sigma_v > \sigma_h$) is determined from small scale fracture tests (micro-fracs) where fluid is injected at a constant rate into either an open-hole interval that has been isolated with straddle packers or a perforated cased-hole. As the bottom hole pressure increases, near wellbore stresses decrease. In the case of the open-hole, the minimum *local* (σ_{min}) stress occurs in the direction of σ_H. Fracturing occurs when the compressive stress is overcome (the wellbore is in tension) and the tensile strength of the rock is overcome. The initial breakdown pressure is:

$$p_b = 3\sigma_h - \sigma_H - p_{pore} + T_{hf} \tag{5}$$

where T_{hf} is the hydraulic fracture tensile strength.

Hydraulic fractures initiate normal to the minimum compressive stress. When a fracture is initiated, the wellbore pressure decreases rapidly and pumping is stopped after a short time. The pressure drops rapidly to a value known as the Instantaneous Shut-In Pressure (ISIP), and continues to decline until fracture closure. Fracture closure behavior is used to estimate σ_h for open-hole micro-fracs. The ISIP exceeds σ_h caused by the back-stresses due to fluid leak-off and the pressure to keep the fracture open. Fracture closure pressure may be difficult to detect due to changing system stiffness. Equation 5 can be used to estimate σ_H if T_{hf} is known. Some use laboratory estimates for T_{hf}; however, subsequent "micro-frac" tests are preferable to estimate T_{hf} [Bredehoft]. After fracture closure, fluid is reinjected for multiple reopenings and closings. It is assumed that T_{hf} is zero for these reopenings and the difference between p_b and the reopening pressure is equal to T_{hf} enabling estimation of σ_H.

Open-hole micro-fracs are costly and the packers used alter the wellbore stresses to varying extents [Warren]. Additionally, the assumption is made that the near wellbore stresses after the initial crack is induced are identical to the pre-existing stresses, an assumption that is valid for either low pumping rates with low viscosity fluids (the general method used) or for fast pumping and high viscosity fluids [Ratigan]. Cased and perforated holes present completely different problems as the initial breakdown pressure is influenced by the orientation and spacing of the perforations, the quality of the cement job, and the stresses induced by the perforations. T_{hf} can not be generally determined from micro-fracs in cased and cemented holes and the value of ISIP can not be assumed to be equal to σ_h. Analysis of fracture closure in cased hole micro-fracs can lead to an estimate of σ_h and the leakoff coefficient, a measure of porous media filtration properties including permeability, porosity, compressibility, *etc.* required for fracture design.

3 STRESS ORIENTATION

Determination of the orientation of the minimum and maximum horizontal compressive stresses is important for vertical wells that are hydraulically fractured for optimal well placement [Meehan 1989] and for naturally fractured reservoirs with permeability anisotropies related to the current stress state [Meehan 1991]. They are vital for horizontal wells where they relate to wellbore stability and the type of hydraulic fracture that will be created in a stimulation. Because hydraulic fractures will be aligned with σ_H, many of the methods to determine stress direction rely on techniques to determine fracture azimuth.

3.1 Seismic Methods

Seismic methods to detect stress orientation exploit the different properties of compressional and shear waves. Compressional, or p-waves have a sense of motion in the direction of propagation and are sensitive to the acoustic impedance of the material through which they are propagated. Shear, or s-waves, move perpendicularly to the direction of propagation and are sensitive to the shear strength of the media. S-wave velocities are approximately one-half of p-wave velocities for identical materials.

Velocity anisotropy is the dependency of seismic velocity variations on the direction of measurement. Unequal horizontal stresses cause azimuthal anisotropy in the shear waves, polarizing them with the fastest component traveling in the direction of the maximum compressive stress. Shear wave velocities depend both on measurement location and direction of polarization of the shear wave. The direction of the principal stress axes then determines the orientation of the velocity with which polarized shear waves travel.

Shear wave splitting is routinely observed in shear wave Vertical Seismic Profiles (VSP). Small cracks are the most important factor in creating large anisotropies. Open cracks at depth are usually oriented parallel to the maximum compressive stress [Nelmat]. Shear waves at the earth's surface are distorted by phase, amplitude, and mode changes, complicating analysis of velocity anisotropy interpretation from surface measurements. It is easier to interpret shear waves recorded below the surface using three-component geophones.

3.1.1 Tri-Axial Borehole Seismic

Tri-Axial Borehole Seismic(TABS) [Dobecki] is a technique which monitors microfracture events associated with a small hydrofracturing operation. A seismic tool is clamped below the perforations while a fracture treatment is conducted without proppant. The tool contains three orthogonally oriented geophones or accelerometers. Downhole orientation is determined by detonating explosives in shotholes or other wells, or by gyroscopic equipment. The Fenton Hill "Hot Dry Rock" experiment conducted by Los Alamos National Laboratories [Albright] exhibited substantial microseismic activity in the fairly narrow seam of rock enclosing the fracture. Microseismicity continued for several hours after pumping had ceased, permitting the mapping of fracture azimuth, height and length.

3.2 Methods Requiring Cores

Coring methods are generally limited to lengths of 30–90 feet. Data scatter, residual stresses, and scale dependency obscure interpretation of core data. Accuracy of orientation is generally limited to the precision of the orienting tool. Difficulties in orientation techniques include core spinning and other translations of the scribe mark, core jamming and breakage, and missing pieces. Core determined orientation of fracture azimuth may be unduly influenced by local heterogeneities, faults, or structural features. On the other hand, multiple tests in a single formation are possible, costs are only marginally higher than for conventional coring, and analysis is straightforward.

3.2.1 Strain Relaxation

Whole cores can be cut at virtually *in situ* strain conditions, depending on the depths, pressures, and rock properties involved. Freshly pulled cores continue to relax from their initial stresses for several hours following recovery. Anelastic Strain Recovery (ASR) [Blanton], [Teufel] is a technique for measuring these relative strains. Three or more pairs of sensitive strain gauge transducers are mounted on a solid ring spaced around the oriented core, measuring the strain response. Cores containing natural fractures may yield erroneous strain relaxation stress directions.

Differential Strain Analysis (DSA) [Strickland] is a related core technique that involves repressuring a sample of oriented core above its initial pressure and recording oriented strain data as pressure is reduced. This method produces results similar to strain relaxation if irreversible changes in the rock have not occurred. Cores recovered from abnormally pressured reservoirs are likely to have such irreversibilities.

3.3 Log Methods

3.3.1 Borehole Televiewers

Although television-type cameras can be used at shallow depths with clear fluids in the hole, most reports of detection of fracture azimuth have been made with the borehole acoustic televiewer (BATV) [Pasternak], [Plumb 1986]. The BATV is an ultrasonic tool with a rotating high resolution transducer measuring reflected sonic travel time and amplitude. Clear and detailed representations of the borehole walls can be obtained by this technique. Spalling (borehole ellipticity) can be observed, and a fairly detailed map of vugs and natural fractures can be obtained.

Wellbore breakouts can predict wellbore stress concentrations [Zoback 1985]. Breakouts are the result of localized shear failure around a wellbore due to horizontal compression. The result is spalling on both sides of the hole corresponding to the least horizontal stress. Presuming that this stress is less than the vertical compressive stress, a hydraulically induced fracture will propagate normal to the observed spalls. Zoback and Zoback demonstrated excellent correlation between the stress-induced spalls and independent measures of stress orientation [Zoback 1988].

3.3.2 Dipmeters and Other Caliper Logs

Four- and six-arm dipmeter logs are the simplest micro-resistivity devices that may be used to detect breakouts and fractures. If spalling is observed continuously over several intervals, dipmeter results may be used to detect stress azimuths. Unfortunately, it is not always possible to conclude that borehole ellipticity, as determined by caliper logs such as the dipmeter, is in fact due to stress-induced wellbore breakouts. Field observations have indicated that the maximum hole elongation direction may result from keyseats of tensile failures. Televiewer logs may be analyzed to confirm the nature of the breakouts observed by dipmeters.

The dipmeter tool has been used to identify natural fractures and to locate potential breakouts. One study [Brown] presented results from 50 wells in the East Texas Basin's relatively deep Cotton Valley formation. Only 1% of the observed "washouts" were not in the general direction of the dominant N35W/S35E trend observed in the other wells studied. Although the data presented are excellent, the paper's conclusion that hydraulic fractures will be generated in the same direction as the wellbore breakouts is incorrect.

3.3.3 Formation MicroScanner

The Formation MicroScanner (FMS) and Formation MicroImager (FMI)[1] are extensions of dipmeter technology which substitute a dense array of microresistivity devices for those on the dipmeter tool. This array provides a strip image of the wellbore with partial coverage of the wellbore. This tool provides detailed spatial resolution in addition to conventional dipmeter data. Plumb and Luthi [Plumb 1985] contrasted the relative strengths and weaknesses of the FMS and the borehole televiewer.

Identification of natural fractures by dipmeters and BATVs in vertical wells is difficult as sub-vertical fractures tend to parallel the wellbore; their intersection forming an acute angle that may be difficult to detect or to differentiate from induced fractures. Svor and Meehan [Meehan 1991a], [Svor] discuss the improved aspects of fracture detection and quantification of results from oriented micro-resistivity devices in horizontal wells.

3.4 Direct Observation Methods

Direct techniques provide good results for determining fracture (and stress) azimuths but are often difficult to apply and are unduly influenced by local wellbore conditions. Direct observations generally require an open-hole section to be tested and hydraulically fractured. A micro-frac in an open hole section may require altering casing setting plans, thus adding substantially to costs. Local heterogeneities may influence the fracture azimuth and make the value ascertained from these techniques unreliable. Some field results show significant influences from nearby faulting and structurally related effects, especially when the difference between the maximum and minimum horizontal stresses is not large.

3.4.1 Impression Packers

Impression packers have malleable rubber jackets covering metal elements that can be expanded mechanically or hydraulically. This method provides a visual estimate of

---

[1] Marks of Schlumberger

the fracture azimuth, height, and aperture in the open hole section. The created fracture leaves a mark which may be indistinguishable from gouges or preexisting hairline fractures; mechanical difficulties are common.

3.4.2 Fracture Overcoring

Fracture overcoring [Daneshy] is a direct technique in which a hydraulic fracture is created in the open hole just prior to drilling the interval of interest. It is assumed that the fracture will extend below the drilled section, "ahead of the bit". An oriented whole core is then run to recover the fractured, but as yet undrilled, section of hole. The success rate of the method is reported to be about 50 %.

3.4.3 Downhole Extensometer

The downhole extensometer [Lin] is a device combining two six-arm calipers between downhole packers and high-resolution pressure monitoring devices. Pre- and post-fracture wellbore deformations can be measured in pumping tests. In principal, stress magnitudes, directions, and *in situ* static rock properties can be determined from this test. There have been only a few commercial tests; most successful tests are for shallow wells.

3.5 Other Methods

Numerous pressure analysis methods to detect hydraulic fracture azimuth have been described in the literature. Meehan *et al* [Meehan 1991], [Meehan 1988] review these methods including the economic issues associated with estimating fracture azimuth. Sensitive tiltmeters are commercially available for the detection of the orientation of shallow hydraulic fractures [Wood]. Regional geology and mapping of faults provide an excellent tool for initial estimation of stress anisotropy. Local interpretation of the magnitude and azimuth of horizontal stresses must ultimately be resolved with geologic interpretations.

4 ROCK MECHANICS ISSUES FOR HORIZONTAL WELLS

The advent of modern horizontal drilling constitutes one of the most significant petroleum extraction technology improvements in the last twenty years. The actual drilling of horizontal and highly deviated wellbores dates back more than fifty years. Modern horizontal drilling dates back only ten years with the advent of MWD (Measurement While Drilling) and advances in downhole mud motors and other specialized tools. These tools either evolved from "conventional" directional drilling or were created for the much tighter turning radii required for onshore applications of horizontal drilling. As recently as 1988, only a few hundred horizontal wells had been drilled; this number now exceeds 10,000. All of the rock mechanics issues which are important in vertical wells are important in horizontal wells. Additional rock mechanics concerns arise in drilling, formation evaluation, completions, and stimulation of horizontal wells.

4.1 Drilling Issues for Horizontal Wells

Initially, the principal technical challenges associated with horizontal wells centered around the actual drilling of the wellbore. Actual results of thousands of horizontal wells indicate surprisingly few wellbore failures during drilling. As in vertical wells, the primary drilling failures are shear failures at low mud weights and tensile failure at high mud weights [Aadnoy], [Bradley]. Vertical wells are drilled parallel to σ_v and have σ_h and σ_H as the principal stresses normal to the wellbore. Horizontal wells initiate vertically and must be deviated 90° with an azimuth that can be controlled. Horizontal wellbores drilled *parallel* with σ_H have σ_h and σ_v as the principal stresses normal to the wellbore. When $\sigma_v > \sigma_H > \sigma_h$, this will result in a large stress differential, indicating a large possibility for wellbore failure. Hydraulic fractures (for wells drilled parallel to σ_h) will parallel the wellbore, this direction will also be relatively unstable with respect to tensile failure. The actual stability of wellbores depends on the relationship of *each* of the stresses to which the wellbore is exposed. For horizontal wells this will change as the well is deviated from vertical to horizontal.

In situ stresses are not the only factors involved in drilling "gage" holes (where the created diameter is equal to the bit diameter). Hole "washouts" can be caused by chemical interactions with the formation, poor hydraulics (excessive velocities leading to erosional failures), or poor filtrate control (leading to extensive fluid leakoff into the formation). Reduced hole diameters can be caused by plastic flow of the rock (characteristic of salts and soft shales) and often lead to high values of torque and drag with the accompanying risk of being unable to move the drill pipe.

Torque and drag are higher in the case of horizontal well drilling because of the more complex path of the drillpipe involving multiple points of contact with the wellbore, the effects of gravity in the deviated section, and poor cuttings transport that may cause accumulations of previously drilled material to form dunes on the low side of the wellbore. While there are clearly many situations where horizontal wellbores may be *less* stable than vertical wells, the drilling of horizontal wells is generally accompanied by improved analysis and planning of mud programs and hydraulics. Numerous authors have presented guidelines for estimating the "safe" mud weights to minimize the risks of borehole collapse or tensile failure as a function of the wellbore azimuth, inclination, and stress regime.

4.2 Sand Control and Alternatives Issues for Horizontal Wells

The problems associated with gravel packing horizontal holes are fundamentally identical to those encountered in gravel packing highly deviated ($> 60°$) wells but more egregious. Excess sand pumped in a conventional vertical well gravel packing operation falls into the rathole below the pay interval. No rathole is present in horizontal wells as the lowest gravitational points are the low side of the hole and any local minima in the well trajectory. This can degrade sand transport and result in a failure to pack the low side of the wellbore. Gravity settling can further complicate placing sand at the top of the interval. Extremely long intervals are exceptionally difficult to pack, with the large volumes and long pumping times straining every part of the job from gel stability to the size of blending equipment.

Thin fluids and low density or neutral buoyancy sands may be considered as potential alternatives for such large jobs. Other fluids such as foams and emulsions may be effective. Calcium sand consolidation (silicate) or other consolidation methods may be feasible or attractive. Prepacked screens are an extremely attractive alternative when applicable.

As in vertical wells, almost all gravel pack operations ultimately lead to decreased well Productivity Index (PI) with a corresponding cost in time delays to recover reserves, increased need for compression or artificial lift, or additional wells. Depending on the direction the well is drilled with respect to the minimum and maximum horizontal compressive stresses, rock properties, wellbore operating conditions, and the magnitudes of all the compressive stresses, it may be possible to completely avoid the need for sand control (or significantly delay it) with respect to a comparable vertical well. Increased productivity from horizontal wells results in decreased pressure drops for a given flow rate and decreased fluid velocities due to fluid influx from a much larger total area. Lower velocities decreases drag forces on the sand particles proportional to velocity-squared. Additionally, water coning with horizontal wells can be avoided or delayed, further decreasing the likelihood of sand production. Methods to estimate the maximum sand-free rates can be improved for horizontal wells utilizing routinely available information from cores, open hole and production logs, *etc.* Determining wellbore stability criteria as a function of well azimuth and dip needs to be simplified to the point of being routinely used. There is a simple method to safely evaluate the need for prepacked screen in slotted liner applications that hold the potential for significant cost reductions and PI improvements. This technique involves combining sliding sleeves and pre-packed screens in an initial completion. If the opened sleeve doesn't result in sand production, the pre-packed screens were obviously not necessary. Subsequent completions of similar wells can be made without the pre-packed screens and the "test well" completion can be pulled at the next opportunity. If sand production is observed, the sleeve is closed and production continues through the screens. If sand production eventually occurs, a workover to install sand control equipment will be necessary.

4.3 Hydraulic Fracturing Issues for Horizontal Wells

Many horizontal wells require stimulation either as part of the initial completion or at a later date. Hydraulic fracture stimulations are indicated both in low permeability formations, laminated zones which require propped vertical fractures to communicate between layers and improve k_v/k_h, naturally fractured reservoirs (connecting disparate fracture sets and/or clean up fracture fill), or thin zones with poor stress barriers. In very low permeability reservoirs, it may be desireable to create a series of hydraulic fractures that are orthoganol to the wellbore. These *transverse* fractures may make economic sense when the drainage area of vertical wells is low and the cost of drilling the vertical section is large. Fractures that parallel the horizontal well are referred to as *longitudinal* fractures and may create one or more fractures that are equivalent to an extremely long vertical fracture.

Much has been written about hydraulic fracturing in deviated and horizontal boreholes because the mechanics of the jobs and the propagation characteristics vary so significantly from those associated with vertical wells [Weng]. In vertical wells, wide planar fractures can usually be created that have large areas of intersection with the wellbore; this minimizes excess frictional pressures while pumping and fluid flow related pressure drops while producing. When large near-wellbore friction (tortuosity) exists, placement of "proppant slugs" can reduce or eliminate this effect in vertical wells [Cleary 1993].

Near-wellbore friction can be extremely high in deviated and horizontal wells. Cased and cemented wellbores drilled in the direction of σ_h will have fractures that ultimately are transverse to the wellbore direction. If even modest perforated intervals exist, multiple fracture initiations can occur near the wellbore, interacting and competing for fluid. Wellbore effects will tend to initiate the fractures parallel to the wellbore. This requires the fracture(s) to turn into the direction of σ_H. Such turning results in higher pumping friction, decreased width, and serious risks of premature screenouts. Proppant slugs have had some success in mitigating these effects [Johnson], [Kogsboll]. Transverse wells may experience a "choke effect" related to restricted contact with the wellbore.

Wells drilled in a direction normal to σ_h will have hydraulic fractures that tend to parallel the wellbore. However, cased and cemented wellbores essentially repel propagating fractures and multiple, interfering hydraulic fractures may be created with large perforated intervals. However, fracture initiation and propagation pressures for longitudinal fractures tend to be significantly lower than for transverse wells.

Either case (and all the intermediate ones) tend to encourage small perforated intervals, testing for and minimization of near-wellbore tortuosity, and minimizing pad sizes for optimal proppant placement.

Some remaining issues for hydraulic fracturing horizontal wells include:

- What are the pumping rate effects on the rate of turning if fracture direction changes?
- What are the appropriate fracture diagnostics for fractures propagating from a horizontal well?
- What are the design and execution considerations for micro-fracs?
- How do we determine stress barriers, leakoff, and fracture height growth when the wellbore is only in one interval?
- What is the "right" stress to use in a heterogeneous reservoir?

Acknowledgements: I appreciate the support of my employer, Union Pacific Resources as well as the Stanford University Horizontal Well Project. I also appreciate assistance from Prof. Michael J. Economides and Dr. Pavel Peška.

References

"Stability, Failure, and Measurements of Boreholes and Other Circular Openings," U. S. Natl. Comm. for Rock Mechanics Geotechnical Board, Washington, D. C. (1993).

Aadnoy, B. S. and Chenevert, M. E.: "Stability of Highly Inclined Boreholes," *SPE Drilling Engineering* (Dec. 1987) No. SPE 16052, 364–374.

Albright, J. N. and Pearson, C. F.: "Acoustic Emissions as s Tool for Hydraulic Fracture Location: Experience at the Fenton Hill Hot Dry Rock Site," *SPE Journal* (August 1982) 525–530.

Blanton, T. L.: "The Relation Between Recovery Deformation and In Situ Stress Magnitudes," paper SPE 11624 presented at the 1983 SPE Symposium on Low Permeability Gas Reservoirs, Denver, CO, March.

Bradley, W. B.: "Failure of Inclined Boreholes," *Transactions of the ASME* (December 1979) **101**, 232–239.

Bredehoeft, J. D., Wolff, R. G., Keys, W. S., and Shuter, E.: "Hydraulic fracturing to determine the regional in situ stress field, Piceance Basin," *Colorado Geol. Soc. Am. Bull.* (1976) No. 87, 250–258.

Brown, R. O., Forgotson, J. M., and Forgotson Jr., J. M.: "Predicting the Orientation of Hydraulically Created Fractures in the Cotton Valley Formation of East Texas," paper SPE 9269 presented at the 1980 55th Annual Meeting of the Society of Petroleum Engineers, Dallas, TX, September.

Cleary, M. P.: "Discussion of SPE 25890 – Comparison Study of Hydraulic Fracturing Models: test Case - GRI - Staged Field Experiment No.3," (1994).

Cleary, M. P. et al.: "Field Implementation of Proppant Slugs to Avoid Premature Screen-Out of Hydraulic Fractures With Adequate Proppant Concentration," paper SPE 25892 presented at the 1993 SPE/DOE Regional Low Permeability Symposium, Denver, CO, April.

Cleary, M. P., Wright, C. A., and Wright, T. B.: "Modeling evidence for major changes in hydraulic fracturing design and field procedures," paper SPE 21494 presented at the 1991 SPE Gas Technology Symposium, Houston, TX.

Daneshy, A. A., Slusher, G. L., Chisolm, P. T., and Magee, D. A.: "In-Situ Stress Measurements During Drilling," *Journal of Petroleum Technology* (August 1986) No. 13227, 891–898.

Dobecki, T. L.: "Hydraulic Fracture Orientation Using Passive Borehole Seismics," paper SPE 12110 presented at the 1983 58th Annual Meeting of the Society of Petroleum Engineers, San Fransisco, CA.

Economides, M. J. and Nolte, K. G.: *Reservoir Stimulation*, Schlumberger Educational Services, Houston, TX (1987).

Warren, E. W.: "Packer-induced stresses during hydraulic well fracturing," *J. Energy Resourc. Technolo.* (1981) No. 103, 336–343.

Handren, P. J., Jupp, T. J., and Dees, J. M.: "Overbalance Perforating and Stimulation Method for Wells," paper SPE 26515 presented at the 1993 Annual Technical Conference and Symposium, Houston, TX, October.

Hubbert, M. K. and Willis, D. G.: "Mechanics of Hydraulic Fracturing," *Trans., AIME* (1957) **210**, 153–168.

Johnson, D. E., Wright, T. B., Tambini, M., Maroli, R., and Cleary, M. P.: "Real-Data On-Site Analysis of Hydraulic Fracturing Generates Optimum Procedures for Job Design and Execution," paper SPE 25920 presented at the 1993 SPE/DOE Regional Low Permeability Symposium, Denver, CO, April.

Kogsboll, H. H., Pitts, M. J., and Owens, K. A.: "Effects of Tortuosity in Fracture Stimulation of Horizontal Wells – A Case Study of the Dan Field," paper SPE 26796 presented at the 1993 SPE Offshore Europe Conference, Aberdeen, Scotland, Sept.

Lin, P. and Ray, T. G.: "A New Method to Determine In-Situ Stress Directions and In-Situ Formation Rock Properties During a Microfrac Test," paper SPE 26600 presented at the 1993 68th Annual Technical Conference and Symposium, Houston, TX, Oct.

Meehan, D. N.: "Optimization of Fracture Length and Well Spacing in Heterogeneous Reservoirs," paper SPE 21717 presented at the 1991 SPE Production Operations Symposium, Oklahoma City, OK.

Meehan, D. N.: "A Method for Evaluating the Necessity for Sand Control in Horizontal Wells," (1993).

Meehan, D. N., Horne, R. N., and Aziz, K.: "The Effects of Reservoir Heterogeneity and Fracture Azimuth on Optimization of Fracture Length and Well Spacing," paper SPE 17606 presented at the 1988 SPE International Meeting on Petroleum Engineering, Tianjin, China, November.

Meehan, D. N. and Svor, T. R.: "Quantifying Horizontal Well Logs in Naturally Fractured Reservoir–II," paper SPE 22792 presented at the 1991 Annual Meeting of the Society of Petroleum Engineers, Dallas, TX, October.

Meehan, D. N.: *Hydraulically Fractured Wells in Heterogeneous Reservoirs: Interaction, Interference, and Optimization*, PhD dissertation, Stanford U., Stanford, CA (June 1989).

Nelmat-Nasser, S. and Horii, H.: "Compression-Induced Non-Planar Crack Extension with Applications to Splitting, Exfoliation, and Rockburst," *J. Geophys. Res.* (1982) **87**, 6805–6821.

Newberry, B. M., Nelson, R. F., and Ahmed, U.: "Prediction of Vertical Hydraulic Fracture Migration Using Compressional and Shear Wave Slowness," paper SPE 13895 presented at the 1985 SPE/DOE Low Permeability Gas Reservoir Symposium, Denver, CO, May.

Pasternak, E. S. and Goodwill, G. P.: "Applications of Digital Borehole Televiewer Logging," 24th Annual SPWLA Symposium (June 1983).

Plumb, R. A. and Hickman, S. H.: "Stress-Induced Borehole Elongation: A Comparison Between the Four Arm Dipmeter and the Borehole Televiewer in the Auburn Geothermal Well," *Journal of Geophysical Research* (1985) **90**, 5513–5521.

Plumb, R. A. and Luthi, S. M.: "Application of Borehole Images to Geologic Modeling of an Eolian Reservoir," paper SPE 15487 presented at the 1986 61st Annual Meeting of the Society of Petroleum Engineers, New Orleans, LA, October.

Ratigan, J. L.: "The Use of the Fracture Reopening Pressure in Hydraulic Fracturing Stress Measurements," *Rock Mechanics and Rock Engineering* (1992) No. 25(4), 225–236.

Shlyapobersky, J.: "Energy analysis of hydraulic fracturing," Proceedings of the 26th U. S. Symp. on Rock Mechanics, Rapid City, SD (1985) 539–546.

Strickland, F. and Ren, N.: "Predicting the In Situ Stress for Deep Wells Using the Differential Strain Curve Analysis," paper SPE 8954 presented at the 1980 Unconventional Gas Recovery Symposium, Pittsburgh, PA.

Svor, T. R. and Meehan, D. N.: "Quantifying Horizontal Well Logs in Naturally Fractured Reservoir–I," paper SPE 22634 presented at the 1991 Annual Meeting of the Society of Petroleum Engineers, Dallas, TX, October.

Teufel, L. W.: "Prediction of Hydraulic Fracture Azimuth from Anelastic Strain Recovery Measurents of Oriented Core," Proceedings of the 23rd U. S. National Rock Mechanics Symposium (1982).

Valkó, P. and Economides, M. J.: "Fracture Height Containment with Continuum Damage Mechanics," paper SPE 26598 presented at the 1993 SPE Annual Technical Conference and Symposium, Houston, TX, October.

Warpinski, N. R.: "Invited paper: Rock Mechanics issues in completion and stimulation operations," Rock Mechanics Proceedings of the 33rd U. S. Symposium, J. R. Tillerson and W. R. Wawersik (eds.), A. A. Balkema (June 1992) 375–386.

Warpinski, N. R., Moschovidis, Z. A., Parker, C. D., and Abou-Sayed, I. S.: "Comparison Study of Hydraulic Fracturing Models: Test Case - GRI-Staged Field Experiment No.3," paper SPE 25890 presented at the 1993 , Denver, CO.

Weng, X.: "Fracture Initiation and Propagation From Deviated Wellbores," paper SPE 26597 presented at the 1993 68th Annual Technical Conference and Symposium, Houston, TX, Oct.

Wood, M. D.: "Method of Determining Change in the Subsurface Due to Applications of Fluid Pressure to the Earth," U. S. Patent 4,271,696 (June 1981).

Wright, T. B., Aud, W. W., Cipolla, C., Meehan, D. N., F.Perry, K., and Cleary, M. P.: "Identification andComparison of True Net Fracturing Pressures generated by Pumping Fluids with Different Rheology into the Same Formations," paper SPE 26153 presented at the SPE Gas Technology Symposium, Calgary, Alberta, Canada, June, 1993.

Zoback, M. D., Moos, D., Mastin, L., and Anderson, R. N.: "Well Bore Breakouts and In Situ Stress," *J. Geophys. Res.* (1985) **90**, 5523–5530.

Zoback, M. D. and Zoback, M. L.: *Neotectonics*, The Geological Society of North America, Boulder, CO (*in press* 1988) Ch. State of Stress in North America:The Geology of North America.

Rock Mechanics, Nelson & Laubach (eds) © 1994 Balkema, Rotterdam, ISBN 90 5410 380 8

Discrete fracture approaches for oil and gas applications

W. Dershowitz & P. LaPointe
Golder Associates Inc., Seattle, Wash., USA

ABSTRACT: Rock Mechanics techniques are becoming increasingly important in oil and gas reservoir exploration and development. The discrete fracture approach is one of the most promising of these, since it provides a more realistic geological description of in situ conditions. This paper provides an overview of recent developments in continuum and discontinuum rock mechanics approaches for oil and gas applications, and presents a number of brief case studies of applications of discrete fracture approaches.

1 INTRODUCTION

It has been estimated that reservoirs in which natural fractures dominate reservoir performance will contribute more than 40 billion barrels to the world's oil supply (McNaughton and Garb, 1975). In addition, joints and faults play a very important role in secondary recovery in many conventional clastic reservoirs, such as Prudhoe Bay or various North Sea fields. Efficient and profitable production of hydrocarbons from fractured reservoirs in all stages of recovery requires that reservoir development account for the effects of compartmentalization, flow anisotropy and heterogeneity induced by faults or connected joint systems.

For many years, reservoir engineers have conceptualized a fractured reservoir as a dual-contunuum (Aguilera, 1980; van Golf-Racht, 1982) A model of this type is shown in Figure 1. In this dual continuum, the matrix and the fractures are represented by separate continua with distinct properties and very regular geometry. This regularity takes the form of matrix blocks whose faces are mutually perpendicular, and are completely isolated from one another by the fracture continua. The fractures are also simplified. They can have up to three, mutually perpendicular orientations, and each set has uniform fluid flow properties and very regular spacing and sizes. This extreme geometrical simplification is required in order to numerically solve the complex differential equations that are used in the simulation codes to capture such physical effects as multi-phase flow, imbibition, wellbore skin effects and residual saturation. In order to more accurately capture the physics of flow in a petroleum reservoir, the dual-continuum modeling approach has had to over-simplify the reservoir's flow geometry.

Rock mechanics provides a perspective that is very different from the dual-continuum model prevalent in the oil industry. Experience in underground mining, tunneling and

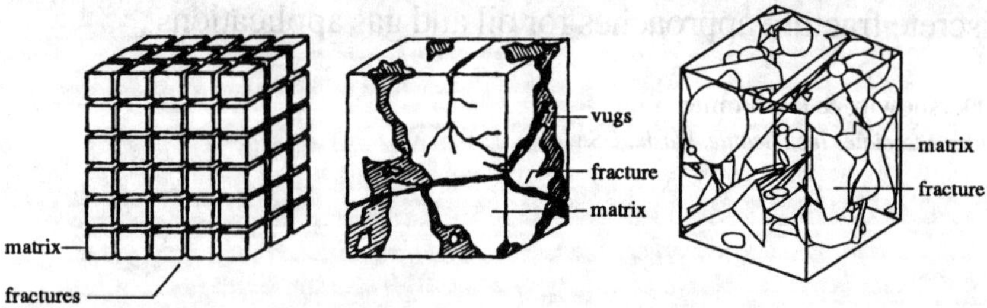

Conventional Dual-Porosity Model
Actual Reservoir
3D Discrete Fracture Model

Figure 1. Discontinuum Approaches.

construction has demonstrated that fractures often play a significant role in the mechanical and hydraulic behavior of rock masses. In many cases, it is the spatial distribution of the fractures and the way in which they forms networks that controls stability (Einstein, 1983) and fluid flow (Gnirk, 1992). As a result, characterization and modeling often focuses on realistically representing the spatial geometry of joints and faults and the way in which they form networks. This paper describes a number of significant recent developments in the use of distinctly rock mechanics approaches to oil and gas exploration and production problems in fractured reservoirs.

2 FLOW BEHAVIOR OF FRACTURED RESERVOIRS

Fracture control of oil and gas reservoirs has been recognized for decades (e.g. Regan and Hughes, 1947). Since fractures are often much more permeable than the matrix, well tests or production from fractured rocks typically show responses different than would be expected in an unfractured porous medium. Among the non-porous media behavior commonly found in fractured reservoirs are:

1. Heterogeneous Connectivity: In a porous media or porous media-equivalent, the spatial distribution of pressures or flow vectors should be a continuous function of distance and direction in response to well production. However, in fractured reservoirs, it is not uncommon for non-continuous large scale flow behavior to take place. This effect is commonly seen in the asymmetric response in interference tests and at larger scales. In an active fractured reservoir play (Figure 2), new horizontal wells were being completed while others were being produced. When one of the new wells was brought into production, existing wells several kilometers distant showed large drops in production within two days, while intervening and neighboring wells were virtually unaffected.

2. Compartmentalization: In order to estimate the effects of infill drilling or for water floods, it is necessary to determine how well the reservoir is connected. In a fractured reservoir, faults and joints can be conduits or barriers to flow. The spatial pattern of the fractures enters into this problem in two ways. First, the fractures may form networks that are not extensive and only drain a relatively small and possibly uneconomic volume of hydrocarbon from the reservoir (Figure 3). This is the situation in many carbonate reser-

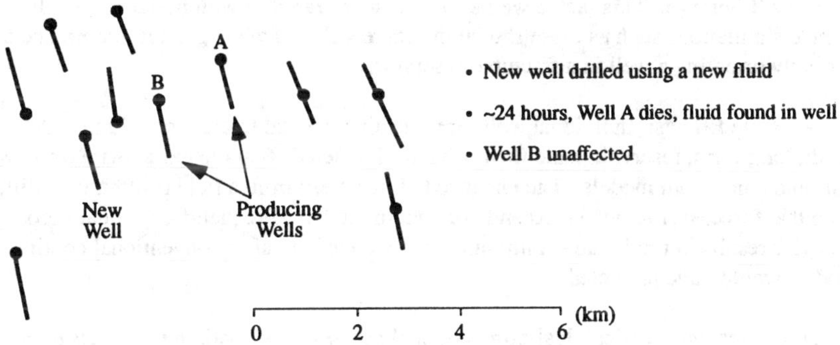

Figure 2. Assymmetric Response, Large Scale.

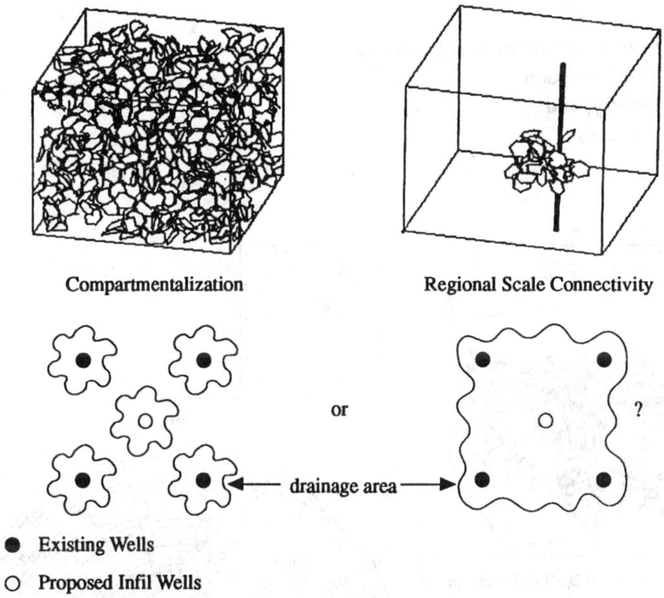

Figure 3. Reservoir Compartmentalization.

voirs with negligible matrix permeability. Secondly, faults may dissect a fractured or clastic reservoir into a series of blocks that are not connected and are at different pressures. If they act as barriers, then development costs will increase, floods will not be as efficient, and ultimate recoveries from the field will probably be less than projected.

3. Well Orientations: In terms of access to the reservoir, well orientation in a clastic unfractured reservoir is not important. Issues of gravity drainage, penetrating the most reservoir interval, or coning may be the most important considerations in orientation. In a fractured reservoir, well orientation can significantly impact production rates and ultimate

recovery (Figure 4). This fact is well known to those familiar with horizontal wells. Simple simulations such as those shown in Figure 4 illustrate the greater importance of correctly orienting a well in a fractured reservoir.

4. Scale Effects: Individual wells are sensitive to local fracture conditions. As a result, long-term, reservoir-scale flow behavior is often difficult to extrapolate from well tests and continuum models. The enhanced oil recovery from a field is often very different from that forecast from well tests and continuum models. Frequently, the dual porosity behavior results in production continuing at low rates long after conventional continuum models would have predicted.

In many cases, the classical dual-porosity, dual permeability continuum models do not even qualitatively reproduce observed reservoir flow behavior, despite the sophistication of the physics and chemistry contained in their formulation. Several new approaches, originating in the fields of rock mechanics and geotechnical engineering, are proving useful in these situations:

Enhanced Porous Continuum Approaches
Discontinuum Approaches
Fractal Geometry, and
Percolation Theory

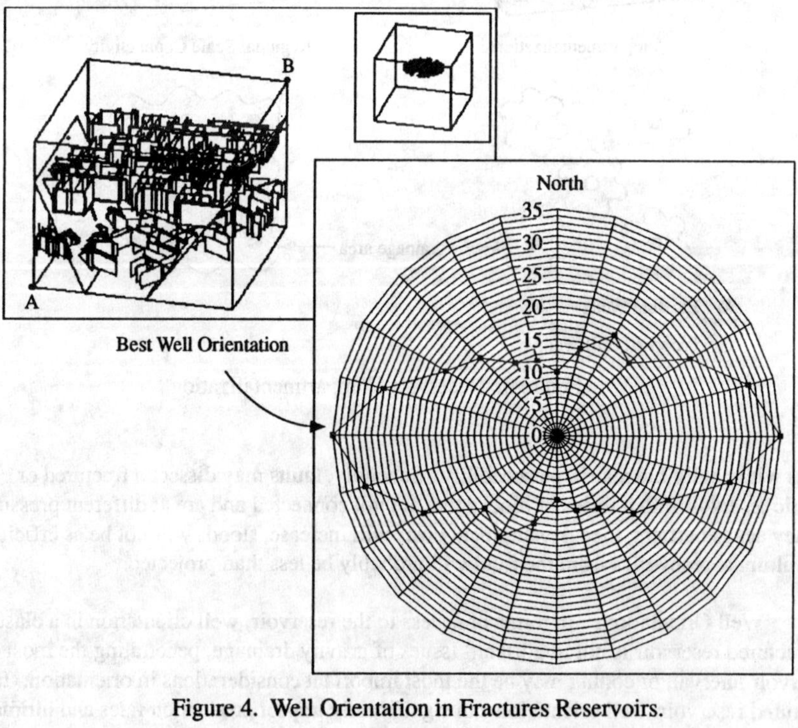

Figure 4. Well Orientation in Fractures Reservoirs.

The following sections summarize some recent development in each of these areas. In addition, applications of the discontinuum approach are highlighted.

3 ENHANCED POROUS CONTINUUM APPROACHES

Rock mechanics research is now being carried out to adapt the more conventional reservoir simulation approaches using rock mechanics techniques. Over the past decade, reservoir engineers have extended techniques developed in the field of rock mechanics to reservoir characterization and simulation. Conventional dual continuum models replace a volume of rock with all of its heterogeneity by an anisotropic continuum that approximates the same bulk response to applied boundary conditions. These approaches generally represent the multiple porosity behavior through interacting continua or multiple storage terms. However, in most cases they do not model heterogeneous connectivity

Perhaps the most widely used technique for approximating the spatial distribution of reservoir properties is through Geostatistical simulation, a technique originating in the mining industry and developed initially for geotechnical and mining applications. Geostatistical simulation is widely used in the petroleum industry now to assign values of permeability and porosity to clastic reservoir models. In the geostatistical continuum approach, the heterogeneity found in fractured rocks is replaced by a heterogeneous continuum, with heterogeneity described by a correlation length and semi-variogram. Geostatistical methods have not been used to describe fracture patterns for fractured reservoir description in the oil industry, since the conventional finite difference dual continuum models are not able to utilize this type of input. However, geostatistical generation of two-dimensional fracture patterns was known to the rock mechanics community at least a decade ago (La Pointe, 1981) for the purposes of calculating effective block-scale rock mass stiffness values, and as a seeding function for discrete fracture models, and could open up a new way of simulating fractured reservoirs using conventional simulators.

Continuum approaches, though widely used in the simulation of conventional clastic reservoirs, do not seem adequate for many fractured reservoir situations. The problem comes from the fact that conventional continuum simulators can only accommodate a few explicit faults without creating a model far too large for solution. If faults at a scale smaller than the simulation region are important, then they are often too numerous to explicitly incorporate. Yet, they create significant non-continuum flow behavior of the type previously described.

There have been attempts to overcome the numerical problems of explicitly representing a realistically large number of faults in a conventional continuum formulation. A promising continuum model which is capable of reproducing observed heterogeneous connectivity has been developed by Intera/ECL (Intera, 1993). The ECLIPSE model utilizes ìnon-neighbor connectionî to simulate heterogeneous connectivity, while maintaining a conventional continuum formulation for the average rock mass (Figure 5). This approach has been used on a variety of fracture dominated reservoirs, to match observed behavior .

4 DISCONTINUUM APPROACHES

Rock mechanics approaches for oil and gas production have been adapted from mining, underground construction, and environmental applications. Increasingly, rock mechanics approaches are being developed within the industry. The most attractive of these approaches are based on Discontinuum approaches similar to those used in rock mechanics applications in tunnel, slope stability and waste repository applications:

Discrete Fracture Approaches: The discrete fracture approach (Dershowitz, 1984; Long, 1984; Robinson, 1984) models the geometry of discrete features such as fractures and faults explicitly, but generally does not model the geometry of the intact rock (Figure 1c).

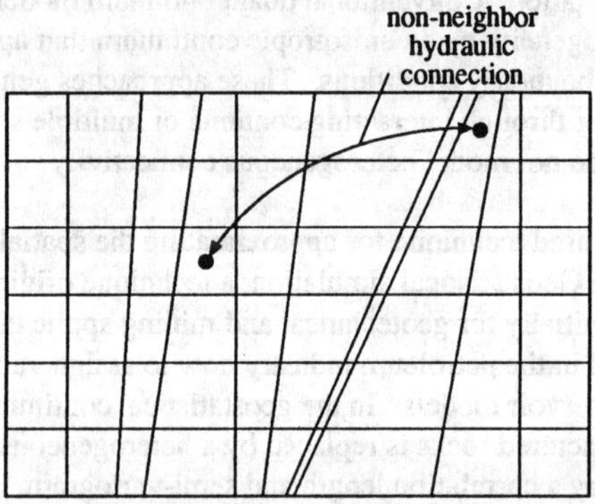

Figure 5. Eclipse Non-Neighbor Connections.

Discrete Fracture Dual Porosity: The discrete fracture dual porosity approach extends analysis to include a ìsecondaryî porosity defined by the rock mass and the smaller fractures not included explicitly in the model (Figure 5). The MAFIC discrete fracture dual porosity simulator (Miller, 1992) is illustrated in Figure 6. This simulator assumes that the primary porosity is fracture porosity, and solves the Galerkin finite element equation 1 for flow and transport in the network of fractures. The model includes two types of nodes to

Equation 1:

$$\sum_{m-1}^{N}\left[\int_R (T_{nm}\overline{\nabla}\xi_n \cdot \overline{\nabla}\xi_n dR)h_m\right] + \sum_{m-1}^{N}\left[\int_R (S_{nm}\,\xi_n\,\xi_n dR)\frac{dh_m}{dt}\right] - \int_R q\xi_n dR \qquad n = 1,2,\ldots N$$

where: T = fracture transmissivity
S = fracture storativity (dimensionless)
q = source flux, volume per unit area (L/T)
ξ = linear or quadratic basis function
R = element area (L^2)
h = nodal hydraulic head (L)
t = time (T)
N = number of nodes

Equation 2:

$$Q_{mf} = \alpha A_m K_m(\overline{h_m}-\overline{h_f}) = -S_{sm} V_m (\partial \overline{h_m}/\partial t)$$

where: Q_{mf} = matrix to the fracture flux (L^3/t)
A_m = area of the matrix block (L^2)
V_m = volume of the matrix block (L^3)
K_{sm} = hydraulic conductivity of the matrix block (L/T)
S_m = specific storage of the matrix block (1/L)
H_f = average matrix of the block head (L)
H = average boundary fracture head (L)
t = time (T)
α = a geometric shape factor = $5/r_{mb}$ for a sphere; $3/r_{mb}$ for a slab (1/L)
r_{mb} = the matrix block radius (L)

represent either (a) rock mass storage only, or (b) rock mass flow and transport. In both cases, oil or gas is assumed to flow between fractures and the rock mass coupled based on the relative pressures as shown in equation 2. MAFIC is a single phase (water or gas) flow simulator. However, because it can realistically represent the pattern of conductive and flow-barrier faults and fractures, it is frequently more useful that continuum multiphase models, which cannot reproduced observed heterogeneous connectivity.

4.1 EXAMPLE APPLICATIONS OF DISCRETE FRACTURE STUDIES

Figures 7 through 9 illustrate recent applications of the discrete fracture approach.

Application #1: Ultimate Recovery in a Oil Reservoir:

As part of a larger reservoir study to assess economic worth and to plan development of a newly-discovered oil accumulation in an overthrust setting, the ultimate production for wells in several structural positions was estimated using discrete fracture models. The initial wells revealed the presence of oil accumulations and good initial flow rates. A detailed discrete fracture model was prepared for each of the wells based upon FMS data, core and flow tests. The model was consistent with the structural style and geology of the region, and consisted of isolated brittle layers that were fractured due to thrusting and bending, separated by strata that were unfractured. The spatial extent of connected clusters (Figure 7) was determined for the model for all clusters and for several stochastic realizations. The results showed that, for certain structural settings, clusters had a sufficient probability of being able to access enough matrix storage to make a well economically viable.

Application #2: Optimized Well Orientation:

In this application, a discrete feature model was developed based on known conductive faults, which were determined to be the primary route for oil production to wells. The known large scale faults were determined from 2D seismic profiles and lineament maps. However, it was possible, given the spacing and location of seismic lines, and the scale of the lineament maps, that not all significant faults were detected. The probability of detection is proportional to fault size (La Pointe and Hudson, 1985). This undetected, yet impor-

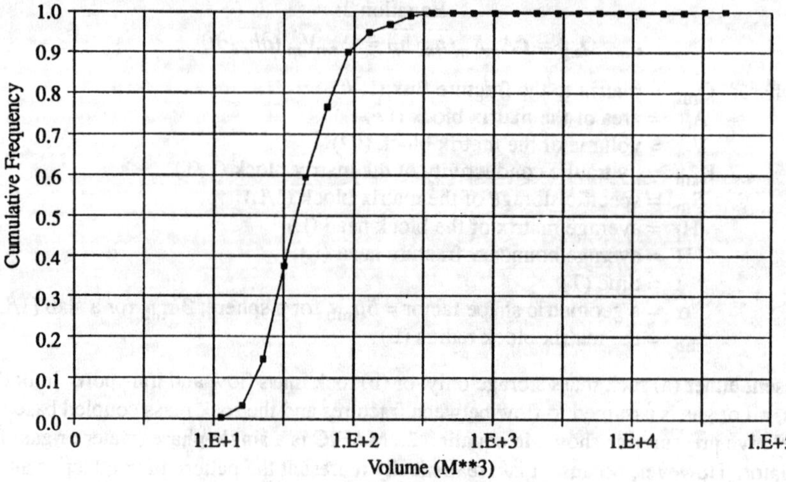

Figure 7. Connected Cluster Analysis.

tant, component was simulated stochastically by extrapolating the fault statistics obtained on the detected faults to faults of smaller, but still significant sizes. Based on these simulations, optimal well orientation were determined. These orientations were specified to maximize the potential drainage volume for horizontal wells was determined, and the potential for destructive interference among wells.

Application #3: Equivalent Continuum Properties:

Standard flow tests to determine cross-flow and identify productive zones were run in three wells. These flows logs when interpreted in the light of FMS data suggested that virtually all of the flow was coming from a seven or eight fractures in the entire 135 meters of open hole. The operator recognized that these well tests were more indicative of local fracturing conditions and fracture geometry than of block scale fluid-flow properties. However, field development required reservoir simulation using blocks of 100 meter size. Discrete fracture models were created to extrapolate from test results to fluid flow properties of these large scale rock blocks,. The well tests, FMS data, seismic profiles and sparse core were used to construct a model that was consistent with all of this data. Stochastic realizations of the model were generated, and 100 m by 100 m by 100 m blocks were extracted from the larger simulation. On each realization, four sides were assigned to be no-flow surfaces (Figure 8). The remaining opposite two faces were each given different, but constant values of head. Flow was then calculated using a discrete fracture finite element code. This set of boundary conditions is analogous to the laboratory scale core floods performed on core to calculate matrix permeabilties, and in a similar way, the effective block scale permeability tensor was determined. Since the calculation subsumes the contributions of both matrix and fractures smaller than the block scale, it was possible to use a single continuum formulation rather than the more cumbersome dual-continuum model.

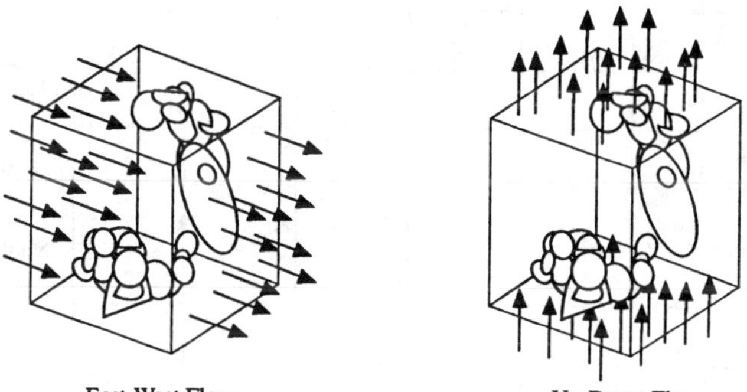

Figure 8. Rock Block Simulation.

5 FRACTAL GEOMETRY:

The fractal approach attempts to represent the complex, heterogeneous fracture system by simple set of indices, the *fractal dimension*. The fractal dimension describes the size and scaling properties of well-connected clusters of objects (Mandelbrot, 1983), such as fracture networks, and how big clusters are spatially related to clusters of other sizes. Since it is the size of these clusters and how close these clusters are to one another that is often important for reservoir engineering, the fractal dimension may describe important properties of the reservoir. This approach assumes that systems of comparable dimension will have comparable hydraulic and connectivity characteristics, as simple numerical experiments demonstrate (Figure 9). Fractal approaches can be combined with either discontinuum or continuum approaches.

Fractals can be used as (a) an index of rock mass behavior (Dershowitz et al, 1991), (b) as an algorithm to generate synthetic rock masses (Sahimi and Yortsos, 1991), or (c) as an approach for extrapolation between scales. In the most common application for oil and gas, FMS data is analyzed to evaluate whether fractures are located according to a fractal model (Figure 10). If they are, then a modified Levy flight model is used to generate the fracture centers for a discrete fracture model (Dershowitz et al., 1993)

6 Percolation Theory

Percolation theory, like the fractal approach, attempts to define indices which can be used describe the connectivity and effective properties of heterogeneously connected systems. Significant work in this area has recently been summarized by Berkowitz and Balberg (1993)

Since heterogeneous connectivity is a key concern in development of reservoirs in fractured rock, it would be nice if a theoretical solution were available for prediction of large scale connectivity based on simple geological and geophysical measurements. Percolation theory was developed to satisfy this need - unfortunately, it has not yet been useful. In general, *percolation theory* solutions are, in fact, discrete fracture network Monte Carlo simulation results for idealized, unrealistic patterns of conductors. As a result, for practical purposes, it may be preferable to utilize discrete fracture modeling directly, since discrete fracture

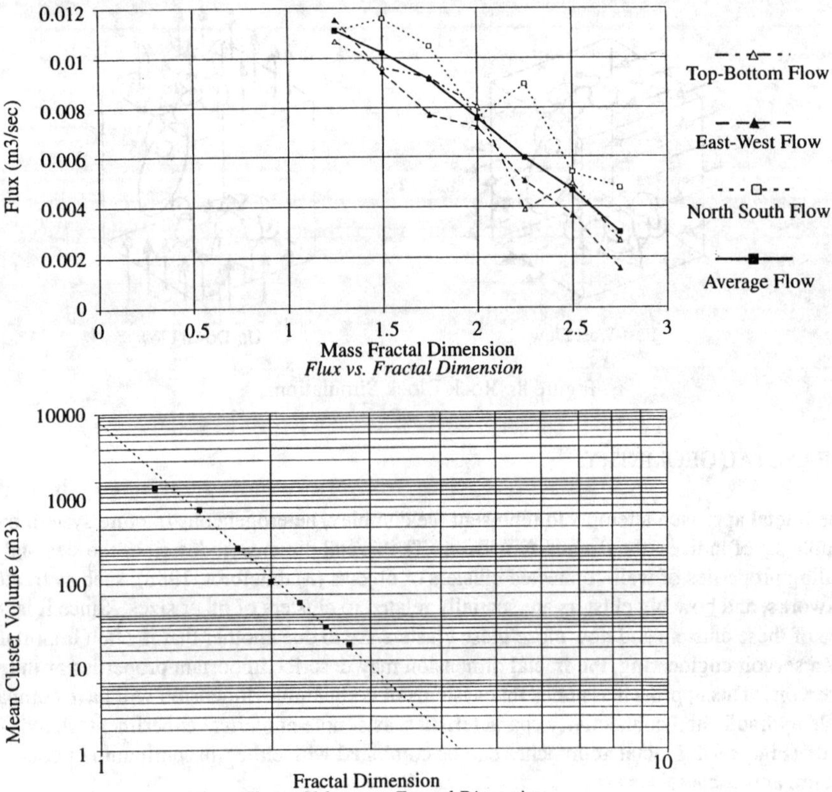

Figure 9. Fractal Dimension and Hydraulic Behavior.

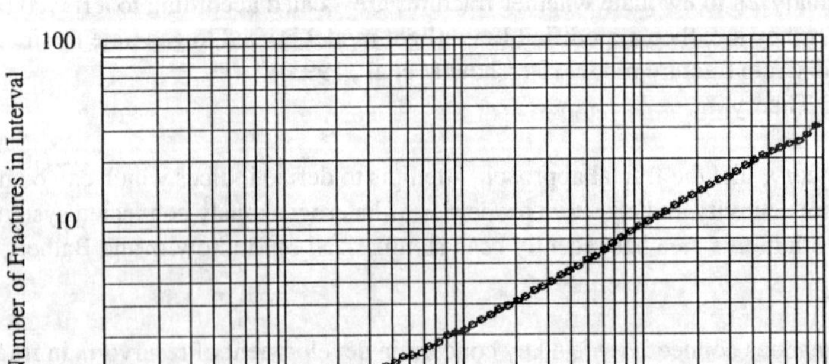

Figure 10. Fractal Mass Dimension Calculated from FMS Data.

Table 1.

Measure	*Definition*	*Application*
Percolation Probability	Probability of connection between locations at given distance R	Location of Wells
Percolation Length	Distance between locations between locations for given probability of connection	Location of Wells
Intersection Intensity	Intersections between features per unit volume	Design of Hydrofracing
Network Volume	Volume of fracture porosity directly connected to a given well interval	Estimation of EOR
Network Radius	Radius of sphere with volume equal to volume of rock mass containing fracture networks connected to a given well interval	Estimation of EOR
Intersection Length	Total length of traces formed by intersections between fracture per unit volume	Design of Hydrofracing

modeling at least utilizes geologically realistic geometries.
Several *connectivity measures* have been developed based on percolation theory, and these have seen some application in oil and gas reservoir development. A series connectivity measures is summarized in Table 1. Of these measures, the percolation probability, the probability of hydraulic connection between points at a given location, is perhaps the most useful. However, it is most useful when derived from discrete fracture Monte Carlo simulations of realistic geometries rather than based on idealized patterns of orthogonal pipe conductors. Another useful index is the *network volume*, which is the total volume and radius of the fractures connected to sources (such as wells) or given size.

6 CONCLUSIONS

Effective development of petroleum reservoirs requires maximizing recovery while minimizing cost. Efficiency and economy requires both more geologically realistic descriptions of the rock mass, as well as the use of methods that can exploit better descriptions. While geologists have long recognized the complexity of natural fracture systems, engineers have lacked ways to incorporate their descriptions into reservoir simulations. Problems that confronted the rock mechanics community preceded similar problems in the oil industry, so that much work in rock mechanics has been assimilated by the oil industry in recent years, particularly in the area of fractured reservoirs. Just as Geostatistics before it, discrete fracture model is now becoming a powerful tool in the oil industry for solving problems in fractured reservoir exploration and development.

REFERENCES

Aguilera, R. 1980. Naturally Fractured Reservoirs. PennWell Publishing Company, Tulsa.

Berkowitz, B. , and I. Balberg, 1993. Percolation Theory and its Applications to Groundwater Hydrology. Water Resources Research, Vol 29, pp 775-794.

Dershowitz W., K. Redus, P. Wallmann, P. LaPointe, and C-L. Axelsson, 1992. The Implication of Fractal Dimension in Hydrogeology and Rock Mechanics. SKB Technical Report 92-17, SKB, Stockholm.

Dershowitz, 1984, Rock Joint System. Ph.D. Dissertation, Massachusetts Institute of Technology, Cambridge, MA.

Dershowitz, W., G. Lee, J. Geier, S. Hitchcock, and P. LaPointe 1993. User Documentation: FracMan Discrete Feature Data Analysis, Geometric Modelling, and Exploration Simulations. Golder Associates, Seattle.

Einstein, 1983. The Effect of Discontinuity Persistence on Rock Slope Stability. International Journal of Rock Mechanics and Mining Sciences, Vol 20, No. 5, pp 227-236.

Gnirk, P. , 1993. OECD/NEA International Stripa Project, Overview Volume II, Natural Barriers. SKB, Stockholm.

Intera, 1993. The ECLIPSE 100 Family of Integrated Petroleum Engineering Software. Intera Information Technologies Petroleum Production Division, Henley, U.K.

LaPointe and Hudson, 1985. Characterization and Interpretation of Rock Mass Joint Patterns. GSA Special Paper 199. Geological Society of America, Boulder.

LaPointe, P., 1981. Analysis of the Spatial Variation in Rock Mass Properties through Geostatistics. Proceedings, 21st US Symposium on Rock Mechanics, pp 570-580.

Long, 1984. Investigation of Equivalent Porous Medium Permeability in Networks of Discontinuous Fractures. Ph.D. Dissertation, University of California, Berkeley.

Mandelbrot, B., 1983. The Fractal Geometry of Nature - Updated and Augmented. WH Freeman, New York.

McNaughton, D. and F. Garb, 1975. Finding and Evaluating Petroleum Accumulations in Fractured Reservoir Rock. Exploration and Economics of the Petroleum Industry, Vol. 13, Matthew Bender and Company.

Miller, 1994. User Documentation, MAFIC Matrix/Fracture Hydraulic Interaction Code with Solute Transport. Golder Associates Inc., Seattle.

Regan, L. and A. Hughes, 1947. Fractured Reservoirs of Santa Maria District, California. Bulletin of the American Association of Petroleum Geologists. Vol 33, No 1, pp 32-51.

Rouleau, 1984. Statistical Characterization and Numerical Simulation of a Fracture System — Application to Groundwater Flow in the Stripa Granite. Ph.D. Dissertation, Waterloo, Canada.

Sahimi, M. and Y.C. Yortsos, 1991. Applications of Fractal Geometry to Porous Media: A Review. SPE 20476, Society of Petroleum Engineers.

van Golf-Racht, T., 1982. Fundementals of Fractured Reservoir Engineering. Elsevier, New York.

Rock Mechanics, Nelson & Laubach (eds) © 1994 Balkema, Rotterdam, ISBN 90 5410 380 8

Rock engineering and civil construction

Arturo A. Bello Maldonado
Mexico City, Mexico

Dear Colleagues of the Rock Mechanics Community :

It is my privilege to be here thanks to the kind invitation from the Organizing Committee of this First North American Rock Mechanics Symposium; I specially want to thank Dr. Priscilla P. Nelson for her communications about the theme of my talk, giving me the opportunity to summarize my personal vision of Rock Mechanics, based on the work I have done during the last 22 years.

I also want to congratulate our colleagues of the U.S.A., the Canadian, the Mexican and the International bodies involved in the Rock Mechanics field who have made a very big effort in organizing this First NARM Symposium during the first year of applicability of the NAFTA and overpassing with much success the scope and covering of the two previous Binational symposia on Rock Mechanics, the first one held in Mexico City in 1983 and the other one in the U.S.A. in 1986.

MY PERSONAL VISION OF ROCK MECHANICS

My first involvement in rock mechanics took place 22 years ago, when the ICA Group, the mexican construction corporation in which I collaborated during the same lapse, was in charge of the construction of the Central Outfall of the Deep Sewage System of Mexico City, a 68 kilometer tunnel being driven through many different geological formations, including hard soils and soft and hard rocks. Many designers and consultants were involved in the job to define the construction procedures and the support systems for the tunnel, which included steel arches, anchor bolts and, for the first time in the mexican civil construction industry, the shotcrete.

The working group I belonged to at that time, was in charge of the geotechnical instrumentation programme established to evaluate the behaviour of the tunnel; very soon we realized that the design methods for the steel arches and for the anchor bolts systems were mostly rules "of thumb" and to the shotcrete, some consultants and suppliers of the required equipment were assigning extraterrestrial characteristics and magic properties. The milestone references for the support systems design were the works by Terzaghi, Obert and Szechy and some considerations started to be given to the ideas and concepts supporting the so called "New Austrian Tunnelling Method". A lot of books and many papers were reviewed trying to find out something that could help us to improve the design of the support systems, but nothing of practical significance was found. In an effort trying to rationalize the design of the support systems making use of anchor bolts and shotcrete, we developed a method for this purpose and published it in the Third International Congress of the ISRM, held at Denver, Colorado, USA, in 1974, (Bello & Serrano, 1974); that method included an adopted and not very well supported hypothesis on the geometry of an arch formed above the opening, but at that time we had nothing better at hand.

The same tunnel, the Central Outfall, faced us with another unsolved problem : that of the evaluation of the quantity of water flowing through the subsoil masses into the tunnel; we measured the highest unit discharge even reported in the technical relevant literature, up to 4,000 liters per second per kilometer, accumulating a total discharge close to 1,700 liters per second to be pumped up, from depths up to 250 meters. No designer or consultant involved in the job was able to predict the future behaviour of the water flow to the tunnel. Making observations of the water inflow, measuring the unit discharges at intervals of time and distance within the tunnel and following the approach dictated by the Scientific Method of Investigation, we developed a mathematical model to predict the "Seepage towards tunnels" and we presented it at the Second International Congress of the IAEG, held at Sao Paulo, Brazil, in 1974, (Bello, 1974).

Those two were our first papers on Rock Mechanics issues and we really believed and hoped, that some retrofeed was to be obtained by presenting them at the international meetings of the technical communities involved in those matters. As far as I know, we never received any comment or suggestion about them.

Later on, we had the opportunity to apply to mining excavations our knowledge on Rock Mechanics and we knew about the problems faced when making really large underground openings : some of them more than 100 meters in width and a few of them, but very significative, with no support at all. This was our chance to finish the development, and prove the validity of a "Simplified Method for the stability analysis of large underground openings", which we presented at the international symposium "Rockstore 77", sponsored by the ISRM and held at Stockholm, Sweden, in 1977, (Bello, 1977). The method keeps the concept of an arch developed above the underground opening, a concept that has been proposed by several authors and has been used extensively, but, in order to avoid an arbitrary definition of the geometry of the arch, a mechanical principle widely used in the Structural Engineering, was applied to determine the geometry of the arch : the Principle of the "Minimum Deformational Work". I do know, and some of my colleagues whom I have had the opportunity to coordinate in our jobs, also do know, that this simplified method allow us to understand, to design and to predict the behaviour of underground excavations, some of them even at the failure stage, and several others that have had stability for dozens of years without support of any kind, like our mexican Grutas de Cacahuamilpa, a natural cavern with more than 100 meters in width.

Through the years we have had to develop simplified models and methods to solve some of the geotechnical problems we have found in our professional practice when we realized that no reasonable or satisfactory solution was available at the time we faced those problems; that was possible because we were involved in jobs big enough to afford the investigation work required to develop those models and methods; in my view, BIG JOBS ARE VERY GOOD UNIVERSITIES : they allow us to develop our knowledge on those matters we feel an improvement is necessary, but of course, we must be lucky to overcome the several types of barriers that are always arised when such a proposal is presented to the people involved in such jobs.

We have been using our simplified models and methods in many different projects : our model for the water flow into tunnels was extended to the case of a finished permeable tunnel and for the case of an aquifer with variable head, when we faced those problems, (Bello, 1983). Our simplified method for stability analysis has been applied to a variety of tunnels and mining excavations, to help in the design of them, (Bello, 1984), or to anticipate the future behaviour of some others, e.g. (Bello, 1985). We know that these simplified models and methods give satisfactory results in the practice of Rock Mechanics applied to civil and mining works if the mechanical properties we must use, are really representative of the rock masses involved and this is another very important aspect in the Rock Mechanics practice : we all know very well that rock masses mechanical behaviour is dictated by the geometric-mechanical properties of one or several of the sets of discontinuities always present in the rock masses and we all also know that the only reasonable way to obtain reliable values of the mechanical properties of rock masses is by means of field tests including a volume of the mass at least of the same order of magnitude of the volume we anticipate will be affected by the works we want to have done on the rock masses. But even to design the field tests or to establish a preliminary design of the project in hand, there is a need to predict in a less expensive and less time consuming way, the approximate values of the mechanical properties of the rock mass to be dealt with; that is why we recently published our simplified methods to get an "Approximate determination of the strength of a rock mass", (Bello & Ayala, 1993) and also an "Approximate determination of the deformability of a rock mass", (Bello and Porras, 1993); both of these simplified methods were developed during several of the last years; they are based on the properties of the rock between the discontinuities of the rock mass and on some hypothesis about the influence of the geometry and nature of the discontinuities of the mass.

SUMMARY AND CONCLUSIONS

I have described what has been, and it is, my vision of Rock Mechanics and, I must say, it is also my vision of other disciplines of the Geomechanical Engineering and of the running of consulting and construction civil engineering firms, which was my responsibility in the last 20 years :

We need simplified models and methods to understand, to design and to predict the behaviour of the rock masses on which we build structures and facilities; not simple, but simplified models and methods, in the same way as they are established and used in other matters of the Civil Engineering : the structural, the hydraulic and even in the management of firms.

We also need an abrupt improvement of the tools and procedures we use, and on the interpretations we made when running field sampling and testing programmes and geomechanical instrumentation :

- Perhaps, new tools based on the cutting capability of water jets at high pressure, could help us to obtain bigger and less disturbed samples from boreholes, without disadvantages in terms of cost. Also, devices of this kind could produce more neatly cutted and again, less disturbed samples for field testing.
- Deformations measurements could be more easily done and at a more con-

venient time if new devices, maybe of the optical type, could be developed and integrated into an automatic data acquisition system appropriate to be installed within tunnels.

We should take more advantage of the back analysis of those case histories involving failure or instability conditions.

In any case, we should prepare ourselves to efficiently make adjustments of our designs and construction procedures, mainly at the early stages of construction, but also, all along the construction time, as it is not possible or even convenient, to try to anticipate every geotechnical condition to be found while making underground works.

We, the rock mechanics community, should compromise to actively participate in the improvement of the present knowledge about the possibility to predict the occurrence of earthquakes and to anticipate the effects they could induce on structures, not only of the Civil Engineering type, but also on the social ones.

I think it is a fortunate situation that our field of work, the Rock Mechanics, like many other disciplines of the human activity, is still open for investigation at the job sites.

I thank you very much for your kind attention. Muchas Gracias.

REFERENCES

Bello M.,A.A. & Serrano L.,F. 1974. Measurements of the behaviour of grouted bolts used as reinforcing elements for the support of underground openings. Proc. 3rd. Int. Congress ISRM, Denver, Colorado, USA. p. IIB-1189. U.S. National Academy of Sciences.

Bello M.,A.A. 1974. Seepage towards tunnels. Proc. 2nd. Int. Congress IAEG, Sao Paulo, Brazil. p. VII,15.1-15.10. Associacao Brasileira de Geologia de Engenharia.

Bello M.,A.A. 1977. Proc. Int. Symp. ISRM, Rockstore 77, Stockholm, Sweden. p. 289-294. Pergamon Press.

Bello M.,A.A. 1983. After construction seepage towards tunnels in variable head aquifers. Proc. 5th. Int. Congress ISRM, Melbourne, Australia. p. B111-B117. Australian Geomechanics Society.

Bello M.,A.A. 1984. Rock mechanics design concepts for post-pillar cut-and-fill underground mines. Proc. Int. Symp. ISRM, Cambridge, England. p.89-96. British Geotechnical Society.

Bello M.,A.A. 1985. Prediction of stresses and subsidence progress for coal mines with continuos miners in a longwall layout. Proc. Int. Symp ISRM, Zacatecas, Mexico. p. 229-244. Sociedad Mexicana de Mecânica de Rocas, A.C.

Bello M.,A.A. & Porras L.,A. 1993. Determinación aproximada de la deformabilidad de una masa rocosa. Proc. IV Reunión Nacional de Mecânica de Rocas, Taxco, Gro. p. 45-61. Sociedad Mexicana de Mecânica de Rocas

Bello M.,A.A. & Ayala M.,R. 1993. Determinación aproximada de la resistencia de una masa rocosa. Proc. IV Reunión Nacional de Mecânica de Rocas, Taxco, Gro. 2nd. volume, in press. Sociedad Mexicana de Mecânica de Rocas, A.C.

Rock Mechanics, Nelson & Laubach (eds) © 1994 Balkema, Rotterdam, ISBN 90 5410 380 8

Competitive at depth: Re-engineering and hardrock mining process

Malcolm Scoble
McGill University, Montreal, Que., Canada

ABSTRACT: Canadian hardrock mines recognize the need to radically improve competitiveness in increasingly deeper reserves. The interaction between mining rock mechanics and the new information and automation technologies is examined as a means to re-engineer the mining process at depth. This will be dependent upon advances in mining methods, geosensing, machine intelligence, and intelligent planning and control systems.

Introduction

This lecture is a personal view of issues that are perceived as key to the advancement of Canadian hardrock mining. In 1990, 62% of Canada's 212 mines were underground hardrock mines. 47 were totally dependent upon blasthole methods and 26 employed a mix of blasthole and other methods, e.g. shrinkage, cut and fill, Hendricks et al (1992). Through the 80's this sector had achieved significant improvements in safety, productivity and costs via workforce rationalization and improved management techniques, in response to recessionary pressures. New mining technology also played an important part: many mines converted to new bulk mining methods. This was facilitated by developments in equipment, such as the ITH drill and the remotely operated LHD. Mining rock mechanics research also evolved and contributed over the period, with a major effort emerging in rockburst-prone ground. In practice, rock mechanics and mining engineering together began to come to terms with strategic planning and tactical ground control for bulk mining at moderate depth. This capability now needs to be extended to deeper deposits and more highly-stressed ground. Exploration success has been limited in recent years; unfortunately the new discoveries have tended to be at depths beyond 1km, such as recent finds in the Sudbury area. Meanwhile, the majority of existing mines approach old age and struggle to extend their underground reserves, invariably by deepening. The challenge then relates to competitiveness in increasingly more difficult mining conditions associated with deeper, though not necessarily richer, reserves. This predicament is faced by most of the established base and precious metal mines in the following areas: Sudbury, Thompson, Flin Flon, Timmins, Kirkland Lake, Hemlo, Red Lake, Yellowknife, Val D'Or, Chibougamau and Bathurst.

In the 90's it became apparent that reduced commitment and success in exploration, together with increased foreign competition had placed Canada's mining future in serious jeopardy. In response, some mines recognized the need to seek competitiveness more urgently. They have observed the recent emergence of information and automation technologies and their potential to re-engineer the mining process. This lecture reviews the status of mining rock mechanics research relating to Canadian hardrock mines and the particular issue of depth. It then examines the progress made to date in automation technology and some current visions of the mine of the future. An attempt is made to evaluate the issues facing the integration of mining rock mechanics with information and automation technologies to re-engineer for competitiveness in deep mining.

Bulk Mining to Moderate Depth.

Mining technology advanced significantly in the last decade as several mines changed to bulk methods, where widths permitted, for aspects of safety, productivity and economics. This change occurred as orebodies were followed to depth, e.g. Campbell Red Lake, Brunswick Mining & Smelting, Inco's Thompson and Sudbury mines, or as new mines, principally at Hemlo, were launched. These have typically tended to encounter ground control problems at moderate depth, approaching say 1km. Dilution and recovery became recognized as important production control parameters, Ashcroft (1991). The product of rock mechanics research and experience continues to be integrated effectively into sound mine design and ground control practice in such mines. A good example would be the William's Mine at Hemlo, Canada's largest gold producer, where once in production.... *as mining progressed, basic design parameters such as stope dimensions, sublevel interval and sequencing were refined to compensate for changing ground conditions. Effective ground control is maintained by using appropriate support systems and backfilling practices, and by controlling blast vibration damage. Computer modelling was calibrated to observed conditions and measured stress levels. It's predictions were combined with accumulated geotechnical data to identify areas where new mining methods would need to be considered...* , Bronkhorst et al (1993).

Cablebolting has tended to dominate the rock support research in Canada, as evidenced at the International Rock Support Symposium in Sudbury in 1992. Publication of a manual and software for support design in hardrock mines recently marked the completion of a major collaborative project at Toronto, Queen's and Laurentian Universities, Hoek et al (1993). New support materials are also beginning to appear, e.g. composite tendons (Khan and Hassani 1993) and polyurethane coatings (Archibald et al 1992.) Paste backfill also has received particular interest for operational and support reasons, Landriault and Lidkea (1993). Research into numerical modelling has become widespread, as evidenced by the support for the 1st Canadian Symposium on Numerical Modelling in Mining, at Montreal in 1993. Some of the resulting proliferation of software has been integrated at minesite as routine design tools, although empirical stope design methodologies continue to be popular, e.g. Potvin et al (1989). Notable in monitoring research has been the continued development of Time Domain Reflectometry, e.g. for monitoring the long term behaviour of pillars, Aston and Charette (1994). Greater recognition of the need to monitor and control blast damage has also seen the more widespread adoption of integrated blast monitoring programs and controlled blasting.

Mining-Induced Seismicity at Depth

Rock mechanics and mining engineering have succeeded together in developing bulk mining technology to moderate depth. This is now being extended to deeper deposits to contend with the hazards of mining-induced seismicity. Rockbursting first became a mining hazard in Canada in the hardrock mines at Kirkland Lake and Sudbury through the 30's as operations began to mature with extensive stope exposures and depths over 1km. Early mining rock mechanics as a rockburst control strategy owed much to Morrison (1942), based primarily on longwall sequencing and advances in support technology. The early 1980's, however, saw the re-emergence of the hazard in several mines, particularly in Ontario. At the Falconbridge Mine in 1984 a series of three rockbursts, each over magnitude 3, caused damage resulting in four fatalities. These rockbursts provided the first clear evidence of fault-slip mechanisms. In the same summer a magnitude 4 event was recorded at the Creighton Mine. A subsequent Ontario government inquiry led to several rockburst research initiatives.

The Canada-Ontario-Industry Rockburst Project was launched in 1985, aiming to develop monitoring systems capable of capturing complete waveforms and to investigate the causes and mechanisms of rockbursts. Field trials were also used to evaluate methods to reduce the magnitude and frequency of rockbursting, Hedley (1992). An expanded second phase project, the Canadian Rockburst Research Program (CRRP), followed in 1990. This set out to consider: the reduction of mining seismicity using stiff backfill; the determination of guidelines in ground support and control; the improvement of mine layout through full waveform analysis; the automation of waveform data analysis; and the improved understanding of fault-slip rockburst mechanisms.

Monitoring, Causes and Mechanisms: Microseismic monitoring has become an accepted tool in seismically-active mines: some twenty ElectroLab MP250 systems have been installed at Canadian mines to provide real time detection and locations of events. CANMET's contribution to CRRP is the study of the larger mining-induced events, i.e. above magnitude 1, Plouffe et al (1993). It has developed and permanently installed remote-controlled macroseismic systems at four mines. CANMET is also responsible for the operation of 8 regional seismograph stations, to determine the magnitude of small mining-induced events and to provide more detailed information about the source properties of the mining-induced seismic activity. Real time quantitative seismology in mines should become a reality with developments such as the integrated seismic system (ISS), Mendecki (1993). This promises not only a research tool to understand the causes and mechanisms of mining-induced seismicity, but also a management tool to monitor changing mine conditions, in association with other rock mechanics or environmental monitoring for more immediate production control decision analysis.

Rockbursts are classified as strain-energy bursts, pillar-type bursts and fault-slip bursts. The latter are generally responsible for the largest and most damaging events, but have proven to be the most intractable problem, Morrison et al (1993). At the Creighton Mine, the most seismically active of the 16 Sudbury mines, the smaller events tend to cluster around the excavations and be promoted in flurries by mining activity. The spatial and temporal distribution of the larger rockbursts, however, could only loosely be correlated with the location of prominent geological structures and the timing of mining activity. This has led to the consideration of possible chaotic behaviour within a mine system. Recognition that not all deep mines in the area are seismically active has prompted study of both seismic and aseismic mines, Morrison (1993).

Modelling and Mine Design: Numerical modelling has begun to play a central role at several mines in attempting to relate mining activities to seismicity, source mechanisms and the effects on excavations. For example, McCreary et al (1993) integrated a full-waveform monitoring system, a source location algorithm, simplified source-mechanism techniques, a consistent event-magnitude estimation method, 3-D boundary-element modelling and underground observations into studies at the Ansil Mine to define rockmass failure criteria. Based upon this work an optimum sill pillar extraction sequence was selected to minimize damage, maximize recovery and facilitate smoother production. Research needs to develop strategic planning to be more effective in identifying stability hazard domains at an early stage within the appraisal of prospective orebodies. 3-D seismic tomographic imaging in conjunction with microseismic monitoring, for example, has been used with some success to delineate brittle, competent and highly-stressed domains, in contrast to weak, destressed domains lacking in seismicity, Maxwell and Young (1992).

Support: A support design rationale for rockburst-prone ground has been proposed, Kaiser (1993). Work is also attempting to predict rockburst damage potential levels on the basis of ground motion and damage data, Jesenak et al (1993). If rockbursts cannot be prevented then it seems sensible to evaluate where and what kind of damage is expected, together with the anticipated failure violence and how rock support systems could contain such rockburst damage. McCreath and Kaiser (1992) reported preliminary guidelines for support levels. Backfill continues to receive increasing recognition as an important mine support member, e.g. stiff rockfill, paste or high density backfill.

Mine Planning and Practice: Mining methods require to be developed which minimize rockbursting, whilst mining practice, including support, requires to be flexible and responsive to changing ground conditions. Landriault and Oliver (1992) describe a technique developed at the Creighton Mine to create a destress slot which allows bulk mining to progress in the face of rockbursting problems at depth, provided the proper sequencing of panels allows the relaxation zone to be passively extended. Field monitoring was key to evaluating the performance of the technique, comprising microseismic monitoring, borehole camera logging, stress cells, rod extensometers and inclinometer casing. The cost of redrilling, reconditioning and production delays are hazards in addition to safety risks when mining in rockburst-prone ground. The increased efforts in developing hazard assessment techniques for mine design and ground control should assist the integration of rock mechanics into automation development. Brummer et al (1993), for example, describe a Risk-Cost-Benefit Analysis which is likely to improve planning procedures. It uses support costs,

costs associated with the failure of the support or loss of the excavation, and benefits achieved by installing rockburst-resistant support.

Towards Remote and Automated Mining

The mid-80's saw some Canadian companies, principally Inco, Noranda and Falconbridge, developing interests in hardrock mining automation research. The Canadian Centre for Automation and Robotics in Mining (CCARM) in Montreal was formed largely as a result of industry initiatives. Early work was in machine control, vehicle guidance and communications, reflected in the five Canadian Symposia on Mining Automation held upto 1994. The most significant advances have been in underground communications systems, Baiden and Scoble (1991), Saindon (1993). These are the basis for machine control and mine-wide monitoring, e.g. for environment, blasting, rock mechanics, materials handling, and shaft management. In 1975 Inco tested the first radio remote controls for the operation of LHDs in large open stopes. By 1991 it had developed a high capacity telecommunications system, based upon broadband technology, capable of moving voice, data and video information throughout the mine. In 1993 the first production application of the teleoperation of multiple LHDs from surface was demonstrated, Baiden and Henderson (1994). Two LHDs, at the 2000 and 3000 levels of the Copper Cliff North Mine, were operated simultaneously from surface by one miner who controlled the load and dump functions directly, but allowed tramming under automatic guidance.

What are the current visions of the hardrock mine of the future? In 1992 a research team from Inco-Falconbridge-CCARM completed a feasibility study for PRECARN Associates Inc. on hardrock mining automation. This proposed the Mining Automation Plan, a R&D programme at around Can$20million over 5 years. This aimed to be precompetitive research to facilitate longer term automation, principally in the following: geosensing, data modelling and fusion, machine monitoring and control, intelligent mine planning and production control, social and implementation issues, Baiden et al (1993). The Finnish mining industry announced in 1992 its Intelligent Mine Technology Programme, towards real-time process-like control of an automated mine, Seppanen and Pukkila (1993). US$12million of R&D projects over 5years were proposed in the following: real-time resources management and production control; a mine-wide information network; new machines and automation; and production and maintenance automation. A recent South African view saw a combination of several innovative technologies in future deep reef mining: concentrated mining, impact fragmentation, hydro-power, continuous scraping, ground penetrating radar, radiowave tomography, gold analysis, grade prediction depositional simulation software, Geographic Information System for real time 3-D planning and monitoring systems, rockburst control and a portable seismic system, backfilling, and hoisting, Langbridge and Wills (1992). There was a marked absence of reference to the teleoperation or automation of mining machines. A study of Australian industry R&D recommendations for innovative underground metal mining technology identified two important objectives: minimization of ore loss and dilution; and continuous mining systems for selective mining and narrow orebodies, Scott et al (1992). Another vision saw the Canadian hardrock mine set in the year 2050 in a remote, northern location. The mining operation exploited mechanical fragmentation and underground ore sorting and milling. The underground machines were operated via a satellite link from Sudbury, no miners worked underground, Ashcroft (1993).

Automation is likely to be achieved incrementally in the following sequence:

- in the short term, remote mining will be implemented. This will likely emulate the existing bulk mining methods with conventional machines. The technology will probably to be prototyped in shallow, less demanding rockmasses, but it will have immediate appeal to reduce the hazards of rockbursting at depth. There would still be a dependence upon human intervention underground, which would impede any radical change in mining method. This will likely represent an important and lengthy learning period, leading to the resolution of many issues relating to the eventual form of the automated mine. It will be important to develop ergonomics, the study and design of human interaction with machines.
- the second phase will probably see the next generation of machines designed to fit a new mining method and environment This could be the system which is first designed to contend with depth; design options for example could include smaller excavations, faster and less development, less reliance on ground stability, improved sequencing of excavation and backfilling, reduced rockmass damage, etc. This could be where we first see non-explosive rock fragmentation, perhaps within a continuous mining system. We will still see remote control but minimal human intervention and some *islands of automation* in this transition phase, e.g. automated haulage levels, automated production drilling and mucking systems. This might be a more challenging scenario to manage than that of full non-entry automation.
- the ultimate, long term goal would be a fully integrated, automated underground complex: but automation is a continuum from completely manual to autonomous, with varying, intermediate degrees of autonomy. These require sensors and actuators similar to an automated system but without the same sophistication of control software. Partial automation would not be so difficult to achieve selectively. Telerobotics would be an option; this is concerned with the human control of semi-automatic robots and other systems from a distance. Supervisory control would involve the miner communicating goals and constraints to a computer, which in turn, through its own sensors and actuators controls the system. It's measurements of the system state would be relayed to the miner to assist in planning, programming, monitoring, failure detection and learning from experience.

Full non-entry mining would be the most difficult to accomplish, requiring automation of all infrastructure systems, such as maintenance, in addition to automatic failure recovery systems. It would, however, provide an opportunity to radically change the mining method, machinery and ground control philosophy. It would also avoid the safety hazards associated with miners working in the proximity of remotely operated or autonomous machines. Otherwise, the mine design would need to account for occasional human intervention, with the necessary extra infrastructure, constraints and costs. The automated mine could employ radically different processes for exploration, ore delineation, fragmentation, transport and processing. Alternatively, costs could be saved by automating the most repetitive tasks and teleoperating those more difficult to automate, e.g. an automated system might turn control over to a teleoperator when handling oversize in drawpoint mucking or loading explosives.

What will be the benefits and costs of removing miners from underground? One basic incentive has been for the reduction of variable costs, primarily labour and energy, provided that capital expense is not excessive. In studying the economics of underground LHD teleoperation, Baiden (1993) demonstrated significant cost reduction arising from productivity, work quality and throughput time improvement. He drew comparisons with factory automation, comparing underground machine control with Computer Numerically Controlled manufacturing. Ground control costs may be saved if the machines require less protection or induce less rockmass perturbation. Where the environment is too hazardous for any human entry, for example in severe ground conditions, then the only alternative may be a non-entry automated system, provided it is adequately robust.

Interaction: Mining Rock Mechanics, Information and Automation Technologies

Ground control will continue to be an issue in the planning and control of remote and automated mining. Manless mining removes a responsibility for human security, which could present the opportunity for radical change to mine design. Perhaps automation technology can assist in extending the limits of mining into more difficult ground conditions, and particularly to depth in rockburst-prone ground. This may be so if the new methods and machines induce less seismicity and associated damage, or contain damage more effectively; or require less protection. Mining rock mechanics, as well as economics, productivity and quality control issues will govern the feasibility of mine automation technology. Mining rock mechanics will contribute to automation through its contribution to new mining methods and machine design, machine intelligence, and mine planning and production control.

Mining Methods and Machine Design The mining method comprises the size, shape and sequence of excavations; the technique for rock fragmentation, handling and transport, and support. Rock mechanics contribution to method and machine design accounts for geomechanical, as well as production and economic criteria. The mining method and machine design are interdependent. A Canadian industry consortium, HDRK Ltd., have invested Can$27million on 25 R&D projects over 9 years. 70% was devoted to mechanical fragmentation towards that elusive, continuous hardrock mining system, Mathews and Sarin (1993). Other efforts are investigating novel, non-explosive fragmentation such as plasma blasting, Nantel and Kitzinger (1990). These machines should facilitate new mining methods and also be amenable to automation, but their feasibility is still as yet unproven.

Only through a full knowledge of the rockmass response to mining can methods and machine systems be designed to accommodate variations in the geomechanical environment. Ground response will likely govern the remote and automated mining rationale, raising such issues as.... □ should we mine aggressively or conservatively? □ how much can we perturb the host rockmass if machine security replaces human security as the prime concern? □ what types of ground control hazard do we need to protect machines against? □ how do we maintain, protect, recover and repair machines? □ should stopes be concentrated or dispersed? □ mined with large or small machine units, i.e. few or many machine units? □ should extraction rates for individual stopes be high or low? □ is continuity of excavation important? □ what are the criteria governing the distribution of local or regional pillars and the degree of extraction? □ what is the role of backfill in supporting the excavation sequence? □ can we mine differently so that stability is ensured or damage is effectively contained? ...or □ must we find new, radically different methods to contend with ultra-deep environments, e.g. in situ leaching?

Machine design relates primarily to size, shape, capacity, cost, flexibility, manoeuvrability, versatility, power source, sensitivity to ground conditions and work functionality. The trend towards mechanization and higher technology in mining has resulted in steep rises in machine and systems capital investment, which emphasizes a much greater sensitivity to machine breakdown or downtime due to environmental factors, such as lack of accessibility due to ground instability. Lewis(1991) reported on HDRK Ltd. underground trials with a new compact fullface, hardrock tunnel borer, 2.44m diameter. The machine development was suspended, partly because of its inability to deal with poor ground conditions. It had not been designed for installing support without interrupting boring. This is a good example of problems that can arise if the interaction between ground conditions and a machine are underestimated. Attention needs to be given to machine design for reliability, protection from ground instability and shock, and also emergency recovery.

Machine Intelligence Automation is seen to bring the machine into more intimate contact with the rockmass. The ultimate generation of robotic mining machines will need to be able to sense, understand and relate to the rockmass in order to develop autonomy, which implies decision-making capability and planning. We will have to transfer our understanding of rockmass characteristics and behaviour into supervisory systems and eventually machine intelligence for machine autonomy and control. The characteristics of machine intelligence relate to sensory perception, information processing and analysis, communications and work functions, as follows:

□ Sensory perception stems from information gathering and diagnosis:
  ▫▫ machine status... health, working efficiency, security, position
  ▫▫ work performance ...productivity, product quality
  ▫▫ environment ...ground and atmospheric conditions
  Sensing derives from a wide range of technology, for example:
  ▫▫ video, laser, radar, thermal imaging to perceive the environment.
  ▫▫ mechanical, thermal, nuclear, optical, chemical, and electrical sensing to define geological and mineralogical parameters.
  ▫▫ gyros, accelerometers, ultrasonic-laser ranging for positioning-navigation.

□ Information processing and analysis, for example, relates to the ability to :
  ▫▫ characterize ground conditions ▫▫ predict ground response to machine actions
  ▫▫ define subsequent actions from production instructions and machine expertise
  ▫▫ decide support requirements and timing
  ▫▫ plan and control tasks .. work functions .. transport .. navigation and guidance

□ Communication with a controller, e.g. for:
  ▫▫ mine-wide sensor data ▫▫ updating mine and orebody models
  ▫▫ instructions relating to tasks ▫▫ feedback for control

□ Work functions are varied and potentially include:
  ▫▫ perception and navigation ▫▫ move - propel ▫▫ survey and map
  ▫▫ maintain accessibility, e.g. clear route of debris, scale loose ▫▫ install support
  ▫▫ prepare/repair roadbed surface ▫▫ install/maintain infrastructure and sensors

▫▫ drill for exploration, delineation, production, or infrastructure
▫▫ fragmentation ..mechanical or chemical rock breakage ▫▫ load explosives
▫▫ load rock ..control trajectory and digging forces ▫▫ handle oversize
▫▫ transport ... rock, consumables ▫▫ machine maintenance, recovery
▫▫ backfill ..preparation, transport, placement, quality control

Automation of these functions will also be affected by other technology developments, for example: lightweight but capable new materials such as polyurethane or shotcrete lining, composite tendons and paste fill would be more amenable to automated placement and manipulation. Machines can be specific or universal in their functional design. One measure of the LHD's success, for example, relates to its versatile functionality.

Intelligent Planning and Production Control Information technology offers software and hardware developments with the ability to develop comprehensive, integrated information systems that link mine and orebody models with real time operational data, enhanced graphics and engineering tools, and business systems (e.g. accounting, maintenance, inventory, personnel). Advances in mine-wide communication systems have been seen to be able to link underground machines to such information systems for machine monitoring and process control, as well as mine-wide monitoring.

The mine planning process is centred on an orebody model, which is used to resolve the shape, size and sequence of stope and pillar extraction to generate a schedule of tonnage and grade of product with time: i.e. the mine plan. The optimum schedule might be assumed to give the maximum amount of product metal in the shortest time, maximizing the Net Present Value. Complicating constraints, however, include market conditions, the design capacity of the mining method and machine system, and the potential ground response to the plan. The orebody model would be an object-oriented database, established and maintained from exploration and delineation data, potentially from the fusion and interpretation of data from excavation mapping and surveying (e.g. video image analysis, radio wave imaging, seismic reflection, ground probing radar), crosshole tomography, borehole logging (e.g. density, magnetic, susceptibility, gamma, dilatometer logs), and traditional core drilling and assaying. Production blastholes should also provide data to maintain these models by undertaking geophysical logging, surveying, chippings analysis, and ground characterization from pattern recognition of drilling performance parameters, i.e. Monitoring-While-Drilling, Peck et al (1989). Analysis of the models by geostatistics or other techniques provides the ability to define the ore reserve: an estimation of cut-off-grade and the reserve tonnage and grade.

Subcomponents of the orebody model will be the geological and geomechanical models. The establishment of mine-wide communications, machine monitoring and geosensing now promises to improve the precision and detail of the ore and geomechanical characterization to significantly improve the planning process. The mine model would comprise the characteristics of the excavations created to date, plus the mine plan, i.e. the detailed plan in terms of the derived sequence of future excavation geometries over time, with grade, tonnage and cost estimates. This represents the base reference for the production control process. An integrated planning system would facilitate automated and frequent simulation of the performance of mining plans when conditions are observed by the mine-wide monitoring system to change with time. This would involve numerical modelling to evaluate ground response, as well as economic, grade and tonnage criteria. Planning could become more responsive to changing circumstances, perhaps emulating *just-in-time* manufacturing by controlling in situ and broken ore inventories, throughput time, and blending.

Mine production control has long suffered from a lack of timely, precise, reliable and pertinent data. Control could be practiced in real time by matching monitored production performance data against mine plan objectives. This should evaluate the mining process efficiency and the mine environmental management quality:

- □ Mining Process Efficiency relates to machine systems activity,in terms of:
  - ▫▫ volumes and weights (drilled, blasted, stored, or hauled), and throughput time
  - ▫▫ grades and impurities (drilled, broken, in transit or stored), dilution, recovery
  - ▫▫ costs, by machine, workplace and task, production and maintenance
  - ▫▫ backfill distribution ..quantities and quality ▫▫ water balance
  - ▫▫ machine performance, health, availability, reliability,utilization
  - ▫▫ fragmentation size-shape distribution and blast design quality
  - ▫▫ rework, rehabilitation downtime ▫▫ energy consumption, costs
- □ Environmental Management Quality relates to :
  - ▫▫ atmospheric control, e.g ... ventilation flow quantities and quality (dust, pollutants, temperature, humidity) ... waste disposal hazard levels
  - ▫▫ ground control, e.g ... ground response (seismicity, instability, deformation, stress), ...support levels, performance and costs ...machine security ....rockmass damage, loose, overbreak, blast vibration

A new generation of production sensors will be required for the rockmass and fragmented rock. These will require to be economic, reliable, easy to install and robust to survive underground conditions, (including limited access and space, hot-humid-corrosive atmospheres, water, rockfall, blasting vibration and flyrock). Priority sensors for development would include the following:

- □ shape and integrity of all excavations, regardless of accessibility; requiring remote surveying in poorly accessible cavities such as stopes, ore passes.
- □ tonnage, grade and throughput time of broken muck resident in stopes, drawpoints,orepasses, bins, vehicle buckets and bodies.
- □ status of integrity and capacity of ore passes and storage
- □ recognition of geology in fragmentation and excavation faces
- □ rockmass damage and loose rock sensing, Jowsey et al (1993).

Conclusion

Underground communication technology now permits consideration of the potential for integrated information systems for mine planning and production control. It also makes remote mining from surface feasible in the short term. These advances offer significant improvements in safety, productivity and economics, together with the potential to reduce the risks associated with deep mining. Development of remote mining will provide a period for research and accumulation of experience to prepare for the subsequent phase of telerobotics and eventual automation. This will require the development of improved geosensing, new mining methods and intelligent machines. Rock engineering will be part of the multidisciplinary effort required to ensure that remote and automated mining at depth is not only practical but also efficient and economic. The opportunity exists for the Canadian industry to become pre-eminent in the automation of deep hardrock mining.

References

Archibald, J.F., Mercer, R.A. and P. Lausch 1992. The Evaluation of Thin Polyurethane Coatings as an Effective Means of Ground Control. *Proc. Int. Symp. on Rock Support, Sudbury,* Rotterdam: Balkema, 105-115.

Ashcroft, J.W. 1991. Rock Dilution - an Opportunity for Total Quality Improvement. 93rd Ann. Gen. Meeting, Can. Inst. Min. Metall., Vancouver.

Ashcroft, J.W. 1993. Mining into the 21st Century in North America, Plenary Address, Int. Congress on Mine Design, Queen's University.

Aston, T and F. Charette 1994. Installation and Monitoring of Three Abandoned Mine Crown Pillar Sites, Ontario. CANMET Div. Report MRL: 93-020(CL).

Baiden, G.R. and E. Henderson, 1994. LHD Teleoperation and Guidance: Proven Productivity Improvement Tools. 96th Ann. Gen. Meeting, Can. Inst. Min. Metall., Toronto.

Baiden, G. and M. Scoble 1993. Mine-Wide Information System Development at Inco Ltd. *Proc. 1st Int. Symp. Mine Mech. Automation.* Golden: Colorado School of Mines, 3, 1-14.

Baiden, G., Scoble, M. and S. Flewelling 1993. Robotics Systems Development for Mining Automation. *Bull. Can. Inst. Min. Metall.*, 86, 972, 75-77.

Bronkhorst, D., Rheault, J. and G.M. Ley 1993. Geotechnical Principles Governing Mine Design at the Williams Mine. *Proc. Int. Cong. Mine Design, Kingston.* Rotterdam: Balkema, 433-442.

Brummer, R.K., Nose, J-P., Hay, L. and P.K. Kaiser, 1993. Risk-Cost-Benefit Analysis Applied to the Design of Support in Burst-Prone Mines. *Proc. Int. Cong. on Mine Design, Queen's University* . Rotterdam: Balkema, 969-980.

Hoek, E., Kaiser, P.K. and W. F. Bawden 1993. *Design of Support for Underground Hardrock Mines.* Mining Research Directorate, Sudbury, p.227.

Hedley, D.G.F. 1992. *Rockburst Handbook for Ontario Hardrock Mines.* CANMET Special Report SP 92-1E, p. 305.

Hendricks, C., Scoble, M.J., Boudreault, F. and J. Szymanski 1992. Blasthole Stoping: Drilling Accuracy and Measurement.Trans. Inst. Min. Metall., London, 101, A173-186.

Jesenak, P., Kaiser, P.K. and R.K. Brummer 1993. Rockburst damage potential assessment - an update. Rotterdam: Balkema, *Proc. 3rd Int. Symp. Rockbursts and Seismicity in Mines, Kingston.* 81-85.

Jowsey, M., Yu, T., Croxall, J., Pettigrew, M. and N. Fisher 1991. Development of a Calibration Technique for HDRK's Loose Detection System. 93rd. Ann. Gen. Meeting, Can. Inst. Min. Metall., Vancouver.

Kaiser, P.K. 1993. Support of tunnels in burst-prone ground - towards a rational design methodology.. *Proc. 3rd Int. Symp. Rockbursts and Seismicity in Mines, Kingston.* Rotterdam: Balkema, 13-27.

Khan U. and F.P. Hassani 1993. Analysis of a New Flexible and Rigid Composite

Tendons for Mining, *Proc. Int. Cong. on Mine Design, Queen's University*. Rotterdam: Balkema, 1033-1043.

Landriault D. and W. Lidkea 1993. Paste Fill and High Density Backfill, Proc. Int. Cong. on Mine Design, Queen's University . Rotterdam: Balkema,111-118 .

Landriault, D. and P.H. Oliver 1992. The Destress Slot Concept for Bulk Mining at Depth.*Proc. Int. Symp. on Rock Support, Sudbury,* Rotterdam: Balkema, 211-217.

Langbridge, C. and P. Willis 1992. The Mine of the Future. *Mining Survey*, COMRO, Johannesburg, 2, 19-25

Lewis, M 1991. Compact Underground Borer. *Proc. 1st Int. Symp. Mine Mech. Automation*. Golden: Colorado School of Mines, 4, 1-13.

Mathews, K. and D. K. Sarin 1993. Progress on Futuristic Mining Technologies. 95th Ann. Gen. Meeting, Can. Inst. Min. Metall., Calgary.

Maxwell, S.C. and R.P. Young 1992. 3D Seismic Velocity Imaging Using Microseismic Systems at Strathcona Mine and Mines Gaspe. 94th Ann. Gen. Meeting, Can. Inst. Min. Metall., Montreal.

Mendecki, A.J. 1993. Real Time Quantitative Seismology in Mines. *Proc. 3rd Int. Symp. Rockbursts and Seismicity in Mines, Kingston*. Rotterdam: Balkema, 287-295.

Morrison, D.M., Swan, G. and C.H. Scholz, 1993. Chaotic Behaviour and Mining - Induced Seismicity. *Proc. 3rd Int. Symp. Rockbursts and Seismicity in Mines, Kingston*. Rotterdam: Balkema, 233-237.

Morrison, D.M. 1993. Seismicity in Sudbury Area Mines.*Proc. 3rd Int Symp on Rockbursts and Seismicity in Mines, Kingston.* Rotterdam: Balkema, 379-382.

Morrison R.G.K. 1942. Report on the Rockburst Situation in Ontario Mines. *Bull. Can. Inst. Min. Metall.*, 45, 225-272

McCreary, R.G., Grant, D. and V. Falmagne 1993. Source Mechanism, Three-Dimensional Boundary-element Modelling, and Underground Observations at Ansil Mine. *Proc. 3rd Int Symp on Rockbursts and Seismicity in Mines, Kingston*. Rotterdam: Balkema, 227-232.

McCreath, D.R and P.K. Kaiser 1992. Evaluation of Current Support Practices in Burst-Prone Ground and Preliminary Guidelines for Canadian Hardrock Mines. *Proc. Int. Symp. on Rock Support, Sudbury,* Rotterdam: Balkema, 611-619.

Nantel, J. and F. Kitzinger 1990. Plasma Blasting Techniques. *Proc. 3rd Int. Symp. Rock Fragmentation, Brisbane,* 79-90.

Peck, J., Scoble, M. and D. Payton 1989. Drilling Intelligence and Automation. *Jnl. Min. Res. Eng.,* London: Taylor and Francis, 2, 2, 109-122.

Plouffe,M., Mottahed, P., Lebel , D and M. Cote 1993. Monitoring of Large Mining-Induced Seismic Events - CANMET/MRL's Contribution. *Proc. 3rd Int Symp on Rockbursts and Seismicity in Mines, Kingston* Rotterdam: Balkema, 399-403.

Potvin, Y. Hudyma, M.R. and H.D.S. Miller 1989. Design Guidelines for Open Stope Design. *Bull. Can. Inst. Min. Metall.*, 82, 926, 53-62.

Saindon, J.P. 1993. Integrated Mine Telecommunications: a Realistic and Innovative Approach. 95th Ann. Gen. Meeting, Can. Inst. Min. Metall., Calgary.

Scott, A., Gurgenci, H. and M. Slifirski 1992. Report on the Review of Underground Metalliferous Mining and Equipment Priorities, Council Min. Tech. Equ., Austr. Min. Ind. Res. Ass.

Seppanen, P. and J. Pukkila 1993. Intelligent Mine Technology Programme. *Proc. 2nd Int. Symp. on Mine Mech. and Automation, Lulea, Sweden.* Rotterdam: Balkema, 79-82.

Rock Mechanics, Nelson & Laubach (eds) © 1994 Balkema, Rotterdam, ISBN 90 5410 380 8

Structuring of rock engineering in South Africa and critical review of developments up to 1993

J. Nielen van der Merwe
Sasol Coal, Johannesburg, South Africa

ABSTRACT: The structuring of rock engineering in South Africa is briefly outlined. The mining industry is dominated by large groups and rock engineering input is provided by in-house specialists rather than external consultants. Safety related research is predominantly concentrated in a single organisation, funded by an accident related levy and controlled by a committee system with strong industry representation. The South African National Group on Rock Mechanics utilises regional branches to serve specialist needs of its members.

Despite significant advances in rock mechanics and greatly enhanced understanding of rock behaviour, mining methods have not undergone meaningful developments in the last seventy years. However, latest developments are very promising and the stage is set for dramatic breakthroughs.

1 INTRODUCTION

Rock engineering has been an integral part of the act of mining since mining was first practised by man. The fact that it was only recognised as a science in the 1960's certainly does not mean that it only started then. Even fairly advanced rock stabilization techniques have been around for centuries - Aqricola, for instance, mentioned backfilling in his celebrated Re de Metallica. While it is one of the youngest sciences in terms of official recognition, it is one of the oldest in terms of existence and application.

Even today, it is often still referred to as an art rather than a science. Art and science, however, are not necessarily mutually exclusive or at opposite ends of some or other scale. One can be coldly scientific in your approach to art. J.S. Bach's music is not very far from attaching sound waves of specified wavelength and amplitude to a rigidly prescribed sequence.

Popularly, though, we apply the classification to distinguish between those aspects of rock behaviour to which we either can or cannot attach the peculiar form of internationally understood short hand called mathematics. The moment we can attach numbers and symbols to any phenomenon we change it's status from "art" to "science".

Furthermore, rock being an imperfect material given to significant variation in properties, the outcome of rock mechanics predictions are given to at least some variation. This very often leads to a credibility gap, further enhancing the perception that there is a serious proportion of art left in the science.

The practitioner responsible for the application of rock mechanics in the mining industry is the human bridge between the science and the rock environment. Rock cannot be controlled directly - it's behaviour can only be influenced by the form, magnitude and timing of openings created therein and the provision of artificial supports of some or other description. In general, these parameters are not directly controlled by the rock engineering practitioner, but by other people.

The mining industry world wide is characterised by conservatism, and in several instances by an authoritarian management style. Sound knowledge of the science and the characteristics of rock in a given locality are relegated to background requirements of a rock mechanics specialist. With those requirements in place however well, he will still fail if the lacks the ability to convince people of his point of view.

It is not the purpose of this paper to expand on the aspect of human interaction, but it has to be mentioned to place this vital requirement on record. Rather, in the following paragraphs the organisational aspects of rock mechanics in South Africa will be explained, followed by the author's personal views on the progress made to date and where attention is required, not restricted to South Africa.

In this paper, the application of rock mechanics to real life situations is referred to as "rock engineering". By way of distinction the classical study of rock behaviour is referred to as "rock mechanics".

2 STRUCTURING OF ROCK ENGINEERING IN SOUTH AFRICA

2.1 Characteristics of the South African mining industry

The mining industry in South Africa is dominated by less than ten major mining groups, each owning a number of very large mines. Within these groups, there are generally divisions (or seperate companies) for different commodities like gold, coal, diamonds and base minerals.

Historically, mines in South Africa have been sheltered from the cyclic nature of mining which has been apparent in the rest of the world for a long time. In the past, the world bought whatever gold the mines produced at favourable and stable prices. Coal mines predominantly produced on long term contracts to local consumers like electricity generators, oil-from-coal plants and the steel industry. Government regulations restricted coal exports to a large extent.

Recently, however, the gold price not only lost it's stable nature, but dropped to levels not experienced in the past. New coal fired electricity plants are not planned for the immediate future and government restrictions on coal exports have been relaxed. More companies are thus entering the export arena, with its fluctuating nature.

The mining industry as a whole is not as profitable as it used to be, and also has to come to terms with price fluctuations it has not had to deal with before. This has also impacted on rock engineering. Previously, safety was the main thrust, but recently the development of more economical mining methods (not sacrificing safety) has become an important aspect.

2.2 Development of rock engineering

Today, it is not yet a direct statutory requirement to employ a rock engineering specialist on any mine in South Africa. However, there is a statutory requirement on mine managers to ensure safe and healthy working environments, and the appointment of a rock engineering specialist may be seen as a method of conforming to that requirement. In general, most major gold mines have their own rock engineering departments, or have the services of such a department centrally employed by the group available to them.

This development started in the late 1960's when researchers at the Chamber of Mines (a non statutory private member organization) and the then government funded Council for Scientific and Industrial Research (CSIR) demonstrated the value and potential benefit of the application of rock mechanics to mining. True to its group oriented nature, the mining houses employed their own specialists in house.

These predominantly self taught pioneers rapidly proved their worth, and as their services grew in demand, their numbers increased. Group departments developed, followed by single practitioners on specific mines, from which developed fully staffed departments on individual mines.

Being shallower, coal and base metal mines did not feel the need for formal rock engineering until later. Base metal mines are mostly situated in hard rock igneous environments employing mining methods (classical narrow reef stoping) similar to gold mines and are served by gold mine trained specialists with relative ease. By contrast, coal mines use vastly different mining methods (longwalling, pillar extraction) in relatively soft and weak environments at shallow depth (50-250m).

Since the early 1980's, coal mining rock engineering has emerged as a seperate class in the discipline. Base metal mining appears to be following the same route, and a need for additional training in massive mining rock engineering has also been identified.

The provision of rock engineering services in South African mining is thus predominantly based on in house departments with a strong tendency for specialisation according to the mining activity. External consultants do not play a significant role in the overall scene.

The activities of these departments vary according to the culture of their mining houses and the basic mining activity. In general the departments on gold mines deliver a more integrated service. This may include advice at the planning and operational levels, the physical installation of non-standard supports such as wiremeshing and control over back filling operations and rapid yielding hydraulic props. Gold mines employ some 300 practitioners at different levels of formal education and training.

By contrast, coal and base metal mines together employ less than 30 practitioners. They tend to operate more at the planning and design levels and are in general at higher levels of formal education.

2.3 Rock Engineering Training

The aspect of rock engineering training is a thorny issue, having been hotly debated over the last few years. The sad fact is that there is no undergraduate university degree in rock engineering available. Mining engineering is taught at only two universities (Witwatersrand and Pretoria) and neither has a chair in rock mechanics. Both universities offer rock mechanics modules in their mining engineering courses.

The mining houses have internal training courses, upon the successful completion of which candidates qualify for in house Elementary Rock Engineering Certificates. Practitioners are then eligible to sit for the Chamber of Mines Rock Engineering Certificate, offered by the Chamber of Mines of South Africa. The examination basically consists of three parts: basic generic rock mechanics, specialist (hard rock or coal) rock engineering and a specialist practical examination underground. All practitioners, irrespective of their level of formal education, are required to obtain the Chamber Certificate before being allowed to certain levels in the organisations of their mining groups. This rule is voluntary, enforced by mutual agreement in the industry.

The highest qualification is available to university graduates only. The Advanced Certificate in Rock Engineering is obtained by successfully completing the Advanced Rock Engineering Course at the University of the Witwatersrand. It may be converted into a master's degree in engineering by successfully submitting a research project report.

While it is often felt in South Africa that more training opportunities should be made available, the current system satisfies the needs of the mining houses at lower levels while ensuring a uniform and high standard at the upper levels.

A need which has been identified as requiring urgent attention is the provision of basic rock engineering knowledge to production personnel, especially at the lower levels. This will, initially at least, be dealt with in house by the different mining groups.

2.4 Role of the South African National Group on Rock Mechanics (SANGORM)

Being affiliated to the ISRM, SANGORM shares the ideals and goals of the ISRM. The Group was formed in the early 1970's under the chairmanship of Dr Dick Bieniawski. Up until the early 1980's it was a monolithic group.

Then, the coal mining specialists felt the need for a specialist discussion group. In order to prevent fragmentation of the fraternity they formed a specialist branch within SANGORM rather than form a seperate society.

The idea proved to work well, and others soon followed. The majority of the mining areas are now served by the five regional branches of SANGORM. Each branch organises its own technical discussions, visits, etc, and is represented on the Executive Committee of SANGORM.

Over the last decade SANGORM has fallen into the habit of presenting annual symposia. The coming of age of the branches was marked by them being given the opportunity of arranging the national symposium in turn. The Executive Committee, in addition to coordinating branch activities, continues to cater for generic issues and those aspects which are not dealt with by the branches.

3. ROCK ENGINEERING RESEARCH IN SOUTH AFRICA

3.1 Structure of research

South Africa has been extremely fortunate in the quality of researchers who are and have been active at the Chamber of Mines Research Organisation and the CSIR. While most favoured moving on to new challenges once they have made their mark, their heritage remains, and fresh insights have always been provided by new arrivals. The success of the research effort was probably due to the well known magnitude of the rockburst problem which tended to attract only the right people : those who had both the mental ability and the confidence to attack the problem.

Rock engineering research has until recently been predominantly voluntarily funded by the Chamber of Mines members. This ensured both the cohesiveness and the focus of the research effort by a system of industry control over the funds.

The funding mechanism recently changed by the implementation of an accident frequency based levy on all mines for safety related research. Safety related research is now controlled by SIMRAC (Safety In Mines Research Advisory Committee) under the chairmanship of the Government Mining Engineer, but still consisting predominantly of industry representatives. Other research is done on a contract basis.

The Chamber of Mines Research Organisation no longer exists. The facilities have been taken over the CSIR, and with one or two exceptions the researchers elected to also join the CSIR in their new Division called Mining Technology (Miningtek). The vast majority of SIMRAC projects and funds have been allocated to Miningtek. So while the organisational structure has a new appearance, the expertise remained and essentially the same research continues to be done by the same people.

3.2 Research results

3.2.1 Coal

In coal mining, the recurrence of a disaster like the Coalbrook collapse in 1960 was prevented by the development of the widely used Salamon and Munro (1967) formula for predicting pillar strength. The formula has been used now for almost 30 years, during which period over a million pillars have been created. The formula has since been modified by Madden (1986) to cater for squat pillars.

Compared to gold mining, the coal mining strata control challenge is easier to handle. Coal mining is much more mechanized than gold, and less people are exposed to potentially hazardous situations. Also, being shallower and in softer rock environments, the rockburst hazard is not present.

Rock engineering practitioners in coal mining are thus faced with a more positive challenge, which is to use their skills to develop ever more economical safe mining methods, such as crush pillar stooping, van der Merwe and Olivier, (1992), and flexiwall mining, Claeys (1988).

However, over the last three years this rosy picture has been tarnished somewhat by the realization that the constant for coal material strength in the Salamon and Munro pillar strength formula was not constant, van der Merwe (1993). An observation by Salamon (1969), which stated that if more data had been available different strengths for different coal seams in different areas would be found, was proven to be correct.

In at least one mining district, the coal was shown to be 40% weaker than the average constant derived by Salamon and Munro.

The consequence of this was that large pillared areas which had been thought to be stable, are now known to be at risk and so are the overlying structures on the surface. While measures like pillar wrapping and back filling are being employed to stabilize risk areas, van der Merwe (1993b), coal pillar strengths are being re-evaluated on a national scale, now making use of the greatly increased data base since Salamon and Munro's work in the 1960's.

3.2.2 Gold

The attention in this section will be focused on the rockburst problem - in the words of WD Ortlepp, as complex a problem as mankind has ever had to contend with. The attack on the problem is based on three inter related broad fronts : supplying broad guidelines for mining methods and techniques, developing effective support methods and developing a fundamental understanding of the mechanism of seismicity. From the latter aspect follows also the development of analysis tools for the rock engineering specialist and special measures to be implemented underground.

a. Development of broad guidelines

Guidelines have to date been formally provided on six occasions : four commissions chaired by the Government Mining Engineer in 1908, 1915, 1924 and 1964 and two publications by the Chamber of Mines in 1977 and 1988.

The 1908 commission concluded that the source of vibrations felt on surface was not roof collapse, but the shattering or punching of support pillars and mitigated against leaving support pillars. This recommendation was endorsed by the 1915 committee.

The 1924 committee distinguished between strain bursts (minor events) and crush bursts (more severe). To prevent "strain bursts", it was recommended inter alia that faces should not be barred completely (i.e to leave a cushion of fractured material) and to change the face direction.

To prevent "crush bursts" it again recommended that pillars should not be left, and very importantly, recommended that the formation of remnants should be avoided by adopting the longwall mining method. For remnant mining it advocated a rapid rate of face advance, competent, dense support and the appointment of a competent person to point out risk areas and to ensure that special instructions were carried out.

The 1964 commission advised that layouts should be planned such that stress peaks are avoided and in general confirmed the recommendations of the previous committees.

The Industry Guide to the Amelioration of the Hazards of Rockbursts and Rockfalls published by the Chamber of Mines in 1977 was the first to contain solid technical material on rock mechanics, having been written by experts in the field. Apart from being much more specific on several counts, it predominantly differed from the previous efforts only in the sense that it carefully advocated the use of stabilizing pillars to reduce the incidence of rockbursting by reducing the Energy Release Rate.

The 1988 Edition - an update of the 1977 document - was even more detailed, more comprehensive and more scientific in its approach. It provided valuable updates on rock engineering technology, like the addition of the use of the Excess Shear Stress concept and various other technical guidelines, but did not add any new broad guidelines as far as mining layouts are concerned.

b. Support methods

The rapid yielding hydraulic prop was developed in the late 1960's. During the 1980's, more effective elongate timber supports came into use and wiremeshing support in tunnels on burst prone mines was more generally applied.

Lately, backfilling is becoming more popular. Gürtunca and Gay (1993) report on the benefits of backfill both as regional and local support and supply valuable guidelines. The stabilizing pillar debate is not yet over. Hobday and Leach (1992) reported an increase in seismicity after a few years to levels higher than those before the pillars were provided. Adams et al (1989) advocated the use of concrete stabilizing pillars.

During the 1950's panel faces at the ERPM mine were softened by preconditioning blasting - Roux et al (1957). The practice was discontinued due to drilling problems but has lately been revived, Adams et al (1993).

The need for yielding tunnel support has been addressed periodically, following the first plea by Ortlepp et al (1973). Following the development of design criteria, Woijno et al (1987), it is now commercially available in the form of the cone bolt - Woijno (1990) - and is beginning to be used by the industry.

c. Fundamental understanding

The concept of Energy Release Rate was developed in the 1960's and was used for a long time as a rockburst predictor. It is no longer used as rigidly as before, but still plays an important role.

There have been meaningful advances in the techniques and interpretations of seismic observation. Several advanced seismic monitoring systems are in use - most gold mines monitor seismicity routinely. These will not be expanded upon, as very little in the form of usable results have yet emerged for the rock engineer in the practical situation. Stewart and Spottiswoode (1993), however, proposed a model for assessing seismic risk based on the combination of several seismic hazard parameters.

An exciting concept is that of Excess Shear Stress, Napier and Ryder (1986). It is fundamental and simple in nature, based on applying sound and proven principles to the observation that large seismic events tend to occur on fault planes. It spawned the practical trial of preventing large seismic events by inducing controlled slip on fault planes through high pressure fluid injection.

4. EVALUATION OF RESULTS

The reason why opinions regarding the real value of progress will differ is because different people will measure the progress against different values. Being a practitioner, one can only measure against the bottom line practical applicability of results.

4.1 Coal

There is no doubt that the effort to determine different coal strengths for different mining areas will be successful. The approach being adopted is to take into account the factors which are thought to influence the strength of coal pillars, like the intact material strength and the frequency, orientation and properties of discontinuities.

But is this the real problem ? It is known that over time, due to several reasons, pillar strength will deteriorate. The inclination of nature is to create a situation of stability, and true stability in mining is only reached once the cavities are no longer there. Whether nature will repair man's damage by failing the pillars, collapsing the roof or causing the pillars to punch into the floor is a matter of circumstances.

The question is not whether a cavity will fail, but when. Is it not time to express stability in terms of the passage of time rather than a ratio of strength to load, both of which are time dependant anyway ?

Considering the Wilson (1972) concept of pillar behaviour, as modified by Barron and Pen (1992), it is possible to attach mathematics to the process of pillar deterioration. Assuming that pillar sides are not artificially contained, the fact that slabbing results in smaller pillar dimensions implies a continuous process.

The remaining challenge is to determine the rate of deterioration, which will enable the prediction of pillar life. At least one such attempt has been made, van der Merwe (1993). However crude or inaccurate, it at least indicated that it should be possible to quantify the life span of coal pillars.

4.2 Gold

As a general statement, the main areas of advance over the last 50 years has been in the fields of mathematical formulation, terminology, numbers processing and remote observation. Apart from the development of yielding support elements with the ability to absorb seismic energy without sacrificing their integrity as support elements, preciously few new mining methods of real practical value have emerged. The understanding of processes in the rock has improved significantly, but the real implementable value thereof is yet to emerge.

For instance, back filling as a technique is centuries old. The ability to understand its role and engineer its properties to meet specific requirements have been added - the technique has been refined.

The principle of longwalling dates from 1924. Understanding the propagation of fractures around it and having the ability to explain the high incidence of unplanned collapses caused by excessive leads or lags is new. Another refinement.

Similarly, was the 1924 recommendation not to bar faces completely not an intuitive hint at preconditioning ? The 1924 commission also made the first distinction between "strain bursts" (at the face) and "crush bursts" (remote). The 1988 Industry Guide distinguishes between "crush type events" (at the face) and "shear type events" (remote).

The 1924 recommendation that "remnants should be removed with all practicable speed by continuous work before the main weight comes upon it" would today probably be rephrased to read "the rate of face advance in remnant extraction should be such as to prevent being exposed to severe fracturing, the development of which is a time dependant reaction to the instantaneous redistribution of stresses, while not advancing so rapidly as to lose the beneficial effect of moderately crushed rock immediately ahead of the face".

The essential difference is that today's rock engineering practitioner, having the benefit of symbols and numbers being attached to phenomena which have been observed very long ago, also has the potential ability to make quantifiable predictions and lay down engineered guide lines. The quantum leap of creativity evident in the 1924 recommendations have not yet resulted from current engineering abilities. We are still in the process of turning the art into science, and hopefully gearing ourselves for the next wave of creativity.

Of course, there are significant refinements without which mining at current depths would not have been possible. Yielding supports were engineered, designed with a sound and quantifiable understanding of the effects of remote seismicity on man made cavities.

The 1964 recommendation to avoid stress peaks can now be implemented, for the Energy Release Rate concept placed computers in the rock engineer's office and created the demand for ever more sophisticated number manipulation processes. The software providers rose to the occasion.

While rock mechanics in the first half of the century suffered from a lack of suitable terminology, we must today guard against an over supply. We are rapidly entering the nth levels of complexity in the conceptual world, and should take care not to cause confusion by terms like "apparent volume", etc. Communication and agreement on uniformity of terminology are vital. We cannot afford misunderstandings.

With regard to the prevention of rockbursts, face preconditioning and fluid injection into fault planes seem to hold considerable promise. Both are based on at least some understanding of two mechanisms of seismicity.

Preconditioning results in the controlled creation of fractures in the working environment which would otherwise have happened uncontrolled and sometimes with negative consequences. The benefits of this method should not be under estimated. If successful, it has the potential to remove at least one source of rockbursts.

There is still some doubt regarding the success of fluid injection to induce controlled fault slip, and certainly much more work needs to be done in this regard. The principle is sound, but the nagging fear is that one may trigger a major seismic event in this fashion. Should it happen, the argument as to whether it would have happened anyway would be interesting to debate.

An important aspect of the attack against the rockburst problem, prediction, is by far the most difficult to deal with. Successful prediction must meet the requirements of being made well enough in advance to enable management to react by for instance evacuating the effected area or changing a layout and the success rate must be well above ninety percent.

In order for a prediction to be made at all, either the mechanism must be sufficiently well understood for its causes to be identified and measured, or someone must stumble onto a pattern in something by chance.

The latter is a matter of chance, and the problem is serious enough for it not to be ignored. Surely, if enough people investigate enough combinations someone will find something and perhaps develop an explanation later. Newton developed his laws by observing and thinking.

The likely success of the former route is a function of our understanding of the basic laws of nature, and our ability to expand our thoughts into the unknown.

Stripped of its complexities, a rockburst is a reaction of the part of the rockmass we can observe visually to an action that occurred somewhere else. Something caused the action.

In essence, Newton's first law states that things respond to a gradient, or a difference. If we hope to explain seismicity, we need to find the gradient. Something that changes in respect of something else.

Seismicity is associated with mining. Therefore, mining - or the creation of a void - is the cause of the gradient. What are the implications of a void in the crust of the earth?

There are two basic changes - the first is the most obvious: we create the potential for a part of the rock mass to be displaced. The second is the redistribution of loads, which is manipulated mathematically to calculate various stresses, energies, etc.

Michelangelo said that the secret of art is to know how to look. If we assume that science follows after art, in other words that we will turn "art" into "science" by attaching numbers and symbols to events, we can solve the problem by first considering our powers of observation.

We have the capacity to extend our basic observational facilities. We can supplement our senses with mathematics, which we can enhance by using computers. We can see through rock by using various remote sensing techniques. We probably have the required powers of observation.

What we possibly lack is the understanding to know what to look for. Without that understanding, we will have the solution right in our field of vision and not recognise it. If we need to find a gradient, and continue looking for a magnitude, we will never solve the problem. Viewed against this argument, the Excess Shear Stress (with its related and derived concepts) stands out amongst the complexities of conceptual developments with the simple beauty of an African sunrise. It is based on a difference, which in turn is quantifiably caused by mining.

The reason why it is not used in practical mining to control slip, are that we do not know enough about the orientation and characteristics of real faults and are uncertain about some of the engineering aspects. But these are mere challenges, indicators of where development is required. Have the urgency and importance of locating faults and their properties been conveyed to our colleagues in the geological and geophysical worlds?

Perhaps we should study the problem indirectly. In our terms of reference, a seismic event is a sudden occurrence, and many problems with regard to observation are related to its suddenness. Our definitions of "sudden" and "gradual" are determined by how quickly we can observe and react. This is in turn a function of our life span. If man had a life span of a micro second, we would not be aware of seismicity as the change would be distributed over more than a life span. Or, if we had a life span of a million years, seasonal variations in temperature would be registered as a barely noticeable vibration.

The point is that we should not be bluffed by the "suddenness" of a seismic event. We define "sudden", not nature. And the challenge is for us to understand nature, nature will not bend to fit into man's frame of reference.

If we had a life span of a million years, would we not experience creep on a fault plane as a seismic event? Is it not perhaps possible that we can study the mechanism of seismicity by observing creep ? If we cannot accelerate our rate of observation, why not study a similar but slower process first and then mathematically observe the effects of accelerating the motions ? Is creep not but a very long series of microseismic events as granules of rock slide over one another ?

5 CONCLUSION

The major rock engineering challenge in mining in South Africa, and in a number of other parts of the world, is the rockburst problem. It may be fair to state that a large part of the problem has been solved technically by the development of yielding support elements.

This has not yet solved the problem of the rock engineering practitioner, who carries at least part of the responsibility for the practical implementation of the systems. Cost may mistakenly be seen as a barrier. One should not fall into the trap of confusing the price of a support element with the cost of stability.

The real cost of stability includes : the cost of the support materials, the cost of labour, the cost of lost production due to the time taken to install support and the cost of unplanned collapses. In most mining situations, the biggest cost is usually that of unplanned collapses, which means that one can profitably pay more for support materials if the time lost due to unplanned collapses can be reduced.

In the practical mining situation, this argument only carries weight if it is quantified, and only the rock engineering practitioner will have the motivation to do the required simple arithmetic. He should in fact do it.

The rock engineering practitioner providing a service to a mine has the benefits of greatly enhanced fundamental understanding of rock behaviour and the computational facilities to quantify them. The road is wide open to design to known and quantifiable requirements for a wide range of applications.

The most important challenge for practitioners now is to consolidate that which is already known into sound and scientifically based engineering procedure. Likewise, our research oriented colleagues have the challenge to bring the solutions which are blinking on the horizon closer to home. The stage is set for the next wave of creative breakthroughs.

REFERENCES

Adams, DJ, Gürtunka, RG Jager, AJ and Gay, NC (1989). Assessment of a new mine layout incorporating concrete pillars as regional support. 4th Int. Symp. on Innovations in Mining backfill Technology, Montreal, Canada.

Adams, DJ Gay, NC and Cross M (1993). Preconditioning - a technique for controlling rockbursts. 3rd Int. Symposium on Rockbursts and Seismicity in Mines, Kingston, Canada.

Anon (1908). Ophirton Earth Tremors Committee. Dept. of Minerals and Energy Affairs of the R.S.A.

Anon (1915). Witwatersrand Earth Tremors Commission. Dept. of Mineral and Energy Affairs of R.S.A.

Anon (1924). Witwatersrand Earth Tremors Committee. Department of Mineral and Energy Affairs of R.S.A.

Anon (1964). Rockburst Committee. Department of Mineral and Energy Affairs of R.S.A.

Anon (1977). An industry guide to the amelioration of the hazards of rockfalls and rockbursts. Chamber of Mines of South Africa P.R.D. Series no. 216.

Anon (1988). An industry guide to the amelioration of the hazards of rockfalls and rockbursts. Chamber of Mines of South Africa, Ed. JA Ryder.

Barron, K and Pen, Y (1992). A Revised Model for Coal Pillars Proc. Workshop on Coal Pillar Mechanics and Design. U.S. Bureau of Mines IC 9315.

Claeys, WD (1988). Geomechanical Implications of Mini Longwalling at New Denmark Colliery. Proc. Rock Mechanics in Africa, SANGORM Symposium, Swaziland, 1988.

Gürtunca, RG and Gay, NC (1993). Mining with backfill - a review of Rock Engineering investigations into the benefits of backfill during the period 1986 - 1992. 5th Int. Symposium on Mining with Backfill. SAIMM. Ed. HW Glen.

Hobday, HMD and Leach, AR (1992). Some case studies in rock engineering application on Western Deep Levels South Mine. SANGORM Symposium: Impact of Rock Engineering on mining and tunnelling economics, 1992.

Madden, BJ (1986). Squat pillar design in South African Collieries. Proc. SANGORM Symposium. Advances in Rock Mechanics in underground coal mining, Witbank 1989.

Napier, JAL and Ryder JA (1986). Application of Excess Shear Stress concepts to the design of mine layouts SAIMM 86, Colloquium on Mining in the Vicinity of Geological and Hazardous Structures.

Ortlepp, WD, More O'Ferrall, RC and Wilson JW (1973). Support methods in tunnels. Association of Mine Managers, South Africa. Papers and Discussions, 1972/73,

Roux, AJA, Leeman, ER and Denkhaus, (1957). De-stressing. A means of ameliorating rockburst conditions. J.S. Afr. Inst. Min. Metall. Vol 58: 101-119.

Salamon, MDG and Munro, AH (1967). A study of the strength of Coal Pillars. J. South Afr. Inst. Min. Metall. Sept. 1967.

Salamon, MDG. (1969). Present state of rock mechanics in bord and pillar coal mining. Chamber of Mines Research Organization, Research Report no. 2/69, Jan. 1969.

Stewart, RD and Spottiswoode, SM (1993). A technique for determining the seismic risk in deep-level mining. Symp. on Rockbursts and Seismicity in Mines, Canada.

van der Merwe, JN and Olivier, FT (1992). The Development of a Crush-pillar system for High extraction Coal Mining. In Massmin 92, Ed. HW Glen, S.A. Inst. Min. Metall.

van der Merwe, JN (1993). Revised strength factor for coal in the Vaal basin. J. S. Afr. Inst. Min. Metall. March 1993.

van der Merwe, JN (1993b). Preventing Unplanned Land Subsidence at Sigma Colliery, South Africa. Proc. Conf. Mine Subsidence in Urban and Developed Areas, Abandoned Mine Land Division, Wyoming Dept. Env. Quality.

Wilson, AH, (1972). Research Into the Determination of Coal Pillar Size. Part I. An Hypothesis Concerning Pillar Stability, Min. Eng, v. 131, 1972, pp. 409-417.

Woijno, LZ, Jager, AJ and Roberts, MKC (1987). Recommended performance requirements for yielding rock tendons. Proc. SANGORM Symp. Design of Rock Reinforcing : Components and systems.

Woijno, LZ, Jager, AJ and Henderson, NB (1990). New developments in the design and support of tunnels under high stress conditions. Proc. SAIMM Symp. Technical Challenges in Deep Level Mining.

Panel discussion: In-situ stress and fluid flow in fractures

Session developers:
Jane Long, Lawrence Berkeley Laboratory
Richard A. Plumb, Schlumberger Cambridge Research

Rock Mechanics, Nelson & Laubach (eds) © 1994 Balkema, Rotterdam, ISBN 90 5410 380 8

Field evidences on the orientation of forced water flow with respect to the regional principal stress directions

F.H.Cornet
Département de Sismologie, Institut de Physique du Globe de Paris, France

R.Jones
Camborne School of Mines Associates, Rosemanowes, Penryn, Cornwall, UK

ABSTRACT

For a relatively shallow experiment (between 400 and 800 m) both borehole observations and induced seismicity locations indicate that flow occurred in pre-existing fractures inclined with respect to the principal stress direction. For a deep (between 2800 and 3700 m) reconnaissance programme (about 40000 m^3 injected water), two situations have been observed. When the injection pressure remains lower than the minimum principal stress, flow occurs along pre-existing fractures inclined at about 30 ° to the maximum principal stress direction. When the injection pressure is slightly larger than the minimum principal stress, the water leaves the well along pre-existing weakness planes, but the mean flow direction is parallel to that of the maximum principal stress.

1. INTRODUCTION

One of the most difficult problems faced by the development of Hot Dry Rock (HDR) geothermal reservoirs is the determination of the main directions of forced fluid flow. This is of paramount importance for defining the relative location of the wells.

In this paper, we discuss results about the direction of flow as observed during two in situ experiments. The first involves a shallow water circulation (800 m) conducted in the granite rock mass of Mayet de Montagne (central France). The second concerns a large scale reconnaissance test conducted at Soultz (Eastern France) in a 3500 m deep well .

Near the wellbores, the direction of flow has been identified by various combinations of logs. The direction of flow in the far field has been identified by analysis of the induced seismicity location. For the Mayet de Montagne site, particular attention has been devoted to the determination of the regional stress field before, during and after the water circulation tests. For the Soultz site, injection induced fractures orientation provide a good control on the principal stress directions.

2. RESULTS FROM A FORCED CIRCULATION TEST (MAYET DE MONTAGNE)

2.1 *Site Reconnaissance*

Site reconnaissance involved first an analysis of the natural fracture network as observed on various outcrops of the granite within a distance of approximately 2 to 3 Km from the site and data derived from electrical imaging logs (Mosnier, 1982) run in the various holes drilled on site (three vertical, 200 m deep boreholes and two vertical, about 800 m deep wells called INAG III-8 and INAG III-9 located about 100 m apart). In addition to a sub-horizontal fracture set, four different preferential fracture strike orientations have

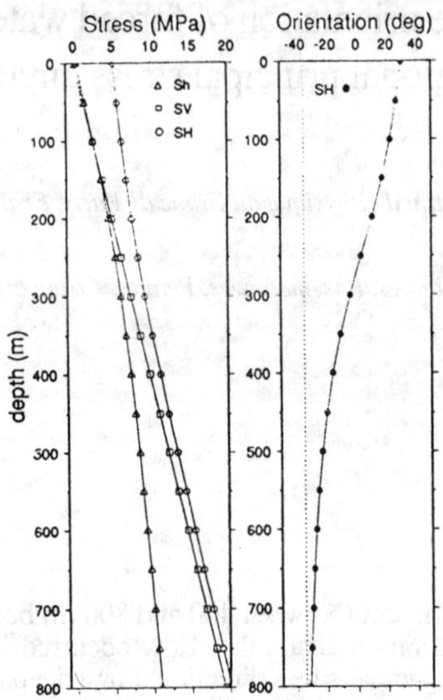

Figure 1. Vertical stress profile obtained with the HTPF method; note the change in stress orientation with depth

been identified (N 20°±20° E, N 60°±10° E, N 100°±10° E and N 155°±10° E); these fracture sets are all sub vertical (within about 20 degree of the vertical direction).
The stress field has been determined by the HTPF method before and after the water circulation experiment (Cornet 1986, 1992). Results show no permanent large scale perturbation induced by the fluid injections. However, locally, the injection has induced some aseismic slip which resulted in local stress heterogeneities (Scotti and Cornet, 1994). At the regional scale (Fig. 1), a 60 ° rotation of the maximum horizontal principal stress direction is observed between the surface (H orientation N 20 ° E) and depths greater than 500 m (H orientation N 40 ° W). Also, while above 200 m the vertical stress is the minimum principal stress, it becomes the maximum principal stress below about 600 m but remains sub-equal to the maximum horizontal principal stress.

2.2 *Identification of fluid flow directions at the boreholes*

The most significant flowing zones at the boreholes have been identified (Bruel and Cornet, 1992) by spinner logs (fig.2) for the injection well (INAG III-9) and by thermal logs for the production well (INAG III-8). In the injection well, the relative importance of each flowing zone (as expressed in percentage of total flow) depends on the injection pressure (and therefore on the injection flow rate). For the production well, the relative significance of each flowing zone depends on both the injection pressure and the effect of stimulations. Indeed the distribution of flow for logs 1 and 3 are slightly different although they have been run with the same injection flow rate (figure 3). In between these two logs, injection had proceeded at larger flow rates (and therefore higher injection pressures) which resulted in some induced seismicity.

For some of the flowing zones a single fracture could be identified on the borehole imaging logs. The corresponding strike and dip are indicated directly on the figures. It may be observed that none of the well identified fractures is parallel to the direction of the maximum horizontal principal stress. However, it has been shown in a previous paper (Cornet and Scotti, 1993) that the large scale geometry of faults and fractures cannot be

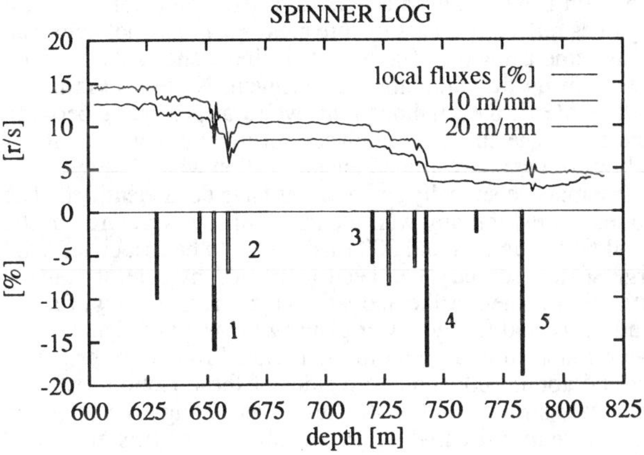

Figure 2. Spinner log recorded in INAG III-9 (16.5 l/s injection flow rate, 11 MPa wellhead pressure). The main flowing zones are indicated with their percentage loss rate. When a single fracture has been observed, its strike and dip are indicated below the graph. At 781 m a fault zone is observed. It takes flow only at low injection rates. For this log, the total flow lost below this depth is shown.

(1) 654 m, N158°E, 90°; (2) 659 m, 2 vertical fracs (N151°E, N164°E); (3) 719 m, N43°E, 82°; (4) 743 m, N38°E, 83°; (5) 781 m, 150°±10°, 70°

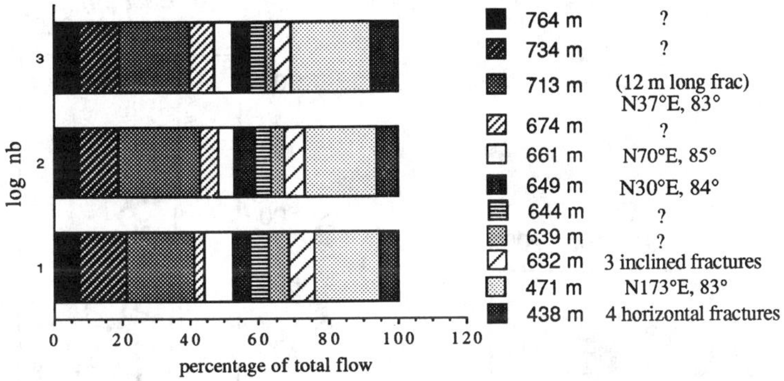

Figure 3. Percentage wise distribution of production flow rate in the well INAG III-8. For each producing depth, the orientation of the associated fracture is given when it has been clearly identified on the electrical imaging log. Log 1 and 3 were obtained when injection in INAG III-9 was equal to 7.5 l/s (9 MPa well head pressure). Log 2 was obtained with a 16.5 l/s injection rate in INAG III-9 (11 MPa wellhead pressure)

derived from borehole imaging logs because their small scale morphology is not representative of their large scale geometry. We will only conclude for now that, locally, the significant flowing zones observed in a fractured rock mass are not all oriented perpendicular to the minimum horizontal principal stress.

2.3 Identification of fluid flow away from the boreholes

The main direction of flow away from the boreholes has been identified by locating induced seismic events (Cornet and Yin, 1992; Cornet and Scotti, 1993). Indeed, it was considered that, since the injection of water induces some increase in local pore pressure, this increase in pore pressure would in turn trigger some seismic activity the location of which would help locate the main flowing zones (Pearson, 1981; Batchelor, 1989).In fact, while induced seismicity does reflect increases in pore pressure, it may not help map all significant flowing zones as demonstrated on figure 4. This figure shows the location of all seismic events observed during the circulation experiment. None occurs in the vicinity of the production well INAG III-8 even though this well was producing between 48 % and 80 % of the injected flow. The absence of induced seismicity in the vicinity of INAG III-8 only reflects the low pore pressure around the production well.

Near the injection well, three main seismically active zones have been identified. The lower zone (below 750 m) does not exhibit any well identified planar structure but the two upper active zones, around 640 m and around 500 m, do seem to be associated with planar structures. Their large scale geometry has been identified by a least squares technique (Cornet and Scotti, 1993). Their strike and azimuth are respectively, for the upper planar zone N 04°E and 61°E, and for the lower planar zone N 174 ° E and 36°E. Thus neither of them has been found to be normal to the regional minimum principal stress direction. Yet they extend about 100 m on both sides of the injection well and cannot be attributed to small scale structures. For this test, even though the injection pressure was larger that the minimum principal stress, the absence of tilts at ground surface (Desroches and Cornet, 1990) together with the large hydraulic impedance observed during injection (Bruel and Cornet, 1992) has been taken as evidence that the mechanical opening of fractures remained limited to the immediate vicinity of the well.

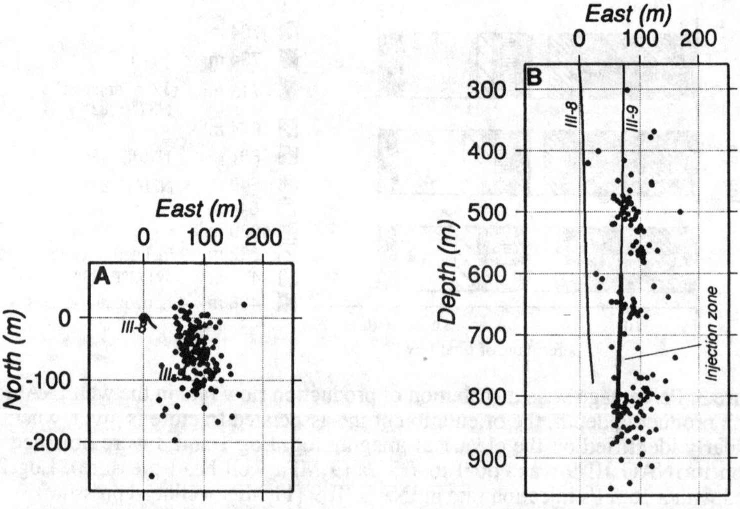

Figure 4. location of microseismic events. (left) Horizontal projection, (right) vertical projection in an East - West plane. Note the absence of events near the production well (INAG III-8)

While the upper zone seems to be associated with only a limited flow according to the spinner logs, the planar structure identified around 640 m absorbed a significant amount of flow (about 41 % of the flow for the 7.5 l/s injection flow rate and 29 % of the flow for the 16.5 l/s injection flow rate). Further, for this zone, the only fracture clearly identified on the borehole imaging log as one of the flowing zones is vertical and striking N 158° E, i.e. with a different strike from that identified by the induced seismicity technique (but roughly aligned with the maximum horizontal principal stress direction). Finally, prior to conducting this hydraulic experiment, a zone of stress heterogeneity had been identified in INAG III-8 around 350 m (Cornet, 1986), i.e. close to the depth at which this borehole is supposed to intersect the upper seismic zone .

It is concluded that, for this granite rock mass, some significant permeable zones exist in directions which are not perpendicular to that of the minimum principal stress. These zones coincide with pre-existing structures the geometry of which cannot be derived simply from borehole imaging logs. Some of them are associated with stress heterogeneities which may be enhanced by the injection of fluid. However, at the regional scale, fluid injections did not alter permanently the stress field .

3. RESULTS FROM A LARGE SCALE INJECTION TEST AT SOULTZ

The site of Soultz is presently being investigated for future development of a HDR Geothermal reservoir. The site is located in the Rhine Graben, about 40 km to the North of Strasbourg (France). A vertical, 3590 m deep well reaches the granite after intersecting about 1400 m of various sedimentary formations. It is cased down to 2850 m. Preliminary hydraulic reconnaissance tests run at very low injection rates have shown that a major fault zone intersected at 3495 m was the only significant hydraulic feature within the open hole section. The mapping of induced seismicity generated by three water injection tests has been used to identify the main flow direction in the granite. During the first test about 25 000 m^3 were injected after the bottom of the well had been sanded off up to 3400 m so as to avoid injecting directly into the fault. Then, during a packer test, about 600 m^3 were injected into the main fault at the bottom of the well. Finally, about 19000 m^3 were injected over the total open hole section .

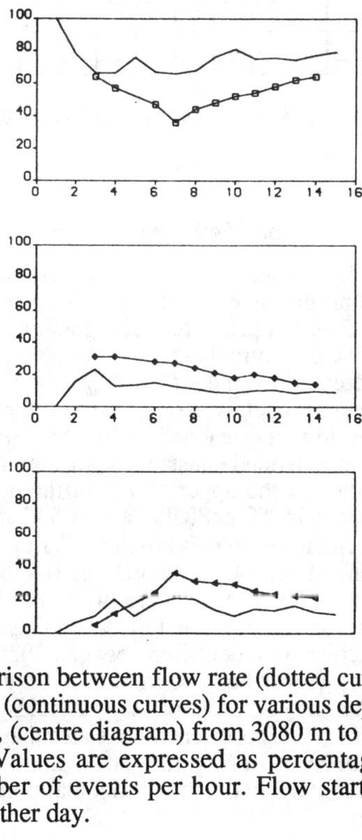

Figure 5.Soultz : Comparison between flow rate (dotted curves) and number of seismic events observed per hour (continuous curves) for various depth intervals : (Top diagram) all depths below 3150 m, (centre diagram) from 3080 m to 3150 m, (lower diagram) all depths above 3080 m. Values are expressed as percentage of total injection rate or percentage of total number of events per hour. Flow started at 6 l/s (day 0) and was increased by 6 l/s every other day.

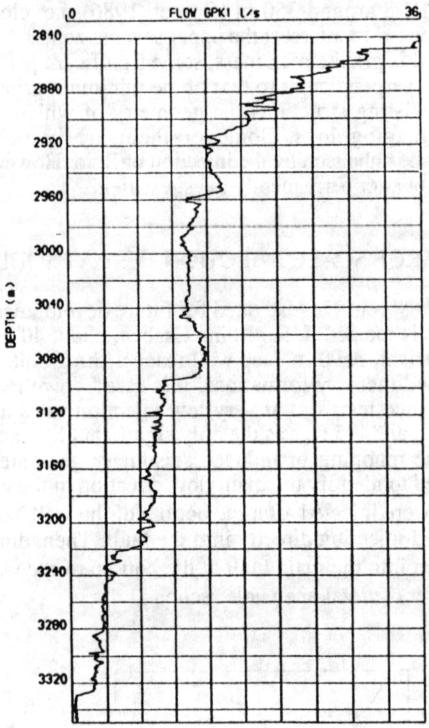

Figure 6. Spinner log recorded when injection proceeded at 36 l/s (day 17) in well GPK1 (Soultz)

3.1 Some hydro-mechanical results from the 25 000 m^3 injection test

Injection started at a 6 l/s flow rate. This flow rate was maintained constant for 48 hours and then was increased progressively to 36 l/s by steps of 6 l/s which each lasted 48 h. A spinner log, combined with a pressure reading and a thermal reading, was run every 24 hours. These logs show that initially the flow involved only a limited amount of fractures. As the flow rate increased from 6 l/s to 18 l:/s, the injection pressure kept increasing linearly and the percentage of flow leaving the well through the deeper fractures increased (figure 5). Once the flow rate reached 18 l/s, the injection pressure stabilised (37.7 ± 0.5 MPa at 2805 m) and remained constant for all subsequent flow rates. Most of the flow increment was lost then in the upper 100 m of the open hole section (fig. 6). This resulted in a cooling by about 115°C at 2850m and 105°C at 3350m.

After the test, a Borehole Televiewer log was run in order to identify newly developed fractures. Just below the casing shoe, a large number of fractures striking about North South but with dip ranging from 60° to 90 ° were observed. Many of them had been already seen on the log run prior to the injection tests when they were less pronounced. A freshly developed vertical fracture was identified between 3025 m and 3058 m, a depth interval where no significant flow loss had occurred according to the spinner logs. It is interpreted as a thermal fracture the orientation of which yields the maximum horizontal principal stress direction (N 175 °E). This orientation is also that of a 30 m high hydraulic fracture developed around 2000 m during the early phases of the reconnaissance program. Thus it is concluded that the maximum horizontal principal stress is oriented N 175°E at least down to 3100 m.

3.2 Mapping the induced microseismicity

During these tests, induced seismicity has been monitored with three four-components accelerometer sondes and a single hydrophone. The accelerometer sondes are grouted in the bottom of boreholes which penetrate through the sediments into the granite by about 20 meters. The hydrophone was deployed at 2100 m in a fourth borehole, also in the granite. The four sensors within the accelerometer sondes give an added degree of redundancy necessary for the long term deployment in this hostile environment (temperatures around 125 °C). The sondes give a resonance free response from a few Hz to over one kHz. The hydrophone's transfer function is flat up to several kHz .

All locations are determined using manually picked P- and S- waves. The picking accuracy is typically better than a millisecond and, along with the geometric configuration of the network, this gives a location error of the order of a few tens of metres for most of the hypocentres. A measured seismic velocity is used in the location scheme. This velocity field is isotropic for both P- and S- waves with station delays of a few milliseconds for the various stations being used. This velocity field was determined by firing three calibration shots with known origin time and co-ordinates at different depth, including the bottom of the well (3540 m).

Altogether, about 20 000 events have been recorded, of which some 14 000 have been located to date. They are presented in figure 7 which shows the plot of all locations projected in a horizontal plane and in a vertical plane oriented N 170 ° E. This figure shows that the induced seismicity lay in a cloud the main striking direction of which is about N 145 ° E, i.e. about 30 ° off the maximum horizontal principal stress direction. Examination of the vertical cross section shows that within the cloud itself, microseismic events are not evenly distributed. In figure 8 we show the horizontal projection of all events observed within three different depth intervals : 2700-2900 m; 2900-3100 m and all events below 3100 m. For the 2700-2900 m depth interval, the cloud is oriented preferentially in the North-South direction, with some local extensions in the NW - SE direction, while for all events observed below 3100 m, the cloud is mostly NW-SE

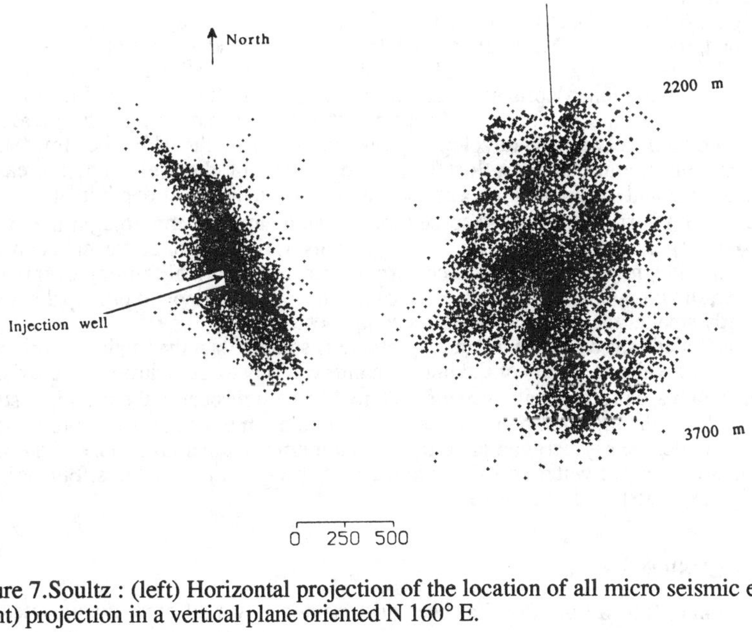

Figure 7.Soultz : (left) Horizontal projection of the location of all micro seismic events; (right) projection in a vertical plane oriented N 160° E.

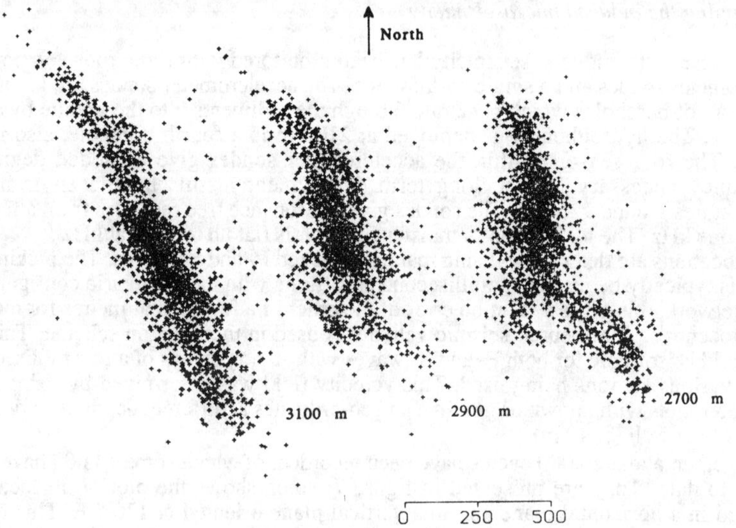

Figure 8. Soultz : Horizontal projection of microseismic events observed in three different depths intervals : (left) all events below 3100 m, (centre) from 2900 to 3100 m; (right) from 2700 m to 2900 m.

oriented. The 2700-2900 m depth interval is precisely that where most of the fluid was lost when the injection became stress controlled. The cloud orientation coincides with the maximum horizontal principal stress direction. Below 3100 m, flow occurred only through pre-existing fractures oriented preferentially about 30° to the maximum horizontal principal stress direction.

On figure 6, we compare the percentage wise number of seismic events observed per day for the three following depth intervals 2200 m (top of cloud)-3020 m ; 3020-3150 m; 3150-3700 m (bottom of cloud). This plot is compared to the relative flow rate lost in the well for the equivalent depth intervals within the open hole section. Initially, the percentage wise amount of fluid flow lost through the lower part of the well increased, and the induced seismicity developed downward. Once the flow became stress controlled, the amount of flow lost in the upper part of the well kept increasing percentage wise and so did the number of seismic events in the upper depth interval.

A 12 stations surface network recorded more than hundred events (magnitude larger than 0.5) during these tests so that the focal mechanisms of the strongest events could be determined. They have been found to correspond mostly to normal faulting events with some strike slip component .This suggests that the maximum horizontal principal stress is only slightly smaller than the vertical stress component.

It is concluded that when the injection pressure is smaller than the minimum principal stress which prevails at the top of the injection interval, flow occurs through pre-existing fractures which are preferentially oriented 20 to 40 ° with respect to the principal stress direction . When the injection pressure reaches the minimum principal stress, pre-existing fractures oriented nearly perpendicular to the minimum principal take most of the flow (up to 70 %) but some water still flows along other pre-existing fractures. Further, the water migrates preferentially upward.

4. CONCLUSION

The direction of forced fluid flow has been documented at two different granite sites, Le

Mayet de Montagne and Soultz. In both instances, pre-existing fractures controlled the direction of outward or inward flow at the wells even though in some instances the injection pressure was larger than the minimum principal stress magnitude. Flow away from the wells was identified by mapping the induced seismicity. This technique precludes the mapping of flow direction in the vicinity of production wells and only outlines zones of large pore pressure. At Soultz, the orientation of the seismic cloud observed within the depth interval where the injection pressure reached a value larger than the minimum principal stress was found to be parallel to the maximum horizontal principal stress direction. In domains where the fluid pressure was lower than the minimum principal stress magnitude, the flow direction was found to be governed by pre-existing structures oriented 20° to 40 ° off the maximum horizontal principal stress direction. At Le Mayet de Montagne, the extension of opened fractures remained limited to the immediate vicinity of the injection well so that the direction of flow was mostly governed by pre-existing structures oriented at an angle with respect to the maximum horizontal principal stress.

ACKNOWLEDGEMENTS

This work was funded by the Commission of the European Communities (DG-XII), by the french CNRS (PIRSEM) and ADEME, and by the U.K. DTI (Renewable Energy Research and Development Program managed by ETSU)

REFERENCES

Batchelor A.S.; The role of the third well in the CSM HDR project; *Hot Dry Rock Geothermal Energy* (R. Parker, edit.); vol.1., pp 185-199; Pergammon Press, 1989

Bruel D. and F.H. Cornet; Forced fluid flow through fractured reservoirs modelling; *Fractured and Jointed Rock Mass* (Cook and Myer, edit.); vol. 3, pp 519-526, Lawrence Berkeley Laboratory report. nb LBL-32379; 1992

Cornet F.H.; Stress Determination from Hydraulic Tests on Pre-existing Fractures, the HTPF method; *Rock Stress and Rock Stress Measurements* (Stephanson ed.), pp. 301-312, Centek Publ.(Lulea) and Balkema; 1986

Cornet F.H.; In situ stress Heterogeneity identified with the HTPF tool; *Rock Mechanics, proc. 33rd U.S. Symp. on Rock Mech.* (Tillerson and Wawersik ed.); pp. 39-49; Balkema (Rotterdam), 1992

Cornet F.H. and O. Scotti; Analysis of Induced Seismicity for Fault Zone Identification; *Int. Jou. Rock Mech. Min. Sc.,* vol. 30, nb 7; pp789-796 ; 1993

Cornet F.H. and J. Yin; Induced Seismicity and Flow Paths associated with forced fluid percolation; *Geothermal Resources Council Transactions*, vol. 16, pp. 287-295; 1992

Mosnier J.; Détection électrique des fractures naturelles ou artificielles dans un forage; Annales Geophys. vol. 38; pp. 537-540; 1982

Pearson C.; The relationship between microseismicity and high pore pressure during hydraulic stimulation experiments in low permeability granitic rock; *Jou. Geophys. Res.*; vol. 86, nb B9, pp 7855-7864; 1981

Scotti O. and F.H. Cornet; In situ evidence for fluid induced aseismic slip events along fault zones; *Int. Jou. Rock Mech. Min. Sc* (in press) 1994 [illegible] de Montrose and Scotia [illegible] controlled the direction of [illegible] even though [illegible]. Injection pressure was greater than the [illegible] pressure. Flow [illegible] from the wells was [illegible]. This technique [illegible] production wells and [illegible] of these [illegible] larger than the [illegible] pressure [illegible]. At [illegible] to the [illegible] of the injection wells [illegible] direction of [illegible] with respect to the [illegible] principal stress.

ACKNOWLEDGEMENTS

This work was supported by the Commission of the European Communities (DG XII) [illegible] Research [illegible] (CRIEPI) and [illegible] Canadian [illegible] Development Programme [illegible].

REFERENCES

[illegible] Proc. [illegible] 1983.

[illegible] and [illegible] in fractured reservoirs [illegible] 1992.

Cornet, F.H. [illegible] Rock Stress and Rock Stress Measurements (Stephansson ed.) [illegible] 1986.

[illegible] stress [illegible] 1982.

Cornet, F.H. and [illegible] for [illegible] 1991.

Cornet, F.H. and [illegible] associated with [illegible] fluid [illegible] 1992.

[illegible] 1987.

[illegible] and their pore pressure [illegible] 1984.

[illegible] slip [illegible].

Rock Mechanics, Nelson & Laubach (eds) © 1994 Balkema, Rotterdam, ISBN 90 5410 380 8

Variations of the least horizontal stress magnitude in sedimentary rocks

R.A. Plumb
Schlumberger-Cambridge Research, Cambridge, UK

ABSTRACT: A multi-basin comparison of least horizontal stress magnitudes in sedimentary rocks has identified the relative importance of geophysical factors influencing intrabasin and interbasin stress variations. The comparison has resolved a global lithology dependence on stress magnitude and has revealed a zone of high horizontal stress at depths less than 1 km. Pore pressure, lithology and geologic structure are the principal factors correlated with intrabasin stress variations whereas pore pressure, the degree of diagenesis and tectonic setting have the greatest influence on interbasin stress variations. Our data indicate that stress variations due to changes in pore pressure are greater than those due to changes in lithology.

1 INTRODUCTION

Knowledge of in-situ stress magnitudes is fundamental to understanding and controlling problems associated with rock deformation encountered during the life of an oil field. Stress magnitudes are required to plan and drill stable wells, to design hydraulic fracture stimulation treatments and to determine the critical draw-down pressure in weak formations (Kunze and Steiger, 1991; Dusseault and Santarelli, 1989; Rosepiler, 1979; Van Eekelen, 1980; Warpinski, 1983; Kry and Gronseth, 1983; Veatch and Moschovidis, 1986; Lacy, 1987; Suman, 1975). As the demand for more efficient oil field operations increases, there are increased demands for solving field scale rock mechanics problems requiring stress magnitudes as inputs. The successful extrapolation of limited in-situ stress measurements to field scale problems requires a better understanding of the processes governing stress variations in sedimentary rocks. Central to this goal is the identification of the factors most likely to affect stress magnitudes in oil fields.

Previous field studies have shown that stress magnitudes vary with depth, pore pressure, lithology and structural or stratigraphic position (Warpinski 1989; Whitehead et al., 1989; Evans et al., 1989; Evans, 1989; Plumb et al., 1991, Teufel et al., 1991). However, on a global basis, only the increase in stress magnitude with depth is well documented (McGarr and Gay, 1978; Brown and Hoek, 1978). To develop better models of in-situ stress, additional global relationships must be established between stress magnitudes and factors related to deformation processes such as: pore pressure, the degree of consolidation, lithology and tectonic setting. The objective of this paper is to identify important geophysical factors correlated with variations of the least principal stress magnitudes in sedimentary rocks and to rank them in terms of their affect on stress magnitudes on a global basis. This is accomplished through analysis of a multi-basin database of stress measurements categorised by lithology, vertical stress and pore pressure.

2 FIELD DATA

This paper summarises over 1,000 least principal stress measurements representing more than 30 oil and gas fields located in 10 countries. The majority of instantaneous shut-in pressure (ISIP) and closure pressure (CP) measurements were made in cased wells located in petroleum producing sedimentary provinces where hydraulic fracturing is required for economic production of hydrocarbons. Porosity in these fields generally ranges from 4% to 7% and permeability ranges from 1 microdarcy to 500 microdarcies. Discussion of this data set is restricted to the least principal stress magnitude as this is the most reliable stress component obtainable from cased-hole hydrofracture measurements (Detournay et al., 1989). Figure 1 identifies the ISIP and CP on a schematic of a typical pressure-time record. Associated with each measurement is an estimate of the vertical stress, constrained by density logs or porosity data, a general indication of lithology (sandstone, shale or carbonate) and an estimate of pore pressure for approximately half of the measurements based on RFT data, well test data or mud weights. In the few cases where density logs or cores were not available from a field, the vertical stress was derived from an assumed value of 0.0226 MPa/m (1 psi/ft).

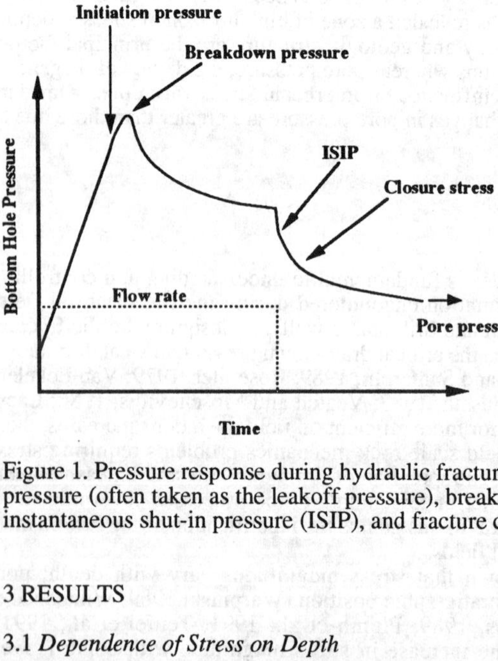

Figure 1. Pressure response during hydraulic fracturing showing fracture initiation pressure (often taken as the leakoff pressure), breakdown, fracture propagation, instantaneous shut-in pressure (ISIP), and fracture closure pressure (CP).

3 RESULTS

3.1 *Dependence of Stress on Depth*

Interbasin comparisons of stress magnitudes are most revealing when ISIP and CP are normalised by the local vertical stress ($R = ISIP/S_v$). Normalising stress measurements by S_v not only allows comparison of measurements from different depths but it preserves basin-specific information on the degree of consolidation and rock strength through the dependence of S_v on porosity. A basin composite plot of R as a function of depth shows that stress data can be divided into two domains (Figure 2). A domain of apparent high stress is observed at depths less than 1 km where R is greater than 1. Stress magnitudes in the other domain, typically at depths in excess of 1 km, are bounded above by the local vertical stress (0.019 MPa/m to 0.027 MPa/m) and below by a normal pore pressure gradient (0.0105 MPa/m, $\lambda \approx .45$). Most measurements in the high stress domain come from the Alberta basin, the Appalachian basin and Japan while the majority of measurements in the database represent geologic settings where ISIP and CP are less than S_v. Outside of the high stress domain ISIP or CP $\approx S_h$ and R becomes R_h (S_h/S_v). A global lithology dependence on R is not observed in Figure 2.

High near-surface stresses have long been recognised in crystalline rocks but are not well documented in sedimentary rocks (Brown and Hoek, 1978; McGarr and Gay, 1978). However, the interpretation of the hydrofracture measurements in the high stress zone at shallow depths is not simple. If new hydraulic fractures were created, fractures should be horizontal and R would equal 1. An R greater than 1 could be an indication that a vertical hydraulic fracture was initiated at the wellbore and the least horizontal stress was measured before the fracture could rotate into a horizontal plane. This situation is most likely when very small fluid volumes are pumped during the test (Evans et al., 1989, Thiercelin et al., 1993). An R greater than 1 might arise if ISIP or CP were over estimated. Although such errors are certainly possible it is unlikely that this explanation would pertain to all near-surface measurements.

The favoured interpretation is that ISIP and CP measure the normal stress acting on pre-existing planes of weakness such as steeply dipping natural fractures, bedding planes or cleavage planes and therefore may not represent a principal stress. If vertical joints, aligned with the contemporary maximum horizontal stress, are opened then ISIP or CP could be a good measurement of S_h. Such joints are prevalent in the upper 1 km in the Appalachian basin (Engelder, 1985; Plumb and Cox, 1987). According to the pre-existing planes of weakness interpretation ISIP and CP are still an indication of high horizontal stress but the relationship to principal stresses is not known. Despite some uncertainty about the magnitude near surface principal stresses it is reasonable to conclude that a zone of high horizontal stress exists at depths less than 1 km in some sedimentary basins. In such basins it appears that a thrust faulting environment in the upper 1 km changes to a normal or strike-slip faulting environment at greater depth.

3.2 *Dependence of Stress on Pore Pressure*

Field studies show that the influence of pore pressure on S_h is not strongly dependent upon

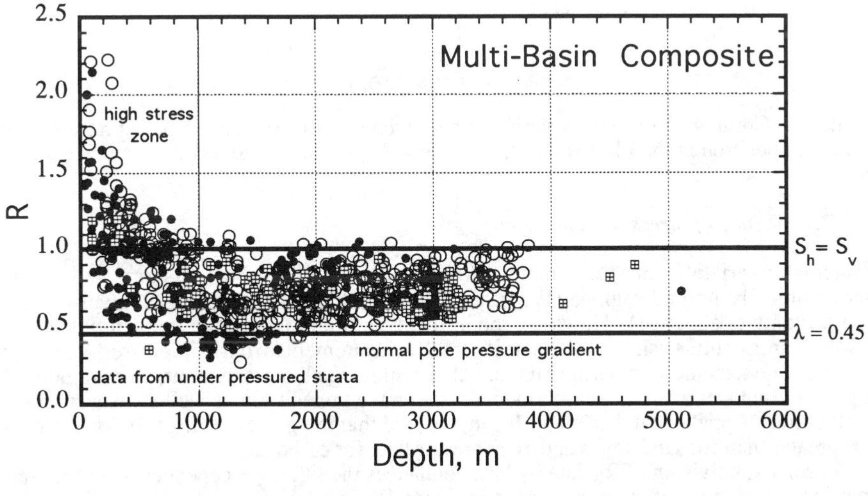

Figure 2. Stress ratio (R) vs. depth for the entire data set categorised by lithology, sandstones (open circles), shales (filled circles), carbonates (squares).

whether the pore pressure changes are spatial or temporal. Our multi-basin data indicate that spatial variations in S_h range from 60% to 80% of the change in pore pressure. This is the range of $\Delta S_h/\Delta P_p$ determined for various lithology groups from measurements made at depths greater than 1 km and averaged over many basins. For example, our global average value for $\Delta S_h/\Delta P_p$ is 0.64 for reservoir sandstones. This average value is similar to values derived by Breckels and van Eekelen (1981) to correct S_h for the effects of spatial changes in pore pressure in young offshore basins. They documented horizontal stress changes ranging from 46% to 56% of the change in pore pressure.

Salz (1977) and Teufel et al. (1991) have documented the effects of temporal changes in pore pressure by comparing stresses measured before and after reservoir pressure declined. In the Vicksburg formation in South Texas, Salz (1977) found S_h in sandstones changes by approximately 50% of the pore pressure change. Teufel et al. (1991) documented changes of S_h in the Ekofisk chalk of approximately 80% of the pore pressure change (Figure 3). Field data suggest that spatial and temporal changes of pore pressure produce a similar change in horizontal stress magnitude and that estimates of $\Delta S_h/\Delta P_p$ are well within the theoretical bounds for both inelastic and poro-elastic deformation processes ($0 < \Delta S_h/\Delta P_p < 1$).

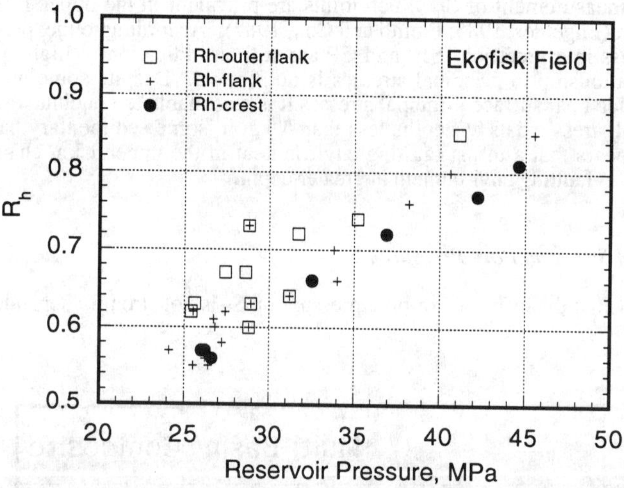

Figure 3. Comparison of stress variations associated with changes in pore pressure and structural position in the Ekofisk field, North Sea, after Teufel et al. (1991).

3.3 *Dependence of Stress on Lithology*

Systematic variations of stress magnitude with lithology have been documented within geographically limited regions (Warpinski, 1989; Whitehead et al., 1989; Evans et al., 1989; Plumb et al., 1991). However, the influence of lithology on stress magnitude is not resolved in a multi-basin comparison if stress measurements from compressed-state and relaxed-state tectonic settings are mixed. After removing all measurements from regions of high horizontal strain, e.g. measurements made in the upper 1 km as well as data from the northeastern Appalachian basin and Japan, we find that the global average R_h for shales is 8% greater than for sandstones and 16% greater than for carbonates.

Further subdivision of R_h data by basin enhances the lithology dependence and reveals interbasin variations of stress magnitude (Figure 4). In relaxed-state basins, R_h for shales ranges from 4% to 15% higher than in sandstones. The Appalachian basin is an example of a compressed-state basin where R_h is greatest in rocks with greater elastic stiffness

(Plumb et al., 1991). Here R_h in carbonates is 40% greater than in sandstones and R_h in sandstones is 20% greater than in shales (cf. (1), Figure 4). The available data suggest that changes in lithology influence stress magnitudes less than changes in pore pressure. Note, however, that lithology dependent stress ratios reported in this study are likely to be low since hydraulic fractures created in most commercial fracturing treatments yield stresses which represent averages over many beds, and many average over different lithologies.

3.4 *Structural and Stratigraphic Influence on Stress*

Changes in structural or stratigraphic position can affect stress magnitudes as much as changes in pore pressure. In the Appalachian basin, Evans (1989) documented a basin-wide discontinuity in the stress ratio which he correlated with the Salina Salt horizon. Above the Salina Salt, R is at least 50% greater than below the salt (above the salt R >1 vs. 0.65 below). Teufel et al. (1991) reported changes in stress with changes in position across the Ekofisk structure. They measured stress magnitudes at the crest of the structure which were approximately 12% lower than on the flanks (Figure 3). In the Ekofisk field, the effect of structural position on S_h is approximately 1/3 of the pore pressure effect.

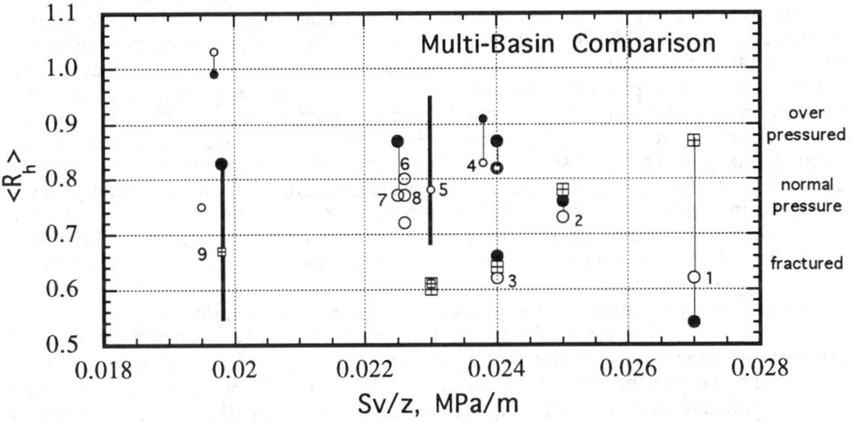

Figure 4. Multi-basin comparison of R_h vs. S_v/z (degree of consolidation) categorised by lithology: field-averaged R_h (small symbols), basin-averaged R_h (large symbols), sandstones (open circles), shales (filled circles), carbonates (squares). Fine vertical lines connect R_h for lithologies from a common region (e.g. 1). Bold vertical lines represent the range of the stress ratio due to temporal changes of pore pressure (e.g. 5,9). Principal references: 1-Appalachian basin (Evans, 1989; Evans et al., 1989, Plumb et al., 1991); 2-Alberta basin (Woodland and Bell, 1988); 3-East Texas basin (Whitehead et al., 1989); 4-Piceance basin (Warpinski, 1989); 5-South Texas (Salz 1977); 6-U.S. Gulf Coast, 7-Brunei, 8-Venezuela (Breckels and van Eekelen 1981); 9-Ekofisk field, North Sea (Teufel et al., 1991).

4 DISCUSSION

A multi-basin comparison of stress magnitudes has shown that pore pressure, lithology, geologic structure and tectonic setting are globally significant factors influencing intrabasin and interbasin stress variations and has identified a zone of high stress in the first km of depth in some sedimentary basins. A comparison of intrabasin and interbasin R_h helps identify which of these factors have the greatest influence on stress magnitudes and helps to assess the influence of tectonic setting and the degree of consolidation on stress magnitudes (Figure 4). Figure 4 shows field-averaged (small symbols) and basin-averaged (larger symbols) values of R_h categorised by lithology plotted against a depth normalised vertical stress to separate locations by their degree of consolidation. Stress measurements from poorly-consolidated low-strength formations plot to the left whereas measurements from well-consolidated high-strength formations plot towards the right. A weak influence of basin consolidation may be seen by considering only sandstone data which show a subtle decrease in the stress ratio as consolidation increases (Figure 4). However, this trend may only be an artefact of interbasin pore pressure differences.

Examples of relaxed-state and compressed-state stress environments are shown in Figure 4. In relaxed-state basins, R_h is lower in high strength rock than in weaker rock. The majority of the data shown in Figure 4 represent relaxed-state environments where R_h is lower in sandstones than in shales. Evidence for compressed-state environments is found in sedimentary rocks at depths less than 1 km and in parts of the Appalachian and Alberta basins. On a basin scale only the Appalachian basin shows an R_h which is higher in sandstones than in shales. This reversal of the stress ratio with lithology is characteristic of a compressed-state environment. However, elastic modulus is more important than lithology for distinguishing the two types of stress environments (Plumb et al., 1991).

Intrabasin stress variations can be strongly influenced by differences in pore pressure, lithology and stratigraphic or structural position. Pore pressure appears to have the dominant effect particularly in flat lying strata from tectonically relaxed basins where reservoir pressures have been heavily depleted. In some cases a coupled pore pressure-lithology effect can not be ruled out owing to the difficulty in determining pore pressure in shales. In basins where strata have been folded, stress variations with changes in structural position can be as important as the variations due to changes in lithology. As in the Appalachian basin, changes of stratigraphic position, with respect to a salt décollement or other potential slip surfaces, can result in stress changes which are as important as those caused by changes in pore pressure. Lithology may have the strongest influence on stress magnitudes in normally pressured fields without structural or stratigraphic complexities.

Interbasin stress variations are linked to pore pressure, degree of consolidation and stress environment. Basin-averaged R_h for sandstones varies by more than 65% among basins. Part of this difference is governed by pore pressure and part is governed by rock strength, e.g. consolidation. The available data suggest that most of the interbasin variations of stress magnitudes in sandstones are dominated by variations in average pore pressure. R_h ranges between 0.75 and 0.80 in young normally pressured sandstones in offshore environments such as the Gulf of Mexico, Lake Maracaibo, and Brunei. Whereas R_h in overpressured sandstones is generally greater than 0.8. For example, contrast the overpressured sandstones South Texas ($R_h = 0.94$) with normally pressured sandstones from East Texas basins ($R_h = 0.62$) (Figure 4).

The lowest R_h values are found in the low-porosity low-permeability sandstones of the East Texas and Appalachian basins and in carbonates where natural fractures are known to contribute to production. For example the North Sea Chalks ($R_h = 0.67$), and East Texas sandstones ($R_h = 0.62$).

5 CONCLUSIONS

A multi-basin comparative analysis of stress variations in sedimentary rock has demonstrated that intrabasin stress variations are related to changes in pore pressure, lithology and structural or stratigraphic position whereas interbasin stress variations are primarily related to changes in pore pressure, tectonic setting and to a lesser extent lithology and the degree of diagenesis.

Pore pressure and lithology are two new globally significant factors resolved using a multi-basin stress database. Typically we find that, pore pressure induced stress changes are greater than stress variations associated with changes of lithology, tectonic setting, or structural or stratigraphic position.

Further improvement in our understanding of in-situ stresses and causes for their variation require better documentation of pore pressure, better characterisation of the rocks where stress is measured, for example, lithology, porosity and elastic modulus and a more complete description of the geologic structure in the vicinity of each stress measurement.

Acknowledgements: I thank the management of Schlumberger for supporting the study of earth stresses and for permission to publish this work. I have benefited greatly from collaborations with many clients and colleagues including: Terry Engelder, Keith Evans, Marc Thiercelin, Norm Warpinski, Bill Whitehead, Larry Teufel, Helen Farrell, Rick Kry and Mark Gronseth.

REFERENCES

Breckels, I. M., and H. A. M. van Eekelen, Relationship between horizontal stress and depth in sedimentary basins, SPE 10336, 56th Annual Fall Technical Conf. San Antonio, Texas, pp. 21, Oct. 5-7, 1981.

Brown, E. T. and E. Hoek, Trends in relationships between measured in-situ stresses and depth, Int. J. Rock Mech. Min. Sci. & Geomech. Abstr., v. 15, p. 211-215, 1978.

Detournay E., Cheng, A. H.-D., Roegiers, J. C. and J. D. McLennan: Poroelastic considerations in in situ stress determination by hydraulic fracturing, *Int. Jour. Rock Mech. Min. Sci. & Geomech. Abstr.*, v. 26, n. 6, p. 507-513, 1989.

Dusseault, M. B., and F. J. Santarelli, A conceptual model for massive solids production in poorly-consolidated sandstones, in Rock at Great Depth, V. Muary and D. Fourmaintraux, editors, Balkema, v. 2, p. 789-798, 1989.

Ervine, W. B., and J. S. Bell, Subsurface in situ stress magnitudes from oil-well drilling records: an example from the Venture area, offshore eastern Canada, *Can. J. Earth Sci.*, 24, p. 1748-1759, 1987.

Engelder, T., Loading paths to joint propagation during a tectonic cycle: an example from the Appalachian Plateau, U.S.A., *J. Structural Geology,* v. 7, p. 459-476, 1985.

Evans, K. F., Appalachian stress study, 3, regional scale stress variations and their relation to structure and contemporary tectonics, *J. Geophys. Res.*, v. 94, p. 17619-17645, 1989.

Evans, K. F., Engelder, T., and R. A. Plumb, Appalachian stress study 1. A detailed description of in-situ stress variations in Devonian shales of the Appalachian Plateau, *J. Geophys Res.*, 94, p. 7129-7154, 1989.

Gretener, P. E., Can the state of stress be determined from hydraulic fracturing data?, *J. Geophys. Res.*, 70, 24, p. 6205-6212, 1965.

Hickman, S. H., Healy J. H., and M. D. Zoback, In situ stress, natural fracture distribution and borehole elongation in the Auburn Geothermal well, Auburn, New York, *J. Geophys. Res.*, 90, 5497-5512, 1985.

Kunze, K. R., and R. P. Steiger, Extended leakoff tests to measure in-situ stress during drilling, Rock Mechanics as a Multidisciplinary Science, Proceedings of the 32nd U. S. Symposium, J. C. Roegiers ed. Balkema, Rotterdam, pp 35-44, 1991.

Kry, P. R., and J. M. Gronseth, In-situ stresses and hydraulic fracturing in the Deep Basin, *J. Can. Petr. Tech.*, p. 31-35, Nov.-Dec. 1983.

Lacy, L. L., Comparison of hydraulic-fracture orientation techniques, Form. Eval., 2, p. 66-76, 1987.

McGarr, A., and N. C. Gay, State of stress in the earth's crust: *Annual Reviews of Earth and Planetary Science*, 6, 405-436, 1978.

Palmer, I. D., and G. M. Kutas, Hydraulic fracture height growth in San Juan basin coalbeds, SPE 21811, Rocky Mnt Reg./Low Permeability Res. Symp., p. 71-88, April 15-17, 1991.

Plumb, R. A., and J. Cox, Stress directions in eastern North America determined to 4.5 km from borehole elongation measurement, *J. Geophys. Res.*, 92, 4805-4816, 1987.

Plumb, R. A., Evans, K. F., and T. Engelder, Geophysical log responses and their correlation with bed-to-bed stress contrasts in paleozoic rocks, Appalachian Plateau, New York, *J. Geophys. Res.*, 96, p. 14,509-14,528, 1991.

Rosepiler, M. J., Determination of principal stresses and confinement of hydraulic fractures in Cotton valley, SPE 8405, 54th Ann. Fall Tech Conf. Sept. 23-26, Las Vegas, NV, pp. 10, 1979.

Salz, L. B., Relationship between fracture propagation pressure and pore pressure, SPE 6870, 52nd Ann. Fall Tech. Conf., Denver, CO, Oct. 9-12, pp. 8, 1977.

Suman, G. O., Unconsolidated sand stabilization through wellbore stress state control, SPE 5717, 50th Ann Fall Tech. Meeting, Dallas, TX, Sept. 28-Oct. 1, pp. 12, 1975.

Teufel, L. W., Rhett, D., and H. E. Farrell, Effect of reservoir depletion and pore pressure drawdown on in situ stress and deformation in the Ekofisk Field, North Sea, in Rock Mechanics as a Multidisciplinary Science, Proceedings of the 32nd U. S. Symposium, J-C. Roegiers, editor, Balkema, p. 63-72, 1991.

Thiercelin, M. J., Plumb, R.A., Desroches, J., Bixenman, P.W., Jonas, J.K., and W.A.R. Davie, A new wireline tool for in-situ stress measurement, SPE 25906, Rocky Mountain Regional/Low Permeability Reservoirs Symposium, Denver, CO, April 12-14, p. 635-646, 1993.

Tsukahara, T., Stress measurements utilizing the hydraulic fracturing technique in the Kanto-Tokai area, Japan, Workshop on Hydraulic Fracturing Stress Measurements, proc., Monterey, CA, Dec. 1-5, 1981, National Academy Press, Washington, D. C., p. 18-27, 1983.

Van Eekelen, H. A. M., Hydraulic fracture geometry: fracture containment in layered formations, SPE 9261, 1980.

Veatch, R. W. and Z. A. Moschovidis, An overview of recent advances in hydraulic fracturing technology, SPE 14085, SPE 1986 International Meeting on Petroleum Engineering, Beijing, China, March 17-20, p. 421-454, 1986.

Warpinski, N. R., Determining the minimum in situ stress from hydraulic fracturing through perforations, *Int. J. Rock Mech. Min. Sci. & Geomech Abstr.*, v. 26, n. 6, p. 523-531, 1989.

Warpinski, N. R., Branagan, P., and R. Wilmer, In-situ stress measurements at DOE's multiwell experiment site, Mesaverde group, Rifle, Colorado, SPE 12142, 58th Ann. Fall Tech. Conf., San Fransico, CA, pp. 12, Oct. 5-8, 1983.

Whitehead, W. S., J. M. Gatens III and S. A. Holditch, Determination of in situ stress profiles through hydraulic fracturing measurements in two distinct geologic areas. *Int. J. Rock Mech. Min. Sci. & Geomech. Abstr.*, v. 26, n. 6, p. 637-645, 1989.

Woodland, D. C. and J. S. Bell, In-situ stress magnitudes from mini-frac records in western Canada, Petr. Soc. CIM, paper 88-39-67, pp. 27, 39th Ann. Tech Meeting, Calgary, June 12-16, 1988.

1 Hydraulic and mechanical properties of discontinuities

Session developers:
Bernard Amadei, University of Colorado
Joe Ratigan, RE/SPEC, Inc.
Jane Geller, Lawrence Berkeley Laboratory

Rock Mechanics, Nelson & Laubach (eds) © 1994 Balkema, Rotterdam, ISBN 90 5410 380 8

The effect of the critical path on fluid flow through a fracture

Laura J. Pyrak-Nolte & Carlo D. Montemagno
University of Notre Dame, Ind., USA

Abstract: Fluid flow through a fracture or fracture network is controlled by the geometry of the network and by the distribution of voids within each fracture. The critical path, i.e., the path of largest apertures across a fracture, dominates the flow of fluid through the fracture. We have directly quantified the importance of the critical path by using Wood's metal porosimetry and methane flow rate measurements. Filling the critical path with Wood's metal reduced flow rates through the fracture by 70% to 98%. X-ray tomographic scanning and photomicroscopy are used to visualize the void spaces in the fracture.

1 INTRODUCTION

Some of richest known gas reservoirs (14 TcF) are located in coal seams. During the production of methane from coal beds, methane and brine flow through a natural fracture network know as cleats which are the primary hydraulic conductors for the coal. An underlying question about the movement of fluids through a fracture or fracture network is how the fracture void geometry affects the permeability of the fracture. Recent modeling of fluid flow through a single fracture suggests that fluid flow through a fracture is dominated by channel flow (Tsang & Tsang, 1987), by the critical path (Pyrak-Nolte et al., 1988) or by preferential flow paths (Yang et al., 1993). The critical path of a fracture is the path consisting of the largest apertures. Through laboratory experiments and computer-aided visualization, we have investigated the geometry of the cleat network and the critical path in fractures in coal. Saturating the critical path with a non-wetting fluid reduces the wetting phase flow rate. We have also examined the relationship between effective cleat porosity and absolute water permeability by imaging the fracture network in coal.

2 COAL CORES

Table 1 lists the dimensions of the coal samples used in this study. Sample #7-2-3 is a whole drill core from the Pittsburgh seam of the Northern Appalachian Basin, Wetzel County, West Virginia. This sample was used to investigate the geometry of the critical path and the effect of the critical path on fluid flow. Coal cores CC and TT are from Seam #1 of the Sundance Pit at the La Plata coal mine and were cored from blocks of coal. Sample CC was drilled perpendicular to the bedding planes and sample TT was drilled parallel to the bedding planes. Samples CC and TT were used to investigate the geometry of the cleat (fracture) network and the effect of the cleat network on the hydraulic properties of the coal.

Table 1. Sample Dimensions			
Sample Number	Diameter (cm)	Length (cm)	Bulk Volume (cm^3)
#7-2-3	5.22	5.01	105.47
CC	8.89	4.41	274
TT	8.89	11.2	694

Table 2. Properties of Wood's Metal (Cerrosafe®)	
Components:	
Bismuth	42.5 %
Lead	37.7 %
Tin	11.3 %
Cadmium	8.5 %
Melting Temperature Range (°C)	70 - 80
Surface Tension (dynes/cm)	485 ± 17
Contact Angle on Coal	130° ± 21°
Density (grams/cm^3)	9.46 ± 0.16
Young's Modulus (GPa)	9.7

Table 3. Experimental Parameters and Results on Coal Sample 7-2-3 from the Pittsburgh Seam of the Northern Appalachian Basin, and Coal Samples CC and TT from Seam #1 of the Sundance Pit at the La Plata Coal Mine, San Juan Basin, New Mexico.

Sample Number	7-2-3	CC	TT
Confining Pressure at Solidification (MPa)	4.5	4.89	5.54
Nitrogen Pressure at Solidification (MPa)	0.41	0.49	0.44
Effective Pressure at Solidification (MPa)	4.09	4.4	5.1
Weight of Wood's Metal Injected (grams)	0.01	2.14	6.19
Connected Void Volume (cm^3)[1]	0.001±2x10-5	0.22 ± 0.005	0.65 ± 0.013
Effective Cleat Porosity (%)[1]	0.001±2x10-5	0.082 ± 0.002	0.094±0.002
Minimum Aperture Size Accessed (μm)[2]	1.47 ± 0.7	1.22 ± 0.59	1.37 ± 0.65

[1] *Error from deviations in the density of Wood's metal and the contact angle of Wood's metal* [2] *Error from deviations in the surface tension*

3 WOOD'S METAL INJECTION

A Wood's metal injection technique (Pyrak-Nolte et al., 1987; Pyrak-Nolte, 1990) is used to measure the effective cleat porosity of a coal core and capture the geometry of the cleat network. The properties of Wood's metal are listed in Table 2. In this technique a low melting point metal is injected into a sample subjected to a confining pressure. The sample is heated to 95 °C and molten metal is injected into the sample using a pressure generator. A nitrogen backpressure is applied during the injection process and maintained during cooling of the sample. After injection, the sample is allowed to cool and the metal to solidify. Gravimetric data and pressure data are used to determine the volume of voids injected, the effective cleat porosity, and the minimum aperture penetrated by the metal. The geometry of the injected flow paths are examined using X-ray tomography (CT) and photomicroscopy. The pressures used during injection of the samples and the results of the metal injection experiment are given in Table 3.

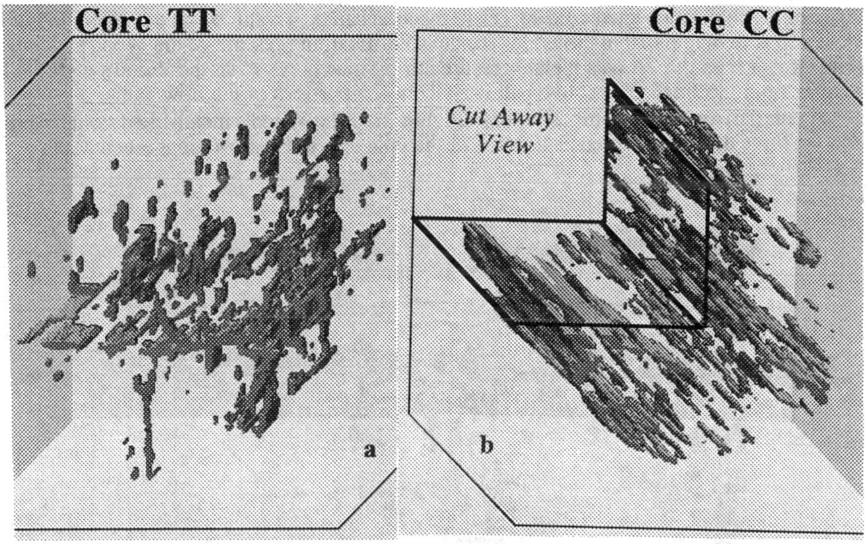

Figures 1 & 2. Three dimensional reconstruction of the cleat networks in coal core TT (a) and coal core CC (b).

4 GEOMETRY OF THE CLEAT NETWORK

The three-dimensional geometry of the cleat or cleat network is reconstructed from the two-dimensional x-ray tomographic (CT) data. The reconstructed fracture networks for coal cores TT and CC are shown in Figures 1 & 2. The fracture network in core TT which was oriented parallel to the bedding planes, and is dominated by a bedding plane fracture which is the horizontal fracture in the Figure 1. The observed fractures in core TT are planar and rectilinear in shape. The fractures in core CC are curvi-linear and show the tortuosity of the cleat network. The geometry of flow paths in the cleats in these coal cores are anisotropic as observed by comparing the reconstructed fracture networks in Figure 1 and Figure 2. The disconnected voids observed in the Figures are a result of the coarse spacing of the CT scans. The scans were taken every 12 mm and 5 mm for cores TT and CC, respectively.

5 GEOMETRY OF THE CRITICAL PATH

Coal core #7-2-3 was injected while subjected to a high confining pressure (4.5 MPa) to examine the geometry of the critical path of the fracture. The critical path of the fracture is the path of largest apertures across the fracture. Figure 3 shows the three-dimensional reconstruction of the cleat in coal core #7-2-3 based on two dimensional CT data. For this sample, the reconstruction shows a single vertical cleat and shows the spatial relationship between the cleat and a bedding plane with high density minerals (Horizontal structure in Figure 3).

Figure 4 is a composite map of the Wood's metal-filled void spaces (represented by black) in the fracture based on photomicroscopy of the metal casts in the fracture. The white regions of the map represent coal to coal contact or inaccessible voids. The direction of metal injection into the core is noted in Figure 4. A swatch of large voids is located in the center of the fracture and is oriented perpendicular to the direction of metal injection. The disconnected void spaces in the map can be a result of (a) the resolution of the composite map (~ 70 microns); or (b) the solidification pressure of the Wood's metal, if it is lower than the Wood's metal pressure during injection so that the Wood's metal retreats from the small connective pathways between portions of the flow paths.

A comparison of the composite map (Figure 3) of the Wood's metal-filled void spaces and the three dimensional reconstruction of the void space (Figure 4) suggests that the

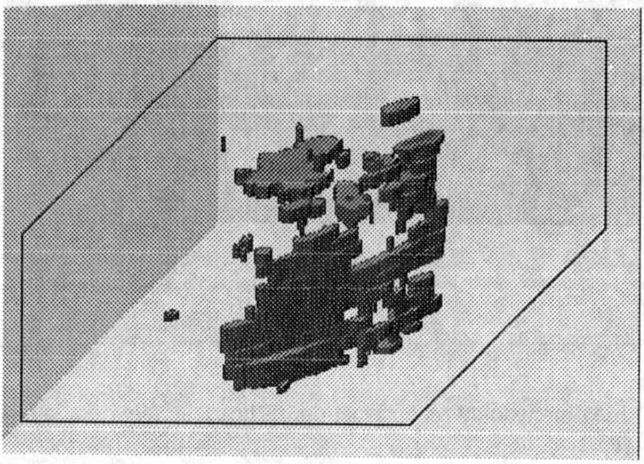

Figure 3. Three dimensional reconstruction of a fracture in coal sample #7-2-3. Horizontail plane features is a region of minerals with high density.

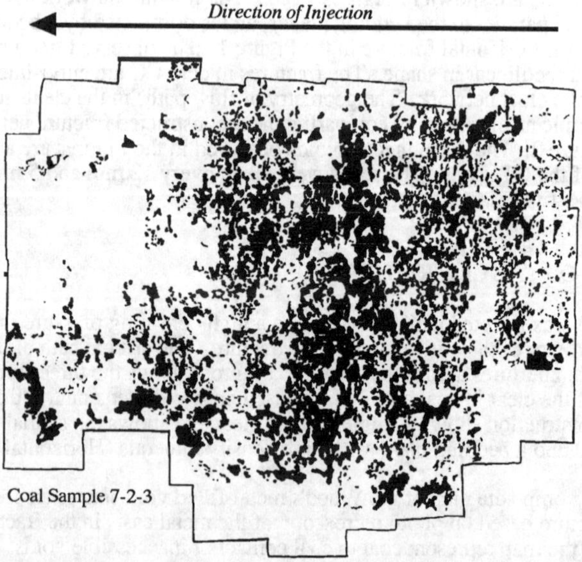

Figure 4. Composite image of the void space in a single cleat in coal sample #7-2-3. Black represents the void spaces. Notice preferential flow path is perpendicular to the direction of the metal injection.

three dimensional reconstruction gives the gross outline of the flow paths. However, the CT scans for this sample were taken every three millimeters while the structure in the composite map is observed to vary on length scales much smaller than that. The three-dimensional reconstruction would be improved by using a smaller spacing between scans.

6 PERMEABILITY-POROSITY PARADOX

Two service laboratories made measurements of absolute water permeability and porosity on samples CC and TT from the San Juan Basin in New Mexico (Pyrak-Nolte, 1992). Each laboratory used the unsteady-state method where the coal core is saturated with brine that is subsequently displaced with humidified helium. The volumes of helium and brine produced are recorded and used in models to determine the relative permeability of the sample (Gash, 1991; Gash et al., 1992). The effective cleat porosity and absolute water permeability are shown in Figure 5. Core CC was oriented perpendicular to the bedding planes (parallel to the face cleats) and core TT was oriented parallel to the bedding planes. Contrary to expectation, the core with the higher permeability (TT) had the smaller porosity (Figure 5).

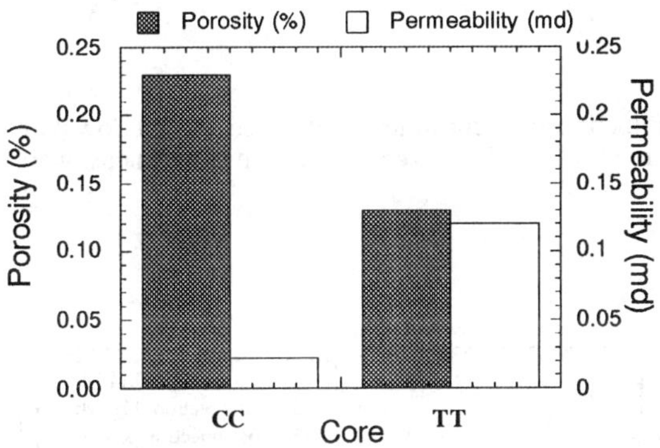

Figure 5. Porosity and absolute water permeability for coal cores CC and TT from the San Juan Basin.

This apparent paradox in the permeability and porosity relationship for these two coal cores was resolved using information from the geometry of the cleats obtained from the Wood' metal injection experiments. The ratio of the effective cleat porosity determined from the Wood's metal method to the effective cleat porosity determined from unsteady-state method gives the fraction of cleat porosity with apertures greater than two microns. This ratio revealed that 36 % of the fracture network porosity in coal core CC consisted of apertures greater than 2 microns. For coal core TT, 76 % of the fracture network porosity consisted of apertures greater than 2 microns. While coal core CC had a higher porosity than coal core TT, the porosity was mainly from small apertures (< 2 microns) which support less flow. Thus, the permeability is controlled by the geometry and interconnectedness of the void spaces within the cleat network structure. Three-dimensional reconstructions (Figures 1 & 2) of the x-ray tomographic images of the fractures illustrate the anisotropy in the fracture network in coal.

7 EFFECT OF THE CRITICAL PATH ON FLOW RATES

Flow measurements were made on coal sample #7-2-3 using methane for a range of confining pressures (0.69 MPa - 4.2 MPa) and a range of pore pressures (0.069 MPa - 0.34 MPa) prior to injection of the sample with Wood's metal. Figure 6 shows that the pre-injection flow rates of methane through the sample (for pore pressures of 0.14 MPa and 0.28 MPa) decrease with increasing confining pressure and increase with increasing pore pressure.

To determine the effect of the critical path on flow through a fracture, coal sample #7-2-3 was injected with Wood's metal while subjected to a stress of 4.5 MPa. The total volume of Wood's metal injected was 0.001% of the total bulk volume of the sample. Apertures as small as 3 microns were penetrated with the metal. Photomicroscopy of the sample has shown that flow paths through the fracture have variable apertures and appear to be spatially correlated. After injection, methane flow rate measurements were made. Figure 6 shows the effect of filling the critical path with a non-wetting phase on the flow rates of the wetting phase (methane). The flow rate is markedly decreased and flow does not occur for confining pressures greater than 2.1 MPa. For purposes of comparison, the ratio of the post-injection flow rate to pre-injection flow rate was taken and is shown in Figure 7 as a function of confining pressure. The post-injection flow rates are observed to be much lower, and were determined to be 70% to 98% lower than the pre-injection flow rates. In addition, the ratio of pre- to post-injection flow rates decreases with increasing stress.

Figure 1. Three dimensional reconstruction of Wood's metal flow paths through a fracture in coal sample 7-2-3. The image is based on two-dimensional x-ray tomographic scans.

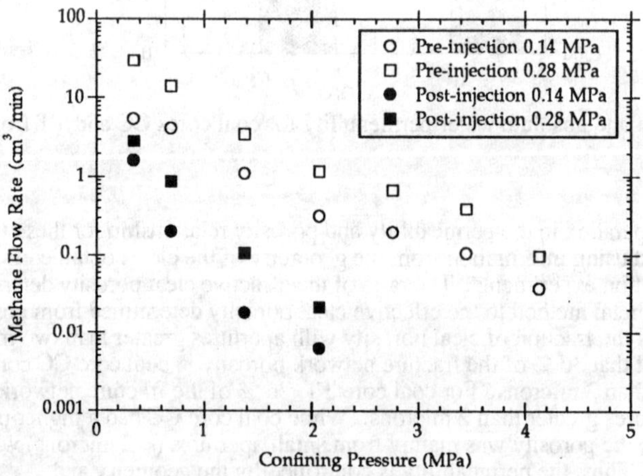

Figure 6. Methane flow rates through coal sample #7-2-3 for pore pressures of 0.14 MPa and 0.28 MPa prior to and after injection of the sample with Wood's metal.

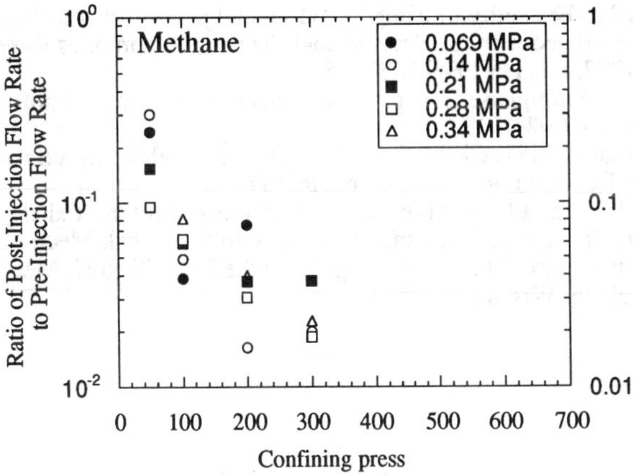

Figure 7. Ratio of post-injection flow rates to pre-injection flows for pore pressures ranging from 0.069 MPa to 0.34 MPa for coal sample #7-2-3.

8 CONCLUSIONS

The dramatic reduction in flow rates is a direct result of obstructing the critical path. The data show that if the critical path of the fracture is saturated with a non-wetting fluid the wetting phase flow rate is greatly reduced. In addition, the effect on the flow rate is not constant as the stress is increased. The reduction in flow rate increases with increasing stress until no flow occurs. These experimental results are significant because they illustrate the dominance of the critical path on fluid flow through a fracture.

Acknowledgments: The authors wish to acknowledge the Gas Research Institute (Contract No. 5092-260-2507), Amoco Production Company, and Argonne National Laboratory. In addition, we thank Matt Owen for constructing the composite photomicrograph.

REFERENCES

Gash, B. 1991. Measurement of 'rock properties' in coal for coalbed methane production. Proceedings of the 1991 SPE Annual Technical Conference & Exhibition. Dallas, Texas, USA. SPE #22909, p221-230.

Gash, B., Volz, R.F., Gary, P., and J. M. Corgan. 1992. The effect of cleat orientation and confining pressure on cleat porosity, permeability and relative permeability in coal. 1992 SCCA Conference paper no. 9224, June, 1992, Tulsa, Oklahoma.

Pyrak-Nolte, L.J., Myer, L.R., Cook, N.G.W. and P. A. Witherspoon. 1987. Hydraulic and mechanical properties of natural fractures in low permeability rock. Proceedings: International Society for Rock Mechanics, 6th International Congress on Rock Mechanics, Montreal, Canada, August 1987, v. 1, Rotterdam, A.A. Balkema p.225-231.

Pyrak-Nolte, L. J., Cook, N.G.W. and D.D. Nolte. 1988. Fluid percolation through single fractures. Geophys. Res. Lett. 15(11): 1247-1250.

Pyrak-Nolte, L.J. 1991. Feasbility of using Wood's metal porosimetry technique to measure fracture void geometry of cleats in coal. Topical Report, Gas Research Institute GRI-91/0373; NTIS PB 92-137678.

Pyak-Nolte, L. J. 1992. Multiphase flow through cleats in coal. Final Report, Gas Research Institute GRI-92/0503.

Tsang, Y. and C.F. Tsang. 1987. Channel model of flow through fractured media. Water Resources Research. v. 23, no. 3, pp.467-479.

Yang, G., Cook, N.G.W., and L. R. Myer. 1994. Analysis of preferential flow paths using graph theory. To appear in the International Journal of Rock Mech Mining Sci. & Geomech. Abstr as part of the Proceedings from the 34th US Rock Mechanics Symposium, Madison, Wisconsin (1993).

Rock Mechanics, Nelson & Laubach (eds) © 1994 Balkema, Rotterdam, ISBN 90 5410 380 8

Fluid flow network and fractal diffusion fronts in fractured rock mass

Yuzhuo Zhang & Yoginder P.Chugh
Department of Mining Engineering, Southern Illinois University at Carbondale, Ill., USA

ABSTRACT: This paper studies fluid flow network and diffusion fronts in two and three dimensions. The fracture system and fluid flow cluster are described with lattice models. A randomly constructed fractal model is used to simulate fractured rock mass, and the fracture distribution is expressed by a fractal measure. The fluid diffusion fronts are given by random walk on fractals. The flow distribution through a fractal object are studied by numerical modeling.

1 INTRODUCTION

Predicting fluid transport in heterogeneous fractured rock masses is one of the most important problems currently studied by geologists and civil and mining engineers. Substantial research has been performed on fluid flow through porous media for which permeability constants are well-defined (Nolte, et al:1989). However, due to the difficulty of knowing the geometry of the heterogeneity on a macroscopic scale, fluid flow through fractured rock is not as well understood. Lately, special interests have been concentrated on fracture networks and the transport properties for the purposes of nuclear waste isolation and other applications.

Methods of studying fluid flow in fractured rock are extraordinarily difficult and diverse. Prior to studying fluid flow in a set of intersecting and interacting fractures, insight has been gained into flow in an isolated fracture[Zhilenkov,1975]. Fluid flow through a single fracture is assumed analogous to laminar flow between two perfectly smooth parallel plates. Flow between two parallel plates representing two walls of the crack can be expressed as:

$$Q=\frac{Wb^3}{12\mu}\frac{dp}{dx}$$

where Q is the flow in units of volume per unit time, dp/dx is the gradient of the hydrostatic pressure, μ is the viscosity and W is the width of the fracture in the direction perpendicular to the flow direction. The above equation is also called cubic law because flow volume is proportional to the cubic power of b. For a natural fracture, however, study has shown that flow depends on larger exponents of aperture than cubic. This may be caused by tortuosity of the flow paths in the real rock fractures which have higher resistance than smooth plates.

In most engineering problems, fluid flow in fractured rock mass is a three-dimensional problem which challenges the traditional description. Fractal geometry, an effective tool exploring complexity of the nature, provides a better avenue for studying flow network and percolation surfaces. The concept of fractal dimensions was introduced by Benoit Mandelbrot(1975), who recognized that many natural phenomena could be described by fractals. One of the common features of fractal objects is that they are self-similar or scale-invariant. For example, a large-scale picture of rock structures shows a fracture cluster with an overall appearance rather similar to another picture the one seen when viewing the rock sample under a microscope. Using fractal notion, Mandelbrot (1983) studied the Bernoulli percolation clusters. Katz and Thompson (1985) studied the fractal properties of sandstone, while Barton and Larson (1985) found that fracture networks may display fractal characteristics. Wheatcraft, et al., (1990) studied fluid flow in fractal porous media. All studies led to the conclusion that fractal concepts are operative over many rock mechanics problems.

The objectives of this investigation were to describe the flow network in fractured rock masses, using fractal concepts combined with numerical analysis.

2 FLUID FLOWING PATHWAYS IN TWO DIMENSIONS

A two dimensional lattice is constructed in Figure 1. The white area in Figure 1 is used to simulate fractures which allow fluid to pass through. If water was put water on the top of the system, continuous pipes would transport the water through the rock mass.

It can be seen from Figure 1 that the percolating pathways exist only when the connection of the pipes exists. In a rock mass with uniformly distributed fractures, percolating paths depend on the fracture density. The critical fracture level at which pathways are likely to be found through the rock mass is called threshold. From the study, we found that the threshold is about 55% . As discussed by Mandelbrot(1977), the critical probability in a Bernouli lattice exists, and the related fractal dimension is:

$$D_c=1+(E-\eta)/2$$

where E is Euclid dimension and η is the mass of the cluster.

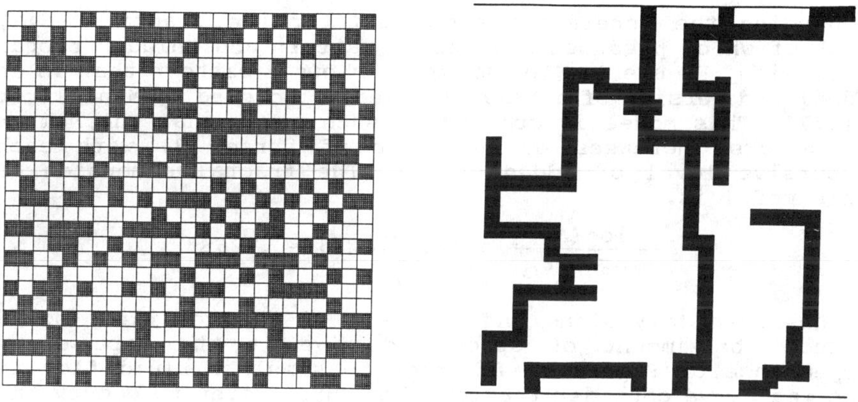

a b

Figure 1. Two dimensional lattice model to illustrate percolating pathways
(a) Percolating pathways exist at 55% occupancy.
(b) Schematic picture of the backbone of the percolation cluster

3 3D LATTICE MODEL AND MENGER SPONGE CONSTRUCTION

Rock masses with fracture systems can be topologically simulated by a 3D fractal medium. A Menger sponge with different fractal dimensions is an ideal model for representing fractured rock with different fractures. Two (2) typical constructions with triangular and square lattice are shown in Figure 2. Also a different choice can be made

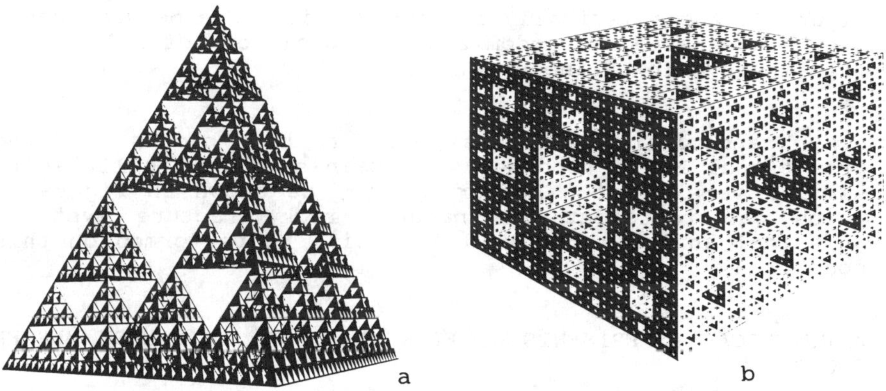

a b

Figure 2. Fractal objects in three-dimensions
(a) A fractal skewed web with D=2.
(b) The Menger Sponge (D=2.7268).
(from B.B. Mandelbrot: The Fractal Geometry of Nature, 1977,1982)

regarding the generating algorithm for the fractal media, each of which produces a distinct pattern and unique fractal dimension. Figure 2a, for example, shows a pattern that is an idealized version of a fractal object discussed by Mandalbrot [1977]. This model is created in such a way that the number of blocks increases by a factor of three (3) with each recursive level of added detail. The fractal dimension of this model is:

$$D=\frac{\log(N)}{\log(1/\varepsilon)}=\log(20)/\log(3)=2.72683$$

Subsequent divisions and removals of the remaining solid reduce the amount of space and give it the more empty appearance of Figure 2. In order to simulate fluid flow in the media, we consider that the fracture system to occupy the solid space, which is assumed to be permeable.

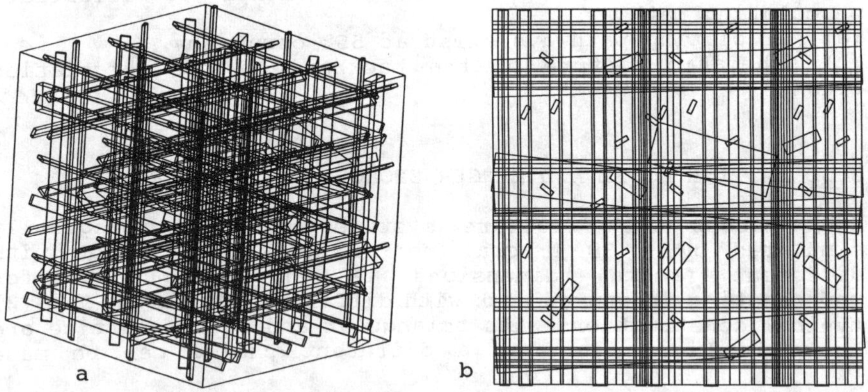

Figure 3. A statistically similar version of a Menger sponge can be generated by randomizing the positions of holes (a) 3D view. (b) Plane view.

By randomizing the position of the blocks in the traditional Menger sponge, we obtain a statistically self-similar version of a Menger sponge, as shown in Figure 3, which can represent the natural rock fracture systems. Numerical analysis of fluid flow will be performed on this model.

4 NUMERICAL EXPERIMENTS OF FLOW TRANSPORT IN RANDOM MENGER SPONGE

If a random Menger sponge is used for a fluid flow distribution, we must make some assumptions regarding how values of hydraulic conductivity are to be assigned. As mentioned above, the filling blocks are taken as rock blocks which have little conductivity, while voids can conduct fluids. As assumed by Wheatcraft[1990], a binary hydraulic conductivity distribution is assigned in which the conductivity in voids is two orders of magnitude higher than that in blocks.

Simulation of fluid flow in fractured rock mass is accomplished numerically using a finite difference scheme to solve the following 3D steady state, continuity equation:

$$\frac{\partial}{\partial x}(K_{xx}\frac{\partial u}{\partial x})+\frac{\partial}{\partial y}(K_{yy}\frac{\partial u}{\partial y})+\frac{\partial}{\partial z}(K_{zz}\frac{\partial u}{\partial z})=f$$

where u is the potential at (x,y,z), K_{xx}, K_{yy} and K_{zz} are hydraulic conductivities in x, y and z directions respectively, and f is the source/sink term at location (x,y,z) as volumetric flux per unit volume. The Dirichlet boundary conditions are applied on the top and bottom faces, while other faces are taken as zero flux.

The example in Figure 4 shows the streamlines and isosurface of the model in Figure 3.

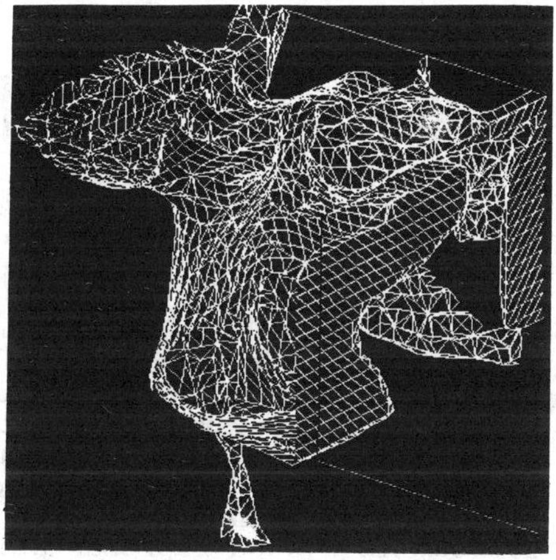

Figure 4. Simulated flowing isosurfaces

5 FLOW FRONTS THROUGH WALK ON FRACTALS

Consider the random motion of a particle placed onto a fractal network of dimension D made of bonds. The walk starts at t=0 and the mass moves with equal probability to any of the nearest neighbor sites belonging to the fractal. The diffusion equation was given by Vicsek(1989)

$$\frac{\partial p(r,t)}{\partial t}\simeq\frac{1}{r^{D-1}}\frac{\partial}{\partial r}(\sigma_0 r^{D-1-\theta}\frac{\partial p(r,t)}{\partial r})$$

where σ is the total conductivity of a shell of $n(r)\sim r^{D-1}$ sites being at a distance r from the origin.

The above equation has the solution

$$p(r,t)=Ct^{-D/(2+\theta)}\exp(-\frac{r^{2+\theta}}{\sigma_0(2+\theta)^2 t})$$

The mean squared distance from the origin

$$L_0^2(t)\sim t^{2/d_w}$$

where $d_w+2=\theta$ depends on the fractal structure.

Diffusion fronts can be obtained by assuming that the concentration range is proportional to the diffusion length.

In fig 5, we show a schematic picture of a simulated sample. A fractured zone was created by an excavation which is bellow an aquifer. The flow paths exist when the diffusion fronts touch the fractured zone.

It is found that the diffusion front is a ideally porous material and the fractal dimension is about 2.5 at its threshold.

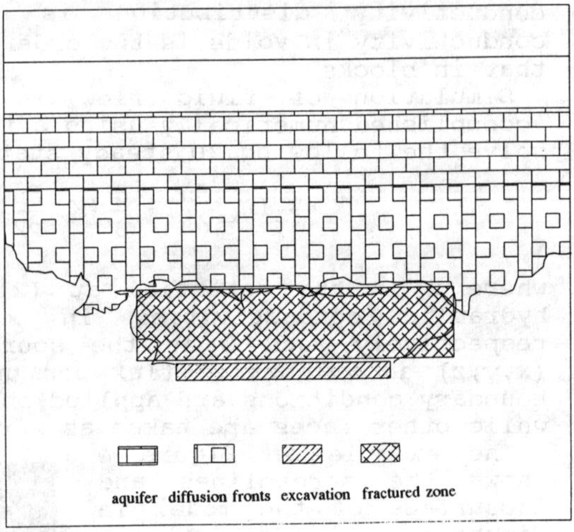

Figure 5. Diffusion fronts and interaction with fractured rock

6 SUMMARY AND CONCLUSIONS

The concept of fractals offers a new viewpoint on quantifying the fluid flow through fractured rock mass. Based on the successful use of fractals to describe rock properties, it appears that it will be a valuable descriptor of the nature of rock fracture systems and their transport properties.

The percolation cluster depends on the density of voids and the threshold is about 55% of occupancies in two dimensions. A random Menger sponger model, with fractal dimension of 2.7268, is a good representation of natural fractures in 3D space. Using finite difference scheme, the flow distribution through this model was simulated, which has inhomogeneous property.

The concept of diffusion fronts is useful to understand the radius of flow percolation within the medium, which may be used to predict water inflow in mining under aquifers. Study found that the diffusion fronts through the fractal objects are very irregular and rough.

REFERENCES

Bunde, A. et al 1991. Fractals and Disordered Systems, Springer-Verlag, p.188-194.

Chernyshev, S.N. 1991. Rock Fractures, Butterworth-Heinemann, p.197-207.

Goff,J.C. Ed 1987. Fluid flow in Sedimentary Basins and Aquifers, Blackwell Scientific Publications, p.201-218.

Kaye,B.H. 1989. A Random walk through Fractal Dimensions, VCH Publishers, p.288-245.

Lee, C.H. 1990. Estimates of Conducting aperture,fracture porosity and fracture permeability in rock masses, Rock

Mechanics Contributions and Challenges: Proceedings of the 31st U.S. Symposium, p.355-362.
Mandelbrot, B.B. 1977. The fractal geometry of nature, W.H. Freeman and Company, p.143-145.
Moon, F. C. 1992. Chaotic and Fractal Dynamics, Jon Wiley & Sons, Inc., p.325-337, 391.
Stevens,R.T. 1989. Fractal Programming in C, M&T Books, p.222-225.
Vicsek, T 1989. Fractal Growth Phenomena, World. Scientific Publishing Co., p.130-135.
Wheatcraft,S.W., et al 1990. Fluid Flow and Solute Transport in Fractal Heterogeneous Porous Media, Ed by Cushman, J.H.,Dynamics of Fluids in hierarchical Porous Media, Academic Press Limited, p.305-325

Rock Mechanics, Nelson & Laubach (eds) © 1994 Balkema, Rotterdam, ISBN 90 5410 380 8

Physically-based models of rock fracture

Carl E. Renshaw & David D. Pollard
Stanford University, Calif., USA

ABSTRACT: The geometric evolution of a fracture set in a controlled laboratory experiment is accurately predicted using a physically-based network simulator (PBNS). The experimental fracture sets were produced by straining a brittle coating applied to an acrylic substrate. The physically-based fracture sets are generated using iterative solutions to the boundary value problem for multiple cracks in an otherwise homogeneous, isotropic, and linear elastic solid subject to uniform remote stress boundary conditions. A preliminary methodology for conditioning the PBNS networks to data from seismic surveys and wellbore flow tests is also presented. The advantage of this methodology is that information concerning the elastic and fracture characteristics of the rocks and the mechanics of the rock fracture process can be combined with other wellbore data to better estimate the geometry of a particular fractured formation.

INTRODUCTION

As fracture networks are an important component of many geological assemblages, an understanding of their geometric characteristics is essential for accurate interpretations of geologic systems (e.g., Rubin and Pollard 1987; Olson and Pollard 1991). Further, since many sub-surface phenomena such as seismic wave propagation and fluid flow and transport are largely controlled by the geometry of the fracture system, our understanding of this geometry directly affects our ability to accurately model many practical problems such as waste isolation, ore deposit genesis, natural resource recovery and aquifer remediation (e.g., Nelson 1985; Garven 1986; Cacas, Ledoux et al. 1990; Murdoch, Losonsky et al. 1990). While our current understanding of the geometry of natural fracture networks is still limited, the systematic nature of most fracture networks strongly suggests that a more complete understanding of the geometry of these networks can be reached using known physical principles (Dey and Wang 1981; Cook 1988; Pollard and Aydin 1988).

We show that the geologic interpretation and simulation of fracture network geometries can be improved by incorporating well established physical principles into models for the generation of natural fracture networks. In particular, we present laboratory and numerical investigations focused on the development of, and appropriate initial and boundary conditions for, a physically-based fracture network simulator (PBNS). A comparison between fracture networks produced in the laboratory and those simulated using the PBNS suggests that the PBNS can accurately predict the evolution of experimentally generated fracture sets.

We also present preliminary results from a methodology we are developing which attempts to condition the networks generated by the PBNS to results from seismic surveys and wellbore flow tests. The advantage of this methodology is that information concerning the elastic and fracture characteristics of the rocks and the mechanics of the rock fracture process can be combined with other types of information to better estimate the geometry of a particular fractured formation. While there are still many issues that need to be addressed, such as the effects of fracture surface roughness, fracture infilling, *in situ* stress and 3D fracture growth, the work presented here is a first step towards improved models of fractured media which are consistent with many different types of data.

In this paper we first review the appropriate boundary conditions for the model by exploring the environment of fracture formation and what happens when fractures intersect each other. We then briefly develop the PBNS and demonstrate its ability to match laboratory data. Finally, we demonstrate the methodology we are developing to condition these networks to various types data by means of an example problem.

ENVIRONMENT OF FRACTURE FORMATION

Field observations indicate that many opening-mode fractures that are significantly longer than the grain-scale of the rock are approximately planar on the scale of the fracture length (Pollard and Segall 1987). In many cases, the development of fracture curvature may have been limited by the ambient stress state during fracture formation (Cotterell and Rice 1980; Olson and Pollard 1989; Cruikshank, Zhao et al. 1991; Renshaw and Pollard 1994, in press). For example, it is well known that in linear elastic materials, fracture propagation paths are sensitive to the ratio between the remote differential stress, or T-stress, and the fracture driving stress. Two mechanically interacting fractures follow straighter paths for greater values of this ratio. We have shown that the degree of curvature that develops between two mechanically interacting fractures is also limited by sub-critical fracture growth and fracture surface roughness (Renshaw and Pollard 1994, in press). In this paper we assume that the fractures propagate along straight paths. Thus we implicitly assume that the fractures are subject to high differential stresses, grow sub-critically and/or have rough fracture surfaces.

A second important assumption of the PBNS is that fracture growth rates are slow enough that they are not limited by the rate of fluid flow from the saturated rock material into the void space created by the growth of the fracture. The growth rate of the fractures may instead be limited by strain rate (Segall 1984) or the combination of a sub-critical stress state and chemical corrosion at the fracture tip (Atkinson 1979; Atkinson 1980; Wilkins 1980; Swanson 1984). An important consequence of this assumption is that the fluid pressure within the fractures during fracture growth, and hence the applied traction along the fracture wall, remains approximately constant and nearly equal to the ambient pore pressure. Results from a complete poroelastic model of fracture growth in saturated linear elastic medium suggest that for isolated fractures having an initial half length of 0.1m, typical maximum propagation rates are on the order of meters per second for fractures in sandstone and meters per day for fractures in granites.

FRACTURE INTERSECTION

Many characteristics of a fracture network (e.g., permeabillity) are sensitive to the extent to which individual fractures from different sets intersect and link together. Qualitatively, we expect that more fracture linkage will lead to more network connectivity, resulting in increased network permeability. We have found that the extent to which individual fractures from two different fracture sets are linked together is enhanced by two processes: (1) *coincident initiation* of fractures from the younger set on older fractures, and (2) younger fractures crossing older fractures rather than terminating at them.

Coincident initiation occurs when a younger fracture initiates from the same flaw that initiated an older fracture (Wu and Pollard 1992). This process enhances fracture linkage since the younger fracture must only intersect *one* additional older fracture in order to link two older fractures. In contrast, if a younger fracture initiates from an isolated flaw, then it must intersect *two* older fractures in order to serve as a link.

Similarly, stress states which allow younger fractures to cross older fractures also enhance fracture linkage. In particular, when younger fractures can not cross older fractures, then a younger fracture that began by coincident initiation can intersect at most three fractures, while a younger fracture that began at an isolated flaw can intersect at most two fractures. In contrast, if the stress environment favors crossing, then a younger fracture can intersect more than three older fractures. Consequently, there is a greater likelihood for increased fracture linkage and more network connectivity.

We have developed a simple criterion which accurately predicts the stress state required for crossing to occur in laboratory samples. The criterion is developed through a first order analysis of the stresses near a fracture almost impinging upon a frictional interface which is oriented normal to the growing fracture. According to this criterion, for crossing to occur the magnitude of the compression acting across the older fracture must be on the order of the tensile strength T_o of the rock (typically a few tens of MPa or less).

PHYSICALLY-BASED MODEL DEVELOPMENT

The PBNS presented here is based on an approximate solution to the boundary value problem for multiple fractures mechanically interacting in an infinite, linear elastic medium. The essential features of the PBNS algorithm may be broken into three steps: 1) a flaw geometry is specified; 2) the flaws are subjected to a constant uniaxial remote tension and the mode I stress intensity factor (SIF), representing the degree of stress concentration at a fracture tip (e.g., Broek 1986), is calculated for each fracture tip; and 3) fracture tips with an SIF greater than a critical value are advanced in a straight line over a distance that is proportional to the SIF. Steps 2 and 3 are repeated in a series of growth iterations to some arbitrary stopping point. Due to the mechanical interaction of the flaws and fractures, some flaws never grow significantly, others grow and then stop and still others grow continuously throughout the entire simulation.

The flaw geometry may be input explicitly or generated randomly. However, in creating a flaw distribution, a minimum length scale for the model is defined. Any length less than the length of the smallest flaw is beyond the limit of resolution of the model. Thus, the PBNS arbitrarily joins any fracture tips that get closer to each other than the minimum flaw length. This joining of close fracture tips implies that the minimum flaw size is on the order of the minimum spacing between the fractures.

Results from some laboratory investigations on sub-critical fracture growth suggest that the rate at which slow fracture growth occurs is proportional to the SIF (e.g., Atkinson 1979; Atkinson 1980). This effect is modeled by specifying the amount each fracture tip is advanced in each growth iteration to be a function of the SIF at that fracture tip. Specifically, the PBNS calculates the amount each tip is advanced, l_{adv}, using:

$$l_{adv} = 2a_o\left(\frac{K_I}{K_I(max)}\right)^{2\alpha} \tag{1}$$

where K_I is the mode I SIF (e.g., Broek 1986). $K_I(max)$ is the maximum SIF for the most intensely loaded fracture in the set at the current growth iteration and α is the *velocity exponent.*

The velocity exponent controls how strongly the propagation velocity of a fracture is related to the energy available for fracture growth, represented by the degree of stress concentration at the fracture tip. When $\alpha = 0$, the growth rate is independent of the degree of stress concentration at the fracture tip. As α increases above zero, the rate at which a fracture grows becomes more and more dependent upon the degree of stress concentration at the fracture tip. Laboratory observations of sub-critical fracture growth (i.e., fracture growth limited by a combination of sub-critical stress and chemical reactions at the fracture tip), suggest that α may be in the range 1-25 (Atkinson 1987).

Figure 1 shows the results from a comparison between a fracture set simulated using the PBNS to an experimentally generated fracture set developed in a brittle coating applied to acrylic (PMMA) sheets (Wu and Pollard 1991). It is assumed that the brittle coating is linear elastic and that the overall change in strain in the coating during the evolution of the fracture set is small enough that, for modeling purposes, the constant strain boundary condition can be approximated as a remotely applied constant stress.

To numerically model the evolution of the experimental fracture set shown in Figure 1, the geometry of the first visible fractures was used as the flaw distribution. Notice, however, that the numerical flaw distribution also contains many members not shown in the experimental distribution. Although initially too small to be visible, these flaws are known to exist because they begin to grow during later stages of the experiment. For modeling purposes, all flaws not visible in the first stage are given the same small initial length and the remotely applied stress and critical SIF are set such that none of these flaws grow at the beginning of the simulation. These flaws only grow when the stresses generated by the surrounding fractures increase the local loading to the point where their SIFs exceed the critical value.

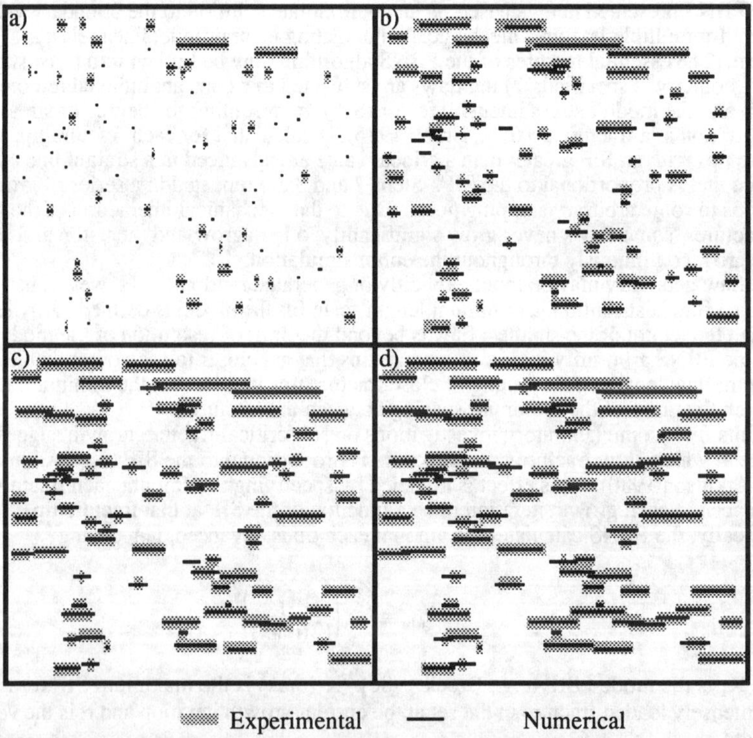

Figure 1. Comparison between experimental and numerical fracture set growth.

The match between the experimental and numerical evolution shown in Figure 1 is quite good in the sense that the differences between the location of the fracture tips in the numerical and experimental sets are small. The value of α found to best simulate the growth of the experimental fracture set was 0.35. Values of α that were either larger or smaller than 0.35 resulted in greater average discrepancies between the locations of the fracture tips. Many of the large discrepancies occur near the perimeter of the network where propagation of experimental fractures is influenced by edge effects not present in the numerical simulation.

FIELD APPLICATION

The advantage of the PBNS over stochastic methods for generating fracture networks is that it allows us to incorporate information gained from fracture mechanics theory into the generation of the network. However, this information is not sufficient to simulate the geometry of a particular fracture system in the field. To do this, the PBNS must be combined with other types of information such as data from pump tests and seismic surveys (i.e., the networks must be *conditioned*). The development of techniques for combining various types of indirect information concerning a particular fracture system into the simulation of the system is important since direct data is often limited. By incorporating several different types of information, the simulation is improved and more accurately reflects the actual system.

We are currently developing a methodology for conditioning the PBNS to field data. To briefly demonstrate our preliminary methodology, consider the hypothetical problem shown in Figure 2. The goal is to simulate the geometry of the fracture network within the unshaded bedded formation underlying the four wells. We assume that the fracture system is two dimensional (i.e., all fractures within the formation are vertical and completely penetrate the formation). We further assume that the velocity exponent and flaw density, defined as:

$$S_p = \frac{1}{A} \sum_i l_i^2 \tag{2}$$

where A is the area of material containing i flaws, each of length of length $2l_i$, are constant throughout the formation and can be determined from the available outcrop exposures. Outcrop exposures also indicate that there are two orthogonal fracture sets, with the fracture set parallel to the fold axis having formed first. The actual fracture system underlying the wells, generated with the PBNS using the known flaw density and velocity exponent, is shown in Figure 3.

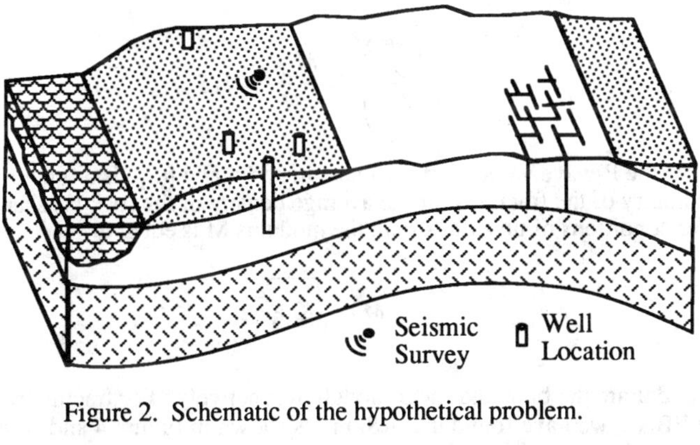

Figure 2. Schematic of the hypothetical problem.

In order to simulate the fracture system under the wells, we condition the PBNS to two types of indirect information gathered from surface investigations. First, we assume that seismic surveys are able to determine the average P-wave velocity through the fractured system in the two directions perpendicular to a fracture set. P-wave velocities through the intact formation are assumed to be known and homogeneous and isotropic. Second, we assume that each well has been pumped while the drawdowns were recorded at the three other wells.

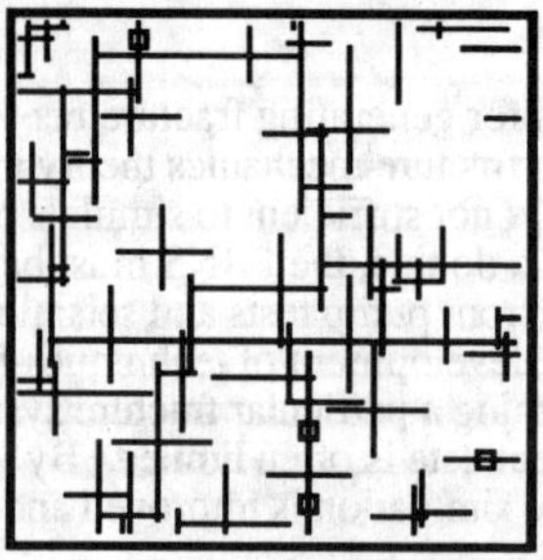

Figure 3. Actual fracure system.

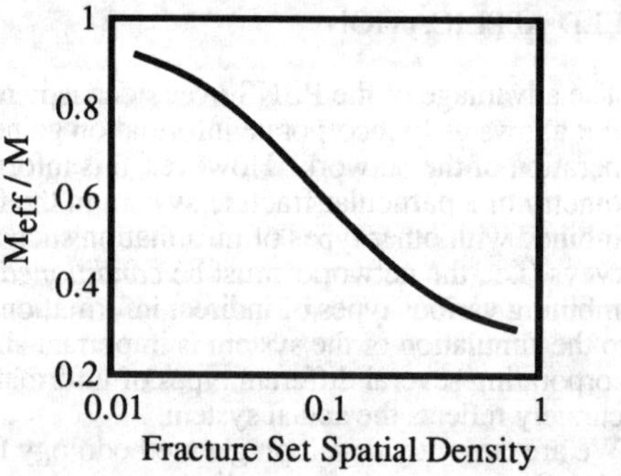

Figure 4. Effective moduli versus S_p.

Seismic conditioning

In order to generate fracture networks using the PBNS which are similar to the network underlying the wells, it is necessary to know each fracture set spatial density. Since it is well known that fracture spatial density is an important control of the effective elastic moduli of fractured solids (e.g., Kachanov 1992), we propose to estimate each fracture set spatial density from the effective elastic moduli of the fractured formation, which are estimated from the effective seismic velocities of the fractured formation.

For example, the relationship between the effective P-wave velocity and the effective modulus M_{eff} of a homogeneous, isotropic elastic solid containing a single set of parallel fractures having a spatial density S_p (determined from equation (2) where A is the area of fracture material containing i fractures 2li in length) is often assumed to be (e.g., Kachanov 1992; Telford, Geldart et al 1990):

$$\left(\frac{V_p^{eff}}{V_p}\right)^2 = \frac{M_{eff}}{M} = f(S_p) \tag{3}$$

where V_p is the P-wave velocity of the intact material and $f(S_p)$ is a function that depends of the geometry of the fractures. If the average density of the rock is assumed constant, then under conditions of uniaxial strain, the modulus M is equal to:

$$M = \sqrt{K + \frac{4}{3}\mu} \tag{4}$$

where K and μ are the bulk and shear moduli, respectively. For fracture sets generated using the PBNS, we have found that $f(S_p)$ is as shown in Figure 4 and is independent of the velocity exponent and flaw density.

Once the spatial density of each fracture set is obtained from the effective seismic velocity and equation (3), they can be combined with the velocity exponent and flaw density to generate a series of *physically-similar* fracture networks using the PBNS. The physically-similar networks are conditioned to the outcrop and seismic data in the sense that the networks have similar velocity exponents, fracture spatial densities and flaw densities as those for the fracture system underlying the four wells. Any differences between two physically-similar fracture networks are only due to differences in the arrangement of flaws used to generate the networks.

Pump test conditioning

Pump tests within a fractured formation give us two important types of information. At the most basic level, each well tells us whether or not a fracture exists at that location. However, if we pump from one well and observe the drawdowns at other wells, we can also determine if two wells are directly connected by the fracture network. Because the effective conductivity between two wells connected by the fracture network is large, wells that are connected will tend to have similar drawdowns. If the wells are not connected, then the effective conductivity between the wells is small and their drawdowns are different.

From Figure 3 we can see that all four wells intersect a fracture and that three of the four are connected by the fracture network. In order to condition the physically-similar networks to this data, we randomly generate a large number (5,000) physically-similar networks and pick out those which happen to match the data. Two such networks are shown in figure 5. By repeating this process it is possible to generate a set of equally-likely realizations of the actual network. Analyses of the flow characteristics of this set of networks provide an estimate of the expected flow and transport characteristics of the actual network.

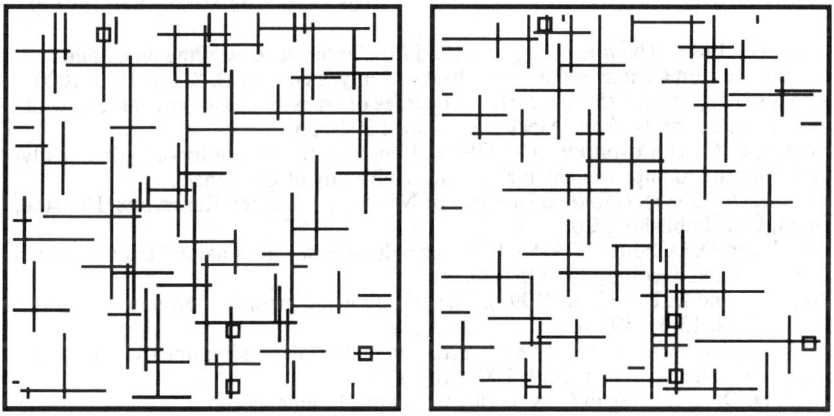

□ Well Location

Figure 5. Two example networks conditioned to the seismic and well data.

SUMMARY AND CONCLUSIONS

A methodology for simulating fracture networks based on the mechanics of fracture formation has been presented. This method is based on the theory of linear elastic fracture mechanics as well as laboratory and numerical experiments designed to better understand the environment of natural fracture formation. As an example application of the model, we simulate realizations of an "unknown" network which are consistent with seismic and flow characteristics of the unknown network. The methodology presented here is still preliminary and further development of the methodology should result in techniques for conditioning the network generation to more types of data. The incorporation of these different types of data into the simulation of the network will result in more accurate simulations and improved models for the behavior of fractured formations.

REFERENCES

Atkinson, B. K. (1979). A fracture mechanics study of subcritical tensile cracking of quartz in wet environments. Pure Appl. Geophys. 117: 1011-1024.

Atkinson, B. K. (1980). Stress corrosion and the rate-dependent tensile failure of a fine-grained quartz rock. Tectonophys. 65: 281-290.

Atkinson, B. K., Ed. (1987). Fracture Mechanics of Rock. Berkeley, Academic Press.

Broek, D. (1986). Elementary Engineering Fracture Mechanics. Boston, Martinus Nijhoff.

Cacas, M. C., E. Ledoux, et al. (1990). Modeling fracture flow with a stochastic discrete fracture network: Calibration and Validation, 2. The transport model. Water Resour. Res. 26(3): 491-500.

Cook, N. G. W. (1988). Natural joints in rock: Mechanical, hydraulic and seismic behavior and properties under normal stress. 29th symposium on Rock Mechanics, Minneapolis, Minnesota,

Cotterell, B. and J. R. Rice (1980). Slightly curved or kinked cracks. Int. J. Frac. 16(2): 155-169.

Cruikshank, K. M., G. Zhao, et al. (1991). Analysis of minor fractures associated with joints and faulted joints. J. Struct. Geol. 13: 865-886.

Dey, T. N. and C. Y. Wang (1981). Some mechanisms of microcrack growth and interaction in compressive rock failure. Int. J. Rock Mech. and Mining Sci. 18: 199-209.

Garven, G. (1986). The role of regional fluid flow in the genesis of the pine point deposit, western Canada sedimentary basin-a reply. Econ. Geol. 81(4): 1015-1020.

Kachanov, M. (1992). Effective elastic properties of cracked solids: critical review of some basic concepts. Appl. Mech. Rev. 45(8): 304-335.

Murdoch, L. C., G. Losonsky, et al. (1990). Report on the feasibility of hydraulically fracturing soil to improve remedial actions. University of Cincinnate.

Nelson, R. A. (1985). Geologic Analysis of Naturally Fractured Reservoirs. Houston, Texas, Gulf Publishing Co.

Olson, J. and D. D. Pollard (1989). Inferring paleostresses from natural fracture patterns: A new method. Geology 17: 345-348.

Olson, J. E. and D. D. Pollard (1991). The initiation and growth of en echelon veins. J. Struct. Geol. 13(5): 595-608.

Pollard, D. D. and A. Aydin (1988). Progress in understanding jointing over the past century. Geol. Soc. Amer. Bull. 100: 1181-1204.

Pollard, D. D. and P. Segall (1987). Theoretical displacements and stresses near fractures in rock: with applications to faults, joints, veins, dikes, and solution surfaces. Fracture Mechanics of Rock. London, Academic Press. 277-349.

Renshaw, C. E. and D. D. Pollard (1994, in press). Are large differential stresses required for straight fracture propagation? J. Struct. Geol. :

Rubin, A. M. and D. D. Pollard (1987). Origins of blade-like dikes in volcanic rift zones. Volcanism in Hawaii. U.S. Geol. Survey Prof. Paper. 1449-1470.

Segall, P. (1984). Rate-dependent extensional deformation resulting from crack growth in rock. J. Geophys. Res. 89(B6): 4185-4195.

Swanson, P. L. (1984). Subcritical crack growth and other time- and environment-dependent behavior in crustal rocks. J. Geophys. Res. 89(B6): 4137-4152.

Telford, W. M., L. P. Geldart and R. E. Sheriff (1990). Applied Geophysics. New York, Cambridge University Press.

Wilkins, B. J. (1980). Slow crack growth and delayed failure of granite. Int. J. Rock Mech. Min. Sci. Geomech. Abstr. 17: 365-369.

Wu, H. and D. D. Pollard (1991). Fracture spacing density and distribution in layered rock masses: Results from a new experimental technique. 32nd U.S. Rock Mech. Symp., University of Oklahoma, Norman, A. A. Balkema.

Wu, H. and D. D. Pollard (1992). Possible secondary fracture patterns due to a change in the direction of loading. Proceedings of the Fractured and Jointed Rock Masses Conference, Lake Tahoe, CA.

Rock Mechanics, Nelson & Laubach (eds) © 1994 Balkema, Rotterdam, ISBN 90 5410 380 8

Modeling of joint compliance using dynamic discontinuous deformation analysis

Wendell O. Miller & M. Eileen Glynn
Geotechnical Laboratory, US Army Engineer Waterways Experiment Station, Vicksburg, Miss., USA

ABSTRACT: A study was conducted using Dynamic Discontinuous Deformation Analysis to model the joint compliance of a model material developed for the Topapah Springs welded tuff located at Yucca Mountain, Nevada. The model used physical properties measured from laboratory tests, a typical or average aperture, and a simulated joint roughness. The degree of joint closure, or remaining aperture, could be controlled by applied stress level. Full closure of the joint was ultimately obtained in the model, and the results compared well with laboratory compliance tests.

1 BACKGROUND

Laboratory and field studies at Yucca Mountain, Nevada have established that several phenomena are unique to, and typical of, rock mass behavior within a fractured volcanic tuff horizon. Models or model components are developed based upon identification of these phenomena. Available information has indicated that the Topapah Springs welded tuff (the previously proposed repository horizon) is moderately fractured with two predominant sets of fractures that are near vertical. The intact rock is relatively stiff and has a high compressive strength. The mechanical response of fractures and sets of fractures is believed to be the key to any model intended to simulate the mechanical behavior of a fractured welded tuff.

Computational models are used extensively in evaluations of potential geologic repositories for disposal of radioactive wastes. With this in mind, the U. S. Army Engineer Waterways Experiment Station (WES) recently performed a small model study of joint compliance using Dynamic Discontinuous Deformation Analysis (DDDA) (Shi 1992). The DDDA concept is a complete and elegant kinematic theory that describes the behavior of discontinuous media. DDDA models a discontinuous medium as a system of interacting blocks, allowing the blocks to move and deform while prohibiting block interpenetration. The purpose of the study was to validate the analytical capabilities of the method using experimental data. This study represented the first attempt at using DDDA to model the mechanical closure of a physical rock joint. Joint stiffness was modelled physically by varying the configuration of the apertures (joint roughness) rather than being modelled as a simple spring with a variable spring stiffness.

2 MATERIAL PROPERTIES

Material property data for the DDDA model were taken from a model material WES developed and characterized for Sandia National Laboratories. This model material simulates the Topapah Springs welded tuff at Yucca Mountain, Nevada. The simulant which was developed was a ceramic brick material with an average unconfined compressive strength of 15,530 psi and a Young's modulus of 2.8 X 10^6 psi. A comprehensive suite of tests was performed on the material which completely characterized the material in terms of physical and mechanical properties. These tests included a large number of joint compliance tests in which single as well as multiple joints were tested. Joint closure was interpreted as the point at which all deformation could be accounted for by the elastic deformation of the intact material. Results of these tests indicated a joint closure stress of 5,133 psi and a stress of 476 psi at 1/2 closure of the joints. The average aperture for the model-material joints, as determined from the compliance tests, was 0.00354 inches. A summary of the material properties used in the model are presented in Table 1.

Table 1

Unconfined compressive strength	15,530 psi
Coefficient of friction	28.67°
Cohesion	5,000 psi
Tensile strength	5,000 psi
Young's modulus	2,856,500 psi
Poisson's ratio	0.19
Unit weight	128.3 pcf
Joint aperture	0.00354 in

3 MODEL GEOMETRY

The model analyzed with DDDA was a model of two 2.375 inch by 4.75 inch ceramic bricks, stacked, with a single joint between them. To facilitate deformations and stress redistributions at and around the joint asperities, these blocks were subdivided into 0.25 inch segments or blocks. The overall model is shown in Figure 1. A simple joint geometry was developed for the model by trial and error. Due to the relatively small aperture of the joint (0.00354 inches), the joint is not visible in Figure 1. The joint geometry selected for the model is shown in Figure 2a, with a 200 to 1 vertical to horizontal exaggeration.

In the trial and error process of selecting a joint geometry it was found that the joint compliance is a direct function of the joint geometry selected. The compliance was, however, easily controlled by varying the number of asperities (contacts), the widths of the contacts, as well as, the asperity heights within the joint. The joint aperture was taken as the maximum open area within the joint, i.e. the center of each void within the joint.

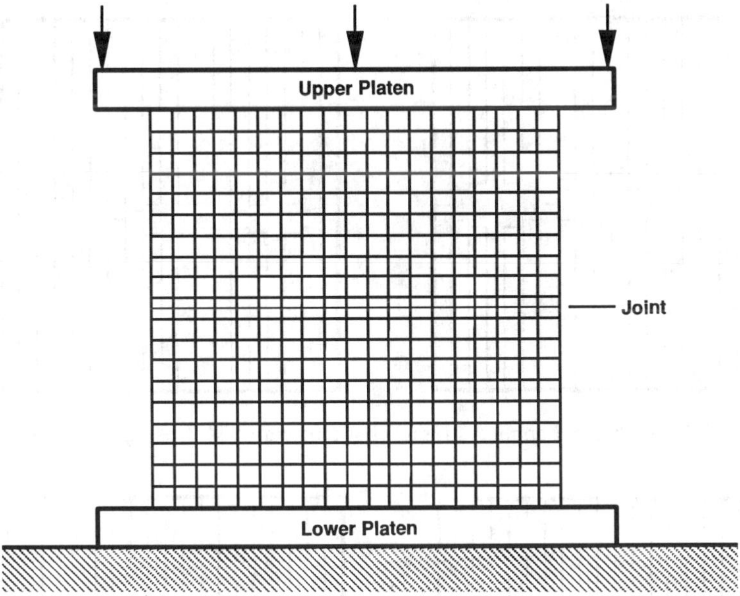

Figure 1. Joint compliance model

4 RESULTS

Stresses were applied to the model incrementally and the deformations within the joint were observed as a function of stress. Total deformation of the model was measured between the loading platens, taking the average of 19 different measurement points across the model. Joint deformation was somewhat more difficult to determine, in that it was variable across the joint. The maximum deformation within the joint occurred at the center of each void (open area), while the minimum occurred at each location where the asperities were in initial contact. In other words, there was a cyclic variation in joint deformation across the joint which varied from minimum to maximum and back to minimum. It was ultimately determined that the average of all deformations across the joint gave a lower value for joint closure than what was actually occurring. As a consequence, joint closure was subsequently defined as the average of those values of maximum deformation that were occurring at the voids. Therefore, joint deformation or closure was determined from the average of 3 measurement points within the joint.

The actual deformation of the joint, and specifically the asperities within the joint, was observed for each incremental stress level. Typical examples of the joint response or configuration at various stress levels are shown in Figures 2b through 2f. As can be seen, the mechanical response of the joint, and specifically the individual asperities within the joint, followed a logical progression, deforming as would be expected. Also, it may be noted that as the stress levels increased, the location of the joint within the figure moved downward. This response was due to the fact that the model was loaded only from the top and elastic deformation of the lower block resulted in the

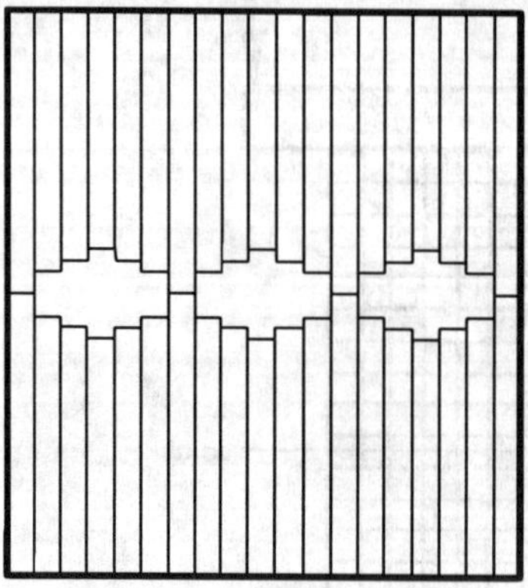

a. Initial joint with 0 psi applied stress

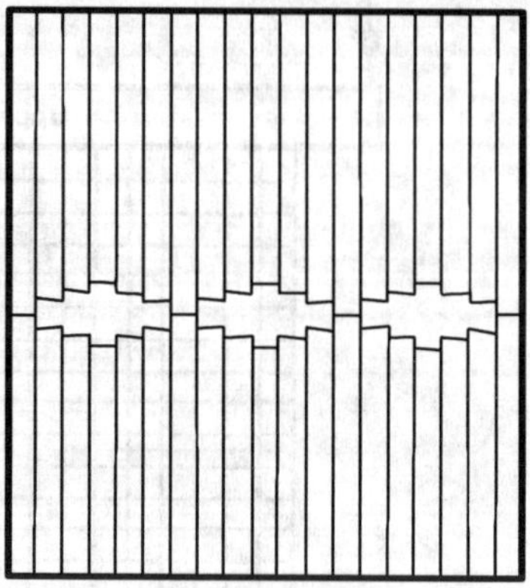

b. Joint with 534.3 psi applied stress

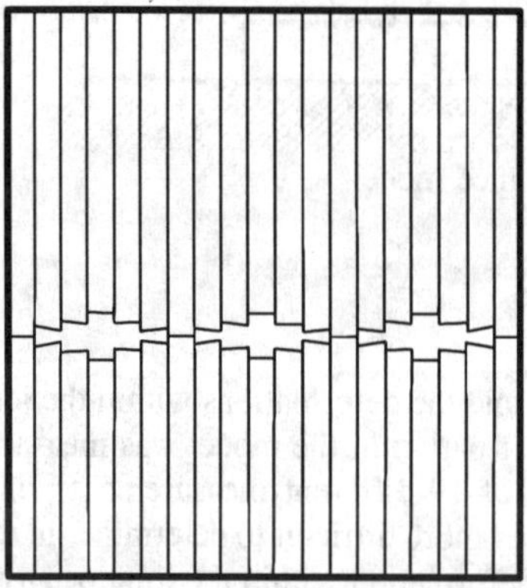

c. Joint with 1,263.2 psi applied stress

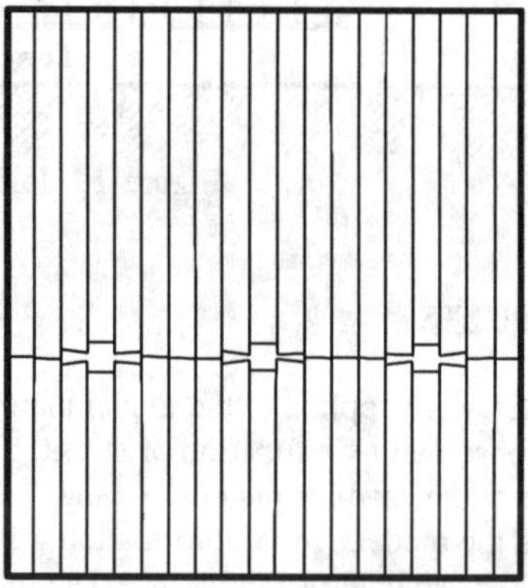

d. Joint with 2,004.6 psi applied stress

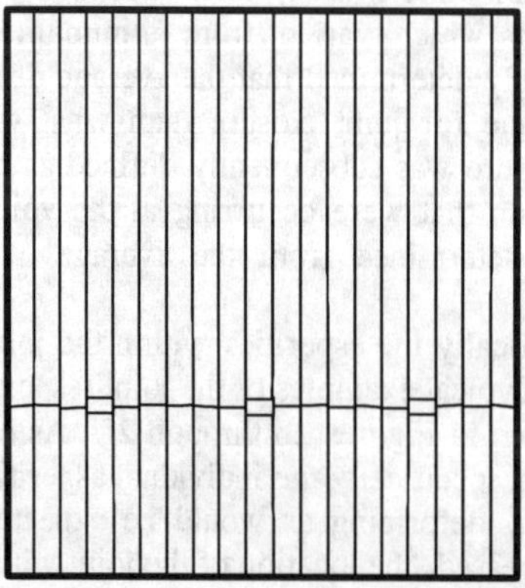

e. Joint with 4,276.4 psi applied stress

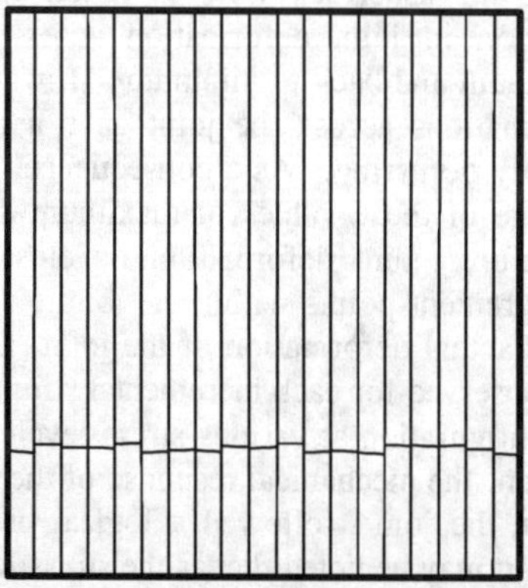

f. Joint with 6,347.4 psi applied stress

Figure 2. Joint opening as a function of applied stress (200 to 1 vertical exaggeration) movement of the joint to lower positions. As closure of the joint occurred, the number of contacts increased, and the percent open area within the joint decreased until total closure of the joint was obtained. This phenomena was contrary to that observed in laboratory tests of the joints, which indicated that at "closure" there remained a relatively significant area of the joints which were not totally closed, or in contact.

A plot of joint compliance, showing joint deformation and total deformation, is shown in Figure 3. For comparative purposes, the joint and total deformations taken from one of the laboratory tests of the material being modelled in this study are also shown in Figure 3. It should be noted that the material properties previously quoted for the model material were average values taken from many suites of tests, and that the laboratory test data plotted in the figure shown does not exactly match those average values. As can be seen, while the results do not track exactly, the same trends are observed for joint compliance. The results presented in this paper are basically the results of one model run using DDDA to model joint compliance. The authors are reasonably confident that minor adjustments to the joint geometry would result in an exact match of the load - displacement characteristics of the joint.

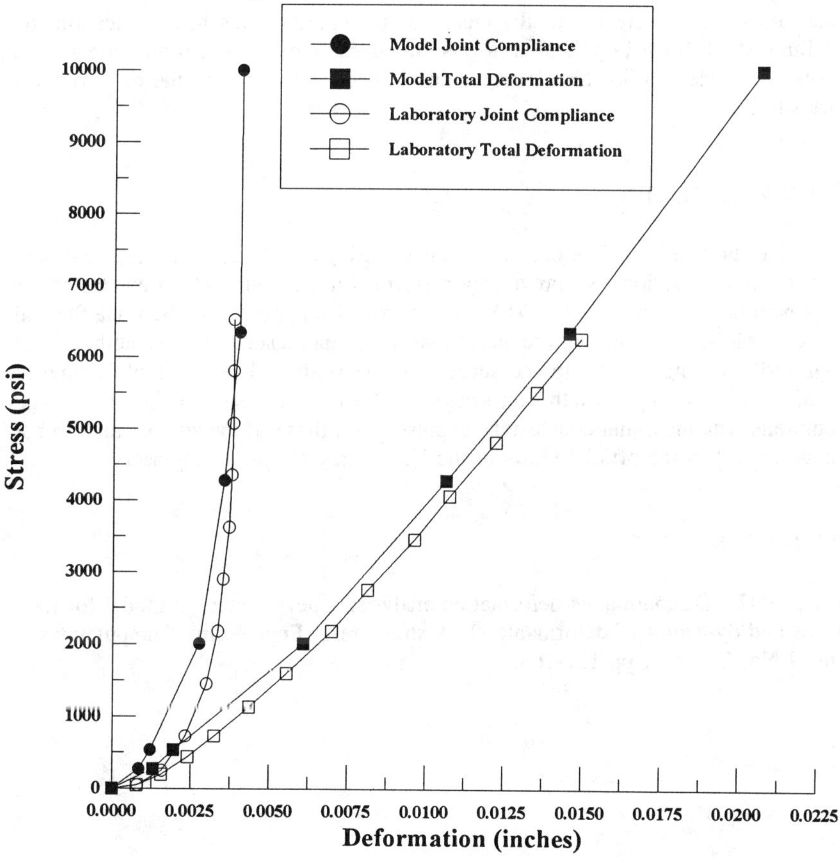

Figure 3. Laboratory and model joint compliance tests results

5 CONCLUSIONS

The results of this one, simple model demonstrate the potential for modelling joint compliance physically rather than with spring constants. The logical extension of the simple joint geometry test would be more complex joint geometries using both continuous and finite joint sets to further demonstrate the capabilities of the DDDA method of analysis. The relative ease with which the appropriate load - deformation characteristics were obtained suggest the potential for establishing standardized joint geometry models which would yield the appropriate joint compliance.

A subject not yet addressed is that of joint shear and sliding friction. The physical closure of the joints as a function of stress resulted in an increase in joint contact area which results in additional shear surfaces. It is intuitively obvious that this would produce a nonlinear increase in shear strength as a function of applied normal load. The realism of such a phenomena is dependent upon the joint geometry. Based upon the success of this quick, rather limited study, the authors are reasonably optimistic that the subject of joint shear using physically modelled joints can be solved in a likewise fashion. What is not currently obvious is how to address the fact that in laboratory tests of the joints, at "closure" there remained a relatively significant area of the joints which were not totally closed or in contact. This has implications on modelling fluid flow in joints at levels of stress above the closure stress. A constitutive model for joint stiffening with closure may represent the best path for future studies.

ACKNOWLEDGEMENTS

The model study described herein was conducted using data from joint compliance and material characterization tests previously performed in a research effort conducted for Sandia National Laboratories by WES. The personal support of Dr. Gen-hua Shi and Ms. Kris McNamara during the performance is appreciated by the authors and recognized as being critical to the success of the study. The Chief of Engineers granted permission to publish this information. However, this study is incomplete and all opinions, findings, and conclusions expressed are those of the authors and do not necessarily reflect the official views of the U.S. Army Corps of Engineers.

REFERENCES

Shi, G. 1992. Discontinuous deformation analysis: a new numerical model for the statics and dynamics of deformable block structures. Engineering Computations, Vol. 9 No. 2, April, pp 157-168

Rock Mechanics, Nelson & Laubach (eds) © 1994 Balkema, Rotterdam, ISBN 90 5410 380 8

An investigation of rock joint models on prediction of joint behavior under pseudostatic cyclic shear loads

Sui-Min (Simon) Hsiung, Amitava Ghosh & Asadul H.Chowdhury
Center for Nuclear Waste Regulatory Analyses, San Antonio, Tex., USA

Commonly used empirical representations of jointed rock behavior under cyclic loading conditions reside in the Mohr-Coulomb, Barton-Bandis, and Continuously-Yielding models. These models were developed based on data obtained under unidirectional loading conditions, and their ability to predict joint performance under cyclic loading conditions has not been tested. Analysis has indicated that these models are not capable of predicting joint behavior observed in the laboratory during pseudostatic shear reversal.

1 INTRODUCTION

In 1987, the United States Congress designated Yucca Mountain, in southern Nevada, as the only site to be characterized to determine its suitability for a high-level nuclear waste repository. The proposed repository horizon is about 300 m beneath the surface, in a densely welded prominently vertically and subvertically jointed tuff. An important phenomenon that could affect the preclosure and postclosure performances of a repository is repeated ground motion due to seismic activities (Kana et al., 1991). The fundamental failure mechanism for an excavation in a jointed rock mass subjected to repetitive seismic loading is the accumulation of shear displacement along joints (Hsiung et al., 1992).

Conditions for joint slip or for sliding of individual blocks from the boundaries of excavations are governed by the shear strengths of the discontinuities concerned. The conditions for joint behavior to be considered explicitly in assessing the response of the jointed rock mass surrounding an opening include: (1) pseudostatic loadings, (2) dynamic loadings, and (3) repetitive episodes of dynamic loadings arising from a series of earthquakes or underground nuclear explosions.

The Mohr-Coulomb, Barton-Bandis, and Continuously-Yielding models are three commonly used empirical representations of rock joint behavior. These models were developed based primarily on data taken under unidirectional pseudostatic loading conditions. The ability of these models to predict joint performance under cyclic pseudostatic loading conditions as well as the second and third conditions listed above has not been tested. This paper assesses these models and their ability to predict joint behavior under cyclic pseudostatic shear loads by comparing with experimental results.

2 LABORATORY OBSERVATION OF JOINT CYCLIC SHEAR BEHAVIOR

Two important, distinct features have been identified through the examination of laboratory-scale direct shear experiments on rock joints of Apache Leap (in Arizona) tuff under cyclic loading conditions; one is that the shear strength upon reverse shearing is smaller than that during forward

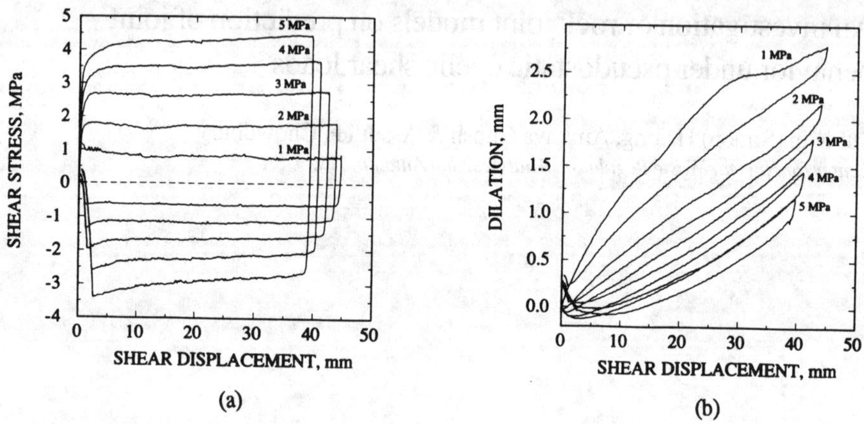

Figure 1. A typical experimental result of joint (a) shear stress response and (b) dilation under pseudostatic cyclic loading condition as a function of normal stress.

shearing, and the other is that the joint dilation resulting from forward shearing recovers during reverse shearing (Hsiung et al., 1993). These two features of joint behavior are independent of the environmental conditions under which the joint is tested. Figure 1(a) shows a typical shear stress response of a rock joint under various normal stress levels. The test sequence followed an ascending order with respect to normal stress. The curve with the 1-MPa normal stress (the first cycle of shearing) illustrates the shear behavior of an originally undamaged (fresh) joint and shows a distinct peak shear strength at the early stage of the shear cycle. The shear strength of the joint gradually reduces to a residual value as shear displacement increases. No distinct peak shear strength was observed for other cycles of shearing. Another feature that can be observed in this figure is that the shear strength during the reverse shearing is smaller than that during the forward shearing. The relation between normal displacement (dilation) and shear displacement for the same specimen is given in Figure 1(b). This figure indicates that joint dilation at various levels of normal stress increases with shear displacement during forward shearing. For reverse shearing, there is some degree of hysteresis in the joint dilation, with the reverse-shearing dilation curve being below that of forward shearing and decreasing towards zero. In general, for repeated shear cycles, the amount of joint dilation decreased with increasing normal stress. The same behavior has also been reported by other researchers (Jing et al., 1992; Wibowo et al., 1992; Huang et al., 1993) for rock joint replicas under cyclic pseudostatic loads.

3 EVALUATION OF MOHR-COULOMB JOINT MODEL

The simplest model for joint strength and deformation is the Mohr-Coulomb model, a linear deformation model that includes a shear failure criterion for a rock joint

$$\tau_o = C + \sigma_n \tan \phi \quad (1)$$

where τ_o is shear strength along the joint, σ_n is normal stress across the joint, C is cohesion, and ϕ is friction angle. Once τ_o is reached, the joint deformation is perfectly plastic. This equation suggests that the joint strength is the same in forward and reverse shearing directions. The joint shear response is governed by a constant shear stiffness K_s

$$\Delta\tau = K_s \, \Delta u_s^{\,e} \quad (2)$$

where $\Delta\tau$ is incremental shear stress and Δu_s^e is an elastic component of the incremental shear displacement. Based on Eqs. (1) and (2), $\Delta\tau$ becomes zero after the condition $|\tau| = \tau_o$ is reached, where τ is the shear stress on the joint.

The Mohr-Coulomb joint model in its basic form does not consider joint wear and dilation behavior. However, the dilation behavior may be added. For example, the dilation may be restricted (that is, the dilation angle ψ is zero) until shear stress has reached the shear strength of the joint, that is, joint dilation starts after the joint begins to deform plastically. Since there is no wear of the joint, joint dilation angle should remain constant with shear displacement. The dilation angle becomes zero after a critical shear displacement is reached. A form of the model has been given by ITASCA Consulting Group, Inc., 1992

(3) $$\text{if } |\tau| < \tau_o, \text{ then } \psi = 0$$

and

(4) $$\text{if } |\tau| = \tau_o \text{ and } |u_s| \geq u_{cs}, \text{ then } \psi = 0$$

where u_s is the joint shear displacement and u_{cs} is the critical shear displacement. Eq. (4) suggests that joint dilation should continue to increase even during reverse shearing.

Figure 2(a) shows the predicted joint shear behavior using the Mohr-Coulomb model compared to the actual experimental results. For clarity, results using normal stresses of 1, 3, and 5 MPa are shown. Results using intermediate normal stresses 2 and 4 MPa are omitted. A friction angle of 40.7° and zero cohesion were assumed for each computation. A shear stiffness, K_s, of 16.65 GPa/m and a dilation angle, ψ, 3.6° were used when the normal stress, σ_n, was 1 MPa. A K_s of 9.18 GPa/m and a ψ of 2.6° were used when σ_n was 3 MPa. For 5 MPa normal stress, 10.14 GPa/m and 2.3° were the values of K_s and ψ, respectively. These values were determined from the experimental results of each cycle. It is clear from the figure that the Mohr-Coulomb joint model simulates the residual shear phenomenon in the forward direction quite well but cannot simulate the peak behavior. As there is no damage accumulation, the predicted strength is independent of shear displacement. Although the Mohr-Coulomb model simulates the 'first-order' shear behavior in the forward direction very well, it overestimates the strength in the reverse direction. This has been observed in each of the sixteen specimens tested.

Figure 2(b) shows the dilation response of the joint when a constant dilation angle of 3.6° was

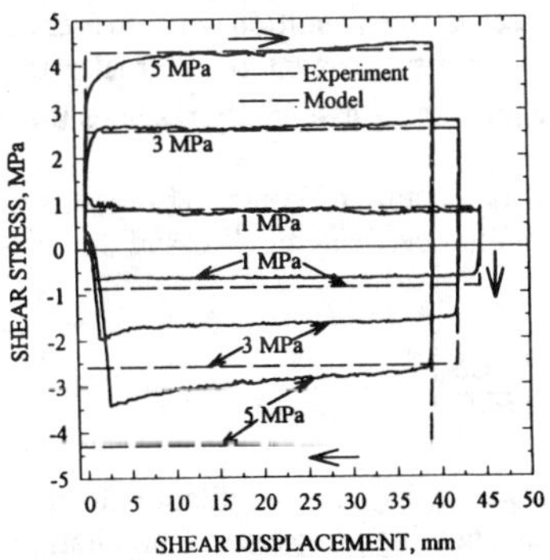

(a)

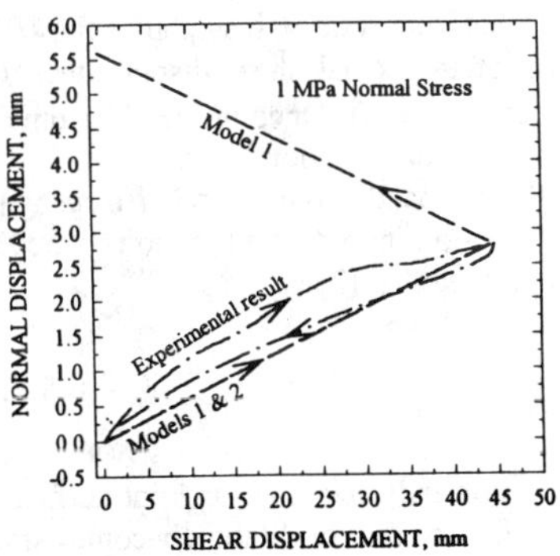

(b)

Figure 2. Comparison of simulated joint (a) shear response and (b) dilation using Mohr-Coulomb model with experimental results.

used with a normal stress of 1 MPa. The figure shows that the Mohr-Coulomb model predicts the dilation behavior in the forward direction reasonably well. In the reverse direction, the dilation curve predicted by Eq. (4), shown as Model 1, does not show any dilation recovery (joint compaction) with increasing shear displacement. But the joint compaction can be simulated numerically if the dilation angle is assumed negative while shearing in the reverse direction. This is shown as Model 2 in Figure 2(b). In this case, the dilation curve for reverse direction overlaps the dilation curve for forward direction, as shown in Figure 2(b).

4 EVALUATION OF BARTON-BANDIS JOINT MODEL

The Barton-Bandis model was proposed with the intent to take into consideration the effect of joint surface roughness on joint deformation and strength. The nonlinear joint strength criterion can be expressed as (Barton et al., 1985)

$$\tau_o = \sigma_n \tan\left[JRC \log_{10}\left(\frac{JCS}{\sigma_n}\right) + \phi_r\right] \tag{5}$$

where *JRC* is joint roughness coefficient, *JCS* is joint wall compressive strength, and ϕ_r is residual joint friction angle. The attrition of the surface roughness or reduction of the *JRC* is represented in a piece-wise linear manner (Barton et al., 1985). When the joint is not sheared, it has a maximum *JRC* value, JRC_{peak}. As the joint is sheared, the *JRC* decreases as a result of attrition of asperities on the joint surface. The peak shear stress (shear strength) is reached at a joint shear displacement u_p that can be calculated using the following equation (Barton and Bandis, 1982)

$$u_p = \frac{L_n}{500}\left(\frac{JRC}{L_n}\right)^{0.33} \tag{6}$$

where L_n is the length of a test specimen.

Based on Eq. (5), the joint shear stress under a constant normal stress depends solely on the *JRC* value. When a joint is sheared in one direction, its roughness is reduced from JRC_{peak} to a value, say JRC_f. If, at this point, the direction of the shear is reversed, the initial shear stress required for the joint to be sheared in the reverse direction is controlled by the JRC_f value, following Eq. (5), which serves as the maximum *JRC* in the reverse direction. In other words, the Barton-Bandis model assumes that the shear strength at initiation of the shear in the reverse direction is equal to the shear strength when the forward shearing was stopped. Also, Eq. (5) suggests that joint wear will stop after the *JRC* becomes zero. This condition will be reached when the ratio of the actual shear displacement u_s to the shear displacement u_p is greater than 100 (Barton et al., 1985). Once the *JRC* becomes zero, the joint shear essentially resumes the Mohr-Coulomb type of behavior.

The Barton-Bandis joint model also recognizes the dilatant nature of joints and suggests that the angle of dilation should be a function of the JRC. The relation between the *JRC* and dilation angle ψ is (Barton et al., 1985)

$$\psi = 0.5\ JRC \log_{10}\left(\frac{JCS}{\sigma_n}\right) \tag{7}$$

This equation indicates that, as joint surface roughness wears, its angle of dilation decreases. In other words, the rate of dilation becomes smaller as joint shearing progresses. The dilation angle will eventually become zero, that is, there will be no further dilation if u_s/u_p becomes greater than 100. Judging from the nature of Eq. (7), ψ is always positive. Joint dilation will continue to increase, although at a gradually slower rate, even after the direction of shear has been reversed.

Figure 3(a) illustrates the shear response of the joint for a JCS of 106.8 MPa, a JRC of 11.9, and a residual friction angle, ϕ_r, of 28.4° with normal stresses of 1, 3, and 5 MPa. Responses for

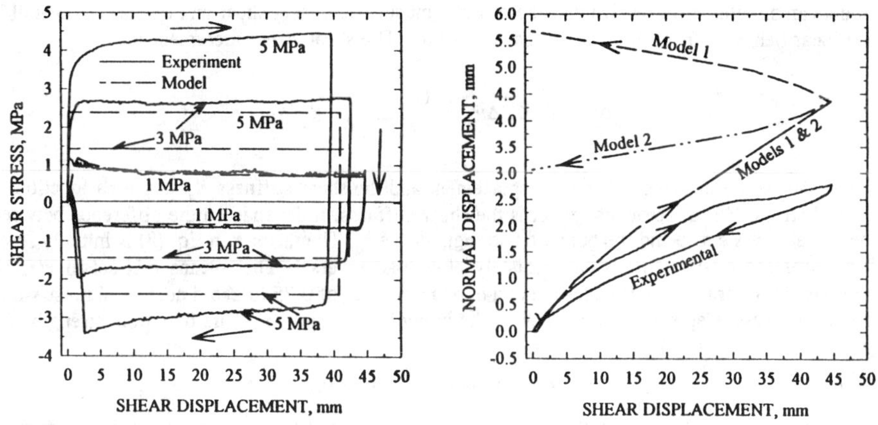

(a) (b)

Figure 3. Comparison of simulated joint (a) shear response and (b) dilation using Barton-Bandis model with experimental results.

intermediate σ_n of 2 and 4 MPa are not shown in the figure for clarity. JRC was back-calculated from Eq. (5) after measuring JCS and ϕ_r independently. The damage accumulation model reduces the JRC value as shear displacement increases. The figure shows that the predicted shear stress in the first cycle is quite close to the actual during forward shearing but overestimates the strength in the reverse direction. The predicted shear strength at the beginning of reverse shearing is exactly the same as the strength at the end of forward shear. This phenomenon is true for all subsequent cycles. This figure also shows that the damage to the joint surface in the Barton-Bandis model accumulates much faster than that observed in actual experiments. As a result, the rough joint surfaces quickly become smooth (i.e., JRC value becomes zero) and the shear response is controlled by the residual friction angle afterwards. Consequently, the strengths achieved in subsequent cycles with higher normal stresses are significantly lower than that observed in actual experiments.

Figure 3(b) shows the dilation of the joint for only the first cycle when the applied normal stress was 1 MPa. The figure shows that the Barton-Bandis model overestimates the dilation in the forward direction. Dilation in the reverse direction was calculated using Eq. (7) and is shown as Model 1. This curve does not show any joint compaction. Contrary to the experimental observation, Model 1 actually shows further increase in dilation while the top block is moved in the reverse direction. Curve designated as Model 2 was calculated assuming negative of the dilation angle in the reverse direction. This curve shows compaction of the joint in reverse shearing but the dilation recovery is far from that observed. The rate of damage accumulation in the Barton-Bandis model is so high that the JRC at the end of forward shear is significantly less than the initial value. As result, the dilation recovery calculated using Eq. (7) while shearing in reverse direction is not sufficient to return the dilation close to zero, as observed in experiments.

5 EVALUATION OF CONTINUOUSLY-YIELDING JOINT MODEL

The Continuously-Yielding model for rock joint deformation was developed by ITASCA Consulting Group, Inc. (ITASCA Consulting Group, Inc. 1992; Cundall and Lemos, 1988). The model can simulate progressive damage of the joint surface during shear and displays irreversible nonlinear behavior from the onset of shear loading. The shear stress increment is

$$\Delta\tau = F\, K_s\, \Delta u_s = \frac{(1 - \tau/\tau_o)}{(1 - r)}\, K_s\, \Delta u_s \tag{8}$$

where Δu_s is the incremental shear displacement and the shear stiffness K_s may be a function of normal stress. F is a factor that governs the shear stiffness. It depends on the difference between the actual stress curve and the bounding strength curve τ_o. The factor r in Eq. (8) is initially zero. It restores the elastic stiffness immediately after load reversal. That means r is set to τ/τ_o (F becomes 1). In practice, r is restricted to a maximum value of 0.75 to avoid numerical noise when the shear stress is approximately equal to the bounding strength, τ_o. The bounding strength, τ_o, is

$$\tau_o = \sigma_n \tan\phi_m \operatorname{sign}(\Delta u_s) \tag{9}$$

where ϕ_m is the friction angle that would apply if the joint is to dilate at the maximum dilation angle and is initially equal to the joint initial friction angle, ϕ_{mo}. As damage accumulates, ϕ_m is continuously reduced according to

$$\phi_m = (\phi_{mo} - \phi)\, e^{\left(-u_s^P / R\right)} - \phi \tag{10}$$

where u_s^P is the plastic shear displacement and ϕ is the basic friction angle of the rock surface. R is a material parameter with a dimension of length that expresses the joint roughness. A large value of R produces slower reduction of ϕ_m and a higher peak stress. The peak is reached when the bounding strength equals the shear stress. After the peak, the joint is in the softening region and the value of F becomes negative.

Based on Eq. (9), joint bounding shear strength under a constant normal stress depends solely on the friction angle, ϕ_m. When a joint is sheared in one direction and its friction angle, ϕ_m, is reduced from its initial value ϕ_{mo} to a value, say ϕ_f, and at this point, the direction of the shear is reversed, the corresponding bounding shear strength in the reversed direction, according to Eq. (9), is controlled by the ϕ_f, which serves as the maximum ϕ_m in the reverse direction. In other words, the Continuously-Yielding model assumes that the maximum bounding shear strength during reverse shearing is the same as the bounding shear strength at the end of the forward shearing process.

The formulation of joint dilation angle in the Continuously-Yielding model is expressed as

$$\psi = \tan^{-1}\left[\frac{|\tau|}{\sigma_n}\right] - \phi \tag{11}$$

Dilation takes place whenever the shear stress is at the bounding shear strength level, and is obtained from the friction angle, ϕ_m (Cundall and Lemos, 1988). Examining Eq. (11) reveals that the joint dilation, as treated in the Continuously-Yielding model, should continue to increase regardless of whether a joint is sheared in the forward or reverse direction. However, the rate of dilation decreases gradually and is governed by the value of ϕ_m. After the bounding shear strength is reached, Eq. (11) can be rewritten as

$$\psi = (\phi_{mo} - \phi)\, e^{\left(-u_s^P / R\right)} \tag{12}$$

Note again the increasing nature of joint dilation as shown in Eq. (12).

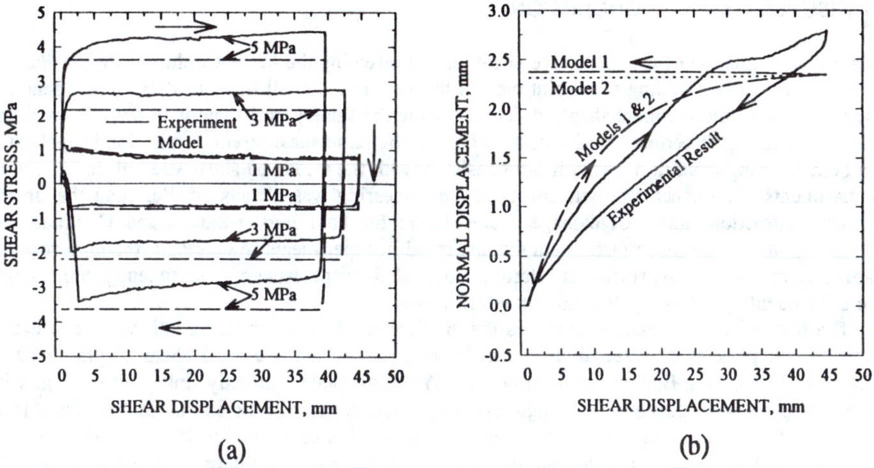

Figure 4. Comparison of simulated joint (a) shear response and (b) dilation using Continuously-Yielding model with experimental results.

Assuming a joint with a constant shear stiffness, K_s, of 19.68 GPa/m, initial friction angle, ϕ_{mo}, of 48.9°, basic friction angle, ϕ, of 36.0°, and roughness parameter, R, of 0.0105 m, the predicted joint shear strength using the Continuously-Yielding model subjected to normal stresses, σ_n, of 1, 3, and 5 MPa is illustrated in Figure 4(a). These parameters were determined using the data from the first cycle of forward shearing only. The damage accumulation model adjusts the ϕ_m value as shear displacement increases.

Figure 4(a) shows that the Continuously-Yielding joint model predicts the shear strength quite well during the first cycle, although it overpredicts the strength during reverse shearing. The predicted shear strength in the reverse direction is almost the same as that in the forward direction. This is a general trend observed in the study. The figure shows another important feature of the joint model. The shear strength predicted by the Continuously-Yielding joint model quickly reaches the residual strength value when compared with the experimental curves. In other words, joint roughness degradation occurs more rapidly in the Continuously-Yielding joint model than in the actual experiment. As a result, numerically the joint becomes two plane surfaces whose behavior is controlled by the basic friction angle following Mohr-Coulomb friction law. Consequently, in subsequent cycles of the experiment, the predicted shear strength is always smaller than the experimental results.

Figure 4(b) shows the dilation response of the joint using the same parameters. For clarity, only the first cycle (σ_n equal to 1 MPa) is shown. The predicted dilation curve in the figure shows overprediction in the beginning and underprediction thereafter. The dilation angle quickly decreases to zero with shear displacement. Due to rapid degradation of the joint roughness, the predicted dilation at the end of forward shearing in the first cycle is smaller than observed experimentally. In the reverse direction, the dilation curve does not return to zero. It stays almost at the same level as the peak value during forward shearing (Model 1). This confirms that the roughness of the joint is completely worn off during forward shearing. Due to the same reason, the dilation recovery is minimal when the dilation angle is taken as negative in the reverse direction (Model 2).

6 DISCUSSION AND CONCLUSION

Careful examination of the respective equations representing the Mohr-Coulomb, Barton-Bandis, and Continuously-Yielding rock joint models has revealed that all three models were essentially developed for unidirectional shearing. They assume the same joint behavior in both forward and reverse shearing directions. With this assumption, the same shear strength criterion is applicable in both shearing conditions. This is not consistent with the measured shear strength in laboratory experiments. The shear strength during reverse shearing was always smaller than that in the forward direction. Rates of damage accumulation for both Barton-Bandis and Continuously-Yielding models are much faster than that observed in experiments. As a result, the joint surfaces are calculated to become smooth at relatively small shear displacement. Consequently, the strength in subsequent cycles is significantly underestimated.

The Mohr-Coulomb model calculates the dilation based on a constant dilation angle. Dilation starts when peak shear strength is reached and ceases when a critical shear displacement is reached. The Barton-Bandis and Continuously-Yielding models directly link joint dilation with shear displacement through the roughness properties. Since a joint continues to wear during shearing, the dilation angle used to calculate joint dilation continues to decrease. The constant decrease of dilation angle implies that the joint will eventually stop dilating when it is completely worn. All three models inherently assume an increase of joint dilation with increasing shear displacement irrespective of the direction of shearing. This is inconsistent with the joint dilation behavior observed in the laboratory.

ACKNOWLEDGEMENTS

The authors thank Dr. Wesley C. Patrick and Dr. Budhi Sagar for their valuable advice and suggestion during preparation of this paper. This report was prepared to document work performed by the CNWRA for the U.S. Nuclear Regulatory Commission (NRC) under Contract NRC-02-93-005. The activities reported here were performed on behalf of the NRC Office of Nuclear Regulatory Research, Division of Regulatory Applications. This report is an independent product of the CNWRA and does not necessarily reflect the views or regulatory position of the NRC.

REFERENCES

Barton, N.R., and S.C. Bandis. 1982. Effects of block size on the shear behavior of jointed rock. *23rd U.S. Symposium on Rock Mechanics*. Berkeley, CA.

Barton, N.R., S.C. Bandis, and K. Bakhtar. 1985. Strength, deformation and conductivity coupling of rock joints. *International Journal of Rock Mechanics and Mining Sciences & Geomechanics Abstracts* 22(3): 121-140.

Cundall, P.A., and J.V. Lemos. 1988. Numerical simulation of fault instabilities with the continuously-yielding joint model. *Proceedings of the 2nd International Symposium on Rockbursts and Seismicity in Mines*. Minneapolis, MN: University of Minnesota.

Hsiung, S.M., D.D. Kana, M.P. Ahola, A.H. Chowdhury, and A. Ghosh. 1993. Laboratory Characterization of Rock Joints. Report CNWRA-93-013. San Antonio, Texas: Center for Nuclear Waste Regulatory Analyses.

Hsiung, S.M., W. Blake, A.H. Chowdhury, and T.J. Williams. 1992. Effects of Mining-Induced Seismic Events on a Deep Underground Mine. *PAGEOPH* 139(3/4): 741-762.

Huang, X., B.C. Haimson, M.E. Plesha, and X. Qiu. 1993. An Investigation of the Mechanics of Rock Joints--Part I. Laboratory Investigation. *International Journal of Rock Mechanics and Mining Sciences & Geomechanics Abstracts* 30(3): 257-269.

ITASCA Consulting Group, Inc. 1992. *UDEC Universal Distinct Element Code Version 1.8 Volume I: User's Manual*. Minneapolis, MN: ITASCA Consulting Group, Inc.

Jing, L., E. Nordlund, and O. Stephansson. 1992. An Experimental Study on the Anisotropy and Stress-Dependency of the Strength and Deformability of Rock Joints. *International Journal of Rock Mechanics and Mining Sciences & Geomechanics Abstracts* 29(6): 535-542.

Kana, D.D., B.H.G. Brady, B.W. Vanzant, and P.K. Nair. 1991. *Critical Assessment of Seismic and Geomechanics Literature Related to a High-Level Nuclear Waste Underground Repository.* NUREG/CR-5440. Washington, DC: U.S. Nuclear Regulatory Commission. 161.

Wibowo, J.T., B. Amadei, S. Sture, and A.B. Robertson. 1992. Shear Response of a Rock Joint Under Different Boundary Conditions: An Experimental Study. *Conference of Fractured and Jointed Rock Masses*. Preprints. June 3-5, 1992. Lake Tahoe, California.

Rock Mechanics, Nelson & Laubach (eds) © 1994 Balkema, Rotterdam, ISBN 90 5410 380 8

The mechanics of progressive shear failure at all scales in rock masses

Guy Archambault, Alain Rouleau, Réal Daigneault & Rock Flamand
Centre d'Études sur les Ressources Minérales, Université du Québec à Chicoutimi, Que., Canada

ABSTRACT: The progressive shear failure development and evolution in brittle to brittle-ductile rock masses behavior is characterized by non-linear, inhomogeneous stress and strain distributions through the media in which various fracture modes intervene in sequence during simple shear displacement. The progressive failure of zones under simple shear deformation is progressive in at least two ways and is influenced by the normal stress magnitude and dilatancy. The progressive failure extends to all scales and is responsible for a marked shear strength scale effect. This paper demonstrates how the rotational and non-coaxial strain characterizing zones submitted to simple shear deformation influence the sequential development and evolution of tension and shear fractures under mode I and II of fracture propagation and their subsequent behavior with shear displacement. The behavioral complexity of a such zone during progressive failure is rather difficult to model, because shear displacement measurements integrate all kinds of variables and phenomena into one single parameter. Stability of workings in rock masses may demonstrate problems if such modes of deformation and failure are induced in the rock media.

1 INTRODUCTION

The progressive shear failure evolution, in brittle to brittle-ductile rock mass behavior at all scales, is characterized by non-linear, inhomogeneous stress and strain distributions through the media, the development of en-echelon tension gashes and shear bands under modes I, II and III of fracture propagation at pre- and peak deformation. This is followed by a post-peak deformation phase localized in the failure zone: a fault or shear zone (Lo 1972). Inside the failed zone, a progressive interlinking of structural discontinuities can be observed to form braided or anastomosed networks during post-peak deformation (Fig. 1). The phenomenon produces lens-shaped domains (Fig. 1) of low stress and strain gradients separated by anastomosing domains of very high gradients at each scale (Gapais et al. 1987; Engelder 1987), where most of the deformation in the rock mass concentrated and brittle failure occurs by propagation of tension fractures and shear bands (Archambault et al. 1990 and 1992).

Field evidence shows that these patterns of progressive failure are generated by the development of several families of shear bands and fractures (Fig. 1). In all cases, shear lenses are formed in the post-peak deformation phase by the interaction of these discontinuities in an anastomosed pattern (Fig. 2). This rock mass softening mechanism, by a braided or anastomosed fracture pattern and shear bands from microscopic to megascopic scale, is responsible for the shear strength scale effect in the rock mass (Archambault et al. 1993). This scale effect is caused by progressive continuing damage to the rock mass from small to large scale (Fig. 3).

In this paper, a major factor that contributes to the progressive failure and the scale

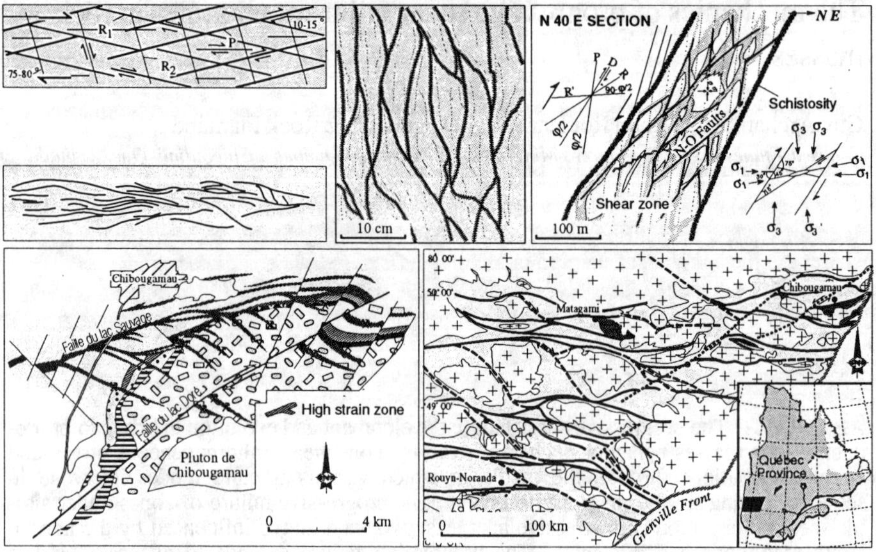

Fig. 1. Schematic illustrations and field case observations of anastomosed patterns at all scales.

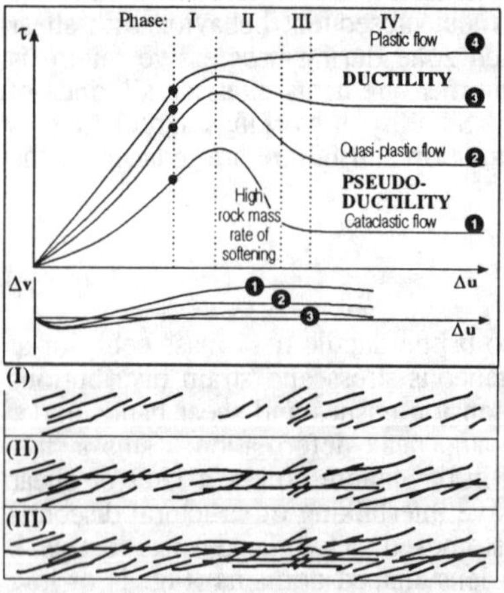

Fig. 2. Schematic representation of the development and evolution of anastomosed patterns of failure: a) stress-strain relationship; b) illustration of patterns.

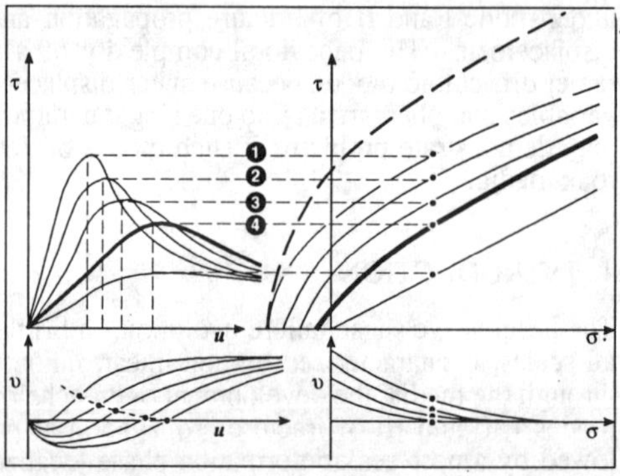

Fig. 3. Illustration of the shear strength scale effects in rock masses: a) the shear stress-shear displacement-dilatancy relationships are schematically given for increasing volume of a rock mass (1 to 4), with b) their corresponding failure envelops and dilatancy curves in relation with the normal stress.

effects on rock mass shear strength is discussed. Specifically it deals with the influence of the progressive heterogeneous non-coaxial and rotational strain which plays an important role on the sequential development and post-peak behavior of tension and/or shear fractures during the progressive deformation of shear zones submitted to simple shear.

2 HOMOGENEOUS PROGRESSIVE SIMPLE SHEAR DEFORMATION

The deformation within zones of rock masses submitted to simple shear are rotational, non-coaxial and progressive with increasing shear displacement (Fig. 4) and are heterogeneous across these zones (Fig. 5) where they show an increasing strain gradient from the borders to the center. The analysis of this problem must be done on the basis of finite strain parameters defined in the following terms (Ramsay and Huber 1983; Jaeger and Cook 1979; Ramsay 1967) for large elastic deformations: i) the infinitesimal extension of a line is denoted by $e = \delta l/l_0$; ii) while the quadratic extension of a line used in finite strain analysis is defined by $\lambda = (l_1/l_0)^2 = (1 + e)^2$; iii) and the shear strain $\gamma = \tan\psi$, where ψ is the angular shear strain or angular deflection of perpendicular lines.

The displacement of any initial point P(x,y) to its final position P'(x',y') in simple shear is given by the following strain matrix (Ramsay and Huber 1983):

$$(1) \qquad \begin{Bmatrix} x' \\ y' \end{Bmatrix} = \begin{bmatrix} 1 & \gamma \\ 0 & 1 \end{bmatrix} \begin{Bmatrix} x \\ y \end{Bmatrix} \quad \text{or} \quad \begin{Bmatrix} x \\ y \end{Bmatrix} = \begin{bmatrix} 1 & -\gamma \\ 0 & 1 \end{bmatrix} \begin{Bmatrix} x' \\ y' \end{Bmatrix}$$

These are the Lagrangian and Eulerian equations for any line in the zone under simple shear, not parallel to shear displacement direction and making an angle α with x axis, that will undergo changes in length and angle during progressive shear displacement (Fig. 4a). Specific material lines may be submitted either to elongation or contraction depending on their initial position before simple shear deformation, but lines parallel to the shear displacement direction show no finite longitudinal strain. Similarly for the angular changes of these lines during shear deformation, imply variations in shear strain for each of the lines except for those parallel to shear displacement direction. Tabulations of the variations in extension (e) and shear strain (γ) for various lines orientation (Ramsay and Huber 1983) show that none of the variations are linear. So even if the media behave in an isotropic, homogeneous, linear elastic manner, the progressive finite strain (λ,γ) of material lines, resulting from simple shear, demonstrate a non linear behavior attributed to the rotational and non coaxial characteristics of the deformation.

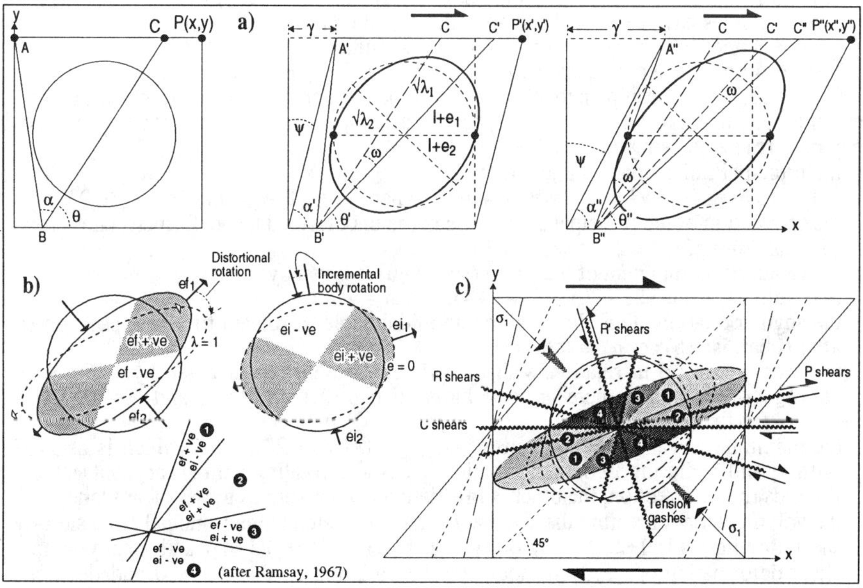

Fig. 4. Progressive finite strain under simple shear with increasing shear deformation.

Simple shear deformation may be considered as progressive with increasing shear displacement (Fig. 4a). Some directions show progressive increases in extension with increase in shear displacement (line BC, Fig. 4a), whereas others (line AB, Fig. 4a) show a complex sequence of shortening followed by stretching. As proposed by Ramsay and Huber (1983), the problem may be solved by dividing the deformation into two parts: the longitudinal finite strain (e_f) recording the total deformation and an incremental infinitesimal longitudinal strain (e_i) recording the changes in extension or shortening. Looking back on line AB (Fig. 4a), in the early stages the total and incremental deformations are negative (e_f - ve; e_i - ve) where ve is the variation of extension. As shear displacement proceeds, the rate of shortening decreases and when A'B is perpendicular to shear displacement direction it ceases to shorten. At this position the deformation is e_f - ve; $e_i = 0$) and following this the line starts to increase in length: at first still shorter than initially (e_f - ve; e_i + ve), then it passes the stage of initial length (e_f, e_i + ve) to culminate with increasing total and incremental deformations (e_f + ve; e_i + ve).

The concept of progressive displacement and deformation provides the key for the understanding of observed structural features occurring in the progressive failure of zones under simple shear deformation, even for the elastic behavior and brittle failure regimes of rock masses. Ramsay and Huber (1983) have discussed this concept of progressive displacement extensively in which a finite displacement field results from displacement increments at a define displacement rate. On this basis, a general progressive deformation process has been proposed using a superposition of the finite and infinitesimal-strain ellipses (Fig. 4b). The superposition of the incremental ellipse on the finite strain ellipse is considered in two parts: a distortional part and a rotational part. The finite-strain ellipse and the incremental infinitesimal-strain ellipse can be divided into sectors according to whether lines have elongated or contracted, and are separated by lines of no longitudinal strain. In general, and particularly in simple shear, there is a rotational component to deformation and the combination of the two ellipses and strain features associated with them is markedly asymmetric; the deformation is said to be rotational and non-coaxial due to a rotation of the principal axes. The superposition of the various sectors (Fig. 4c) of the two ellipses leads to four zones in which lines have had different histories:

- zone 1 lines have expanded but are now contracting (e_f + ve; e_i - ve)
- zone 2 lines have expanded and are still expanding (e_f + ve; e_i + ve)
- zone 3 lines have contracted but are now expanding (e_f - ve; e_i + ve)
- zone 4 lines have contracted and are still contracting (e_f - ve; e_i - ve).

The various relationships between the different angular, shear and extension parameters of a body undergoing simple shear, particularly the principal extensions (λ_1 and λ_2), may be formulated using eq. (1):

- the principal finite strains are given by
  (2) $\lambda_1;\lambda_2 = 1/2\,[\,\gamma^2 + 2 \pm \gamma(\gamma^2 + 4)^{1/2}\,]$ where $\lambda_1 = (1+e_1)^2$ and $\lambda_2 = (1+e_2)^2$;
- the orientation of the principal strain axes in the undeformed and deformed states are:
  (3) $\tan 2\theta = 2/\gamma = 2 \cot \psi$; $\tan 2\theta' = -2/\gamma$;
- the rotational component of the strain ($\omega = \theta' - \theta$) is given by:
  (4) $\tan 2\omega = \tan (2\theta' - 2\theta) = 4\gamma(\gamma^2-4)$; $\tan \omega = \gamma/2$;
- the angular relation of any line making an initial angle α, rotated to a new position α' after shear, is: (5) $\cot \alpha' = \cot \alpha + \gamma$;
- the shear strain (γ) in relation with the principal extensions is given by: (6) $\gamma = (\lambda_1+\lambda_2-\lambda_1\lambda_2-1)$ and if there is no volume change then $\lambda_1\lambda_2 = 1$ and $\gamma = (\lambda_1)^{1/2} - (\lambda_2)^{1/2}$ (so that $\lambda_1 = 1/\lambda_2$);
- the maximum shear strain is given by $\gamma_{max} = (\lambda_1-\lambda_2)/\ 2(\lambda_1\lambda_2)^{1/2}$ which is at ±45° with λ_1 and λ_2 during initial deformation but is also rotating with shear displacement towards the λ_1 direction, the other is invariant and corresponds to the shear plane;
- the volume variation within the shear zone may be related to the principal extension by the following: (7) $1+\Delta_A = (\lambda_1\lambda_2)^{1/2}$ or $1+\Delta_A = (1+e_1)(1+e_2)$ and ideally simple shear deformation proceeds without volume variation so that $\Delta_A = 0$ and develops under plane strain conditions ($e_3 = 0$, $\lambda_3 = 1$). In reality this last is not often the case so volume variation must be integrated into the relationships.

3 IMPLICATIONS ON PROGRESSIVE SHEAR FAILURE

This preceding shows that progressive deformation, resulting from simple shear loading conditions in rock masses attributed to the non-coaxial and rotational strain characteristics of the deformation. The rotational behavior of strain between the principal axes of finite strain may be deduced from one increment of displacement to the other due to rotation of the principal axes. A non-coaxial behavior may be observed between the principal axes of the incremental infinitesimal strain and the cumulative finite strain. The non-coaxial behavior reflects also the non-coaxiality between the stresses/strain rate/incremental infinitesimal strain in regards to the cumulative finite strain (Fig. 5a). The zones of elongation in the strain ellipses (Fig. 4c) indicate that the rock material is submitted to tension stresses in this direction and it may fail under mode I of fracture development, but under the incremental infinitesimal strain condition (phase I, Fig. 2), so that the en-échelon tension fractures will be oriented at around 45° to the shear displacement direction or parallel to the major principal stress (σ_1) (Fig. 5b) if there is no volume variation within the shear zone. They will be submitted to a progressive opening and rotation with the subsequent increments of shear displacement under the cumulative finite strain condition (Fig. 5b).

The zones in contraction in the strain ellipse indicate that the rock material is submitted to high compressive stresses in this direction so that it may fail under mode II of shear fracture development (phase I, Fig. 2), but also under the incremental infinitesimal strain condition, so that en-échelon Riedel shears (R-R') are developed in a conjugate system oriented at 30° with the major principal stress or 15° with the shear displacement direction (Fig. 5b). Also, another type of shears (P shears) appears symmetric but in oppposed

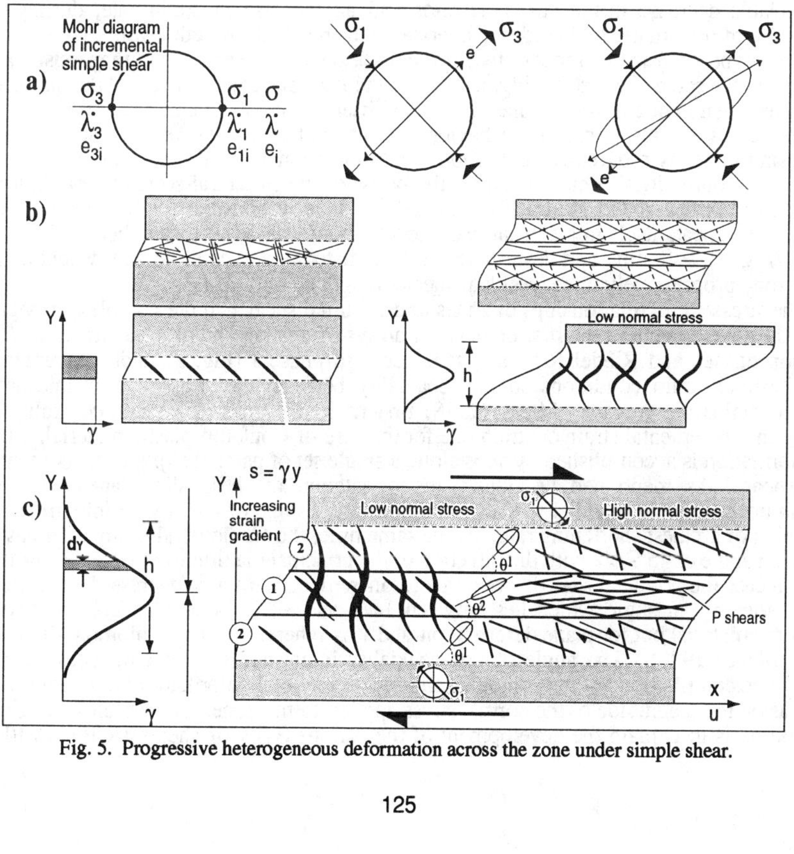

Fig. 5. Progressive heterogeneous deformation across the zone under simple shear.

dip orientation to Riedel-R shears (phase II, Fig. 2) in a thrust direction (Fig. 5b). Both are developed under Rankine conditions: active state R-R' shears and passive state P-P' shears following Mohr-Coulomb failure criteria. With subsequent shear displacement the R shears tend to rotate in an opposite direction to the tension gashes, while the R'-P-P' shears rotate in the same direction as the tension fractures. The P-P' shears usually appear after the R-R' shears (phase III, Fig. 2) in a subsequent stage and forming an anastomosing pattern with the earlier formed R-R' shears (Archambault et al. 1990, 1992, 1993; Mandl 1988; Tchalenko 1970).

A third family of shears appear in the final stage: the direction or C shears (phase IV, Fig. 2), parallel to the shear plane, the only invariant, no-infinitesimal and no-finite longitudinal extension orientation ($e_1 = e_2 = 0$ or $\lambda_1 = \lambda_2 = 1$) in simple shear. It also corresponds with one of the planes of maximum shear strain (cumulative finite and incremental infinitesimal) the other being rotated with shear displacement and with the plane of maximum shear stress. This orientation in the zone mobilizes both the maximum shear stress and strain (cumulative and incremental) and the failure plane is constrained to this direction. But, because of applied stress state in the zone under shear deformation and internal friction characterizing the rock material, it fails at lower shear stress under failure mode II or a combination of modes I and II. The rock material soften progressively in the zone, under minimized energy conditions, and proceeds at residual strength (phase IV, Fig. 2) under bulk pseudo-ductile deformation by friction plasticity (Mandl 1988). These C plane shears may result from a combination of the earlier R and P shears after their rotation during shear displacement or by the creation of new shear planes during the softening process in the post-failure phase (Fig. 2). All these sequential shears, after their initiation under incremental infinitesimal strain, are submitted to progressive rotation and shear deformation, under the cumulative finite strain, and the principal stress orientation is progressively modified inside the zone under simple shear. There is also an appreciable decrease in the stress deviator and an increase in strain rate, during the softening of rock material, in order to operate on the residual strength level.

Both type of fracture modes (tension and shear) may develop simultaneously and coexist in the same shear zone (Fig. 6a) giving incongruous coexistence and this has been observed even in ductile shear zones and field observations of several cases are reported in the literature. This ambiguous behavior comes from the cumulative finite strain which with simultaneous progressive extension and contraction force the development of the mode I discontinuities. This constitutes the basis for the structural softening mechanism reported in Archambault et al. (1990 and 1993). It is responsible for the scale effects because it repeats itself in a self similar process from one scale to another in the rock mass (Fig. 3), softening and destroying the rock material at each scale and widening the fault zone progressively under residual shear strength conditions (Fig. 5).

The stress/strain relationships of zones under simple shear is quite complex, it begins with the development of the first en-echelon arrays of tension and/or shear fractures (the tension gashes and Riedel R-shears) that occur in phase I (Fig. 2) under incremental infinitesimal strain conditions so that coaxiality between the stress and infinitesimal incremental strain prevails. Mandl (1988) first noted the problem of non-coaxiality of stress and incremental strain or strain rate for the case of Coulomb-plastic material, when a deformation is accomplished by slips along a single set of parallel slip planes as in simple shear. As mentioned previously the orientation of these slip planes with the maximum shear rate (parallel to shear direction) at an angle of 45° (by definition) to the direction of greatest shortening rate. At the same time, the Coulomb slip direction has to make an angle of $45° - \phi/2$ with the direction of the principal maximum stress. So, in this special case the principal axes of strain rate (or incremental strain) and stress deviate from each other by an angle which lies between $+\phi/2$ and $-\phi/2$ for mechanically perfectly isotropic material but for plane deformation and it is inherent to the Coulomb's slip concept (Mandl 1988). The angle of non-coaxiality is not constant and depends on the respective contributions of the conjugate slip sets to the total deformation, to dilatancy as well as on the magnitude of the normal stress applied to the zone. Nevertheless, it gives a good basis to explain the development of the various types of shears (R-P-C) within shear zone at an angle with the shear direction. Once developed, the shears evolve with the cumulative rotational finite strain.

Another type of progressive deformation may characterize simple shear loading conditions in rock masses as illustrated in figure 5. It may be attributed to inhomogeneous strain within a wide zone, either by expansion of a smaller zone (Fig. 5b) or mobilized as such. The first possibility implies a progressive migration of the zone walls under simple shear with a concentration of the strain gradient towards the center of the zone as illustrated in figure 5b. The other possibility may also develop under constrained conditions and in this case the concentration of the strain gradients may occur anywhere in the zone. Both possibilities are observed in the field and in the experimental study (Fig. 6a). Various zone thicknesses were studied to evaluate the effect of progressive deformation on the mechanics of failure in these zones (Fig. 6) and the previous observations regarding the structural morphology of the fracture development during simple shear deformation. The results (Fig. 6b) show quite a dispersion on the Mohr diagram that depends partly on the analog material but above all on the loading path used, on the dilatancy properties of the material, on the normal stress magnitude and on the failure modes.

Simple shear stress and total shear displacement measurements under various normal stress values and normal displacements on the shear zone (for dilatancy) integrate all the previous variables in the same four parameters without distinction, including even heterogeneous strain developed in wide zones under simple shear (Fig. 5). The measured data do not show the complexity of the various aspects of the progressive simple shear deformation, the resulting non-linear and non-coaxial behavior of the stress-strain relationships and do not permit an adequate evaluation of the failure process. Only measurements done

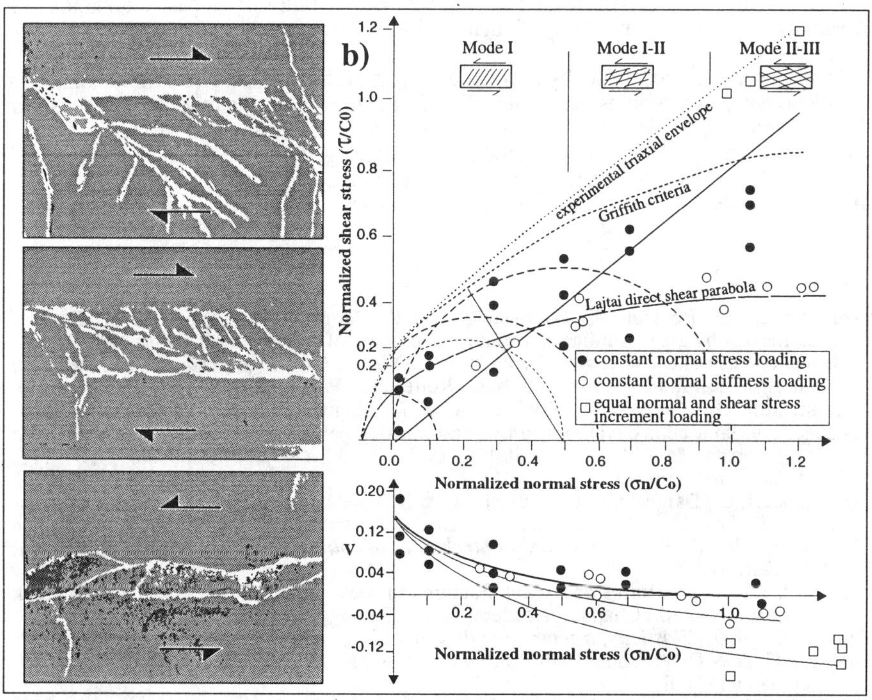

Fig. 6. Structural morphology of discontinuities and shear a) on tests specimen in the sheared zone after failure and b) Mohr diagram on the shear strength results under three different confitions of loading and the dilatancy-normal stress corresponding to these loading conditions.

for punctual strain give some information regarding the deformation of zone under simple shear, which is the object of the experimental work done in this research project. There are still two other important variables to integrate in this strain approach of progressive deformation and failure in zones under simple shear loading: the dilatancy characteristics and the influence of normal stress. The influence of the latter on the rupture morphology in direct shear tests with applications to natural fault features was already evaluated (Petit 1988), but the first has been less studied. This will be the subject of a further paper which will also look into the stress/strain relationships, showing non-linear behavior and its repercussion on the mechanics of progressive shear failure in including the effects of inhomogeneous shear strain for wide zones under simple shear.

4 CONCLUDING REMARKS

It was demonstrated that rotational and non-coaxial strain affects the initiation, development and evolution of tension and/or shear fractures in a zone submitted to simple shear and the progressive deformation and failure resulting from such a loading mode, which are common in geotechnical problems in rock masses. The scale effect on rock mass shear strength induced by this type of progressive deformation and failure is quite evident and underlined in the discussion. Also, the problem of rotational and non-coaxial strain has an influence on the stress path followed during loading and was discussed in a previous paper (Archambault et al. 1992). There are still other types of progressive deformation and failure induced by non-homogeneous deformation superposed on non-coaxial and rotational strain to be examined in wider failure zones. Other problems arise regarding the resulting non-linear and non-coaxial relationships which are pointed out in some studies on this type of failure in rock masses. The stability of slopes, foundations or openings may develop instability if such modes of deformation are induced in the rock media.

This paper show the difficulty to solve even part of the two first phase of initiation and development of the progressive failure (Fig. 2) induced in rock masses by simple shear loading and deformation, either from the strain and stress/strain point of view. Unfortunately, simple shear displacement measurements integrate into single measurements all the variables in the same parameter and do not permit an adequate evaluation of the behavior and development of the failure process in such zones.

REFERENCES

Archambault, G., Rouleau, A., Daigneault, R. & R. Flamand 1993. Progressive failure of rock masses by a self similar anastomosing process of rupture at all scales and its scale effect on their shear strength. In A. Pinto da Cunha (ed), *Proc. 2nd Int. Workshop on Scale Effects in Rock Masses:* 133-142. Rotterdam: Balkema.

Archambault, G., Flamand, R., Rouleau, A. & R. Daigneault 1992. Mechanics of shear zones and fault belts development by anastomosing patterns of fractures at all scales. In J.R. Tillerson & W.R. Wawersik (eds), *Proc. 33rd U.S. Symposium:* 101-110. Rotterdam: Balkema.

Archambault, G., Daigneault, R., Rouleau, A. & O. Tavchandjian 1990. Mechanics of shear zones and fault belts development by anastomosing patterns of fractures at all scales. In Rossmanith (ed), *Proc. Int. Conf. on Mechanics of Jointed and Faulted Rock:* 197-204. Rotterdam: Balkema.

Engelder, J.T. 1987. Joints and shear fractures in rock. In Fracture Mechanics of Rock (B.K. Atkinson, ed.) Chap. 2. Academic Press Geology Series, pp. 27-80.

Jaeger J.C. 1969. *Elasticity, fracture and flow* 3rd ed. London: Methuen.

Jaeger, J.C. & N.G.W. Cook 1979. *Fundamentals of rock mechanics* 3rd ed. London: Chapman and Hall.

Lo, K.Y. 1972. An approach to the problem of progressive failure. *Can. Geotech. J.* 9: 407-429.

Mandl, G. 1988. Mechanics of tectonic faulting: Models and basic concepts. Developments in structural geology, R, Elsevier. 407 p.

Petit, J.P. 1988. Normal stress dependent rupture morphology in direct shear tests on sandstone with applications to some natural fault surface features. *Int. J. Rock Mech. Min. Sci. & Geomech. Abstr.* 25(6): 411-419.

Ramsay, J.G. 1967. *Folding and fracturing of rocks.* New York: McGraw-Hill.

Ramsay, J.G. & R.H. Graham 1970. Strain variation in shear belts. *Can. J. Earth Sci.* 7: 786-813.

Ramsay, J.G. & M.I. Hubber 1983. The techniques of modern structural geology: Vol. I - Strain analysis (307 p.) and Vol. 2 - Folds and fractures (595-638). London: Academic Press.

Tchalenko, J.S. 1970. Similarities between shear zones of different magnitude. *Bull. Geol. Soc. Am.* 81: 1625-1640.

Rock Mechanics, Nelson & Laubach (eds) © 1994 Balkema, Rotterdam, ISBN 90 5410 380 8

Development of bedding plane faults and fracture localization in a flexed multilayer: A numerical model

M.L.Cooke & D.D.Pollard
Stanford University, Calif., USA

ABSTRACT: Geologic folds are not comprised of a single flexed layer but of a heterogeneous stack of rock strata. If layer interfaces slip upon flexure, the frictional contact between layers dictates that the fold deform inelastically. This is numerically modeled using the boundary element method. The model is loaded in flexure and then unloaded in to explore both elastic and inelastic deformation of the fold. The curvature maxima are larger for frictionally slipping than for bonded interfaces and occur limbward of the crest and trough of the fold. The inelastically deformed fold has straight limbs and concentrated hinges after unloading.

The presence of bedding plane faults within a flexure may play an important role in the localization of fractures. Fractures are likely to be associated spatially with regions of greater curvature and with large slip gradients on bedding plane faults.

1 INTRODUCTION

Fracture networks in folded sedimentary strata can have a significant effect on the flow of fluids throughout the geologic structure. Joints within folds can provide either conduits for migration of hydrocarbons into other portions of the structure or provide conduits for hydrocarbon escape to the surface. Similarly, fractures within aquifers may control contaminant transport. The study of fracture development in folded strata is a first order problem for understanding the flow of fluids through reservoirs and aquifers.

Geologic folds are not comprised of a single flexed layer but of a heterogeneous stack of rock strata. During flexure, the layers may slip along bedding contacts. To accurately analyze jointing of flexed rock strata, the mechanical behavior of interfaces between layers should be understood (Kemeny and Cook 1985). To this end, we have investigated a mechanical model of sedimentary rock flexure to determine the stress fields that could produce fractures in geologic folds.

Sedimentary rock within the upper 10 km of the earth's crust is thought to deform elastically for relatively small strains and duration of loading (Scholz 1990). Thus, to first order, the formation of some geological folds can be modeled as the flexural deformation of linear elastic layers. The heterogeneous nature of geologic strata can be modeled as layers bounded by interfaces that are weak in shear. Mechanical models show that if layers slip along interfaces when loaded in flexure, folds will have greater amplitude and sharper hinges than folds with bonded layer contacts (Chapple and Spang 1974; Feund 1979; Koch, et al. 1981).

The frictional resistance (resistance to slip) of the interface can play an important role in the deformation of the stack of layers. Analytical solutions with zero frictional

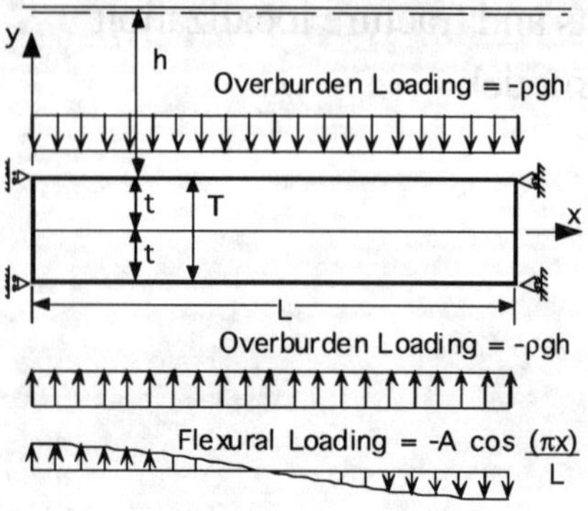

Figure 1 - Model configuration.

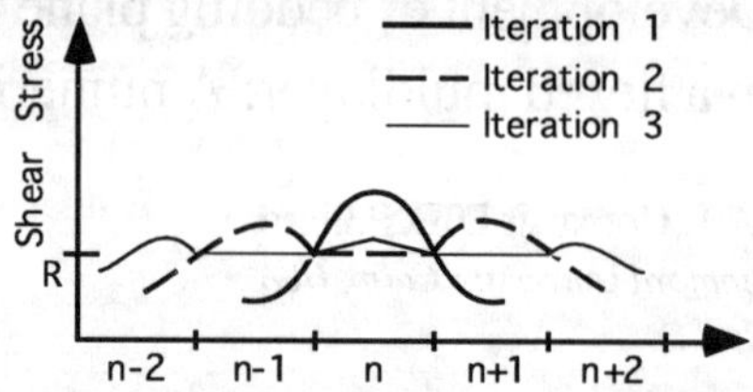

Figure 2 - Iteration 1: shear stress on element n exceeds R and n slips. Iteration 2: slip on n increases shear stresses on n-1 and n+1. Shear>R for n-1 and n+1 and they slip. Iteration 3: shear stresses on n, n-2 and n+2 are increased. In this fashion slip distribution converges.

resistance yield freely-slipping layers which can support no shear stress (Feund 1979; Koch, et al. 1981). Other analytical solutions include modeling interfaces as thin viscous layers (Chapple and Spang 1974). The benefit of numerical models is that more complex (and perhaps more realistic) boundary conditions can be investigated. The potential for slip can be evaluated anywhere along the interface and the stress field can be found for any slip configuration. Numerical models may be readily unloaded to investigate inelastic deformation of the structure. Roth et al. (1982) used the finite element method to investigate flexural slip during compression of a preformed fold.

This paper uses a two-dimensional numerical model to investigate the flexural behavior of a stack of two elastic layers with a frictional interface. In the following section, the two-layer model is described along with its relevance to geologic structures. Section 3 describes the numerical method and special elements used to define frictional interface behavior. The model is tested by comparing results with analytical solutions for bonded and freely-slipping interfaces. In section 5 the results of frictional analysis are presented and discussed. Section 6 discusses potential fracture distributions based on the stress field throughout the two-layer fold and slip distribution along the interface.

2 DESCRIPTION OF MODEL

The two-dimensional numerical model utilizes the boundary element method (BEM) to investigate slip along layer interfaces. The benefit of BEM compared to other methods such as finite element (FEM), is that only the layer boundaries and interfaces need to be discretized. The stress and displacement field throughout the system are determined from the tractions and/or displacements prescribed along the boundary elements.

The basic model includes two layers bounding a slipping interface,Fig. 1. The model has length, L, total thickness, T, and is located at a depth, h, from the earth's surface. The interface lies between two layers of equal thickness, t. The boundary conditions relative to the x, y coordinate system shown in Fig. 1 are:

(1) $u_x=0;\ \sigma_{xy}=0;$ along: $x=0$ and $x=L$, $-t\leq y\leq t$.

(2) $\sigma_{yy}=-\rho gh;\ \sigma_{yx}=0;$ along: $y=t$, $0\leq x\leq L$.

(3) $\sigma_{yy}=-A\cos(\pi x/L)-\rho gh;\ \sigma_{xy}=0;$ along: $y=-t$, $0\leq x\leq L$.

Normal displacements and shear tractions are fixed along the sides of the multilayer (Eq. 1). The weight of rock overlying the model is applied to both the upper and lower surfaces. The thickness of the model is assumed small compared to the depth of burial so that the compressional stress due to weight of the overburden can be considered constant through the thickness of the model. A sinusoidal loading function is also applied to the base of the multilayer to induce flexure.

This two-dimensional model can represent a three-dimensional geologic fold if the fold is cylindrical: length of the fold axis much larger than distance across fold and no changes in geometry or boundary conditions along the axis. The flexural loading schematically shown in Figure 1 will raise the left half of the structure and lower the right half, forming a monocline. The fundamental results from this basic model can be used to investigate the deformation of a series of anticlines and synclines simply by extending the monocline perpendicular to the fold axis using symmetry.

3 NUMERICAL METHOD

The boundary element code, FRAC2D developed by Thomas (Liang 1993), was used with some special augmentations for this problem. FRAC2D uses the displacement discontinuity method which is based on the TWODD program by Crouch and Starfield (1983). The boundaries of the layer are discretized into equal-length line elements, each associated with a normal and shear displacement discontinuity. The displacement discontinuity methd is described in detail in Crouch and Starfield (1990).

Special fault boundary elements which can accommodate inelastic slip are used to model the monocline interface (Crouch 1979; Crouch and Starfield 1990) . The fault elements are described in detail by Crouch (1979) and Crouch and Starfield (1983). The elemental mechanical behavior is described by 4 parameters: shear stiffness, K_S, normal stiffness, K_N, cohesion, c, and coefficient of friction, μ.

The normal and shear stiffnesses of the fault elements describe the elastic behavior. A fault stiffness much greater than the material stiffness of the neighboring layers produces a low displacement discontinuity. For the fault elements a relatively high shear stiffness (K_S >> Shear Modulus) mimics bonded behavior while a negligible shear stiffness approximates a freely slipping fault. Until the fault element slips, the displacements across the fault are linearly related to the stresses on the element by the stiffness.

The Mohr-Coulomb friction criterion is used to determine if an interface element slips. The criterion relates the shear stress required for slip to the frictional resistance (Jaeger and Cook, 1979). Slip occurs under compressive (negative) normal stress conditions if

(4) $$|\sigma_{yy}| \geq c-\mu\sigma_{yy} \text{ only if } \sigma_{yy}<0.$$

The frictional resistance, R, is expressed in terms of cohesion, c, the normal stress across the fault, σ_{yy}, and the coefficient of friction, μ, of the fault. If on a fault element $\sigma_{yx}>R$, the element will slip until the shear stress after slip is just less than the frictional resistance of the element. At this point, the stress state for the system must be recalculated because slip on one element may change the stress state on neighboring regions along the fault. The shear stress on elements adjacent to the slipped element may then exceed their frictional resistance and slip as well, Fig. 2. After successive iterations, $\sigma_{yx}<R$ for every fault element and the displacement discontinuities converge to the final solution for the loading step. Slip along fault elements is inelastic.

4 MODEL TESTING

4.1 Description of analytical model

The validity of numerical models can be tested by comparison of numerical results with analytically derived solutions. By changing the shear stiffness of the interface elements, our numerical results approach the analytical solutions derived from thin plate theory for stacks of both freely slipping and bonded layers.

The governing equation of thin plate theory relates the transverse loads applied, P, to the deflection of the plate, u_y, by the flexural rigidity of the plate, D.

(5) $$D\, \partial^4 u_y(x)/\partial x^4 = P(x)$$

The x-axis lies in the plane of the plate along the middle (neutral) surface and the y-axis perpendicular to the plate and parallel to applied loading and deflection. For two-dimensional plates the flexural rigidity depends on the Young's Modulus, E, and the plate thick-ness cubed, that is $D=12/ET^3$. An expression for the plate deflection can be found by integrating Eq. 3 and solving for the four constants of integration using boundary conditions. We will compare plate deflections from the numerical and analytical solutions.

The plate configuration is similar to Fig. 1 except that the loading is only flexural and there is no overburden loading, that is $P(x) = -A \cos(\pi x/L)$. Using the boundary conditions described in Fig.1 the deflection of a single layer of thickness T is:

(6) $$u_y(x) = -12\, L^4/\pi^4 E T^3\, A \cos(\pi x/L) \quad \text{for } 0 \le x \le L.$$

Table 1 defines the symbols and give their values used to compute the deflection shown in Fig. 3. Eq. 6 describes the deflection of a bonded interface.

The deformation of a stack of two plates each of thickness t bounding a freely slipping interface can be analyzed similarly. The deflection of the entire stack is expressed as:

(7) $$u_y(x) = -6\, L^4/\pi^4 E t^3\, A \cos(\pi x/L) \quad \text{as } 0 \le x \le L.$$

Using parameter values of Table 1, the analytical solution for deflection of a freely slipping stack is presented in Fig 3.

Table 1: Model Parameters

Parameter Description	Symbol	Value
Length	L	1000 meters
Stack Thickness	T	200 meters
Layer Thickness	t	100 meters
Young's Modulus	E	50 GPa
Poisson's Ratio	ν	0.25
Loading Amplitude	A	10 MPa
Depth to top of stack	h	1000 meters
Overburden Density	ρ	2600 kg/m^3
Shear Stiffness	K_S	1000 GPa
Normal Stiffness	K_N	1000 GPa
Cohesion	c	0.0 Pa
Coefficient of Friction	μ	0.6 (frictional slip)
Frictional Resistance	R	0.6 σ_{xx}

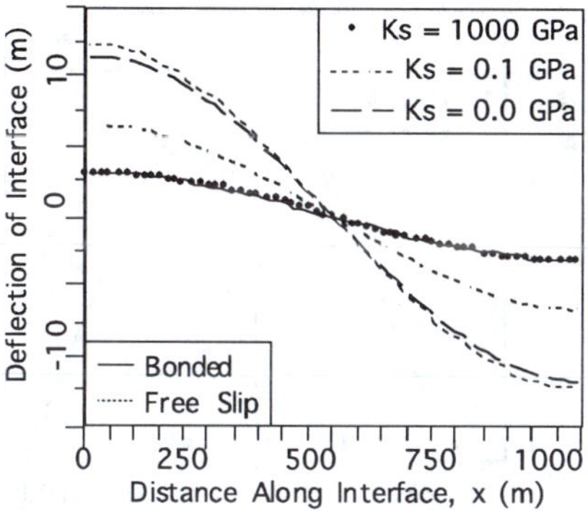

Figure 3 - Comparison of numerical and analytical data.

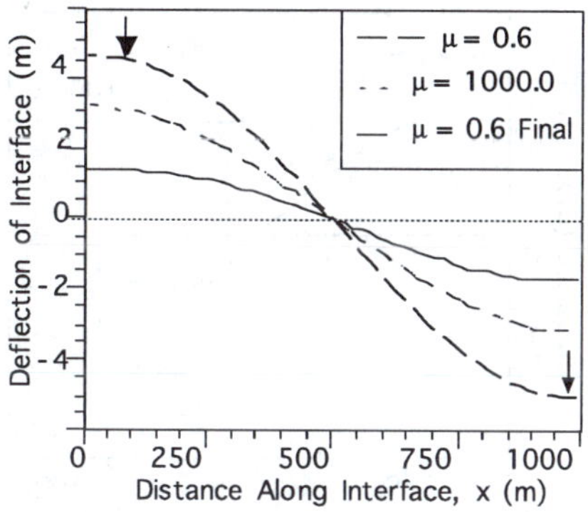

Figure 4 - Deflection at maximum flexural loading and after unloading.

4.2 Numerical results and comparison

In the numerical model the shear stiffness of interface elements can be altered to approximate bonded, and freely-slipping contacts. For this comparison, the interface elements are not allowed to slip due to friction; the deformation is due to elastic element stiffness. The boundaries and interface were discretized into 204 equal length elements. Requisite model parameters were taken from Table 1.

A bonded contact is approximated by assigning $K_S = K_N = 1000$ GPa, 20 times greater than E of the layers. Displacement discontinuities across the interface elements are on the order of 0.02 mm. Deflection of the top and bottom surfaces of the model lie within 2% of the interface deflection so we only show profiles for y=0. The deflection profile of $K_S = K_N = 1000$ GPa closely matches the analytical result for a bonded layer, Fig 3. We can quantify the match by considering the maximum deflection, $u_y(x=0, y=0)-u_y(x=1000, y=0)$. The max. deflection for the $K_S= 20$ E is only 3.6% greater than the analytical solution for perfectly bonded interfaces. Larger stiffnesses will decrease the difference between the numerical and analytical deflections.

The freely-slipping interface is approximated with $K_S = 0.0$ GPa. Normal displacements across the interface are constrained by assigning $K_N = 1000$ GPa. Slip is greatest around x=500 m and approaches zero at x=0 and x=1000 m. The max. displacement discontinuity is 3.6 m. The max. deflection of $K_S = 0.0$ GPa, is 5.8% less than the analytical solution for a perfectly slipping interface. As the number of elements used to discretize the boundaries is increased, the difference between the numerical and analytical solution decreases. Deflection for the case $K_S = 100$ GPa is shown to lie between the two end member cases of perfectly bonded and free slip.

5 MODEL RESULTS

Deformation of elastic systems is independent of loading history; however, deformation of a stack of elastic layers with frictionally slipping contacts is inelastic. We apply a monotonic increase in flexural load followed by a monotonic decrease in flexural load to zero. The monotonic decrease in flexural loading to zero permits separate investigation of the inelastic and elastic deformation occurring within the structure. A constant overburden loading is applied to the model for realistic frictional behavior of the interface.

As elements along the interface slip, convergence of the interface element displacement discontinuities is completed between successive iterations when the $(R-\sigma_{yx}) < 0.1\% \ \sigma_{yx}$

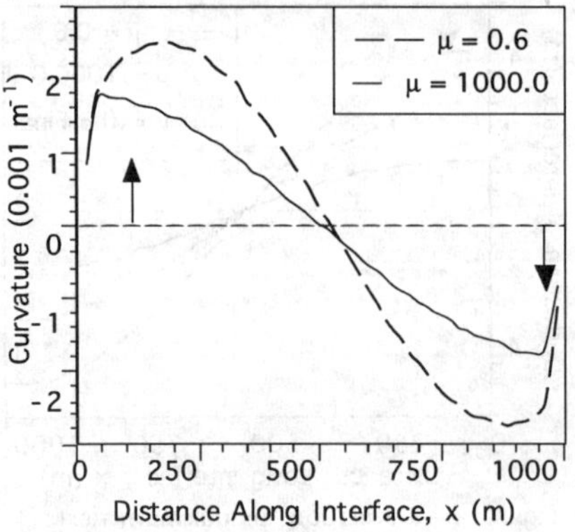

Figure 5 - Curvature at max. flexural load.

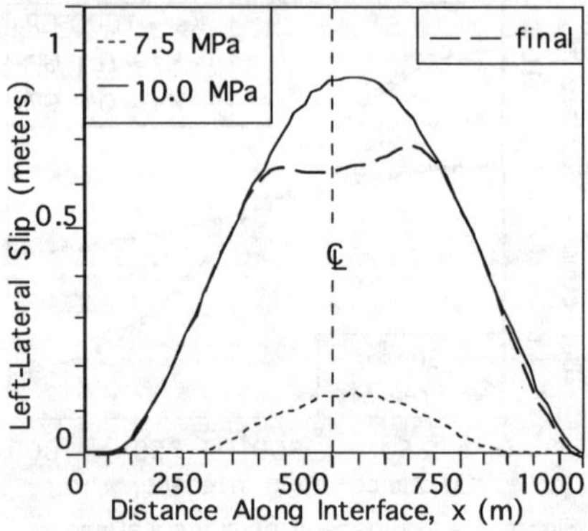

Figure 6 - Slip distribution during loading and after unloading.

for every slipped element. The model is discretized into 204 equal-length boundary elements; this extent of discretization matches the analytically solutions well.

The model was loaded to a maximum flexural loading amplitude of 10.0 MPa and then unloaded to 0.0 MPa in an equal number of steps. Elastic deformation does not depend on the number of steps taken, but inelastic behavior may. The change in the maximum deflection for 4 and 8 loading steps was about 5%. For this paper, we will describe the deformation of the multilayer loaded and unloaded in 8 steps.

5.1 Monotonic Loading

The deflection of the model with frictional interface elements is shown in Fig 4. A coefficient of friction of 0.6 is reasonable for geologic materials (Byerlee 1978). The case of a frictionally bonded interface is investigated by assigning μ=1000. We will compare the deformation of the frictional slipping interface with the elastic deformation of the frictionally bonded interface.

The interface elements that experienced slip lie between the two arrows on Fig. 4. The bedding plane faults occur along the limb of the fold, as expected. The slip patch is not centered on the monocline but extends to the right. The flexural loading increases relative tension on the interface in the region $500 \leq x \leq 1000$ m and lowers the frictional resistance. The presence of frictional slip increases the max. deflection, steepens the fold and tightens fold hinges; in agreement with previous analytical models of bedding plane slip.

The maximum curvature of the monocline is greater for the frictionally slipping (μ=0.6), varies with location along the interface and is asymmetric about the interface midpoint, Fig. 5. The point of max. curvature to the right of the midpoint ($x \approx 900$ m.) is located closer to the sides of the model than the point of max. curvature to the left of the midpoint ($x \approx 200$ m). The curvature for μ=1000 is nearly linear away from the sides of the model and symmetric about the interface midpoint. Maxima occur at the crest of the anticline and the trough of the syncline.

The first two steps of loading elastically deform the interface elements; no slip occurs. At the third loading step, 7.5 MPa, about 30 elements slip left-laterally near the center of the interface (Fig. 6). The offset has a maximum of about 0.13 m at $x \approx 550$ m and tapers to zero at the ends of the slip patch ($x \approx 250$, 850 m.).

The slip distribution at 10.0 MPa loading shows a similar shape. The maximum offset of 0.84 m is reach at $x \approx 525$ m. The distribution is asymmetric because of the decreased frictional resistance of elements to the right of the interface midpoint.

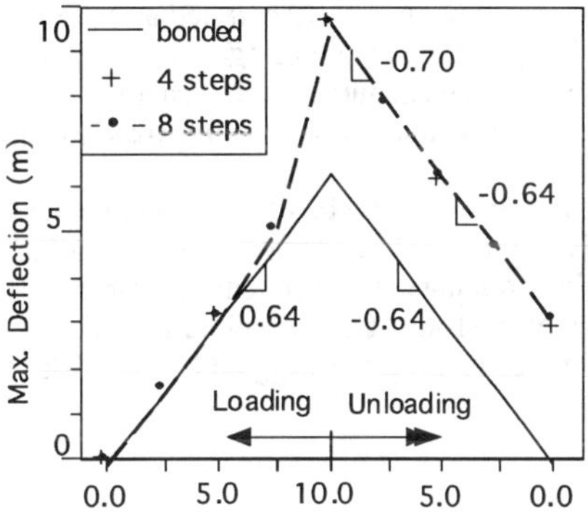

Figure 7 - Max. deflection for elastic and inelastic cases.

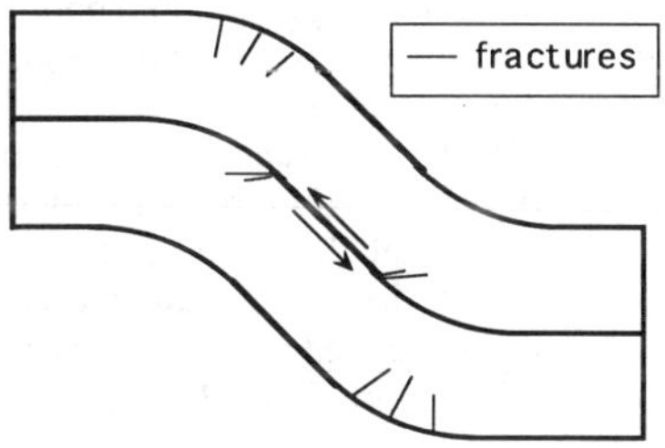

Figure 8 - Schematic distribution of fractures along the monocline. Some are related to fold curvature while others are related to bedding plane slip.

5.2 Monotonic Unloading

The multilayer is monotonically unloaded from the maximum flexural load (10 MPa) to zero while the overburden load is kept constant. The bonded case (μ=1000) shows linear elastic behavior with a constant ratio of maximum deflection to load amplitude of 0.64 for loading and -0.64 for unloading, Fig. 7. The case of μ=0.6 shows inelastic behavior during loading past 5 MPa. At that point, the interface elements begin to slip and inelastic deformation accumulates.

During unloading, the behavior of the multilayer with a frictional interface is nearly linear elastic. The ratio of max. deflection to load amplitude of the first unloading step is -0.70 while the remaining steps have a slope of -.064. During the first step of unloading the middle 250 meters of the interface experiences right-lateral offset, depicted as a decrease in left-lateral offset in Fig. 6. The ratio of the max. deflection to load amplitude for the first step of unloading is -0.70. The first unloading step contained more than 90% elastic strain release and less than 10% inelastic right-lateral slip. For the last 3 unloading steps, the slope of Figure 7 is -0.64 and the remaining elastic strain accumulated in loading is released leaving the inelastic deformation due to slip as evidence of the loading history.

The deflection of the monocline with a frictionally slipping interface after unloading, is shown on Fig. 4. The max. deflection for the unloaded monocline is about 3.1 m and most of the deformation occurs within the central 850 meters. The 150 meters to either side of this central region have approximately zero slope. The central 200 meters are nearly linear with a slope of 0.5%. The hinge sections of the fold are about 200 meters wide. Rather than a broad fold with constant curvature, the fold shape can be described as roughly three connected linear sections.

6 DISCUSSION OF FRACTURE GROWTH

The distributions of slip along the interface differs from the elliptical distributions for uniformly loaded mode II fractures. Because the slip approaches zero more gradually at the ends of the slipping patch, the concentration of stresses there would be less than for the case of a uniform stress drop. It is unlikely that joints would form at the tips of a slip patch. This remains to be investigated by computing the actual values of K_{II}.

Changes in the frictional behavior along the interface may cause abrupt changes in the distribution of slip. For example at the end of a shale seam in an otherwise massive sandstone. Where the slip gradient is high, stresses will be more concentrated and the greater tensile stresses are more likely to induce failure of the neighboring rock. The change in material property at one point along the interface encourages the formation of many cracks initiating at the same location. Thus, the distribution of fractures may not be constant along the fold.

Fold curvature increases the layer parallel stresses encouraging bedding perpendicular fracture propagation. Our model suggests the distribution of fractures caused by bending will not be constant along the length of a monocline, but have greatest density near x≈200 m and x≈900 m; limbward of the crest and trough of the fold. The overall pattern of fractures will be related to the curvature of the monocline during loading and to the distribution of slip on interfaces, Fig. 8.

7 CONCLUSIONS

Although each layer is elastic, a stack of layers with frictional interfaces will deform inelastically leaving residual deformation after unloading. For the same loading, geometry and elastic properties of the layers, the deflection for a fold with frictionally slipping interface is greater than that for a frictionally bonded interface. The maximum curvatures of the layers are larger for frictional slipping interfaces and occur limbward of the crest and trough of the fold.

By removing the flexural load, we found that the residual inelastic deformation of frictional interfaces between two stacked layers forms a fold with straighter limbs and tighter hinges than elastic models. For sinusoidal loading distributions, anticlines and synclines would have flat tops and bottoms, more like a kink fold.

The presence of bedding plane fault within a flexure may play an important role in the localization of fractures. The distribution of fractures along the fold, will be concentrated at points of material change along bedding planes. Fractures density will also increase to either side of the crest and trough where fold curvatures are greatest.

REFERENCES

Byerlee, J. 1978. Friction of Rocks. *Pageoph.* 116: 615-626.

Chapple, W.M., and J.H. Spang 1974. Significance of layer-parallel slip during folding of layered sedimentary rocks. *GSA, Bull.* 85: 1523-1534.

Crouch, S.L. 1979. Computer simulation of mining in faulted ground. *J.S. African Inst. of Mining and Metallurgy*. 79: 159-173.

Crouch, S.L., and A.M. Starfield 1990. *Boundary Element Methods in Solid Mechanics*. Boston: Unwin Hyman.

Feund, R. 1979. Progressive strain in beds of monoclinal flexures. *Geology*. 7: 269-271.

Jaeger, J. C. and N.G.W. Cook 1979. *Fundamentals of Rock Mechanics*. New York: Chapman and Hall.

Kemeny, J., and N.G.W. Cook 1985. Formation and stability of steeply dipping joint sets. 26th U. S. *Symp. on Rock Mechanics*. 26: 471-478.

Koch, F.G., A.M. Johnson, and D.D. Pollard 1981. Monoclinal bending of strata over laccolith intrusions. *Tectonophysics*. 74: T21-T31.

Liang, J. 1993. Frac2D v. 1.0 User's Manual. *Rock Fracture Project: Stanford Univ.*

Roth, W.H., J. Sweet, and R.E. Goodman 1982. Numerical and physical modeling of flexural slip phenomena and potential for fault movement. *Rock Mechanics, Suppl.* 12: 27-46.

Scholz, C.H. 1990. *The mechanics of earthquakes and faulting*. New York: Cambridge University Press.

Rock Mechanics, Nelson & Laubach (eds) © 1994 Balkema, Rotterdam, ISBN 90 5410 380 8

Experimental observations of static and dynamic properties of joints

G.N. Boitnott
New England Research, Inc., White River Junction, Vt., USA

1 INTRODUCTION

The importance of joints in the attenuation of high frequency seismic waves originates from the fact that the mechanical properties of rock joints contrast greatly in comparison to the mechanical properties of intact rock. Efforts to incorporate fractures in high frequency wave propagation models have primarily been based on the model of *Schoenberg* [1980], where an individual joint is treated macroscopically as a displacement discontinuity [see *Pyrak-Nolte et al.,* 1990]. When an elastic wave encounters a displacement discontinuity, part of the energy is reflected and part transmitted. The transmission coefficient is a function of the stiffness of the displacement discontinuity (the joint stiffness), the frequency of the elastic wave, and the acoustic impedance of the host material. While the displacement discontinuity appears to fit experimental data quite well at ultrasonic frequencies, questions arise due to the fact that the dynamic stiffnesses are consistently higher than quasi-static stiffnesses measured using the same sample.

In this study, joint closure and ultrasonic wave transmission experiments are conducted on an artificial joint in glass. Glass was chosen because it is a simple, homogeneous elastic material, a characteristic which greatly simplifies the task of separating the effects of deformation of the matrix from those of the joint. In previous work on natural joints in granite [*Pyrak-Nolte et al.,* 1990], measurements based on ultrasonic transmission coefficients indicated that the dynamic stiffness was considerably greater than quasi-statically measured stiffnesses. They found compressional wave stiffnesses to be 2 to 5 times greater than quasi-static unloading stiffnesses, with a high degree of variation from one sample to the next, as well as from one normal load to the next. In addition, compressional stiffnesses were found to be greater than shear stiffnesses by as much as a factor of 5 (typically near 2), again with considerable variation from sample to sample.

The problem lies in the fact that these observations seem inconsistent with developed linear elastic contact models which have been successful at predicting quasi-static mechanical properties of joints in both normal compression and shear. Two hypotheses to explain these observations and the resulting discrepancy with developed models are examined. The first is that hysteresis in the deformation of the joint (i.e. loading path dependence) leads to discrepancies between dynamic and quasi-static stiffness. The second is the model of *Pyrak-Nolte and Nolte* [1992], where heterogeneity of the stiffness within the plane of the joint produces an apparent frequency dependent dynamic stiffness.

The artificial joint tested here was used as a means of minimizing these two potential effects, and thus producing a more ideal displacement discontinuity. Owing to the fine scale of the roughness, the joint exhibits little hysteresis in deformation and thus the quasi-static stiffness is well constrained independent of the loading history. Additionally, in comparison to natural joints, the joint tested here is thought to have more uniform mechanical properties at the scale of the ultrasonic wavelengths, minimizing the potential frequency dependence of the stiffness resulting from heterogeneity.

2 EXPERIMENTAL RESULTS

2.1 Background

The experiments were conducted on a servo-controlled uniaxial apparatus operated in load feedback from a foil strain gage load cell mounted above the sample. Three spring loaded displacement transducers are mounted on rings which are attached to the sample on either side of the joint. The gage length between the rings is 28.6 mm. Two ultrasonic transducers mounted on 6.35 mm thick titanium plates are housed in the upper and lower end plates. Each transducer consists of a 22.2 mm diameter stack with one compressional and one shear PZT5A piezo-electric crystal along with an absorptive backing. The center frequency of each crystal is 700 kHz.

Right cylinders of optical quality BaK1 glass, nominally 7.62 cm in diameter were used for the construction of the samples. Starting with 8.9 and 5.1 cm long cores provided by Schott Glass Technologies Inc., one end of each cylinder was lapped with 60 grit silicon carbide powder to produce nominally flat surfaces with short wavelength random roughness. This lapping procedure is commonly used in the development of artificial joints for testing linear elastic contact models (see for example *Brown and Scholz,* [1985, 1986]). It was believed that such a simple joint would make it easier to understand the differences between static and dynamic joint stiffness and the relationship with contact mechanics models.

2.2 Joint Closure Experiments

An example of a joint closure experiment on the artificial glass joint is shown in Figures 1a and 1b. The joint closure, plotted along the abscissa in Figure 1b, is the displacement which is attributed to the presence of the joint (i.e. the deformation of the intact glass between the rings has been removed). The main features to notice in Figure 1b are that the overall stiffness increases rapidly with normal load and that the stiffness upon reversal of loading direction increases with respect to the stiffness just prior to a loading reversal, resulting in cusps in the stress-displacement relation at the loading reversals. While there appears to be some permanent closure with each successive cycle, the deformation is primarily elastic (recoverable), exhibiting little permanent deformation.

In Table 1, we present the joint stiffness just before (K_l) and just after (K_u) each reversal in loading direction at each stress maximum. Here we see that the joint stiffness after a loading reversal is no more than 1.4 times that of the joint stiffness prior to the loading reversal (a relatively small effect compared to similar measurements on tension fractures in rock [see *Boitnott,* 1992]). Results such as those in Table 1 are important here because the amount of stiffening after a load reversal is an important parameter needed for accessing the loading path dependence of the quasi-static stiffness.

Table 1

Load at Reversal MPa	K_l MPa/μm	K_u MPa/μm	K_l/K_u -
10	2.4	3.5	0.7
10	2.6	3.6	0.7
10	2.9	3.9	0.7
5	1.5	1.6	0.9
5	1.5	1.6	0.9

2.3 *Dynamic vs. Quasi-static Stiffness*

The displacement discontinuity model of *Schoenberg* [1980] and *Pyrak-Nolte et al.* [1990] predicts that the transmission coefficient is a function of the stiffness of the displacement discontinuity (the joint stiffness), the frequency of the elastic wave, and the acoustic impedance of the host material. The transmission coefficient for a shear or compressional wave with normal incidence is given by

$$T = \frac{2K/Z}{-i\omega + 2K/Z} \tag{1}$$

where K is the joint stiffness (for shear or compression), ω is the frequency of the incident wave, and Z is the seismic impedance of the host material. The validity of such a model has been tested by *Pyrak-Nolte et al.* [1990] who found that the functional form of equation (1) did a good job of quantitatively describing the spectrum of the transmitted energy for ultrasonic frequencies in the band 200 kHz to 1.0 MHz.

Here, we use time-domain waveform matching to determine the joint stiffness. In the time domain, the transfer function defined in equation (1) results in an impulse response function of the form

$$A(t) = 2\frac{K}{Z}e^{-\left[2K(t-t_0)/Z\right]} \quad ; \quad t \geq t_0 \tag{2}$$

Use of time-domain waveform matching includes both the amplitude and phase response of equation (1), and thus is a more robust (restrictive) test of the displacement discontinuity model than comparison of amplitude spectra.

The same experimental configuration was used for ultrasonic tests as was used in the quasi-static closure experiments. A number of calibration experiments were performed on a solid sample of identical glass to assess the reproducibility of the waveforms. The compressional wave amplitudes through a 14 cm long solid piece of BaK1 glass were reproducible to ± 5% and the S wave amplitudes to ± 10%.

Once the reference waveforms were determined, the reference waveforms were used as the forcing function and convolved with the impulse response function (equation 2) to yield the predicted waveforms for a joint with specified stiffness. Example computations of expected waveforms as a function of stiffness are shown in Figures 2a. Here we see that the transmitted amplitude is roughly linearly related to joint stiffness under these conditions. Thus, the 10 to 20 percent uncertainty in the reference amplitudes will lead to similar errors in the inferred joint stiffness.

Measured waveforms as a function of normal load are shown for comparison in Figure 2b. Here we see good correlation between the measured and theoretical waveforms, making it relatively easy to constrain the joint stiffness by type-curve matching. Long portions of the waveforms are well matched, indicating little contamination from extraneous reflections. The ultrasonically derived compressional stiffnesses are plotted in Figure 3a along with the quasi-static stiffness from the closure curve. Here we see the ultrasonically derived stiffnesses are greater than the quasi-statically measured values, consistent with the observations of *Pyrak-Nolte et al.* [1990].

Ultrasonically derived compressional and shear wave stiffnesses for multiple experiments on the same sample are shown in Figure 3b. As would be expected, the shear stiffnesses are less than the compressional stiffnesses, and follow the familiar linear increase with normal load. The ratios of the compressional stiffness to the shear stiffness were near 1.5, less than that found for natural fractures by *Pyrak-Nolte et al.* [1990], but still measurably greater than the value of 1.2 predicted by quasi-static elastic Hertzian contact theories (see discussion).

3 DISCUSSION

3.1 Predictions of Contact Theory

The contact of two rough surfaces has generally been described based on the consideration that surfaces come into contact over only a fraction of the nominal area of the joint. A number of constitutive models have been developed and experimentally tested which predict the macroscopic mechanical properties of the joint from the material properties of the solid and the surface topography (or equivalently the aperture distribution) [*Greenwood and Williamson,* 1966; *Swan,* 1983; *Brown and Scholz,* 1985, 1986; *Hannan,* 1988; *Boitnott et al.*, 1992]. These models all belong to a class of model referred to as non-interacting contact models (NICM), reflecting the fundamental assumption that each asperity undergoes the same displacement during deformation. As a result of this assumption, the total load resisting deformation is simply the sum of the forces contributed by each contacting asperity. Using realistic distributions of asperity heights, the models predict many of the fundamental properties of joints, including the approximately linear dependence of the joint stiffnesses (both normal and shear) with load.

A popular asperity model has been the use of the elasticity solutions of *Hertz* for the contact of two linear elastic spheres [see *Mindlin,* 1949]. The ratio of normal to shear compliance for a Hertzian contact turns out to depend only on the Poisson's ratio of the material, independent of load and size. Thus any NICM for which the Hertz solution is used will predict the ratio of macroscopic normal to shear stiffnesses to be independent of load and dependent only on the Poisson's ratio of the matrix. For the glass used here, the Poisson's ratio was measured to be $\nu = 0.25$, yielding a predicted $K_p/K_s = (2-\nu)/(2-2\nu)$ of 1.2 .

In the experiments reported here, we do see that the stiffness increases approximately linearly with load (see figure 3) and that, as a consequence, the ratio K_p/K_s is constant independent of the load. However, while both of these observations are qualitatively consistent with Hertzian contact theories, quantitative discrepancies exist. First, we must explain the discrepancy between quasi-static stiffnesses and dynamic stiffnesses. Second, the results here, as well as those of *Pyrak-Nolte et al.* [1990], consistently show K_p/K_s greater than the predicted 1.1 to 1.2 for reasonable Poisson's ratios. Interestingly, however, the results of *Sherif and Kossa* [1991], who measured shear and normal stiffness of joints in steel at frequencies on the order of 1 *kHz* do find $K_p/K_s \approx 1.2$ independent of load. This suggests dynamic effects may be responsible for the high K_p/K_s at ultrasonic frequencies.

3.2 Effects of Hysteresis

This suite of experiments tests the notion that loading path dependence may be the dominant cause of much of the consistent difference between measured values of dynamic and static stiffnesses. The argument follows that proposed by *Walsh* [1965] to explain the difference between static and dynamic moduli of rocks. *Haupt et al.* [1992] and *Martin and Haupt* [this issue] have shown that the strain history effect is one of the dominant variables (in comparison to dynamic effects) controlling attenuation and apparent modulus dispersion for dry, intact rock, and thus we might expect that the same would be true for a single joint as well.

In the results of this study, as well as those of *Pyrak-Nolte et al.* [1990], the stiffnesses needed to model the observations were significantly higher than would be expected from measurements of quasi-static stiffnesses on the same samples. While results using natural joints in rock could easily be contaminanted by hysteresis effects, for the case of the glass joint reported here the sample exhibited little hysteresis. Despite this fact, the dynamic stiffness was clearly greater than the quasi-static stiffness. Thus, the results shown here seem to speak against the strain history explanation.

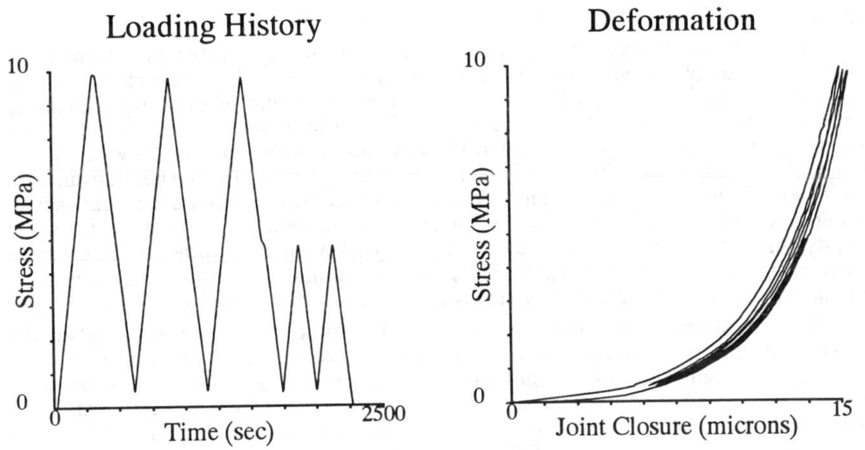

Figure 1: (a): Time history of the normal load for a joint closure experiment (gbad20) on a artificial joint in BaK1 glass. (b): Normal load versus joint closure for the experiment shown in (1a).

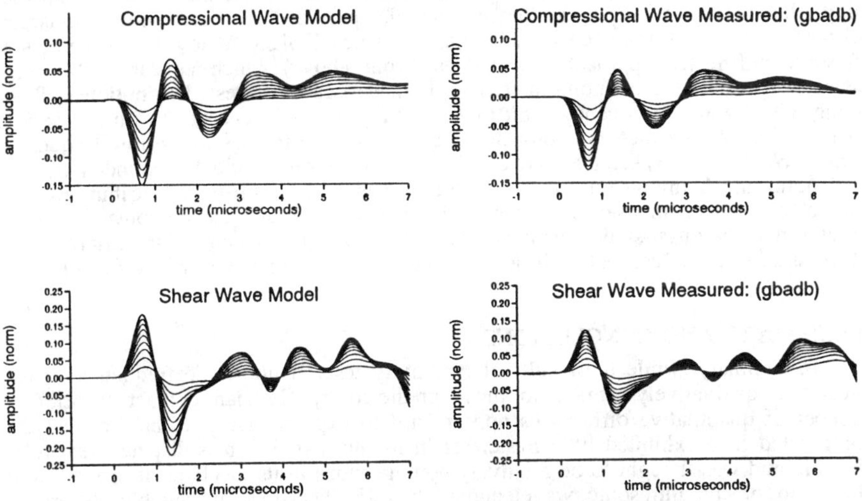

Figure 2: (a) Predicted compressional (top left) and shear (bottom left) waveforms for propagation across a sample with a single joint with joint stiffness ranging from 1 $MPa/\mu m$ to 9 $MPa/\mu m$ in 1 $MPa/\mu m$ increments. The waveforms were computed by convolving the reference compressional waveforms with the impulse response function (equation 2). The amplitudes are normalized by the peak to peak amplitude of the reference waveform. (b) Measured compressional (top right) and shear (bottom right) waveforms for the jointed glass sample as a function of normal loads from 1 MPa to 10 MPa in 1 MPa increments. Note the similarity to the modeled waveforms in (a).

3.3 *Effects of a Heterogeneous Joint*

In an attempt to explain the difference between static and dynamic stiffness of fractures, *Pyrak-Nolte and Nolte* [1992] offered a model which attempts to include the effects of heterogeneity of the stiffness within the plane of the joint. Assuming that the local transmission coefficients depend on the local static stiffnesses via equation (1) and that regions of different stiffness transmit independently, they developed a model in which the effective dynamic stiffness reflected the mean transmission coefficient. Thus, the macroscopic dynamic stiffness becomes frequency dependent, despite the fact that at the meso-scale (the scale of the acoustic wavelength) the static and dynamic stiffnesses are assumed identical. *Pyrak-Nolte and Nolte* [1992] go on to show that such a model does a better job of fitting the observed transmission coefficients than the simple homogeneous displacement discontinuity.

Two questions arise concerning their model. First, can the model explain the observation that the macroscopic K_p/K_s is greater than the local (mesoscopic) K_p/K_s (i.e. can we assume the contact mechanics prediction of $K_p/K_s \approx 1.2$ at the meso-scale and yield $K_p/K_s > 1.5$ at the macro-scale)? Second, is the model quantitatively consistent with the measured difference between static and dynamic moduli? In Figure 4 we show computations of K_p and K_p/K_s vs frequency for a variety of stiffness distributions, using the measured acoustic impedances of the glass ($Z_p = 1.69 \cdot 10^7 kg/sec \cdot m^2$, $Z_s = 0.97 \cdot 10^7 kg/sec \cdot m^2$). In each case, the mean stiffnesses of $\bar{K}_p = 1.25\bar{K}_s = 5.0\ MPa/\mu m$ were used (for computational reasons we used 1.25 instead of 1.2). It is found that as the frequency is increased, the dynamic stiffnesses approached (from below) the mean stiffnesses ($\bar{K}_p$,$\bar{K}_s$) for all distributions studied. In addition, as K_p and K_s approach $\bar{K}_p$ and $\bar{K}_s$ with increasing frequency, K_p/K_s approaches (from above) the assumed meso-scale K_p/K_s. Thus the model is capable of producing elevated K_p/K_s at the macro-scale, provided the macroscopic dynamic stiffnesses have not approached their high frequency limits. At low frequencies, the dynamic stiffnesses approach low values (from above) which appear to be consistently less than the harmonic means of the meso-scale stiffness distributions. This, along with the high frequency limit of $\bar{K}_p$ and $\bar{K}_s$, seems inconsistent with the notion that the dynamic stiffnesses approach the quasi-static stiffnesses in the low frequency limit. We would expect that the quasi-static stiffnesses would be bounded by the arithmetic and harmonic means of the local (mesoscopic) stiffness distributions. In fact, as a result of imposed displacement boundary conditions, elastic contact theories assume that the quasi-static stiffnesses equal the arithmetic means of the microscopic stiffnesses, in direct conflict with these predictions of the dynamic theory for joints.

4 SUMMARY AND CONCLUSIONS

In summary, while the results of laboratory tests at ultrasonic frequencies have properties qualitatively similar to those predicted by Hertzian contact theories, a number of quantitative differences appear hard to explain. Despite the fact that the joint tested here exhibited little hysteresis in its quasi-static stress-displacement relationship and was thought to be relatively homogeneous in its mechanical properties at the scale of the ultrasonic wavelengths, there is clearly a measurable difference between static and dynamic stiffnesses and unexpectedly high values of K_p/K_s. Strain history dependence of quasi-static deformation does not appear to explain the discrepancy between dynamic and quasi-static stiffnesses. The frequency dependent model of *Pyrak-Nolte and Nolte* [1992] has the potential of explaining the results (even in the context of Hertzian contact models), however quantitative discrepancies concerning the low frequency (quasi-static) limit of the dynamic stiffnesses require explanation. Of course, we should not assume *a priori* that the predictions of Hertzian contact theories are reasonable. In addition, some of the observed discrepancies between the measured values and those predicted by contact theory may result from dynamic effects not yet considered. In any case, more experimental data, particularly as a function of frequency, are needed to address these questions.

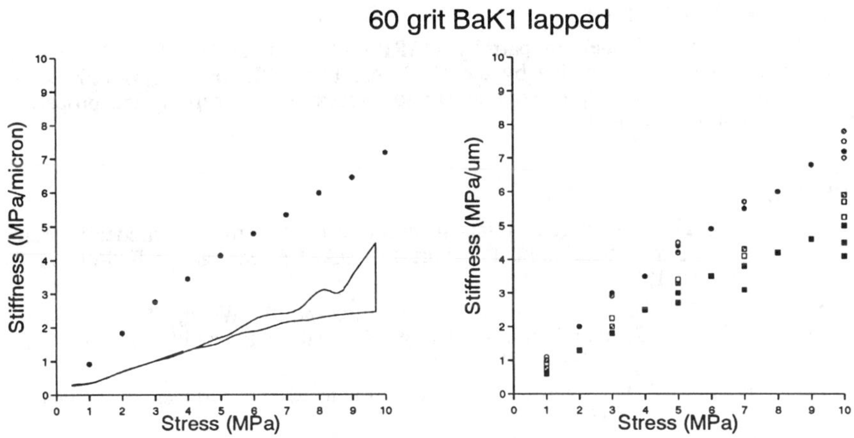

Figure 3: (a) Comparison of ultrasonically derived compressional stiffnesses (circles) with the quasi-statically measured stiffness (solid line) for the 60 grit lapped joint in glass. The ultrasonically derived stiffnesses are systematically higher than the quasi-statically measured values.
(b) Inferred compressional (circles) and shear (boxes) wave stiffnesses from four different experiments on the jointed glass sample (gbad).

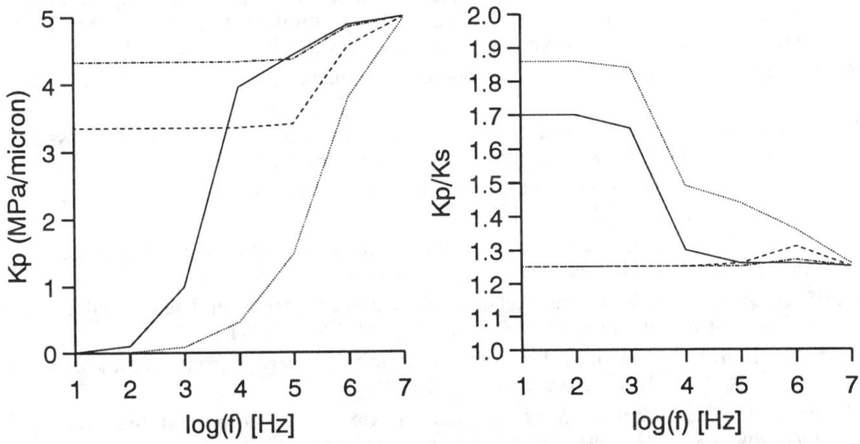

Figure 4: Computed dynamic K_p and K_p/K_s as a function of frequency using the model of *Pyrak-Nolte and Nolte* [1992]. Results for four different meso-scale distributions of K_p and K_s are shown. In each case the mean of the meso-scale K_p distribution was specified to be 5 $MPa/\mu m$. The local values of K_s were enforced to be equal to $K_p/1.25$. The distributions for K_p were: solid line - Gaussian with mean of 5 $MPa/\mu m$ and standard deviation of 1.25 $MPa/\mu m$; dotted line - Exponential distribution where the probability of K_p went as $e^{-K_p/\bar{K}}/\bar{K}$; dashed line - bimodal distribution with even probability between two values 2.5 $MPa/\mu m$ and 7.5 $MPa/\mu m$; dot-dashed line - band-limited constant distribution of values between 2.5 $MPa/\mu m$ and 7.5 $MPa/\mu m$.

5 ACKNOWLEDGMENTS

This work was funded in part by DARPA-PL contract #F29601-91-C-DB26. Additonal support was provided by the U. S. Air Force PL/GPEH. I would like to thank Martin Smith for helpful comments and suggestions concerning the properties of displacement discontinuities.

6 REFERENCES

Boitnott, G. N., The role of rock joints in non-linear attenuation in moderate strain amplitude regimes Semi-annual Report for DARPA contract # F29601-91-C-DB26, August 1992.

Boitnott, G. N., N. Yoshioka, R. Biegel, C. Scholz, and W. Wang, Micromechanics of rock friction: 2, quantitative modeling of initial friction with contact theory, *J. Geophys. Res., 97,* 8965-8978, 1992.

Brown, S. R., and C. H. Scholz, Closure of random elastic surfaces in contact, *J. Geophys. Res., 90,* 5531-5545, 1985.

Brown, S. R., and C. H. Scholz, Closure of rock joints, *J. Geophys. Res., 91,* 4939-4948, 1986.

Greenwood, J. A., and J. Williamson, Contact of nominally flat surfaces, *Proc. R. Soc. London Ser. A, 295,* 300+, 1966.

Hannan, S. S., An experimental study of fracture closure in elastically and non-elastically deformable rocks, Ph.D. thesis, Dep. of Geol., Univ. of Toronto, Toronto, Ont. 1988.

Haupt, R. W., R. J. Martin, III, X. M. Tang, and W. Dupree, Elastic moduli and seismic wave attenuation in dry and saturated rock: modulus dispersion and attenuation. Final Report PL-TR-92-2199(II), Phillips Laboratory, Air Force Material Command, Hanscom Air Force Base, MA., July 1992.

Mindlin, R. D., Compliance of elastic bodies in contact, *J. Appl. Mech., 16,* 259-268, 1949.

Pyrak-Nolte, L. J., L. Myer, and N. Cook, Transmission of seismic waves across single natural fractures, *J. Geophys. Res., 95,* 8617-8638, 1990.

Pyrak-Nolte, L. J., and D. D. Nolte, Frequency dependence of fracture stiffness, *Geophys. Res. Let., 19,* 325-328, 1992.

Schoenberg, M., Elastic wave behavior across linear slip interfaces, *J. Acoust. Soc. Am., 68,* 1516-1521, 1980.

Sherif, H. A., and S. S. Kossa, Relationship between normal and tangential contact stiffness of nominally flat surfaces, *Wear, 151* 49-62, 1991.

Swan, G., Determination of stiffness and other joint properties from roughness measurements, *Rock Mech. Rock Eng., 16,* 19-38, 1983.

Walsh, J. B., The effect of cracks on the uniaxial elastic compression of rocks, *J. Geophys. Res., 70,* 399+, 1965.

Rock Mechanics, Nelson & Laubach (eds) © 1994 Balkema, Rotterdam, ISBN 90 5410 380 8

Seismic geotomography: Amplitude versus velocity in consideration of joint aperture and spacing

Chikaosa Tanimoto & Kiyoshi Kishida
Department of Civil Engineering, Kyoto University, Sakyo, Japan

ABSTRACT : *Both laboratory and field tests show that the seismic amplitude tomography (SAT) owns higher capability in sensing joint properties and distribution than the available velocity tomography (SVT). Experimental comparisons between amplitude and velocity are illustrated concerning joint parameters such as aperture, joint roughness, joint frequency, and dilation under various normal and shear stresses. Then, two kinds of mapping by SVT and SAT which were obtained from the investigation of the dam site are discussed in terms of joint orientation and distribution.*

1. INTRODUCTION

Geotomographic technique is widely applied to various fields so as to improve its abilities in mapping a section of earth interior. Its capability, however, in rock engineering has not been clarified sufficiently, that is, a constructed tomographic image does not help our understanding on joint characteristics by itself. It is because the propagation velocity of a seismic wave remains being as an overall index of the dynamic behaviour of a rock mass without clarifying any characteristics of rock mass. The first author initiated the study on the relationship between seismic behaviour and mechanical properties of jointed rock in association with the research for the magnitude of loosening event during the construction of the long suspension bridge in late 1970s [Tanimoto & Ikeda, 1983].

Our experimental study in laboratory on seismic behaviour has been carried out with respect to joint orientation, joint frequency, aperture, roughness, infilling material, alteration, moisture content, etc. under normal and shear stresses [Tanimoto, et al., 1994a]. To speak of the representation of joint roughness, Barton's JRC has been conveniently employed, but the determination of a JRC-value has to be estimated visually with referential profiles given by Barton et al. [Barton, et al., 1977]. For the detailed discussion on the experimental result, it is necessary to establish any other scientific way to determine the magnitude of roughness directly. We propose a method to measure joint roughness by means of the three dimensional, noncontact profiler (TOK-3DRP) and the maximum entropy method (MEM). MEM is the method which is applied to the wave analysis as well as the fast fourier transform (FFT). The results concerning the behaviours of amplitude have shown remarkable distinctions in sensitivity under various conditions from those of velocity. They support the advantage of our approach which combines the amplitude tomography (SAT) with the velocity one (SVT).

In field, first of all, it is emphasized that reliable joint survey must be executed in undisturbed state. Our use of Borehole Scanner (BSS) is going quite well, and provides very valuable information on the above-mentioned joint characteristics [Tanimoto, et al., 1992]. In 1992 and 1993 we carried out joint observation through BSS and seismic survey in the exploratory adit and along the boreholes at a certain dam site. It is concluded that the SAT map corresponded much more reasonably with the observed joint distribution than the SVT map [Tanimoto, et al., 1994b].

2. EXPERIMENTAL STUDY ON THE RELATIONSHIP BETWEEN SEISMIC BEHAVIOUR AND MECHANICAL PROPERTY

2.1 Determination of rock joint roughness

When rock substance is hard enough, e.g. its compresstve strength is higher than 50 MPa, most of mechanical properties of jointed rock are strongly affected by the characteristics of joints and their distribution. They cause the reduction of shear strength, unfavorable deformation, and change of permeability. Roughness of joint is one of major parameters in the determination of mechanical properties of rock mass.

Although Barton's JRC is very practical in field, it is difficult to estimate the magnitude of joint asperity so quantitatively as to avoid human error and to seek high accuracy in laboratory test. It was pointed out as a common subject at Rock Joint meeting in Loen [1990] what JRC means.

Analysis of fractal dimension has been popularly employed in the determination of rock joint roughness. Maetz [1990] showed that the correlation between JRC and fractal dimension is as high as 0.973 in reliability, representing the primary order of approximation, and its fractal dimension is between 1.00 and 1.02. There are some other papers applying fractal dimension to roughness [Carr, et al., 1987,Turk, et al., 1987, Muralha & Grassa, 1990]. Nevertheless, we do not think that the application of fractal dimension to joint roughness is practical, because the range of fractal dimension corresponding to JRC (from 1 to 20) is too narrow to distinguish JRC rating, namely, its resolution is too low to represent the roughness of joint surface.

Then, we have newly proposed our own method. It is as follows : Considering the rock joint roughness, which is expressed in the relation of asperity vs. shear-directional displacement, as a series of waves in the amplitude vs. time relation, wave forms corresponding to various roughnesses were analyzed in terms of power spectrum by FFT (Fast Fourier Transform) [Cooley & Tukey, 1965] and MEM (maximum Entropy Method) [Burg, 1967]. MEM is better than FFT, for the former provides more stable result than the latter even in case of highly various numbers of sampling data. The former is very convenient to be applied to roughnesses of natural joints. The measurement of asperities is materialized by the noncontact-type laser-scan micrometer which is situated on the three dimensional movable stage 3-D Roughness Profiler (TOK-3DRP). Both operation and data acquisition are fully automated by a micro computer. We call the entire system 'Ms-method'.

The power spectrum vs. frequency relation analyzed by MEM for various JRC ranges is shown in Fig.1. The maximum values of power spectra for all ranges of JRC appear in the low frequency range for any cases, and values of power spectra in the range of higher than 1 Hz are lower than 1/100 of the maximum values. It means that, assuming the maximum value (amplitude) to be 1 mm, the amplitude of a wave whose frequency is higher than 1 Hz becomes smaller than 0.01 mm. Hence, we can neglect waves in the range of higher than 1 Hz, and we may set the upper limit of frequency at 1 Hz. We define our concept of power spectrum moment, Ms, by the product of an 'area (As)'

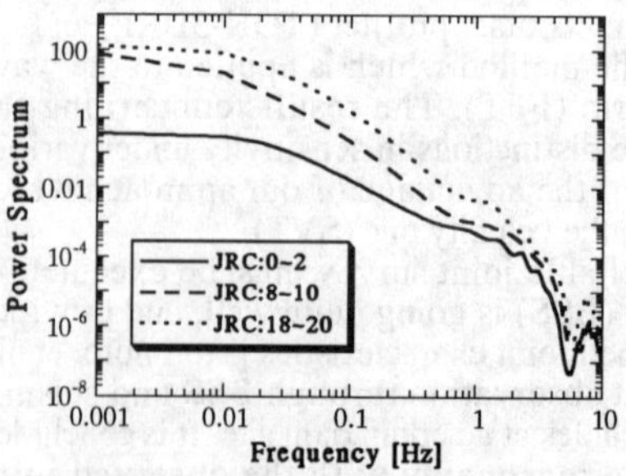

Fig.1 Power Spectrum ~ Frequency relation by MEM for representative JRC ranges

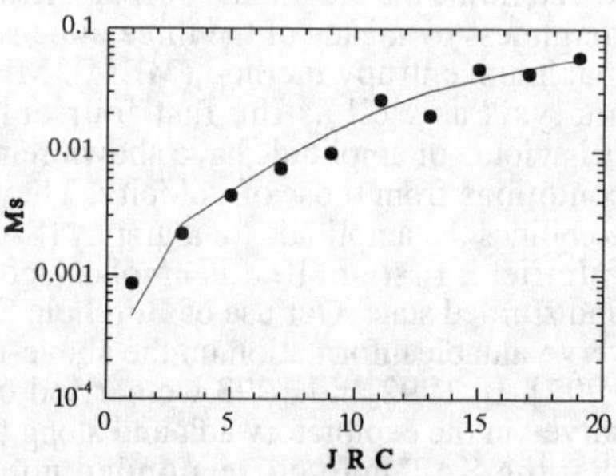

Fig.2 Ms ~ JRC relation

surrounded by a power spectrum curve, the upper (1 Hz) - and the lower (0.001 Hz) limits and an 'abscissa (f_G)'. The relationship between Ms and JRC is as shown in Fig.2. It is very clear that there exists a good relationship and Ms has high resolvability for the estimation of joint roughness in practical use.

2.2 SAT versus SVT in laboratory test

Based on the authors' past research aiming at clarifying the loosening behaviour of jointed rock, it is generally concluded that, when the magnitude of joint aperture is less than 0.2 mm or confining pressure at joints varies in the range of higher than 2 MPa, the velocity change does not correspond with the influence of existing joints, but the change of amplitude of propagating wave can reflect the effect of joint, showing the clear reduction of energy (amplitude) at joints [Tanimoto, et al., 1983, 1994a].

Cylindrical specimens of intact rock obtained from the sites, where seismic surveys were carried out, and plaster/cement cylinders of various lengths were used for laboratory tests. The rock types used in the study were : sandstone, rhyolite, shale, granite, granite-prophyry, and slate. 86 pieces of specimens were employed totally. For the test subject to normal loading the samples were constructed by stacking individual cylinders of the rock or plaster/cement to produce a "jointed" sample with joints normal to the axial direction. Filter paper was used to produce the desired joint aperture and moisture content. For the test onto natural joints, samples including a single or multiple parallel fracture planes were carefully recovered, and joint planes were replicated with hard mortar or plaster so as to revive the same roughness as a prototype of natural joint.

Several examples of experimental set-up are shown in Fig.3. A seismic event (a single pulse of 10 μsec for each event) was generated by a AE-sensor (NFCD made) tightly contacted to the end of the sample. An identical sensor was used to pick up the signal at the opposite end of the specimen. Through the laboratory test we have observed the remarkable distinction between velocity (Vp) and amplitude (A) (or amplitude ratio A/A_0, where A_0 is an amplitude obtained through intact rock) in both normal and shear loadings as shown Fig.4 to 9. Fig.4 shows the comparison between Vp and A/A_0 in the relation between contact pressure (at a joint), Pc, and joint frequency, n. A/A_0 increases depending on n as Pc becomes high, while Vp converges at the constant level regardless of joint frequency. The Vp ~ Pc ~ n relation does not make a noticeable sense. To speak of the roughness (Ms) ~ Pc relation as shown in Fig.5(a) and (b), the common tendency which was obtained in Fig.4 is recognized. The variation of roughness is not reflected clearly in the Vp ~ Pc relation. It should be noted that amplitude ratio (A/A_0) becomes smaller as roughness is higher.

When aperture is wider than 1 mm, both velocity and amplitude can correspond with the change of contact (confining) pressure as shown in Fig.6. But, the amplitude owns higher capability of sensing the reduction of confining pressure still. It is very useful to the discussion on the effects of support elements in tunnelling or large underground excavation works.

Shear displacement along joint is of great concern to us, because it causes a comparatively aggressive reduction of shear resistance against excavation and it is not easy to be known. As shown if Fig.7, Only the amplitude may suggest a magnitude of shear displacement in the primary stage which is subjected to rather high change in amplitude for the low range of displacement such as 1 mm or so. Loosening is related to this low range

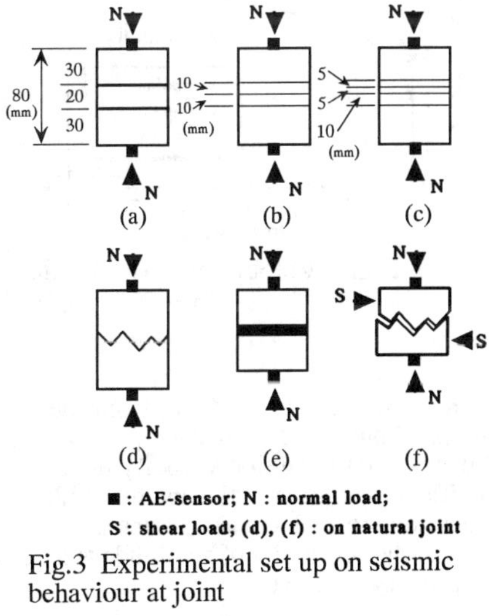

Fig.3 Experimental set up on seismic behaviour at joint

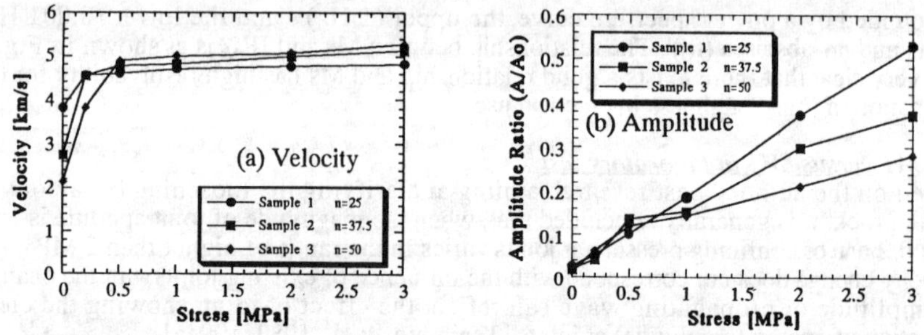

Fig.4 Velocity vs. Amplitude on joint frequency under normal stress (n = 25, 37.5, and 50 [1/m])

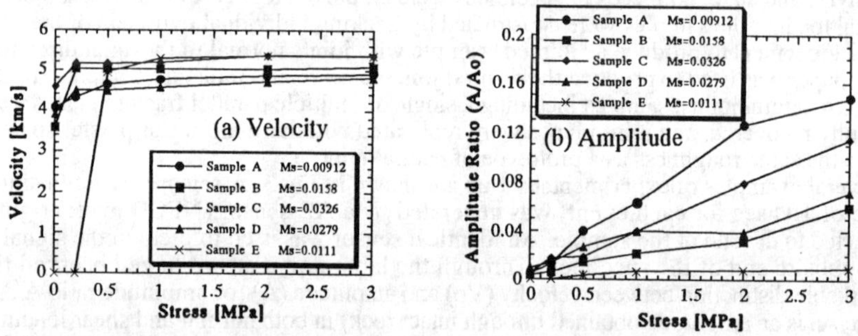

Fig.5 Velocity vs. Amplitude on joint roughness under normal stress (Ms = 9.12, 11.1, 15.8, 27.9, and 32.6 x 10^{-3})

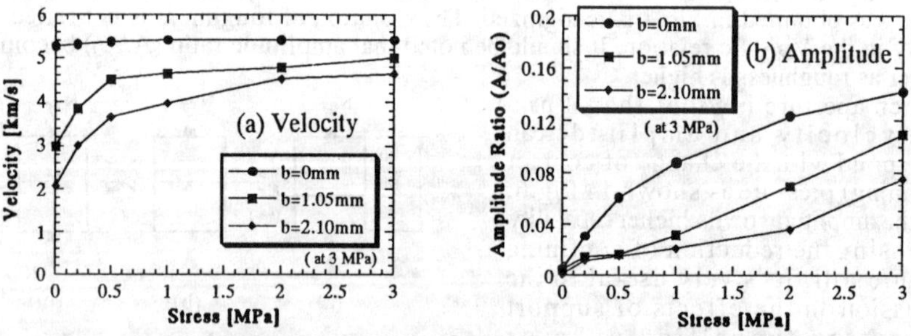

Fig.6 Velocity vs. Amplitude on aperture under normal stress (b = 0, 1.05, and 2.10 mm at 3 MPa)

of shear displacement. Also, from the same point of view, the change of aperture of joint(s) should be detected adequately in excavation work. From the results from Fig.8 and 9, we can see that velocity does not show any change, but amplitude suggests significant variation in the aperture (dilation) ~ roughness ~ confining pressure relations under both normal and shear stresses.

Thus, the sensitivity of amplitude is much higher than that of velocity. It is greater than our primary expectation.

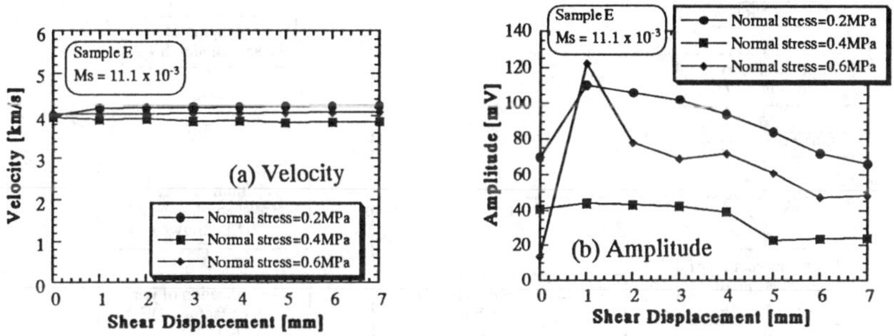

Fig.7 Velocity vs. Amplitude on shear displacement under normal and shear stresses (Ms = 11.1 × 10^{-3}, Pc = 0.2, 0.4, and 0.6 MPa)

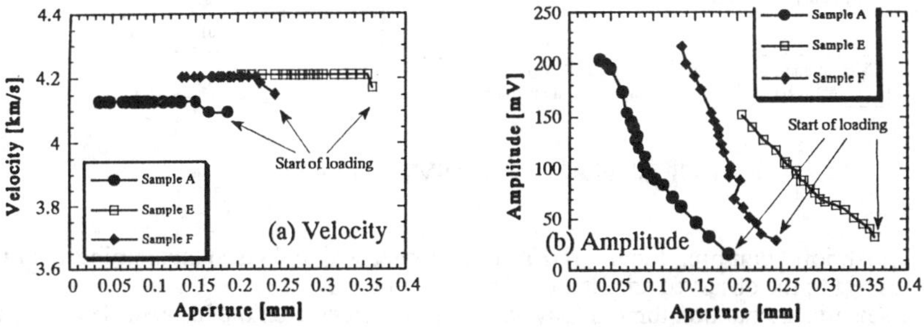

Fig.8 Velocity vs. Amplitude on aperture and roughness under normal stress (Ms = 9.12, 11.1, and 7.79 × 10^{-3} for A, E, and F)

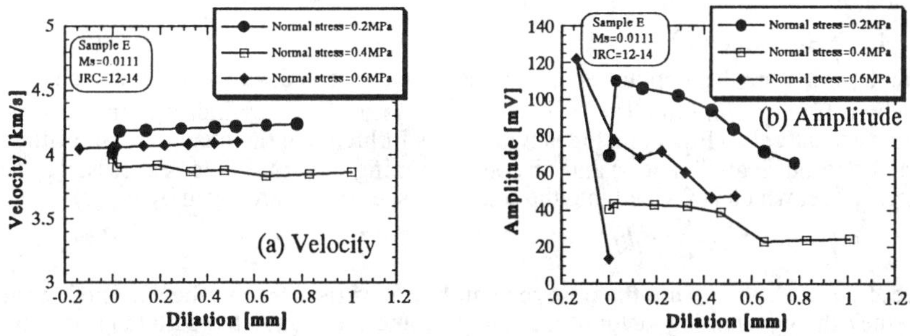

Fig.9 Velocity vs. Amplitude on dilation under normal and shear stresses (Ms = 11.1 × 10^{-3}, Pc = 0.2, 0.4, and 0.6 MPa)

3. AMPLITUDE (SAT) VS. VELOCITY (SVT) TOMOGRAPHY IN FIELD

3.1 Algorithm of SAT

The seismic velocity tomography has widely become popular among Japanese rock engineers as well as the resistivity tomography. Fig.10 shows a flow chart for the constructions of SVT and SAT [Itoh, et al., 1983, Scales, 1987, Vidale,1988, Qin, et al., 1992, Tanimoto, et al., 1994a & b]. Specific feature of our SAT concept is how to

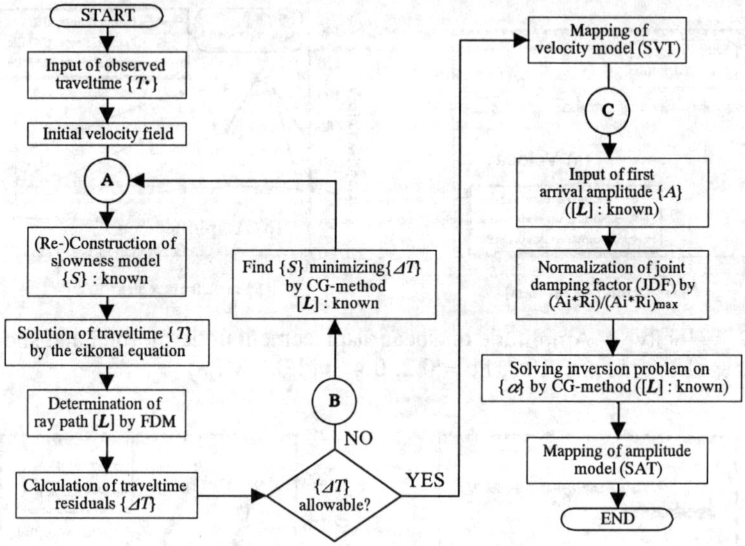

Fig.10 Flow chart for SVT and SAT

distinguish joint damping factor (JDF) from geometrical and viscous damping factors with respect to the observed amplitudes of propagating waves.

The seismic attenuation tomography (SAT) is constructed on the basis of SVT result, that is the ray paths. The SAT method is proposed by the authors as follows:

Let us consider the observed wave which penetrates through a certain medium of l in thickness and α in damping factor. When a medium is homogeneous and continuous, the relation of amplitude between a projected and received waves is generally represented in the form of

$$A = \frac{k}{l} A_0 \cdot exp\,(-\alpha l) \tag{1}$$

where A_0 and A are the amplitudes of a projected and received waves, and k is a constant given by diffusion condition (k=1, $(1/\sqrt{\pi})$, $(1/2\sqrt{\pi})$ for planar, cylindrical, and spherical waves, respectively). Eq.(1) includes two causes influencing the change of amplitude : namely, geometrical- (length-) and viscous- damping factors. Waves expressed in the discrete model which is divided into the square meshes (cells) are given by Eq.(2).

$$A_i = \frac{k}{L_i} A_0 \cdot exp\,(-\sum_j \alpha_j l_{ij}) \tag{2}$$

where A_i, L_i, α_j, and l_{ij} are the obseved amplitude of the i-th ray, the summed length along the i-th ray, damping factor of the j -th cell, and the segmented path length of the i -th ray in the j-th cell, respectively.

As ray paths have been already known through the SVT analysis, Eq.(2) is changed to

$$A_i \cdot L_i = k\, A_0 \cdot exp\,(-\sum_j \alpha_j l_{ij}) \tag{3}$$

where k and A_0 are constant regardless of ray path and cell. Then, by taking a ratio of Eq.(3) to the maximum value of a product of $A_i{*}L_i$, $(A_i{*}L_i)_{\max}$, Eq.(4) is given by

$$\frac{A_i \cdot L_i}{(A_i \cdot L_i)_{max}} = exp\left\{-\sum_j (\alpha_j l_{ij} - \alpha_j l_{mj})\right\} \tag{4}$$

where l_{mj} is a segmented path length along the ray which showed the maximum value of $A_i{*}L_i$. Eq.(4) is expressed in logarithmic form,

$$-ln \frac{A_i \cdot L_i}{(A_i \cdot L_i)_{max}} = -\sum_j (l_{ij} - l_{mj})\alpha_j \qquad (5)$$

The left side of Eq.(5) becomes a constant e_i, and also the value of $(l_{ij}-l_{mj})$ is known. The calculation of damping factor distribution α_j attributes to the same problem as in SVT. Any iteration procedure is not necessary.

It should be noted that the damping (or attenuation) factor distribution obtained by Eq.(5) does not include any influence given by the attenuation due to travel length, namely it is determined independently from travel length. It means that values of α_j for respective cells express the relative magnitude of propagating energy attenuation to a common amount (a reference scale) of $(A_i{*}L_i)_{max}$, corresponding with the magnitudes of joint frequencies and joint apertures. The inner (viscous) damping effect in rock substance has been eliminated, and now the damping factor matrix [α] corresponds only with rock joints. A value of α_j whose unit is [1/m] may become bigger than 1.0 in some cases. The bigger an α-value (JDF) becomes the more an attenuation at discontinuities in rock mass takes place.

3.2 Application to the site for the underground powerhouse

The feasibility study for a new underground powerhouse is being carried out in the central part of Houshu Island. The objective site is situated 70 m deep beneath the mountain, which is subjected to the late Cretaceous cauldron movement. Basement rocks are granite porphyry, rhyolite and slate. These rocks contain many joints dipping at steep angles, with comparatively high joint frequency, and with wide apertures.

Four boreholes of 70 m deep were prepared for the investigation by the geotomography technique. The number of shot- and receiver-points are 70 for each and they are installed at a 1 m interval, respectively. It means that a number of meshes (or cells) for the geotomographic discrete model is (70 x 70) and 4,900 ray paths were analyzed. Energy

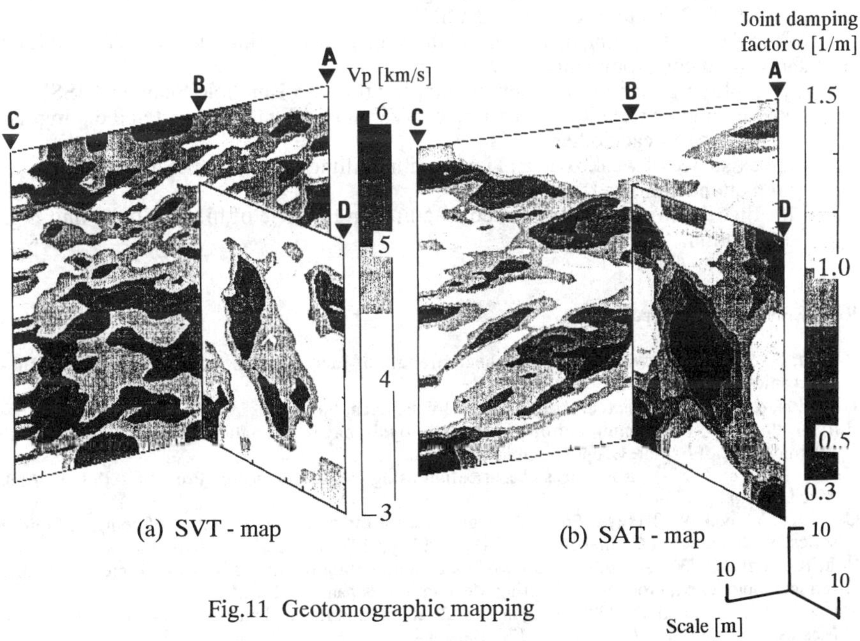

Fig.11 Geotomographic mapping

source at each shot point was the detonation of 2 pieces of instantaneous EB-caps, and the data logger and the receivers employed in the survey were McSeis-16000 (OYO made) and Hydro-phones (ITI made).

Fig.11(a) and (b) show the velocity distribution (SVT) and the joint damping factor distribution (SAT) maps for section A-C, and B-D.

Adding the joint frequency (n) map to these two tomographic maps, let us compare the appearances of various rock conditions in the respective maps, which are classified into three classes. They are i) good (H), ii) medium (M), and iii) poor (L) classes.

In comparison of three different maps concerning section A-C, Zone-c (right, bottom) and -e (left, middle) show good agreements each other regarding of Class-M and -L. Zone-b (right, middle) and -d (left, top) show the agreement between SVT and n maps, and also there are remarkably good agreements between SAT and n maps except the case of Zone-b.

Judging from the results of the comparison among SVT, SAT and n maps, the magnitude of velocity does not always correspond with the conditions of rock joint.

As well as the agreements in the rock classification, the sensitivity to (capability of detecting) joint orientation was discussed. The propagating behaviour of projected waves is very spatial, and it is highly recommended to carry out seismic survey in three dimensional manner as much as possible. This project is still in progress.

4. CONCLUSION

It is shown that the seismic geotomographic technique based on the attenuation behaviour of traveling waves can provide more detailed information concerning rock joint through the both considerations of laboratory experiment and application to field.

It is concluded as follows :

1) When a number of joints with smooth fracture planes and thin apertures increases, the amplitude distribution (SAT) corresponds much better than the velocity one (SVT).
2) Concerning the variation of joint roughness and the change of normal and shear stresses, SAT is more sensitive than SVT.
3) When there exist filling materials and thick apertures at joints, both of SAT and SVT show good correspondence.
4) By employing the joint frequency mapping through Borehole Scanner (BSS), the geotomographic technique consisting of SVT and SAT may suggest and distinguish following cases each other :
   - the existence of weak zone and joints with infillings in the case of low velocity and high attenuation, and
   - the distribution of tightly jointed fractures in the case of high velocity and high attenuation.

REFERENCES (* : in Japanese)

Barton, N. & V. Choubey (1977) : The shear strength of rock joints in theory and practice, Rock Mechanics, Vol.10, pp.1-54.

Barton, N. & O. Stephansson (ed.) (1990) : Rock Joints, Loen, Balkema.

Burg, J. P. (1968) : Maximum entropy spectral analysis, paper presented at the 37th Annual Int'l Meeting, Soc. of Explor. Geophys., Oklahoma.

Carr, J. R., et al. (1987) : Rock mass classification using fractal dimension. Proc. of 28th U.S. Rock Mech. Symp., pp.73-80.

Cooley, J. W. & J. W. Tukey (1965) : An algorithm for the machine calculation of complex Fourier series, Mathematics of Computation, Vol.19, No.90, pp.297-301.

Itoh, K., et al. (1983) : A model for approximate computation of first arrival traveltimes in seismic refraction modeling, Proc. of 1983 Spring Meeting SEG Japan, pp.11-12.*

Maetz, N. H. & J. Franklin (1990) : Roughness scale effects and fractal dimension, Scale Effects in Rock Masses, Pinto da Cunha (ed.), pp.121-126, Balkema.

Muralha, J. & Charrua-Grassa, J. G. (1990) : An experience on the application of fractal theory to basic shear strength studies, Rock Joints, N. Barton & O. Stephansson (ed.), pp.667-671, Balkema.
Qin, F., et al. (1992) : Finite-difference solution of the eikonal equation along expanding wavefronts, Geophysics, Vol.57, No.3, pp.478-487.
Scales, J. A. (1987) : Tomographic inversion via the conjugate gradient method, Geophysics, Vol. 53, No. 2, pp.179-185.
Tanimoto, C. and K. Ikeda (1983) : Acoustic and mechanical properties of joited rock, Proc. of 5th ISRM Congress, Melbourne, pp.A15-18.
Tanimoto, C., et al. (1992) : Immediate Image and Its Analysis of Fractured/Jointed Rock Mass Through the Borehole Scanner. Int'l Symp. on Fractured and Jointed Rock Masses, Lake Tahoe, Vol.1 of 3, pp.225-232.
Tanimoto, C., et al. (1994a) : Fundamental Study on Seismic Geotomography in consideration of Amplitude Attenuation at Discontinuities, Proc. of the 9th Japan Symp. on Rock Mech., ISRM-Japan, pp.325-330.*
Tanimoto, C., et al. (1994b) : Seismic Attenuation Geotomographic Technique in Designing a Large Underground Cavern, The 1994 ISRM Int'l Symp..
Turk, N., et al. (1987) : Characterization of rock joint surfaces by fractal dimension, Proc. of 28th U.S. Rock Mech. Symp.,pp.1223-1236.
Vidale, J. (1988) : Finite-difference calculation of traveltimes, Bull. Seis. Soc. Am., Vol. 78, No. 6, pp.2062-2076.

Rock Mechanics, Nelson & Laubach (eds) © 1994 Balkema, Rotterdam, ISBN 90 5410 380 8

Guided waves propagation along a fracture with contact

Koji Nagano
Muroran Institute of Technology, Japan
Hiroaki Niitsuma
Tohoku University, Sendai, Japan

ABSTRACT: We measured guided wave trapped in a subsurface artificial fracture in Higashi-Hachimantai Hot Dry Rock model field, Japan. The subsurface fracture was produced at about 370 m depth in an intact welled tuff layer by hydraulic fracturing. We used a downhole air gun as a wave source. Because energy of the guided waves concentrates in a fluid layer, a hydrophone was used to detect the guided waves. We determine the arrival of guided wave by an analysis of tube wave, to which the guided wave converts at an intersection of a borehole and the subsurface fracture.

We found that travel time of the guided wave increased corresponding to pressurization of fracture. This phenomenon cannot be explained by the simple parallel wall fracture model. Observed velocity decrease is thought to be caused by effects of actual properties of the subsurface fracture; contact of asperities of the fracture surfaces and elastic variation of the rock due to pressurization.

1 INTRODUCTION

The guided waves trapped in a subsurface fracture are potentially utilized to characterize the fracture because the propagation characteristics of the waves depend not only on the elastic behavior of rock but also on the configuration of the fracture.

Chouet (1986) analyzed numerically a transient response of a fluid-filled crack to an impulsive pressure in the fluid. He reported that a velocity of one mode of the waves propagating at the crack is always lower than the fluid velocity. Ferrazzini and Aki (1987) studied analytically the waves trapped in the crack where the fluid layer is sandwiched between half-space elastic media. They found that a very slow wave propagates in the crack. Hayashi and Sato (1992) also made a numerical study on the guided waves behavior in interface with contact.

We have measured the guided waves in Higashi-Hachimantai Hot Dry Rock model field, Iwate Prefecture, Japan. A subsurface single fracture was produced by hydraulic fracturing in the field. Three dimensional configuration of the fracture was evaluated by a cross hole seismic measurement (Niitsuma and Saito 1991) and three dimensional tectonic stress near the fracture was determined by a hydraulic fracturing method (Hayashi et al. 1985). We show detected waveform data corresponding to the guided wave which propagates along the subsurface fracture. Moreover, distinctive wave behavior, velocity decrease of the guided wave, was observed when the subsurface fracture was pressurized.

2 FIELD MEASUREMENT OF THE GUIDED WAVES

A field experiment to detect the guided waves was carried out at Higashi-Hachimantai

Hot Dry Rock model field in Iwate prefecture, Japan (Niitsuma 1989). Figure 1 is an illustration of the field. An artificial subsurface fracture was created at 369.0 m depth in F-1 well by the hydraulic fracturing. During the hydraulic fracturing, 40 mesh sand proppant was injected. Formation of the zone in which the artificial fracture exists is intact welded tuff, and core samples proves that there is no significant joint or crack. EE-4 well was drilled into the fracture after the fracturing and it intersected the artificial fracture at 358.2 m depth. Distance between the intersecting points in F-1 and EE-4 wells is about seven meters. Niitsuma and Saito (1991) estimated the three dimensional configuration by the triaxial shear-shadow measurement and showed a diameter of the fracture was about 60 m. Relationship between the crack aperture and wellhead pressure in this artificial fracture was estimated by a transmissibility test (Hayashi and Abé 1989). The crack aperture was estimated to be 0.08 mm without pressurization and 0.2 mm at 3.1 MPa in wellhead pressure. The velocities of the compressional and the shear waves in the field are about 2710 m/s and 1730 m/s, respectively (Sato et al. 1986).

We carried out experiments using this artificial single fracture to detect the guided waves. Figure 2 illustrates the guided waves measurement. An air gun (Bolt, DHS-5500), which was modified for downhole use, was used as a wave source.

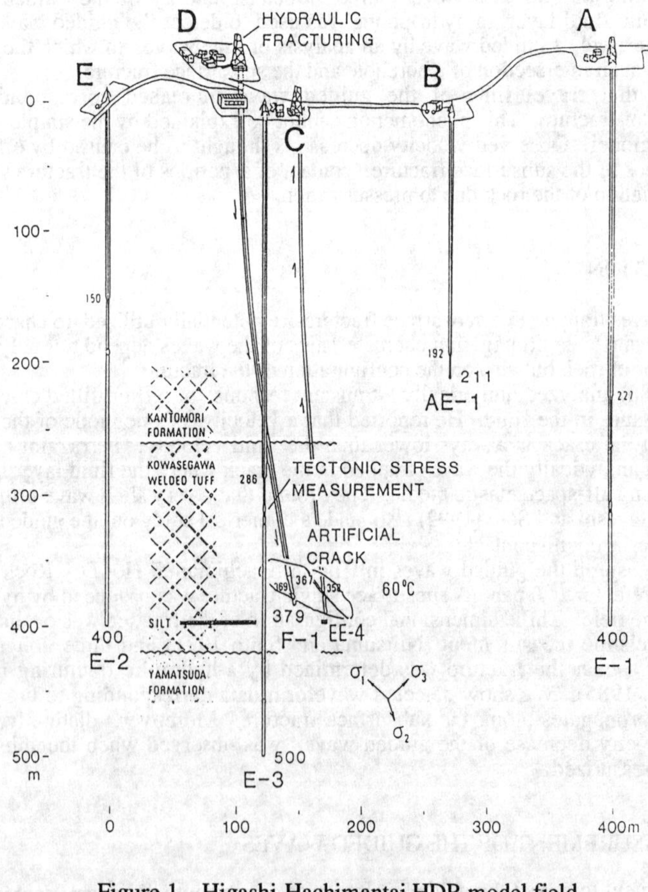

Figure 1. Higashi-Hachimantai HDR model field.

A hydrophone was employed to measure the guided waves.

The wave source was set above the intersecting point of F-1 at 367m. Wave energy of the air gun is mainly distributed in the frequency range of 1 Hz - 300 Hz.

The hydrophone was suspended in the water near the intersection of the borehole and the fracture. Because the guided wave energy is concentrated in the fluid layer rather than the rock layers, the hydrophone, which is sensitive to the fluid motion, is suitable to detect the guided waves. The hydrophone was installed at 300.0m, 310.0m, 330.0m, 350.0m, 358.2m, and 359.0m depth in EE-4 so that the tube waves, to which the guided waves converted in the borehole, could be identified. The hydrophone can detect a pressure wave in frequency range from 1 Hz to 10 kHz.

We also measured propagation characteristics of guided waves by the hydrophone, which was installed at the intersection, changing the fracture aperture. Wellheads of injection and production wells were shut with wireline lubricators in order to pressurize the subsurface fracture. The wellhead pressure was increased by 0.1 MPa per one minute and the waves were shot at 30 s intervals. When the wellhead pressure reached 3.2 MPa, we shut the wellhead. After the shut-in the wellhead pressure was naturally decreased due to permeation into the rock.

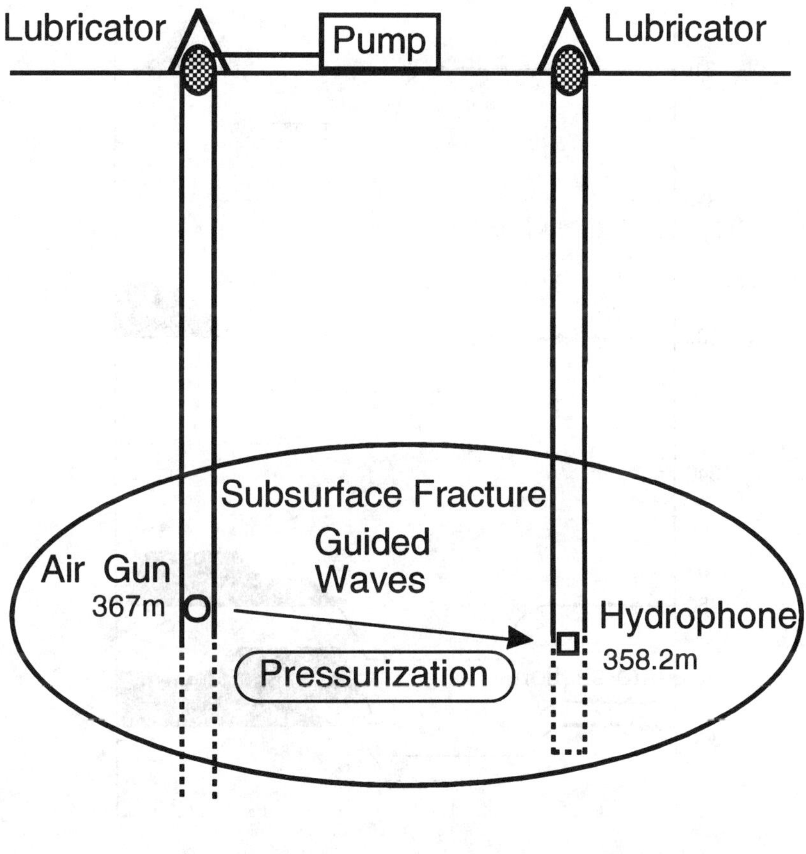

Figure 2. Guided waves measurement. Because energy of the guided waves is concentrated in a water layer, a hydrophone is suitable to detect the guided waves.

3 WAVEFORM DETECTED IN THE FIELD MEASUREMENT

Waveform data detected by the hydrophone which was installed in several locations in the borehole is shown in Figure 3. The waveforms were saturated because gain of an amplifier in the hydrophone was too high to detect the tube waves. Both up-going and down-going tube waves were clearly detected. These tube waves were generated at the intersection of the borehole and the fracture. It is, therefore, concluded that wave corresponding to the tube wave was the guided wave which propagated along the subsurface fracture.

The waveform data detected during a pressurization of the subsurface fracture is

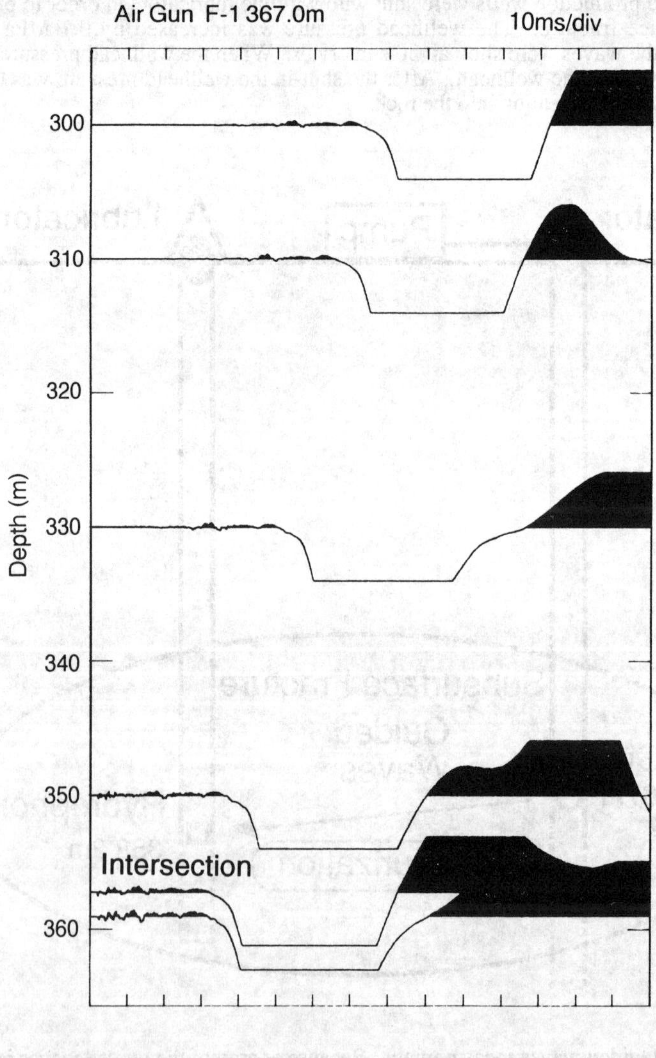

Figure 3. Waveform detected by the hydrophone. Both an up-going and down-going tube waves, which was generated at the intersection, were observed.

shown in Figure 4. In this experiment the gain of the amplifier in the hydrophone was adjusted to detect the guided waves. The guided wave, which has been determined by the above tube wave analysis, is indicated by an arrow in the figure. The phase of the guided wave was delayed corresponding to the wellhead pressure. Furthermore, the guided wave data showed a dispersive property. When the wellhead pressure was higher, the low frequency component velocity was lower than that of the high frequency component.

The amplitude was affected by the pressurization as well as the velocity. A calibration test of the downhole air gun proved that the air gun shot a same wave energy when the

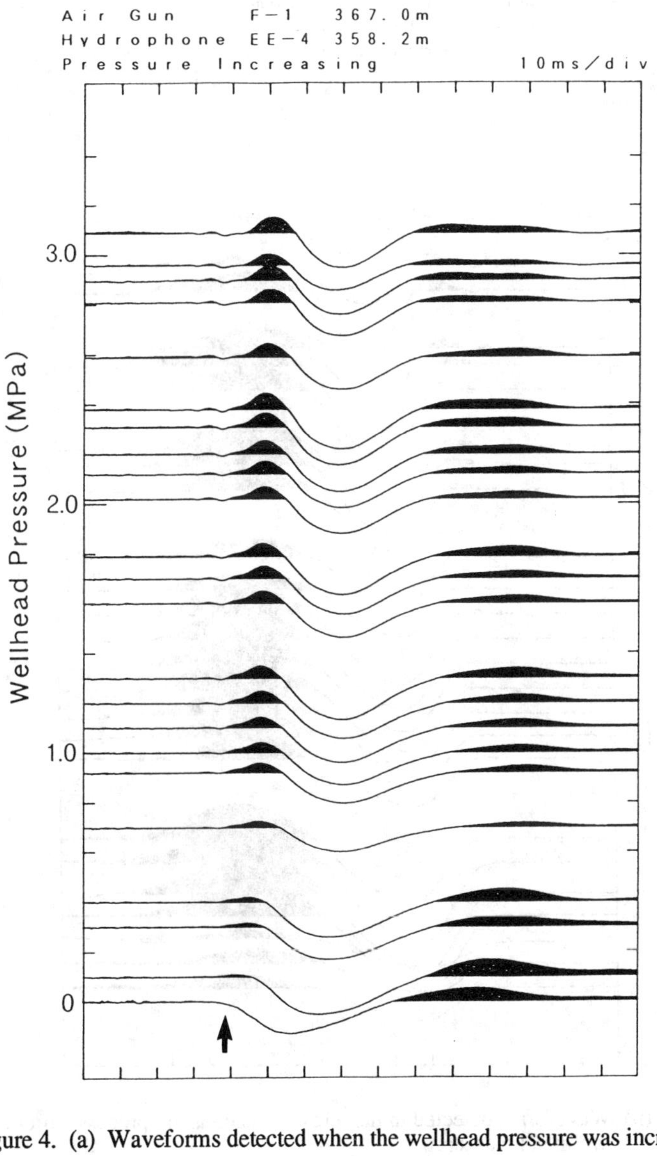

Figure 4. (a) Waveforms detected when the wellhead pressure was increased.

wellhead pressure was changed from 0 MPa to 3.0 MPa. The amplitude in the pressure decreasing process was larger than that in the pressure increasing process even though the wellhead pressure was identical.

4 DISCUSSION

Characteristics of the guided waves trapped in a fluid-filled fracture was studied analytically by Ferrazzini and Aki (1987). They modeled the fluid-filled fracture as a

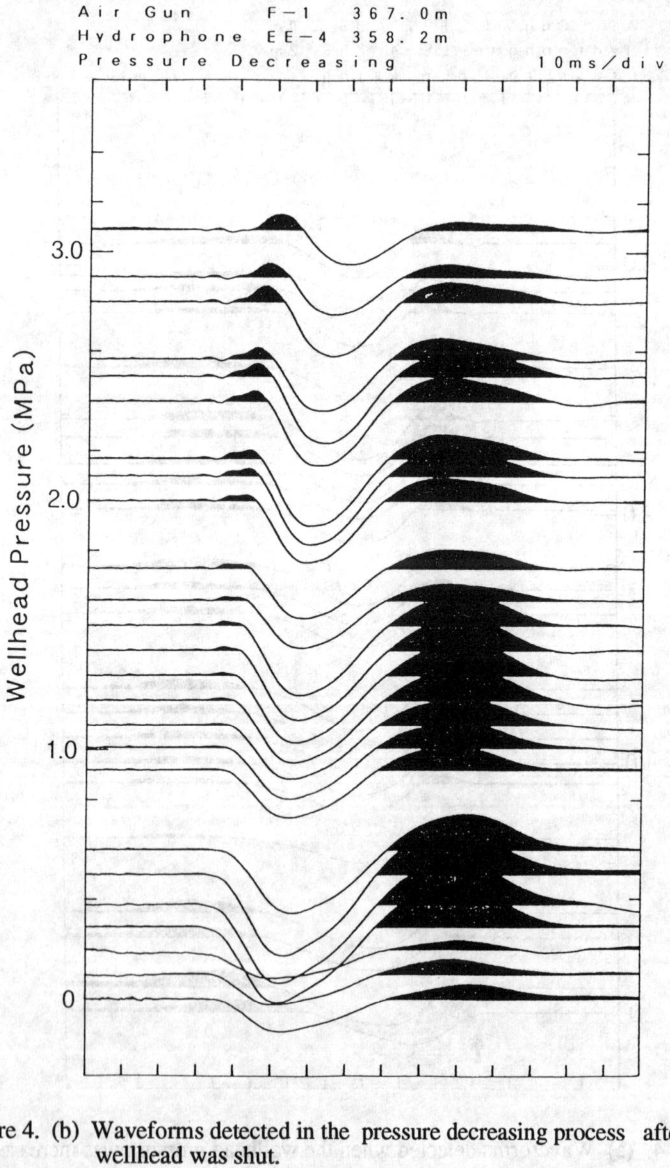

Figure 4. (b) Waveforms detected in the pressure decreasing process after the wellhead was shut.

fluid layer lying between two elastic half-spaces, neglecting the contacts between crack surfaces. The velocity of the symmetric fundamental mode of the guided waves is slower than the fluid velocity. One of the important results of their analysis is that velocity of the fundamental symmetric mode monotonously increases as the crack aperture increases and/or the wavelenght decreases.

The dispersion observed in the figure 4 seems to be explained by the theoretical result by Ferrazzini and Aki (1989). However, the velocity monotonously increases in their analysis when the crack aperture increases. The guided wave velocity decrease observed in the pressurization experiment cannot be explained by their simple model.

Hayashi and Sato (1992) introduced the crack stiffness which models the crack contacts in their analysis of the guided waves propagation. In their analysis the crack stiffness can be represented as a function of the specific contact area. This analysis predicts that the velocity decreases when the contact area decreases. The velocity variation detected in the guided waves measurement is in qualitative agreement with the prediction by the Hayashi's analysis.

5 CONCLUSION

We have described the results of measurement of the guided wave which propagates along the artificial single fracture. The guided wave arrival has been determined by the tube wave analysis. The delay of the guided wave was observed depending on the wellhead pressure when the subsurface fracture was pressurized. The simple parallel wall crack model cannot explain this result. The effect on guided waves behavior of contact of fracture surfaces has been pointed out to explain the detected velocity variation.

REFERENCES

Chouet, B. 1986. Dynamics of a fluid-driven crack in three dimensions by the finite difference method. *J. Geophys. Res.* 91:13967 - 13992.

Ferrazzini, V. and Aki, K. 1987. Slow waves trapped in a fluid-filled infinite crack: implication for volcanic tremor. *J. Geophys. Res.* 92:9215-9223.

Hayashi, K., Shoji, T., Niitsuma, H., Ito, T. and Abé, H. 1985. A new in-situ tectonic stress measurements and its application to a geothermal model field. *Trans., Geother. Res. Couc.* 9:99-104.

Hayashi, K. and Abé, H.1989. Evaluation of hydraulic properties of the artificial subsurface system in Higashihachimantai geothermal model field. *J. Geothermal Research Society of Japan.* 11:203-215.

Hayashi, K. and Sato, K.1992. A theoretical study of AE traveling through a fluid-filled crack with application to characterization of a geothermal reservoir crack. In T. Kishi, K. Takahashi and M. Ohtsu (eds.), - *Proc. 11th. Int. AE Sym., Fukuoka, 26-29 October 1992* :423-430. Tokyo: Japan NDI.

Niitsuma, H.1989. Fracture mechanics design and development of HDR reservoirs -concept and results of the Γ-project, Tohoku University, Japan. *Int. J. Rock Mech. Min. Sci. & Geomech. Abstr.* 26:169-175.

Niitsuma, H. and Saito, H.1991. Evaluation of the three-dimensional configuration of a subsurface artificial fracture by the triaxial shear shadow method. *Geophysics.* 56:2118-2128.

Sato, M. Nakatsuka, K., Niitsuma, H. and Yokoyama, H. 1986. Calibration of downhole AE measuring system by detonation test. In K. Yamaguchi, K. Aoki and T. Kishi (eds.), - *Proc. 8th. Int. AE Sym., Tokyo, 21-24 October* 1986:389-395. Tokyo: Japan NDI.

Rock Mechanics, Nelson & Laubach (eds) © 1994 Balkema, Rotterdam, ISBN 90 5410 380 8

Numerical modelling of the hydro-mechanical behavior of single fractures

Marte Gutierrez & Nick Barton
Norwegian Geotechnical Institute, Oslo, Norway

ABSTRACT: A numerical modelling approach for studying the hydraulic and mechanical properties and behavior of single fractures is presented. Fractures are represented by a series of straight line contacts which are obtained from digitized profiles of the fracture surfaces. Each contact is given simple mechanical and hydraulic properties, and the response of the fracture is obtained by integrating the response of all contacts along the fracture length. The distinct element code UDEC is used to implement the proposed approach. The proposed model is used to obtain basic understanding of fracture behavior at the micro- (or asperity) scale.

INTRODUCTION

The behavior of fractured rock masses is of considerable interest in such areas as nuclear waste disposal, hydrocarbon recovery, underground storage of fluids and contaminant transport. Fractures affect both the mechanical and hydraulic behavior of a rock mass. Mechanical and hydraulic behavior are coupled in the sense that fracture conductivity is dependent on fracture deformation and, conversely, fluid pressure in the fracture affects mechanical response.

In recent years, there has been a number of research works on the behavior of fractures. There are two main approaches to the quantitative description and modelling of the properties and behavior of fractures: 1) the empirical approach, and 2) the theoretical/numerical approaches. Empirical models are formulated by correlating physical data and properties with observed behavior. The most simple approach is probably that based on the fracture coefficients JRC (Joint Roughness Coefficient) and JCS (Joint Compressive Strength) established by Barton (1976) for rock joint surfaces. On the other hand, theoretical models tended to be either complicated or only applicable for simple fracture surface geometries (e.g., parallel plate, sawtooth and sinusoidal fracture profiles).

What is currently lacking is a basic understanding of fracture behavior at the micro or asperity scale. Although mechanical and hydraulic testing of fractures can be routinely done, observation of the deformation and fluid flow within the fracture is difficult. Yet such observations are essential especially when progressive damage of the fracture surface needs to be quantified. An experimental program specifically to observe deformation and asperity damage in direct shear testing of a fracture was recently reported by Pereira and de Freitas (1993). However, the experiments were performed on artificial fracture surfaces with a simple sawtooth profile. Another approach is to measure asperity deformation on transparent analogues of the fracture surface by the use of transmitted light microscopy.

An alternative approach to the understanding of fracture behavior is presented here. The applicability of the proposed method is demonstrated. Using the proposed approach, some basic aspects of fracture behavior in the asperity scale are highlighted and analyzed.

NUMERICAL MODELLING APPROACH

The numerical modelling proposed in this paper consists of studying the mechanical and hydraulic behavior of single fractures during direct shear loading. Figure 1 gives a layout of the basic model analyzed in this paper. The fracture is modelled as a sequence of straight line segments or contacts and the interactions of all contacts in a fracture determine the fracture behavior. The top and bottom fracture surfaces constituting a fracture are digitized as a series of straight line segments, and two opposing straight line segments constitute a contact.

The basic element of the model is the contact. Each contact is assumed to be smooth and given simple mechanical and hydraulic properties (Fig. 1). The behavior of the whole fracture is then obtained by integrating the individual contact response over the length of the fracture. For the mechanical behavior, each contact is assumed to behave as an elasto-plastic Mohr-Coulomb joint with either constant or stress-dependent elastic normal and shear stiffnesses k_n and k_s, a constant slip friction angle which is equal to the base friction angle ϕ_b, and zero angle of dilation. For the hydraulic behavior, each contact is given a smooth parallel plate aperture e equal to the opening between the two straight line fracture surfaces making the contact. The behavior of the fracture is also dependent on the behavior of the asperities which in turn depends on the intact rock properties. Three possible intact rock behaviors that can be modelled are: rigid (fracture deformation occurs only due to sliding along contacts), elastic, or elasto-plastic. The model shown in Fig. 1 is subjected to the typical loading conditions used in direct shear tests. In addition, fluid flow into the fracture maybe analyzed by the introduction of different fluid pressures at the two ends of the fracture.

Each contact is further characterized by its location, length, and orientation given by the angle the contact makes from the horizontal. These parameters will not only vary with fracture geometry but will also change during normal and shear loading. Asperity deformation and damage will also change the contact geometry. The smallest size of each contact is dependent on the resolution of the device used to profile the fracture surfaces. The number and size of contacts that can be handled depend only on the available computational resources.

The ability to keep tract of the evolution of the contact geometry and the creation and deletion of contacts is the most crucial requirement of the proposed modelling procedure. This requirement is easily met by using Cundall's distinct element computer code UDEC (Universal Distinct Element Code, Itasca; 1991). UDEC is specifically designed for modelling problems involving relatively large fracture displacement which require frequent contact updating. UDEC handles both mechanical and hydraulic calculations and both calculations can be made in a coupled manner. Asperity deformation can be modelled by using one of the several intact rock constitutive models available in UDEC. However, it is also possible to write a new code to handle the contact logic required in the procedure proposed in this paper.

DETAILS OF THE MODEL

The UDEC model analyzed in this paper consists of a 100 mm long and 20 mm thick fractured block (Fig. 1). The lower half of the block is fixed and prevented from moving in both the horizontal and vertical directions while the upper half is subjected to a constant vertical stress σ_n and a constant horizontal velocity. Different fluid pressures p_1 and p_2 are introduced at the two ends of the fracture.

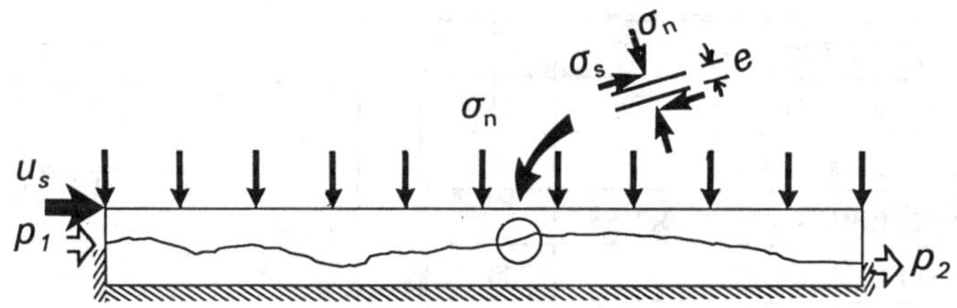

Fig. 1 Fracture subjected to direct shear loading

Typical digitized 100 mm fracture profiles with Joint Roughness Coefficients of 5, 13 and 19 as shown in Fig. 2 were analyzed. The fracture surfaces are assumed to be initially matched although this is not a limitation of the method. The 100 mm fracture length is digitized with approximately 1 mm intervals. The effects of the size of contacts in the response of the model will be investigated in future studies. The other input required in the simulations are given in Table 1. The analyses were carried out for two sets of intact rock models: elastic and Mohr-Coulomb elastoplastic.

The properties giving the most important effects on the results of the simulations are the contact stiffnesses given by k_n and k_s. These parameters determine the time increments that will give stable calculations. The normal stiffness k_n determines the amount of contact overlap that can occur during loading. The contacts cannot be made too stiff by using high values of k_n and k_s, otherwise the stresses cannot propagate through the contacts and contact stresses will be higher than those in the intact rock. High contact stiffnesses also slow down the calculations considerably. On the other hand, contacts cannot be made too soft which could cause contact interpenetration or overlap. The k_n and k_s values used in the simulations are based on the rule-of-thumb given by Itasca (1991) that the contact stiffnesses should not be more than 10 times the equivalent stiffnesses of the intact rock in the direction normal to the contact. This rule insures that the contact deformations and stresses are consistent with those of the intact rock.

The apparent contact stiffness is given as

$$k_s = k_n = \frac{10}{l}\left(K + \frac{4}{3}G\right) \tag{1}$$

where K and G are the bulk and shear moduli of the material and l is the average contact length. The initial hydraulic aperture e_o corresponds to a zero normal stress (i.e., open contact). The hydraulic aperture e as a function of normal deformation u_n is calculated as

$$e = e_o + u_n \tag{2}$$

The minimum hydraulic aperture is given by e_r. Since the contacts are assumed smooth, the mechanical aperture E is the same as the hydraulic aperture e.

Table 1 - Intact rock and contact properties used in the simulations.

ROCK ELASTIC PROPERTIES	
Bulk modulus, K (GPa)	13.3
Shear modulus, G (GPa)	8.0
ROCK STRENGTH PROPERTIES	
Angle of friction, ϕ (°)	30
Dilation angle, ψ (°)	0
Cohesive strength, c (GPa)	1.0
CONTACT PROPERTIES	
Normal stiffness, k_n (GPa/mm)	193.3
Shear stiffness, k_s (GPa/mm)	193.3
Base friction angle, ϕ_b (°)	30
Initial aperture, e_o (mm)	$2 \cdot 10^{-3}$
Residual aperture, e_r (mm)	$1 \cdot 10^{-6}$

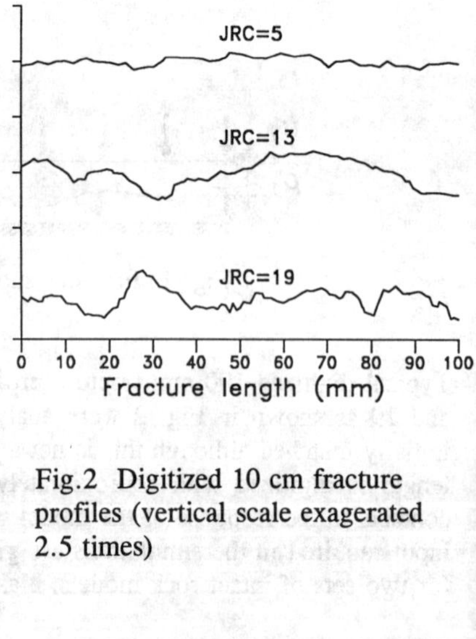

Fig.2 Digitized 10 cm fracture profiles (vertical scale exagerated 2.5 times)

RESULTS AND DISCUSSIONS

Several results of the analysis can be presented and used to analyze how the fracture behaves. Typical results are shown in Fig. 3 in terms of the displacements and stresses that are developed in the model. The results shown are for a rock with elastic rock properties and a fracture surface with JRC = 5 (see Fig. 2) subjected to a shear displacement u_s of 0.5 mm and a constant normal stress σ_n of 1 MPa. For this shear displacement fracture dilation is already evident. The principal stress plot shows the stress concentration close steep asperities and the formation of tensile stresses. The maximum compressive stress is about 38 times higher than the applied normal stress and tensile stresses of as high as 13 MPa was obtained.

Figure 3 shows that some of the contacts overlap or inter-penetrate due to high contact stresses, and plots of contacts which have completely opened up. Most "uphill" contacts are at the shear limit (σ_s = 0.57 MPa), while most "downhill" contacts are opened. These plots show what percentage of the fracture surface carries the applied load.

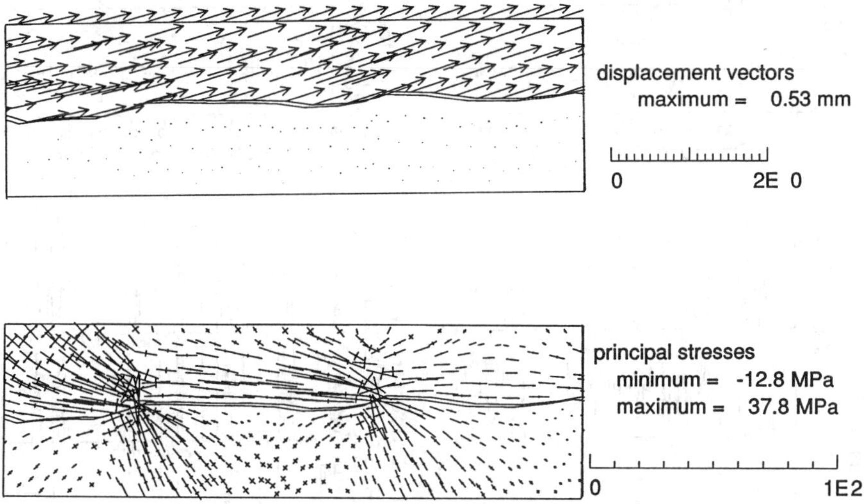

Fig. 3 Typical results of UDEC calculations

Analysis of contact data along the fracture length are shown in Figs. 4 which gives the variation with shear deformation of: (a) contact length l_i, (b) contact angle α_i , (c) contact aperture e_i, and contact stress ratio $r_i = \sigma_{si}/\sigma_{ni}$. The data are for a JRC=5 fracture with elasto-plastic intact rock behavior and an applied normal stress of 1 MPa. The initial contact lengths typically vary from 0.5 mm to 1 mm before shearing. There is a general tendency for contact lengths to become smaller during shearing (Fig. 4a). Contacts tend to smooth out with shear loading as can be seen in the decrease in variation in contact angles in Fig. 4b. Figures 4c illustrates that contacts are either "closed" (with $e_i = e_r = 1 \cdot 10^{-6}$ mm) or opened (with $e_i = e_o = 2 \cdot 10^{-3}$ mm), and only very few contacts have apertures between e_r and e_o. The number of opened contacts increases with shear displacement. The stresses are irregularly distributed among the different contacts (Fig. 4d) with some contacts yielding earlier than the others. The applied load to the fracture is carried only by a few contact points. Although detailed correlations between contact data have not yet been made, there appears to be a good coincidence of high contact stress ratios r_i with small contact lengths and high contact angles.

The behavior of the whole fracture can be obtained by integrating the contact data such as those shown in Fig. 4 over the whole fracture length. For instance, the total normal and shear stresses applied to the fracture can be calculated as:

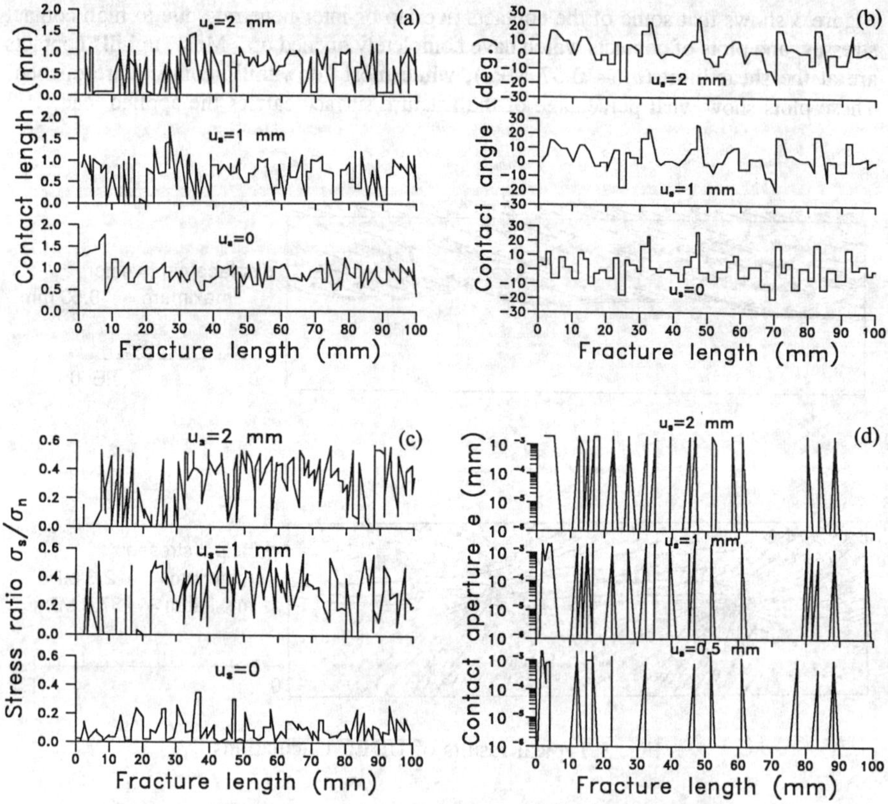

Fig. 4 Variations of contact data with shear displacement
(a) length, (b) angle, (c) stress ratio, and (d) asperity

$$\sigma_n = \frac{1}{L}\sum_{i-1}^{N}\left(\sigma_{ni}\cos\alpha_i - \sigma_{si}\sin\alpha_i\right) l_i \tag{3}$$

$$\sigma_s = \frac{1}{L}\sum_{i-1}^{N}\left(\sigma_{ni}\sin\alpha_i + \sigma_{si}\cos\alpha_i\right) l_i \tag{4}$$

where i is the contact number, N is the total number of contacts, and L is the total length of the fracture.

The normal stress calculated from Eq. (3) was found to be very close to the applied normal stress of 1 MPa, indicating that the model is numerically correct. The variation of shear stress with applied shear displacement is shown in Fig. 5a for both fractures in elastic and elasto-plastic rocks. The results show the important effects of the intact rock behavior. A fracture in elastic rock shows a peak response followed by some softening while the fracture in elasto-plastic rock shows a continuing increase in shear stress with shear displacement. The peak shear stress for the elastic rock is close to 0.7 MPa which is consistent with the value given by the Barton-Bandis model of $\sigma_n \cdot \tan(\mathrm{JRC} \cdot \log(\mathrm{JCS}/\sigma_n) + \phi_b)$ assuming JCS $\approx$ 10 MPa. For the plastic rock, there seems to be very little contribution of roughness since the peak shear stress is only about 0.56 MPa. Some stick and slip response were obtained for both types of rocks as the asperities ride one peak to another.

The fracture dilation (calculated from displacement of the upper block) for the two types of rocks are shown in Fig. 5b. As expected, dilation is higher for the elastic rock than the elasto-plastic rock. In both cases dilation starts only after about 0.15 mm of shear displacement has been applied. The angle of dilation is not constant especially for the elastic rock but slightly decreases with shear displacement, as predicted by the Barton-Bandis model.

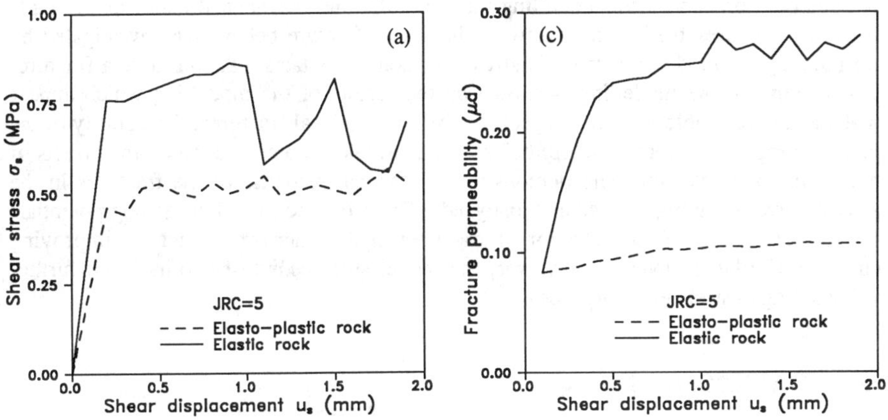

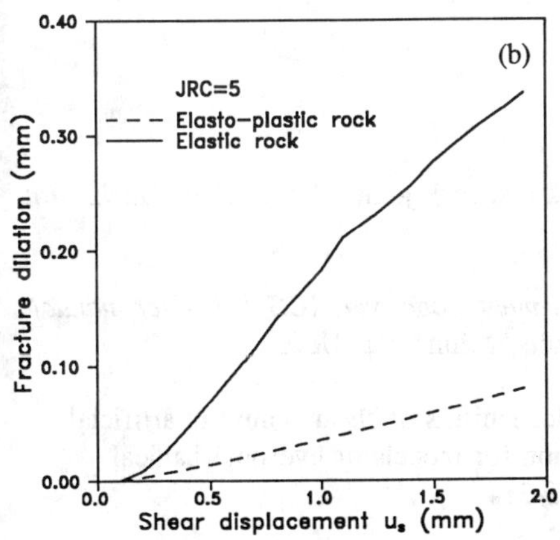

Fig. 5 Calculated fracture response during shear loading (a) stress strain curve, (b) dilation, and (c) permeability

The fracture permeability is calculated as average of all contact permeabilities k_i which is equal to the parallel plate permeability $e_i^2/12$. The average fracture permeability is given as:

$$k = \frac{L}{\sum_{i=1}^{N}\left(\frac{l_i}{k_i}\right)} \tag{5}$$

The variation of fracture permeability (in μd) as a function of applied shear displacement is shown in Fig. 5c for both elastic and elasto-plastic rocks. Due to dilation, fracture permeability increases with shear displacement.

CONCLUSIONS

A numerical procedure for modelling the hydraulic and mechanical behavior of single fractures was presented. The micro-mechanics of fracture behavior is investigated by formulating the model in terms of individual fracture contacts that constitute a fracture. The validity of the modelling is shown by the results of the modelling which appear realistic and resemble in many respects the behavior of real fractures. The ability of the model to replicate typical fracture behavior and to take into consideration the effects of the intact rock behavior were demonstrated. Typical response of the fractures in the asperity level were highlighted and analyzed. The proposed model has strong potentials for further studies of the behavior of fractures in the micro-scale and for improving currently available models. However, the model still needs to be refined and further validated against experimental results.

ACKNOWLEDGMENT

The study presented in this paper was partially funded by the Norwegian Research Council (Norges Forskninsgråd).

REFERENCES

Barton, N. 1976. The shear strength of rock and rock joints. *Int. J. Rock Mech. Min. Sci. Geomech. Abstr.* 13: 255-279.

Itasca. 1991. *UDEC - Universal Distinct Element Code, ver. ICG 1.7; User manual* Itasca Consulting Group, Inc., Minneapolis, Minnesota, USA.

Pereira, J.P. and de Freitas, M.H. 1993. Mechanisms of shear failure in artificial fractures of sandstone and their implication for models of hydromechanical coupling. *Rock Mech. Rock Eng.* 26:195-214.

2 Borehole stability and hydraulic fracture

Session developers:
Jean-Claude Roegiers, University of Oklahoma
Thomas Boone, Imperial Oil Resources, Ltd

Rock Mechanics, Nelson & Laubach (eds) © 1994 Balkema, Rotterdam, ISBN 90 5410 380 8

Hydraulic fracture orientation and pressure response during fracturing of producing wells

M.G. Mack & J.L. Elbel
Schlumberger Dowell, Tulsa, Okla., USA

ABSTRACT: Poroelastic effects cause the stress field around a producing well to differ from the initial stress field. The pressure response during a hydraulic fracture treatment may be misinterpreted due to this effect, leading to incorrect estimates of fracture geometry. In addition, the stress field around a producing well in anisotropic or heterogeneous formations may be such that the fracture orientation differs from that which would occur if the well were fractured prior to production. Several examples show how these effects may be estimated.

1 INTRODUCTION

Hydraulic fracture treatments are usually performed on new wells under conditions of uniform pore pressure. However, treatments are often performed on wells which have been producing, and have pore pressure variations in the reservoir. These variations are not usually considered when treatments are designed or simulated, although they may affect both the fracture orientation and the pressure responses during the treatments.

The effect of pore pressure changes on hydraulic fracture orientation has been recognized by several authors, including Avasthi *et al.* (1991), who reported that the stress orientation changed from layer to layer, due to production, in the McElroy field of the Permian Basin. They also noted that the production response to waterflood operations was more rapid at producers in one direction than another, which would be consistent with a preferred flow direction (caused by permeability anisotropy or induced fractures). In the current work, we investigate how production in formations having permeability and elastic property anisotropy may affect subsequent fracture orientation. Bruno and Nakagawa (1991) reported laboratory experiments in which fractures grew towards areas of higher pore pressure. Detournay and Boone (1993) challenged Bruno and Nakagawa's claim that "the orientation of any pore pressure gradient field influences fracture extension." Our simulations show that, for typical field scenarios, the direction of higher pore pressure gradient corresponds to the minor principal plane (i.e., perpendicular to the minimum principal stress), which is the direction of fracture initiation and propagation.

2 TREATMENT PRESSURE RESPONSE

The models used to design and simulate hydraulic fracture treatments vary in complexity from simple analytical models in which one-dimensional fluid flow and a fixed fracture height is assumed, to highly complex ones in which full three-dimensional fracture growth and fluid flow are included. The models currently used for fracture treatment design typically assume a planar fracture lying in the minor principal plane, and that the stress field is

Table 1 Simulation Parameters for Base Case

Reservoir Rock Parameters		
Young's modulus	4.35e6 psi	30 GPa
Poisson's ratio	0.2	0.2
permeability	0.01 md	1e-14 m^2
porosity	0.2	0.2
stress ($\sigma_{xx} = \sigma_{yy} = \sigma_{zz}$)	7250 psi	50 MPa
reservoir zone height	65.6 ft	20m
Reservoir Fluid Parameters		
gas compressibility	3.4e-5 psi^{-1}	5e-8 Pa^{-1}
viscosity	0.02 cp	2e-5 kg/m/s
reservoir pressure	6500 psi	44.8 MPa
Production Configuration		
well radius	6 in	0.15m
bottom-hole producing pressure	2000 psi	13.8 MPa
Fracture Treatment Parameters		
Fluid volume	50000 gall	189.4 m^3
Injection rate	25 BPM	6.63e-2 m^3/s
n'	0.7	0.7
K'	0.02 lb $sec^{n'}/ft^2$	1 $Pa.s^{n'}$
leakoff coefficient	0.002 $ft/min^{1/2}$	8e-5 $m/s^{1/2}$

uniform around the well. This assumption is reasonable for typical treatments of new wells because the fracture length is many times larger than the well diameter. Around producing wells, however, the stress field may be modified tens to hundreds of feet into the reservoir. A previous paper (Elbel and Mack 1993) showed that the observed treating pressure during a fracture treatment of a recently-produced layer would rise more rapidly than that in an initial fracture. The stress in the produced layer was assumed to be uniform, but at a lower level than the initial stress, so that added confinement was provided by the increased stress contrast with the surrounding layers. In this section, we examine the effect of the production-induced horizontal variation in the stress-field on the pressure response observed during hydraulic fracturing of such wells.

The P3D hydraulic fracture simulator described by Mack *et al.* (1992) has been used in this work. Although it can be used to model height growth and lateral elastic coupling, it is used here in a constant-height mode without coupling, to emphasize that the effects described are independent of fracture height growth models. It has recently been extended to account for the variation of stress in the horizontal direction.

Consider an unfractured well, isotropic and homogeneous in both permeability and elastic properties, which has been producing for five years prior to a fracture treatment. The reservoir is 20m thick, and is bounded above and below by layers with the same elastic properties but very low permeability. (The parameters of both the well and the fracture treatment are given in Table 1). The effect of production on the stress and pore pressure fields in the reservoir were calculated using a finite difference simulator (Cundall 1976). which includes fully-coupled stress and fluid flow analysis. The production-induced pore-pressure and $\sigma_{\theta\theta}$ fields prior to the treatment are shown in Figure 1. It should be noted that although the production from an isotropic reservoir decreases all three principal stresses, the decrease perpendicular to a radial vertical plane through the well ($\sigma_{\theta\theta}$) is

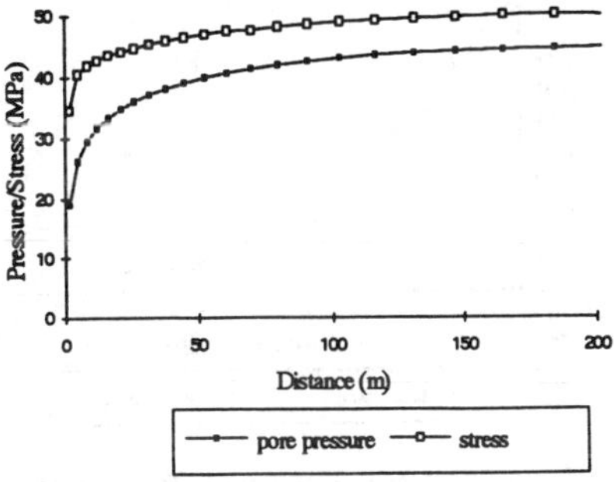

Figure 1 Pore pressure and stress distribution around a producing well

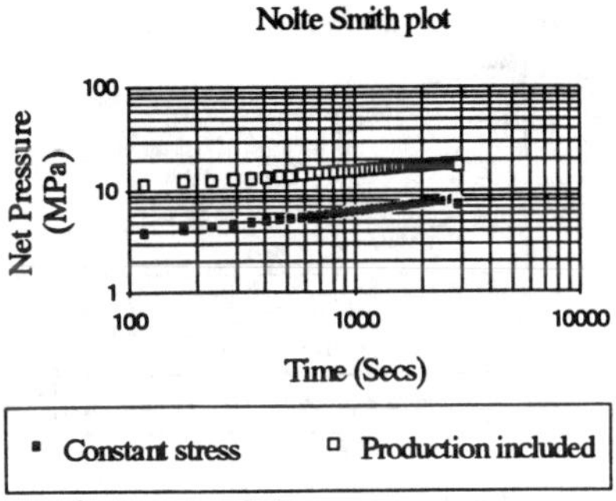

Figure 2 Nolte Smith plot with and without production effects

smaller than that on the perpendicular vertical plane (σ_{rr}) and the horizontal plane (σ_{zz}) through the same point. Thus, depending on the initial stress differences, production may alter the stress sufficiently to result in the propagation of a fracture which is initially horizontal. The stress field analysis presented here is not coupled to a fracture mechanics simulator, so it is not possible to estimate when or how rapidly the fracture will turn out of the initial vertical plane. Instead, when simulating the fracture treatment, we have assumed that the initial stress distribution was such that, even after production, the orientation of the minor principal plane was unchanged.

Net pressure is defined as the downhole measured treating pressure at the mouth of the fracture minus the minimum principal stress or closure stress, σ_{clos}. The closure stress is usually assumed to be constant throughout the formation, but after production it is not, unless the well is shut in for a long time prior to the treatment. We define p_{net} with reference to the near-wellbore stress which would be measured just prior to the fracture treatment. The fracture treatment was simulated under the conventional assumption of uniform closure stress, and then with the variation in stress included. The net pressure histories from these two simulations are compared in Figure 2. Note that if net pressure were defined relative to the initial stress state, it would be negative at the beginning of the treatment simulated here, and remain below that measured in a treatment on a new well. Negative net pressures have been observed in the field and reported by Elbel *et al.* (1984). In practice, the field engineer would probably simply reduce the closure stress to obtain an initial net pressure perceived as realistic.

During the post-production treatment, the net pressure rises rapidly as the fracture passes through the near-wellbore region with the lowest pore pressure and most rapidly varying stress field so that, apart from a very short initial period, the pressure record can be represented by a straight line on the conventional Nolte-Smith plot (Nolte and Smith 1981). The magnitude of the net pressure predicted when production-induced stress changes are accounted for is significantly greater than that predicted by assuming a uniform stress field, although the slope is only slightly smaller. The small difference in slope (about 0.16 vs. 0.23) might well go unnoticed in field practice, or be attributed to height growth, but the significant difference in magnitude of the net pressure (a factor of two) could erroneously be attributed to excess pressure at the fracture tip. Numerous authors (e.g., Shlyapobersky *et al.* 1988, Jeffrey 1989, Gardner 1992, SCR 1993) have discussed such tip effects and proposed explanations for them. If the "excess" net pressure is modeled by a tip effect, fracture width will be significantly increased over the entire length of the fracture, resulting in a reduced fracture length. The effect of the pressure increase due to prior production is to alter the width only in a relatively small region over which the stress changes significantly, and not to materially affect the predicted fracture length. In the

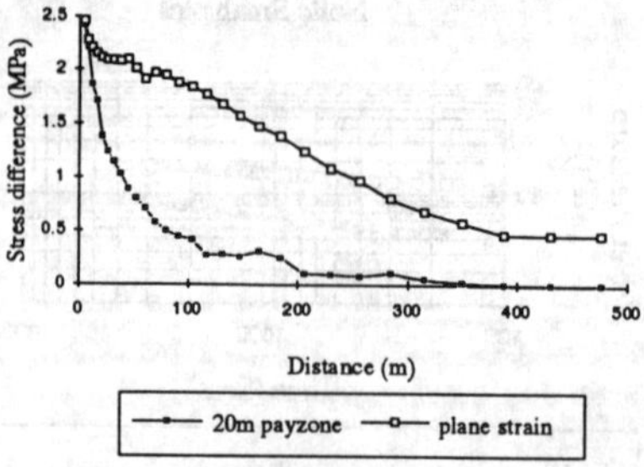

Figure 3 Stress difference for finite reservoir thickness and plane-strain solution

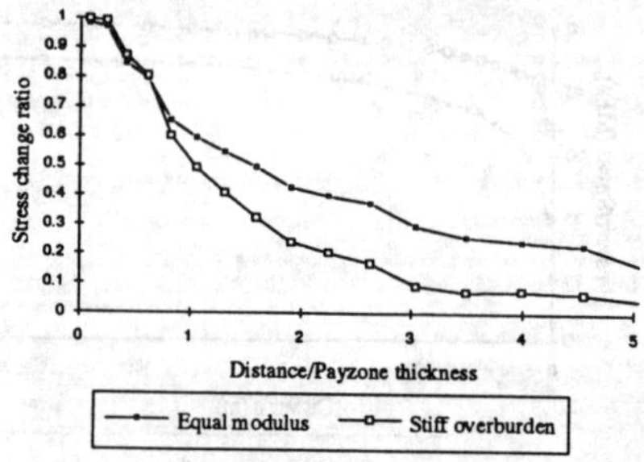

Figure 4 Ratio of stress difference to plane strain solution

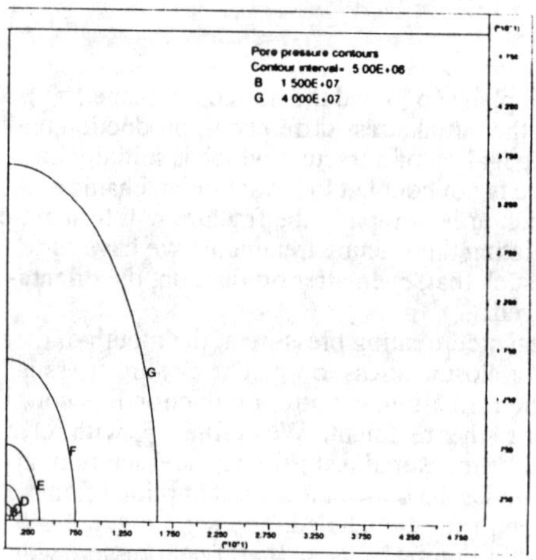

Figure 5 Pore pressure distribution around well

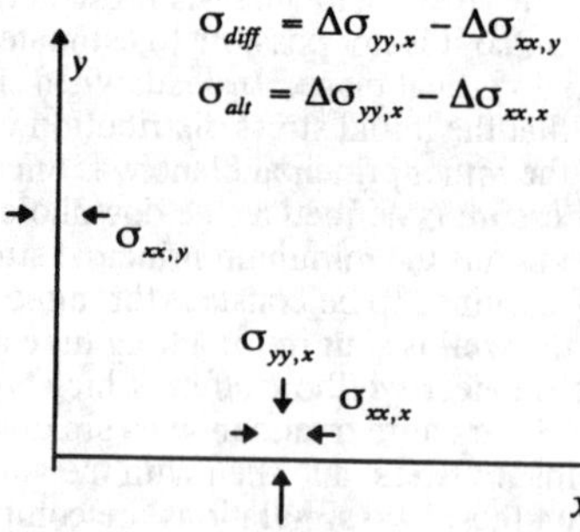

Figure 6 Definition of σ_{alt} and σ_{diff}

above example, the fracture length differed by less than 10% when production effects were included.

3 HYDRAULIC FRACTURE ORIENTATION

As previously noted, the preferred direction of fracture propagation is in the minor principal plane. Several authors (e.g., Avasthi *et al.* 1991, Shuck 1977, Warpinski and Branagan 1989) have discussed the stress alteration in the vicinity of wells under injection or production. Experimental evidence (Bruno and Nakagawa 1991) for such effects has also been reported. In previous work (Elbel and Mack 1993) a two-dimensional plane strain analysis of stress redistribution around a well which had been fractured and then produced was provided. It was shown that the change in stress was sufficient to change the orientation of the minor principal plane by 90^0, hundreds of feet from the fracture. In this section, we extend that analysis to reservoirs of finite height, and also investigate the effects of permeability and elastic property anisotropy and heterogeneity.

3.1 *Three-dimensional effects*

It is well-known that the "radius of influence" of an effect is proportional to a typical dimension of the problem. In the case of a reservoir of finite height but large areal extent intersecting a well of small radius, such as the problem described in Section 2 above, the representative dimension of the problem is the reservoir height.

We investigated the effect of altering the reservoir height in the example in Section 2, by determining the difference between the result of a plane strain analysis (in the horizontal plane) and an axisymmetric analysis, in which the finite reservoir height is considered. To obtain the same results as a plane-strain analysis, the overburden was replaced by the limiting case of an infinitely thick reservoir (with a boundary condition of no motion in the *y*-direction, but free movement in the *r*-direction). In the limit of an infinitely thin reservoir, the overburden will prevent any motion of the reservoir, so the stress change will be isotropic, and proportional to the pore pressure change. Figure 3 shows the magnitude of $\sigma_{\theta\theta} - \sigma_{rr}$ as a function of distance from the well for the 20m reservoir and the infinitely thick zone (the plane strain solution). Figure 4 shows the ratio of the two solutions. Simulation showed that this ratio was independent of time and reservoir height (provided the dimension was scaled according to reservoir height as shown in Figure 4). Figure 4 also shows the effect of an overburden with infinite modulus, which provides the most rapid falloff relative to the plane strain solution. The results presented in the following sections are all obtained from two-dimensional analyses in the horizontal plane. They can be corrected by multiplying the indicated stress distributions by the ratio shown in Figure 4.

3.2 *Permeability Anisotropy*

Production reduces the pore pressure in a reservoir, which causes changes in stress, especially near the well. In the presence of horizontal permeability anisotropy, the reduction in pore pressure extends further into the reservoir in the direction of the higher permeability. This leads to a greater reduction in the closure stress near the well in the direction of lower permeability than in the perpendicular direction. Depending on the initial stress state, this difference between the reductions in closure stress in the two directions may be sufficient to alter the direction of fracture initiation by 90^0.

We define σ_{diff} as the difference between the reduction in production-induced closure stress for a fracture propagating in the *x*-direction and that for a fracture in the *y*-direction, as shown in Figure 5. σ_{diff} near the well thus represents the anisotropy in the initial stress which can be overcome by production-induced stress changes, causing a change in the direction of fracture initiation .

The preferred direction of fracture propagation once initiation has occurred is determined by the full stress field around the crack tip. Mode I and Mode II components combine to determine the direction of propagation, and rate of turning if the orientation of the minor principal plane at the crack tip changes as the fracture extends further into the formation. In the example considered here, the *x*-direction is the preferred direction of fracture initiation after production. The orientations of the principal stress directions on this plane remain 0^0 and 90^0, although the relative stress magnitudes may be reversed. Define "stress alteration" (σ_{alt}) as $\Delta\sigma_{yy} - \Delta\sigma_{xx}$ on the *x*-axis, so that positive values of σ_{alt} represent an increased preference for propagation in the *x*-direction, and negative values an increased preference for turning into the perpendicular direction, as shown in Figure 5. The actual direction of propagation is determined by the superposition of the initial stress state and σ_{alt}.

The example of section 2 is re-analyzed with a reduced permeability in the *x*-direction. Figure 6 shows the pore-pressure distribution after 5 years of production, Figure 7 shows the change in stress perpendicular to the x and y-axes ($\Delta\sigma_{yy}$ and $\Delta\sigma_{xx}$ respectively), and Figure 8 shows the distribution of σ_{diff}. It is clear that a significant preference may be developed for the *x*-direction, but that this preference is reversed about 10m from the well.

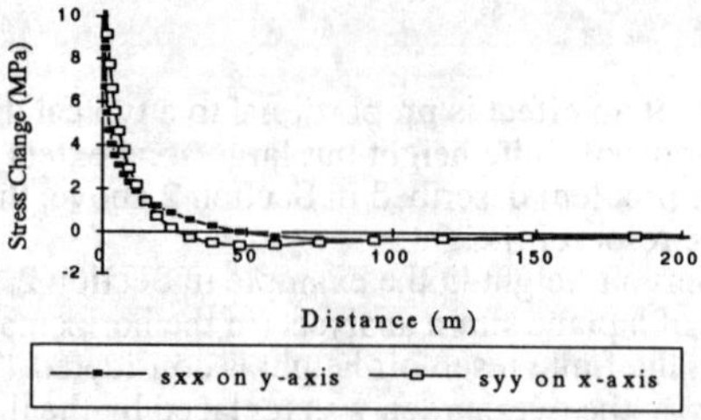

Figure 7 Stress change with anisotropic permeability

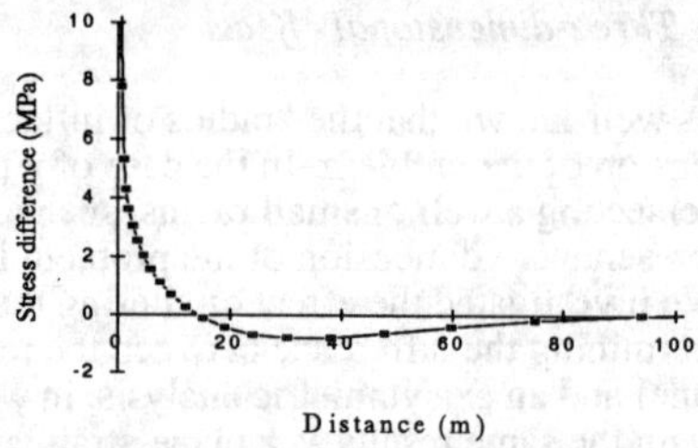

Figure 8 σ_{diff} with anisotropic permeability

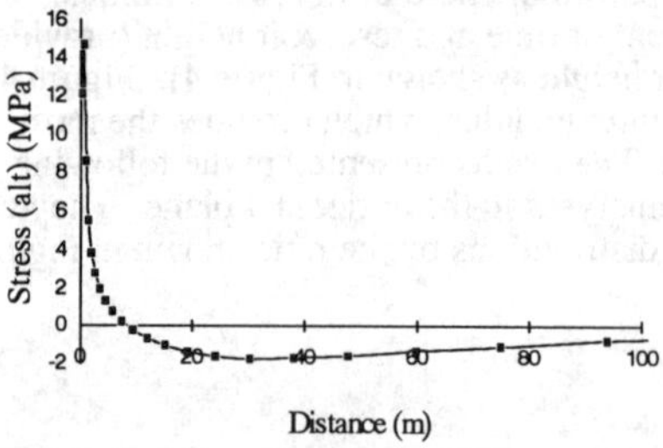

Figure 9 σ_{alt} with anisotropic permeability

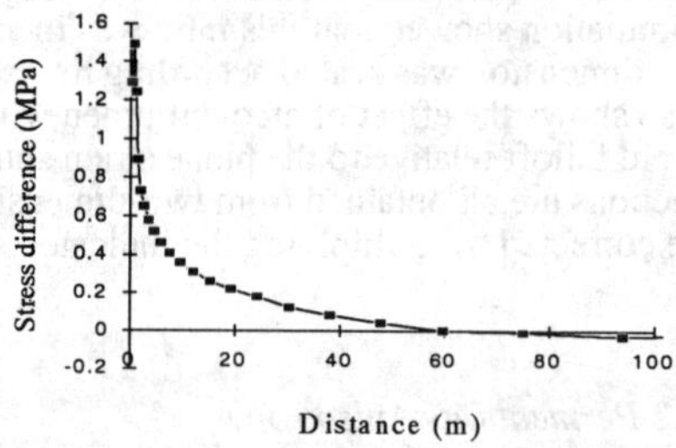

Figure 10 σ_{diff} with anisotropic modulus

Once the fracture is established in the x-direction, however, we need only consider how the stress components on that plane, represented by σ_{alt}, affect fracture propagation.

Figure 9 shows that σ_{alt} dies off rapidly with distance from the well in this case, mainly because the pore pressure changes have not penetrated very far into the reservoir. If the permeabilities were higher, or the production time longer, the tendency to turn during the treatment could be postponed until the fracture is longer. Nevertheless, in this case, even 10m of propagation prior to turning allows fracture penetration into regions with pore pressures which are only observed three times as far in the other direction. If significant fracture penetration in the *x*-direction is desired, the fracture treatment should be postponed until later in the life of the well.

It can be observed that the effect of anisotropic permeability is to enhance the likelihood of fracture propagation in the direction from which less fluid has been extracted. This implies that the fracture will enter the region which has been produced less, and also that after the treatment flow towards the fracture will be in the direction of high permeability.

3.3 *Modulus Anisotropy*

A similar analysis was performed with an anisotropic elastic model. The modulus in the *y*-direction was lower than in the previous examples by a factor of two. The distribution of σ_{diff} (Figure 10) shows that the added preference for reorientation is only about one tenth as large in this case as it was in the case of a five-fold permeability anisotropy, and the σ_{alt} distribution (Figure 11) shows that the turning tendency is only reduced within a very small distance from the wellbore. It also reaches the maximum magnitude in the opposite direction (about 2 MPa) (i.e., to favor turning) within 20m of the well. These two factors

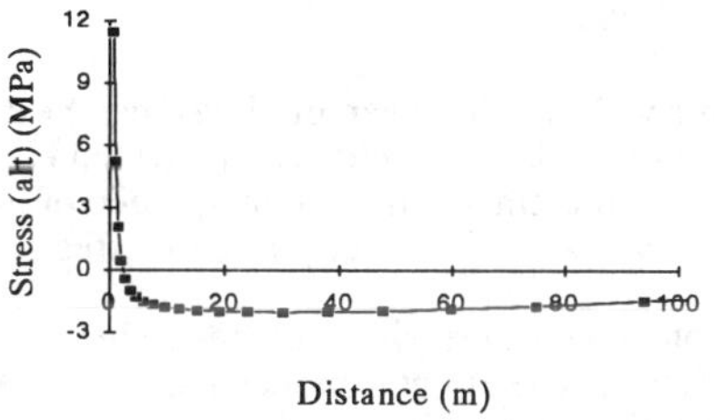

Figure 11 σ_{alt} with anisotropic modulus

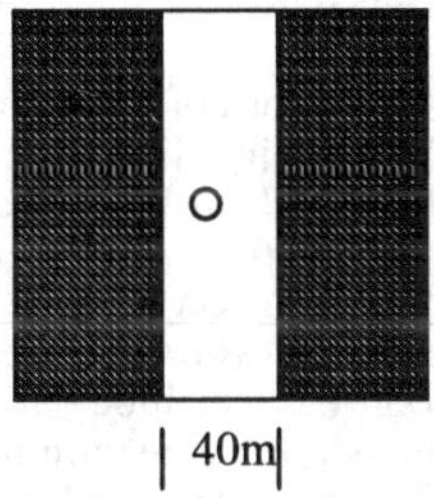

Figure 12 Geometry of permeability heterogeneity

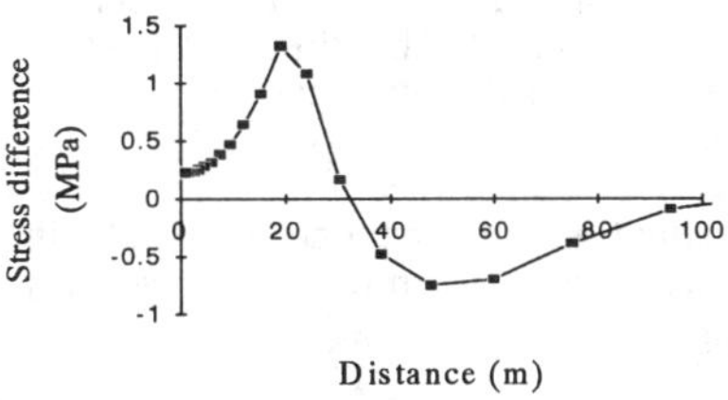

Figure 13 σ_{diff} for well in permeable channel

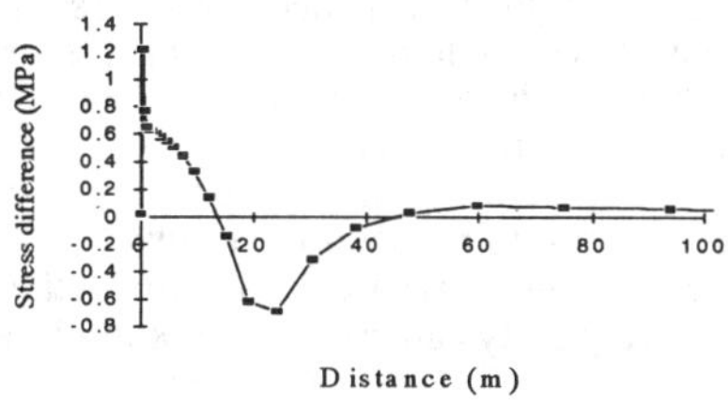

Figure 14 σ_{diff} for well near high-permeability region

combine to suggest that production in the presence of modulus anisotropy has significantly less effect on fracture orientation than permeability anisotropy.

3.4 *Permeability And Modulus Heterogeneity*

There are numerous different scenarios under which permeability and modulus heterogeneity can develop. Based on the observation in the previous section that modulus has a small effect, only permeability heterogeneity is considered in this section. One fairly common example of such heterogeneity is a channel. Typical channels are roughly sinusoidal and have a channel width of about one tenth of the meander length (Leopold and Wolman 1957). Meander lengths of three or four hundred meters are not uncommon.

Two cases in which a well is drilled near a sharp change in permeability were modeled. The first case is identical to that in Section 2, except that the permeability is reduced by a factor of five 20m away from the well as shown in Figure 12, representing a well drilled in the center of a 40m wide channel. In the second case, the permeability is *increased* by a factor of 5 in the same region. Channel curvature is neglected in both cases. In the first case, the inability of fluid to dissipate in the lower permeability zone exerts a tensile stress in the high permeability zone. This results in a preference for fracturing across the channel, as illustrated in Figure 13. The preference is so small that even small local variations in the initial stress state are likely to override the production-induced effect. The magnitude of σ_{alt} is also so small that the fracture could rapidly turn away from the initial direction. If longer production times were considered, the pore pressure changes would have propagated further from the well, and a larger effect would be observed. In the second case, the increase in permeability allows flow-induced stresses to dissipate more quickly, resulting in a *decreased* tendency to prefer the cross-channel direction, as shown in Figure 14. Once again, however, the effect is so small that it is unlikely to be of any practical benefit.

4 CONCLUSIONS

The pressure record of a fracturing treatment in a well that has been produced can be represented by a straight line on the conventional Nolte-Smith plot, with a slope which is only slightly different from that observed when a new well is fractured. The magnitude of the net pressure is significantly higher, however, and can be misinterpreted as a tip effect, leading to an incorrect estimate of fracture length.

The effect of stress redistribution calculated from a two-dimensional plane-strain analysis can be modified to include three-dimensional effects. The effect is reduced to about 40% of the plane-strain value within about three reservoir heights of the well.

Production in the presence of permeability anisotropy may result in subsequent fracture propagation in the direction from which less fluid has been extracted, which is advantageous for subsequent production. Modulus anisotropy has a small effect on stress reorientation, and is unlikely to result in significant changes in fracture initiation and propagation compared with fracturing a virgin well. In the examples in this paper, permeability heterogeneities, such as channels, have a small effect on fracture orientation. It is expected that this effect would be significant if the analysis were continued to later times. Any effect which produces a large-scale anisotropy in pore-pressure distribution can be expected to produce correspondingly large anisotropy in the stress field, which may result in a different orientation of a hydraulic fracture treatment than would be obtained if the well were fractured prior to production. Production after hydraulic fracture treatments, especially in low permeability formations, and production from wells in areas of permeability anisotropy and heterogeneity can cause such pore-pressure changes.

ACKNOWLEDGMENT

The authors wish to thank Schlumberger Dowell for permission to publish this paper.

REFERENCES

Avasthi, J.M., R.C. Nolen-Hoeksema & A.W.M. El Rabaa 1991. In-situ stress evaluation in the McElroy field, West Texas. *SPEFE* 6:301-309.

Bruno, M.S. & F.M.Nakagawa 1991. Pore pressure influence on tensile fracture propagation in sedimentary rock. *Int.J.Rock Mech.Sci. & Geomech. Abstr.* 28:261-273.

Cundall, P.A. 1976. Explicit finite difference methods in geomechanics. *Proc EF Conference on Numerical Methods in Geomechanics* 1:132-150.

Detournay, E. & T.J. Boone 1993. Discussion of "Pore pressure influence on tensile fracture propagation in sedimentary rock." *Int.J.Rock Mech.Sci. & Geomech. Abstr.* 30:323.

Elbel, J.L., R.L. Howard, G.R. Talley and B.H. McLaughlin 1984. Stimulation study of Cotton Grove formation. . *J. Pet. Tech.* 36:1199-1205.

Elbel, J.L. and M.G. Mack 1993. Refracturing: Observations and Theories. *Proc. SPE Production Operations Symposium*:521-531. Richardson:SPE.

Gardner, D.C. 1992. High fracturing pressures for shales and which tip effects may be responsible. *Proc 67th Annual SPE Tech. Conf. and Exhibition*:879-893. Richardson:SPE.

Jeffrey, R.G. 1989. The combined effects of fluid lag and fracture toughness on hydraulic fracture propagation. *Proc. Low Permeability Reservoir Symp.*:269-276. Richardson:SPE.

Leopold, L.B. & M.G. Wolman 1957. River channel patterns - braided, meandering and straight. *USGS Professional Paper* 282:34-85.

Mack, M.G., A.R.Piggott & J.L.Elbel, 1992. Numerical representation of multilayer hydraulic fracturing. *Proc 33rd U.S. Symp. Rock Mechanics.* Rotterdam:Balkema.

Nolte, K.G. & M.B. Smith 1981. Interpretation of fracturing pressures. *J. Pet. Tech.* 33:1767-75.

Shlyapobersky, J., G.K. Wong & W.W. Wolhaug 1988. Overpressure calibrated design of hydraulic fracture stimulations. *Proc 63rd Annual SPE Tech. Conf. and Exhibition*:133-148. Richardson:SPE.

Shuck, L.Z. 1977. Method for selectively orienting induced fractures in subterranean earth formations. *U.S. Patent 4005750.*

Warpinski and Branagan Altered Stress Fracturing, *J. Pet. Tech.* 41:990-997.

SCR Geomechanics Group 1993. On the modelling of near tip processes in hydraulic fractures. *IntJ.Rock Mech.Sci. & Geomech. Abstr.* 30:1127-1134.

Rock Mechanics, Nelson & Laubach (eds) © 1994 Balkema, Rotterdam, ISBN 90 5410 380 8

Hydraulic fracture from the interface of a cased wellbore

B.J.Carter, P.A.Wawrzynek & A.R.Ingraffea
Cornell University, Ithaca, N.Y., USA

H.Morales
Schlumberger Dowell, Sugar Land, Tex., USA

ABSTRACT: Hydraulic fracture initiation and propagation from a cased and perforated wellbore is strongly dependent on the influence of the steel casing, the cement between the casing and the rock, and the strength of the steel–cement and cement–rock interface bonds. 2D and 3D simulations are presented to illustrate the effects of the casing on both the stress in the rock and the subsequent fracture propagation. The wellbore pressure necessary to induce tension in the rock can be reduced by adding perforations to the model. A casing reduces the stress transfer into the rock; however, poorly bonded interfaces between the casing and rock allow greater stress transfer into the rock. Fluid pressure in the interface changes the tangential stress distribution in the rock, creating a local increase in stress at the ends of the pressurized section and a reduction in stress over the pressurized area. A casing can alter the fracture propagation trajectory and delay fracture reorientation, depending on the strength of the interface bonds.

1 INTRODUCTION

Hydraulic fracture from a cased and perforated wellbore can involve 1) fluid flow into the perforation, 2) possible failure of the rock–cement or cement–steel interfaces, leading to slip on and fluid flow into the interfaces, 3) hydraulic fracture initiation in the heterogeneous rock mass, 4) arbitrary, 3D fracture propagation in the rock, and 5) possible complex interactions and coalescence of the multiple cracks.

Laboratory tests (Weijers and de Pater, 1992; Behrmann and Elbel, 1991; Hallam and Last, 1990; Daneshy, 1973) and analytical and numerical studies (Li, 1991; Rogiers and Detourney, 1988) have been done to study hydraulic fracture, leading to some interesting observations. If the casing is cemented into the borehole after loading and the perforations are not aligned with the in situ stresses, the fracturing fluid often penetrates the interface between the rock and the casing (Behrmann and Elbel, 1991; Weijers and de Pater, 1992). The process of creating perforations often induces a zone of residual compression in the rock (Warpinski, 1983; Cosad, 1992), leaving the perforation filled with debris and restricting fluid flow into the perforation. The perforating process also can destroy the cement between the rock and the steel casing close to the perforation (Economides, 1992). The original cementing is often poor and incomplete due to residual water, oil or mud on the rock or casing walls, borehole breakouts or fractures, and other factors, such as the applied pressure and poor heat dissipation (Bol *et al.*, 1991; Nelson, 1990).

Fractures either initiate from the perforations or directly from the borehole wall (Behrmann and Elbel, 1991; Yew and Li, 1987). These fractures are aligned with the near-wellbore stresses initially, but quickly reorient themselves with respect to the far-field stresses during propagation. Different types of fractures initiate depending on the orientation of the borehole with respect to the in situ stresses (Aguilera et al., 1991). If the borehole is perpendicular to the minimum compressive stress, then transverse (aligned with the borehole axis) fractures form. If the borehole is parallel to the minimum compressive stress, radial fractures are created; some of these fractures start as transverse, but change to radial as they propagate away from the wellbore (Economides, 1992). If the borehole is inclined with respect to the in situ stresses, the initial fractures are inclined as well.

In this paper, 2D and 3D simulations are used to model hydraulic fracture propagation from both uncased and cased wellbores. A number of 2D models were previously analyzed (Carter et al., 1994) and a summary of the results is first presented. Then, a series of 3D simulations is presented, using a horizontal wellbore, both with and without casing, perforations and interface cracks. A poorly bonded interface is modeled by putting cracks in the interface between the rock and casing. The crack surface consists of a small area around the perforation, and fluid in the interface is modeled by pressurizing the crack faces. The change from transverse to radial crack is discussed also.

2 FRANC2D ANALYSES

A number of 2D analyses were done using FRANC2D (Carter et al., 1994), a finite element code written specifically for modeling crack propagation (Wawrzynek, 1987). These analyses concentrated on the effects of the casing and interface behavior on hydraulic fracture initiation and propagation. The plane strain, 2D model is shown in Figure 1. The perforation is not modeled, although in several cases a perforation is assumed so that fluid can pass from the inside of the steel casing to the rock–cement interface. At the same time, it is assumed that there is no flow into the rock-portion of the perforation, corresponding to the case of a plugged perforation.

The tangential stress near the borehole perimeter is shown for two different cased models, a perfectly bonded (Figure 2a) and a slipping interface (Figure 2b) model. For the cased–perfectly bonded model, the stress in the rock is fairly constant, at about one third of the applied wellbore pressure and there is no tension zone. With slip elements (using a coefficient of friction equal to 0.33) in both the rock–cement and steel–cement interfaces, the stress in the rock and in the concrete is almost identical and is only a fraction of the stress in the casing. There is more stress transferred into the rock for this model than for the perfectly bonded model due to the weaker interfaces. The stress in the rock is slightly less than that in the cement and, again, there is no tension in the rock. These models do not correspond exactly to the field situation, where the hole is drilled into the pre-loaded rock and the casing is then cemented into the hole. However, the effect of the casing in reducing the fluid-induced tangential stress in the rock near the borehole perimeter should be similar in both situations (Li, 1991).

If the interface is weak and flow into the perforation is restricted, it is possible that fluid will flow into the interface rather than into the perforation. In the next set of analyses, a constant fluid pressure is applied to varying lengths of the rock–cement interface. Two cases are modeled, with a perforation centered at 90° and 45° locations, where the angle is measured from the positive x-axis (see Figure 1).

In the first case, a perforation at 90°, four pressure distributions are shown (Figure 3a). These correspond to lengths spanning 0°, 22°, 112° and 180° along the interface, centered at 90°. A wellbore pressure of 69 MPa produces a small zone of tension in the tangential stress distribution, at 90°, without pressurizing any of the interface. When a 22°-span of the interface is pressurized, the tangential stress at 90° is depressed (more compressive, with a dip in the typical bell-curve), but there are two zones of higher tension on either side of the pressurized length. If a hydraulic fracture does not start in the rock and more of the interface becomes pressurized (112°-span), then the maximum tension returns to the 90° location. A fully pressurized interface produces maximum tension at 90° and leads to a stress distribution similar to that of an uncased model.

The next model assumes a perforation at 45°. A pressurized length of interface, spanning 22° and centered about the 45° location, depresses the tangential stress over the length of the pressurized interface. The stress becomes more tensile immediately after

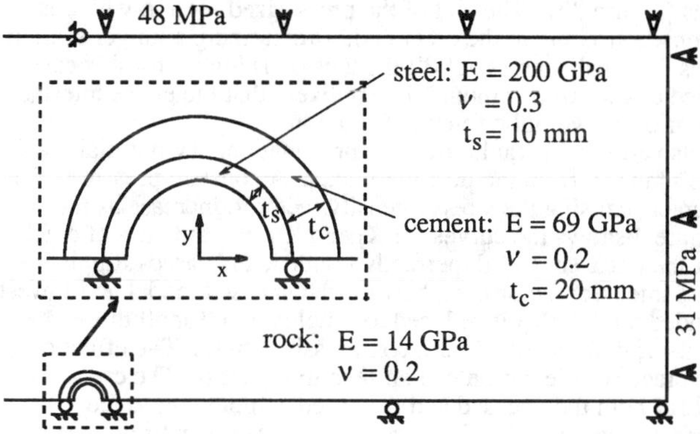

Figure 1. Portion of 2D model, showing boundary conditions and loading, with a close-up of the steel casing and cement bond.

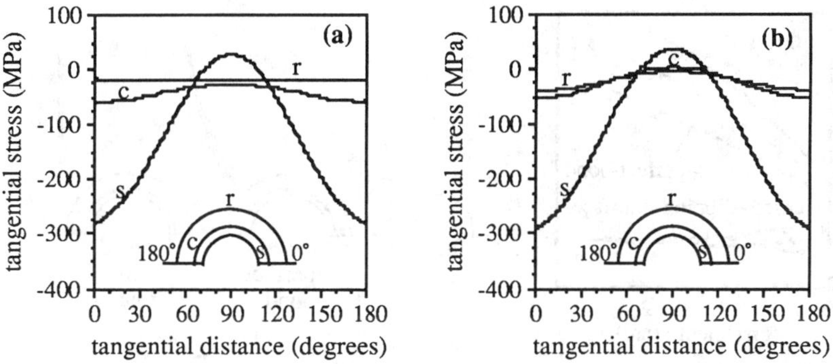

Figure 2. Tangential stress inside the perimeter of the steel (s), concrete (c) and rock (r) for (a) the cased model with a perfectly bonded interface and for (b) the cased model with a slipping interface. Wellbore pressure is 51.7 MPa. Tension is positive.

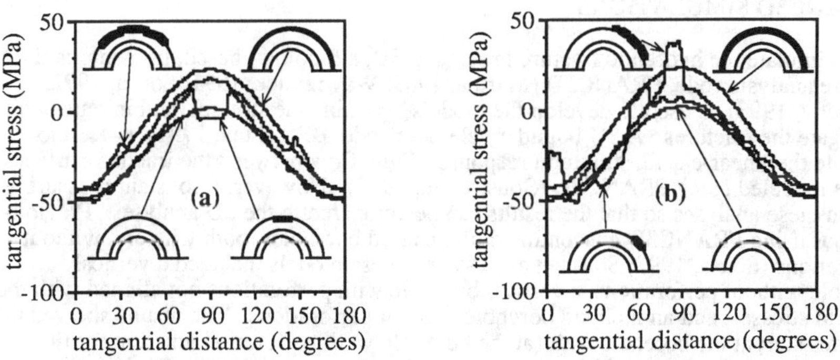

Figure 3. Tangential stress inside the perimeter of the rock for the cased model with a slipping and pressurized interface. A perforation is assumed at 90° (a) and at 45° (b). The span of pressurized interface is indicated by the thick lines. The wellbore and interface pressure is 69 MPa.

the pressurized section ends (Figure 3b). The end of the pressurized section which is closer to 90° has a higher tensile stress than the other end. Pressurizing a longer section (67°-span), further increases the tensile stress at 90°; the tension is higher for this case than for the fully-pressurized case. For this model, it is unlikely that the entire interface would become pressurized prior to hydraulic fracture initiation.

The interface behavior also affects hydraulic fracture propagation. Two initial crack locations are modeled at 45° and 90° from the positive x-axis. A constant pressure of 69 MPa is applied inside the crack, causing the stress intensity factor to increase as the crack length increases. Figure 4 shows the curves for K_I and K_{II} as a function of crack length for the 45° crack for the uncased, cased–perfectly bonded, and cased–slipping interface models. The stress intensity factor is highest for the uncased model and lowest for the cased–perfectly bonded model. K_{II} is reduced to roughly zero for all three models as the crack aligns itself with the maximum compressive stress. The effect of the casing and interface on crack re-orientation can be seen in Figure 4. The crack quickly changes to vertical for both the uncased and the cased–slipping interface models. For the cased–perfectly bonded model, the crack still was not entirely vertical when the analyses were stopped. The crack at 90° begins and remains in the preferred orientation with K_{II} remaining near zero. The plot of K_I versus crack length for the 90° crack is similar to that of the 45° crack.

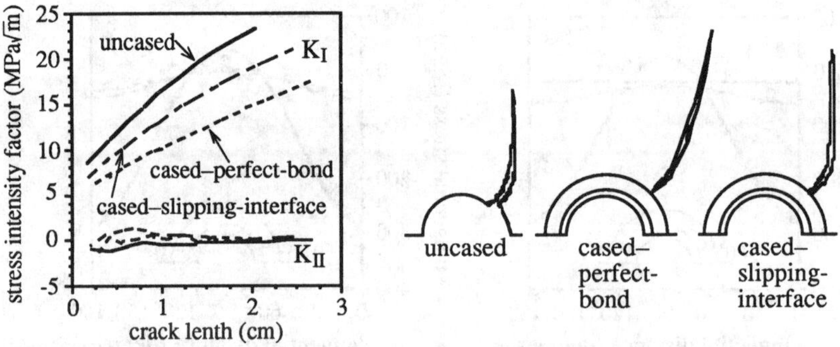

Figure 4. Stress intensity factors (Mode I and II) plotted against crack length for a wellbore and crack pressure of 69 MPa and the deformed-crack propagation trajectories. The initial crack is at 45°.

3 FRANC3D SIMULATIONS

To simulate the hydraulic fracture process in 3D, a topology-based, user-interactive, fracture analysis code, FRANC3D (Martha, 1989; Wawrzynek, 1991; Sousa, 1992; Potyondy, 1993), is used to develop the models, examine the stresses, and initiate and propagate the fractures. A 3D boundary element code, BES (Lutz, 1991), is used to evaluate the linear-elastic structural response. Fluid flow between the fracture surfaces can be modeled using FRANC3D (Sousa et al., 1993). However, a constant pressure is used in these analyses so that the results can be compared to the 2D analyses. Previous analyses using FRANC3D concentrated on uncased boreholes, both with and without perforations (Sousa, 1992; Sousa et al., 1993). These models included a vertical borehole without perforations, a vertical borehole with perforations not aligned with the far-field stresses, and an inclined borehole without perforations. The results showed that fractures start at the borehole wall (at the borehole wall–perforation junction for the perforated model), change orientation to align themselves with the far-field stresses, and merge as they propagate.

A number of models are analyzed (Table 1), starting with a horizontal, uncased and unperforated wellbore (Figure 5). The applied far-field loading is shown in Figure 5;

the minimum compressive stress is aligned with the borehole axis. The steel casing in models 4–8 is perfectly bonded to the rock. A weakened interface is modeled by including an area of interface crack around the perforation. The interface crack is pressurized to simulate fluid flow in the interface. The far-field stress is the same for all models, while the wellbore pressure is increased until there is tension in the rock. The full lengths of the wellbore and the perforations are pressurized.

There are three zones in the rock where tension is first induced (Figure 7). One zone (z-1) is located on the top and bottom of the borehole, at the junction of the borehole and perforation if a perforation is aligned with the y-axis. The second zone (z-2) is located within the perforations, a short distance from the borehole perimeter. The third zone (z-3), which is not shown in Figure 7, consists of the area around the interface crack.

Due to limitations of computer time and memory, only preliminary results could be obtained from these models. The meshes are rather coarse, especially for the cased models. However, some general observations can be made from the analyses. The model number, the applied wellbore pressure, and the zone where tension is first induced are given in Table 2. The addition of perforations to the uncased model reduces the wellbore pressure that is required to induce tension in the rock. As in the 2D analyses, a casing reduces the stress transfer to the rock, requiring an increased wellbore pressure to induce tension in the rock. Pressurizing the interface crack around the perforation reduces the wellbore pressure that is needed to induce tension. However, the tension in zone z-3 is more likely to induce further interface cracking rather than the desired hydraulic fracture in the rock mass.

Table 1. Model description

Model	Description
1	uncased, no perforations
2	uncased, four perforations aligned with the x and y axes (Figure 6a)
3	uncased, four perforations offset by 45° from the x and y axes
4	cased, four perforations aligned with the x and y axes
5	cased, four perforations offset 45° from the x and y axes (Figure 6b)
6	cased, interface crack, one perforation aligned with +x axis (Figure 6c)
7	cased, interface crack, one perforation offset 45° from the +x axis
8	cased, interface crack, one perforation aligned with the +y axis

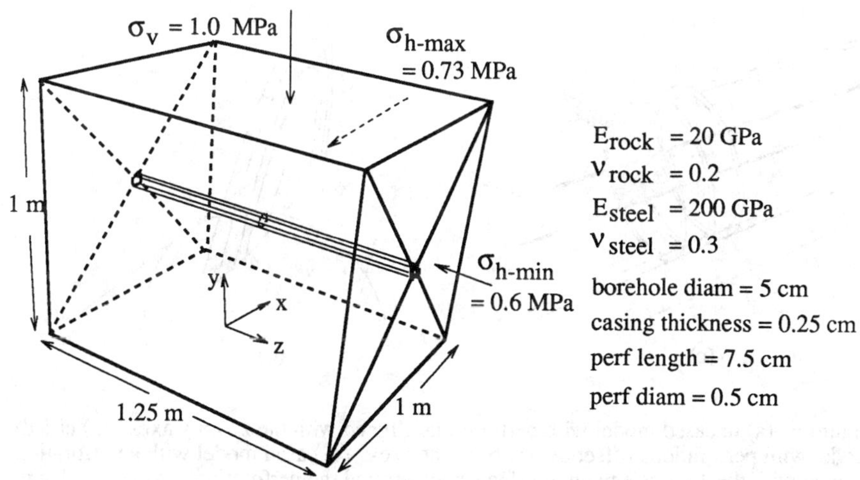

Figure 5. Base 3D model with applied far-field loading. Properties and dimensions for the casing and perforations are given for reference with Figure 6.

The uncased, unperforated borehole (model 1) and the models with perforations aligned with the y axis (models 2 and 4) have tension in zone z-1; this will induce an initial transverse crack. A transverse crack is created for model-1 (Figure 6d). The model is perfectly symmetric, thus the crack propagates in the initial plane and is not influenced by the load applied in the x-direction. If the crack symmetry is perturbed slightly (Figure 6d), the crack starts to change to a radial crack, wanting to propagate perpendicular to the minimum compressive stress. The models which have four perforations offset by 45° (models 3 and 5) can be used to promote radial crack initiation and propagation. These models have tension in zone z-2 for all perforations. Cracks should start from all perforations. A radial crack should form as the individual cracks propagate both away from and towards the borehole wall and finally merge.

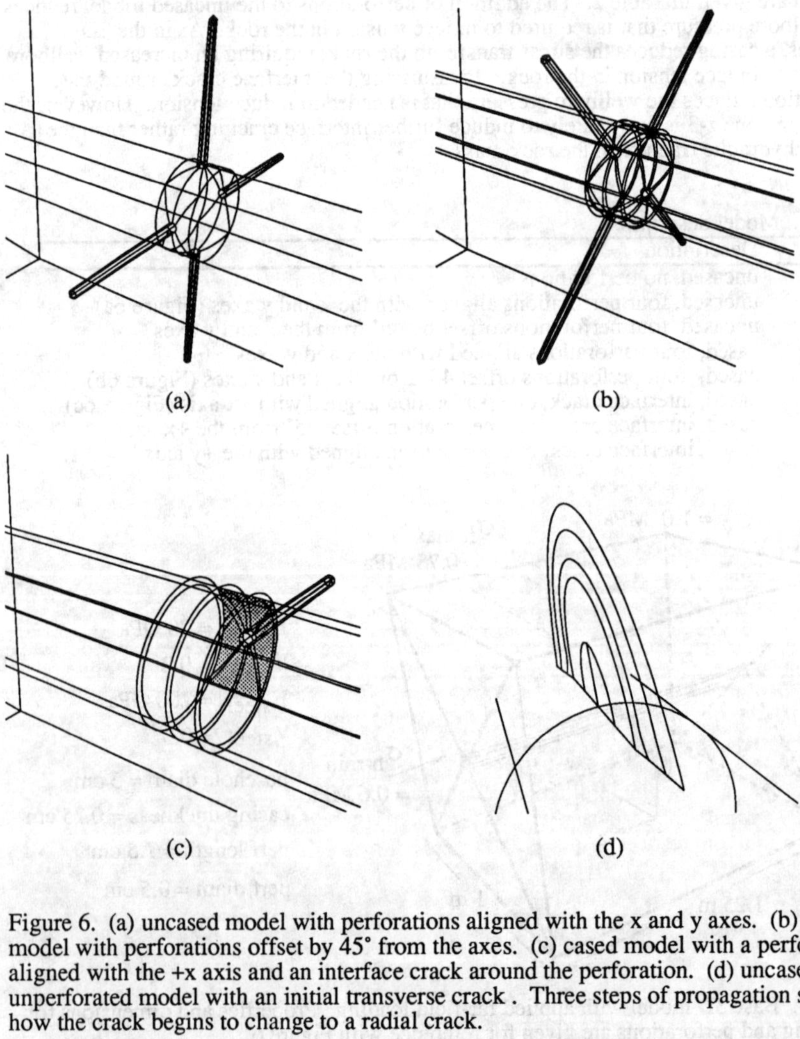

Figure 6. (a) uncased model with perforations aligned with the x and y axes. (b) cased model with perforations offset by 45° from the axes. (c) cased model with a perforation aligned with the +x axis and an interface crack around the perforation. (d) uncased and unperforated model with an initial transverse crack . Three steps of propagation show how the crack begins to change to a radial crack.

Table 2. Model, applied pressure and induced tension.

Model #	Pressure (MPa)	Tension zone
1	1.3	z-1
2	1.0	z-1
3	1.1	z-2
4	1.3	z-1
5	1.4	z-2
6	1.2	z-3
7	1.2	z-3
8	1.1	z-1 and z-3

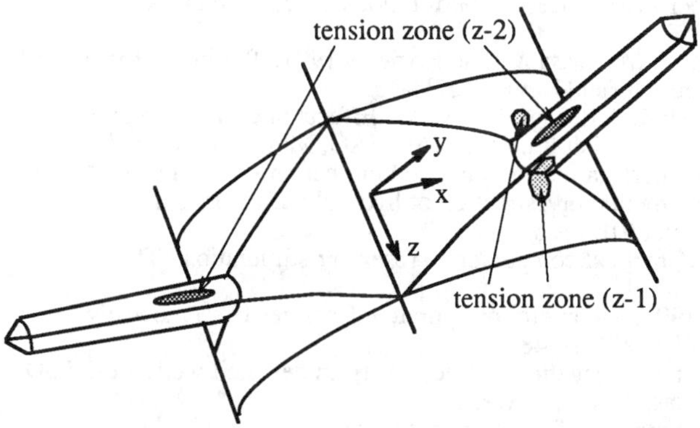

Figure 7. Zones in the models where tension is first induced.

4 SUMMARY & CONCLUSIONS

The results from this study are quite preliminary. The 3D models, especially, require further analysis, including initiation and propagation of cracks for all models. However, some conclusions can be made from these simulations.

1) Pressurized perforations, modeled in 3D, reduce the wellbore pressure needed to induce tension in the rock.

2) A casing reduces the stress in the rock near the wellbore perimeter compared to the uncased model.

3) Slip on the interface allows more stress transfer than a perfectly bonded interface.

4) Pressure in the interface changes the stress distribution in the rock depending on the amount of interface that is pressurized, and reduces the wellbore pressure required to induce tension in the rock. In general, as the amount of pressurized interface increases, the tangential stress in the rock becomes more tensile.

5) A perfectly bonded casing inhibits crack propagation more than a casing with a slipping interface; both inhibit propagation more than the uncased model.

6) The perfectly bonded casing also affects the crack propagation trajectory, delaying reorientation with respect to the far-field stress.

These conclusions show the importance of modeling the interface behavior as well as the casing when simulating hydraulic fracture. This is especially true if the perforations are not aligned with the in situ stresses. If the cementing is poor, fluid can flow into the interface and fractures can start from the interface surface rather than from the perforations. This could affect any measurements or calculations based on the interpreted break-down pressure or pressure-time curves, the fracture orientation and shape, and the transport of fluid and proppant into the fracture.

ACKNOWLEDGMENTS

The authors would like to acknowledge the other members of the Cornell Fracture Group who were involved in developing the software used in these analyses. Dr. Carter is supported by a Post-Doctoral Scholarship from the Natural Sciences and Engineering Research Council of Canada. This research was made possible through the continued funding and support of Schlumberger Dowell.

REFERENCES

Aguilera, R., G.M. Cordell, G.W. Nichol, J.S. Artindale, M.C. Ng, & G.A. Runions, 1991. Horizontal Wells. Houston: Gulf Publishing Co.

Behrmann, L.A. & J.L. Elbel 1991. Effect of perforations on fracture initiation. JPT, May:608-615.

Bol, G., H. Grant, S. Keller, F. Marcassa & J. de Rozieres 1991. Putting a stop to gas channeling. Schlumberger Oilfield Review, 3:35-43.

Carter, B.J., P.A. Wawrzynek & A.R. Ingraffea 1994. Effects of casing and interface behavior on hydraulic fracture. 8th Int. Conf. IACMAG, Morgantown, WV.

Cosad, C. 1992. Choosing a perforation strategy. Schlumberger Oilfield Rev., 4:54-69.

Daneshy, A.A. 1973. Experimental investigation of hydraulic fracturing through perforations, JPT, October:1201-1206.

Economides, M.J. 1992. A practical companion to reservoir stimulation. The Netherlands: Elsevier.

Hallam, S.D. & N.C. Last 1991. Geometry of hydraulic fractures from modestly deviated wellbores. JPT, June:742-748.

Li, Y. 1991. On initiation and propagation of fractures from deviated wellbores. PhD Thesis, University of Texas, Austin, Texas.

Lutz, E.D. 1991. Numerical methods for hypersingular and near singular boundary integrals in fracture mechanics, PhD Thesis, Cornell University, Ithaca, New York.

Martha, L.F. 1989. Topological and geometrical modeling approach to numerical discretization and arbitrary fracture simulation in three dimensions, PhD Thesis, Cornell University, Ithaca, New York.

Nelson, E.B. (ed.) 1990. Well Cementing. The Netherlands: Elsevier.

Potyondy, D.O. 1993. A software framework for simulating curvilinear crack growth in pressurized thin shells. PhD Thesis, Cornell University, Ithaca, New York.

Rogiers, J-C. & E. Detourney, 1988. Considerations on failure initiation in inclined boreholes. 29th U.S. Rock Mechanics Symposium:461-468.

Sousa, J.L. 1992. Three-dimensional simulation of near-wellbore phenomena related to hydraulic fracturing from a perforated wellbore. PhD Thesis, Cornell University.

Sousa, J.L., B.J. Carter, & A.R. Ingraffea, 1993. Numerical simulation of 3D hydraulic fracture using Newtonian and power-law fluids. 34th U.S. Rock Mechanics Symposium, Madison, Wisconsin.

Warpinski, N.R. 1983. Investigation of the accuracy and reliability of in situ stress measurements using hydraulic fracturing in perforated cased holes. 24th U.S. Symposium on Rock Mechanics:773-786.

Wawrzynek, P.A. 1987. Interactive finite element analysis of fracture processes: an integrated approach. Theor. and Appl. Fract. Mech., 8:137-150.

Wawrzynek, P.A. 1991. Discrete modeling of crack propagation: theoretical aspects and implementation issues in two and three dimensions. PhD Thesis, Cornell University, Ithaca, New York.

Weijers, L. & C.J. de Pater 1992. Hydraulic fracture reorientation in model tests. Formation Damage Control Symposium, Lafayette, SPE Paper 23790.

Yew, C.H. & Y. Li 1987. Fracturing of a deviated well. 62nd Tech. Conf. & Exhib., Dallas, Texas, SPE Paper 16930:509-517.

Rock Mechanics, Nelson & Laubach (eds) © 1994 Balkema, Rotterdam, ISBN 90 5410 380 8

The influence of constitutive behavior on the fracture process zone and stress field evolution during hydraulic fracturing

Kunsoo Kim & Cunying Yao
Henry Krumb School of Mines, Columbia University, N.Y., USA

ABSTRACT: The nature of the fracture process zone varies from material to material depending upon the micro-structure, the constitutive behavior, the boundary condition and the loading rate to which the material is subjected. We investigated the micro-mechanics of the process zone development during hydraulic fracturing using a new 2-D dynamic network model which incorporates several types of material constitutive relationships. The modeling results suggest that the fracture pattern, the fracture branching behavior and the breakdown pressure are influenced considerably by the constitutive relationship of the material. The crack tip stress field and its evolution during the fracture propagation are also found to be informative in interpreting the process zone development.

1 INTRODUCTION

It is well known that during fracture propagation the fracture process zone develops in the vicinity of the advancing fracture tip (Hoagland et al 1973, Kobayashi and Fourney 1978, Kim and Mubeen 1981, Wawersik and Stone 1989). The significance of the process zone has drawn renewed attention recently in the secondary recovery of petroleum production by hydraulic fracturing method. Sophisticated hydraulic fracturing simulators, which model the fracturing process numerically either in 2-D or 3-D (Morale and Abou-Sayed 1989, Advani et al, 1990, Clearly et al 1990), have been developed to optimize the benefit of the hydraulic fracturing treatment. In these models, the influence of the process zone on the fracture propagation is assumed to be practically negligible. Recently, indirect field observations in well bore treatment operations (Vinegar et al 1992) have suggested that the contribution of the process zone may be significant in fracture propagation and must be taken into consideration for prediction of the fracture growth. It has been suggested that the process zone is a contributing factor to the discrepancy between the model prediction and the actual observations of the hydraulically induced fractures(Warpinski and Teufel 1987, Warpinski 1992). In these predictive models used currently in the industry, the fracture process zone is largely neglected due to the difficulty of realistically modeling the multiple fracturing phenomenon despite the observations made in the laboratory and field tests (Shlyapobersky and Chudnovsky 1992, Huang et al 1991, Labuz et al 1987, and Heuze et al 1990).

We employed the Dynamic Lattice Network Model in an attempt to gain an insight into the fracture process zone which develops during HF process. The modeling analysis has been performed to simulate the hydraulic fracturing in rocks with various constitutive behaviors ranging from linear elastic to strain softening behavior. The results indicate that the material constitutive behavior affects the fracture initiation and propagation significantly and thus the process zone development, and that the simplistic premise that rock fails in an elastic brittle fashion in tension may apply to limited, but not general cases.

2 MODELING APPROACHES

The Dynamic Lattice Network Model, a 2-D triangular lattice network model (Huang 1992), is used to describe the hydraulic fracturing process in rock. The model consists of particles interconnected by springs. The particles have mass m which is determined by the rock density. Connecting any two adjacent particles is a spring which is characterized by its spring constant K, natural length L_0, and threshold lengths L_{max} and L_{min} beyond which the spring snaps. K is determined by Young's modulus and Poisson's ratio of the rock. L_{max} and L_{min} define rock's compressive and tensile strength respectively. For each particle an equation of motion governs its displacement vector **X**.

$$m\frac{d^2\mathbf{X}}{dt^2}+\alpha\frac{d\mathbf{X}}{dt} = \textit{forces on springs} + \textit{any boundary force} \tag{1}$$

where α is the damping factor representing an inner energy dissipation process. The forces on the springs are determined by the general Hooke's law or any constitutive law we prescribe to the springs:

$$f_{ij} = f(\varepsilon_{ij}^{k}, \varepsilon_{ij}^{k(p)}) \tag{2}$$

where subscripts ij indicate the particle's number and superscript k indicates the spring's number connected with that particle. ε^{k}_{ij} denotes the spring's total strain and $\varepsilon^{k(p)}_{ij}$ is its inelastic strain. The boundary forces, the confining stress and the hydraulic pressure, are applied to the particles on the external boundaries of the mesh, and on the borehole interior and the induced fracture surfaces respectively.

For the given in situ stress field and the hydraulic pressure, a set of equations(1) can be solved by a finite difference scheme at every time step. At any given time, if either L_{max} or L_{min} of a spring is exceeded, that spring breaks irreversibly to initiate a new fracture or extend an old one and no longer exerts any force. The forces are then redistributed dynamically through the mesh, and the above equations are solved again. Thus, the evolution of this dynamic system is determined, and the fracture initiation and propagation process are simulated.

Due to the tensile stress concentration at the crack tip, some weaker springs that are not on the fracture surface may break to form microcracks that are part of the fracture process zone. Some of them grow and join the main fracture, while others remain isolated beyond the main fracture. In this model, the hydraulic fluid is allowed to flow with full pressure to the new main fracture surface instantaneously. No water pressure exists inside the isolated microcracks.

In this analysis, a particular emphasis is placed on the influence of the material constitutive behavior on the fracture process(Yao et al 1993). Four types of constitutive relationships have been used to describe rocks, ranging from linear elastic to strain-softening behavior. The following are the basic parameters used for the model:

Network size :	200 x 200 particles
Spring natural length L_0 :	100 μm
Spring constant :	K_C: 4.97 - 5.05 x 10^9dyne/cm (comp.)
	K_T: 2.98 - 3.03 x 10^9dyne/cm (tension)

Spring Threshold lengths:	L_{min}: 99.5 ± 0.1 μm (comp.)
	L_{max}: 100.05 ± 0.01 μm (tension)
Particle mass m :	3 μg

It should be noted that the spring constants and the threshold lengths are all randomly assigned in the given ranges with a uniform distribution, to account for the inherent variabilities in rock property. The time step length used for the finite difference scheme is 4 x 10^{-10} seconds.

The network with the above parameters represents a rock sample of 2cm by 2cm in size and 3g/cm^3 in density. Young's modulus is about 70 GPa for compression and 42 GPa for tension, except for the case of a linear elastic relationship described below where 70 GPa is used for both compression and tension. Poisson's ratio of 0.25 has been used throughout. The maximum strain failure criterion is used for all analyses. The mean maximum strain is 5.0 x $10^3\mu\varepsilon$ in compression, and 0.5 x $10^3\mu\varepsilon$ in tension.

3 CONSTITUTIVE RELATIONSHIPS USED FOR THE MODELING ANALYSIS

The prime objective of this modeling analysis is to gain an insight into the influence of material constitutive behavior on the fracture process. In the analyses we use four types of constitutive equations in lieu of the general expression given in Equation (2). These relationships are graphically illustrated in Figure 1.

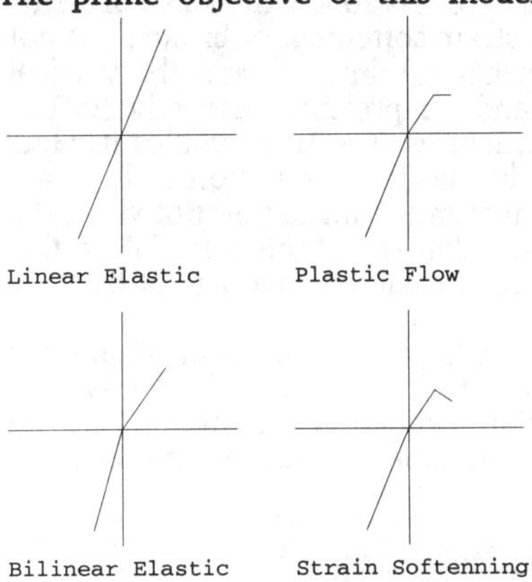

Figure 1. The constitutive models

Model 1. Linear Elastic Model : Springs behave linear elastically until they break. For each spring, The spring constant or stiffness, K, is equal to K_C which varies randomly in the range from 4.97x10^9 to 5.05 x 10^9 dyne/cm both in compression and tension.

Model 2. Bilinear Elastic Model : Springs follow the linear elastic constitutive law both in compression and tension, but the spring constants differ in compression and tension. The spring constant in tension, K_T, is 60 % of that in compression, K_C.

Model 3. Elasto-Plastic Model : Springs follow the bilinear elastic relationship as in the bilinear model, until they reach the yielding point in tension when they start behaving perfectly plastically until the moment of failure.

Model 4. Elastic-Strain Softening Model : Springs follow the bilinear relationship until the tensile yielding point when they start showing the strain softening behavior. The spring constant in the descending portion of the force-displacement curve in tension is one half the spring constant K_T in the ascending portion of the curve.

4 MODELING RESULTS

4.1 Fracture and fracture process zone

We carry out a series of numerical simulations of hydrulaic fracturing process using the models described above. The boundary conditions are imposed on the external boundaries to simulate the far field stresses existing at depth: 7.5 MPa and 19.75 MPa in vertical and horizontal directions in the figures, respectively. A circle with a radius of 10 grid-point is dug out in the center of the mesh to model a borehole. Then, a prescribed level of hydraulic pressure is applied on to the borehole boundary. As the springs respond to the applied borehole pressure, some of the springs on the boundary of the borehole which exceed the threshold level snap. Then, the stresses are redistributed dynamically around the borehole with the newly created fractures and the process continues as the time step increases.

We present several snapshots of the fracture patterns taken at 8,000 time steps for four different types of constitutive relationships (Figures 2 - 5). The plot (a) in each figure is a part of the network mesh with the fracture system inside it. They illustrate the initiation and propagation of the fractures through the mesh and their effects on the rest of the mesh. The fractures initiate slightly off the horizontal plane because the borehole in the lattice network is made of triangles and thus the circumference of the borehole is not precisely circular. The plot (b) in each figure shows the fracture pattern with its process zone which also presents the springs which are in plastic deformation, in strain softening, or broken but not connected to the main fractures. The small horizontal lines "-", and the vertical lines "l" represent the springs failed in tension and compression, respectively. The figures show that all the fractures are tensile fractures. The tiny bubbles in plots (c) of figures 4 or 5 are the springs which are in plastic deformation or in strain softening stages, but not yet broken. From the micromechanical point of view, the plastic deformation is the result of the permanent damage of the material. In this sense, those plastic deformation zones should be considered as a part of fracture process zone as well.

This graphic presentation provides the development of main fractures accompanied by the process zones of different nature. The major fractures vary quite distinctly depending on the constitutive relationships of the material. And the characters of the fracture process zones are also unique for different materials.

4.2 The tensile stress field evolving around the fracture tip

The stresses in all springs change continually during the propagation of the hydraulic fracture. The stresses in the neighborhood of the main fractures connected to the borehole undergo the most drastic changes and they affect directly the growth of the fractures and the development of the process zones. We compute the stress changes at each time step in all cases. Due to the space limitation we present only one case which describes the evolution of the stress field at the main fracture in Figure 6.

In this figure we plot tensile stresses only for the case of the bilinear model. As the time step advances, the tensile stress zones on each side of the fracture increase in size. The darker areas at the immediate vicinity of the main fractures are the zones of highest stress concentration. This high stress concentration leads to the failure of springs and the main fractures advance with process zone development. It is very clear from this figure that the fracture initiation and propagation are both related to the tensile stress concentration. The shape and size of the tensile stress concentration change as the fracture propagates.

The results suggest that the fracture branching is attributed to tensile stress concentration deviating from the main fracture direction. During the fracture propagation, the highest tensile stress can occur at an angle other than $\theta=0$,

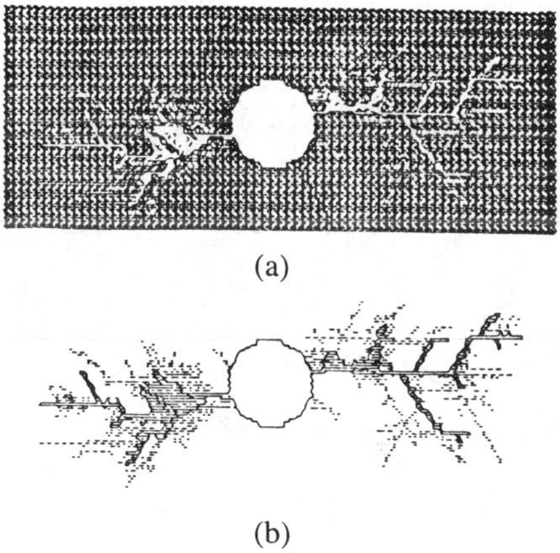

Figure 2. Fracture pattern for linear elastic model

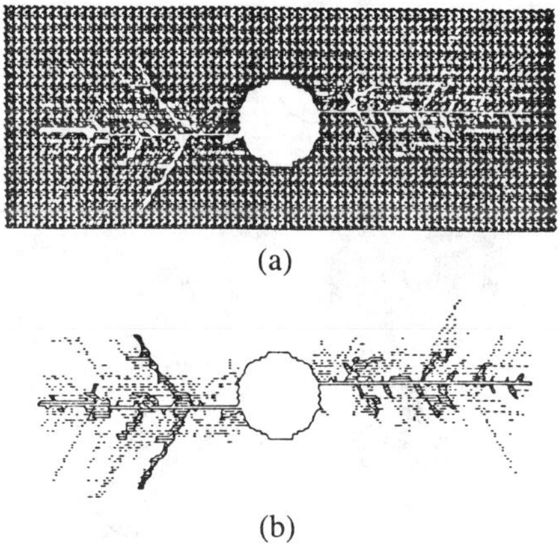

Figure 3. Fracture pattern for bilinear elastic model

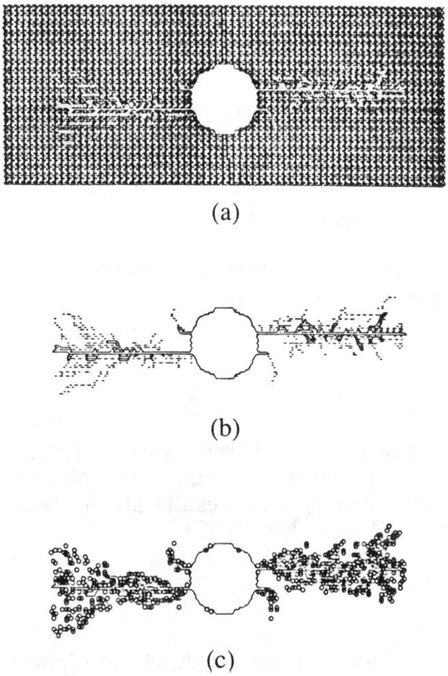

Figure 4. Fracture pattern for elasto-plastic model.

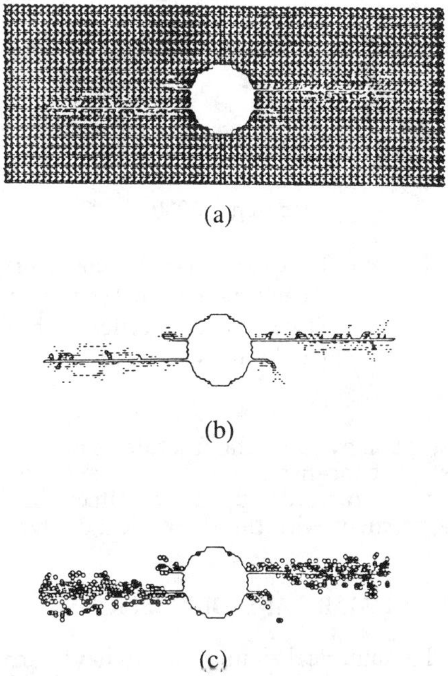

Figure 5. Fracture pattern for strain-softening model.

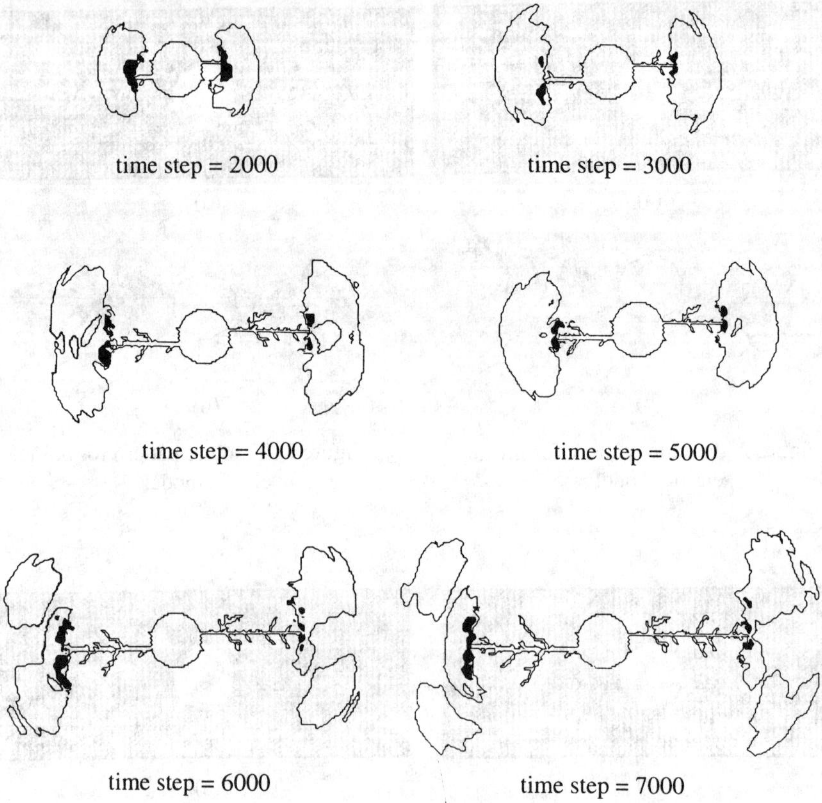

Figure 6. The evolution of the tensile stress concentration at the main crack tips for the bilinear model. The contours are plotted for the normal tensile stresses in the vertical direction of the figure. The dark areas indicate the highest tensile stress concentration.

which may cause the fracture to bifurcate. The resulting morphology of the tensile stress concentration affect the evolving fracture patterns. The directional change of the fractures appears to affect the fracture velocity. The results are in basic agreement with the theoretical analysis(Yoffe 1951, Baker 1962).

5 SUMMARY AND DISCUSSIONS

The numerical simulation we have carried out demonstrates, as shown in Figures 2-6, that the fracture process zone develops during the hydraulic fracturing process in a unique way depending on the material constitutive relationship. We define the fracture process zone as the area around the main fracture which consists of fracture branchings, incipient fractures, and the plastic deformation area in the plastic cases. The effect of material constitutive behavior on the fracture process zone is quite pronounced. The fracture branching and the incipient fractures develop most intensively along the main fracture in the linear elastic case. For the bilinear elastic case, the fracture process zone is still quite pronounced, but slightly less than the linear elastic case. The borehole pressures needed for fracture propagation in the bilinear case and linear case are 30.0 MPa and 37.5 MPa respectively.

In the elasto-plastic and the strain softening models shown in Figure 4 and 5, the breakdown pressure is 22.5 MPa and 18.75 MPa, respectively. The fracture branching becomes much less pronounced in comparison with the elastic brittle cases. The incipient fractures subside also. The fracture process zone including the plastic deformation area is pronounced in the elasto-plastic model and becomes less pronounced in the softening model. From Figures 2 to 5 the process zones show a progressively decreasing trend.

The modeling results also indicate that the fracture velocities vary depending upon the constitutive models. We are currently endeavoring to study the interactions between the fracture velocity and the fracture branching behavior, and the fracture tip singularity and the fracture velocity. These questions must be addressed before we can develop a numerical simulator that can satisfactorily predict the hydraulic fracturing process.

REFERENCES

Advani, S.H., Lee, T.S. and Lee, J.K. 1990. Three-dimensional modeling of hydraulic fracture in layered media. *ASME Jour. Energy Resources Technol,* Vol. 112, pp. 1-9.

Baker,B.R. 1962. Dynamic stresses created by a moving crack. *J. Appl. Mech.* E29: 449-458.

Clearly,M.P., Barr,D.T. and Willis,R.M. 1990. Enhancement of realtime hydraulic fracturing methods with full 3-D simulation. *SPE paper No. 17713.*

Heuze,F.E, Shaffer,R.J., Ingraffea,A.R. and Nilson,R.H. 1990. Propagation of fluid driven fractures in jointed rock. Part 1 - development and validation of methods of analysis. *Int.J. Rock Mech. Min. Sci. & Geomeh. Abstr.* 27: 243-254.

Hoagland,R.G., Hahn,G.T. and Rosenfield,A.R. 1973. Influence of microstructure on fracture propagation in rock. *Rock Mechanics.* 5: 11-106.

Huang,J-I. 1992. Modeling dynamic fracture growth with an elastic network, *Geophy. Res. Letters* 19: 1855-1858.

Huang, J-I and K. Kim, 1993, Fracture process zone development during hydraulic fracturing, *Int. Jour. Rock Mech. Min. Sci. & Geomech. Abst.* 1295-98

Huang,W.L., Kunin,B. and Chudnovsky,A. 1991. Kinematics of damage zone accompanying curved crack. *Int. J. Fracture.* 50: 143-152.

Kim,K. and Mubeen,A. 1981. The relationship between differential stress intensity factor and the crack growth rate in cyclic tension in westerly granite. *ASTM, Spec. Tech. Pub.* 745: 157-168. Eds. S.M. Friedman and E.R. Fuller.

Kobayashi,T. and Fourney,W.L. 1978. Experimental characterization of the development of the micro-crack process zone at a crack tip in rock under load. *Proc. 19th U.S. Symp. on Rock Mech.*: 243.

Labuz,J.F., Shah,S.P. and Dowding,C.H. 1987. The fracture process zone in granite: evidence and effect. *Int. J. Rock Mech. Min. Sci. & Geomech. Abstr.* 24: 235-246.

Morale,R.H. and Abou-Sayed,A.S. 1989. Microcomputer analysis of hydraulic fracture behavior with a pseudo-three-dimensional simulator. *SPE Production Engineering.* 4: 69-74.

Shlyapobersky,S. And Chudnovsky,A. 1992. Fracture mechanics in hydraulic fracturing, *Proc. 33rd U.S. Symp. Rock Mech.*: 827-836.

Vinegar,H.J., Willis,P.B., deMartini,D.C., Shlyapobersky,J., Deeg,W.F.J., Adair,R.G., Woerpel,J.C., Fix,J.E. and Sorrells,G.G. 1992. Active and Passive seismic imaging of a hydraulic fracture in diatomite. *Jour. Pet. Tech.*: 28-34.

Warpinski,N.R. 1992. Rock Mechanics issues in completion and stimulation operations. *Proc. 33rd U.S. Symp. on Rock Mechanics.*: 375-386.

Warpinski,N.R. and Teufel,L.W. 1987. Influence of geologic discontinuities on hydraulic fracture propagation. *Jour. Pet. Tech.*: 209-220.

Wawersik,W.R. and Stone,C.M. 1989. A characterization of pressure records in inelastic rock demonstrated by hydraulic fracturing measurements in salt. *Int. J. Rock Mech. Min. Sci. & Geomech. Abstr.* 26: 613-627.

Yao,C., Kim,K. and Huang,J-I. 1993. Simulation of hydraulic fracturing process with non-linear constitutive relationships, *EOS Trans* 74: 293. Amer. Geophys. Union.

Yoffe,E.H. 1951. The moving Griffith crack. *Philosophical Magazine*, Series 7, 41: 739.

Rock Mechanics, Nelson & Laubach (eds) © 1994 Balkema, Rotterdam, ISBN 90 5410 380 8

Shearing motion of dense proppant material in hydraulic fracturing

Dezhang Lin & Jean-Claude Roegiers
School of Petroleum and Geological Engineering, University of Oklahoma, Norman, Okla., USA

1 INTRODUCTION

Proppant transport and placement are critical parts of hydraulic fracturing treatments. Such particulate flow consists of two interacting materials i.e., proppants and the carrying fluid. Two-phase flow mechanics requires specification of stresses for each phase, plus a relation expressing the interaction of both materials.

The character of the interparticle forces in the dilute and dense regions is essentially different. The constitutive equations for dilute flow specialize the drag to include Stokes and Faxen forces, and the lift to accommodate the slip-shear force and the disturbance-shear force. While in dense proppant flow the solid fraction is high, it is more difficult for them to move with respect to one another. In the state where maximum packing has been reached, the particles are so closely interlocked that only rigid motion is possible. A locking phenomenon occurs.

Here we review briefly the approach of Passman, Nunziato and Bailey (1986) to clarify the origin of the momentum equations arising in the dense proppant flow. In addition, the constitutive equations for the momentum transfer are formulated to account for the viscous transfer arising from the highly dense particles.

In order to try to fully understand the problem, we have performed parameter studies and numerical analyses. To solve the Poiseuille flow problem in the fracture channel we have used a spectral method which is known to produce accurate results. The proppant concentration profiles, shear stress distribution, and velocity profiles for fluid and proppant are given. The explanation of the results is provided.

2 EQUATIONS OF BALANCE

The rheological properties of suspensions are often determined under the assumption that the material in the test device is homogeneous. However, it is well known from both direct observations and single-particle analyses that, in certain flow configurations, particles migrate across the mean fluid streamlines. It is desirable, then, to develop models with which to explore the complex interactions of the fluid, particles, and the boundaries of a given device.

Toward this end, we present here results based on the continuum theory of mixtures, a structure that provides field equations for fluid and particle velocities as well as the particle concentration.

We consider the behavior of a two-phase mixture of small proppant particles in a linearly viscous fluid. At every spatial position, x, occupied by the mixture and at every time, t, each constituent is assigned a material density, $\gamma_a(\mathbf{x}, t)$, a = p(proppant), f(fluid), which represents the mass of constituent, a, per unit volume occupied by the constituent.

In application to multiphase flow, particle (proppant) phase with density, ρ_p, and velocity, $\mathbf{v_p}$, suspended in fluid phase with mass density ρ_f and velocity $\mathbf{v_f}$.

Both the disperse proppant and the continuous fluid phase are assumed to be incompressible; thus, the densities γ_p, γ_f are material constants. It is also useful to define the partial density, $\rho_a(\mathbf{x}, t)$, which is related to the material density by

$$\rho_a = \phi_a \gamma_a. \tag{1}$$

Here $\phi_a, 0 \leq \phi_a \leq 1$ is the volume fraction of constituent a and indicates the proportion of the total volume occupied by the constituent. Clearly, ϕ_a is a measure of the local volume changes that result from the relative motion of the phases and we require that

$$\Sigma \phi_a = \phi_p + \phi_f = 1 \tag{2}$$

where Σ denotes the sum. It is evident that the volume fraction of either phase may vanish at any $\mathbf{x}$ and t as long as they both do not vanish simultaneously. Thus, particle-free zones in the flow may exist, in which case $\phi_f = 1$.

The density of the mixture, ρ, is defined by

$$\rho = \phi_p \gamma_p + \phi_f \gamma_f. \tag{3}$$

Notice that, although the constituents are assumed to be incompressible, the mixture density will vary with time and position in the flow field except when the particles are neutrally buoyant.

The motion of each phase of the mixture is assumed to be as smooth as required and we assign to each phase a velocity field $\mathbf{v_a}(\mathbf{x}, t)$. The velocity gradient with respect to the spatial position, $\mathbf{x}$, is

$$\mathbf{L_a} = \nabla \mathbf{v_a} \tag{4}$$

and the symmetric part of $\mathbf{L_a}$ is called the rate of deformation tensor, $\mathbf{D_a}$

$$\mathbf{D_a} = \frac{1}{2}(\mathbf{L_a} + \mathbf{L_a^T}) \tag{5}$$

and the deformation rate of the fluid is

$$\mathbf{D_f} = \text{sym grad } \mathbf{v_f} \tag{6}$$

The velocity of the mixture, $\mathbf{v}$, is defined in terms of the total momentum

$$\rho \mathbf{v} = \rho_p \mathbf{v_p} + \rho_f \mathbf{v_f}. \tag{7}$$

Next we wish to consider the general equations of motion for each of the phases and for the mixture. We make two important assumptions (see Rajagopal et al (1986)):

(i) Each phase of the mixture behaves as if it were a single material except that it interacts with the other phase.

(ii) The equations of motion for the mixture are the same as those for a single material and result from summing the equations of motion for the individual phases.

Assumption (i) asserts that balance of mass and momentum for each constituent take the form

$$\dot{\rho_a} = -\rho_a \nabla \cdot \mathbf{v}_a \tag{8}$$

$$\rho_a \dot{\mathbf{v}}_a = \nabla \cdot \mathbf{T}_a + \rho_a \mathbf{b}_a + \mathbf{m}_a^+. \tag{9}$$

In Eq. (9), $\mathbf{T_a}$ is the symmetric partial stress tensor, $\mathbf{b_a}$ is the external body force (e.g. gravity), and $\mathbf{m}_a^+$ is the momentum transfer resulting from the forces acting on the disperse phase due to the presence and motion of the continuous fluid phase.

For our subsequent discussion, it is convenient to introduce the extra stress $\mathbf{T}_a^*$

$$\mathbf{T}_a^* = \mathbf{T}_a + \phi_a p_a \mathbf{1} + \mathbf{S} \tag{10}$$

where p_a is the isotropic pressure associated with each phase. $\mathbf{S}$, the equilibrated stress, is introduced to determine the evolution of granular material.

$$\mathbf{S} = 2\alpha_a \nabla\phi_a \otimes \nabla\phi_a \tag{11}$$

Then since the material density of each phase is constant, Eqs. (13) and (14) can be written

$$\dot{\phi_a} = -\phi_a \nabla \cdot \mathbf{v}_a \tag{12}$$

$$\rho_a \dot{\mathbf{v}_a} = -\nabla(\phi_a p_a) + \nabla \cdot \mathbf{T}_a^* + \rho_a \mathbf{b}_a + \mathbf{m}_a^+. \tag{13}$$

Along with appropriate constitutive equations for the stresses $\mathbf{T}_a^*$, pressures, p_a, and momentum transfer, $\mathbf{m}_a^+$, the equations of motion Eqs. (12) and (13) form the basis of our analysis of fluid-particle flows.

Before considering our second assumption, we must highlight some aspects of the momentum transfer, $\mathbf{m}_a^+$. Usually $\mathbf{m}_a^+$ is interpreted as a momentum exchange representing the momentum a phase may gain or lose from the other phase as a result of, say, drag forces.

$$\mathbf{m}_p^+ = -\pi \nabla \phi_p - D(\mathbf{v}_p - \mathbf{v}_f) \tag{14}$$

$$\mathbf{m}_f^+ = -\pi \nabla \phi_f - D(\mathbf{v}_f - \mathbf{v}_p) \tag{15}$$

As a consequence of assumption (ii), the equations of motion for the mixture are obtained by summing Eqs. (8) and (9)

$$\dot{\rho} = -\rho \nabla \cdot \mathbf{v} \tag{16}$$

$$\rho \dot{\mathbf{v}} = \nabla \cdot \mathbf{T} + \rho \mathbf{b} \tag{17}$$

3 FIELD EQUATIONS

The governing differential equations of the theory we have developed are obtained by substituting the constitutive equations into the equations of balance. In order to make these equations tractable, we introduce further simplifications. These are justified in terms of the boundary-value problems we will subsequently consider.

We restrict our attention to *isothermal* motions, i.e., $\theta = \theta_0$ for all $\mathbf{x}$ and t. Thus, the temperature ceases to appear as a parameter in constitutive equations, and the equations of energy and entropy need not be considered.

For the equations of conservation of mass, we have

$$\frac{\partial \phi_p}{\partial t} + div(\phi_p \mathbf{v}_\mathrm{p}) = 0 \tag{18}$$

$$\frac{\partial \phi_f}{\partial t} + div(\phi_f \mathbf{v}_\mathrm{f}) = 0. \tag{19}$$

The appropriate equations for balance of linear momentum are then

$$\begin{aligned} \rho_p(\frac{\partial v_p}{\partial t}) = -\nabla(\phi_p p_p) - 2div(\alpha_p \nabla \phi_p \otimes \nabla \phi_p) + \\ div(\mu_p[\nabla \mathbf{v}_\mathrm{p} + \nabla \mathbf{v}_\mathrm{p}^\mathrm{T}]) + \gamma_\mathrm{p} \phi_\mathrm{p} \mathbf{b}_\mathrm{p} - \pi \nabla \phi_\mathrm{p} - \mathrm{D}(\mathbf{v}_\mathrm{p} - \mathbf{v}_\mathrm{f}) \end{aligned} \tag{20}$$

$$\begin{aligned} \rho_f(\frac{\partial v_f}{\partial t}) = -\nabla(\phi_f p_f) - 2div(\alpha_f \nabla \phi_f \otimes \nabla \phi_f) + \\ div(\mu_f[\nabla \mathbf{v}_\mathrm{f} + \nabla \mathbf{v}_\mathrm{f}^\mathrm{T}]) + \gamma_\mathrm{f} \phi_\mathrm{f} \mathbf{b}_\mathrm{f} - \pi \nabla \phi_\mathrm{f} - \mathrm{D}(\mathbf{v}_\mathrm{f} - \mathbf{v}_\mathrm{p}). \end{aligned} \tag{21}$$

Forms convenient for computation are obtained by writing the equation for the fluid component and the difference of the equations for the two components, the interface pressure, π, and the configuration pressure, P_c.

$$\pi = -(p_f - \lambda_f) - 2div(\alpha_f \nabla \phi_f), \tag{22}$$

and

$$P_c = p_p - p_f + 2div[(\alpha_p + \alpha_f)\nabla \phi_p]. \tag{23}$$

The pressure in the solid is the pressure in the fluid plus the configuration pressure:

$$\mathbf{T}_\mathrm{p} = -\phi_\mathrm{p} \mathrm{p}_\mathrm{p} \mathbf{I} = -\phi_\mathrm{p}(\mathrm{p}_\mathrm{f} + \mathrm{P}_\mathrm{c}). \tag{24}$$

For uniform spherical particles the maximum value of ϕ_p, ϕ_m corresponds to the hexagonal close-packed configuration and has a value of approximately $\phi_m = 0.74$.

Experiments on shearing motions also indicate that there is another value of ϕ_p called the critical column fraction, ϕ_c, above which shearing of the material will cause expansion, below which it will cause contraction. For uniform spheres this corresponds to a simple cubic lattice, so $\phi_c = 0.52$ approximately.

This may be visualized as a 'locking' phenomenon, i.e., as the solid particles come closer together, they become more closely interlocked, so that it is more difficult for them to move with respect to one another. In the state where the maximum packing has been reached, the particles are so closely interlocked that only rigid motion is possible. On this basis we choose

$$\mu_p = \frac{\bar{\mu_p} \phi_p^2}{\phi_m - \phi_p)^2}. \tag{25}$$

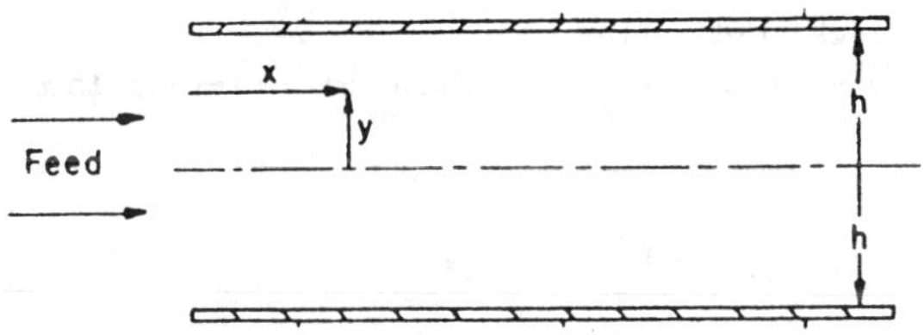

Figure 1: Fracture channel coordinate system.

Similarly, we take

$$\alpha_p = \frac{\bar{\alpha_p}}{(\phi_m - \phi_p)^2}. \tag{26}$$

For the viscosity of the fluid, we let

$$\mu_f = \phi_f^2 \bar{\mu_f}. \tag{27}$$

We also take the drag coefficient D to be a constant.

By Eqs. (26) it gives

$$P_c = (\phi_p - \phi_c)[\phi_p(\alpha_p\gamma_p - \alpha_f\gamma_f) + \alpha_f\gamma_f]. \tag{28}$$

Thus, P_c has the same sign as $\phi_p - \phi_c$. However, we expect that when the particles are in contact, which occurs when $\phi_p > \phi_c$, P_c will be positive, when the particles are not in contact, $\phi_p < \phi_c$, P_c will vanish. These conditions are satisfied if

$$P_c = \begin{cases} (\phi_p - \phi_c)[(\alpha_p\gamma_p - \alpha_f\gamma_f)\phi_p + \alpha_f\phi_f] & \text{for } \phi_p \geq \phi_c \\ 0 & \text{for } \phi_p < \phi_c \end{cases} \tag{29}$$

$$\lambda_f = \begin{cases} \gamma_f\alpha_f\phi_a[\phi_f + \phi_c - 1] & \text{for } \phi_p \geq \phi_c \\ 0 & \text{for } \phi_p < \phi_c \end{cases} \tag{30}$$

4 PROPPANT SEGREGATION IN POISEUILLE FLOW

Consider now plane Poiseuille flow (Figure 1) of the mixture. We assume rectilinear flow of both phases, such that $\mathbf{v_f} = (\mathrm{v_f(y)}, 0, 0)$, $\mathbf{v_p} = (\mathrm{v_p(y)}, 0, 0)$, and $\phi_a = \phi_a(y)$ only. The flow will be assumed to be symmetrical about the midplane of the channel of half-thickness h.

The flow is subject to a negative pressure gradient $\partial p/\partial x$, where p is the mixture pressure

$$p = (1 - \phi_p)p_f + \phi_p p_p. \tag{31}$$

We are interested in steady solutions of the form

$$\mathbf{v_p} \sim (\mathrm{v_p(y)}, 0, 0), \mathbf{v_f} \sim (\mathrm{v_f(y)}, 0, 0) \tag{32}$$

and $\phi_p, \phi_f.p_p, p_f, \pi$, function of y only.

We use a prime (') to denote the differentiation with respect to y.

$$\phi_p + \phi_f = 1 \tag{33}$$

$$2[(\alpha_p + \alpha_f)\phi_p^{'}]^{'} + p_p - p_f = P_c \tag{34}$$

$$\pi = -(p_f - \lambda_f) - 2[\alpha_f \phi_f^{'}]^{'} \tag{35}$$

$$[2\alpha_p(\phi_p^{'})^2 + \phi_p p_p]^{'} + \pi\phi_p^{'} + \gamma_p\phi_p = 0 \tag{36}$$

$$[2\alpha_f(\phi_f^{'})^2 + \phi_f p_f]^{'} + \pi\phi_f^{'} + \gamma_f\phi_f = 0. \tag{37}$$

The velocity field can be solved by the following two equations

$$(\mu_p v_p^{'})^{'} - \phi_p \frac{\partial p_p}{\partial x} - D(v_p - v_f) = 0 \tag{38}$$

$$(\mu_f v_f^{'})^{'} - \phi_f \frac{\partial p_f}{\partial x} - D(v_f - v_p) = 0. \tag{39}$$

Once Eqs. (33)-(37) are solved, Eqs. (38) and (39) may be viewed as linear equations with variable coefficients.

The component of nonzero stresses in the solid granular proppant are given by

$$T_{xx} = T_{zz} = -\phi_p p_p \tag{40}$$

$$T_{xx} = -\phi_p p_p - 2\alpha_p(\phi_p^{'})^2 \tag{41}$$

$$T_{xy} = T_{yx} = \phi_p \mu_p v_p^{'} \tag{42}$$

with the stresses in the fluid having similar forms. Thus, this theory exhibits normal stresses.

We note that specifying the normal stresses, T_{yy}, at the wall surface and the centerline specify a relation among the volume fraction and pressure.

Once the above problem is solved, the remaining equations are second order ϕ_p and ϕ_f and first order in p_p and p_f, so six boundary conditions are needed.

Thus, first we choose to specify the volume fraction ϕ_p and ϕ_f at the wall; then we consider that the condition of vanishing proppant and fluid stresses on the centerline and at the wall are satisfied.

$$T_p yy(0) = 0, T_f yy(0) = 0 \tag{43}$$

$$T_p yy(h) = 0, T_f yy(h) = 0. \tag{44}$$

The equations for velocity profiles are separated from the equations for the pressures and volume fractions. The remaining equations for velocity field are second order and linear in v_p and v_f, so four boundary conditions are needed to get the solutions. No-slip boundary conditions are specified at the wall and the symmetry condition of vanishing shear stress on the centerline are needed.

at the channel wall

$$v_p(h) = 0, v_p(h) = 0, \tag{45}$$

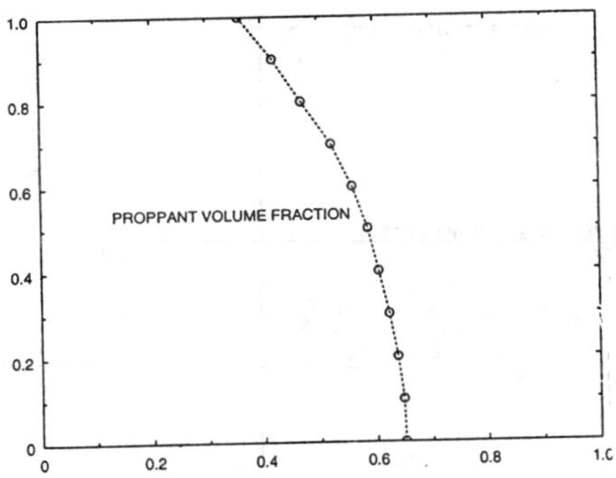

Figure 2: Proppant volume fraction profile.

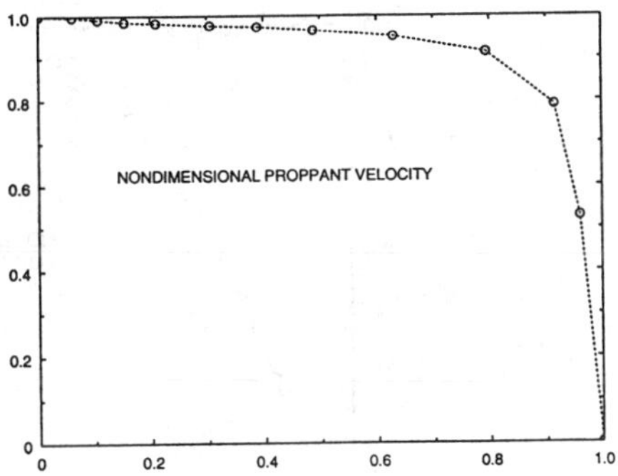

Figure 3: Proppant velocity profile.

on the center line

$$v_p'(0) = 0, v_f'(0) = 0. \tag{46}$$

The Poiseuille flow problem is a two-point boundary-value problem of some difficulty. To solve it, we have used a spectral method which is known to produce accurate results.

5 NUMERICAL RESULTS

In order to investigate the behavior of the coupled, nonlinear system, a simple transformation reduces the original second-order system of equations to a system of four, first-order, ordinary differential equations.

We choose to first investigate the case with the estimated parameters corresponding to the sand-water mixture, $\bar{\alpha}_p = 0.003lbft/s^2, \alpha_f = 0.0025lbft/s^2, D = 200lb/ft^2s, \alpha_p = 0.2ft^2/s^2, and\ \alpha_f = 0.1ft^2/s^2$. We take $\phi_p(h) = 0.36$, an assumed volume fraction at the wall. The density of sand $\gamma_p = 137.2lb/ft^3$ and the density of water $\gamma_f = 62.2lb/ft^3$. The viscosity of sand is chosen as $\mu_p = 485.8lb/fts$ and the viscosity of water $\mu_f = 0.00076lb/fts$.

Typical results for the proppant concentration profile is illustrated in Figure 2. In the vicinity of the wall boundary if the proppant volume fraction is low a layer of high shear rate occurs, while if the proppant volume is high a locking phenomenon occurs. There is a tendency for particles to accumulate in regions of low shear rate.

The proppant velocity appearing to be the same as the fluid velocity in Figure 3, is in fact slightly less than the fluid velocity. In the wall region of the low proppant volume fraction is a region of high shear rate for the proppant and the fluid, the velocities of the two constituents being virtually indistinguishable because of the high drag coefficient.

The effect of changing the values of drag coefficient, D, is not so obvious. The proppant velocities for all three cases are so close to each other, although, in fact decreasing the value of D decreases the proppant velocity.

In Figure 4 it is shown that the channel wall region of low proppant volume fraction is a region of high shear stress. A boundary layer forms near the wall.

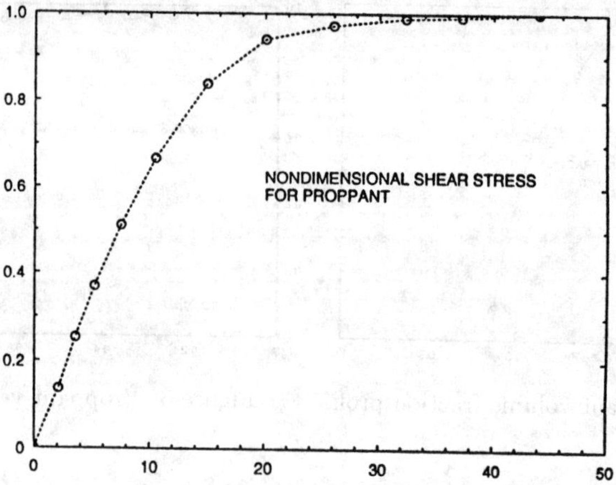

Figure 4: Nondimensional shear stress profile.

6 CONCLUDING REMARKS

We studied the constitutive equations of the saturated mixture of a linearly viscous fluid and a high concentration granular material. Using the data available for a mixture of natural angular sand with water, we obtained a solution in which the solid particles tend to concentrate themselves near the centerline of the flow and the velocity profile is nonlinear, with the shearing highest in the region of low particle concentration, i.e., near the bounding fracture walls. These results are dependent on boundary conditions; however, they are in qualitative agreement with experimental results on high-concentration suspensions and analytic results on single-particle migration.

The viscosity of the flowing proppant particles is a function of volume fraction with a locking phenomenon, i.e., as the proppants come closer together, they become more closely interlocked, so that it is more difficult for them to move with respect to one another. In the state where the maximum packing has been reached, the particles are so closely interlocked that only rigid motion is possible.

The constitutive equations have been formulated specifically for very concentrated proppants. We performed a number of calculations wherein ϕ_p had very low values at the channel boundaries. The results obtained were not in agreement with accepted phenomena for dilute proppant (see Passman (1990)).

REFERENCES

Passman, S.L., J.W. Nunziato and P.B. Bailey 1986. Shearing motion of a fluid-saturated granular material. J. Rheology. 30(1), 167-192.

Rajagopal, K.R., M. Massoudi and J.M. Ekmaan 1986. Mathematical modeling of fluid-solid mixture. in Recent Developments in Structured Continua, D. De Kee and P. N. Kaloni, eds., vol. 2, 236-248.

Passman, S.L. 1990. Stress in dilute suspensions. in Two Phase Flows and Waves, D.D. Joseph and D.G. Schaeffer, eds., 80-87.

Rock Mechanics, Nelson & Laubach (eds) © 1994 Balkema, Rotterdam, ISBN 90 5410 380 8

Behaviour and stability analysis of a wellbore embedded in an elastoplastic medium

Panos Papanastasiou, Marc Thiercelin & John Cook
Schlumberger-Cambridge Research, Cambridge, UK

David Durban
Faculty of Aerospace Engineering, Technion, Haifa, Israel

ABSTRACT: A large strain elastoplastic analysis is presented for a wellbore embedded in Mohr-Coulomb and Drucker-Prager media. Material behaviour is modelled by a non-associated deformation theory with strain hardening. The constitutive equations were fitted on true stress-strain data from triaxial tests on Castlegate sandstone. A comparison is presented between predictions obtained from the two models and experimental data from hollow cylinder tests.

1 INTRODUCTION

Most of the existing work on wellbore stability is based on elastic/ brittle models that are calibrated with data taken from triaxial compression experiments. Failure is assumed to occur when the stresses reach the elastic limit that is usually set at the peak of the experimental stress-strain curve. The assumption of linear elastic behavior up to failure leads to underpredictions of the stresses required to fail a borehole, and to excessive stress concentrations at the borehole wall. These classical design procedures lead to an overestimation of the drilling fluid density. However, too high a mud density leads to other problems such as a reduction in the rate of penetration, and it may lead to the fracturing of upper sections with possible mud loss risks. Therefore, the range of mud-density within which a driller can operate is narrow and must be determined by non-conservative models.

Lower values of the minimum mud density required for wellbore stability are predicted by elastoplastic models. Most of the existing analytical work is based on elastic-perfectly plastic models and small strain analysis (e.g Wang and Dusseault 1991). Although the use of numerical techniques allows more realistic constitutive laws to be implemented (Papanastasiou 1990), analytical or semi-analytical solutions are still valuable since they can provide quick solutions in real time.

Axially symmetric plane strain deformation patterns for radially stressed hollow cylinders, of arbitrary thickness, are examined within the framework of large strain plasticity. Invariant, non-associated deformation-type theories are formulated for the pressure sensitive Mohr-Coulomb and Drucker-Prager models with arbitrary strain hardening and accounting for an initial hydrostatic state of stress. The analysis centers on the effective stress as the independent variable with universal stress histories as a natural outcome. With the Mohr-Coulomb model, presented in section 2, we arrive at a single first order differential equation, while for the Drucker-Prager solid, presented in section 3, an algebraic constraint supplements the governing differential equation. The importance of this work lies in the indroduction of strain-hardening into a complete large strain analysis. Similar work is presented by Durban and Fleck (1994) for the problem of a spherical cavity expansion in a Drucker-Prager solid.

Triaxial tests on Castlegate Sandstone have been used to calibrate the constitutive parameters of the two models in section 4. We present the comparison between predictions obtained from the two models and experimental data from hollow cylinder tests in section 5 and finally outline the main conclusions.

2 MOHR-COULOMB MODEL

Consider a cylindrical cavity embedded in a pressure-sensitive elastoplastic medium and subjected to monotonically increasing external pressure p. Assuming a plane-strain axially-symmetric deformation pattern we may expect the stress components to satisfy (with standard notation),

$$\sigma_\theta < \sigma_z < \sigma_r \tag{1}$$

The plastic potential ϕ and effective stress σ_e for the Mohr-Coulomb constitutive model are given by

$$\phi = \eta\, \sigma_r - \sigma_\theta \qquad\qquad \sigma_e = \mu\, \sigma_r - \sigma_\theta \tag{2}$$

where (η, μ) are material parameters. Associative behaviour is obtained when $\eta = \mu$ and the Tresca solid is recovered with $\eta = \mu = 1$.
Normality and the principle of plastic power equivalence

$$\sigma_r \dot{\epsilon}_r^p + \sigma_\theta \dot{\epsilon}_\theta^p = \sigma_e \dot{\epsilon}^p \tag{3}$$

dictate the plastic strain rates

$$\dot{\epsilon}_r^p = \eta \frac{\sigma_e}{\phi} \dot{\epsilon}^p \qquad\qquad \dot{\epsilon}_\theta^p = -\frac{\sigma_e}{\phi} \dot{\epsilon}^p \qquad\qquad \dot{\epsilon}_z^p = 0 \tag{4}$$

The total plastic strain ϵ^p is a function of the effective stress known from experiment σ_e.

Assuming that the elastic branch of the constitutive relations is Hookean, we arrive at the plane strain ($\dot{\epsilon_z} = 0 \quad \dot{\sigma_z} = \nu(\dot{\sigma_r} + \dot{\sigma_\theta})$) rate formulation

$$\dot{\epsilon_r} = \frac{1}{E}[(1-\nu^2)\dot{\sigma_r} - \nu(1+\nu)\dot{\sigma_\theta}] + \eta \frac{\sigma_e}{\phi} \dot{\epsilon^p} \tag{5}$$

$$\dot{\epsilon_\theta} = \frac{1}{E}[-\nu(1+\nu)\dot{\sigma_r} + (1-\nu^2)\dot{\sigma_\theta}] - \frac{\sigma_e}{\phi} \dot{\epsilon^p} \tag{6}$$

where (ν, E) are the usual elastic constants.

For the associated Mohr-Coulomb model, with $\sigma_e = \phi$, relations (5) and (6) are holonomic and can be exactly integrated along the loading history. Since we expect a nearly proportional path in stress space, we proceed with a deformation-type theory in which (5), (6) are replaced by the finite strain versions

$$\epsilon_r = [(1-\nu^2)\Sigma_r - \nu(1+\nu)\Sigma_\theta] + \eta \frac{\Sigma}{\Phi} \epsilon^p + (1+\nu)(1-2\nu)P_0 \tag{7}$$

$$\epsilon_\theta = [-\nu(1+\nu)\Sigma_r + (1-\nu^2)\Sigma_\theta] - \frac{\Sigma}{\Phi} \epsilon^p + (1+\nu)(1-2\nu)P_0 \tag{8}$$

Here we have introduced the non-dimensionalized notation

$$\Sigma_r = \frac{\sigma_r}{E} \qquad \Sigma_\theta = \frac{\sigma_\theta}{E} \qquad \Sigma_z = \frac{\sigma_z}{E} \qquad \Sigma = \frac{\sigma_e}{E} \qquad \Phi = \frac{\phi}{E} \qquad P_0 = \frac{p_0}{E} \tag{9}$$

and we account for the background hydrostatic stress ($\sigma_r = \sigma_\theta = \sigma_z = -p_0$). Notice that while no definite yield point has been introduced so far, the plastic branch is activated only beyond yield. Thus $\epsilon_p = 0$ for $\sigma_e < \sigma_Y$ though the yield stress σ_Y can be taken as zero.

Denoting by r the original undeformed radius and by u the radial displacement, we have the finite strain components

$$\epsilon_r = \ln(1 + \frac{du}{dr}) \qquad \epsilon_\theta = \ln(1 + \frac{u}{r}) \tag{10}$$

which may be combined to produce the compatibility relation

$$d\epsilon_\theta = (e^{\epsilon_r - \epsilon_\theta} - 1)\frac{dr}{r} \tag{11}$$

Radial equilibrium requires that

$$\frac{d\sigma_r}{d(r+u)} + \frac{\sigma_r - \sigma_\theta}{r+u} = 0 \tag{12}$$

or with the aid of (9) and (10),

$$d\Sigma_r = (\Sigma_\theta - \Sigma_r)e^{\epsilon_r - \epsilon_\theta}\frac{dr}{r} \tag{13}$$

Eliminating the radial coordinate between (11) and (13), and taking the dimensionless effective stress Σ as the independent variable, results in

$$(1 - e^{\epsilon_\theta - \epsilon_r})\Sigma_r' - (\Sigma_\theta - \Sigma_r)\epsilon_\theta' = 0 \tag{14}$$

where the prime denotes differentiation with respect to Σ. Now, equation (14) can be reduced to a single O.D.E. for Σ_r in terms of Σ. From (2)

$$\Sigma_\theta - \Sigma_r = (\mu - 1)\Sigma_r - \Sigma \qquad \Phi = \Sigma + (\eta - \mu)\Sigma_r \tag{15}$$

It follows, from (7) and (8), that

$$\epsilon_\theta - \epsilon_r = (1+\nu)[(\mu - 1)\Sigma_r - \Sigma] - (1+\eta)\frac{\Sigma}{\Phi}\epsilon^p \tag{16}$$

$$\epsilon_\theta = (1+\nu)[-(1-\nu)\Sigma + (\mu(1-\nu) - \nu)\Sigma_r] - \frac{\Sigma}{\Phi}\epsilon^p + (1+\nu)(1-2\nu)P_0 \tag{17}$$

are known functions of (Σ_r, Σ).
Inserting the derivative ϵ_θ' from (17) in (14), along with (15) and (16), we arrive at the governing equation

$$\Sigma_r' = -\frac{[\Sigma + (1-\mu)\Sigma_r][\alpha_1 + \frac{\Sigma}{\Phi}\epsilon^{p'} - \beta_1\frac{\Sigma_r}{\Phi^2}\epsilon^p]}{\exp\{-\alpha_2[\Sigma + (1-\mu)\Sigma_r] - (1+\eta)\frac{\Sigma}{\Phi}\epsilon^p\} - 1 - [\Sigma + (1-\mu)\Sigma_r][\alpha_3 + \beta_1\frac{\Sigma}{\Phi^2}\epsilon^p]} \tag{18}$$

where $\alpha_1 = 1 - \nu^2 \qquad \alpha_2 = 1 + \nu \qquad \alpha_3 = (1+\nu)[\mu(1-\nu) - \nu] \qquad \beta_1 = \eta - \mu$

A second equation is obtained from (18) and (13), using (15) and (16), viz

$$\ln r' = -\frac{\exp\{-\alpha_2[\Sigma + (1-\mu)\Sigma_r] - (1+\eta)\frac{\Sigma}{\Phi}\epsilon^p\}}{[\Sigma + (1-\mu)\Sigma_r]}\Sigma_r' \tag{19}$$

The governing equations (18) and (19) are supplemented with the boundary data

$$\Sigma_r = -P_0 \quad at \quad r = a, \quad \Sigma_r = -P \quad at \quad r = b \tag{20}$$

for increasing external pressure ($P > P_0$), or

$$\Sigma_r = -P \quad at \quad r = a, \quad \Sigma_r = -P_0 \quad at \quad r = b \tag{21}$$

for decreasing internal pressure ($P < P_0$), where (a, b) are the inner and outer undeformed radii, respectively. The background hydrostatic pressure P_0 is kept constant in both cases. Available numerical procedures can now be employed to solve the governing system of equations. The independent variable here is the effective stress Σ and a standard integration routine would begin, for example in the case of increasing external pressure, the integration from the external boundary where the values of (Σ_r, Σ) are known. The integration will stop at the point where the radial stress equals the internal pressure $\Sigma_r = -P_0$. Repeated integrations with increasing values of P will determine the entire loading history. The circumferential stress component is then obtained from (2) while the spatial stress profiles, with respect to the deformed radial coordinate $(R = r + u)$, are provided using (17) and the second of (10) to give

$$\ln \frac{R}{r} = -\alpha_1 \Sigma + \alpha_3 \Sigma_r - \frac{\Sigma}{\Phi} \epsilon^p + (1 + \nu)(1 - 2\nu) P_0 \tag{22}$$

3 DRUCKER-PRAGER MODEL

The plastic potential ϕ and effective stress σ_e are here defined by

$$\phi = (\frac{3}{2}\, \tilde{s}\tilde{s})^{1/2} + \frac{1}{3}\eta(\tilde{\sigma}\tilde{I}) \qquad \sigma_e = (\frac{3}{2}\, \tilde{s}\tilde{s})^{1/2} + \frac{1}{3}\mu(\tilde{\sigma}\tilde{I}) \tag{23}$$

where $\tilde{\sigma}$ is the Cauchy stress tensor, $\tilde{s}$ the stress deviator, (η, μ) are material parameters, defined differently from those in the Mohr-Coulomb model.

Following a procedure similar to that for the Mohr-Coulomb model, the deformation theory relations for the Drucker-Prager model are

$$\epsilon_r = \Sigma_r - \nu(\Sigma_\theta + \Sigma_z) + \frac{\Sigma}{\Phi}\epsilon^p \left[\frac{3(\Sigma_r - \Sigma_h)/2}{\Sigma - \mu \Sigma_h} + \frac{1}{3}\eta \right] + (1 - 2\nu) P_0 \tag{24}$$

$$\epsilon_\theta = \Sigma_\theta - \nu(\Sigma_z + \Sigma_r) + \frac{\Sigma}{\Phi}\epsilon^p \left[\frac{3(\Sigma_\theta - \Sigma_h)/2}{\Sigma - \mu \Sigma_h} + \frac{1}{3}\eta \right] + (1 - 2\nu) P_0 \tag{25}$$

$$\epsilon_z = \Sigma_z - \nu(\Sigma_r + \Sigma_\theta) + \frac{\Sigma}{\Phi}\epsilon^p \left[\frac{3(\Sigma_z - \Sigma_h)/2}{\Sigma - \mu \Sigma_h} + \frac{1}{3}\eta \right] + (1 - 2\nu) P_0 \tag{26}$$

where $\Sigma_h = (\tilde{\Sigma}\tilde{I})/3$ is the dimensionless hydrostatic stress. The dimensionless deviator $\tilde{S}$ obeys

$$(\frac{3}{2}\, \tilde{S}\tilde{S})^{1/2} = \Sigma_e - \mu \Sigma_h = \sqrt{\frac{3}{2}} [(\Sigma_r - \Sigma_h)^2 + (\Sigma_\theta - \Sigma_h)^2 + (\Sigma_z - \Sigma_h)^2]^{1/2} \tag{27}$$

In plane strain deformation $\epsilon_z = 0$ and (26) can then be solved for Σ_z in terms of Σ and Σ_h. To this end we employ the simple identity $\Sigma_r + \Sigma_\theta = 3\Sigma_h - \Sigma_z$ and insert it in (26) to obtain

$$\Sigma_z = \frac{[6\nu\Phi\Sigma_h - 2\eta\Sigma\epsilon_p/3 - 2(1-2\nu)\Phi P_0](\Sigma - \mu\Sigma_h) + 3\Sigma_h\Sigma\epsilon_p}{2(1+\nu)\Phi(\Sigma - \mu\Sigma_h) + 3\Sigma\epsilon_p} \tag{28}$$

where from (23) and (27) $\Phi = \Sigma + (\eta - \mu)\Sigma_h$

Next, we solve for $\Sigma_r + \Sigma_\theta$

$$\Sigma_r + \Sigma_\theta = \frac{[6\Phi\Sigma_h + 2\eta\Sigma\epsilon_p/3 + 2(1-2\nu)\Phi P_0](\Sigma - \mu\Sigma_h) + 6\Sigma_h\Sigma\epsilon_p}{2(1+\nu)\Phi(\Sigma - \mu\Sigma_h) + 3\Sigma\epsilon_p} \tag{29}$$

Equation (27) can be written as

$$(\Sigma_\theta - \Sigma_r)^2 = 4(\Sigma - \mu\Sigma_h)^2/3 - 3(\Sigma_z - \Sigma_h)^2 \tag{30}$$

Combining this with (28) gives the stress difference $(\Sigma_\theta - \Sigma_r)$ in terms of (Σ, Σ_h), namely

$$\Sigma_\theta - \Sigma_r = -(\Sigma - \mu\Sigma_h)\left[\frac{4}{3} - 3(\frac{2(1-2\nu)\Phi(\Sigma_h + P_0) + 2\eta\Sigma\epsilon_p/3}{2(1+\nu)\Phi(\Sigma - \mu\Sigma_h) + 3\Sigma\epsilon_p})^2\right]^{1/2} \tag{31}$$

Here we have assumed that $\Sigma_r - \Sigma_\theta > 0$ which is appropriate for externally pressurized cavities. The upshot of the analysis so far is that all three stress components are expressible as algebraic functions of the effective stress Σ and the hydrostatic stress Σ_h. The axial component is given by (28) while Σ_r and Σ_θ follow from (29) and (31).

Turning now to equilibrium (13) and compatibility (11) requirements we have from (24) and (25),

$$\epsilon_r - \epsilon_\theta = -[(1+\nu) + \frac{3\Sigma\epsilon_p}{2\Phi(\Sigma - \mu\Sigma_h)}](\Sigma_\theta - \Sigma_r) \tag{32}$$

which is an explicit function of (Σ, Σ_h) by (31). Similarly, equation (25) can be written again as an explicit function of (Σ, Σ_h),

$$\epsilon_\theta = [(1+\nu) + \frac{3\Sigma\epsilon_p}{2\Phi(\Sigma - \mu\Sigma_h)}](\Sigma_\theta - \Sigma_h) + \eta\Sigma\epsilon_p/(3\Phi) + (1-2\nu)(\Sigma_h + P_0) \tag{33}$$

In order to facilitate the integration of (14) we rewrite the $(\Sigma_\theta - \Sigma_h)$ term in (33) as this avoids later differentiation of (31)

$$\Sigma_\theta - \Sigma_h = L_1 - \Sigma_r \tag{34}$$

Here $L_1 = (\Sigma_r + \Sigma_\theta) - \Sigma_h$ is a known function of (Σ, Σ_h) by (29). In the differential form ϵ_θ is written

$$d\epsilon_\theta = \frac{\partial\epsilon_\theta}{\partial\Sigma}d\Sigma + \frac{\partial\epsilon_\theta}{\partial\Sigma_h}d\Sigma_h + \frac{\partial\epsilon_\theta}{\partial\Sigma_r}d\Sigma_r \tag{35}$$

Substituting this in (14) gives the governing equation

$$\left[1 - e^{\epsilon_\theta - \epsilon_r} - (\Sigma_\theta - \Sigma_r)\frac{\partial\epsilon_\theta}{\partial\Sigma_r}\right]\frac{d\Sigma_r}{d\Sigma} - (\Sigma_\theta - \Sigma_r)\frac{\partial\epsilon_\theta}{\partial\Sigma_h}\frac{d\Sigma_h}{d\Sigma} = (\Sigma_\theta - \Sigma_r)\frac{\partial\epsilon_\theta}{\partial\Sigma} \tag{36}$$

This equation can be solved in conjuction with the expression for Σ_r obtained from (29) and (31). The partial derivatives of ϵ_θ can be evaluated without difficulty from (33) for any hardening characteristic.

To further simplify the analysis let us introduce the notation

$$L_2 = \Sigma_r + \Sigma_\theta \qquad L_3 = \Sigma_\theta - \Sigma_r \qquad \Sigma_r = (L_2 - L_3)/2 \tag{37}$$

Along the same lines write (32) as

$$\epsilon_\theta - \epsilon_r = \left[(1+\nu) + \frac{3\Sigma\epsilon_p}{2\Phi(\Sigma - \mu\Sigma_h)}\right] L_3 = L_4 \tag{38}$$

which defines L_4. The governing equation (36) is now

$$\left[1 - e^{L_4} - L_3\frac{\partial\epsilon_\theta}{\partial\Sigma_r}\right]\frac{d\Sigma_r}{d\Sigma} - L_3\frac{\partial\epsilon_\theta}{\partial\Sigma_h}\frac{d\Sigma_h}{d\Sigma} = L_3\frac{\partial\epsilon_\theta}{\partial\Sigma} \tag{39}$$

with Σ_r given in (37).

Once the variation of Σ_r and Σ_h has been determined we can find the radial profiles by integrating (13) in the form

$$\frac{dr}{r} = \frac{e^{L_4}}{L_3}d\Sigma_r \tag{40}$$

The stress components $(\Sigma_\theta, \Sigma_z)$ are easily evaluated from (28) and (29) and the radial displacement is found from (33).

4 MATERIAL PARAMETERS

Material parameters were determined for Castlegate sandstone (a weak sandstone) using triaxial compression tests. It was found that the elastic constants are $E = 8.1$ GPa and $\nu = 0.35$. The angle of internal friction was found to be $\phi_i = 32.6^o$ and an associated flow-rule with the dilation angle $\psi = \phi_i$, was assumed because the rock showed pronounced dilatancy. These values give the following parameters for the Mohr-Coulomb model

$$\mu = \eta = \frac{1+\sin\phi_i}{1-\sin\phi_i} = 3.329 \tag{41}$$

Because of diffuse and localized bifurcations, standard laboratory tests have the tendency to yield apparent values for the material softening which are unrealistic and sometimes not even physically meaningful. Therefore, because of the absence of reliable data for true material softening, the calibration of effective stress σ_e as a function of plastic strain strain, ϵ^p, was based on interpolation of the more reliable pre-peak data (Figure 1). The fitted curve is given by the power law function

$$\epsilon^p = B(\sigma_e - \sigma_Y)^A \tag{42}$$

where for the Mohr-Coulomb model, $\sigma_Y = 25$ MPa is a conventional threshold value of the effective stress defining the state of initial yield; $A = 3.547$ and $B = 9.58927 \;\; 10^{-8}$.

The corresponding parameters for the Drucker-Prager outer bounding cone are:

$$\mu = \eta = \frac{6\sin\phi_i}{3-\sin\phi_i} = 1.311 \tag{43}$$

$\sigma_Y = 14$ MPa, $A = 3.547$ and $B = 1.307 \;\; 10^{-6}$.

For the Drucker-Prager inner bounding cone are:

$$\mu = \eta = \frac{6\sin\phi_i}{3+\sin\phi_i} = 0.912 \tag{44}$$

$\sigma_Y = 9.8$ MPa, $A = 3.547$, and $B = 6.79568 \;\; 10^{-6}$.

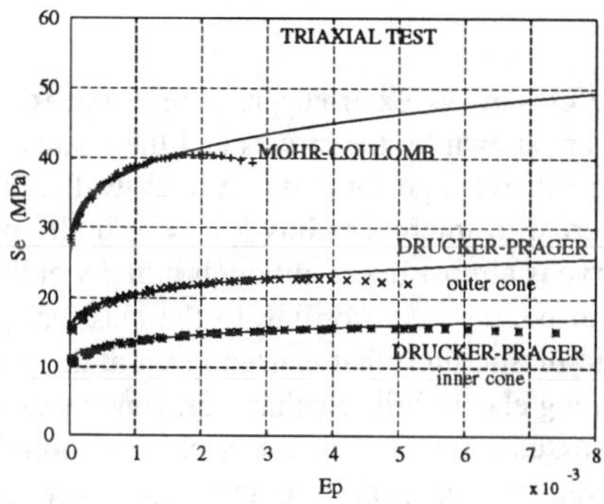

Figure 1: Calibration of effective stress. Points are experimental data for test at confining pressure 10 MPa. Continuous lines are the fitted models from eqn. (42)

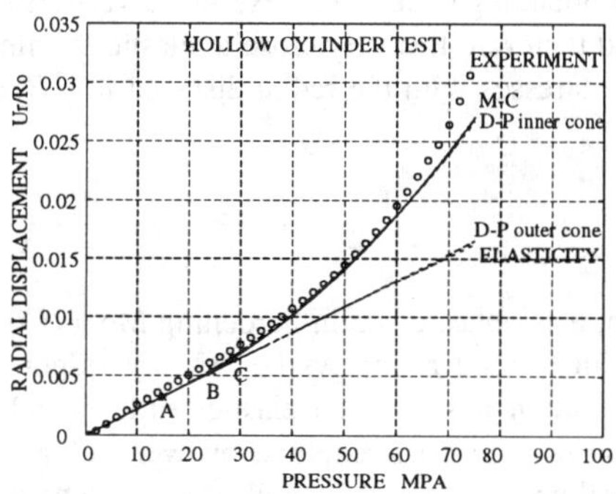

Figure 2: Hole closure vs external pressure (internal pressure p_0 = 2 MPa)

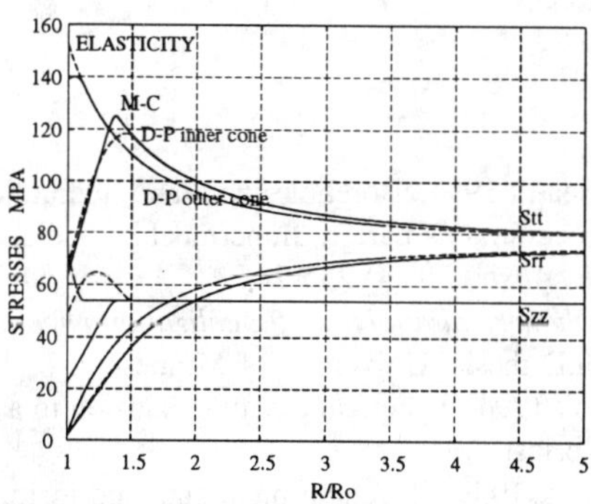

Figure 3: Radial variation of stresses (internal pressure p_0 = 2 MPa, external pressure p=75 MPa)

5 RESULTS

Figure 2 shows the hole closure vs external pressure curve for a thick-wall cylinder. The experimental points are shown by the circles and the results of the different models by the lines. The experiments were performed in a thick-wall cylinder of internal radius $a = 0.5$ cm, ratio of external to internal radius $b/a = 5$, under plane strain conditions. The initial part of the curve is almost linear although initial yielding at the borehole wall takes place quite early (at point A). According to the classical procedure, if failure is associated with the elastic-plastic limit that is usually set at the peak of the experimental stress-strain curve, assuming elastic-brittle behaviour, rupture should take place at point B. Another option is to assume that failure takes place at point C where the element near the borehole wall reaches the peak strength. However, the thick wall-cylinder reaches a state of higher deformation under increasing external pressure without the appearance of any signs of failure. It is obvious that the Mohr-Coulomb (M-C) and Drucker-Prager (D-P) inner bounding cone predict results close to the experiment. The slight deviation from the experimental data at higher load is due to the fact that the fitted curve on the triaxial experimental data shows continuously hardening behaviour. The Drucker-Prager outer bounding cone model exhibits a very stiff behaviour resulting to small scale yielding and therefore lies very close to elastic solution. Finally, Figure 3 shows the variation of the stresses with the radial distance $\rho = R/a$ from the borehole axis.

6 CONCLUSIONS

The presented large-strain, elastoplastic, strain-hardening Mohr- Coulomb and Drucker-Prager models predict with a good accuracy the primary response of the thick wall cylinder experiment. Prediction based on an elastic/brittle model leads to significant underestimation of the hollow-cylinder strength. However, before a final assessment is made of the failure of hollow cylinder and stability of wellbores the presented work must be supplemented with failure criteria which based on the prediction of localization of deformation (Papanastasiou 1990), since it is now well-recognized that failure in geomaterials is accompanied by localization of deformation. It is also known that hollow cylinder strength is scale-dependent (Haimson and Herrick 1989), and this effect must be examined in future models.

REFERENCES

Wang, Y. and M.B. Dusseault 1991. Borehole yield and hydraulic fracture initiation in poorly consolidated rock strata - part 1. Impermeable media, *Int. J. Rock Mech. Min. Sci. & Geomech. Abstr.* 28 (4):235-246.

Papanastasiou, P. 1990. *Numerical analysis of localization phenomena with application in deep boreholes.* Ph.D thesis, University of Minnesota.

Durban, D. and N.A. Fleck 1994. Spherical cavity expansion in a Drucker-Prager solid. *J. Applied Mech.* (in print).

Haimson, B. and C. Herrick 1989. Borehole breakouts and insitu stress. *Proc. Dri lling Symp. at ETCE, ASME, Houston, TX* 22: 17-22.

Rock Mechanics, Nelson & Laubach (eds) © 1994 Balkema, Rotterdam, ISBN 90 5410 380 8

Modeling of episodic failure leading to borehole breakouts in Alabama limestone

Courtney G. Herrick & Bezalel C. Haimson
Geological Engineering Program, Department of Materials Science and Engineering, University of Wisconsin, Madison, Wis., USA

ABSTRACT: Based on our experimental observations, we developed a numerical model of borehole breakout growth in Alabama limestone. For given states of far-field stress breakout shape and dimensions were obtained by simulating the episodic nature of the spalling process and utilizing a rock strength-size effect determined for boreholes of differing radii. These geometric characteristics of the simulated breakouts correctly matched our experimentally produced ones. The results suggest a means to estimate one of the principal in situ horizontal stresses when the other principal stress is known and breakout dimensions are logged.

1 INTRODUCTION

Borehole breakouts are important in situ stress indicators. It has been demonstrated repeatedly that breakout orientation around a borehole is aligned with the least in situ horizontal stress σ_h (e.g., Bell and Babcock, 1986; Plumb and Cox, 1987). Less understood is the relationship of breakout characteristics, such as failure mechanism or breakout dimensions, to the magnitudes of the prevailing stresses. The advantages of using borehole breakouts to infer the in situ state of stress result from their widespread occurrence in deep holes, and our ability to detect their existence, dimensions, and orientation from borehole logging (e.g. four-arm dipmeter, sonic televiewer). In boreholes where hydraulic fracturing stress measurements may be difficult to conduct, especially at great depths and under high temperatures, breakouts may be the only source of in situ stress information.

Our research has been aimed primarily at assessing the potential for utilizing borehole breakout dimensions, the angular span θ_b and the maximum depth r_b, to estimate the principal horizontal stress magnitudes. In this paper we present a simple numerical model based on our experimental observations of breakout development in Alabama limestone. The objective of the numerical simulation is to establish a correct relationship between breakout dimensions and the far-field horizontal principal stresses. There are two working models (Zoback et al., 1985; and Zheng et al., 1989) which relate breakout dimensions to the in situ stress regime. For a given state of stress and rock strength characteristics, these models yield identical breakout widths but substantially underestimate and overestimate, respectively, the expected breakout depths. The numerical simulation presented here reaffirms the breakout widths obtained by the previous models, but significantly improves the accuracy of the predicted breakout depths.

2 REVIEW OF IMPORTANT EXPERIMENTAL RESULTS

We have conducted breakout initiation tests on prismatic specimens (13x13x17 cm^3) of Alabama limestone in which three unequal principal stresses were first applied, and then a central borehole (R = 1.1 cm) was drilled in a direction perpendicular to the plane containing σ_h and σ_H lines of application (Haimson et al., 1991). The experimental procedure simulated more closely real field conditions than previous tests in which loading was applied after the hole had already been drilled. The vertical stress was applied by a hydraulic ram; the horizontal stresses were provided through a polyaxial cell (Haimson and Herrick, 1989). A diamond coring bit was driven through the entire specimen. Water was used as the coolant under just enough flow to insure removal of the rock debris. Drill advancement was accomplished by means of rack-and-pinion gears in an attempt to produce minimal bit pressure.

Four series of tests were carried out for preset magnitudes of σ_h (14, 21, 28, and 35 MPa) and σ_v (= σ_h + 7 MPa). Within each series, the magnitude of σ_H was varied from test to test so as to cover the entire range of stresses leading to borehole breakouts. Following are the major experimental results which were subsequently used in the development of the numerical model relating breakout dimensions to far-field principal stresses:

- Triaxial tests revealed that Alabama limestone follows closely the linear Mohr-Coulomb criterion, with an angle of internal friction $\phi = 18°(\pm 2°)$, and a cohesion of 15.5 (±1) MPa.
- Analysis of the tangent modulus of elasticity at 50% of the peak axial stress in triaxial tests shows that this material property is not pressure-dependent in Alabama limestone. The modulus is E = 28 GPa. The Poisson's ratio is $\nu = 0.27$.
- Breakouts in Alabama limestone always initiate at the borehole wall along the σ_h-springline.
- Photomicrographs of thin sections reveal the mechanism of breakout formation. It commences with extensile microfracturing behind the borehole wall along and around the σ_h-springline. Some microcracks extend into one another and create large extensile fractures subparallel to the borehole wall. Others that form away from the σ_h-springline coalesce in an en echelon pattern creating shear fractures that delineate the ends of the slab. With increased stress, the fractures open and the slab is detached from the borehole wall (Figure 1).
- Breakouts in Alabama limestone form as a result of a succession of spalling events. The rock slabs detached during each event are roughly crescent-shaped. The span subtended by the initial failure episode remains, for the most part, unchanged as spalling continues. Deepening of the breakout occurs by the detachment of progressively smaller slabs behind the preceding ones. Eventually, spalling ceases as the breakout attains a stable configuration. The entire spalling process appears to remain brittle as evidenced by a lack of plastic effects. Figure 2 shows a stage after

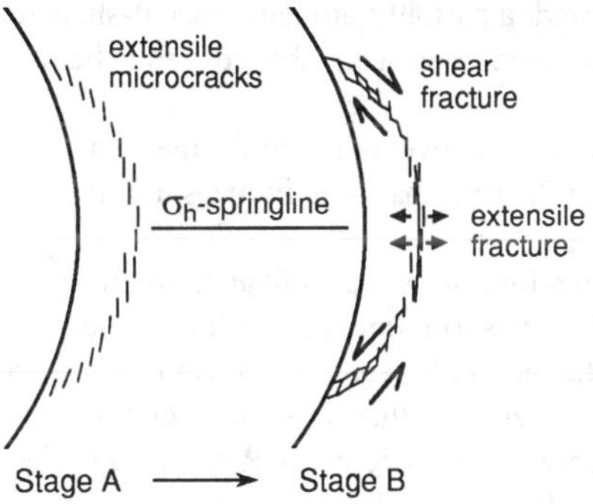

Figure 1. Schematic of extensile and shear mechanisms leading to spalling.

Figure 2. Thin section showing breakout process in Alabama limestone.

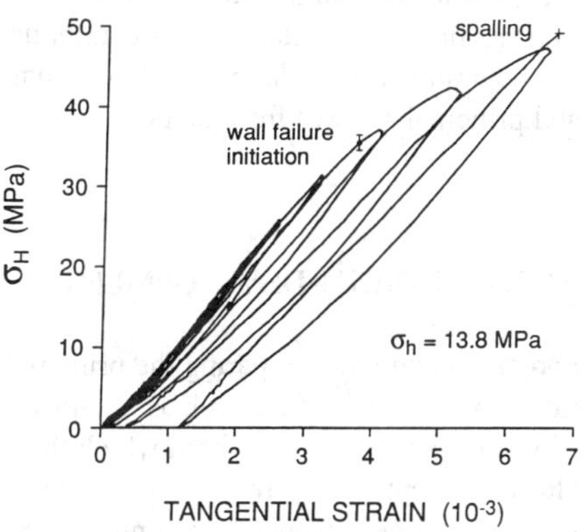

Figure 3. Borehole tangential strain behavior along the σ_h-springline as a function of σ_H.

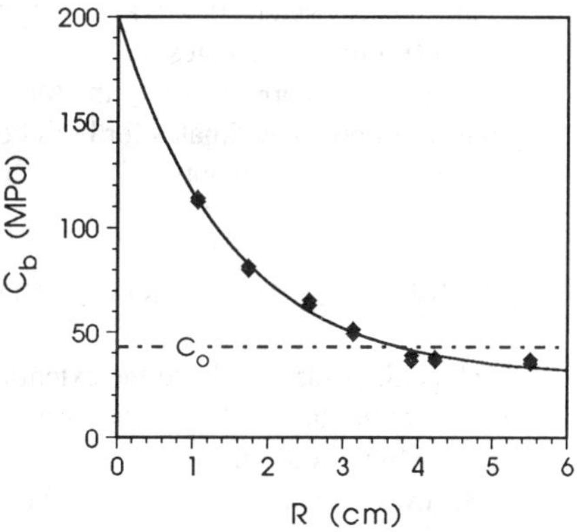

Figure 4. Variation of borehole compressive strength C_b with borehole radius R.

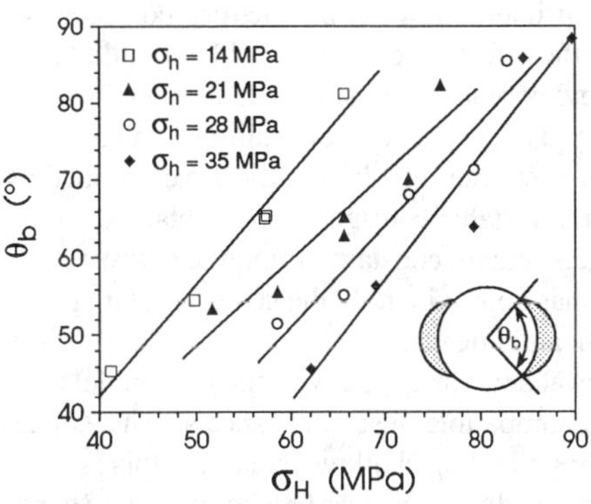

Figure 5. Experimental results of θ_b as a function of σ_H for different levels of σ_h.

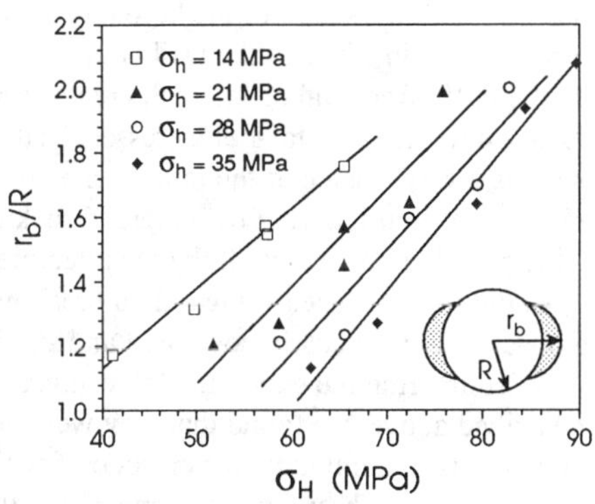

Figure 6. Experimental results of r_b/R as a function of σ_H for different levels of σ_h.

the first episode of spalling has already occurred: a partially broken crescent-shaped slab is practically detached from the wall, and another crescent-shaped slab is being formed behind it.

- For given σ_h and σ_v borehole failure begins at the σ_H level required for the first breakout episode to initiate. The formation of the first spall slab induces nonlinear stress-strain behavior (Figure 3).
- The borehole compressive strength of Alabama limestone C_b, equivalent to the tangential stress at the borehole wall (along the σ_h-springline) at which breakouts initiate, is not a constant. It is a function of the borehole radius as shown in Figure 4.
- Measurements of induced breakout dimensions revealed that for each level of σ_h, separate linear relationships could be fitted between both θ_b and r_b/R and the applied σ_H (Figures 5 and 6). These experimental results suggest that by mapping the dimensions of breakouts, one of the two horizontal principal stresses could be assessed if the other stress is inferred independently, say from hydrofracturing shut-in pressure results, or from other indicators.
- Plotting r_b/R versus θ_b it was noted that the experimental data lie within a narrow band (Figure 7a), suggesting that more than one pair of horizontal stresses can induce a breakout of approximately the same size. The implication is that it is unlikely that we could obtain estimates for both horizontal principal stresses from mapping the dimensions of breakouts.

3 NUMERICAL SIMULATION OF BOREHOLE BREAKOUT DEVELOPMENT

Zoback et al. (1985) modeled the extent of the breakout zone by comparing the principal stresses around the borehole to the triaxial compressive strength of the rock as obtained from the Mohr-Coulomb criterion. A contour line enclosing the zone within which the stresses are sufficient to bring about failure yields a crescent-shape breakout that is wide and relatively flat-bottomed. Zoback et al. consider only one episode of spalling, assuming that the initial breakout is final. Zheng et al. (1989) conducted a numerical simulation of the borehole breakout phenomenon in which the episodic nature of the spalling process was taken into account. When a set of far-field stresses create tangential stresses within a sector of the borehole wall sufficient to overcome the rock compressive strength, a thin sliver of the wall bounded by the affected sector is considered failed and spalled off. Removal of the spall creates a new borehole geometry, which is again exposed to the same far-field stresses, leading to a new episode of spalling. The next spall is always inside of the failed zone but is of shorter length. Iteration is continued by removal of thin flakes of overstressed rock until a stable configuration is obtained. Zheng et al. assume that all the rock parameters remain constant during the entire episodic failure. Their typical final breakout has approximately the same angular span as Zoback's, but is acutely dog-eared and significantly deeper.

In comparing the experimental results in Alabama limestone with the two models described above, we found that they were not compatible in several aspects. The Zoback et al. model is limited to one episode of spalling yielding shallow breakouts; this is contrary to our observations of episodic failure leading to deeper breakouts. The Zheng et al. model assumes that rock fails one thin sliver at a time and that the strength of the rock is constant, resulting in very deep breakouts; in our experiments Alabama limestone failed in thicker crescent-shaped chunks that left successively shorter borehole sectors of decreasing radii of curvature and enhanced strength, leading to less deep breakouts.

Thus, we undertook our own numerical modeling in which we attempted to replicate as closely as possible our experimental results listed in the previous section. We employed the plane stain, linear elastic boundary element method (Crouch and Starfield, 1983) to model a borehole drilled vertically through rock under a constant in situ stress state. In this model the numerically obtained stresses around the borehole are compared to the Mohr-Coulomb strength in order to determine the extent of a breakout zone. The borehole compressive strength is selected from our experimentally determined relationship between borehole-wall failure initiation and radius of curvature (Figure 4). For example, for the initial breakout the radius of curvature is 1.1 cm, and the borehole compressive strength used in the Mohr-Coulomb criterion is 116 MPa. The angle of friction, the modulus of elasticity and the Poisson's ratio are those given above. Linear interpolation between boundary elements and field points is used to determine the initial spalling extent given by the breakout width θ_{b1} and maximum depth r_{b1}. A circular arc, as an approximation to the actual shape of the failure zone, is fit through the coordinates of θ_{b1} and r_{b1} delineating the initial episode of spalling.

The borehole is redrawn to remove the initial spall (Figure 8b) and the numerical analysis is again performed using the new geometry. The stresses behind the first failure zone obtained from this run are again compared to the Mohr-Coulomb criterion. However, the criterion is now modified to account for the rise in borehole compressive strength resulting from the decrease in radius of curvature (Figure 4). A new episode of spalling occurs within the existing failed arc, producing a new depth and width. Iteration is continued until a stabilized breakout is obtained (compare Figure 8a with Figure 8b). Each episode covers a progressively smaller area that lengthens the breakout by a decreasing amount relative to the previous spalling episode. A breakout is considered stable when the increase in depth with each iteration is less than 0.3 mm (1/3 to 1/2 grain diameter). This cutoff was chosen because it is the smallest distance observed between adjacent fractures, or between a free surface and a fracture, based on our thin-section analysis. The initial breakout width does not change throughout the simulation.

4 RESULTS AND DISCUSSION OF THE NUMERICAL SIMULATION

Using our numerical model, we attempted to simulate the experiments in Alabama limestone described above. Numerical simulations were conducted at four levels of σ_h (14, 21, 28, and 35 MPa). The maximum horizontal stresses selected were equally spaced between levels not leading to breakouts and levels resulting in breakouts of maximum size for the σ_h applied. The results of the numerical simulations for $\sigma_h = 21$ and 28 MPa in the form of θ_b and r_b/R as a function of σ_H are presented in Figure 9.

Generally, the higher the tangential stress at the borehole wall in the direction of σ_h, the larger the simulated breakout in terms of both angular span θ_b and normalized breakout depth r_b/R. The increase in both θ_b and r_b/R with σ_H for a given level of σ_h is approximately linear within the stress range tested, similar to the experimental results. Our model, as well as those by Zoback et al. and Zheng et al., yield the same θ_b since breakout spans are determined by the first episode of spalling and the same borehole

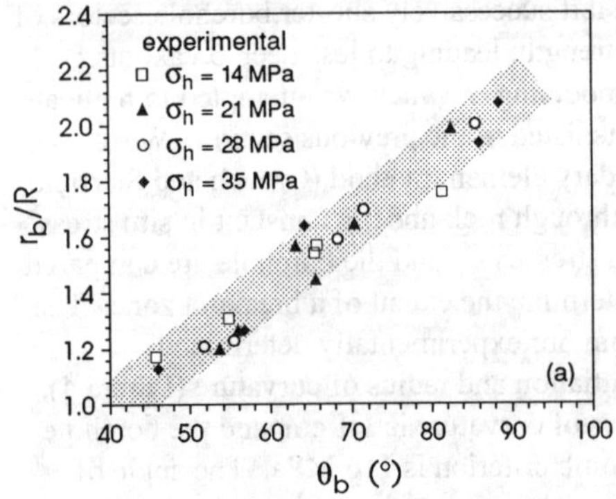

Figure 7. (a) Experimental results of r_b/R as a function of θ_b for different far-field stresses.

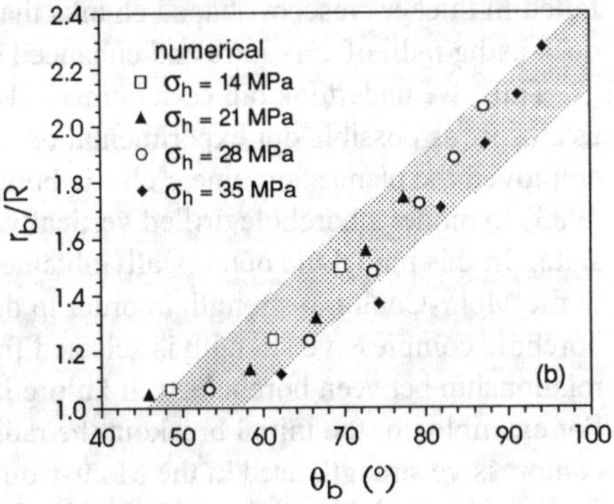

Figure 7. (b) Numerical results of r_b/R as a function of θ_b for different far-field stresses.

Figure 8. (a) Experimental breakout shape and size for σ_h = 28 MPa and σ_H = 83 MPa.

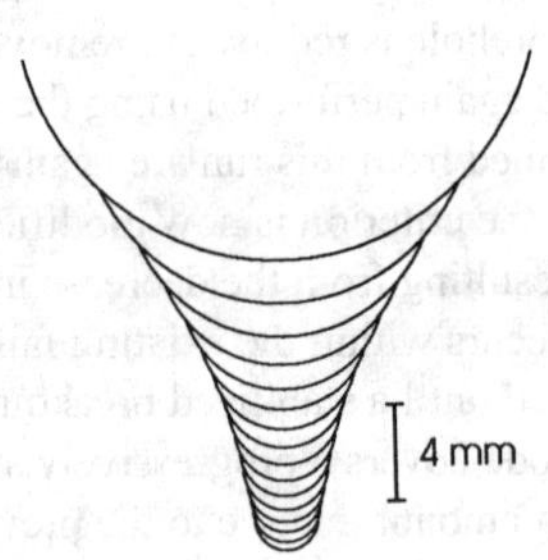

Figure 8. (b) Model breakout shape and size for σ_h = 28 MPa and σ_H = 83 MPa.

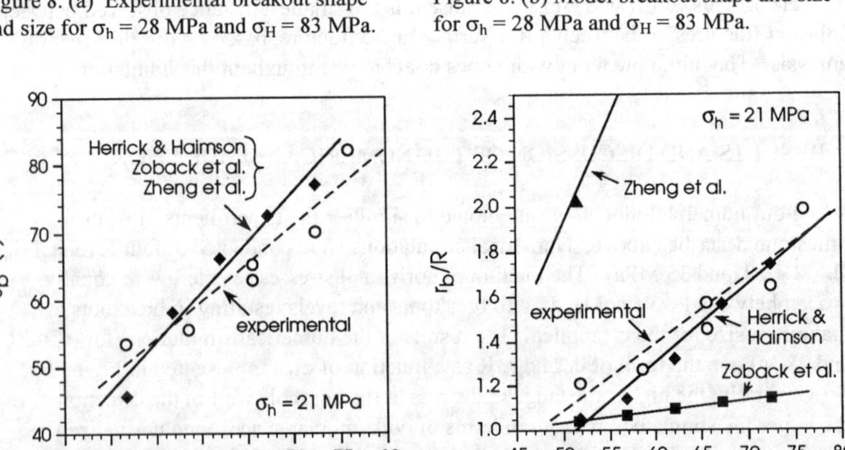

Figure 9. (a) Experimental breakout dimensions compared to those predicted by three models (σ_h = 21 MPa).

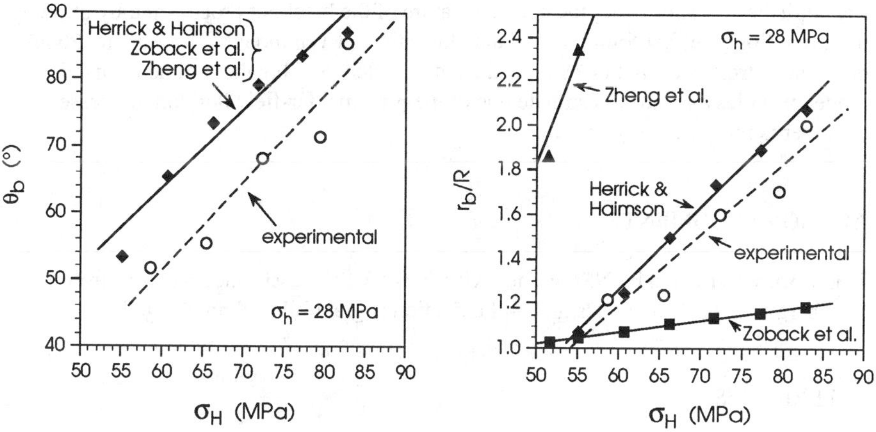

Figure 9. (b) Experimental breakout dimensions compared to those predicted by three models ($\sigma_h = 28$ MPa).

compressive strength. On the other hand, our numerical model, which is based on borehole compressive strength variation with radius, predicts normalized breakout depths r_b/R that are surprisingly close to the experimental results and yield significantly better fits than those obtained by either Zoback et al. or Zheng et al. (Figure 9).

Simulated breakouts resulting from our model are dog-eared and relatively deep, similar to those experimentally produced. A comparison between a modeled and an actual breakout for the same far-field stress state is shown in Figure 8. The similarity is quite remarkable.

We also plotted the modeled breakout depth versus breakout width for all the far-field stresses applied (Figure 7b). As in our experimental study, the results of which are juxtaposed in the figure, the numerically generated breakout dimensions lie within a narrow band. This suggests that different far-field stress conditions can result in breakouts of similar dimensions, and hence that knowledge of both angular span and maximum depth of a breakout is not sufficient to uniquely determine the two far-field principal stresses. The numerical results lie very close to the experimental ones in Figure 7, attesting to the accuracy of our simulation.

An important implication of the results shown in Figure 9 is that if one or both of the breakout dimensions and one horizontal principal stress are known (typically this will be σ_h), it should be possible to obtain an estimate of the other principal stress (σ_H).

5 CONCLUSIONS

A simple numerical method was used to model our laboratory breakout test results in which boreholes were drilled through blocks of Alabama limestone subjected to three unequal principal far-field stresses. The input parameters and material behavior attributes were based on experimental evidence. In general, the method yields realistic breakout shapes and depths. The improvement over previous breakout models was accomplished by simulating the episodic nature of the breakout process and by utilizing the strength-size effect found in Alabama limestone. Even though it appears that both horizontal stresses cannot be estimated from knowledge of breakout dimensions, the model provides a means to estimate one of the principal far-field horizontal stresses if the other is independently known.

ACKNOWLEDGMENTS

This work was funded by NSF grant EAR-9004688. The assistance received from R. Sandberg with equipment design and fabrication is gratefully acknowledged.

REFERENCES

Bell, J.S. and E.A. Babcock. 1986. The stress regime of the western Canadian Basin and implications for hydrocarbon production. *Bull. Can. Petrol. Geol.* 34: 364-378.

Crouch, S.L. and A.M. Starfield. 1983. *Boundary Element Methods in Solid Mechanics*. London: George Allen & Unwin.

Haimson, B.C. and C.G. Herrick. 1989. Borehole breakouts and in situ stress. In *Proc. Drilling Symposium 1989*, J.C. Rowley (ed.), Houston: 17-22. New York: ASME.

Haimson, B.C., M.Y. Lee, and C.G. Herrick. 1991. Recent advances in in-situ stress measurements by hydraulic fracturing and borehole breakouts. In *Proc. 7th Congr. Int. Soc. Rock Mech.*, W. Wittke (ed.), Aachen. vol. 3: 1737-1742. Rotterdam: Balkema.

Plumb, R.A. and J.W. Cox. 1987. Stress directions in eastern North America determined to 4.5 km from borehole elongation measurements. *J. Geophys. Res.* 92: 4805-4816.

Zheng, Z., J. Kemeny, and N.G.W. Cook. 1989. Analysis of wellbore breakouts. *J. Geophys. Res.* 94: 7171-7182.

Zoback, M.D., D. Moos, L. Mastin, and R.N. Anderson. 1985. Well-bore breakouts and in situ stress. *J. Geophys. Res.* 90: 5523-5530.

Rock Mechanics, Nelson & Laubach (eds) © 1994 Balkema, Rotterdam, ISBN 90 5410 380 8

Borehole breakout interpretation in the Gulf Coast, offshore Louisiana

A. Zerwer & N. A. Yassir
Department of Earth Sciences, University of Waterloo, Ont., Canada

ABSTRACT: A study of breakouts was carried out for 8 wells in the Viosca Knoll area of the Gulf Coast to determine the orientation of the maximum horizontal principal stress. Breakout interpretation in a region of low stress anisotropy and low sediment strength is difficult because of the weakness of the stress signature. However, it is shown that the use of visual analyses of the data using trajectory and polar plots highlights reliable data. A general NE-SW maximum horizontal principal stress orientation was obtained for the region.

1 INTRODUCTION

Borehole breakouts are wellbore elongations that form in response to unequal horizontal stresses. Breakouts, systematically studied since the late seventies (Bell and Gough 1979), have repeatedly proved to be an effective method of mapping horizontal stresses in sedimentary basins. Much of the data required for breakout analysis can be economically obtained from uncomputed four arm dipmeter logs which are routinely run by petroleum companies in most wells.

Breakout formation depends on the magnitude of stress anisotropy and on material strength. The Gulf Coast is a tectonically passive, deltaic basin containing thick sequences of overpressured shales and sands which are frequently soft with little strength. Consequently, it is not obvious whether there is sufficient horizontal stress anisotropy to cause borehole ovalisation and, if so, what the strength of the stress signature is. Examination of numerous dipmeter logs from wells located offshore from Louisiana indicates that borehole ovalisation does, indeed, occur in Gulf Coast sediments.

The purpose of this paper is to address some of the inherent problems encountered when defining breakouts in this type of environment and to illustrate that breakout analysis can be used to resolve principal horizontal stress orientations for the Gulf Coast region. An alternative methodology is introduced taking into account ovalisations that may not be categorized as "classic" breakouts but which may, nevertheless, give a true indication of the stress field. Borehole breakouts from 8 wells (Figure 1) were examined using four arm dipmeter logs. These wells were located in the Viosca Knoll region in an area near the Mississippi - Alabama shelf along the shelf break. This is part of a larger study conducted at the University of Waterloo on mapping of the horizontal stress regime in the Gulf Coast region.

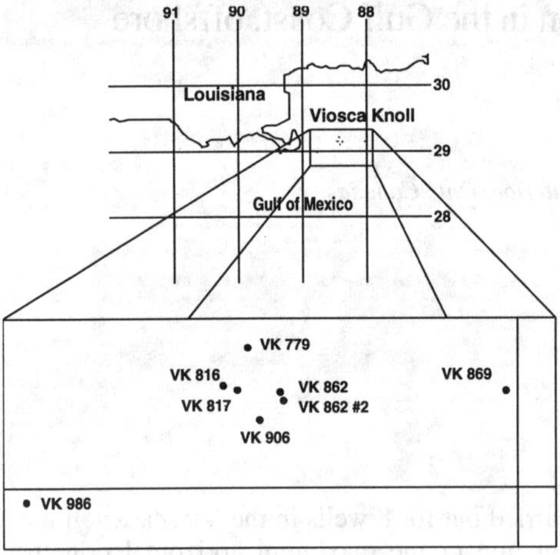

Figure 1. Location map of Viosca Knoll

2 BREAKOUT IDENTIFICATION

The dipmeter tool rotates while moving up the wellbore measuring the resisitivity with four spring-loaded pads. When the tool encounters an elongation in the wellbore, the caliper pairs are differentially extended and, at the same time, the tool becomes "stuck", so that a constant azimuth is recorded. This orientation is assumed to be perpendicular to the maximum principal stress. A number of criteria were established in order to distinguish breakouts from drilling-induced or other unreliable ovalisations (Plumb and Hickman, 1985; Springer, 1987; Bell and Caillet,1992), including,

i) The dipmeter log must show tool rotation above and below the breakout zone and must stop rotating within the breakout zone.
ii) The extended caliper pair must be larger than the hole diameter while the opposite pair remains close to the hole diameter.
iii) The extended caliper pair must show a rapid increase and decrease of caliper size within the breakout zone.
iv) The breakout orientation should not follow the azimuth of borehole deviation or be offset from it at a consistent angle.
v) The wellbore deviation must be within 5° from vertical.
vi) The minimum length is 30 cm.

Where the material is soft and/or the stress anisotropy is subtle, however, slightly different criteria from the above may have to be applied to obtain a stress signal. For example, if a shallow breakout has formed, it may not affect the tool rotation, yet still be a reliable indicator of stress orientation. This is the type of problem encountered in the Gulf Coast and a re-assessment of some of the above criteria was necessary to minimize the elimination of potentially useful data.

2.1 *Caliper Extensions*

For the "ideal" breakout, one pair of calipers should have an extension larger than bit size while the other pair should be close to bit size. Key-seating, hole wear caused by tool drag, normally shows a below-gauge reading on one caliper pair. In reality, this anomaly may not always be present and differentiating the "ideal" breakout from a key-seat can be difficult. Plumb and Hickman (1985), Springer (1987) and Bell and Caillet (1992) suggest that the larger diameter extension measured by the dipmeter should show an abrupt increase at the beginning of a breakout rather than a gradual increase. For example, a key-seat may develop in a small bend in the wellbore. When the dipmeter approaches it, the caliper extension will gradually increase to some maximum value then decrease as it passes the zone of maximum wear. As discussed below, this criterion is not necessary because other, more definitive, methods can be used to eliminate the key-seating effect. Finally, a washout is recognized by extension on both caliper pairs. Again, it is argued that the data should not be automatically discarded if the calipers are still differentially extended; this could indicate a breakout which has "washed out" (Dart and Zoback, 1989). Abrupt larger caliper extensions and in-gauge smaller extensions are therefore strong indicators of the "classical" breakout, but this may be inadequate in areas such as the Gulf of Mexico where stress indicators are more subtle due to soft sediments and possibly weak stress anisotropy.

2.2 *Tool Rotation*

Tool rotation need not cease in a breakout zone. Santarelli and Brown (1989) showed that breakout shape is highly dependent on rock type. Shallow, curved breakouts can form as easily as deep triangular "dog ears". Where sediments are soft, the breakout may not be well defined, in which case, tool rotation can be expected within the breakout zone. Even in zones where the "dog ear" shape is present, the dipmeter tool may not stop rotating as shown in Figure 2.

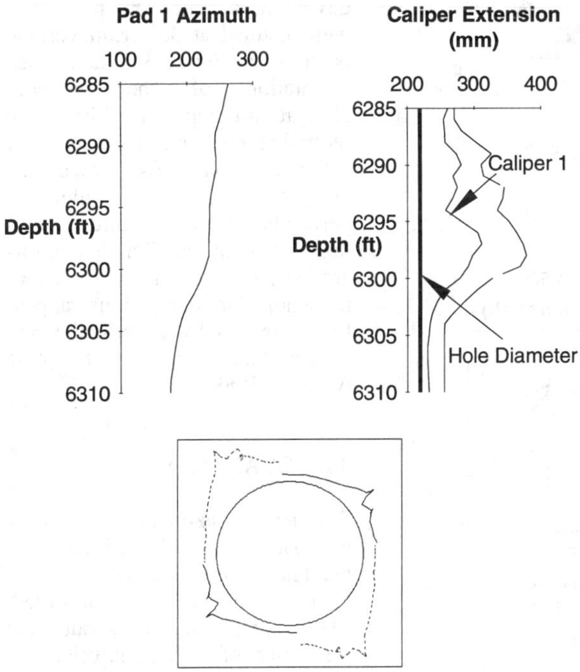

Figure 2. Polar plot from VK 906 #1 illustrating tool rotation throughout a breakout zone.

Polar plots, which were utilized by Brereton and Evans (1987), facilitate the estimation of breakout orientation as well as give their approximate shape, assuming that the tool is centered in the hole. Caliper extensions are plotted with respect to their azimuth, constantly monitored by pad one. Figure 2 shows that the dipmeter tool continues to rotate from 270 to 170°, but still defines an elongation which looks like a breakout.

2.3 *Wellbore Deviation*

Several important factors must be considered with regards to breakouts and wellbore deviation (Guenot, 1989; Yassir and Dusseault, 1991). Key-seating becomes an important effect because softer sediments are more susceptible to tool wear at lower borehole deviations. It is demonstrated here that, in an area like the Gulf Coast, deviations of <5° can still result in key-seating effects. Boreholes are usually not perfectly vertical but rather spiral downward. Plumb and Hickman (1985) regarded elongations which systematically coincided with the high side of the borehole as key-seating effects. Drill stem wear can also occur at a consistent angle to the high side of the hole. To minimize such an effect, Bell and Caillet (1992) rejected breakouts oriented within 30° of the hole azimuth.

Mastin (1988) further notes that the tectonic regime has a strong effect on the occurrence of breakouts, even in near-vertical wells.

To illustrate the effect of hole deviation, a trajectory plot of a well inclined at 38° from vertical is shown in Figure 3. The average orientation of the wellbore elongation is represented by a line plotted at the depth of the center of the elongation. As shown, the elongations are both parallel and perpendicular to the direction of hole deviation. Parallel elongations are key-seating effects whereas perpendicular elongations appear to be formed by a vertical stress concentration on the borehole (Guenot, 1989).

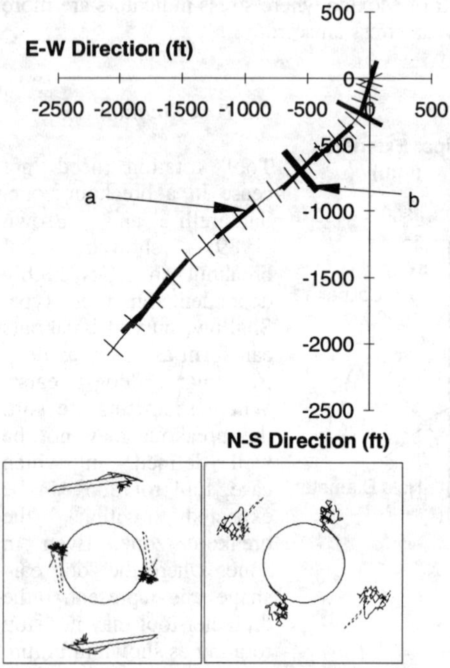

a b

Figure 3. Deviation effects observed in a well inclined at 38° from vertical (VK 862).

3 DATA ANALYSIS

The length, azimuth, amount of extension and hole azimuth are tabulated for each zone where caliper extension is observed. These are possible breakouts and the subject of further analysis.

3.1 *Plotting Procedures*

Two types of plot are used to initially examine the quality of the data, the trajectory and polar plots. As previously discussed, elongations which consistently plot parallel, or near-parallel to the direction of the wellbore on a trajectory plot are likely key-seating effects. Polar plots are examined for zones of caliper extension to study the shape of the elongation and, if possible, to determine the breakout orientation. Breakouts usually appear as rough shaped bulges extending past the borehole radius as shown in Figure 4. Key-seating effects, illustrated in Figure 3a, usually have smooth contours in the extended zone. After visual inspection of the two types of plot, the unreliable data are removed from the data set.

Figures 4 and 5 show orientations of ovalisation from three Viosca Knoll wells illustrating the variety of data encountered. The plots show all the data including key-seats. The breakout orientations shown in the trajectory plot of Figure 4 follow two directions: NE-SW and NW-SE (others are key-seats). Data from these illustrate the difficulty in pin-pointing the stress directions the methodology of which is described below.

The polar plots in Figure 4 illustrate a variety of wellbore elongations, some of which allow tool rotation (Figure 4a), and some which do not (Figures 4b and 4c). A potential washout zone (Figure 4d), is possibly a reliable indicator of stress orientation as it does not follow the well trajectory.

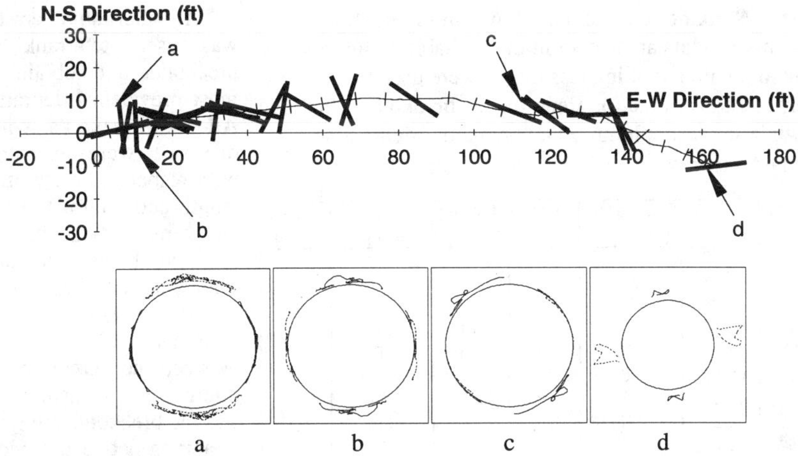

Figure 4. Trajectory and polar plots from VK 986. a. significant tool rotation: b. c. little tool rotation: d. washout zone.

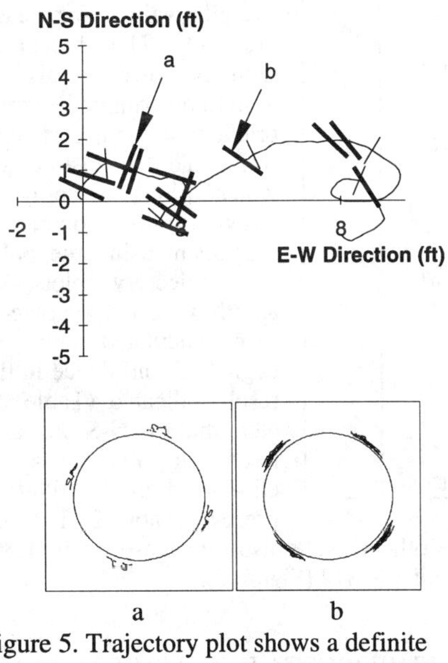

Figure 5. Trajectory plot shows a definite NW-SE trend in breakout orientation (VK 817).

The breakouts shown in Figure 5 have a consistent NW-SE direction with a few exceptions oriented 90° to the trend. The polar plot of the breakout perpendicular to the trend (Figure 5a) has the characteristic rounded shape previously observed, whereas the breakout in the second polar plot (Figure 5b) is not obvious and could possibly be confused with a key-seat. Both are probably reliable breakouts, one because of the distinctive nature of the polar plot (Figure 5a) and the other because of the unmistakable trend (Figure 5b). The origin of the NE-SW oriented breakouts is unknown, but may be fracture-related.

3.2 *Ranking Breakouts*

It is clear that the world stress map ranking of breakouts (Zoback and Zoback, 1991) is inappropriate for a region which seems to be more influenced by basin geometry than crustal movements.

The data show that: a. although some

wells had a reasonably strong stress signature (eg. Figure 5), others had a complex signature (eg. Figure 4), possibly dictated by lithological variations; b. the breakouts tended to be small in length and width. This implies that the stress anisotropy in the area is weak, but that it is possible to obtain a signal by careful handling of the data.

Table 1. Ranking of breakouts shown in Figure 4. Data without key-seats are shown in the 'Reliable' column. Only breakouts with a ranking less than 9 were used to determine the orientation trend for the breakouts which is shown in the last column. Final breakout orientation obtained was 114°.

Length (ft)	Width (inches)	Orientation (degrees)	Reliable	Ranking (sum)	Orientation (ranking<9)	Chosen Trend
71	0.90	77				
28	0.53	72				
12	0.85	174	✓	11	----	
51	0.69	9	✓	10	----	
146	1.04	178	✓	7	178	
15	0.68	76	✓	9	----	
18	0.55	18	✓	12	----	
199	1.29	116	✓	5	116	✓
26	1.09	114	✓	7	114	✓
38	0.72	106	✓	7	106	✓
11	1.23	73	✓	7	73	
50	1.40	68	✓	5	68	
22	0.65	67	✓	9	----	
20	0.55	64	✓	9	----	
19	0.41	7	✓	12	----	
9	0.67	101	✓	9	----	
9	1.05	109	✓	8	109	✓
22	1.80	113	✓	8	113	✓
60	1.24	106				
17	0.75	25	✓	11	----	
28	0.59	8	✓	11	----	
15	0.53	124				
24	0.46	159				
24	0.67	15				
30	0.71	129	✓	8	129	✓
34	1.23	114				
12	0.35	128				
13	0.47	134				
14	0.66	115				
38	0.80	89				
19	0.37	163				
15	1.00	156				
13	2.97	82	✓	6	82	

An alternative system was used to rank the breakouts and obtain the most prevalent orientation. All the breakouts within one well were normalised with respect to orientation, length and width. Initially, an average of the variable (eg. length) is calculated. The values are then divided by the average. Four intervals are calculated between the maximum and minimum measurements and the breakouts are given a ranking between 1 and 4, in decreasing order of reliability. Summing the number given by each variable results in an overall value for each breakout. The breakouts with the lowest totals are used to determine the stress orientation for a particular well. The decision on which filter to apply is based on the operator's judgement using the polar and trajectory plots. A length-weighted average is then calculated and the degree of confidence in the result indicated (Table 2). Note that the N-S breakout trend in Figure 4 is eliminated in the filtering process (Table 2) because these elongations are much smaller than the others. Note also that these ovalisations correspond with the the anomalous trend in well VK 817 (Figure 5).

4 PRINCIPAL STRESS ORIENTATIONS IN THE VIOSCA KNOLL AREA

Eight wells were examined, all with deviations less than 5° except for one well where the hole deviation reached 38°, which was discarded. Final principal stress directions from the remaining wells are shown in Figure 6. Four wells, 816 #1, 817#2, 906 #1 and 986 #2 appear to follow a NE-SW stress direction whereas the other three wells show a NW-SE orientation (Figure 6 and Table 2).

Table 2. Maximum horizontal stress (S_{hmax}) orientations and confidence ratings.

Well	Lat.	Long.	Shmax	Confidence
VK 779	29.21383	88.43314	150	High
VK 816	29.1567	88.4709	38	Medium
VK 817	29.1498	88.4493	24	High
VK 862#2	29.14468	88.38078	160	Low
VK 869	29.14825	88.03434	142	Low
VK 906	29.1047	88.4145	63	High
VK 986	28.98207	88.77527	24	Medium

The maximum horizontal principal stress orientation for an extensional regime, as in offshore Louisiana, would be expected to follow the orientation of the shelf break, in a NE-SW direction. Four of the wells agreed with this direction while the other three had perpendicular orientations.

A number of geological factors may contribute to this stress signature in the Viosca Knoll area. Figure 6 shows that there are many faults somewhat parallel to the shelf break with a few exceptions perpendicular to this direction (Kindinger, 1989). The faulting is mainly related to downslope movement of sediment and to diapiric activity. Also shown in Figure 6 are a number of salt or shale diapirs present in the area. Faults and diapirs can cause stress deflections due to contrasts in elastic moduli (eg. Bell and Lloyd, 1989).

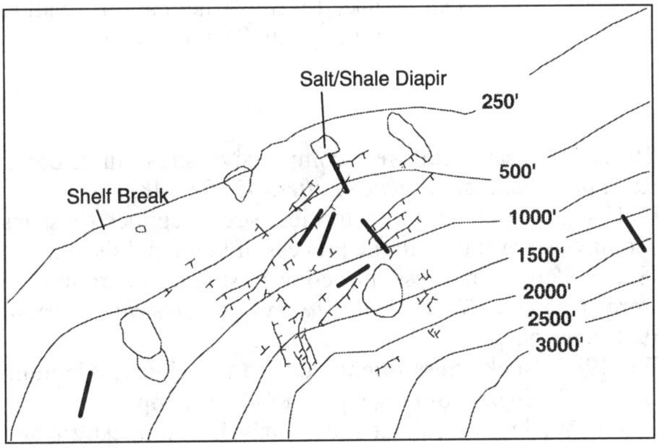

Figure 6. Maximum horizontal stress orientations. Structural and bathymetry data from Kindinger (1989).

It is interesting, however, that even on the well scale, the breakouts do not have a random distribution, but follow the geometry of the shelf break (parallel or orthogonal). In these areas the stress anisotropy might be low enough to cause 90° shifts in the maximum principal stress orientation (Bell and Caillet, 1992). Alternatively, in localized areas where isotropic stress conditions might exist, the formation of breakouts may be controlled by material properties or natural fracture orientations. Figure 7 illustrates, however, that the breakouts in all the wells do, in fact, have a strong NW-SE trend, corresponding to the geometry of the shelf.

5 SUMMARY AND CONCLUSIONS

Determining stress orientations, particularly in the Gulf Coast is a difficult task. Conventional means of choosing zones of caliper extension combined with no tool rotation is insufficient for the kind of breakouts observed in the Gulf of Mexico. Trajectory and polar plots proved to be effective in removing key-seating effects and extracting reliable breakout zones. Remaining breakouts were subsequently sorted to determine the principal horizontal stress orientations.

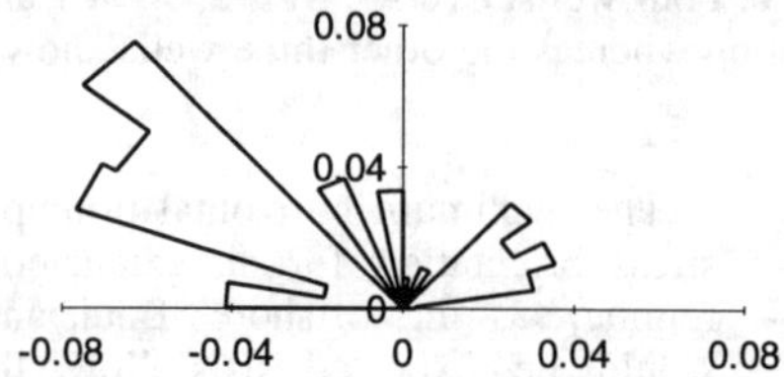

Figure 7. Rose diagram of reliable breakout orientations plotted with respect to length/total length for all breakouts. (Note that orientations are at 90° to S_{hmax})

Breakouts analyzed from wells, 816 #1, 817 #2, 906 #1, and 986 #2, located in the Viosca Knoll area, give a NE-SW maximum principal stress orientation. Due to low sediment strength, low stress anisotropy or diapirism, the other wells, 779 #1, 862 #2, and 869 #1, display perpendicular stress orientations. The breakouts in all the wells, however, have a NW-SE trend, consistent with the shelf geometry.

ACKNOWLEDGEMENTS

Our thanks to BP America for funding and providing data for this project.

REFERENCES

Bell, J.S. and Gough, D.I. 1979. Northeast-southwest compressive stress in Alberta: Evidence from oil wells. *Earth and Planetary Science Letters* 45: 475-482.

Bell, J.S. and Caillet, G. 1992. Using oil well data to determine stress regimes - lessons from the Aquitaine Basin, Southwestern Basin. In the process of being published.

Bell, J.S. and Lloyd, P.F. 1989. Modelling of stress refraction in sediments around the Peace River Arch, Western Canada. *Geological Survey of Canada, Current Research, Part D, Paper* 89-1D: 49-54.

Brereton, N.R. and Evans, C.J. 1987. Rock stress orientations in the United Kingdom from borehole breakouts. *British Geological Survey report* RG87/14: 36p.

Dart, R.L. and Zoback, M.D. 1989. Wellbore breakout stress analysis within central and eastern continental United States. *Log Analyst* Jan-Feb: 12-23.

Guenot, A. 1989. Borehole breakouts and stress fields. *Int. J. Rock Mech. Min. Sci. & Geomech. Abstr.* 26:3/4 185-195.

Kindinger, J.L. 1989. Upper Pleistocene to Recent shelf and upper-slope deposits of offshore Mississippi-Alabama. *GCSSEPM Foundation Seventh Annual Research Conference Proceedings*: 163-174.

Mastin, L. 1988. Effect of borehole deviation on breakout orientations. *J. Geoph. Res.* 93: 9187-9195.

Plumb, R.A. and Hickman, S.H. 1985. Stress induced borehole elongation: A comparison between four arm dipmeter and the borehole televiewer in the Auburn geothermal well. *J. Geoph. Res.* 90: 5513-5521.

Santerelli, F.J. and Brown, E.T. 1987. Failure of three sedimentary rocks in triaxial and hollow cylinder compression tests. *Int. J. Rock. Mech. Sci. & Geomech. Abstr.* 26: 401-413.

Springer, J.E. 1987. Stress orientations from well bore breakouts in the Coalinga Region. *Tectonics* 6:5 667-676.

Yassir, N.A. and Dusseault, M.B. 1991. Stress Trajectories in southwestern Ontario using wellbore breakout orientations. *Proc. of 32nd U.S. Rock Mech. Symp.* 83-92.

Zoback, M.D. and Zoback, M.L. 1991. Tectonic stress field of North America and relative plate motions. *The Geology of North America* , Decade Map vol. 1, *Neotectonics of North America*, edited by: B. Slemmons et al. 339-366.

Rock Mechanics, Nelson & Laubach (eds) © 1994 Balkema, Rotterdam, ISBN 90 5410 380 8

Experimental modeling and numerical analysis of well casing loading by rock creep

Deng Jingen, Huang Rongzun, Wang Kangping & Chen Mian
Department of Petroleum Engineering, University of Petroleum, Chang Ping, Beijing, People's Republic of China

ABSTRACT In this paper, the constitutive equations of rock creep are described, and the external well casing loading due to rock creep under nonuniform in—situ stress is calculated by using finite element method. The external loading distribution and its time—dependent relationships, and the effect of the value of in—situ stress, the rheological characteristics of the muddy shale and rock salt formations, as well as the elastic and geometric parameters of well casing and cement mantle are also studied. The results would have a great value for the design of well casing in creep formations.

1 INTRODUCTION

In the development of oil and gas fields , the abnormal collapse of well casing is often encountered and is reported in the Gulf of Suez of Egypt, North Dakota and Wyoming of the U. S. A. , North Sea of the U. K. , West Germany and Russia [1] [2] [3]. Begining from the early 1980's, in Daqing, Jilin and Yumen oil fields of the P. R. C. , as a result of enhanced recovery by high pressure waterflooding, the injected water flowed into the muddy shale formations, and caused oil well casing collapse. Besides, serious well casing collapse was also encountered in salt formations of Zhongyuan, Jianghan and Tarim oil fields. The authors suggestthe clay and rock salt formations exhibit creep characteristics and under the action of in—situ stress, rock creep brings about a nonuniform additional loading on well casing which is not considered in casing design. Investigation of the additional loading characteristics would be helpful in revealing the mechanism of this abnormal well casing collapse .

2 CONSTITUTIVE MODELS

For some viscoelastic materials, e. g. muddy shale and rock salt, strain is related not only to stress, but also to time, and it can be divided into two parts —— elastic strain not relating to time and creep strain relating to time. For convenience, the viscoelastic strain can be expressed in the form:

(1) $$\varepsilon = \varepsilon_e + \dot{\varepsilon}_c \cdot t$$

where ε is total strain, ε_e is elastic strain, $\dot{\varepsilon}_c$ is rate of steady—state creep, t is time.

In Eq. (1), the initial creep stage is neglected. From Eq. (1), it is clear that so long as the steady—state creep rate of the formation $\dot{\varepsilon}_c$ is known, the creep equation of viscoelastic rock is determined. Usually, the steady—state creep rate has a nonlinear relation with the stress and is also affected by other factors. For muddy shale the creep rate is mainly affected by its water content, and by temperature for rock salt. The results of the triaxial creep tests by the authors show that the constitutive equation for these rocks may be written as follows[4][5]:

(2) For muddy shale: $\dot{\varepsilon}_{ij} = \sigma_e^{N-1} \cdot \left(\frac{1+\mu_c}{\eta_0}\right) \cdot e^{\frac{1}{W_0 - W}} \cdot S_{ij}$ $\qquad (i,j=1,2,3)$

(3) For rock salt: $\dot{\varepsilon}_{ij} = \frac{sinh(B\sigma_e)}{\sigma_e} \cdot A \cdot e^{-\frac{Q}{RT}} \cdot (1+\mu_c) \cdot S_{ij}$ $\qquad (i,j=1,2,3)$

where $\sigma_e = [(\sigma_1-\sigma_2)^2+(\sigma_2-\sigma_3)^2+(\sigma_3-\sigma_1)^2]^{1/2}/\sqrt{2}$, $\sigma_1, \sigma_2, \sigma_3$ is principal stresses, S_{ij} is deviator stress (MPa), $\dot{\varepsilon}_{ij}$ is steady—state creep rate (1/s), μ_c is ratio of lateral to axial creep rate , W_0 is clay water content(%), T is absolute temperature (K), R is ideal gas constant ,Q is activation energy of rock salt (mol/cal), N, W , A, B, $\eta 0$ is rheological parameters determined by nonlinear regression of creep test data. The rheological parameters of the tested muddy shale and rock salt are listed in Table 1. By comparing Eq. (2) and Eq. (3), the rock creep model can be written in a general form:

Table 1. Rheologic Parameters of the muddy shale and Salt Rock

Rheologic Parameters of the muddy rock				
Sample Number	W	N	η_0	W_0
Daqing NO. 31	5.48 6.24 7.26	1.59 1.48 1.35	1.29×10^7	8.5
Daqing NO. 44	4.51 5.81 6.75	1.38 1.20 1.11	2.9×10^6	9.0

Rheologic parameters of the Salt rock			
Sample Number	A	B	Q
Zhong yuan S3—2	0.046	0.34	13.5×10^3
Zhong yuan S3—4	66.82	0.69	22.7×10^3
Talimu 3th Pedigree	42.0	0.61	20.0×10^3

(4) $\dot{\varepsilon}_{ij} = \sigma_e^{N-1} \cdot \left(\frac{1+\mu_c}{\eta}\right) \cdot S_{ij}$

where η is the viscocity coefficient and N is the stress nonlinear exponent:

(5) For muddy shale: $\eta = \eta_0 \cdot e^{-\frac{1}{W_0 - W}}$

(6) For rock salt : $\eta = \frac{1}{A} \cdot e^{\frac{Q}{RT}}$; $\quad N = \frac{ln\ sinh(B\sigma_e)}{\sigma_e}$

It can be seen from Eq. (4) that the constitutive models for the creep ofmuddy shale and rock salt can be expressed by one general form, and the same method can be used to calculate casing loading by rock creep forr these two type of formations. In the following analysis, only the muddy shale creep parameters is considered, the result can be also applied to rock salt.

3 CASING LOADING DUE TO ROCK CREEP

The in—situ stresses consist of over—burden stress and tectonic stress, Before drilling, the rock formations are in a state of relative equilibrium. When a well is drilled, the stressed rock in the borehole is removed and replaced with drilling mud pressure, the stresses arround the borehole will be redistributed and rock creep will be induced. Upon installation of casing, the external load of the casing is the borehole mud column pressure only. After the cement mantle is formed, deformation due to rock creep is restricted by the casing. Therefore the external loading of the casing will increase continuously with time. If the in—situ stresses are nonuniform, the external loading consists of two parts———radial and tangential loading.

From the analysis of the above section, the constitutive equations of rock creep are nonlinear, and if in—situ stresses are nonuniform there is no way to find out the analytical solution of casing loading. For this reason the 2—D finite element method is used to solve this problem. The problem is simplified as a plane strain problem shown in Fig1 (because of the symmetry, only one quarter of the cross section is considered), and a finite element program to compute external loading of the casing by rock creep is developed By using viscous initial strain method. In this program the 8 nodals mapped elements are used. The program has been checked by comparing the result of solving the problem of a cylindrical thick wall tube subjected to uniform internal pressure with the analytical solution. Its relative error is less than 2 %. The casing external loading due to creep of clay formation in 1,000 meters depth in Daqing oil field was computed, where the calculation parameters are taken from the oilfield as follows: $a = 6.213\ cm, b = 6.985\ cm, c = 11.875\ cm, \sigma_{h1} = 27.5\ MPa, \sigma_{h2} = 20.9\ MPa P_i = 13\ MPa, E_1 = 2 \times 10^5\ MPa, E_2 = 1 \times 10^4\ MPa, \mu_1 = \mu_2 = 0.25, E_3 = 2 \times 10^3\ MPa, \mu_3 = 0.3, \eta = 9.32 \times 10^4 (MPa \cdot hour)$.

Supposing the casing is in close contact with the cement mantle and there is no slip between them, the calculation results are shown in Fig. 2, where the 6 curves represent radial external loading on the casing located at θ=0°, 18°, 36°, 54°, 72°, 90°against the maximum horizontal in—situ stress, respectively. From Fig. 2 it is obvious that at the initial time (when the cement mantle is just formed) the radial external loading of the casing is equal to the mud column pressure in the borehole, and the external loading by rock creep increases with the lapse of time hereafter. The increaserate of the external loading is changes at different times and at different positions of casing due to nonuniform in—situ stresses. The distributions of radial external loading of the casing at three given times are shown in Fig. 3. From Fig. 3, it is clear that the loading distributions are non—uniform and change with time. After a certain time, the final radial loading tends to be a steady value. Moreover, besides the radial loading, a tangential loading on the casing has also been observed, which increases with the lapse of time too. The distribution of the tangential loading is also nonuniform and is dependent upon time. Fig. 4 shows a distribution of the final tangential loading (steady state value), which can be expressed approximately as a sinusoid. In the direction of θ=45°, the tangential loading reaches its maximum value.

Therefore , under the action of nonuniform in—situ stresses ,both radial and tangential nonuniform loadings may be induced by rock creep. The combination of these nonuniform loadings may bring about enormous danger for the well casing, which must be considered in casing design for these special cases.

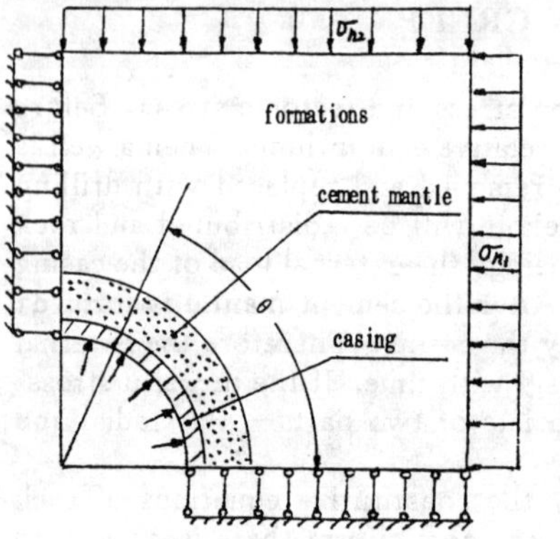

Fig. 1 The calculation section of the external loading of the casing

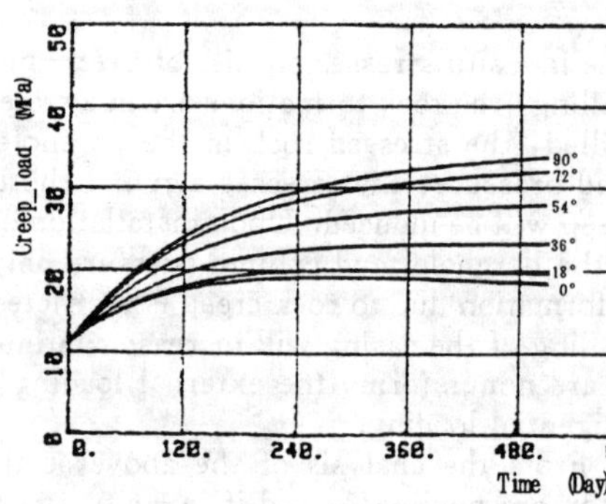

Fig. 2 The curve of the external loading of the casing

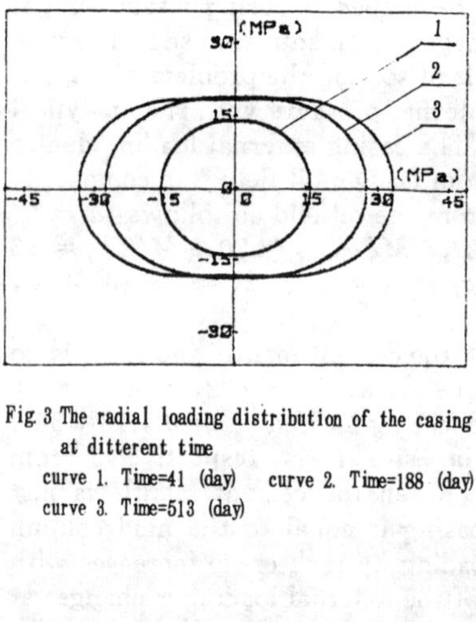

Fig. 3 The radial loading distribution of the casing at ditterent time
curve 1. Time=41 (day) curve 2. Time=188 (day)
curve 3. Time=513 (day)

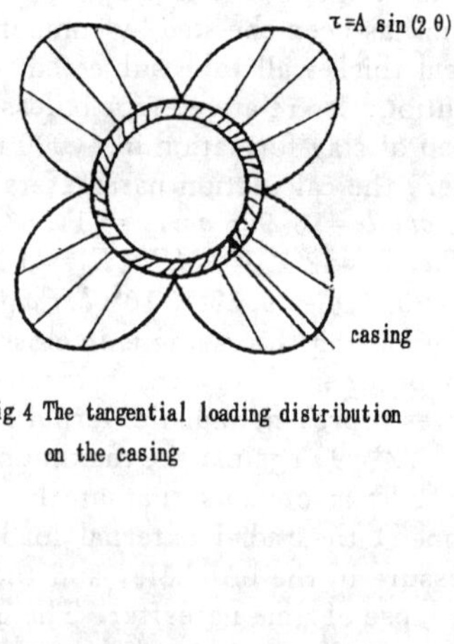

Fig. 4 The tangential loading distribution on the casing

4 FACTORS AFFECTING CASING LOADING

4.1 *INFLUENCE OF IN—SITU STRESSES*

The variation of radial external casing loading with the magnitude and nonuniformity of the in—situ stresses would differ greatly. Fig. 5 shows the distributions of radial loading on well casing under four types of in—situ stresses. It is seen from Fig. 5 that the value and the degree of nonuniformity of external casing loading by rock creep increase with the in — situ stress and the degree of its nonuniformity. For convenience, let:

$$k_1 = \frac{P'_0 + P'_{90}}{\sigma'_{h1} + \sigma'_{h2}} \qquad k_2 = \frac{P'_0 - P'_{90}}{\sigma'_{h1} - \sigma'_{h2}} \tag{7}$$

where $P'_0 = P_0 - P_i$, $P'_{90} = P_{90} - P_i$, $\sigma'_{h1} = \sigma_{h1} - P_i$, $\sigma'_{h2} = \sigma_{h2} - P_i$ and P_0, P_{90} is represent final steady values of the radial casing external loading in the direction of 0°, 90° respectively, σ_{h1}, σ_{h2} is maximum and minimum horizontal in-situ stress respectively, P_i is casing internal fluid pressure, In Eq. (7) k1 represents the magnitude of the radial casing loading; k2 the its nonuniformity. k2=0 indicates that the distribution of casing external loading is uniform; $|k2|>1$ indicates that the degree of nonuniformity of the external loading is greater than that of the in-situ stresses; $k2<0$ and $k2>0$ indicate that the position the maximum external casing loading is in the direction of minimum in-situ principal stress and the direction of maximum in-situ principal stress respectively.

The values of k1, k2 obtained from numerical computation under four types of in-situ stresses are shown in Table 2. These data show that k1 and k2 have approximately the same value for the given four in-situ stresses, where $k1 \approx 1.15$, $k2 \approx -2.90$. It is said that under given calculation conditions the value and the degree of nonuniformity of the radial casing loading are both greater than those of in-situ stresses, and the position of maximum casing loading and is in the direction of minimum in-situ stress (because $k2<0$). k1 and k2 are determined by the elastic and geometric parameters of the formation--cement mantle--casing system.

Table 2. Effect of in situ stress on the external Loading

NO	IN-SITU STRESS			P_0	P_{90}	K_1	K_2
	σ_{h1}	σ_{h2}	P_i				
1	22.5	20.9	11	21.0	25.6	1.15	−2.88
2	27.5	20.9	11	16.6	35.8	1.15	−2.90
3	32.5	20.9	11	12.4	45.9	1.16	−2.89
4	24.7	18.7	9.9	14.7	32.2	1.15	−2.91

4.2 *INFLUENCE OF CASING WALL THICKNESS ON LOADING*

The distributions of the external radial loading by rock creep of 5 1/2″ API standard casing with different wall thickness are shown in Fig. 6 (the calculation parameters are the same as before). It can be seen that the wall thickness affects both the value and distribution of the radial loading significantly. For the API casing with ordinary wall thickness, the loading reachs the maximum final steady value at $\theta = 90°$, and the shape of its distribution is likely fitted to a Cassini's ellipse with concave waist.

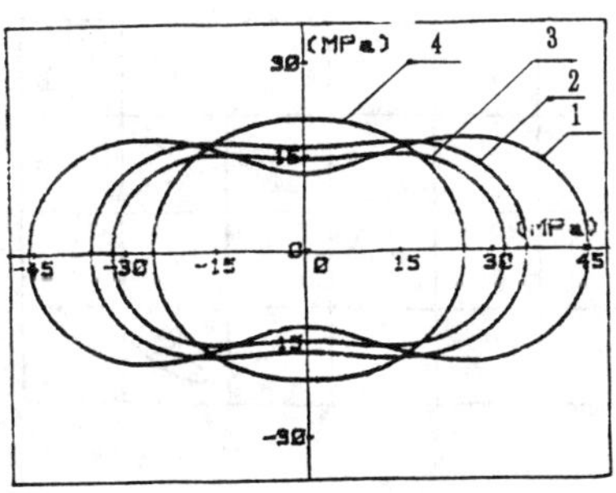

Fig 5 The radial loading distribution on the casing at different in-situ stresses

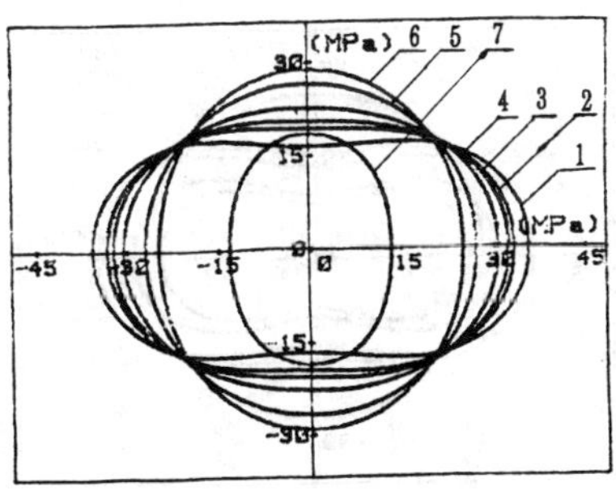

Fig 6 The radial loading distribution on the casing at different casing thickness

The nonuniformity of the external radial loading distribution decreases with the increase of casing wall thickness, and the shape of loading distribution changes from an ellipse with concave waist into an ellipse with convex waist. When the casing wall thickness reaches a specific value (such as t=16.0 mm in Fig. 6), the radial loading distribution becomes approximately uniform (turn to a circular shape). However, the nonuniformity of radial loading distributionwill increase again as the casing wall thickness further increases and the radial loading, reversedly, reaches final maximum steady value at $\theta=0°$, not at $\theta=90°$. The values of k1 and k2 of casing with different wallthickness are listed in Table 3. From these data, the degree of nonuniformityof casing radial loading distribution can be determined.

4.3 *INFLUENCE OF ELASTIC MODULUS OF CEMENT MANTLE*

The formation creep loading acts on the well casing through the cement mantle, which must play a role in partially resisting formation loading. The distributions of casing radial loading by rock creep for the case of different cement elastic modulus E2 are shown in Fig 7. The data were obtained by finite element calculation. The value of k1 and k2 may be seen in Table 4 and the relationships between k1,k2 and E2 are shown in Fig 8. From Fig. 8 it can be seen that with the

Table 3. Efect of casing thickmess on the external Loading

NO.	thickness h(mm)	P_0	P_{90}	K_1	K_2
1	7.72	16.6	35.9	1.15	−2.90
2	9.17	19.5	33.5	1.17	−2.33
3	9.70	20.7	32.5	1.17	−2.12
4	10.54	22.7	30.8	1.18	−1.79
5	12.09	26.6	27.3	1.20	−1.23
6	16.0	29.1	25.1	1.21	−0.11
7	20.0	18.7	13.3	1.22	−.61
$\sigma_{h1}=27.5$MPa, $\sigma_{h2}=20.9$MPa, $P_i=11.0$MPa					

Table 4. Effect of cement mantle's elastic modulu on the external Loading

NO.	Elastic Modulu E_2(MPa)	P_0	P_{90}	K_1	K_2
1	5×10^2	25.9	27.4	1.19	−0.23
2	5×10^3	16.6	35.8	1.15	−2.89
3	1×10^4	14.4	35.8	1.07	−3.24
4	5×10^4	9.84	29.8	0.67	−3.03
5	1×10^5	8.4	25.3	0.44	−2.56
$\sigma_{h1}=27.5$MPa, $\sigma_{h2}=20.9$MPa, $P_i=11.0$MPa					

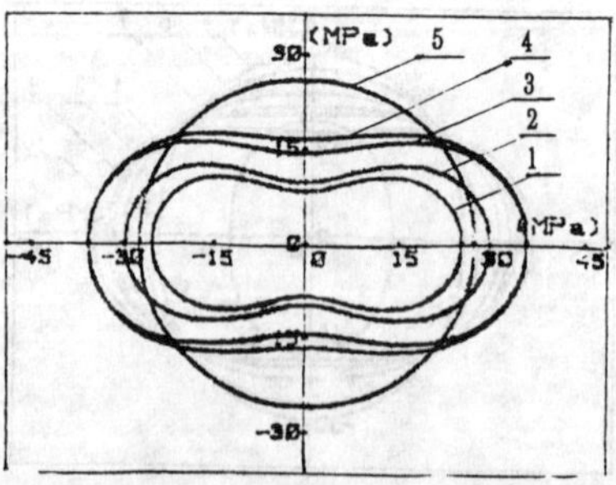

Fig 7 The radial loading distribution on the casing at different cement mantle elastic modulu

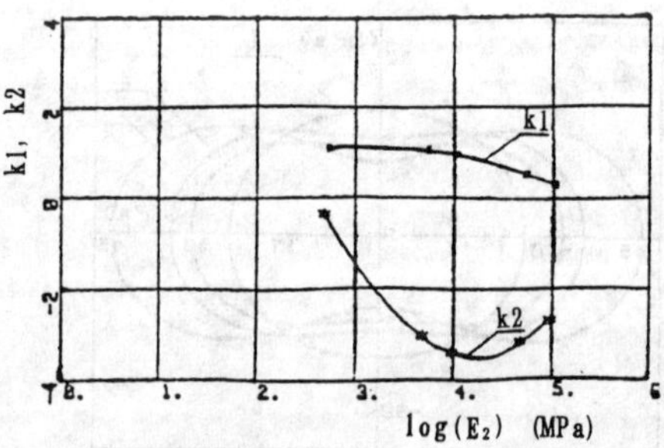

Fig 8 The curve of k1, k2 VS the cement mantle elastic modulu

increase of the elastic modulus of cement mantle, the magnitude of the external casing loading changes only little, but the degree of its nonuniformity increases greatly. That is k1 decreases little and k2 increases greatly. While E2 increases further, both k1 and |k2| decrease, the external casing loading decreases, that is good to protecting casing.

The above analysis shows that when $E2<5,000$ MPa, the nonuniformity of the loading becomes smaller for smaller E2, but its magnitude increases only a little. In addition, when $E2>20000$ MPa, the magnitude of the loading and its nonuniformity decrease with the increase of E2, they are all good to preventing casing collapse. Therefore it is suitable to let $E2<5000$ MPa or $E2>50000$ MPa in engineering.

4.4 *EFFECT OF OTHER FACTORS ON EXTERNAL CASING LOADING*

There are many other factors that affect the external casing loading, such as the formation creep characteristics, the cement mantle thickness and the elastic modulus of the casing. The calculation results show that rock creep characteristics affects only the time while the external casing loading reaches a steady value, but do not affect the steady value itself. Since the creep rate of muddy shale increases greatly with the increase of its water content, therefore in injection oilfields with high pressure, absorbing of water by muddy shale may result in a lot of casing abnormal collapse in a short period.

The cement mantle thickness has a little effect on the external casing loading, and because both it and the elastic modulus of the casing and the formation can hardly be changed, there is no use to study their effect on the loading in practice.

5 CONCLUSION

In this paper, the creep constitutive equation of rock creep (rock salt and muddy shale are established and the external casing loading by rock creep and its distribution are studied by means of finite element method. The conclusions are summarized as follows:

(1) Under nonuniform horizontal in—situ stress, a nonuniform radial andtangential loading will be caused by rock creep, and it increases with time, eventually tends to be a steady maximum value. Generally the magnitude and the nonuniformity of its steady value are larger than and are proportional to thatof in—situ stresses.

(2) With the increase of the water content of muddy shale, the timewhile the loading reaches a steady value becomes greatly shorter, it will cause earlier casing collapse.

(3) Both the magmitude and the nonuniformity of the external casing loading decrease greatly with the increase of the casing wall thickness. Therefore the increase of the casing wall thickness not only makes casing itself stronger, but makes the external loading on casing smaller, it is double sure in preventing casing collapse.

(4) The external casing loading is closely related to the elastic modulus of cement mantle, it can be made less harmful by changing the elastic modulus, but the change of the cement mantle thickness has only little effect on the loading.

The above conclusions are supported by the data of large scale true triaxial simulation tests, and are consistent with the situation of casing collapse in the oilfields. The casing strength under nonuniform loading caused by rock creep are studied by simuiation tests, and the results show that nonuniform loading is likely to cause casing collapse. This will be reported in a succeeding paper.

REFERENCES

[1]. C. Marx A. H. El—sayed " Evaluation of Collapse Strength of Cemented Pipe—in—Pipe Casing Strings" , *SPE/IADC*13432, 1985

[2]. Burkowsky, M. " Cemented Pipe—in—Pipe Casing Strings Solved Field Problems " *World Oil*, 1981, P134—147

[3]. Pattillo, P. D. "How Amoco Solved Casing Design Problems in the Gulf of Suez" *Petroleum Engineering International*, 1981,10 *P*86—112

[4]. Huang Rongzun, Zhou Zuhui, Deng Jingen " The Creep Characteristics of Shale formation and the Analysis of its Loaing On the oil well Casing " , *Proceeding of* 6*th ISRM Inter. Conf.* 1987 , *P*137, Monteral, Canada

[5]. Huang Rongzun, Zhou Zuhui, Deng Jingen " A Study of Viscoelastic—Plastic Behavior of Rock Salt and its Application In Petroleum Engineering " *Rock at Great Depth*,*ISRM —SPE*, 1989, *P*827—834 France

3 Rock engineering for underground excavations

Session developer:
Birger Schmidt, Parsons Brinckerhoff Quade & Douglas

Rock Mechanics, Nelson & Laubach (eds) © 1994 Balkema, Rotterdam, ISBN 90 5410 380 8

Prediction of boom tunnelling machine excavation rates

R.J.Fowell
Department of Mining and Mineral Engineering, University of Leeds, UK
G.Richardson & M.J.Gollick
Dosco Overseas Engineering Ltd, Tuxford, UK

ABSTRACT: The paper describes how instrumented laboratory cutting tests conducted on a full scale boom tunnelling research rig and monitored field trials have led to the development of a method of predicting excavation rates for existing machines or quantifying the specification of machines for a desired excavation rate.

The limitations of compressive strength as a sole determinator of cuttability are illustrated and the advantages of an instrumented cutting test which allows the cutting motor size and boom forces to be determined are given. The importance of providing adequate boom or hold in force is demonstrated.

1 INTRODUCTION

The performance of boom tunnelling machines depends on the intact rock strength, the properties of the rock mass, (Fowell and McFeat-Smith, 1976) the abrasive properties of the strata to be excavated (Johnson 1986) and the machine specification in terms of cutting head design and the forces the machine may generate. The influence of rock mass characteristics are in the main related to the discontinuities present, their frequency and orientation for most applications. If the machine can excavate the massive ground at the desired rate without relying on discontinuities the application is satisfactory. This paper concentrates on the prediction of cutting performance in massive conditions. A factor may be introduced to account for the rock mass effect.

To determine how the cutting and hold in forces acting on the cutting head relate to cutting head design and excavation rate, a full size instrumented laboratory based boom tunnelling machine research rig was constructed. The details of the rig have been given elsewhere [Speight and Fowell, 1987, and Fowell, 1988] so only a brief description will be given in this paper.

The cutting relationships for a number of rock materials were determined and the results were compared to the cutting force components recorded during a standardised instrumented cutting test. Using the functions obtained for each rock type the advance per revolution for the head may be predicted and hence cutting performance. An illustration is provided to demonstrate the method.

TABLE 1 Comparison between Rock Salt and Coal Cutting Characteristics

	Salt	Coal
Unconfined compressive strength (MPa)	28.70	27.20
Cerchar abrasivity number	Nil	Nil
Density (t/m^3)	2.14	2.40
Mean Cutting Force (kN)	1.40	0.31
Mean Normal Force (kN)	0.46	0.19
Cutting Tool Wear (mm/m)	Nil	Nil
SPECIFIC ENERGY (MJ/m^3)	10.52	4.72

2 COMPRESSIVE STRENGTH AS A PREDICTOR OF PERFORMANCE

Unconfined compressive strength (UCS) is often used as the sole predictor of cutting machine performance. Relationship have been established figure 1 between compressive strength and excavation rate in rock types that have a known history of machine applications. Usually this is rock associated with coal seams or the weaker sedimentary rocks such as sandstone, siltstone and mudstone. When the range of application has to be extended into other rock materials compressive strength alone is not always a good predictor of performance. Evaporite rocks, breccias and chalk containing flints are materials that do not follow the established relationships for compressive strength and excavation rate.

Table 1 shows the UCS and cutting properties for a coal and a rock salt. Though the UCS is similar the specific energy recorded under standardized conditions is not. The specific energy and force components quoted were obtained from a full sized instrumented drag tool cutting test (ICT). Figure 2

A fixed composition tungsten carbide chisel geometry tool, is used to take 5mm deep cuts along the outside of a borehole core or rock block sample. The mean cutting force and the volume of rock excavated give the specific energy. As the rock is the only variable in the test the specific energy is considered to represent the cuttability of the rock and relates to the size of cutting motor required for excavation of unit projected cutting head area, provided the depth of cut per tool is around 5mm.

The instantaneous cutting rate is determined by the 'hold in' force available at the cutting head. If this is insufficient the tools only take a shallow depth of cut which results in inefficient cutting.

The normal force component from the ICT test gives an indication of the boom force required for a given production rate. This value is obtained from relationships obtained from a laboratory instrumented boom tunnelling machine research rig described in the next section, and field trials in other materials.

3 INSTRUMENTED BOOM TUNNELLING RESEARCH RIG

The basis of the rig was a boom from a Dosco Mk IIA roadheader powered by a 36 kW thyristor controlled motor, providing variable head rotational speeds. The

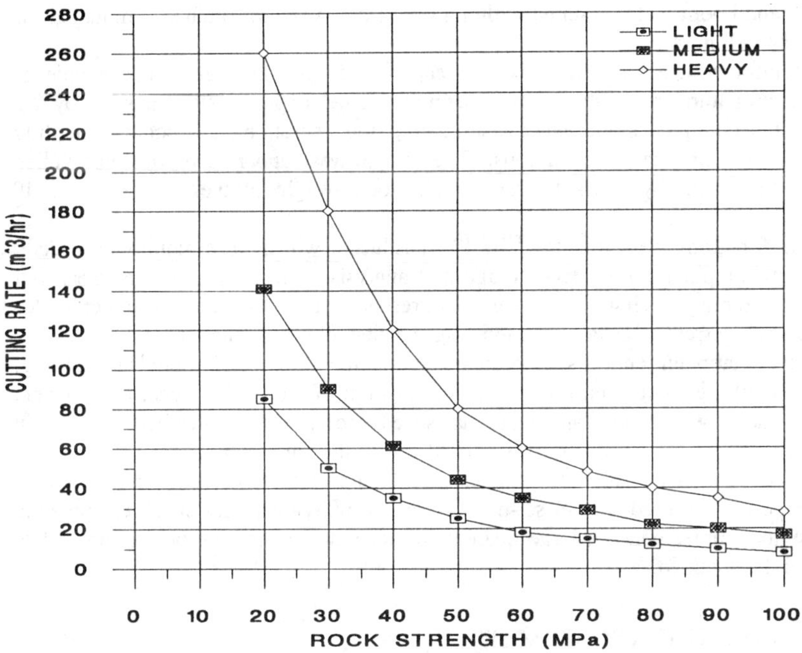

Figure 1. Typical ICR curves for axial roadheaders.

Figure 2. Instrumented cutting test.

pedestal and boom were mounted within a steel framework in which was mounted the sample box. Figure 3.

The reinforced steel sample box was capable of holding a 20 tonne sample of artificial or natural rock, representing a mock tunnel approximate 3m wide by 2m high. Cuts could be made exactly simulating the operation of a boom tunnelling machine under production conditions. The sample was supported on rails and pulled towards the boom by hydraulic rams. The total weight of the rig was some 40 tonnes.

The rig functions were measured by 13 transducers which were monitored directly by a microcomputer based data storage and analysis system. Parameters measured included sumping, arcing and lifting pressures and displacements and cutter motor voltage and current. Data reduction was carried out in a matter of minutes via interactive computer analysis programs. Parameters calculated included sumping, arcing and lifting forces and speeds, spatial position of the cutting head, cutting rate, cutting head power, torque and rotational speed and arcing rate per head revolution (Advance/Rev.). Data analysis was carried out at the rig location.

Individual cuts and associated analysis took approximately five minutes with a similiar period required for cut set-up. Some 200 individual trial cuts were possible in a single sample, hence a large quantity of accurate roadheader performance data was rapidly obtained.

4 EXPERIMENTAL PROGRAMME

The principal aim of the trials was to investigate the relationships between the monitored parameters for a range of rock materials and cutting head designs. A complementary programme to investigate the benefits of high pressure water jet assistance was also undertaken (Fowell, 1988).

For each cutting speed and advance per revolution the specific energy, torque and boom force was determined. Figure 4 shows a typical relationship between specific energy against advance per revolution.

Boom force is the arcing or lifting force at the cutting head, analogous to normal or hold in force for the instrumented cutting test. Torque and boom force was normalised by expressing them per unit area swept by the head during arcing or lifting. Such normalisation oversimplifies the actual geometric conditions during cutting, however it provides a simple and valuable means of comparison for prediction rate purposes.

All parameters were related to advance per revolution of the head across or up the face. This is analogous to the depth of cut taken in single tool experiments.

Figure 5 shows a typical relationship between torque per unit area and advance per revolution. The form of the boom force per swept area against advance per revolution relationships obtained for a concrete and a sandstone are given in figurre 6, along with those obtained for a salt and a coal from monitored field trials.

5 PREDICTION OF INSTANTANEOUS CUTTING RATE

Three main parameters are used to predict instantaneous cutting rate (ICR) which is the cutting rate of the machine when undertaking a traversing or under cut with the

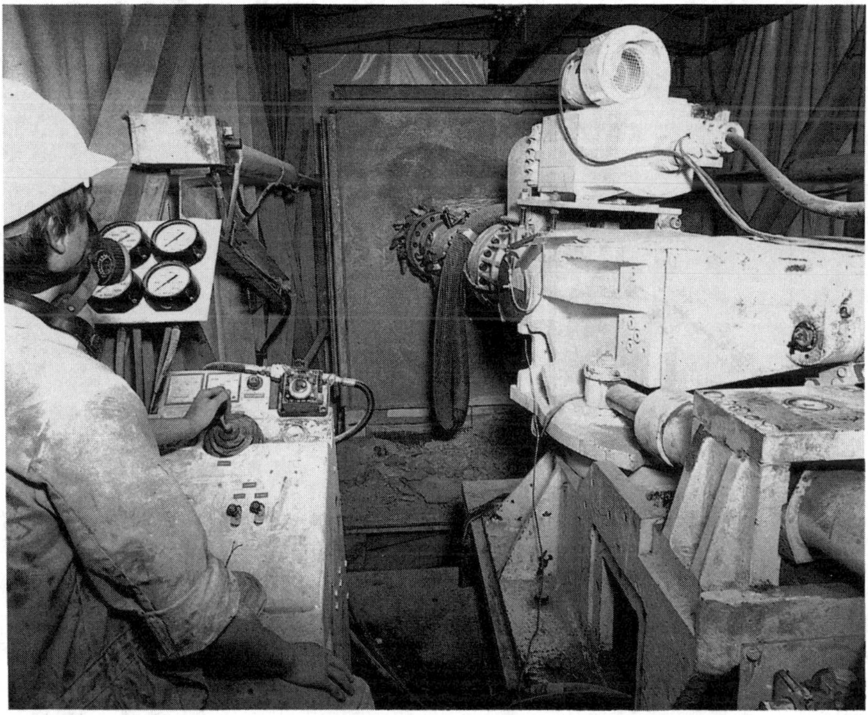

Figure 3. Instrumented boom tunnelling machine test rig.

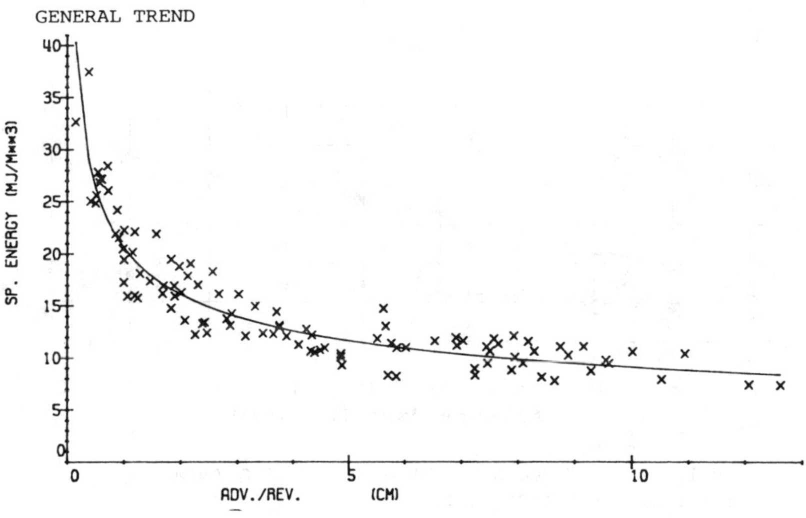

Figure 4. Typical specific energy versus advance per revolution relationship.

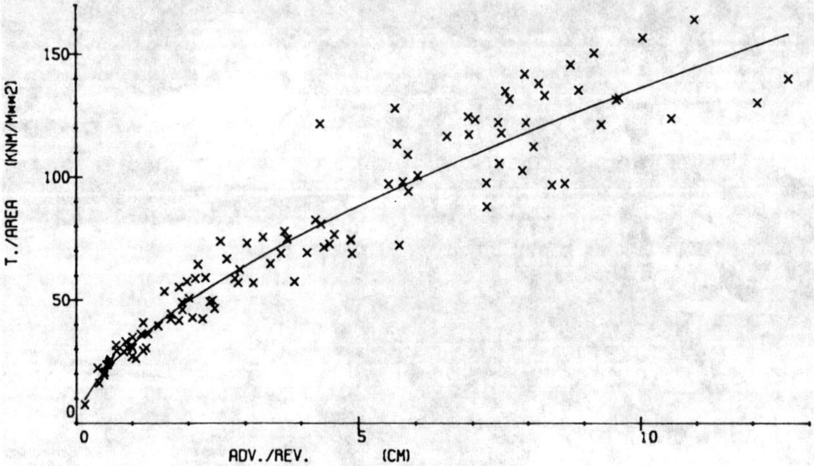

Figure 5. Typical torque per unit area versus advance per revolution relationship.

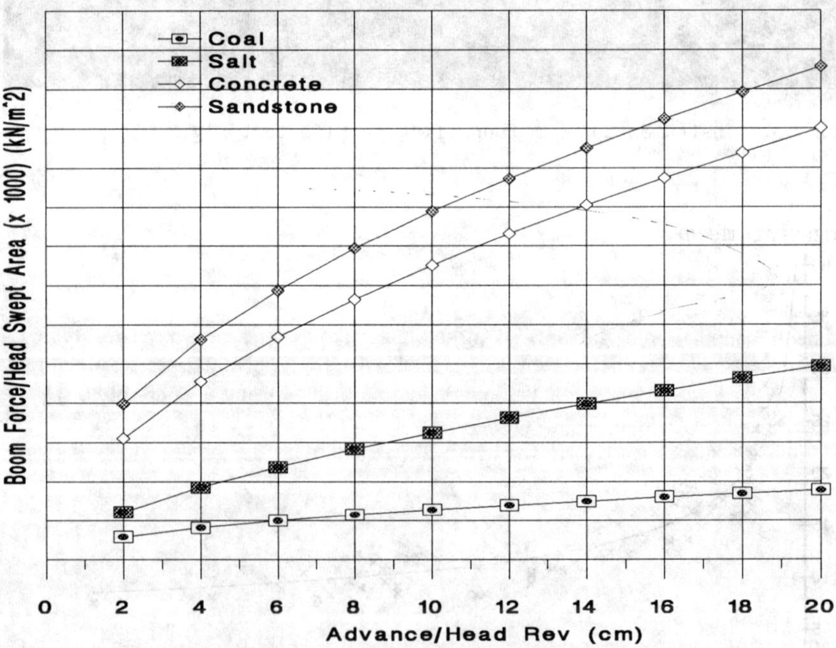

Figure 6. Typical boom force per unit area versus advance per revolution relationships.

cutting head. A factor dependent on site conditions, F_1, may be used to convert the ICR into a cutting and loading rate (CLR) which takes into account sumping, boom manoeuvring, loading, trimming etc., and gives overall maching performance.

The factors are:

1. Cutting head contact or swept area (SA) in m^2. For a traverse cut this is the profile area of the head projected in the direction of boom movement.
2. The head rotational speed in revolutions per minute (rpm).
3. Pick penetration or advance per revolution (A/R) in centimetres, which is determined by the available boom force (BF) kN. The relationships similar to figure 6 for the material under consideration are used to obtain the advance per revolution. The values of the constants in the relationship are obtained from the ICT force components.
4. The torque for the advance per revolution determined in (3) is a fourth factor and used to check that there is sufficient power available in the cutter motor. The torque is determined from the related boom torque per swept area versus advance per revolution relationship for the rock under consideration. The power is given by the torque multiplied by the angular velocity of the head.

The first three parameters are combined in the following equation to give the ICR in m^3/h.

(1) $$SA \times A/R \times RPM \times \frac{60}{100} = ICR\ m^3/h$$

Example: To determine the excavation rate for a machine operating in hard coal.
Head contact swept area (SA) = $1.0m^2$
Head rotational speed = 52 rpm
To determine the advance/revolution of the head. Using the advance/revolution characteristic for this coal, figure 4, which may be expressed as:

(2) $$\frac{\text{Boom Force (BF)}}{\text{Swept Area (SA)}} = 41.0\,(A/R)^{0.49}$$

where the Boom Force is expressed in kN and A/R in centimetres per revolution.

For an available boom force of 130 kN this gives and advance per revolution of 10.5cm. The excavation rate is given by equation (1) and using a value of 0.6 for the factor F_1 gives a CLR of 196.6 m^3/h.

Field trials subsequently showed close agreement with the predicted cutting and loading rate.

6 DISCUSSION

Though there are other factors to take into account related to cutting head design such as number of starts, line spacing and tool cutting tip geometry the proposed method has been found to be accurate for initial design of heads and machine specification for a particular rock and far more informed than relying on UCS values on their own.

The method can be used to predict production rates for existing machines or for specifying the boom force and head area required for a given excavation rate.

CONCLUSIONS

1 The available boom force which is related to weight of machine is the important limiting factor governing boom tunnelling machine excavation rate.
2 The cutting and normal force components from a single full scale drag tool instrumented cutting test have been related to the cutting motor power and the boom force respectively.
3 A method of predicting excavation rates for existing machines and the specification of machines for a given excavation rate has been provided.

ACKNOWLEDGEMENTS

The authors wish to express their gratitude to Dosco Overseas Engineering Ltd for support of the laboratory studies and permission to use information gained during field trials.

The opinions expressed in the paper are those of the authors' and not necessarily those of Dosco Overseas Engineering Ltd.

Acknowledgement is made to Drs H.E. Speight and A. Waggott former students of the University of Newcastle who built and developed the instrumented full scale boom tunnelling machine research rig.

REFERENCES

Fowell, R.J. and McFeat-Smith, I. 1976. Factors influencing the cutting performance of a selective tunnelling machine. In Proc. Tunnelling' 76, ed. M.J. Jones, IMM, London, pp 301-309.

Fowell, R.J., Waggott, A. and Anderson, I. 1988. Full scale boom tunnelling machine waterjet trials at pressures up to 140 MPa, 9th Int. Symp. on Jet Cutting Technology, Sendai, Japan, BHRA, Cranfield, UK. pp 323-339.

Johnson, S.T. and Fowell, R.J. 1986. Compressive strength is not enough - Assessing pick wear rates for drag tool equipped machines, Proc 27th US Rock Mechanics Symp, AIME, pp 840-845.

Speight, H.E. and Fowell, R.J. 1984. The influence of operational parameters on roadheaders productivity and efficiency with particular reference to cutting pick wear. In Conf. Tribology in Mineral Extraction: War on Wear, I.Mech.E., pp 175-182.

Rock Mechanics, Nelson & Laubach (eds) © 1994 Balkema, Rotterdam, ISBN 90 5410 380 8

Geotechnical parameters and tunnel boring machine performance at Goodwin Tunnel, California

J.E.O'Rourke, J.E.Springer & S.V.Coudray
Woodward Clyde Consultants, Oakland, Calif., USA

ABSTRACT: Two methods of predicting tunnel boring machine (TBM) penetration rates in rock are discussed in relation to site conditions and TBM performance for a 4.3 m diameter and approximately 4.8 km length hard rock tunneling project in northern California. One method relies on methodology for the extrapolation of penetration rates predicted by a TBM manufacturer based on special rock sample tests performed in its laboratory. Another method is based on the analysis of actual TBM performance in relation to rock index properties, particularly Total Hardness, H_t, and the Field Penetration Index, R_f, and the definition of the relationship between them.

1 INTRODUCTION

The Goodwin Tunnel Project was constructed in 1990 and 1991. The tunnel conveys water from the Stanislaus River northwest beneath a foothill range of the Sierra Nevada to a canal system supplying water to Stockton, California. The project includes approximately 4.8 km of 4.3 m diameter machine-bored tunnel, 0.4 km of drill-and-blast decline and incline tunnel, and a slide gate intake structure on the bank of a diversion pool behind Goodwin Dam. The dam is a buttressed twin arch concrete dam, almost 30 m high, built in the 1920s. It serves as a control structure diverting water from the Stanislaus River to two irrigation canals. This paper discusses design predictions and construction performance for the Tunnel Boring Machine (TBM), as related to rock properties of the metavolcanic, and metasedimentary deposits encountered.

2 GEOLOGY

The topographic relief is locally high and steep, with canyons that have been cut into Jurassic metamorphic rocks by westerly-draining rivers. The Jurassic rocks are unconformably overlain by interlayered Tertiary alluvium and volcanic flows and by Holocene alluvial, talus, and fill deposits. Geologic mapping and seismic refraction surveys established that the tunnel route was crossed near the intake end by a major, buried paleovalley, and then by another minor tributary paleovalley further along the route. Elsewhere along the route, bedrock outcropped, or was close to the surface.

The tunnel was driven through the Upper Jurassic Gopher Ridge Volcanics (Clark,

1964) which consists of mafic and felsic massive, porphyritic, and pyroclastic rocks. These rocks have undergone regional metamorphism, and have a strong foliation. Near the inlet structure, these rocks include a jointed, highly foliated stretch-pebble metaconglomerate. Foliation in these metavolcanic rocks generally strikes N 30° W to N 50° W and is nearly vertical. One vertical joint set is approximately parallel to the foliation and two other sets are approximately orthogonal to it. The tunnel bore is roughly normal to the foliation.

3 GEOTECHNICAL INVESTIGATION

The geotechnical investigation included core sampling a sufficient number of drill holes along the proposed tunnel alignment to adequately define the geology and rock mass properties needed for design of the project, and for estimating construction costs. The required number and locations of the drill holes were influenced by the complexity of the geology along the tunnel route.

Two drill holes were started by others in earlier investigations for a feasibility study. An additional 8 holes were drilled by the tunnel designer, Woodward Clyde Consultants (WCC). The drill hole depths ranged from 30 m to 130 m. Core sampling in the drill holes was required within the depth interval from 5 tunnel diameters above to 2 tunnel diameters below the preliminary selection of the tunnel invert grade. Some of these core samples were sent to the laboratory of Robbins Co., a TBM manufacturer located in Seattle, for testing to help predict the performance of a TBM in this rock. Other samples were taken to the design engineer's laboratory, for additional rock properties testing.

4 TESTING AND DESIGN

Selected rock core samples were tested in the Robbins Company laboratory and the WCC laboratory. Densities averaged 2.79 g/cm^3 and varied from 2.65 to 2.98 g/cm^3. The unconfined compressive strength (UCS) of 32 separate samples from five of the holes varied from 26 to 332 MPa with a mean of 137 MPa. Modified Taber abrasion tests and Schmidt Hammer Rebound tests were performed by WCC on nine samples from three holes. Table 1 presents results of the rock core test results from three of the 5 holes, including UCS tests performed by WCC and Robbins, and the rock core index tests performed by WCC. Other rock core tested by Robbins included samples with UCS up to 332 MPa. Robbins also performed some of their own non-standard tests of rock hardness on their inventory of samples. The test procedure includes measurement of a hardness category using a rebound technique, and includes identifying the quartz content of the sample. Their samples had quartz contents of up to 25 percent.

The drilling logs and core sample evaluations provided the conventional basis for classifying the rock mass conditions into four categories of ground types, requiring varying degrees of excavation support, ranging from occasional rock bolts, to shotcrete, to bolts and shotcrete, to steel supports and shotcrete. However, the prediction of TBM penetration rates using the available data was a more uncertain task. The rock index properties had been measured with a view toward using them to estimate and predict the TBM penetration rate. The standard rock hardness indices shown in Table 1, H_a,

H_r, and H_t, have been correlated with some past success to TBM performance in specific sedimentary rocks. An overview of past research and applications of rock core index properties are given in O'Rourke et al. (1986). However, the core drilling advance rates had demonstrated a wide range of rock conditions, including some very hard rock, and it was generally concluded by WCC that the available correlations of index properties with construction performance predictions would not be applicable to the present hard rock conditions. Robbins, however, were able to use their test data on strength and hardness, and their own empirical data base on TBM performance, to predict the TBM penetration rates in rock representative of the core samples supplied to them. Their predictions were a function of both the UCS, and the relative percentage of quartz present in the rock core samples. WCC decided to use and extrapolate the Robbins prediction data to the overall rock conditions at the site using the total core sample inventory from the drilling program.

TABLE 1. LABORATORY TEST DATA

Boring No.	UCS (MPa)	H_a	H_r	H_t
2	166	9	42	127
2	138	11		
2	194	8	41	120
2	108	9		
2	152	9	40	117
2	71	11		
3	82	2	30	40
3	70	5		
3	103	4	33	68
3	90	8		
3	122	2	34	52
3	95	7		
3	81			
6	30	7	25	68
6	28	9		
6	26	3	14	24
6	26	3		
6	32	6	26	65
6		7		

H_a - Modified Tabor Abrasion Hardness: H_r - Schmidt Hammer Rebound Hardness: H_t - Total Hardness.

Based on Robbins' criteria, WCC established three major classifications of rock penetration. The classification categories were arrived at by comparing drilling performance and rock core characteristics to the unconfined compressive strengths and quartz contents measured on selected core samples in the laboratory. It was assumed that the characterized rock mass would be similar to laboratory measured properties of rocks with similar mineralogy and textures. The classification categories arrived at were as follows:

Classification Categories	Description	Estimated Unconfined Compressive Strength and Quartz Content
I	Hard porphyritic metavolcanic rock and meta-felsite.	175 MPa to 275 MPa High quartz content.
II	Aphanitic metamorphic rock and slightly schistose metamorphic rock.	70 MPa to 175 MPa Low to moderate quartz content.
III	Highly schistose metamorphic rock.	Less than 70 MPa Low to moderate quartz content.

This rock is relatively massive with characteristic fabric, mineralogy, and joint patterns, but with gradational variations as well as zones of concentrated weakness, such as shear zones and weaker rock (eg. talc and serpentine). Based on these classifications, the drilling data and core for the holes drilled along the TBM tunnel alignment were examined.

Fig. 1 shows the resulting predictions by WCC for ranges of TBM penetration rates. A range of low, average, and high predicted Penetration Rates are shown for each rock hardness classification. The estimated proportion of total tunnel length represented by each rock hardness classification is shown in percentage next to the classification category, I, II, and III.

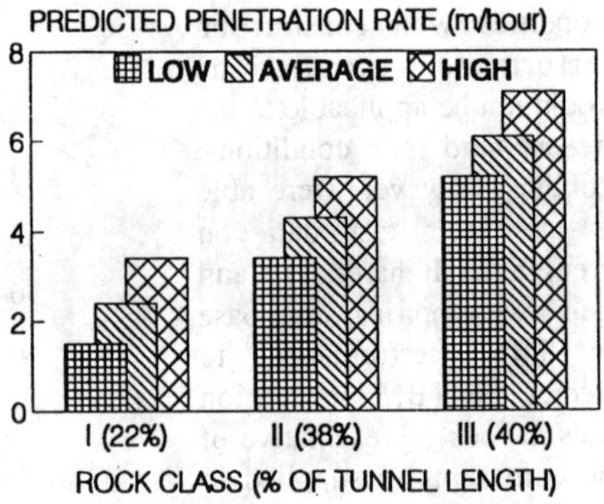

Figure 1. Predicted penetration rates for hardness classification categories.

5 TUNNEL BORING MACHINE (TBM) AND CONSTRUCTION OPERATIONS

The TBM for Goodwin Tunnel was a refurbished Atlas Copco, Jarva Mark 12/1400. It was outfitted by Robbins with a new 4.3 m cutterhead equipped with 29 disk cutters (432 mm diameter). The TBM was rated at 746 kW (1000 Horsepower) nominal (4 electric drive motors of 186 kW (250 hp) each), with a drive torque of 450 kN m (4 Million lb/in), and a cutter head thrust of 6.23 MN (1,400,000 lb).

The tunnel was excavated on a 5 day per week, 24 hour per day basis with the necessary support and maintenance activities performed on Saturdays. The diesel locomotives with 3 to 4 muck cars (depending on cycle time) transported the muck out of the tunnel to a disposal area. The actual penetration rates achieved by this TBM over the total length of tunnel, including all rock classification types, ranged from 1 to 5 m/hour, and the mean rate achieved was approximately 2.5 m/hour. This rate is less than would be predicted from Fig. 1; however, the TBM was an older refurbished machine with a slightly larger cutter head and designed to use fewer cutters than specified by Robbins.

6 ANALYSIS OF TBM PERFORMANCE

It will be apparent that there is significant data scatter in the following figures. Some of the scatter is probably attributable to the absence of any automatic, real-time recording of parameters onboard the TBM, and the uncertainties inherent in daily, manual recording of all parameters by the tunnel inspector. For example, there were occasional failures of one of the four electric motors during 20% of the tunnel drive, and it would be difficult to quantitatively account for the decreased average cutter thrust in the manual recordings for those periods. There were even times when the TBM operating clock failed in the latter part of the tunnel drive. This necessitated reconstructing the TBM operating time and down time by using the clocks on the conveyor system and the hydraulic ram system. Nevertheless, it is felt that the large amount of data collected over the total tunnel drive satisfactorily averages and portrays the TBM operational parameters.

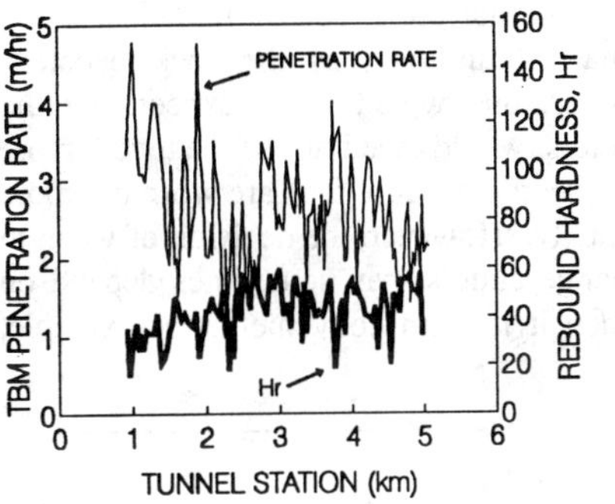

Figure 2. TBM Penetration rate and H_r over the length of the tunnel drive.

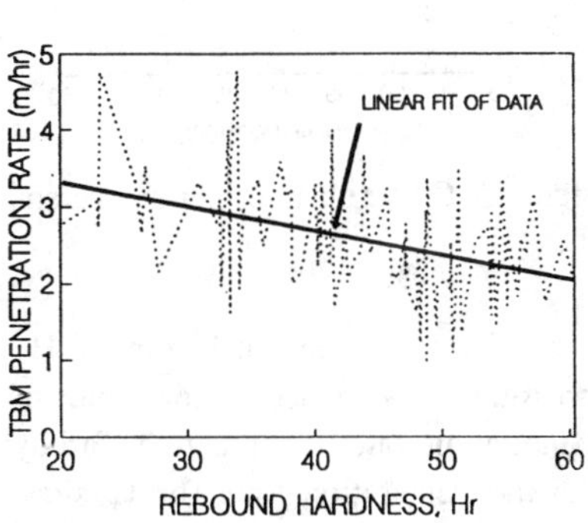

Figure 3. TBM Penetration rate versus H_r.

There is also obvious scatter in the field data obtained for the rock index property of Rebound Hardness, H_r, taken in situ on the tunnel walls throughout the tunnel drive. Measurements were made with a Schmidt Hammer, which is a hand-held, spring-loaded impact device. The spring mechanism is first cocked, the impact seat is held against the rock surface, and the rebound of the slide hammer after striking the rock is registered on an index scale. Although the Schmidt Hammer is typically used on rock core samples held in a cradle, a number of other investigators also have reported its use on in situ rock in tunnel and mine walls (O'Rourke el al., 1986). The in situ rock measurement was reported to generally yield higher index values than laboratory index values for rock core from the same location. This is probably due to the confinement pressure acting on the in situ rock in the tunnel wall. Also, the degree of data scatter was reported to be larger for measurements on weak rock types in situ, compared to laboratory measurements on the same rock type. This is probably due to the inherent better quality of rock samples that are able to be tested in an unconfined condition in the laboratory.

Fig. 2 shows the TBM Penetration Rate and the Rebound Hardness, H_r, recorded over the length of the tunnel drive. The inverse relationship of Penetration Rate to H_r could be shown a little more clearly if moving averages of each data set were plotted. This would tend to smooth the curves. However, close inspection of this raw data will generally show the "mirror" image of one curve to the other. A clearer view of the inverse relationship can be seen in Fig. 3, where Penetration Rate is plotted as a function of H_r. A linear regression fit of the Penetration Rate data has been shown in addition to the raw data to better demonstrate the trend.

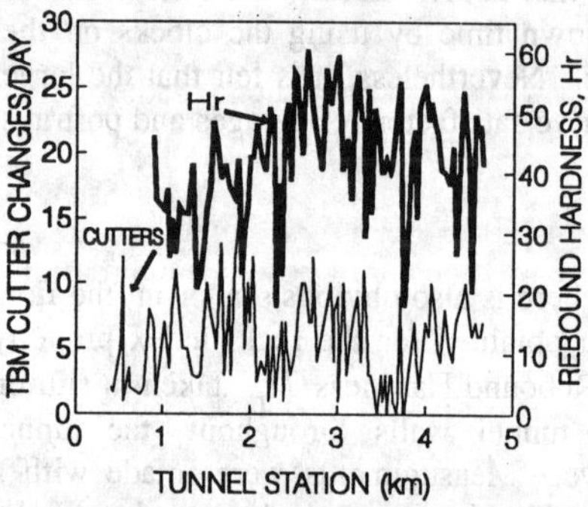

Figure 4. Cutter changes per day and rebound hardness over the length of the tunnel drive.

Fig. 4 shows the number of worn TBM cutters changed per day over the length of the tunnel drive. The Rebound Hardness measurement is again shown on this plot. There is a proportional relationship between these two parameters, although there is much more "noise", or outside factors, that are probably included in the data. Intuitively, as the rock becomes harder, one would expect that the cutters would need to be changed more frequently, because they are worn out more quickly. However, the decision of when to change cutters may sometimes depend on when it is most convenient; for example, when other maintenance on the TBM happens to be needed. Such events would produce "noise" in the true relationship of the data. In any case, the relationship of Penetration Rate with H_r is shown in Fig. 5, where the raw data, and a linear regression fit of the data, are included.

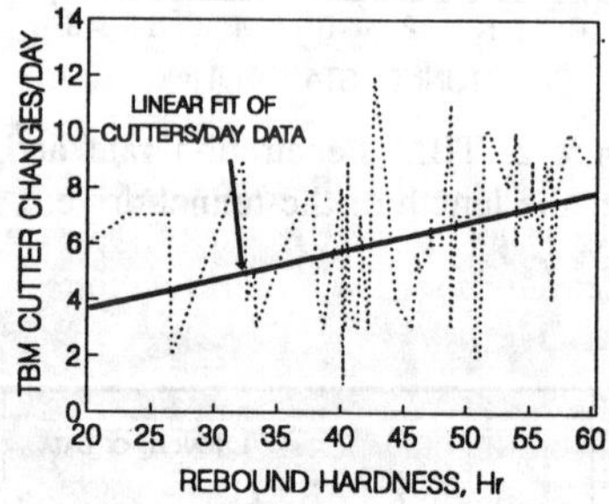

Figure 5. Cutter changes per day versus H_r.

Fig. 6 shows the interrelationship of the dependent variable, Penetration Rate with the independent parameters, TBM Cutter Changes/Day, and H_r. The qualitative trends, and not exact, quantitative relationships, are emphasized in this figure. In general, the relative effects of Abrasion Hardness, H_a (Cutter Changes/Day), and H_r, on the Penetration Rate are as expected for a plot of these components of Total Hardness, H_t, against Penetration Rate. To illustrate, Tarkoy (1983) showed significant correlations of H_t with the Penetration Rate for 13 sites, where H_t was defined as follows:

(1) $$H_t = H_r + H_a^{1/2}$$

Abrasion Hardness, H_a, is measured in the laboratory (see Table 1), but it is a time-consuming effort involving abrasive test grinding of sawed rock core disks and measurement of resulting disk weight loss. The wearing of TBM cutters on the rock face is an analogous process, and the Cutter changes/day parameter in Fig. 6 provides data that should be representative of the H_a index. A comprehensive and detailed study on TBM performance in shales, sandstone, and limestone formations by Nelson et al. (1983) showed a strong correlation of a "normalized" Penetration Rate vs. H_t over a broad range of H_t values:

(2) $$R_f(\text{kN/mm}) = 5.95 + (0.18)H_t$$

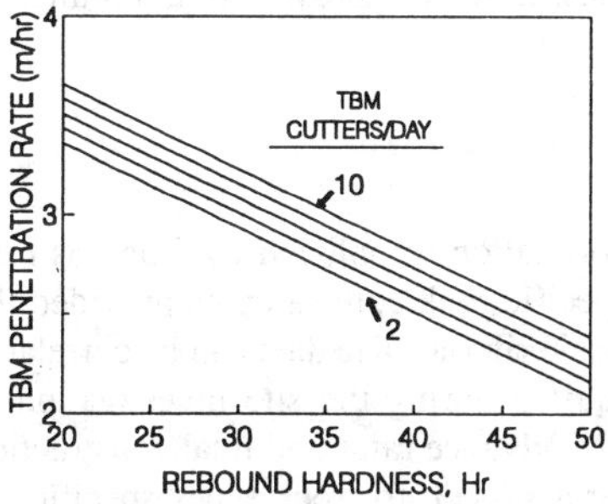

Figure 6. TBM penetration rate as a function of cutter changes per day and H_r.

The Field Penetration Index, R_f, normalizes the performance of different TBMs, operated in the same rock types. The R_f is the thrust force per cutter divided by the cutter head penetration into rock for one revolution of the cutter head. Since it appears that the Goodwin Tunnel TBM performance is indeed sensitive to both of the H_t components, a plot of the R_f vs. H_t values compared to the Nelson et al. (1983) relationship would be of value. Some H_t data are directly available from the laboratory data in Table 1, and could be used with a TBM penetration rate selected from a tunnel section located near the core hole to compute a value of R_f. However, this would only provide a few data points. Alternately, a linear regression analysis of the relationship of H_t to H_r using the laboratory data can provide a reasonably good correlation between the parameters ($r^2 = 0.73$), which will allow the H_r data taken along the tunnel drive to be used to calculate related H_t values vs. tunnel station. This procedure was followed and the Field Penetration Index was then computed from the Goodwin Tunnel TBM thrust/cutter and the Penetration Rate data for tunnel stations. The resulting relationship between H_t and R_f is shown in Fig. 7. The equation of the fitted line for the Goodwin data is:

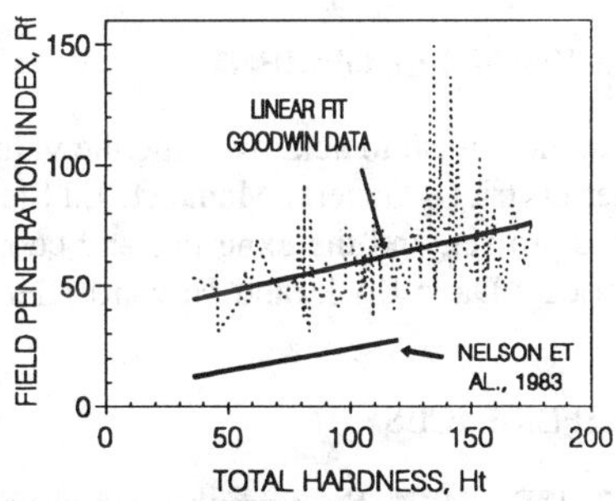

Figure 7. Relationship of R_f and H_t for Goodwin Tunnel and TBM data from Nelson et al. (1983).

(3) $$R_f\ (\text{kN/mm}) = 36 + (0.23)H_t$$

It is obvious in Fig. 7 that the statistical significance of the fitted equation for predicting individual R_f values anywhere along the analyzed Goodwin Tunnel data set is low. However, such a trend line has greater than usual value when there is judged to be a significant amount of noise and random event effects influencing individual data points. It also has significant validity for analysis when the data set is large and spans a broad range of the independent variable. In Fig. 7, a comparison of the trend line with the other referenced data set seems reasonable and useful. Note that the relationship of a linear regression fit of R_f to the calculated H_t data approximately parallels the Nelson et al. (1983) relationship. The R_f for the Goodwin Tunnel rock is significantly higher, however. This shows the greater resistance of the harder metamorphic rock to penetration by a TBM, as compared to the shales, sandstones and limestone sequences of the sedimentary rock formations in the other referenced data set. cementing, and other rock property characteristics, may be represented as families of (sub)parallel lines on the R_f vs. H_t plot.

7 CONCLUSIONS

The rock core testing and empirical data base correlations employed by Robbins may be useful for predicting penetration rates from specific rock core samples provided the characteristics of the particular TBM meet Robbins' criteria. The data can be correlated by the engineer to a core sample inventory acquired during the site investigation to make an overall, project estimate for average TBM advance rates and total construction time and costs. These predictions can be further assessed, for rock types specifically investigated by Nelson et al. (1983) and the Goodwin Tunnel Project, by measuring Total Hardness, H_t, and using the relationship between H_t and the Field Penetration Index, R_f, documented by the two projects.

8 ACKNOWLEDGEMENTS

The authors wish to acknowledge the valued participation and support of Stockton East Water District's General Manager, Ed Steffani, in the design and construction activities for the project, and the expertise and cooperation of Dillingham Construction of North America's Dave Bock, and Howard Klein, representing the construction contractor.

9 REFERENCES

Clark, L.D., 1964, Stratigraphy and structure of part of the western Sierra Nevada metamorphic belt, California: U.S. Geological Survey Professional Paper 410, 70p.

Nelson, P., T. D. O'Rourke, and F. H. Kulhawy, 1983, Factors Affecting TBM Penetration Rates in Sedimentary Rocks, Proceedings, 24th U.S. Symposium on Rock Mechanics, Texas A&M, College Station, TX, pp. 227-237.

O'Rourke, J. E., P. H. Rey, E. Alviti, and C. C. Capps, 1986, Rock Index Properties for Geoengineering in the Paradox Basin, Technical Report No. BMI/ONWI-579, published by Office of Nuclear Waste Isolation, Battele Memorial Institute, Columbus, OH, 198p.

Tarkoy, P. J., 1983, Truth, Judgement, Responsibility, Integrity, Conflict and Resolution in Exploration and Heavy Construction, Proceedings, Rapid Excavation and Tunneling Conference, Chicago, IL.

Rock Mechanics, Nelson & Laubach (eds) © 1994 Balkema, Rotterdam, ISBN 90 5410 380 8

Rock mechanics conditions in circular tunnels through Austin Chalk

R.F.Cook
The PB/MK Team, Dallas, Tex., USA

ABSTRACT: Twenty Four kilometers of tunnel were driven as part of the construction for the Superconducting Super Collider Project, primarily through Austin Chalk. During the drive, local failures of ground were encountered. The factors controlling the failures have included high horizontal in situ stresses, a regional geologic structure with major faults trending NNE, the presence of minor faulting, the mineralogical composition of shaley layers in the chalk, strength variations in the chalk, ground water, tunnel shape and orientation, and the type of excavation equipment used. Each factor alone has generally not been sufficient to cause instabilities but in combination they have led to a number of mechanisms resulting in instabilities in the tunnel through the chalk.

1. INTRODUCTION

The Superconducting Super Collider, as originally planned, included a 87 km long tunnel ring in Ellis County, North East Texas (see Figure 1). Prior to termination of the project, about 24 km of tunnel with a diameter between 4.5 and 5 m was driven, mainly through the Austin Chalk but with about 6.5 km through Eagle Ford Shale. The tunnel through the chalk was temporarily supported during construction with rock dowels. The tunnel in the shale was fully lined due to the low strength of the shale and its susceptibility to deterioration on contact with air and moisture. This paper identifies the various factors that have contributed to conditions in the excavations in Austin Chalk, and describes the mechanisms that have produced the local instabilities and the observations that can predict failures.

2. GEOLOGY

The project site is on the eastern margin of the Texas Craton with the underground structures located in sedimentary rocks ranging in age from Paleozoic (Upper Cretaceous Units) to Quaternary (alluvium deposits). A geologic profile showing the sequence of Taylor Marl, Austin Chalk, and Eagle Ford Shale formations through which the tunnel passes is shown in Figure 2. The conditions anticipated in the tunnel drives were presented by Nelson and Lundin (1990), and are described in Geotechnical Design Summary Reports for N20 to N25, (1991); for N25 to N40, (1992); and, for N40 to N55, (1992).

The Austin Chalk is a light to medium grey, fine to very fine grained, slightly to moderately fossiliferous highly calcareous mudstone, with horizontal to subhorizontal bedding. The chalk is composed of an average of 85 percent calcium carbonate and contains thin beds of calcareous shale, argillaceous beds and occasional bentonite layers. The interbeds can be rich in montmorillonite, illite and bentonite. The Austin Chalk has been divided into three members in the area (Norton,1965; Allen and Flanigan, 1986): the lower chalk, the middle marl and the upper chalk. The lower and upper chalk have similar lithologies, and consist of massive 0.6 to 1.5 m beds of chalk, interbedded with 0.3 to 0.6 m beds of marl. The middle member consists of 0.6 to 1.5 m beds of marl interbedded with 0.3 to 0.6 m beds of chalk. The middle chalk contains a distinctive bentonite layer, generally less than 0.5 m thick, and referred to as the "Marker Bed", that is about 20 m above the contact with the underlying Eagle Ford Shale.

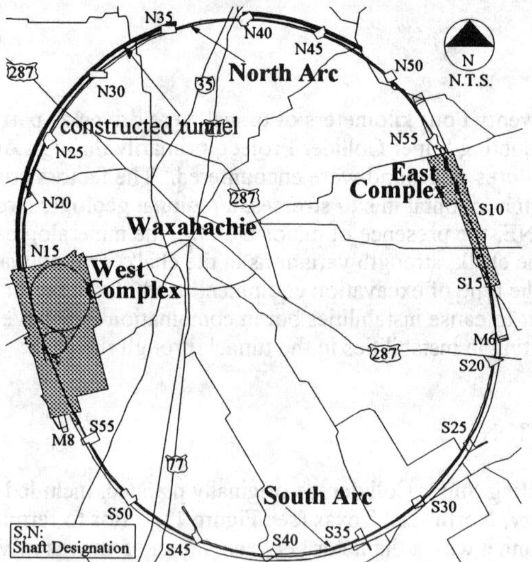

Figure 1 Layout of Superconducting Super Collider

The project site is crossed by the Balcones Fault System, a system of dip-slip normal faults showing both vertical and horizontal displacements. The faults generally strike east-northeast, dip at about 70 degrees to the east, and have displacements of less than 30 m. The faults occur on the order of tens of thousands of meters apart and produce horst and graben structures in the Austin Chalk.

A more random orientation of discontinuities than expected based on the regional geology was observed in the tunnels. Nance (1993) has shown that this is due to the presence of not only the NNE trending primary fault system but a secondary system of minor faults running perpendicular. The secondary system exhibited a lack of continuity due to shaley layers within the chalk reducing its propagation.

The vertical stress in the chalk is equivalent to the overburden pressure. The horizontal/vertical in situ stress ratio, K_O, varies from 1.5 and 2.2 with the maximum value oriented at about N35 degrees E, subparallel to the trend of the regional fault system (Kim and Schmidt, 1992).

Although the Austin Chalk is saturated at tunnel depths, its low porosity assures that the rock is not a major source of water inflow to the tunnels except occasionally at fault zones where interconnectivity to the surface alluvial deposits provides a source for water. Tunnel inflows have averaged about 10,000 liters/day/km once the high initial inflows have ceased.

3. GEOTECHNICAL PARAMETERS

The chalk varies from a very soft material (where weathered) to a moderately hard rock (fresh) with a uniaxial compressive strength of about 1.5 MPa. The permeability of the massive chalk is low, on average 10^{-6} cm per second. The massive chalk contains non plastic material (LL of 36%, PI of 19%). However, samples taken from shaley layers in the N30 to N35 section of the tunnel have indicated that layers showing deterioration contained plastic (CH) materials.

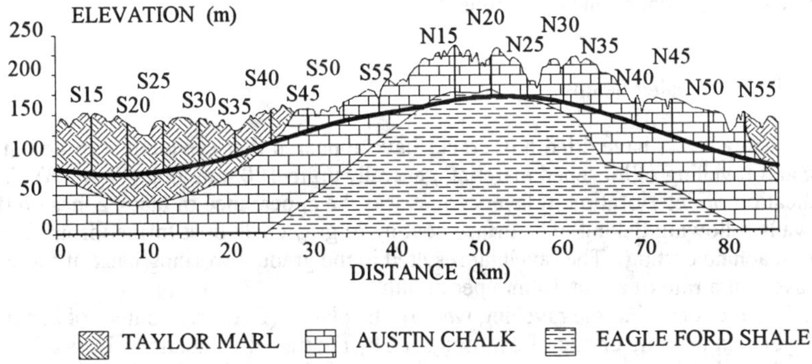

Figure 2 Geologic Profile Along Collider Ring

The Eagle Ford Shale and Taylor Marl provide an indication of the swell potential for the shaley seams in the chalk since they have similar composition and have been subject to the same geologic changes. Based on these results (Olsen et al,1993), swell pressures are in the range of 0.15 MPa to 0.5 MPa.

4. STRESS CONDITIONS IN TUNNEL

The stress concentrations at the crown, invert and springline for a circular opening are shown in Table 1 for the extreme values of K_o found on the project. These stresses are not sufficient to cause failure in the massive chalk for the range of tunnel depths encountered on the project. However, weaker shaley layers occuring in the crown or invert may fail.

Table 1. Stress Concentration Factors for Circular Tunnel

Orientation of Far Field Horizontal Stress	N 55 W - S 55 E	N 35 E - S 35 W
Horizontal stress/Vertical Stress (K_o)	1.5	2.2
Tangential Stresses at Crown and Invert	3.5	5.6
Tangential Stresses at Springline	1.5	0.8

Where the tunnel is perpendicular to the maximum horizontal far field stress, low compressive stresses occur at the springline. Under these conditions, discontinuities can open up and, in massive chalk, shrinkage or tensile cracks can develop.

5. OBSERVED CONDITIONS IN AUSTIN CHALK

Ground support sufficient to ensure safety during construction was installed in the 17.5 km of tunnel through Austin Chalk. This has allowed the rock mechanics performance of the chalk to be observed before the tunnel was permanently stabilized. A number of different types of localized rock failures have occurred in the tunnel. The factors that have controlled the failures have included in situ stress conditions in the rock, the regional geology, the strength properties of the chalk, and inhomogeneities within the chalk formation. Although they did not materially affect tunnel progress, the failures did require an assessment of actual tunnel conditions.

5.1 Ravelling of Shaley Seams

As some shaley layers have dried out, cracks have appeared that eventually resulted in the spalling and ravelling of loosened material, generally up to 100 mm long (Figure 3). The first indication of deterioration of layers zones was the appearance of black zones on the tunnel walls following flaking of a surface coating of grey chalk dust formed from fines left after machine cutting. The ravelling resulted in the gradual working back of these shaley layers at a rate of about 10 mm per month.

The main factors causing the ravelling were i). the plastic (CH) clay content of some shaley layers, and ii). the drying of exposed rock by the tunnel ventilation. Loss of moisture from the shales resulted in volume decreases, that produced shrinkage cracks. Further deterioration was aggravated by moisture content changes in the tunnel air that caused cyclic dampening and drying out of the rock walls with associated expansion and contraction of shaley material.

Figure 3 Ravelling of Shaley Seams

Figure 4 Longitudinal Tensile Fractures in Chalk

5.2 Fractures Induced by Excavation

Tensile fractures running longitudinally along the tunnel have developed above the springline in some sections of tunnel (Figure 4). The fractures were generally about 3 m long and up to 3 mm wide. They continued en echelon along the length of the tunnel, generally located between 60 and 120 degrees from the tunnel crown. They did not appear immediately the tunnel was excavated but were observed some weeks after excavation.

The fractures developed in sectors of the tunnel that are perpendicular to the direction of the greatest horizontal far field stresses (K_o of 2.2). These stress conditions provided for relatively low compressive tangential stresses on the tunnel walls. With time, the exposed chalk surface dried out, leading to volume contractions which were apparently sufficient to overcome the tensile strength of the chalk (1.5 MPa) and resulted in the fracturing. The lack of jointing within the rock mass probably contributed to fracture development as volume contraction could not take place within the rock material without affecting its integrity.

5.3 Block Fallouts from Shoulder (failure surface through intact chalk)

Block fallouts have occurred from the tunnel shoulder (Figure 5). The upper surfaces for the blocks were usually defined by horizontal shaley seams, generally less than 75 mm thick, exposed at locations between 30 and 45 degrees from the tunnel crown. The failure surfaces were generally steeply inclined and sometimes concoidal shaped. The fallout cross sections were about 150 to 300 mm deep and about 150 to 500 mm high. Each fallout was generally between 1 to 4 m long. However, where conditions favored the fallouts, they have tended to run continuously for hundred's of meters until the stratigraphy changed. The shale exposed on the upper surfaces bounding the fallouts often appeared to have dried out, but in some cases, absorption of moisture and softening of the shale layers was observed. The fallouts did not occur immediately upon excavation of the tunnel but generally took between one and four weeks to develop.

The main factor contributing to the splitting of the intact chalk appears to be extensional strains developing in the softer shaley seam once the excavation had been opened up. The strains were caused by squeezing of the shaley layers either by the compressive stresses developing about the tunnel, or by the expansion of bentonite or smectite rich shale absorbing moisture from the chalk. Despite the low permeability of the chalk, the opening has acted as a sink and moisture has flowed towards the opening. The shaley seams can absorb moisture since the natural moisture contents is no longer in equilbrium with the changed stress regime. As the shales take in moisture, they swell, or where confined, start to exert swell pressures. These expansive forces have led to splitting of the intact chalk.

5.4 Block Fallouts from Walls (failure surfaces defined by natural structural features)

Block fallouts from the walls have occurred with the upper surfaces again, defined by thin horizontal shale seams but with the back failure surfaces defined by natural fractures in the chalk (Figure 6). The natural fractures were generally high angled and oriented within 20 degrees of the tunnel alignment. Often, the fractures forming the sliding or failing surface were slickensided and black coated with evidence of moisture. These failures have occurred relatively soon following excavation but may develop gradually over weeks or months particularly if the failure surfaces deteriorate with contact with moisture.

Figure 5 Wall Fallout
(failure surface through intact chalk)

Figure 6 Wall Fallout
(failure surface defined by natural feature)

High angled fractures (greater than 70 degrees) lying subparallel to the direction of the tunnel provide the typical conditions for this failure. Sometimes, the failure of an intact rock slab by compression or buckling as the result of stresses being concentrated by the natural feature occurred. Failure was often associated with moisture or areas of seapage that led to softening or swelling of the clays and reduced the shear strength on the joint surfaces.

5.5 Failure of Shaley Layers in Crown

As weak shaley seams were exposed in the tunnel crown, they immediately broke up and back to overlying non argillaceous chalk. This was generally 50 to 500 mm above tunnel crown. Where shaley seams were first exposed in the roof, a stippled grey/black effect developed due to localized undulations in the chalk/shaley layer boundaries. Where shaley layers remained covered by a thin layer of competent chalk, the confinement of the shale was not always sufficient to prevent deterioration and the shale often broke up and voids developed in the roof. Such drummy roof conditions generally took from days to months to develop.

5.6 Heave of Tunnel Invert

In some areas where the Eagle Ford Shale lay within 3 m of the tunnel invert, floor heave has occurred. Generally, the heave was less than 100 mm and was associated with open fractures crossing the tunnel and introducing water into the underlying shale. The mechanism causing the heave has been expansion of the Eagle Ford Shale which has a high montmorillonite content and is therefore extremely moisture sensitive. The expansion of the shale has, in some cases, fractured the overlying chalk, easing the hydraulic pathway to the shale. The source of water was both natural and from construction activities.

5.7 Wedge Fallouts

Intersecting joints encountered in the tunnel, have resulted in wedge development. Wedge fallouts were not common since fracture frequency in the chalk was low. However, orthogonal joint sets have occasionally combined with the tunnel geometry to produce wedge failures.

5.8 Block Fallouts from Fault Zones

Where the tunnel alignment crossed major fault zones, broken ground was encountered. The major faults occurred, on average, about five thousand meters apart. Fault zones often intersected the tunnel over lengths of up to 30 m with poor conditions on one side of tunnel but with sound rock exposed in the other tunnel wall. The fault zones occasionally proved to be sources of significant water inflows to the tunnels (up to 1000 liters per minute) when first encountered but flows reduced within a few weeks to nominal flows. Some rock falls occurred immediately upon excavation while other blocks worked loose with time.

The presence of groundwater in the fault zones undoubtedly contributed to deterioration. It washed out fines, thereby loosening rock blocks and also lubricated rock surfaces making conditions more favorable for slippage. Since the fault zones were often supplied by the surface alluvial deposits, weather conditions could influence ground water inflows.

Figure 7 Floor Heave

Figure 8 Wedge Fallout

5.9 Equipment induced Failures

Failure of the chalk did occur under the application of the gripper pads used to provide the reaction against the tunnel walls as the tunnel boring machines were moved forward. The pads could apply pressures up to 25 MPa that were sufficient to crush the chalk (compressive strength of 15 MPa) on occasion. High pad pressures were applied when moist conditions were encountered and the development of high frictional resistance was required to move the machines forward.

5.10 Bentonite Marker Bed Deterioration

The Marker Bed has deteriorated rapidly upon exposure to air. The bed is about 500 mm thick and composed of bentonite (100<LL<150%, 50<PI<150%) which makes it extremely moisture sensitive. It was encountered over approximately 600 m length of tunnel where the 100 mm thick bentonite layer resulted in crown fallouts and large spalls from the walls.

6. CONCLUSIONS

The tunnel drive for the Supercollider has provided an opportunity to observe the performance of excavations in Austin Chalk prior to their permanent stabilization. Observations have shown that localized failures can occur in these tunnels.and were particularly associated with the regional geology, the moisture sensitive nature of layers within the chalk, and moisture conditions including wetting/drying cycles. Ground instabilities did not materially affect tunnel progress, and were stabilized temporarily by dowels, minestraps and mesh but required further support including protection of exposed surfaces from air and moisture in order to permanently stabilize conditions.

ACKNOWLEDGEMENTS: The SSC is operated by Universities Research Association, Inc for the U.S. Department of Energy. The PB/MK Team, a joint venture of Parsons Brinckerhoff Quade & Douglas Inc. and Morrison Knudsen Corp., is responsible for the design and construction management of the conventional surface and underground structures. Photographs were provided by PB/MK Design Team Engineers.

REFERENCES

Allen P. M. and W. D. Flanigan, 1986, Geology of Dallas, Texas, United States of America, Bulletin of the Association of Engineering Geologists, Volume XXIII.

Geotechnical Design Summary Report, N25 to N40 Tunnel Basic Construction, 1992, prepared by the PB/MK Team for the Universities Research Association Inc., Superconducting Super Collider Laboratory, Texas.

Geotechnical Design Summary Report, N40 to N55 Tunnel Basic Construction, 1992, prepared by the PB/MK Team for the Universities Research Association Inc., Superconducting Super Collider Laboratory, Texas.

Geotechnical Design Summary Report, N20 to N25 Tunnel Basic Construction, 1991, prepared by the PB/MK Team for the Universities Research Association Inc., Superconducting Super Collider Laboratory, Texas.

Kim, K and B. S. Schmidt, 1992, Characterization of the in situ stress for the design of the Superconducting Super Collider Interaction Hall, International Society of Rock Mechanics Symposium on Rock Characterization, Chester, England.

Nance, S, 1993, Personal Communication.

Nelson, P. P. and T. K. Lundin, 1990, Geotechnical Characterization and Construction Methods for Superconducting Super Collider Tunnel Excavation, Unique Underground Structures Symposium.

Norton, G. H., 1965, Surface Geology of Dallas County. In Dallas Geological Society, The Geology of Dallas: The Dallas Geological Society, Dallas, TX.

Olsen, R. E., J. Lai, and S. Hsu, 1993, Measurement of Rock Properties for the Superconducting Super Collider Project, University of Texas, Austin, Report to the PB/MK Team.

4 New developments in blasting

Session developers:
Dale Preece, Sandia National Laboratories
H. P. Rossmanith, Technical University Vienna

Rock Mechanics, Nelson & Laubach (eds) © 1994 Balkema, Rotterdam, ISBN 90 5410 380 8

The use of probability and risk in blast design

Tom Kleine & Bill Roberds
Golder Associates, Seattle, Wash., USA

Alan Cameron
Golder Associates, Sudbury, Ont., Canada

Bill Forsyth
Golder Associates, Vancouver, B.C., Canada

ABSTRACT: The concepts of probability and risk can be applied to blast designs and the expected results of these designs. The natural variability in some blast design parameters, its impact on blast performance predictions and the use of probability in blast design are discussed.

Measurements can be made of the variability of blast design parameters and combined with estimates of our lack of knowledge of these parameters. These are expressed in terms of frequency distributions. From these data probability distributions for the expected results of the blast (i.e. maximum ground vibration, flyrock range, fragmentation size distribution, extent of damage etc.) can be calculated. This approach determines the uncertainty in achieving the desired results from the variability of outcomes.

If the probability of a successful blast (i.e. "reliability" of the design) can be determined it can be weighed against the associated cost of implementing the design and the likely cost of failure. This leads to more informed decisions based on quantifiable success, quantifiable design parameters and estimates of the probability of failure with associated costs, particularly in critical situations. The cost of failure includes such things as reduced production due to oversized muck through to closure of operations due to excessive vibration or flyrock.

1 INTRODUCTION

The optimisation of blasting must consider the total mining system because blasting is just the excavation and preparation of the ore and waste rock for further handling. The resulting rock size distribution controls productivity of the mine and the blast results have a strong influence on recovery, support requirements and dilution. In addition, blasting causes an environmental impact in terms of ground vibration, airblast, dust and flyrock. The environmental impact of blasting can have a large impact on the mining operation, through enforced reductions in productivity, or even closure.

Blast design parameters are commonly used to predict ground vibration levels, muckpile fragmentation size distribution, extent of damage and maximum flyrock distance. Historically, blast design and prediction have only considered single value estimates for parameters (e.g. delay detonators fire exactly at their designed delay time or explosive charge weights are constant from hole to hole) and have not acknowledged the fact that during implementation a range of values for parameters will exist. This is far from reality as the data show (Figure 1). Unsurprisingly, predictions of blast results are often inaccurate due to the inhomogeneity of the rock mass, scatter in the design

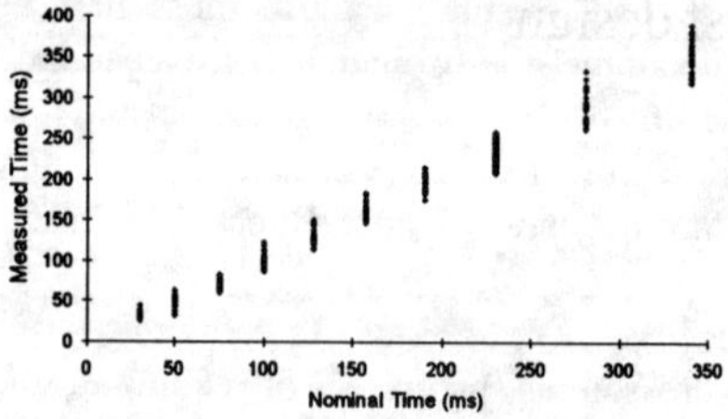

Figure 1a Delay detonator firing times

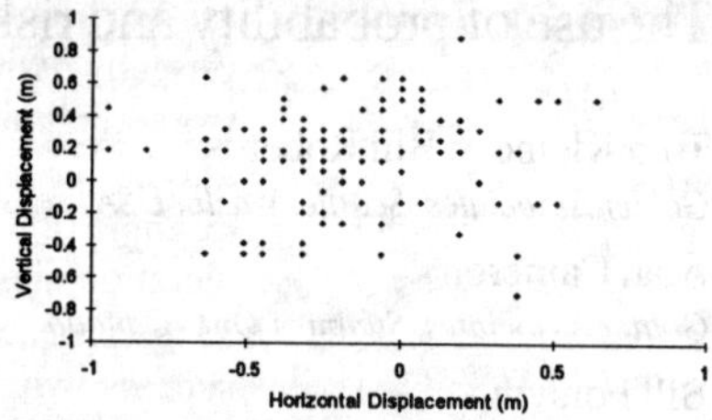

Figure 1b Blasthole alignment

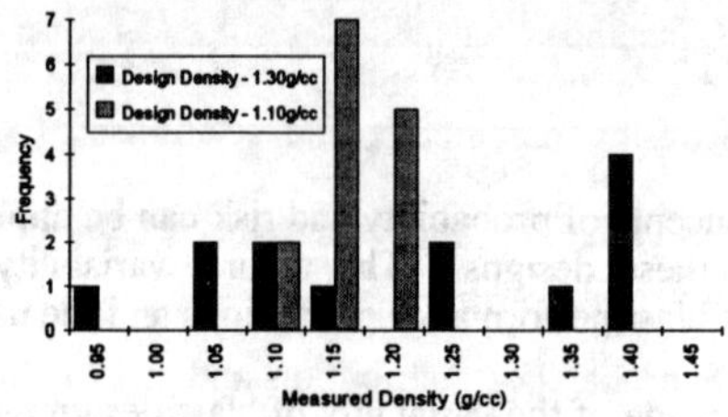

Figure 1c Explosive density

Figure 1 Examples of variations in blast design parameters

parameters when the design is implemented and errors in the empirical equations. The result is the adoption of overly conservative designs with corresponding high costs. By incorporating knowledge of parameter variability and uncertainty into the prediction of results it is possible to estimate the probability of achieving success for single blasts and over several blasting periods.

Once the variability of the design parameters is known and our uncertainty in their value quantified then the probability of achieving a particular result and the probability distribution for a range of results can be determined. The risk of not achieving a desired result is obtained from this probability distribution. These analyses are presented in detail along with examples.

The reliability of a blast design is a function of the cost of implementing the blast, the risk of not achieving the required result and the cost associated with failure to achieve this result. These factors can be linked such that the potential profitability (i.e. reliability) of a blast is determined and the distribution of profitability for a series of blasts is estimated.

It is useful to define

Probability: as the relative likelihood of a particular result. For example the probability of a random charge weight that is 10 kg or more over design might be 20%.

Probability distribution: as the set of probabilities, one per result, for every possible result. For a small number of possible results, such as rolling a fair dice, it is just a set of six equal numbers. For a very large number of possible results, such as the weight of a particular charge in a blast, it is better to draw a graph or use an equation.

Probability of failure: as the probability of a "bad" result. Failure must first be carefully defined. For example, the probability of exceeding a vibration limit might be 5.0%.

Risk: as the probability of failure and its consequences. For example, the probability of mine closure due to excessive vibration might be 5.0%.

Reliability: as the probability of success, or not failing. In economic terms success would be expressed as profit. For example, the probability of the value of broken ore exceeding the cost of the blast and clean-up might be 65%. Alternative definitions are based on achieving a minimum level of profitability, for instance a $1 million profit or a return of 10% on investment.

The steps in applying concepts of probability to blast design are:

- establish the design parameters and relationships that affect the expected result,
- understand the variability and uncertainty of the design parameters and the uncertainty of the equations,
- define failure,
- determine the probability distribution and risk associated with the required result,
- establish the income/cost function associated with the range of possible results,
- determine the reliability of the design.

2 VARIABILITY IN BLASTING

Blast design parameters include such things as; hole diameter, hole pattern dimensions, explosive density, explosive charge length, explosive energy, initiation sequence and initiation timing. These parameters determine the explosive charge weight per delay, spatial distribution of the explosive charges and the time between initiation of the explosive charges. Blast results in terms of ground vibration, flyrock, airblast, fragmentation, damage and muckpile shape are all influenced by these parameters and can be incorporated in an assessment of the risk involved in not achieving the desired results. Figure 2 shows an example of the relationships between some of the more important blast design parameters.

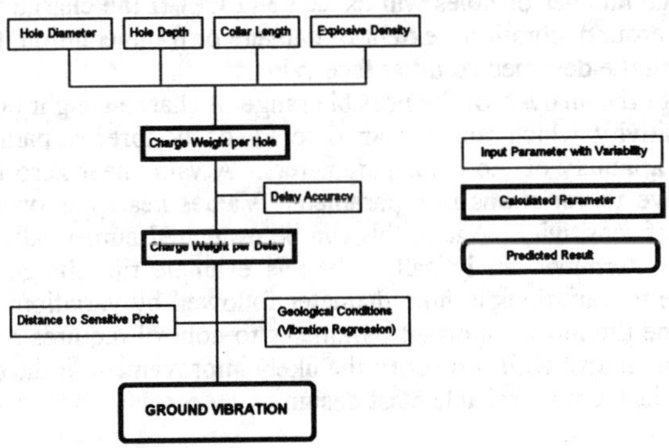

Figure 2 Example of some relationships between blast design parameters

The development of a blast design and the prediction of the results of the blast usually assumes that the design parameters are constant at their design value. During the implementation of a blast design (i.e. during the drilling operation, explosive charging and tying-in) there are a large number of factors which result in the actual blast having a range of charge weights per delay, an irregular explosive charge distribution and a range in times between the firing of explosive charges.

To implement an analysis of blast variability it is necessary to start with information of the variability of individual and independent parameters. Figure 1 demonstrates some observed variations in delay scatter, hole locations and explosive density. With numerical simulations it is possible to combine frequency histograms and equations to describe the variation of design parameters and their effect on numerous outcomes. The simple example of charge weight variability and its effect on blast vibration was chosen with which to demonstrate the concepts and their application (Figure 2).

The effect of worn drilling equipment on charge weight per unit length of hole can be a 10% reduction in hole diameter which causes a 19% reduction in charge weight per unit length of hole. Worn deck bushings or bent drill rods could cause an increase in hole diameter. A 10% increase in hole diameter corresponds to a 21% increase in charge weight per unit length of hole.

Other factors which affect the charge weight per delay are the density of the explosive and the length of the explosive charge. The density of bulk explosives can deviate significantly from design and/or can exhibit variability as shown in Figure 1. In addition to this, the depths of blast holes usually vary within a range of ± 0.5 m and the collar length varies within a similar range. These factors all contribute to variability in the mass of the explosive charges, which can be evaluated if the frequency distributions of the controlling factors (i.e. blasthole diameter, explosive density and explosive charge length) are known. There is little uncertainty in this relationship.

An analysis of the possible range of explosive charge weight per hole for a typical blasting situation is shown in Figure 3. It was assumed that the blast design parameters were independent (i.e. unrelated) and normally distributed in a single random hole. If we also assume that the holes are independent, then the uncertainty in charge weight for a single hole becomes the variability among many holes. While it is possible to use analytical expressions for a problem of this simplicity, in general it is necessary to adopt the numerical approach used in the example. The variability of the actual blast design parameters used in the analysis are also shown in the figure.

Results from the analysis show a wide range of charge weights (i.e. ±100 kg) with a mean of 314.6 kg, which compares with the designed charge weight of 314.2 kg. In fact, there is a probability of 16% that a random charge weight will exceed 341 kg. Again, assuming independence among the blast holes, this means that the charge weight in 16% of a large number of holes will exceed 341 kg. If the charge weight per hole is used to control ground vibrations, extent of damage or fragmentation, then care must be taken in accepting the designed result at face value.

Part of the analysis of the possible range of charge weight per hole is an assessment of the sensitivity, which ranges from -1 to +1, of the forecast parameter (i.e. charge weight) to the variability of the input parameters. A value near zero indicates that the result is insensitive to variations in a parameter. Values near plus or minus one indicate a high degree of sensitivity. Again, this can be estimated numerically or for simple problems it can be determined analytically. In this example the charge weight per hole is most sensitive to variations in hole diameter followed by variations in explosive density. To determine the most important parameter to control requires a knowledge of the cost of the extra control with respect to the likely improvement in the blast result. Extra control should yield a more reliable blast design.

Parameter	Mean	Standard Deviation	Sensitivity
Avg Hole Diameter (mm) per hole	200.00	5.00	0.59
Avg Explosive Density (g/cc) per hole	1.25	0.06	0.58
Hole Depth (m)	12.00	0.25	-0.36
Collar Length (m)	4.00	0.25	0.35

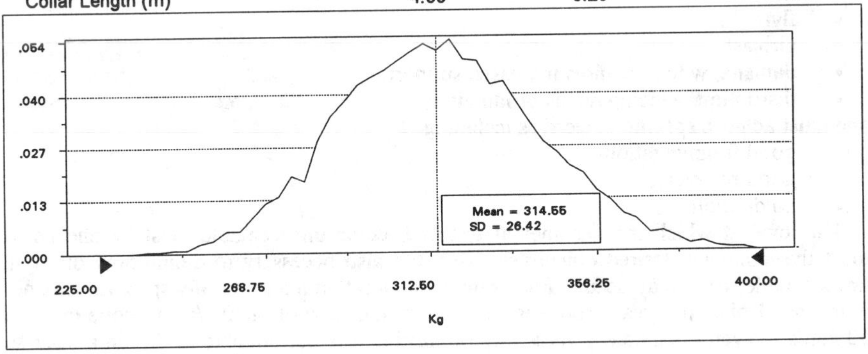

Figure 3 — Forecast variability of charge weight per hole

Other questions that can be answered are the probability of exceeding a certain charge weight in, say, 10 blasts of 200 charges each. This question typically arises when blasting practices are evaluated. It is a slightly more complex analysis. Assuming independence, one could start by choosing 200 charge weights at random from the distribution in Figure 3 and determining the peak weight. This would be one possible peak weight. The probability distribution for the peak charge weight of a 200 charge blast has an average that is higher than the average charge weight per blast, and a standard deviation that is smaller.

The cumulative "extreme value" distribution can be evaluated assuming that individual charges are independent and identically distributed (Benjamin & Cornell 1970):

$$F_Y(y) = [F_X(y)]^n \tag{1}$$

Where:

$F(y)$	cumulative distribution (probability of being less than y)
y	charge weight
F_X	probability of a single random charge weight
F_Y	probability of multiple charge weights (single blast)
n	number of charges per blast

Using the peak probability equation it is possible to find the approximate distribution for the peak charge weight for any number of charges per blast. The equation is applicable to other analyses of maxima or minima of repeatedly sampled distributions, such as boulder size or peak vibration — in fact, whenever best or worse case scenarios are being analysed. However, it is important to note that a fundamental assumption in this equation is that each variable (e.g. single charge weight) is independent. When it is applied to multiple blasts this means that blast practices are assumed to remain constant. In practice it is often possible to improve poor designs by observing actual blast results.

3 RISK ASSESSMENT AND BLASTING

Risk is a function of what is considered unacceptable and how likely that unacceptable result is. Blasting may produce numerous detrimental results including:

- high vibration,
- flyrock,
- airblast,
- damage, which requires increased support,
- disruption, which reduces productivity,

and must achieve specific objectives including:

- good fragmentation,
- good recovery,
- no dilution.

The level at which the detrimental effects become unacceptable must be chosen as must the minimum desired objectives. But it is also necessary to define how often an unacceptable result may occur. For example, a vibration limit typically specifies that not more that 1 blast per year produces a vibration in excess of 18 mm/s and none exceeds 50 mm/s. Given the inherent variability of blasting it is not possible to design a blast so that it never produces an unacceptable outcome. However, in the life of the mine there might be 1000 blasts, running at 100 per year. This translates to two risk points, one at (18 mm/s, 1.0 %), the second at (50 mm/s, 0.1 %) if we allow one blast to exceed 50 mm/s in 10 years.

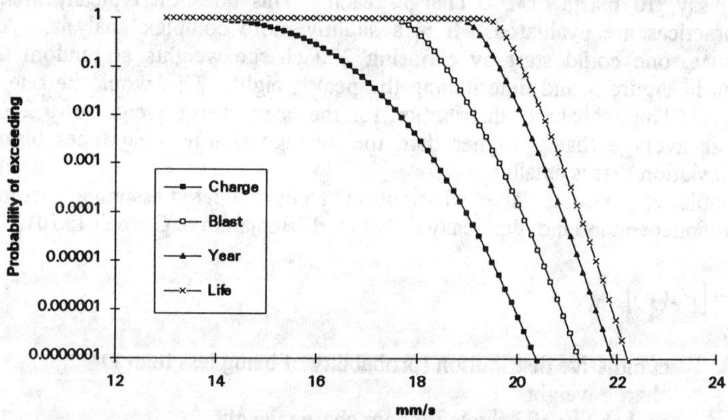

Figure 4 — Peak Vibration Risk Graph

Figure 4 shows the cumulative probability distribution of peak vibration for a single random charge, which was determined from a scaled distance relationship for vibration and the variability analysis in Figure 3. No uncertainty was assumed in the calculation of the vibration. The effect of model uncertainty may be included relatively easily as an error parameter that alters the calculation by a random amount.

The cumulative distribution shows the probability of not exceeding a particular vibration level. Figure 4 also shows the combined probability of exceeding a vibration level during a 200 hole blast, a 100 blast (1 year) period and a 10 year (mine life) operation. The curves were obtained from Equation 1, assuming a normal distribution for the distribution of peak vibration from a single random charge (Figure 3). In this example there is a significant risk that a blast will exceed the limits during one year despite easily meeting the apparently more stringent requirement.

4 RELIABILITY OF A BLAST DESIGN

A design's reliability is a measure of the trade-off between the blast's cost and its benefit. This can be expressed in terms of the net return, in which case the benefit must be measured in the same units as the cost. (i.e. If blasting costs are expressed in dollars the benefit must be also.) The blasting cost must include not only the cost of implementation but also the cost penalties for unacceptable results, which can be expressed as decreased benefits or as direct penalties that are independent of the benefits and can in fact exceed the benefits. For example, excess damage introduces a direct cost penalty for extra support whereas bad fragmentation may result in a lower quality product that has less value. In short, every combination of outcomes produces a benefit from which is subtracted the total cost of that outcome.

This information is combined with the probability density that describes the likelihood of, say, bad fragmentation to calculate the net-return probability density function. The calculation may be complicated by having more than one blast outcome that yields the same net-return. The *reliability graph* (Figure 5) shows the probability of exceeding any hypothetical profit level. The *simple reliability* ($R_{\$>0}$) is the probability of not making a loss. Reliability graphs (like risk graphs) may be calculated for multiple blasts and multiple operating periods, although in this case it is not the maximum profit but the average profit that must be used.

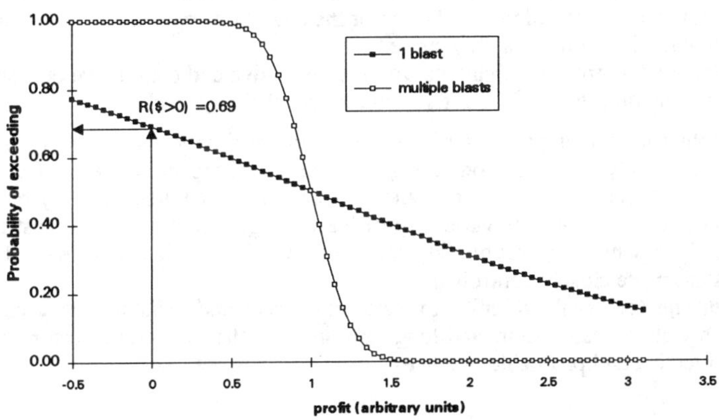

Figure 5 — Typical Profit Reliability Graph

Assuming independence among blasts, the probability of exceeding the average net-return per blast decreases as the number of blasts increases, while the probability of not achieving the average net return decreases. Simply, after many attempts (multiple blasts) the overall outcome (and hence return) will be close to the average outcome with very little likelihood of deviation.

The implication is that it is possible to tolerate a certain error or uncertainty provided that there are enough attempts (at blasting) to even out the bad performers with good performers. The analysis quantifies the risk (probability of making a loss) after a certain number of attempts. It provides a basis for evaluating alternative strategies in terms of the risk of losing money (or achieving a lesser return) in the short term (i.e. cash flow per blast or per period) versus the long term higher return (per year). It may be possible to develop a blasting strategy that, due to less blasting control, is cheaper to implement though it has higher variability. Providing that it returns the same or higher mean profit (at greater risk) and with sufficient resources it might be better to use this more risky short term strategy because short term losses can be survived. In the longer term the seemingly risky strategy may be more reliable.

5 CONCLUSIONS

There are three broad steps to applying probability, risk and reliability concepts in blasting.

Understand the variability and uncertainty:

- Determine the parameters and relationships to predict the outcome.
- Determine the variability and uncertainty in the parameters and relationships.
- Calculate the probability distribution of the outcome.
- Determine the sensitivity of each parameter.

Determine the risk:

- Define the criteria for failure or success.
- Determine the consequences of failure.
- Calculate the risk curve for comparison with risk points

Determine the reliability:

- Define a penalty function for the variation of an outcome.
- Calculate the probability distribution of the amount of return.
- Calculate the profit reliability curve.
- Calculate the profit reliability curve for consecutive and cumulative outcomes.
- Compare long term results for various profit reliability levels.

The simple analysis of parameter variability and sensitivity can be conducted very quickly using readily available computing tools and desktop computers. Even if the second and third steps to examine risk and reliability are not pursued, every competent engineer should understand the variability of blast design outcomes for a blast's critical objectives. The sensitivity information alone can yield rapid improvements if critical parameters are more closely controlled.

Although the technical difficulty of calculating risk and reliability is greater, the analyses can yield valuable short and long term insights that can guide blasting to more rigorous, objective and profitable practices.

6 ACKNOWLEDGMENTS

The authors wish to thank Golder Associates for permission to use the variability data and scaled distance relationship, which were taken from various technical studies conducted by that company.

REFERENCES

Benjamin & Cornell, 1970. *Probability, Statistics and Decision for Civil Engineers*, McGraw Hill, **3.3.3**, pp. 271-85.

Rock Mechanics, Nelson & Laubach (eds) © 1994 Balkema, Rotterdam, ISBN 90 5410 380 8

Survivability of shotcrete near blasts

D.R.McCreath
School of Engineering, Laurentian University, Sudbury, Ont., Canada
D.D.Tannant
Geomechanics Research Centre, Laurentian University, Sudbury, Ont., Canada
C.C.Langille
INCO Ltd, Ontario Division, Copper Cliff, Ont., Canada

ABSTRACT: Reinforced shotcrete maintains its functionality and sustains only minor damage when exposed to peak particle velocities of 1.5 to 2 m/s (60 to 80 in/s) due to nearby blasting. Mesh reinforced shotcrete performs better than steel fibre reinforced shotcrete under very severe dynamic loading conditions. Mesh reinforced shotcrete can continue to retain broken rock even when extensively cracked, whereas fibre reinforced shotcrete loses this ability and may itself become a safety hazard.

INTRODUCTION

In Canadian hardrock mines, shotcrete is finding increasing application as one element of standard support systems used in production areas (Morrison & Beaudry 1993). Historically, shotcrete was generally used only where extra safety or security was required, such as in permanent crusher stations or in personnel refuge stations, and was viewed as a high-cost additional item. Recent experience suggests that shotcrete may play a valuable and cost-effective role in standard ground support systems within production areas because of its ability to withstand repeated dynamic loading from near-by blasting or from moderate levels of rockburst activity. The most common ground support systems in production areas of hardrock mines in Ontario and Quebec are a combination of mechanical rockbolts, resin-grouted reinforcing bars, and welded-wire mesh screening. However, this support is prone to severe damage when installed close to production blasts. In critical areas of a mine, such as bottomsills, drawpoint troughs and brows, and topsills, this may lead to recurring requirements for rehabilitation of the support, adding substantially to the costs, delays, and risks associated with mining.

This paper first summarizes some recent evidence, based largely on empirical observations in operating mines in the Sudbury (Ontario) area, which indicates that inclusion of mesh-reinforced shotcrete as an element of a support system significantly increases the toughness and survivability of the system under conditions of intense dynamic loading, and which suggests that mesh-reinforced shotcrete may remain partially functional when subjected to peak particle velocities as high as 2 to 6 m/s (60 to 80 in/s). The paper then presents some data from a series of full-scale tests undertaken at the Bousquet mine in Quebec which substantiate the previous observations regarding the effectiveness of shotcrete under conditions of dynamic loading. Comparative performance of fibre-reinforced and mesh-reinforced shotcrete is also addressed.

OBSERVATIONS FROM SUDBURY AREA MINES

Since 1989, Inco Ltd. has been investigating the use of shotcrete for support in critical production areas of their Sudbury area mines, most notably the Frood-Stobie Complex, Creighton Mine and North Mine. An early use of mesh-reinforced shotcrete was in the topsill drifts and final sub-level cave cross-cuts for recovery of the sill pillar at Frood Mine at a depth of about 400 m (1,200 ft), where conventional support systems had required constant rehabilitation due to blast damage. Despite relatively poor installation practice due to lack of experience, the results were sufficiently encouraging to suggest that wider use could be made of shotcrete in areas prone to blast damage (Morrison & Beaudry 1993).

Over the last several years, trials have been underway with welded-wire reinforced shotcrete in various production areas at 2,200 m depth (7,200 ft) in the Creighton Mine (Langille & Burtney 1992). The damaging effects of continual blasting associated with Vertical Retreat Mining (VRM), the cumulative effects of multiple low-energy rockbursts, the corrosive nature of the high-sulphide groundwater, and the fractured nature of the pillars between stopes resulted in rapid deterioration of the conventional rockbolt, cable bolt and welded-wire screen support systems, requiring constant rehabilitation of these systems. As indicated in Figure 1, ground support in both the topsills and the bottomsills is subjected to continual blast loading as the VRM stope is mined, and in the case of the bottomsill drawpoint, the initial blasting occurs within only a few metres of the drawpoint brow. Due to high induced stresses and effects of deliberate destress blasting when excavating development drifts, there is generally an envelope of broken rock around these tunnels. Despite the use of welded wire mesh with closely spaced resin-grouted rebar support, and the installation in the brow of two rings of cable bolts comprising 6.4 m (21 ft) long twin 7-strand grouted cables, loss of ground in the brow and damage to the bottomsill support was a continuing problem, resulting in rehabilitation costs and delays.

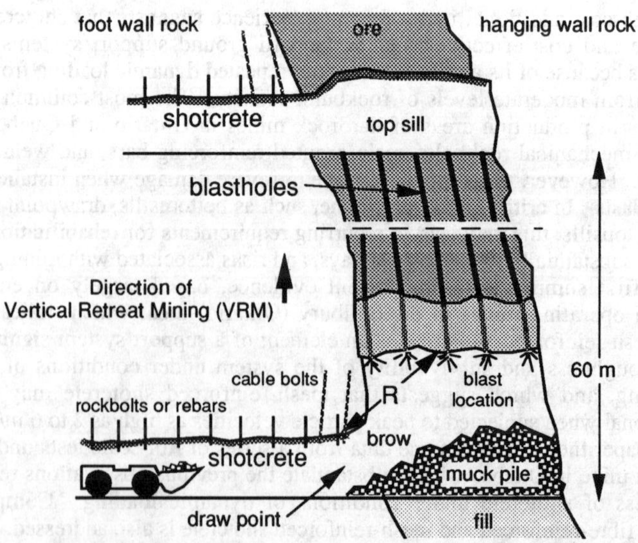

Figure 1 Shotcrete support in top and bottom sills for vertical retreat mining

In 1990-91, welded-wire reinforced shotcrete was used in the development and rehabilitation of selected bottomsills on the 7200 level, installed after the rock had been deliberately fractured using heavy destress blasting to minimize the danger of violent strain-type rockbursts. In the first trials, the shotcrete was placed over top of the regular support pattern of welded-wire screen with 2.4 m mechanical bolts on a 1.2 m pattern, split-spaced with a 1.2 m pattern of 2.4 m resin-grouted dowels. Rings of 6.4 m long cable bolts were placed in the drawpoint brows. In recent trials, the cable bolts have been omitted. Because shotcrete was only used in selected bottomsills, it was possible to directly compare the performance of shotcreted and non-shotcreted areas under similar conditions. Observations by the ground control staff at the mine showed clearly that the shotcrete had greatly increased the survivability of the support systems, and had either decreased or eliminated the need for rehabilitation. Despite cracking of the shotcrete, no shotcrete slabs were displaced or ejected by the blasting, and hammer sounding of the shotcrete did not indicate significant increases in drumminess as mining progressed.

Although no direct measurements were made of the peak particle velocities (ppv) to which the shotcrete had been subjected, it was possible to gain an approximate idea of the likely range of ppv values which the shotcrete had survived. Conventional square-root and cube-root scaling laws were used to predict the ppv levels at the location of the drawpoint brows (Nicholls et al. 1971; Ambraseys & Hendron 1968). The initial VRM blasts occurred at a distance of approximately 3 m (10 ft) from the brow, with a charge per delay of 60 kg (125 lbs). A range of values were used for the site constants and attenuation factors required in the scaling laws, based both on published values and on values back-calculated from the peak particle velocity data recorded at a series of triaxial geophones which had been installed in other areas of the mine as part of a seismic monitoring system. Based on these calculations, it was estimated that peak particle velocities of between 2 m/s (80 in/s) and 6 m/s (235 in/s) had been experienced, and survived, by the shotcrete at the drawpoint brows.

An initial trial was recently completed at Inco's Stobie Mine to further clarify and quantify the behaviour of shotcrete under blast loading, and to investigate the possibility of using mesh reinforced shotcrete as a primary support system, without the use of bolts or dowels. Figure 2 shows, schematically, a longitudinal section of a sub-level cave mining slot at the 2400 level. The test drift was supported with welded-wire mesh reinforced shotcrete with the mesh attached to the back and walls with 0.2 m long (9 in) Hilti nails. No rockbolts or dowels were installed. Instrumentation consisted of five geophones (3 mounted to the rock and 2 mounted to the shotcrete) and six 5-pin tape extensometer convergence stations. Five blasts, involving one or two rings of blastholes per blast, were taken over a period of four weeks resulting in increasing levels of vibration as the blasts approached the test sections.

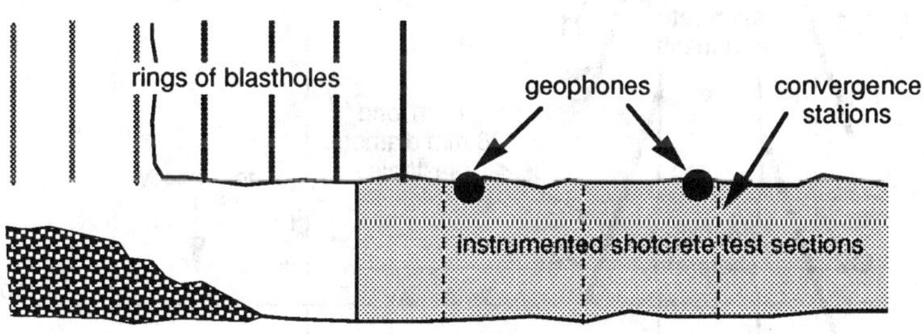

Figure 2 Longitudinal section (schematic) showing shotcrete test sections at Stobie Mine 2400 foot level

The results of this initial boltless shotcrete trial are somewhat inconclusive due primarily to problems with the instrumentation (a difficult issue in the environment of an operating mine) and application of the shotcrete during development. On a tentative basis, it does appear that the use of shotcrete alone is not as effective as a composite bolt-dowel-shotcrete system, which is not unexpected. The maximum ppv recorded in the shotcrete during the blasts was 0.4 m/s (15 in/s), which is substantially lower than the previously noted values.

A second, more comprehensive trial is planned for Stobie mine during 1994 to evaluate shotcrete and mesh attached with different holding elements including short, 0.5 m (18 in) closely spaced Split-Set Stabilizers, 0.2 m (9 in) Hilti bolts and standard resin grouted rebar. The effect of blast vibrations on these support combinations will be determined and the applicability of mesh reinforced, boltless shotcrete as a primary support system will be assessed.

BOUSQUET FIELD EXPERIMENT

In order to more specifically evaluate the performance of shotcrete-based support systems subject to nearby blasts, a carefully controlled and monitored field experiment was conducted at the Bousquet #2 gold mine in northern Quebec. Figure 3 shows the location of six test sections in highly foliated sericitic schist in the footwall of the mine at the 6-1 level, about 700 m below surface. Welded-wire mesh reinforced shotcrete and steel fibre reinforced shotcrete were monitored at distances ranging from 1 to 5 m from single blastholes loaded with ANFO. The shotcrete was used in conjunction with mechanical rockbolts and grouted tubular (pipe) bolts. The shotcrete was only sprayed on the walls, which were thoroughly washed, and was applied in a 75 to 80 mm nominal thickness. In all cases a bulk bag, dry-mix method was used.

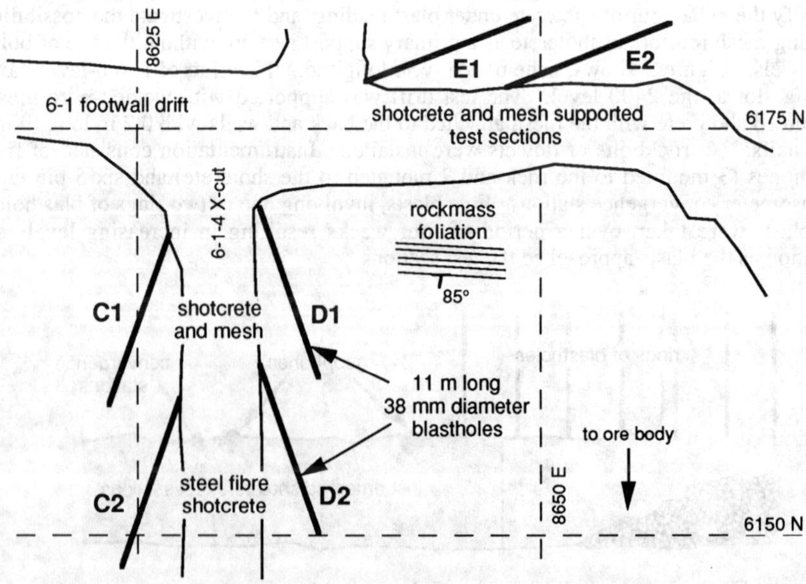

Figure 3 Test sections and blastholes at shotcrete monitoring site

Each test section contained a 38 mm diameter blasthole that was collared at mid-height of the drift wall, drilled horizontally, and inclined at 20° to the wall (Fig. 3). The inclination of the borehole relative to the drift wall was designed to generate a variation in the blast loading intensity along the drift. Particular attention was paid to the selection of instrumentation to measure ground motions during the test blasts. Oyo uniaxial long travel geophones (McSeis-101 LT, 8 Hz) were mounted both to the surface of the shotcrete and to the rock surface beneath the shotcrete layer. Tannant et al. (1994) describe the response of the shotcreted test sections in greater detail.

Results

Careful field observations coupled with mapping and measurements formed the basis for characterizing the extent and nature of the damage caused by the blasts, i.e., the performance of the shotcrete. The relative performance of the shotcrete was based on comparisons of the volume of rock and shotcrete ejected, observations of the fracturing in the test panels, and the degree of drumminess behind the shotcrete.

The use of reinforced shotcrete reduced the volume of rock ejected during the blasts by a factor of 1.5 to 6.5 when compared to similar blasts conducted in a cross-cut that was not supported with shotcrete or mesh. In addition, shotcrete and mesh shortened the length of the ejected zone. It was noted that the condition of the rockmass and more specifically the orientation of the dominant foliation relative to the shotcrete layer controlled to a substantial degree the extent of damage and the volume of rock and shotcrete ejected during essentially identical blasts.

Figures 4 and 5 show a summary of typical data and observations from test sections C2 and D1 respectively, where section C2 was supported using steel fibre reinforced shotcrete and bolts, and section D1 was supported with welded-wire mesh reinforced shotcrete and bolts. Measured values of peak particle velocity are indicated, along with the contours of predicted ppv values based on the scaling relationship by Holmberg and Persson (1979). These figures also show the area over which the shotcrete was ejected, the tonnage of rock ejected, the fracture patterns in the remaining shotcrete, and the locations of geophones and failed bolts.

Reliable measurements of peak particle velocity are difficult to obtain very close to a blast hole, due to the response characteristics of available instrumentation as well as the natural variability in the response of the rock–support system, thus there is considerable variability in the measured ppv values at different locations along the test sections. Nevertheless, several tentative conclusions can be drawn from the data indicated in Figures 4 and 5. First, the area of ejected shotcrete and the volume of ejected rock was about twice as high for the fibre-reinforced section (C2) than the mesh-reinforced section (D1). Second, there is good general correspondence between the measured and predicted ppv's, except at locations within a few centimetres of the edge of the ejection zones (e.g., Fig. 4). Third, the results indicate that both types of shotcrete were able to survive (i.e., remain in place) at ppv values in the order of 1.5 to 2 m/s (60 to 80 in/s). However, in terms of maintaining functionality by continuing to effectively retain broken rock in the drift wall, despite fracturing of the shotcrete, differences were noted between the two types of shotcrete, as discussed below. Fourth, it is evident that the mesh-reinforced shotcrete was able to survive at smaller distances from the blast, i.e., at higher ppv values. Based on the predicted ppv contours as shown, it is suggested that the mesh-reinforced shotcrete was actually able to survive ppv's up to about 3 m/s (120 in/s), compared to about 2 m/s (80 in/s) for the fibre-reinforced shotcrete.

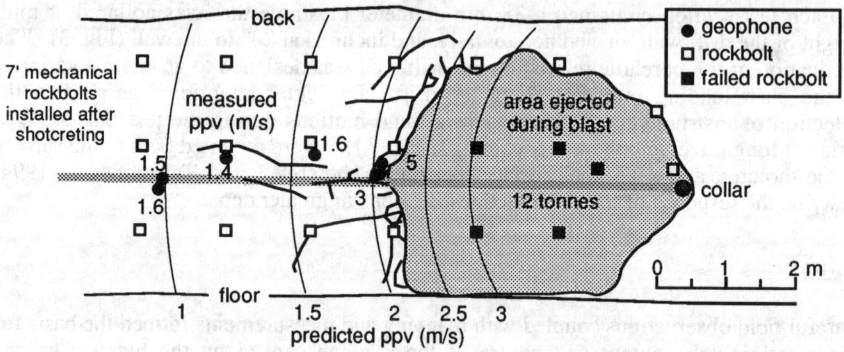

Figure 4 Section C2 supported with steel fibre reinforced shotcrete and mechanical rockbolts showing ejected area, shotcrete fractures, instrumentation, and rockbolts

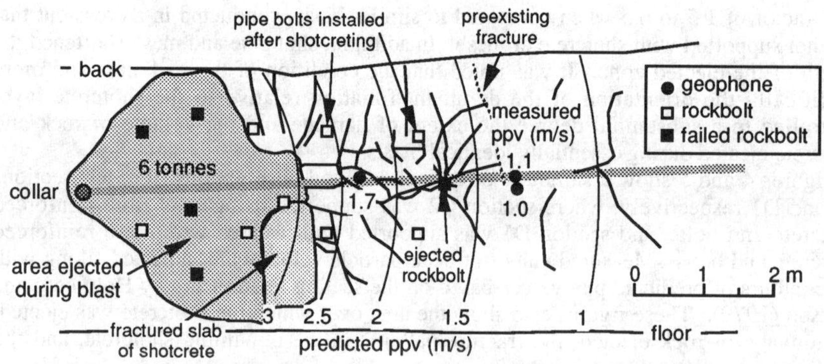

Figure 5 Section D1 supported with welded-wire mesh reinforced shotcrete and rockbolts showing ejected area, shotcrete fractures, instrumentation, and rockbolts

The ppv values at which the shotcrete support systems were able to survive were higher than those which we expected based on previous work by others on blast damage thresholds. In general, new fracturing of rock or concrete is expected at ppv's of about 0.5 m/s (20 in/s), with complete break-up of rockmasses expected to occur at ppv's in the order of 2 m/s (80 in/s) (Dowding 1985; Atlas Powder Co. 1987). Thus, the reinforced shotcrete systems were able to survive well beyond the threshold ppv levels at which rock fracturing would be anticipated.

Although it would appear that both the fibre-reinforced and the mesh-reinforced shotcrete were able to withstand high peak particle velocities, there was an important difference in the behaviour of the two materials in the region near the ejection zone, where the remaining shotcrete existed in a severely cracked condition, in terms of the functionality of the two systems. While there were fewer large fractures in the fibre-reinforced than in the mesh-reinforced shotcrete, where large fractures did occur the fibre-reinforced material did not provide an effective retaining function, and actually posed a safety hazard in some areas. The reason for the difference is related to the greater toughness of the mesh-reinforced system under conditions of large displacement.

When fractures wider than about 3 to 5 mm (0.1 to 0.2 in) were created, the short steel fibres were broken or pulled out of the shotcrete, losing their reinforcing function and allowing a mosaic of disconnected shotcrete plates or panels to be formed. Coupled with the fact that the intense dynamic loading reduced or eliminated the effective shotcrete–rock adhesion, due either to direct debonding or to formation of slabbing fractures in the rock, the disconnected shotcrete panels provided little or no effective retaining function.

Under the same conditions, however, the wire-mesh reinforced shotcrete was significantly more robust. Even with the formation of large open cracks of 5 to 10 mm (0.2 to 0.4 in), which often caused the breakage of several strands of the mesh, some of the wire strands remained intact and provided continuity and connection between the otherwise separated plates of cracked shotcrete. In addition, the wire mesh provided stronger and tougher connections between the shotcrete and the bolts. Overall, the wire-mesh reinforced shotcrete system continued to provide a measure of functionality in retaining the broken rock, and thus maintaining a "ground arch" around the opening by acting as the type of "supermesh" which is desirable for conditions of continued dynamic loading from blasting or rockbursts, as described by McCreath and Kaiser (1992).

The role of rockbolts in the performance of shotcrete systems under intense dynamic loading is not well understood. This issue is currently under investigation, as noted previously, and the results will be reported at a later stage. However, several tentative observations from the Bousquet tests are of interest. Clearly the bolts play a role in resisting the radial ejection of material. In addition, in a hard rock environment under high stresses, it appears that the bolts may also play an important role in helping to effectively mobilize the composite ground arch (comprising the broken rock and the shotcrete), by helping to maintain intimate contact and a high tangential shear strength between the shotcrete and the rock elements. This mechanism is also under investigation. In terms of these two general functions of the bolts, the Bousquet tests indicated that both the mechanical (ungrouted) rockbolts and the grouted pipe bolts tended to act as stiff elements under dynamic loading, resulting in displacement incompatibilities between the bolts and the shotcrete which caused fracturing in the shotcrete to radiate from the bolt locations. In all cases, bolts which were installed through the shotcrete after it was placed seemed to perform better in terms of resisting radial ejection and in terms of maintaining interface contact than bolts which were installed prior to shotcreting.

CONCLUSIONS

Welded-wire mesh reinforcement provides substantial toughness to shotcrete in the post-fracture region, i.e., after the shotcrete has been cracked and then subjected to further dynamic loading and/or deformations. This observation is consistent with the results of static laboratory tests of shotcrete panels recently reported by Kirsten (1993). Even when the shotcrete is highly fractured, and several sections of wire strand are broken, the remaining wires serve to interconnect the shotcrete "panels". The mesh also ties the panels back to the rockbolt holding elements, and effectively reinforces the shotcrete-bolt connections. Overall, the key function of maintaining safe access to an area under conditions of extreme dynamic loading and/or large deformations is well served by a support system which includes mesh reinforced shotcrete.

The performance of steel-fibre reinforced shotcrete is less acceptable under the extreme loading and deformation conditions found in these mining situations. The primary advantage of the weld-mesh reinforcement is in its more effective action as a net of catenary wires at extensive deflections, once the shotcrete is substantially cracked, whereas the steel fibres pull out of the shotcrete under these conditions.

ACKNOWLEDGEMENTS

This work was funded by the Mining Research Directorate as part of the Canadian Rockburst Research Program. The assistance and cooperation of staff from the Bousquet #2, Creighton, and Stobie mines is gratefully acknowledged. We are also grateful for the contributions by GRC staff.

REFERENCES

Ambraseys, N.N. and Hendron, A.J. 1968. Dynamic behaviour of rockmasses. in *Rock Mechanics in Engineering Practice*. Wiley and Sons.

Atlas Powder Company. 1987. *Explosives and Rock Blasting*. Field Technical Operations, Dallas, Maple Press.

Dowding, C.H. 1985. *Blast Vibration Monitoring and Control*. Prentice-Hall Inc., Englewood Cliffs, N.J.

Holmberg, R. and Persson, P.A. 1979. Design of tunnel perimeter blasthole patterns to prevent rock damage. *IMM Proc., Tunnelling*, London, 280-283.

Kirsten H.A.D. 1993. Equivalence of mesh- and fibre-reinforced shotcrete at large deflections. *Can. Geotech. J.*, 30: 418-440.

Langille, C.C. and Burtney, M.W. 1992. Effectiveness of shotcrete and mesh support in low energy rockburst conditions at Inco's Creighton mine. *Rock Support in Mining and Underground Construction*, Balkema, 633-638.

McCreath, D.R. and Kaiser, P.K. 1992. Evaluation of current support practices in burst-prone ground and preliminary guidelines for Canadian hardrock mines. *Rock Support in Mining and Underground Construction*, Balkema, 611-619.

Morrison, D.M. and Beaudry, F. 1993. Application of shotcrete for ground support at Inco Mines. *Shotcrete for Underground Support VI*, A.S.C.E., 173-178.

Nicholls, H.R., Johnson, C.F. and Duvall, W.I. 1971. Blasting vibrations and their effects on structures. *US Bureau of Mines Bulletin* 656.

Tannant, D.D., McDowell, G.M. and McCreath, D.R. 1994. Shotcrete performance during simulated rockbursts. *Integral Approach to Applied Rock Mechanics, IV South American Congress on Rock Mechanics*, Santiago.

Rock Mechanics, Nelson & Laubach (eds) © 1994 Balkema, Rotterdam, ISBN 90 5410 380 8

Effect of 'crack-like' defects on rock fracture

Dal Sun Kim
University of California, San Diego, Calif., USA

ABSTRACT: Experiments were designed to test the hypothesis that the energy to crush explosively damaged rock is less than that for intact rock. It has been found that ultrasonic velocity and density of "crack-like" defects are closely related to the magnitude of energy dissipation in the form of new fracture surface energy. Crack-like defects increase as shock energy dissipated in rock material increases. Measurements following shock loading indicate that ultrasonic velocity in rock partially recovers with time. It appears that compressive strength is independent of shock-treatment. Energy efficiency required for creating cracks in less competent (sedimentary) rock may be much lower than for more competent (igneous) rock. Shock induced damage on fragmentation is measurable at the levels of explosive energy delivered by the split "Hopkinson" pressure bar device.

1 INTRODUCTION

Rocks can be generally identified as aggregates of mineral particles. These mineral particles may be held together by a bonding material (matrix) at the grain boundaries or by interlocking of grains. Thus, the apparent strength of rock depends on the combined local strength of the matrix and the contact area between the grains (Farmer, 1968). It is known that rock materials contain numerous crack-like defects (or discontinuities) defined by mismatching contact areas along grain boundaries. The terminology "crack-like" defects (or damage) as used in this paper includes defects at the microscopic scale generated at the time of rock formation or subsequent stress application.

Since the energy efficiency in comminution (crushing and grinding) of rock materials is substantially low (National Materials Advisory Board, 1981), the integration of improved mineral extraction technology with understanding of rock failure mechanism may lead to cost reductions in mining operation. Rock breakage is an energy-intensive process and quite inefficient. Considerable effort has been devoted to optimizing the design of machinery used in mineral processing and to understanding breakage physics of rock. Complicated and ubiquitous failure processes hinder complete understanding of rock failure. Furthermore, the effects of blast-induced damage on conventional comminution have not received much attention. The objective of this research is to investigate the effect of shock-induced crack-like defects on rock properties and subsequent breakage efficiency for petrologically distinct rock materials under laboratory conditions.

2 DAMAGE VARIABLE THEORY

Griffith (1921, 1924) proposed that a glass-like elastic solid is inherently filled with microscopic cracks which behave as stress raisers under an externally applied traction force. A drawback of Griffith theory is that it is difficult to define the stress field for anisotropic cracks distributed in a solid material. Multiple cracks create a complex state of stress and make quantitative calculation impractical. It is also reported that elastic and inelastic deformation accompany crack propagation, and microcracks are zoned around the crack tips in rock material. This zone is called the fracture process zone (FPZ) which is experimentally defined by many researchers (Hoagland et al., 1973; Labuz, 1985; others). Because of countless sites of crack-like damage and resulting computational impracticality of fracture mechanics, the concept of damage mechanics may be more appropriate for cracked materials.

Kachanov (1986) first proposed damage theory for assessing metallic material damage during creep deformation. Damage in a material is accumulated by initiation and growth of crack-like defects with an increase in externally applied load. Change in Young's modulus or compressibility of rock can reflect internal damage development (Walsh and Brace, 1966; Lemaitre, 1984). The ratio of Young's moduli in intact (E_o) and cracked (E_n) solids can be approximated by a crack density (ζ)

$$\text{(1)} \qquad \frac{E_n}{E_o} \simeq \exp(-\alpha\,\zeta) \ ,$$

where α is a constant depending on the Poisson's ratio (Bruner, 1976). According to ASTM (C769-80), P-wave velocity (V_p) can be approximately equal to the bar velocity ($C_b = \sqrt{E/\rho} \simeq V_p$). Then, the ratio of P-wave velocities yields

$$\text{(2)} \qquad \left\{\frac{V_p'}{V_p}\right\}^2 = \frac{E_n}{E_o} \ .$$

Kim (1993) first proposed that the damage variable can be experimentally determined by measuring ultrasonic velocities

$$\text{(3)} \qquad \omega = 1 - \left\{\frac{V_p'}{V_p}\right\}^2 \ , \qquad (0 \leq \omega \leq 1)$$

where V_p and V_p' are P-wave velocities for a pre- and postshock rock specimen, respectively. Postshock specimens were obtained by employing the split Hopkinson pressure bar (SHPB) designed to measure the energy loss in the specimen. The energy (W) associated with a stress wave is given by

$$\text{(4)} \qquad W = \frac{A_b\,C_b}{E_b} \int_o^{t_o} \sigma^2 \, dt \ ,$$

where A_b, C_b and E_b are the pressure bar cross-sectional area, the bar velocity and the Young's modulus, and σ the stress at time (t). Employing equation (4), the energy (W_L) dissipated in the postshock specimen is quantified by

$$\text{(5)} \qquad W_L = W_I - (W_R + W_T),$$

where W_I, W_R and W_T are the incident, reflected and transmitted energy. The detailed experimental procedure and data analysis can be found elsewhere (Kim, 1993).

3 EXPERIMENTAL TECHNIQUES

Quartz monzonite (QM), quartz latite porphyry (QL), diopside wollastonite (DW) and subarkosic siltstone (SS) were chosen for the study because they possess petrologically distinct textures. Preparation of cylindrical specimens, measuring a nominal 3.190cm in diameter with a length-to-diameter (L/D) ratio of 2.0, complied with ASTMs (D2938-79, D3148-80) and ISRM suggestion (1979). Petrographic analyses were used to identify mineral composition and texture of each material (Kim and McCarter, 1994; Kim, 1994). In case of QL specimens, a portion of its data will be reported in this paper.

The explosively driven SHPB apparatus was constructed for this study (Kim, 1993). The SHPB consists of the incident and transmitting bars. Postshock condition was made by placing specimens in the gap of the two bars and detonating an explosive charge placed at one end of the incident pressure bar. The shock energy delivered to all specimens was carefully held below the threshold necessary to develop visible surface crack in the cylindrical specimens. This was done to facilitate measurement of mechanical parameters such as strength and ultrasonic velocity for damaged material.

Nineteen QM, 15 QL, 10 DW, and 21 SS specimens for unconfined uniaxial compressive strength (UUCS) tests were subjected to shock loading in the SHPB. The amount of PETN necessary to reach the threshold of visible damage was determined as follows: 0.90g for QM, 1.30g for QL, 0.75g for DW, and 0.80g for SS. The UUCS tests were made within a few hours after shock loading. This procedure was followed to reduce effects on the test results caused by changes in environmental factors (temperature and humidity) and possible time dependent material behavior. P-wave velocity measurements (ASTM D2845-83) were performed on both pre- and postshock specimens and ω was determined by using equation (3). The P-wave velocity for the preshock condition was made immediately before placing the specimen in the SHPB. The postshock condition was made within two to three minutes after detonating the explosive.

A laser scanning confocal imaging microscope (LSIM) was used to quantify crack-like damage in the intact and postshock specimens. A total of 19 samples were prepared -- one intact sample of each of the three rock types, five postshock samples of QM, five postshock samples of DW, and six postshock samples of SS. Postshock specimens were prepared by subjecting cylindrical samples to different levels of force. All specimens were sectioned (in longitudinal direction) into rectangular plates (nominal 6.699cm x 3.181cm x 0.530cm). The sectioned plates were impregnated with low-viscosity epoxy mixed with fluorescent dye. After overnight curing in a vacuum jar, the impregnated specimens were mounted on sample holders using sealing wax and then ground flat on a semi-autofeeding grinding machine. Care was taken to remove only the excess epoxy, provide a uniform surface, and expose the dye-penetrated crack-like defects. The fluorescent dye provides a high-contrast crack image relative to areas not penetrated by the dye. The LSIM provides a numeric value proportional to the dyed area divided by the total area in each measurement. This ratio, expressed as a percentage, represents the relative crack density. Additional details concerning the LSIM technique can be found elsewhere (Kim, 1993 and 1994).

Time-dependent properties in QM, DW, and SS were characterized by measuring the postshock P-wave velocity over a period of six months. Five representative specimens were chosen from each rock type and shocked using different amounts of explosive (0.90g-PETN for QM, 0.75g for DW and 0.80g for SS specimens). The postshock P-wave velocity was then measured at intervals ranging from one hour immediately following shock loading up to 180 days near the end of the test period. Effects of temperature and moisture on the results were minimized by maintaining near constant ambient temperature and humidity.

Fifteen intact and 15 postshock samples for each of the four materials were prepared and crushed using the impact apparatus (Kim, 1993). The equivalent PETN-load for the postshock samples was 0.90g for QM, 1.30g for QL, 0.75g for DW, and 0.75g for SS, respectively. The resulting crushed product for each rock material was subjected to sieve analyses. The particles retained on the 0.371, 3, 4, 6, 8, 10, and 14 screen (Tyler mesh) were counted for both intact and postshock samples of each rock type.

Table 1. Summary of engineering properties of intact and postshock specimens.

Rock Type	Compressive Strength (MPa)		P-wave Velocity (km/s)		PETN
	Intact	Damaged	Preshock	Postshock	(g)*
Quartz-Monzonite	118.9 {34/13.8}	112.5 {19/16.4}	3.4 {19/0.1}	2.6 {19/0.1}	0.90
Quartz-Latite Porphyry	125.6 {15/25.0}	117.7 {13/17.1}	4.4 {13/0.1}	4.2 {13/0.1}	1.30
Diopside-Wollastonite	63.0 {10/10.3}	60.6 {10/11.8}	4.9 {10/0.5}	3.8 {10/0.7}	0.75
Subarkosic-Siltstone	50.0 {28/4.5}	48.6 {21/4.3}	3.3 {21/0.1}	2.8 {21/0.1}	0.80

{Number of samples / Standard deviation} * Includes 0.4 g cap base charge.

4 EXPERIMENTAL RESULTS AND DISCUSSION

Figure 1 shows the number of specimens in each interval plotted against UUCS of intact and postshock specimens (QM). The mean compressive strength for each material is presented in Table 1. The table shows a relative reduction in strength of 6% for QM, 7% for QL, 4% for DW, and 3% for SS. Although these differences between intact and damaged materials may appear insignificant, there is a trend towards lower strength in damaged rocks relative to the intact rocks. The results for the four materials chosen for this investigation appear to agree with the findings of these researchers (Olson et al., 1973; Holmberg and Mäki, 1981). Based on these results, it appears that compressive strength is independent of shock-treatment.

Figure 2 shows the number of specimens (QM) in each interval plotted against P-wave velocity. As indicated, postshock P-wave velocity was significantly decreased. A similar decrease was reported by Roblee and Stokoe (1989).

Reduction of P-wave velocity strongly suggests that considerable internal damage was accumulated in the specimen during the passage of the wave. The damage variable (ω) is plotted as a function of relative density of crack-like defects in Figure 3. For QM and SS, the damage variable appears to increase as the density of the crack-like damage increases.This trend suggests that ω is related to the increase of crack-like defects. The correlation between these two quantities is not as clear for DW. Petrographic thin section analysis clearly discloses that DW material is extensively weathered and deformed. Ratio of P-wave velocities for QM and SS is plotted as a function of relative LSIM crack density (Figure 4), using equations (1) and (2) with $\alpha \simeq 0.112$. The internal crack-like damage increases with an increase in energy dissipation. In other words, the moduli decrease exponentially as crack density increases. Open symbols designate the crack density for intact samples. The velocity ratio near zero for intact specimens does not imply that these samples are free of crack-like damage.

Figure 5 shows an interesting time-dependent effect on the change in P-wave velocity. Each data point represents average values for five postshock specimens of each rock type. Open symbols designate the preshock P-wave velocities, while black symbols indicate the postshock P-wave velocities. It is evident that the velocity was reduced by 28% for QM, 43% for DW, and 16% for SS immediately following the shock. Measurements confirmed that the postshock length of the specimen was the same as the original length and that ambient conditions (temperature and humidity) remained relatively constant. The plots clearly show an increase in velocity with the passage of time. No substantial change in P-

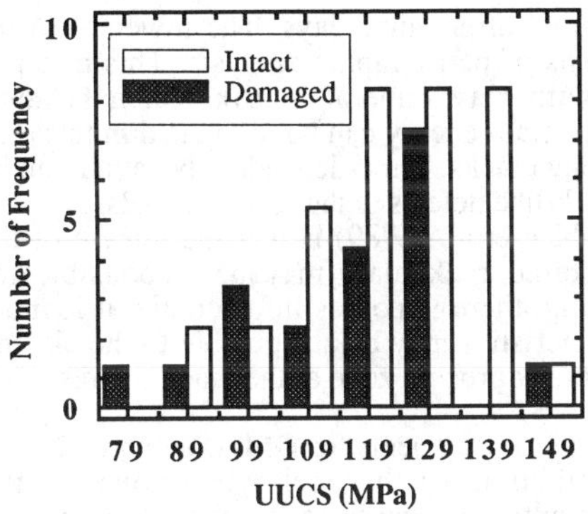

Figure 1. Compressive strength of QM.

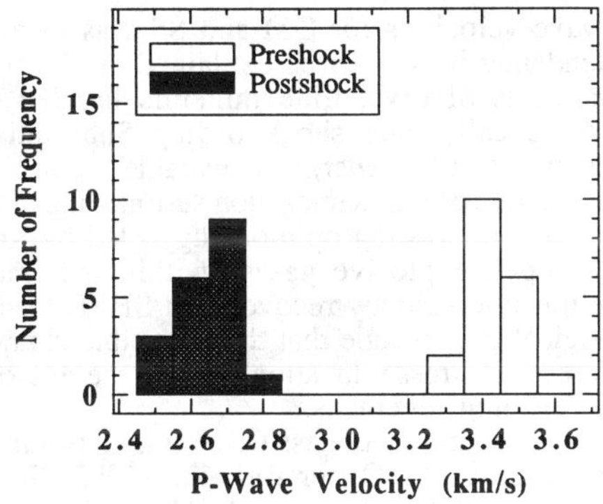

Figure 2. P-wave velocity of QM.

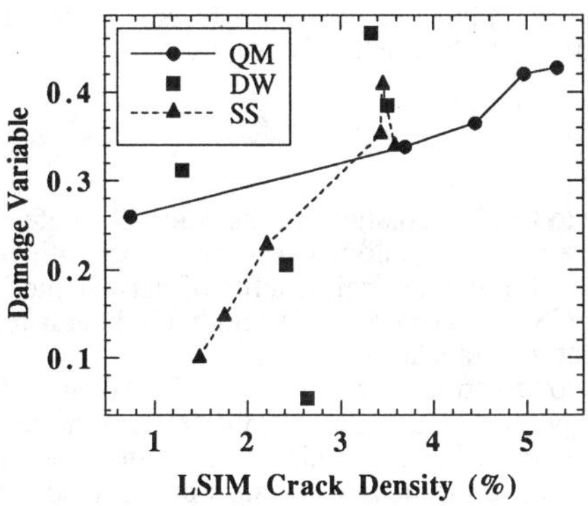

Figure 3. ω vs. crack density.

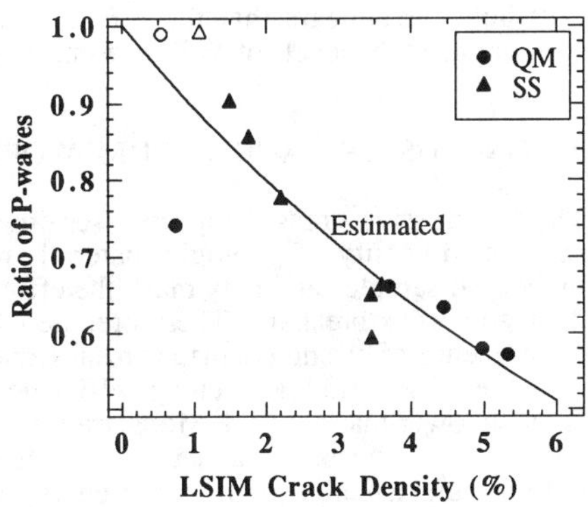

Figure 4. Decrease in P-wave velocity.

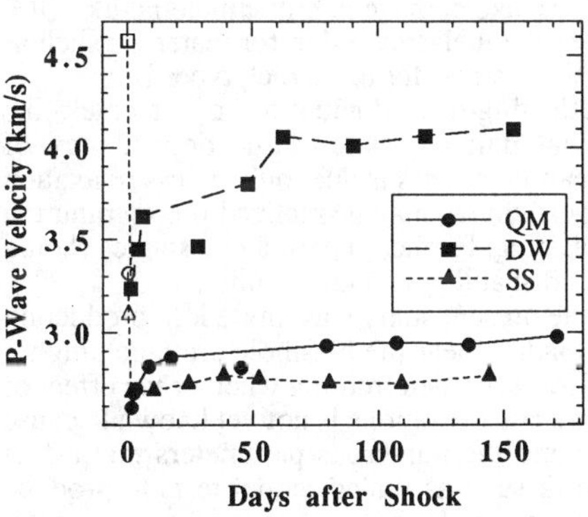

Figure 5. Time dependency of P-wave.

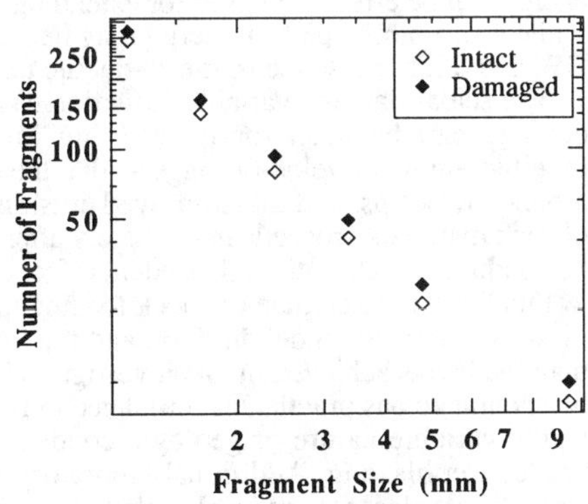

Figure 6. Counted fragments of QL.

wave velocities for QM and SS was measured after thirty days. The lower recovery tendency in SS can be explained on the basis of petrographic analysis. This material consists of crystalline minerals bonded with clay mineral matrix which behaves inelastically under shock loading. Substantial strain energy can be dissipated in this clay matrix, and less energy is available for creating cracks. Time-dependent behavior can be indicative of stress relaxation surrounding crack-like defects in the rock materials.

One interesting opinion suggested by Roblee et al. (1989) is that the release of the "trapped explosive gases" within the fractured rock mass may be responsible for ultrasonic velocity recovery. In SHPB loading, there is no gas infusion and it is more logical to conclude that the reversible characteristics may be attributable to the closing stress (or stress relaxation) around the microcrack process zone at the crack tip, resulting in "healing" of cracks.

The cumulative mass fraction and its number of fragments for QM, DW, and SS were reported by McCarter and Kim (1993). In addition, another rock type of quartz latite porphyry (QL) was selected and tested. The result shows the same tendency (Figure 6). In all cases, more particles in the size range of 1 mm to 10 mm were counted for postshock relative to intact rock materials. This observation was consistent throughout the experiments. Although it may be difficult to draw a statistical inference that a significant effect of shock treatment on the change in rock breakage exists between intact and postshock, it appears that the effect of shock-induced damage on fragmentation is measurable at the levels of explosive energy delivered by the SHPB device.

5 CONCLUSIONS AND FUTURE WORK

The compressive strength appears insensitive to local fluctuations of the crack-like defect. Inherent diversity of geologic materials results in a random variation in strength of individual samples and may mask the effort to know potential benefits of blast-induced damage in rock breakage. Therefore, the UUCS may not provide a reliable indicator for the influence of blasting damage on subsequent rock breakage.

In the present study, the energy efficiency for creating new surface area in DW and SS is much lower than for QM. More strain energy must be dissipated at the fracture process zone when the surrounding material around the cracks behaves inelastically. If the material deforms inelastically, extra strain energy is needed to advance cracks and extend the inelastic zone. It is well known that brittle-to-ductile transition can occur right before and immediately after reaching the maximum failure stress for some rocks. Goodman (1980) reports that brittle-to-ductile transition stress (σ_{bd}) for sandstone and limestone is much less than granite ($\sigma_{bd} >> 100$MPa for granite). This result implies that most of the input energy can be effectively used for extending crack-like damage in brittle material like QM, while a substantial part of energy may be lost at the inelastic region for materials such as DW and SS, because the brittle-to-ductile transition stress for these rock types is low.

The scalar damage variable (equation 3) reflecting local fluctuations of the crack-like damage can be used for quantifying internal damage state. Time dependency of longitudinal wave velocity suggests that the healing process is due to the stress relaxation around crack tips, and that improved crushing efficiency may be realized if comminution of rock materials proceeds immediately after blasting. Further experimental studies should be conducted such as time-dependent effect on crushability and grindability.

Finally, the magnitude of shock loading in the present study was limited. In production blasting, the magnitude, duration and rate of loading near the blasthole are much higher than the levels achieved in this investigation. However, determining whether the effect of blast damage has practical significance in field scale operations is not yet known because of the variable nature of geologic conditions and the numerous parameters that affect production blasting. Additional laboratory work such as scaled-model tests to produce stress levels closer to the level of that near a detonating blasthole can be performed.

ACKNOWLEDGMENT

This paper is prepared at the University of California, San Diego, while the author is working as a Postgraduate Research Engineer in the Department of Applied Mechanics and Engineering Sciences. The author would like to particularly thank my supervisor, Prof. Sia. Nemat-Nasser, for his support.

REFERENCES

Annual Book of ASTM Standards, Subject Index, Alphanumeric List 1990. ASTM Designations: C769-80, D2845-83, D2938-79, D3148-80.

Bruner, W.M. 1976. Comment on 'seismic velocities in dry and saturated cracked solids' by Richard J. O'Connell and Bernard Budiansky. *JGR* 81(14): 2573-2576.

Farmer, I.W. 1968. *Engineering properties of rocks*. London: E. & F.N. Spon Ltd.

Goodman, R. E. 1980. *Introduction to rock mechanics*. John Wiley & Sons.

Griffith, A.A. 1921. The phenomena of rupture and flow in solids. *Phil. Trans. R. Soc. Lond.* A 221: 163-198.

Griffith, A.A. 1924. The theory of rupture. In C.B.Biezeno & J.M.Burgers (eds), *Proc. First Int. Cong. Appl. Mech.* Delft, 55-63.

Hoagland, R.G., G.T. Hahn & A.R. Rosenfield 1973. Influence of microstructure on fracture propagation in rock. *Rock Mech.* 5: 77-106.

Holmberg, R. & K. Mäki 1981. Case example of blasting damage and its influence on slope stability. In C.O.Brawner (ed), *Third Int. Conf. Stability in Surface Mining*, Vancouver, British Columbia Canada, 773-793.

ISRM Suggestions 1979. Determining water content, porosity, density, absorption and related properties and swelling and slake-durability index properties. *Int. J. Rock Mech. & Min. Sci. & Geomech. Abstr.* 16(2): 141-156.

Kachanov, L.M. 1986. *Introduction to continuum damage mechanics.* Martinus Nijhoff Publishers.

Kim, D.S. 1993. *The effect of shock-induced damage on comminution of rock materials.* Dissertation, University of Utah, Salt Lake City, Utah, USA.

Kim, D.S. & M.K. McCarter 1994. Quantitative assessment of extrinsic damage in rock materials. Submitted to *Rock Mech. & Rock Eng.*

Kim, D.S. 1994. Experimental technique of rock fractography with the laser scanning confocal imaging microscope. To be published elsewhere.

Labuz, J.F. 1985. *A study of the fracture process zone in rock.* Dissertation, Northwestern University, Evanston, Illinois, USA.

Lemaitre, J. 1984. How to use damage mechanics. *Nucl. Eng. Des.* 80: 233-245.

McCarter, M.K. & D.S. Kim 1993. Influence of shock damage on subsequent comminution of rock. In H-P.Rossmanith (ed), *Proc. 4th Int. Symp. Rock Fragmentation by Blasting–Fragblast-4*, Technical University of Austria, Vienna, Austria, 5-8 July 1993: 63-70. Rotterdam: Balkema.

National Materials Advisory Board 1981. *Comminution and Energy Consumption*, Report of the Committee on Comminution and Energy Consumption, Publication NMAB-364, Washington, D.C.: National Academy Press.

Olson, J.J., R.J. Willard, D.E. Fogelson & K.E. Hjelmstad 1973. Rock damage from small charge blasting in granite. *US, BurMines, Rep. Invest. 7751.*

Roblee, P.P.N. & K.H. Stokoe II 1989. Changes in seismic measurements with blast induced blasting: a field experiment. *Proc. 30th U.S. Rock Mech. Symp.*, West Virginia University, Morgantown, 19-22.

Walsh, J.B. & W.F. Brace 1966. Elasticity of rock: a review of some recent theoretical studies. *Felsmechanik* 4(4): 283-297.

Rock Mechanics, Nelson & Laubach (eds) © 1994 Balkema, Rotterdam, ISBN 90 5410 380 8

Rock fragmentation results for a simulated underground leaching stope

Farrokh Djahanguiri
Denver Research Center, US Bureau of Mines, Colo., USA

Mark S. Stagg & Rolfe E. Otterness
Twin Cities Research Center, US Bureau of Mines, Minneapolis, Minn., USA

John M. Kemeny
Department of Mining and Geological Engineering, University of Arizona, Tucson, Ariz., USA

ABSTRACT: In situ stope leach mining is part of a clean mining research program designed to increase the domestic supply of metals for the United States. The U. S. Bureau of Mines (USBM) is developing this innovative mineral recovery technology to help preserve the ecosystem. The USBM is applying a systems approach to develop mineral recovery methods that minimize environmental impact by selectively extracting and recovering targeted minerals. The motivation for this approach derives from the conviction that we can no longer afford the environmental or economic cost of moving large volumes of rock to recover small amounts of metal. To accomplish a clean mining system, one must maximize the solutions contact with ore minerals and also maintain control of the leach solutions. This can be achieved by performing in situ fragmentation to obtain relatively uniform rock particle size to increase rock mass permeability, and also minimize blast damage to the perimeter of the excavation. The USBM is investigating the feasibility of in situ stope leach mining, where leaching solutions are passed through rubblized ore. Rubblizing blocks of ore by blasting for subsequent underground in-stope leaching requires special applications of blasting techniques.

This paper presents the results of in situ fragmentation of a simulated cylindrical stope which is 1.8 m in diameter and 6.1 m high. The blast design combined conventional and smoothwall blasting techniques. Three drop-raise and smoothwall blasts were used to fragment the simulated test stope.

The average powder factor for the stope fragmentation was 2.6 kg per metric ton, compared to 1.10 kg per metric ton for routine drift development work at the mine. All the material from the first blast, 14.2 metric tons, was sieved. The resulting distribution was compared to the distribution predicted by empirical equations. The best fit was found with a USBM developed equation based on over 50 sieved, reduced-scale (1- to 2-m) high wall blasts. Comparing the broken rock distribution data for round 1 with the photo image processing data for rounds 1, 2, and 3 shows that using photo image processing is cost effective and reasonable when compared to screening and weighing.

1 INTRODUCTION

The U.S. Bureau of Mines is currently researching the feasibility of incorporating in situ stope leaching into active underground metal mines. This modified in situ leaching technique proposes to recover metals from ore by creating underground rubblized stopes, leaching the stopes to dissolve the target mineral, and collecting and processing the metal-bearing solution for metal recovery.

A USBM field experiment at the Colorado School of Mines Experimental Mine (Edgar Mine) in Idaho Springs, Colorado, has been designed to evaluate rock blasting techniques, to compare size distribution data with that predicted by empirical equations, and to correlate applications of photo image processing techniques of rock pile photos with data from screen analysis.

2 SOLUTION CONTROL TEST SITE DESCRIPTION

The simulated leaching stope fragmentation tests were made at the Edgar Mine, located in Idaho Springs, Colorado, at 2,300 m elevation. The depth of cover at the mine varies from zero at the portal to 200 m at the test site. The mine is located above the water table and is essentially dry.

The structural geology of the area is complex, with many local variations. Six major joint systems orientations form 350 discontinuities in the drifts surrounding the underground research site. Strike of the most predominant joints are N.W.

The most important structural feature at the site, mapped during blasthole drilling, is a fracture zone dipping roughly to the S-S.E. In this zone, water loss during drilling was common in most of the holes. The fracture zone is not continuous but cuts through only the S.E. quarter of the rock mass.

Predominant rock types at the stope site are quartz biotite gneiss, pegmatitic granite, and hornblende schist. Quartz zones are abundant in the gneiss and define the banding. The schist zones often define the foliation orientation structurally because they are relatively weak. Average attitudes of the foliations over the total area mapped are N.70°E. with a dip of 75° to the N.W.

3 STOPE FRAGMENTATION

Mining routinely uses blast fragmentation to develop orebodies and extract ores. Stope leaching fragmentation is unique and has two primary functions: First, to uniformly fragment rock at an acceptable size to maximize the permeability of the rock mass for leaching, and second, to minimize blast damage around the stope to prevent leaching fluids from migrating into the surrounding rock mass or adjacent stopes. Another design consideration is resource recovery by maintaining rock mass strength between the stopes so that stopes are situated as close as possible compatible with minimum pillar width. However, since optimum fragmentation is desired, a closely spaced drill pattern and high density explosives are necessary. Therefore, controlled blasting that minimizes blast damage is required to prevent damaging large zones of surrounding rock.

Controlled blasting is applied to specific requirements in construction, nuclear waste disposal, and mining engineering projects. In construction engineering, controlled blasting has been used successfully to minimize overbreak and seepage, to improve diggability, and to improve roof and wall stability and ventilation in tunneling projects (Cyr 1991; Day et al. 1982; Revey 1991).

In addition to structural and safety considerations, controlled blasting is also routinely used worldwide in mines and tunneling projects to control ore dilution and rock size (Can. Min. J. 1991-a,-b; Fletcher et al. 1990; Revey 1991). Controlled blasting, similar to the stope fragmentation presented in this paper, minimized overbreak and prevented ore dilution and water migration through ore pillars in the Lockerby Copper-Nickel Mine in Sudbury, Ontario (Can. Min. J. 1991-a,-b).

4 BLAST DESIGN

A full-face blast design was used to fragment the simulated leaching stope. (Figure 1). The stope was excavated using three separate drop-raise and smoothwall blast rounds. The lower round was shot first; the middle, second; and the uppermost third. The first round was 2.1 m deep, and the second and third rounds were each 1.8 m. The blast design had one 10-cm-diameter burn cut hole, twenty-nine 3.8-cm-diameter blast holes, and seven additional 3.8-cm-diameter perimeter holes containing no explosives. The seven open holes were included to protect the existing fracture zone in the SE section of the rock mass.

All blast holes were drilled the full 6.1 m raise depth from the top of stope, and the rounds were loaded from the top of the stope. Two types of explosives were used for fragmentation: Powerditch 750[1] high-energy dynamite in the two inner circles of blastholes, and Kleen-Kut E ammonia dynamite in the two outer circles of blastholes.

The weaker Kleen-Kut E was used to control overbreak around the perimeter of the stope.

The burn-cut was loaded with six sticks of Powerditch 750, and the trim and buffer holes were loaded with three 24-inch-long sticks of Kleen-Kut E. Blastmaster LP nonelectric detonators were used, and a detonation cord was attached to the entire length of trim and buffer hole charges. After the 29 holes were initially loaded, extra explosive was added to any holes that were low so that the top of the explosive column in each hole created a plane equidistant from the top of the stope.

In the second round, 26 holes were loaded like the first round except that each hole was stemmed completely to the top with sand. Three blast holes were unusable as a result of rock displacement from the initial blast. In the third round, the 21 holes that remained unplugged earlier were loaded as in the second round.

In the blast sequence, the main blast holes were fired in sequence outward from the center of the round. The smoothwall and trim holes were then fired on the longest delay to bring down rock between the perimeter and the main blast holes (Revey 1990). The powder factor for each round (1, 2, and 3) was 3.3, 2.5, and 1.9 kg/mt, respectively, and gradually decreased because the number of loaded blast

---

[1]Reference to specific product does not imply endorsement by the USBM

holes for rounds 2 and 3 was less due to the plugging of the blast holes, and the reduction in depth of blast holes as a result of previous round blastings.

The term powder factor is loosely defined in the content of this study and refers to a combined total weight of explosives used per metric tons of fragmented rock per blast round. However, a more accurate method for measurement of performance of the two explosives needs to be developed based on energy released, detonation velocity, and explosive density.

Photographic assessment of the fragmmented rock after blasting was an important part of the field work. After each blast round, fragmented rock was exposed along the axis of the drift and at the stope, due to confinement. The fragmented rock was expanded into a limited space. Naturally bigger boulders can be expected to accumulate at the perimeter of a fragmented rock pile due to gravitational and rotational forces. Therefore, photos taken after each round blast may present fragmented rock sizes which are not a true representative for image analysis purposes. It might have been better to take additional pictures at various stages of the rock handling, screening and disposing process, so that the final results of the image analysis studies could have been refined and better results achieved. Unfortunately, due to schedule limitations and manpower, only limited field activity was possible.

The broken rock from the first blast round, 14.2 m tons, was collected and screened at the surface. Rocks larger than 22.7 cm were screened by hand using square plywood templates, and the plus-15.3 cm and plus-7.6 cm sizes were separated with a small square-grid grizzly. The remaining rock was passed through a mechanical vibrating screen to sort it into the plus-3.8 cm, plus-1.9 cm, plus-0.95 cm, and minus-0.95 cm sizes. For round 1, 63% of screened and weighed rock was under 7.6 cm, and 21% was between 7.6 cm and 15.3 cm. Overall, 84% of the fragmented rock was under 15.3 cm.

5 ANALYSIS OF STOPE FRAGMENTATION DATA USING PHOTO IMAGE PROCESSING

Photos of each rock pile were taken after blasting and before the removal of the rock. These photos were studied using image processing techniques. Photo image processing uses computerized image processing and statistical programs to analyze photographic images of rock fragments taken in the field to estimate the rock size distribution of the fragments (Kemeny et al. 1993; Kemeny 1994).

The image processing analysis of the photos was carried out as follows. First, each photo was scanned into the computer and scaled. A total of nine photos were scanned from rounds 1, 2, and 3. Secondly, the individual fragments in each digital image were delineated, and the area and shape of each fragment were calculated. Third, the areas of each fragment were corrected for the angle between the normal to the rock pile and the axis of the camera. Fourth, statistical algorithms were used to estimate the volume and sieve size of each fragment. The statistical analyses were carried out in terms of probabilities, in order to take into account the effects of fragment overlap (this makes a fragment appear to have a smaller sieve size than it actually has) and fragment orientation in the pile (this could make a fragment appear to have a larger sieve size than actual since the sieve size for a fragment

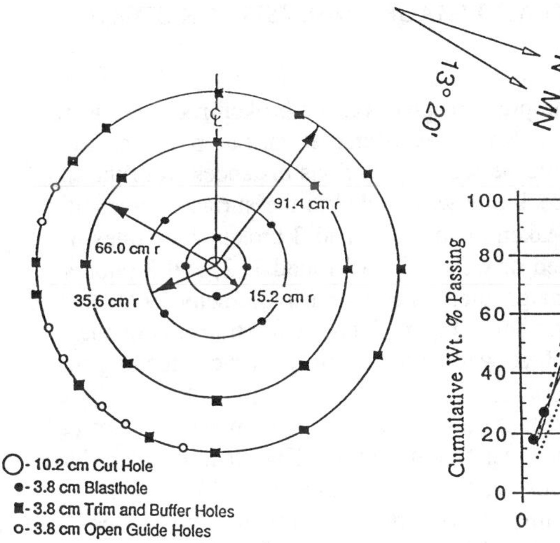

Figure 1. Full-face stope blast design layout.

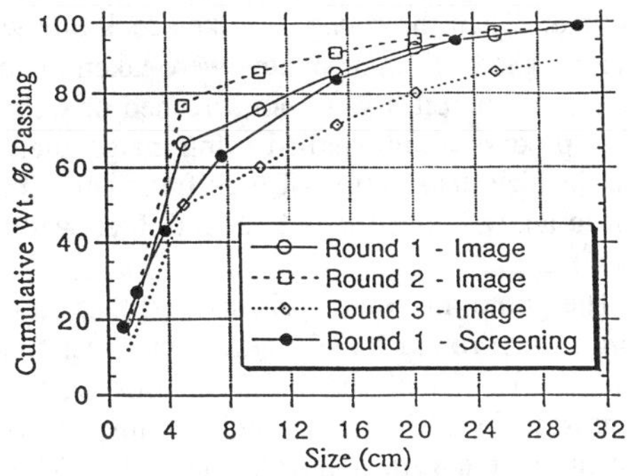

Figure 2. Comparison of screening and image analysis.

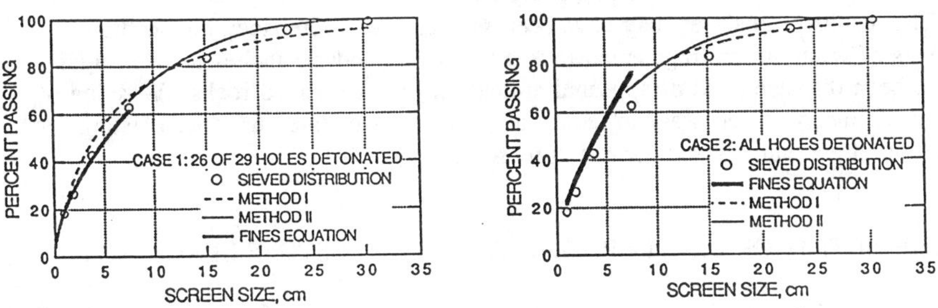

Figure 3. Predicted distribution versus sieve data for stope blast, case 1 and 2.

does not depend on the largest dimension of the fragment). An algorithm was also used to combine the fragment information from images taken at different scales. Fragments too small to be delineated were considered as fines. In this study fragments less than 2 inches in size were not delineated. The percentage of the total pile containing particles smaller than 2 inches was estimated from the area of the non-delineated region (total area minus the sum of all the fragment areas). The functional form for the size distribution of the fines was not known and was assumed to be linear. Details of the procedure are described in Kemeny, et al. (1993) and Kemeny (1994).

6 COMPARISON OF SCREENING AND IMAGE ANALYSIS FOR STOPE FRAGMENTATION DATA

After blasting round 1, photos of the pre- and post-washed broken rock pile were taken at the lower drift, with 1-ft wooden scales located at various positions on the broken rock pile so that the pile could be scaled at various distances from the camera. Then the round 1 broken rock pile was hauled to the surface, screened, and weighed. Similar photos were taken of rounds 2 and 3 broken rock piles in place, but the piles were not screened or weighed. All round 1, 2, and 3 photos were processed and studied using image analysis techniques (Kemeny 1994) to find out the size distribution of rock fragments. Figure 2 presents the results of the image analysis of rounds 1, 2, and 3, as well as the result from the screening of round 1.

The photo image analysis showed that round 2, with a lower powder factor (2.5 kg/mt) than round 1 (3.3 kg/mt), had finer fragmentation. Two factors might have caused this: First, recorded data from accelerometers, high-speed films, and cartridges found in the broken rock pile showed that some round 1 explosives did not fire as planned. Second, rounds 1 and 3 free surfaces were subjected to fracturing from the conventional blasting used during routine lower drift and stope site development; the round 2 rock mass was more confined. If the second shot fired exactly as planned, it should have the fragmentation expected from the given powder factor. Figure 2 shows that the round 1 image analysis compares reasonably well with the screening analysis.

The results of analyses may have been subjected to discrepancy due to the effects of washing the fragmented rock pile. Rocks under 2 inches in size might have been displaced and disappeared around bigger fragmented rocks. Washing of the rock piles was necessary to reduce accumulations of fine dust after blasting for ease in taking pictures and for health reasons.

7 PREDICTION OF BLAST FRAGMENTATION FOR STOPE LEACHINGS

Several existing empirical equations can predict the fragmentation that results from bench blasts (Cunningham 1987; Lin 1988). These equations were developed for multi-hole bench blasts where timing and shot geometry were combined to provide finer fragmentation from the interaction of stress-wave and gas-pressure between adjacent boreholes. These conditions differ from underground blasting where long delays produce, in effect, a series of single-hole blasts. Very little sieved data from underground blasting have been published to evaluate the applicability of these equations, or to determine underground empirical guidelines (Haller 1972).

However Stagg, et al., (1994) have analyzed the fragmentation data from a stope site to assess the application of the bench-blast derived equations to underground blasting.

The Bureau-developed equations divide the fracture process into two regions: (1) the radial fracture network propagating out from the borehole, which accounts for most of the finer sizes, and (2) the rest of the bench, where larger fragments are formed, (Stagg, et al. 1994)

Three relationships were developed: (1) a curve for the finer sizes, (2) a curve for average size (K_{50}), and (3) a curve for the 80% passing size (K_{80}). Two different methods were used to find the K_{50} and K_{80} points. Method I is a power fit developed to compare with other empirical equations. Method II involved separating the shot data into high- and low-powder factor groups and finding a linear fit to the data. Method II reduced error appreciably for bench blasts. The data from previous surface bench blasts, including single-hole shots, were reviewed to determine how to apply the equations appropriately. Stagg, et al., (1994) discusses the USBM and other empirical equations further.

Either Method I, or Method II combined with the fines equations could be used to predict the total size distribution. The predicted distribution for the current underground blasting was determined for two scenarios: case 1, based on explosive performance determined by accelerometer and high-speed camera data which indicated that 26 of 29 holes fired; and case 2, based on all holes firing as planned.

For applying the USBM equations to underground blasting, charges with similar burden and explosives are grouped. Then the volume of rock which will be fractured by each group of rounds and its proportion to the total shot are calculated from the blast design geometry. Overall size distribution can then be determined. In Figure 3, which shows both case 1 and case 2, the one-curve approach using a Weibull distribution (Method I) appears as effective as the two-curve approach. Also of note is the predicted distribution for case 2 which has finer fragmentation than the screened data.

8 CONCLUSION

The simulated in situ stope fragmentation project conducted at the Edgar Mine is unique because there is very little experience to draw upon for in situ stope leaching. Preliminary results of this field investigation confirm that uniform rock mass fragmentation can be achieved, although ongoing borehole geophysical investigations might verify the actual extent and approximate magnitude of blast damage to the surrounding rock mass.

The screen analysis of round 1 compared reasonably well with the photo image processing for the same round. The results of studying the broken rock size distribution through photo image processing show that the process is cost effective and yields reasonable results when compared to the screening and weighing for round 1 broken rock. Also, the empirical fragmentation prediction equations developed from surface bench blasts by the USBM can be modified and applied to underground rounds as well. The fragmentation distribution predicted by those equations matched well the sieved data after correcting for the effect of undetonated holes.

The two empirical methods developed by the USBM proved to be nearly equally accurate when applied to this underground round.

REFERENCES

Canadian Mining Journal, 1991-a, Controlled blasting, September/October pp. 33-34.

Canadian Mining Journal, 1991-b, Feeding a hungry mill. September/October pp. 47-55.

Cunningham, C.V.B. 1987. Fragmentation estimations and the Kuz-Ram model-four years on. *Second Internat. Symp. on Rock Fragmentation by Blasting*. (Ed. W.L. Fourney and R.D. Dick), Society of Experimental Mechanics, Bethel, CT 06801 USA. pp. 5-487.

Cyr, R., 1991, *Drop raise drilling with conventional longhole equipment*, Presented at The World Boring Association Seminars at Laurentian University, September, Sudbury, Ontario.

Day, P.R. and W.K. Webster 1982, *Controlled blasting to minimize overbreak with big boreholes underground*, C1M Bulletin, pp. 112-121.

Fletcher, L.R., M.N. Pils, and V.J. Stachura 1990, Overbreak control in the White-Pine underground copper mine, <u>*Skilling's Mining Review*</u>, Vol. 79, No. 34, August, pp. 4-7.

Haller, H.F. 1972. *Interrelationship of in situ rock properties, excavation method, and muck characteristics*. Annual Technical Report No. 1. Advanced Research Projects Agency, pp. 59.

Kemeny, J.M., Devgan, A., Hagaman, M.R., and Wu X. et al. 1993, Analysis of rock fragmentation using digital image processing, <u>*Journal of Geotechnical Engineering*</u>, Vol. 119, No. 7, pp 1144-1160.

Kemeny, J.M. 1994, A practical technique for determining the size distribution of blast benches, waste dumps, and heap-leach sites, submitted for publication to Mining Engineering.

Lin, N.S. 1988. *New hard rock fragmentation formulas based on model and full-scale tests, licentiate thesis*, Lulea Univ. Tech., Lulea, Sweden, 138 pp.

Revey, G.F. 1991. Controlled blasting at the Hanging Lake Tunnels. *Proceedings of the 17th Conference on Explosives and Blasting Techniques*. Soc. of Explosive Eng., Las Vegas, NV., Vol. 1, pp. 145-155.

Revey, G.F., 1990. Blast design proposed for USBM 6 ft diameter drop raise, an internal report submitted to Bureau of Mines, 15 pp.

Stagg, M.S., Otterness, R.E., and Djahanguiri, F. 1994. Prediction of Blast Fragmentation of Underground Stopes for In Situ Leaching, <u>*20th Annual Conference on Explosives and Blasting Techniques*</u>, Austin, Texas. (In press)

Rock Mechanics, Nelson & Laubach (eds) © 1994 Balkema, Rotterdam, ISBN 90 5410 380 8

Numerical simulation of delay blasting by the dynamic lattice network model

John Song & Kunsoo Kim
Henry Krumb School of Mines, Columbia University, N.Y., USA

ABSTRACT: Delay blasting has been practiced for several decades in mining and quarrying operations and is considered a mature technology, but with the advent of precision detonators capable of achieving delays in the order of microseconds, further research is required to assess the efficacy of using such precise delay times for common applications such as fragmentation and controlled excavation. For a preliminary assessment, we have simulated the explosives induced fracturing process using the Dynamic Lattice Network Model for processes involving varying delay times including simultaneous detonations.

The discrete model is composed of particles and springs with randomly assigned stiffnesses to reflect the material property variations and their influence on the initiation and propagation of fracture in rocks; moreover, the model is capable of simulating the different stages of fracture including crack branching. In the simulation, the springs are assumed to respond to explosive induced stresses in a linear elastic and brittle manner and the material response is expressed in terms of particle dislocations and spring elongations and compressions. Failure occurs when a spring is stretched or compressed beyond a randomly assigned ultimate limit which is based on the material's ultimate strength in tension and compression. Upon failure, the stresses are dynamically redistributed and the fracture process is continued.

The emphasis of this modeling effort is on the study of the fracture phenomenon under dynamic loading, especially the extent of fracturing and branching, and the analysis of the influence of delay time in multiple hole blasting on fragmentation. Of particular interest is the stress wave interaction and its influence on the fracture propagation, and the resultant fracture pattern.

1 INTRODUCTION

Although the technology of blasting is well developed, its numerical simulation is still a subject for research. Recent efforts in modeling the process have included a finite element approach using a constitutive damage model (Thorne et al 1990) and a discrete element model (Preece and Burchell 1993) that uses spherical elements and accounts for the motion induced by gas flow.

The finite element model as described by Thorne et al (1990) simulates stress wave induced fragmentation in an isotropic material weakened by the presence of randomly distributed microcracks. The dynamic fracture process is associated with the propagation and interaction of these microcracks under tensile loading and fragmentation is attributed to the continual accumulation of damage caused by microcracking. The model itself is based on a system of constitutive relationships derived for a cracked medium and it incorporates effective elastic moduli to relate stress and strain, and the influence of the strain rate on the fracturing process. This model has been applied to the study of cratering and the authors, comparing actual blast experiments with the numerical simulations, report moderate success.

The other modeling approach is the use of discrete elements to follow particle motions caused by a blast. Preece and Burchell (1993) describe one such model called the Distinct Motion Code which uses spherical elements to discretize a rock mass and adopts the concept of a packing angle to address the influence of bedding in sedimentary rocks on rock motion. The packing angle determines the different degrees of element conformity from loose- packing to close-packing. Sample simulations indicate differences in rock motion for different packing angles with closer packing resulting in more material bulking in fragmentation. This model is currently being validated by field experiments.

The Dynamic Lattice Network Model described in this paper is also a discrete element model, but the emphasis of this model is on the micromechanics of fracture initiation and fracture propagation. In the model, the rock mass is discretized into particles and springs and the explosives induced fractures which indicate large particle displacements and spring failures are simulated at different stages in the dynamic process.

Using the Dynamic Lattice Network Model, a numerical simulation is performed to determine the influence of delay time in blasting with an emphasis on the study of the fracture phenomenon, especially on the extent of fracture propagation and crack branching. The simulation is performed on Charcoal Gray Granite whose properties are given in table 1. Three basic cases are considered in the parametric analyses on a block of material constrained on all sides. The first case involves the simultaneous detonation of explosives of equal strength at two opposite faces of the block. The second and third case, similarly configured as the first case, involve the staggered detonation of explosives with the second charge detonating at a delay equal to the time of travel for the elastic wave generated by the firstly detonated hole to traverse half the length of the block and to traverse the full length of the block, respectively. The purpose of studying these cases is to study the effects of delay time on the resulting fragmentation.

The influence of delay time on fragmentation has been investigated in physical experiments performed on various media by several researchers. Anderson et al (1985) have conducted full scale blasting experiments in a 45-ft granite bench using delay intervals between holes ranging from 1 ms/ft to 6 ms/ft. They report that the finest overall fragmentation is attained using 2 ms/ft delays for massive rock and conclude that the optimal delay time lies in the intermediate range and that using shorter or longer delays results in coarser fragmentation. As a possible explanation for these observations, they postulate that using a short delay time is insufficient to allow the stress wave emanating from the first hole to fully debilitate the rock mass before the second hole is detonated and that using longer delays may cause fragmented rock to separate from the bench before the second detonation and therefore sever wave communication between the strain waves for the separated fragments.

Swift et al (1978) have conducted tests on polystyrene under confined conditions using simultaneous and sequential detonations. They observe that simultaneous detonations tend to produce fractures propagating in the plane of the blastholes, with crack communication occurring between the two holes. For a delay time comparable to the wave transit time between the two holes, they find that the fractures produced in the second hole are strongly oriented in the direction of wave propagation with a near elimination of fractures oriented perpendicular to this direction. They also find that using such a delay reduces the damage at the second hole.

Stagg and Rholl (1987) find in test blasts in a single-row bench of horizontally bedded limestone that using a short delay time, .33 ms/ft, produces coarser fragmentation than when using a longer delay of 1 to 8 ms/ft. They conclude that when using short delay times the stress wave fracturing around the blasthole is incomplete before the second hole is detonated; thus, poor fragmentation results.

Having conducted this literature survey for actual test experiments conducted on delay blasting, we have conducted a numerical analysis to simulate the delay blasting using the Dynamic Lattice Network Model.

2 DESCRIPTION OF MODEL

The Dynamic Lattice Network Model (Huang, 1992) is composed of triangular lattices which form the basic network elements. There is a particle whose mass is determined by the rock density at each lattice site and two adjacent particles are connected by a spring which is characterized by its spring constant, natural length, and threshold level beyond which the spring snaps. The network has two elastic parameters, Young's Modulus and Poisson's Ratio. The parameters need not be uniform; they can have statistical distributions to reflect the inhomogeneous nature of rocks. Each particle is governed by Newton's equation of motion and the forces exerted on a particle are the Hookean forces from the springs connecting the particle. The particle is governed by the following equation:

$$m_i\ddot{x}_i = f_i - a\dot{x}_i + \Sigma k_{ij} \left(|x_i - x_j| - l_{ij}\right)(x_j - x_i) / |x_j - x_i|$$

where: m_i = particle mass, x_i = position of particle i,
$\dot{x}_i$ = velocity of particle i, $\ddot{x}_i$ = particle acceleration
f_i = external force applied to particle i
a = damping constant, k_{ij} = spring constant,
l_{ij} = spring length

For particles on the model boundaries, the additional forces can applied externally. Boundary forces are applied in two different ways to simulate the blasting operations frequently practiced; one with line charges on the boundaries of the block, and the other with charges placed in circular boreholes. As the forces are applied to the respective boundaries, the particles and springs respond to these applied forces in a dynamic fashion. The positions of all particles as functions of time are determined by solving a set of coupled second order ordinary differential equations.

When a spring on the borehole wall or on the model boundary is stretched beyond the threshold level as the forces are applied, the spring is irreversibly broken to initiate a fracture. The gas generated by blasting is allowed to instantaneously migrate into that newly created void and continue to exert the same pressure on the fracture surface. After the forces are dynamically redistributed, more springs on the fracture surface can break, consequently the fracture growth continues.

Due to the stress concentration near the crack tip, some weaker springs that are not on the fracture surface can break, forming microcracks that are part of the fracture process zone. Some of these microcracks can grow into the main fracture; some will remain isolated.

A more detailed discussion of the Dynamic Lattice Network Model can be found in earlier papers by Huang and Kim (1993), Song and Kim (1994), and Kim and Yao (1994).

3 MODELING ANALYSES

The computer simulation has been performed on a 200 X 200 particle mesh which has equivalent material properties as Charcoal Gray Granite whose reported properties are summarized in Table 1. The strength and elastic properties are assigned stochastically in the model within the ranges of material property variations. The constitutive behavior of the material is assumed to be linearly elastic

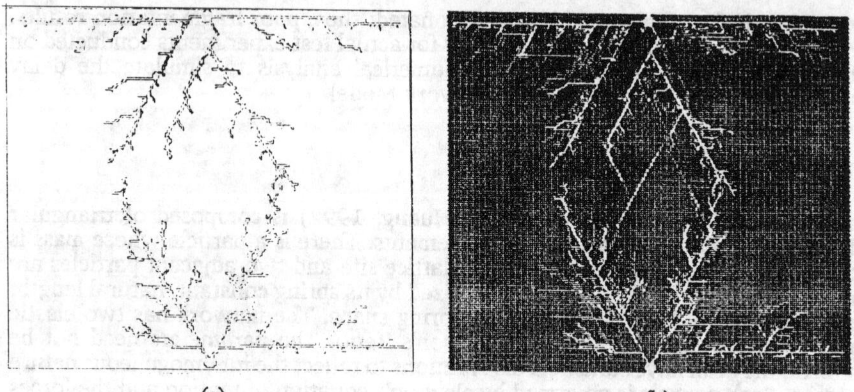

(a) (b)

Figure 1. Fracture patterns for simultaneous blasts after 10,000 time steps: (a) radial fracture pattern (b) complete mesh showing incipient fractures or areas of weakness in which a spring or springs have failed but have not yet formed into a crack.

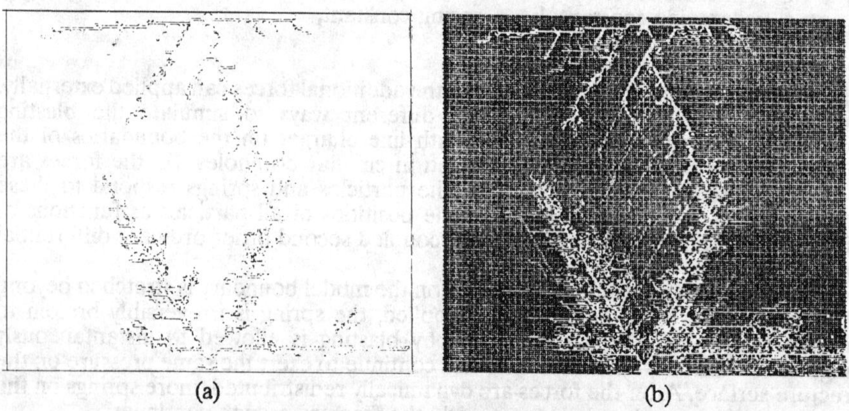

(a) (b)

Figure 2. Fracture patterns for a delay case. The top hole detonates 5,000 time steps after the bottom hole. (a) radial fracture propagation (b) complete mesh at 15,000 time steps.

Table 1. Material Properties for Charcoal Gray Granite

Dynamic Young's Modulus	6,450 - 7,020 kpsi
Tensile Strength	839 - 1,053 psi
Density	169.8 pcf
P-wave Velocity	15,090 ft/s
Poisson's Ratio	.25

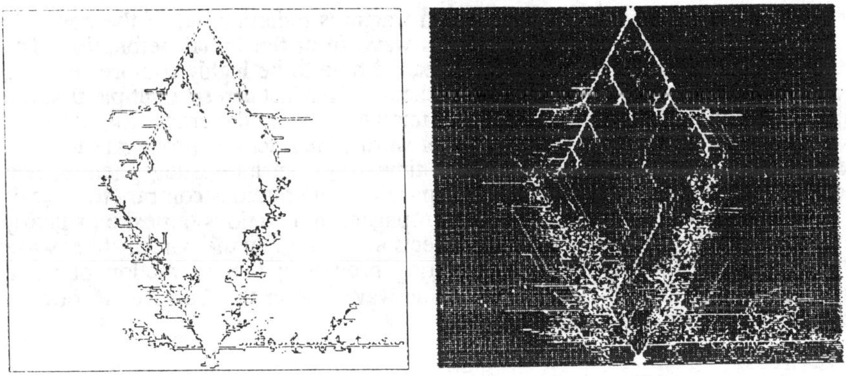

(a) (b)

Figure 3. Fracture patterns for a delay case in which the top hole detonates 10,000 time steps after the bottom hole. (a) radial fracture propagation (b) complete mesh at 20,000 time steps.

and brittle and failure is assumed to occur when a maximum strain is attained.

The simulations have been performed on blocks with fixed boundary conditions with no confining stresses applied at the boundaries. In all the cases, the effective pressure from each detonation is assumed to last 10,000 time steps. As previously outlined above, three cases are studied which simulate the effects of delay time on the fracture initiation and propagation.

The first case involves the simultaneous detonation of charges at opposite ends of a block. The numerical simulation shows multiple fracture initiation from the boreholes and much crack branching as the fractures propagate. The majority of the fractures propagate in the plane of the hole placements. Furthermore, the branching becomes more pronounced and indicates a preferred direction towards the axis of wave propagation at later time steps. This result indicates the influence of the stress waves as the wave generated at the opposite face is transmitted through the damage zone created by the passage of the initial stress wave. The cracks branch in the radial direction towards the impinging wave because under the influence of the passing wave that direction is that of the highest stress. As can be observed from the fracture pattern generated, the block is well fractured, but the overall fragmentation will be characterized by large pie shaped fragments.

The next case considered is one in which the second hole is detonated when the stress wave from the first detonation travels half the distance to the second hole. In this case, the fractures emanate from the second hole similar to that for the simultaneous case, but when the wave front from the first hole interacts with the stress wave from the delayed charge, the fracture propagation is slowed and the region around the second hole is strengthened. Furthermore, the crack branching is inhibited in directions contrary to the axis of wave propagation and the fractures show a tendency to propagate towards a preferred orientation,i.e. the direction of the wave front. Consequently, less damage is incurred by the second explosion. At later times, when the wave front from the delayed detonation enters the damaged region produced by the first explosion, the weakened rock is pulverized and the region around the first hole exhibits great damage which suggests that the region around the first borehole is rubbilized.

The last simulation is for the case where the delay time corresponds to the wave to traverse the distance between the two holes. The modeling results show a pronounced rubbilization around the firstly detonated hole and little damage around the subsequently fired hole with two main fractures radiating out of the borehole in the direction of the propagating stress wave from the initial detonation. The stress wave causes the area around the second hole to be highly compressed and, as a result, the immediate region is strengthened. This fact arrests multiple fracture initiation found in the previous cases examined and causes the cracks to propagate in the direction of the stress wave; in other words, the fractures propagate towards the direction of highest stress. For later time steps, an interesting phenomenon occurs; the cracks start to branch and propagate in orientations contrary to the axis of wave propagation; some cracks even propagate in directions opposite or nearly opposite to the initial direction. These effects are caused by the partial stress wave reflections that occur in the late fracturing processes; the interaction of these reflected waves with the transmitted elastic waves cause the fractures to deviate from their initial course of propagation.

4 SUMMARY

In this parametric study, we have examined the influence of delay time on the fracturing processes induced by blasting using the Dynamic Lattice Network Model. Though the model still needs to be validated against experimental data, initial comparisons between the numerical simulation results with past observations from actual blasting experiments are encouraging. For instance, the modeling indicates that using short delay times will produce a well propagated fracture pattern, but the overall fragmentation will be characterized by large, coarse material. This finding confirms the observations made by Stagg and Rholl (1987) and Anderson et al (1985), on the use of short delays; they conclude that the stress wave from the firstly detonated hole must be allowed sufficient time to damage the rock before the second hole is detonated for optimal fragmentation.

Furthermore, experimental tests conducted on polystyrene by Swift, et. al., under confined conditions and using various delays compare remarkably well with the results obtained using the Dynamic Lattice Network Model. For the simultaneous detonation case, the simulation shows the propagation of radial cracks in the plane of the two holes with near convergence between two fractures propagating from separate holes as found by Swift et al (1978) in actual laboratory experiments on polystyrene. For a longer delay corresponding to the time needed for the stress wave to travel the distance between the two holes, they find a definite fracture orientation in the radial direction of the impinging stress wave and a reduction of damage around the second hole as a result of the compressional wave from the first hole suppressing fracture initiation at the second hole except in the direction of the propagating wavefront. In the numerical simulation for this case, the fracture initiation at the second hole is constrained by the zone of compression around this hole created by the stress wave from the first explosion to initiate in a direction of this wavefront. Therefore, fewer fractures develop around the second hole and crack branching and propagation is stifled. Confirming observations by Swift et al (1978) the damage is minimized at the second hole and the fractures are generally oriented in the direction of the propagating wavefront. However, for late time steps, the simulation does indicate peculiar crack behavior due to the wave interaction of the reflected waves from the boundaries.

In terms of overall fragmentation, the simulation shows that using longer delays will result in finer material, produced mostly in the region of the first hole, but the damage produced around the second hole will be considerably less than that produced by the same charge detonating alone. In the simultaneous detonation case, the damage will be more controlled and characterized by radial cracks propagating into the material from both holes. Hence, the damage will be better distributed in the rock than for the longer delays, but the fragments will be larger and fewer fines will result from the blasts.

As we have demonstrated, the Dynamic Lattice Network Model is a capable tool for blast design. Presently, the modeling is performed on a small mesh of particles and springs to keep the computational requirements at a minimum and yet be able to observe the actual physics behind the fracturing process. Future development will require a larger mesh and a better approximation for the fluid flow into the cracks.

REFERENCES

Anderson, D.A., Winzer, S.R., & Ritter, A.P., "The Relationships Between Rock Structure and Delay Timing, and Fragment Size and Distribution in Explosively-Loaded Rock", *Fragmentation by Blasting, 1st Ed.,* Society for Experimental Mechanics, pp. 41-62, 1985.

Barker, D.B., Fourney, W.L., & Dally, J.W., "The Influence of the Stress Waves on Explosive Induced Fragmentation-Borehole Crack Network," *Proceedings of the 19th U.S. Symposium on Rock Mechanics,* Stateline, NV, pp. 144-150, 1978.

Huang,J-I., "Modeling dynamic fracture growth with an elastic network", *Geophy. Res. Letters,* Vol. 19, pp. 1855-1858, 1992.

Huang, J-I., & Kim, K., "Fracture Process Zone Development During Hydraulic Fracturing," *Int. J. Rock Mech. Min. Sci. & Geomech. Abst.,* pp. 1295-1298, 1993.

Kim K. & Yao, C., "The Influence of Constitutive Behavior On Fracture Process Zone And Stress Field Evolution During Hydraulic Fracturing," *Proceedings First North American Symposium on Rock Mechanics,* Austin, TX. 1994.

Kutter, H.K., & Fairhurst, C., "On the Fracture Process in Blasting", *Int. J. Rock Mech. Min. Sci.,* Vol. 8, pp. 181-202, 1971.

Pearsson, P.A., Holmberg, R., & Lee, J., Rock Blasting and Explosives Engineering, CRC Press, 1994.

Preece, D.S., & Burchell, S.L., "Variation of Spherical Element Packing Angle and Its Influence on Computer Simulations of Blasting Induced Rock Motion", *Proceedings of the 1st International Conference on Discrete Element Methods,* MIT, Cambridge, MA, 1993.

Preece, D.S., & Knudsen, S.D., "Computer Modeling of Gas Flow and Gas Loading of Rock in a Bench Blasting Environment", *Proceedings of the 33rd U.S. Symposium on Rock Mechanics,* Santa Fe, NM, pp. 295-303, 1992.

Song, J., & Kim, K., "Dynamic Modeling of Stress Wave Induced Fracture", *Proceedings the 8th International Conference of the Association for Computer Methods and Advances in Geomechanics.* (1994)

Stagg, M.S., Otterness, R.E., & Siskind, D.E., "Effects of Blasting Practices on Fragmentation", *Proceedings of the 33rd U.S. Symposium on Rock Mechanics,* Santa Fe, NM, pp. 313-322, 1992.

Stagg, M., & Rholl, S.A., "Effects of Accurate Delays on Fragmentation for Single-row Blasting in a 6.7m(22-ft) Bench", *Second Int. Symp. on Rock Fragmentation by Blasting,* Society of Exp. Mech., Bethel, CT, pp. 210-223, 1987.

Swift, R.P., Schatz, J.F., Durham, W.R., Hearst, J.R., & Kusabov, A., "Effect of Simultaneous and Sequential Detonation on Explosive-Induced Fracture", *Proceedings of the 19th U.S. Symposium on Rock Mechanics,* Stateline, NV, pp. 235-242, 1978.

Taylor, L.M., Chen, E., & Kuszmaul, J., "Microcrack-induced Damage Accumulation in Brittle Rock Under Dynamic Loading", *Computer Methods in Applied Mechanics and Engineering,* Vol. 55, no. 3, pp. 301-320, 1986.

Thorne, B.J., Hommert, P.J., & Brown, B., "Experimental and Computational Investigation of the Fundamental Mechanisms of Cratering", *Fragblast'90,* Brisbane, Australia, pp. 117-124, 1990.

Winzer, S.R., & Ritter, A.P., "Role of Stress Waves and Discontinuities in Rock Fragmentation", *Fragmentation by Blasting,* 1st Ed., Society for Experimental Mechanics, pp. 11-23, 1985.

5 Rock cutting and TBM performance simulation

Session developer:
David Summers, University of Missouri-Rolla

Rock Mechanics, Nelson & Laubach (eds) © 1994 Balkema, Rotterdam, ISBN 90 5410 380 8

A boreability model for TBM's

Nils-Olof Sundin & Stefan Wänstedt
Division of Mining Engineering, University of Luleå, Sweden

ABSTRACT: The Boreability Index (I_B) is a measure of the full face boreability of the rock mass. The index I_B can be calculated in two ways depending on the application. In a boreability model the index I_B used is the quotient between net penetration rate and the two machine parameters cutterhead revolution speed and thrust per disk cutter. In a prediction model I_B consists of data from an indentation test (modified Handewith test) and a factor for joint and weakness planes.

By applying the boreability model to measurement while boring data, it is possible to detect important changes in the rock mass during the boring process, e.g. fracture zones and different rock types. This model maybe also can be a helpful tool when mapping the tunnel and in planning of reinforcements after boring. In our opinion I_B is more sensitive measure of the rock mass parameter than the net penetration rate.

1 INTRODUCTION

The prediction of tunnel boring machine (TBM) performance in various surroundings is of great importance for contractors and other companies involved in tunnelling business. Even during the project there is a continuous need to monitor changes in the rock quality, so that the machine can be stopped before it gets into major problems.

Variations in the rock mass affect the penetration rate of TBM to a higher extent than conventional drilling and blasting. The penetration rate of a TBM has been observed to vary with as much as a factor of 10, depending on rock type and fracturing of the rock mass. Thus, TBM projects require a more thorough geological investigation of the site than conventional drilling and blasting projects. The TBM-technology differs from drilling and blasting in three other points as well, namely regarding system technology, work organisation and distribution of costs. Since full face boring is a continuous operation, the personnel must work well together to ensure that operation and maintenance of the system is performed satisfactorily.

For several years the Division of Mining Engineering at the University of Luleå has evaluated a number of Swedish and European TBM projects. A prediction model for boring time and a boreability model has been developed at the division. The prediction model for calculating boring time has continuously been developed during this time, taking in consideration how the TBM system, work organisation and geology influences the drifting (Sundin, 1991). In this case the boreability index is calculated from an indentation test on small rock samples and a factor for weakness planes in the rock mass.

In the boreability model the index I_B is calculated from real boring data. Three different boring factors are used to calculate the boreability: net penetration rate, cutterhead revolution speed and thrust per cutter.

In this paper four different evaluated TBM-projects are discussed. Two of them are used to illustrate the boreability index in the production stage of tunnelling. The remaining two are used to demonstrate the use of the boreability index in the prediction model of boring time.

2 BOREABILITY MODEL

The Boreability Index (I_B), a measure of the full face boreability of the rock mass, is calculated from equation (1) below.

$$(1) \quad I_B = \frac{1}{F}\frac{npr}{RPM}\frac{1000}{60} \quad \text{[mm/kN]}$$

In the equation *npr* [m/h] is the net penetration rate, *RPM* [rev/min] the revolution speed of the head and F [kN] the load per disc cutter. The calculated values of I_B are smoothed before presentation, to enhance their readability. Figure 1, shows raw and smoothed I_B-values, calculated from data in shift reports along a part of a TBM tunnel.

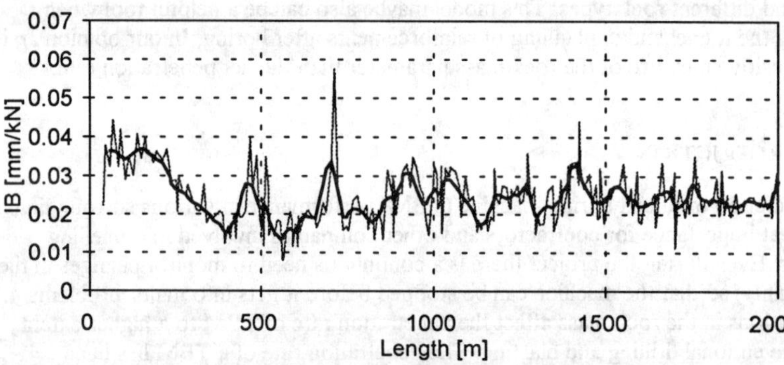

Figure 1 An example of the index I_B along part of a tunnel, calculated from shift reports.

The only data available to evaluate TBM-projects are usually the shift reports, which are compiled by the TBM operator. During the compilation the amount of uncertainty varies with the motivation of the personnel. In some cases, analogue printers have been used to improve the recording of boring parameters. Due to the format of data, it was time consuming to analyse and present the results. Following the development of electronics it is now possible to record and evaluate boring data digitally. We suggest that this method should be called Measurement While Boring (MWB).

2.1 *Case 1*

This tunnel was constructed during 1985 and 1986 for a hydropower project. The main rock types in this area (figure 2) are gneissic granites with varying grain sizes, fine (F), medium (M) and coarse (C), and compressive strengths between 120 and 200 MPa. Other rock types are amphibolite and gabbro (G). Some tectonics has occurred in the area resulting in narrow mylonite zones (My). The average strike of the structures and fractures is approximately perpendicular to the direction of the tunnel, with dips varying from 40 to 60 degrees. Refraction seismics along part of the tunnel indicated slow velocity zones (3200-4000 m/s) from about 430 m to 650 m in the tunnel.

In figure 2, the boreability index I_B for the tunnel, calculated from boring data from shift reports, is presented together with the mapping of the rock classes in the tunnel. In this case different levels of I_B fit correlate well to different kinds of rock classes. Peaks indicate zones with higher a frequency of weakness planes.

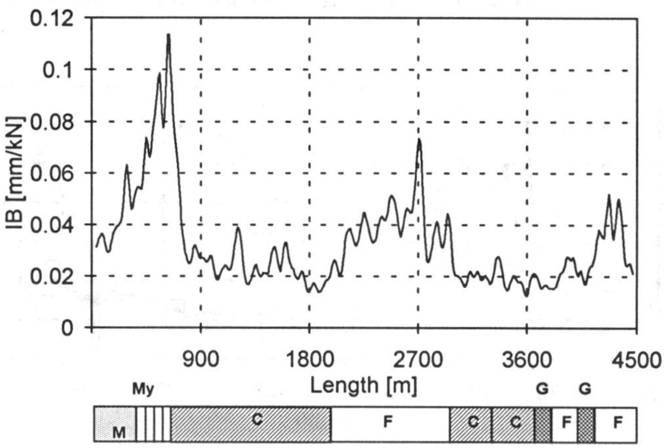

Figure 2. The boreability index I_B for the first tunnel, calculated from shift reports. The manual mapping of the rock mass is plotted in the bottom of the figure.

The TBM got stuck in a fracture zone after 450 m of boring. This resulted in a delay of more than one month. Data from the shift reports indicated an increase in the penetration rate and a decrease in the thrust force, 2 to 3 working shifts before the problem occurred. Note the high I_B values for that zone.

Only 5 % of the tunnel required any kind of reinforcements. In drill-and-blast tunnels in the same area almost 70 % needs to be reinforced.

2.2 *Case 2*

This is also a hydropower tunnel, presently under construction in the Swedish Caledonians. The diameter of the tunnel is 6.5 m. At the writing of this paper, 2 km of tunnel remains to be bored of a total of 11 km. The rock mass in the area consists mainly of gneissic rock types of varying composition, and mica schists.

The TBM (Atlas Copco Mk 27) is equipped with sensors that continuously, at an interval of 20 seconds, records digital information from different parts of the machine

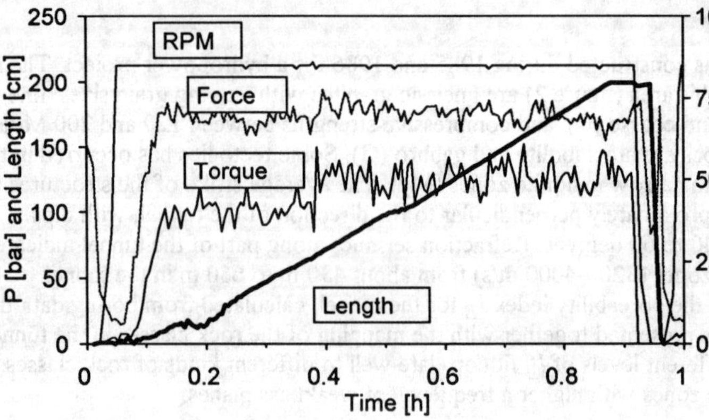

Figure 3. Data recorded from the TBM during one stroke length of boring.

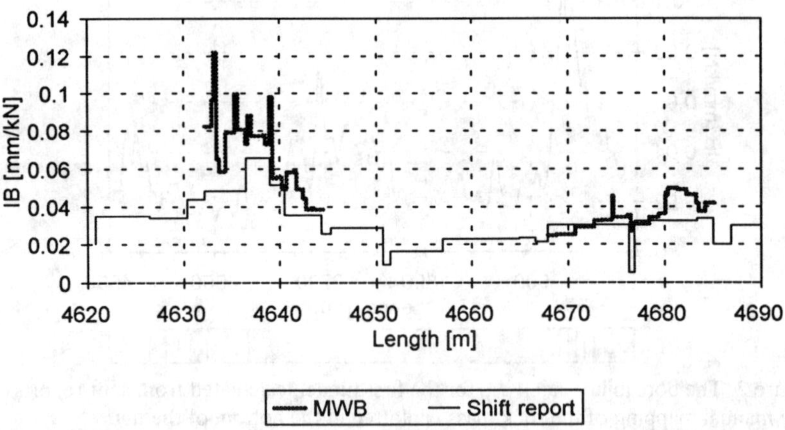

Figure 4. Recorded data compared to information obtained from the shift reports.

system. This is the first time one can use a TBM to record MWB data. An example of the data can be seen in figure 3, which displays four different parameters recorded during one stroke length.

The data has not been recorded on a routine basis since the software is still being developed. At present data exists from approximately 500 m of the tunnel.

The boring had to stop when a 20 m wide water bearing fracture zone was encountered at 4632 m in the intake tunnel. The tunnelling could continue after 5 weeks of intensive grouting, both from the surface and in the tunnel. Unfortunately no MWB data were recorded at the time when they bored into the water bearing zone. The recording of MWB data was resumed when the boring continued. Shift reports, on the other hand, exist from the entire tunnel. In figure 4, the I_B calculated from MWB-data is compared to the corresponding values calculated from shift reports. The fracture zone has not been mapped.

The figure shows that the MWB data provides a much more sensitive analysis of the boreability than the shift reports. Both curves, though, indicate clear differences in I_B when boring in "soft" rock (4620-4640 m) and "hard" rock (4650-4690 m).
In figure 5, I_B is compared to *npr*. One can see that the boreability index I_B has a higher resolution regarding rock properties. For instance, the level of *npr* does not change significantly when drilling in "hard" or "soft" rock. The I_B, on the other hand, clearly shows the difference in hardness.

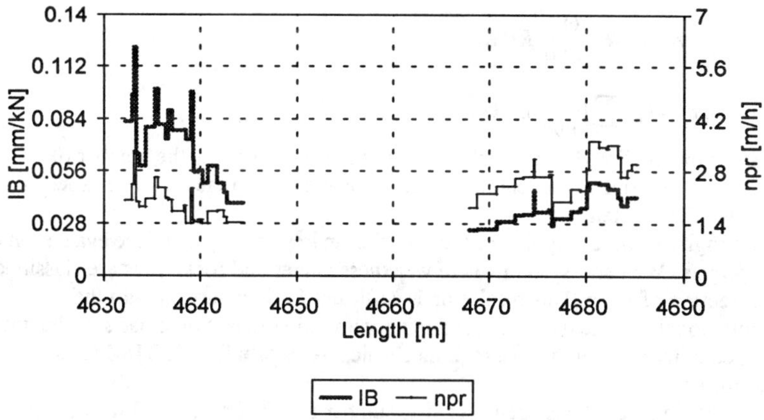

Figure 5. A comparison of the boreability index I_B and the net penetration rate *npr*.

In table 1, we have made an attempt to classify the boreability in different classes. The classification is based on our experience from tunnelling in crystalline rocks, e.g. granite and gneiss.

Table 1. Classification of the boreability of the rock based on the index I_B

I_B (mm/kN)	**Boreability class**
< 0.01	very low
0.01 - 0.02	low
0.02 - 0.04	medium
0.04 - 0.05	high
> 0.05	very high

3. PREDICTION MODEL FOR TBM TIME

The prediction of TBM project time is performed by dividing the total tunnel length in independent homogeneous zones and studying certain parameters that affect the TBM performance. The effect of each zone is estimated in consideration to how the thrust per cutter (*F*), utilisation (*U*), rock distribution and fracturing influence the TBM performance.

In the prediction model the boreability index (I_B) is based on an indentation rock test and a factor for weakness planes in the rock mass, see equation (2). The test is a modified version of the Handewith indentation test (1972) and gives a penetration index (P_{BI}). The factor for weakness planes (k_J) used in the equation, is based on empirical data from a large number of evaluations of TBM-projects (Johannessen, 1988).

Three main equations are used for the calculation of boring time:

2) $$I_B = P_{BI}\ k_J$$

3) $$npr = I_B\ F \frac{60}{1000} RPM$$

4) $$sh_t = k_{nx} \sum (\frac{1}{npr} h_{sh}\ U)$$

where sh_t is the predicted total number of shifts of tunnelling, h_{sh} the number of working hours per shift, l the length of each zone and k_{nx} is a factor for unexpected shutdowns of the system.

A thorough geologic mapping of the tunnel is an important part of the evaluation of a TBM project. We use a system where weakness planes and rock types are classified for each section of 10 m. This resolution is sufficient for comparison with the prediction model and the data from preliminary investigations. Some parts of the tunnel are mapped in more detail, where rock mechanical tests or full scale TBM tests have been performed.

The prediction model has been used in a number of TBM projects. Two of them from the Stockholm area are presented here. The same TBM and back-up system (BUS) were used in both projects, but the equipment was modified before the second tunnel was bored, based on the experience from the first tunnel.

Dominant rock types in the Stockholm area are granite and gneiss. In table 2, some rock mechanical data from rock samples are listed. The samples originate from core holes drilled in the tunnel.

Table 2. Rock mechanical data for the dominant rock types in the Stockholm area. (Tollerup, 1993)

Rock Type	Granite	Gneiss
Uniaxial Compressive Strength (MPa)	150-220	65-180
Point load (MPa)	5-9	1-7
Fracture toughness ($MN/m^{1.5}$)	2.2-3.0	2.3-3.3
Wear CAI	3.0-5.5	1.9-5.9

3.1 *Case 3*

A 6.5 km long tunnel was constructed 1987-1989 to transport sewage from a heating plant, beneath Stockholm city to the Baltic Sea. In the heating plant the sewage is cooled which results in the release of energy (about 120 MW) to heat nearby houses.

An entirely new constructed Tunnel Boring Machine (Atlas Copco FORO 900 S/3.5 m) was used. The TBM was only 6 m long to enable the drilling of injection holes close to the front. In the prognosis the tunnel was divided into 7 zones, based on penetration

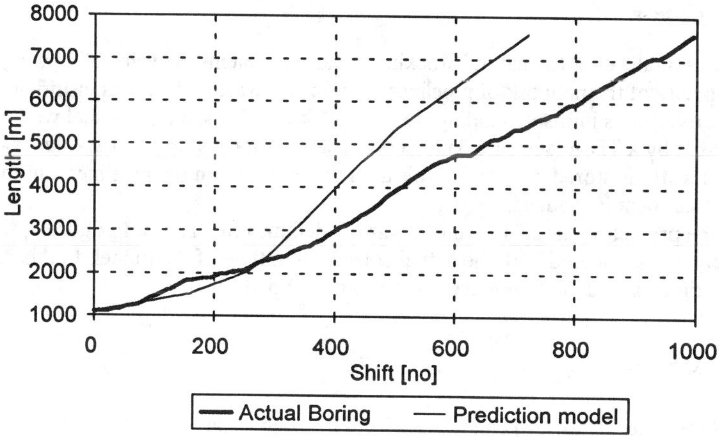

Figure 6. Predicted time compared to actual boring time for Case 3 in the Stockholm area.

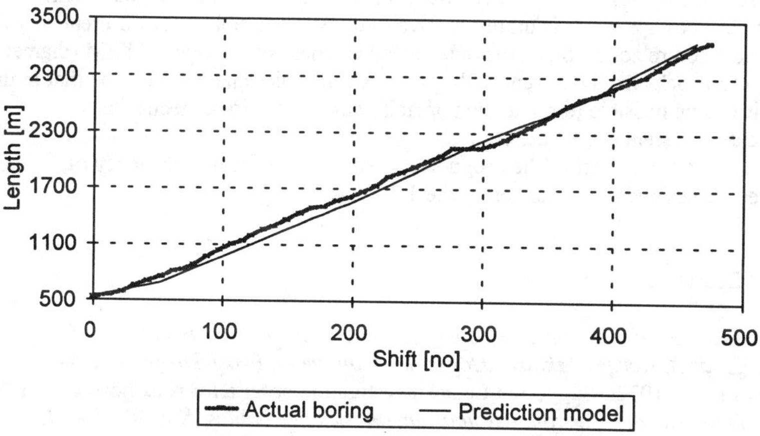

Figure 7. Predicted time compared with real boring time for Case 4 in the Stockholm area.

tests and preliminary investigation data. The performance of the TBM was evaluated and in figure 6 the prognosis is compared to the outcome.

The prognosis underestimated the boring time with about 25 %. The reason for the deviation is that the estimated factor for fracturing (k_J) was too high. Also, large problems with the TBM and BUS resulted in a low utilisation for boring, in fact much lower than the prognosis.

3.2 *Case 4*

Heavy rain falls over the city of Stockholm cause problems in cycles of 5 years, since the capacity of the network of pipelines for rain and waste water was insufficient. To reduce overflows into surrounding rivers and lakes, a 2.7 km long tunnel was excavated by a TBM (see case 3) to serve as a temporary storage of surplus water. The surplus water is stored in the reservoir until the pressure on the pipelines and the waste water treatment is reduced.

In the prediction model the tunnel was divided into five zones. In figure 7, the prediction is compared with the actual construction time of the tunnel. In this case the boring time exceeded the predicted time by only 2.5 shifts.

4. CONCLUSIONS

The Boreability Index (I_B) can be used both for prediction work, during the actual boring and as an aid for the evaluation of the tunnel. The index I_B increases the possibility to detect changes in the rock mass of importance during the drifting, rock type boundaries, variations in the frequency of weakness planes, fracture zones etc. The index has been observed to increase before weak zones are entered. If the index is monitored continuously the TBM can be stopped before the machine gets stuck.

Compared to npr the I_B is more sensitive to classify boreability for different rock types and fracture zones. Since the index is independent of the type of TBM (diameter, power, etc.) rocks from different TBM projects can be compared. We hope that in the future it will be possible to relate the boreability index to a rock mechanical classification system (Q, RMR, etc.).

The most difficult part of the prognosis work is to predict the frequency of weakness zones and the utilisation of the TBM.

5. REFERENCES

Johannessen, O. 1988. Full face boring of tunnels. (in Norwegian) *Institutt for Anleggsdrift, Norges Tekniske Högskole, Trondheim, Prosjektrapport* 1-88.

Handewith, H. 1972. Suggested tunnel investigation criteria for rock boring machines. *The 8th Canadian symposium on rock mechanics, Nov 30 - Dec 1, University of Toronto.*

Sundin N.O., Lindqvist, P.A. 1983. Comparison of predicted and encountered TBM performance for the Wöllatunnel. *Luleå University, Sweden, Dept of Mining and Underground Construction.* TULEA 1983:41.

Sundin N.O., 1991. Prediction model for fullface boring. (in Swedish) *Luleå University, Sweden, Dept of Mining and Underground Construction. Licentiate Thesis.* 1991:08 L.

Tollerup, L. 1993. Stockholm builds a "Snake Tunnel" to avoid flooding and improve water quality. *Tunnelling and Underground Space technology*, Vol. 8, No. 4. pp.425-427.

Rock Mechanics, Nelson & Laubach (eds) © 1994 Balkema, Rotterdam, ISBN 90 5410 380 8

Development of a mechanized excavation system performance simulation package

Yousof Abd Al-Jalil, Priscilla P. Nelson & Christopher Laughton
Department of Civil Engineering, The University of Texas at Austin, Tex., USA

ABSTRACT: This paper presents the design and development of a package, based on Monte Carlo simulation techniques, that can model the operation of mechanized underground excavation systems. The primary use of the package is to estimate time and cost required to complete a given tunnel, in addition to predicting equipment utilization. A data base containing performance records from more than 450 tunnel boring machine (TBM) projects is used to derive a ground classification system, predictive equations for various performance parameters, and probability density functions of various components of mechanized underground excavation systems. Analysis of various possible excavation system configurations is achieved by using rate of failure and delay information for each comprising subsystem. The estimation of time to complete a given tunneling project can be done in different ways depending on available project information.

1 INTRODUCTION

The objective of this research is to create a tool that can be used by different authorities involved in TBM-based excavation systems, to evaluate the performance of these systems. These systems include all the components necessary to perform tunnel excavation: from the cutting face at the tunnel heading to last point of muck disposal. The scope of the research is limited to TBMs with single disc cutters, and to tunnels that are excavated in rock. Data were compiled from different projects completed over the last ten years. Primary reliance was placed on 30 projects with the most complete information. Most of these were excavated in sedimentary rock.

The problem is to estimate time (and cost, if cost data are available) required to excavate a given tunnel in a given geologic setting. This task includes estimating machine utilization, penetration rate, cutter wear and the rate of cutter replacement. A major application of the results is the selection the most efficient excavating system for the given job, while identifying expected ground-related problems and pinpointing weak links in the excavation process. The three major variables are: machine, geology, and management. Machine includes both the TBM itself and its back-up system. Geology includes an assessment of intact rock properties and rock mass characteristic variability. Management includes the experience of contractor, crew, and how the maintenance program is carried out.

In the data collection phase, data collected can be split into two broad groupings. The first data group is at project level (so-called "global data"). This "global" set includes the average utilization, average project advance rate, average penetration rate for each geologic unit in the drive, and some cutter wear information. In addition, general information such as diameter and length of tunnel, model of TBM, year of completion, and purpose of tunnel construction is entered. The second data set is much more detailed. For each project, information was compiled regarding machine specifications, site investigation reports, mapped tunnel walls, shift reports, and cutter replacement information. Shift reports also vary in level of detail, with some detailed projects having records of delays and mining time logged with 5-minute resolution.

The first set of data is compiled in a data base, accessed to exercise the ability to estimate average performance of new tunnels similar to those in this data base. This data base can be sorted according to the criteria set by package user. Such criteria may include machine power, tunnel diameter, type of rock unit(s), excavation system configuration, and contractor experience. The second and more detailed set is used to derive the distributions of penetration rate and utilization per ground class, time to repair (TTR) and time between failure (TBF) for various subsystems including all breakdowns, and the development of the ground classification system (Abd Al-Jalil *et al.*, 1993).

In the following sections, a description of methodology of analysis, and how the package can be used in practice are given.

2 METHODOLOGY OF ANALYSIS

The backbone of the package is the Monte Carlo simulation program. This is a modified version of program SIMSUPER (Salazar and Einstein, 1986). There are two types of simulation; geology simulation, and construction simulation. Inputs to geology simulation are parameters describing both intact rock and rock mass characteristics. The output is a classification of ground along tunnel alignment. The construction simulation input is the method of construction for each ground class. It is not uncommon that several ground classes share one construction method. However, one or more performance parameters may assume different values for different ground classes.

In this analysis of case histories, a new ground classification system has been developed by the third author. This system takes into account the intensity of fracturing in rock mass, rock strength, and water inflow. The ground class identification was done by monitoring the impact of changing ground characteristics on the performance of TBM. Basic output from this analysis is a set of modifiers, or weights, to be applied to the distribution of utilization and penetration rate achieved in "best ground class". These modifiers are used later in the construction simulation.

The second part of this analysis focuses on modeling performance parameters at different levels. Results from this part can be summarized as follows:

1) A set of parameters is generated to model advance rate, utilization, penetration rate, and cutter replacement. This set is divided into two subsets: the first subset contains sets of parameters derived from each case history separately; and the second subset is based on all case histories that have common features (in terms of machine, management, and geology). These sets are grouped by ground class. One approximate classification is done by sorting data points according to the ground support installed.
2) A set of parameters is generated to model the various activities and delays encountered in the excavation process.
3) A set of equations is generated to predict cutter wear, and transition delays (which occur when construction method is changed due to changing ground support). These are derived using regression analysis techniques.
4) A set of probabilities and delay durations for specific ground-related problems are generated which can be identified consistently from case histories for the different ground classes.

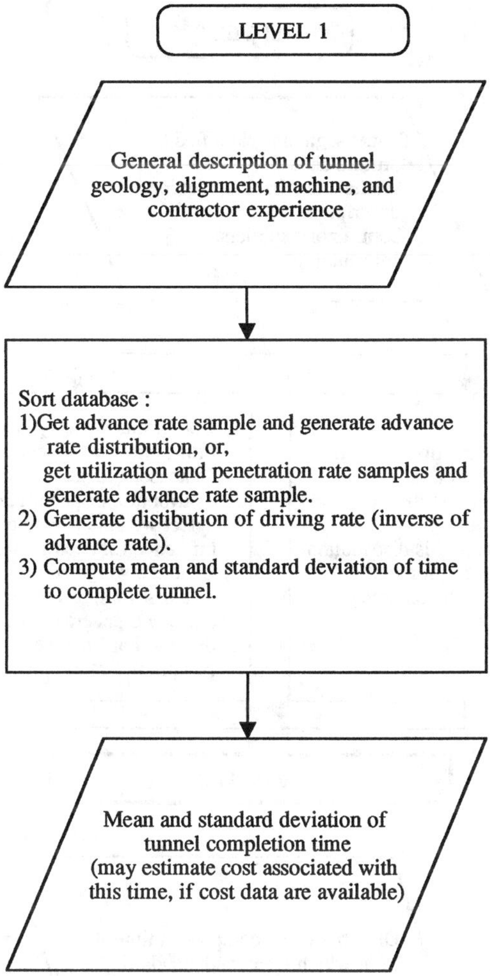

Figure 1 Level 1 prognosis for tunnel excavation by TBMs

The first step in using this package is to make use of site investigation data to generate many realizations of ground class distributions along a proposed tunnel. This step quantifies the uncertainty in ground class along the tunnel. A "strip" map of tunnel geology is generated, with stations indicating start and end of each ground class. Next the user selects parameters, from sets 1 - 4 in the listing above, to represent various components of the excavation process. This is done for each construction method. The first task is estimation of a "basic" penetration rate for the "best ground class", which can be regarded as representative of unfractured rock. The distribution of penetration rate for various other ground classes can be generated from the basic penetration rate, employing the corresponding penetration rate modifier. Users have to link each ground class to a unique construction method. The program is then run through many simulations of production, resulting in an evaluation of the time to complete tunnel (TTCT) and cost to complete tunnel (CTCT), if cost data are available. This process is repeated for each

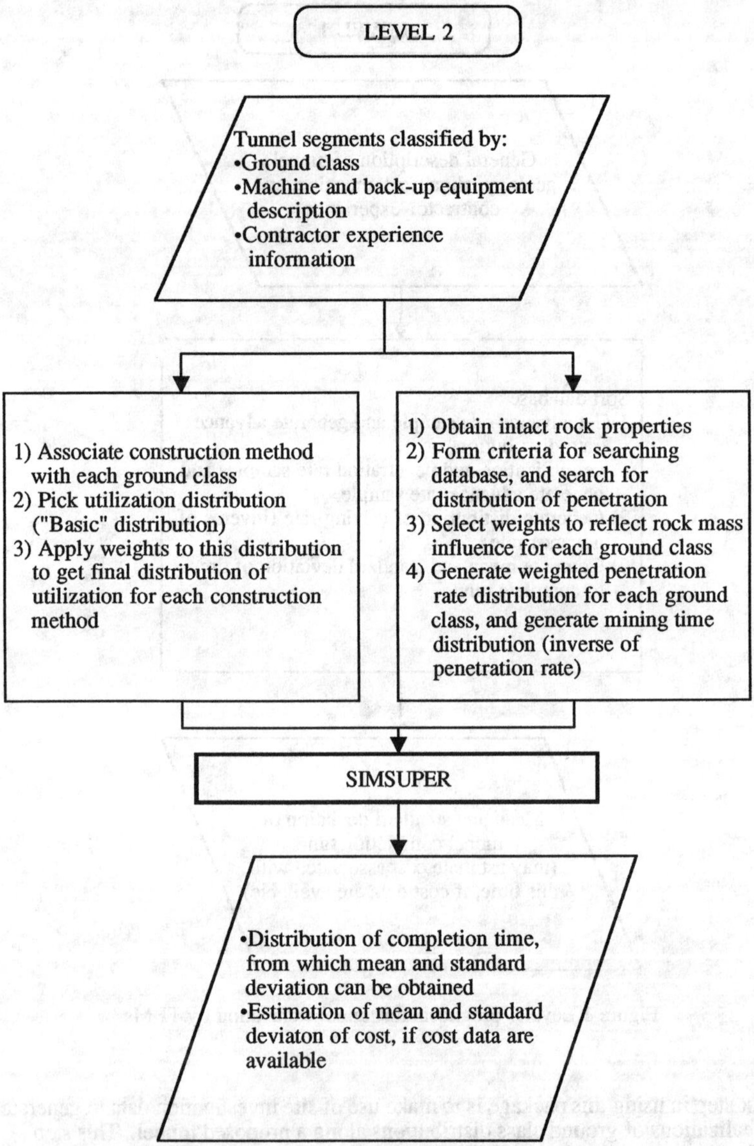

Figure 2 Level 2 prognosis for tunnel excavation by TBMs

geology realization. The last step is to combine all TTCT and CTCT from different geologic realizations into one sample, and compute sample mean and standard deviation. Of course, many excavation system configurations can be tried, so that users can select the optimum system configuration.

It should be noted that all these calculations are for the production phase of the project. The project start-up time and learning curve can be estimated separately, and a separate data base module for these is included in the package. Time and cost associated with scheduled maintenance, holidays, and site-specific special items of construction can be estimated discretely and added directly to TTCT and CTCT.

3 LEVELS OF SIMULATION

There are four levels of prognosis of the tunnel excavation process. These are depicted in Figs. 1 through 4. Since data exist at different levels of detail, simulations are performed at levels parallel to that of the data. The less detailed level, Level 1 (Fig. 1) with project mean penetration rate and utilization, or the Level 2 (Fig. 2) with penetration rate and utilization data on a daily or shift basis, can be used for the preliminary estimation of TTCT and CTCT.

For a more extensive analysis, information on time between failure and time to repair, together with ground related problems and ground classifications, can be used in the detailed estimation of TTCT and CTCT at the bidding stage of a project using a Level 3 (Fig. 3) or Level 4 (Fig. 4) approach. Here, the construction activities are defined for each construction method through a network. This network is based on the critical path method, and due attention should be paid to whether activities are linked in series or parallel. Delays are associated with activities or subsystems. The Level 4 simulation, which is reliability-based, can be used to study the performance of different equipment assemblages for a proposed project. In this case, a very good data set should be used. Use of this most detailed level can only be justified if there has been a very detailed site investigation and ground classification, and if there is a reliable set of data for the proposed excavation system and good historical records for crew and management experience.

4 DEVELOPMENT OF SIMULATION PROGRAM

The development of SIMSUPER, as of the time of writing, consists of the following:

1) Defining new functions to account for TBM clock-, shift time-, and distance-based delays, in addition to representing delay duration.
2) Coding and testing random number generators. This suite of subroutines can generate random numbers from the following distributions:
    - 2.1) Exponential
    - 2.2) Shifted exponential
    - 2.3) Truncated normal
    - 2.4) Lognormal
    - 2.5) Triangular
    - 2.6) Beta
    - 2.7) Weibull
    - 2.8) Empirical cumulative distribution function (ECDF)

   These distributions are standard, and interested readers are referred to Law and Kelton (1982).
3) Adding cutter wear estimation subroutine. This takes into account the impact of abrasive rock units on the rate of cutter wear .

Verification of the program includes many steps. The first step is to check that the random number generators give samples of random numbers that have means and standard deviations very close to the population from which these random numbers are sampled. Each generator is tested separately, and the relative error, defined as the difference between population mean and sample mean divided by the population mean, is less than 1% for sample size of 10,000 random numbers.

The program is then tested for the production phase of an actual project in the data base. For example, with 30 simulations the mean TTCT was 97.91 days, compared to that for

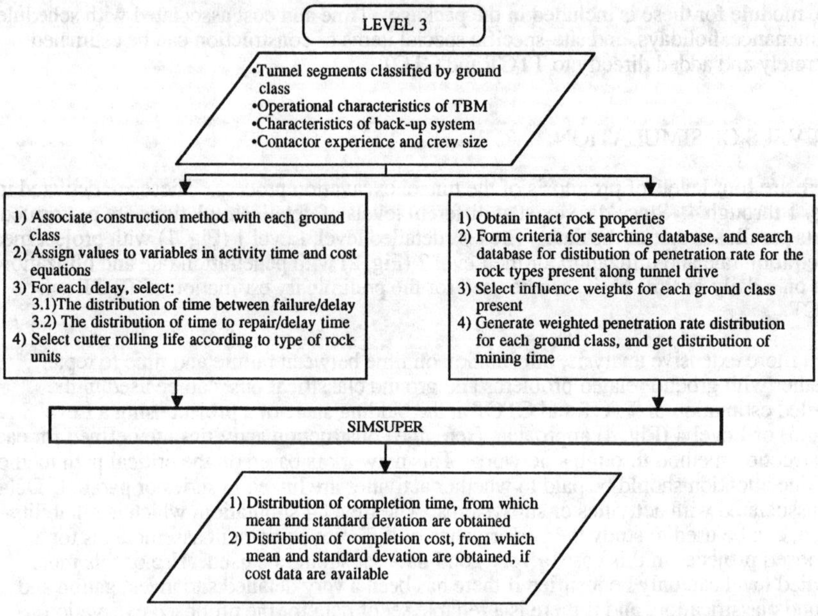

Figure 3 Level 3 prognosis for tunnel excavation by TBMs

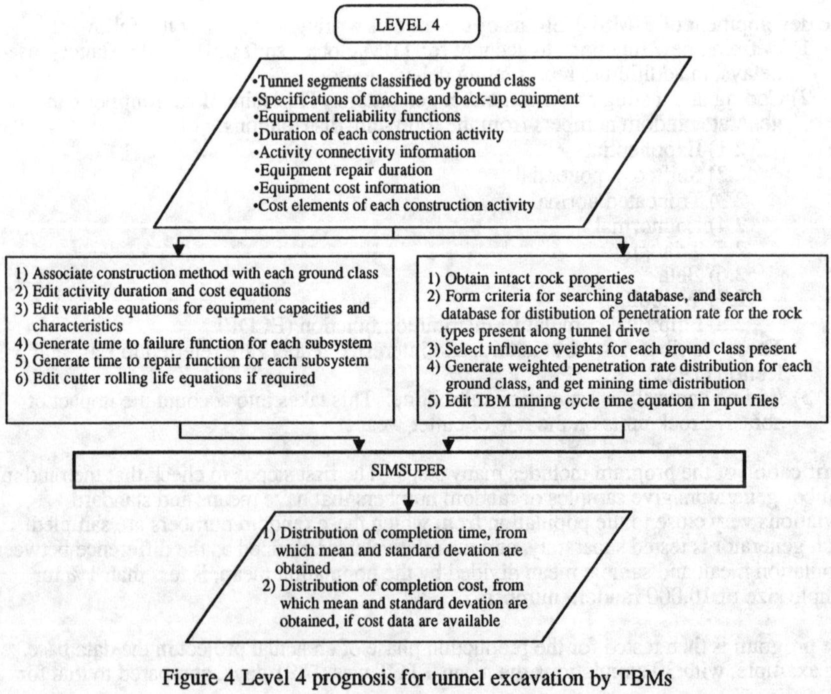

Figure 4 Level 4 prognosis for tunnel excavation by TBMs

the actual project TTCT which was 96.04 days. This corresponds to less than 2 percent relative error. The actual project duration is within 0.70 standard deviations from the mean TTCT obtained by simulation. Improvement of accuracy can be achieved with more simulations, but computer time increases also.

5 FUTURE DEVELOPMENT

The next step is to modify the input and processing of site investigation data - within the program - to classify ground along the tunnel. The modules of penetration rate prediction and user interface need to be developed to make the program more user-friendly. The final module will use all public domain accessible information and relationships to predict penetration rate. In a subsequent phase of research, a new predictive tool will be developed which will be based on a fundamental analysis of the process of disc indentation and chipping, developed from a solid and fracture mechanics perspective.

ACKNOWLEDGMENTS

We would like to thank the National Science Foundation, Division of Design and Manufacturing Systems and the Division of Geomechanical, Geotechnical and Geo-Environmental Systems for sponsoring this project. We also would like to thank Prof. Herbert Einstein and Mr. Vijaya Halabe, of Massachusetts Institute of Technology, for their help in modifying the program SIMSUPER. The authors also thank the international tunneling community for their support in providing access to case history information.

REFERENCES

Abd Al-Jalil,Y., P. P. Nelson and C. Laughton 1993.TBM performance analysis and rock mass impacts. *Proc. Mine Mechanization and Automation.* Almgren, Kumar and Vagenas (eds.). Balkema, Rotterdam.

Law, A. M. and W. D. Kelton 1982. *Simulation modeling and analysis.* McGraw-Hill Book Co. New York.

Salazar,G. F. and H. H. Einstein 1986. SIMSUPER5: tunneling construction simulation. *Proc. Fourth Conference on Computing in Civil Engineering*, Boston, MA, ASCE, 461-467.

Rock Mechanics, Nelson & Laubach (eds) © 1994 Balkema, Rotterdam, ISBN 90 5410 380 8

Indentation fracture mechanics and rock disc cutting

X. L. Zhao & J.-C. Roegiers
The University of Oklahoma, Norman, Okla., USA

E. K. S. Passaris
The University of Newcastle upon Tyne, UK

ABSTRACT: An attempt was made in this paper to apply the principles of indentation fracture mechanics to disc cutting. A total of 25 disc cutters with edge angles varying between 60° to 100° and diameters ranging from 100 mm to 200 mm was investigated. Following the indentation tests, cutting tests were carried out. As the same disc cutter was used in cutting tests, the data generated by the indentation tests could easily be transferred to those of cutting tests via a fracture mechanics model. Good agreements between theoretical analyses and experimental results were obtained.

1. INTRODUCTION

The microcracking which occurs at the contact loading of brittle solids was first studied by Hertz in 1881. The scientific evaluation and engineering application of such phenomena is the indentation test, now widely adopted as a standard indicator of material "hardness." Some publications qualitatively describe the indentation process; while others propose models of the rock failure under the indenter which involve various stress-strain relationships.

Lindqvist et al. (1984) observed the indentation fractures using a Scanning Electron Microscope and found that: (a) cracks initiated mainly at the edge and corners of wedges, or in the interior of the rock; (b) stable fracture appearing dominated the progressive penetration process; and, (c) all such cracks were much longer than the actual wedge penetration depth. Lindqvist and Lai (1983) also indicated that the crushed material could act as an extension to the cutting tool and had the effect of distributing the load over a wider area, thereby reducing its effectiveness.

In most cases, the fracturing process preceding failure by an indenter may be divided into four stages (Cook et al., 1984): (a) closure of microcracks; (b) linear elastic deformation; (c) stable crack growth; and, (d) unstable crack propagation. Paul and Sikarskie (1965) used the Mohr-Coulomb failure criterion to evaluate the orientation of the plane of failure and the load associated with shear failure due to the penetration of a symmetrical wedge-shaped indenter. Basically, the force transmitted to the solid rock depends on two parameters; namely, the geometry of the indenter and the angle of friction which describes the indentation between the surface of the indenter and the rock.

Following the development of fracture mechanics, the indentation test was adopted as a practical method for determining the fracture toughness and its applicability has been assessed via a number of tests on ceramics, glasses and rocks (Marion, 1979; Atkinson and Avdis, 1980). Fracture mechanics models of indentation were formulated by Lawn and Fuller (1975) and Swain and Lawn (1976).

Generally speaking, the indenters can be of any geometry; for example, Robson (1983) and McGann (1984) used the pencil indenters on small rock samples which simplified the core-grooving techniques for predicting the performance of pick cutting. An edge indenter was used by Fauvel (1981) to investigate the mechanism of rock breakage by disc cutter. Therefore, in order to predict field boreability from laboratory indentation tests, two major factors should be considered (Ozdemir et al., 1977): (a) the geometry of the indenter; and, (b) the proper translation of indentation results to the situation of rock cutting. Instead of the indenters which are normally used to predict the performance of a field machine, Ozdemir et al. (1977) used disc type cutters and found that the ratios of penetration induced for the static and rolling cases fluctuated between 1.8 and 2.1; the average being exactly two. This is simply due to the fact that the idealized rolling disc has only half of the rejected areas in the thrust direction compared to the static situation.

In this paper, a total of 25 disc cutters indentation experiments with edge angles varying between 60° to 100° and diameters ranging from 100mm to 200mm are reported. Following the indentation tests, cutting tests were carried out and a theoretical model was built up based on fracture mechanics concepts. The comparison of theoretical analyses and experimental data lead to some new conclusions.

2. THEORETICAL BACKGROUND

2.1 Contact stress

The contact stress, or contact pressure, is defined by:

$$\sigma_0 = \frac{P}{A} \tag{1}$$

where σ_0 is the mean contact pressure; A is the projected area; and P is the applied load.

Because of the complex nature of the strain field produced by an indentation, analytical solutions have been limited to rigid ideal plastic or ideal elasto-plastic materials. Tabor (1951) expressed the relationship of the mean contact stress and yield stress of a material, σ_y, in simple compression as:

$$\sigma_0 = C\sigma_y. \tag{2}$$

Though the constraint factor, C, depends upon the geometry of the indenter, it approximately equals three for all of the indenters considered. The origin and characteristics of the constraint factor, C, are indeed the most important considerations in indentation hardness testing.

Johnson (1970) considered the indenter to be encased in a hemispheric core of radius and assumed that a hydrostatic uniform pressure exists throughout the core, which is equal in magnitude to the applied stress, σ_0:

$$\frac{\sigma_0}{\sigma_y} = \frac{1}{\sqrt{3}}[1 + \ln(\frac{4E}{3\pi\sigma_y \tan\frac{\theta}{2}})] \tag{3}$$

where θ is the angle of an indenter and E is Young's modulus.

2.2 Indentation fracture mechanics

As the crack configuration during indentation presents a complex elasto-plastic problem, and in order to avoid most of the analytical complexities associated with the general indentation fracture problem, a simple fracture mechanics analysis is very helpful.

Swain and Lawn (1976) considered the problem as shown in Figure 1 in which P is the applied load acting over a line, L, i.e., $P = P_L L$ with P_L being a line force per unit length and c is a characteristic crack length. Recalling Griffith-Irwin's energy-balance condition for equilibrium cracks, a relationship between the crack length, c, and the stress intensity factor, K_I, can be written as:

$$K_I = \frac{P_L}{\sqrt{\pi c}\tan\frac{\theta}{2}}. \tag{4}$$

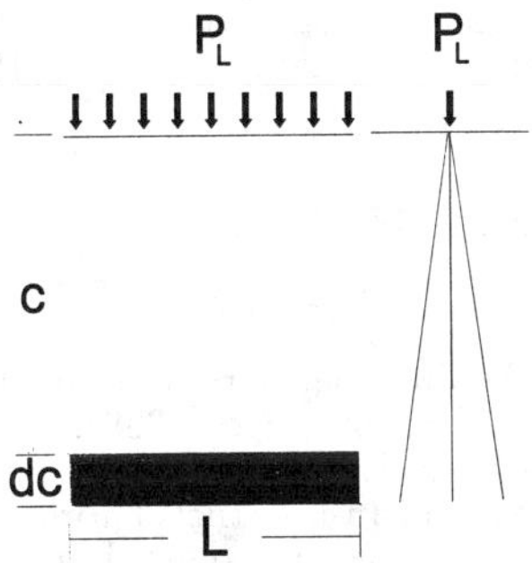

Figure 1. Indentation crack for line-force configurations.

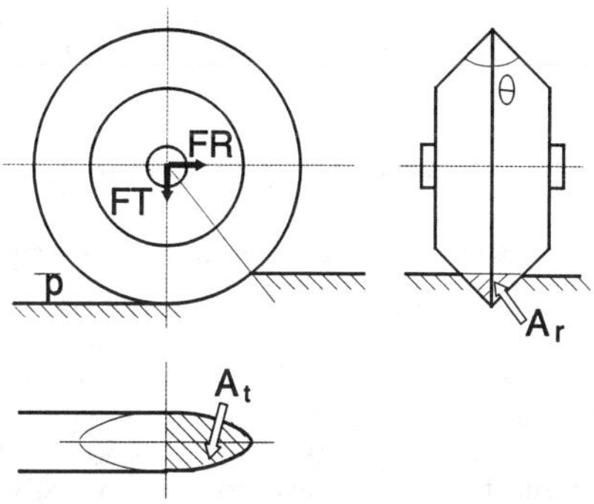

Figure 2. Contact areas of a disc with rock.

2.3 Analysis of unrelieved cutting forces

It is widely accepted that a hydrostatic stress formed under the indenter is responsible for crushing (Hartman, 1959). Under the indenter, stresses forming the crushed zone are larger than those forming the chip. Also such stresses are formed in advance of those forming a chip (Maurer, 1966). In such a case, the thrust force (refer to Figure 2) is:

$$FT = \sigma_0 A_t \tag{5}$$

where A_t is the projected area of the crushed zone which can be approximated as:

$$A_t = \frac{4}{3}p\tan\frac{\theta}{2}\sqrt{Dp - p^2}. \tag{6}$$

2.4 Analysis of relieved cutting forces

The above formulae are only applicable to the unrelieved cutting conditions. In fact, the spacing distance plays a very important role in rock cutting. At small (s/p) ratios, cracks propagating from one groove interact with the cracks produced by indentation of the cutter situated in the neighboring groove, and chip formation happens at lower cutter forces than would be required for chip release from grooves spread further apart. At (s/p) ratios larger than some critical value, grooves are too far apart for the interaction to occur, and chips form at applied force levels which are independent of further increases in groove spacing.

From equation (4)

$$P_L = \sqrt{\pi c}K\tan\frac{\theta}{2}. \tag{7}$$

If the disc is rolling, the total penetration force may be obtained by

$$FT = \int_0^{\sqrt{Dp-p^2}} P_L dx. \tag{8}$$

If the cutting is conducted within the critical regions of s/p ratios, and assuming that the length of cracks increase approximately linearly with x (Sanio, 1985), then

$$c(x) = \frac{s}{\sqrt{Dp - p^2}} x. \tag{9}$$

Incorporating the above equation into Equation (8) and integrating leads to:

$$FT = \frac{2}{3} K_I \sqrt{\pi s(Dp - p^2)} \tan \frac{\theta}{2}. \tag{10}$$

3. ARRANGEMENT AND PROCEDURES OF EXPERIMENTS

3.1 Rock samples and apparatus description

Two rock samples, one Springwell Sandstone and one Whinstone, were used in this investigation. The mechanical properties of rocks were given by Zhao (1989).

The machine used in the indentation and cutting tests is a modified shaping machine with a forward stroke of 800 mm and a maximum in-line thrust force of 50 kN. A rock specimen of 500×500×300 mm in size can be accommodated on the machine table and lowered and laterally traversed with respect to the cutting tool. The details of the rig were given by Tecen (1982) and Moses (1985).

The indenters used on the small cutting rig were machined from blanks of high carbon, high chrome, non-distorting steel and subsequently heat treated. Twenty five disc cutters were used in the indentation tests; while only one disc cutter with 150 mm diameter and 60° edge angle was used in the cutting tests.

3.2 Test procedures

The thrust load was produced by raising the bed with a manual screw with a deformation transducer mounted between the disc holder and the rock block. Useful information concerning the failure process can be gained by recording the force-displacement curve during the indentation of rock. The standard load vs penetration curve, finally, can be drawn based on the data from UV chart recorder. A number of representative force vs. penetration curves were obtained. Generally they were similar to the one depicted in Figure 3 where p_m is the depth of penetration of the indenter into the rock, F_m is the measured force, and (p_{mj}, F_{mj}) represents the jth point of chip formation. A set of (p_{mj}, F_{mj}) values was collected from the tests for each indenter. The average pressure was obtained by:

$$\sigma_0 = \frac{1}{n} \sum_{j=1}^{n} \frac{F_{mj}}{A_{mj}} \tag{11}$$

where A_{mj} is the contact area.

The first approach for determining the contact area assumed the shape of the area to be a pair of parabolas, giving:

$$A_{mj} = \frac{8}{3} p_{mj} \sqrt{Dp_{mj} - p_{mj}^2} \tan \frac{\theta}{2} \tag{12}$$

where θ is the edge angle of the disc cutter.

The variables studied in the cutting experiment were penetration and spacing. After shaving the block level with a shaving tool, the bed was unbolted and raised so that the disc tool just came into contact with the level block surface and a zero reading was taken on the deformation dial gauge. The bed was then raised and locked to the required depth of cut; thereafter series of cuts were made cross the rock surface. At the end of each cut, the tool was reversed and the debris was collected and weighed.

4. RESULTS AND DISCUSSION

4.1 Contact stress

The results of indentation tests, in terms of the ratios of contact pressure/yield stress, σ_0/σ_y, with different disc cutters for Springwell Sandstone and Whinstone, are listed in Table 1. The theoretical analysis based on equation (3) is also given in the Table. Note that for rock materials, $\sigma_y = \sigma_c$. More stable results were found in the tests with disc cutters which have either a diameter of 200 mm or an edge angle of 100°. The indentations, using a disc cutter with a smaller diameter and edge angles, give (σ_0/σ_y) ratios which are higher than those obtained by theoretical analysis. The values of the ratios of (σ_0/σ_y) fall as the disc diameter increases from 100 mm to 200 mm. This is equally true as the disc edge angle increases from 60° to 100°.

The effective angle of an indenter plays a very important role in hardness testing. Most indenters used in engineering practice are nearly blunt and the constraint developed is essentially elastic. Unless the friction is very large, the indenters with an effective angle should be expected to show a small upward flow. In fact, the more blunt the indenter, the less will be the upward flow, and the less will be the influence of friction on the contact pressure or so-called hardness.

For metal materials, nearly all practical indenters perform with some upward flow. However, it is fortunate that the constraint factor, C, is relatively insensitive to the mode of action (that is, whether flow or elastic constraint pertains). With respect to rock materials, the flow is formed early in the process and then developed so fast as to form a chip. Consequently, the force is reduced and further increase of the force causes another chip, and so on. This makes the analysis of such a flow even more difficult. To simplify the problem, only the loads at which the chips are formed are considered in this study and the influence of the upwards flow in this case was not taken into account.

Table 1. Experimental and theoretical results of indentation tests

θ	D	σ_0/σ_y S.s.		σ_0/σ_y W.s.		D	σ_0/σ_y S.s.		σ_0/σ_y W.s.	
degree	mm	exper.	theor.	exper.	theor.	mm	exper.	theor.	exper.	theor.
60	200	3.031	3.554	3.478	3.037	125	4.943	3.554	3.524	3.037
70	200	–	–	2.461	2.926	125	–	–	2.922	2.926
80	200	3.054	3.338	2.710	2.821	125	4.418	3.338	2.738	2.821
90	200	–	–	1.564	2.720	125	–	–	2.750	2.720
100	200	2.758	3.136	1.858	2.619	125	2.741	3.136	2.260	2.619
60	175	3.684	3.554	3.088	3.037	100	5.586	3.554	3.884	3.037
70	175	–	–	2.750	2.926	100	–	–	3.787	2.926
80	175	3.336	3.338	3.462	2.821	100	3.978	3.338	3.044	2.821
90	175	–	–	2.680	2.720	100	–	–	2.966	2.720
100	175	2.421	3.136	1.682	2.619	100	3.092	3.136	2.630	2.619
60	150	4.466	3.554	3.393	3.037					
70	150	–	–	2.446	2.926					
80	150	3.864	3.338	3.042	2.821					
90	150	–	–	2.345	2.720					
100	150	3.047	3.136	1.660	2.619					

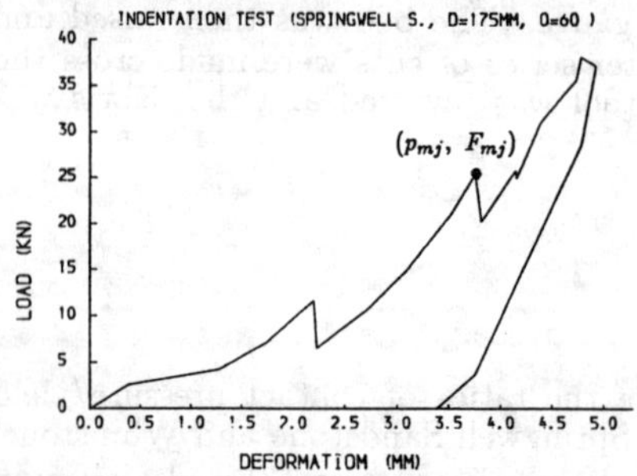

Figure 3. Force vs. penetration curve.

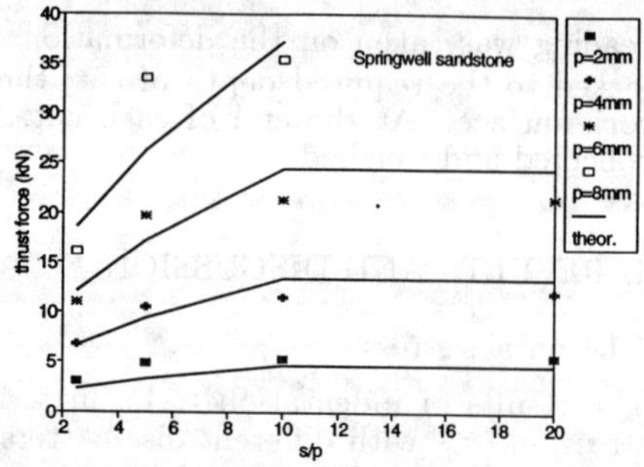

Figure 4. Results of rock cutting tests (Springwell sandstone).

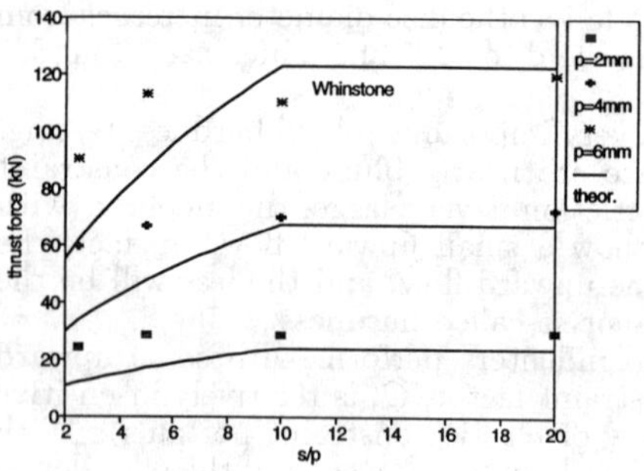

Figure 5. Results of rock cutting tests (Whinstone).

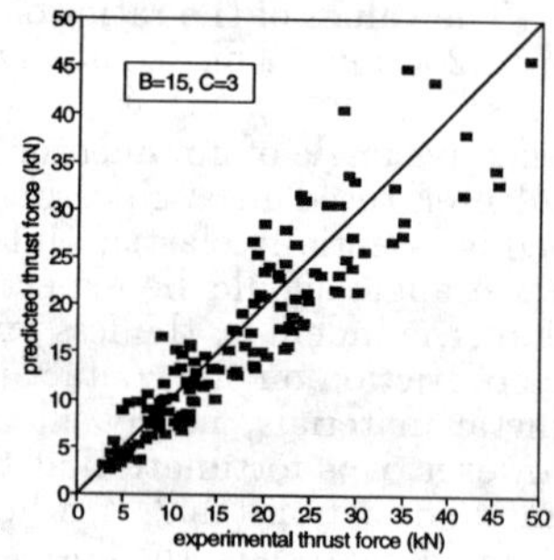

Figure 6. Comparison studies of experimental data and theoretical results.

4.2 Disc cutting

According to the reported laboratory tests (Ozdemir et al., 1977; Snowdon et al., 1982), the effect of spacing is limited when the ratios of spacing and penetration, s/p, approach a critical value, B, e.g., $s = Bp$. Considering single pass tests on a smooth rock surface, this study has revealed a critical value of (s/p) ratio to be about 10, and for single pass tests on a previous cutting surface (see Snowdon et al., 1982) to be about 20. For multiple pass tests, the critical value could be more than 50, but equal to 20 when the independent thrust forces are the same as the thrust force in relieved cutting (Ozdemir et al., 1977). In order to determine the influence of (s/p) ratios on cutting forces, one may compare the models for relieved and unrelieved cutting and consider the situation when the (s/p) reaches its critical value. So, from equations (5) and (10), one has

$$K_I = \frac{2}{\sqrt{B\pi}} \sigma_0 \sqrt{p}. \tag{13}$$

Thus, equation (10) can be rewritten as:

$$FT = \frac{4}{3\sqrt{B}} \tan\frac{\theta}{2} \sqrt{sp}\sqrt{Dp - p^2}\ \sigma_0. \tag{14}$$

The contact stress obtained in the indentation tests can be readily used in prediction of disc cutting. Figures 4 and 5 show that agreements between predicted cutting forces and the experimental results are fairly good. Furthermore, using equation (2), the predicted cutting forces can be obtained based on the material's uniaxial strength. A comparison study of the theoretical results and the experimental data provided by several other workers (Roxborough and Phillips 1975; Bilgin, 1977) is illustrated in Figure 6, showing satisfaction.

4.3 Disc cutter design

If there are two groups of disc cutters with different edge angles, θ_1 and θ_2, based on equation (14), the ratio of thrust forces may be written as:

$$\frac{FT_{\theta_1}}{FT_{\theta_2}} = \frac{\tan(\theta_1/2)}{\tan(\theta_2/2)} \frac{\sigma_{\theta_1}}{\sigma_{\theta_2}}. \tag{15}$$

From Table 1, one may find that the contact stress for the disc cutters with a 60° edge angle is about double in the value of the cutters with a 100° edge angle. Applying equation (15), for example, with a diameter of 150 mm, a 100° edge angle cutter gives almost the same thrust force as that given by a 60° edge angle cutter. A similar tendency can also be found when the diameter of a cutter increases from 100 mm to 200 mm; but at a smaller scale. As the tool wear is directly related to the contact stress on the tip of the cutters, these findings suggest that for a TBM excavating in hard rock formations, the disc cutter with a larger edge angle and a larger diameter would be more effective. Additional benefits, such as reduction in cutter numbers and longer life of the cutters, may also be obtained.

5. SUMMARY

The indentation test is a useful method for predicting the performance of TBMs. In this study, the disc cutters were used as a type of indenter. This has advantages that (a) the geometry of the indenter is identical to that of a TBM tool; and (b) the result given by the indentation tests can be easily transferred to those of spaced cuts. The mechanism of rock cutting, in such a case, is more similar to the penetration of a disc cutter into rocks. Fracture mechanics analysis provides a new way of understanding rock breakage by the disc cutter.

The contact stress for the disc cutter with a 60° edge angle is about double that of the cutter with a 100° edge angle. As the tool wear is directly related to the contact stress on the tip, this finding suggests that for a TBM excavating in very hard rock formations, the disc cutter with a larger edge angle would be more effective. Furthermore, some other benefits may also be obtained to reduce the total costs of TBM operation.

ACKNOWLEDGEMENTS: The experiments were carried out at The University of Newcastle upon Tyne, UK. The assistance given by the technicians in the Dept. of Civil Engineering is gratefully acknowledged. The authors thank Carla Cates for her help with this manuscript.

REFERENCES

Atkinson, B.K. and V. Avdis. 1980. Fracture mechanics parameters of some rock forming minerals determined using an indentation technique. Int. J. Rock Mech. Min. Sci. and Geomech. Abstr., 17: 383-386.

Bilgin, N. 1977. Investigations in to the mechanical cutting characteristic of som medium and high strength rocks. Ph.D. thesis, University of Newcastle upo Tyne.

Cook, N.G.W., M. Hood and F. Tsai. 1984. Observations of crack growth in har rock loaded by an indenter. Int. J. Rock Mech. Min. Sci & Geomech. Abstr 21: 97-107.

Fauvel, O.R. 1981. Implications of laboratory rock cutting for the design of a tunn boring machine cutter head. Ph.D thesis, University of Newcastle upon Tyne.

Hartman, H.L. 1959. Basic studies of percussion drill. Trans. AIME(Mining TP478: 68-75.

Johnson, K.L. 1970. The correlation of indentation experiments. J. Mech. Phy Solids, 18: 115-126.

Lawn, B.R. and E.R. Fuller. 1975. Equilibrium penny-like cracks in indentatio fracture. J. Mat. Sci., 10: 2016-2024.

Lindqvist, P.-A. and H.H. Lai. 1983. Behavior of crushed zone in rock indentatior Rock Mech. & Rock Eng., 16: 199-207.

Lindqvist, P.-A., H.H. Lai and O. Alm. 1984. Indentation fracture development i rock continuously observed with a scanning electron microscope. Int. J. Roc Mech. Min. Sci & Geomech. Abstr., 21: 97-107.

Marion, R.H. 1979. Use of indentation fracture to determine fracture toughness Fracture Mechanics Applied to Brittle Materials, ASTM STP 678, Freiman S.W ed., ASTM, 103-111.

Maurer, W.C. 1966. The state of rock mechanics knowledge in drilling. Proc. 8t U.S. Symp. on Rock Mech.

McGann, E.A. 1984. An investigation into rock hardness using a cone indente method. M.Sc Thesis. Dept. of Geotechnical Engineering, University of Newcas tle upon Tyne.

Moses, D.M. 1985. High pressure water jet assisted disc cutting. M.Sc Dissertation Dept. of Geotechnical Engineering, University of Newcastle upon Tyne.

Ozdemir, L., R. Miller and F.D. Wang. 1977. Mechanical tunnel boring predictio and machine design. Annual Rept. NSF IRA-770199, Colorado School of Mines Golden.

Paul B., D.L. Sikarskie.1965. A preliminary theory of static penetration by a rigi wedge into a brittle material. Trans. of the Soc. of Min. Eng. of AIME, 372-382 Dec.

Peng, S. and A.M. Johnson. 1972. Crack growth and faulting in cylindrical speci mens of Chelmsford granite. Int. J. Rock Mech. Min. Sci. & Geomech. Abstr. 9: 37.

Robson, J.A. 1983. Indentation testing as an aid to excavating performance predic tion for boom-type tunneling machines. Dissertation for postgraduate diploma Dept. of Geotechnical Engineering, University of Newcastle upon Tyne.

Roxborough, F.F. and H.R. Phillips. 1975. The mechanical properties and cutting characteristics of selected rock formations. Report to the TRRL. University o Newcastle upon Tyne.

Sanio, H.P. 1985. Prediction of the performance of disc cutters in anisotropic rock Int. J. Rock Mech. Min. Sci. & Geomech. Abstr., 22: 153-161.

Snowdon, R.A., M.D. Ryley and J. Temporal. 1982. A study of disc cutting in selected British rocks. Int. J. Rock Mech. Min. Sci. & Geomech. Abstr., 13 311-319.

Swain, M.V. and B.R. Lawn. 1976. Indentation fracture in brittle rocks and glasses Int. J. Rock Mech. Min. Sci. & Geomech. Abstr., 13: 311-319.

Tabor, D. 1951. The Hardness of Metals. Clarendon, Oxford.

Tecen, O. 1982. High pressure water jet assisted drag tool cutting of rock materials Ph.D Thesis, Univ. of Newcastle upon Type.

Zhao, X.L. 1989. A fracture mechanics study of unassisted and water jet assisted rock disc cutting. Ph.D. Thesis, Department of Civil Engineering, University of Newcastle upon Tyne.

Rock Mechanics, Nelson & Laubach (eds) © 1994 Balkema, Rotterdam, ISBN 90 5410 380 8

Observation and simulation of rock indentation fracture

X.C.Tan, P.-A. Lindqvist & S.Q. Kou
Department of Civil and Mining Engineering, Luleå University of Technology, Sweden

ABSTRACT: Rock indentation experiments in granite and marble are carried out using a hemispherical indenter. A thorough investigation of the indenting process is made and quantitative data of rock fragmentation and fracture obtained. A two dimensional micro-mechanics constitutive model is applied to simulate the actual median fracture system. The model simulation reveals quantitatively the effects of loading force, rock mechanical properties and micro fractures on the median crack system.

1 INTRODUCTION

Rock indentation is the basic process in rock disintegration and excavation by mechanical means. A sound understanding of rock fragmentation mechanisms and the quantification of the influences of various factors, such as loading characteristics, rock properties and tool configurations, will help in the design of mechanical tools and equipment to enhance the efficiency and productivity of rock fragmentation.

Rock indentation failure is characterised by the high rates of change of stresses and strains, and extreme confinement in the vicinity of the indenter. Rock beneath the indenter undergoes severe compaction, crater formation, crushing and crack propagation.

A number of researchers, e. g. Wagner and Schümann (1971), Hood (1977), Gregory (1979), Cook et al (1984), Kumano and Goldsmith (1982), Lindqvist et al (1984) and, Pang and Goldsmith (1990) made indentation tests on rocks with different indenter shapes, sizes and loading forces. The indenter types included cylinder, wedge, cone and hemispherical shapes. The sizes of the indenters used in the rock indentation tests range from 5 to 20 mm in diameter. Since rock is not transparent and behaves differently under various confining conditions, the processes of rock fracture and fragmentation were not fully explored.

In our rock indentation experiments, systematic investigations were made of rock failure and fracture processes.

The modelling of rock fracture and failure caused by indentation and cutting actions were made by a number of researchers, such as Sikarskie and Altiero (1973), Lundberg (1974), Gregory (1979), Zuech et al (1983), Pang et al (1989), Wijk (1989) and Whittaker et al (1992). They dealt with the major rock disintegration phenomena namely the cratering process. A complete view of rock damage and fragmentation by indentation is lacking. Models of the damage and fracture in the rock subsurface were developed by Kumano et al (1982), Cook et al (1984) and Pang (1990). Failure criteria of tensile and compression were applied to predict the failure and fracture zones. The results predicted by the rock indentation models did not satisfactorily agree with the experimental results. The use of conventional strength theories, which prevail mainly in rock mass, seem to be inadequate for describing the rock indentation damage.

In our work, the modelling of the subsurface median crack propagation in rock is carried out using the fracture mechanics approach. The effects of the influencing factors, such as, loading force, indenter shape and rock properties are examined.

2 ROCK INDENTATION TESTS AND OBSERVATIONS

2.1 Experimental set-up

In the rock indentation experiments, two displacement controlled rock mechanics testing machines, Instron 0-100 kN and Instron 0-4,500 kN, were used for different loading ranges. A hemispherical indenter with a spherical radius of 13.2 mm, commonly used in rock drilling and boring, was used. For each kind of rock, the rock specimens were indented at different loads from about 10 to 100 kN. A constant rate of indentation, 0.05 mm/s was used for all the experiments. The force-indenter displacement curves were also recorded during the tests.

2.2 Rock types and specimen preparation

Two types of rock, Bohus granite and Ekeberg marble were tested. The Bohus granite is a medium grained rock (mean size around 1.8 mm) with large individual grains of potassium feldspar up to 1 cm in size and quartz grains of up to 3.6 mm. Ekeberg marble is a fine to medium grained, well recrystallized carbonate that has comparatively similar sized grains of equal dimensions. The rock is very homogeneous.

The values of important rock mechanics parameters are shown in Table 1 (Vutukuri, et al, 1983).

The rock specimens tested in the experiments had a diameter of 116 mm and a height of 85 mm. Each specimen was cemented by a prefabricated fast setting cement, named

Tab. 1 Rock mechanical parameters

Mechanical properties	Density ρ (kg/m^3)	Uniaxial. compressive strength, σ_c (MPa)	Young's modulus, E (GPa)	Poisson's ratio, ν	Fracture. toughness, (K_{IC}/K_{IIC}) (MPam$^{1/2}$)	Mean grain size (mm)
Bohus granite	2634	152.8	64.1	0.31	1.0/0.90	1.8
Ekeberg marble	2848	160.4	84.4	6.35	0.7/0.66	0.3

Tab. 2 Testing program

Granite

Test No.	Maxim Force	No. of Views
1	35.9	2
2	41.9	2
3	44.8	4
4	70.3	6
5	86.8	8
6	102.0	11

Marble

Test No.	Maxim Force	No. of Views
7	15.0	4
8	27.4	4
9	29.4	4
10	38.9	6
11	61.3	12

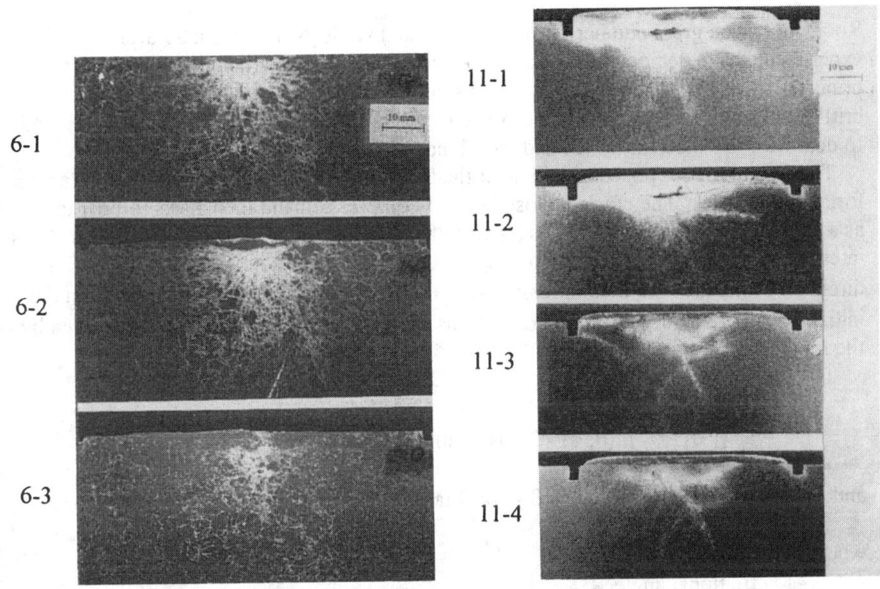

Fig. 1 Photos of typical slabs of granite and marble

Betec, in a steel tube. The steel tube confined the rock specimen in a manner similar to the confinement of a semi-infinite rock mass.

The testing program is shown in Tab. 2

After testing, half of the specimen was sectioned into 2 to 5 slabs about 3 mm thick. The two sides of each slab were treated with a fluorescence penetrant and photographed in black light and natural light. The craters, crushed zones and fractures of each rock specimen were mapped.

2.3 Rock indentation observation

Typical sets of observations in granite and marble are shown in Fig. 1. Craters, crushed zones and different crack types are formed. A cracked zone of many small cracks also developed in granite. They are median cracks, radial cracks, or possibly cone cracks. Cracks parallel to the rock surface, classified as side cracks were developed. Side cracks participated to form craters with a partly flat bottom, see Photo 6-2 in Fig. 1. A complicated crack system with a prism type crack was formed (see Photos 11-3 and -4 in Fig. 1).

3 MODELLING OF ROCK INDENTATION FRACTURE

3.1 Description of a micro-mechanics constitutive model

Rock contains a great number of micro cracks and voids. Numerous laboratory experiments have demonstrated that cracks propagate towards the direction of the major principal stress under compression. Axial splitting is the main mechanism of failure in brittle rocks by compression. A micro-mechanics constitutive model has been developed to deal with this mechanism (Li and Nordlund, 1993).

The fracture is supposed to initiate at the tips of the initial micro cracks (Fig. 2 (a)) to form a wing crack. The crack is closed due to compression and then starts to propagate as a shearing crack when the stress intensity factor K_{II} at the tip exceeds the critical stress intensity factor K_{IIC}. While the two wings extend, they propagate gradually toward the direction of the maximum compressive stress σ_1 by the driving force F generated on the initial crack surface due to σ_1 and σ_2. At this stage the crack propagation is governed by the law of tensile fracture, see Fig. 2(b).

The driving force is given by

$$F = a_o \ [(\sigma_1 - \sigma_2)\, f(\theta, \phi) - 2 \cos\theta \tan\phi] \qquad (1)$$

and $$f(\theta, \phi) = (1 + \cos 2\theta)(\sin\theta - \cos\theta \tan\phi). \qquad (2)$$

where, a_o = original crack length,
ϕ = frictional angle,
θ = acute angle between the normal of an original crack and the direction of σ_1.

When the load is larger than the critical value, the wing cracks will develop. The length of crack propagation is given by

$$L = \frac{1}{4\pi}\left[\frac{\sqrt{K_{IC}^2 + 4F\sigma_2} - K_{IC}}{\sigma_2}\right]^2 \qquad (3)$$

In indentation field, the development of original cracks at a number of representative points will be investigated. The cracks are located at r/a= 0, 1, 2, 3 along three radial lines, $\psi = 0°$, 30° and 60° (Fig. 3). The Hertzian solution for the hemispherical indenter and the rock material (Hertz, 1882 and, Lawn and Wilshaw, 1975) is used. The results are symmetric with respect to the z axis.

Parameters and assumed magnitudes applied in the model calculations for the two kinds of rock are:

$$a_o = 0.1 \sim 0.2 \text{ mm}, \quad \phi = 30^o, \quad \theta = 45^o + \phi/2 = 60^o$$

The mechanical parameters used in the model are given in Tab. 1.

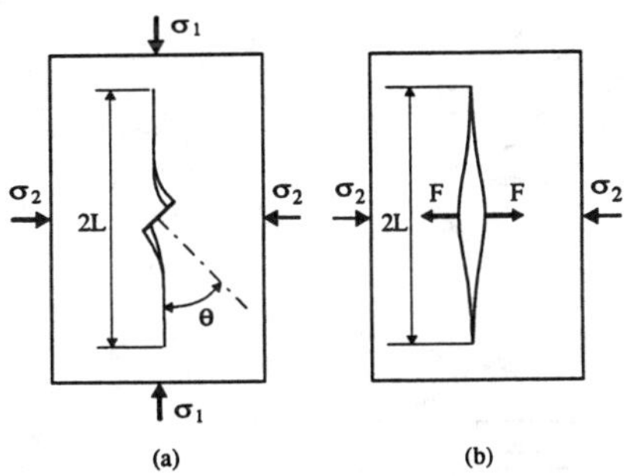

Fig. 2 Development of wing crack in the micro mechanics model (after Li and Nordlund, 1993)

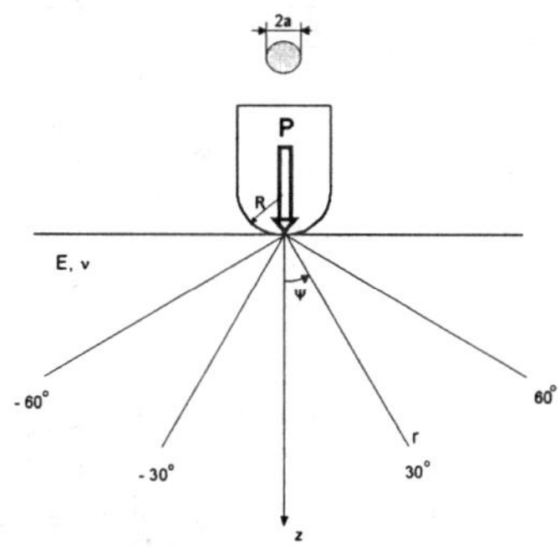

Fig. 3 The indentation coordinates

3.2 Indentation fracture simulations

The principal stresses, σ_1, σ_2 of Hertz field in the vertical plane containing the z axis are applied to the model. The fractures in the axial symmetric plane will be simulated. The critical loads for initiating the original cracks to wing cracks are calculated for the representative points along the three radial lines (Fig. 3).

The relationships between the crack length and the loads in the indentation field are demonstrated in Figs. 4 and 5. It can be seen that the cracks on the centre line develop slower when near the indenter (r/a=1), but faster at some distance outside the confining zone for $r/a \approx 2$). These form the median crack system with a penny-like front.

The relationships between the crack length and the loads in the indentation field indicate that the stress distributions determine the fracture pattern. The magnitudes of σ_1 and σ_2 increase linearly as the load P increases. The locations of original cracks at different r/a and ψ have a different σ_1/σ_2 ratio. The critical load for the initiation and propagation of cracks is greater away from the rock indentation contact. For points very close to the indentation centre, where both principal stresses are compressive, no crack initiation occurs. At some critical distance from the origin centre, maximum crack propagation will be achieved. The maximum length of crack propagation is different for the original crack with different ψ coordinates. The larger the $|\psi|$ is, the smaller the maximum length. Therefore, a circle band of median fracture front forms.

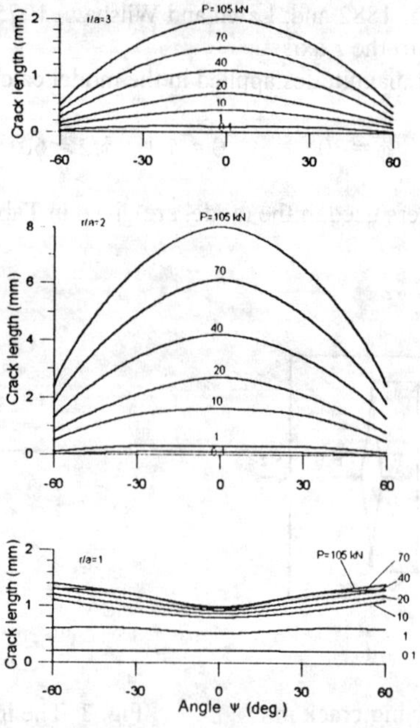

Fig. 4 Cross section of fracture development at different distances from the indenting centre

Since the cracks in the middle obtain longest propagation, they may intersect with the fractures outside, if their wing cracks are on, or very close to, the same trajectory of principal stress σ_1 and the sum of their propagation lengths are larger than their distances. Therefore, a set of long and short cracks will be generated.

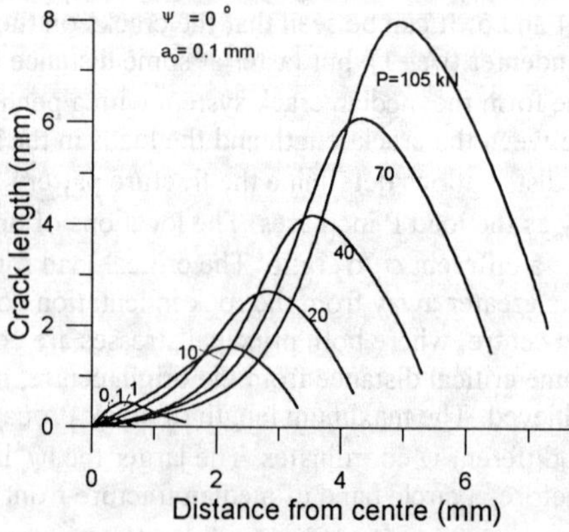

Fig. 5 Fracture development along z axis

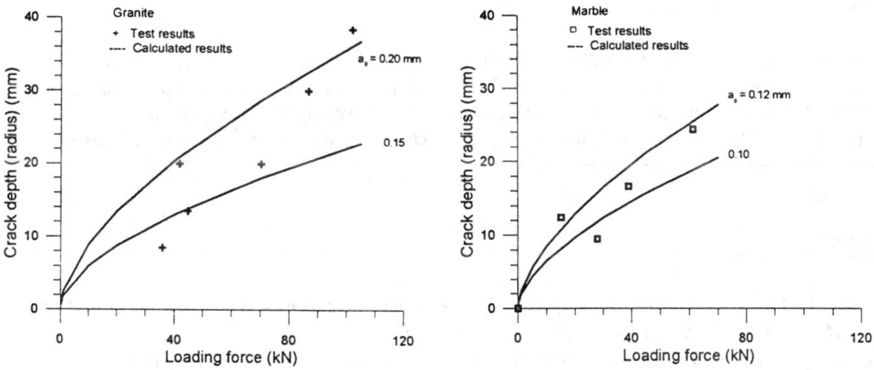

Fig. 6 Simulation of fracture depth for granite and marble

We simulate the depth of the fractured zone representing the median crack boundaries. The simulated fracture depths for the two rock types are shown in Fig. 6. We predict in a similar manner that the fracture boundaries in other radial lines are smaller than that in the z axis.

4 DISCUSSIONS AND CONCLUSIONS

Rock indentation fractures consist of side cracks, radial cracks, cone cracks and median cracks. Side cracks and radial cracks interact to form the flat craters. They are believed to be the main rock removal mechanisms in percussive drilling and boring. The static results are not expected to deviate much from those of dynamic loading, since the loading rate of actual drilling is not significantly high (Lawn and Wilshaw, 1975).

Both the observations and the model simulation of rock indentation fractures indicate that the sub-surface median cracks initiate outside the crushed zone and develop roughly along the radial direction with a penny shaped front around the indenting centre in the vertical plane.

The micro-mechanics constitutive model is useful in the study of brittle failure and sub-surface fracture. It reveals quantitatively the effects of indentation stress (indenter shape), original micro cracks and rock mechanical properties on rock damage by indentation. In our simulation, only the original cracks in the direction of maximum shear stress are considered. They require minimum critical loads for crack initiation. Our results present the most serious development of fracture. In addition to these cracks, many micro cracks at different directions will be initiated. The simulated crack length ($2a_o$) 0.4 mm for granite is close to the microscope observation of the void and fault in the rock. In comparison to its grain size, the value for marble, about 0.2 mm, seems rather large. Since marble is more ductile than granite, the fracture may not be fully described by brittle splitting. Other properties, such as, plasticity, might need to be considered.

A change in the indenter shape will influence the stress field, the indenter confinement and hence, the crack pattern. To increase the efficiency of rock fragmentation, less confinement should be created and a hemispherical indenter is preferred to flat or truncated indenters.

ACKNOWLEDGEMENTS

The financial support of the National Swedish Board of Technical Development for the experiments is gratefully acknowledged. The development of crack discrimination methods by Prof. M. Montoto and Dr. L. M. Suárez del Río at the University of Oviedo (Spain) is highly appreciated. The authors would also like to thank Dr. Chunlin Li for constructive discussions on the application of the constitutive model.

REFERENCES

Cook, N. G. W. et al 1984. Observation of crack growth in hard rock by an indenter. *Int. J. Rock Mech. Min. Sci. & Geomech. Abstr.* 21(2):97-107.

Gregory, E. K. 1979. Factors influencing the performance of full face hard rock tunnel boring machines, *Technical report UMTA-CA-06-0122-79-1 (for Urban Mass Transportation Administration)*, Washington DC..

Hertz, H. 1882. On the contact of elastic solids (in Ger.). *J. reine und angewandte Mathematik* 92:156-171.

Hood, M. 1977. Phenomena relating to the failure of hard rock adjacent to an indentor. *J. of South African Inst. of Min. and Metallurgy* Dec.:113-123.

Kumano, A. & Goldsmith, W. 1982. An analytical and experimental investigation of the effect of impact on coarse granular rocks. *Rock Mech.* 15:67-97.

Lawn B. & Wilshaw R. 1975. Review indentation fracture; principles and applications, *J. Mater. Sci.* 10:1049-1081.

Li, C. & Norlund E. 1993. Deformation of brittle rocks under compression —with particular reference to micro cracks. *Mech. of Mater.* 15:223-239.

Lindqvist P-A., et al 1984. Indentation fracture development in rock continuously observed with a canning electron microscope. *Int. J. Rock Mech. Min. Sci. & Geomech. Abstr.* 21(4):165-182.

Lundberg, B. 1974. Penetration of rock by conical indenters. *Int. J. Rock Mech. Min. Sci. & Geomech. Abstr.* 11:209-214.

Pang, S. S. et al 1989. A force-indentation model for brittle rocks. *Rock Mech. & Rock Engng.* 22:127-148.

Pang, S. S. & Goldsmith, W. 1990. Technical note: Investigation of crack formation during loading of brittle rock. *Rock Mech. & Rock Engng.* 23:53-63.

Sikarskie, D. L. & Altiero, N. J. 1973. The formation of chips in the penetration of elastic-brittle materials. *J. Appl. Mech.* 40(3):791-798.

Vutukuri, V. S. et al 1983. Influence of physical properties of rock and rock-like material on blastability in crater and slab blasting — A literature and model study. *Report GF (Swedish mining research foundation) 8221*, Luleå, Sweden.

Wagner H. & Schümann E. R. H. 1971. The stamp-load bearing strength of rock —an experimental and theoretical investigation. *Rock Mech.* 3/4:185-207.

Wijk, G., 1989. The stamp test for rock drillability classification. *Int. J. Rock Mech. Min. Sci. & Geomech. Abstr.* 26(1):37-44.

Whittaker, B. N. et al 1992. *Rock Fract. Mech. — principles, design and applications.* Amsterdam: Elsevier.

Zeuch D. H. et al 1983. Subsurface damage development in rock during drag-bit cutting: Observations and model predictions. *24th U. S. Symp. on rock Mech.*:733-742.

Rock Mechanics, Nelson & Laubach (eds) © 1994 Balkema, Rotterdam, ISBN 90 5410 380 8

A new approach for simulation of the rock cutting process

M. Kadiu
Paris School of Mines, CGES, France

ABSTRACT : Modeling of rock-cutting tool interaction has historically incorporated the assumption that the cutting process can be presented by the maximum and average values of the cutting forces. This assumption places considerable limitations on the model's ability to simulate the instantaneous behavior of a cutting head as the dynamic modeling of this one. Recent studies have shown that robotics in rock cutting machines require a good understanding of the forces generated on the cutting head. The approach as presented in this paper provides the capability to simulate the rock cutting process. It shows that rock breakage by a mechanical tool contents either determinist and uncertain aspects. The determinist aspect is closely linked with average values of the mechanical characteristics of the rock matrix whereas the discontinuities and other structural features of the rock mass introduces the uncertain aspect in the rock cutting process. The variation of the tool forces required to penetrate the rock is simulated using the Random Boolean Functions. An extention towards rock-cutting head interaction allows to predict the instantaneous behavior of the cutting head. Here, interaction is based on the instantaneous forces simulated for each cutting tool in contact with the rock.

1 INTRODUCTION

The application of robotics in underground works such as mining and tunnelling activities has been very limited in the past. Despite interesting works carried out on this issue by different research teams the practical industrial use of robotics in rock cutting machines (roadheaders, tunnelling...) is virtually not existant (Fuentes, 1993). The use of classical robotic analysis in rock cutting is faced with two major problems :

- The uncertainity of external variables such a rock strength and in general forces applied on cutting head, which are considered as being unpredictable.
- The lack of data regarding the dynamic characteristics of the machine, which usually is different for almost every machine.

The above problems have a great influence on all topics related with automation or robotics. In special, the modeling of the cutting action of the machine and the reaction of the rock is more complicated. However, it is well known that the instantaneous behavior of the cutting head is a function of the forces applied on each cutting tool being in contact with the rock, with tool arrangements on the cutting head, with the head kinematic values, and with the machine stability. Thus, if the instantaneous forces required by each tool are known during the cutting process then the dynamic modelling of the rock-cutting head interaction would be resolved, assuming the stiffness of it is high.

CGES-Research in the rock cutting field has been carried out over 15 years together with the French Coal Board Research Organisation (CERCHAR) and French or foreign companies of Europeen countries. Theoritical and experimental analysis of the rock cutting process, its efficiency, cutting head modeling, vibrations levels etc..., were the objects of the several project works. The knowledge gained allows to resolve a major part of the problems encountered in the rock excavation machines technology.

This paper is part of recent development in rock cutting machines research at CGES. The first part deals with a statistical analysis of the rock cutting process. The second part describes the model used for the simulation of the cutting process as an comparaison between the theoritical and the experimental results. An application on the rock-cutting head interaction is presented in the third part of this paper.

2 ANALYSE OF THE ROCK CUTTING PROCESS

The rock cutting process is generally described by the variation of three components of the cutting forces F_c, F_n, and F_l during tool penetration in rock mass. It's widely accepted that the rock cutting process has a cyclic character and every cycle contents of four phases (figure 1) :

a) The initiation phase: Forces are almost equal to zero.
b) The penetration phase: Forces increase irregulary.
c) The chip breaking phase: F_c and F_n obtain their maximal values.
d) The phase of the chip removing: Forces go drastically down to zero.

Most of the authors (Deliac, 1986) have assumed that four parameters define the cutting cycle: maximal and average values of the cutting forces, the length or the period of the cycle, and the specific energy. Thus, historically the principle problem in mechanical rock breaking was to establish the relationships linking rock parameters with tool caracteristics and the depth of cut.

The models describing tool-rock interaction (proposed during last decade or before) allow therefore to predict the evolution of the cutting forces with the depth of cut. These models, build on the Coulomb/Griffith criteria of rock failure (Lebrun, 1978; Atkison, 1980), calculate the maximal cutting and normal forces by integration of the stress on the chipping area. Moreover, chipping shape is assumed to be almost spheric and, theoritical explications about low variation of the value of the rock failure angle, α, are not done yet.

The experimental results carried out by several research laboratories have shown a good merger with theoritical predictions. Despite of this, the real structure of the cutting process remains greatly unknown.

Figure 1.b shows a light disagreement between length of cycle and the maximal value of the cutting force: Shorter cycles have sometimes higher maximum value than those having a longer. Moreover, the length of cycles seems sufficiently dispersed whereas the maximum value is more stabilized. The maximal value of the cutting cycle appears to be independant of its length as well as from one cycle to another. These remarques introduce immediatly the following questions :

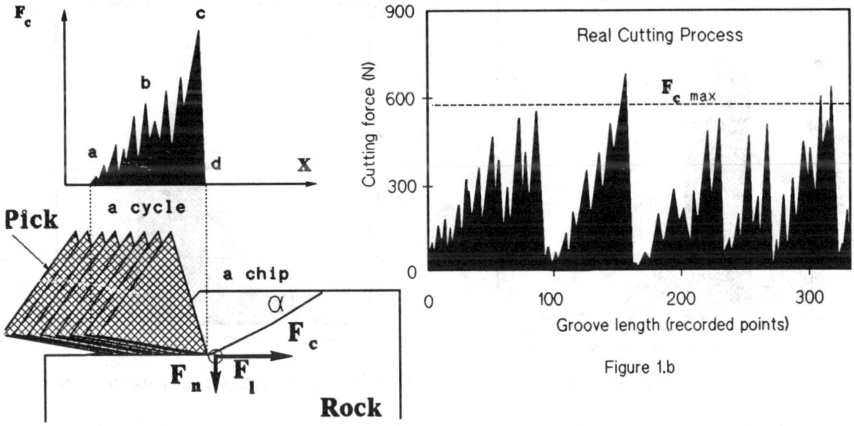

Figure 1.a

Figure 1.b

Figure 1 : View of the rock cutting process with a mechanical tool.

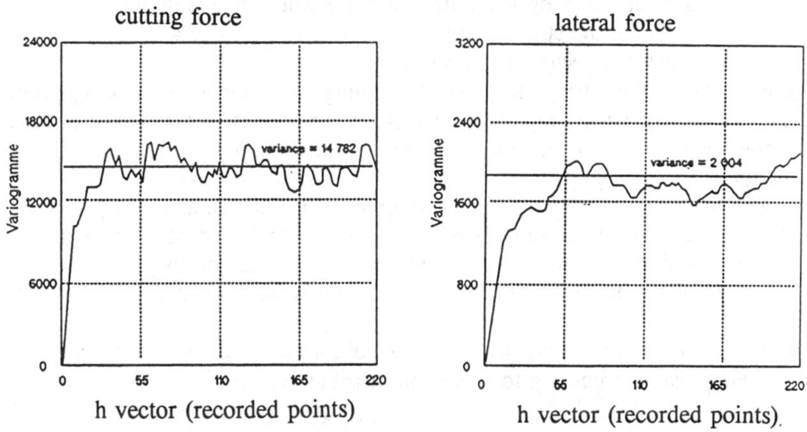

Figure 2 : Variogramme of the cutting process.

- What's the relation between the length of the cutting cycle and it's maximum value?
- Which are the parameters describing the real structure of the cutting cycle ?

In order to answer these questions we consider the cutting forces, F_c, F_n and F_l as are regionalized variables according to the groove length, x. Two mathematical models are generally used to describe this type of phenomenom :

a) The histogramme of the extreme values of the phenomenom.

b) The variogramme, calculated by the following equation :

$$\Gamma(h)=\frac{1}{2}\int_{x\in S}[f(x+h)-f(x)]^2dx \qquad (1)$$

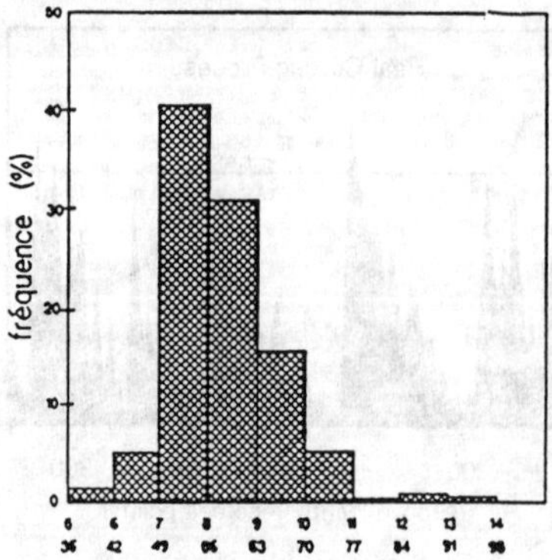

Cycle lengths (recorded points)

Figure 3: Histogramme of cycles length.

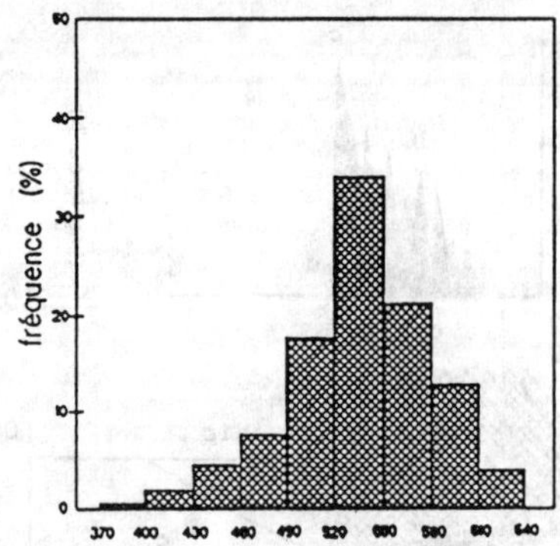

Cutting force (maximum values, N)

Figure 4 : Cutting forces histogramme.

where : f(x) being the value of the function at point x∊S;
S being the geographic limited space where f(x) is defined;
dx being the current point of the geographic space and
h being a vector of this space[1].

Figure 2 shows the variogramme of the cutting force for a real cutting process, Figures 3 and 4 respectively represent the histogrammes of the length of the cycles and of the maximum values of the cutting forces. These functions fournish the following information:

- The behavior at origine of the variogramme is very continuous, proving that cutting process has a regular structure (nugget effect is equal to zero); the value of the cutting force in a cycle at any moment depends on its preceding value.
- The range of the variogramme is equal to the average value of the length of the cycles.
- The cycles of the cutting process are independant; their length fluctuate around average value according to a Poisson distribution law :

$$P_{(x)} = \frac{\epsilon^{-m} m^{x}}{x} \tag{2}$$

where : x and m respectively are the length of the cycle and the esperance matematique.

- The maximal cutting force has a Gaussian distribution law with a low standard deviation compared to its average value.
- The sill of the variogramme is equal to the variance of the cutting force, proving the stability of the average value of the cutting force.

This analysis shows that cutting process is, first of all, a stationary phenomenom. The average of the maximum values for several cycles is stable; it depends only on the tool and rock matrix charcteristics, and on the depth of cut. The variation of the cutting force from a cycle to another can be explained by rock hetérogeneity, e.a. structural

[1]) This nomenclatura is adapted from a space with 1 dimension.

features or discontinuities of the rock mass. The variation of the force within a cycle (in the penetration phase, only) is provoked by the successives removes of the small chips. This aspect of cutting cycles can be interpreted as a multitude of cutting cycles where the depth of cut varies from zero to h. Moreover, in contrast to existing correlation between average values of the cycle length and maximum force, no relationship can be established between the individual length of cycles and their maximum value (Kadiu, 1991).

3 SIMULATION OF THE CUTTING PROCESS

When the rock nature and the cutting tool are known the cutting models fournish generaly the evolution of the forces with the depth of cut (in average). The only values known about the cutting cycle (for a fixed depth) are the averages of the length and the maximal cutting force. The simulation of the cutting cycle raises therefore the question: How to give the force value on each point looking at maximal force and its support[2]?

This problem is known under the name : **support changement by the maximum.** The mathematical models better adapted for resolving this kind of problem are the Random Boolean Function, R.B.F. and the Random Sequentiel Function with Markov's Jump, R.S.F.M.J. (Serra, 1989; Chatrou, 1989). The R.B.F. are generaly applied on Boolean assembly having a Poisson distribution law. They have been introduced by Jeulin (1982) and successful used on the study of surface roughness as well as image analysis (Prêteux, Schmitt, 1983). In fact, the tests justifying the application of the R.B.F. to a process simulation are very complex and introduce some difficult operations, in particular, when the data are not joined.

The detailed presentation of R.B.F. is done by Serra (1989) and Jeulin (1982). The construction of the R.B.F. passes through three steps (figure 5) :

i) It gives a Poisson processus in R^n.
ii) It introduces at every Poisson point a "primary function" or a "grain". This one is defined by a geometrical figure independant on every point.
iii) The R.B.F. is obtained by the external shadow (shape) of the primary functions, i.e. the support on a point.

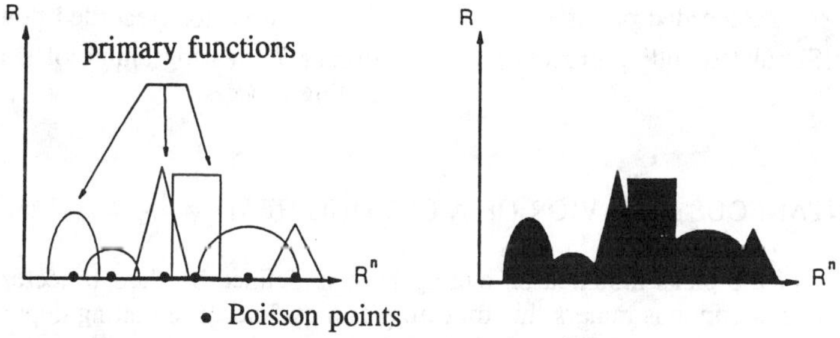

Figure 5 : Construction of the R.B.F.

[2])the length of the cycle

The base of the primary function is the support of the maximal value. This obeys a Poisson law, whereas the maximal value follows the law of maximum values of the processus.

In addition to other tests realized at CGES with the rock cutting experimental database, the use of the R.B.F. is build, on one hand, on the variogramme behavior and, on the other hand, on the Poisson ditribution of the cycle lengths. Moreover, it's assumed that the penetration phase of the cutting cycle is nothing else then a multitude of small cycles. The cutting depth of this cycles is considered as being variable, and both the average value of maximal force and the length of the cycle are replaced by mobile averaging. The choice of the primary function has to be compatible to the real phenomenon. For the cutting process the triangle shape is best adapted as the primary function.

Figure 6 shows the simulation of the cutting force, F_c, for the same conditions that those corresponding to the experimental cutting process represented in Figure 1.b. Figure 7 represent the variogramme of the simulated cutting process.

The use of the R.B.F. to the rock cutting simulation seems to be confirmed. Several simulations realized by Kadiu (1991) have shown the validity of this approach for any other type of pick and for the lateral force. The normal force, F_n, is at every point of the cycle proportional to the cutting force.

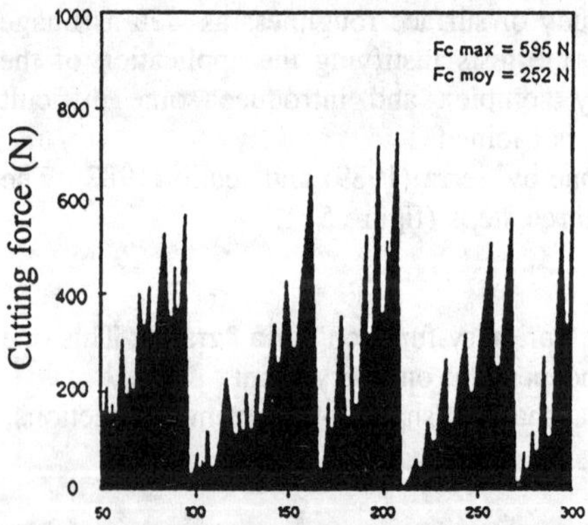

Figure 6 : Simulated cutting process.

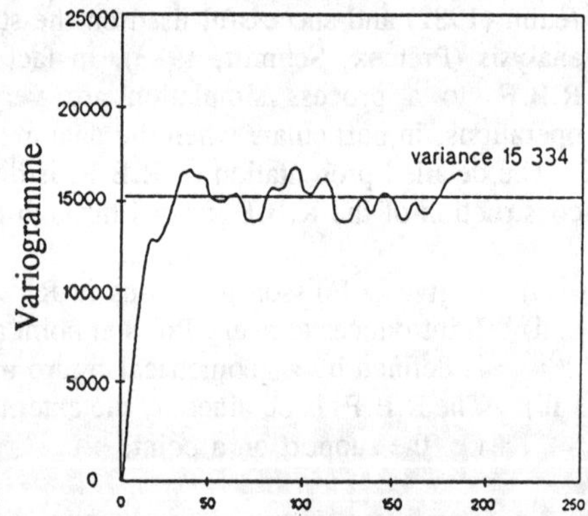

Figure 7 : Variogramme of simulated cutting process.

4 INSTANTANEOUS BEHAVIOR OF A CUTTING HEAD

The work of the picks mounted on cutting head is defined by their trajectory. For roadheaders or continuous miners this trajectory is a cycloid. The cutting depth varies from zero to a maximum value and decrease after to zero value. Thus, the picks encounter three cutting modes : Unrelieved cut, Relieved cut and Deepening cut. The simulation of the forces applied on the picks at every moment has to consider this alternance of the cutting mode.

For tunnelling machine or PDC drag bits the picks or PDC cutter trajectory is a helicoid. The depth of cut during one rotation is constant, generaly equal to penetration rate. The tools encounter only the relieved cutting mode. The instantaneous behavior of the machine is nothing else then momentarily some of the forces applied on the cutting tools removing the same rock area (depth).

In reality, the alternance of the cutting mode during cutting head rotation does't change the nature of the cutting process. The parameters influenced of this are the average values of the maximal forces and of the cycle lengths. Theoritical and experimental study have shown that relationships between average values of forces and different cutting modes exist in the practice (Deliac, 1986; Kadiu, 1991).

To avoid the problem of the machine reaction, it's assumed that the stiffnes of this one is very high. Consequently, the trajectory of the tools during the simulation is considered unchangeable and the same of these theoritical. In laboratory condition the test rigs ensure these conditions.

The curves represented in Figure 8 concern the experimental results and theoritical simulation obtained for a clearance ring of a cutting drum. The conditions used for the simulation and the experiments are the same, such that illustrated in the table 1.

Table 1 : Conditions of the simulation.

Type of the pick : point attack pick	Advandced speed : 13.2cm/min
Number of the pick : 18	Rotation speed : 20 tr/min
Rock : Vosges sandstone	Reduced scale 1/4

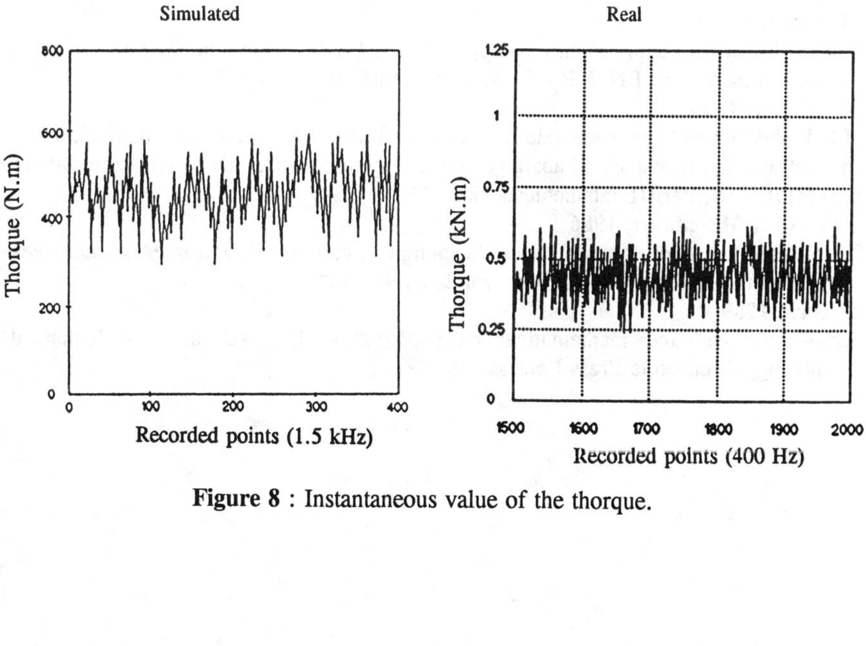

Figure 8 : Instantaneous value of the thorque.

CONCLUSION

The use of R.B.F. for the simulation of the rock cutting process allows to determine the instantaneous forces applied on the cutting head (rock reaction). The dynamic modelling of a cutting head can therefore be simulated looking at dynamics equations of the system and its stifness.

REFERENCES

B.K. Atkison, V.Advis, 1980
"Fracture mechanics parameters of some rock forming minerals determined using an indentation technique", Int. Journal of Rock Mechanics & Mining Sciences (Technical Notes), Vol. 17, pp. 383-386.

J.M.Chatrou, 1989
"Modélisation probabiliste de gisements de nodule polymétallique", Doctoral dissertation, ENSMP/CG, Paris/Fontainebleau, pp. 63.

E.P.Deliac, 1986
"Optimisation des machines d'abattage à pics", Doctoral dissertation, ENSMP/CGES, Paris/Fontainebleau, pp. 502.

J.L.Fuentes-Cantillana, A.Rodriguez, C.Fernandez, J.J. Sanchez, 1993
"Robotic modelling and controll of roadheader boom", Second International Symposium on Mine Mechanization and Automation/Lulea/Sweden, June 1993, pp. 231-238.

D.Jeulin, 1982
"Morphologie mathématique et propriétés physiques des agglomères de minerai de fer et du coke métallurgique", Doctoral dissertation, ENSMP/CG, Paris/Fontainebleau, pp. 93.

M.Kadiu, 1991
"Etude du travail du pic incliné & application à l'optimisation du disque rouilleur", Doctoral dissertation, ENSMP/CGES, Paris/Fontainebleau, pp. 208.

M.Lebrun, 1978
"Etude théorique et experimentale du travail de l'abattage mécanique; application à la conception de machines d'abattage et de creusement", Doctoral dissertation, ENSMP/CGES, Paris/Fontainebleau, pp. 143.

F.Prêteux; M.Schmitt, 1986
"Fonctions Booléenes et modélisation du spongieux vertrébral", 2nd International week of the Electronic Image, april 1986, vol.2, pp.476-481.

J.Serra, 1989
"Image analysis and mathematical morphology", vol.2, Advanced mathematical morphology, Academic Press London, pp. 297.

Rock Mechanics, Nelson & Laubach (eds) © 1994 Balkema, Rotterdam, ISBN 90 5410 380 8

The effect of surface-active chemical agents on rock cutting with shear/drag bits

Alexander V. Staroselsky
Mechanical Engineering Department, Massachusetts Institute of Technology, Cambridge, Mass., USA

Kunsoo Kim
Henry Krumb School of Mines, Columbia University, N.Y., USA

ABSTRACT: The effect of surfactant on rock cutting is evaluated using the fracture mechanics approach. Experiments are carried out to elucidate the mechanics of the surfactant assisted fracture development during rock cutting. The mathematical model developed based on the observation hypothesizes that two types of fractures are created by cutting action under the influence of surfactant, namely shallow surface cracks created mechanically and deeper fractures induced by the aid of surfactant. This model permits an estimate of the surfactant influence on cutting force.

1 INTRODUCTION

Rock drilling is an essential part in mineral and energy resource extraction processes. Rock drilling technology, however, remains an extremely inefficient process in terms of energy utilization. It typically utilizes less than a few percent of the total energy input into the bit for rock breakage. Thus, even an incremental improvement in the technology can result in substantial cost savings for the energy and mineral extraction industry. Drilling is accomplished by either crushing and/or chip formation through the application of force to the exposed surface of rock with an aid of drilling fluid. It is a complicated process and is often masked by many concurrent events. We attempt to gain an insight on the mechanics of chip formation under the influence of surfactant using experimental and analytical means.

2 BACKGROUNDS

Since Rebinder's pioneering work (Rebinder et al, 1948), a number of investigators confirmed the beneficial effects of surface-active chemical additives in drilling fluids. They report that the use of certain surface charge modifiers as drilling fluid additives can improve the drilling rate as much as two to three fold in some hard rock types; the optimum drilling rate is achieved when a zero point of charge on the surface of the rock is maintained (Westwood and Goldhein, 1970, Westwood et al, 1974, Engelmann et al, 1987). They explain the improvement based on the changes in rock strength, surface hardness, friction at the bit-rock interface, crack development and propagation. Rebinder and Shchukin (1973) explain that molecules of the surfactant form diffusion double layers on the surfaces of the cracks, which play the role of a shield and prevent closure of the crack edges. It is also reported that the formation of double layers of electrically charged particles and a build-up of osmotic pressure in the cracks and pores of as much as 0.1 to

10 atm could promote the fracture process in some rock types (Paul, 1988). More recently, it has been suggested that the change of electro-catalytic reactions at the crack tip has an influence on the crack activities during the chip formation process based on lab testing of fracture toughness under the various types of chemical additives (Akram and Karfakis, 1991).

A substantial body of knowledge exists toward the phenomenological understanding of the effect of surfactant on drilling as stated in the foregoing paragraph. Yet, the actual mechanism causing the change in the drilling rate is only qualitatively understood and the combination of the chemical and mechanical destructive process has been under-exploited for drilling. In this investigation we attempt to assess quantitatively the effect of surfactant on rock cutting through experimental and analytical approaches.

3 EXPERIMENTAL

Experiments have been conducted to determine the effect of surfactant on the basic fracture properties of rock to establish baseline data, and then to measure its effect on rock cutting. The surfactant effect on rock property is measured in terms of reduced surface energy which is accomplished by the standard fracture mechanics tests. The tests help select the surfactant types which yield pronounced effects on rock cutting. Four types of surfactants are tested on five types of sandstone specimens in the tests.

3.1 Fracture Property Tests

The first series of tests was carried out to determine the values of K_{Ic}, the mode I fracture toughness of five types of rock specimens in four different surfactant environment. During the early experiments it has been found that the K_{Ic} values, determined on sandstone specimens with a saw-cut notch of 0.5 mm width, are 1.5 - 2 times higher than the corresponding values determined on specimens with a natural edge crack. This effect is observed to be more pronounced under the influence of surfactant.

After this observation we attempt to develop a sample preparation method which creates a sharp notch. The development of an edge crack in rock specimens by cyclic loading (Kim and Mubeen 1982) is too time consuming, and it has a drawback of altering the microstructure of the specimens during the load cycling. A mechanical apparatus is devised and used to induce a sharp crack in the specimen. The tests are conducted using the standard four-point bending scheme. The values of K_{Ic} have been determined using the following equation (Nisitani and Mori, 1985):

$$\mathbf{K}_{IC} = \frac{3Pl}{tw^2}\sqrt{\pi a}\;F(\alpha) \tag{1}$$

where a : Length of the initial crack; *w* and *t* : Height and thickness of the sample respectively; *P* : Critical load; and
$\alpha = a/w,\ F(\alpha) = 1.12 - 1.12\alpha + 3.74\alpha^2 + 3.87\alpha^3 - 19.05\alpha^4 + 22.55\alpha^5$

The K_{IC} values obtained from the fracture property tests in the five types of fine grained sandstones treated with aqueous solutions of surfactants are presented in Table I.

Table I, Summary of K_{Ic} Values & Coefficients of Variations Obtained from Fracture Property Tests with & without surfactant on 5 types of sandstones

Rock Types	Surfactant Types	K_{Ic} MPa $m^{1/2}$	Coeff. of Variation	Rock Characteristics
Sandstone 1	Dry	0.82	10	
	SF-1	0.65	31	Co : 66.1 MPa
	SF-2	0.57	20	To : 6.2 MPa
	SF-4	0.67	25	Fine Grain
Sandstone 2	Dry	1.33	8	
	SF-1	1.00	28	Co : 100.6 MPa
	SF-2	0.92	32	To : 23.9 MPa
	SF-3	1.12	11	Medium Grain
Sandstone 3	Dry	1.05	7	
	SF-2	0.58	12	Co : 86.0 MPa
	SF-3	0.78	24	To : 9.1 MPa
	SF-4	1.47	15	Fine - Medium Grain
Sandstone 4	Dry	0.86	8	
	SF-1	0.74	24	Co : 85.3 MPa
	SF-2	0.72	30	To : 9.9 MPa
	SF-3	0.79	13	Fine - Medium
	SF-4	0.80	16	Grain
Sandstone 5	Dry	1.19	7	
	SF-1	1.11	20	Co : 192.7 MPa
	SF-3	1.15	15	To : 10.2 MPa
	SF-4	1.02	18	Fine Grain

The pH and concentration levels of the four types of surfactant used are summarized in Table II.

Table II. Summary of the pH and concentration levels of surfactant used for the investigation

Surfactant Types	pH	Concentration Level
SF-1 (Sulfonic acid, NaCl, Foam Generator)	8.75	0.3%
SF-2 (Sulfonic acid, NaCl)	9.0	0.2%
SF-3 (Carboxy Methyl Cellulose, Na_2CO_3)	10.5	0.15%
SF-4 (Katamin, Na_2CO_3)	10.2	0.15%

The change in the fracture toughness values under the influence of surfactant is attributed to the process zone evolving near the crack tip, which depends on the electrical charges of active ions: cationic or anionic activities. Surfactant SF-3 is a high-molecular compound and thus its penetration into the specimen occurs slowly. The molecules of the anion-active surfactant, SF-1, are much smaller and penetrate more easily to the crack tips. When the solution is fed directly into the crack, the most pronounced effect (20%- 25%) is obtained using an anionic surfactant with electrolyte additives. A cationic surfactant is effective for a sandstone with closely-packed quartz grains and an increase of fracture toughness is observed when this solution is introduced into sandstones with irregular packing. The increase in the coefficients of variation is ascribed to the material behavior changes in different moisture contents which cause a change in material properties from elastic-brittle to inelastic.

3.2 Bench Scale Rock Cutting Experiments

The role of surfactant in the formation of the process zone has been experimentally proven by Arzimovich (1985). He conducted a cutting experiment on a block of rock with and without surfactant. With the aid of a special florescent solution added to the drilling fluid, he was able to observe the pattern of surfactant-induced fractures in the block specimen under ultra-violet light after each bit pass. The surfactant-induced damage zone was found to be as much as two to three times bigger than the damage zone created without the use of surfactant. This test proved unambiguously that the surfactant promoted the formation of a damage zone which influenced the drilling.

In our bench-scale tests, the first cutting pass which removes a layer of chips is made while surfactant is fed into the rock specimen. Subsequent cutting passes are done without feeding surfactant. A decrease in the cutting force is observed which also indirectly indicates the size of the damage zone induced by surfactant effects (Staroselsky et al, 1991). In comparison with cutting dry rock samples, a 45 - 55% decrease in the resistance to cutting is observed. The cutting force indices even decrease when the third layer of chips is removed (line 1 on Figure 1). The size of the damage zone after removal of the first layer of chips is found to be about two chip thicknesses.

The experiments also indicate that the ratio, $(Q/K_{Ic})^2$, remains almost constant (where Q is the thrust force applied to the bit and K_{Ic} is the mode I fracture toughness of the rock being cut) under surfactant influence. This suggests that the increase in the damage zone is not attributable to the decrease in K_{Ic} alone under the influence of surfactant. In order to explain this observation, we consider a model with two types of induced flaws; the shallow surface cracks developed by the mechanical action of the bit and the deeper fractures developed under the influence of surfactant. The roles of these mechanically and chemically induced flaws in the rock being cut and the degree of their influence on the cutting process are analyzed.

4. ANALYTICAL APPROACH

Based on the experimental observations described in the foregoing section we develop a mathematical relationship which permits us to estimate the surfactant effects quantitatively.

4.1 Damage Zone Development With the Aid of Surfactant

In our analysis we consider the cutting tool as a rigid bit which moves quasi-statically along the surface of a half-plane which has shallow surface cracks as illustrated in Figure 2. The deeper fractures are induced with the aid of surfactant.

Friction is generated at the bit-rock interface as the bit moves on the rock surface separating chips in front of it. A compression zone is formed in rock in front of the bit. The fracture in front of the bit is filled with liquid and is compressed by the load applied by the bit as shown in Figure 3.

The stresses in the rock decrease with depth into the half-space. The fluid pressure is uniformly distributed over the entire wet fracture surface. For this reason, the stresses on the banks of the fracture near the free surface are compressive, while those near the tip of the fracture are tensile. The boundary

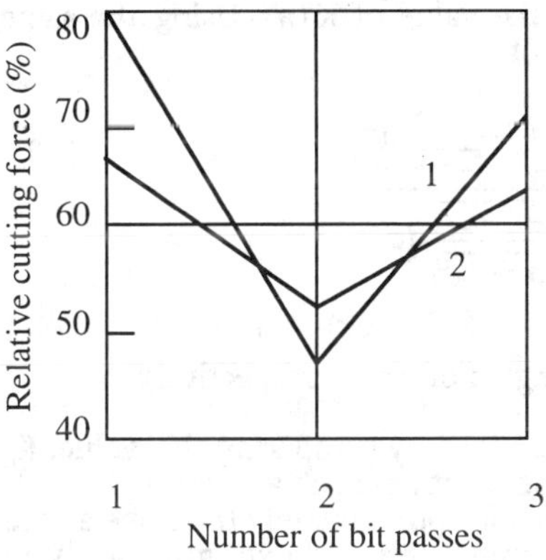

Figure 1. A comparison between experimental and analytical results

conditions for this problem are:

$$\sigma_n = \sigma_x - P_0,\ \tau_{nt} = 0\ ;\ d \le y \le b$$
$$\tau_{xy} = \sigma_y = 0, \qquad y = 0. \qquad (2)$$

where σ_x is the normal stress induced by the bit in the half-space; P_0 is the fluid pressure in the fracture.

In formulating the boundary problem we substitute the bit load for the stresses it produces. The fracture is closed to a depth d due to compression. The closing condition is $K_I(d)=0$, since the stresses are finite at this point. The value of the pressure P_0 is determined and the fracture length a is found from the equilibrium condition $K_I(a)= K_{Ic}$. The surfactant molecules are adsorbed on the fracture surfaces and electric double-layers are formed which prevent displacement of the liquid from the crack. Physically, this problem is somewhat analogous to hydraulic fracturing.

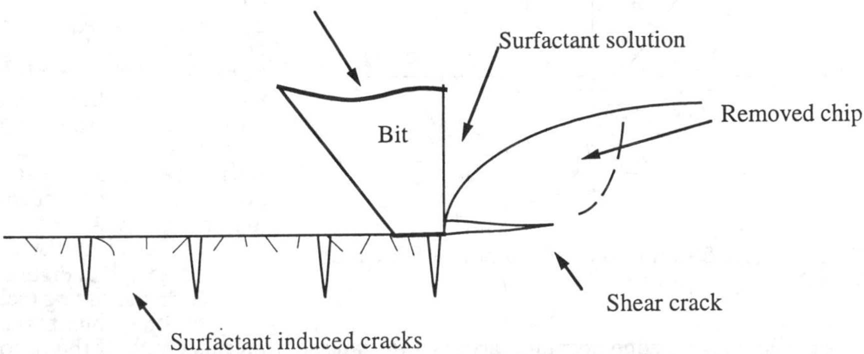

Figure 2. Mechanically induced shallow surface cracks and deeper fractures induced by the cutting action of the bit with the aid of surfactant

This problem can be solved using asymptotic methods. A rough estimate shows that the size of the closed zone d is much less than the fracture length a. So we can obtain the asymptotic expression for $K_I(d)$ under the condition d/a approaching zero. After that we equate this expression to zero. The results of this analyses are given below:

$$K_I(a) = \sqrt{2ca^{c-1}/\pi^3}\int_0^a \frac{P_0(x) - \sigma_x}{\sqrt{a^c - t^c}}dt, \qquad (3)$$

where $c=2\pi^2/(\pi^2 - 4)$. The stress σ_x is determined by the contact problem from the theory of elasticity (Galin, 1961).

If we know the value P_0 we can determine the value of $K_I(a)$. Using the same asymptotic method we obtain the expression of:

$$P_0 = \frac{\int_0^a \sigma_x \ln\frac{a^{c/2}+\sqrt{a^{c}-t^{c}}}{a^{c/2}-\sqrt{a^{c}-t^{c}}}dt}{\int_0^b \ln\frac{a^{c/2}+\sqrt{a^{c}-t^{c}}}{a^{c/2}-\sqrt{a^{c}-t^{c}}}dt} \tag{4}$$

where b and a are the initial and the final lengths of crack, respectively.

The length of an equilibrium fracture is determined by comparing this value, K_I, with the fracture toughness of the rock, K_{Ic}. The length of such a fracture depends only on K_{Ic} and the friction coefficient, ρ, which is approximately the same as the ratio of the shearing and thrust forces. The comparison is shown graphically in Figure 4.

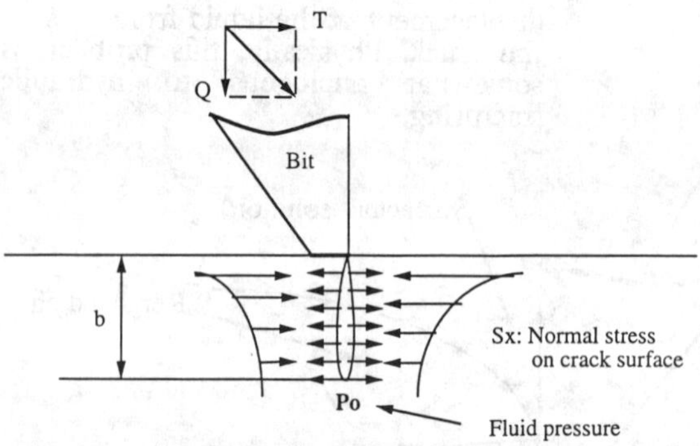

Figure 3. Edge fracture filled with surfactant in front of the bit

The horizontal line in this figure correlates with the fracture toughness index. The curves are determined by the right hand side of Equation 3. The stable damage zone is formed if the crack growth $a - b$ after each cutting pass is approximately equal to the removed chip thickness. If the length b is too small, the crack will be closed by compression and there will be no crack propagation. The critical minimum crack length can be estimated from the graph in Figure 3. As the cutting tool becomes blunt, the size of the damage zone becomes larger as we can see from this graph. If the ratio a/l is constant and l increases, then a should also increase. Thus, from this point of view, the blunter the cutting tool becomes, the larger the chip sizes become.

We have shown that the fractures in the damage zone develop under the bit when we use surfactant solutions. Without surfactant, we observe only very shallow cracks induced mechanically. The average size of the surfactant induced fractures in the damage zone is up to 2.5 times the removed chip thickness. The fractures propagate mostly in front of the bit. These fractures are filled with surfactant which facilitate the subsequent cutting passes. This initial estimate of the size of the damage zone provides a reasonably close value to the experimental observation.

4.2 Influence of the Damage Zone on Cutting Force

After estimating the length of the fracture in the damage zone we attempt to

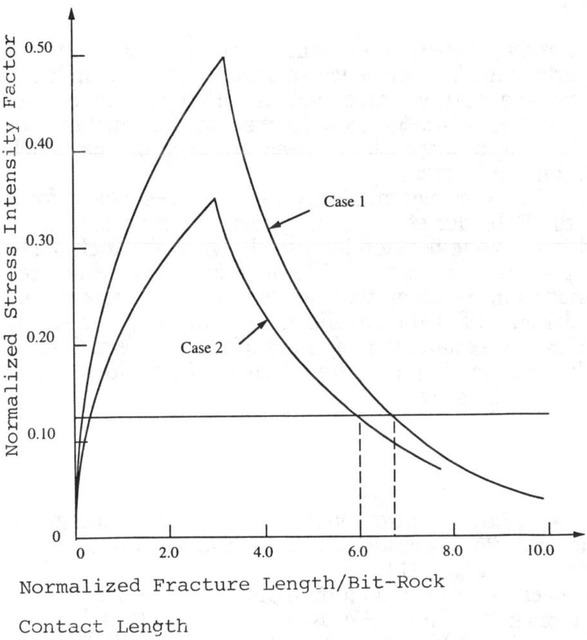

Figure 4. Normalized stress intensity factors vs. normalized fracture length / bit-rock contact length

assess the influence of these fractures on the cutting forces. The surfactant filled fracture located in front of the shearing fracture which forms a chip, transmits normal stresses, but not tangential stresses, since the friction between its surfaces decrease significantly due to the effect of surfactant. The fracture in the damage zone is located in the stress field created by the shearing fracture, which is approximately parallel to the free surface of the rock being cut.
This fracture propagates in a stable fashion. Immediately before the shearing of the chip occurs (when the fracture is open) K_I reaches K_{Ic}. The estimates of this value can be obtained by Equation (5) given below:

$$T = \frac{1}{\eta} K_{IC} \sqrt{2h}\, H\, f(\lambda, Y); \qquad \lambda = \frac{a}{h}; \quad Y = \frac{b}{h}. \tag{5}$$

where η = the coefficient depending on the cutting tool shape; h = chip thickness; H = bit width; $f(\lambda, Y)$ = function of the mutual effect of the surfactant-filled fractures, whose explicit expression can be obtained using the J-integral method.

The results of the calculation using Equation (5) are plotted in Figure 1 (line 2). This figure displays the good agreement between the experimental and analytical results. The estimate of this function and its minimum value range approximately from 0.5 to 0.9 depending on the size of the fractures in the damage zone and other associated parameters. This analysis supports that the Rebinder effect is attributed to two processes, namely a decrease of fracture toughness and an effect arising from the interaction of the surfactant-filled fracture with the bit loading. The surfactant aided rock cutting can thus be described as combined effects from chemical action and the mechanical action by bit loading during the rock cutting process which results in the facilitation of the chip formation process.

5 SUMMARY AND DISCUSSIONS

The cracks become significantly deeper when surfactant is used than the comparable case without surfactant. The presence of these surfactant induced fractures appreciably reduces the cutting force and, accordingly, the energy requirement for the cutting process. A 45% - 55% decrease in the resistance to cutting in comparison with cutting of dry rock has been observed on the bench scale testing with surfactant on sandstones.

A mathematical model based on experimental observation provides a basis for a quantitative explanation of the Rebinder effects during shear/drag bit cutting of rock. The fractures in the damage zone develop intensively when the rock being cut is treated with an appropriate surfactant. The lengths of the fractures developing with the aid of surfactant reach as much as two times the thickness of the chips removed. This model quantifies the beneficial effect of using surfactant in rock drilling, and provides an explanation of the surfactant effect from a mechanistic view point. This explanation is a step forward in exploiting the surfactant for improving drilling efficiency.

REFERENCES

Akram, M. and Karfakis, M. 1991, Rock fracture toughness and fracture energy in aqueous chemical environments. *Rock Mechanics as a Multidisciplinary Science,* Roegiers (ed.) Balkema, Rotterdam. pp. 211-220.

Arzimovich, G.V. 1985, The mechanical physical foundation of the design of rock drilling and excavation equipments. "*Nauka*" *Novosibir sk,* 265 p. [Russian]

Engelmann, W.H., Watson, P.A., Tuzinski, P.A. and Pahlman, J.E. 1987, Zeta potential control for simultaneous enhancement of penetration rates and bit life in rock drilling. *US Bureau of Mines RI 9103*, 18 p.

Galin, L.A. *Contact problems in the theory of elasticity.* Translated from the Russian. I.N. Sneddon (ed.) 1961, Dept. of Mathematics, School of Physical Sciences and Applied Mathematics, North Carolina State College.

Kim, K. and A. Mubeen 1982, Relationship between Differential SIF and Crack Growth Rate in Cycle Tension in Westerly Granite, *Amer. Soc. for Testing and Materials, Spec. Tech. Pub.*, Vol. 745, pp. 157-168.

Nisitani H. and Mori K., 1985, Influence of supporting condition on stress intensity factors for single-edge-crack specimens under bending, *Tech. Rep.*, Kyushu Univ., 58, #5.

Paul, A.D., 1988, Electrocatalytically induced liberation of mineral matter from coal, *Ph.D.Thesis*, Virginia Polytech. Institute & State University, Blacksburg, VA.

Rebinder, P.A., Shchukin, E.D. 1973, Surface phenomena in solids during deformations and fracture. *Progress in Surface Sc.*, Vol. 3, pp. 97-5.

Rebinder, P.A., Schreiner, L.A. and Zhigach, K.F. 1948, *Hardness reducers in Drilling. A Physico-Chemical Method of Facilitating Mechanical Destruction of Rocks Drilling*. Translated by CSIRO, Melbourne, Australia, 163p.

Staroselsky, A.V., Shobolova, L.P., Edelshtein, O.A. 1991, Effect of surface-active media on the degree of mechanical crushing of rocks. *Doklady Acad. Nauk USSR (Proc Acad Sc.)(Physical Chemistry)* Vol. 317, No 5 pp. 1150-1153.

Westwood, A.R.C. and Goldhein, D.L. 1970, Mechanism for Environmental Control of Drilling MgO and CaF monocrystals. *J. of the American Ceramic Soc. Vol.* 53, No. 3, pp. 143-147.

Westwood, A.R.C., Macmillan, N.H. and Kalyoneu, R.S. 1974, Chemomechanical phenomena in hard rock drilling. *Trans. Metal. Soc. ASME*, vol. **256**, pp.106-111.

6 In-situ stress

Session developers:
David Yale, Mobil Research and Development
Douglas Schmitt, University of Alberta, Edmonton

Rock Mechanics, Nelson & Laubach (eds) © *1994 Balkema, Rotterdam, ISBN 90 5410 380 8*

In-situ stress predictions and measurements in an unconsolidated sandstone formation, the Lower Frio, East Texas

G. Ramos, K. Katahara, R. Keck & M. Batzle
ARCO Exploration and Production Technology, Plano, Tex., USA

ABSTRACT: Field tests at a depth of 5000 ft, in the Lower Frio Formation (Jasper County, East Texas), demonstrated that core testing, log interpretation, and micro-fracturing techniques for *in situ* stress determination need to be modified when investigating unconsolidated sandstones. Anelastic relaxation-strains, though < 200 microstrains, can be measured within the first 7 hours. Laboratory differential strain analysis had high failure rates, but provided reliable strain ratios. Shear-wave velocity anisotropy was measured from reinforced whole-cores with a new system of polarized transducer-mounts that allowed simpler travel-time determination and sample rotation. The measured core strain anisotropy had a consistent preferential orientation with an azimuth of N00E-N40E. A small number of borehole elliptical cross sections were elongated NW-SE, suggesting a S_2 azimuth of NE-SW. The mapped East-West strike of nearby gravity faults suggested a far-field in *situ* maximum horizontal stress azimuth of N60E-N90E. Downhole stress tests in cased holes were performed after packers got stuck in open-hole micro-fracture tests. Stress magnitudes in the sand derived from logs agreed with downhole micro-fracture tests if *in situ* stresses are assumed to be at the limit of mechanical stability. These demonstrate the difficulty in core testing, log interpretation and stress testing techniques when applied to unconsolidated formations.

1 INTRODUCTION and BACKGROUND

ARCO has undertaken a major field research program to demonstrate the feasibility of fracturing for industrial solids waste disposal. This project, called the Deep Well Treatment and Injection Program (DWTI), has recently completed field operations in Jasper County, Texas. It involved drilling three vertical, 5000-ft deep wells shown in Figure 1: FTD #1 is the core and logging pilot well, converted later into a monitor well; FTD #2 is the injection/fracturing well; and FTD #3 is a monitoring well. FTD #1 is the first of the three, the only core well, and provided the engineering data for locating the other two wells and designing the hydraulic fracture treatment. In order to accurately determine the initiation, growth, containment, and orientation of the hydraulic fracture, the monitor wells, which contain micro-seismic geophones, must be strategically located. Since fracture azimuth and containment are functions of *in situ* stresses $S_1 > S_2 > S_3$, then *in situ* stress determination was paramount and critical to the success of the injection program.

The original project plan included numerous techniques for *in situ* stress and fracture azimuth determination. Implementation of this plan was hampered by difficult drilling conditions brought about by poor rock quality. Anelastic strain relaxation (ASR) was used at the rig-site, followed by laboratory differential strain curve analysis (DSCA) and velocity anisotropy (VELAN). In addition to the standard logging suite, borehole televiewer (BHTV) and dipole sonic tools were run. Extreme washouts limited the use of multi-arm calipers and other pad tools. High permeabilities and wellbore instability complicated the downhole stress tests or hydraulic microfractures (commonly referred to as *microfracs*).

The success of the *in situ* stress measurement program was instrumental in creating a well-contained, massive, hydraulic fracture. For the main massive fracture, a total of 3 million

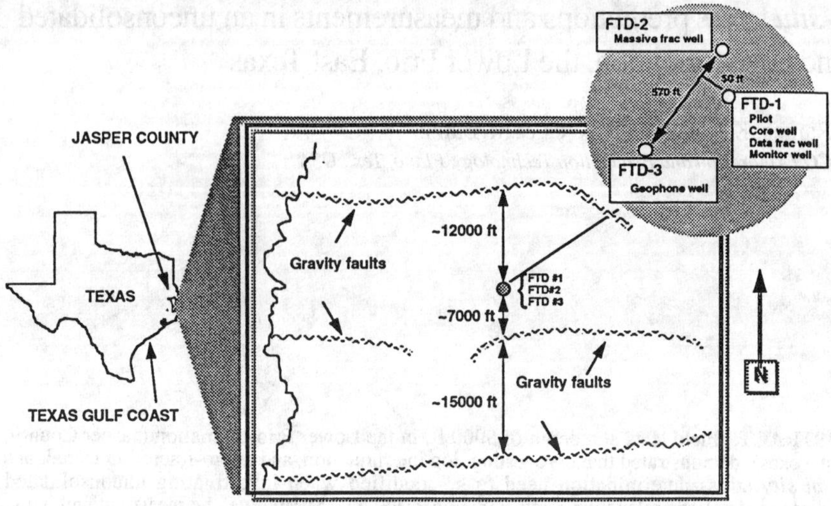

Figure 1. Location of FTD test wells, Jasper County, Texas

pounds of sand, 1 million pounds of bentonite and 50,000 bbls of water was injected in FTD #2 over a period of five days. Fracture growth was monitored with temperature logs, surface tiltmeters and hydraulic impedance testing. A new state-of-the-art micro-seismic data acquisition and locating system was developed which performed on-site processing and real-time event detection and location.

1.1 *Geological setting*

The target sand is the Lower Frio Sand, located at a depth of 4500 feet in Jasper County, Texas. The Frio is one of a number of major Oligocene-Miocene progradational wedges of coastal plain sediments in the Texas Gulf Coast Basin. It consists of a series of fluvial, deltaic and strandplain depositional systems. The Lower Frio is from 4430' to 4585', bounded above by a 130 foot-thick shale, and below by a series of shale/sand sequences that extend down to the Yegua sand at 6500'. This interval was selected due to its low dip, shallow depth, large thickness, wide areal extent, and excellent confining layers. However, behind these ideal features are the less desirable aspects of a shallow target: its low strength and high permeability. Testing on FTD #1 pilot core confirmed that the average static Young's elastic modulus of the sandstone is 250,000 psi with cohesive (shear) strengths less than 500 psi. The elastic moduli of the shale are about half that of the sandtones but the shear strength of the shale is about twice that of the sand.

In this part of the Texas Gulf Coastal Plain, S_2, the *in situ* maximum horizontal stress or S_{hmax}, is considered to be roughly parallel to the shoreline (Zoback and Zoback, 1980). The coast is only 60 miles to the south of DWTI, striking N50-80E. Seismic surveys on the deeper Yegua reveal normal faults whose general strike is N70-90E, but locally curved along NNE, Figure 1. Based on local structure, the expected S_2 azimuth is N50-90E.

2 CORE TESTING

2.1 Anelastic Strain Relaxation (ASR)

The anelastic strain relaxation (ASR) method has gained wide acceptance because of its relatively low cost, simplicity and the quick turn-around offered by several contractors (See for example Blanton, 1983, Blanton and Teufel, 1983, Yale et al, 1992). Since only the horizontal

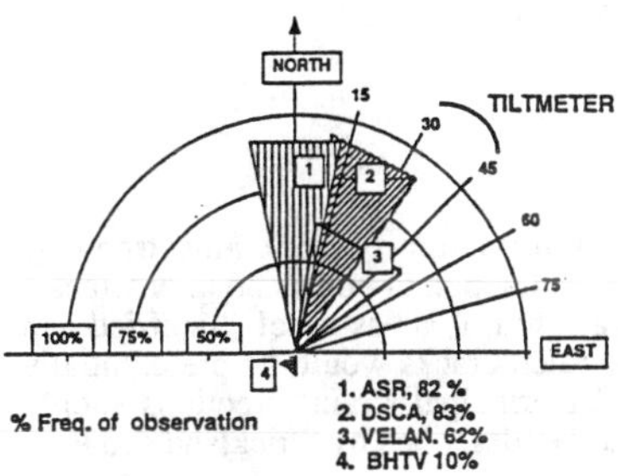

Figure 2. Diagram of dominant trends of anisotropy from core and log

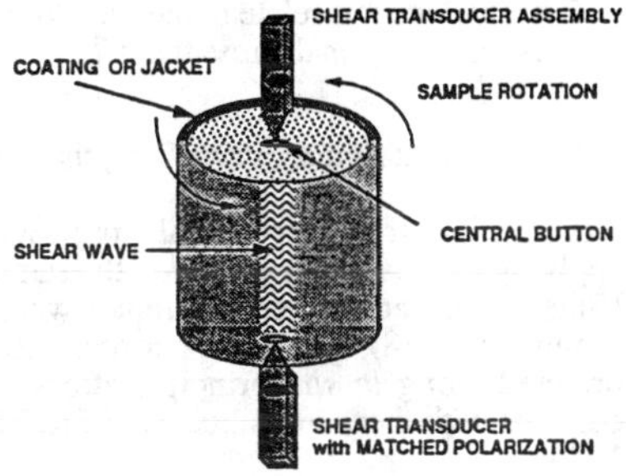

Figure 3. Shear wave velocity anisotry test system.

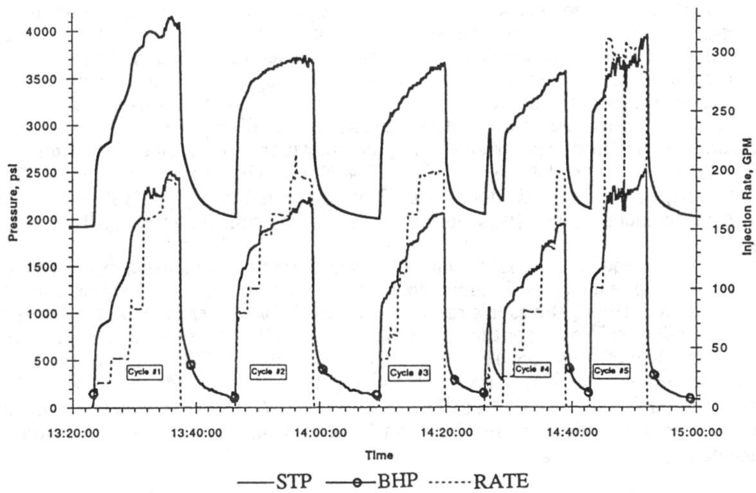

FIGURE 4. Typical job plot for a cased-hole hydraulic microfracture, showing surface treating pressure (STP, solid line), bottomhole pressure (BHP, line with circles), and injection rate (dashed line)

stress azimuth was desired early in the program, the selected ASR equipment (described by Toronto and Owen, 1988) was considered flexible enough for hard as well as soft formations.

Sample selection conditions were generally poor. Sample retrieval times were long (5.5 to 7 hours) and competent core intervals were rare. Common problems were the absence of measurable strain suitable for calculation of S_2 orientation and lack of scribe marks or orientation survey data. Trip-time on this rig ranged from 3 to 6 hours depending on the crew and conditions. The core came out reamed and under-gauge, with faint or no scribe marks. All 17 ASR samples produced time-strain curves, but only 11 had reliable scribe references. Most samples contracted in one or more directions and three samples expanded on all three channels. Some samples failed during the tests. It was common to observe transducer embedment into the soft matrix and temperature-dependent deformations.

Field plots of strain versus elapsed time indicated that anelastic strain, typically less than 200 microstrains, was complete within 6 to 10 hours after core retrieval. Of 11 successful ASR tests, 9 tests gave S_2 azimuths within N31W to N21E. The maximum horizontal strain orientation rose diagram (Figure 2) showed a dominant North-South trend. The consistency of the measured azimuth was good but questions remained unanswered:

1. Is this an effect of rock fabric instead of stress relaxation ?

2. How does this relate to the East-West strike of nearby gravity faults ?
3. Is this a regional stress trend ?

2.2 Differential Strain Curve Analysis (DSCA)

Differential strain curve analysis provides additional information on rock anisotropy by applying hydrostatic pressure on the sample, forcing the contraction of pores and microcracks. If the microcracks in such samples were generated as a result of stress-relief (Siegfried and Simmons, 1978), e.g. during coring and ASR, then the microcracks would be preferentially oriented along *in situ* principal stresses. Ideally, DSCA principal strain directions should coincide with those measured by ASR. The method is further described by Strickland and Ren, 1980.

The 11 ASR samples could not be tested with the conventional DSCA because the poorly cemented rock was not suitable to epoxy-bonding of the strain gauge-rosettes used conventionally. Instead of cutting a cube required for the complete strain tensor, DSCA was performed on a circular plug or whole-core (Ramos and Rathmell, 1989). Its horizontal faces, if firm enough, were bonded with foil-type strain rosettes. Otherwise, a jacket, pre-instrumented with three LVDTs at 45 degrees separation, was used. A pair of LVDTs was also used for axial displacements. LVDTs were preferred because of deformations exceeding 3%, well beyond the capabilities of bondable resistance gauges. The major difficulties encountered were sample disintegration during sample preparation and testing, poor gauge bonding on weakly cemented rock, resistance-gauge embedment into pore spaces, and large strains exceeding 3%. Consequently, the results were widely scattered and some were discarded.

Only six samples gave meaningful results for S_2 interpretation. The measured S_2 azimuth of 5 samples are 12, 13, 23, 23, and 40 degrees (from True North) with an average of N25E (Figure 2). When two adjacent cubes of siltstone from 4605' were tested, one gave N13E and the other N82E. The same inconsistency was observed on the adjacent plugs tested for velocity anisotropy.

Poisson's ratios are required in order to convert principal strains to principal stresses, using the same relationships in reducing ASR strains to stresses. (See for example Blanton, 1983, and Blanton and Teufel, 1983). Poisson's ratio ν is not available for each sample because the DSCA test to 4000 psi (net) confining pressure practically “destroyed” the sample by breaking inter-particle cementation. Thus, the value assumed for ν is 0.21, based on the average of the static tests on the Frio Sands. From 10 DSCA tests on samples with no scribe-lines, the average principal strain ratios are $\varepsilon_1:\varepsilon_3$ of 2.0 and $\varepsilon_1:\varepsilon_2$ of 1.56. Converting these strain ratios, using the assumed ν, the principal stress ratio $S_3:S_1$ is 0.74 and $S_2:S_1$ is 0.81. Therefore, the implied in-situ stress gradients are:

S_1	S_{vert} gradient	= 0.90 psi/ft = 0.90 psi/ft	(from the density log)
S_2	S_{hmax} gradient	= 0.81 * 0.90 = 0.73 psi/ft	
S_3	S_{hmin} gradient	= 0.74 * 0.90 = 0.66 psi ft	

2.3 *Shear wave velocity anisotropy* (VELAN)

Many investigations have shown that oriented cracks will lower velocity and induce anisotropy (e.g. Anderson et al., 1974, Sayers et al., 1990). Many of the FTD #1 cores were so weak that a layer of quick setting epoxy was also applied to all surfaces for strength and to retard drying. Small aluminum contact buttons are first attached to the center of each end of the whole-core sample. The transducer assembly consists of a piezo-electric crystal mounted in a conical punch, shown in Figure 3, which allowed sample rotation in increments of 10 degrees. A shear wave was transmitted from one transducer and received by a similar transducer with the same motion polarity. By rotating the sample, shear velocity can be determined at any azimuth. Other schemes have been developed that detect anisotropy by the amount of converted or rotated waves received (e.g. Yale et al., 1992). However, this technique of velocity measurement is a simpler procedure and the entire process takes only about 20 minutes on a prepared sample. A check is provided by the technique since two velocity cycles should be seen as an anisotropic sample is rotated through a complete 360 degree rotation. Errors due to mounting, changes in coupling during rotation, or the measurement apparatus itself will generally appear as a single cycle.

Based on the average of 5 samples, the lowest speed (maximum travel time) was along N35E ± 15 degrees, (Figure 2). Orientations for 3 other samples were not so clear because they were not strongly anisotropic. A strong shear wave velocity anisotropy in a sample yielded a travel-time difference (the difference between minimum and maximum travel-period) of 2 to 6 microseconds. The 3 samples with time-differences less than 2 microseconds were considered isotropic.

3 STRESSES FROM MICROFRACS

The magnitudes of S_2 and S_3 were measured downhole with low-volume hydraulic fracturing, or microfracs. For S_2 or fracture orientation, a tiltmeter array was used during FTD #1 microfrac and real-time micro-seismic monitoring system was installed for the FTD #2 massive fracture.

The open-hole microfracs were attempted in the lower bounding shale and in the target sand interval, performed with a set of inflatable open-hole packers to provide 18' interval isolation (Truby and Keck, 1994). This packer was selected over side-wall anchor type packers due to concerns about the integrity of the unconsolidated sandstone formation. A 30lb/1000-gal guar gel was used for all the open-hole microfracs. Given the high permeability and unconsolidated sandstone formation, it was determined that a wall-building fluid (other than drilling mud) would be more appropriate. Typically a series of five injections were performed. The first was at matrix rate to determine the leakoff rate. The second was to break the formation. The remaining three cycles were for fracture re-opening and short extension, with pressure declines used for observation of fracture closure pressure. Injection rates ranged roughly from 5 to 20 gpm, for periods of about two to ten minutes, with total volume of approximately 200 gallons. After attempting eight open-hole microfracs, the packer assembly became stuck, but was successfully fished. However, due to additional wellbore washouts in the target sand and upper shale, it was believed that the chances of achieving adequate zonal isolation from the primary cement job were quite low.

It was decided to obtain additional stress information from cased-hole microfracs rather than risk further open-hole packer runs in FTD #1. A total of six cased-hole microfracs were attempted: one in the lower shale , three in the sand interval, and two in the upper shale. Thc first two were pumped down casing; the remaining were pumped down tubing using a retrievable packer and bridge plug assembly. This assembly included quartz memory bottom-hole pressure gauges. A 30 lb/1000-gal gel was used throughout. Due to the 1 darcy sand permeability, the sand interval microfracs were conducted at much higher injection rates than those in the shales. The sand interval procedure typically consisted of a series of three to five step-rate tests to assure fracture extension was achieved in each injection cycle. The rate steps typically were 0.5, 1.0, 2.0, 4.0 and 5.0 BPM, with an occasional 7.0 BPM step. The total volume pumped was on the order of 100 bbls. Figure 4 shows results of one these injections, showing injection rate, surface and bottom-hole pressures, and casing pressure. Figure 5 log shows the depth and measured S_2 (denoted as $\mathbf{S_{2ufrac}}$) and S_3 ($\mathbf{S_{3ufrac}}$) for FTD microfracs. Results from these stress tests showed that there was roughly 750 psi stress contrast between the target sand and the bounding shales, which indicates excellent fracture containment was possible in this formation. The tiltmeter survey of FTD #1 microfracs indicated a N30E fracture, in apparent close agreement with core predictions. However, the micro-seismic monitoring of the longer and more massive hydraulic fracture of FTD #2 showed a N60E fracture azimuth, more in line with S_2 azimuth of the gravity faults 7000 ft from the well (Figure 1).

4 LOG-DERIVED STRESSES

Having estimated the azimuth of S_2 from core tests and tiltmeters, and the magnitudes of S_2 and S_3 from microfracs, the remaining unknown component needed to complete the stress tensor is S_1. In addition, a continuous vertical profile of these stresses was desired. Thus, a comprehensive log processing was performed, and briefly summarized below.

4.1 *Stress magnitudes*

The density log was integrated to give the vertical stress or S_1. The vertical stress gradient over the zone of interest was about 0.89 psi/ft. Pore pressure was assumed to be hydrostatic and was estimated by integrating pore water density from the surface. The water density was obtained from the resistivity and porosity by inverting Archie's law in clean sands, and assuming NaCl as the only dissolved solid. The pore pressure gradient in the zone of interest is about 0.44 psi/ft.

A conventional log analysis solved for clay, sand and water volumes. The relative volumes, density, and travel time logs were then used to solve Gassmann's equation for the rock skeleton elastic moduli as described by Greenberg and Castagna (1992).

The conventional model for oil-field stresses assumed uniaxial vertical strains and a poro-elastic medium, with $S_1 = S_{vert}$ and $S_3 = S_{hmin}$:

$$S_{hmin} = \left(\frac{\nu}{1-\nu}\right)(S_{vert} - \alpha P) + \alpha P \qquad (1)$$

where α is the Biot poro-elastic coefficient and P is the pore pressure. Track 2 in Figure 5 presents two stress estimates, one ($S_{h,eu}$) using the acoustic values for ν and α, and another ($S_{h,ed}$) using the drained skeleton values. Horizontal stresses from the open- and cased-hole microfracs, also shown in Figure 5, indicated a much greater stress contrast between the sand and the shales than do either $S_{h,eu}$ or $S_{h,ed}$. Equation (1) was also tried assuming α=1 with drained and with undrained values. None of these elastic models reproduced the directly-measured stress contrasts, essentially because ν from the sonic logs is about the same in both the sands and the shales. Aside from its conceptual shortcomings (Perkins, 1976; McGarr, 1988), poro-elastic equation (1) did not apply in this case.

A Mohr-Coulomb model proposed by Perkins (1976) assumed that stresses are on the shear failure envelope. The log's failure envelope was estimated using empirical relations in sandstone data bases. This stress estimate, plotted as $S_{h,mc}$ in track 3 of Figure 5, was highly uncertain in the shales because of lack of strength data for clay-rich rocks. Since there are growth faults through these zones (Figure 1), the stress may be limited by friction rather than fault initiation. In the absence of laboratory data on friction, the model used a Mohr-Coulomb internal friction angle estimated from an empirical porosity-friction angle correlation. This gives the curve labeled $S_{h,fn}$ in Figure 5. Both $S_{h,mc}$ and $S_{h,fn}$ predicted a significant sand-shale stress contrast, but $S_{h,fn}$ gave more reliable stress contrast.

A soil mechanics model may be applicable because of the poorly-consolidated nature of the sediments. This model assumed uniaxial plastic compaction and that K_0, the ratio of horizontal to vertical stress, is a function only of clay content:

$$K_0 = 0.4 + 0.5V_{cl} \qquad 0.0 < V_{cl} < 0.4 \qquad (2a)$$

$$K_0 = 0.6 + 0.167(V_{cl} - 0.4) \qquad 0.4 < V_{cl} < 1.0 \qquad (2b)$$

where V_{cl} is the clay volume fraction. Equation (2) was chosen to give rough agreement with the observations of Karig and Hou (1992) and Karig (1993) on sands and clays. This stress estimate is shown as $S_{h,ko}$ in track 4 of Figure 5. Given the uncertainties involved, $S_{h,ko}$ agreed fairly well with the microfrac S_{hmin} data.

In the lower shale, Figure 5 indicates that S_2 from microfracs (S_{2ufrac}) is about the same as S_1, the vertical stress (S_{vert}). This implied that the uniaxial vertical strain assumption is incorrect. For the soil mechanics model it is straightforward to show that if $S_{hmin} = S_{vert}$ then

$$S_{hmin} = S_{vert}(4K_0 - 1)/(K_0 + 2) \qquad (3)$$

This is $S_{h,k1}$ in track 4 of Figure 5. This model gave better agreement with the microfrac sand-shale stress contrast than $S_{h,ko}$.

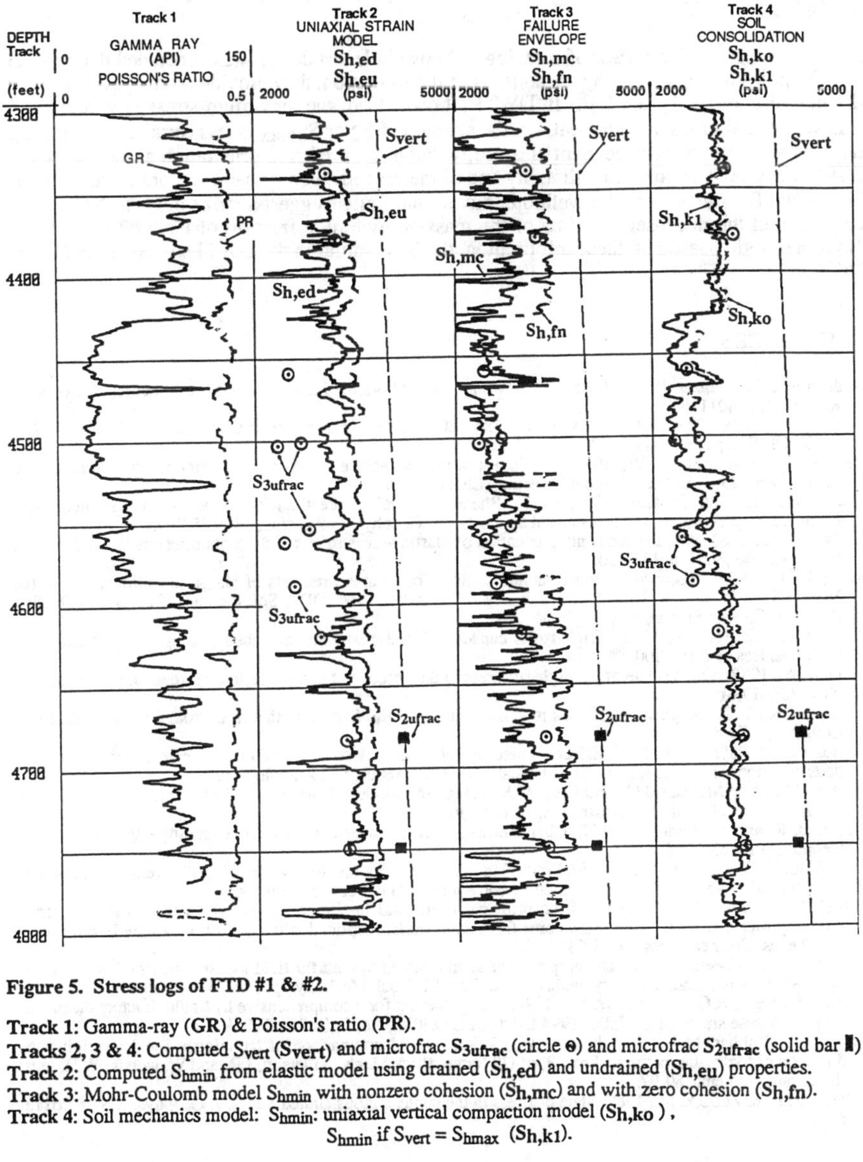

Figure 5. Stress logs of FTD #1 & #2.

Track 1: Gamma-ray (GR) & Poisson's ratio (PR).
Tracks 2, 3 & 4: Computed S_{vert} (Svert) and microfrac S_{3ufrac} (circle ⊙) and microfrac S_{2ufrac} (solid bar ▮)
Track 2: Computed S_{hmin} from elastic model using drained (Sh,ed) and undrained (Sh,eu) properties.
Track 3: Mohr-Coulomb model S_{hmin} with nonzero cohesion (Sh,mc) and with zero cohesion (Sh,fn).
Track 4: Soil mechanics model: S_{hmin}: uniaxial vertical compaction model (Sh,ko),
S_{hmin} if $S_{vert} = S_{hmax}$ (Sh,k1).

A borehole imaging tool was used to map borehole ellipticity, with the objective of locating stress-induced breakouts. A circumferential acoustic scanning tool (BHTV) was preferred in soft formations because it has no pads that may get stuck. The BHTV yielded high-resolution ultrasonic images of the wall and the borehole shape indicated that the hole was generally washed out. In a few places, the cross-section looked as if breakout-outs had been present but were being eroded. The major axes were oriented roughly NW-SE, implying S_2 to be NE-SW. Less than 10% of the imaged interval provided discernible borehole ellipticity, and this gave the lowest frequency when compared to the core-derived azimuths (Figure 2).

5 CONCLUSIONS

After a comprehensive review of core, log and downhole test data, it was concluded that in spite of difficulties brought about by the softness of the formation, they provide reliable estimates for the magnitudes of S_1 and S_3. In DWTI's Lower Frio, the maximum stress is vertical and follows normal lithostatic gradient, but the azimuth of S_2 or S_{hmax} is less certain. Core testing has revealed the stress-dependent anisotropy that has a preferred orientation along the NNE-SSW direction. The borehole ellipticity and tiltmeters indicated a near-wellbore S_2 azimuth of N30E-N50E. Away from the wellbore, the normal faults suggest a strike close to N60E ± 20 degrees and the microseismic data of the massive hydraulic fracture of FTD #2 indicated a N60E azimuth. Based on these information, the S_2 azimuth in the DWTI site is within N20E-N60E.

REFERENCES

Anderson, D.L., Minster, B., and Cole, D., 1974, The effect of oriented cracks on seismic velocities, J. Geophys. Res., v.79, p. 4011-4015.

Blanton, T.L., 1983, The relation between recovery deformation and insitu stress magnitudes, SPE/DOE 11624, SPE/DOE Symposium on Low Permeability, Denver, CO,

Blanton, T.L. and Teufel, L.W., 1983, A field test of the recovery method of stress determination in Devonian shale. SPE 12304, Eastern Regional Meeting, Champion PA,

Greenberg, M. L. & Castagna, J.P., 1992. Shear-wave velocity estimation in porous rocks: theoretical formulation, preliminary verification and applications. Geophys. Prospecting. 40:195-209.

Hudson, J. A., 1981, Wave speeds and attenuation of elastic waves in materials containing cracks, Geophys. J. Roy. Ast. Soc., v.64, p. 133-150.

Karig, D. E., 1993. Reconsolidation tests and sonic velocity measurements of clay-rich sediments from the Nankai Trough, in I. A. Hill, A. Taira & J. V. Firth (eds.), Proc. ODP, Sci. Results, 131. Ocean Drilling Program, College Station, TX.

Karig, D. E. & Hou, G., 1992. High-stress consolidation experiments and their geologic implications. J. Geophys. Res. 97:289-300.

McGarr, A., 1988. On the state of lithospheric stress in the absence of applied tectonic forces. J. Geophys. Res. 93:13,609-13,617.

Perkins, T.K., 1976, Application of rock mechanics in hydraulic fracturing theories, Proc. Seventh World Pet. Congress, 3:75-84

Ramos, G. and Rathmell, J., 1989, Effects of mechanical anisotropy on core strain measurements for *in situ* stress determination, SPE 19593, Proc. SPE Annual Tech. Conference, San Antonio, TX

Sayers, C.M., Van Munster, J.G., and King, M.S., 1990, Stress-induced ultrasonic anisotropy in Berea Sandstone, Int. J. Rock Mech. Min. Sci. Abstr., v.27 p. 429-436.

Siegfried, R. and Simmons, G., 1978, Characterization of Oriented Cracks with differential strain analysis, J. Geophys. Res, 83:1268

Strickland, F. G. and Ren, N.K., 1980, Predicting *in situ* stresses for deep wells using differential strain curve analysis, SPE 8954, SPE Symposium on Unconventional Gas Recovery, Pittsburgh

Teufel, L.W. 1984, Determination of the principal horizontal in-situ stress directions from anelastic strain recovery measurements of oriented cores from deep wells: Application to the Cotton Valley Formation of East Texas, Geomechanics, 57:55-63

Toronto, T. and Owen, L, 1988, Development of an automated system for field measurements of *in situ* stress orientation by the strain recovery method, Proc. 29th US Rock Mech. Symp, 743-744.

Truby L.S., Keck, R.G. and Withers, R.J. 1994, Data gathering for a comprehensive hydraulic fracture diagnostic project: A case study, SPE 27506, 1994 IADC/SPE Drilling Conf., Dallas, TX

Yale, D.P., Stubhar, M.K., El Rabaa, A. W.., 1992, A field comparison of techniques for determining the direction of hydraulic fractures, Proc. 33rd US Symp. Rock Mech., Tillerson, J.R. and Wawersik, W.R. (eds) Rotterdam, Balkema, 89-97.

Zoback, M.L. & Zoback, M., 1980, State of stress in the conterminous United States, J. Geophys. Res., 85:6113-6156.

Rock Mechanics, Nelson & Laubach (eds) © 1994 Balkema, Rotterdam, ISBN 90 5410 380 8

The influence of near surface faults on in situ stresses in the Canadian Shield

N. Chandler & D. Martin
AECL Research, Pinawa, Man., Canada

ABSTRACT: The influence of reverse-dip displacement along low-dipping faults at the Underground Research Laboratory (URL) was assessed by constructing numerical models using both finite difference and distinct element methods. The URL was constructed by AECL Research to provide a realistic environment for in situ studies into aspects of the deep geological disposal of nuclear fuel waste. Fracture zones which intersect the URL shaft at depths of less than 300 m have a significant effect on the magnitudes of the measured horizontal in situ stresses. The numerical models discussed in this paper demonstrate that reverse-dip displacement along these near-surface faults is one factor leading to the observed near-surface reduction in stresses. The models also suggest that the horizontal stresses below the deepest fracture zone do not increase significantly with depth.

INTRODUCTION

In situ stress will play an important role in the design of a vault for nuclear fuel waste disposal. It has implications for both damage to the rock at the surface of excavations and the hydraulic transmissivity of existing fractures and fracture zones (Chandler and Martin 1990). Experience at the Underground Research Laboratory (URL) has demonstrated that it is not always possible to obtain reliable measurements of in situ stress using traditional methods (i.e., hydraulic fracturing and overcoring). Prior to excavating a shaft it may be necessary to make a reasonable assessment of in situ stress conditions based on measurements made elsewhere in the Canadian Shield, and on sound engineering judgement. The simple numerical model presented in this paper is an example of a tool that could be used for making assessments based on engineering judgement.

THE UNDERGROUND RESEARCH LABORATORY

AECL Research constructed the URL for the purpose of examining elements of the concept for deep geological disposal of nuclear fuel waste. The assessment of stress determination methods is one element of the research program at the URL. One result of the in situ stress program is a well characterized in situ stress field to a

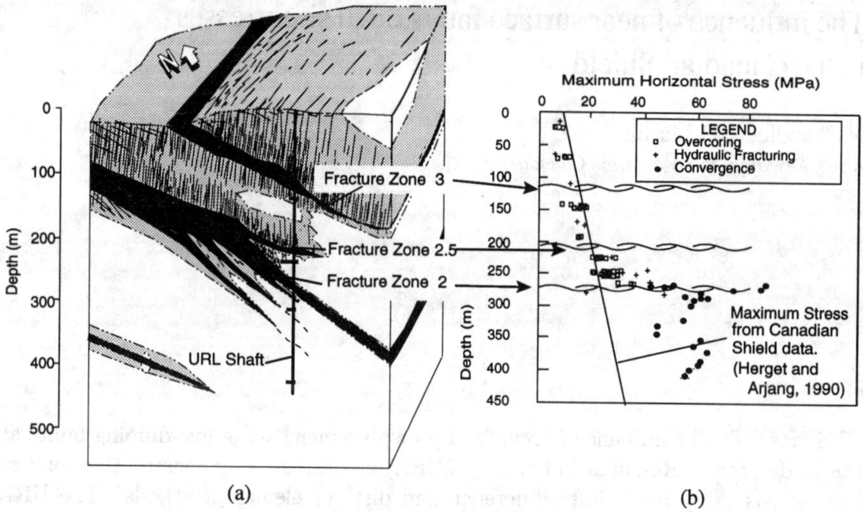

Figure 1: Fracture zones at the Underground Research Laboratory (a) and their influence on measured stresses (b) with a comparison to stresses measured in the Canadian Shield.

depth of 500 m (Martin 1990). Within the volume of rock immediately surrounding the URL, there exist two major fracture zones with a number of associated splays (Figure 1(a)). These fracture zones are thrust faults, along which reverse-dip displacement has occurred. Fracture Zone 2, which intersects the URL shaft at a depth of approximately 280 m, is the most predominant fracture zone at the URL. More than 7 m of differential displacement has been observed along Fracture Zone 2 (Everitt et al. 1990). The horizontal stresses measured at the URL can be grouped into three domains, as is evident in Figure 1(b) (Martin and Chandler 1993). Horizontal stresses above Fracture Zone 2.5 are similar to the average of stresses measured elsewhere in the Canadian Shield at similar depths (Herget and Arjang 1990). Below Fracture Zone 2, the horizontal stresses measured at the URL are anomalously high compared to the average Canadian Shield stresses, while the stresses in the domain between Fracture Zones 2.5 and 2 are intermediate in magnitude. Also, the direction of the maximum horizontal stress rotates by 90°, from the strike of the fault to its dip direction below Fracture Zone 2 (Figure 2). To determine if these observations could be replicated, a simple numerical model of the volume of rock surrounding the URL was developed.

Although overcoring and hydraulic fracturing stress measurements have not been successful at depths below Fracture Zone 2, the in situ stresses have been determined with reasonable confidence using convergence measurements. The maximum horizontal stress is 55 MPa, with a horizontal stress ratio of about 1.2. These values have been supported by the results of under-excavation tests (Kaiser et al. 1990), reopening of subhorizontal hydraulic fractures (Haimson et al. 1993), and doorstopper overcoring tests (Corthésy and Gill 1993). This stress condition was used in the calibration of the numerical models.

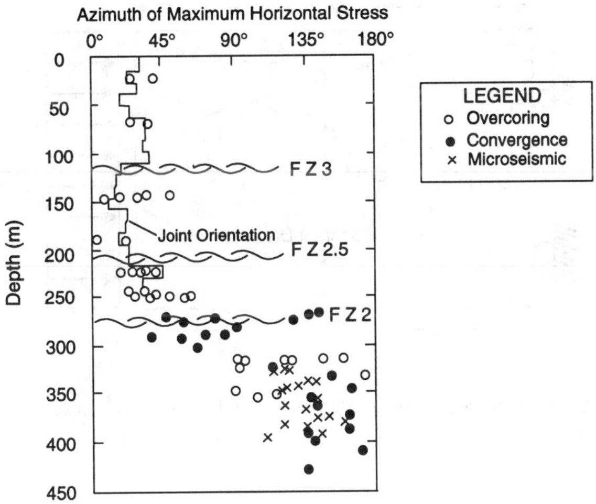

'igure 2: The measured orientation of the maximum horizontal stress at the URL. Note the 90° rotation across Fracture Zone 2.

MODELING THE URL STRESS FIELD

Numerical modeling was used to examine the influence of the faults on the in situ tresses in rock along the URL shaft. Two software packages, both produced by tasca Consulting Group Inc., were used in the analysis, FLAC and UDEC. FLAC s a finite difference code, whereas UDEC makes use of the distinct element method. When the same material properties and boundary conditions were specified, both nodels produced similar results. A 2-km-long by 1.5-km-deep section of rock was nodelled assuming plane strain conditions. Fracture Zones 2 and 3 were considered o be simple linear structures, dipping at 25° and separated vertically by 180 m. The boundary conditions were as shown in Figure 3. The right-hand edge of the nodel is a constant-displacement boundary, which satisfies the condition of continuity with the surrounding rock. The horizontal distance between the boundaries was decreased until stresses at the 420-m depth were representative of measured values, approximately 55 MPa. In the absence of faults this displacement condition will produce a horizontal stress (σ_h) which increases linearly with depth (z) by the "Poisson " effect:

$$\frac{\partial \sigma_h}{\partial z} = \frac{\nu}{1-\nu}\gamma z$$

where ν is Poisson's ratio and γ is the rock unit weight.

Figure 4 compares the dip direction (northwest-southeast) stresses measured at the URL with the horizontal stresses obtained from the UDEC model. Also shown in the figure is the out-of-plane or strike direction stresses from the UDEC model. Both the strike- and dip-direction stresses are assumed to be known at the lower boundary of the model (48 and 55 MPa respectively). The dip-direction stresses are the horizontal stresses calculated by the model and the strike-direction stresses are the out-of-plane stresses that satisfy the condition of plane strain.

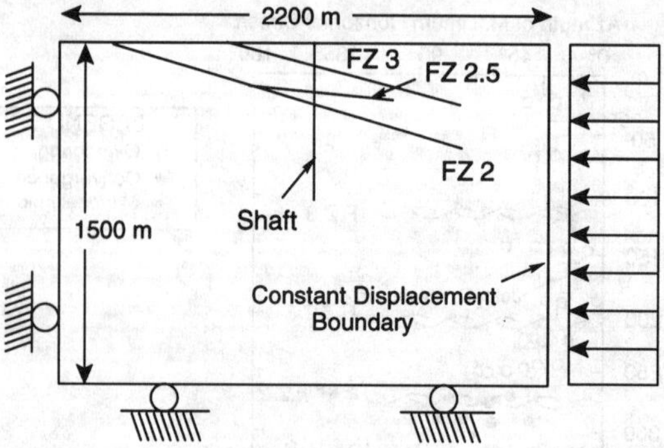

Figure 3: The boundary conditions imposed in the UDEC and FLAC numerical models of the URL.

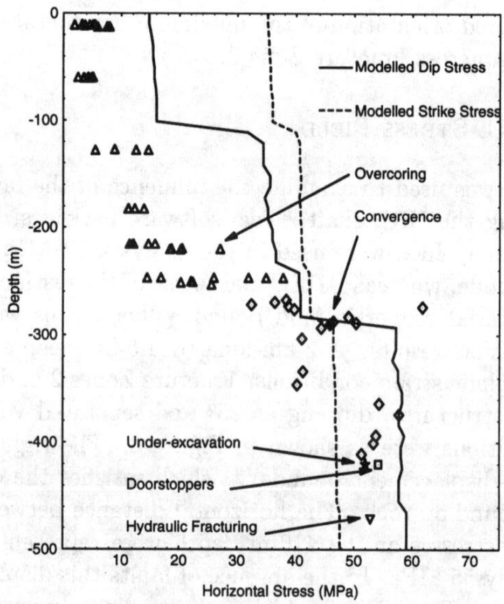

Figure 4: The horizontal stresses calculated using UDEC and the horizontal stresses measured in the dip direction of Fracture Zone 2.

Two important observations should be noted. Modelled reductions in stress with decreasing depth resulting from slip along the faults are comparable to the measured stresses. The model suggests that the stresses between fracture zones and below the deepest fracture zone remain essentially constant with depth. The second observation is that the direction of the maximum horizontal stress rotates from the dip direction below Fracture Zone 2 to the strike direction above the fracture zone; this is precisely what occurs in the field (Figure 2).

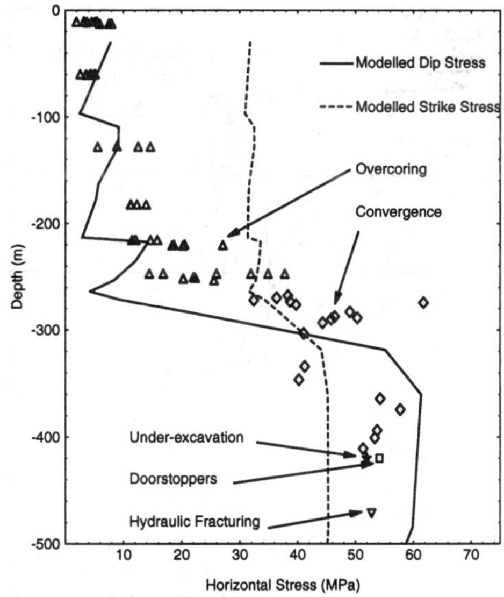

Figure 5: The horizontal stresses calculated using FLAC and the horizontal stresses measured in the dip direction of Fracture Zone 2.

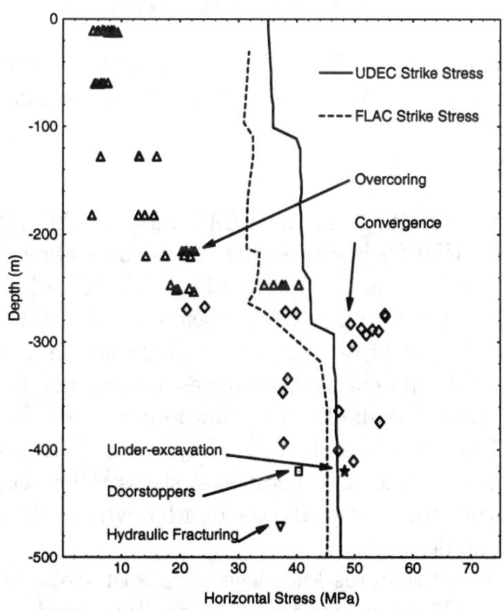

Figure 6: The out-of-plane stresses calculated using both UDEC and FLAC and the horizontal stresses measured in the strike direction of Fracture Zone 2.

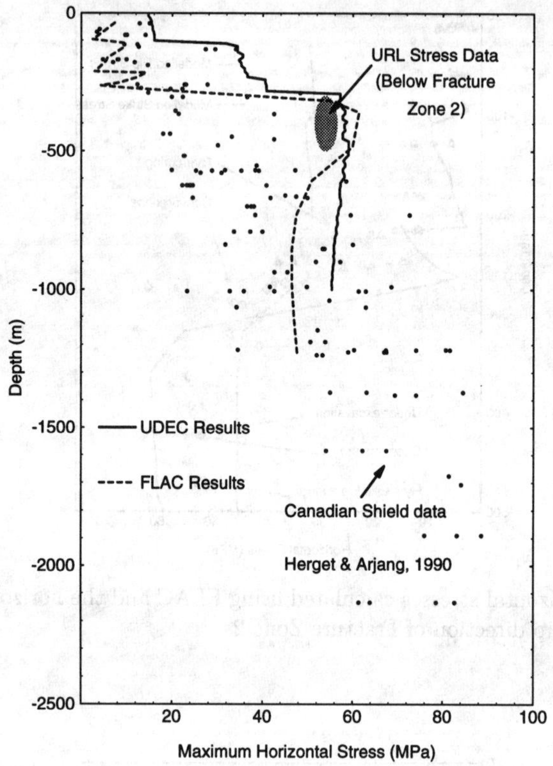

Figure 7: The horizontal stresses from the numerical models compared to the maximum horizontal stresses measured elsewhere in the Canadian Shield (Herget and Arjang 1990).

Figure 5 shows the results from the FLAC model. The results are similar to stresses calculated by UDEC, however, the stress drop above Fracture Zone 2 is greater in the FLAC model. The results from both UDEC and FLAC imply that the observed dip direction stress drop with decreasing depth across the fracture zones can be explained by reverse-dip displacement along the fault surfaces. However, both models overestimate the horizontal stresses in the strike direction at shallower depths (Figure 6). Since the models are two-dimensional, the fracture zones are depicted as planer features when, in fact, they are dish shaped. The curvature of the fracture zones should lead to additional stress relief occurring in the strike direction of the fracture zones at shallower depths, which can not be simulated in the two-dimensional models.

This particular model requires knowlege of in situ stress conditions at depths below Fracture Zone 2. If other sites are to be similarly modelled, this information will not necessarily be available. However, the range of horizontal in situ stresses reported by Herget and Arjang (1990) at depths below 1000 m can be used to calibrate the model. This method of analysis can provide useful information regarding possible stress rotations and stress variations with depth resulting from dip displacement along known fractures or fracture zones.

IMPLICATIONS WITH RESPECT TO STRESSES IN THE CANADIAN SHIELD

The model presented in this paper is geometrically very simple, and the results are not surprising. Stress relief caused by fault slip, one could argue, is intuitive without the effort of constructing a numerical model. However, the results of the model provide one viable explanation for the stress variations observed at the URL. Perhaps the most important implication is that the stresses below Fracture Zone 2 should not increase substantially with depth. A question commonly asked is "Are these stresses at the URL representative of stresses measured elsewhere in the Canadian Shield?" In Figure 1 the measured stresses at 420 m are high with respect to Herget and Arjang's (1990) average major principal stress. However, the modelled stresses nearer the base of the model (1500 m depth), compare favourably with the range of Herget and Arjang's published data for the maximum principal stress (Figure 7). The model, therefore, suggests that URL stresses are representative of Canadian Shield stresses.

Does this model have implications with respect to vault design? The model demonstrates that one factor leading to reduced in situ stresses at near-surface depths is reverse slip displacement along low-dipping faults. The disposal vault site selection process may include sites with low-dipping faults similar to those at the URL. The magnitude of in situ stresses at the potential sites will be one of the factors influencing the selection of the location and depth of a vault. The analysis presented in this paper demonstrates that simple numerical simulations can be developed to provide insight into the variation of near-surface stresses with depth, provided reasonable assumptions can be made with respect to deeper in situ stress magnitudes. This type of analysis can be carried out to support measured in situ stress data, or to provide insight into local in situ stresses when traditional in situ stress determination methods are not entirely successful.

ACKNOWLEDGEMENT

The experimental program at the Underground Research Laboratory is part of the Canadian Nuclear Fuel Waste Management Program which is funded jointly by AECL Research and Ontario Hydro under the auspices of the CANDU owners group.

REFERENCES

Chandler, N. A. and C. D. Martin (1990). Implications of *in situ* stress on vault design in granite. In R. S. Sinha (Ed.), *Proc. Int. Symp. on Unique Underground Structures*, Denver, Colorado, pp. 65:1–19. CSM Press, Denver.

Corthésy, R. and D. E. Gill (1993). Doorstopper stress measurements at the 420 Level of the URL - Final Report. AECL Technical Report TR-M-013, École Polytechnique, Montréal.

Everitt, R. A., A. Brown, C. C. Davison, M. Gascoyne, and C. D. Martin (1990). Regional and local setting of the Underground Research Laboratory. In R. S. Sinha (Ed.), *Proc. Int. Symp. on Unique Underground Structures*, Volume 2, Denver, Colorado, pp. 64:1–23. CSM Press, Denver.

Haimson, B. C., M. Lee, N. A. Chandler, and C. D. Martin (1993). Estimating the state of stress from sub-horizontal hydraulic fractures at URL, Manitoba. In B. C. Haimson (Ed.), *Proc. 34th U.S. Rock Mechanics Symp.*, Volume II, Madison, Wisconsin, pp. 585–588.

Herget, G. and B. Arjang (1990). Update on ground stresses in the Canadian Shield. In G. Herget (Ed.), *Proc. Stresses in Underground Structures*, Ottawa, Ontario, pp. 33–47. Energy, Mines and Resources Canada.

Kaiser, P. K., D. Zou, and P. A. Lang (1990). Stress determination by back-analysis of excavation induced stress changes – A case study. *Rock Mechanics and Rock Engineering 23*(3), 185–200.

Martin, C. D. (1990). Characterizing *in situ* stress domains at the AECL Underground Research Laboratory. *Can. Geotech. J. 27*, 631–646.

Martin, C. D. and N. A. Chandler (1993). Stress heterogenity and geological structures. *Int. J. Rock Mech. Min. Sci. & Geomech. Abstr. 30*(7), 993–999.

Rock Mechanics, Nelson & Laubach (eds) © 1994 Balkema, Rotterdam, ISBN 90 5410 380 8

Measurement of in-situ initial stress and excavation-induced stress changes in the vicinity of underground opening

Akio Tamai, Tsutomu Kaneda & Tetsuji Mikami
Civil Engineering Technical Department, Obayashi Corporation, Tokyo, Japan

ABSTRACT: A private underground research laboratory has been constructed by Obayashi Corporation to study various aspects related to the design and construction of underground openings. As part of the comprehensive research programme, a representative state of in-situ initial stress of the site has been determined by stress measurements at a number of points by means of the conical-ended borehole technique. Following the measurements of initial stress, a tunnel with a cross section of 4m-high and 5m-wide was excavated in the rock mass. The stress changes have been measured at many locations in the rock mass during the advance of the tunnel. The distribution of stress changes around the opening is calculated and presented here. In order to evaluate the stress response to excavation, to use the monitored results as an aid for future design process and, also, to validate an appreciation of the numerical analysis for the subject, the three-dimensional finite element analysis was performed, taking the excavation sequences into consideration. As a results, although the predictions by the analysis generally agree with the monitored results, the stress redistribution obtained from in-situ measurements is much more various than that of the numerical estimation. The results from this study demonstrate that the stress changes induced by excavation would be highly influenced by the joints in a rock mass. Those, also, suggest both the propriety and the limitation of numerical analyses applied for the predictions of stress redistribution.

1 INTRODUCTION

The state of in-situ initial stress is an important design parameter for the construction of underground openings. The stress changes induced by the excavation of an opening are related to the initial stress field. The new stress distributions resulting from excavation are particularly important if they are likely to cause failure. In order to estimate the stress redistribution, e.g. for the design of permanent works underground, numerical analyses such as the finite element method are usually used. Many reports have been made in respect of the measurements of in-situ initial stress and then, using the results, the numerical estimation of the stress redistribution caused by excavation has been carried out. There are, however, not so many cases measuring stress changes during advance of a tunnel at many locations around an opening. It is, therefore, thought that enough discussion has not been made with regard to the propriety of numerical analyses to predict stress redistribution.

The purposes of this study are to monitor the stress response to tunnel excavation in the wide area, to compare the results with the predictions by the three-dimensional finite element analysis and to evaluate the capability of current stress change monitoring technique. Although it is an ultimate purpose for us to improve the constitutive model of the rock mass, only the process for the in-situ measurements and the comparison between the results in the site and those from the numerical analysis are presented here.

2 DESCRIPTION OF THE TEST SITE

The test site is located in a metal mine at a depth of 175m in Paleozoic gneiss. The site has been occupied by Obayashi Corporation as a private underground research laboratory in order to study issues related to the design and construction of underground openings in rock. In accordance with our comprehensive research programme, various kinds of research projects concerning the mechanical behaviours and hydraulic characteristics of rock mass have been carried out. The location and profile of the site are shown in Fig. 1 and Fig. 2. The geotechnical properties show that the uniaxial compressive strengths range from 38 to 172 MPa with a modulus of 27 to 81 GPa and a Poisson's ratio in the order of 0.1 to 0.4. No major fracture zones or faults appear around the test site. The rock is however essentially discontinuous with some visible joints or natural fractures. The discontinuity spacing of the rock mass is 0.2m on the average.

A drift for investigation and monitoring, shown in Fig. 3, was excavated by drilling and blasting. The area surrounded by the drift is about 30m wide and 40m long. The rock mass is an object for detailed investigation.

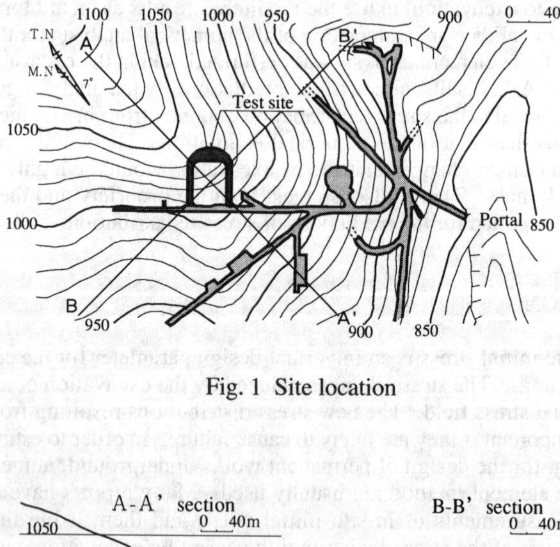

Fig. 1 Site location

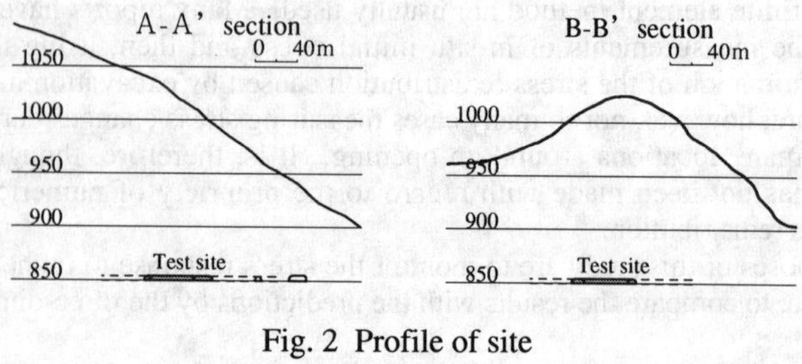

Fig. 2 Profile of site

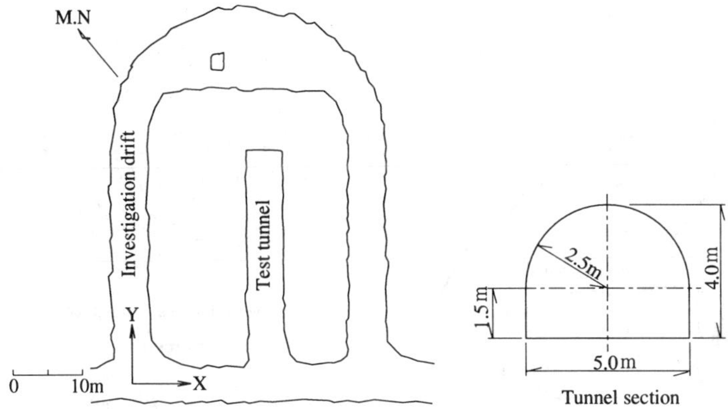

Fig. 3 Plan of test site and tunnel section

3 INITIAL STATE OF IN-SITU STRESS

The initial state of in-situ stress near the research site has been determined by stress measurements by means of the conical-ended borehole technique. This is one of the stress relief methods and enables one to determine the complete state of stress from a set of measurements in a single borehole. The conical-ended borehole technique is the modification of the hemispherical-ended borehole technique presented by Sugawara et al. (1985), and both techniques, developed by Sugawara and others of Kumamoto University in Japan, are based on a principle very similar to that of the doorstopper system, apart from the shape of the borehole bottom. The induced strains on the hemispherical or conical bottom surface of a borehole have been analyzed by the theory of elasticity and the arrangement of strain gauges was theoretically examined to obtain a high accuracy of stress determination. On the basis of these theoretical studies, an axisymmetrical arrangement has been proposed with the 16-element spherical or conical strain cell. Using the strain cell with 16 gauges, the three-dimensional stress tensor in rock is determined by measuring the strains that occur on the bottom surface of a single borehole when the stresses are relieved by overcoring. Although these techniques are not popular throughout the world, they are now in common use in Japan, because they are methods currently available for stress measurements in jointed rock formations with joint spacing less than 0.4m. The principles and practice of these methods are presented in details by Sugawara and Obara (1993) and Nakayama et al. (1993).

Using two horizontal boreholes drilled from investigation drift walls, the stresses in the rock mass were measured at 29 points with the distances from the walls ranging from 1.0m to 15.0m. Fig. 4 shows the locations of stress measuring points. As the initial state of in-situ stress is disturbed in the vicinity of an opening, a representative state of initial stress in the domain was determined by the results of 13 sets of stresses measured at the distance more than 9m from drift walls. The magnitudes and the orientations of the principal stresses are presented in Fig. 5 using the stereographic projection of lower reference hemisphere. The results indicate that the vertical stress component is nearly equal to the value estimated from the weight of the superincumbent strata and the directions of the principal stresses reflect the topographical features.

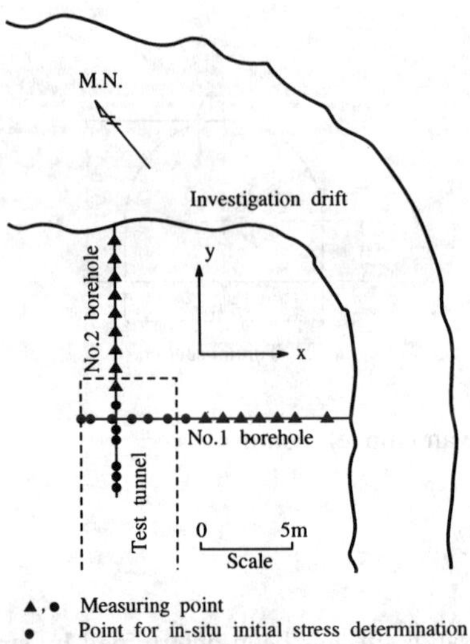

Fig. 4 Measuring points of initial stresses

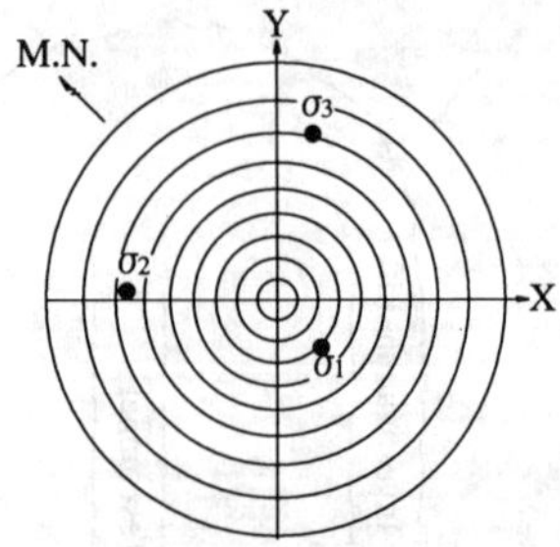

principal stresses

σ_1 = 5.04MPa
σ_2 = 1.88MPa
σ_3 = 1.22MPa

stress components

σ_x = 2.13MPa
σ_y = 1.92MPa
σ_z = 4.12MPa
τ_{yz} = -1.33MPa
τ_{zx} = 0.74MPa
τ_{xy} = -0.60MPa

Fig. 5 In-situ initial stress

4 MEASUREMENTS OF STRESS CHANGES

After the measurements of in-situ stress, a test tunnel was excavated in the rock mass. The tunnel has the same cross section of 4.0m-high and 5.0m-wide as the investigation drift's. To monitor the stress changes induced by the excavation, strain measurements were carried out using the same strain cell with 16 strain gauges as used for the conical-ended borehole technique. Seven strain cells, one in front of the excavation face and the others near the side walls of a tunnel, were installed at the end of each horizontal borehole ahead of the excavation. The locations of each cell are shown in Fig. 6. The distances from each instrument to the wall of the opening (i.e., after the excavation) are varied from 1 to 9 metres.

In stress change monitoring, the stresses calculated from associated strains at each cell location should change from the initial state to new states as the excavation face advanced. The strains were continuously recorded until the excavation had passed the instrument locations by about 10m. Drill and blast excavation for about 1.0m advance was employed to change the initial state of stress gradually. Unfortunately, two of the strain cells, No.5- and No.7-cell, could not monitor the strain changes properly, probably due to the inadequate installation.

According to the monitored results, some of the strains changed sharply at around the passage of excavation face. This indicates the stress response to excavation. It was however thought that the monitored strains included time dependent changes or drifting. The distributions of stress changes around the opening are presented afterward.

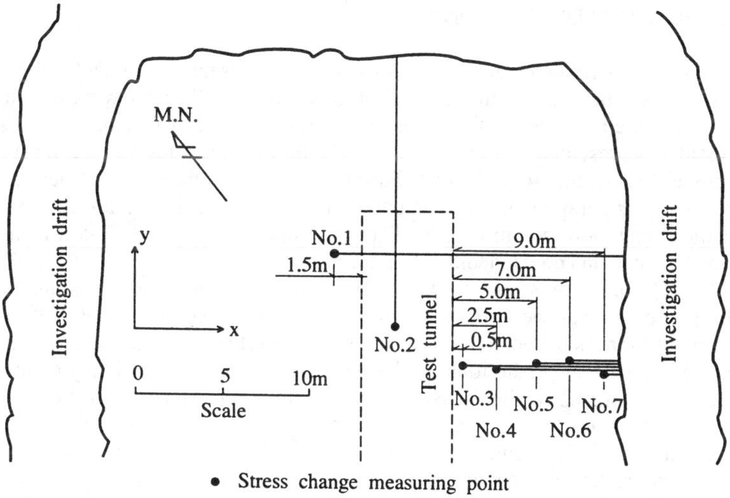

Fig. 6 Locations of stress change measurements

5 THREE-DIMENSIONAL NUMERICAL ANALYSIS

In order to compare the in-situ results with predictions by numerical analyses, the three-dimensional finite element analysis was performed, taking the excavation sequences into consideration. In the analysis, the rock mass excavated is assumed as a continuous, homogeneous, isotropic and linear elastic medium. The geotechnical parameters used for the FEM analysis were decided from in-situ and laboratory test results. A Young's modulus is 10.3 GPa and a Poisson's ratio is 0.23. The in-situ stress, described before, is used as the initial state of stress in the analysis and it is assumed to be uniform in the domain. A numerical model is illustrated in Fig. 7.

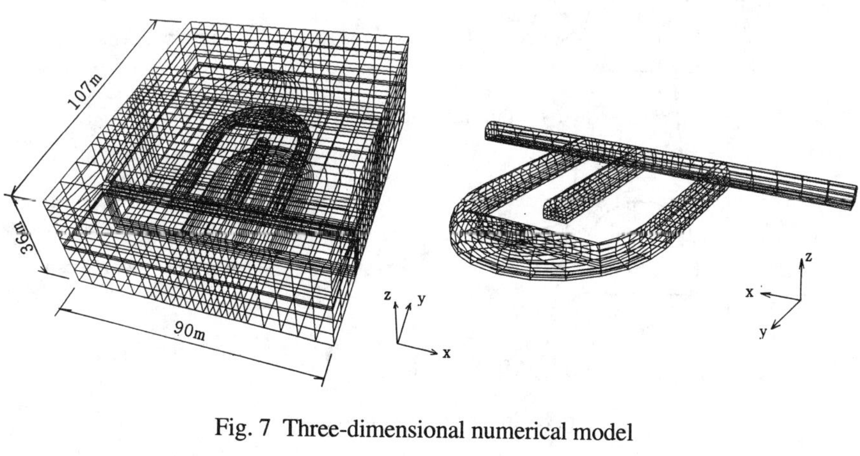

Fig. 7 Three-dimensional numerical model

6 COMPARISON AND DISCUSSIONS

For the calculation of stress changes, the quality of each measurement must be checked carefully. Although the assessment of monitored strains is still under way, the stress changes are estimated provisionally. The comparison between measurements and analytical results is presented in Fig. 8 and Fig. 9. Fig. 8 shows the two-dimensional principal stresses in a section with the distribution of major discontinuities in boreholes. The changes of stress components are compared at each strain cell location against the advance of the excavation face in Fig. 9. The close blasting to the No.3-cell location damaged the strain cell and concluded the monitoring.

As shown in figures, the stress changes estimated by numerical analysis are small because the stresses used as the initial state in the analysis are relatively low. Although there are many differences between measurements and analytical results, especially in the magnitudes, the numerical predictions are likely to agree with the tendency of in-situ stress changes. According to the measurement results, stress changes by excavation extend to wide area. Therefore, in consideration of loose zones around a new tunnel, the size of an opening will be estimated larger than a real tunnel size. Stresses do not concentrate, but are relieved by excavation in the vicinity of an opening. This is thought because of the influence of pervasive discontinuities around the opening. In future, careful assessment and processing for measured strains should be carried out again. Then, using the numerical model of stress changes, in-situ initial state of stress will be estimated from the stress change measurements as Zou and Kaiser (1990) proposed.

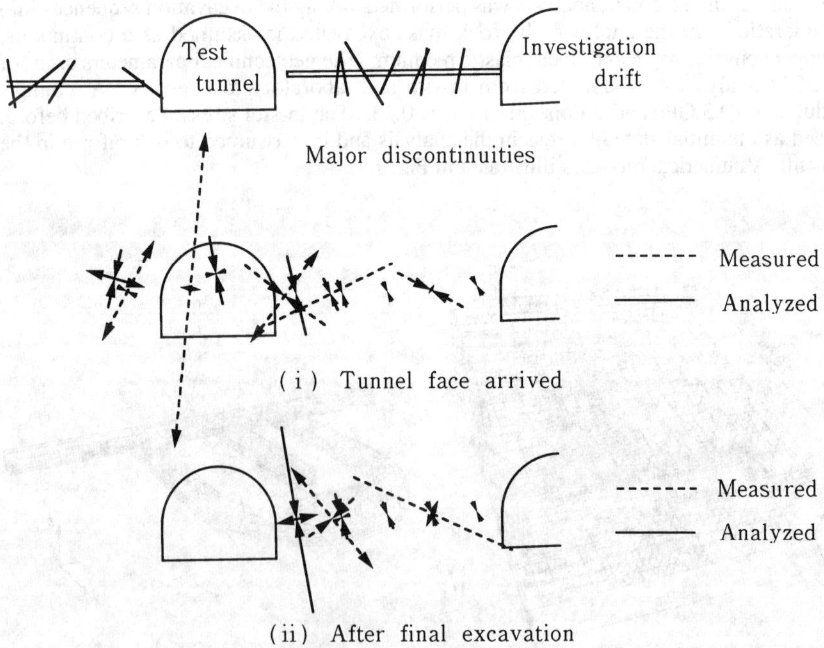

Fig. 8 Stress changes in section

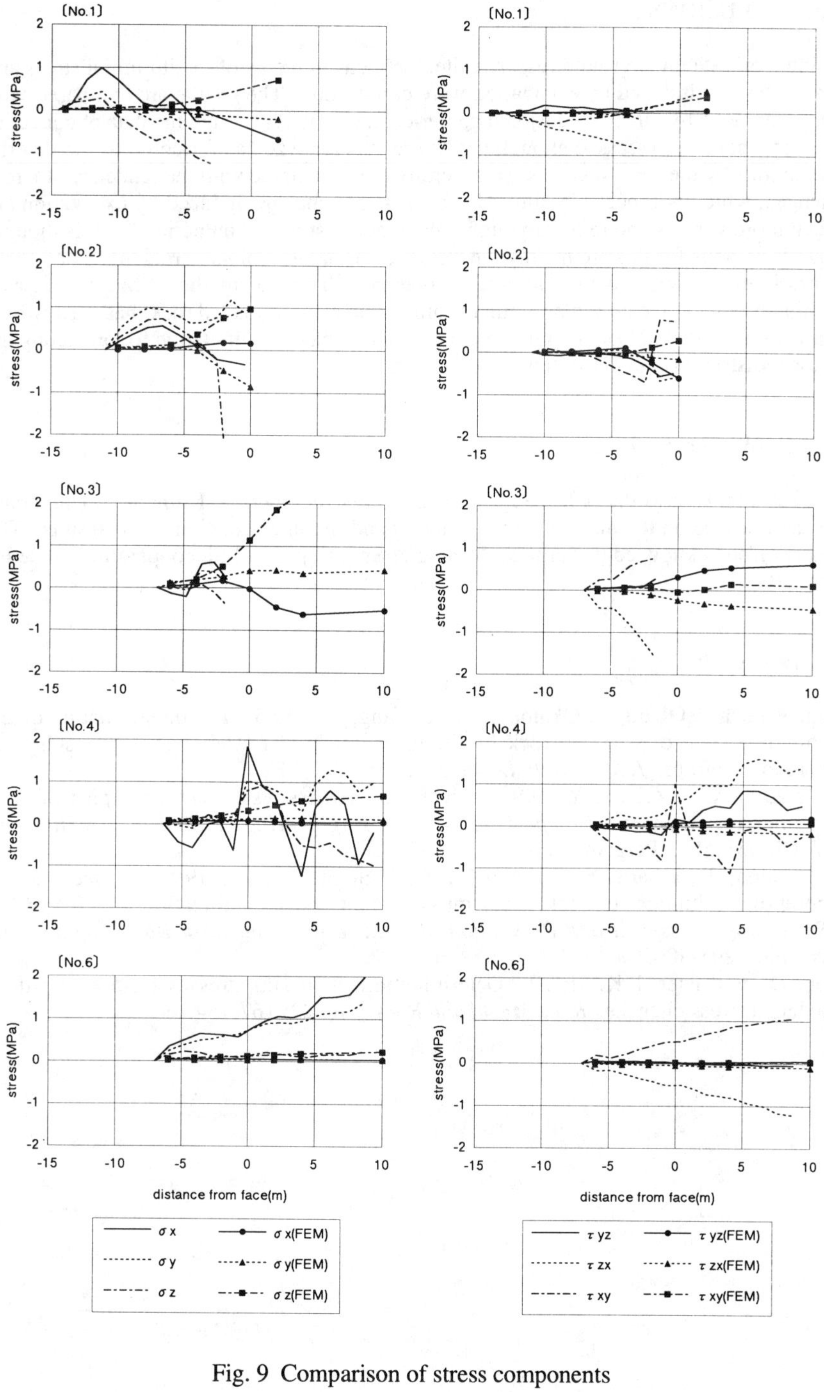

Fig. 9 Comparison of stress components

7 CONCLUSIONS

Using a private underground research site, the measurements of in-situ initial stress and stress changes induced by excavation were carried out. The results are compared with the predictions by three-dimensional numerical analysis. Although the data obtained in-situ have not been assessed enough yet, the results analyzed so far demonstrate that the predictions by the FEM analysis are generally likely to agree with the tendency of stress changes, which will occur in-situ. However, stress changes induced by excavation are much more various and relatively higher than those estimated numerically. It is thought that the reasons for this are mainly the influence of natural joints in rock mass and loose zones induced by excavation around an opening. This means the FEM analysis as a continuum medium can be a preliminary simulation even in jointed rock mass. However, this also suggests that continuum numerical analyses have the limitation for predicting the stress redistribution in discontinuous rock mass.

ACKNOWLEDGEMENT

The authors wish to thank Dr. Sugawara and other members of Kumamoto University for their advice on the measurements of stress and the interpretation of the results. The authors also acknowledge Kamioka Mining & Smelting Co., Ltd. cooperating in the in-situ tests.

REFERENCES

Sugawara, K. , Obara, Y. Okamura, H. & Wang, Y. 1985. The determination of the complete state of stress in rock by the measurement of strains on a hemispherical borehole-bottom. *J. Min. Metall. Inst. Jpn.*: 101: 277-282.

Sugawara, K. & Obara, Y. 1993. Measuring rock stress: Case examples of rock engineering in Japan. *Comprehensive Rock Engineering (Edited by J.A.Hudson).* Vol.3: 533-552. Pergamon Press.

Nakayama, T., Obara, Y., Sakaguchi, K. & Sugawara, K. 1993. Conical-ended borehole technique for rock stress measurement and its applications. *Proc. of Int. Symp. on Assessment and Prevention of Failure Phenomena in Rock Engineering,* Istanbul, 295-300.

Zou, D. & Kaiser, P.K. 1990. Determination of in situ stresses from excavation-induced stress changes. *Rock Mech. and Rock Eng.* 23: 167-184.

Rock Mechanics, Nelson & Laubach (eds) © 1994 Balkema, Rotterdam, ISBN 90 5410 380 8

First application of the RPR method to field measurements

Robert Corthésy, Maria Helena Leite, Guang He & Denis E.Gill
Department of Mineral Engineering, École Polytechnique, Montréal, Que., Canada
Paul Germain
Aur Resources, Val d'Or, Que., Canada

ABSTRACT: The RPR method for the doorstopper is a new concept in stress measurement interpretation models, as it allows to estimate an extra stress component by using the strain recovery curve geometry. This paper explains the principle on which the RPR method is based and reports the latest developments related to it. Results from laboratory stress measurement simulations allow the validation of the method and field measurement results using the RPR method are compared to results obtained with the CSIR triaxial cell.

1. INTRODUCTION

In situ stress measuring techniques are numerous, each of them having their own advantages and inconveniences. Techniques based on the recovery principle (usually called overcoring techniques) are the most commonly used. They can be divided in two groups, the first including triaxial cells, CSIR and CSIRO, which allow the calculation of the complete 3D stress tensor from a single measurement, without prior knowledge of the principal stress orientation. The second group includes borehole deformation gauges and CSIR doorstopper cells, which require a minimum of three non parallel boreholes to calculate the 3D stress tensor, when the orientation of the principal components is not known *a priori*. Contrary to what is generally believed, the two last techniques, when used in a single borehole do not allow the estimation of three far field stress tensor components, since each measurement yields three independent equations containing four unknowns, these being the three components of the far field stress tensor in a plane perpendicular to the borehole axis, σ_x, σ_y and τ_{xy} and the stress component parallel to this axis, σ_z. In this paper the x axis is horizontal, the z axis is the borehole axis and the y axis is normal to the xz plane. In order to determine the stress components in a plane perpendicular to the hole axis from a measurement in a single borehole, the far field stress component parallel to the hole axis must be known. Such is the case for conventional hydraulic fracturing stress measurements, where σ_1 is assumed to be vertical, parallel to the borehole and equal to the overburden pressure. Under these circumstances, borehole deformation gauges and doorstopper cells also yield the complete 3D stress tensor from a single measurement, but it should be emphasized here that σ_1 is not necessarily vertical and exactly equal to the overburden pressure, which in the authors view, greatly limits the applicability of hydraulic fracturing as a stress measuring technique.

This paper reports improvements to the original RPR method (Corthésy et al., 1994) which allows the calculation of σ_z and consequently σ_x, σ_y and τ_{xy} from a single modified doorstopper stress measurement.

2. ADVANTAGES AND INCONVENIENCES OF THE DOORSTOPPER TECHNIQUE

In highly fractured rock masses or when the stress field induces core discing while drilling proceeds (Corthésy et al., 1993a), only in situ stress measuring techniques requiring a short length of overcoring can be used. The doorstopper cell stress measuring technique (Leeman, 1965) is the only currently used stress measuring method showing this advantage.

The doorstopper has, as well, the following advantages: (1) loading the recovered core on which the doorstopper is glued in a biaxial Hoek type cell and conducting subsequently, on the same core, a diametral loading perpendicular to the direction of the major principal strain measured in the biaxial test, give the deformability parameters required in the stress calculation model, which ever it is; such a procedure reduces both the rock heterogeneity and the experimental errors effects to a minimum (Leite et al., 1993); (2) a simple stress calculation model allowing to take into account anisotropy combined with non linearity (Corthésy et al., 1993b) is available. Some stress measuring techniques allow strain or displacement recording while stress relief is in progress. The strain or the displacement recovery curves can then be used as a diagnose tool to evaluate the quality of the measurements by comparing them with theoretical strain recovery curves. This has been first suggested by Blackwood (1978). The standard doorstopper cell stress measuring technique does not have this advantage. In order to obtain strain recovery curves, to monitor rock-cell temperature variations and data acquisition stability while performing in situ stress measurements, a series of simple and inexpensive modifications to standard doorstopper cells have been devised at the Rock Mechanics Laboratory of École Polytechnique de Montréal (Corthésy et al., 1990). Concerning the inconveniences of the standard doorstopper technique, it requires that measurements be performed in three differently oriented boreholes in order to calculate the complete 3D stress tensor. The same holds for borehole deformation gauges but, the doorstopper will only require one hole diameter, which makes it less expensive in terms of drilling costs. As will be shown, using modified doorstopper cells and the RPR method, reduces the number of holes required from three to two, bringing the drilling costs, cell costs and delays very close, if not lower, than what they are for CSIRO and CSIR triaxial cell stress measuring techniques.

3. THE RPR METHOD

The RPR method is described in detail in Corthésy et al. (1994). As this paper focuses on the validation and application of the technique, it will only be briefly explained.

The following equations relate the far field stress components to the stresses measured at the borehole bottom.

$$\overline{\sigma_x} = H_1\sigma_x + H_2\sigma_y + H_3\sigma_z \tag{1}$$

$$\overline{\sigma_y} = H_1\sigma_y + H_2\sigma_x + H_3\sigma_z \tag{2}$$

$$\overline{\tau_{xy}} = (H_1 - H_2)\,\gamma_{xy} \tag{3}$$

where $\bar{\sigma}_x$, $\bar{\sigma}_y$ and $\bar{\tau}_{xy}$ are the stress components at the borehole end and σ_x, σ_y, σ_z and τ_{xy} are the far field stress components. H_1, H_2, H_3, are the stress concentration factors which are a function of the Poisson's ratio and which can be obtained from the following empirical relationships (Rahn, 1984):

$$H_1 = 1.32 + .08\nu\,(1 + \nu) \tag{4}$$

$$H_2 = -0.12 + 0.18\nu\,(1 + 2.55\nu) \qquad (5)$$

$$H_3 = -1.09(0.33 + \nu) \qquad (6)$$

If σ_z could be obtained, it would then be possible to calculate the stress components σ_x and σ_y in equations 1 and 2. The information necessary to determine σ_z, comes from the strain recovery curve and consists in finding the relationship that exists between the Recovered to Peak Ratio, RPR, and the Stress Ratio, SR. RPR(ν) is given by,

$$RPR = \frac{D_r}{D_p} \qquad (7)$$

where D_r is the mean recovered strain invariant in the measurement plane and D_p is the mean peak strain invariant obtained during the stress relief drilling as shown in Figure 1. The doorstopper cell having four strain gauges spaced at 45° from each other, two strain invariants are available, and the mean strain invariant should be used in equation 7. Since the RPR is a ratio, wether it is calculated with the mean strain invariant or the average strain from the four strain gauges does not make any difference. As for a given stress state, the stress concentration factors are known to be a function of the Poisson's ratio only, it is obvious this will also be the case for the RPR parameter. The stress ratio SR is given by,

$$SR = 2\frac{\sigma_z}{\sigma_x + \sigma_y} \qquad (8)$$

where σ_z is the far field stress component parallel to the hole axis and the sum σ_x + σ_y is the far field stress invariant in the measurement plane.

The relationship between SR and RPR gives the fourth equation required to completely determine the four unknowns in equations 1 to 3, without the need for extra strain gauges, boreholes or complicated modelling. It only requires using information already available on the strain recovery curve.

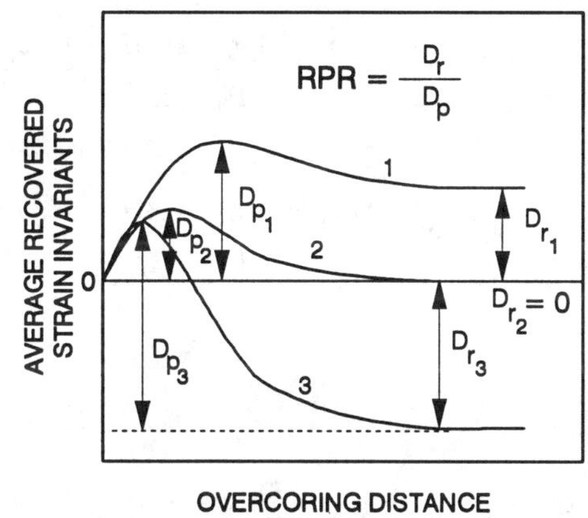

Figure 1: Method for calculating RPR

THE SR-RPR RELATIONSHIPS

Calculating strain recovery curves for doorstopper stress measurements using an analytical solution is extremely difficult. As powerful finite element codes are available on micro computers, it was found more practical to simulate stress measurements using numerical models and to validate the results by using physical models in the laboratory.

The COSMOS/M finite element code from Structural Engineering was used to obtain the strain recovery curves for different load cases and Poisson's ratios. Axisymetric analyses were considered sufficient as the ratio σ_x/σ_y was found to have no effect on the RPR parameter. This has been demonstrated by 3D finite element modelling, by physical modelling and also analytically by Corthésy et al. (1994). Combining various SR and ν values allows to plot the graph in Figure 2a and 2b, which can be used for calculation purposes. SR values for Poisson's ratios other than the ones plotted can be obtained by interpolation.

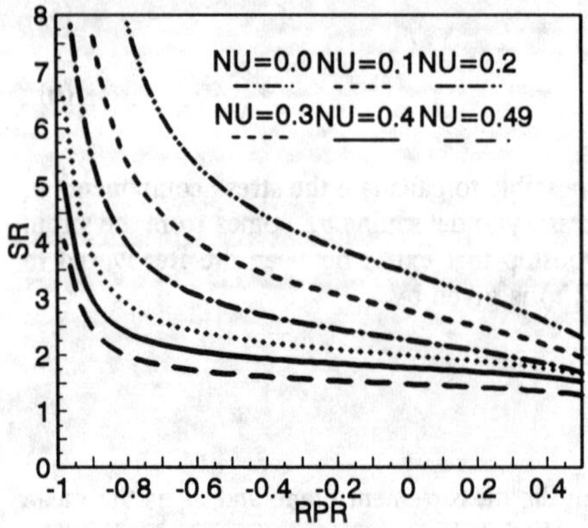

Figure 2a: RPR-SR relationships

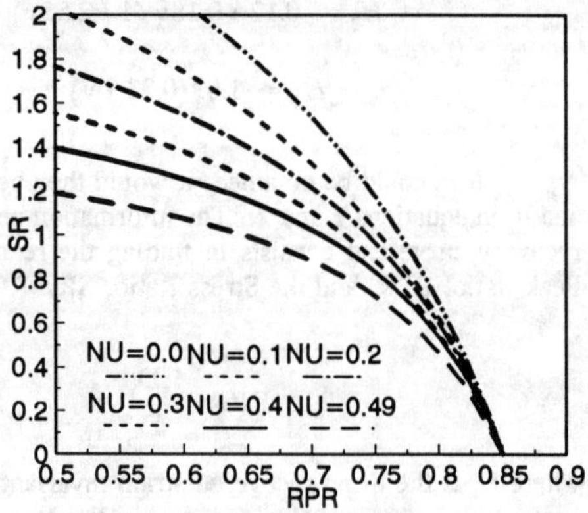

Figure 2b: RPR-SR relationships

CALCULATING THE AXIAL STRESS COMPONENT

From equations 1, 2, 3 and 8, the following system of simultaneous linear equations can be solved for σ_x, σ_y, τ_{xy} and σ_z.

$$\sigma_y = \frac{H_1\overline{\sigma_y} - H_2\overline{\sigma_x} - H_3(H_1 - H_2)\sigma_z}{H_1^2 - H_2^2} \tag{9}$$

$$\sigma_x = \frac{H_1\overline{\sigma_x} - H_2\overline{\sigma_y} - H_3(H_1 - H_2)\sigma_z}{H_1^2 - H_2^2} \tag{10}$$

$$\tau_{xy} = \frac{\overline{\tau_{xy}}}{H_1 - H_2} \tag{11}$$

$$\sigma_z = \frac{SR\ (\bar{\sigma}_x + \bar{\sigma}_y)}{2[(H_1+H_2)+(H_3\ SR)]} \tag{12}$$

One problem that arises when using equation 12, is that the numerator may equal zero. This is the case when $\bar{\sigma}_x = \bar{\sigma}_y = 0$ (RPR=0). This may happen, for example, in a vertical drill hole in undisturbed sedimentary rocks with a Poisson's ratio of about 0.35, where the in situ stresses are only cause by gravity loading. In such a situation, the horizontal stress components are equal and their effect, combined with the vertical stress component, induce a null stress state and consequently null recovered strains at the borehole bottom. σ_z must then be calculated independently from the recovered strains. Another problem with equation 12 is shown in Figure 3. The multiplication factor of $(\bar{\sigma}_x + \bar{\sigma}_y)$ in equation 12 is plotted for different SR and ν. The discontinuity in the curves represents the case where the denominator in equation 12 equals zero for certain combinations of ν and SR. Even when the denominator is not zero, the graph shows that small inaccuracies in the determination of SR may lead to erroneous values of σ_z, where the slope of the multiplication factor becomes steep. When faced with such

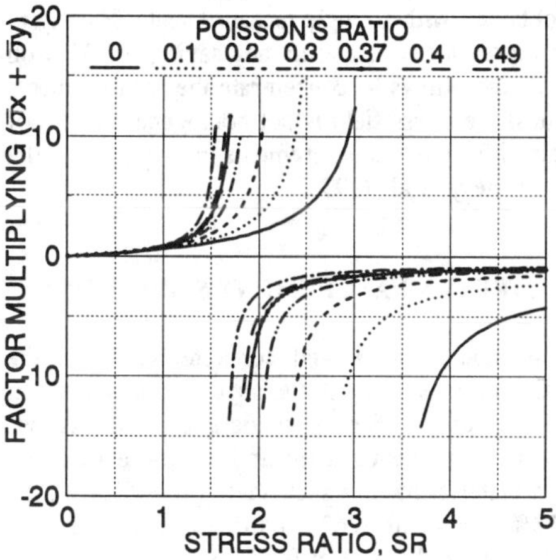

Figure 3: Graphical representation of Eq. 12

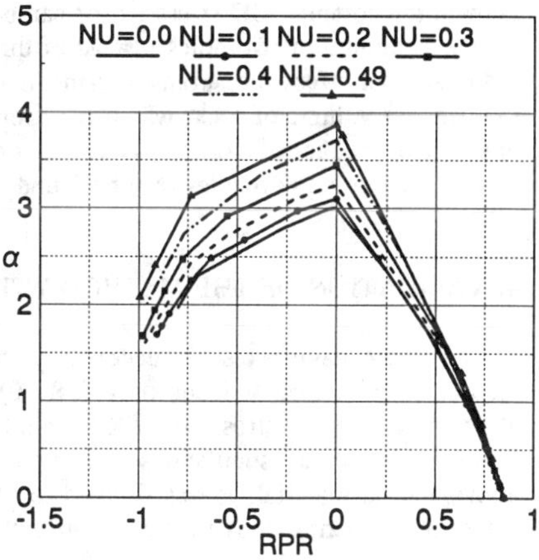

Figure 4: α-RPR relationships

situations (null numerator or ν-SR combinations in the steep slope areas of Figure 3), an alternate solution may be used. It consists in relating σ_z directly to the peak strain invariant without using equation 12. This requires the direct introduction of the Young's modulus of the rock in the relationship. The following equation may then be used,

$$\sigma_z = \alpha\,(\overline{\epsilon}_{px} + \overline{\epsilon}_{py})\;E \tag{13}$$

where α is a function of the Poisson's ratio and the RPR parameter as shown in Figure 4. $\overline{\epsilon}_{px}$ and $\overline{\epsilon}_{py}$ are the peak strains measured in μ strains. The Young's modulus, E, is expressed in MPa and the resulting stresses in Pa. The relationships between σ_z and the peak strain invariants were obtained by finite element analyses. Equation 13 applies for the critical zones identified in Figure 3. It should be used instead of equation 12 in such circumstances.

To illustrate the use of equation 13, let us suppose a measurement is performed in a vertical borehole in gravitational stress field. If one measures a peak strain invariant of $\overline{\epsilon}_{px} + \overline{\epsilon}_{py} = 350$ μ strains and the recovered strains are null (RPR=0), four stress components can still be calculated from a single measurement. Suppose the rock has a Poisson's ratio of 0.35 and a Young's modulus of 7 GPa. The factor α = 3.585 is obtained on the curve in Figure 4 for a Poisson's ratio of .35 and an RPR of 0. For the example given, as the recovered strains are null, RPR = 0, which for ν =0.35, leads to SR = 1.88, as determined from Figure 2b. From equation 13, one has σ_z = 3.585 * 350 * 7000 = 8.78 MPa. Once σ_z is known, σ_x and σ_y can easily be estimated using equation 8, leading to $\sigma_x = \sigma_y$ = 4.67 MPa. This corresponds to the particular case mentioned previously in a gravitational stress field. Having a vertical stress of 8.78 MPa gives a horizontal stress of $\nu/(1-\nu)$ * 8.78 = 4.72 MPa. The difference of 0.05 MPa is related to the fact that the Poisson's ratio corresponding to null stresses at the hole bottom in a gravitational stress field is not exactly 0.35, and also to the fact that the linear interpolation between Poisson's ratios of 0.30 and 0.40 is not very accurate as the relationship between α and ν is not linear.

3D STRESS TENSOR

Having four independent equations per hole, it is now possible to calculate the 3D stress tensor without prior knowledge of the principal stress orientation, with only two holes instead of the three holes required until now. If one principal stress orientation (not intensity) is known, the complete 3D stress tensor can be obtained with a single measurement. Having to drill a maximum of two holes instead of three represents time and money savings of 33% on a doorstopper stress measurement campaign. It also allows to concentrate the measurements in a smaller volume of rock, which is important if the stress field is not homogeneous. Proof of the independence of 6 of the 8 equations obtained from two measurements in two non parallel or non orthogonal boreholes, can be found in Corthésy et al. (1994).

4. VALIDATION OF THE METHOD WITH STRESS MEASUREMENT SIMULATIONS

After having tested successfully the method on rock and concrete blocks in the laboratory with biaxial loads only (SR=0) (Corthésy et al., 1994), an experimental setup allowing to vary the stress ratio SR from 0 to 5 was used to further validate the RPR method. Since the numerical simulations were performed by supposing a linear isotropic elastic and homogeneous material, it was decided to use a material having such ideal properties. Ciba Geigy 3969 casting epoxy resin mixed with 12% weight of 3269-1 hardener was used. The epoxy had a Young's modulus of 6.9 GPa and a Poisson's ratio of 0.37, as determined from uniaxial compression tests following ASTM standard D-3148. The model consisted in a 152 mm diameter by 355 mm long cylinder in which a 38 mm diameter borehole was drilled up to mid length. These dimensions allowed to consider the hole small in comparison with the dimensions of the cylinder which could adequately simulate an infinite medium. A standard doorstopper was used to monitor the borehole bottom strains but its diameter was reduced to 25 mm on a lathe so the stress relief drilling could be done with a 31 mm inner diameter bit. In order to be able to simulate a wide range of loading conditions with a single cylinder, the stress relief drilling process was done by increments of 1 mm at a time, between which the cylinder would be loaded biaxially in a triaxial cell up to a stress of 6.9 MPa and the axial stress would then be applied gradually up to a stress of 34.5 MPa, allowing to obtain SR ratios ranging from 0 to 5. During the loading process, the strains were continuously monitored with a multichannel data acquisition system. The strain recovery curves were obtained for a wide range of loading conditions by plotting the measured strains as a function of the drill bit advance.

The applied stresses could then be recalculated using equations 12 and 13. Figure 5 shows the results obtained by comparing the applied to the calculated stresses using both equations. When approaching the critical zone with a $\nu=0.37$ and SR=2, equation 13 gives better results, whereas outside this zone, equations 12 and 13 give comparable results, although one should avoid using equation 13 for RPR values smaller than -0.75 and greater than 0.50. The small differences observed between applied and calculated stresses arise from the fact that the curves in figures 2a, 2b and 4 were plotted from a limited number of points. More finite element analyses would allow a better definition of the relationships plotted in these figures.

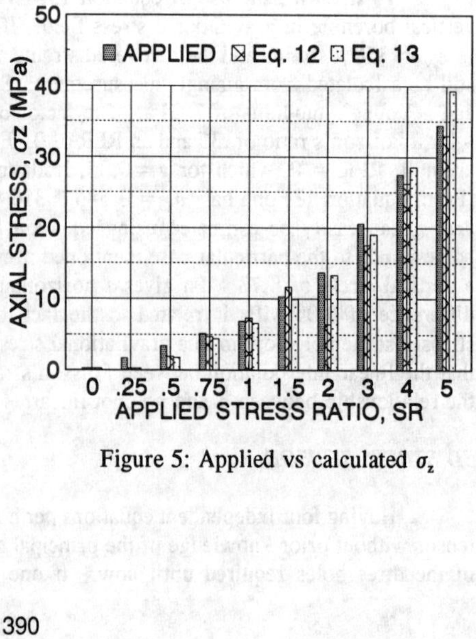

Figure 5: Applied vs calculated σ_z

5. FIRST FIELD MEASUREMENTS USING THE RPR METHOD

The RPR method was first tested in the field at the AUR-Louvicourt mine project in Val d'Or Québec, at a depth of 650 m. In order to test the results obtained, two parallel horizontal holes were drilled. A 75 mm hole (NX) was instrumented with modified doorstopper cell, and a 100 mm hole was instrumented with an AECL modified CSIR triaxial cell. The results obtained are summarized in Table 1. It should be added that the doorstopper measurements were interpreted using a model that accounts for the anisotropy and non linear elastic behaviour found in many rock types (Corthésy et al., 1993b) whereas the triaxial cell measurements were interpreted using a conventional stress calculation model.

Table 1

	sigma horizontal σ_x (MPa)	sigma vertical σ_y (MPa)	sigma // hole σ_z (MPa)	shear stress τ_{xy} (MPa)
Triaxial cell AECL modified	21.96	7.96	22.18	.59
Doorstopper RPR method	19.92	13.70	21.85	-.79

What comes out from these tests is that there is a very good match between the horizontal stress components and the stress components parallel to the hole axis (the one estimated using the RPR method). The only discrepancy is found for the vertical stress component. If one calculates the vertical stress from the overburden pressure, the value should be approximately 17.7 MPa considering a unit weight of .027 MN/m^3. On the other hand, regression analyses performed on stress measurements in the Abitibi mining district in Québec (Gill, 1990), indicate that the vertical stress component has a gradient of .02 MPa per meter, which would give a 13 MPa stress intensity. The doorstopper measurement gives a vertical stress intensity very close to this one. In the authors view, these results do not validate or invalidate the method, as only laboratory tests in a controlled environment can be used for this purpose. The results in Table 1 simply illustrate the applicability of the method to field measurements.

6. CONCLUSIONS

The paper showed that through improvements that can be done very easily, rapidly and at a relatively low cost, strain recovery curves, temperature and equipment stability monitoring can be obtained for the doorstopper cell. These curves give directly the total strains to be fed into the stress calculation models and they can be used as a diagnose tool. But the main advantage one gets from these strain recovery curves, is that they allow the calculation of four stress components from only one measurement in a single borehole, and the complete 3D stress tensor can be obtained from measurements performed in two non parallel or non perpendicular boreholes. Nevertheless, if one wishes to conduct measurements in three boreholes although now, only two are necessary, the extra equations given by the RPR method should be used to increase the quality of the calculated stresses.

Moreover, the RPR method is very simple to apply and the laboratory validation it has undergone indicates it is a tool that can be trusted. Improvements to the original calculation method of σ_z have been presented. These allow a more accurate evaluation of σ_z when certain combinations of ν and SR prevail. Although more field measurements will have to be performed with the RPR method to develop more confidence in its potential, the first results obtained in the field, presented in this paper, are very encouraging. Another field test program is already underway where three boreholes are used, giving the possibility to combine holes two by two, using the RPR method, and all three holes at a time with the conventional method, in order to calculate a 3D stress tensor. Comparing the results using two and three holes will give another indication of the accuracy of the RPR method when used in the field. Finally, the influence of anisotropy and non linear stress-strain relationships on the RPR method are currently being investigated.

ACKNOWLEDGEMENTS

The authors are grateful to the Natural Sciences and Engineering Research Council of Canada (grants CRSNG OGP0089752 and OGP0003412) for financing part of this research.

REFERENCES

Blackwood, R.L.,(1978). Diagnostic stress relief curves in stress measurements by overcoring. Int. J. Rock Mech. Min. Sci., Vol. 15, pp. 205-209.

Corthésy, R., Gill, D.E., Nguyen, D. (1990), "The modified doorstopper cell stress measuring technique", Proc. of the Conf. on Stresses in Underground Structures: Determination, prediction and monitoring, Ottawa, Canmet Editor, pp. 23-32

Corthésy, R., Gill, D.E., Leite, M H., Thompson, P. (1993a), Stress measurements in high stress zones using the modified doorstopper technique, Canadian Geotechnical Journal, Vol. 30, no 6, pp. 991-1002.

Corthésy, R., Gill, D.E., Leite M.H., (1993b). An integrated approach to rock stress measurement in non linear elastic anisotropic rocks, Int. J. Rock Mech. Min. Sci. and Geomech. Abstr., Vol. 30, no 3, pp. 395-411.

Corthésy, R., Leite, M.H., He, G., Gill, D.E. (1994), The RPR method for the doorstopper technique: Four or six stress compoenents from one or two boreholes. In print, Int. J. Rock Mech. Min. Sci.

Gill, D.E., (1990). Unpublished progress report submitted to the Ground Control Committee of the Québec Mining Association, 2 p.

Leeman, E.R.,(1965). The measurement of stress in rock, Part II: Borehole rock stress measuring instruments. Proc. Symp. on rock mechanics and strata control in mines, South African Inst. Min. Met., pp. 285-317.

Leite, M.H., Corthésy, R., Gill, D.E., (1993). Stress and strain measurement scales - Influence of micro-heterogeneity, Proceedings of the 24th APCOM, Montréal, Canada, pp. 277-284.

Rahn, W., (1984). Stress concentration factors for the interpretation of doorstopper stress measurements in anisotropic rocks. Int. J. Rock Mech. Min. Sci., Vol. 21, pp. 313-326

Rock Mechanics, Nelson & Laubach (eds) © 1994 Balkema, Rotterdam, ISBN 90 5410 380 8

In-situ stress and hydraulic fracture orientation in the mid-continent area, US

D.P.Yale
Mobil Research and Development Corporation, Dallas, Tex., USA

T.C.Ryan
Mobil Exploration and Producing U.S., Liberal, Kans., USA

ABSTRACT: Knowledge of the direction of hydraulic fractures can impact reservoir management in fracture stimulated reservoirs. Fracture direction can affect where wells are placed for optimum drainage (especially in-fill wells), how fracture treatments are designed, and how well patterns for EOR floods are designed. In-situ stress direction is especially important in the drilling of horizontal wells in order to determine the optimum direction of the wellbore with respect to wellbore stability and intersection of the wellbore with hydraulic and natural fractures. Reliable, cost-effective methods for determining fracture direction can aid production engineers and reservoir engineers alike in optimizing production from fields stimulated through hydraulic fracturing and/or developed with horizontal wells.

This paper details a study of in-situ stress and hydraulic fracture direction in the Hugoton, Panoma, and South Cutter fields in Kansas and Oklahoma. Data were collected in 6 wells across the field in the Chase, Council Grove, and Chester formations to determine in-situ stress and hydraulic fracture orientation. The azimuths of hydraulic fractures (determined from tiltmeter surveys and passive borehole seismic measurements) compared well with the orientation of maximum horizontal in-situ stress (determined from shear acoustic anisotropy and strain relaxation measurements).

Of the six wells studied, four suggest an azimuth of maximum horizontal stress between 90° and 98°. The data suggest that the east-west regional compressive stress trend for the southern mid-continent area (Zoback, 1992b) continues at least as far north as mid-Kansas. The orientation of stress in the other two wells appears to be affected by local faulting in the area. The data will be used to help choose in-fill well locations to maximize drainage and to determine the most advantageous orientation of horizontal wells.

1 INTRODUCTION

Because hydraulic stimulation fractures open normal to the least principal stress, most fractures are vertical and propagate in the direction of the maximum horizontal stress. This direction therefore influences the design of hydraulic fracture treatments, the drainage patterns around fractured wellbores, and the stability of horizontal or highly deviated wells. Knowledge of the direction of the in-situ stress is therefore important for optimal drilling, effective well completions, and efficient reservoir management.

The drainage pattern around fractured wells is highly anisotropic (see Figure 1). In-fill drilling is most often done to access undrained portions of the reservoir. However, placement of in-fill wells without regard to fracture direction can result in overlapping drainage areas and still leave large areas of the reservoir undrained. Reservoir performance can be optimized by accounting for fracture direction in the placement of in-fill wells (see Figure 1). Premature water or steam breakthrough during EOR floods can be due to fluid movement along fractures from an injector to a producer if the injectors and producers lie along a line which is near parallel to the fracture orientation. Characterizing the orientation of stress in the reservoir allows the engineer to forestall premature breakthrough and enhance sweep efficiency by changing the pattern of injecting and producing wells so that the fractures can act as line injectors and help fluid sweep across the field (i.e. injectors and producers lie along a line which is perpendicular to the fracture direction).

Wellbore stability in horizontal or highly deviated wells is enhanced by drilling the well parallel to the minimum horizontal stress direction. This allows the widest margin in mud weights between well collapse due to under-balanced drilling and well fracturing due to overbalanced drilling. Horizontal well stimulations also are controlled by the direction of the maximum horizontal stress. If an engineer wishes to create several transverse fractures in the wellbore, the horizontal well must be drilled parallel to the minimum horizontal stress direction

2 DESCRIPTION OF STUDY

This study was undertaken to determine the hydraulic fracture orientation for optimal in-fill well placement and horizontal well orientation. The permeability in Chase and Council Grove carbonate sequences and the Chester sandstone is low, and most wells are hydraulically fractured to make them economic. Because of the anisotropic drainage pattern around fractur stimulated wells (see Figure 1) and the close well spacing (80 or 40 acres), placement of in-fill wells to produce undrained portions of the reservoir depend on knowledge of the hydraulic fracture orientation. In addition, since horizontal wells in general will be completed with transverse (i.e. normal to the well direction) fractures, horizontal wells should be drilled in the direction of the minimum horizontal stress.

The location of the six wells is shown in Figure 2. The wells are in southwest Kansas and extreme northwest Oklahoma. The area is characterized by very gently dipping (less than 1°) strata at the formations of interest with very little faulting. Locations of two faults in the area are shown in Figure 2.

3 STRESS AND FRACTURE ORIENTATION METHODS

In-situ stress and hydraulic fracture orientation were determined from tiltmeter surveys, borehole microseismic measurements, shear acoustic anisotropy measurements, recovery strain

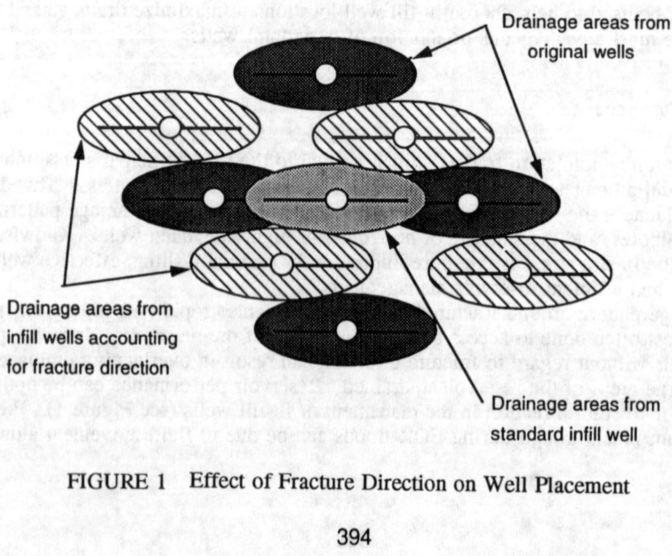

FIGURE 1 Effect of Fracture Direction on Well Placement

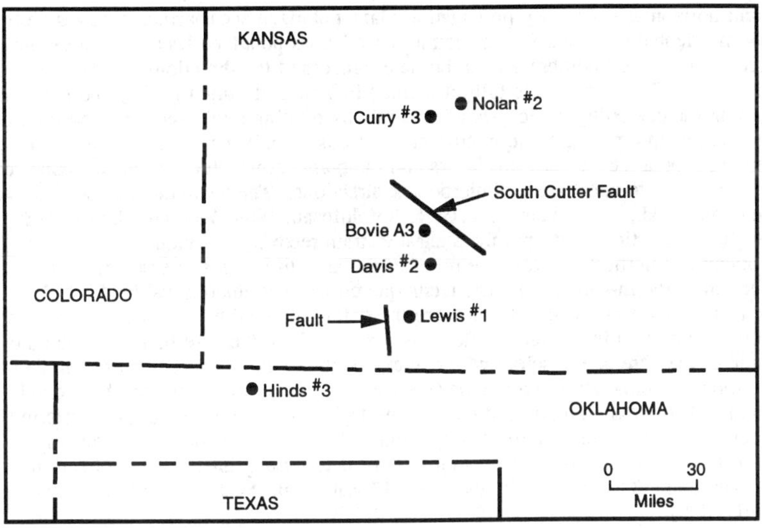

FIGURE 2 Well Locations for This Study

measurements and from the orientation of natural fractures. The first two techniques determine hydraulic fracture orientation as the fracture is created. The last three techniques predict fracture orientation by measuring the orientation of the maximum horizontal stress from tests on oriented core.

3.1 *Hydraulic fracture orientation methods*

The borehole microseismic technique is based on the observation that microseismic events apparently continue to occur several hours after cessation of pumping a fracture treatment due to the "closure" of the fracture (Schuster, 1978). The technique, describe in Schuster (1978) and Sorrels and Mulcahy (1986), utilizes an oriented three component geophone placed in the wellbore near the location of the hydraulic fracture to record the microseismic events after a fracture treatment has ceased. By polarization analysis of these events, a coherent pattern of signals emerges to define a fracture plane, giving both strike and dip. The geophone is usually oriented using a check shot at a known azimuth away from the wellbore.

The tiltmeter survey technique utilizes precision tiltmeters to sense the deformation of the surface due to the creation of the fracture at depth. Use of tiltmeter surveys is generally limited to shallow wells (<3000 ft depth) due to the inability to get good surface deformations from hydraulic fractures created deeper. A full description of the tiltmeter technique is given in Evans (1983) and Evans and Holzhausen (1983). For the Nolan #2, Hinds #3, and Curry #3 wells in this study, sixteen tiltmeters were arranged in a roughly 1000 foot circle around the wellbores tested. Surface tilts were measured and data analyzed using the method of Evans (1983).

3.2 *Stress orientation methods for prediction of fracture orientation*

The shear acoustic anisotropy technique for in-situ stress direction (Yale, 1987) is based on the premise that microcracks are generated in a core sample due to the relief of in-situ stress during coring. The relief of an anisotropic stress field leads to an anisotropic distribution of microcracks with more microcracks having strikes perpendicular to the direction of the maximum horizontal stress. By propagating plane polarized shear acoustic waves vertically through an oriented core sample between a pair of cross-polarized shear acoustic transducers, the direction of these microcracks can be determined and the direction of maximum horizontal stress inferred. The technique is fully described in Yale and Sprunt (1989) and is similar to determining the crystallographic axis of minerals by rotating a thin section between cross-polarized lenses in a petrographic microscope. In this case, however, a core sample is rotated between cross-polarized shear transducers and the points at which the acoustic signal reaches "extinction" (goes to zero) are the microcrack directions. The technique has been proven very reliable (Yale et. al., 1993; Yale et. al.1991; Rafolofosan, 1989; Yale and Sprunt, 1989).

As in shear acoustic anisotropy, the anelastic strain recovery technique relies on the development of microcracks from the relief of the anisotropic in-situ stress field to determine the direction of the in-situ stress. The technique utilizes extremely sensitive LVDT (linear voltage differential transducers) to measure the deformation with time of the core (in a number of different azimuths) immediately after it is removed from the core barrel. Assuming the expansion of the core is ellipsoidal with one axis vertical, the measured strains can be resolved into the three principal strain directions (vertical, horizontal minimum, and horizontal maximum). The maximum stress direction and hydraulic fracture direction is inferred to be the direction of maximum horizontal expansion. The technique, originally developed by Teufel (1983), has received widespread usage for determining in-situ stress direction (Teufel, 1984; Blanton and Teufel, 1983; Lacy, 1987; El Rabaa and Meadows, 1986; Ramos and Rathmell, 1989)

The oriented core from several of the wells was inspected for natural and coring induced fractures and the strike of these fractures in the horizontal plane was determined. Frequently the strike of natural fractures is oriented similarly to the strike of hydraulically induced fractures if the fractures are natural hydraulic fractures (i.e. no shear motion detected in the fracture plane) and the stress field direction at the time the fractures were formed is the same a the current day stress field direction. Drilling or coring induced fractures have been used mo recently (Ramos and Rathmell, 1989; Laubach et. al., 1992) as an indication of present day maximum horizontal stress direction and therefore hydraulic fracture direction.

4 RESULTS

The stress and fracture orientation data are summarized in the following table. The data in the table are divided by well and by stress/fracture orientation technique. The well average is a weighted mean of the data from the different techniques. High quality data was given twice th weight of fair quality data. Poor quality data was not included in the averages. The formatio tested is shown in parentheses below the well name. The formations tested are the C(ouncil) G(rove) carb(onates), the Ch(ase) carb(onates), and the Ch(ester) sand(stone). The data in eac table cell are presented as data mean ± standard deviation with the number of measurements shown in parentheses.

Data from four of the wells show the maximum horizontal stress and hydraulic fracture orientation coincide and are oriented between 90° and 98° azimuth. In the Hinds #3 well, the acoustic anisotropy, recovery strain and core fracture orientation agree well with each other . The recovery strain data in this well is rotated 20° to the north of the rest of the data but only five cores were tested. In the Nolan #2 well, the tiltmeter survey results agree well with the passive borehole seismic interpretation of the fracture direction. In addition, these data are similar to the stress and fracture orientation found in the nearby Curry #3 well where the predicted fracture direction from acoustic anisotropy agrees with the fracture orientation measured with the tiltmeter survey. It is interesting to note that the stress/fracture orientation i Chase carbonates at the northeast end of the study area (Curry and Nolan wells) is essentially the same as the orientation in the deeper Council Grove carbonates in the southwestern portio of the study area (Hinds well) and in the much deeper Chester sandstone in the central portio of the area (Davis well).

SUMMARY OF FRACTURE DIRECTION MEASUREMENTS
Hugoton/Panoma/South Cutter Fields, Southwest Kansas/Oklahoma

Method	Hinds #3 (CG carb.)	Nolan #2 (Ch carb.)	Curry #3 (Ch carb.)	Davis #2 (Ches sand)	Bovie #A3 (Ches sand)	Lewis #1 (Ch carb)
Acoustic Anisotropy	101°±16 (18)		94°±14 (19)	92°±22 (40)	130°±20 (30)	351°±11 (8)
Recovery Strain	82°±9 (5)					353°±23 (5)
Borehole Seismic		90°±8 (2)				
Tiltmeters	105°±6 (1)	90°±12 (1)	102°±6 (1)			
Core Fractures	95°±21 (9)					360°±5 (8)
Well Average	98°	90°	98°	92°	130°	354°

Tiltmeters results from the Curry #3 and Hinds #3 are excellent and from the Nolan #2 fair. This is somewhat surprising considering that the depth of the formations (near 3000 feet in all cases) is near the limit of where tiltmeters are generally used. The surface tilt characteristics of the fracture from all three wells show that the induced fractures were essentially vertical (dip angles in excess of 86°). Often in shallow reservoirs the least principal stress is vertical causing hydraulic fractures to be horizontal. The tiltmeter data shows that this is not the case in these reservoirs.

The acoustic anisotropy results from the five wells which were tested are very good, with slightly higher than usual scatter in the two wells penetrating the Chester sandstone. Yale et. al. (1993) and Laubach et. al. (1992) have suggested that the scatter in core based stress orientation techniques will be higher in areas of small differences between the minimum and maximum horizontal stress.

The apparent stress orientation in the area of the Lewis and Bovie wells is markedly different from the other four wells. Acoustic anisotropy results show a very strong southeast (130° azimuth) orientation in the Bovie well and acoustic anisotropy, recovery strain, and natural fracture data all suggest a north-south orientation near the Lewis well. These two wells, however, are both close to faults (see Figure 2). In the case of the Bovie well, the nearby South Cutter fault has an orientation of 135° azimuth. The Lewis well is near a smaller north-south trending fault.

5 DISCUSSION

The stress and fracture orientation in the Hinds #3, Nolan #2, Curry #3 and Davis #2 wells suggest that the maximum horizontal in-situ stress is oriented roughly east-west over a wide area of southwest Kansas and extreme northwest Oklahoma. The data show that the stress orientation is consistent with depth between 2500 and 6500 feet subsurface. This data is consistent with the general east-west compressive trend noted by Zoback (1992b) for the southern mid-continent region of the U. S. (see Figure 3). With the exception of one earthquake focal mechanism plot, Zoback's (1992a) extensive World Stress Map compilation does not include any stress orientation data further north than central Oklahoma (see Figure 3). The stress orientation data in this study, therefore, shows that the east-west mid-continent compressive trend extends at least as far north as mid-Kansas (see Figure 3).

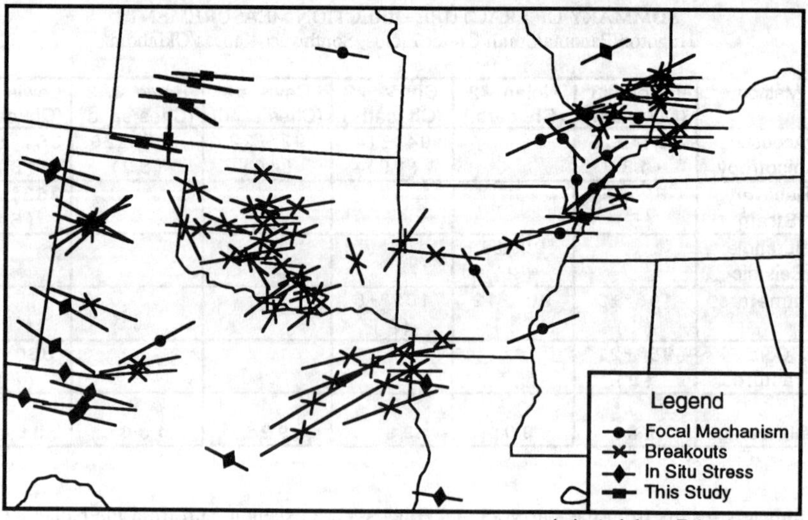

FIGURE 3

Orientation of Maximum Horizontal Stress, Mid-Continent Area, U. S.

5.1 *Rotation of stress orientation by local structure*

In the case of wells Lewis #1 and Bovie #A3, the local stress orientation appears to have been rotated by the presence of the nearby faults (see Figure 2). A number of authors have reporte on the apparent rotation of the local stress field near faults (Barton and Zoback, 1993; Yale et al., 1991; Aleksandrowski et. al., 1992). Laubach et. al. (1992) and Yale et. al. (1993) have suggested that the rotation of the orientation of the regional stress field near geologic structure and faults is often indicative of low horizontal stress contrast (i.e. the difference between the minimum and maximum horizontal stress). The tectonics of the mid-continent region (Zoba 1992b) and the magnitude and scatter in the acoustic anisotropy results from this study strongly suggest that the study area is a region of low horizontal stress contrast. This gives us additional confidence that the stress orientation has been rotated in the vicinity of the Lewis # and Bovie #A3 wells due to the presence of the nearby faults.

Laubach et. al. (1992) and Yale et. al. (1993) have also pointed out that rock fabric can sometimes affect core based measurements of stress orientation. Since only core based stress orientation techniques (acoustic anisotropy and recovery strain) were used in the Lewis #1 an Bovie #A3 wells, macro and microscopic rock fabrics were analyzed to determine if they may have affected the measurements. Macroscopic fractures were observed only in the Lewis #1 core but those fractures trended north-south and confirm the acoustic anisotropy and recover strain data. Yale et. al. (1992) and El Rabaa and Meadows (1986) have shown that fractures can sometimes rotate acoustic anisotropy and recovery strain data. However, they rotate the apparent stress orientation perpendicular to the macrofractures and therefore should not have influenced the north-south stress orientation found in the Lewis #1. Microscopic analysis of selected samples from both wells shows no mineral fabric which could have been responsible for the apparent measured stress orientation. This further suggests that the stress field was locally rotated in the vicinity of the Lewis #1 and Bovie #A3 wells.

5.2 *Comparison of stress orientation techniques*

This study also gives an opportunity to compare several of the stress orientation determination techniques currently used in the industry. The data quality and accuracy from all the techniques was good to excellent in the formations tested. The fact that the various techniques agree well with each other and across the area gives confidence in the techniques.

From a field engineers point of view, the shear acoustic anisotropy technique stands out from the other three techniques tested. In planning fracture treatments, horizontal well orientations, and in-fill well placements, it is convenient to know the orientation of the hydraulic fractures before a hydraulic fracture is created. This gives the predictive techniques (acoustic anisotropy and recovery strain) more value. In addition, both of these techniques are less expensive that the tiltmeter and passive borehole seismic techniques. In addition, it is often operationally more convenient to utilize a technique which is laboratory based and can be done at any point in the life of the core (acoustic anisotropy) than using a technique which requires well site testing of the core immediately upon removal from the wellbore (recovery strain).

However, there are several caveats to these conclusions. First off, it is always better to utilize two or more stress orientation techniques in any given field since no technique is always correct. In addition, although they are not predictive, the tiltmeter and passive borehole seismic techniques are closer to the "ground truth" than the core based techniques. We also should note that recovery strain technique is useful not only for stress orientation but often can be used to determine the magnitudes of the in-situ stresses as well.

6 CONCLUSIONS

- This study suggests that the regional maximum horizontal stress orientation and therefore hydraulic fracture direction is roughly east-west in southwestern Kansas and northwestern Oklahoma.

- In this area, local faulting can affect the orientation of stress field. Data suggest that the horizontal stress contrast in the area is low.

- Tiltmeter surveys, passive borehole seismic measurements, shear acoustic anisotropy measurements, and recovery strain measurements all work well in this area to determine stress orientation. From a field engineers point of view, shear acoustic anisotropy is the most useful.

- Knowledge of the orientation of in-situ stress and hydraulic fractures is needed for in-fill drilling decisions, hydraulic fracture treatment planning, and for optimal horizontal well orientation.

7 ACKNOWLEDGMENTS

The authors would like to acknowledge the help of J. Michael Rodriguez and Ron Moore in conducting the laboratory measurements. We would also like to thank Mobil Research and Development Corporation and Mobil Exploration and Producing U. S. for granting permission to publish this material.

8 REFERENCES

Aleksandrowski, P., Inderhaug, O.H., and Knapstad, B. (1992) "Tectonic structures and wellbore breakout orientation", in *Proceedings of the 33rd U. S. Symposium on Rock Mechanics,* Tillerson and Wawersik (ed.), Balkema, Rotterdam.

Barton, C. B. and Zoback, M. D. (1993) "Stress perturbations associated with active faults penetrated by boreholes: Evidence for near-complete stress drops and a new technique for stress magnitude measurement", *Jour. Geophys. Res.,* in press.

Blanton, T. L. and Teufel, L. W. (1983) "A Field Test of the Strain Recovery Method of Stress Determination in Devonian Shale", SPE paper 12304, presented at Eastern Regional SPE Meeting, Champion, Pennsylvania.

Daneshy, A. A., Slusher, G. L., Chisholm, P. T., and Magee, D. A. (1986) "In-situ Stress Measurements during Drilling", Jour. Pet. Tech., August 1986, pp.891-898.

El Rabaa, A. W. and Meadows, D. L. (1986) "Laboratory and field application of the strain relaxation method", presented at the 56th California Regional SPE Meeting, Oakland, California, March, SPE paper #15072.

Evans, K. (1983) "On the development of shallow hydraulic fractures as viewed through the surface deformation field: Part I - Principles", *Jour. Pet. Tech.*, February 1983, pp. 406-410.

Evans, K. and Holzhausen, G. (1983) "On the development of shallow hydraulic fractures as viewed through the surface deformation field: Part II - Case histories", *Jour. Pet. Tech.*, February 1983, pp. 411-420.

Lacy, L. L. (1987) "Comparison of Hydraulic Fracture Orientation Techniques", *SPE Formation Evaluation*, March 1987, pp. 66-76.

Laubach, S. E., Clift, S. J., Hill, R. E., and others, (1992) "Stress directions in Cretaceous Frontier Formation, Green River Basin, Wyoming", in *Wyoming Geological Association Forty-Third Field Conference Guidebook*, Mullen (ed.), Pioneer Printing, Cheyenne, Wyoming.

Rasolofosan, P. N. J., and Zinzner, B. (1989) "Effect of in-situ relaxation on velocity anisotropy of ultrasonic P and S waves in rock core samples", in *Rock at Great Depth*, ed. Maury and Fournointraux, Balkema, Rotterdam.

Schuster, C. L. (1978) "Detection within the Wellbore of Seismic Signals created by Hydraulic Fracturing", SPE paper 7448, presented at 53rd Annual SPE Meeting, Houston, Texas.

Sorrels, G. G. and Mulcahy, C. C. (1986) "Advances in the Microseismic Method of Hydraulic Fracture Azimuth Estimation", SPE paper 15216, presented at the SPE Unconventional Gas Technology Symposium, Louisville, Kentucky.

Teufel, L. W. (1983) "Determination of In-situ Stress from Anelastic Strain Recovery Measurements of Oriented Core", SPE paper 11649, presented at the SPE/DOE Joint Symposium on Low Permeability Gas Reservoirs, Denver, Colorado.

Teufel, L. W. (1984) "Determination of the Principal Horizontal In-situ Stress Directions from Anelastic Strain Recovery Measurements of Oriented Cores from Deep Wells: Application to the Cotton Valley Formation of East Texas", *Geomechanics,* vol. 57, pp.55-63.

Ramos, G. G. and Rathmell, J. J. (1989) "Effects of Mechanical Anisotropy on Core Strain Measurements for In-situ Stress Determination", presented at 1989 Society of Core Analysts Technical Conference, San Antonio.

Yale, D. P. (1987) "Method and apparatus for measuring the mechanical anisotropy of a material", U. S. Patent #4,713,968.

Yale, D. P. and Sprunt, E. S. (1989) "Prediction of Fracture Direction using Shear Acoustic Anisotropy", *The Log Analyst,* March-April 1989, pp.65-70.

Yale, D. P., Nieto, J. A., Rodriguez, J. M. R., and Moore, R. D. (1991) "In-situ Stress Direction from Shear Acoustic Anisotropy - A Case History from the Southern North Sea", presented at 14th Annual European SPWLA Conference, London, England, December, 1991.

Yale, D. P., Strubhar, M. K., and El Rabaa, A. W. (1993) "Determination of Hydraulic Fracture Direction, San Juan Basin, New Mexico", presented at SPE Production Operations Symposium, Oklahoma City, March, SPE paper #25466.

Zoback, M. L. (1992a) "First and second-order patterns of stress in the lithosphere: The World Stress Map Project", *Jour. Geophys. Res.*, vol. 97, pp.11703-11728.

Zoback, M. L. (1992b) "Stress field contraints on intraplate seismicity in eastern North America",*Jour. Geophys. Res.*, vol. 97, pp.11761-11782.

Rock Mechanics, Nelson & Laubach (eds) © 1994 Balkema, Rotterdam, ISBN 90 5410 380 8

The use of thermal strain relaxation of core to determine stress directions

N.A. Yassir & D.G. Robertson
Department of Earth Sciences, University of Waterloo, Ont., Canada

ABSTRACT: Thermal strain relaxation tests were carried out on core samples from southwestern Ontario and from the Paris Basin to assess the feasibility of the method in orienting horizontal stresses. The strain response for both sets of core was anisotropic and repeatable. Maximum horizontal stress orientations obtained for the Paris Basin core was NE-SW, corresponding to measurements using the differential strain curve analysis and anelastic strain recovery techniques. The value obtained for southwestern Ontario was ESE-WNW. The results from both areas are orthogonal to the known directions. This is discussed in the light of the influence of in-situ microcracks on core-based stress orientation techniques.

1 INTRODUCTION

Stress determination continues to be one of the most important topics of research in geomechanics. It has applications in numerous areas of study including directional drilling and fracture strategy in the petroleum industry,
rock squeeze in tunnels, structural geology and neotectonics.

The most reliable current methods of measuring stress magnitude and direction in situ are expensive, especially offshore. A great deal of research was therefore launched over the last two decades on the use of core to determine stress, cheaply and efficiently. The most prominent methods are briefly described below,

1.1 Direct observation

1. Core discing (Paillet and Kim 1987) is frequently observed, where retrieved core splits perpendicular to its axis, the surface of the discs being curved. Measurements on the orientation of the high side of the curve show that it is parallel to the minimum horizontal stress.
2. Centreline fracturing (Kull 1984, Dyke 1989) is also observed, where core shows a fracture running down its centre that seems to be consistently parallel to the maximum horizontal stress.

The methods described below are based on the assumption that the core was originally isotropic and that the release of stresses that were acting on the core results in microcracks that open against the maximum horizontal stress, i.e. parallel to the minimum horizontal stress.

1.2 Wave-properties

1. Sonic wave velocities (Yale and Sprunt 1989). Measurements of wave velocity show that core has anisotropic velocity. This was correlated with the presence of microcracks, so that the minimum velocity is perpendicular to those and therefore parallel to the maximum horizontal stress.

2. Shear wave birefringence (Crampin 1985, Yale and Sprunt 1989). Sonic shear waves are observed to split when going through microcracks so that there is a birefringence effect which further aids the determination of maximum stress.

1.3 Strain measurements

1. Anelastic strain recovery, ASR (Voight 1968, Teufel and Warpinski, 1984). This method requires the use of core within two days of its retrieval. The strain relaxation is measured in the core and the maximum strain direction is correlated with the maximum stress that had been acting on the material (i.e. the maximum stress is where the microcracks open the most). Attempts are made to correlate the strain with stress magnitude, with varying degrees of success. One of the reasons could be the above-mentioned simple assumptions which do not take previous rock fabric into account. Also, core contraction is sometimes observed in one or more direction(s) (Butterworth *et al.* 1991, Perreau *et al.* 1989), and its cause is poorly understood.

2. Differential strain curve analysis, DSCA (Strickland and Ren 1980, Dyke 1989). Older, but well-preserved core can be used in this method which requires that the core be cut into cubical samples and instrumented then placed in a cell and repressured hydrostatically. Again, maximum strain is correlated with maximum stress direction. This method is also used to estimate stress magnitudes.

3. Thermal strain relaxation (TSR) is the topic of this paper. It had hitherto only been tentatively attempted by a few other workers (Lacy 1987, Butterworth *et al.* 1991, Hagan, *pers. comm.* 1993). It is based on the assumption that core will "remember" the stresses that had acted on it, and that it will expand preferentially, showing maximum expansion in the direction of maximum stress.

There has been no systematic work to date on the thermal recovery method as a viable alternative to other core uses for stress orientation. The advantages of developing this method are that it does not require fresh core as with the ASR method, nor does the core need to be cut and re-stressed at high pressures as with the DSCA method.

This paper discusses the use of thermal strain recovery on core from the Paris Basin and southwestern Ontario.

2 SAMPLES AND METHODOLOGY

Two different core types were used in this study:

1. 4" oriented limestone core from the Paris Basin, France (well B, Perreau *et al.* 1989), provided by Elf Aquitaine.

2. 1.5" unoriented carbonate core from southwestern Ontario, Toronto region, (Pickering, OGS-83-3, 34 to 385.5m) provided by the Ontario Geological Survey.

The core sample was cut so that its top and bottom edges were flat. Three strain gauges were then glued to the sides at 120° from each other. The sample was placed in a special oven which was heated at 2°C/min intervals until a maximum of 100°C was reached. The rate of heating and the gauge type were compatible with the coefficient of thermal expansion of limestone (Clark 1966, Gangi, 1987). Two thermocouples, placed at the top and bottom of the sample, monitored the temperature difference between the core and the air in the oven. The core was placed on an insulating surface to allow the heat to migrate from the sides of the core towards the centre. Strain measurements were taken as soon as the top and bottom thermocouples were within one degree of the target temperature. Measurements were recorded at 10°C intervals. Maximum and minimum strain orientations were calculated following the method in Perreau *et al.* (1989).

Effects of moisture, temperature range and heating cycles were investigated in the preliminary study. The samples were dry and tests did not show sensitivity to moisture changes. The strain response to temperatures below 70°C tended to be homogeneous. Above this temperature, contraction was sometimes observed in one direction (Figure 1); this was also observed for the Elf core by Perreau *et al.* (1989).

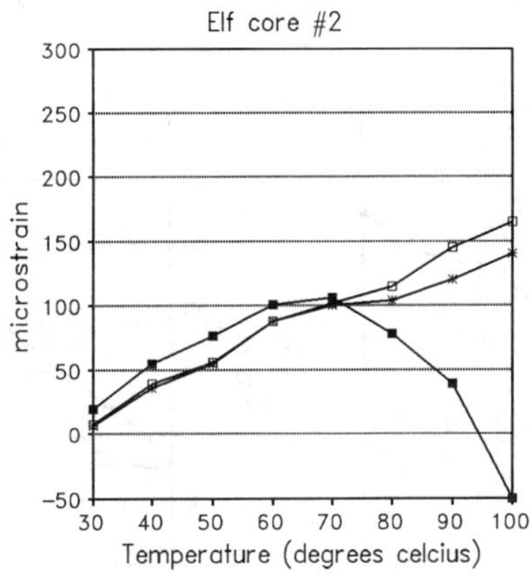

Figure 1. The thermal strain response in three orientations of Elf core #2.

The effect of cycling on the materials tested was also found to be small, giving repeatable results in later cycles. Figure 2 illustrates the variation in response over 4 cycles of heating of the same sample (Elf core #19). Interestingly, although there was variation in the amount of strain recorded, the orientations obtained were repeatable, with the exception of the last cycle (Table 1). Note again that the results in Figure 2 and Table 1 show the effect of temperatures greater than 70°C. Only data below 70°C were therefore used in computing the stress orientations in this study.

Table 1. Orientation of maximum strain on Elf core sample #19 at 10°C intervals. The calculated average angle is repeatable for temperatures below 70°C.

Temp (°C)	30	40	50	60	70	80	90	100	<70°	>70°
Cycle 1	64	61	63	62	59	55	51	40	62	48
Cycle 2	65	62	61	61	59	55	48	41	62	49
Cycle 3	58	63	64	67	67	69	68	68	64	68
Cycle 4	62	68	69	71	74	76	89	73	69	79

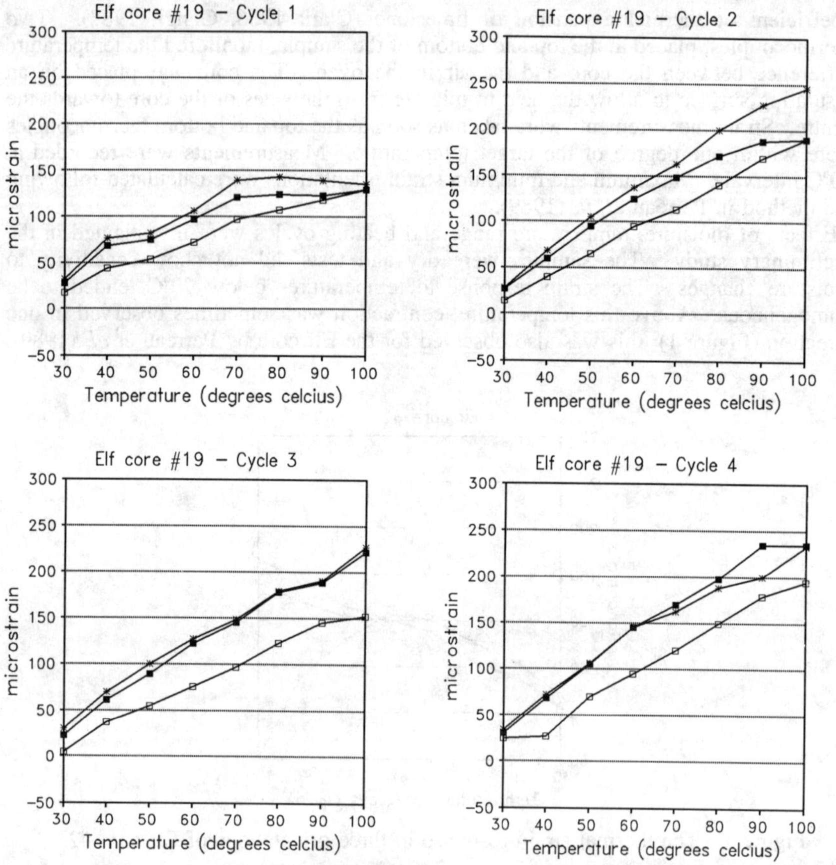

Figure 2. The thermal strain response in three orientations of Elf core #19 over four cycles. Note that the first cycle has less strain associated with it. The anisotropy of strain response is repeatable except in the fourth cycle at higher temperatures (see Table 1).

3 RESULTS

3.1 Elf Core

The results for the three Elf core samples are presented in Table 2.

Table 2. Average orientation obtained for maximum strain of Elf core (<70°C). Degree of reliability: H - high; M- medium.

Sample	1	2	19
Orientation	47 H	57 M	62 H

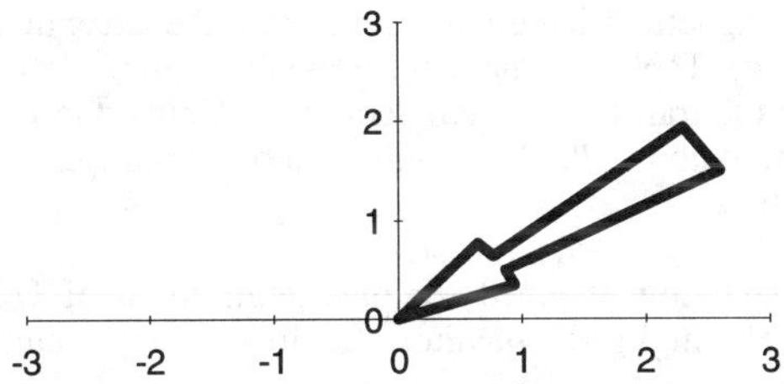

Figure 3. Rose diagram of maximum strain orientations for the Elf core. Data include DSCA and ASR results obtained by Perreau *et al.* (1989).

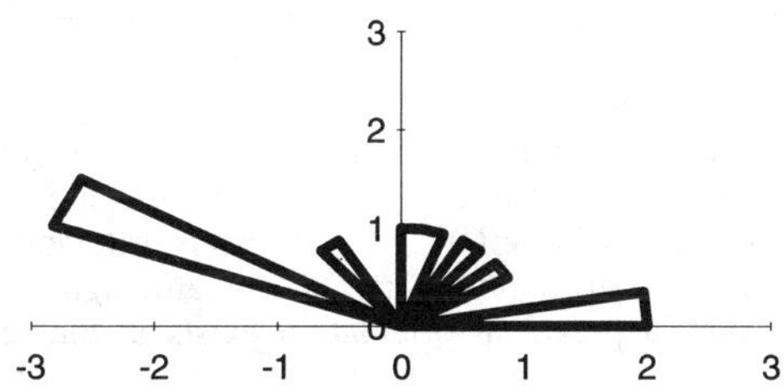

Figure 4. Rose diagram of all maximum strain orientations for the OGS core.

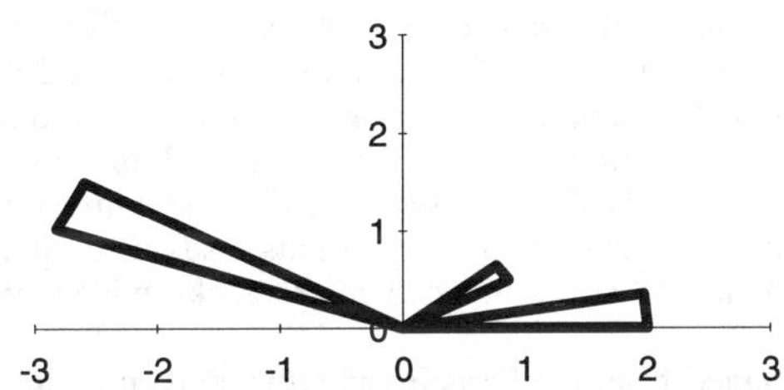

Figure 5. Rose diagram of medium and high reliability orientations for the OGS core.

The tests gave maximum strain directions in the NE quadrant (Figure 3). Interestingly, these results correspond well to test results obtained by Elf Aquitaine (Perreau *et al.*, 1989) on core samples from the same well using anelastic strain recovery and differential strain curve analysis (Figure 3). The angles of maximum recorded strain from north are approximately 50° and 58° respectively.

The results suggest that the orientation of the maximum horizontal stress is NE-SW. However, hydraulic fracturing tests in the Paris Basin indicate that the maximum stress is oriented NW-SE (Perreau, 1989), i.e. the samples are expanding in a direction orthogonal to the expected orientation.

3.2 OGS Core

This core had to be oriented using a cryogenic magnetometer. There was a good level of confidence in the magnetic orientation results, with the exception of samples # 34 (medium) and 278 (low), Table 3. The other tests given low reliability ratings (Table 3) showed cross-over of strains at low temperatures. Unlike the Elf core, no change in response was observed above 70°C, but only data below that temperature were used.

Table 3. Average orientation obtained for maximum strain of OGS core (<70°C). Degree of reliability: H - high; M - medium; L - low. Sample number is depth (m).

Sample	34	56.5	75	87.5	203	278	291.5	343	357.5	3855
Orientation	83 M	116 H	144 L	32 L	80 H	115 L	8 L	116 H	59 H	14 L

The orientations of maximum strain of all the data (low to high reliability) are presented in Figure 4 and the medium to high reliability data are presented in Figure 5. The dominant orientation is ESE-WNW. Again, this is orthogonal to the perceived orientation of maximum stress in southwestern Ontario, although there is considerable debate as to whether such a prevalent orientation exists in the region (Yassir and Dusseault, 1991).

4 DISCUSSION AND CONCLUSIONS

Core-based techniques rely on the premise that core develops microcracks upon unloading which are oriented parallel to the minimum stress and which open in the direction of the maximum principal stress. In that case, maximum strain directions should be the same as maximum stress directions. Both sets of core gave an anisotropic strain response to heating, however, with maximum expansion orthogonal to the expected orientation of maximum stress. This leads us to question whether the core is responding to in situ stress-induced microcracks which override effects of unloading.

The orientations obtained from the Paris Basin (Elf) core, although at 90° to the true orientation, corresponded with those from other techniques (ASR, DSCA; Perreau *et al.*, 1989), which is an encouraging result for the TSR method. Perreau *et al.* (1989) suggest that the discrepancy is caused by a natural fracture pattern in the area, oriented parallel to the maximum stress.

It is difficult to assess the results for the southwestern Ontario (OGS) core because the stress signal in the region is complex and there is little agreement on whether or not it is directional (Yassir and Dusseault, 1991). It is worth noting, however, that data from hydraulic fracturing and overcoring within the same depth range in Pickering and in Darlington, a site close to Pickering, gave a maximum stress orientation in the NE quadrant (Adams, 1987), orthogonal to the trend obtained in this study.

Another possible explanation for the 90° anomaly observed in our tests is thermal expansion causing closure of microcracks (Gangi 1987), which may explain the observed negative strain response in some samples (Figure 1). However, this does not explain the same orientation being obtained in the DSCA method, which requires core compression.

Although a great deal of valuable research has been carried out on core-based techniques, it is clear that there are some areas which merit closer attention. These include,

1. effects of lithology, and fabric on microcrack propagation during unloading (Charlez *et al.*, 1986; Gangi, 1987; Dyke, 1989).
2. effects of moisture; and
3. effects of sample shape. DSCA, for example, requires cubical samples; other techniques can be carried out on cylindrical core.

These issues become particularly pertinent when stress magnitude is being estimated by core-based techniques.

The preliminary results obtained in this study are encouraging, but too few to draw strong conclusions. More research is needed to develop TSR into a stress orientation technique.

ACKNOWLEDGEMENTS

This project was funded by a government of Canada URIF matching grant. The assistance of Drs. Fred Santarelli, Agip (formerly Elf) and Sebastian Bell, Geological Survey of Canada, is gratefully acknowledged. Thanks also to Dr. David Symonds at the University of Windsor for orienting the Ontario core.

REFERENCES

Adams, J. 1987. Canadian crustal stress data, a compilation to 1987. *GSC Open File Report* 1622.

Butterworth, S. R., Chroston, P. N., Davenport, C. A. and others 1991. Anelastic strain recovery of rock core from the English Midlands. Rock Mechanics as a Multidisciplinary Science, *Proc. 32nd U.S. Symp. Rock Mech.*, Roegiers (ed.): 55-62.

Charlez, Ph., Hamamdjian, C. and Despax, D. 1986. Is the micro-cracking of a rock a memory of its initial state of stress? *Proc. Int. Symp. on Rock Stress and Rock Stress Measurement*, Stockholm: 341-350.

Clark, S. P. 1966. Handbook of physical constants. *Memoir Geol. Soc. America* 97.

Crampin, S. 1985. Evaluation of anisotropy by shear-wave splitting. *Geophysics* 50: 142-152.

Dyke, C. G. In situ stress indicators for rock at great depth. PhD thesis, University of London, U.K.

Gangi, A. F. 1987. The variation of the thermal expansion coefficient of cracked and porous rocks with pressure and temperature. *Proc. 28th U.S. Rock Mech. Symp.*, Farmer *et al.*, (eds.): 391-398.

Kull, T. Coring and core analysis. PETEX, Austin, Texas. (From Dyke 1989).

Lacy, L. L. 1987. Comparison of hydraulic-fracture orientation techniques. *SPE Formation Evaluation*, March 1987: 66-76.

Paillet, F. L. and Kim, K. 1987. Character and distribution of borehole breakouts and their relationship to in-situ stresses in deep Columbia River Basalts. *J. Geophys. Res.* 92 (B7): 6223-6234.

Perreau, P. J., Heugas, O. and Santarelli, F. J. 1989. Tests of ASR, DSCA and core discing analyses to evaluate in situ stresses. *SPE paper* 17960: 325-336.

Strickland, F. G., and Ren, N-K, 1980. Predicting the in-situ stress state for deep wells using differential strain curve analysis. *Proc. 21st U.S. Rock Mech. Symp.*: 523-533.

Teufel, L. W. and Warpinski, N. R. 1984. Determination of in situ stress from anelastic strain recovery measurements of oriented core: comparison to hydraulic

fracture stress measurements in the Rollins sandstone, Piceane Basin, Colorado. *Rock Mechanics in Productivity, Protection*, Chapter 19: 176-185.
Voight, B. 1968. Determination of the virgin state of stress in the vicinity of a borehole from measurements of a partial anelastic strain tensor in drill cores. *Felsmechanik V. Ingenieureol* 6: 201-215.
Yale, D. P. and Sprunt, E. S. 1989. Prediction of fracture direction using shear acoustic anisotropy. *The Log Analyst*, March-April 1989: 65-70.
Yassir, N. A. and Dusseault, M. B. 1991. Stress trajectories in southwestern Ontario using wellbore breakour orientations. Geological Applications of Wireline Logs, *Proc. 21st U.S. Rock Mech. Symp.*, Roegiers (ed.): 83-92.

7 Mechanical properties of intact rock – Laboratory testing

Session developer:
C. Derek Martin, AECL Research

Rock Mechanics, Nelson & Laubach (eds) © 1994 Balkema, Rotterdam, ISBN 90 5410 380 8

Material-dependent size effect of hollow cylinder stability: Theory and experiment

P.J. van den Hoek & D.-J. Smit
Shell Research B.V., Rijswijk, Netherlands

M. Khodaverdian
Terra Tek Inc., Salt Lake City, Utah, USA

ABSTRACT: It has been experimentally observed in studies carried out on Castlegate and Berea sandstones that the size effect of hollow cylinder (HC) strength can depend on the material being tested. For tight Berea sandstone, a much weaker decrease in the HC strength was observed with increasing hole size (8 - 180 mm) than for friable Castlegate sandstone. This was especially noticeable for larger holes (30 - 180 mm). This paper presents the results of a theoretical and experimental study that aimed to explain these observations. Model calculations, incorporating splitting failure near the borehole wall, spall buckling and elasto-plastic hardening, reproduce the experimental results well for Castlegate sandstone and for small (8 - 30 mm) holes in Berea sandstone. However, large (30 - 180 mm) holes in Berea sandstone are not satisfactorily explained with this model.

HC strength tests with a large ratio of outer to inner diameter (6:1) and different hole sizes (16.5 and 39 mm) showed stable (but size- and rock-type-dependent) breakout structures. In all Castlegate samples and in Berea samples with a small hole (16.5 mm), these breakouts had a pointed cusped structure, whereas in Berea with a larger hole (39 mm), they had a flat-bottomed 'dog-ear' structure. Both the theoretical and experimental results suggest extensile failure for Castlegate and small-hole Berea, but shear failure for large-hole Berea. These different failure mechanisms are explained in terms of different spall strengths against buckling.

1 INTRODUCTION

The hollow cylinder (HC) collapse test is an important way to assess open-borehole or perforation stability during drilling, completion and production. Routine HC collapse tests are usually performed on samples with outer diameters between 25 and 50 mm and with outer-to-inner diameter ratios of between 3 and 4 (Antheunis, *et al.*, 1976a and 1976b). The HC collapse strength (i.e., the external load needed to cause a sample to fail without an inner-hole pressure being applied) is used as a measure for initial borehole or perforation failure in the field.

It is questionable whether the collapse of 8 to 15 mm holes is a good measure of the collapse of a 250 mm borehole. Size effects are commonly known to play a role in

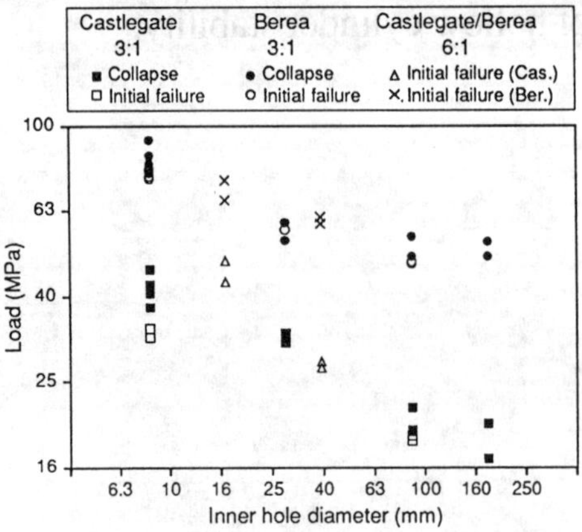

Fig. 1 Hollow cylinder collapse strength and initial failure: Experimental results

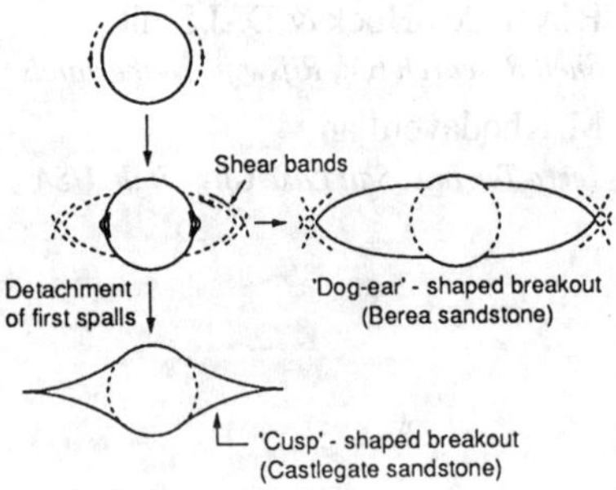

Fig. 2 Possible breakout shapes

failure, e.g., Brazilian tensile strength tests (Jaeger and Cook, 1971; Bazant *et al.*, 1991) and uniaxial compressive strength tests (Jaeger and Cook, 1971). The general trend is a decrease in the measured strength with increasing sample size. A size effect of HC strength was observed in limestone (Haimson and Herrick, 1988) as well as in friable (Kooijman *et al.*, 1991) and tight (van den Hoek *et al.*, 1992) sandstone. In all these experiments, HC strength decreased with increasing hole size. Moreover, the experiments on sandstone showed a material-dependent size effect: for friable (Castlegate) sandstone, the HC strength exhibits a much stronger reduction with increasing hole size than tight (Berea) sandstone. These results are summarised in Fig. 1.

It has been proposed that the observed difference between friable and tight sandstones may be due to different failure mechanisms (van den Hoek, *et al.*, 1992): in friable Castlegate sandstone, the formation and growth of borehole breakouts is governed by tensile failure, whereas in tight Berea sandstone it is determined by shear failure. These two mechanisms have different characteristic shapes associated with their failure, which are schematically depicted in Fig. 2. After initial localisation (either by splitting (e.g., Lee and Haimson, 1993) or shear, (e.g., Song and Haimson, 1993)), spalls start to be removed from two diametrically opposite breakouts (Fig. 2). In the case of extensile (splitting) failure, a stable cusp structure is finally formed (Maury, 1987; Zheng *et al.*, 1989; Ewy and Cook, 1990; Lee and Haimson, 1993); see Fig. 2. In this case, the breakout boundary has a convex curvature which ends in a pointed tip. Alternatively, if the removal of spalls is more difficult (e.g., because of inner-hole support pressure (Maury, 1987; Ewy and Cook, 1990)), shear bands may be formed (Song and Haimson, 1993), which then determine the final breakout shape ('dog-ear') with a concave curvature and a much less pointed tip than in the case of a cusp structure. Since shear failure is generally less size dependent than spalling failure, this may explain the difference observed between the failure in Berea and Castlegate sandstones.

This paper presents results of a theoretical and experimental study aimed at explaining the differences observed between failure strengths in friable Castlegate sandstone and tight Berea sandstone. Model calculations were performed to assess the extent to which splitting (tensile) failure (localisation) and buckling near the borehole wall in combination with elasto-plastic rock hardening prior to failure can account for the observed material-dependent size effect. Earlier modelling of the HC size effect concentrated on shear localisation using bifurcation analysis (Papanastasiou *et al.*, 1989; Tronvoll *et al.*, 1993). Experiments were performed using HC samples having a large ratio (6:1) between their outer to inner diameters (OD:ID) to observe growth of stable breakout structures without total collapse of the sample due to the proximity of the outer-boundary (Kooijman *et al.*, 1991; van den Hoek, *et al.*, 1992). The inner-hole diameters were 16.5 and 39 mm. The model calculations reproduce the experimentally observed size dependence of the HC strength for Castlegate sandstone well. For Berea sandstone, only the observed size dependence for smaller holes is reproduced by the calculations. For larger holes, the model gives a stronger size effect than experiment. These results suggest that in the case of larger holes of Berea sandstone, failure may not be of a mere splitting nature. This idea is supported by experimental observations of initial failure and progressive breakout growth for 'small' (16.5 mm) and 'large' (39 mm) holes, where cusped structures were observed for Castlegate and small-hole Berea sandstones, and dog-eared structures for large-hole Berea sandstones.

Sections 2 and 3 below discuss the modelling and experimental results, respectively.

2 MODELLING

An energy-balance argument is used to estimate the size-dependence of the HC initial failure load. Initial failure is assumed to start when it becomes energetically favourable to buckle a spall and create a splitting (tensile) crack parallel to the inner-hole wall; see Fig. 3. The onset of failure by this mechanism requires that the elastic energy of the spall exceeds the strain energy associated with the buckling stress. Below, we first present an "elastic-brittle" analysis of initial failure, i.e., failure in the absence of elasto-plastic rock hardening. In this case, a simple analytical expression can be derived for the failure load. The calculation is subsequently extended to include elasto-plastic rock deformation prior to splitting failure.

2.1 *Elastic-brittle analysis*

Consider a HC sample with inner radius a and outer radius b (Fig. 3). The sample is loaded from the outside by a pressure p with the inner-hole pressure at zero. For a certain value of p, a splitting crack parallel to the inner-hole wall is created at radius αa ($\alpha > 1$) following buckling of the rock slab (spall) between a and αa (see Fig. 3). The critical (buckling) stress σ_{cr} at which the spall starts to buckle is given by (Jaeger and Cook, 1971)

$$\sigma_{cr} = \frac{\pi^2}{12} E' \left[\frac{(\alpha - 1)a}{L}\right]^2, \qquad (1)$$

where E' is the plane strain Young's modulus and L is the length of the spall (see Fig. 3). Note that this stress only depends on HC sample dimensions through the aspect ratio of the spall (i.e., its thickness to length ratio). Experiments suggest that this aspect

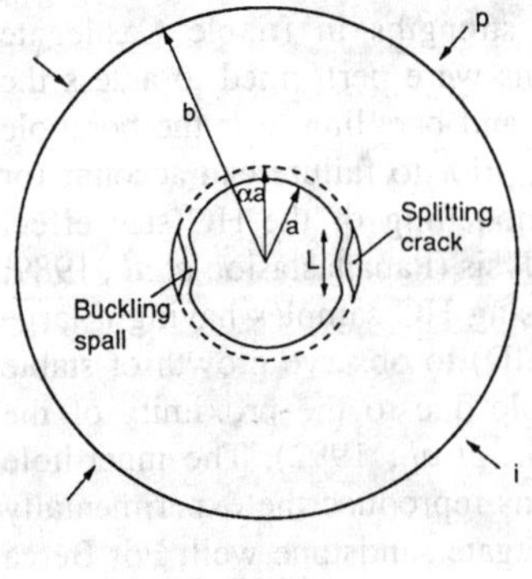

Fig. 3 Buckling and splitting crack growth in a hollow cylinder sample.

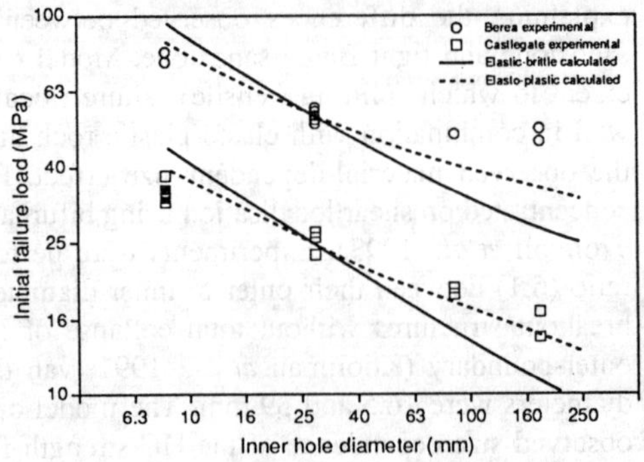

Fig. 4 Initial failure for a hollow cylinder test: Computed and observed size effect

ratio remains approximately constant with increasing HC sample size (Kooijman, 1991; van den Hoek, 1992). Therefore, it seems reasonable to assume σ_{cr} to be size-independent.

We now perform an energy-balance analysis to evaluate the outside pressure p_{cr} at which enough strain energy is available to cause buckling of the spall and growth of a splitting crack. For a purely elastic rock, the result is

$$p_{cr} = \left\{ \frac{1}{4}\left(1-\lambda^2\right)^2 \sigma_{cr}^2 + \frac{\alpha\left(1-\lambda^2\right)\left(1-\lambda^2\alpha^2\right)}{\alpha^2-1}\,\frac{K_{Ic}^2}{a} \right\}^{\frac{1}{2}}, \tag{2}$$

where $\lambda = a/b$ and K_{Ic} is the fracture toughness of the rock. In the derivation of Eq. (2), a radially symmetric distribution of the stresses and strains has been assumed for simplicity. From Eq. (2), one can see that for small HC sample sizes, the initial failure stress p_{cr} behaves as (sample size)$^{-1/2}$, whereas for very large samples, it levels off to a constant, non-zero value.

Figure 4 shows p_{cr} calculated according to Eq. (2) as a function of inner-hole diameter for Castlegate and Berea sandstone. In these calculations, it has been assumed that, for $a \to \infty$, initial failure takes place if the inner-hole stress exceeds the unconfined compressive strength (UCS), implying $\sigma_{cr} \approx$ UCS in Eq. (2). This corresponds to the elastic-brittle limit as observed in tunnels (Hoek and Brown, 1980). The value of α has been derived from the experimentally observed slab thickness, $\alpha \approx 1.06$. The fracture toughness K_{Ic} has been determined by calibrating Eq. (2) to experimental observations (Kooijman, 1991; van den Hoek, 1992) of initial failure for a = 14 mm. This gives $K_{Ic} \approx$ 2.5 MPa·min$^{1/2}$ for Berea and $K_{Ic} \approx$ 1.5 MPa·min$^{1/2}$ for Castlegate sandstone, which are reasonable values for rock. In Fig. 4, Eq. (2) is compared with experimental values of HC initial failure (copied from Fig. 1). For cases where no measured initial failure data were available, these values have been extrapolated from measured HC total collapse strengths, using an average experimental figure of 85% for the ratio: *initial failure to total collapse strength* (Kooijman, 1991; van den Hoek, 1992). As can be seen, the elastic-brittle formula of Eq. (2) gives a size dependence that is too strong for the range of inner-hole diameters of interest. Improvements in this result by taking elasto-plastic rock deformation into account are studied below.

2.2 *Elasto-plastic analysis*

In this analysis, we use a constitutive relationship based on Mohr-Coulomb with friction hardening as proposed by Vermeer and De Borst (1980). Rock strength parameters (cohesion and friction angle) have been derived from triaxial strength tests. The same energy-balance analysis as described in Section 2.1 is carried out numerically. The results are given in Fig. 4. Compared to the elastic-brittle results, the elasto-plastic results give a clear improvement. For Castlegate sandstone, the theoretical results give a good reproduction of the experiments. For Berea sandstone, however, there is only such agreement for smaller holes (up to about 30 mm). These results suggest that in the case of Castlegate sandstone, localisation may be of a mere splitting nature, whereas both splitting (small holes) and shear (large holes) localisation may prevail in Berea sandstone, which would lead to an overall weaker size dependence.

3 EXPERIMENTAL

HC tests were performed to check, in more detail, the possibility of different failure mechanisms (splitting and shear) operating in Castlegate and Berea sandstones. Two different hole sizes were used: 16.5 mm and 39 mm. The samples had a large OD:ID ratio (6:1) in order to minimise the influence of the outer boundary on breakout growth (Kooijman, 1991; van den Hoek, 1992).

The samples were loaded hydrostatically (0.2 MPa/min) up to about 60 MPa (Castlegate) or 80 MPa (Berea), when a stable breakout shape appeared to have formed which did not change much with further loading. During testing, breakout growth was monitored using an endoscope. After each test, the borehole of each sample was filled with a fast-setting grout (hydrocal) and the samples were cut with a saw perpendicular to the sample axis at several places along the cylinder. The grout served to avoid further damage to the borehole and preserve the breakout geometry during sawing.

The points of initial failure as observed with the endoscope are plotted in Fig. 1 as a function of hole size, together with results taken from Kooijman (1991) and van den Hoek (1992). Note that these references used an OD:ID ratio of 3:1, in which case total collapse usually takes place at pressures about 15% above the initial failure. The results show that with a 6:1 ratio, initial failure starts at somewhat higher external loads than with a 3:1 ratio. This obviously makes sense, since a decreasing OD:ID ratio will result in decreasing hole stability. Furthermore, it can be seen that 6:1 samples exhibit a similar size effect of initial failure as 3:1 samples.

Figures 5a and 5b show post-test breakout shapes for small holes (16.5 mm). A clear cusp shape can be seen in Fig. 5a (Castlegate sandstone). In Fig. 5b (Berea sandstone), one of the breakout 'wings' (the left one) is clearly cusp shaped, indicating a splitting failure mechanism. The right 'wing' in Fig. 5b has a less clearly defined shape: it may be either a cusp or a 'dog-ear'. Figures 5c and 5d show the post-test breakout shapes for large holes (39 mm). In the case of Castlegate (Fig. 5c), these shapes are still

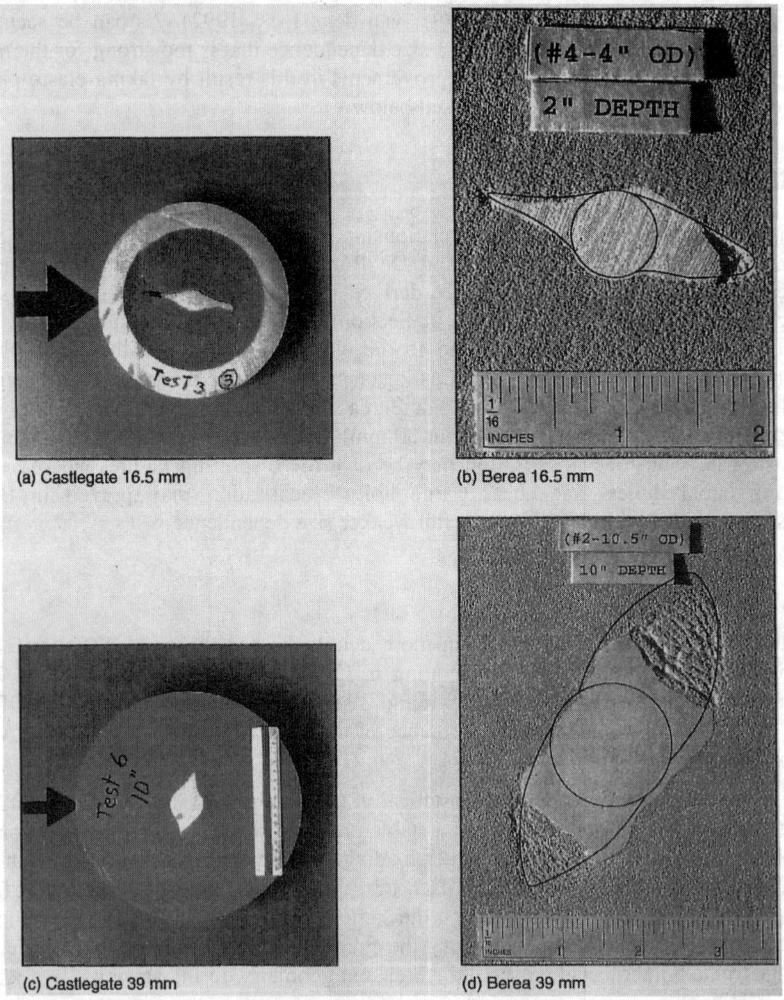

Fig.5 Observed breakout shapes

pointed (cusp), but in the case of Berea (Fig. 5d), two large, dog-ear-shaped breakouts can be identified around the borehole, with arrays of shear bands in the adjacent rock.

The above results indicate that for friable Castlegate sandstone in the absence of an inner-hole support pressure, breakout generation appears to be governed by a tensile (splitting) failure mechanism, leading to cusp-shaped breakouts (Figs. 5a and 5c). In tight Berea sandstone, however, breakout generation adjacent to 'small' boreholes may be governed by tensile failure (Fig. 5b), but adjacent to 'large' boreholes it appears to be governed by shearing (Fig. 5d). These results are consistent with the previously discussed differences between model calculations assuming tensile failure and experimentally observed points of initial failure for large-hole Berea samples.

The difference between Castlegate and small-hole Berea on the one hand and large-hole Berea on the other may be understood from differences in spall strength. In the large-hole Berea samples, a large number of spalls came loose from the breakouts upon relief of the external pressure, indicating that they had been tightly pinched between the breakout walls during loading. This was not observed in Castlegate and small-hole Berea samples. In the large-hole Berea case, these tightly pinched spalls provide a certain inner-hole support pressure, thus favouring a shearing mechanism, which is in line with earlier experimental results (Ewy and Cook, 1990). In the case of small Berea holes, the spalls appear to be more easily crushed, since their thickness is equal to only a small number of grain diameters. As a result, they cannot provide inner-hole support pressure. In Castlegate sandstone, the spalls are also crushed because of the generally friable nature of the rock.

CONCLUSIONS

A previously observed material-dependent size effect of HC stability (friable Castlegate versus tight Berea sandstone) can be explained in terms of different failure mechanisms at the borehole wall, i.e. tensile vs. shear failure. For borehole sizes in the tested range (8 - 180 mm), Castlegate appears to fail in a tensile mode. Model calculations assuming tensile (splitting) failure gave a good reproduction of the experimental results for initial failure at the borehole wall. HC experiments on Castlegate sandstone with a large OD:ID ratio showed development of stable, cusp-shaped breakouts, which is also an indication of tensile failure. In the case of Berea sandstone, there appears to be a transition from tensile failure to shear failure at a hole diameter of about 30 mm. Model calculations based on tensile failure can only reproduce observed trends of initial failure up to this hole size. Large OD:ID experiments indicated that cusp-shaped breakouts formed for small holes, but dog-ear shaped ones for large holes. In the latter case, tightly pinched spalls appear to provide inner-hole support pressure, thus favouring a shear failure mode.

ACKNOWLEDGEMENTS

The authors are grateful to the Shell Internationale Research Maatschappij for permission to publish this paper.

REFERENCES

Antheunis, D., Vriezen, P.B., Schipper, B.A. and van der Vlis, A.C., 1976. Perforation collapse: Failure of perforated friable sandstones, *Paper SPE 5750, Proc. SPE European Spring Meeting*. Amsterdam, April 8-9, 1976.

Bazant, Z.P., Kazemi, M.T., Hasegawa, T. and Mazars, J., 1991. Size effect in Brazilian split-cylinder tests: measurement and fracture analysis, *ACI Mater. J.* 88:325 May-June, 1991

Ewy, R.T. and Cook, N.G.W., 1990. Deformation and fracture around cylindrical openings in rock - II. Initiation, growth and interaction of fractures, *Int. J. Rock. Mech. Sci. & Geomech. Abstr.* 27:409.

Haimson, B.C. and Herrick, C.G., 1988. Borehole breakouts and in situ stress, *Proc. 12th Ann. ASME Energy-sources Techn. Conf. Drilling Symp.*:17 Houston .

Hoek, E. and Brown, E.T., 1980. *Underground excavations in rock,* The Institute of Mining and Metallurgy, London: Chapter 7.

Jaeger, J.C. and Cook, N.G.W., 1971. *Fundamentals of rock mechanics* (Chapman and Hall, London,) Ch. 7, and references therein.

Kooijman, A.P., van den Elzen, M.G.A. and Veeken, C.A.M., 1991. Hollow cylinder collapse: measurement of deformation and failure in an X-ray CT scanner and observation of size effect, in "Rock mechanics as a multidisciplinary science", *Proc. 32nd US Symposium on Rock Mechanics*: 657. Ed. J.C. Roegiers Rotterdam: Balkema.

Lee, M.Y. and Haimson, B.C., 1993. Laboratory study of borehole breakouts in Lac du Bonnet granite: A case of extensile failure mechanism, *Proc. 34th U.S. Symposium on Rock Mechanics*, Madison, Wisconsin, 28-30 June 1993.

Maury, V., 1987. Observations, recherches et resultats recents sur les mecanismes de ruptures autour de galeries isolees, *Proc. Sixth Int. Congr. on Rock Mech., Montreal*:1119 (in French) Rotterdam: Balkema.

Papanastasiou, P.C., Vardoulakis, I.G. and Santarelli, F. J., 1989. Modelling of borehole breakouts, *Proc. 12th Ann. ASME Energy-sources Technol. Conf. Drilling Symp.* (Houston, 1989): 49.

Song, I. and Haimson, B.C., 1993. Laboratory study of borehole breakouts in Cordova Cream limestone: A case of shear failure mechanism, *Proc. 34th U.S. Symposium on Rock Mechanics*, Madison, Wisconsin, 28-30 June 1993.

Tronvoll, J., Papamichos, E. and Kessler, N., 1993. Perforation cavity stability: Investigation of failure mechanisms, *Proc. Int. Symp. on Soil Mechanics, Athens*, September 20-23, 1993.

van den Hoek, P.J., Kooijman, A.P., Kenter, C.J., Khodaverdian, M., Hyland, C.R., and McLennan, J.D., 1992. Size-dependency of hollow cylinder collapse strength, *SPE 24800*.

Vermeer, P.A. and De Borst, R., 1984. Non-associated plasticity for soils, concrete and rock, *Heron* 29, 3.

Zheng, Z., Kemeny, J. and Cook, N.G.W., 1989. Analysis of borehole breakouts, *J. Geophys. Res.* 94:7171.

Rock Mechanics, Nelson & Laubach (eds) © 1994 Balkema, Rotterdam, ISBN 90 5410 380 8

Quantifying drilling-induced damage in samples of Lac du Bonnet granite

C. Derek Martin
AECL Research, Pinawa, Man., Canada

ABSTRACT: The laboratory properties of Lac du Bonnet granite indicated that either the *in situ* properties were changing with depth or that the samples were changing with depth. Comparison of P-wave velocities in the samples and *in situ* confirmed the latter, i.e., drilling-induced damage was affecting the laboratory properties. The strength of the damaged samples indicated that drilling-induced damage affected the cohesion of the material, not the frictional properties, and that the cohesion loss cannot be recovered by applying a confining stress.

INTRODUCTION

It is generally recognized that soils and softer rocks, i.e., shales and porous sandstones, are susceptible to drilling-induced damage and that this process affects their laboratory properties. However, little effort has been devoted to quantifying the effects of drilling-induced damage, i.e., sample disturbance, on brittle rocks such as granite.

The process of drilling a core sample from a stressed medium induces a stress concentration at the sampling point. When this stress concentration is sufficient, "microcracking", and/or "core discing", (an extreme form of drilling-induced damage) is observed (Obert and Stephenson 1965; Dyke 1989). *In situ* stress magnitudes generally increase with depth, consequently, one would suspect that samples of the same rock obtained at increasing depths should contain a higher density of microcracks. Thus the process of drilling-induced damage, which could ultimately lead to core discing, should be progressive and a function of the stress environment, i.e., the sampling depth. Geotechnical properties of laboratory specimens are affected by crack density/damage. In this paper we examine the evidence and the effects of stress-induced microcracks/damage, caused by sample drilling, on the laboratory properties of Lac du Bonnet granite samples obtained at depths ranging from near-surface to 1000 m.

SITE DESCRIPTION

The Underground Research Laboratory (URL) is owned by AECL Research and situated in southeastern Manitoba, about 100 kms northeast of Winnipeg. The URL is located within the Lac du Bonnet granite batholith, which is considered to be representative of many granitic intrusions of the Precambrian Canadian Shield. The URL shaft intersects two major thrust faults that dip about 25 to 30° southeast (Figure 1). These faults are

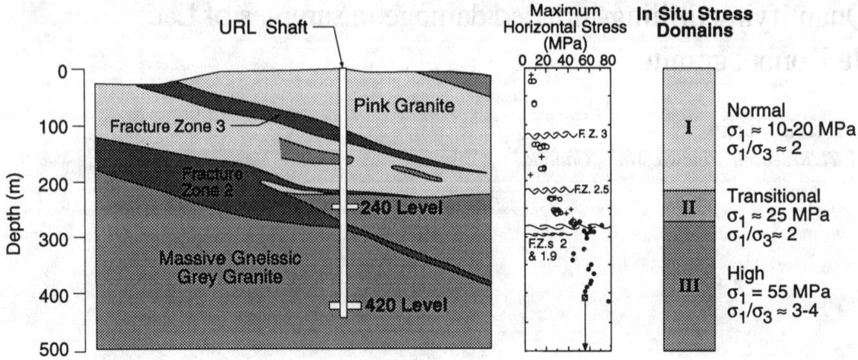

Figure 1: Geological section through the Underground Research Laboratory showing three stress domains.

referred to as Fracture Zone 3 and Fracture Zone 2 and their splays as Fracture Zones 2.5 and 1.9. The rock mass from the surface to Fracture Zone 2.5 is a mixture of pink and grey granite, and contains two or more regular joint sets, which decrease in persistence with depth. Below Fracture Zone 2 the rock is sparsely fractured grey granite, with no prominent joint sets.

The fracture zones divide the rock mass into several stress domains (Figure 1). Stress Domain I, which extends from the surface to Fracture Zone 2.5, exhibits stress magnitudes that are considered normal for the Canadian Shield for depths up to 200 m (Herget and Arjang 1990). In this domain, the *in situ* deformation modulus is about 30 GPa, and the maximum horizontal stress ($\approx\sigma_1$) increases with depth to about 20 MPa just above Fracture 2.5.

Stress Domain II serves as the transition from Stress Domain I to Stress Domain III. In this domain, the rock mass contains very few joints and σ_1 is about 25 MPa. Within Stress Domain III the rock mass contains no joints and the *in situ* deformation modulus is about 60 GPa. Stress measurements within this domain have been carried out to a depth of 550 m and show that there is little variation of the maximum horizontal stress magnitude with depth (Figure 1). The magnitude of σ_1 shown in Figure 1 for Stress Domain III is $\approx$55 MPa, and there is no evidence at present to suggest that it increases significantly below 550 m. The ratio of σ_1/σ_3, in this domain, is quite high. At the 420 Level this ratio is about 4, whereas in Stress Domain I, the ratio of σ_1/σ_3 is about 2.

Evidence for drilling-induced damage

Core discing

The most significant form of visual evidence for drilling-induced damage is core discing. Core discing has been observed at the Underground Research Laboratory in core samples ranging in diameter from 18 mm to 1.25 m. The thickness of the discs varies from a few mm to several cm depending on *in situ* stress magnitudes, e.g., a borehole drilled from an underground opening on the 420 Level displayed the thinnest discs near the opening. As the borehole advanced away from the stress concentration near the opening, the thickness of the discs increased. It should be noted that core discing is observed only in Stress Domain III and near fracture zones at the Underground Research Laboratory.

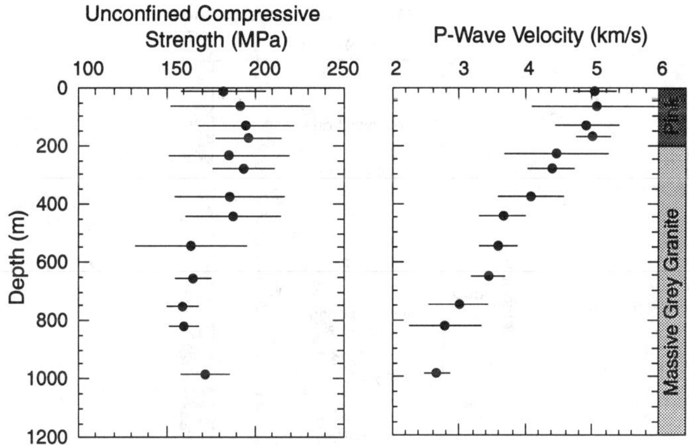

Figure 2: Laboratory properties versus depth, modified from (Jackson, Lau, and Annor 1989).

Mechanical properties versus depth

Samples were collected from various near-vertical boreholes drilled in the Lac du Bonnet batholith near the Underground Research Laboratory and were tested to establish the unconfined compressive strength and P-wave velocity (Figure 2). All samples were tested at ambient room temperature and humidity with a standard loading rate of 0.75 MPa/s (Jackson et al. 1989). It is evident from Figure 2 that either the *in situ* rock is changing with depth or that the microcrack density in the samples is simply increasing with depth. Based on P-wave velocities measured *in situ*, and discussed later, the latter is true.

P-wave velocity is sensitive to crack density, and hence is a good indicator of drilling-induced damage. In order to establish that the P-wave velocity trend observed in the laboratory samples in Figure 2 is present *in situ*, P-wave velocities from several *in situ* velocity borehole surveys carried out at the Underground Research Laboratory were compiled (Figure 3). The velocity surveys were carried out using a variety of techniques: standard borehole logging tool, cross-hole geophysics and acoustic emission/microseismic (AE/MS) survey. Figure 3 provides no evidence to suggest that the microcrack density *in situ* increases with depth. Consequently, it can be concluded that the samples taken at depth are disturbed.

Modulus

Young's modulus is very sensitive to crack density. The tangent moduli of samples from Stress Domain I and Stress Domain III were determined at confining stresses ranging from 0 to 60 MPa (Figure 4). The moduli were measured when the samples were loaded to about 50% of their ultimate load. The samples from Stress Domain I show no dependence of the modulus on confining stress, whereas the samples from Stress Domain III show a strong pressure-dependent modulus, as described by Santarelli and Brown (1987). However, the granite from Stress Domains I through III is all part of the same batholith and shows no significant variation in mineralogy or grain size. In addition, the samples were selected from the main mass of homogeneous granite, and hence it is concluded that the stress-

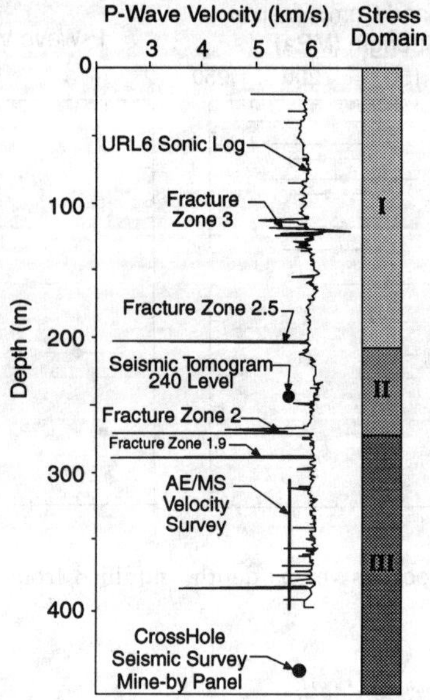

Figure 3: P-wave velocities *in situ.* The sonic log for borehole URL 6 shows a constant P-wave velocity with depth, except where fracture zones are encountered.

dependent modulus for the samples from Stress Domain III is an expression of the increased drilling-induced damage in the sample.

The moduli of the samples from Stress Domain III in Figure 4 remain somewhat lower than the results from Stress Domain I, even at confining pressures above 20 MPa. This would imply that the moduli of the damaged samples cannot be recovered by simply applying confining stress.

Compressive strength

To estimate the rock strength based on laboratory testing, it becomes important to establish whether or not drilling-induced damage affects the strength envelope. The samples were first divided into three groups, corresponding to the three *in situ* stress domains established at the Underground Research Laboratory: Stress Domain I (0 to 200 m depth), Stress Domain II (240 Level) and Stress Domain III (420 Level). The σ_1 stress magnitudes increased from ≈10 to 20 MPa in Stress Domain I to ≈55 MPa in Stress Domain III, suggesting that the stress-induced microcracking in the samples should increase with depth. The samples from Stress Domain I were further subdivided into two groups (pink and grey) to determine if the colour would indicate different mechanical properties. The samples from all domains were also screened for lithological uniformity to ensure that only samples of the medium-grained homogeneous granite typical of the main mass of granite at the Underground Research Laboratory were analyzed.

Triaxial test results for the four sample groups were analyzed using the Hoek-Brown failure criterion. The Hoek-Brown failure criterion is expressed in terms of σ_1 and σ_3 by

$$\sigma_1 = \sigma_3 + \sqrt{m\sigma_3\sigma_c + s\sigma_c^2} \qquad (1)$$

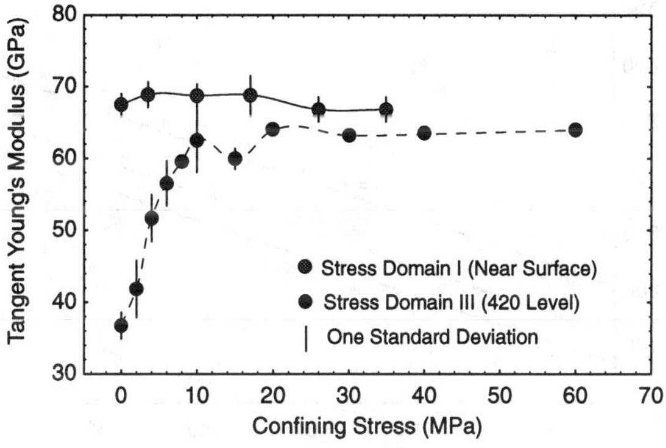

Figure 4: Tangent Young's modulus versus confining stress.

where σ_1 is the stress at failure, σ_3 is the confining stress, σ_c is the uniaxial compressive strength of intact rock and, m and s are empirical constants.

The Hoek-Brown failure envelopes for the four groups are summarized in Figure 5. As shown in Figure 5, there is essentially no difference between the two groups from Stress Domain I, indicating that the colour of the samples has no bearing on the strength. Note in Figure 5 that the samples from the 240 Level show a slight reduction in strength at low confining stresses, but that the entire strength envelope is reduced for samples from Stress Domain III regardless of the confining stress. Based on the relative shapes of these failure envelopes, it appears that the drilling-induced damage process is one that reduces the cohesion of the highly disturbed samples to about 70% of that of the undisturbed samples, but the frictional component of the failure envelopes is not affected.

This hypothesis of reduced cohesion is in keeping with the strength predicted from the sliding crack model of Cook (1965), which was shown by Martin (1993) to be given by

$$\sigma_{1_f} = 2\sqrt{\frac{2\alpha G}{c\pi(1-\nu)}}\tan\left(45+\frac{\phi}{2}\right) + \sigma_3\tan^2\left(45+\frac{\phi}{2}\right) \quad (2)$$

where α is the fracture surface energy, G is the shear modulus, c is the crack half-length, ν is Poisson's ratio, and ϕ is the angle of friction. Martin (1993) used the sliding-crack

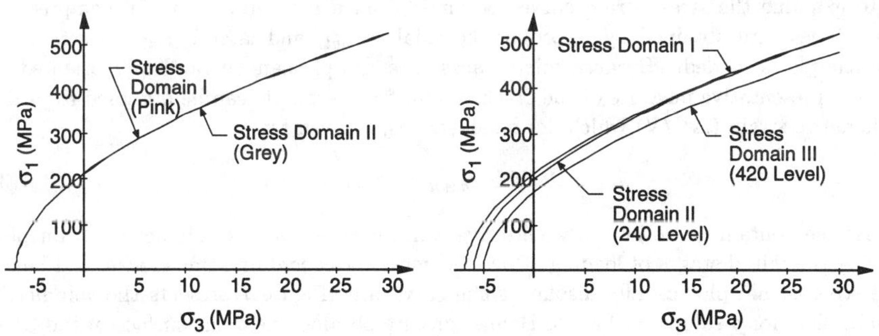

Figure 5: Hoek-Brown failure envelopes. Note the loss in cohesion for samples from Stress Domain III.

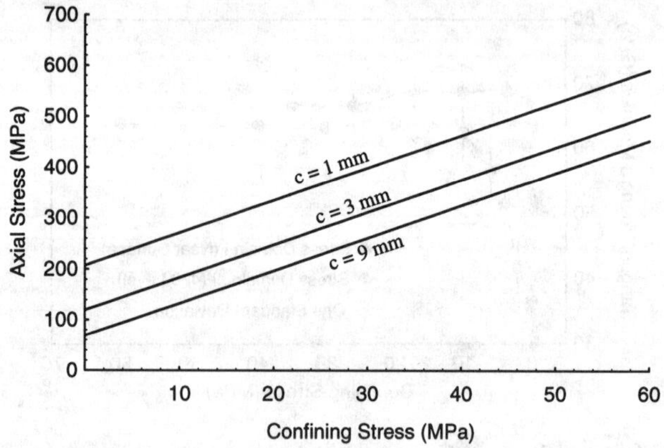

Figure 6: Predicted failure envelopes for Lac du Bonnet granite with different crack lengths using a sliding crack model. Note the increase in crack length only affects the cohesion and not the friction.

model to demonstrate that crack extension only affects the cohesion of the material and not the frictional properties. For example, if a crack length of 1 mm is used for the average grain size and the crack length is increased to 3 mm, approximately a 40% reduction in strength (cohesion) is realized (Figure 6). It is also worth noting that a further 3-fold increase in crack length to 9 mm only results in a further decrease in cohesion of about 25%. Thus initial crack extension is significant, even as little as 1 or 2 mm.

DISCUSSION

The data presented in the preceding sections illustrate that the laboratory properties of Lac du Bonnet granite can be correlated to the various stress domains established for the URL and that the properties measured in the laboratory are not representative of the *in situ* properties even when the rock mass *in situ* is massive and homogeneous. Thus, it would be extremely useful to know when the properties of a sample taken at depth are likely to be affected by drilling-induced damage.

One of the simplest methods to determine the possible extent of sample disturbance is to examine the stress-strain curves obtained from a laboratory uniaxial compression test. These are obtained by recording the axial (ϵ_{axial}) and lateral ($\epsilon_{lateral}$) strains as the sample is loaded. Because microcracks in a sample can be randomly oriented, a more representative measure of the crack volume in the sample can be obtained from the volumetric strain (ΔV/V) which, for small strains, is given by

$$\Delta V/V \simeq \epsilon_{axial} + 2\epsilon_{lateral}. \tag{3}$$

A sample containing no cracks would give a linear stress versus volumetric strain plot during the initial stages of loading. However, most rocks contain some cracks, and hence the stress-strain plot usually displays some curvature. Figure 7 presents the volumetric strain plots for a sample of Lac du Bonnet granite obtained near the surface of the URL and a sample taken from Stress Domain III at a depth of 420 m. Figure 7 illustrates the significant crack closure at low stresses associated with the deeper sample suggesting that this sample, relative to the near-surface sample, contains a significantly greater volume of

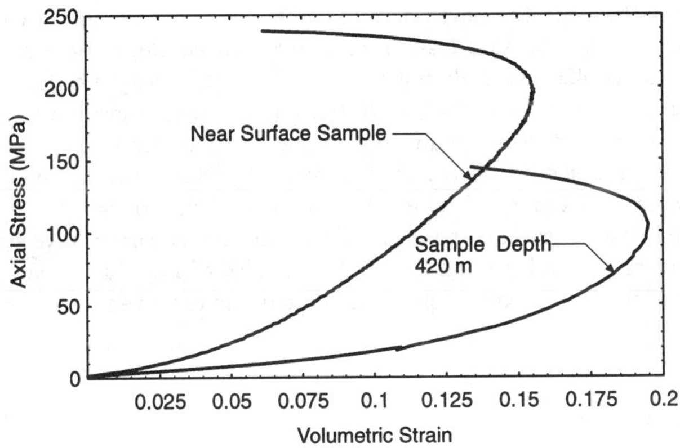

Figure 7: Measured volumetric strain for a near-surface sample and a sample taken from Stress Domain III at a depth of 420 m. Note the large amount of crack closure in the early stages of loading for the deep sample.

cracks. This simple comparison may be used as a first step in determining whether or not drilling-induced damage is present in samples taken at depth.

An alternative to the stress-strain approach is to consider under what compressive stresses cracking occurs. Griffith (1924) introduced the concept that local tensile stresses could exist in a material subjected to all-around compression if the material contained flaws. Intact rocks contain numerous flaws, e.g., grain boundaries, cleavage, vugs, microcracks, and hence should develop "Griffith" cracks if subjected to sufficient stress concentrations. The stress required to initiate cracking usually occurs at about 30 to 50% of the peak strength as reported by Brace *et al* (1966) and Bieniawski (1967) for a variety of rock types. Martin (1993) demonstrated that cracking is also first observed in Lac du Bonnet granite at about 30 to 40% of the failure load. This crack-initiation stress can be approximated using the Griffith theory of fracture given by

$$(\sigma_1 - \sigma_3)^2 = 8\sigma_t(\sigma_1 + \sigma_3) \tag{4}$$

where σ_t is the tensile strength of the rock. In Equation 4, crack initiation at $\sigma_3 = 0$ occurs when the maximum applied stress (σ_1) is 8 times the tensile strength (σ_t). Hence, for undisturbed Lac du Bonnet granite, with a $\sigma_t = 8$ MPa, stress-induced cracking should occur at about 64 MPa. This is in reasonable agreement with the crack initiation stress of ≈70 MPa determined by Martin (1993) for Lac du Bonnet granite at the Underground Research Laboratory.

Conclusions

The preceding discussion raises the important issue that drilling-induced damage in a brittle rock can be a significant problem when trying to establish *in situ* rock properties at depth. The results from our testing program show that the strength, deformability, and P-wave velocity decrease as the samples are drilled from rock at increasingly high *in situ* stresses, i.e., increasing depth. The uniaxial compressive strength was reduced by nearly 30%. What is most significant is that the cohesion of the damaged samples cannot be recovered by applying a modest (10 MPa) confining stress, and if the damage is severe, as in the samples from the 420 Level, no reasonable (30 MPa) amount of confining stress can compensate for the cohesion loss. Hence, strength envelopes of damaged samples, i.e., samples from the 420 Level, do not reflect the *in situ* strength of intact rock. Thus in order to estimate the *in situ* strength of the Lac du Bonnet granite at the 420 Level of the Underground Research Laboratory, samples with a minimum amount of stress-induced crack volume should be tested, i.e., samples from Stress Domain I. It is also important to note that damage begins when the maximum stress concentration at the point of sampling is greater than about 1/3 of the unconfined compressive strength. Also, the stress-dependent modulus observed in the samples of Lac du Bonnet granite from Stress Domain III is only an expression of drilling-induced damage rather than an intrinsic property.

ACKNOWLEDGEMENTS

Funding for this work was provided by the Canadian Nuclear Fuel Waste Management Program jointly funded by AECL and Ontario Hydro under the auspicies of the CANDU Owners Group.

REFERENCES

Bieniawski, Z. T. 1967. Mechanism of brittle fracture of rock, Parts I, II and III. *Int. J. Rock Mech. Min. Sci. & Geomech. Abstr. 4*(4), 395–430.

Brace, W. F., B. Paulding, and C. Scholz 1966. Dilatancy in the fracture of crystalline rocks. *Journal of Geophysical Research 71*, 3939–3953.

Cook, N. G. W. 1965. The failure of rock. *Int. J. Rock Mech. Min. Sci. & Geomech. Abstr. 2*, 389–403.

Dyke, C. G. 1989. Core discing: Its potential as an indicator of principal *in situ* stress directions. In V. Maury and D. Fourmaintraux (Eds.), *Proc. ISRM-SPE International Symposium Rock at Great Depth, Pau*, Volume II, pp. 1057–1064. A.A. Balkema, Rotterdam.

Griffith, A. A. 1924. Theory of rupture. In *First International Congress on Applied Mechanics, Delft*, pp. 55–63.

Herget, G. and B. Arjang 1990. Update on ground stresses in the Canadian Shield. In G. Herget (Ed.), *Proc. Stresses in Underground Structures, Ottawa*, pp. 33–47.

Jackson, R., J. S. O. Lau, and A. Annor 1989. Mechanical, thermo-mechanical and joint properties of rock samples from the site of AECL's Underground Research Laboratory, Lac du Bonnet, Manitoba. In *Proc. 42nd Canadian Geotechnical Conference, Winnipeg*, pp. 41–49.

Martin, C. D. 1993. *Strength of massive Lac du Bonnet granite around underground openings.* Ph. D. thesis, Department of Civil & Geological Engineering, University of Manitoba, Winnipeg, Manitoba, Canada.

Obert, L. and D. E. Stephenson 1965. Stress conditions under which core discing occurs. *Transactions of the Society of Mining Engineers of AIME* (6), 227–235.

Santarelli, F. J. and E. T. Brown 1987. Performance of deep wellbores in rock with a confining pressure-dependent elastic modulus. In G. Herget and S. Vongpaisal (Eds.), *Proc. 6th Int. Soc. Rock Mech. Congress, Montreal*, Volume 2, pp. 1217–1222. A.A. Balkema, Rotterdam.

Rock Mechanics, Nelson & Laubach (eds) © 1994 Balkema, Rotterdam, ISBN 90 5410 380 8

Dynamic stress mapping utilizing ultrasonic tomography

T.E. Scott, Jr, Q. Ma & J.-C. Roegiers
School of Petroleum & Geological Engineering, University of Oklahoma, Norman, Okla., USA
Z. Reches
School of Geology & Geophysics, University of Oklahoma, Norman, Okla., USA

ABSTRACT: Two indentation experiments were conducted to generate acoustic tomographic velocity imaging of a dynamically developing complex stress field. The indentation tests were conducted on dry Berea sandstone cores with successive loading increments up to 110 MPa being applied to a hardened steel indenter. A series of 20 acoustic sensors were arranged either in vertical or horizontal planes to tomographically image cross-sectional planes parallel and perpendicular to the indentation direction. The indentation generates a strongly heterogeneous stress field with the highest stresses located in the rock directly beneath the indenter. The regions of high stresses, delineated by an ABAQUS finite element solution, correspond to the tomographically mapped regions of high velocity. These results suggest that tomographic acoustic velocity changes resulting from perturbations in elastic stress states can be successfully mapped in porous reservoir rocks both in laboratory tests and in the field environment.

1 INTRODUCTION

Ultrasonic and seismic tomography represents an exciting new technique for monitoring stresses during engineering operations. Kormendi et al. (1988) conducted a seismic tomographic survey to map out the intensity of stresses in an active mine during long-wall mining of a coal seam. Their results clearly indicated a marked increase in acoustic velocity as the face was cut back into the mine. The increased stresses resulted from a decrease in cross-sectional area of that load bearing section of the mine.

The use of acoustic velocity as a possible stress indicator finds its basis in many laboratory rock mechanics studies. For example Nur and Simmons (1969) and Lo et al. (1986) noted that the application of hydrostatic stress led to an increase in acoustic compressional and shear waves velocities - a factor attributed to the closure of pre-existing microcracks and pore spaces. Velocity changes in rocks subjected to triaxial stresses are more complex. Studies, such as those of Sayers et al. (1990), Sammonds et al. (1989), Fjaer et al. (1989), and Scott et al. (1993), indicate that acoustic velocities may increase or decrease depending upon the directions of applied load and the degree of microcracking.

These types of laboratory studies involve measurements of velocity and stresses along coaxial pathways. The pattern generated by in-situ stress fields or created during engineering projects are more complex. Typically these have stress isobars which are curved, the stress fields are heterogeneous, and the stress field changes in response different loading conditions. See Gough (1969) for an example of the stress field induced beneath an artificial lake.

Santamarina (1992) suggested that tomographic imaging of acoustic velocities beneath a foundation or footing could provide an image of stress concentrations in soils. An 'indentation' type of experiment, similar to that proposed by Santamarina (1992), is developed in this paper for imaging the stress field in lithified rock samples. The experiment involves creating a complex elastic, heterogeneous stress field during indentation into Berea sandstone (Figure 1) and then tomographically correlating the observed velocities to the changes in stresses.

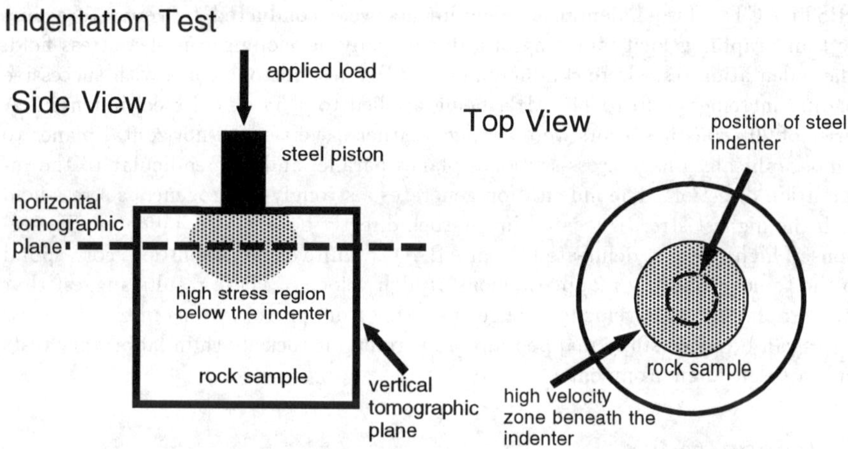

Figure 1: Schematic of the indentation experiment.

2 EXPERIMENTAL PROCEDURE

The Berea sandstone utilized in these indentation tests had a porosity of 18% and a permeability of 150 md. Samples were cut as right circular cylinders with a length and diameter of 10.4 cm. Samples were indented normal to a weakly defined bedding plane fabric with a hardened steel cylinder 2.54 cm in diameter and 5.08 cm in length. IBM-PC's were used to acquire the stress-strain data and acoustic waveforms from a Tektronix 2440 digital storage oscilloscope through a Hewlett-Packard 3488a VHF switchbox (Figure 2). The ultrasonic pulse transmission transducers were constructed in-house from 1 MHZ PZT-5 piezoelectric crystals with a diameter of .15 cm. These crystals were mounted in epoxy housing and glued to the sample with conductive epoxy.

Two samples of Berea sandstone were tested, each with a different configuration of the acoustic sensor array. The first sample was tested with the 20 acoustic sensors mounted around the circumference of the core sample (perpendicular to the

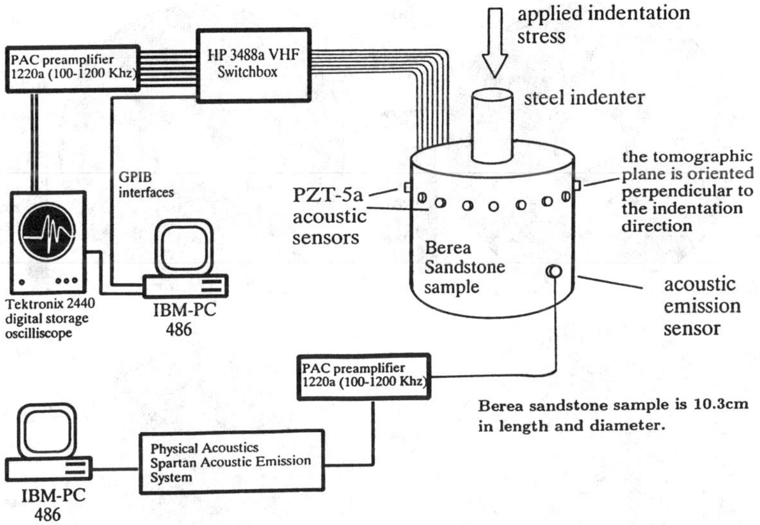

Figure 2: Data acquisition system for acoustic tomography.

the top of the core sample and was designed to crosscut the high stress region just below the indenter. The second sample had the 20 acoustic sensors mounted in a vertical plane with 4 sensors on both the top and bottom faces and 6 sensors on the lateral faces. Sensors beneath the sample were built into a special base platen to prevent damage during loading.

Each experiment was conducted with an incremental step loading procedure which allowed a constant load to be maintained during the 10 minute period required to collect acoustic pulse transmission data along each of the 130 raypaths. The Simultaneous Iterative Reconstructive Technique (SIRT) was used to invert the acoustic velocity in each of the 97 individual pixel elements in the sample cross-sectional area. Each pixel had a resolution of .95 cm.

3 EXPERIMENTAL RESULTS

3.1 Horizontal tomographic imaging

Eight acoustic velocity tomograms have been generated for stresses of 0, 6.9, 20.6, 48.3, 69.0, 89.7, 110.3 MPa, and the unloaded post-test condition (Figures 3a to 3h). The initial undeformed core sample (Figure 3a) has acoustic velocities which average around 2300 m/sec. As the indentation stress is increased to 20.6 MPa the acoustic velocities in the central part of the core sample show a large increase from 2300 m/sec to 3000 m/sec (Figure 3c). This increase is significantly above the velocity variations present in the initial undeformed sample and can only be associated with the application of load to the indenter. Further increases in applied stresses result in a marked intensification of the central high velocity zone beneath the indenter. At an indentation stress of 48.3 MPa stress the velocity increases to

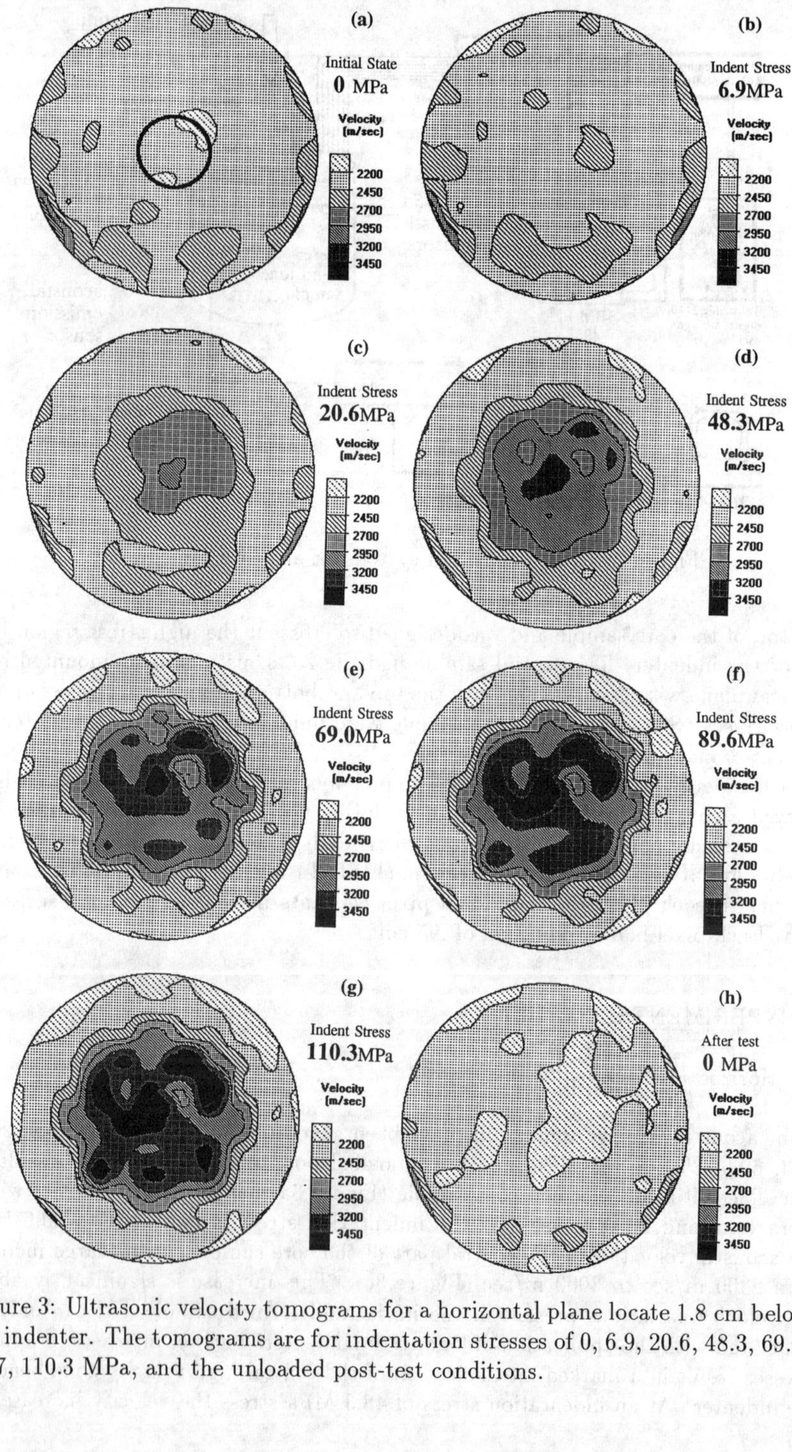

Figure 3: Ultrasonic velocity tomograms for a horizontal plane locate 1.8 cm below the indenter. The tomograms are for indentation stresses of 0, 6.9, 20.6, 48.3, 69.0, 89.7, 110.3 MPa, and the unloaded post-test conditions.

3300 m/sec, and at the maximum applied stress of 110.3 MPa it has achieved 3550 m/sec. Velocity in the central portion of the sample increased by about 50% (from the initial value of 2300 m/sec to 3550 m/sec) at the highest indentation stress level.

One interesting observation is that there is a decrease in the rate of velocity change as indentation stresses are increased during each successive loading step. For example, between indentation stresses of 6.9 MPa and 20.6 MPa the velocities in the center of the sample increase from 2800 m/sec to 3100 m/sec (Figures 3a and 3b). However, from 89.6 MPa to 110.3 MPa the velocities increase only from 3500 m/sec to 3550 m/sec (Figures 3f and 3g). This is attributed to the fact that there are considerably less available open microcracks and pore spaces which can close in response to an incremental increase in the applied stress at the highest stress levels.

The margins of core have velocities nearly equal to, or slightly lower than, the velocities evidenced in the deformed state. Acoustic velocities on the outer 1-2 cm rim have velocities which range around 2200 to 2500 m/sec at all levels of indentation (Figures 3a through 3f).

The high velocity zone develops in the center of the sample as predicted by the ABAQUS finite element solution for the indentation test (Figure 4). The position of the horizontal tomographic plane on the stress solutions does not, at this time, allow a determination of which aspect of deformation stresses (such as shear stress, mean stress, or normal stress) is most affecting acoustic velocities. All profiles indicate that centralized high stress regions will develop under the indenter.

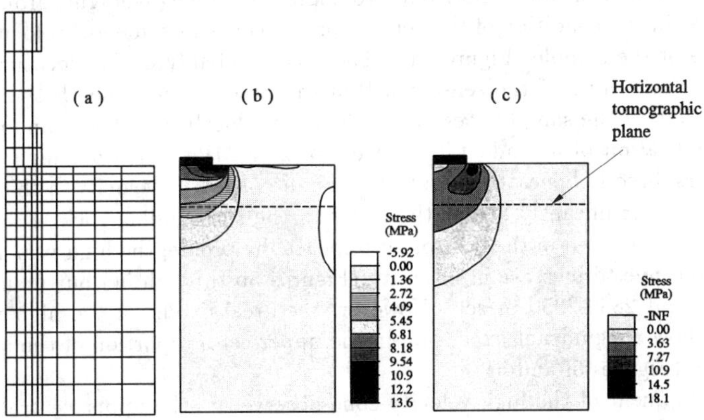

Figure 4: (a) ABAQUS finite element grid pattern; (b) contoured mean stress values under the indenter; (c) shear stress values under the indenter.

The application of indentation stresses appears to be nearly elastic. The evidence for elastic nature of the indentation in these experiments is based on three observations. First, the acoustic velocities between the initial and final unloaded states are very similar (Figures 3a and 3h). The acoustic velocities in the unloaded state are slightly lower than the initial state, possibly indicating that some slight damage has occurred. However, this change is very small when compared to the intense velocity changes resulting from the high stresses applied during indentation. Second, the load-displacement curve has a morphology typical of stress-strain curves which are elastic with some hysteresis but it is nearly completely recoverable. It does not show any evidence of a break in slope typical of yielding associated with dilatancy or microcracking. Third, the acoustic emission sensors did not record any intense acoustic emission activity typical of dilatancy or large scale microcracking in the indented sample.

3.2 Vertical tomographic imaging

A second sample of Berea sandstone was prepared to allow tomographic imaging of velocities in the vertical plane paralleling the indentation direction. Tomographic imaging of this vertical plane is much more difficult than the horizontal plane due to the irregular array of the acoustic sensors. The irregular array is required by (1) the presence of the steel indenter between sensors, (2) the square geometry of the acoustic array, and (3) the difficulty in placing sensors at the base of the sample. In spite of these difficulties a high velocity zone, located just beneath the steel indenter, did appear during loading.

The indentation experiment for vertical tomography was carried out with exactly the same experimental procedures as the previously discussed horizontal tomography test. Figures 5a through 5h present the vertical tomographic profiles at 0, 6.9, 20.7, 48.3, 69.0, 89.6, 110.3, and unloaded condition.

The initial velocities recorded in the vertical tomography test have values similar to those observed in the initial horizontal tomography profile (averaging around 2300 m/sec). The initial condition of the vertical profile contains some low velocity zones at the base of the sample (Figure 5a). These are probably due to location errors created by positioning of the sensors within the base platen beneath the sample. The upper half of the sample does not evidence any localized velocity anomalies.

Upon application of an indentation stress of 20.6 MPa a distinct high velocity zone appears directly beneath the indenter (Figure 5b). The velocity in this region, 2850 m/sec, is significantly higher than that in the surrounding rock. In contrast to the pattern observed in the horizontal tomography profile, the high velocity zone does not continue to increase in intensity (Figures 5b through 5f) but maintains a velocity around 2830-2950 m/sec. However, the areal extent of the high velocity zone does show a significant expansion in the upper central portion of the indented core sample (Figures 5f and 5g).

The development of the high velocity zone observed in the upper central half of the core sample corresponds to the position where the intense stresses are created during indentation (compare to the ABAQUS solutions in Figure 3). It is interesting to note that the acoustic velocities observed in the vertical profile do not match the intensities of the horizontal profile. The cause for this is unclear but it is important to remember that there is no reason to assume that acoustic velocities in the vertical and horizontal tomographic planes should have equal intensities. The deformations at any pixel will be subject to differing components of the triaxial stresses and the research of Sayers (1990) and Scott et al. (1993) have shown that this results in anisotropic acoustic velocities.

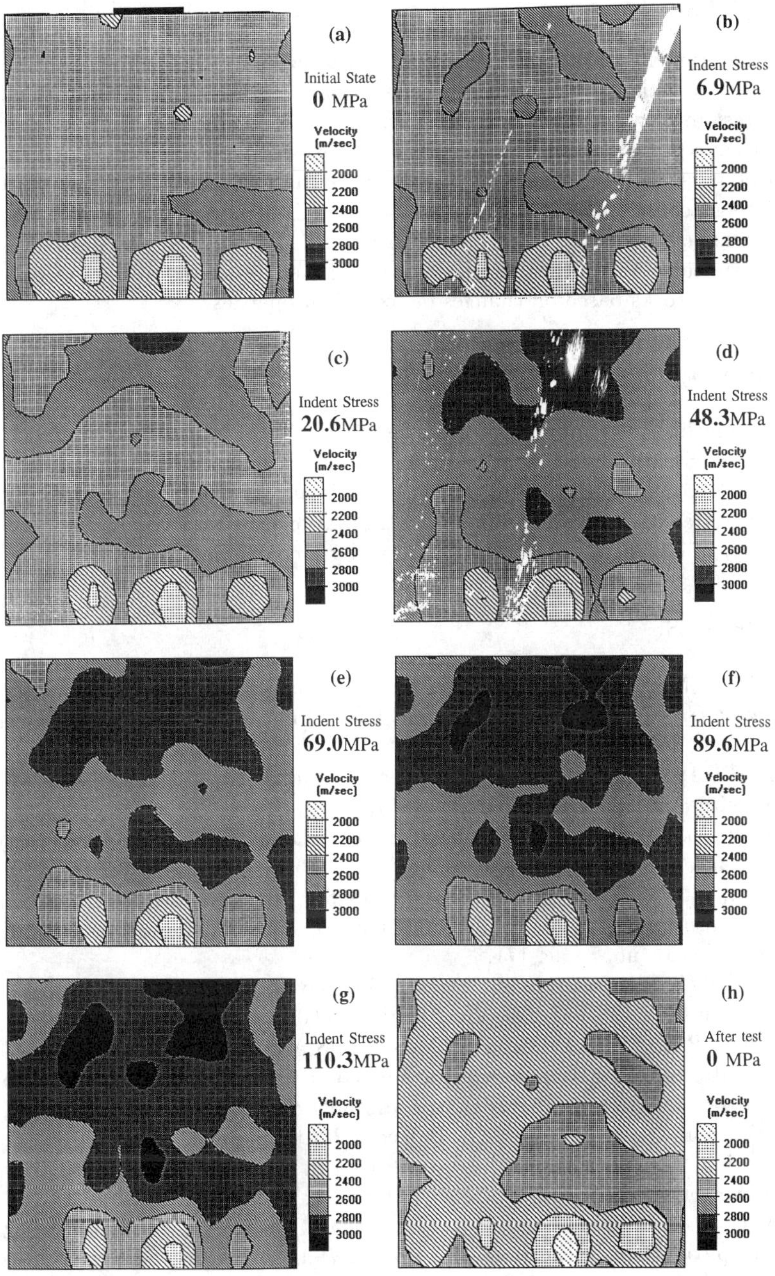

Figure 5: Ultrasonic velocity tomograms for the vertical profile. The tomograms are for indentation stresses of 0, 6.9, 20.6, 48.3, 69.0, 89.7, 110.3 MPa, and the unloaded post-test conditions. Note that the velocity scale is smaller than that of the horizontal profile (Figure 3) to enhance detail.

4 CONCLUSION

These experiments demonstrate the success of utilizing acoustic tomography to dynamically track the development of an elastic stress field during a laboratory indentation test into Berea sandstone. The primary observation of the study is that a qualitative correlation can be made between zones of high acoustic velocities and the regions calculated to have high stresses. Future research will concentrate on correlating acoustic velocity and stress data obtained in the triaxial compression experiments of Scott et al. (1993) to values generated during the tomographic imaging experiments. A quantitative relationship can then be derived to allow calculation of stresses in rocks based on changes in acoustic properties.

ACKNOWLEDGEMENTS

We would like to thank Younane Abousleiman, Mike Batzle, Ray Brown, Louis Boldt, Dave Stearns, and Roger Young for helpful discussions. Thanks also to Peter Lemmon, Tony Haas, and David Snyder for assistance in the laboratory. The Rock Mechanics Research Center is funded by the National Science Foundation and the University of Oklahoma consortium of oil and gas companies.

REFERENCES

Fjaer, E., R.M. Holt, & A.M. Raaen. 1989. Rock mechanics and acoustics, in *"Rock at Great Depth"*, Maury, V. and Fourmaintraux, D. eds., A.A. Balkema, Rotterdam, 355-362.

Gough, D.I. 1969. Incremental stress under a two-dimensional artificial lake, Canadian Journal of Earth Science 6, 1067-1075.

Kormendi, A., T. Bodoky, L. Hermann, L. Dianiska, & T. Kalman, 1988. Seismic measurements for safety in mines, Geophysical Prospecting 34, 1022-1037.

Lo, T.-w., K.B. Coyner, & M.N. Toksoz, 1986. Experimental determination of elastic anisotropy of Berea sandstone, Chicopee shale, Chelmsford granite, *Geophysics*, vol. 51, no. 1, 164-171.

Nur, A., & G. Simmons, 1969. Stress-induced velocity anisotropy in rock, An experimental study, Jour. Geophys. Res., 74, 6667.

Sammonds, P.R., M.R. Ayling, P.G. Meredith, S.A.F. Murrell, & C. Jones, 1989. A laboratory investigation of acoustic emission and elastic wave velocity changes during rock failure under triaxial stresses, in *"Rock at Great Depth"*, Maury, V. and Fourmaintraux, D., eds., A.A. Balkema, Rotterdam, 233-240.

Santamarina, C. 1992. Imaging the state of stress in particulate media, 29th Meeting of the Society of Engineering Science (Abstract).

Sayers, C.M., J.G. van Munster, & M.S. King, 1990. Stress-induced ultrasonic anisotropy in Berea Sandstone, *Int. J. Rock Mech. Min. Sci. & Geomech.* Abstr., vol. 27, no. 5, 429-436.

Scott, T.E., Q. Ma, & J.-C. Roegiers, 1993. Acoustic velocity changes during shear enhanced compaction of sandstone, Proc. 34th U.S. Symposium on Rock Mechanics, University of Wisconsin, 29-32.

Rock Mechanics, Nelson & Laubach (eds) © 1994 Balkema, Rotterdam, ISBN 90 5410 380 8

Tensile properties of rocks in four-point beam tests under confining pressure

Ram Weinberger, Ze'ev Reches & Amir Eidelman
School of Geology and Geophysics, University of Oklahoma, Norman, Okla., USA (Presently: Department of Geology, Hebrew University, Jerusalem, Israel)

Thurman E. Scott
School of Geology and Geophysics, University of Oklahoma, Norman, Okla., USA

ABSTRACT: The strength and elastic properties of rocks were measured with a four-point beam device placed inside a pressure vessel. The experiments were conducted with Tennessee sandstone, Indiana limestone, and Berea sandstone. The tensile Young's modulus is nonlinear and best represented by $\sigma_t = A\,\varepsilon_t^{B}$, where σ_t, ε_t are the tensile stress and tensile strain and A and B are constants. B ranges from 0.56 for tests without confinement to 0.85-0.9 for tests with confinement of 10 MPa or more. The tensile strength depends only slightly on the confining pressure and it ranges from -8.8 MPa to -5.1 MPa. The yielding envelope agrees with the parabolic shape predicted by Griffith. The fractures observed under 20 MPa of confinment are invariably tensile with no indication for transition to shear fractures (faults).

INTRODUCTION

The mechanical properties of rocks strongly depend on the sign of applied loads: tensile versus compressive. For example, the tensile strength of rocks and rock-like materials is 8-10 times smaller then their compressive strength (Griffith, 1921). Therefore, the tensile strength and tensile elastic moduli, should be determined under tensile stress conditions. Probably the most comprehensive series of experiments in the tensile regime was published by Brace (1964). He used a "dog bone" design for his samples and loaded them in a standard triaxial cell under extension conditions in which the confining pressure was larger then the axial stress. Brace could also generate axial tension on his samples. Similar loading conditions were used in the experiments of Griggs and Handin (1960).

It is experimentally difficult to determine the tensile properties of rocks under confining pressure, and tensile strength and tensile moduli are usually determined under room pressure (e.g., plate splitting, Brazilian test, four-point beam). We measured the tensile strength and tensile elastic properties of three rocks under confining pressure by using a four-point device placed inside a pressure vessel (Fig. 1). Four-points beam tests are a simple and effective method to measure the tensile properties of metals and rocks (e. g., Nadai, 1950; Oldroyd, 1971). Experiments with the four-point device generated excellent agreement with the results of uniaxial tests for metals (Yokoyama, 1988). In the present tests the beam samples were deformed under confining pressure and provided the tensile properties of the rocks for polyaxial stress states.

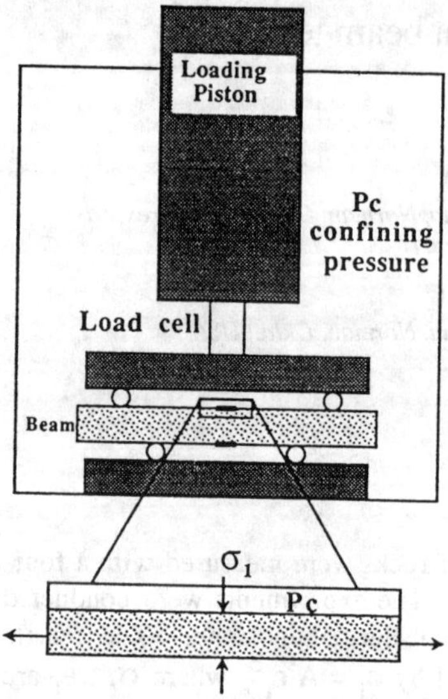

Figure 1. Schematic presentation of the experimental setting that includes a four-point device and a beam inside a pressure vessel. Note the strain gauges on top and bottom of the beam. σ_1 and σ_3 are the maximum and minimum compressive stresses.

EXPERIMENTAL APPROACH

The four-point bend test is a relatively simple method to simultaneously determine the compressive and tensile elastic moduli. Yokoyama (1988) derived a useful formulation for the stress-strain relations in a four-point bending test. His analysis utilizes the experimental measurements of the axial load, P, the strains at the top and bottom of the beam, and the geometry of the device and the beam. The derivations of Yokoyama (1988) lead to the following values of σ_t and σ_c, the maximum tensile stress and the maximum compressive stress along the beam (fiber stresses),

$$\sigma_t = \frac{dM(\epsilon_t + \epsilon_c) + 2M(d\epsilon_t + d\epsilon_c)}{bh^2 d\epsilon_t} \qquad \sigma_c = \frac{dM(\epsilon_t + \epsilon_c) + 2M(d\epsilon_t + d\epsilon_c)}{bh^2 d\epsilon_c}$$

where $M = 0.5\,P\,(L_t - L_c)$ is the bending moment of the beam (L_t and L_c are the spacing of the pairs of the loading points for the tensile side and for the compressive side, respectively); ε_t and ε_c are the tensile and compressive strains measured at the top and bottom of the deformed beam, respectively (fiber strains); b is the beam width; h is the beam height; dM, $d\varepsilon_t$ and $d\varepsilon_c$ are the increments of the moment and the strains during the experiment (the differentials between two consecutive steps in the experiment). We used the last set of equations to calculate the stress-strain curves for both tension and compression from the experimental data of P, ε_t and ε_c.

Sample Preparation. Twenty experiments were conducted with three rock types: Tennessee sandstone of 6% porosity (eight tests), Indiana limestone (eight tests), and Berea sandstone of 17-19% porosity (four tests). Samples were machined into 15.0 cm long rectangular beams with cross sectional dimensions of 1.8 cm by 1.8 cm. The surfaces were grounded parallel to within 0.001 cm under dry conditions with an 80 grit aluminum oxide wheel. Two strain gauges were mounted parallel to the long axis of the beam surfaces for monitoring the axial strain (Fig. 1). The beams tested under confining pressure were jacketed with heat shrink polyolefin tubing.

Experimental Setting. We used a four-point beam device that was designed and built in the Halliburton Laboratory, University of Oklahoma, Norman, by Bjornen (1994) following Yokoyama (1988). Unconfined tests were run on a MTS 319 load frame. Tests under confining pressure were performed on the servo-controlled MTS Model 315 load frame with an integral MTS 138 MPa (20,000 psi) pressure vessel. The axial load P, was measured with a sensitive internal load cell and the beam parallel strains were measured with the stain gauges at top and bottom (Fig. 1). The benchmarking with aluminum (6061) beam yielded accuracy smaller than 1% in the strain measurements.

Data acquisition was controlled by an IBM-PC 486. We used a Hewlett-Packard 3497a for data acquisition/control unit. It acquired parameters for storage at 1 second intervals during the test. The parameters were recorded as voltage values of the axial load, axial stroke, confining pressure, tensional strain(s) and compressional strain(s).

The experimental conditions are listed in Table 1. All samples were room dry and were deformed at room temperature. The confining pressure ranges from 5-20 MPa, with one experiment under 50 MPa. A typical experiment lasted 30-50 minutes under constant stroke rate. The strain rates monitored from the strain gauges were approximately 10^{-6} sec^{-1}.

RESULTS

Elastic Moduli

The tensile and compressive stresses within the beams were calculated from the measured axial load, P, and the strains measured at the top and bottom of the beam, ε_t and ε_c. Representative stress-strain curves are shown in Fig. 2 for the three tested rocks. Almost all curves are relatively smooth in their initial part, up to a tensile strain of 0.0001-0.0003. At larger strain values, both σ_t and σ_c curves become crooked and the intensity of the irregularities increases with the strain increase (Fig. 2). The smooth, initial curves indicate elastic deformation whereas the subsequent stress irregularities indicate local microcracking. In all experiments the compressive stress-strain curves are generally linear and the tensile stress-strain curves are generally nonlinear (Fig. 2). Accordingly, we fit linear coefficients to the compressive curves, and the relations of $\sigma_t = A\ \varepsilon_t^B$ to the tensile curves (Table 1). All curve fittings were with correlation coefficients of $r > 0.98$.

The compressive Young's modulus for Tennessee sandstone is 19,000 to 40,000 MPa, with some consistent increase with confinement (Table 1), in good agreement with those of Scott and Nielsen (1991). The compressive Young's modulus for Indiana limestone is 30,600 ± 3,900 MPa with no dependence on the confining pressure. The

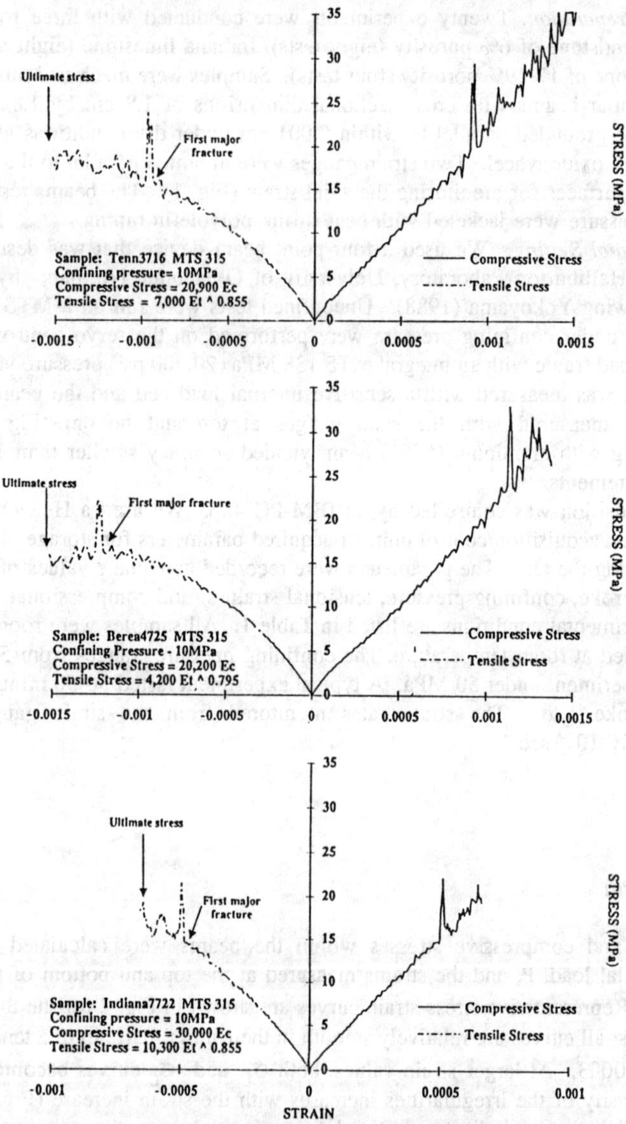

Figure 2. Stress-strain curves of three tests under confining pressure of P_c= 10 MPa. The tensile and compressive stresses σ_t and σ_c are plotted versus the corresponding strains ε_t and ε_c , tensile curve on the left (negative strain) and compressive curve on the right (positive strain). Note that while tensile stresses are negative, both tensile and compressive stresses are plotted as absolute values. The principal stresses at the surface of the convex side of the beam are: $\sigma_1 = P_c$ and $\sigma_3 = -(\sigma_t - P_c)$; the principal stresses at the surface of the concave side of the beam are: $\sigma_1 = (\sigma_c + P_c)$ and $\sigma_3 = P_c$. The experimental data and the coefficients of the elastic moduli appear in the lower left part of each figure.

Table 1. Experimental conditions and results.

Test #	Confining pressure (MPa)	Compressive modulus (MPa)	A-Tensile slope (MPa)	B-Power of tensile curve	Tensile strain at yielding	Tensile Strength (MPa) Lower (3)	Upper (4)
INDIANA LIMESTONE							
24A	0.1	26,600	4,000	0.82	0.00041	-5.3	-5.3
24B	0.1	29,100	3,600	0.80	0.00041	-6.2	-6.2
3720	5	34,900	8,000	0.86	0.00054	-5.9	-5.9
8722	5	28,600	23,500	0.99	0.00038	-3.6	-7.4
7722	10	30,000	10,300	0.86	0.00048	-4.6	-8.2
5722	20	37,200	22,000	0.94	0.00053	2.6	-5.9
9803	20	28,200	16,000	0.91	0.00100	-4.6	-4.6
4716	10	N/A	N/A				
Mean (1)		30,657			0.00054	-3.9	-6.2
STD (1)		3,882			0.00021	3.0	1.2
BEREA SANDSTONE							
24A	0.1	10,200	170	0.56	0.00109	-3.2	-3.8
6810	5	26,100	6,400	0.83	0.00069	-8.6	-9.8
4725	10	20,200	4,200	0.80	0.00121	-6.3	-6.9
5803	10	27,500	4,800	0.80	0.00105	-7.4	-9.5
Mean		21,000			0.00101	-6.4	-7.5
STD		7,864			0.00022	2.3	2.8
TENNESSEE SANDSTONE							
24A	0.1	22,600	350	0.56	0.00113	-7.1	-7.1
24B	0.1	16,300	1,100	0.66	0.00081	-9.4	-9.4
5722	5	25,000	5,000	0.85	0.00066	-8.4	-15.6
8803	5	22,800	3,200	0.75	0.00095	-10.0	-14.1
3716	10	20,900	7,000	0.86	0.00093	-5.8	-9.1
9803	10	24,000	4,900	0.79	0.00072	-4.3	-9.1
6722	20	31,500	12,500	0.87	0.00093	-5.4	-8.1
7727	50	40,400	14,600	0.86	0.00099	14.1	-18.2
Mean (2)		25,438			0.00088	-7.2	-10.4
STD (2)		7,387			0.00016	2.2	3.2

(1) Not including experiment 4716 (2) Not including experiment 7727
(3) First major fracture (4) Ultimate stress

modulus for the Berea sandstone range from 10,000 MPa to 27,000 MPa with slight dependence on the confining pressure (Table 1). The tensile modulus is represented by the constants A and B. The Berea and Tennessee sandstones have similar A and B. A ranges from ≈ 1,000 MPa at room pressure to ≈ 14,000 MPa at 50 MPa; B ranges from 0.56 for room pressure to 0.85-0.9 for tests with confinement of 10 MPa or more. A and B of Indiana limestone has no clear relation to the confinement (Table 1).

Tensile Strength

The deformed beams were first examined with a binocular, thin sections were prepared from selected regions, and these sections were analyzed with a petrographic microscope (Fig. 3). We traced the fractures on two surfaces to obtain their 3-D shape: the top surface which is the surface with the maximum tensile stress (A, B in Fig. 3), and the sides of the sample (C in Fig. 3).

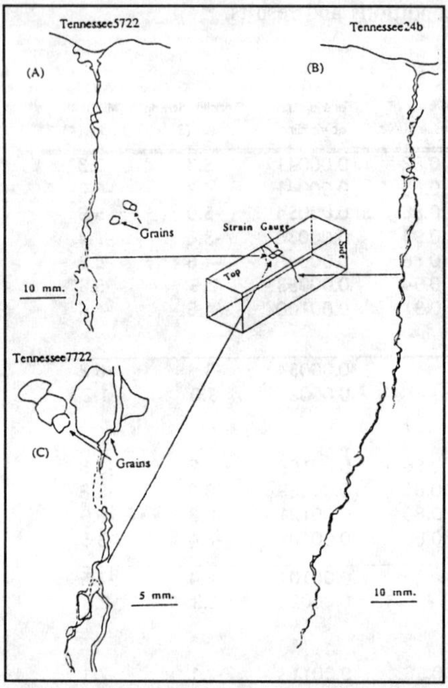

Figure 3 (above). Micrographic view of some tensile fractures in three tests of Tennessee sandstone marked on the figure. Note the crooked shape of the fractures as they bypass the quartz grains.

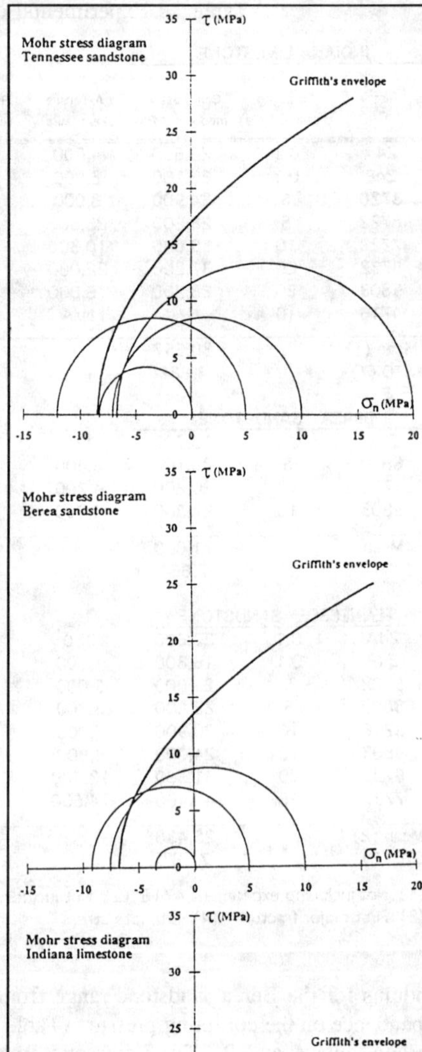

Figure 4 (right). The σ_1 and σ_3 magnitude during the tensile yielding in the present tests plotted on Mohr diagrams. The tensile strength is the mean value of the lower and upper bound for the given confining pressure (see text). Solid, heavy lines are the apparent yielding envelopes calculated according to Griffith. Rock type marked on figures.

Almost all strain-stress curves display irregular curves of σ_t and σ_c for tensile strain larger then 0.0001 to 0.0003; the irregularities increase with increasing strain (Fig. 2). We interpret these irregularities as local, stable fracturing. These fractures initiate at the tensile surface of the beam, where the tensile stresses are the highest, and they propagate into the inner part of the beam. These early fractures grow stably, and they were arrested as they approach the neutral surface of the beam (Fig. 3). Eventually, the sample fails at the ultimate stress (Fig. 2).

Therefore, it appears that the rock beams display two modes of tensile yielding: local, stable fracturing during the early stage, and unstable, crosscutting fracturing during the final stage. We regard the tensile stress of the first major local fracturing (marked in Fig. 2) as the lower bound of the tensile strength and the ultimate stress as the upper bound. The first major fracture is associated with a major stress change as shown in Fig. 2. Thus, for each test we derived a lower and upper bound on the tensile strength (Table 1) and we used the mean value of these bounds as the tensile strength. We found that the tensile strength, S_t, is -8.8 ± 3.1 MPa for Tennessee sandstone, -5.1 ± 2.5 MPa for Indiana limestone and -6.9 ± 2.4 MPa for Berea sandstone (tensile stress is negative) (Table 1).

These results are plotted on Mohr diagrams (Fig. 4). We traced the parabolic yielding envelope of Griffith, $\tau^2 = 4\, S_t\, (\sigma_n + S_t)$ (heavy curves in Fig. 4). This curve fits the stress circles and the observation that the fractures are tensile (normal to maximum tension in Fig. 3). It should be noted that the tensile part of the yielding envelope is usually extrapolated, and only seldom is it bounded by experimental data, for example by Brace (1964). This figure also shows that the tensile strength does not depend on σ_1, the confining pressure. This observation is in good agreement with the experimental results of Brace (1964, Table 4) for several other rock types.

Another interesting problem is the enigmatic transition from tensile fractures (joints) to shear fractures (faults). The present observations indicate that within the range of tested confining pressures (up to 20 MPa), all fractures are tensile fractures with no indication to "hybrid" or transitional fractures. This result is in agreements with the conclusions of Griggs and Handin (1960) and Brace (1964).

CONCLUSIONS

A. It is feasible to run four-point bending tests under confining pressure. This design can provided the tensile strength of the tested rocks and their tensile elastic moduli for confining conditions.

B. The tensile elastic modulus is invariably nonlinear and could be fitted by the form $\sigma_t = A\, \varepsilon_t^{B}$. Microcracking initiates after strain of approximately 0.0003. In the present beam configuration, this early microcracking is stable and does not cause global yielding, and it drives the stress-strain curves into irregular shape.

C. The tensile strength of the three tested rocks apparently does not depend on the confining pressure.

D. The yielding envelope in the tensile regime of the Mohr diagram seems to agree, in general, with the parabolic shape predicted by Griffith's theory and it does not agree with the predictions of Coulomb criterion.

E. The fractures observed under 20 MPa of confining pressure are invariably tensile in character with no indication for transition to shear fractures (faults).

ACKNOWLEDGEMENTS

The advise, help and suggestions of several people made this work possible. In particular, Lou Boldt who suggested to use bending device inside a pressure vessel and Kevin Bjornen who kindly provided the four-point device built by him. We also wish to thank Carla Cates, David Snyder, Karen Hogan, Peter Lemmon and Tony Haas for their technical support and advise.

The study was conducted as part of the Rock Mechanics Consortium, University of Oklahoma, Norman. It was originally submitted as Report 22A of the Consortium.

BIBLIOGRAPHY

Bjornen, K., 1994. Epoxyl-lastic material used in large scale fluid loss test. M.Sc. Thesis, Univ. of Oklahoma (in preparation).

Brace, W. F., 1964, Brittle fracture of rocks, in William, R. J., ed., *State of stress in the earth's crust*: New York, US, Elsevier, p. 111-174.

Griffith, A. A., 1921. The phenomena of rupture and flow in solids, *Philos. Trans. R. Soc. London Ser. A*, 221, 163-198.

Griggs D. and Handin, J., 1960. Observation of fracture and a hypothesis of earthquakes, *GSA Mem*., 79, 347-362.

Nadai, A., 1950. *Theory of flow and fracture of solids*, McGraw-Hill, New York, pp. 353-359.

Oldroyd, P. W. J., 1971, Tension-Compression stress-strain curves from bending tests, *J. of Strain Analysis*, 6, 286-292.

Scott, T. E. and Nielsen K.C., 1991, The effects of porosity on the brittle-ductile transition in sandstones, *J. Geophy. Res*., 96, 405-414.

Yokoyama, T.A., 1988. A microcomputer-aided four-point test system for determining uniaxial stress-strain curves, *J. of Testing and Evaluation*, 16, 198-204.

Rock Mechanics, Nelson & Laubach (eds) © 1994 Balkema, Rotterdam, ISBN 90 5410 380 8

Determination of the microcrack tensor in rock: Evaluation of coring induced damage

D.R.Schmitt & Y.Y.Li
University of Alberta, Edmonton, Alb., Canada

ABSTRACT: Strains measured on the surfaces of rectangular prisms machined from low porosity, low permeability rock cores are used to determine that portion of the porosity comprised of small low aperture cracks. In these measurements, which are similar to the the differential strain analysis technique, the linear microcrack strains are the difference between the observed response of the strain gage and the estimated intrinsic material strain. These crack strains may be displayed in a tensor form and principal directions and magnitudes calculated, the trace of the crack strain tensor further yields the microcrack porosity of the rock. Two example experiments in a weakly foliated granitic gneiss and a sedimentary dolomite indicate that these crack principal crack strains are not necessarily aligned with the in situ horizontal and vertical principal stresses as might be anticipated if the crack damage resulted solely from stress relief mechanisms after coring.

1. INTRODUCTION

Damage within rock cores removed from the earth is manifest by the existence of small low aperture microcracks. These microcracks result from the transient stresses locally applied to the rock during the act of drilling and coring and from poorly understood stress relief mechanisms which commence immediately after the core is cut. Both processes depend to some degree on the direction and magnitude of the in situ principal stresses. For example, both macroscopic discing and petal fractures, the orientations of which provide information on principal stress directions are the end result of the coalescence of microscopic cracks resulting from geometric stress concentrations and the weight of the drill string. Further, present stress relief models suggest that the severity of microcrack damage in a given orientation is roughly proportional to the magnitude of the normal stress parallel to this orientation. Consequently, determining the orientations and densities of these microcracks may in some cases be useful for aiding in delineation of stress directions with previously oriented core, or alternatively, orienting the core if obtained in regions where the stress fields are well understood.

The damage microcracks contribute to the bulk porosity of a rock as measured under standard room conditions. In many crystalline cores these microcracks may form nearly all of the measurable porosity. In low porosity and permeability sedimentary cores determination of the microcrack porosity is necessary for an accurate economic evaluation of the reservoir. If these damage induced microcracks form a substantial portion of the total porosity measured in the core then the amount of fluid in place may be overestimated. For this reason there may be cases where determining the fraction of the porosity attributable to cracks becomes important.

Both the stress and porosity determinations require some means of separating the effects of microcracks from the bulk behaviour of the rock. In this paper we describe a modification to the well known differential strain analysis technique in which the estimated response of the rock forming material is removed from the bulk observations to yield a microcrack strain tensor. The principal crack strains provide a measure of the crack porosity; the directions of these principal crack strains may provide further insight into the source of the damage. Below follows a brief explanation of the experimental technique and data analysis. Results for measurements in crystalline and sedimentary rocks are also shown.

2. EXPERIMENTAL METHOD AND DATA ANALYSIS

The method used is similar to the differential strain analysis technique described by Simmons et al (1974) and more details of the present analysis may be found in Schmitt and Li (1994). Briefly, a square oriented prism of rock is machined from a core. After cutting the sample is well dried and the surfaces made smooth and flat with fine grained emery papers before cleaning with acetone. Strain gages (MicroMeasurements Group CEA-06-250UR-350) are then applied to the sample using a special strain gage adhesive according to the configuration shown in Figure 1. The z axis is always taken parallel to the vertical axis of the core and the x and y axes are aligned with respect to the foliation of the rock if it exists. The entire sample is then sealed in a castable urethane compound which provides an impermeable jacket.

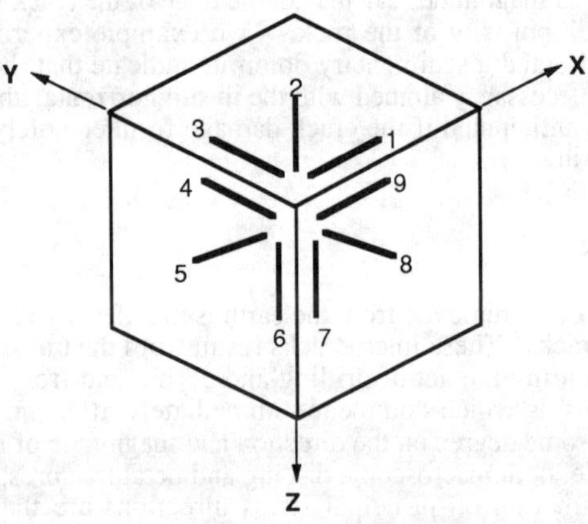

Fig.1 Configuration of strain gauges

Each strain gage forms one arm of a Wheatstone bridge completed by three stable precision resistors, these bridges are activated by an input voltage of 10.000 ± 0.003 Volts. The sample is then placed in a pressure vessel and the voltage outputs of the bridges, typically in the range of a few millivolts, were recorded using a digital voltmeter with 100 nV resolution. These bridge voltages were converted to strains using well known equations for an unbalanced bridge. As the response of these strain gages are temperature dependent and especially since the co-efficients of thermal expansion of the gages and the rock are not easily matched, care was taken to ensure that the room temperature remained constant to within 1° C and measurements were only taken after waiting for equilibration the temperature in vessel fluid after a change in pressure. A fused quartz standard of known compressibility containing a strain gage was also placed in the vessel in order to monitor the stability of the experimental conditions and to provide a baseline for pressure corrections to the measurements allowing calculation of the absolute strain magnitudes (Siegfried and Simmons (1978)).

A set of typical observed strain curves for a weakly foliated granitic gneiss is seen in Figure 2a. As is well known (Brace (1965)) the low pressure strain response is nonlinear due to the progressive closure of the microcracks. Once the microcracks are closed a linear strain regime indicative of the intrinsic solid strains is observed. The compressibilities of the minerals of pressure dependent and strictly speaking there will be some curvature to the observed intrinsic strain response but this effect is sufficiently small over the pressure range studied to 200 MPa to be unnoticeable. As a result, the intrinsic linear compressibility λ_k for the material beneath each strain gage k may be independently estimated by fitting a straight line to the strain ε_k at higher pressures such that one may predict the intrinsic strain $\varepsilon_k{}^i$ via

$$(1) \quad \varepsilon_k{}^i = -\lambda_k P$$

where P is the hydrostatic confining pressure. Note that here a dilation is assumed positive. Siegfried and Simmons (1978) provide arguments to show that the total observed strain $\varepsilon_k{}^o$ at any pressure is the sum of those portions due to the intrinsic strain $\varepsilon_k{}^i$ and the microcrack strain $\varepsilon_k{}^c$ such that the latter may simply be calculated by

$$(2) \quad \varepsilon_k{}^c = \varepsilon_k{}^o - \varepsilon_k{}^i.$$

In the experiments there are nine separate strain gages at a variety of directions with respect to the co-ordinate frames of Figure 1. As a result the relationships between the observed crack strains and the co-ordinate axes based strain tensor will be (Timoshenko and Goodier (1970))

$$(3) \quad \varepsilon_k{}^c = l^2\varepsilon_{xx}{}^c + m^2\varepsilon_{yy}{}^c + n^2\varepsilon_{zz}{}^c + 2lm\varepsilon_{xy}{}^c + 2ln\varepsilon_{xz}{}^c + 2mn\varepsilon_{yz}{}^c$$

where the $\varepsilon_{ij}{}^c$ (i,j = x, y, z) are the components of the strain tensor. The nine observed crack strains consequently provide an overdetermined system of equations which may be solved using any least squares method to provide the six independent $\varepsilon_{ij}{}^c$. The principal crack strains and their directions may then be determined in the standard fashion (e.g. Fung (1969)). Finally the crack porosity is calculated by taking the absolute value of the sum of the three principal crack strains.

3. RESULTS AND DISCUSSION

Figure 2 shows the results of measurements on the gneiss taken from the crystalline basement in Northern Alberta. The nine observed strains in Figure 2a display a linear behaviour above the confining pressures of 100 MPa. Each of the strain curves above this pressure yielded intrinsic linear compressibilities which ranged from 5.3 X 10^{-3} GPa^{-1} to 6.5 X 10^{-3} GPa^{-1} close to the values expected from the rock's mineral composition. These compressibilities were used in equations 1 and 2 to determine the crack strains as a function of confining pressure (not shown) which in turn were used to calculate the crack strain tensor and finally the principal crack strains (Figure 2b) and their directions (Figure 2c). The crack porosity for this sample is small at 8.2 X 10^{-4} (0.082%). Note that the three stars in Figure 2c represent to eigenvector directions calculated at pressures above 100 MPa and the various data points leading up to these stars are the changes in the principal directions with increasing pressure. The pole positions of the eigenvectors of the principal crack strain directions show that the

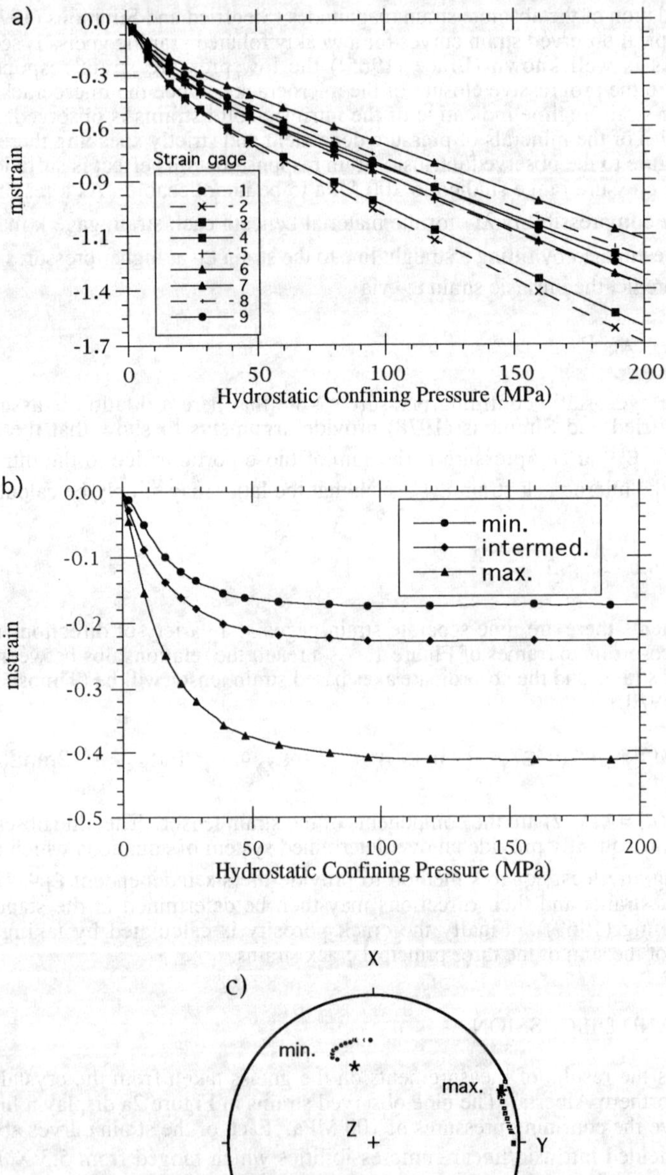

Fig.2 The pricipal crack strains and microcrack strain principal axes of a granite core. a) The observed strain; b) the principal microcrack strains; and c) the principal axes of microcrack strains.

maximum strains in the sample occur in a nearly horizontal direction 20° counter-clockwise from the y axis. This direction is roughly perpendicular to the plane of the near vertical foliation within the sample and may suggest that this largest strain results from opening of microcracks along grain boundaries.

The two lesser principal crack strains occur at orientations different from the horizontal and vertical directions. These orientations are not related to the texture of the sample as there are no obvious lineations of the mineral grains. Their positions on opposite sides of the vertical also differs from that expected if the strains were solely the result of stress relief expansion of the cores with respect to the principal stress directions, one of which is here expected to be vertical on the basis of the depth of the sample and on the subtle surface topography in the region.

Figure 3 contains strain data acquired from a dolomite core obtained from a depth near 4700 m. The raw strain curves for this dolomite in Figure 3a display nonlinearity to the highest pressures of 200 MPa. This result was unexpected as it was thought that any microcracks should be closed above pressures equivalent to the overburden near 120 MPa predicted on the basis of an average density of the overlying rocks of 2500 kg/m^3. This remnant nonlinear strain behaviour above these pressures may be representative of the higher aperture, pre-existing porosity within the sample. If all the microcracks within the test piece are indeed closed at higher pressures, then the porosity would tend to increase the effective intrinsic compressibility of the material as a whole as predicted by various theoretical studies (e.g. Zimmerman (1984)).

This nonlinear response makes estimation of the intrinsic material compressibility problematic as there are no unique analytic formulations which allow calculation of both the compressibility and the change in the compressibility with pressure. To calculate the crack strains, however, it was assumed that the microcracks were closed above a pressure of 140 MPa and to obtain a linear compressibility to use in equation 2 a linear fit was made to the data above this confining pressure. This is justified only on the basis that the strain curves have very little curvature at these higher pressures. Higher order polynomial modelling to these data are not ruled out but the physical validity of such fits in representing the true intrinsic material strain is unknown. With the linear fit, however, one may be confident that the estimated intrinsic material strains will be a minimum with the result that the crack strains derived with equation 2 will be the upper bound.

The tangent intrinsic compressibilities estimated by least squares fitting of a model line to the higher pressure data yields values ranging from 7.2×10^{-3} GPa^{-1} to 9.5×10^{-3} GPa^{-1} which indicate bulk compressibilities of 22.5×10^{-3} GPa^{-1} to 28.5×10^{-3} GPa^{-1}. These bulk compressibilities are nearly twice that expected for single crystal dolomite itself (Schmitt and Li (1994)). This observation is consistent with, but cannot be uniquely predicted by, the bulk dry porosity of the dolomite sample of 5.3%.

The principal crack strains shown in Figure 3c indicate a crack porosity within the dolomite of 7.9×10^{-3} (0.79%). This value is in excellent agreement with an earlier test conducted on another prism of rock machined from the same core (Schmitt and Li (1994). This represents about 15% of the total porosity within the sample. As in the gneiss above, the directions of the principal crack strains are not aligned with the vertical or horizontal directions. This cannot be related to any visible fabric to the sedimentary rock which is extremely fine grained and has a homogeneous appearance to the naked eye. These directions too, are not in agreement with the anticipated vertical and horizontal principal stress directions in the region.

4. CONCLUSIONS

In both of the experiments described above the principal crack strain directions were not aligned with the vertical and horizontal directions. Aligned crack strains might be anticipated if crack damage within the core pieces resulted only from stress relief

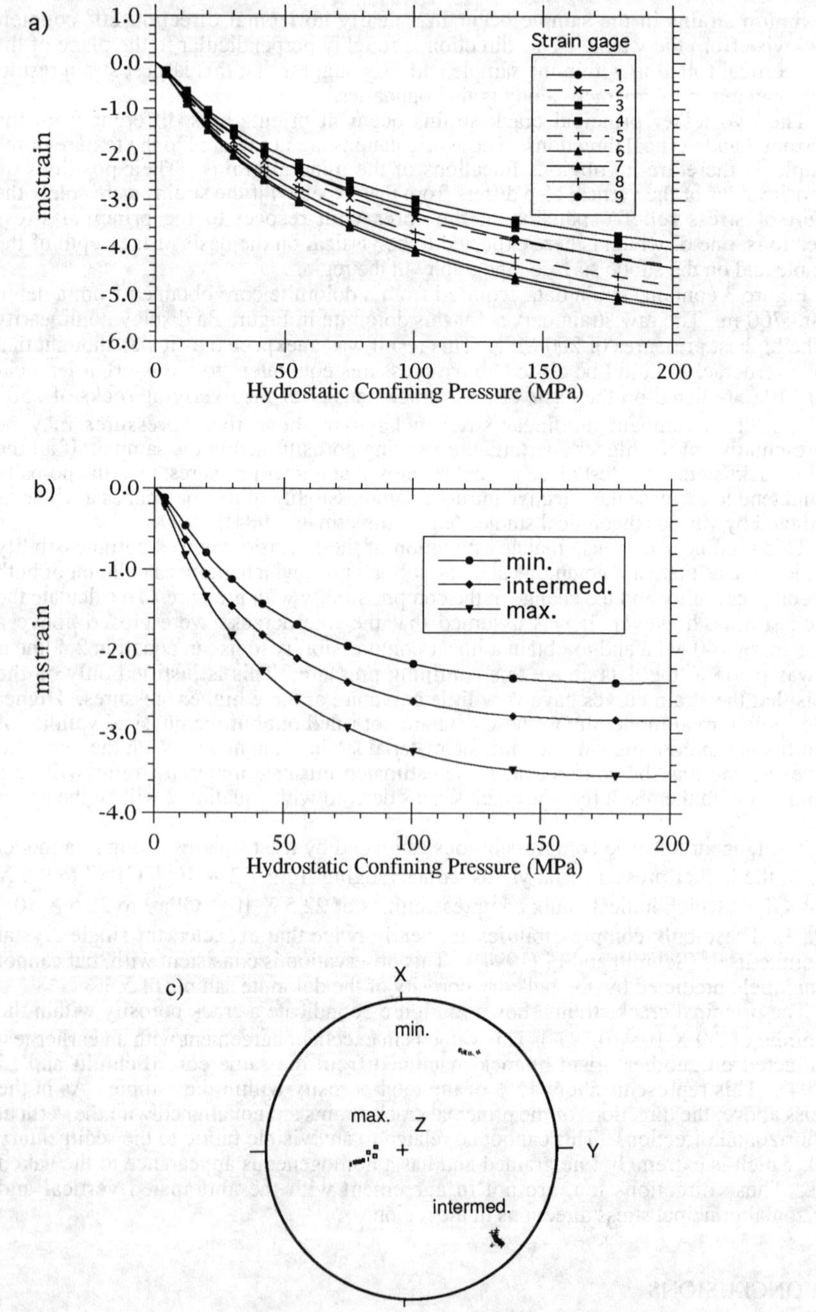

Fig.3 The principal crack strains and principal axes of a dolomite core. a) The observed strains; b) the principal microcrack strains; and c) the principal axes of microcrack strains.

expansion in the directions of the principal stresses and proportionally with the principal stress magnitudes. A possible explanation for this discrepancy is that some of the microcrack damage in these cores may result from the concentrated stresses applied to the rock during the coring operation. These concentrated stresses are known to produce visible core discing and centerline petal fractures in many cores, and the crack stain orientations observed here may be representative of incipient microcrack damage resulting from such effects. Other sources of microcrack damage could be internal stresses generated by differences in thermal expansion co-efficients between the various mineral grains forming the rock.

Future work will concentrate on examining the microcrack structures using optical and electron microscopy to see if there is any correlation the observed strain results and the density and orientations of microcracks in the rock. Other work will focus on further experimental tests including tests for heterogeneity and compatibility of the observed strain tensors.

Acknowledgements: Mr. B. Madu greatly aided in the development of the pressure vessel system and the machining of the rock samples. This work supported by the National Science and Engineering Research Council of Canada. Equipment for this work was purchased with grants from the LITHOPROBE Alberta Basement Transect and from Shell Canada Resources.

REFERENCES

Brace, W.F. 1965. Some new measurements of linear compressibilities of rocks, *J. Geophys. Res.* 70:391-398.

Fung, Y.C. 1969. *A First Course in Continuum Mechanics*. Englewood Cliff: Prentice-Hall.

Schmitt, D.R. & Y.Y. Li 1994, A high pressure technique for determining the microcrack porosities of damaged brittle materials, submitted *Can. J. Phys.*

Siegfried, R. & G. Simmons 1978, Characterization of oriented cracks with differential strain analysis, *J. Geophys. Res.* 83:1269-1278.

Simmons, G., R.W. Siegfried, & M. Feves 1974. Differential strain analysis: A new method for examining cracks in rocks. *J. Geophys. Res.* 79:4383-4385.

Timoshenko, S.P. & J.N Goodier 1970. *Theory of Elasticity*. New York: McGraw-Hill.

Zimmerman, R.W. 1984, Elastic moduli of a solid with spherical pores: new self-consistent method.

Rock Mechanics, Nelson & Laubach (eds) © 1994 Balkema, Rotterdam, ISBN 90 5410 380 8

Modelling crack propagation in compression

A.V. Dyskin
The University of Western Australia, Perth, W.A., Australia
L.N. Germanovich, K.K. Lee & L.M. Ring
The University of Oklahoma, Norman, Okla., USA
A.R. Ingraffea
Cornell University, USA

ABSTRACT: Fairhurst and Cook's (F-C) 2-D model of extensively growing crack in uniaxial compression, consisting of a straight crack wedged at the center by a couple of concentrated forces is investigated, both experimentally and theoretically. Such a crack alone grows in a stable manner. The presence of a free surface parallel to the crack amplifies its growth such that the crack becomes unstable after reaching a certain critical length. Although compression parallel to a straight crack does not normally make the crack open, the presence of the free surface is found to provide a case when the crack can grow. This happens if the compression on the sample top concentrates between the crack and the free surface, whereas the balancing compression at the bottom is distributed over the whole sample width. The crack propagates mostly parallel to the free surface as a mode I crack until the separated layer becomes slender and buckles. This may be a mechanism of rock chipping. In the 3-D case when a single inclined initial crack cannot grow extensively, the F-C crack is shown to be a reasonable model of growth of large tensile cracks driven towards the compression direction by certain combinations of interacting initial cracks.

1 INTRODUCTION

A typical mechanism of fracture in brittle rocks under uniaxial compression is based on stable growth of cracks initiated from various pre-existing defects. After initiating, a crack is maintained in the open state by the initial defect, which virtually makes it a mode I (tensile) crack. In many situations the crack grows in the presence of a free surface, e.g., in a mechanism of skin rock burst (e.g., Dyskin and Germanovich, 1993), or in rock cutting (e.g., Cherepanov et al., 1987). In these cases the crack additionally suffers shear loading which is supposed to direct the crack propagation towards the free surface. However experimental observations suggest that this change of the crack path does not always occur: rather the crack keeps growing parallel to maximum compression. This paper considers some aspects of crack propagation sub-parallel to a free surface. The consideration is predominantly based on 2-D experiments and models with some insight into the 3-D situation.

2 F-C MODEL OF 2-D CRACK GROWTH IN UNIAXIAL COMPRESSION

In 2-D when an initial crack extends through the full thickness of the plate, it is known that two secondary tensile cracks dynamically branch from both ends of the initially inclined crack (fig. 1a). Their further extension proceeds in a quasi-stable manner towards the major axis of compression requiring an increase in the applied stress to maintain this process (e.g., Brace and Bombolakis, 1963; Hoek and Bieniawski, 1965; Ingraffea and Heuze, 1980; Ashby and Hallam, 1986).

Let the initial crack be of length $2a$, inclined with respect to the compression direction at an angle α . Suppose also that the faces of the initial crack are in contact. Then, the initial crack forms an inclined contact area and crack faces can slide with respect to each other. It is this sliding that actually drives the growth of secondary branches.

The simplest model for such a crack at a developed stage of its growth (fig. 1b) was proposed by Fairhurst and Cook (1966); hereafter the F-C crack. This is a rectilinear crack opened by a pair of concentrated forces which simulate the action of the inclined contact area (the former initial crack). These forces originate from the sliding of the opposite lips of the contact area (dotted lines in fig. 1b). As only brittle rocks are considered here, it is assumed that the crack grows due to tensile stress concentrations at its tips, i.e., in Mode I. Therefore, only horizontal projections, F, of these forces tending to produce normal opening of the crack have to be taken into account.

To assess the magnitude of these forces (per unit thickness), assume the average normal stress acting through the area to be equal to $\sigma \sin^2\alpha$, i.e., to the normal stress which acted on the area prior to the growth of the secondary cracks. This assumption, that the influence of the secondary cracks on the average stress is negligible, is widely accepted (e.g., Horii and Nemat-Nasser, 1986; Dyskin and Salganik, 1987; Ashby and Sammis, 1990). In the absence of friction, the force, F, is then equal to the horizontal projection, $2a\sigma \sin^2 \alpha$, of the normal force produced by this average stress. If friction is present, F may be estimated by (cohesion neglected),

$$F = 2a\sigma\beta(\alpha), \quad \beta(\alpha) = \sin^2 \alpha \cos\alpha(1 - \tan\alpha \tan\mu). \qquad (1)$$

Here, σ is the magnitude of the applied compressive stress and μ is the friction angle.

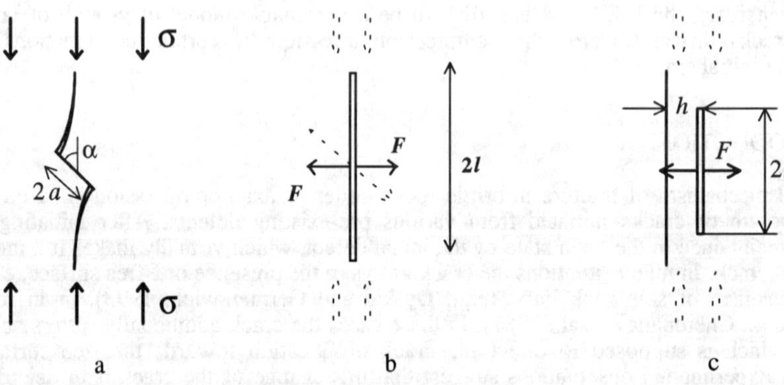

Figure 1. 2-D crack propagating in compression in brittle material (a) and F-C crack located far (b) and close (c) to the free surface.

According to linear fracture mechanics, crack growth is controlled by the stress intensity factor (SIF) characterizing the stress singularity at the crack tip. The SIF for the crack loaded by forces (1) can be calculated based on the known solution (e.g., Murakami, 1987):

$$K_I = \frac{2a\sigma\beta(\alpha)}{\sqrt{\pi l}} \qquad (2)$$

where $2l$ is the total crack length (Fig. 1b). By using the simplest criterion of crack growth $K_I = K_{Ic}$, where K_{Ic} is rock fracture toughness, the length can be expressed in the following form

$$l = C\sigma^2, \quad C = \frac{1}{\pi}\left(\frac{2\beta(\alpha)a}{K_{Ic}}\right)^2 \tag{3}$$

It is seen that the crack grows stably with the length increasing as the second power of the applied stress.

3 INFLUENCE OF A FREE SURFACE ON STABILITY OF CRACK GROWTH

Consider a crack growing near a free surface. Since high compressive stress acts parallel to the surface, the developed crack will grow parallel to the surface as well (Fig. 1c). The F-C crack problem near a surface has been solved numerically (Germanovich and Grekov, 1994) and asymptotically (Dyskin and Germanovich, 1993). The values of the SIFs when the surface is free are plotted on Fig. 2. The minimum for K_I at $l_{cr} \approx 0.47h$ indicates the transition from stable to unstable crack growth. The presence of the free surface adds mode II to this originally mode I crack model. However, according to Fig. 2, the values of K_{II} are considerably smaller than K_I up to the moment of unstable crack growth.

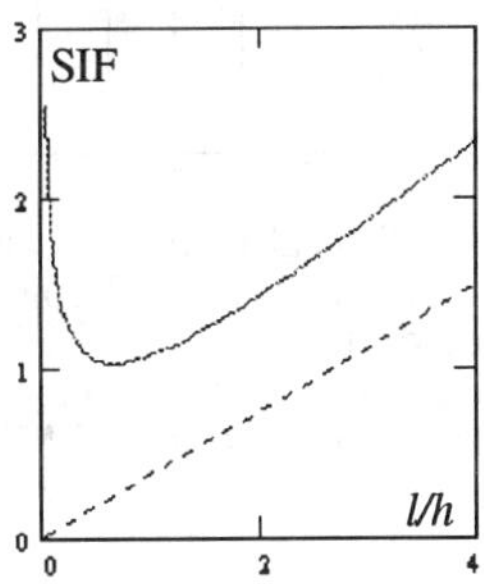

Figure 2. Dimensionless SIFs, $K_I h^{1/2}/F$ (solid line) and $K_{II} h^{1/2}/F$ (dotted line) vs. the normalized crack length, l/h.

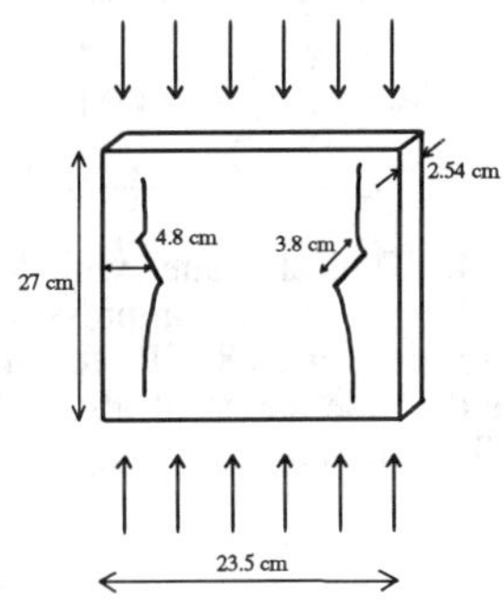

Figure 3. Typical test with near surface cracks. Right crack (45°) propagated unstably at 21 MPa. Left crack (30°) propagated at 37.3 MPa. They stopped due to the lateral compression near the loading platens.

To study free surface influence on a real branching crack (Fig. 1a) experiments were conducted on Tennessee Sandstone plates with single inclined slits located near the boundary and filled with thin plates of zinc to insure side contacts (Fig. 3-5). In all experiments the crack grew unstably from the very beginning (the figures show trajectories of the dynamic crack propagation). It is likely that after the initial unstable (jump-like) phase of branching the resulting crack length reaches the critical one (minimum in Fig. 2), such that the branches have no chance to stop. This phenomenon is even more pronounced for edge cracks (Fig. 4). Although, their growth is dynamic from the very beginning, a quasi-static model can be useful to show that again the crack cannot stop propagating under load control once initiated. Another application of quasi-static modeling is sub-critical stable crack propagation.

The simplest model of such a crack is similar to F-C-crack, Fig. 6. The force, F, could also be estimated by eq. (1). Important point is though that in this case we have to account explicitly for stresses acting along the separated rock layer (Fig. 6) since they create moments about the crack tip and may influence its growth. Due to the sliding of crack faces the stress p acting on the top of the layer should be assumed to be smaller than the applied compressive stress σ. This is discussed below in more detail and shown that the Mode I SIF strongly increases if such a crack grows; hence the crack cannot stop propagating even under load control.

The tests shown in Figs. 4, 5 are similar to the ones performed by Nemat-Nasser and Horii (1982) with Columbia resin. However, the result for the curvilinear boundary (Fig. 5) is different: the crack grows vertically rather than following the boundary geometry as in Nemat-Nasser and Horii's tests. The reason for this could be in the dynamic fashion of

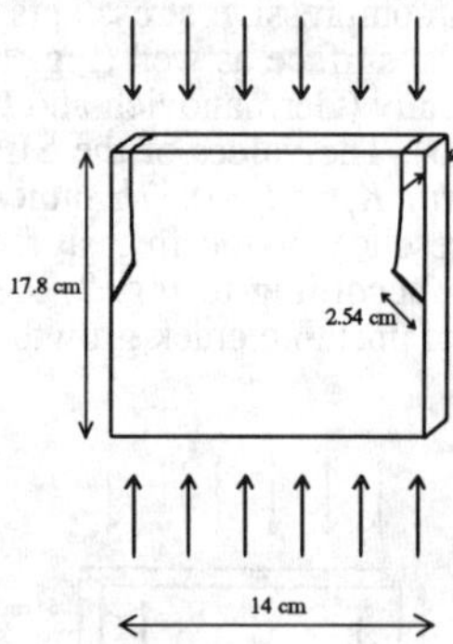

Figure 4. Typical plate with edge slits subjected to uniaxial compression. Cracks started to grow at 18.8 MPa for the 45° slit, right, and at the stress of about 32.8 MPa for the 30° slit, left.

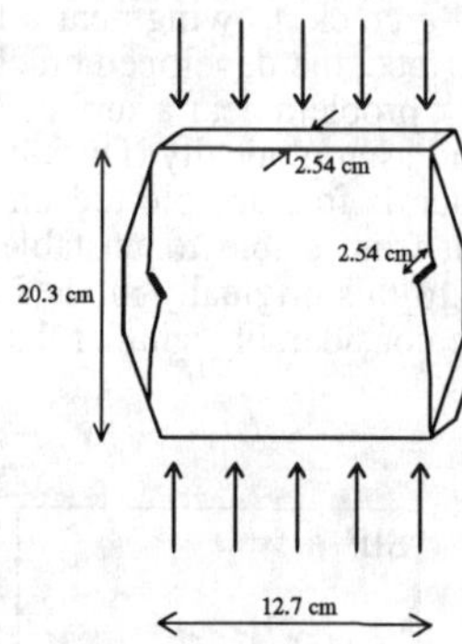

Figure 5. Typical plate with curvilinear boundaries under uniaxial compression. Right crack (45°) propagated unstably at 34.2 MPa. Left crack (30°) propagated unstably at 28.7 MPa.

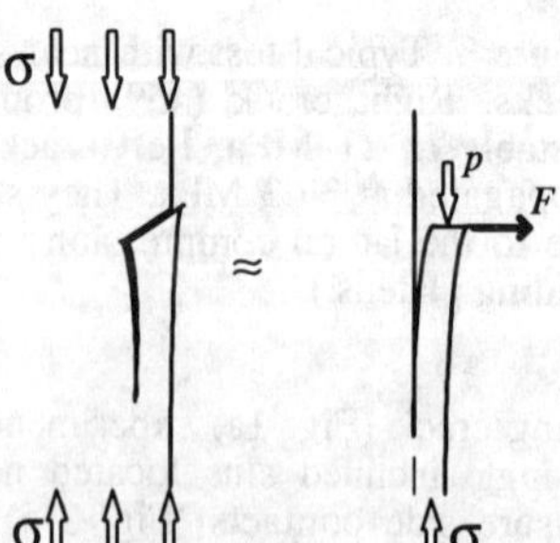

Figure 6. F-C-type model for an edge crack in compression.

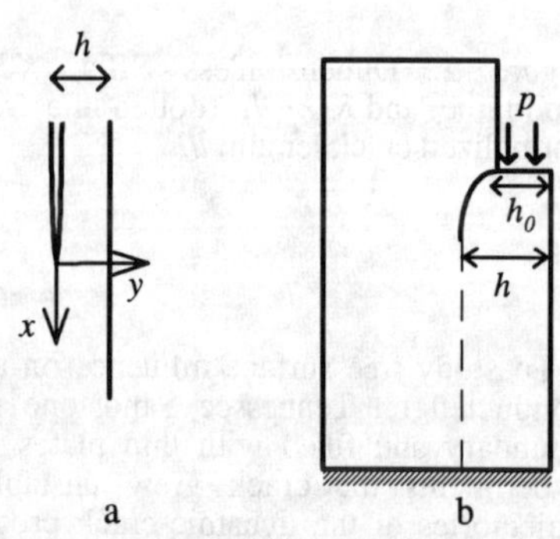

Figure 7. Semi-infinite crack in the half-plane (a) and crack growth near free surface under compression of the layer between the crack and the boundary (b; solid line was observed in the test).

crack propagation, with inertia preventing the crack from changing its path. The same reason could contribute to the straight crack path in the second type of tests (Fig. 4). Note, that in analogous tests on PMMA (Adams and Sines, 1978) this crack immediately propagated towards the free surface. Another reason for this difference was probably in the absence of the contact in Nemat-Nasser and Horii's and Adams and Sines' tests with *empty* slots. The presence of contact in our tests led to the passage of the strong axial compression through the initial crack. This compression could direct the cracks vertically.

4 PATH OF CRACK GROWTH IN THE PRESENCE OF FREE SURFACE

Crack growth in compression near a free surface is, in principle, controlled by three major factors: (1) the wedging action of the initial inclined crack which drives the growing crack in mode I; (2) the presence of the free surface which adds mode II and; (3) the presence of high compressive stress acting approximately along the crack. Including the last point, we recognize that although the compression parallel to the crack does not affect the stress intensity factors, since the real crack process zone has a finite size (not infinitesimally small), high external compressive stresses can be comparable with the crack-generated stress concentration and thus may affect the crack path. However, as the first approximation we use conventional analysis and assume that crack propagate in that direction where $K_{II}=0$

Numerical modeling of extensive crack propagation, especially near a free surface often requires considerable computer power, human effort, and is not always free from algorithm instability. That is why simple analytical or semi-analytical models, supported by numerical calculations, can be of assistance, at least for preliminary analysis. A simple approximate model based on the analytical solution for a semi-infinite crack parallel to a free surface (Zlatin and Khrapkov, 1986) is described here. For the semi-infinite crack situated at distance h from the free surface and subjected to a load with the given forces, (N, T), and moment, M, the SIFs in the coordinate set shown in Fig. 7a have the form

$$\begin{pmatrix} K_I \\ K_{II} \end{pmatrix} = K_M M h^{-3/2} + K_N N h^{-1/2} + \left(K_T - \frac{1}{2} K_M \right) T h^{-1/2} , \tag{4}$$

where

$$K_M = \begin{pmatrix} 1.932 \\ -1.506 \end{pmatrix}, \quad K_N = \begin{pmatrix} 1.951 \\ 0.032 \end{pmatrix}, \quad K_T = \begin{pmatrix} 0.4346 \\ 0.5578 \end{pmatrix} \tag{5}$$

Here (N, T) is the total force owe to the stresses acting on the line of crack continuation. M is the total moment of these stresses about the crack tip. They can be expressed through the total force and moment of the applied load by considering the equilibrium of the layer being separated, i.e., the beam (plate) on the right of the crack in Fig. 7a.

For example, Fig. 7b depicts "chipping" a rock surface by applied compression, acting parallel to a free surface. Such chipping is typical, for instance, for drilling process (see also Cherepanov et al., 1987). The bottom and left side of the figure 7b represent the remote boundaries. Therefore, for this type of loading, the force is locally applied to the small portion of material which is supported at infinity (Fig. 7). If the bottom is loaded uniformly and the sample width is much greater than the thickness, h, of the layer between the crack and the free surface, the pressure on the bottom boundary is very low and a moment is created by the pressure, p, on the top of the layer between the crack and the free surface (this moment acts only within the layer adjacent to the free surface; the whole plate is in equlibrium). This moment alone can open the crack and make it grow. The SIFs for this case will have the form:

$$\begin{pmatrix} K_I \\ K_{II} \end{pmatrix} = ph_0 \left[h^{-1/2} K_T + \frac{1}{2} K_M h^{-3/2} (h - h_0) \right] \tag{6}$$

The criterion K_{II}=0 imposes that the layer should eventually be of the thickness $h\approx 3.86h_0$. However, this model cannot predict the crack path and it is not clear how quickly this final thickness is reached. Fig. 8 shows the results of numerical modeling of the crack trajectory with the FRANC code (Wawrzynek and Ingraffea, 1991). It is seen that the crack trajectory rapidly becomes parallel to the boundary with the distance approximately $3.8h_0$. This demonstrates the applicability of the considered simplified model which also gives acceptable values for SIFs that differ less then by 10% from those given by FRANC.

Cherepanov et al. (1987) introduced a model geometrically similar to Fig. 7. However, they assumed that the crack was of the shear type. As one can see from the above analysis and numerical modeling, in reality the crack is Mode I. Therefore, the "slip toughness," K_{IIc}, obtained by Cherepanov et al. (1987) may be interpreted as Mode I fracture toughness K_{Ic} provided that compression does not additionally influence fracture toughness (see above).

The results obtained with FRANC show that the applied force (pressure p in Fig. 7) has to be slightly increased to propagate the crack until the separated layer thickness becomes almost constant. After that, this force remains of about the same magnitude, the crack keeps growing untill the separated layer buckles. Therefore, the magnitude of the force (per unit thickness) required for the rock chipping can be estimated by equating fracture toughness, K_{Ic}, to K_I, obtained from eq. (6) for h=3.86h_0:

$$T_{\max} = p_{\max} h_0 = 1.71 \cdot K_{Ic} \sqrt{h_0} \tag{7}$$

This is only 20% higher than obtained by Cherepanov et al (1987), i.e., all their experimental results can be explained by the model suggested here. Note again the important conceptual difference: we use Mode I crack propagation in compression and correspondingly Mode I fracture toughness rather then shear crack growth and "slip toughness" K_{IIc} in order to explain plate-like rock pieces produced by certain chipping regimes.

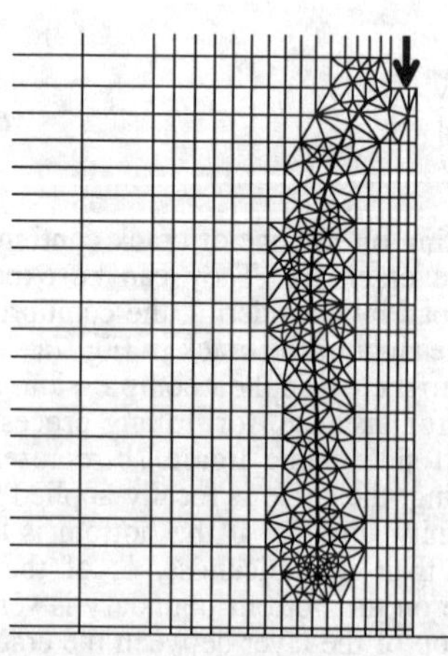

Figure 8. FRANC calculations of the crack trajectory near free surface.

Figure 9. The largest wing cracks developed from single laser-induced cracks in PMMA close to the free surface (Germanovich and Lee, 1993).

The considered result was produced by the special kind of boundary conditions at the top and bottom of the sample. Although compression along a straight crack does not normally make the crack open, the presence of the free surface is found to provide a case when the crack can grow. This happens if the compression on the sample top concentrates between the crack and the free surface, whereas the balancing compression at the bottom is distributed over the whole sample width as shown in Fig. 7b. If balancing compression stresses were concentrated only on the layer part of the sample bottom (like in Fig. 6 but with balanced loads $p=\sigma$), the moment about the crack tip created by the pressure, p, from the top would be balanced by the moment from symmetrically applied pressure from the bottom. Furthermore, due to the equlibrium of the separating layer and the absence of the wedge force, F, in. eq. (4), the shear force $T=0$, normal force $N=0$, and the moment $M=0$. According to (4), this load creates $K_I=0$ and $K_{II}=0$. Such a crack could not grow. After applying the wedge force, F, the SIFs have the form: $K_I=(1.932L/h+1.951)F$; $K_{II}=(-1.506L/h+0.032)F$, where L is the crack length. It is seen that in this case the wedge force, is necessary to make the crack grow.

Using this consideration for the loading shown in Fig. 6, we see that, asymptotically, when the crack size, L, is large and separating layer thickness, h, remains constant, the main term for the Mode I SIF, as follows from (4), comes from the moment, M created by the wedge force, F. Then, $K_I \sim Lh^{-3/2}F$. Since in the tests shown in Fig. 4 the beam thickness was about the same during the crack propagation, K_I was strongly increasing during the test (proportionally to L), which means that crack could not stop and had to propagate dynamically.

5 3-D MECHANISM OF CRACK GROWTH

The above study has been based on the 2-D concept which conventionally presumes the possibility of extensive growth of a single crack under uniaxial compression. While a 2-D approach might be acceptable for a process such as rock chipping (Fig. 7 and 8), growth of an isolated crack in the compression field (Fig. 1) requires special consideration. Such a crack is driven by a single pre-existing defect; inclined crack with contacting faces is the strongest source of extensive crack growth (Dyskin et al, 1993).

In reality, pre-existing cracks in rock are mostly internal and three dimensional. Direct experiments (Dyskin et al, 1993; Germanovich and Lee, 1993) with a variety of transparent (casting resin, boro-silicate glass, plexiglass) and cementicious materials show that a single internal disk-like crack is able to sprout wings at most to an extent of the order of the crack dimensions (Fig. 9). The presence of a free surface insignificantly increases the wing size.

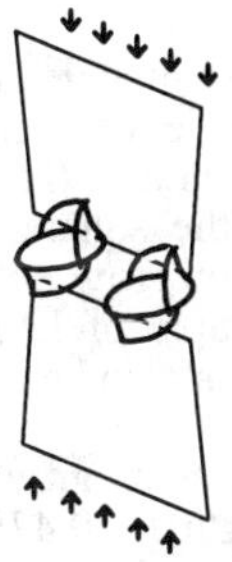

Figure 10. Sketch of tensile cracks produced by wings sprouted from horizontally aligned coplanar cracks (Dyskin et al, 1994).

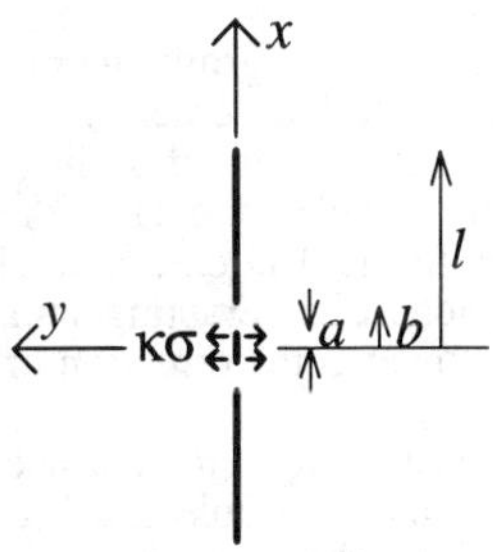

Figure 11. A 2-D model of extensive 3-D crack growth.

However, for samples with two initial cracks there are special arrangements when the cracks starting to grow independently, as single cracks, and eventually produce large tensile cracks towards the direction of compression (Fig. 10). The main feature of these arrangements is that the sprouted wings should be subcoplanar. Obviously, the reason for the appearance of the large cracks is in the corresponding tensile components of the remote stress disturbance produced by the sprouted wings.

Therefore, instead of the conventional kinked crack (Fig. 1a) underpinning the 2-D concept of extensive crack growth, a straight crack driven by the remote tensile stresses generated by certain groups of wing cracks should be discussed. This situation may be approximately modeled by considering the central vertical cross-section of a couple of tensile cracks in a plane-strain approximation. This model is shown in Fig. 11; the small tensile crack in the centre simulates the driving arrangement of wing cracks. This crack is opened by the stress $\kappa\sigma$, where σ is the magnitude of the compressive load; κ is a factor accounting for the crack density in the arrangement. The corresponding remote tensile stress on the x-axis produced by the center crack has the form

$$\Delta\sigma_y = \frac{\kappa a^2 \sigma}{2x^2} + o\left(\frac{a^2}{x^2}\right). \tag{8}$$

This stress field is assumed to be applied to the large cracks. The back influence of the large cracks is neglected, since the central crack models the wings whose opening is determined by the powerful external compressive load and therefore should not be sensitive to the stress disturbances caused by the large cracks.

It can be shown that in the asymptotics of long cracks ($l \gg b$, see Fig. 11) the stress intensity factor at the outer ends of the large cracks has the form:

$$K_I = \frac{a\kappa\sigma}{\sqrt{\pi l}\,(4+\ln l/b)} \tag{9}$$

It is seen that the difference between (9) and (2) is in a logarithmically varying factor. This means that even in the 3-D case the F-C crack is a reasonable approximation. Perhaps this explains why the 3-D model of dilatancy and fracture (Dyskin et al, 1991; Germanovich et al, 1993) based on expression (2) gives a good agreement with experimental data.

It could be concluded that these special arrangements of wing cracks approximately act as concentrated forces wedging the large straight crack in the centre (Fig. 1b).

6 CONCLUSION

In 2-D, extensively growing cracks in uniaxial compression are modeled by the F-C crack which is a straight crack wedged at the centre by a couple of concentrated forces simulating the action of sliding faces of the initial inclined crack. Such a crack alone grows in a stable manner with the length proportional to the square of applied load. Although in the 3-D case a single inclined initial crack cannot grow extensively, certain combinations of interacting initial cracks can produce a couple of large tensile cracks growing towards the direction of compression. This couple can also be modeled by a F-C crack.

The presence of a free surface parallel to the crack amplifies its growth such that the crack becomes unstable after reaching the critical semi-length $l_{cr} \approx 0.47h$, were h is the distance from the surface. This theoretical conclusion is qualitatively valid for real 2-D (through) cracks: if an initial inclined crack is situated close to the free boundary it starts to grow unstably from the very beginning.

Compression along a straight crack does not normally make the crack open hence wedging is required. However, in the presence of a near parallel free surface, there are cases when the compression alone can drive the crack. In particular, this happens if the compression on the sample top concentrates between the crack and the free surface, whereas the balancing compression at the bottom is distributed over the whole sample width. Then the crack is opened by the moment acting on the layer between the crack and the surface. This may be a mechanism of rock chipping. After the short initial stage of growing into the rock the crack then propagates parallel to the free surface as a mode I crack until the separated layer becomes slender and buckles.

ACKNOWLEDGMENT

The authors are thankful to J.-C. Roegiers, K.B. Ustinov, and P.A. Wawrzynek for useful discussions. The authors are also grateful to E.A. Haas, P.M. Lemmon, and D. Snyder for their assistance with the tests described in this paper.

REFERENCES

Adams, M. & Sines, G. 1978. Crack extension from flaws in a brittle material subjected to compression. *Tectonophysics* 49: 97-118.

Ashby, M.F. & Hallam, S.D. 1986. The failure of brittle solids containing small cracks under compressive stress states. *Acta Metallurgica* 34: 497-510.

Brace, W.F. & Bombolakis, E.G. 1963. A note on brittle crack growth in compression. *Journal of Geophysical Research* 68: 3709-3713.

Cherepanov, G.P., Vorozhtsov, M.I., & Eigles, R.M. Rock cutting. *Soviet Physics-Doklady* 32(9): 728-730.

Dyskin, A.V. & Germanovich, L.N. 1993. Rock burst model based on cracks growing near free surface. Rockbursts and Seismicity in Mines 93. R.P. Young (ed.). *Proceedings of 3rd International Symposium*: 169-173. Kingston, Ontario, Canada.

Dyskin, A.V., Germanovich, L.N., & Salganik, R.L. 1991. A mechanism of deformation and fracture of brittle rocks. *Proceedings of the 32nd U.S. Symposium on Rock Mechanics*:181-190. A.A. Balkema Publishers:Rotterdam, Brookfield,

Dyskin, A.V., Germanovich, L.N., & Ustinov, K.B. 1993. On the cracks growing from a pre-existing pore in compression. *International Journal of Fracture* 62: 307-324.

Dyskin, A.V. & Salganik, R.L. 1987. Model of dilatancy of brittle materials with cracks under compression. *Mechanics of Solids* 22 (6): 165-173.

Dyskin, A.V., Jewell, R.J., Joer, H., Sahouryeh, E. & Ustinov, K.B. 1993. Experiments on 3-d crack growth in uniaxial compression. *The University of Western Australia. Geomechanics Group:* Report G1116 (Submitted to *Intern. J. Fracture*).

Dyskin, A.V., Germanovich, L.N., Jewell, R.J., Joer, H., Krasinski, J.S., Lee, K.K, Roegiers, J.-C., Sahouryeh, E. & Ustinov, K.B. 1994. Study of 3-d mechanisms of crack growth and interaction in uniaxial compression. *ISRM News Journal* (in print).

Fairhurst, C. & Cook N.G.W. 1966. The phenomenon of rock splitting parallel to the direction of maximum compression in the neighborhood of a surface. *Proc. First Congress International Society for Rocks Mechanics*. Lisbon.

Germanovich, L.N., Dyskin, A.V., & Tsyrulnikov, M.N. 1993. A model of brittle rock deformations under compression. *Mechanics of Solids* 28(1).

Germanovich, L.N. & Lee, K.K. 1993. Study of three-dimensional crack initiation and propagation in transparent brittle material subject to compression. Part II. Development of experimental techniques. *Rock Mechanics Research Center at The University of Oklahoma*: Report RMRC-93-14.

Germanovich, L.N. & Grekov, M. A. 1994. Influence of free surface. In: Germanovich, L.N., Dmitriev, A.P., & Goncharov, S.A. *Rock Fracture Thermomechanics*, Gordon & Breach Publishers, London, New York, etc. (in print).

Hoek, E. & Bieniawski, Z.T. 1965. Brittle fracture propagation in rock under compression. *International Journal of Fracture* 1: 137-155.

Horii, H. & Nemat-Nasser, S. 1986. Brittle failure in compression: splitting, faulting, & brittle-ductile transition. *Phil. Trans. R. Soc. Lond.* A 319:337-374.
Ingraffea, A.R., & Heuze 1980. Finite element models for rock fracture mechanics. *Int. J. for Numer. & Analit. Methods in Geomechanics* 4: 25-43.
Wawrzynek, P.A., & Ingraffea, A.R. 1991. Discrete modelling of crack propagation: theoretical aspects & implementation issues in two & three dimensios. *School of Civil & Enviromental Engineering, Cornell University*: Report 91-5.
Murakami, Y. 1987. *Stress intensity factors handbook.* Vol. 1, Pergamon Press, Oxford, New York.
Nemat-Nasser, S. & Horii, H. 1982. Compression-induced nonplanar crack extension with application to splitting, exfoliation, & rockburst. *Journal of Geophysical Research.* 87(B8): 6805-6821.
Zlatin A.N. & Chrapkov A.A. 1986. A semi-infinite crack parallel to the boundary of the elastic half-plane. *Soviet Physics-Doklady.* 31(12): 1009-1010.

8 Static and dynamic properties of intact rock

Session developers:
Mukul Sharma, The University of Texas at Austin
Michael Batzle, ARCO Exploration and Production Technology

Rock Mechanics, Nelson & Laubach (eds) © 1994 Balkema, Rotterdam, ISBN 90 5410 380 8

Static and dynamic mechanical properties of carbonates

D.P. Yale
Mobil Research and Development Corporation, Dallas, Tex., USA

W.H. Jamieson, Jr
Mobil Exploration and Producing U.S., Liberal, Kans., USA

ABSTRACT: An understanding of the mechanical properties of petroleum reservoirs is important for drilling, well stimulation and horizontal well development. These mechanical properties can be measured in the laboratory from triaxial tests ("static" tests) or they can be estimated from wireline measurements of compressional and shear acoustic velocities and rock density ("dynamic" tests).

There can significant differences between static and dynamic mechanical properties in most rocks. Static measurements on cores are much more indicative of the mechanical properties of the reservoir than the dynamic results, however, the information from acoustic well logs covers much more of the reservoir than core measurements and is less expensive. This makes it important to be able to translate the "dynamically" derived mechanical properties to the static values which better represent the reservoir.

A comparison of static and dynamic mechanical properties was carried out on 85 core samples from two wells in the Chase and Council Grove carbonate sequences of the Hugoton and Panoma fields, Kansas. The purpose of the study was to characterize the mechanical properties of the different facies and calibrate the dynamic mechanical properties so acoustic well logs from other wells can be utilized more effectively in determining areal and lithologic variation in mechanical properties of the field.

Results of the tests show that Young's modulus correlates strongly with lithofacies and porosity and that the carbonate sequences can be separated into six "mechanical facies", defined by their modulus-porosity trends and their lithology. Poisson's ratio is less sensitive to lithology but correlates with porosity. The dynamic Young's moduli are 15% to 40% higher than the static values. Dynamic and static Poisson's ratios correlate well with each other for liquid saturated samples.

Dynamic to static transforms based on lithofacies were developed from these data to correct the acoustic log derived mechanical properties to static values appropriate for reservoir deformations. The corrected mechanical properties along with the trends observed between mechanical properties, lithofacies, and porosity allow for the design of more effective hydraulic fracture treatments by utilizing accurate values for the reservoir's mechanical properties and their variability.

1 PURPOSE OF STUDY

Information on the rock mechanical properties of petroleum reservoirs influences decisions on drilling, completion, and stimulation of the well. Variability of properties with depth and areal extent control the height and length of stimulation fractures and what regions of the wellbore will be stable or unstable, especially in horizontal wells. Poisson's ratio, Young's modulus, and the Biot alpha parameter are all important input into the design of hydraulic fracture treatments.

This study investigates the dynamic and static mechanical properties of carbonate and clastic sequences of the Chase and Council Grove Groups in the Hugoton and Panoma fields in Kansas. The samples cover the porosity and lithofacies range of the hydrocarbon producing intervals and the siltstones separating them. The purpose of the study was to characterize the mechanical properties of the different reservoir formations so that hydraulic fracture treatments and horizontal well completions could be more effectively designed. Transforms between static and dynamic mechanical properties were developed so that acoustic logs could be used throughout the field to characterize the areal variability in mechanical properties.

Differences between static and dynamic mechanical properties are caused by the "anelasticity" of reservoir rocks and the differing effects of pore fluid saturation on static and acoustic measurements. The high frequency of dynamic measurements lead the pore fluid to support some of the elastic loading of the rock (undrained conditions) whereas the low frequency of static tests allow the pore fluid to drain out of the regions under stress (drained conditions) . Wide differences in the frequency and loading conditions of static and dynamic tests allow static tests to sample the elastic and anelastic portions of the rock deformation where acoustic measurements sample only the elastic response.

2 GEOLOGIC FRAMEWORK

2.1 *Structural and Stratigraphic Setting*

The Hugoton and Panoma fields are within the tectonic feature termed the Hugoton Embayment which is a northwest extension of the Anadarko Basin. The Hugoton field is one of the largest gas fields in the U.S., stretching 175 miles from just north of Amarillo, Texas, through the Oklahoma panhandle to about 20 miles north of Garden City, Kansas; the width of the field varies from 20 to 50 miles. It produces gas from the Lower Permian Chase Group whereas the Panoma field (which is much smaller and is within the geographic boundaries of the Hugoton field) produces gas from the Lower Permian Council Grove Group which underlies the Chase Group. The focus of this study are those parts of the field which lie in southwest Kansas. The structural configuration of the Hugoton field area is gradual dip from west to east of less than one degree. Faulting is not present over a majority of the field.

2.2 *Chase and Council Grove Group Lithologies*

The Chase and Council Grove Groups have a composite thickness of 650 feet in the study area. The two groups consist of a repetitive sequence of cycles which typically begin with a clastic-rich marine carbonate and terminate with a "red bed" siltstone. The Chase Group consists of six of these cycles and the Council Grove consists of a minimum of seven of these cycles. The carbonate portion of each cycle ranges from 5 to 70 feet in thickness and the siltstone portions range from 5 to 20 feet in thickness.

Diagenesis has significantly altered the lithology of the Chase and Council Grove Groups. Dolomitization is pervasive in the upper half of the Chase Group and diminishes downward toward the base. Dolomite is less common in the carbonate interval of the Council Grove. Diagenesis has also altered the siltstone intervals, which have been extensively cemented by dolomite and/or calcite to the point where many of the siltstones contain more than 20% carbonate. Another product of diagenesis in these groups is anhydrite replacement of carbonate material, which is common in both the Chase and Council Grove. Some intervals within the Chase contain over 20% anhydrite. All of these diagenetic products have a significant impact on the mechanical properties of the Chase and Council Grove reservoir intervals.

3 EXPERIMENTAL METHODS

3.1 *Acoustic velocities and dynamic mechanical properties*

All measurements were done on 1 inch diameter by 2 inch long vertical core plugs. Before mechanical or acoustic testing, porosity and grain density of the samples were measured under ambient (no stress) conditions.

The acoustic velocities (compressional and shear) of the samples were then determined at 1500 psi laboratory confining pressure and without pore fluid ("drained" conditions). Frequencies of the acoustic measurements were approximately 700 kHz for the P-wave and 400 kHz for the S-wave. The system was calibrated using aluminum and steel plugs of known velocity. Dynamic-drained bulk modulus and shear modulus were calculated from the shear and compressional velocities and the bulk density of the sample (Jaeger and Cook, 1976).

Young's modulus and Poisson's ratio were calculated from the elasticity equations and the bulk and shear moduli.

To facilitate comparison with acoustic log data, dynamic-undrained moduli were calculated from the dynamic-drained measurements. Part of the difference between static and dynamic log derived mechanical properties is the fact that compressional velocities are higher in fully liquid saturated rock (Gassmann, 1951) than in gas saturated rock. This higher compressional velocity (and higher moduli and Poisson's ratio) is due to the pore fluid supporting part of the elastic load as the wave passes (called "undrained" conditions). The frequency of static measurements is such that pore fluid can move away from the stressed zone and does not support any of the load even in fully liquid saturated cases (called "drained" conditions).

The equation which relates the "drained" dynamic bulk modulus, K_d (i.e. dry or gas saturated case) to the "undrained" or water/mud saturated dynamic bulk modulus, K_s is:

$$K_d = (K_s*(1/K_m + \phi/K_w - \phi/K_m) - 1) / ((1/K_m + \phi/K_w - \phi/K_m) + K_s/K_m^2 - 2/K_m)$$

where K_w and K_m are the bulk moduli of pore fluid (water/mud) and rock matrix respectively. This is taken from Geertsma's (1961) analysis of Gassmann's (1951) original formulation. The shear modulus is not affected by the pore fluid since the shear strength of all fluids is zero.

3.2 *Static Mechanical Properties*

Static mechanical properties were done on the same core samples as the acoustic tests. The measurements were made in a triaxial (true biaxial) cell by recording axial and volumetric strain as the sample was loaded axially under constant confining stress without pore fluid. A confining stress of 1500 psi was used for all samples to simulate approximate reservoir stress conditions. Axial differential stresses were taken to 5000 psi at strain rates of 2 to 8 X 10^{-6} per second. Axial and volumetric strain were measured over the increasing and decreasing stress portions of the test to eliminate any hysteresis effects. Tangent Young's moduli and Poisson's ratio were then calculated and averaged over the increasing and decreasing stress cycles. The system was calibrated using an aluminum slug of know properties before the tests and after every set of 20 core tests.

4 RESULTS

4.1 *Static mechanical properties and mechanical facies*

As shown in Tables 1a and 1b and in Figure 1, the various lithofacies appear to group together into six "mechanical facies". These "mechanical facies" are groups of geologically similar facies which have similar mechanical properties and show a linear trend in Young's modulus versus porosity. These six mechanical facies are "dolostones w/anhydrite", "dolostones", "limestones", "siltstones w/dolomite", "siltstones", and "mudstones". The mechanical facies have very different Young's moduli versus porosity trends.

The type, weight fraction, and distribution of matrix and cement material appears to control which mechanical facies a sample falls into. For example, "dolostones w/anhydrite" have higher moduli than the "dolostones" since anhydrite is much stiffer than calcite or dolomite. Similarly the "dolostones" have higher Young's modulus than the "limestones" since dolomite is stiffer than calcite. There also appears to be a minimum amount of "stiffer" matrix material or cement which is needed to give the rock a higher modulus. Analysis of the XRD (X-ray

Table 2

Facies	Anhydrite	Dolomite	Calcite	Quartz	Clay
Dolostones w/anhyd	>8%	>40%			<15%
Dolostones		>30%		<10% quartz and clay	
Limestones		<30%	>50%	<10%	
Siltstones		<25%		>40%	>15%
Siltstones w/dolo cem		>30%		>30%	<15%
Mudstones		~20%	~50%	~15%	~15%

Table 1a

Mechanical Facies	Depth (ft)	Porosity (fraction)	Vp (ft/s) (dry)	Vs (ft/s) (dry)	Matrix Density (g/cc)	Poisson's Ratio			Young's Modulus (psi)			Dynamic/Static Ratio		
						(undrain) (dynam)	(drained) (dynam)	(drained) (static)	(undrain) (dynam)	(drained) (dynam)	(drained) (static)	Youngs (undrained)	Youngs (drained)	Shear
dolo w/anhyd	2506.6	0.035	18604	10405	2.91	0.288	0.272	0.252	1.06E+07	1.04E+07	9.18E+06	1.15	1.14	1.12
dolo w/anhyd	2557.2	0.057	18913	10763	2.88	0.273	0.261	0.301	1.08E+07	1.07E+07	8.79E+06	1.23	1.21	1.25
dolo w/anhyd	2557.6	0.070	17574	10294	2.86	0.264	0.239	0.236	9.59E+06	9.40E+06	7.54E+06	1.27	1.25	1.24
dolo w/anhyd	2567.0	0.084	18922	10829	2.85	0.268	0.256	0.242	1.05E+07	1.04E+07	8.13E+06	1.29	1.27	1.26
dolo w/anhyd	2566.9	0.086	18598	10659	2.85	0.269	0.255	0.260	1.01E+07	1.00E+07	7.41E+06	1.36	1.35	1.35
dolo w/anhyd	2596.6	0.104	17832	10584	2.81	0.250	0.228	0.246	9.49E+06	9.32E+06	7.69E+06	1.23	1.21	1.23
dolo w/anhyd	2596.7	0.122	17851	10746	2.81	0.244	0.216	0.243	9.56E+06	9.34E+06	7.41E+06	1.29	1.26	1.29
dolo w/anhyd	2710.7	0.229	14161	8422	2.84	0.263	0.226	0.316	5.28E+06	5.13E+06	4.72E+06	1.12	1.09	1.17
dolo w/anhyd	2711.1	0.247	13992	8251	2.84	0.266	0.233	0.303	4.97E+06	4.84E+06	3.86E+06	1.29	1.25	1.32
dolostone	2695.0	0.038	17855	10198	2.76	0.281	0.258	0.239	9.52E+06	9.35E+06	7.51E+06	1.27	1.24	1.23
dolostone	2695.1	0.039	17891	10324	2.69	0.269	0.250	0.249	9.43E+06	9.29E+06	7.55E+06	1.25	1.23	1.23
dolostone	2506.3	0.093	15763	9564	2.82	0.255	0.209	0.241	7.92E+06	7.63E+06	5.16E+06	1.53	1.48	1.52
dolostone	2506.6	0.094	15888	9782	2.80	0.244	0.195	0.181	8.14E+06	7.82E+06	5.38E+06	1.51	1.45	1.44
dolostone	2503.6	0.102	15194	9126	2.90	0.263	0.218	0.239	7.38E+06	7.12E+06	6.10E+06	1.21	1.17	1.19
dolostone	2503.7	0.108	14875	9123	2.90	0.252	0.199	0.236	7.26E+06	6.95E+06	6.05E+06	1.20	1.15	1.18
dolostone	2532.3	0.134	14421	8967	2.82	0.245	0.185	0.272	6.59E+06	6.27E+06	5.51E+06	1.19	1.14	1.22
dolostone	2532.6	0.138	14867	9225	2.82	0.240	0.187	0.283	6.92E+06	6.62E+06	5.52E+06	1.25	1.20	1.30
dolostone	2663.8	0.144	13438	8314	2.83	0.254	0.187	0.256	5.66E+06	5.37E+06	4.58E+06	1.24	1.17	1.24
dolostone	2676.5	0.155	13981	8569	2.73	0.254	0.199	0.264	5.72E+06	5.46E+06	4.36E+06	1.31	1.25	1.32
dolostone	2663.5	0.170	13059	7926	2.84	0.266	0.189	0.281	5.05E+06	4.82E+06	4.16E+06	1.22	1.16	1.23
dolostone	2719.8	0.207	12554	7777	2.79	0.252	0.189	0.350	4.52E+06	4.29E+06	3.60E+06	1.26	1.19	1.42
dolostone	2715.2	0.250	11594	7255	2.90	0.254	0.178	0.380	3.87E+06	3.63E+06	2.87E+06	1.35	1.27	1.48
limestone	2518.5	0.034	16356	9444	2.71	0.288	0.250	0.233	8.10E+06	7.86E+06	5.94E+06	1.36	1.32	1.31
limestone	2652.5	0.043	14421	9210	2.70	0.262	0.156	0.215	7.46E+06	6.83E+06	5.24E+06	1.42	1.30	1.37
limestone	2652.3	0.052	15093	9388	2.70	0.257	0.185	0.223	7.65E+06	7.21E+06	5.85E+06	1.31	1.23	1.27
limestone	2518.8	0.058	15576	9102	2.71	0.281	0.241	0.289	7.30E+06	7.07E+06	5.58E+06	1.31	1.27	1.32
limestone	2509.2	0.059	15240	9335	2.71	0.259	0.200	0.248	7.54E+06	7.19E+06	5.62E+06	1.34	1.28	1.33
limestone	2641.5	0.067	16049	10002	2.71	0.233	0.182	0.248	8.39E+06	8.05E+06	6.10E+06	1.38	1.32	1.39
limestone	2644.8	0.074	15678	9329	2.71	0.265	0.226	0.231	7.44E+06	7.21E+06	5.93E+06	1.26	1.22	1.22
limestone	2644.5	0.079	15469	9138	2.71	0.270	0.232	0.231	7.13E+06	6.92E+06	6.00E+06	1.19	1.15	1.15
limestone	2641.7	0.080	16349	9395	2.71	0.277	0.253	0.256	7.56E+06	7.42E+06	6.25E+06	1.21	1.19	1.19
limestone	2743.3	0.082	14721	9073	2.74	0.252	0.194	0.268	6.99E+06	6.66E+06	5.51E+06	1.27	1.21	1.28
limestone	2755.3	0.087	15007	9163	2.71	0.260	0.203	0.260	7.05E+06	6.73E+06	5.02E+06	1.40	1.34	1.40
limestone	2743.6	0.094	14234	9156	2.74	0.227	0.147	0.261	6.88E+06	6.43E+06	4.88E+06	1.41	1.32	1.45
limestone	2763.1	0.102	16225	9259	2.70	0.280	0.259	0.227	7.16E+06	7.04E+06	5.84E+06	1.23	1.21	1.18
limestone	2755.6	0.107	14792	9133	2.72	0.250	0.192	0.216	6.81E+06	6.49E+06	5.09E+06	1.34	1.27	1.30
limestone	2762.5	0.118	15983	8916	2.70	0.293	0.274	0.272	6.59E+06	6.50E+06	5.63E+06	1.17	1.15	1.15
limestone	2762.3	0.119	15913	9045	2.70	0.283	0.261	0.273	6.72E+06	6.61E+06	5.66E+06	1.19	1.17	1.18
limestone	2822.0	0.134	14490	8221	2.69	0.294	0.263	0.276	5.49E+06	5.35E+06	4.33E+06	1.27	1.24	1.25
limestone	2582.4	0.135	13801	8299	2.71	0.268	0.217	0.304	5.52E+06	5.29E+06	4.25E+06	1.30	1.25	1.34
limestone	2582.6	0.140	13867	8562	2.71	0.236	0.192	0.291	5.69E+06	5.48E+06	4.44E+06	1.28	1.24	1.34
limestone	2822.3	0.141	14990	8501	2.69	0.290	0.263	0.323	5.80E+06	5.68E+06	4.42E+06	1.31	1.29	1.35
limestone	2858.4	0.209	12178	7450	2.71	0.267	0.201	0.347	4.07E+06	3.85E+06	3.18E+06	1.28	1.21	1.36
limestone	2858.6	0.211	12068	7411	2.71	0.266	0.197	0.328	4.00E+06	3.78E+06	3.23E+06	1.24	1.17	1.30

Table 1b

Mechanical Facies	Depth (ft)	Porosity (fraction)	Vp (ft/s) (dry)	Vs (ft/s) (dry)	Matrix Density (g/cc)	Poisson's Ratio (undrain) (dynam)	Poisson's Ratio (drained) (dynam)	Poisson's Ratio (drained) (static)	Young's Modulus (psi) (undrain) (dynam)	Young's Modulus (psi) (drained) (dynam)	Young's Modulus (psi) (drained) (static)	Dynamic/Static Ratio Youngs (undrained)	Dynamic/Static Ratio Youngs (drained)	Dynamic/Static Ratio Shear
mudstone	2867.8	0.030	14311	9126	2.72	0.278	0.157	0.190	7.57E+06	6.85E+06	5.45E+06	1.39	1.26	1.29
mudstone	2868.3	0.031	13968	8993	2.71	0.281	0.146	0.213	7.34E+06	6.56E+06	4.93E+06	1.49	1.33	1.41
mudstone	2881.4	0.043	14462	8991	2.68	0.276	0.185	0.245	7.13E+06	6.62E+06	5.09E+06	1.40	1.30	1.37
mudstone	2880.3	0.044	14458	9038	2.67	0.273	0.179	0.198	7.14E+06	6.62E+06	5.17E+06	1.38	1.28	1.30
mudstone	2508.8	0.057	13027	8524	2.72	0.254	0.126	0.201	6.29E+06	5.64E+06	3.81E+06	1.65	1.48	1.58
silt w/dol	2631.3	0.080	14580	9007	2.77	0.235	0.191	0.254	6.88E+06	6.64E+06	4.99E+06	1.38	1.33	1.40
silt w/dol	2774.8	0.092	12741	8126	2.72	0.244	0.157	0.236	5.46E+06	5.08E+06	5.44E+06	1.00	0.93	1.00
silt w/dol	2497.6	0.097	13858	9120	2.76	0.215	0.118	0.172	6.79E+06	6.25E+06	4.38E+06	1.55	1.43	1.50
silt w/dol	2497.3	0.098	13747	9151	2.76	0.208	0.102	0.261	6.79E+06	6.19E+06	4.71E+06	1.44	1.31	1.50
silt w/dol	2390.0	0.100	12283	8354	2.73	0.231	0.070	0.274	5.69E+06	4.94E+06	5.01E+06	1.14	0.99	1.18
silt w/dol	2390.2	0.100	13050	8654	2.76	0.231	0.108	0.253	6.17E+06	5.55E+06	4.87E+06	1.27	1.14	1.29
silt w/dol	2780.6	0.136	12906	8282	2.72	0.224	0.150	0.243	5.32E+06	5.00E+06	4.26E+06	1.25	1.17	1.27
silt w/dol	2780.3	0.140	12620	8181	2.72	0.220	0.138	0.230	5.15E+06	4.80E+06	3.60E+06	1.43	1.33	1.44
silt w/dol	2781.3	0.150	11752	7814	2.72	0.217	0.104	0.248	4.62E+06	4.19E+06	3.65E+06	1.27	1.15	1.30
silt w/dol	2638.2	0.176	12972	8104	2.71	0.236	0.180	0.276	4.88E+06	4.66E+06	3.36E+06	1.45	1.38	1.50
silt w/dol	2638.5	0.180	12993	8042	2.70	0.242	0.189	0.277	4.79E+06	4.59E+06	3.31E+06	1.45	1.39	1.49
siltstone	2629.6	0.050	13057	8605	2.66	0.229	0.116	0.141	6.20E+06	5.63E+06	4.17E+06	1.48	1.35	1.38
siltstone	2629.1	0.051	12845	8781	2.69	0.209	0.061	0.158	6.42E+06	5.64E+06	4.12E+06	1.56	1.37	1.49
siltstone	2629.8	0.065	11948	8660	2.69	0.184	-0.054	0.189	6.02E+06	4.82E+06	3.97E+06	1.52	1.21	1.52
siltstone	2631.4	0.073	13343	8722	2.77	0.207	0.127	0.291	6.35E+06	5.93E+06	4.19E+06	1.52	1.42	1.62
siltstone	2383.7	0.085	12641	7959	2.73	0.244	0.172	0.181	5.30E+06	5.00E+06	3.79E+06	1.40	1.32	1.33
siltstone	2630.0	0.086	11944	8018	2.70	0.220	0.090	0.186	5.21E+06	4.65E+06	3.34E+06	1.56	1.39	1.52
siltstone	2630.2	0.088	12131	8032	2.71	0.223	0.110	0.180	5.26E+06	4.77E+06	3.43E+06	1.53	1.39	1.48
siltstone	2762.6	0.094	12939	8599	2.70	0.199	0.104	0.222	5.84E+06	5.38E+06	3.78E+06	1.54	1.42	1.57
siltstone	2616.1	0.096	11111	7599	2.68	0.227	0.061	0.250	4.63E+06	4.01E+06	3.42E+06	1.35	1.17	1.38
siltstone	2691.7	0.096	12630	8776	2.73	0.168	0.033	0.189	5.99E+06	5.30E+06	4.38E+06	1.37	1.21	1.39
siltstone	2616.5	0.097	11461	7842	2.69	0.215	0.060	0.228	4.88E+06	4.26E+06	3.90E+06	1.25	1.09	1.26
siltstone	2514.7	0.097	12202	8170	2.73	0.208	0.094	0.210	5.35E+06	4.84E+06	3.65E+06	1.47	1.33	1.47
siltstone	2382.0	0.099	12637	8401	2.73	0.202	0.104	0.228	5.62E+06	5.16E+06	3.58E+06	1.57	1.44	1.61
siltstone	2624.1	0.105	11753	7924	2.73	0.211	0.083	0.249	5.00E+06	4.48E+06	3.19E+06	1.57	1.40	1.62
siltstone	2578.7	0.114	11746	7853	2.72	0.214	0.096	0.211	4.86E+06	4.39E+06	3.70E+06	1.31	1.19	1.31
siltstone	2514.3	0.114	11441	7928	2.72	0.195	0.038	0.201	4.88E+06	4.24E+06	3.61E+06	1.35	1.18	1.36
siltstone	2691.7	0.115	12141	8164	2.70	0.200	0.087	0.269	5.16E+06	4.67E+06	3.45E+06	1.50	1.36	1.58
siltstone	2671.0	0.117	10594	7334	2.67	0.224	0.040	0.207	4.17E+06	3.55E+06	3.32E+06	1.26	1.07	1.24
siltstone	2762.9	0.118	12047	7987	2.71	0.212	0.108	0.239	4.98E+06	4.55E+06	3.22E+06	1.54	1.41	1.58
siltstone	2624.3	0.120	12919	8101	2.77	0.231	0.176	0.306	5.30E+06	5.06E+06	3.47E+06	1.52	1.46	1.62
siltstone	2383.4	0.122	10789	7585	2.73	0.200	0.011	0.197	4.46E+06	3.76E+06	2.71E+06	1.64	1.39	1.64
siltstone	2484.0	0.125	12140	7835	2.73	0.226	0.143	0.319	4.84E+06	4.52E+06	3.52E+06	1.38	1.28	1.48
siltstone	2484.3	0.128	11936	7861	2.72	0.216	0.117	0.265	4.80E+06	4.41E+06	3.47E+06	1.38	1.27	1.44
siltstone	2579.0	0.131	10849	7666	2.70	0.191	0.001	0.195	4.43E+06	3.72E+06	3.16E+06	1.40	1.18	1.41
siltstone	2671.3	0.135	11017	7440	2.68	0.220	0.081	0.254	4.22E+06	3.74E+06	3.42E+06	1.24	1.10	1.27
siltstone	2493.3	0.168	10343	6812	2.70	0.243	0.117	0.243	3.50E+06	3.14E+06	2.06E+06	1.70	1.52	1.70

diffraction) results suggests that about 30% dolomite is required to increase the modulus of the limestone or siltstone significantly but less than 10% anhydrite is required to increase the modulus of a dolomite.

Comparison of these mechanical facies with the XRD data show strong correlations between the mechanical facies and the percent of certain minerals in the samples. Table 2 summarizes the ranges of mineral compositions found in the various mechanical facies.
Some rocks which mechanically fall into the "dolostones" category have >10% anhydrite but their high clay content (20%) lowers their modulus. Similarly some rocks which fall into the "limestone" mechanical facies have >30% dolomite but have high quartz or clay percentages (>15%) which lowers their modulus. Also some dolomite cemented siltstones with >30% dolomite and high clay contents (25%) have moduli into the "siltstone" category. Figure 1 also shows a small set of rocks which are geologically described as "mudstones". Their mineral content places them as borderline "limestones" to "dolostones" but their texture or distribution of minerals in the rocks causes them to behave mechanically much more like "siltstones".

4.2 *Static and dynamic Young's modulus*

A comparison of the static measurements and dynamic calculations of Young's modulus are presented in Tables 1a and 1b and in Figure 2. This shows that the dynamic-undrained moduli (which are representative of properties calculated from acoustic logs) are 15% to 70% higher than the static moduli. Even after correcting for the pore fluid, the dynamic-drained moduli are still 10% to 40% higher than the static moduli. From the data in Figure 2 we can derive transforms to correct acoustic log derived Young's moduli to static moduli. We find that the transform depends on lithology. A 0.79 factor corrects the limestone and dolostone log derived moduli to static values, a 0.73 factor corrects the siltstones with dolomite cement and a 0.68 factor corrects the siltstones and mudstones to static values.

We note that the difference between the drained and undrained dynamic moduli is not great. The effect of the pore fluid on acoustic velocities depends on the ratio of the bulk modulus of the rock frame to the modulus of the pore fluid. Since the bulk modulus of the rock frame for most of the samples is more than 10 times the bulk modulus of water, there are only small drained to undrained differences.

4.3 *Static and dynamic Poisson's ratio*

Figure 3 shows static Poisson's ratio as a function of porosity and lithology. There is no systematic difference in Poisson's ratio between the various carbonate facies, however, there does appear to be a slight linear trend between porosity and Poisson's ratio for the carbonates.

Tables 1a and 1b and Figure 4 show the relationships between static and dynamic Poisson's ratio. Unlike Young's modulus there is significantly more scatter in this relationship. In addition, there is a much larger difference between the drained and undrained dynamically calculated Poisson's ratio. Figure 4 shows that the dynamic-drained (dry) Poisson's ratio are significantly lower that the static (dry) Poisson's ratios.

Since liquid saturation in the undrained case increases the dynamic-undrained Young's modulus without affecting the shear modulus, the dynamic-undrained Poisson's ratio is higher (see Figure 4) than the dynamic-drained Poisson's ratio. It is coincidental that the difference between dynamic-drained and undrained is nearly equal to the difference between static and dynamic-drained. This leads to the dynamic-undrained Poisson's ratios being a fair match to the static values of Poisson's ratio. Because of the liquid saturation in the near wellbore area, acoustic logs yield dynamic-undrained Poisson's ratios, which in this case match fairly well with the static values (see Figure 4). However, note that the range of dynamic-undrained values is smaller than the range of static values for Poisson's ratio. This suggests that the log derived values will show smaller differences in Poisson's ratio between zones than static measurements.

4.4 *Reasons for static and dynamic differences*

A number of studies have shown the differences between static and dynamic mechanical properties (Howarth, 1987; Jizba and Nur, 1990; King, 1967; Montmayeur and Graves, 1986; Nieto and Yale, 1991; Tutuncu and Sharma, 1992; Yale, 1993). Most of the studies show that the dynamic moduli (Young's, shear, and bulk) are higher than the static values. There is less consistency in the correlations between static and dynamic Poisson's ratios. The two reasons for the differences are 1) liquid saturation effects (Gassmann differences) and 2) frequency

FIGURE 1

Young's Modulus versus Porosity and Lithology

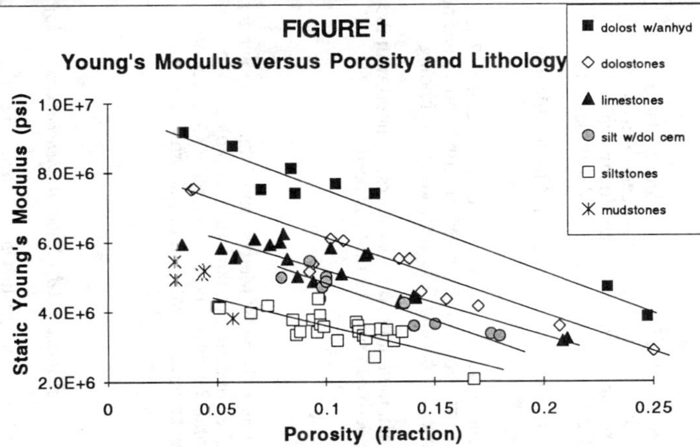

FIGURE 3

Poisson's Ratio versus Porosity

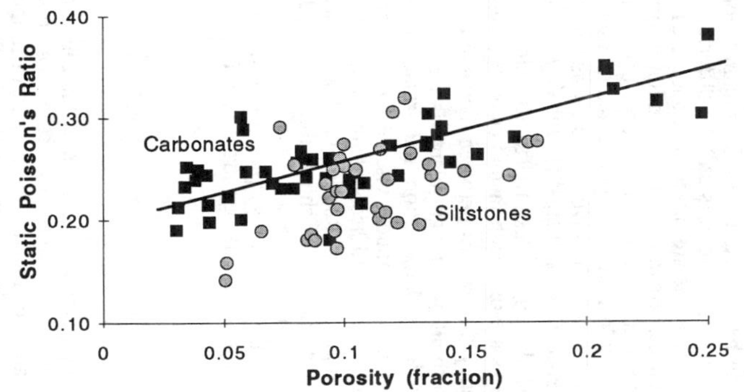

FIGURE 2

Dynamic/Static Young's Modulus Ratio

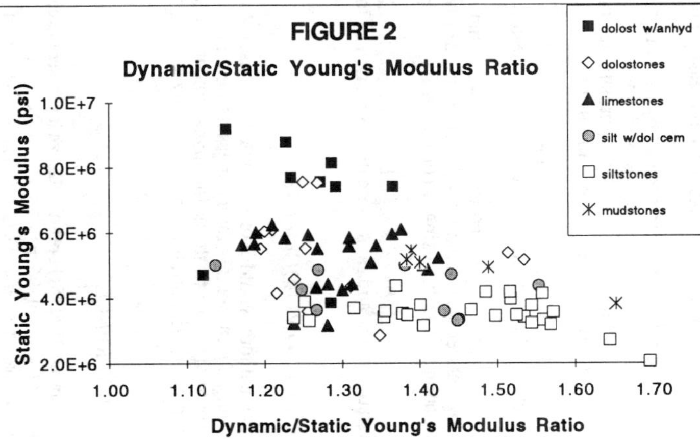

FIGURE 4

Static versus Dynamic Poisson's ratio

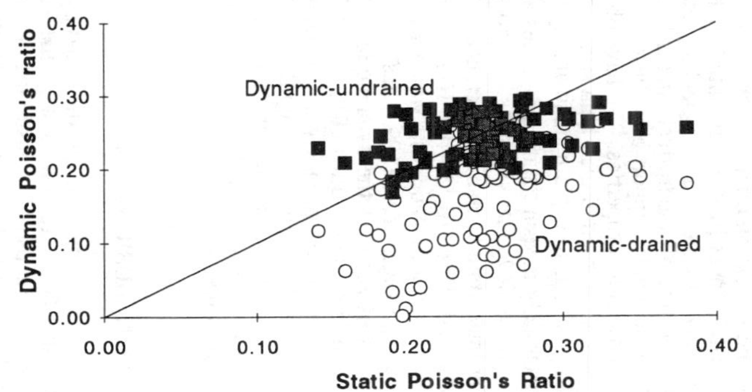

and loading condition differences between the two measurements (intrinsic differences). The liquid saturation effects (undrained to drained) can be corrected with Gassmann theory.

Differences in strain amplitude between static and dynamic tests and the differing effects of microcracks on acoustic velocities and static measurements are most often quoted as responsible for the intrinsic static/dynamic differences. However, little quantitative evidence has been shown to support how these mechanisms cause the observed differences between static and dynamic properties. We have observed that very low strain amplitude static tests actually yield lower moduli than high strain amplitude static tests (the inverse should be true if strain amplitude is to explain static and dynamic differences). However, it is probably impossible to accurately measure static properties at acoustic strain amplitudes (10^{-5} to 10^{-6} strain).

We believe that viscoelastic factors (microcracks are one such factor) cause the bulk of the intrinsic static/dynamic differences. The frequency differences between dynamic (10^4 to 10^5 Hz) and static (10^{-4} to 10^{-2} Hz) are large enough to allow for significant viscoelastic deformation in static tests which would not occur in the deformation time frames of acoustic wave passage. Yale (1993) has shown that frequency independent viscoelasticity (constant Q factor of 20) over the dynamic to static frequency band could account for dynamic/static Young's modulus ratios of up to 2. Yale (1993) suggests that although most viscoelastic mechanisms are probably not frequency independent, his model shows that very moderate viscoelasticity can cause significant "dispersion" in dynamic and static moduli due to the enormous frequency differences. Strain amplitude differences may play a role in a viscoelastic model as certain viscoelastic mechanisms may operate at the higher strain amplitudes of static measurements.

5 CONCLUSIONS

• Measurement of the static and dynamic core mechanical properties of the Chase and Council Grove reservoirs show that correlations can be developed from core measurements to improve the use of acoustic logs for the determination of mechanical properties in the Hugoton and Panoma fields. Static values are more representative of the reservoirs properties, but dynamic values from logs are more easily obtained.

• The formations represented by the core can be divided into six "mechanical" facies. We find that each of these facies is characterized by different mechanical property trends. Within any given "mechanical facies", porosity has the strongest control on the Young's modulus. Poisson's ratio appears fairly insensitive to lithofacies but correlates weakly with porosity.

• Only 10% and 30% of anhydrite and dolomite respectively are needed to significantly increase the modulus of the rock in these formations.

• There is a good correlation between static and dynamic-undrained Young's moduli. The log derived (dynamic-undrained) values can be corrected to static, reservoir condition values of Young's modulus using a correction factors between 0.68 and 0.79 depending on lithology.

• Although there is significant scatter, the dynamic-undrained Poisson's ratios trend with the static values and no systematic correction to log derived Poisson's ratios is needed.

6 ACKNOWLEDGMENTS

The authors would like to thank J. Michael Rodriguez, Tony B. Mercer, and John Tanigowa for their help in the laboratory measurement and in the collection of core for this project. We would also like to thank the managements of Mobil Research and Development Corporation and Mobil Exploration and Producing - U. S. for granting permission to publish this paper.

7 REFERENCES

Gassmann, F. (1951) über die elastizität poröser medien, *Vierteljahrsschrift der Naturforschencher Gesellschaft in Zürich*, Vol. 96, p. 1-23.

Geertsma J. and Smit, D. C. (1961) Some aspects of elastic wave propagation in fluid-saturated porous solids, *Geophysics*, Vol. 26, p. 169-181.

Jaeger, J. C. and Cook, N. G. W. (1976) *Fundamentals of rock mechanics*, Chapman and Hall, London.
Jizba, D., and Nur, A., (1990) Static and dynamic moduli of tight gas sandstones and their relation to formation properties, Gas Research Institute Report 90/0169.
King, M. S., (1969) Static and dynamic elastic moduli of rocks under pressure, in *Proceedings of the 11th U. S. Rock Mechanics Symposium*, Soc. Mining Eng. of AIME.
Montmayeur, H., and Graves, R. M., (1986) Prediction of static elastic/mechanical properties of consol-idated and unconsolidated sands from acoustic measurements: Correlations, presented at 61st Annual Technical Conference and Exhibition of the Society of Petroleum Engineers, New Orleans, LA, SPE paper 15644.
Nieto, J. A., and Yale, D. P., (1991) Integration of core and downhole acoustic measurements - Shear and compressional, presented at Second European Core Analysis Symp. (SPWLA), London.
Tutuncu, A. N., and Sharma, M. M., (1992) Relating static and ultrasonic laboratory measurements to acoustic log measurements in tight gas sands, presented at 67th Annual Technical Conference and Exhibition of the Society of Petroleum Engineers, Washington, D. C., SPE paper 24689.
Yale, D. P., (1993) A comparison of static and dynamic rock mechanical properties, presented at 63rd Society of Exploration Geophysicists Annual Meeting and International Exposition, Washington, D. C.

Rock Mechanics, Nelson & Laubach (eds) © 1994 Balkema, Rotterdam, ISBN 90 5410 380 8

Static and dynamic elastic moduli in granite: The effect of strain amplitude

R.J. Martin, III & R.W. Haupt
New England Research, Inc., White River Junction, Vt., USA

INTRODUCTION

Understanding and modeling seismic wave propagation from nuclear explosions require information on the behavior of rocks in the source region under appropriate loading conditions. Within the source region the rocks are loaded in uniaxial strain; during unloading damage occurs, permanently altering the rocks. Rock properties measured at the laboratory scale constrain the models. Static and dynamic elastic moduli and attenuation as a function of mean stress and strain amplitude are incorporated into forward models to determine particle velocity and displacements and time histories in the nearfield.

An underground nuclear explosion in granite was detonated near Semipalatinsk as part of a Joint Verification Experiment with the USSR. In order to model the range of possible responses and further our understanding of the relationship between static and dynamic moduli, a suite of experiments was performed on five granites representative of the source region.

EXPERIMENTAL PROCEDURE

Five low porosity, granitic rocks were tested in hydrostatic compression and uniaxial strain with simultaneous compressional and shear wave velocity measurements. The granites were collected from Mount Katahdin, Maine; Nickerson Lake, Maine; Raymond, California; Westerly, Rhode Island; and an NRDC borehole near Semipalatinsk, Kazakhstan. The rock compositions were similar, but the mean grain size varied from 0.7 to 1.1 mm. Detailed data is presented only for Sierra White granite (SWG) from Raymond, California. The physical properties of SWG are bulk density, 2.638 g-cm^{-3}; porosity, 0.9%; and mean grain diameter, 0.7 mm. The composition is microcline, 44%; quartz, 38%; albite, 2%; biotite, 11%; and muscovite, 5%.

Each specimen was a ground right circular cylinder 25.4 mm in diameter and 40 ±6 mm in length jacketed with 0.13 mm thick copper. The specimens were instrumented with a minimum of two strain gages epoxied to the jacket to measure strain parallel and normal to the core axis.

The experiments were conducted in a servo-hydraulic apparatus. Each specimen was loaded in hydrostatic compression to 120 ±10 MPa. At selected pressures the sequence was interrupted for compressional and shear wave velocity measurements. Subsequently, a uniaxial strain test was performed to a mean stress of approximately 250 MPa. The radial strain was maintained constant and the confining pressure allowed to increase in response to an increasing axial stress. At selected points along the loading path, the experiment was interrupted for compressional and shear velocity measurements.

This work was funded by the Department of the Air Force, Geophysics Laboratory, Hanscom Air Force Base, Massachusetts, under Contract No. F19628-89-C-0097.

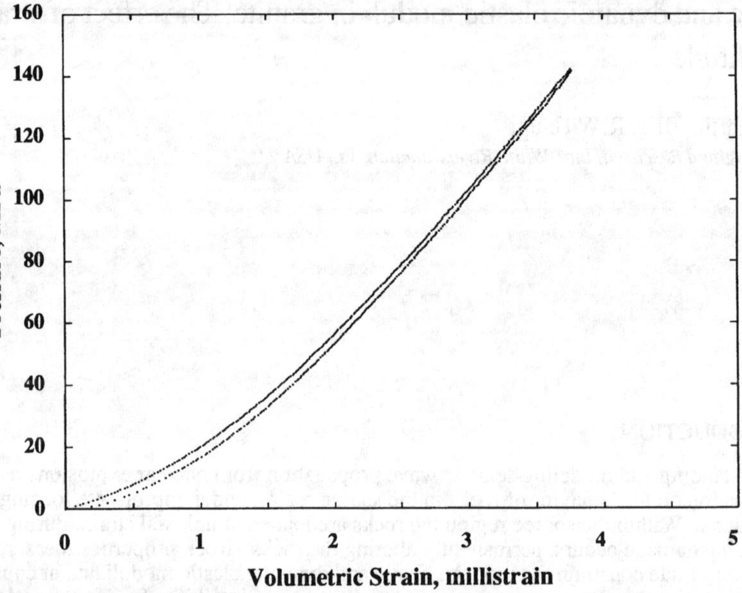

Figure 1: Pressure is plotted as a function of volumetric strain for the hydrostatic compression of a dry Sierra White granite specimen.

One experiment was performed in uniaxial compression on a large specimen of SWG (155 mm in diameter and 510 mm in length) to study the influence of strain amplitude on Young's modulus. The specimen was instrumented with three LVDTs to monitor the axial displacement. First, the specimen was cyclically loaded to a strain of 2×10^{-4}. During the second loading cycle the test was interrupted for low strain amplitude load-unload cycles; peak to peak strains were on the order of 2×10^{-6}.

A specimen of 50.8 mm in diameter and 50.8 mm in length was cored from the same block as the large specimen. Compressional and shear wave velocities were measured on the specimen subjected to an axial stress of 0.8 MPa. Young's modulus was computed from these data.

EXPERIMENTAL RESULTS

Static loadings of SWG in hydrostatic compression and uniaxial strain are shown in Figures 1 and 2 respectively. The characteristics of both data sets are similar. The stress (pressure) versus strain curves are concave upward, and there is a pronounced hysteresis on unloading. The data collected for the other granites exhibit the same behaviors.

Figures 3 and 4 show the compressional and shear wave velocities measured at selected points along the loading curves for hydrostatic compression and uniaxial strain. The compressional and shear wave velocities increase with increasing pressure and mean stress. The change in velocity is greater for hydrostatic compression than for uniaxial strain. At stresses in excess of 60 MPa, the velocities for both boundary conditions approach the same values.

In Figure 5, the bulk modulus is plotted as a function of confining pressure for the five rocks studied. Bulk modulus increases with increasing confining pressure; the changes are greatest below 20 MPa. At 100 MPa, Katahdin and Sierra White granite exhibit

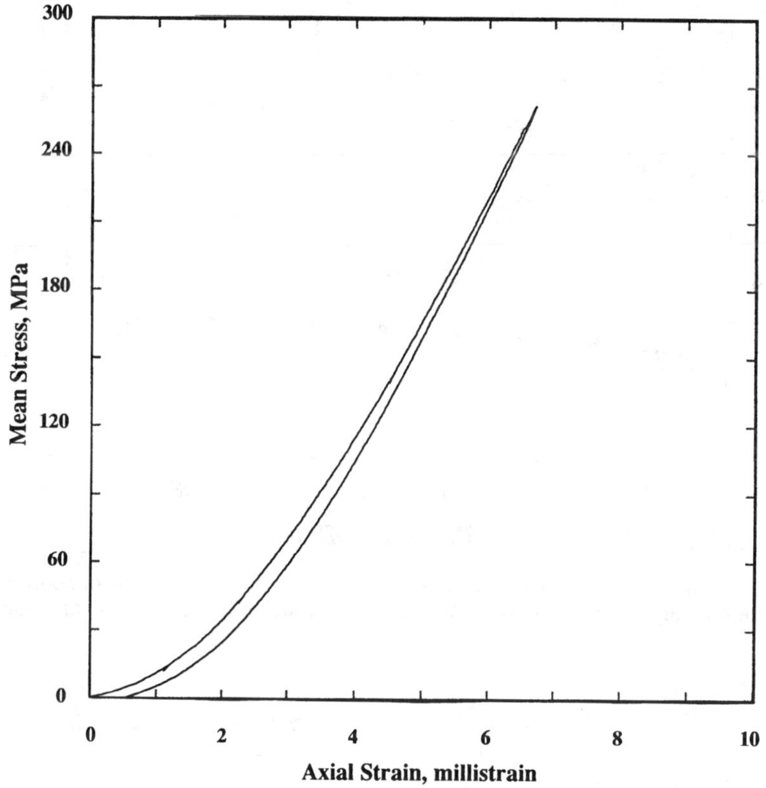

Figure 2: Mean stress is plotted as a function of axial strain for the uniaxial strain of a dry Sierra White granite specimen.

lower bulk moduli than the Westerly and Nickerson Lake granite.

The static and dynamic bulk moduli for SWG loaded in hydrostatic compression are shown in Figure 6. The dynamic moduli exceed the static moduli at all pressures. At low pressure, the ratio of the dynamic to static moduli is nearly a factor of two. The static and dynamic Young's moduli obtained under uniaxial strain conditions are shown in Figure 7. The dynamic moduli exceed the static moduli for mean stresses below 150 MPa; at 150 MPa the moduli are nearly equal.

To explore the strain amplitude dependence on Young's modulus in more detail, a uniaxial compression experiment was conducted on a large specimen of SWG. These data are shown in Figure 8. Young's modulus is plotted as a function of axial strain. The solid line indicates the instantaneous Young's modulus computed from the slope of the stress-strain curve. The open circles represent Young's moduli determined from small strain amplitude excursions at various points along the stress-strain curve. The X indicates the dynamic modulus computed from compressional and shear wave velocities at a strain of 2×10^{-5}. The Young's modulus for monotonic loading remains nearly constant for strains below 3×10^{-6}. As the strain on the specimen increases, the modulus decreases and reaches a minimum near 6×10^{-5}. With continued loading the granite stiffens. In contrast, the moduli determined from small strain amplitude measurements are nearly equal to the instantaneous modulus at strain amplitudes up to 5×10^{-6} and

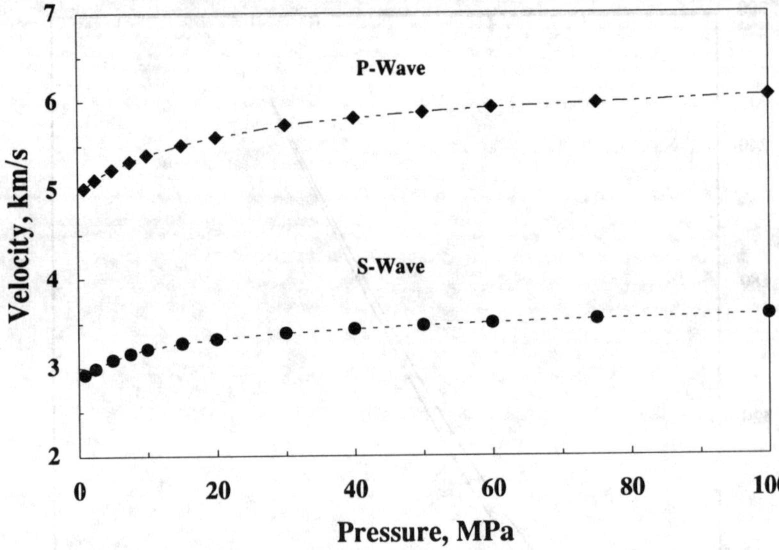

Figure 3: Compression and shear wave velocities are plotted as a function of confining pressure for a dry specimen of Sierra White granite tested in hydrostatic compression.

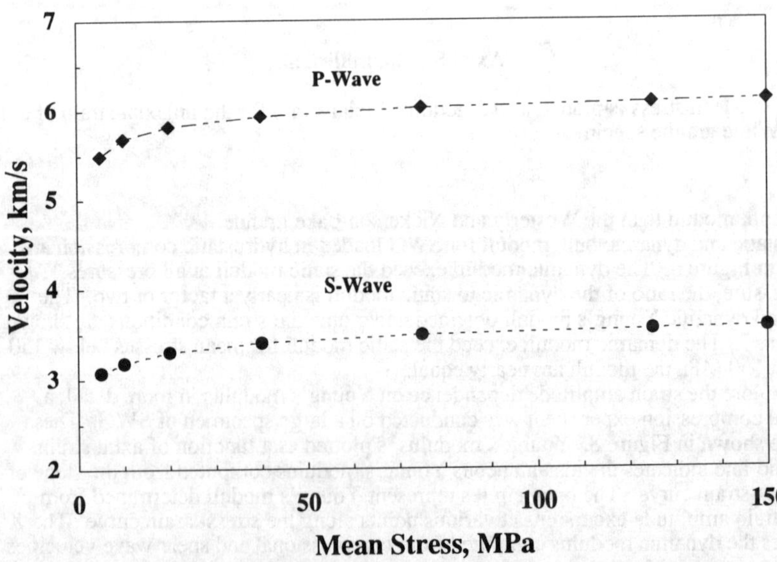

Figure 4: Compressional and shear wave velocities are plotted as a function of mean stress for a dry specimen of Sierra White granite tested in uniaxial strain.

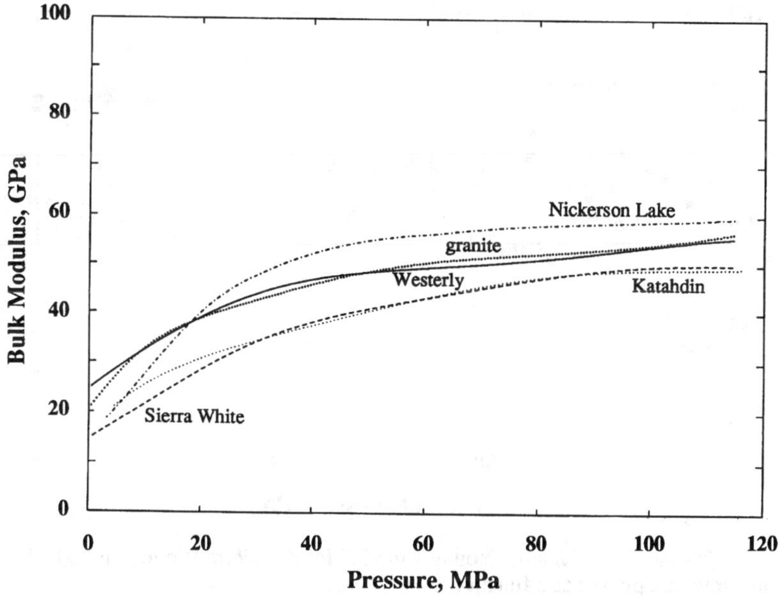

Figure 5: Bulk modulus is plotted as a function of confining pressure for five representative granite specimens.

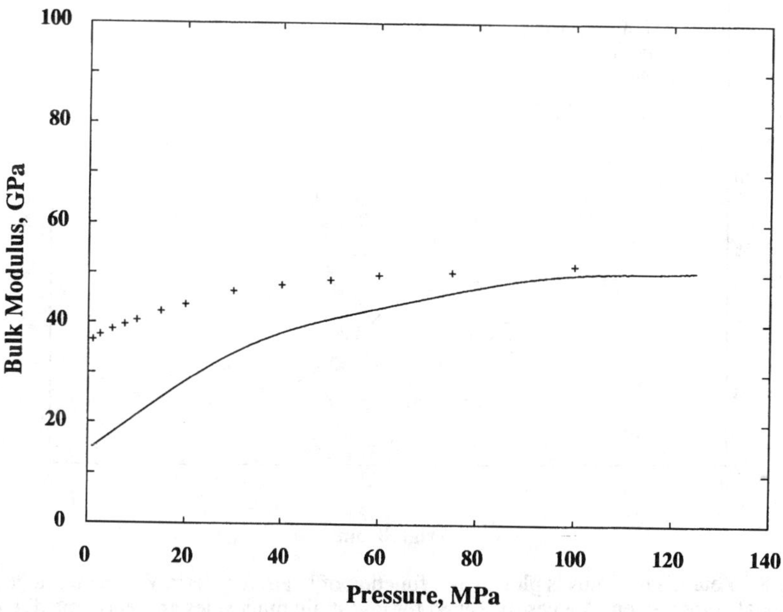

Figure 6: The static and dynamic bulk moduli for dry Sierra White granite measured in hydrostatic compression are plotted as a function of confining pressure.

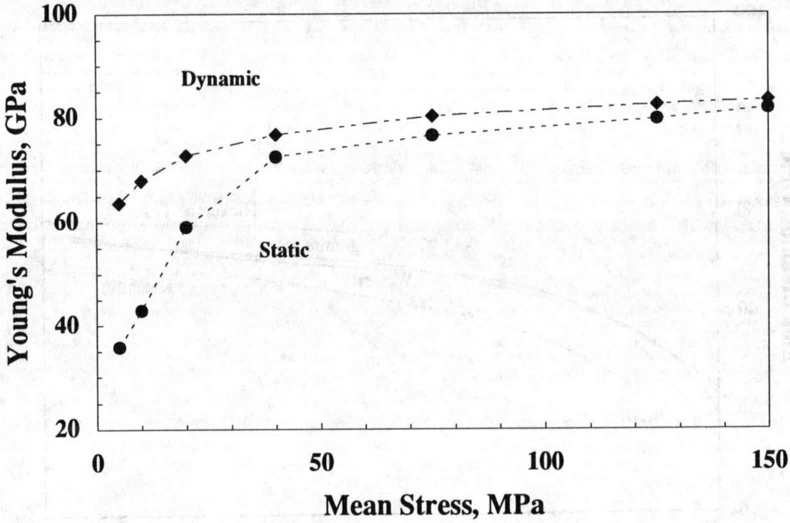

Figure 7: The static and dynamic Young's moduli for dry Sierra White granite tested in uniaxial strain are plotted as a function of mean stress.

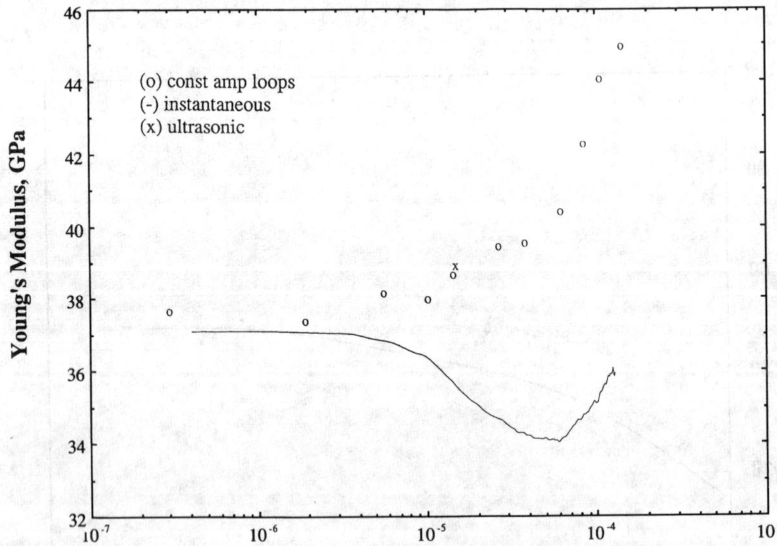

Figure 8: Young's modulus is plotted as a function of strain for Sierra White granite tested in uniaxial compression. Values measured for low strain minicycles are represented with open circles.

increase with increasing load. At a strain of 6 x 10^{-5} the instantaneous modulus experiences a low near 34.4 GPa, whereas the small strain amplitude modulus is in excess of 39 GPa.

DISCUSSION

The data presented above exhibit a consistent trend: bulk modulus and Young's modulus increase with increasing confining pressure or mean stress. The stress dependencies for the static measurements are greater than those for the dynamic. Since both static and dynamic measurement techniques are sampling the same rock, the reasons for the discrepancies need to be explored.

First we consider the differences in the moduli of the granites tested statically in hydrostatic compression. The rocks are very compressible at low confining pressures and become less compliant with increasing pressure. Walsh (1965) demonstrates that the decrease in compressibility with pressure is attributable to the closing of low aspect ratio, elliptical cracks, the lower the aspect ratio the greater the change at low pressures. Cracks close completely at high pressure, and the compressibility approaches the Voight-Reuss average modulus for an uncracked rock.

The low pressure bulk moduli of the granites range from 15 to 25 GPa. With increasing confining pressure, the moduli increase and asymptotically approach a high pressure value of 50 to 60 GPa at 100 MPa. Since the compositions and mean grain diameters of the granites are similar, the reason for the differences in the bulk moduli between the granites is most likely related to the crack morphology.

At low pressure, the bulk moduli vary greatly with pressure. The greatest pressure sensitivity is observed for the Nickerson Lake granite (NLG) and the lowest for the SWG, Westerly, and the granite from Kazakhstan. The pressure dependencies are a direct reflection of the crack morphology, the crack length, and the manner in which they close with pressure. The data for the Nickerson Lake specimen suggest that the cracks are low aspect ratio (less than 10^{-3}) that close by 80 MPa.

In contrast, SWG may have cracks with higher aspect ratios, nonelliptical cracks, and higher closure pressures. Mavko and Nur (1978) analyze the effect of nonelliptical cracks on compressibility. They consider tapered cracks and cracks that make contact over a portion of the crack surface during closure. The latter condition gives rise to multiple crack spectra. The difference in bulk modulus between SWG and NLG at high pressures may be due to a multiple of crack populations that require pressures greater than 120 MPa to completely close.

O'Connell and Budiansky (1974) show that seismic wave velocity for a rock with elliptical cracks is proportional to $N\langle a^3\rangle$, where N is the number of cracks per unit volume of length a. As crack lengths decrease, presumably due to stress, the velocities increase. The three models are consistent with our observations, but fail to account for the differences in static and dynamic moduli. Static and dynamic properties of granite depend on crack density and the distribution of crack lengths, yet under the same state of stress the computed elastic moduli differ, particularly at low stresses. Simmons and Brace (1965), Cheng and Johnson (1981), and Jizba (1991) have reported similar differences. Several explanations have been forwarded to explain these discrepancies. Two frequently sighted possibilities for crystalline rocks are frequency and strain amplitude. Seismic wave velocities are measured at frequencies between 400 and 700 kHz and strains of 10^{-7} or less. In contrast, the static measurements are conducted at a much lower loading rate, less than 10^{-3} Hz and at strain amplitudes of 10^{-4} and greater. For dry specimens, the effect of frequency on Young's modulus is small when the strains for the static measurements are less than 10^{-6}, but Young's modulus decreases with increasing strain (Spencer, 1981; Winkler, 1986; and Haupt et al, 1992). Therefore, it seems plausible that strain due to crack closure, not frequency, controls the static and dynamic behavior of low porosity crystalline rocks.

The data in Figure 8 indicate that crack closure is not a linear function of stress. The low strain amplitude moduli are consistent with the crack models. With increasing stress, crack lengths decrease and modulus increases. For the moduli computed for monotonic loading, as the cracks close the modulus decreases and then increases. At a strain of 10^{-4} the low strain amplitude modulus is 23% greater than that derived from the monotonic loading curve. The differences may be due to the way cracks close. At low strains, cracks are stiff due to high surface energy at crack tips and asperities in contact on the crack surfaces. As the stress increases, the surface forces are overcome and slip occurs on the asperities in contact, resulting in larger strains and lower moduli. Dynamic and static low strain amplitude moduli are high because the small strains do not cause slip on contact points, and surface forces play an important role.

CONCLUSIONS

The discrepancy between static and dynamic moduli in low porosity granitic rocks is strain dependent. At low strains the static and dynamic moduli are equal. At strains greater than 5×10^{-6}, the stiffness of the cracks decreases, and Young's modulus decreases. This effect may be due to slip on the contacts between the surfaces. Surface forces at crack tips are important for small strain excursions, less than 5×10^{-6}, experienced in seismic wave propagation and small strain amplitude cyclic loading contributing to the higher dynamic moduli. Differences in static bulk moduli for granites of similar compositions and grain sizes at pressures below 120 MPa are attributable to differences in microcrack population and crack geometry. One granite, NLG, appears to have a single low aspect ratio crack population that closes at pressures below 100 MPa.

REFERENCES

Cheng, C.H. and D.H. Johnston 1981. Dynamic and static moduli. Geophys. Res. Let. 8:39-42.

Haupt, R.W., R.J. Martin, III, X. Tang, W.J. Dupree and R.H. Price 1992. Modulus dispersion and attenuation in tuff and granite. Proceedings of the 33rd U.S. Symposium on Rock Mechanics. 899-908.

Jizba, D.L. 1991. Mechanical and acoustical properties of sandstones and shale. Thesis submitted to the Department of Geophysics, Stanford University.

Mavko, G.M. and A. Nur 1978. The effect of nonelliptical cracks on the compressibility of rocks. J. Geophys. Res. 83:4459-4468.

O'Connell, R. and B. Budiansky 1974. Seismic velocities in dry and saturated cracked solids. J. Geophys. Res. 79:5412-5426.

Simmons, G. and W.F. Brace 1965. Comparison of static and dynamic measurements of compressibility of rocks. J. Geophys. Res. 70:5649-5656.

Spencer, J.W. 1981. Stress relaxations at low frequencies in fluid-saturated rocks: Attenuation and modulus dispersion. J. Geophys. Res. 86:1803-1812.

Walsh, J. 1965. The effect of cracks on the compressibility of rock. J. Geophys. Res. 70:381-389.

Winkler, K. 1986. Estimates of velocity dispersion between seismic and ultrasonic frequencies. Geophysics. 51:183-189.

Rock Mechanics, Nelson & Laubach (eds) © 1994 Balkema, Rotterdam, ISBN 90 5410 380 8

A thermodynamic basis for static and dynamic scaling laws in the design of structures in rock

B.T. Brady
US Bureau of Mines, Denver Research Center, Colo., USA

ABSTRACT: Brittle fracture is modeled as a phase change that exhibits the characteristics of a thermodynamic system at its critical point. Universality and critical scaling are shown to exist in brittle materials at incipient failure. The law of corresponding states, which applies to fluid and magnetic systems, applies to brittle fracture. Several critical exponents are calculated and compared with observation. The Griffith theory of fracture is shown to correspond exactly to Landau mean field theory for continuous phase transitions. Applied to failure of geotechnical materials, the widely used empirical Hoek-Brown type failure criterion when written in shear stress (τ) versus normal stress (σ) space is derived using the commonly used critical exponent of $\beta \approx 0.33$.

1 INTRODUCTION

Despite extensive research performed on the mechanisms of brittle fracture during the past several decades, results have not substantially contributed to our understanding of the physical processes involved leading up to and including failure. For example, we have not been particularly successful in predicting earthquakes and mining-induced failures. We have not developed a rational design methodology for geotechnical structures in rock. Nor, for that matter, have we completely eliminated the problem of catastrophic failures in metallic or concrete structures. Rather, there is an acute lack of a coherent theoretical link between experimental studies and the analysis and modeling of fracture. The problem is that we simply do not have an understanding of the fundamental physics of the nonlinear processes involved in how cracks form and interact in solids. Its solution will require an understanding of the detailed microphysics of how atomic and molecular bonds are formed and broken. The beginnings of a solution to this problem is proposed in this article.

Brittle fracture is formulated as a continuous phase change that exhibits the characteristics of a thermodynamic system at its critical point. In particular, fracture of brittle materials is modeled as a tricritical thermodynamic phenomenon. Tricritical refers to critical phenomena in a three-phase system: solid material, solid plus microcracks, and solid plus macrocrack(s). Formation of the macrocrack(s) for example would denote failure of a laboratory sample into two pieces. A macrocrack represents the endpoint of the microcrack coalescence process within the nucleation zone of what will become the macrocrack. Universality and critical scaling are shown to exist in brittle materials near incipient failure. In the following discussion, the theory is applied to geotechnical materials. However, the theoretical approach should hold for all brittle materials. Treatment of brittle fracture as a continuous phase change suggests a gauge theoretic formulation of dissipative processes may yield realistic equations of state and may profoundly alter the way we view and approach the subject of dissipative processes in materials.

2 EXPERIMENTAL CONSTRAINTS ON MODELS OF BRITTLE FRACTURE

Power law relationships, the hallmark of scale invariant (fractal) processes and critical phenomena, are of crucial importance in seismology and condensed matter physics (Bak et al., 1988; Hirata, 1987; Hirata et al.,1987; Kanamori and Anderson, 1975; Paterson, 1978). In seismology, important parameters exhibiting power law behavior are seismic magnitude and moment, fault dimension, and stress drop determined from earthquakes, mine failures, and laboratory experiments. In condensed matter physics, power law relationships are observed before phase changes. Laboratory studies of rock fracture also show power law relations emerge during the fracturing process. For example, Hirata (1987) observed that bursts of acoustic emissions decayed exponentially during the primary and secondary creep deformation phases of the experiments but tended to decay by a power law during the tertiary phase leading to catastrophic failure. Hirata et al. (1987) observed that the spatial distribution of acoustic emission, while random during the secondary creep phase, gradually becomes organized to become fractal as creep deformation progresses into the tertiary deformation phase.

Since long-ranged interactions are required to yield power law behavior, results of laboratory and large scale failure investigations in mine failures and earthquakes indicate that both temporal and spatial long-ranged interactions among cracks (joints, faults) increase as a rock mass evolves to failure. Self-similar phenomena are observed in larger scale phenomena, such as earthquakes and mine failures. Power law behavior is observed in the structure of fractured samples of low-carbon steels. These observations suggest a self-similar or scale invariant (fractal) structure of crack formation and growth exists that is becoming self-organized during the fracture process. If this is valid, then brittle fracture of materials may be modeled as a continuous phase change and the mathematical power of critical phenomena theory and the renormalization group can be brought to bear on this important unsolved problem (Stanley 1971).

3 A THERMODYNAMIC MODEL OF BRITTLE FRACTURE

Symmetries play a great role in science. Discovery of symmetries is fundamental if one is to understand the intricacies of the laws of nature; for example, knowledge of a system's symmetry often allows us to greatly simplify the search for a solution to a given problem. Important symmetries include, for example, Galilean invariance, Poincaré invariance, and gauge invariance. Problems with one type of symmetry may have solutions with different and lesser symmetries. For example, during the evolution of a physical system through a phase transition, it may pass from a symmetric to a less symmetric state; its symmetry is said to be broken at the transition. As with systems exhibiting power law behavior near phase transitions, broken symmetries are viewed as hallmarks of phase transitions. Of interest to the materials scientist, formation of a crack is an example of a "broken" symmetry. Galilean invariance is spontaneously broken once a crack forms.

Several comments on applying the phase transition construct to fracture are in order. Phase transformations are commonly perceived to pervade the whole structure. Cracks, however, are two-dimensional entities and while it may trouble one to consider them as phase transitions, we should recall that many interesting systems of lower dimensionality exist that exhibit phase transitions; for example, one dimensional or quasi two-dimensional systems, such as threads, polymers, solids made up of chains of atoms or molecules with only very weak coupling between the chains. Two-dimensional or quasi two-dimensional systems like films, absorbed phases, or solids made up of weakly coupled planes, such as mille-feuilles pastry, are also examples. In this category we also include the transitions between surface phases that, strictly speaking, are two-dimensional phase transitions, even though their effects extend through a small but finite volume. With regard to fracture, experiment has shown that before the transition to tertiary deformation, cracks are distributed throughout the sample volume (dimensionality $d \approx 3$). Following the transition, these cracks begin to cluster, and at a point within the cluster, the microcracks coalesce into a $d \approx 2$ structure (macrocrack) at failure.

The prototype energy function in systems, described by a "density" defined at all points in space, for example, fluid/gas density or magnetization density, is the Landau-Ginzburg (LG) potential. This potential has the form (Stanley, 1991)

(1) $$V = \int d^{d}x\left[\frac{\lambda}{2}(\nabla\psi)^2 + \frac{1}{4}\psi^4 + \frac{1}{2}a\psi^2 + b\psi\right],$$

where $(\nabla\psi)^2 = (\partial\psi/\partial x_1)^2 + ... + (\partial\psi/\partial x_d)^2$ and d is the spatial dimension. The density ψ is called an order parameter.

Reduced to its essence, the LG model is as follows. ψ, is introduced to describe the state of a physical system. The total energy of the system is expressed as the sum of a kinetic energy term [$\sim (d\psi/dt)^2$] and a potential energy term $V(\psi)$. This potential depends on external control parameters as well as the order parameter ψ. The system ground state is determined by minimizing the sum of these two terms. In the static case ($d\psi/dt = 0$), the ground state is determined by the value of ψ which minimizes $V(\psi)$. The Landau-Ginzburg potential is chosen to be an even function of ψ ($b = 0$) and "a" is a control parameter. This potential has a unique minimum at $\psi = 0$ when "a" ≥ 0, and two equally deep minima at $\psi = \pm\sqrt{(-a)}$ when "a" < 0 with a continuous phase transition occurring as "a" goes through zero. The LG phase transition is often used to describe continuous or second-order phase transitions. In this case the control parameter "a" is identified with $(T_c - T)$ in the neighborhood of the critical temperature. Experimentally, continuous phase transitions are subject to difficulties not readily apparent from the above equation. For example, a continuous phase transition will simply disappear under an arbitrarily small symmetry-breaking perturbation ($b \neq 0$) in the potential function. The phase transition may reappear at a far-distant point as a first-order transition. Note that this behavior has a natural interpretation with the context of elementary catastrophe theory.

A non-zero value of the order parameter corresponds to a breaking of a symmetry. For a ferromagnetic transition, the order parameter is M, the homogeneous magnetization density; for a liquid-gas transition, the order parameter is the density difference between the two phases. When the order parameter has a discontinuity at the transition, we are dealing with a first-order transition following Landau; when there is no such discontinuity, the transition is second-order or continuous, such as at the critical point (CP). Close to a CP, one observes quantities obeying power laws with exponents that are not integers. Thus, just below T_c, the order parameter is observed by falling proportionally to $(T_c - T)^{\beta}$ and in the magnetic case, specific heats, correlation length (a measure of fluctuations in the order parameter near the CP) and susceptibility all fall the same, but with differing exponents. These exponents, termed critical exponents, are defined near a given point, and have been measured with sufficient accuracy that clearly suggest they are not integers. What is remarkable is that such behavior, though singular, has very simple characteristics. One observes that the critical exponents exhibit simple relations called scaling laws, whose degree of universality is even greater. Thus, amid complexity, a simplicity arises.

The LG potential is generally taken to be the most general one-parameter family of even functions. When the conjugate parameter to the order parameter ψ, say b, is zero, the family can only exhibit continuous phase transitions. The symmetry of the potential is broken when $b \neq 0$ and the transition becomes first-order. Breaking the symmetry allows more general perturbations to occur. Since V is an integral over space, it clearly increases as the volume of the system. Thus, in the large volume limit, the largest weight in the configuration average is associated with the configuration of lowest energy. Since $(\nabla\psi)^2$ always costs energy because it is positive, the minimum energy configuration has $\psi = M$ for the ferromagnetic (uniaxial magnet) [or $\psi = \Delta\rho = \rho_{liq} - \rho_{gas}$ for the liquid-gas system], a constant that satisfies the condition $\partial V/\partial\psi = 0$. With the identification that the critical point divides $M = 0$ from $M \neq 0$ (for $H = 0$), where M and H represent the magnetization density and the applied magnetic field, we must have $a \propto (T - T_c)$, up to a general constant and neglecting terms of order $(T - T_c)^2$. Equation 1 then becomes the Landau-Ginzburg or mean field equation of state relating magnetization density M to T and H. It can be written in the scaling form $H/M^{\delta} = F[(T - T_c)/M^{1/\beta}]$, where $\delta = 3$, $\beta = 0.5$.

Based on extensive experimental evidence (Paterson, 1978), brittle fracture will be modeled as a three-phase system that includes solid material, microcracks in the solid matrix, and finally, the macrocrack that represents the final stage of the fracture process. The macrocrack denotes coalescence of the microcracks. The phase transition is characterized by separation of the solid into two pieces. In general, each microcrack that forms in the solid matrix locally breaks the translational and rotational symmetry of the system (solid) and if each microcrack is a phase change, then one must view the fracture problem as a true polycritical or N-critical system and the local tensile stress at which a given microcrack forms as the critical point for that microcrack. Ordinary and tricritical points are the simplest cases of polycritical points. However, for ease of argument, we shall assume all microcracks can be grouped into one classification class. Thus, for simplicity, we assume a tricritical system. In this context, the application of a shearing stress (strain) to an otherwise undistorted lattice can be viewed as breaking the original symmetry of the lattice. For the tricritical system, the order parameter will be chosen to be the shear stress.

As its name is meant to suggest, a tricritical point is a point where three critical lines meet. In a typical phase diagram, one can visualize in addition to the temperature T and the magnetic field H (coupled to the order parameter M), another field h (coupled to a variable m). We can compare this situation to an ordinary critical point, where two phases, say A and B, of a system become identical. Similarly we can consider a tricritical point as a point where three phases, A, B, and C become identical. Thus, in an enlarged parameter space, the coexistence line and the critical point are replaced by three first-order surfaces (coexistence surfaces for A and B, B and C, or C and A), and by three critical lines where A = B, B = C, or C = A; finally the tricitical point is the point where A = B = C. A tricritical point differs from an ordinary critical point in the number of relevant fields. It has an additional field h. At points on the line L_1 the fluctuations in the order parameter M diverge, inducing fluctuations in the variable m conjugate to h. Consequently, the effect of the field h is to shift the critical point in question but without changing its character, except in the case of the tricritical point, where there is competition between the fluctuations of m and M. Beyond the tricritical point any small variation in h inhibits the fluctuations and results in a first-order transition with a discontinuity in the order parameter.

In LG theory, an expression like

(2) $$V(\psi;a,b,c,d)=\frac{1}{6}\psi^6+\frac{1}{4}a\psi^4+\frac{1}{3}b\psi^3+\frac{1}{2}c\psi^2+d\psi$$

yields a continuous transition for $a > 0$, a first-order for $a < 0$ and a tricritical point for $a = 0$. If "a" depends on, say pressure, we can eventually encounter a tricritical point. The characteristic dimensionality for tricritical systems is $d_c = 3$. Thus, regarding tricritical points, our three-dimensional world is precisely at the limit of LG theory. This limiting case entails logarithmic corrections to the theory, a not uncommon occurrence. The classical tricritical exponents are $\alpha = 1/2$, $\beta = 1/4$, $\delta = 5$ and differ from the classical bicritical exponents ($\alpha = 0$, $\beta = 1/2$, $\delta = 3$). In the following, we briefly discuss the β critical exponent. The remaining exponents and an in-depth discussion of the above model will be published elsewhere.

To apply the above constructs to fracture, we take the order parameter to be the shear stress τ. Then we can write the functional relationship between τ and temperature as $\tau \propto (\Delta T)^\beta$. It can be shown that ΔT is proportional to $(t_c - \sigma)^2$. We find

(3) $$\frac{\tau}{t_c}=B\left(\frac{t_c-\sigma}{t_c}\right)^{2\beta},$$

where t_c is the theoretical tensile strength of the material, τ and σ are the local shear and normal stress, and B is a numerical constant. In terms of the far-field values, the same relationship holds, with the exception that the tensile strength is replaced by t_o, the measured strength of the medium. Note that the convention is taken that tension is positive. The classical or mean-field equation is then

(4) $$\frac{\tau}{t_o}=A\left(\frac{t_o-\sigma}{t_o}\right)^{0.5},$$

where A is a constant equal to 2.0 by the condition of limiting circles for parabolic equations (Ode, 1962). Equation 4 with the classical or mean field value of $\beta = 1/4$ corresponds to the Griffith equation expressed in shear stress-normal stress space. The exponent β for thermodynamic systems at criticality is observed to be about 0.33 ± small correctional terms (Standley, 1971). Modeling brittle fracture as a critical phenomena, one would expect the Mohr envelope for materials at incipient failure can be written

$$\frac{\tau}{t_o} = A\left(\frac{t_o - \sigma}{t_o}\right)^{0.66}, \tag{5}$$

where A is a material constant whose value, while on the order of 2.0, must be found experimentally. It is relatively simple to show that this equation is a statement of the law of corresponding states applied to brittle materials.

4 EXPERIMENTAL RESULTS AND ANALYSIS

Table 1 illustrates the applicability of equation 5 to data on shear strength of undisturbed rock masses as a function of rock type and its rock mass rating (RMR) (Brady and Brown, 1985; Bieniawski, 1976). The failure envelopes in τ-σ space were constructed from envelopes determined in σ_1, σ_3 space using standard mathematical procedures (1962; Ode, 1960). Several important conclusions result from this table and equation 5.

1. Rock mass behavior near incipient failure, is independent of the physical characteristics and the chemical makeup of the rock mass. This observation shows universality and that the law of corresponding states is satisfied near the phase change (fracture/macrocrack) boundary. The law of corresponding states is the result of observation that data, properly normalized and scaled fall on the same curve.
2. The theory predicts that a universal scaling hypothesis is applicable to all materials that belong to the same universality class.
3. The subject of critical phenomena and its powerful mathematical apparatus can be brought to bear to the subject of brittle fracture with the provision that one can view t_o as the critical point for the rock mass phase change.
4. The classical Griffith theory of fracture corresponds exactly with the classical Landau mean field theory of critical point thermodynamic systems with $2\beta = 0.50$. In particular, the empirical Hoek-Brown theory of failure for geological materials results using the commonly measured critical exponent of $2\beta = 0.684 \pm 0.054$. In table 1, the equations for the shear strength of undisturbed rock masses are statistically identical within a confidence interval of 95% and independent of material type and rock mass quality, which is

$$\frac{\tau}{t_o} = A\left(1 - \frac{\sigma}{t_o}\right)^{2\beta}, \tag{6}$$

where $A = 2.466 \pm 0.58$, $2\beta = 0.684 \pm 0.054$, and t_o refers to the tensile strength of the rock mass. Experience gained by using the empirical Hoek and Brown failure criterion showed that the estimated rock mass strengths for underground excavations were reasonable.
5. One can readily decide the functional relationship between the m and s values as a function of the RMR and the tensile strength of the rock mass. The result is where t_o is the tensile strength of the intact in situ rock mass and t_o(RMR) is the

$$\log_e \frac{t_o(RMR)}{t_o} = 0.13(RMR-100), \quad (7)$$

tensile strength of the in situ rock mass for RMR $<$ 100. Equations 6 and 7 can be used to decide the shear strength of the rock mass provided spatial variations and temporal variations in the order parameter are neglected. Of course, any other appropriate rock mass rating system could be used; the only requirement is the variation of m and s with the rating factor.

5 DISCUSSION

The theoretical argument presented in this article is that there may be great simplicity in describing the observed complexity of material behavior at incipient failure. The assumption that material near incipient failure can be treated as a tricritical system near criticality has a basis in experiment and theory. Several results can be drawn from this study. First, rock mass strength, properly scaled and normalized, is independent of the physical characteristics of the rock mass and its chemical makeup. Second, this condition may place design of engineering structures in rock on a rational basis much like design of concrete and metal engineering structures. Third, a universal scaling hypothesis is applicable to all materials at incipient failure. Such materials would include concrete, ceramics, and any other materials exhibiting isotropy, spatial homogeneity, and stationarity. These latter constraints can be easily relaxed and taken into account. Fourth, the physics of critical phenomena and the renormalization group and their powerful mathematical apparatus can be brought to bear on the subject of brittle fracture of materials when t_o, the tensile strength of the material, represents a critical point in parameter space. Five, the classical or Landau formulation of critical phenomena, when applied to brittle fracture, leads unambiguously to the well-established Griffith theory of fracture. Using the tenets of critical scaling exponents and their universality leads to the Hoek-Brown empirical failure criterion expressed in (τ, σ) space.

REFERENCES

Bak, P., Tang, E., and K. Wiesenfeld 1988. Phys. Rev. A, v. 38, pp. 364-374.

Brady, B. H. G., and E. T. Brown 1985. Rock Mechanics for Underground Mining, George Allen and Union, Lon., 527 pp.

Bieniawski, Z. T. 1976. in Exploration for Rock Engineering, Z. T. Biewiawski ed., Balkema, pp. 97-106,

Hirata, T. 1987. Jour. Geophys. Res., v. 92, pp. 6215-6221.

Hirata, T., Satoh, T., and K. Ito 1987. Geophys. J. R. Astron. Soc., v. 90, pp. 369-374.

Kanamori, H., and D. L. Anderson 1975. Bull. Scis. Soc. Amer., v. 65, pp. 1073-1095.

Ode, H. 1960. GSA Mem. 79, Rock Deformation, pp. 293-322.

Paterson, M. S. 1978. Experimental Rock Deformation - The Brittle Field, Springer-Verlag, New York, 254 pp.

Stanley, H. E. 1971. Introduction to Phase Transitions and Critical Phenomena, Oxford U. Press, New York, 308 pp.

TABLE 1.—Strength of Undisturbed Intact and Jointed Disturbed Rock Masses (Modified from Hoek and Brown)
(t_o is the tensile strength of the rock mass)

ROCK TYPE - ROCK MASS QUALITY	Fine Grained Polymineralic Igneous Crystaline Rocks (Andesite, Dolerite, Diabase, and Rhyolite)	Coarse Grained Polymineralic Igneous and Metamorphic Crystaline Rocks (Amphibolite, Gabbro, Gneiss, Granite, Norite and Quartz Diorite)	Carbonate Rocks with Well Developed Crystal Cleavage (Dolomite, Limestone, and Marble)	Arenaceous Rocks with Strong and Poorly Developed Crystal Cleavage (Sandstone and Quartzite)	Lithified Argillaceous Rocks (Mudstone, Siltstone, Shale, and Slate)
Intact Rock Samples Laboratory Size Rock Specimens Free From Structural Defects (RMR – 100)	$\frac{\tau}{t_o} = 2.572\left(1 - \frac{\sigma}{t_o}\right)^{0.696}$	$\frac{\tau}{t_o} = 3.155\left(1 - \frac{\sigma}{t_o}\right)^{0.705}$	$\frac{\tau}{t_o} = 1.598\left(1 - \frac{\sigma}{t_o}\right)^{0.658}$	$\frac{\tau}{t_o} = 2.400\left(1 - \frac{\sigma}{t_o}\right)^{0.692}$	$\frac{\tau}{t_o} = 1.937\left(1 - \frac{\sigma}{t_o}\right)^{0.677}$
High Quality Rock Mass Tightly Interlocking Undisturbed Rock with Unweathered Joints Spaced at 3m. (RMR – 85)	$\frac{\tau}{t_o} = 3.107\left(1 - \frac{\sigma}{t_o}\right)^{0.705}$	$\frac{\tau}{t_o} = 3.808\left(1 - \frac{\sigma}{t_o}\right)^{0.712}$	$\frac{\tau}{t_o} = 1.933\left(1 - \frac{\sigma}{t_o}\right)^{0.679}$	$\frac{\tau}{t_o} = 2.913\left(1 - \frac{\sigma}{t_o}\right)^{0.702}$	$\frac{\tau}{t_o} = 2.339\left(1 - \frac{\sigma}{t_o}\right)^{0.692}$
Good Quality Rock Mass Fresh to Slightly Weathered Rock, Slightly Disturbed with Joints Spaced at 1-3m. (RMR – 65)	$\frac{\tau}{t_o} = 2.381\left(1 - \frac{\sigma}{t_o}\right)^{0.698}$	$\frac{\tau}{t_o} = 4.112\left(1 - \frac{\sigma}{t_o}\right)^{0.707}$	$\frac{\tau}{t_o} = 2.124\left(1 - \frac{\sigma}{t_o}\right)^{0.669}$	$\frac{\tau}{t_o} = 3.164\left(1 - \frac{\sigma}{t_o}\right)^{0.695}$	$\frac{\tau}{t_o} = 2.552\left(1 - \frac{\sigma}{t_o}\right)^{0.683}$
Fair Quality Rock Mass Several Sets of Moderately Weathered Joints Spaced at 0.3 -1m. (RMR – 44)	$\frac{\tau}{t_o} = 3.659\left(1 - \frac{\sigma}{t_o}\right)^{0.691}$	$\frac{\tau}{t_o} = 4.478\left(1 - \frac{\sigma}{t_o}\right)^{0.700}$	$\frac{\tau}{t_o} = 2.308\left(1 - \frac{\sigma}{t_o}\right)^{0.662}$	$\frac{\tau}{t_o} = 2.862\left(1 - \frac{\sigma}{t_o}\right)^{0.688}$	$\frac{\tau}{t_o} = 2.785\left(1 - \frac{\sigma}{t_o}\right)^{0.675}$
Poor Quality Rock Mass Numerous Weathered Joints Spaced at 30-500mm with Some Gouge Filling/Clean Rock Waste (RMR – 23)	$\frac{\tau}{t_o} = 4.455\left(1 - \frac{\sigma}{t_o}\right)^{0.676}$	$\frac{\tau}{t_o} = 5.376\left(1 - \frac{\sigma}{t_o}\right)^{0.686}$	$\frac{\tau}{t_o} = 2.965\left(1 - \frac{\sigma}{t_o}\right)^{0.646}$	$\frac{\tau}{t_o} = 3.340\left(1 - \frac{\sigma}{t_o}\right)^{0.672}$	$\frac{\tau}{t_o} = 3.449\left(1 - \frac{\sigma}{t_o}\right)^{0.655}$
Very Poor Quality Rock Mass* Numerous heavily weathered spaced less than 50mm with gouge filling/waste rock with fumes (RMR – 3)	$\frac{\tau}{t_o} = 2.923\left(1 - \frac{\sigma}{t_o}\right)^{0.684}$	$\frac{\tau}{t_o} = 3.962\left(1 - \frac{\sigma}{t_o}\right)^{0.684}$	$\frac{\tau}{t_o} = 1.428\left(1 - \frac{\sigma}{t_o}\right)^{0.684}$	$\frac{\tau}{t_o} = 2.435\left(1 - \frac{\sigma}{t_o}\right)^{0.684}$	$\frac{\tau}{t_o} = 1.902\left(1 - \frac{\sigma}{t_o}\right)^{0.684}$

*Exponent 2β is taken to be the mean of exponents for all rock types and rock mass qualities from intact to poor.

Rock Mechanics, Nelson & Laubach (eds) © 1994 Balkema, Rotterdam, ISBN 90 5410 380 8

Strain amplitude and stress dependence of static moduli in sandstones and limestones

A. N. Tutuncu, A. L. Podio & M. M. Sharma
Department of Petroleum Engineering, The University of Texas at Austin, Tex., USA

ABSTRACT: A series of uniaxial stress cycling experiments have been performed on dry Berea, Bandera, Boise, Ohio and Pecos sandstones, Carthage, Leuders and Glen Rose limestones and Austin Chalk in order to investigate effects of strain amplitude and stress on the static moduli. A servo-controlled rock mechanics test system has been used for the uniaxial stress cycling tests. Tests are conducted at a frequency of 10^{-2} Hz frequency at different levels of strain amplitude and average stress. The static Young's moduli in sandstones shows a decreasing trend as strain amplitude is increased. Poisson's ratios increase with increasing strain. As the average stress is increased, sandstone samples yield higher Young's moduli whereas the opposite is observed in limestones. Data is presented to show that the dry static moduli are very sensitive to the strain amplitude and applied stress.

1 INTRODUCTION

Stress cycling measurements have been extensively used for determining fatigue limits of building foundations and soils and for wear of metals. The experiments reported in the literature usually consist of measuring the number of cycles to failure under either compressive or tensile cyclic loading conditions. There are numerous studies on the effects of the rate of loading, frequency of vibration and anisotropy of the samples on failure. The application of cycling tests to earth sciences has been introduced within the past two decades when significant differences between the static and dynamic moduli were observed in laboratory core measurements. This difference arises primarily due to differences in measurement frequency (10^6 Hz vs. 10^{-2} Hz) and strain amplitude (10^{-7} vs. 10^{-3}). The presence of clays and microcracks in the rock, the amount and type of fluid in the pore space, the petrophysical properties of the samples tested and variations in magnitude of the effective stress all control the magnitude of the difference between the two moduli. The effects of frequency, presence of clay, saturation, and fluid type have been investigated both experimentally and theoretically by the authors (Tutuncu and Sharma, 1992; Tutuncu et al., 1994). As pointed out by Tutuncu and Sharma (1992), at low frequencies such as in the cycling tests, moduli are expected to stay constant at constant stress in both dry and saturated rocks. On the other hand, at fixed stress levels, a reduction in moduli is commonly reported in fatigue studies. Haimson and Kim (1971) performed cyclic uniaxial compression tests on white Tennessee marble and their data indicated a definite weakening of the rock when the maximum applied stress is 75-100 % of the compressive strength. The value of the tangent Young's modulus per cycle in their tests decreased as the number of cycles increased.

Haupt et al. (1992) measured the effects of loading frequency, strain amplitude and saturation on elastic moduli and attenuation in welded tuff and Sierra White granite. They used four different laboratory techniques (cyclic loading, resonant bar, waveform inversion and ultrasonic measurements) for determining Young's modulus and extensional wave attenuation at frequencies ranging from 0.01 Hz to 1 MHz. Their modulus and attenuation data in room dry samples remained frequency independent while frequency dependent attenuation and modulus dispersion were observed in saturated samples. They found the properties of tuff to be independent of strain amplitude in both dry and saturated conditions. On the other hand, they observed a decrease in modulus with increasing strain amplitude, the attenuation also increased remarkably as strain amplitude increased from $5x10^{-7}$ to $5x10^{-4}$ for granite. Brace and Jones (1971) compared the uniaxial strain data in plane shock and static loading for Westerly granite and Solenhofen limestone. They found that for granite and limestone, the stress-strain relation was almost identical and strains were elastically recoverable at low and intermediate stresses. At high stresses permanent deformation took place in limestone.

The main objective of our study is to understand the strain amplitude and stress dependence of the moduli in various dry sandstone and limestone samples with a wide range of porosities and permeabilities. The petrophysical properties of the samples used in the study are listed in Table 1.

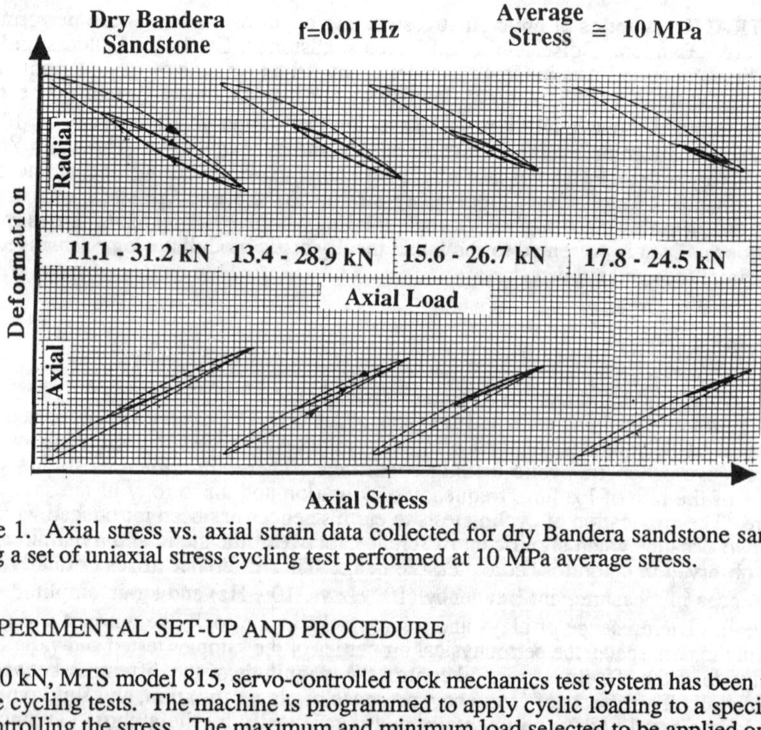

Figure 1. Axial stress vs. axial strain data collected for dry Bandera sandstone sample during a set of uniaxial stress cycling test performed at 10 MPa average stress.

2 EXPERIMENTAL SET-UP AND PROCEDURE

A 4450 kN, MTS model 815, servo-controlled rock mechanics test system has been used for the cycling tests. The machine is programmed to apply cyclic loading to a specimen by controlling the stress. The maximum and minimum load selected to be applied on the sample along with the frequency of the loading-unloading process is entered into the program as an input and the number of cycles that can be recorded is unlimited. Ultrasonic arrival times are simultaneously recorded at both maximum and minimum loads in order to determine the corresponding dynamic moduli at the same stresses as the static moduli and changes in the waveform are also monitored to provide information on whether any microfracturing takes place during the tests.

Both the axial and radial deformations are recorded using 350 ohm foil CEA type strain gauges connected into a four arm Wheatstone bridge circuit to provide temperature compensation. The strain gauges are bonded vertically and horizontally to approximately the mid-length of the specimen on the opposite sides for recording the axial and radial deformations, respectively. 5-cm diameter room dry cylindrical core samples are used for the tests. The length of the samples varies from 6.2 to 10.1 cm. The ends of the cores are ground flat and parallel to within $\pm 2.5 \times 10^{-3}$ cm.

A heaversine loading program is used to obtain sinusoidal variations in load during each cycle. Cyclic frequencies of 0.001, 0.005, 0.01, 0.2 and 0.5 Hz have been studied for a few samples tested and the results indicate that at the low frequencies the frequency is not an important factor controlling the variations in static elastic moduli. Therefore, test results are reported here for only the 0.01 Hz.

The tests are performed at least at two average stress levels which are less than or equal to 50% of the compressive strength of the samples studied (for most samples 10 MPa and 20 MPa have been selected) and at each average stress level, three or four tests are performed to investigate the role of strain amplitude at constant average stress. Once a test is completed at a certain differential stress level, the minimum axial load is increased about 2 kN while the maximum axial load is decreased by the same amount in order to keep the average stress constant and the cycling test is continued at the new differential stress level. When a set of tests is completed, the identical procedure is repeated for the next selected average stress value. Tests for all samples have been repeated at least twice in order to check the repeatability of the static data and there is no indication of permanent plastic deformation of the samples.

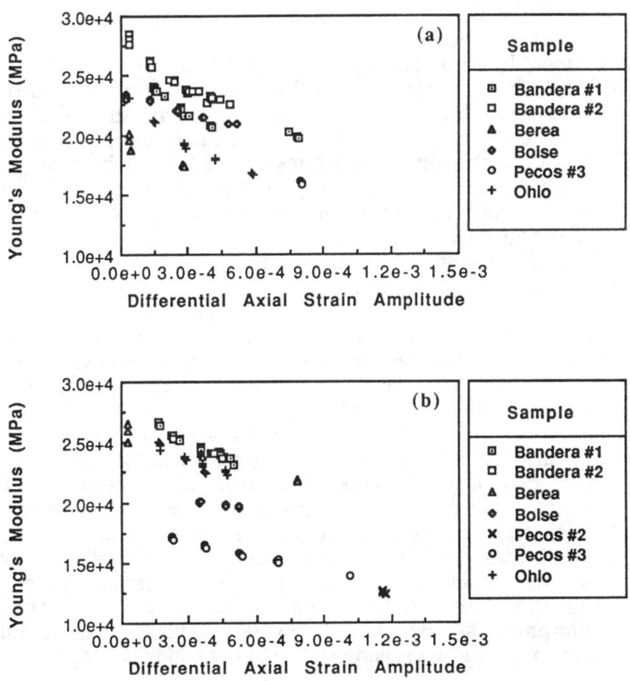

Figure 2: Static Young's moduli as a function of differential axial strain amplitude in five sandstone samples at average stresses of 10 MPa (a) and 20 MPa (b).

An example of the stress-strain plots obtained with the procedure described above for a dry Bandera sandstone sample at 10 MPa average stress is shown in Figure 1. In a viscoelastic medium (such as sedimentary rocks), a sinusoidally varying uniaxial stress results in a proportionate sinusoidally varying strain, differing only in magnitude and phase angle. This phase shift or lag is observed as a hysteresis loop on a stress-strain plot. The area within the hysteresis loop is proportional to the energy loss during a stress cycle. This energy loss divided by the maximum energy stored during the cycle has been used as a parameter (attenuation) for characterizing a lossy solid. It is clear from Figure 1 that the smaller the differential stress, the smaller the hysteresis loop, i.e. attenuation.

3 EXPERIMENTAL RESULTS FOR THE STRAIN AMPLITUDE AND STRESS DEPENDENCE OF STATIC MODULI

It is well known that materials without pores and cracks such as aluminum and steel do not display hysteresis on their stress-strain diagrams, while porous and/or fractured materials such as sedimentary rocks do. One of the earliest proposed mechanisms for explaining the difference between the behavior of these two types of materials is frictional sliding between crack surfaces and at the grain boundaries, Walsh (1966), Mavko (1979), White (1966), Winkler (1979), Winkler et al. (1979), Stoll (1989). In order to examine whether or not frictional sliding is the dominant factor in cycling tests and to investigate the role of stress and strain amplitude on static moduli, a series of uniaxial stress cycling tests have been conducted on room dry Berea, Bandera, Pecos, Ohio and Boise sandstones, Carthage, Leuders and Glen Rose limestones and Austin Chalk samples. The variation of static Young's moduli and Poisson's ratios have been monitored and attenuations from the hysteresis loops recorded have been calculated at 10 MPa and 20 MPa average stress.

Static Young's moduli are obtained by plotting the axial strain calculated at each loading and unloading cycle separately as a function of axial stress and fitting a first degree polynomial through the data points. The Young's modulus obtained from this procedure is similar to the tangent modulus. The same procedure is followed for calculation of static Poisson's ratio (radial strains are plotted as a function of axial strain and the slope of a linear polynomial fit through data points gives Poisson's ratio). The loading cycles for all samples are in general linear while slightly non-linear curves are observed for the unloading data.

The dynamic moduli are always greater than the static moduli at low stresses. The ratio of dynamic to static Young's moduli approaches 1 as stress is increased.

Static Young's moduli of all five sandstones have been plotted as a function of strain amplitude at these two stress levels in Figure 2. All plots have been prepared using the loading data. The corresponding static Poisson's ratio data are shown in Figure 3. It is evident that as strain amplitude increases, static Young's moduli decrease whereas static Poisson's ratios increase slightly. The softening observed causes the Young's modulus to decrease linearly with strain amplitude. The limestone data also shows that Young's moduli decrease linearly with strain amplitude (Figures 4 and 5). The difference in the stiffness of the samples is also a function of the difference in their porosities. One important point to note here is that for sandstones, as the average stress is increased from 10 MPa to 20 MPa, static Young's moduli increase whereas they decrease for the limestones tested. It is believed that in both sandstones and limestones at low stresses, closure of existing microcracks and variations in the separation distances of grain asperities govern the variations in static moduli. The softening of rock at higher stress values may be interpreted as frictional sliding taking place at the grain contacts. Permanent deformation or yielding is another possibility which is suggested by Haupt et al. (1992).

4 DISCUSSION

Haimson (1977) reported experimental results for dry Tennessee marble, Indiana limestone, Berea sandstone and Westerly granite under uniaxial cyclic compression and tension. He used stress controlled cycles at 1 Hz frequency and simultaneously recorded microseismic activity and observed large hysteresis in the first cycle followed by decreasing loops in the next few cycles, narrowing of the loop to an almost constant shape in the following groups of cycles and reopening of the loops in the last cycles prior to failure. Acoustic emission increased only as failure was approached. He determined the mechanism of cyclic fatigue as microfracturing.

In the uniaxial stress cycling tests performed in this study, since the maximum axial stress applied was less than the 50% of the compressive strength of the rocks studied and no variation was observed in the ultrasonic waveforms during the tests, the microfracturing is not a realistic explanation for the trends obtained.

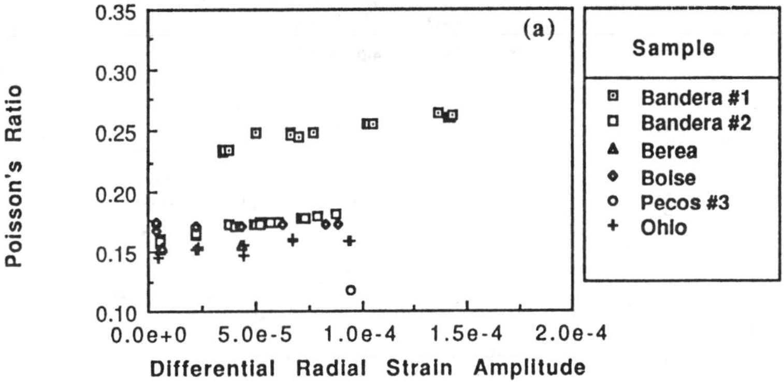

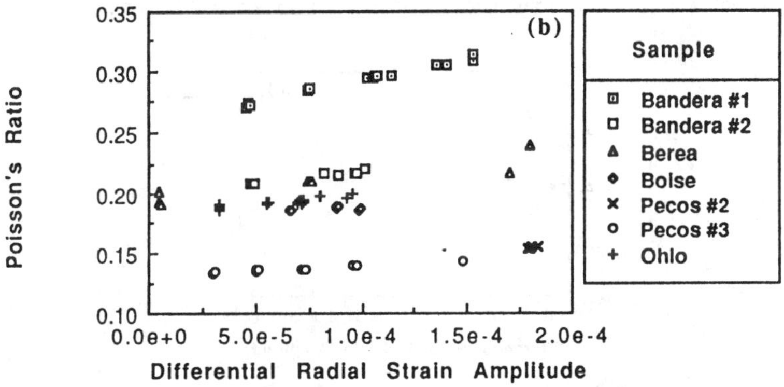

Figure 3. Static Poisson's ratios as a function of differential radial strain amplitude in five sandstone samples at average stresses of 10 MPa (a) and 20 MPa (b).

Attenuation is calculated from two different hysteresis loops for each test. The first cycle is between the minimum and maximum stresses which we will call "small cycles". When attenuation for the first loading and last unloading (large cycles) are calculated, larger attenuations are obtained (See Figure 1). This is due to larger differential axial stresses applied on these cycles which results in larger strains. The static moduli and attenuations have been plotted as a function of strain amplitude and stress for the small and large cycles for Ohio sandstone and Leuders limestone samples in Figures 6 and 7. It is evident that attenuation tends to increase with differential strain amplitude (or differential stress).

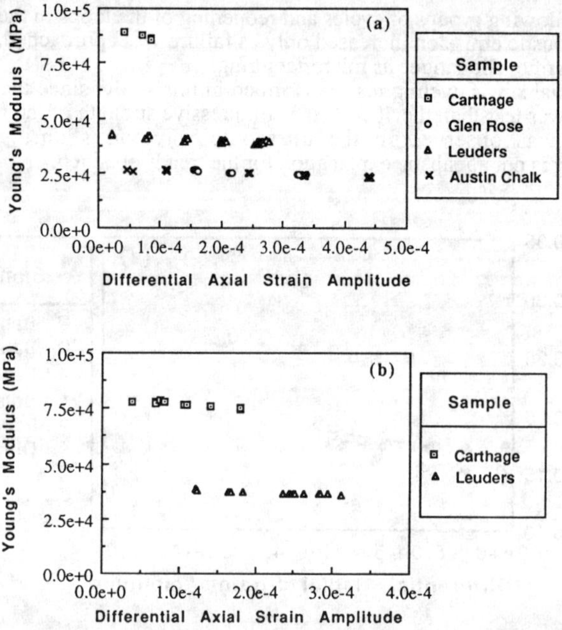

Figure 4. Static Young's moduli as a function of differential axial strain amplitude in four carbonate samples at average stresses of 10 MPa (a) and 20 MPa (b).

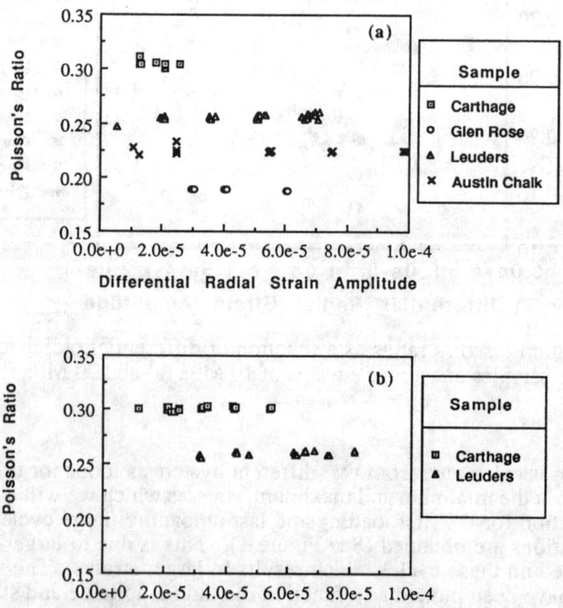

Figure 5: Static Poisson's ratios as a function of differential radial strain amplitude in four carbonate samples at average stresses of 10 MPa (a) and 20 MPa (b).

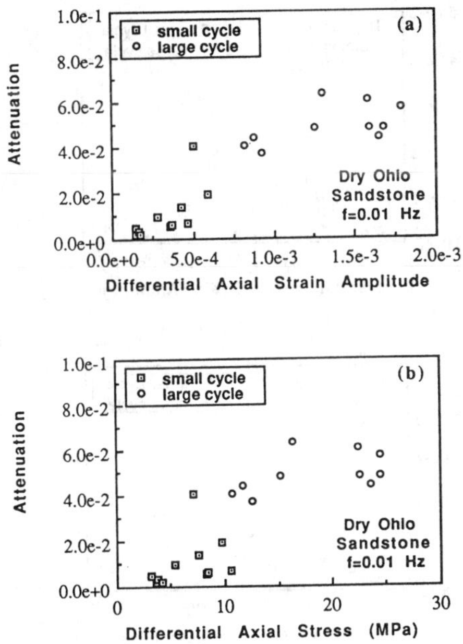

Figure 6. Variation of attenuation with strain amplitude (a) and stress (b) in small and large cycles in dry Ohio sandstone.

5 CONCLUSIONS

Static moduli and attenuation are found to be strong functions of strain amplitude and stress in sandstones and limestones. As strain amplitude increases, static Young's moduli decrease whereas static Poisson's ratios increase. Attenuation also increases almost linearly with strain amplitude.

In addition to frictional sliding an intrinsic loss mechanism is believed to be responsible for the strain amplitude dependence of the static moduli and attenuation and the differences between static and dynamic moduli in dry rocks. Work is underway to quantitatively describe this intrinsic loss mechanism in granular sedimentary rocks.

Table I. Petrophysical properties of the samples tested in this study.

Sample	Porosity (%)	Permeability (md)	Density (gr/cm^3)
Sandstones			
Bandera	18	7.9	2.20
Berea	21	500.0	2.13
Boise	24	1500.0	2.10
Pecos	18	0.7	2.13
Ohio	16	0.1	2.24
Limestones			
Carthage	2	< 0.1	2.66
Glen Rose	20	6.8	2.52
Leuders	18	1.0	2.21
Austin Chalk	20	< 0.2	2.16

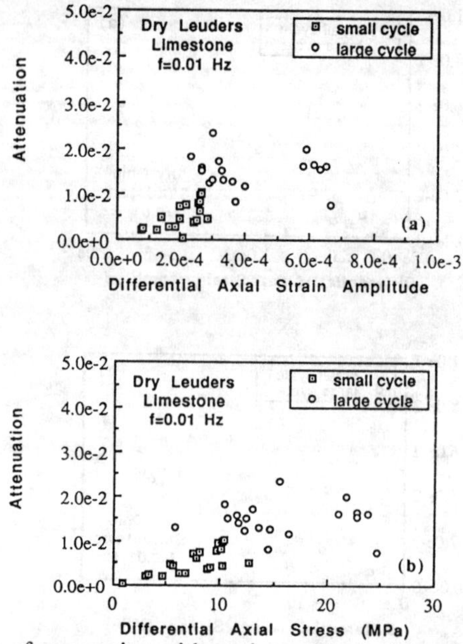

Figure 7. Variation of attenuation with strain amplitude (a) and stress (b) in small and large cycles in dry Leuders limestone.

REFERENCES

Brace W.F. and Jones A.H. 1971. Comparison of uniaxial deformation in shock and static loading of three rocks. J. Geophysical Res. 76: 4913-4921.

Haimson B.C. 1977. Effect of cyclic loading on rock. Dynamic Geotechnical Testing. ASTM Special Publication 654: 228 - 245.

Haimson B.C. and Kim C.M. 1971. Mechanical behavior of rock under cyclic fatigue. Stability of Slopes. Proc. 13th Symp. Rock Mech.: 845-863.

Haupt R.W., Martin III R.J., Tang X, Dupree W.J. and Price R.H. 1992. Modulus dispersion and attenuation in tuff and granite. Proc. 33rd U.S. Symp. Rock Mech : 899-908.

Mavko, G. 1979. Frictional attenuation : an inheritent amplitude dependence. J. Geophys. Res. 84:4769-4775.

Stoll, R.D. 1989. Sediment Acoustics: Lecture notes in Earth Sciences 26. Springer Verlag.

Tutuncu, A.N. and Sharma, M.M. 1992. The influence of fluids on grain contact stiffness and frame moduli in sedimentary rocks. Geophysics 57:1571-1582.

Tutuncu, A.N., Podio A.L. and Sharma, M.M. 1994. An Experimental Investigation of Factors Influencing Compressional and Shear Wave Velocities and Attenuations in Tight Gas Sandstones. Geophysics 59: 77-86.

Walsh, J.B. 1966. Seismic wave attenuation in rock due to friction. J. Geophys. Res. 71: 2591-2599.

White, J.E. 1966. Static friction as a source of seismic friction. Geophysics 31: 333-339.

Winkler, K. 1979. The effects of pore fluids and frictional sliding on seismic attenuation. Ph.D. Dissertation. Stanford University.

Winkler, K., Nur, A. and Gladwin, M. 1979. Friction and seismic attenuation in rocks. Nature 227: 528-531.

Rock Mechanics, Nelson & Laubach (eds) © 1994 Balkema, Rotterdam, ISBN 90 5410 380 8

Effects of strain amplitude on the static and dynamic nonlinear deformation of Berea sandstone

L.B.Hilbert, Jr, T.K.Hwong & N.G.W.Cook
Department of Materials Science and Mineral Engineering, University of California, Berkeley, Calif., USA
K.T.Nihei & L.R.Myer
Earth Sciences Division, Lawrence Berkeley Laboratory, Calif., USA

ABSTRACT: A series of carefully controlled static compressive loading and ultrasonic pulse transmission tests on room-dry Berea sandstone were performed to develop a fundamental understanding of the micro-mechanisms that result in nonlinear dynamic and static stress-deformation relationships of granular rocks. Young's modulus and Poisson's ratio were calculated as a function of axial stress from static uniaxial strain and uniaxial stress tests. Dynamic Young's modulus were computed as a function of axial stress from compressional (P) and shear (S) wave velocities measured during the ultrasonic pulse transmission tests. We found that measured dynamic moduli match initial tangent moduli computed from small-strain, small-stress amplitude cycles executed during a static uniaxial strain test. Further, the nonlinear stress dependence of both P- and S-wave velocities can be reproduced from these tangent moduli. These results show that nonlinear static and dynamic properties of a granular rock, such as Berea sandstone, are due to the strain amplitude of the loading, rather than to frequency-dependent dynamic loading effects. By applying appropriate effective medium models based on the micromechanics of microcrack closure and grain elasticity we have related the nonlinear dynamic response of a granular rock to the nonlinear behavior under static loading. Such a relationship between nonlinear static and dynamic properties will be helpful in understanding the nonlinear behavior of granular rocks under high strain amplitude waves.

1. INTRODUCTION

The closure of rock intra-granular cracks and grain boundaries, referred to generally as microcracks, under increasing applied compressive stress results in a well-known nonlinear relationship between stress and strain. This nonlinear load-deformation relationship results in a nonlinear dependence of elastic properties on applied stress. In addition, microcracks cause the velocities of elastic waves in rocks to have a nonlinear stress-dependence.

Although a great deal of literature exists on theoretical models for the behavior of cracked and porous solids (Walsh, 1965a, b, c, O'Connell and Budiansky, 1974, Bruner, 1976, Kemeny and Cook 1985) there is little experimental data available with which to establish a relationship between the commonly observed static and dynamic nonlinear behavior (Cook and Hodgson, 1965). To investigate the micromechanics of nonlinear behavior of clastic rocks, laboratory experiments on Berea sandstone were performed. Static stress-strain and ultrasonic pulse transmission tests were performed concurrently.

2. EXPERIMENTAL INVESTIGATIONS

A cylindrical sample of Berea sandstone 50.8-mm in diameter and 42.4-mm in height was used for all tests. The sample used in the tests had a density of 2100 kg/m^3 and a porosity of 22%. Two types of static loading tests were executed: a uniaxial stress test and a uniaxial strain test. Changes in specimen length and diameter as well as applied axial and radial loads were monitored using a personal computer and custom laboratory software. Dynamic pulse transmission tests were also conducted concurrently with the static loading tests.

In each test, static axial load was increased to a maximum of about 25 MPa and then the sample was unloaded, forming a "major" stress cycle. During the static major stress cycle, loading was periodically stopped and a "minor" stress cycle with an amplitude of 1.5-3.0 MPa was executed. Plots of static axial stress versus axial strain for these tests are shown in Fig. 1. The major stress-strain cycles exhibit considerable hysteresis. However, there is little hysteresis observable in the minor stress-strain cycles where the maximum change in axial strain during unloading or loading was ~0.00013.

The P- and S-wave velocities computed from the travel times measured during the ultrasonic pulse transmission tests are shown in Fig. 2. It can be seen that both wave velocities have a nonlinear dependence on axial stress. Moreover, the velocity data also exhibit a dependence on loading direction.

3. MICROMECHANICS OF STATIC AND DYNAMIC DEFORMATION

Under static conditions, the static instantaneous Young's modulus and Poisson's ratio are given by:

$$E_{static} = \frac{d\sigma_z}{d\varepsilon_z} \quad \text{and} \quad \nu_{static} = \frac{d\varepsilon_r}{d\varepsilon_z} \tag{1}$$

Values of the dynamic material properties computed from the results of dynamic pulse transmission tests can be calculated from the P- and S-wave velocities using the following equations for elastic waves in an isotropic medium:

$$\rho V_s^2 = \mu_{dyn} \quad \text{and} \quad \rho V_p^2 = K_{dyn} + (4/3)\mu_{dyn} = \left[E_{dyn}(1-\nu_{dyn})\right] / \left[(1+\nu_{dyn})(1-2\nu_{dyn})\right] \tag{2}$$

where E_{dyn}, K_{dyn}, μ_{dyn}, and ν_{dyn} are the Young's modulus, bulk modulus, shear modulus, and Poisson's ratio calculated from a dynamic test. The dynamic Poisson's ratio can be calculated from the ratio of the wave velocities:

$$\nu_{dyn} = \frac{1}{2}\left\{\left[(V_s/V_p)^2 - 2\right] / \left[(V_s/V_p)^2 - 1\right]\right\} \tag{3}$$

Dynamic elastic material properties calculated from the P-wave and S-wave velocities are shown in Fig. 3. The hysteresis exhibited in the wave velocities manifests itself in the dynamic elastic material properties as well. The moduli approach an asymptotic value as axial stress increases, closing properly oriented open cracks and leaving only the pores persistently open (under hydrostatic stress conditions all cracks would close). As long as stresses are below the stress required for new crack creation or crack extension, the rock stiffens with increasing stress and the moduli asymptotically approach values for a porous rock with no cracks. The asymptotic values of Young's, bulk, and shear moduli are approximately 27 GPa, 17 GPa, and 11 GPa, respectively.

Since a dynamic value of Young's modulus has been determined as a function of axial stress, a "dynamic" stress-strain curve can be determined by calculating a "dynamic" axial strain corresponding to the applied stress. Such a dynamic stress-strain curve is compared to the static stress-strain curves in Fig. 1. As can be seen, each of the three curves have different stress-dependent slopes, or stiffnesses. The uniaxial stress curve has the lowest stiffness and the dynamic curve has the highest stiffness. In the uniaxial stress test, the sample was free to deform radially. Radial deformation is prohibited in the uniaxial strain test, and thus the resistance of the sample against axial deformation is greater, resulting in a higher stiffness. Both static tests subject the sample to large strains, relative to the dynamic test. The large static strains mobilize frictional sliding between microcracks, which results in a lower stiffness, that is, lower resistance to deformation, relative to the dynamic tests. Observable attenuation due to frictional sliding has been shown to occur under strain amplitudes above on the order 10^{-6} (Mavko, 1979). In our ultrasonic pulse tests, the strain amplitude was less than 10^{-6}, thus deformation during the passage of a strain wave results only from elastic response of a grain with no concomitant frictional sliding. The stiffness reflected in the dynamic stress-strain relationship is therefore a function of only the properties of the porous structure, which has a higher stiffness than that which would include an additional component due to frictional sliding.

Static and dynamic Young's moduli are compared in Fig. 4. Three curves are plotted: static Young's moduli computed from the static uniaxial strain test major stress cycle using Eqn. 1; Young's moduli computed from the initial slope of the minor stress cycles in the static uniaxial strain test; and dynamic Young's moduli computed from the wave velocities. It can be seen that the Young's moduli calculated from the initial unloading portions of the minor stress cycles agree well with the dynamic Young's moduli. These two moduli reflect the stiffness of the porous rock structure sampled during small strain amplitudes in which frictional sliding is not mobilized. At low axial stresses, the moduli computed from the wave velocities and the static minor stress cycles are both higher than the Young's moduli computed from the static major stress cycle. This is a result of mobilization of grains or sliding between microcracks under the larger strains in the major stress cycle, which contributes greater deformation and thus lower stiffness than in dynamic tests.

It is of interest to note that during the unloading phase of the major stress cycle, the computed moduli are close to the dynamic moduli. In has been stated previously that during the passage of an ultrasonic strain wave the strain amplitude is so small that the response of the rock is the elastic response of the porous structure with no frictional sliding on microcracks. That the moduli from the unloading portion of the static major stress cycle matches the dynamic value indicates that the static unloading response has no frictional sliding either. This is indeed the case. During the loading phase of the major stress cycle, strains (or stress) are large enough so that eventually frictional sliding occurs between microcracks. However, when the axial load is reversed during the unloading cycle, the grains are "locked" in place until the axial load is reduced to the point at which the elastic response of the grains is sufficient to overcome the frictional resistance between the grains. This also explains difference in slopes in the loading and unloading phases of the static stress test and also accounts for the hysteresis in the cycle. The micromechanism of frictional sliding between microcracks also plays a fundamental role in the radial deformation of the rock structure, which effects the value of Poisson's ratio, as will be discussed below.

Poisson's ratios as calculated from the static test data and from the ultrasonic wave velocities are plotted in Fig. 5. Again three curves are plotted in the figure: static values from the major stress cycle in the uniaxial stress tests; values calculated from changes in strain during a minor stress cycle; and dynamic values calculated from the wave velocities. Poisson's ratio calculated from the minor stress cycles agrees with that calculated from the major stress-strain curve during the unloading path. This is expected because microcracks are frictionally "locked" in place. In fact, it can be seen that Poisson's ratio is constant for nearly the entire unloading phase, indicating that the grains continue to be frictionally locked and the Poisson's ratio reflects the deformation associated with the porous rock structure, with no frictional sliding.

It might be expected that the dynamic Poisson's ratio should be the same or close to the static values during unloading, since in the unloading case the grains are also frictionally "locked" in place and the radial and axial deformations are a result of only the porous rock, with no attendant frictional sliding. However, Fig. 5 shows that the dynamic values of Poisson's ratio are the highest of all. The reason for this is the effect of stress-induced anisotropy under non-hydrostatic loading. P- and S-wave velocities depend on the microcrack contact stiffness (Ita, et al., 1993, Nihei, 1992). Under a hydrostatic state of stress all microcrack contact surfaces are, for all practical purposes, isotropically stressed; but under uniaxial stress conditions, contact surfaces are preferentially compressed only in the longitudinal direction and cracks perpendicular to the axial direction are "loose" or remain open. Thus, in the uniaxial stress or strain cases, the S-wave velocity will be lower than in the hydrostatic (or isotropically stressed) case, since these waves sample transverse stiffness. Application of Eqn. 3 to the uniaxial stress case will result in higher Poisson's ratio than for the hydrostatic stress case, since decreasing S-wave velocities will result in increasing dynamic Poisson's ratio. It is also recognized that presence of a bedding plane can result in significant anisotropy of any or all measured values. The Berea sandstone sample used in the work described in this paper did not exhibit any visible evidence of a bedding plane.

4. ANALYSIS USING EFFECTIVE MEDIUM THEORY

The objective of the following analysis is to show that the material properties of this rock as measured under dynamic loads can be computed knowing only properties determined under static loading. Also, it will be shown that shear and compressional wave velocities can be computed from the static tangent moduli determined from the initial unloading portion of the minor stress cycles performed in the static uniaxial strain tests.

The mineral composition of the Berea sandstone grains used in this study is: 0.88 quartzite, 0.05 calcite, and 0.07 feldspar. Zimmerman (1984) computed the intrinsic elastic moduli of Berea sandstone grains by computing the elastic moduli of the three-phase mixture according to the average of the Voigt and Reuss values for the mixtures.

The resulting moduli for the grains are shown in Table 1. Since these values refer to the intrinsic properties of the sandstone grains, a subscript of "i" is used.

The addition of pores should make the stiffness of a solid equivalent to the asymptotic value of Young's modulus, which is the high-stress value in which microcracks are closed and pores are open. For sandstone, one may be tempted to assume that the pores can be approximated by spherical voids which are imbedded in a simply connected matrix of material with the elastic properties of the grains, a geometry for which there are many effective medium models (Nemat-Nasser and Hori, 1993). The results of application of various effective medium models are shown in Table 2, in which a subscript "p" denotes a porous material. The values shown in Table 2 for models with spherical pore shapes are all much higher than the asymptotic values indicated in Fig. 3. This is because spherical pores are much stiffer than the actual pores in the sandstone. Since actual rock has pores that are connected to form tortuous tubular pathways, it might be expected that a model with tubular pores of circular cross section could provide a more favorable approximation. Material properties were computed for a model with a dilute concentration of cylindrical pores with and without interactions. For interacting aligned circular cylindrical pores under plane stress, the self-consistent theory gives (Nemat-Nasser and Hori, 1993):

$$\mu_p/\mu_i = [1+3\phi]/[1+\phi(\kappa-2)], \quad E_p/E_i = 1-3\phi, \quad \nu_p/\nu_i = 1-(3-1/\nu_i)\phi \tag{4}$$

where ϕ is the porosity and for plane stress $\kappa = (3-\nu)/(1+\nu)$. Material properties computed from these equations for tubular pores are also shown in Table 2. It is seen that the values from the non-interacting pores are still higher than the experimentally-determined values. However, the values determined from the self-consistent model are very close to the asymptotic values from the experimental results. This indicates that the circular cylindrical pores more closely approximate the overall effects of the porous structure of Berea sandstone. In addition, the Poisson's ratio calculated using the above equation for the self-consistent method was ν_p=0.195, which agrees well with the values shown in Fig. 5 for the unloading portion of the static tests and the minor stress cycles. This validates the proposal that the grains are "locked" in place, since the calculated value based on the effective medium model does not allow for grain sliding.

There is a further stress-dependent reduction in the elastic moduli from the poro-elastic value. This modulus reduction is due to the closing of microcracks as stress is increased. A modified self-consistent method for modelling a solid containing randomly oriented, penny shaped cracks has been proposed by Bruner (1976). Bruner gives the following approximate solutions from his model:

$$E_{eff}/E_p = e^{-(16c/9)} \quad \text{and} \quad \nu_{eff}/\nu_p = e^{-(16c/10)} \tag{5}$$

where c is the crack density. Although Zimmerman (1984) gives exact solutions for the original nonlinear differential equations, it is more practical to use the above approximations. The subscripts "eff" indicate that these are properties of the effective medium, including both pores and cracks.

It has not proved to be practical to attempt to determine the shape distribution of all cracks in the rock sample. Instead, Eqns. 5_1 and 5_2 were used to determine the crack density c from the ratio of moduli E_{eff}/E_p. This was done by using for E_{eff} the values of the initial tangent moduli to the minor stress cycles in the static uniaxial strain tests

Table 1. Elastic properties of Berea sandstone grains (after Zimmerman, 1984).

K_i (GPa)	μ_i (GPa)	E_i (GPa)	ν_i
39.5	31.6	73.8	0.19

Table 2. Effective elastic properties for porous material and no microcracks.

Model	Pore shape	K_p (GPa)	μ_p (GPa)	E_p (GPa)
Non-interacting	sphere	27.4	21.6	51.4
Self-consistent	sphere	22.1	17.5	41.5
Modified self-consistent	sphere	25.5	20.1	47.7
Kuster-Toksoz	sphere	25.3	19.9	47.3
Non-interacting	tubular	23.5	18.8	44.5
Self-consistent	tubular	13.3	10.6	25.1

Table 3. Calculations for effective moduli and wave velocities.

Axial stress (MPa)	**Young's modulus (GPa)**	**Crack density**	**Poisson's ratio**	**Shear modulus (GPa)**	**Bulk modulus (GPa)**
3.7	17.2	0.253	0.13	7.6	7.8
8.7	22.7	0.097	0.167	9.7	11.4
13.6	25.3	0.036	0.184	10.7	13.4
18.7	26.5	0.01	0.192	11.1	14.4
14.4	26.7	0.006	0.193	11.2	14.5
9.5	25	0.043	0.182	10.6	13.1
4.2	15.6	0.308	0.119	7	6.8

and for E_p the value of 25 GPa calculated previously for circular cylindrical pores. This resulted in the crack density as a function of axial stress. After determining the crack density *c*, ν_{eff} was determined at each axial stress using the value of ν_p computed previously. Now having computed E_{eff} and ν_{eff} as functions of axial stress, the shear and compressional wave velocities can be calculated using Eqns. 2_1 and 2_2, along with the standard equations relating E and ν to K and μ. The results of these calculations are shown in Table 3 and the calculated wave velocities are plotted along with the measured wave velocities in Fig. 6. The agreement between the predicted and calculated wave velocities is seen to be excellent. In addition, the crack densities shown in Table 3 agree well with the range of crack densities determined by Sayers and van Munster (1991).

5. SUMMARY AND CONCLUSIONS

Detailed static stress-strain and dynamic ultrasonic pulse transmission tests were performed concurrently on a sample of room-dry Berea sandstone with the objectives of developing a better understanding of the effects of micro-structure of granular rock on the nonlinear load-deformation relationships of static and dynamic material

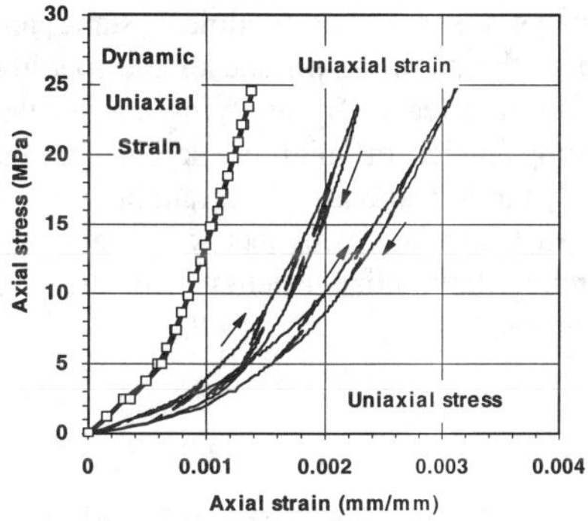

Figure 1. Stress versus strain plots.

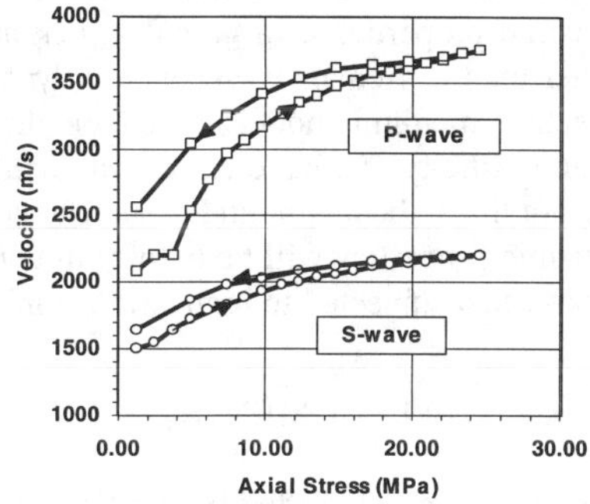

Figure 2. Wave velocities from ultrasonic pulse tests.

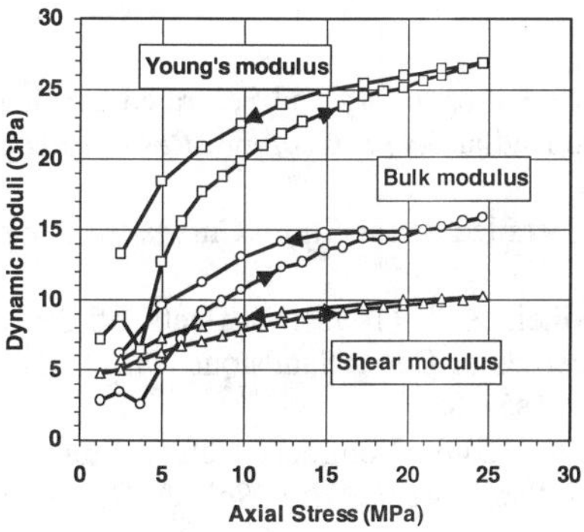

Figure 3. Dynamic moduli computed from wave velocities.

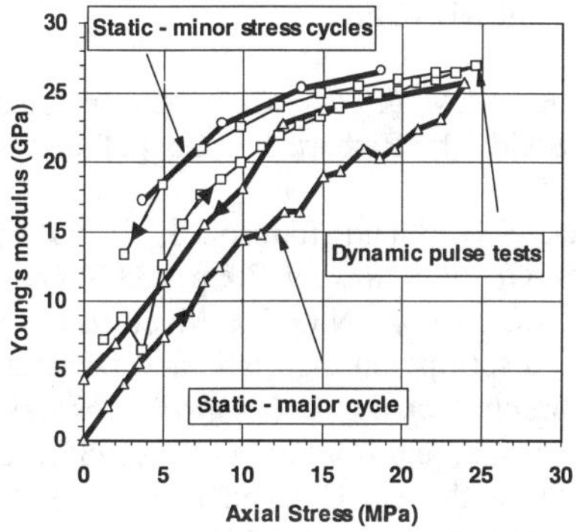

Figure 4. Dynamic and static Young's moduli.

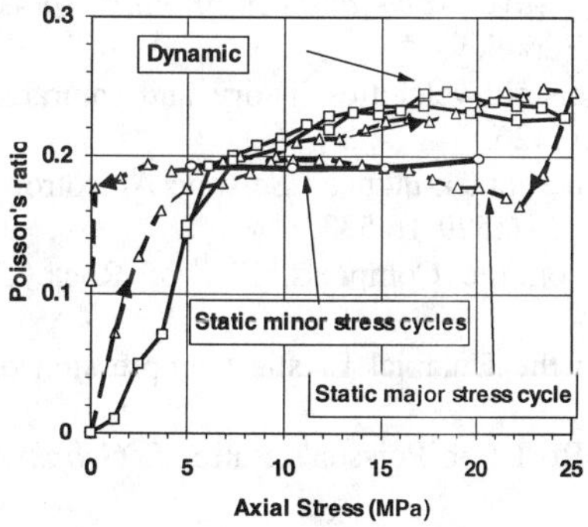

Figure 5. Dynamic and static Poisson's ratio.

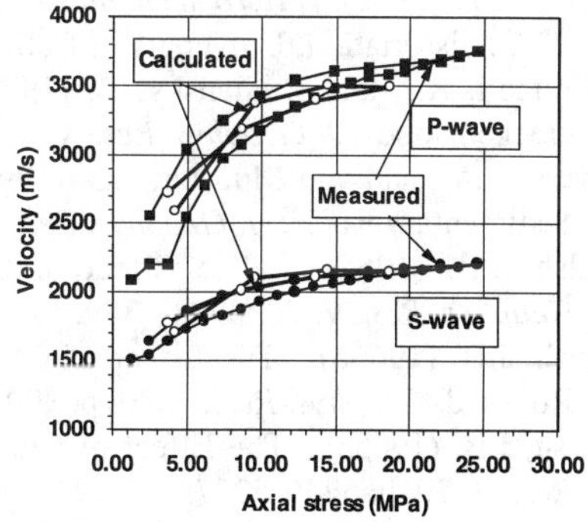

Figure 6. Computed and measured wave velocities.

properties. Our experimental results and analyses show that nonlinear static and dynamic properties of a granular rock are due to the strain amplitude of the loading, rather than to frequency-dependent dynamic loading effects. By applying appropriate effective medium models based on the micromechanics of microcrack closure and grain elasticity, we have related the nonlinear dynamic response of a granular rock to the nonlinear behavior under static loading. Such a relationship between static and dynamic properties will be helpful in understanding the nonlinear behavior of granular rocks when subjected to high strain amplitude waves.

6. ACKNOWLEDGMENTS

This work was supported by the Director, Office of Energy Research, Office of Basic Energy Sciences, Engineering and Geosciences Division, of the US Department of Energy under Contract No DE-AC03-76SF00098.

7. REFERENCES

Bruner, W.M. (1976). Comment on "Seismic Velocities in Dry and Saturated Cracked Solids" by Richard J. O'Connell and Bernard Budiansky, *J. Geophys. Res.*, v. 81, p. 2573-2578.

Cook, N.G.W. and Hodgson, K. (1965). Some Detailed Stress-Strain Curves for Rock, *J. Geophys. Res.*, v. 70, p. 2883-2888.

Ita, S.L., Cook, N.G.W., Myer, L.R., and Nihei, K.T. (1993). Effects of Stress Anisotropy on the Static and Dynamic Properties of Berea Sandstone, *Int. J. Rock Mech. Min. Sci. & Geomech. Abstr.*, v. 30, p. 785-788..

Kemeny, J. and Cook, N.G.W. (1985). Effective Moduli, Nonlinear Deformation and Strength of a Cracked Elastic Solid, *Int. J. Rock Mech. Min. Sci. & Geomech. Abstr.*, v. 23, p. 107-118.

Mavko, G.M. (1979). Frictional Attenuation: An Inherent Amplitude Dependence, *Geophysics*, v. 84, p. 4769-4775.

Nemat-Nasser, S. and Hori, M. (1993). *Micromechanics: Overall Properties of Heterogeneous Materials*, North-Holland, Amsterdam.

Nihei, K.T. (1992). *Micromechanics of seismic wave propagation in granular rocks*, Ph.D. dissertation, University of California, Berkeley.

O'Connell, R.J. and Budiansky, B. (1974). Seismic velocities in dry and saturated cracked solids, *J. Geophys. Res.*, v. 79, 5412-5426.

Sayers, C.M. and van Munster, J.G. (1991). Microcrack-induced Seismic Anisotropy Sedimentary Rocks, *J. Geophys. Res.*, v. 96, p. 16,529-16,533.

Walsh, J.B. (1965a). The Effect of Cracks on the Compressibility of Rock, *J. Geophys. Res.*, v. 70, p. 381-389.

Walsh, J.B. (1965b). The Effect of Cracks on the Uniaxial Elastic Compression of Rocks, *J. Geophys. Res.*, v. 70, p. 399-411.

Walsh, J.B. (1965c). The Effect of Cracks in Rocks on Poisson's Ratio, *J. Geophys. Res.*, v. 70, p. 5249-5257.

Zimmerman, R.W. (1984). *The Effect of Pore Structure on the Pore and Bulk Compressibilites of Consolidated Sandstones*, Ph.D. dissertation, University of California, Berkeley.

Rock Mechanics, Nelson & Laubach (eds) © 1994 Balkema, Rotterdam, ISBN 90 5410 380 8

Relationship between static and dynamic rock properties in welded and nonwelded tuff

R.H. Price
Sandia National Laboratories, Albuquerque, N.Mex., USA
P.J. Boyd, J.S. Noel & R.J. Martin, III
New England Research, Inc., White River Junction, Vt., USA

1 INTRODUCTION

An integral part of the licensing procedure for the potential nuclear waste repository at Yucca Mountain, Nevada, involves accurate prediction of the in-situ rheology for design and construction of the facility and emplacement of the canisters containing radioactive waste. The data required as input to successful thermal and mechanical models of the behavior of the repository and surrounding lithologies include bulk density, grain density, porosity, compressional and shear wave velocities, elastic moduli, and compressional and tensile strengths. In this study a suite of experiments was performed on cores recovered from the USW-NRG-6 borehole drilled to support the Exploratory Studies Facility (ESF) at Yucca Mountain. USW-NRG-6 was drilled to a depth of 1100 feet through four thermal/mechanical units of Paintbrush tuff. The thermal/mechanical stratigraphy was defined by Oritz et al (1985) to group rock horizons of similar properties for the purpose of simplifying modeling efforts. The tuff samples from USW-NRG-6 have a wide range of welding characteristics (usually reflected in sample porosity) and a small range of mineralogy and petrology characteristics. Generally samples are silicic ashfall tuffs that exhibit large variability in their elastic and strength properties (see Price and Bauer, 1985). The four thermal/mechanical units are designated: TCw, a relatively low porosity welded tuff; PTn, a non-welded very porous tuff sometimes having a high percentage of zeolites and clay; TSw1, a lithophysal rich and vapor-phase altered welded tuff; and TSw2 (the potential repository horizon), a welded tuff with relatively few lithophysal cavities and vapor-phase altered zones.

These data constitute a substantial data set to establish correlations between the fracture strength, bulk properties, and elastic properties of the tuff. The emphasis in this study is to collect as much data as possible, in a more or less routine way, on the potential repository horizon and surrounding thermal/mechanical units.

The data collected in the study have both long and short term uses. In the short term, the data are used to characterize the bulk and mechanical properties of the various tuff units for the design of the ESF. On a longer range, these data, combined with specimens tested from other boreholes, will be used to explore the lateral and vertical variability of the tuff to support repository design and performance calculations.

In an effort to characterize the tuffs at Yucca Mountain as thoroughly as possible, compressional and shear wave velocities were routinely measured on all specimens. The velocity data were used in conjunction with the density measurements to compute the dynamic elastic moduli of the rock. These values are compared with the static moduli to infer as much as possible about the pore structure of the rock. A licensed repository will

This work was performed under the auspices of the U.S. Department of Energy (DOE), Office of Civilian Radioactive Waste Management, Yucca Mountain Site Characterization Project, under Contract No. DEAC04-76DP900789.

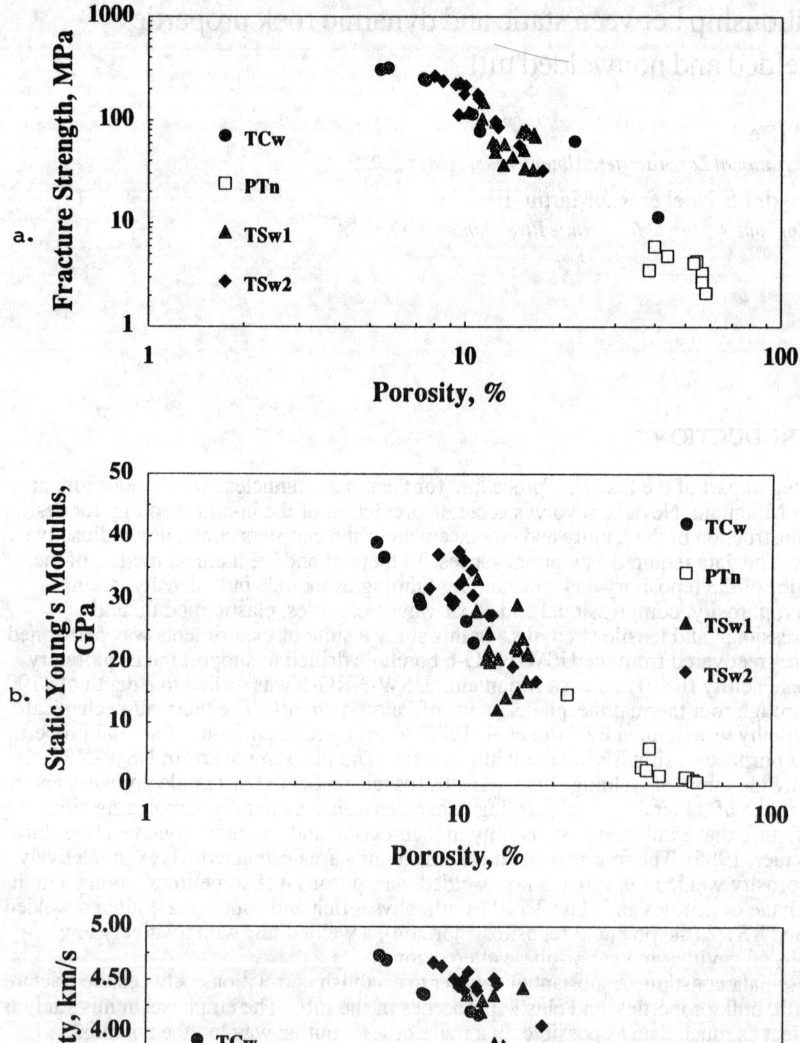

Figure 1: Fracture strength (a), static Young's modulus (b), and P-wave velocity (c) are plotted as a function of porosity for specimens of USW-NRG-6 tested in uniaxial compression.

be in existence for thousands of years, and it is important to know as much as possible about the variability in the rocks and their properties for prediction of the properties of the surrounding rock mass. The most remotely measured geophysical parameter is compressional wave velocity. Therefore, it would be useful to establish a scaling law that incorporates fracture strength, Young's modulus, and compressional wave velocity. Fracture strength and Young's modulus are important parameters for modeling long term stability behavior of the rock mass at Yucca Mountain.

2 EXPERIMENTAL PROCEDURE

Right cylindrical specimens of tuff 50.8 mm in diameter and 101.6 mm in length were prepared from forty-eight cores recovered from borehole USW-NRG-6. A single CT (X-ray computerized tomography) scan was performed along the specimen axis. Each specimen was dried at 110°C for a minimum of ten days, and the dry bulk density was determined. The specimen was pressure and vacuum saturated with distilled water. Compressional and shear wave velocities were measured for both dry and water saturated conditions.

Unconfined compression tests to failure were performed at a constant axial strain rate of 10^{-5} s^{-1} at room temperature. The specimen was instrumented with LVDT based transducers to monitor the axial and radial deformation. From these, data strains were calculated. Static Young's modulus and Poisson's ratio were computed from the stress and strain data between 10 and 50% of the fracture strength.

3 EXPERIMENTAL RESULTS

The results of the experiments are summarized in Figures 1a, b, and c. Following the convention of Price and Bauer (1985), fracture strength, Young's modulus, and dry compressional (P) wave velocity are plotted as a function of porosity. Figure 1a shows a decrease in fracture strength with increasing porosity. The data are distinguished according to the individual thermal/mechanical units. The strength of the tuff ranged from 314 to 1.0 MPa as the porosity increased from 5% to near 58%. The strongest specimens were from thermal/mechanical unit TCw, a low porosity welded tuff. In contrast, PTn, a non-welded porous tuff, exhibited the lowest strengths. The potential repository horizon, TSw2, exhibited strengths between 56 and 236 MPa for porosities between 9 and 18%.

Figure 1b shows the change in static Young's modulus as a function of the specimen porosity. The modulus ranges from near 40 GPa to less than 1 GPa. Young's modulus decreases with increasing porosity. Although the trend is distinct, there is significant scatter in the observed modulus at each porosity, in some cases as much as a factor of two.

Figure 1c shows the dry compressional wave velocity as a function of porosity. The velocity decreases with increasing porosity. The scatter is consistent with that observed for other parameters. Dry P-wave velocities were chosen for detailed consideration for two reasons: shear wave data was not obtainable on all specimens, and compressional and shear wave velocities for saturated specimens were difficult to measure.

4 DISCUSSION

Price and Bauer (1985) reported that Young's modulus and fracture strength for tuffs in the vicinity of Yucca Mountain are related to functional porosity. Functional porosity is defined as the measured porosity plus the volume fraction of montmorillonite. Strength and modulus decreased with increasing functional porosity. Although there were strong

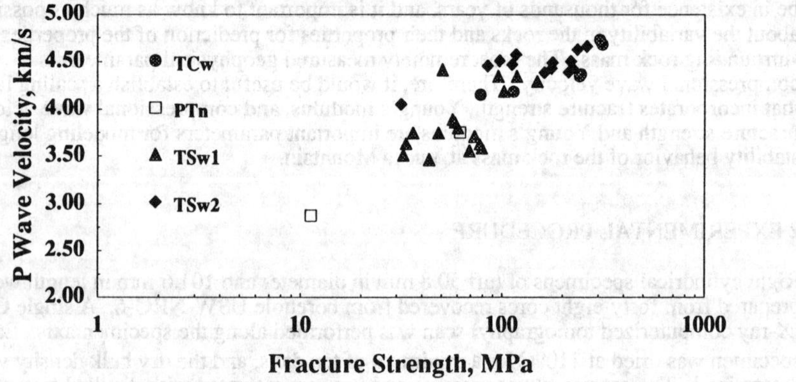

Figure 2: P-Wave velocity is plotted as a function of fracture strength for specimens of USW-NRG-6 tested in uniaxial compression.

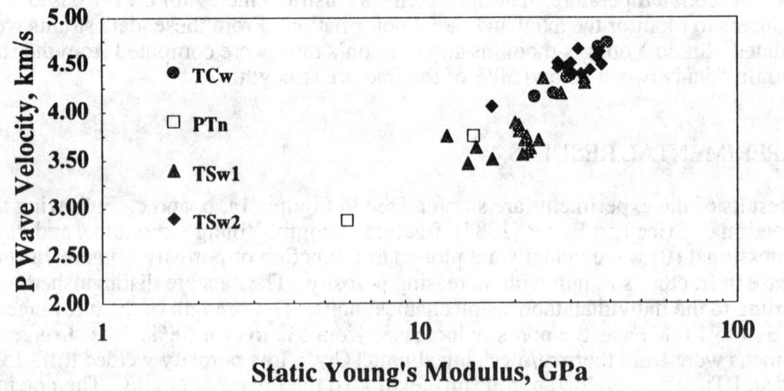

Figure 3: P-Wave velocity is plotted as a function of static Young's modulus for specimens of USW-NRG-6 tested in uniaxial compression.

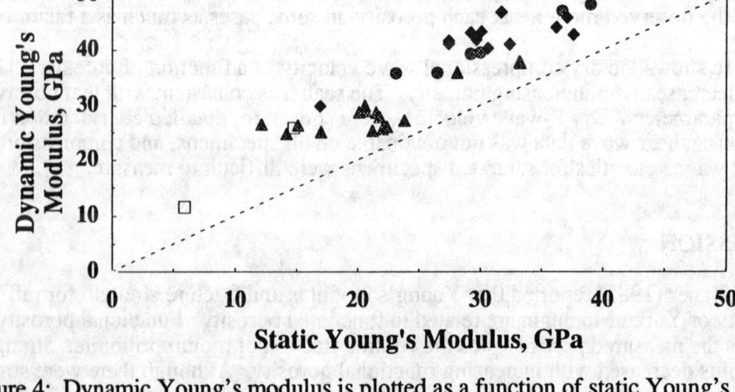

Figure 4: Dynamic Young's modulus is plotted as a function of static Young's modulus for specimens of USW-NRG-6 tested in uniaxial compression.

dependencies of fracture strength and modulus on porosity, significant scatter at a fixed porosity was observed. Frequently the fracture strength differed by as much as a factor of two or three for a given porosity. A similar effect was noted for static Young's modulus. We have plotted the recently collected data as a function of porosity rather than functional porosity in Figures 1a, b, and c because detailed chemical composition of each specimen has yet to be determined.

The relationship between P-wave velocity and porosity is not surprising and forms a useful link between the static and dynamic properties of the rock. Since the goal of establishing a relationship between the static and dynamic properties is to determine the spacial distribution of static properties throughout the potential repository, the P-wave velocity has been plotted as a function of fracture strength and as a function of static Young's modulus in Figures 2 and 3, respectively. Finally, the relationship between static and dynamic Young's modulus is shown in Figure 4. The ratio of dynamic to static moduli is very instructive in establishing the microstructure of the tuff.

The analysis of the data has concentrated on two aspects: first, the relationship between the static and dynamic moduli, and second, the relationship between dynamic properties, such as compressional wave velocity and fracture strength. Specifically, it is important to understand why the dynamic moduli are consistently greater than the static moduli and the reason for the scatter in fracture strength and moduli at a fixed porosity.

Both the static and dynamic moduli have been measured on a wide variety of rocks (e.g. Simmons and Brace, 1965; Cheng and Johnston, 1981; Jizba, 1991). Generally, for a specific rock type, the dynamic moduli exceed the static moduli. For specimens from USW-NRG-6, the ratio of the dynamic to static moduli ranges from 1.13 to 2.17.

The elastic properties and fracture strength of most brittle rocks have been modeled in terms of the pore size and geometry. Elastic constants, as well as compressional and shear wave velocities, have been shown to be related to the micro-crack density, crack length, and the shape of the pores (e.g. Walsh, 1965; O'Connell and Budiansky, 1974; Mavko and Nur, 1978). Each of these models shows that with increasing crack length and crack density, the bulk modulus, Young's modulus, and compressional wave velocity decrease. While these models establish the dependence of elastic properties on porosity, they do not offer insight into the differences between static and dynamic elastic moduli widely reported and present in the tuffs at Yucca Mountain.

The fracture strength of brittle rocks are strongly controlled by the initial damage. Damage typically consists of micro-cracks and irregularly shaped pores that act as stress concentrators and loci for crack initiation and propagation (e.g. Costin, 1983; Sammis and Ashby, 1986; Holcomb, 1993). Given that both the elastic and deformational properties are controlled by the distribution and size of cracks and pores, the correlations between static and dynamic rock properties as shown in Figures 2, 3, and 4 are reasonable.

In many studies a single rock type with well defined properties, such as granite or sandstone, has been studied in detail to examine the effect of one variable on the static and dynamic properties and/or strength. At Yucca Mountain, a wide variety of tuffs are represented. While the chemical composition of the tuffs is relatively uniform, the differences reside in the porosity, the pore geometry and distribution, and the alteration products within the tuff. For example, there are lithophysal and vapor-phase altered zones in TSw1 and TSw2. The objective of collecting dynamic properties of the tuff is to ultimately correlate them with the mechanical properties and to use the empirical relationships to assess the lateral variability in each lithologic unit throughout Yucca Mountain. As we have seen, the fracture strength and elastic properties of tuff can be related, at least to the first order, to porosity. Within a given lithology, even specimens with nearly the same porosity exhibit properties that can differ by as much as a factor of two.

The CT scans allow us to examine the shape and distribution of pores in the specimen. In some cases the pores are relatively large and easily resolved. The data from four specimens representing end members for the ratio of dynamic to static Young's modulus (E) were selected for analysis. Two TSw1 specimens recovered from 427.0 and 488.0

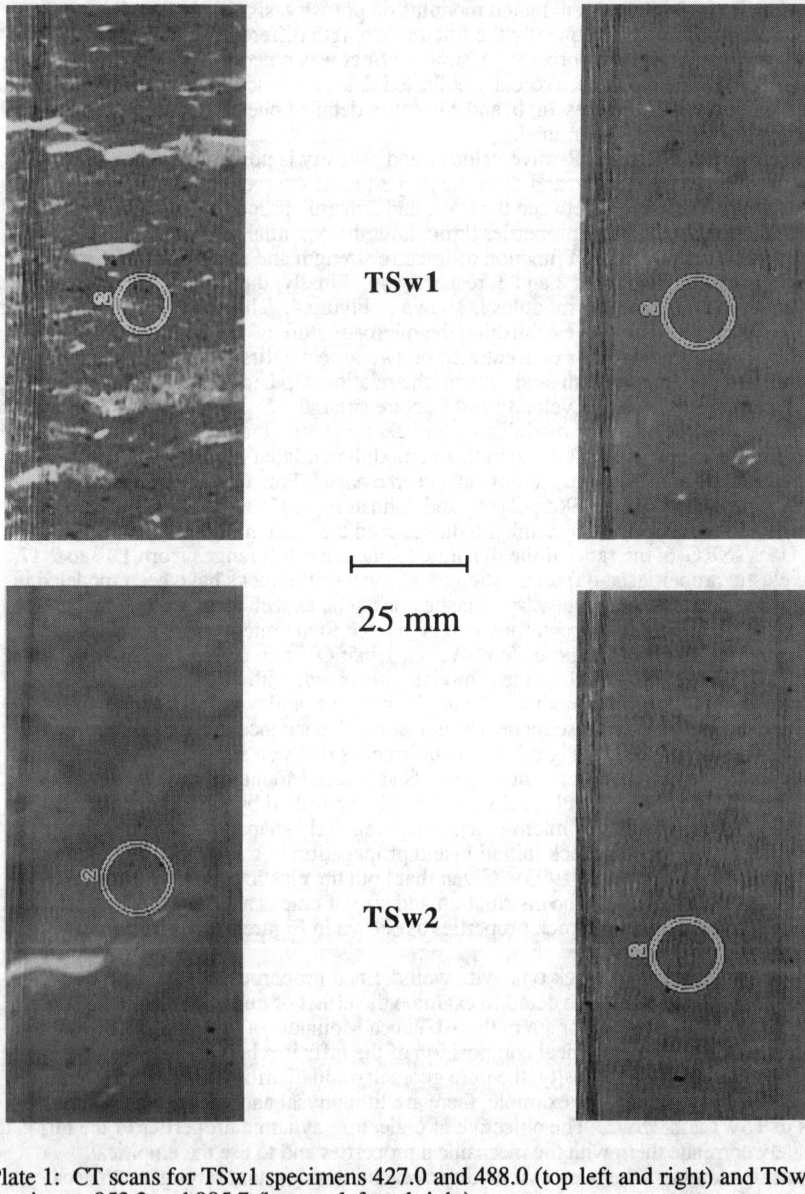

Plate 1: CT scans for TSw1 specimens 427.0 and 488.0 (top left and right) and TSw2 specimens 953.2 and 985.7 (bottom left and right).

feet and two TSw2 specimens recovered from 953.2 and 985.7 feet were analyzed. The properties measured on each of these specimens are shown below. Although the differences in porosity for the specimens within each unit are relatively small, there are large differences in the fracture strength, compressional and shear wave velocities, and static and dynamic Young's modulus.

The CT scans for the four specimens are shown in Plate 1. The lighter the shading the lower the density of the material represented in the scan. In order to quantitatively assess the relative proportion and distribution of pores within each scan, image analysis has been performed. We are looking for the volume fraction of the pores, the largest contiguous pore, and the mean size of the pores. For the TSw1 specimens (Plate 1), the volume fraction of low density material in 427.0 is approximately fourteen times greater than for 488.0. Zones of low density material are larger by a factor of nearly twenty. The difference in porosity between the two specimens is small, but the differences in the distribution and characteristic length of the pores are large. From a qualitative point of view, if the strength and the dynamic properties of the rock are related to the size and shape of the pores, specimen 427.0 should be weaker and exhibit lower moduli and velocities than 488.0.

Two specimens from TSw2 are shown in Plate 1. The number of low density zones in these two specimens is not as pronounced as those for TSw1. Specimen 953.2 is characterized by large, light colored zones representing vapor-phase altered regions. A relatively large lithophysal zone is apparent in the upper right quadrant of the specimen. In contrast, the vapor-phase altered regions for 985.7 are much smaller and not as pronounced. The large percentage of weak, low density, vapor-phase altered zones in 953.2 contributes to the low strength, velocity, and static modulus.

Bulk and Mechanical Properties Data for Four Specimens from USW-NRG-6

Unit	TSw1	TSw1	TSw2	TSw2
Depth, ft	427.0	488.0	953.2	985.7
Porosity, %	14.1	10.5	17.6	9.9
Grain Density, gcm^{-3}	2.527	2.504	2.548	2.554
P-Velocity, $km\ s^{-1}$	3.463	4.299	4.054	4.499
S-Velocity, $km\ s^{-1}$	2.228	2.684	2.408	2.812
Strength, MPa	43.4	149.4	31.6	177.3
Static E, GPa	14.2	33.3	16.9	37.6
Dynamic/Static E, GPa	1.74	1.13	1.77	1.14

The variations in pore structure for the specimens shown from TSw1 and TSw2 support the differences in P-wave velocity, Young's modulus, and fracture strength. As lithophysae and vapor-phase altered zones increase in size and abundance, the elastic moduli and fracture strength decrease. This is one of several wells that have been and will be drilled at Yucca Mountain in support of the ESF design. As the data set increases and image analysis is performed on CT scans from each specimen, a relationship between statistical properties of the pore structure and the mechanical properties of tuff will be developed. It is clear that it will not be possible to give an unequivocal explanation for the differences between the static and dynamic Young's modulus. In a previous study (Haupt et al, 1992), Young's modulus was measured as a function of strain amplitude and frequency for a specimen of TSw2. In the dry state there was no dispersion in Young's modulus at frequencies between 10^{-2} and 10^{6} Hz. At strains between 10^{-6} and 10^{-5}, Young's modulus was constant for a fixed loading frequency of 10^{-2} Hz. These results suggest that the pores in welded tuff are elliptical to spherical and do not result in a significant difference between the static and dynamic moduli. With saturation, however, the dynamic moduli increases and the static moduli should decrease. In this case we are comparing dry dynamic moduli with saturated static moduli. If all the pores were elliptical and spherical, one would anticipate the static and the dynamic moduli would be very near equal. The fact that the differences are large in some cases indicates that there are a number of effects influencing the behavior of the rock; most likely these include strain amplitude, chemical weakening of the bonds along grain boundaries, particularly in specimens of PTn, and the effect of strain amplitude on the closure of cracks and pores in some units.

5 CONCLUSION

A large data set has been collected on specimens recovered from borehole USW-NRG-6. Analysis of the results of these experiments show that there is a correlation between fracture strength, Young's modulus, compressional wave velocity, and porosity. Additional scaling laws relating static Young's modulus with compressional wave velocity and fracture strength with compressional wave velocity are promising. Since there are no other distinct differences in material properties, the scatter that is present at each fixed porosity suggests that the differences in the observed property can be related to the pore structure of the specimen. Image analysis of CT scans performed on each test specimen are currently underway to seek additional empirical relationships to aid in refining the correlations between static and dynamic properties of tuff. Once these studies are complete, it may be possible to examine, in more detail, mechanisms to explain the differences between static and dynamic moduli and refine the interpretation of the dependence of fracture strength and static Young's modulus on acoustical measurements, such as compressional wave velocity.

REFERENCES

Cheng, C.H. and D.H. Johnston 1981. Dynamic and static moduli. Geophys. Res. Let. 8:39-42.

Costin, L.S. 1983. A microcrack model for the deformation and failure of brittle rock. J. Geophys. 88:9485-9492.

Haupt, R.W., R.J. Martin, III, X. Tang, W.J. Dupree and R.H. Price 1992. Modulus dispersion and attenuation in tuff and granite. Proceedings of the 33rd U.S. Symposium on Rock Mechanics. 899-908.

Holcomb, D. 1993. General theory of the Kaiser effect. 34 U.S. Symposium on Rock Mechanics, University of Wisconsin-Madison. 2:423-430.

Jizba, D.L. 1991. Mechanical and acoustical properties of sandstones and shale. Thesis submitted to the Department of Geophysics, Stanford University.

Mavko, G.M. and A. Nur 1978. The effect of nonelliptical cracks on the compressibility of rocks. J. Geophys. Res. 83:4459-4468.

O'Connell, R. and B. Budiansky 1974. Seismic velocities in dry and saturated cracked solids. J. Geophys. Res. 79:5412-5426.

Oritz, T.S., R.L. Williams, F.B. Nimick, B.C. Whittet, and D.L. South 1985. A three dimensional model of reference thermal/mechanical and hydrologic stratigraphy at Yucca Mountain, Southern Nevada. SAND84-1076, Sandia National Laboratories Albuquerque, NM.

Price, R.H. and S.J. Bauer 1985. Analysis of elastic and strength properties of Yucca Mountain tuff, Nevada. Research and Engineering Applications in Rock Masses. Proceedings of 26th U.S. Symposium on Rock Mechanics, Eilleen Ashworth (ed) 1:89-96.

Sammis, C.G. and M.F. Ashby 1986. The failure of brittle porous solids under compressive stress states. Acta metall. 34:511-526.

Simmons, G. and W.F. Brace 1965. Comparison of static and dynamic measurements of compressibility of rocks. J. Geophys. Res. 70:5649-5656.

Walsh, J. 1965. The effect of cracks on the compressibility of rock. J. Geophys. Res. 70:381-389.

9 Uncertainty and reliability in rock engineering

Session developer:
P.H.S.W. Kulatilake, University of Arizona

Rock Mechanics, Nelson & Laubach (eds) © 1994 Balkema, Rotterdam, ISBN 90 5410 380 8

Application of the Unit-Cell Method to determine the optimum tunnel orientation

J.S. Kuszmaul
Department of Geology and Geological Engineering, University of North Dakota, Grand Forks, N.Dak., USA

ABSTRACT: The Unit-Cell Method is a technique developed to describe the relationship between joint spacing, excavation width, and keyblock size. The method is summarized and then applied in design. The application considered involves a rock mass with five joint sets. After considering the multiple keyblocks possible, the optimum tunnel orientation is determined.

1 INTRODUCTION

The design of excavations in blocky rock masses must take into account the likelihood of the formation of dangerous blocks (keyblocks) on the excavation periphery. An example of such a keyblock is shown in Figure 1; the keyblock in this case lies above a tunnel of rectangular cross section and is defined by the intersection of three joints. Methods for determining the geometry of keyblocks require information about the geometry of the joint sets (orientation, spacing, and persistence), the excavation profile, and the mechanical properties of the joint sets. These properties will determine whether a particular combination of joints constitute an unstable keyblock. Conventional techniques (for example, Goodman 1989) can be used to evaluate the stability of a specific keyblock formed by excavation.

To determine how a rock mass that has been thoroughly investigated will respond to a proposed excavation requires a predictive design tool. Rock mass classification systems have served (and continue to serve) this role in many cases. However, a number of alternative design approaches have been proposed that attempt to make better use of detailed joint set orientation data and joint spacing information. These include approaches based upon Monte-Carlo simulation, such as Chan (1986) and Hoerger (1988), that employ a conceptual model of the jointed rock mass applied in a large number of realizations to simulate much of the full range of keyblock possibilities. Other methods have been proposed that assume a particular model of jointed rock masses in an attempt to avoid some of the computationally intensive tasks of simulation (McCullagh and Lang 1984 and Tyler 1991).

One method for determining the effect of joint orientation alone is Goodman and Shi's Block Theory (Goodman and Shi 1985). Goodman and Shi (1987) have previously demonstrated how Block Theory might be used as a predictive tool for

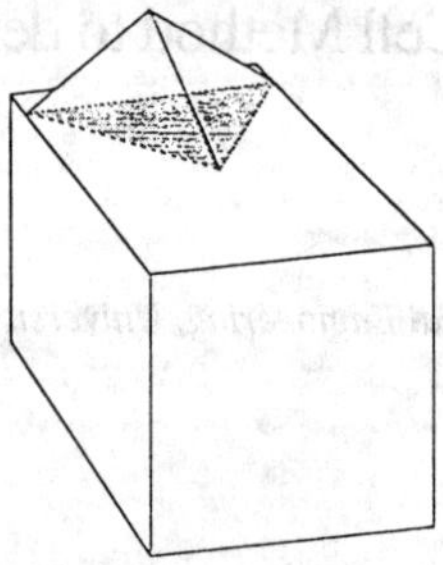

Figure 1 An example of the type of keyblock being studied in this case, which are those above tunnels of rectangular cross section.

determining the optimum tunneling orientation within a rock mass. This design approach is based upon the assumption that the maximum keyblock possible from a particular joint combination is the appropriate basis for comparing design alternatives. Joint spacing, a rock-mass feature not originally considered in Block Theory, should be recognized as an important influence on the size of keyblocks likely to form.

The recently developed Unit-Cell Method (Kuszmaul 1993) attempts to incorporate a model of joint spacing (in rock masses with widely spaced joints) to determine the probability of the formation of a keyblock of any size between zero and the maximum keyblock size. This method holds potential in determining the optimum excavation geometry or the excavation support requirements. The Unit-Cell Method has been developed only in a simplified model and has yet to incorporate variations in orientation within a joint set, limited joint persistence, or the mechanical properties of joints. But the method has resulted in an analytical solution rather than a numerical approximation for the effect of joint spacing. Thus, it offers the possibility of allowing the incorporation of the important realistic variations in joint properties using reliability analysis (Der Kiureghian 1990).

2 DESCRIPTION OF THE UNIT-CELL METHOD

The method has been presented previously in a simplified two-dimensional form (Kuszmaul and Goodman 1992) and as a three-dimensional solution (Kuszmaul 1993), but has yet to be fully explored in application. One application of the Unit-Cell Method is in comparing multiple joint combinations to determine which pose the greatest threat to excavation stability. Knowing which combinations are likely to yield large keyblocks would assist in excavation design. One design parameter that must be considered is the tunneling direction. This paper introduces the application of the Unit-Cell Method in determining the optimum orientation for tunneling.

The Unit-Cell Method is based on a simplified model of a rock mass in which all joints within a set will be assumed to have the same orientation. Joint spacing within the set is measured along the normal to the joints and is described by a probability density function (pdf). No particular pdf for joint spacing is assumed as it has been shown that the details of the pdf are unimportant, because the result of the Unit-Cell

Method depends only upon the mean spacing. The method has been restricted to rock masses with widely spaced joints. Joints are assumed to be infinitely persistent (a conservative assumption). Mechanical properties of the joints have been considered in other analyses of keyblock stability, but the present formulation of the Unit-Cell Method does not consider these stabilizing effects. Instead, the method has been applied only to keyblocks that can form in the roof or crown of a tunnel of rectangular cross section. Only those keyblocks formed beneath three intersecting joints are considered. In this simplified model of a rock mass, the Unit-Cell Method examines the relationship between keyblock size, joint spacing, and excavation width.

To describe the probability of keyblock formation, it is necessary to introduce the parameter x, which is the fraction of the maximum keyblock size that poses concern,

$$x \equiv \frac{\text{keyblock size of interest}}{\text{maximum keyblock size}} \ . \tag{1}$$

Thus x varies from 0 (when a keyblock of any size is a concern) to 1 (when only the maximum keyblock is a concern for design). In this method, failure is defined as the existence of a keyblock of any size larger than size fraction x within the excavation cross section at a randomly selected point along the length of the tunnel. The probability of such a failure then can be described as

$$p_f(x) = C^*(1 - 3x^{2/3} + 2x) \ , \tag{2}$$

where the constant C^* is a function of the joint set orientation, mean spacing of the joint sets, and the excavation width and orientation. The coefficient C^* is found by evaluating an analytical expression obtained from the Unit-Cell Method. The combination of three joint sets within a rock mass divide the rock mass up into parallelepipeds. These parallelepipeds are called unit cells. Because of the similar geometry among parallelepipeds, it is sufficient to study the interaction of a single unit cell with the proposed excavation. The complete derivation of this solution has been presented (Kuszmaul 1993) where it was also validated by comparing the analytical equation to results of numerical simulation. The definition of the coefficient C^* has been shown to be

$$C^* = \frac{\text{volume of keyblock-forming region}}{\text{mean volume of unit cells}} \ . \tag{3}$$

This equation involves operations on three-dimensional vectors that are most easily accomplished using a computer or programmable calculator (a copy of such a program was given in Kuszmaul 1993).

3 DEMONSTRATION OF THE UNIT-CELL METHOD

To demonstrate the results of the Unit-Cell Method, the case of a rock mass with three joint sets is considered; the joint-set properties are given in Table 1. A 5-m wide, rectangular tunnel will be driven through the rock mass, with the axis of the tunnel aligned in the North-South direction. The application of Equations 2 and 3 in this case yields the probability of failure distribution shown in Figure 2. This result indicates that the most likely condition in a cross section of the tunnel selected at random is the formation of no keyblock at all. The probability of finding a keyblock

Table 1 Rock Mass Properties Used in Initial Demonstration of the Unit-Cell Method			
Joint Set Number	Dip Direction	Dip	Mean Joint Spacing
1	60°	60°	10 m
2	180°	60°	10 m
3	300°	60°	10 m

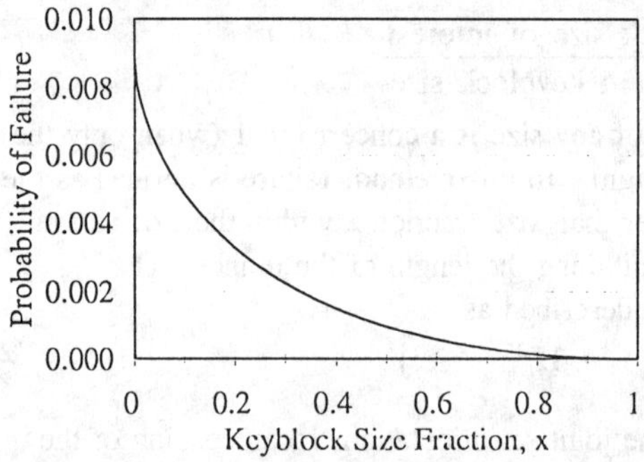

Figure 2 The probability of failure function for the tunnel through the rock mass given in Table 1. Keyblock size is found by multiplying the size fraction, x, by the maximum keyblock size, 9.02 m^3.

larger than any size fraction x can be found using this figure. It shows that the probability of keyblock formation decreases sharply to zero as the size of interest approaches the maximum keyblock size.

To validate this new method, it is necessary to compare its results to the results of keyblocks observed in the field. This is not yet possible because of the simplified model of a rock mass considered for developing the method. Realistic joint properties such as variations in joint orientation, limited joint persistence, and joint strength must be considered by this new technique before it can be compared to field conditions. While that research continues, it is worth demonstrating that the method holds the potential for a significant contribution to excavation design.

One application of the Unit-Cell Method involves applying Equations 2 and 3 to a jointed rock mass where a tunnel of a chosen width is to be sited, but the orientation of the tunnel is to be selected so as to minimize the deleterious effect of the jointing. An example of this type of application is in a rock with the same properties as those given in Table 1 and an excavation width of 5 m. In this application, however, the tunnel orientation has yet to be determined. The probability of keyblock formation will be examined in all directions to determine the optimum orientation. The results of this analysis are shown in Figure 3. To use this figure the designer must first select a level of risk from among the probability contours shown in Figure 3. For the selected level of risk, the contour shows the keyblock size that must be expected (given by the radial coordinate of the curve) for each tunnel direction (given by the azimuth direction measured clockwise from North). In this example, a tunnel axis oriented North-South or 60° from that axis (in either direction) results in the

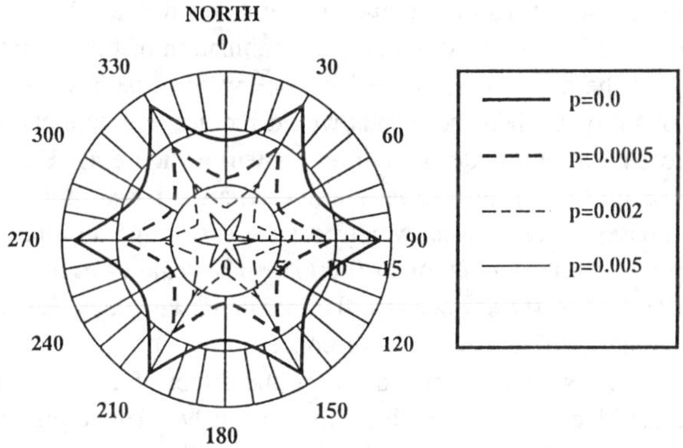

Figure 3 The design curves to determine the optimum tunnel orientation at various levels of probability of failure. The radial coordinate of the plot gives the size of the keyblock (in m^3) to be expected in the direction of interest.

minimized risk of large keyblock formation. If the outer curve in Figure 3 is used alone to determine the optimum tunnel orientation, the results would match the analysis possible with conventional Block Theory. The additional information resulting from the application of the Unit-Cell Method yields the inner contours.

4 THE UNIT-CELL METHOD AND MULTIPLE KEYBLOCK TYPES

The information provided by the Unit-Cell method can also be used to determine the effect of multiple keyblock types that may be present within the same rock mass. To demonstrate this ability, a sample set of rock mass properties was prepared as shown in Table 2. It is not possible yet to use real rock-mass data, because joint spacing must currently be large relative to the excavation width to avoid an overconservative solution. In this problem, a tunnel 2.5-m wide is to be excavated. If the orientation of the tunnel is to be selected so as to minimize the support requirements, the Unit-Cell Method can assist by determining the likelihood of various keyblock types.

Table 2 Rock Mass Properties Used in Unit-Cell Method Demonstration Involving Multiple Keyblock Types

Joint Set Number	Dip Direction	Dip	Mean Joint Spacing
1	45°	63°	15.8 m
2	187°	55°	7.8 m
3	285°	48°	8.7 m
4	150°	61°	12.7 m
5	358°	57°	5.8 m

For this particular set of data there are five joint sets that result in five keyblocks of the type of interest (those formed under the combination of three joints in the crown of the tunnel). If the orientation of the keyblock were to be based solely on a particular keyblock type, then the results would require five applications of the Unit-Cell Method to determine the design curves shown in Figure 4. For each of the five keyblocks, there are two design curves: an outer curve based on conventional Block Theory (which bases design on the worst possible case), and an inner curve associated with a probability of failure contour of 0.0005. The scale of each of the diagrams shown in Figure 4 are different because the maximum keyblock size varies significantly among the five keyblocks of concern.

The design in this case is to be based not on the effect of a single type of keyblock, but on all five keyblocks being possible simultaneously. This represents a structural series system in which there are multiple modes of failure (Madsen et al. 1986). In this case a failure could occur because a keyblock of any type (or two or more key-blocks of different types) occur within a selected tunnel cross section. To determine the exact probability of failure in this case is geometrically complex. It is possible, however, to approximate the probability of failure using upper and lower bounds on the size of keyblock possible in each direction. To find the lower bound, the five design curves are plotted on the same scale. In each direction, the outermost curve becomes the minimum curve used to represent the lower bound of the net effect of all keyblock types. To find the upper bound, the five design curves are summed in each direction to produce the outer limits on probable keyblock sizes. In this particular case, the results for these bounds are shown for two cases. The first case, given in Figure 5a, is the result for the application of Block Theory alone. The second case, given in Figure 5b, is the result using a probability of failure of 0.0005 in interpreting the results of the Unit-Cell Method. In the first case, the difference between the upper and lower bounds on the keyblock size is noticeable, but it would not change the determination of the optimum orientation for tunneling. These results indicate that an East-West tunnel axis is to be avoided. The likely keyblock sizes remain reasonably small (less than 2 to 4 m^3) for tunnels oriented between 120° and 230° measured clockwise from North. The design curves shown in Figure 5b reveal that results of the Unit-Cell Method create tighter bounds on the probability of keyblock formation. This allows a suitable orientation to be selected with more confidence. Again these results indicate that tunnel directions within 10° of East and possibly near 60° East of North are to be avoided, while nearly any other tunnel orientation is unlikely to lead to large keyblocks.

5 CONCLUSIONS

The Unit-Cell Method has been shown to be useful in selecting the optimum tunnel orientation within a rock mass. Simple bounds on the series system were possible for the case of multiple keyblock types, but the bounds are close enough to determine dangerous tunneling directions. The Unit-Cell Method will only be useful for design after it incorporates realistic joint properties, but it can already assist in understanding the effect of joint spacing on the sizes of likely keyblocks.

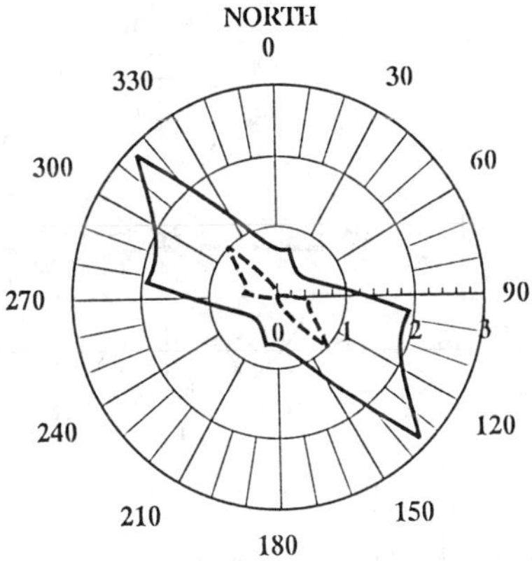

Keyblock Formed by Sets 1, 2 & 3

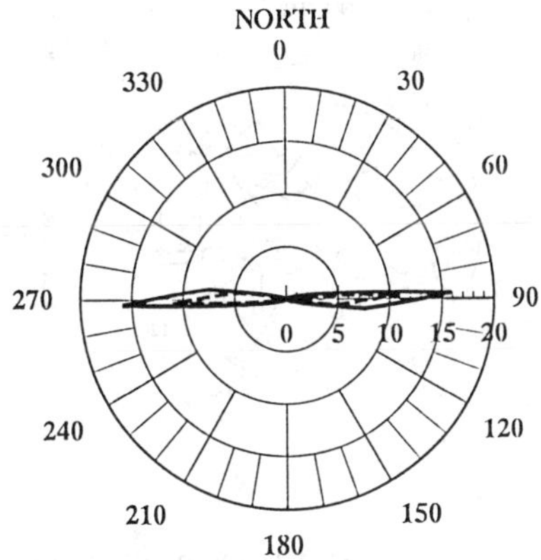

Keyblock Formed by Sets 2, 4 & 5

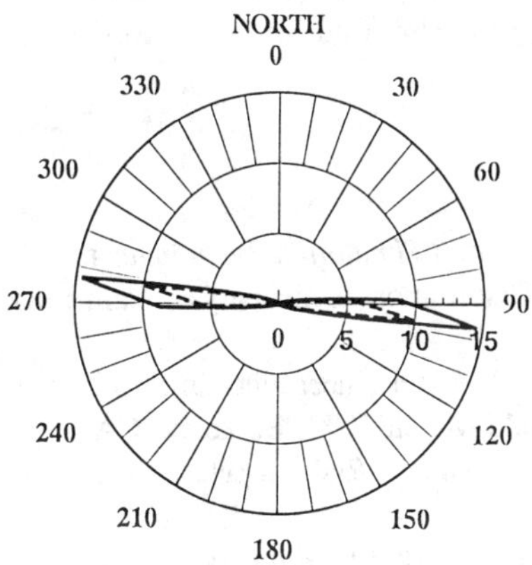

Keyblock Formed by Sets 1, 2 & 5

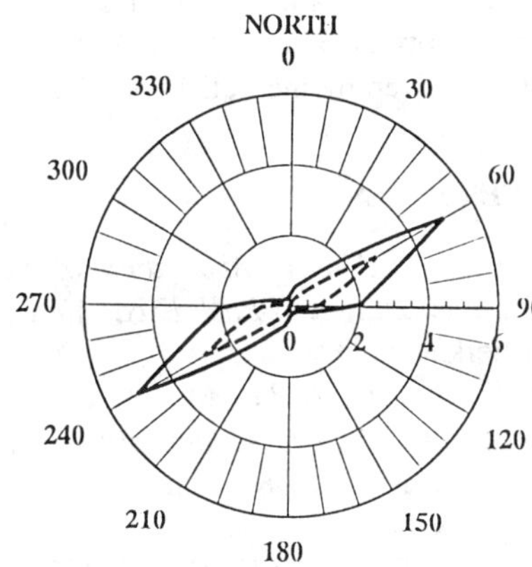

Keyblock Formed by Sets 3, 4 & 5

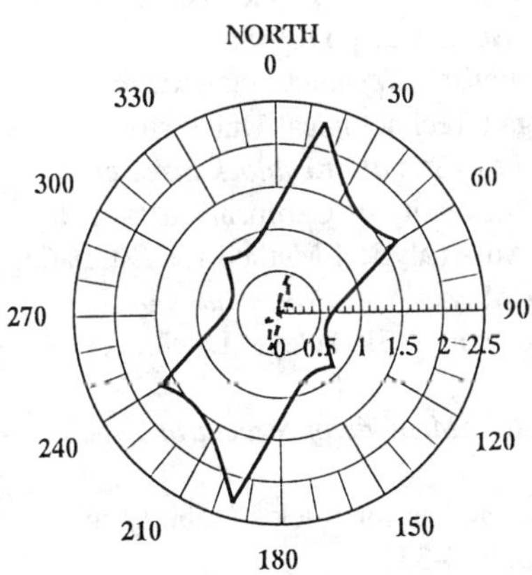

Keyblock Formed by Sets 1, 3 & 4

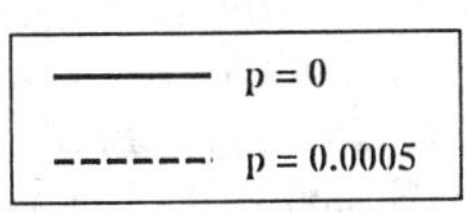

Figure 4 The design curves to determine the optimum tunnel orientation if design were based solely on the possibility of one of the five individual keyblock types considered. Keyblock size is given by the radial coordinate in m^3.

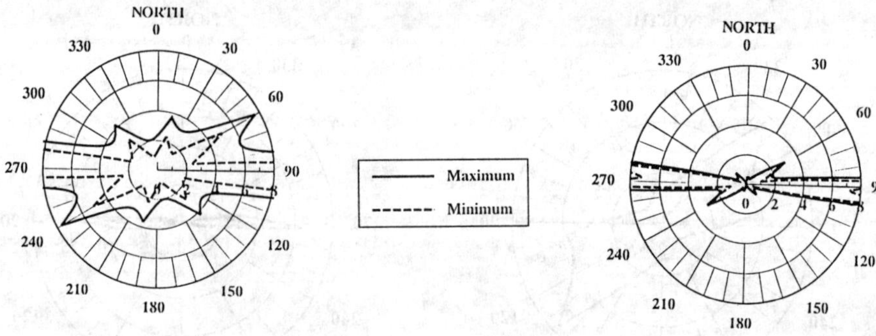

(a) (b)

Figure 5 The maximum and minimum bounds on the design curves for determining the optimum tunnel orientation based on (a) the bounds on a probability of failure of 0 (obtained from Block Theory) and (b) the bounds on a probability of failure of 0.0005 (obtained using the Unit-Cell Method). Keyblock size is given by the radial coordinate in m^3.

REFERENCES

Chan, L.-Y. (1986). *Application of Block Theory and Simulation Techniques to Optimum Design of Rock Excavations*, Ph.D. Dissertation, University of California at Berkeley.

Der Kiureghian, A. (1990), "Bayesian Analysis of Model Uncertainty in Structural Reliability," in *Proceedings of the 3rd IFIP*, held March 1990, Berkeley, CA.

Goodman, R. E. (1989). *Introduction to Rock Mechanics*, 2nd ed. (New York: John Wiley & Sons).

Goodman, R. E. and G.-H. Shi (1985). *Block Theory and Its Application to Rock Engineering* (London: Prentice-Hall International, Inc.).

Goodman, R. E. and G.-H. Shi (1987). "The Application of Block Theory to the Design of Rock Bolt Supports for Tunnels," *Felsbau* 5:2, pp. 79-86.

Hoerger, S. F. (1988). *Probabilistic and Deterministic Keyblock Analyses for Excavation Design*, Ph.D. Dissertation, Michigan Technological University.

Kuszmaul, J. S. (1993). *A Probabilistic Method to Estimate Keyblock Sizes in Underground Excavtions*, Ph.D. Dissertation, University of California at Berkeley.

Kuszmaul, J. S. and R. E. Goodman (1992). "An Analytical Method for Estimating Keyblock Sizes in Excavations in Jointed Rock Masses," in *Proceedings of the Fractured and Jointed Rock Masses Conference*, vol. 1 (Berkeley: Lawrence Berkeley Laboratory), pp. 26-33.

Madsen, H. O., S. Krenk, and N. C. Lind (1986). *Methods of Structural Safety* (Englewood Cliffs, NJ: Prentice-Hall, Inc.).

McCullagh, P. and P. Lang (1984). "Stochastic Models for Rock Instability in Tunnels," *J. R. Statist. Soc.*, Series B, 46:2 pp. 344-352.

Tyler, D. B. (1991). *A Probabilistic Key Block Analysis for the Prediction of Tunnel and Stope Support*, Ph.D. Dissertation, Cambore School of Mines.

Rock Mechanics, Nelson & Laubach (eds) © 1994 Balkema, Rotterdam, ISBN 90 5410 380 8

A test is worth a thousand guesses – A paradox

Glenn A. Nicholson

Geotechnical Laboratory, US Army Engineer Waterways Experiment Station, Vicksburg, Miss., USA

ABSTRACT: Within the U.S. Army Corps of Engineers (USACE) results from laboratory shear strength tests are an important ingredient in the often complex process of selecting shear strengths judged to be representative of insitu strengths. Variability in test results on seemly similar rock specimens are frequently accounted for by increasing conservatism in parameters selected for design. The reason for conservatism lies in the fact that designers do not have a feel for the source of variability. Toward this end, this paper addresses some of the more important finding of a recently completed interlaboratory direct shear testing program. A total of five USACE laboratories participated in the program. Tests were conducted on uniform Berea Sandstone specimens with three different types of pre-sawn and lapped shear surfaces, as well as two different specimen nominal diameters of 10 cm and 15 cm. The test data were analyzed for variability within and between laboratories.

1 INTRODUCTION

Practitioners in rock mechanics frequently refer to process from which design shear strengths are chosen as a selection process. Hence, rock engineers think in terms of selecting design shear strengths rather than precisely determining strengths. The selection process relies heavily on an understanding of the many factors influencing the strength of rock as well as engineering judgment tempered with experience. Nevertheless, in most cases the central basis of the selection process is centered around test results from laboratory shear strength tests performed on rock specimens carefully chosen to reflect the upper and lower bounds of insitu strengths judged to be representative of a given potential failure mode.

Within the U.S. Army Corps of Engineers (USACE) shear strengths selected for design are based, in part, on the results of direct shear tests conducted on specimens whose shear surfaces are planes of weakness, (i.e., bedding planes, joints, etc.). From time to time various USACE District and Division offices have observed seemingly large variations in direct shear test results obtained from specimens of the same rock type and similar shear surface conditions but tested at different USACE laboratories. These variations were in addition to variations observed within individual laboratory test results.

2 VARIATION SOURCES

For tests on supposedly identical specimens, potential sources for test result variations may be grouped into three general categories: (1) the human factor; (2) the test equipment used (including equipment calibration); and (3) to some extent, the environment (humidity and temperature). Reason would dictate minimal variations in test results in cases were all tests were conducted by the same operator using the same equipment and the same environment controlled (heated/cooled) room. Such cases are typical within each USACE laboratory. However, each potential source changes from laboratory to laboratory and as such contribute appreciably to the variability of test results. Variations in test results, both within and among laboratories can be controlled in so far as possible, by ensuring that laboratory technicians are properly trained and follow prescribed standardized test procedures and that the test equipment is properly maintained and calibrated. However, of the five USACE laboratories that routinely perform direct shear tests on rock specimens, all use different types of shear equipment. For this reason, variations in equipment was a prime suspected cause of observed variations in tests results among USACE laboratories.

3 TESTING PROGRAM

In order to quantify the magnitude of variations both among and within the USACE testing laboratories, the USACE Waterways Experiment Station (WES) developed and implemented an interlaboratory direct shear testing program. All five USACE laboratories that routinely perform direct shear tests on rock participated in the program. The USACE Division laboratories that took part included: the Mississippi River Division; the North Pacific Division; the Ohio River Division; and the South Atlantic Division. The WES also participated. To ensure anonymity, each laboratory is assigned a number, one through five, for the purpose of this paper.

A testing program designed to examine variability of test results must, in so far as possible, exercise control of extraneous sources of variability. The testing program closely followed the guidance for conducting an interlaboratory study contained in the American Society for Testing and Materials (ASTM), E 691 (American Society for Testing and Materials, 1992). However, E 691 describes a Standard for conducting an interlaboratory study to determine the precision of a test method. As such, the standard requires a minimum of six participating laboratories. Since only five laboratories participated in this study, no direct conclusions should be made with respect to the results of this study and the likely precision of the USACE direct shear test method.

4 TEST SPECIMENS

All tests were performed on "standard" specimens prepared at the WES from a uniform Berea Sandstone. Specimens consisted of two sizes, 10 cm and 15 cm nominal diameters. For each size, three different standard shear surfaces were tested: (1) a smooth pre-sawn and lapped surface parallel to the direction of the shear stress (horizontal sliding); (2) a smooth pre-sawn and lapped surface inclined up 4.0 degrees with respect to the direction of the shear stress (upslope sliding); and (3) a smooth pre-sawn and lapped surface inclined down 4.0 degrees with respect to the direction of the shear stress (downslope sliding). The latter two types of shear surfaces were intended to simulate dilatation and compression processes commonly observed when testing rough natural joints.

5 TEST PROCEDURES

The tests were conducted in accordance with the USACE standard procedures for conducting direct shear test on rock specimens contained in the Rock Testing Handbook, RTH-203. (U.S. Army Engineer Waterways Experiment Station, 1989). In addition, each laboratory received a detailed test protocol that supplemented the RTH-203 standard procedure. Specifically, the protocol described how to handle the test specimens and the format for test data.

Each laboratory conducted three suites of tests with each suite consisting of three individual tests subjected to a constant normal stress of 0.192 MPa, 0.767 MPa or 1.533 MPa. There was one exception to normal stress test levels. The direct shear equipment used by the laboratory listed as laboratory 3 in the following discussions could only generate a maximum normal stress of 1.15 MPa as the upper end stress level for the 15 cm diameter specimens. Specimens were tested in an air dried condition at room temperature. The as tested specimen moisture content ranged from 0.10 to 0.80 per cent (by weight) with an average of approximately 0.25 percent. In all, 54 tests were conducted by each laboratory for a total of 270 tests.

Because shear test on smooth shear surfaces seldom produce a clearly defined peak shear stress, particularly for the case of upslope sliding, failure was defined as the shear stress corresponding to 6.35 mm of shear displacement. The time from initiation of a given test to failure was specified as 35 minutes plus or minus 5 minutes.

6 TEST RESULTS

The test protocol required that each laboratory provide, in part, a complete record of shear stress versus shear displacement as well as the shear stress corresponding to failure for each test. The author was responsible for analyzing all test data. Although there are a number of ways of displaying the results, perhaps the most useful for engineering purposes is to determine the angle of friction for each suite of test using a line of best fit approach. Lines of best fit determined from the least squares best fit approach generally resulted in small cohesion intercepts (usually less than plus or minus 0.05 MPa) as illustrated in Figure 1. Theoretically, shear test conducted on smooth shear surfaces should not produce cohesion intercepts. In this respect, an equation was developed from the two so called normal equations associated with the least squares linear regression approach. (e.g. see Davis 1973). This equation assumes that line of best fit passes through the origin (i.e., cohesion = 0) and solves directly for the slope of the line as follows:

$$b = \sum_{i=1}^{n} (X_i\, Y_i) \,/\, \sum_{i=1}^{n} (X_i)^2$$

where b is the slope (arctan ϕ) of the line of best fit, X_i is the independent variable (i.e., normal stress), and Y_i is the dependent variable (i.e., shear stress). Figure 2 provides a plot summarizing the friction angle values calculated from the test results.

An examination of Figure 2 indicates a number of disturbing trends. The more significant trends are as follows. (1) Theoretically there should be no difference in test results between 10 cm and 15 cm diameter specimens since there are no scale effects associated with smooth shear surfaces. Yet for horizontal sliding friction angle values obtained from laboratories 1 and 4 indicated that friction values for the 15 cm diameter specimens were, on the average, 4.5 degrees higher than valves obtain from 10 cm diameter specimens. (2) Comparison of all results from among laboratory tests indicate that friction values for horizontal and downslope sliding varied by 12.5 degrees while upslope sliding varied by 10.5 degrees. (3) Variation for within laboratory results ranged from a minimum of less than 1 degree for 10 cm diameter specimens subjected to horizontal sliding to a maximum of 8.7 degrees for 15 cm diameter specimens subjected to down sloped sliding. Both the minimum and maximum spread in results were obtained by the same laboratory (i.e., laboratory 3). Visually observed trends in within laboratory variability appear to be independent of specimen size or type of shear surface (i.e., horizontal, upslope or downslope sliding).

The above observations say little about statistical predictions in variability. In terms of statistical trends in variability two terms become important; repeatability and reproducibility. According to

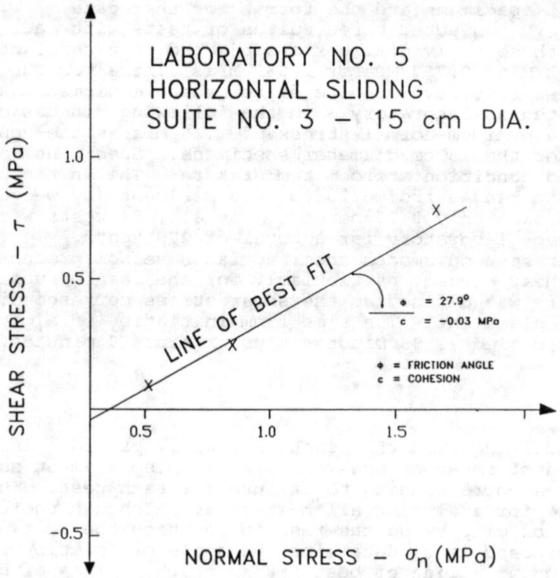

Figure 1. Typical plot of shear stress versus normal stress and the least squares line of best fit.

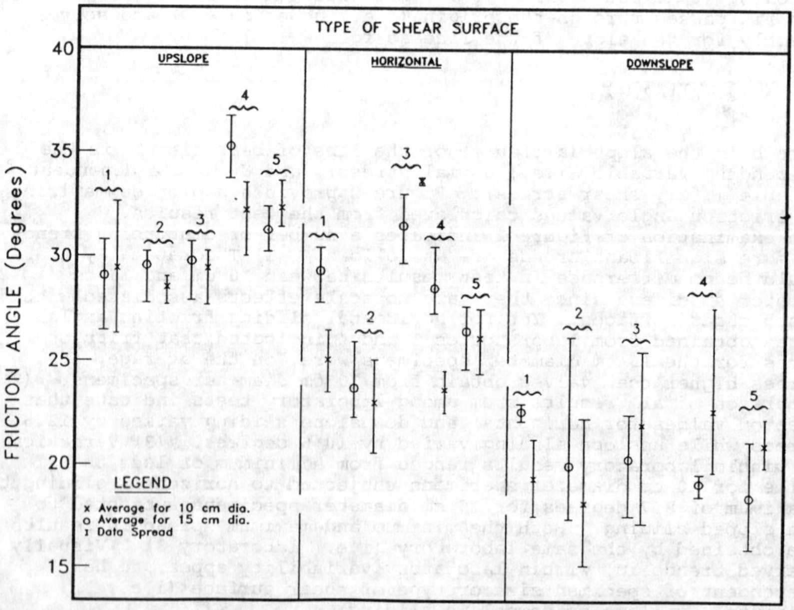

Figure 2. Plot of the average and data spread boundaries for friction angles versus type of shear surface according to testing laboratory.

ASTM E 691 repeatability concerns the variability between independent test results obtained within a single laboratory in the shortest practical period of time by a single operator with a specific set of test apparatus using test specimens taken at random from a single quantity of homogeneous material obtained for the study. Reproducibility deals with the variability between single test results obtained in different laboratories, each of which has applied the test method to tests specimens taken at random from a single quantify of homogenous material obtained for the study. In this respect, repeatability pertains to variability within laboratories where as reproducibility pertains to variability among laboratories. To examine trends in repeatability and reproducibility the author used software obtained from ASTM (ASTM 1990) to calculate the repeatability and reproducibility 95 percent probability limits denoted r and R, respectively. The calculated limits along with the average friction angle valves for all results of a given shear surface type and specimen size are summarized in Table 1. Table 2 summarizes the average friction angles for each laboratory according shear surface type and specimen size. Because of the limited number of participating laboratories and limited number of tests conducted by each laboratory, the summaries offered in Tables 1 and 2 should be considered as only general trends in likely test results.

7 SHEAR EQUIPMENT

All shear devices are similar in that they provide a means for applying and measuring a shear force and normal force parallel and perpendicular, respectively, to an intended shear surface. However,

Table 1. Summary of repeatability and reproducibility limits as well as average friction angle values for all laboratories according to type of shear surface and specimen size.

	Upslope sliding		Horizontal Sliding		Downslope sliding	
	10 cm	15 cm	10 cm	15 cm	10 cm	15 cm
Average friction angle (degrees) $\bar{x}$	30.30	30.90	26.47	28.04	20.61	20.45
Repeatability limits(degree) r	1.90	1.69	1.58	1.58	3.46	2.68
Reproducibility limits (degrees) R	2.44	2.97	4.27	2.91	3.40	2.68

Table 2. Summary of average friction angles for each laboratory according to shear surface type and specimen size.

	Average friction angles (degrees)					
Laboratory	Upslope sliding		Horizontal sliding		Downslope sliding	
No.	10 cm	15 cm	10 cm	15 cm	10 cm	15 cm
1	29.39	28.98	25.12	29.37	19.05	22.38
2	28.27	29.20	23.67	24.53	18.37	20.87
3	29.10	29.87	33.53	31.27	22.20	20.93
4	32.22	35.36	23.94	28.55	22.43	19.29
5	32.34	31.08	26.11	26.48	20.98	18.80

different details of construction dealing with the manner in which shear and normal loads are applied and measured led to the potential for variations in stress distribution over the shear surface and/or errors between the actual applied loads and the measured loads. Nicholson (1983) describes the problems associated with designing and fabricating a direct shear device suitable for testing rock specimens. The more important provisions that all direct shear devices should contain are listed as follows:

(1) The line of action of the shear force should pass through the centroid of the intended shear surface.
(2) The normal load system should be capable of maintaining a constant normal stress.
(3) The normal load should be applied to the specimen in such a manner as to ensure a uniform distribution of the normal stress.
(4) A frictionless system should be provided to allow the bottom half of the specimen to move as the shear load is applied.
(5) The shear and normal loads should be directly measured (e.g., as opposed to measuring hydraulic jack pressures).

Of the various devices used in this study, only the equipment used by laboratories 3 and 5 meet all five provisions. The equipment used by laboratories 1, 2 and 4 were deficient in at least one of the five provisions.

8 DISCUSSION

One of the more important provisions of any direct shear device used to test natural discontinuities is the ability to maintain a constant normal load. Most normal load systems apply the normal load by means of a hydraulic ram. As such, the ram must be able to track with the specimen as it dilates or compresses. In this respect, such systems must overcome a certain amount of system inertia which in most cases requires manual adjustment of the hydraulic pressure. Upslope and downslope sliding tests conducted in this study would require almost constant manual adjustment thus leading to the potential for greater variability in results than might be expected for the horizontal sliding tests.

An inspection of Table 1 indicates that the within laboratory repeatability limits (r values) were only slight higher for upslope sliding versus horizontal sliding. However, downslope r values were greater than the horizontal r values by factors of 2.2 and 1.7 for 10 cm and 15 cm diameter specimens, respectively. It is interesting to note that the same trends do not exist for the among laboratories reproducibility (R values). In fact, the greatest variability, indicated by the R values, occurred for the horizontal sliding tests on 10 cm specimens.

One of the central issues of this study centered around which test device(s) produced the best results. In this sense, best results implies the least amount within laboratory variability as well as most accurate results. Theoretically, there should have been no difference in the test results between the 10 cm and 15 cm specimen diameters. An inspection of Figure 2 indicates that laboratory 5 results contain the least amount of variability (at least marginally). In other words, there appears to be slightly less variation in the maximum and minimum friction angles for both specimen sizes tested.

To answer the question of accuracy first requires that the true value of the parameter being measured is known. Of course the true value of any parameter relating to a geologic material is seldom, if ever, known. Initially, the author intended to compare the average values from each laboratory with the overall average values from all laboratories. However, considering the variability of the average values recorded in Table 2, such an exercise would have little physical meaning.

9 CONCLUSIONS

The maximum variation in friction angles of 12.5 degrees observed from among laboratory test results is excessive. In addition, trends in the reproducibility limits indicate that even under the best test conditions, where variability is controlled in so far as possible, we can measure simple sliding friction angles to within only three to four degrees among testing laboratories. The disturbing issue is that in the real world where tests are likely to be conducted on natural discontinuities with random shear surface roughnesses, the variability is likely to be much greater.

Variability in test results has certain implications with respect to design. The minimum acceptable factors of safety imposed to satisfy sliding stability criteria of gravity structures or slopes is directly related to uncertainty associated with the geotechnical engineers ability to predict representative shear strengths along potential planes of instability. At least part of this uncertainty is attributed to the issues of test result variability. Designers frequently perceive test result variability as an indication of prototype strength variability and, as a result, designs are based on the weakest link in the chain concept (i.e., shear strengths selected for design are frequently based on lower bound test values). Reductions in the variability of test results will lead to less conservatism and hence more economical structures. In this respect, the findings of this study will likely lead to standardization of direct shear equipment used to test rock specimens within the Corps of Engineers.

ACKNOWLEDGEMENTS

This study was made possible through funding provided by the U.S. Army Corps of Engineers, Office of the Chief of Engineers, under the Civil Works Research and Development Program. The study was conducted at the U.S. Army Engineer Waterways Experiment Station, Geotechnical Laboratory, Soil and Rock Mechanics Division.

REFERENCES

American Society for Testing and Materials. 1992. Designation: E 691, Standard practice for conducting an interlaboratory study to determine the precision of a test method, published by ASTM, Philadelphia, PA.

American Society for Testing and Materials. 1990. Interlaboratory data analysis software, E 691 user's guide, ASTM, Philadelphia, PA.

Davis, J.C. 1973. "Statistics and data analysis in geology," John Wiley and Sons, Inc., New York, NY.

Nicholson, G.A. 1983. Insitu and laboratory shear devices for rock: a comparison, technical report GL-83-14, U.S. Army Engineer Waterways Experiment Station, Vicksburg, MS.

U.S. Army Engineer Waterways Experiment Station. 1989. "Rock testing handbook," RTH-203, Vicksburg, MS.

Rock Mechanics, Nelson & Laubach (eds) © 1994 Balkema, Rotterdam, ISBN 90 5410 380 8

Probabilistic analysis of progressive block failures

Charles Stone & Dae Young
Michigan Technological University, Houghton, Mich., USA

ABSTRACT: In order to simplify the analysis of complex key block problems, a matrix approach can be employed to analyze the stability of individual points in a rock mass in the vicinity of an underground excavation. This technique involves a three-dimensional grid of connected, discrete nodal points representing the region around the excavation. The grid is assembled with standard finite element procedures, employing an eight-noded cubic element with 28 connection bars.This approach is an effective way to relax some of the restrictive assumptions of the traditional key block theory. Because of the ease with which this approach can handle the probabilistic analysis of progressive block failures, it may be employed in the development of block caving models. This will facilitate the efficient solution of realistic problems involving complex excavations and joint structures with a general three-dimensional and point specific analysis.

1 INTRODUCTION

The traditional approach to key block analysis, as it has been developed by Goodman and Shi, Warburton, and others has been hindered by the restrictive assumptions of infinite, perfectly planar, deterministic joint surfaces. These assumptions have been required in order to achieve workable solutions to design problems. However, they make analysis of complex three-dimensional key block problems awkward and time consuming.

In order to simplify the analysis of complex problems, a matrix approach can be employed to analyze the stability of individual points in a rock mass in the vicinity of an underground excavation. This technique involves a three-dimensional grid of connected, discrete nodal points representing the region around the excavation. The grid is assembled with standard finite element procedures, employing an eight-noded cubic element with 28 connection bars. From the grid data a global connectivity matrix is formed which represents the continuity of adjacent points in the rock mass.

Joints are subsequently simulated by whichever joint modelling method is deemed appropriate. Random joint generators utilizing Fisher, Watson, or Bingham distributions are programmed so that a Monte Carlo simulation can be run. During each simulation, every connection in the connectivity matrix is checked to see if any joint in the rock mass separates the two physical locations in that rock mass which are represented by the two nodal points of that particular connection. Any such nodal connections are then removed from the connectivity matrix. The removal of these connections forms independent substructures within the matrix which correspond to the various blocks which are formed in the rock mass by the joint pattern. Subsequently, the nodal points are evaluated by a sorting algorithm based upon graph theory to determine which nodes correspond to particular blocks. The identified blocks are then classified according to whether they are stable or key blocks. A data array is set up to record the outcome of each node during the Monte Carlo simulation from which probabilities of failure are subsequently determined.

2 MESH GENERATION

The first step in this type of analysis is to generate a discrete mesh of nodal points which represent the continuous three-dimensional rock mass surrounding an excavation. This is accomplished with the use of finite element analysis techniques. A three-dimensional, eight-noded cubic finite element is employed. An incidence matrix is established based upon this element. Since each node in the cubic element is considered to be adjacent to each of the other nodes, a value of "1" is given to show that these two nodes are connected in three-dimensional space. The result is an incidence matrix for each element as follows:

$$K = \begin{matrix} 7 & 1 & 1 & 1 & 1 & 1 & 1 & 1 \\ 1 & 7 & 1 & 1 & 1 & 1 & 1 & 1 \\ 1 & 1 & 7 & 1 & 1 & 1 & 1 & 1 \\ 1 & 1 & 1 & 7 & 1 & 1 & 1 & 1 \\ 1 & 1 & 1 & 1 & 7 & 1 & 1 & 1 \\ 1 & 1 & 1 & 1 & 1 & 7 & 1 & 1 \\ 1 & 1 & 1 & 1 & 1 & 1 & 7 & 1 \\ 1 & 1 & 1 & 1 & 1 & 1 & 1 & 7 \end{matrix} \tag{1}$$

Obviously, this is a symmetric matrix. The "7" indicates that there are seven connections between that particular node and other nodes in the cubic element. When the individual elemental matrices are combined into a global connectivity matrix, M', for the representation of a region of rock mass, node numbering becomes important. A mesh generation technique is required to efficiently generate the global connectivity information for a region. The following mesh generation technique can be employed.

Two opposite corners, x_0 and x_1, of a rectangular prismatic mesh are selected, along with the number of divisions d_x, d_y, d_z, along each axis. The number of nodes along an axis are therefore: $(d_x + 1)$, $(d_y + 1)$, $(d_z + 1)$. This forms a global mesh consisting of b_k elements and g_p nodes as follows;

$$b_k = d_x\, d_y\, d_z \quad \text{and} \tag{2}$$

$$g_p = (d_x+1)\,(d_y+1)\,(d_z+1). \tag{3}$$

The g_p nodes are numbered starting with x_0 = node #1 and proceeding with (d_z+1) nodes upwards along the z-axis. Next, (d_y+1) similar sets of nodes are numbered in sequence along the y-axis. Finally, (d_x+1) similar grids are numbered in sequence along the x-axis so that the final node x_1 = node # g_p.

The elements are numbered in a similar manner. Starting with the element whose lower left corner is x_0; d_z individual elements are generated in sequence upward along the z-axis forming a stack of elements. Subsequently, d_y stacks of elements are generated along the y-axis, forming a grid. Finally, d_x grids of elements are generated along the x-axis, forming the complete mesh of b_k individual elements.

As each element is generated in sequence, the correspondence between the element numbering and the node numbering system is determined with reference to the lower left node position of the element. The function providing the node number of the lower left node position for a particular element, n_1 is as follows:

$$(4) \qquad n_1 = i + (((i-1)/d_z) - (d_y((i-1)/(d_z\, d_y)))) + (d_z+d_y+1)((i-1)/(d_z\, d_y)).$$

It should be noted here that the FORTRAN conventions for integer division must be employed in order for this function to return the correct node number.

As the element is generated, "connection bars" corresponding to the 28 connections between the corners of that element are generated and placed in the global connectivity matrix. However, since the global connectivity matrix is symmetric, only those connections in the upper triangular portion of the global connectivity matrix are stored. As each connection for a particular element is entered the position on the diagonal of the matrix corresponding to a particular node is incremented by 1 so that the total number of connections at each node is identified on the matrix diagonal. In some cases a particular connection between two nodes will have been previously established during the formation of a previous element. In these cases the diagonal position on the global connectivity matrix is not incremented. The result is that upon completion of the global connectivity matrix, each node will have either 7, 11, 17, or 26 connections with other nodes depending upon how many elements contain that particular node.

Upon completion of the global connectivity matrix it is easy to see that this is a multiply banded symmetric matrix. This allows for a considerable savings in memory space if those entries which can have only a "0" entry are not stored in memory. Chandrupatla and Belegundu (1991) discuss an algorithm for storing an (n x n) symmetric banded matrix in an (n x n_{bw}) matrix where n_{bw} is the band width. Since many of the diagonals within the bandwidth in the global connectivity matrix are also filled with zeros, this method has been modified to further reduce the storage space required to an (n x 14) matrix. This is accomplished with a function providing potential nodal point positions.

The band width for the global connectivity matrix may be quite large. However, only the 13 diagonal bands which correspond to the possible node positions with node numbers larger than the current node, i, can possibly contain non-zero values. Only the main diagonal and these 13 diagonal bands are stored because the matrix is symmetric and multiply banded. The connectivity data is therefore stored in a two-dimensional (g_p x 14) storage array, M. Upon completion of the global connectivity matrix, the first column $M_{i,1}$, $i=1,g_p$ contains only the values 7, 11, 17, and 26 for the number of connections at node i. All of the other positions in the matrix contain a 1 if there is a connection between the two potential nodal point positions or a 0 if there is no connection. The second column of the storage matrix $M_{i,2}$ $i=1,g_p$ represents the connectivity of the potential nodal point position directly above the node i, along the z-axis. Note that there may be no connection here such as in the case of node number $(d_{vz}+1)$ which is on the top surface of the mesh. Thus, for each of the storage matrix columns j, j=2,14 there is a corresponding node position function $n_p(j)$ which can be used to evaluate a node number for a particular potential node position as shown in Table 1.

Thus, the function $n_p(j)$ returns the number which must be added to the current node number i to get the node number of the node occupying the particular relative position corresponding to the j th column of the connectivity data storage matrix M as indicated above. This method of data compression can be employed as long as the original grid node numbering system is preserved. Therefore, the storage matrix $M_{i,j}$ $i=1,g_p$ $j=1,14$ can be used throughout the analysis.

Table 1: Nodal point position functions

		Node point position function	Potential nodal point position
1			current node position
2	$n_p(2)$	1	directly above the current node +1 along z-axis
3	$n_p(3)$	d_z	+1 along y-axis, -1 along z-axis
4	$n_p(4)$	d_z+1	adjacent to current node +1 along y-axis
5	$n_p(5)$	d_z+2	+1 along y-axis, +1 along z-axis
6	$n_p(6)$	$n_p(10) - d_z -2$	+1 along x-axis, -1 along y-axis, -1 along z-axis
7	$n_p(7)$	$n_p(10) - d_z -1$	+1 along x-axis, -1 along y-axis
8	$n_p(8)$	$n_p(10) - d_z$	+1 along x-axis, -1 along y-axis, +1 along z-axis
9	$n_p(9)$	$n_p(10) - 1$	+1 along x-axis, -1 along z-axis
10	$n_p(10)$	$(d_z+1)\ (d_y+1)$	adjacent to the current node +1 along x-axis
11	$n_p(11)$	$n_p(10) + 1$	+1 along x-axis, +1 along z-axis
12	$n_p(12)$	$n_p(10) + d_z$	+1 along x-axis, +1 along y-axis, -1 along z-axis
13	$n_p(13)$	$n_p(10) + d_z + 1$	+1 along x-axis, +1 along y-axis
14	$n_p(14)$	$n_p(10) + d_z + 2$	+1 along x-axis, +1 along y-axis, +1 along z-axis

3 REMOVAL OF NODES INTERIOR TO THE EXCAVATION

Once the grid has been established, it is necessary to eliminate the node points interior to the excavation surface. This is done by eliminating all connections to that node in the global connectivity matrix and resetting the number of connections to that node to zero. The node numbers of the excavation interior nodes are not eliminated because this would destroy the utility of the nodal point position functions.

Once the nodes interior to the excavation have been removed, it is necessary to identify the nodes which are on the surface of the excavation. For each node on the excavation surface, the "1" indicating a connection to any node interior to the excavation is replaced by the negative of the number of the adjacent excavation plane. This identifies excavation plane equations for later use by the block classification algorithm.

4 JOINT MODELLING

The method of joint modelling for this approach to key block analysis is completely arbitrary. Random joint generators utilizing Fisher, Watson, Bingham, etc. distributions are programmed so that Monte Carlo simulations may be run. Any type of joint may be simulated provided that an appropriate routine is developed to break the connection bars in the global connectivity matrix. However, the use of circular disks in the joint modelling scheme allows for a rather simple algorithm to break nodal connections within the global connectivity matrix.

When a connection bar is broken by a joint plane, the "1" in the storage matrix is replaced by the number of the joint plane. Therefore, the joint plane numbering system always starts with the number two.

The node breaking procedure is one of the most time consuming aspects of this type of analysis. It is therefore important to keep this algorithm as quick as possible.

5 ASSEMBLING BLOCK TYPE INFORMATION

The breaking of node connections subject to the given joint data forms independent substructures within the global connectivity matrix which correspond to separate rock blocks formed within the rock mass. It is therefore necessary to identify which nodes correspond to particular rock blocks. A sorting technique based upon graph theory is employed.

The data storage scheme for block information consists of three arrays. The block nodal point array, BNP, contains a re-ordered list of all of the node numbers. The second array, NBK, is doubly subscripted containing for each block, the position in the BNP array of the first node in the block and the total number of nodes in the block. Together these arrays specify all of the nodes contained in a particular block. A third array contains the number of the block in which a given node is located.

The sorting algorithm employs a main loop which is incremented so that all nodes will be tested for inclusion in a block. This loop controls the block data assembly procedure. The node corresponding to the value of the loop variable is considered as the "root" of a spanning tree. This node is stored in the BNP array in the location indicated by the first subscript of the NBK array. The connectivity matrix is subsequently checked for all connections to both higher and lower node numbers. These "son" nodes, or "branches," are then stored in the BNP array also. As they are being stored, three operations are performed:

1) the number of nodes in the block are tallied,
2) the current position in the BNP array is advanced, and
3) a logical array subscript corresponding to these node points is set to TRUE in order to indicate that the node number has already been stored in the BNP array.

Subsequently, for each of the son nodes, the next generation of son nodes is also collected and stored in the BNP array. As this is being done the position in the BNP array whose sons are being collected is advanced. When this position is equal to the number of nodes in the block plus the position in the BNP array of the first node in the block, the last leaf in the spanning tree of the current block has been collected. The number of nodes for the block is stored in the NBK array. Control returns to the main loop to start collection of the nodes for the next block. If the logical array indicates that the next node has not already been stored in the BNP array then this node is used as the root of the spanning tree for the next block. It should be noted that upon completion of the sorting algorithm the complete block size distribution information is contained in the NBK array, referenced by the second subscript.

The block classification technique involves a three step approach as follows;

1) the storage matrix is analyzed to collect the joint and excavation plane geometry associated with a particular block,
2) the kinematic stability of the block is evaluated with an algorithm based upon Shi's theorem, (Goodman & Shi, 1985) and if kinematically unstable,
3) the mechanical stability is evaluated with Warburton's (1981) algorithm.

For those blocks which are kinematically unstable, this approach involves a re-calculation of the joint intersection vectors for the mechanical stability evaluation. Both Shi's theorem and Warburton's algorithm are important contributions to the key block theory. However a straight forward evaluation with Shi's theorem does not handle the mechanical stability as efficiently as Warburton's algorithm when considered within the scope of this matrix approach. Warburton's algorithm, on the other hand, does not consider the excavation geometry as well as Shi's theorem. Since the number of blocks which are kinematically unstable is small compared to the total number of blocks, this approach, employs their use in sequence. Further research may provide an algorithm which is capable of eliminating the recalculation of the joint intersection vectors.

6 PROGRESSIVE FAILURE

One of the interesting aspects of this type of analysis is that it provides for the evaluation of progressive key block failures. If it is assumed that a key block displaces into the excavation, the next level of blocks which are exposed to the excavation surface become potential key blocks. The process may continue over many levels of failure. With this type of analysis it is very simple to evaluate the successive levels of key block failure around the excavation surface.

Following the assembly of the block data a loop is required to evaluate the desired number of levels of failure. After the classification of the blocks on the excavation surface any blocks which are considered key and their associated nodes must be set to failed key and the nodes on the excavation surface for the next level of failure must be determined. The classification is then run on the next set of blocks exposed to the excavation surface.

7 INCREMENTING PROBABILISTIC RESULTS

The probabilistic results are stored in a triple subscripted array in which the first subscript indicates the node number, a second subscript indicates the level of failure, and a third subscript organizes the following information; the node type, and the number of times the node was classified as "key," "indeterminate," and "stable." Following a Monte Carlo simulation the positional probabilities can be calculated simply by dividing the number of times the position was evaluated as being contained in a key block by the number of simulations.

Block types are stored in a double subscripted array in which the first subscript indicates the block number and the second indicates the level of failure. When a block is classified as a key block, the block and all of its nodes must be set to indicate that it is previously failed for all future levels of failure. Codes used to identify block and node types are shown in Table 2.

Table 2: Node and block types

	Node types		Block types
0	not classified	0	not classified
1	inside the excavation	1	failed on a previous level of failure
2	inside previously failed key block	2	key block on the present level of failure
3	on the excavation surface	3	indeterminate on present level of failure
4	in the grid interior	4	stable block on present level of failure
5	on the exterior surface of the grid		

8 DISCUSSION OF THE ALGORITHM

This algorithm-intensive approach to key block analysis evaluates the positional probability of failure of large numbers of nodal points simultaneously. It is this batch processing of nodal points which is responsible for the efficiency of the algorithm.

The most important aspect of this approach is that the restrictive requirement of infinite, planar, deterministic joint surfaces has been eliminated. The infinite plane assumption leads to a conservative design because more potential key blocks are predicted than actually exist in the rock mass. With this approach, realistic excavations and rock masses can be simulated utilizing any approach to joint modelling. In fact, even very complex concave excavation geometries can be handled as efficiently as convex ones.

The use of Shi's theorem and Warburton's algorithm, in sequence, is an effective way to incorporate the mechanical stability analysis into this matrix approach to key block theory. In addition to this consideration, it is not inconceivable that an algorithm to handle rotational failure mechanisms can be developed.

This method is similar to Goodman and Shi's approach to the traditional key block theory in the sense that only blocks formed at the surface of the excavation are considered as potential key blocks. It is therefore subject to Warburton's (1987) criticism that it may tend to underestimate the number of blocks which may require support. However, since the progressive failure of blocks has been incorporated into the programming, the algorithm also identifies all "neighbor" blocks when it evaluates the next level of failure. Therefore, while it is not immune to some of the dangers of the traditional key block theory, it does provide for a more comprehensive analysis of the interaction of the various blocks in the vicinity of the excavation.

One of the disadvantages of the discrete region approach is that it requires the assembly of a grid of nodal points with finite extent. This leads to the inevitable situation where blocks are formed which contain points on the edge of the grid. In such cases there is insufficient information available to make a determination as to whether these blocks are shaped by the joint surfaces so as to be key blocks. This problem can not be avoided. However, it can be minimized by making the grid large enough that the edge blocks are located far away from the excavation surface.

As previously noted, the progressive failure capability is one of the interesting aspects of this type of analysis. There is a potential here for utilizing this approach in the development of caving models for underground mining operations. The method can be used to predict, in probabilistic terms, the most favorable orientations of the mine layout for ore caving to occur.

As with any new technique much work remains in order to refine and improve the technique. There are many possible ways to develope block classification algorithms. Therefore, some experimentation will be required to develope faster and more efficient methods. It would also be desirable to integrate this approach with the finite element technique. The stresses at the nodal points could be generated with the finite element technique and subsequently the stresses on each of the block faces could be integrated to generate the driving force on the block employed in the mechanical stability evaluation.

9 CONCLUSIONS

This matrix approach is algorithm-intensive, but quite different from the traditional key block approach. It is simple to apply and runs quickly, but may have difficulty classifying blocks on the grid exterior. Because of the ease with which it can handle the probabilistic analysis of progressive block failures, it may be employed in the development of block caving models. This approach, therefore, is an effective way to relax some of the restrictive assumptions of the traditional key block theory. This will facilitate the efficient solution of realistic problems involving complex excavations and joint structures with a general three-dimensional and point specific analysis.

ACKNOWLEDGEMENT

This work was supported by a mining engineering research fellowship provided by the United States Department of Education and Michigan Technological University.

REFERENCES

Chandrupatla, Tirupathi R. and Ashok D. Belegundu 1991. *Introduction to finite elements in engineering*. Englewood Cliffs: Prentice Hall.

Goodman, Richard E. and Gen-hua Shi 1985. *Block theory and its application to rock engineering*. Englewood Cliffs: Prentice-Hall, Inc.

Stone, Charles A. 1994. *A matrix approach to probabilistic key block analysis*. Michigan Technological University Ph. D. Thesis.

Young, Dae S., and Wenbin Zhang 1992. Fragment size distributions and rock joints, *Proceedings of the 33rd U.S. Symposium on Rock Mechanics*. Rotterdam: A. A. Balkema, pp. 775 - 784.

Warburton, P. M. 1987. Implications of keystone action for rock bolt support and block theory, *International Journal of Rock Mechanics*, 24 #5, pp. 283-290.

Warburton, P. M. 1981. Vector stability analysis of an arbitrary polyhedral rock block with any number of free faces, *International Journal of Rock Mechanics*, 18 pp. 415 - 427.

Rock Mechanics, Nelson & Laubach (eds) © 1994 Balkema, Rotterdam, ISBN 90 5410 380 8

Uncertainty, reliability and risk in rock block stability

José Muralha
Laboratório Nacional de Engenharia Civil (LNEC), Lisbon, Portugal

ABSTRACT: In some Rock Engineering projects the stability of rock blocks is the main failure scenario. The relevant parameters required to study the stability of a rock block can be divided in two groups (geometrical and mechanical). The definition of such parameters is subjected to several kinds of uncertainties that lead to the need of using probabilistic approaches in geo-engineering. This paper deals with the determination of the statistical distributions that can be used to model these parameters.

1 INTRODUCTION

In Rock Mechanics problems and studies, statistical calculations for the evaluation of the safety conditions are not usually performed. One of the reasons deals with the fact that rock masses are formed by rock and discontinuities, which are natural materials. In opposition, other construction materials, as steel and concrete, are man-made, and we make them comply with certain specifications guaranteeing that a given strength will be assured. This distinction allows the statistical distributions of the mechanical parameters required to perform probabilistic analyses in concrete and steel projects to be well known, perfectly defined and even controlled.

In Rock Engineering problems, uncertainties are not only related to the materials dealt with. From site characterization, passing through the definition of the failure modes, the determination of the models and parameters that define the properties of the rock and joints, the estimation of the loads, down to the choice of the method to assess safety uncertainty is ,at different levels, present and, in some cases, can only be overcome by expertise and empirical knowledge.

In order to perform any probabilistic analysis the choice of statistical distributions for the rock and joint properties and the determination of their parameters is required. In this particular field, geotechnical problems face major sampling problems. Due to the heterogeneity and to the variability that often occur at the scale of the projects, the number of observations or tests that are usually performed is very limited from a statistical point of view. This adds to the difficulties to simulate field conditions in laboratory tests, mainly due to scale effects, and to the methods that are used to assess safety. In this last step, safety is generally defined by a safety factor that comes out of a somewhat elaborate modelling technique. It is not seldom to find complex mathematical techniques, numerical tools and computer programs being used to assess the safety of a Rock Mechanics problem, based in hypothesis that do not apply fully to the real conditions and that lack of reliable input data. This does not mean that computer programs are to be seen with suspicion, but the generalized use of these powerful tools must be regarded with caution, since the results of a computer code are the direct outcome of a modelling of the real problem and of the input parameters. Concerning this last point, it is frequent to find references where no tests were performed to determine relevant parameters for the input data of elaborate programs.

An important drawback for the use of probabilistic methods in Rock Engineering concerns the difficulty of interpreting and using the results. From the designer's point of view, a probabilistic approach to a geotechnical problem would replace the usual safety factor by a failure probability or a safety index. The interpretation of a safety factor is well born in mind; on the other hand, a failure probability, meaning that following strictly the same conditions a number (even if very low) of failures is to be expected, is a difficult tool to assess safety with. In concrete or steel engineering a large number (millions) of identical beams or slabs can be easily found in the same country; in Geo-engineering it is difficult to find in the whole world ten dam foundations that can be grouped together. This is one of the reasons why building regulations are, already for some decades, risk-based and performance-oriented and in Rock Engineering the first steps are being done nowadays. One of the general aims of Rock Mechanics theoretical research should be to provide the basis of new regulations that relate the performance requirements of geotechnical works with acceptable risks.

Taking into account these ideas it is advisable to implement probabilistic or statistical methods in Rock Engineering whenever possible in order to allow for the reduction of unsatisfactory design procedures and significant cost cutbacks, in other words, to increase safety saving money. Since risk is a function of the likelihoods or probabilities and their respective consequences and costs, probabilistic approaches are the proper tool to perform comparative studies between different solutions. Another domain where probabilistic studies should be employed is in the definition of the most relevant parameters for a given failure scenario of a geotechnical project.

2 PARAMETERS INVOLVED IN ROCK BLOCK STABILITY

The stability of rock blocks is an important issue in Rock Mechanics and one of the areas where probabilistic approaches are more frequent. The safety assessment of rock blocks can be found in rock slopes, created by surface excavations for open pit mining and rail or motorways, and in rock faces in structurally controlled underground excavations, as caverns, for hydropower plants or waste disposal, and tunnels for water supply or transportation networks.

The reason why a large number of probabilistic studies related to the stability analysis of rock blocks (*Priest and Brown 1983, Kulatilake et al. 1985, Muralha 1991, Trunk 1993*) is the fact that the required critical parameters are very few and not difficult to model (*Hoek 1991*). Moreover, no elaborate methods or equations are required for the evaluation of safety, as simple limit equilibrium analyses can be used to determine the failure probability or the reliability index.

In these kind of problems the parameters can be divided in two groups: geometrical and mechanical. The geometrical parameters are the orientation, area, intensity (number of joints per meter or, under certain conditions, the inverse of the spacing) and the trace length of the joint sets. As this is a case where an ultimate limit state (failure), and not a serviceability limit state, is in question, the deformability characteristics of the rock mass is not particularly important. So, the required mechanical parameters are only those needed to define the shear strength of the joint sets.

Several statistical distributions can be used to define the orientation of a joint set. The mean bivariate normal distribution on the tangent plane seems, however, to be the most appropriate, since it is an anisotropic distribution, i.e. the dispersion around the mean attitude is not independent of the direction considered, it is independent from the geological coordinates or the dip or dip direction, and it assumes that a discontinuity perpendicular to the mean attitude can not belong to that set. These reasons seem to be sufficient physical support to employ this distribution, that a good fitting in practical cases along the last two decades tends to validate.

From a collection of geological data the different joints can be grouped in sets and then the 5 parameters defining the bivariate normal distribution (mean strike and mean dip, maximum and minimum dispersions and the angle between the dip direction and the direction of the maximum dispersion) can be evaluated (*Grossmann 1985*). The estimation of these parameters is made using a large number of joints for the most important sets, so, apart from measuring errors, they are statistically quite reliable.

The occurrence of a joint can be regarded as a Poisson process, since it is a rare event, two discontinuities of the same set can not occur in the same place (non-multiplicity) and the location of a joint can be considered independent from the overall network (independence). To characterize these kind of processes, a single parameter exponential distribution is required. The determination of the joint set intensity can be performed using different kinds of observation surfaces (*Grossmann 1988*), but the simplest case refers to the scanline whose formula for the intensity is

$$I = \frac{N}{l \cos\epsilon} \tag{1}$$

where I is the intensity of the joint set, N is the number of joints from the given set that are sampled during the geological survey, l is the length of the scanline and ϵ is the angle between the normal to the mean attitude of the set and the scanline.

The two confidence limits I_L (lower and upper) for the intensity are the two solutions of the following equation (*Grossmann 1993*)

$$\frac{I_L}{I} e^{\frac{I_L}{I}} = \frac{\sqrt[N]{P_L N!}}{N} \tag{2}$$

where P_L is the probability of occurrence for the confidence limits.

Figure 1 was composed by solving equation (2) for an unit intensity (I=1 m^2/m^3) and with 3 different probabilities of occurrence P_L. It plots the upper and lower confidence limits for 3 values of P_L (0.05, 0.025 and 0.01), as a function of the number of discontinuities of the joint set N intersecting the observation surface. As N is always an integer, only discrete functions have to be considered, and consequently the dashed lines are just a helping support for a better reading of the figure.

This figure shows that the lower limit tends to be closer to the expected value than the upper. This is not an unexpected conclusion but just a validation, since positively skewed distributions are foreseen when dealing with Poisson processes. A visual evaluation of the figure allows to state that the estimates of the intensity begin to be

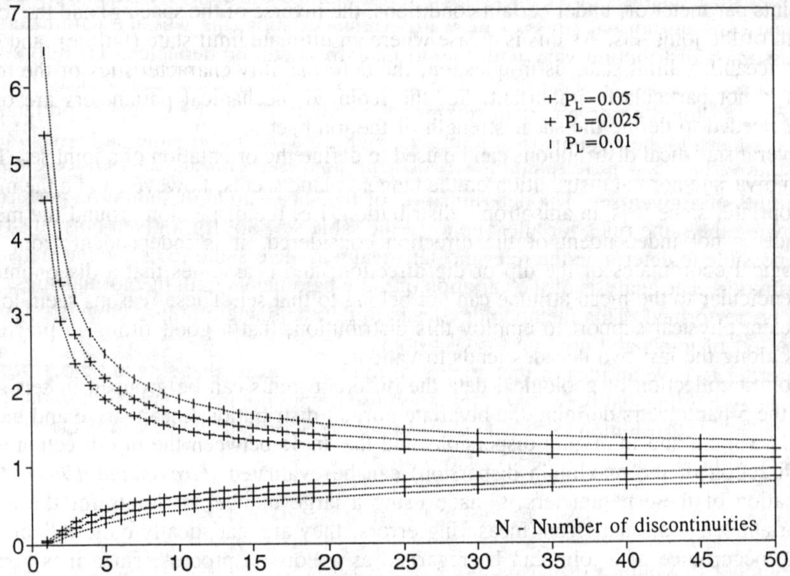

Figure 1 - Confidence limits for the joint set intensity

reliable for samples larger than 10 and for samples larger than 25 our confidence in the estimates does not increase much with the increase of *N*. A last conclusion relates to the relatively small differences, except for low values of *N*, that the probability of occurrence induces, specially when thinking about the practical consequences. However, a 5 times increase in our confidence (from 0.05 to 0.01) in finding the correct value in the interval means a 2 times enlargement of that interval.

In what concerns the modelling of the area of a discontinuity set, a Bessel distribution function is supported by strong physical reasons, since it derives from the product of the two exponential distributions of the trace lengths. Assuming this single parameter distribution, the probability density function of the areas is given by

$$f(A) = \frac{2}{\bar{A}} K_0\left(2\sqrt{\frac{A}{\bar{A}}}\right) \tag{3}$$

where $\bar{A}$ is the mean area and K_0 denotes the modified Bessel function of the 2nd kind and zero order (*Grossmann and Muralha 1987*).

A more elaborate model considers the distribution of the areas of a joint set as a distribution of the areas of parallelograms whose sides follow exponential distributions. This leads to a 4 parameter distribution that allows to establish formulas for the area evaluation and also for the confidence limits for the mean area (*Grossmann 1993*).

In both cases, due to the product of the two exponential distributions, the confidence interval can be expected to be wider than for the intensity.

A final remark should be done in what concerns the geometrical parameters of joint sets. All these ideas are based on the Poisson process hypothesis that refers to rare events. In highly ordered rock masses, where foliation or schistosity are present, this concept does not hold, so, special care should be taken.

A statistical analysis of the shear strength of a joint set should first be performed for a given constant normal stress. Afterwards, remarks about the way the parameters of the assumed distribution vary with the normal stress can be deducted. Therefore, the procedure to assess the shear strength of a joint set based on the results of a group of shear tests should observe certain routines. The joints should be grouped, preferably, in accordance to their geometrical joint sets, and if that is not possible, at least, to their lithology. The shear tests should be performed under the same normal stresses for all the joints of a given set. The determination of the shear strength should be performed following the same rules for all samples of the same set. So, for a given normal stress, it is possible to determine the relevant statistical parameters and to define the statistical distribution. The analysis of the variation of these parameters with the normal stress can then be performed. This methodology leads to define a single failure envelope for the whole set of joints and not one for each joint.

In order to have an idea of the number of joint shear tests that should be performed, the Chebyshev Inequality is very helpful because it does not make any assumption about the statistical distribution. If it is expected that the estimate of the mean is close to the correct value in a large number of cases

$$P(\,|X - \mu\,| < 10^{-k}) \geq 1 - \alpha \tag{4}$$

the number of samples (tests) required is given by

$$n \geq \frac{\sigma^2}{\alpha}\,10^{2k} \tag{5}$$

Considering that real values for the coefficient of variation lie around 0.20 and a unit mean, for simplicity, if the deviation 10^{-k} was not to exceed $\pm 20\%$ of the mean value with a 95% probability (α=0.05), at least 20 samples of the same joint set should be tested. If a higher confidence is to be considered, reducing the value of ε, the number of tests would have to be increased in the same proportion. If the coefficient of variation was to vary, to 0.25 for instance, the number of tests would be of 32 to maintain the same confidence limits.

To show the variability of the shear strength of rock joint sets a few examples referring to dam foundations will be summarized and presented in Table 1. The table displays, for the normal stresses σ_n applied during the shear tests, the values of the mean (arithmetic) τ and median τ_{50} shear strengths, the correspondent standard deviations s_τ and coefficients of variation C_V, the upper τ_l and lower τ_u confidence limits and a general description of the joint set and/or rock lithology and the number of tests performed n. The equations for the confidence intervals around the mean for large and small samples are well known and will be used to assess how reliable is the estimation of the mean. The values presented will relate to a 95% probability (α=0.05).

The table allows to have a general idea about the ranges of the parameters and to take further interesting conclusions. Rows 2 and 3, vertical and horizontal joint sets in granite, are from the same dam foundation and present a case of poor grouping of the joints. Among the samples from the vertical set a relevant percentage was filled with clay and so the shear strength is relatively low and the dispersion is very high (coefficients of variation higher than 30%). In this case, two groups of tests should be analyzed: one with the clay-filled joints and the other with the rest.

Large coefficients of variation are also found in rows 5 and 6 where, for another project, the grouping was made in accordance with the rock lithology, thus showing that

Table 1 - Shear strength statistics. Some examples
(Units are MPa except when referred)

Description	σ_n	τ	S_τ	C_V (%)	τ_{50}	τ_l	τ_u
Horizontal	0.4	0.441	0.119	26.9	0.428	0.357	0.525
joint set	0.8	0.711	0.122	17.1	0.707	0.625	0.797
Shale	1.2	0.976	0.116	11.8	1.003	0.895	1.058
n=10	1.6	1.267	0.177	14.0	1.300	1.143	1.392
Vertical	0.25	0.267	0.090	33.9	0.301	0.211	0.323
joint set	0.50	0.452	0.154	34.1	0.485	0.357	0.548
Granite	1.00	0.798	0.254	31.8	0.856	0.640	0.955
n=33	1.50	1.133	0.343	30.2	1.236	0.921	1.346
Horizontal	0.25	0.362	0.070	19.4	0.377	0.319	0.406
joint set	0.50	0.594	0.113	19.0	0.598	0.524	0.664
Granite	1.00	1.037	0.160	15.5	1.045	0.937	1.136
n=36	1.50	1.478	0.238	16.1	1.458	1.330	1.625
Schistosity	0.3	0.310	0.047	15.1	0.312	0.279	0.342
planes	0.6	0.524	0.061	11.6	0.525	0.483	0.565
Schist	0.9	0.726	0.063	8.7	0.722	0.683	0.769
n=15	1.2	0.925	0.088	9.5	0.932	0.865	0.984
Granite	0.25	0.354	0.109	30.9	0.339	0.283	0.425
n=27	0.50	0.566	0.153	27.1	0.591	0.466	0.665
	1.00	0.973	0.226	23.2	0.978	0.827	1.120
	1.50	1.329	0.283	21.3	1.351	1.145	1.512
Hornfels	0.25	0.287	0.100	34.7	0.262	0.223	0.352
n=29	0.50	0.457	0.156	34.2	0.459	0.355	0.558
	1.00	0.792	0.250	31.5	0.767	0.631	0.953
	1.50	1.099	0.333	30.3	1.037	0.884	1.315
Vertical	0.5	0.405	0.073	18.0	0.389	0.357	0.454
joint set	1.0	0.748	0.145	19.4	0.727	0.652	0.844
Schist	1.5	1.083	0.197	18.2	1.073	0.954	1.213
n=20	2.0	1.405	0.262	18.6	1.397	1.232	1.578

it should be preferably performed using the main joint sets.

The relation between the mean shear strength and the normal stress is almost linear for all cases, thus enabling to use the Coulomb criterion as the failure model for the sets. Prevalently, the standard deviation also increases with the normal stress and this happens, in some cases, following linear trends with high correlation coefficients (over 0.99). Figure 2 presents this last relation for a group of 12 joint sets that comprise the ones presented in Table 1.

Since there is no statistical evidence to state that the median is smaller than the mean, a centred distribution for the shear strength is to be expected. The parameter-free Kolmogorov test was used to verify if the data fitted the most common non-skewed distribution: the normal or gaussian. In all 4×12, cases the K-test showed that the tests

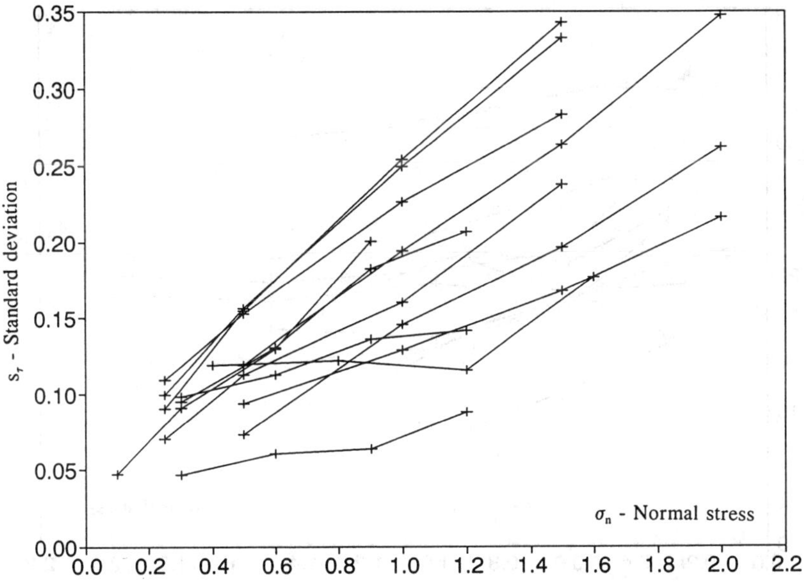

Figure 2 - Standard deviation of the shear strength versus normal stress

results were samples from normal distributions with means and standard deviations corresponding to those presented in Table 1, with high confidence levels. Merging the two linear relations and the normality allows to infer, for any normal stress, the parameters of the respective normal distribution.

In practical terms this model seems to be very useful, enabling to establish reduced strength values, for instance the 5th percentile. However, it yields non-zero probabilities for negative shear strengths. To overcome this restraint other distributions can be used: log-normal, gamma or a truncated normal distribution, for instance.

As Figure 3 shows for the 12 examples previously mentioned, though not very clear, the coefficient of variation reveals a decreasing trend, somewhat asymptotic, with the normal stress.

3 FINAL REMARK

Since failures never occur for the mean values, the use of probabilistic methods in Rock Engineering should be encouraged, in order to, in the future, allow regulations to be based on reliability analyses.

Only statistical approaches enable the comparison between different solutions to solve a problem in terms of safety and costs, i.e., in terms of a risk analysis. For now their application is still restricted to problems such as determining which solution will improve more safety or as to assess the risks behind our empirically well-established safety factors. The use of these methods in rock block stability has already allowed to say that a global safety factor seems not to be adequate for this problem (*Muralha and Trunk 1993*), as the weight of the geometrical parameters is dominant to determine the failure conditions.

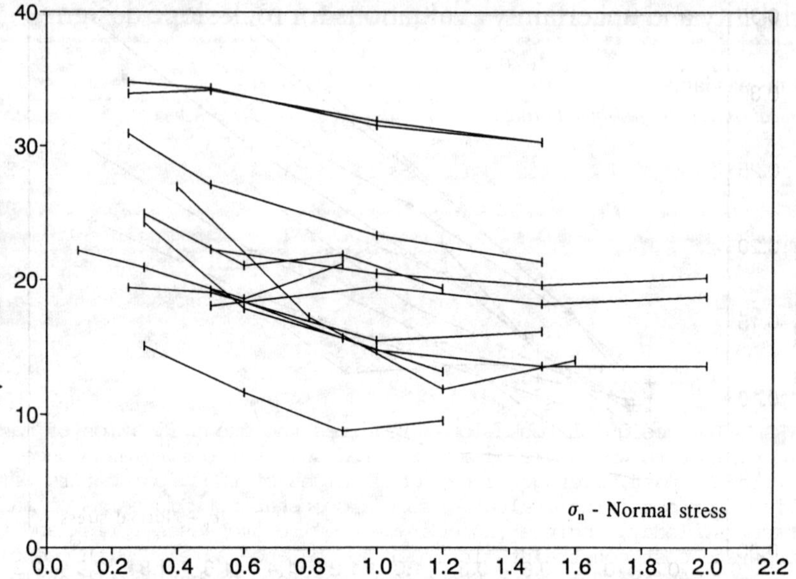

Figure 3 - Coefficient of variation versus normal stress

REFERENCES

Grossmann, N.F. 1977. Contribution to the study of jointing in rock masses (in portuguese). Research Officer Thesis. Lisbon: LNEC.

Grossmann, N.F. 1985. The bivariate normal distribution at the tangent plane to the mean attitude. Int. Symp. Fundamentals of Rock Joints, Bjorkliden. Lülea: Centek.

Grossmann, N.F. 1988. About the joint set intensity. Int. Symp. on Rock Mechanics and Power Plants, Madrid, Spain. Rotterdam: A.A. Balkema.

Grossmann, N.F. 1993. New developments in the in-situ determination of rock mass parameters. Short course on Dam Foundations. Lisbon: LNEC.

Grossmann, N.F. and Muralha, J. 1987. About the mean area of a joint set. 6th Congress of the ISRM, Montreal, Canada. Rotterdam: A.A. Balkema.

Hoek, E. 1991. When is design in rock engineering acceptable? Müller Lecture 1991, 7th Congress of the ISRM, Aachen, Germany, Volume 3. Rotterdam: A.A.Balkema.

Kulatilake, P.H.S.W.; Finley, R.E. and Ghosh, A. 1985. Effect of variability of joint orientation and strength on factor of safety of wedge stability, Int. Symp. Fundamentals of Rock Joints, Bjorkliden. Lülea: Centek.

Muralha, J. 1991. A probabilistic approach to the stability of rock slopes. 7th Congress of the ISRM, Aachen, Germany, Volume 2. Rotterdam: A.A. Balkema.

Muralha ,J. and Trunk, U. 1993. Stability of rock blocks. Int. Symp. Assessment and Prevention of Failure Phenomena, Istanbul, Turkey. Rotterdam: A.A. Balkema.

Priest, S.D. and Brown, E.T. 1983. Probabilistic stability analysis of variable rock slopes, Transc. Inst. Mining and Met. A92:1-92.

Trunk, U. 1993. Probabilistic stability analysis for rock wedges, EUROCK '93, LNEC, Lisbon. Rotterdam: A.A. Balkema.

Rock Mechanics, Nelson & Laubach (eds) © 1994 Balkema, Rotterdam, ISBN 90 5410 380 8

Variability and uncertainty evaluations for rock slope design

C. Dinis da Gama
Technical University of Lisbon, Portugal

ABSTRACT: The design of rock slopes requires the quantification of many field parameters which are characterized by variability and uncertainty. In the past, average or expected values of those parameters were utilized for obtaining evaluations of slope stability in terms of safety factors. Today, engineering design requires to assess risk and to incorporate cost constraints in order to balance safety with economy before appropriate decisions are taken. Therefore, probabilistic analyses of slope stability must be based on a reliable framework involving all relevant factors, thus justifying a sound approach to formulate their variability and uncertainty. A description of these concepts and examples of their implementation are presented.

1. INTRODUCTION

Rock slope design involves the determination of geologic field conditions and geotechnical properties that must be quantified in order to develop reliable engineering projects. As most decisions are taken under conditions of uncertainty, it is advisable to utilize probabilistic rather than deterministic methods of analysis, because these require constant parameters and one must decide to use either average values, or maximum (optimistic) and minimum (pessimistic) values, depending on the adopted hypothesis for design, thus obtaining completely different results.

The application of probabilistic analysis has the merit of quantifying both variability and uncertainty, the first being characterized by the occurrence of random events, which may be calculated by means of probabilities, and the second concept meaning unpredicted events of unknown occurrence probabilities, only to be assessed by engineering feeling and experience.

The most important aspects of variability are:

a) As a result of the geologic setting, the existence of heterogeneities, anisotropies, non-elastic behavior and the presence of discontinuities.
b) As a consequence of observation errors, due to observer difficulties, instrument malfunctions, and environmental problems.
c) Sampling errors of different types, such as those depending od improper sample size, the chosen inference method and reduced sample representativity.

Regarding uncertainties to be taken into account, they include:
a) The applicationn of unexpected or unpredicted loads during slope life.
b) Ignorance about real slope mechanical behavior.
c) Unperfect observations and measurements of slope data.
d) Oversimplifications or approximations of mathematical models used for stability analysis.

Due to these difficulties and restrictions, it is obvious that any reliable method of evaluating slope stability must consider its inherent variability and uncertainty, and so deterministic techniques are not appropriate for the purpose.An additional advantage is the provision of a mathematical framework for the quantification of risk and a definition of optimum slope angles by means of integrating geotechnic and economic criteria.

However, the use of probabilistic techniques is not unanimously accepted by many engineers.To begin with the definition of failure probability of a slope, its meaning varies according to different authors:
a) Coates (1965) proposed that failure probability is no more than the percentage of a certain slope that slides, with respect to its initial stable volume.
b) Shuk(1970)considers that such a probability represents the percentage of slopes that slide among a certain number of geometrically similar slopes.
c) In 1981, Coates stated that it has the meaning of relative likelihood of occurrence of failure in a slope, given a critical geometry and the combined variations in strength, groundwater and dip angles to produce instability.
d) Other authors consider a stochastic variable in a Bernoulli (or binary) process, due to the hypothesis that all slopes are going to slide along the geologic time scale, thus meaning the relative measure of time for each slope to fail with respect to all slopes of the same population.
e) McMahon(1975,1982)regards the collapse of a slope as an event whose probability depends on the probabilities of the dip of sliding plane being greater than a critical value and its length larger than the dimension of a fracture extending from the top to the slope basis. This criterion is also adopted by Call & Savely (1990).

The definition used in this paper considers the probability of slope failure as the frequency of occurrence, in the surface of sliding, of conditions where its shear strength is exceeded by the acting tangential stresses within the slope, thus causing its collapse.

2. PROBABILISTIC DESIGN OF ROCK SLOPES

In view of the difficulties in the acquisition of reliable field variables and the complexity, as well as the subjectivity, of uncertainties influencing rock slope design a step by step method of analysis is suggested. The method is divided in three parts:
a) Formulation of the geomechanical model of the slope under study.
b) Probabilistic computer analysis of slope stability.
c) Selection of optimum slope characteristics by the stochastic benefit/cost criterion.

The first part is formed by a series of surveys involving namely:
- Detailed geotechnical mapping of the rock mass, including their lithologies, the presence and frequency of discontinuities, as well

as weathered zones, and finishing with the division of the rock mass in classes of similar constitution and properties.

- For each discontinuity set, development of statistical analyses on its attitude,spacing,exposed lengths, roughness and infilling materials.
- Gathering of representative rock samples and discontinuity volumes in order to perform laboratory shear and tensile testing, for an appraisal of scale effects on those mechanical properties.
- Whenever possible, backanalysing pre-existing slope failures is recommended in order to evaluate shear strength parameters of the rock mass, which must also be validated by means of bibliographic information.
- Field determination of the positions of the water level within the slope, and their flow characteristics, together with other acting or expected external forces.
- Utilization of conventional (deterministic) methods of slope stability analysis,aimed to calculate safety factors for the various possible sliding mechanisms, in order to identify the most probable ones to occur under the prescribed conditions.

The second part of the method will focus its attention on the application of probabilistic methods of rock slope stability. Upon the use of Monte Carlo techniques, probabilities of collapse are computed for the various slope sectors, in order to obtain a series of relationships between slope height and inclination, water levels and with their corresponding probabilities of failure.

When this information is available the third and last part of the method is performed by means of employing economic criteria for the decision,either using the minimum generalized cost of slope construction or the maximum benefit cost ratio. These criteria are implemented on the basis of the expected costs of slope failure under the various circunstances involved in the probabilistic analysis.

The concepts of generalized cost of slope design and the corresponding benefit cost ratio were presented elsewhere (Gama-1987,1989).The first one C_g represents the sum of C_o (the initial cost of slope excavation and construction) with the expected costs of its failure:

$$C_g = C_o + P_c\,(C_r + C_d)$$

where P_c indicates the probability of slope collapse,C_r the reconstruction costs and C_d the value of damages resulting from its failure.

The variation of the generalized cost with the slope angle is normally following a trend that is depicted in Fig.1, where the optimum slope angle is indicated by means of the corresponding minimum generalized cost.

This notion can be extended if the benefits of slope utilization are estimated, a procedure that is still seldom employed in rock engineering. For example, if a new area is gained upon slope excavation and stabilization their benefits are easily evaluated. In Fig. 2 such a representation is suggested and benefits are linearly rising with the slope angle, while the various costs are indicated (construction, stabilization and expected failure costs).

It can be observed in the bottom of Fig.2 the variation of the benefit cost ratio with slope angles, thus providing a clear hint for the definition of an optimization criterion, leading to the best slope angle.

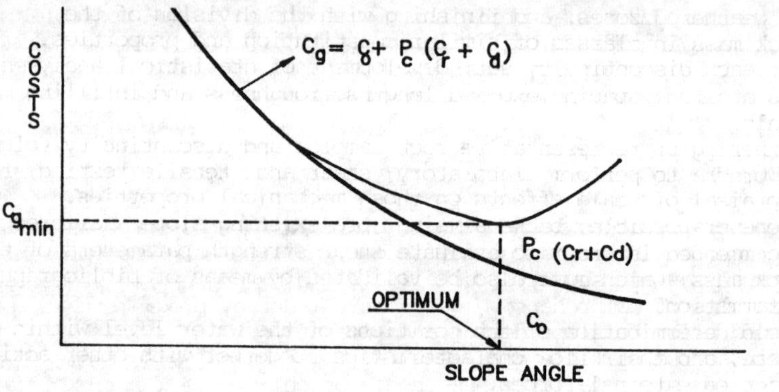

Fig. 1 - Generalized cost vs. slope angle with definition of its optimum (Gama, 1987)

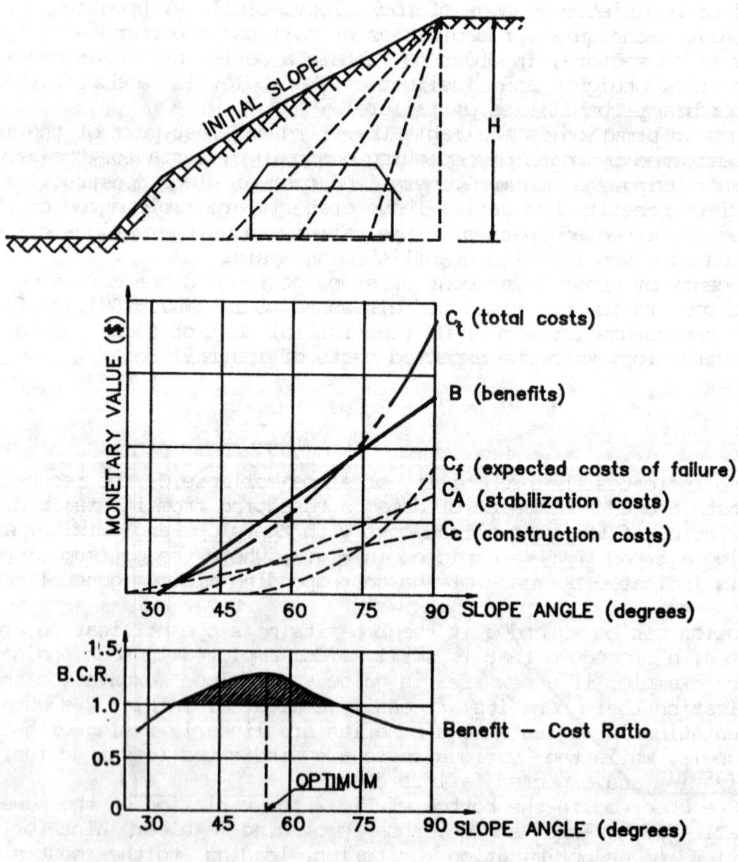

Fig. 2 - Application of the benefit cost ratio for slope design and optimization (Gama, 1989)

It is obvious that the probabilistic approach made possible the integration of geomechanical aspects of slope failure with economic factors influencing its design, therefore implementing a reliable framework on which strategic decisions can be based.

3. FIELD EXAMPLES

It is generally acknowledged that field data for the purposes of slope design is often scarce and unsufficient, particularly when probabilistic analyses are sought. Specially in mining engineering there are few examples of comprehensive geotechnical survey programs that were developed prior to slope design and excavation.
The main source of field data results from backanalysis studies conducted after slope failures happened, not only to look for explanations to these phenomena,but also to establish new design criteria for the remaining slope volumes.That was exactly the source of field information available to the author, allowing the formulation of probabilistic analyses of rock slope stability.
In the first example, a phosphate open pit mine in Brazil was being prepared for exploitation when a large slide took place in the two initial benches, occupying a length of 90 m.
The rock mass is formed by stratified metasiltstones and marls within which occurs a thick lens of phosphate rock dipping around 45° to W. Excavation started at the ore outcrop by means of 10 m height benches with an overall angle of 45°, which proved to be unstable when only two benches were created.After the collapse,that moved around 2,700 m^3, a detailed topographic survey of the failed material was conducted, providing a series of 19 cross-sections,each one 5 m apart to the next, and supplying information about the geometry of the displaced rock volume, as well as the attitudes of the failure surface (Fig.3 shows one of those sections).
Water conditions prevailing at the instant when the sliding ocurred were dry and indicated clearly that most of the movement was along the rock mass stratification, through bedding planes.

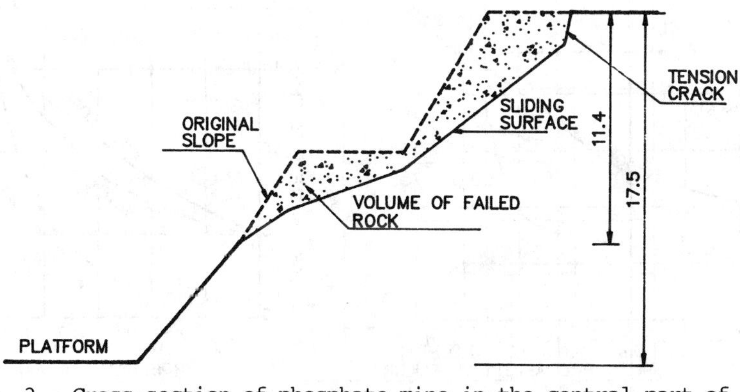

Fig. 3 - Cross-section of phosphate mine in the central part of the collapsed slope

Backanalysis studies were developed for each of the 19 cross-sections, and supplied 19 equations of the type:

$$K_1\ c + K_2\ \tan \emptyset = K_3$$

where c is the cohesion and Ø the friction angle along the plane of sliding, and K_1, K_2, K_3 are constants. A total of 171 pairs of equations were obtained, each one giving the values of (c, Ø) that were afterwards submitted to statistical analysis followed by the application of the Monte Carlo method. Each simulation with this method computed the safety factor of that slope, by the expression:

$$F_s = \frac{W \cos\alpha\ tg\ \emptyset + cL}{W\ \sin\alpha}$$

where W is the weight of displaced rock and α the dip of sliding plane. The shear strength parameters (c, Ø) were taken from their cumulative frequency distributions and according to an intrinsec dependency with respect to any sequence of random numbers.
After 10,000 simulations to evaluate the safety factors, they provided a mean value of 1.09, with a maximum of 1.51 and a minimum of 0.84, mode of 0.87, median of 1.06 and standard deviation of 0.18.
The corresponding probability of slope failure, which represents the frequency of safety factors less than 1, was found to be 37%.
These results were utilized for the design of the remaining open pit slope, which is expected to reach a total depth of 60 m: at this final situation the risk of failure was equated for various overall slope angles (35°, 40° and 45°)and for several water levels at the bottom of

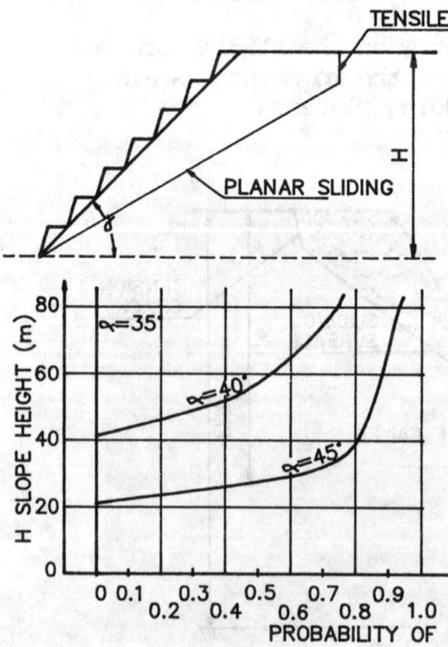

Fig.4-Probabilistic assessment of slope stability against depth of excavation

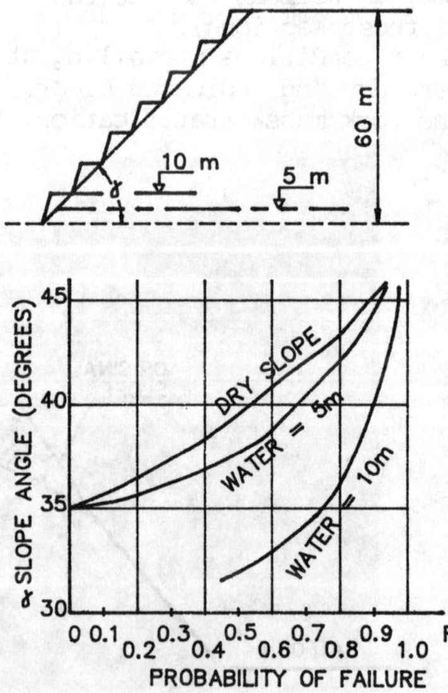

Fig.5-Probabilistic assessment of slope stability against groundwater

the pit. These two cases are represented in Figs.4 and 5,respectively, and they provided information for the mine owner to decide on the basis of an acceptable failure probability, the adopted slope inclination for different depths of the open pit, as well as the magnitude of water drainage required to keep its long range stability.

The second example deals with the stabilization project of a canal that was excavated for the water and tailings drainage of a phosphate concentration plant in Jacupiranga, Brazil. The canal, which is shown in Fig.6, suffered a series of slides during the first rainy season after its completion, leading to flow obstructions at several points. The original slope angles were obtained with a 2(V):1(H) inclination on each of the 3 individual benches,separated by horizontal berms with 2.4m to 4.6m breadths, thus creating an overall slope angle of around 45°.Detailed survey of the failures detected a total of 18 slides,9 on each margin, with essentially rotational and planar mechanisms. Although in both cases there was evidence of water saturation within the slopes,prior to their failure, in the rotational mode it affected mostly the stability of a layer of weathered rock where the dip of discontinuities was against the slope angle.However,in the other side of the canal those joints caused most of the failures by means of a mechanism of planar sliding which was subjected to detailed survey in order to provide information for backanalysis.

Those studies conducted to a set of 12 equations relating c and Ø values of the joint plane that controlled failure. A statistical analysis provided then the frequency distributions that were needed for the application of the Monte Carlo method, with which probabilities of collapse were computed.

The purpose of the study was to recommend solutions for the stabilisation of the overall slope and two procedures were selected:the change of slope geometry and the control of groundwater within it.

The framework for decisions on these alternatives was established by means of a series of graphs containing not only the common stability

Fig.6 - General view of the Jacupiranga canal with several slope failures

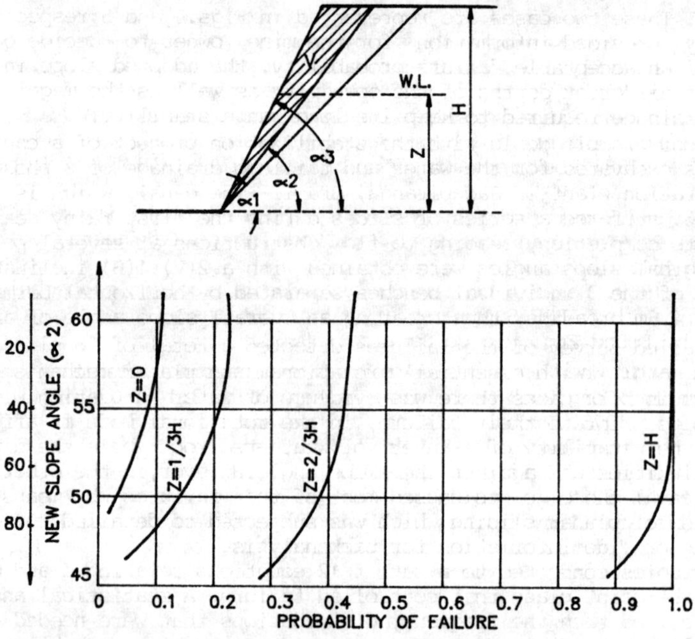

Fig.7 - Slope geometry with joint dip (α_1), new slope angle (α_2), old slope angle (α_3), water level (Z) and slope height (H). Variation of and the costs of excavating volume V with the probability of slope failure, for various water conditions.

evaluations, but also their corresponding costs. Fig. 7 illustrates these variations in the form submitted to the mine owner.
The field work was conducted on the basis of adopting a certain risk of slope instability against a provision for water drainage by means of a series of sub-horizontal boreholes opened at the slope faces. In general the reconstruction costs were chosen for water conditions not higher than half the slope height and proved to be realistic, as far as the subsequent rainy seasons were concerned.

4. CONCLUSIONS

The treatment of rock mass properties when field data are abundant, as it is the case when backanalyses of slope failures are performed, leads to the adoption of probabilistic techniques either for explaining the mechanisms of their instability or to establishing the basis for design of new engineering works. The inclusion of cost constraints into those techniques contributes for the extension of their context, thus providing new insights for Rock Mechanics design practices.

REFERENCES

Call,R.D. & Savely,J.P.1990.Open pit rock mechanics. Chapter 6.8 in Surface Mining, 2nd ed. B.A.Kennedy, editor, AIME.

Coates,D.F. 1965.Planning slopes in shale and other rocks. ASCE Water Resources Engineering Conference. Preprint No.342.

Coates,D.F. 1981 Rock mechanics principles, 3rd ed.CANMET Monograph 874. Ottawa.

Gama,C.D.1987.Sensitivity analysis for failure probability of slopes. VIII Pan-American Conf.Soil Mech.Found.Eng., Cartagena, Vol.3: 441-451.

Gama,C.D. 1989. Benefit cost ratio as a design tool in geotechnical engineering.ABMS Supplementary Contributions. IX Int.Cong.Soil Mech. Found Eng.,Rio de Janeiro: 59-62.

Hoek, E. 1991. When is a design in rock engineering acceptable? Proc. 7 th Int.Cong.Rock Mechanics, Aachen, Vol. 3:1485-1497.

McMahon,B.K. 1975. Probability of failure and expected volume of failures in high rock slopes. Proc. 2nd Aust. New Zealand Conf. Geomechanics. Brisbane: 308-313.

McMahon,B.K.1982.Redesign of the west wall Kanmantoo mine,South Australia.Stability in Surface Mining vol.3, Vancouver. AIME Publ.

Shuk,T.1970. Optimization of slope designed in rock.Proc.2nd Int.Cong. Rock Mechanics, Belgrade, Vol.2, paper 7-2.

[illegible] 1990. [illegible]

Gill, R.L. & Savely, J.P. 1990. Open pit rock mechanics. Chapter 9.8 in Surface mining, 2nd ed.: B.A. Kennedy (editor), SME.

Coates, D.F. 1965. Planning slopes in soils and open rocks. ASCE [illegible] Resources [illegible] Conference Preprint No. 142.

Coates, D.F. 1981. Rock mechanics principles, 3rd ed. CANMET Monograph 874. Ottawa.

[illegible]ra, C.O. 1987. Sensitivity analysis for failure probability of slopes. [illegible] of Pan-American Conf. Soil Mech. Found. Eng., Cartagena. Vol. [illegible] 141-151.

Cena, C. 1989. Benefit cost ratio as [illegible] design tool [illegible] In ARMS [illegible] supplementary Contributions. [illegible] Mech. Road [illegible]: 62.

Hoek, E. 1991. [illegible] design in rock engineering. [illegible] Proc. 7th Int. Cong. Rock Mechanics, Aachen, Vol. 3: 1485-1497.

Kraemer, D. [illegible] 197[illegible]. [illegible] probabilities of failure and expected volume of [illegible] [illegible] 2nd [illegible] Conf. [illegible] Brisbane: 309-313.

McMahon, B.K. 1985. [illegible] Pit Slope Manual [illegible] Vancouver [illegible]

Stacey, T.R. 1970. [illegible] slope designed [illegible] Rock Mechanics, Vol. 2, [illegible] 7-2[illegible]

Rock Mechanics, Nelson & Laubach (eds) © 1994 Balkema, Rotterdam, ISBN 90 5410 380 8

Modeling the effect of disorder on the compressive strength of rock

S.C. Blair
Lawrence Livermore National Laboratory, Livermore, Calif. & Department of Material Sciences and Mineral Engineering, University of California, Berkeley, Calif., USA

N.G.W. Cook
Department of Material Sciences and Mineral Engineering, University of California, Berkeley, Calif., USA

S.A. Daveler
Lawrence Livermore National Laboratory, Livermore, Calif., USA

ABSTRACT: We have used a field theory model to simulate the effect of microscale heterogeneity in grain strength, and stress concentration due to grain shape and loading geometry, on the stress-strain behavior and ultimate strength of rock in compression. A brief description of the model is presented and results of a series of simulations are presented. Ultimate strength for uniaxial conditions had a strong inverse dependence on the geometric parameter but was independent of the width of the site strength distribution and may instead depend on the mean of the strength distribution. Ultimate strength was higher for confined simulations, and was inversely dependent on the range of the geometric parameter, However, an increase in ultimate strength with increasing width of site strength distribution was observed for some ranges of the geometric parameter.

1 INTRODUCTION

Rocks are known to fracture in compression via the formation, extension and coalescence of extensile cracks. These cracks are generally oriented in the direction of the maximum principal stress and as more cracks are formed or existing cracks grow, crack interaction becomes important in determining the direction of fracture. The formation and extension of these cracks under conditions of compression is attributed to local tensions at the microscale that may be caused by heterogeneity in the shape, orientation and strength of grains and other factors (Myer et al., 1992). Moreover, on the microscale rock is not a continuum and many rock types can be thought of as a stochastic assemblage of grains. This stochastic nature of the microstructure affects the interaction of microcracks and other aspects of the behavior when the rock is stressed. In addition, when microcracks form and open, strain energy is stored in the rock that is not recovered upon unloading (Jaeger and Cook, 1976). Because of the discontinuous nature of rock at the microscale, the micromechanics of deformation and failure in rock is not amenable to analysis using a standard continuum approximation.

Recently, Blair and Cook (1993) presented a field theory model for fracture of rock in compression that incorporates the discontinuous nature of rock at the microscopic level. This model uses the concepts of stress decomposition and superposition and the techniques of boundary element analysis (Crouch and Starfield, 1990) and percolation theory (Stauffer, 1985) to form a simple yet powerful method for the study of progressive fracture of rock and other disordered materials. In this paper we briefly review the concept of the model and then present preliminary results of an investigation of the effect of microscale heterogeneity in strength and geometry on macroscopic mechanical behavior in compression.

2 MODEL DESCRIPTION

The field theory model is based on the assumption that the stress field in a rock can be decomposed into an elastic continuum component and a crack interaction component so that the total stress field is given by:

$$\sigma = \sigma_{ec} + \sigma_{ci}. \qquad \text{Eq. 1}$$

Following this decomposition the model uses the boundary element method (BEM) to estimate the total stress and deformation fields in a region of interest as follows. Three problems with identical geometries are set up, corresponding to the three terms in equation 1. Boundary conditions are applied to the elastic continuum problem, while the boundary of the crack interaction problem remains unstressed. The stress field in the total problem is then formed by superposition of the stresses in the elastic continuum and crack interaction problems. Moreover, areas of high stress concentrations where cracking or fracture are expected, are identified and a grid or mesh of sites representing grain centroids is imposed over these selected regions in each problem. Cracks are allowed to form only at these sites and the orientation of any crack that forms is taken to be along the direction of the maximum local principal stress in the elastic continuum problem. The stress free surface of a crack at a site in the total problem is formed by inserting a crack at the corresponding site in the crack interaction problem, and opening the crack with a traction that is equal in magnitude but opposite in sign to minimum local principal stress at that site in the elastic continuum problem. This concept is illustrated in Figure 1.

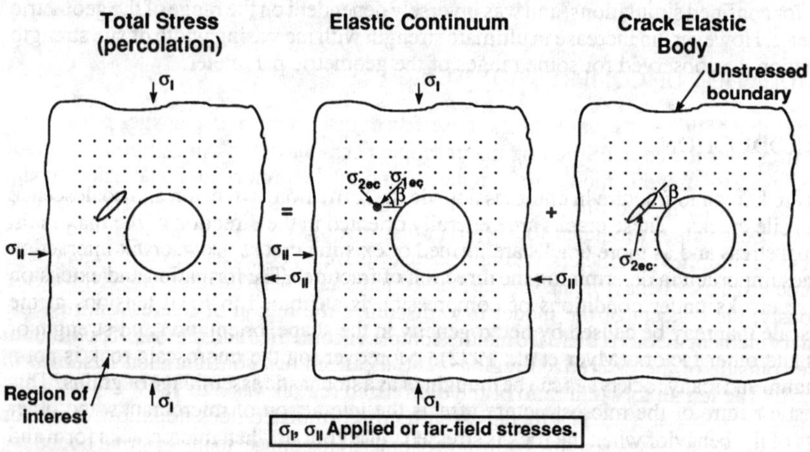

Figure 1. This figure illustrates the stress decomposition used in the field theory model. The total stress field is formed using superposition of the elastic continuum and crack interaction components.

The sites are used as a vehicle to incorporate heterogeneity into the analysis. Disorder due to variations in the shape and loading of grains in a rock is simulated by defining a stress perturbation factor, ρ, that acts on the elastic continuum space in the direction normal to the maximum local principal stress at each site. This stress perturbation is given by

$$\rho = m_1(\sigma_{1ec} - \sigma_{2ec}) \qquad \text{Eq. 2}$$

where m_1 the stress difference multiplier represents a geometric stress concentration factor assigned to each site, and σ_{1ec}, σ_{2ec} are the local principal stresses computed at the site in the elastic continuum. To preserve equilibrium the summation of ρ over the discretized region is defined to be zero. The stress in the elastic continuum problem along the local σ_{2e} direction is given by

$$\sigma_{2ec}^* = \sigma_{2ec} + \rho \qquad \text{Eq. 3}$$

and the total stress, σ_2, at a site is then given by

$$\sigma_2 = \sigma_{2ec}^* + \sigma_{2ci} = \sigma_{2ec} + \rho + \sigma_{2ci} \qquad \text{Eq. 4}$$

where σ_{2ci} is the stress at the site in the crack interaction problem in the direction of σ_{2ec}.

Disorder in grain strengths due to variations in composition and/or weathering are simulated by assigning a tensile strength to each site. A site breaking rule is defined as follows: if stress at a site is tensile and exceeds the tensile strength assigned to the site, the site is said to crack or "break." If the tensile stress exceeds the strength at more than one site, the site with the highest stress is broken. When a site is broken a crack is inserted at the site in the crack interaction problem and a "crack opening" traction with magnitude σ_{2ec}^* is imposed normal to the crack surface. Clusters of cracked sites are identified and the sample is said to fail when a percolating cluster is formed. A percolating cluster is a group of neighboring cracked sites that spans the lattice. The stress field is recalculated after each breaking event and the breaking rule is re-evaluated for the sites that are still unbroken. This process continues at a given stress level until (a) no more sites break, or (b) a percolating cluster is formed. The model is discussed in detail in Blair and Cook (1993).

3 SIMULATION OF COMPRESSION TESTS

The objective of this study was to assess the effect of disorder in a) the tensile strength and b) the stress difference multiplier on macroscopic behavior of rock in compression. Our method was to use the model to simulate standard laboratory uniaxial and triaxial compression tests in two dimensions and then compare the stress-strain behavior and ultimate stress values predicted for different distributions of the site strength and stress difference multiplier parameters. Compression tests were chosen as they are widely used to characterize rock behavior and standard procedures have been developed. As discussed previously, the concept of the model is to discretize the region of interest and use the boundary element method (BEM) to handle complex geometries and regions of continuum stresses outside of the discretized region. The idealized test configuration is shown in Figure 2. The length to width ratio of the discretized region was 3 to 1.

The model as currently formulated does not allow cracks or fractures (clusters of cracks) to connect to the external boundary of the simulated sample. Consequently, the discretized region to be studied is surrounded by a narrow border of uncracked elastic material, and the properties of this border region do not vary as cracks are added in the discretized region. For the simulations of laboratory compression tests the boundary conditions were as shown in Table 1 and Figure 2. Axial stress was applied in steps of 10 MPa.

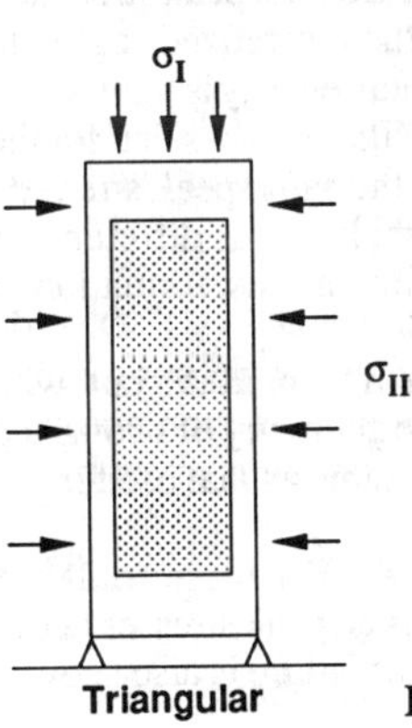

Figure 2. Model geometry used to simulate compression tests

Table 1. Boundary conditions for uniaxial and triaxial simulations.

Location	Type	Normal	Shear
Top	Mixed	Stress: $\sigma_n = \sigma_I$	Displacement: $u_s = 0$
Bottom	Displacement	Displacement: $u_n = 0$	Displacement: $u_s = 0$
Sides	Stress	Stress: $\sigma_n = \sigma_{II}$	Stress: $\sigma_s = 0$

For this study the computations have been carried out with sites in the discretized region located on a triangular grid. This grid was chosen because it is isotropic and commonly used in the statistical modeling community. Trials computed using the following conditions are discussed:

confining pressure:	0.1 MPa., 10 MPa.
stress difference multiplier (m_1)	uniform over: ±0.3, ±0.5, ±0.7
site strength (T_g)	uniform over: 20 ±0.2, 20 ±19.9, bimodal

Variation in site strength was investigated to evaluate the effect of the presence of a population of weak and/or strong sites on the macroscopic behavior. For this study the terms homogeneous and heterogeneous site strength distribution will refer to the distributions 20 ±0.2 MPa and 20 ±19.9 MPa, respectively. The values for the stress difference multiplier are investigated to evaluate how variations in site geometry and site loading will affect macroscopic behavior. The ranges of values used for this parameter are based on the observations of Zheng (1989) and Hadley (1976) and others who have examined grain geometry in rocks tested in compression. For example the multiplier for a site with circular cross section and unit diameter, loaded in compression across a diameter taken parallel with the σ_1 direction would have a multiplier of –0.64, based on formula for the Brazilian geometry.

One of the objectives of this study is to assess how disorder it the rock microstructure affects the stress strain behavior predicted by the simulation. Stress strain behavior is one of the fundamental diagnostic measures of macroscopic rock behavior in compression, and results for ten trials with homogeneous site strength distributions and with m_1 in ±0.5 are shown in Figure 3. It is important to note that the values of stress and strain shown are those computed after each cracking event and are plotted sequentially. This figure shows linear behavior up to a peak stress (or ultimate strength) that is identical for all the trials. This section of the curve reflects the fact that as load was applied, no cracks were formed at axial stress levels less than the peak stress. Nonlinear behavior is shown after the peak stress and most of the trials show class I behavior with energy going into the discretized region to produce increased axial strain. One trial shows class II behavior where energy is transferred out of the region and into the uncracked side border regions. This figure shows that for the homogeneous site strength distribution, all ten trials had exactly the same peak strength. The ensuing sequence of crack formation that eventually led to fracture was different for each trial, but the peak strength was the same. For each of these simulations, formation of the first crack led to fracture of the sample.

Next, we consider the case of uniaxial compression with m_1 in ±0.5, but with heterogeneous site strength distribution. For these trials the relative geometry of strong and weak sites and their spatial correlation with the m_1 values was the same as that used in the homogeneous site strength trials discussed above.

Stress strain curves for unconfined trials are shown in Figure 4. For these trials we observe nonlinearity at axial stress greater than 25 MPa. The observed peak stresses occur over a wide range of axial stress values and a wider range of strain at failure is also shown.

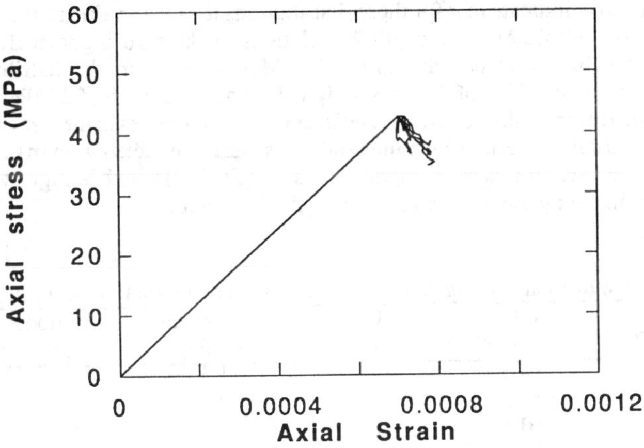

Figure 3. Stress strain curves predicted for ten simulations of uniaxial compression tests using homogeneous site strength distributions and stress difference multiplier in the range ±0.5. For all simulations initial cracks formed at the same axial stress.

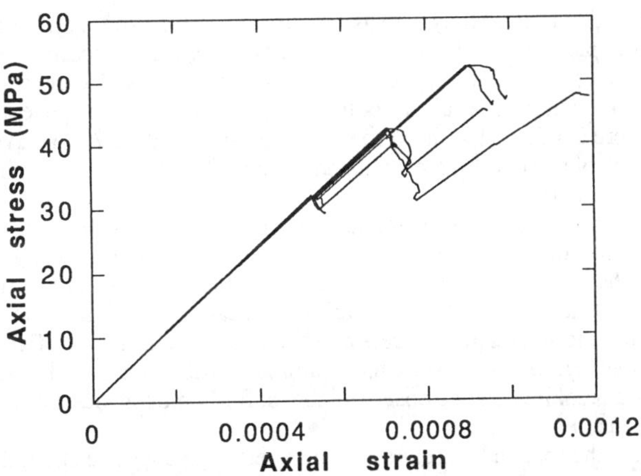

Figure 4. Stress strain plots for uniaxial simulations of compression tests using heterogeneous site strength distribution and m_1 in the range ±0.5. Variation in peak stress and strain at failure is greater than for simulations using homogeneous site strength distributions.

Strain softening behavior is shown for all trials and several show sequences of strain softening–strain hardening–strain softening as is sometimes observed in laboratory tests of rock samples. Some of the trials show class II behavior, and others show strain occurring at constant stress.

The peak or ultimate strength observed during compression test is another measure of the macroscopic behavior of rock. Average values of this parameter for the homogeneous and heterogeneous trials discussed above are 42.7 MPa and 42.9 MPa respectively (see

Table 2). This may indicate that for these distributions the mean value of the site strength distribution may control the peak strength. Simulations conducted using normal distributions of site strength values produced similar results. Moreover, a bimodal distribution of site strengths was used with 25% of the sites assigned a tensile strength of 2 MPa. and 75% of the sites assigned a strength of 26 MPa. The mean value of the sites in the array was 20 MPa as in the above trials. A series of ten uniaxial trials were conducted with m_1 in ±0.5. The average ultimate stress for these trials was 48.5 MPa (Table 2) which is slightly higher than the values predicted by the uniform site strength distributions.

Table 2. Mean ultimate strength for ten simulations at a variety of conditions.

Pc (MPa)	M1 (range)	20±0.2 (MPa)	20±19.9 (MPa)	Bimodal (see text)
0.1	±0.3	74.8	71.5	
0.1	±0.5	42.7	42.9	
0.1	±0.5	-	-	48.5
0.1	±0.7	31.4	34.1	
10	±0.3	133.1	159.0	
10	±0.5	84.4	102.0	
10	±0.7	68.6	68.9	

A second objective of this study is to assess the effect of the stress difference multiplier, m_1. on macroscopic behavior. This parameter is designed to incorporate the effect of variation in the shape and loading geometry of a grain in a rock. To investigate the effect of m_1, a series of ten uniaxial trials for both homogeneous and heterogeneous site strength distributions were conducted with m_1 in the ranges ±0.3, and ±0.7. The average ultimate strengths for each of these series are tabulated in Table 2, and plotted in Figure 5 along with the results for m_1 in ±0.5.

Figure 5 shows that for uniaxial simulations the average ultimate strength has a strong inverse dependence on the range of m_1, with ultimate strength decreasing as the range in m_1 increases. The dominant effect of m_1 is due to the fact that it is a direct multiplier of the stress difference and reducing it's range (about a mean of zero) reduces the range of the stress perturbation term at a given stress condition, making it more difficult for cracking to occur. Moreover, when all sites have similar tensile strengths, higher stresses are required initiate breaking, and as observed above, for homogeneous strength cracking of a single site leads to failure.

The results for the uniaxial simulations using heterogeneous site strength distributions an d different ranges of m_1 indicate that as discussed above, the ultimate stress is determined by the mean site strength, at least over this range of m_1 values. We interpret to mean that once a certain stress level is reached the crack interaction and geometric factors dominate the behavior. However more analysis needs to be done to fully evaluate this behavior.

To assess the effect of confining pressure on ultimate strength, the above trials were repeated using the same site strength and stress difference multiplier values at each site, but with the confining pressure set to 10 MPa. Values of average peak strength determined for the various sets of trials are listed in Table 2 and plotted in Figure 5. The data for trials at 10 MPa show that, as expected, increasing the confining pressure increases the ultimate strength at all conditions. Moreover, both homogeneous and heterogeneous trials show inverse dependence of ultimate strength on the range of m_1 that is stronger than that displayed by the uniaxial simulations. The results for confined simulations also show that unlike the uniaxial simulations, for m_1 in the ranges ±0.3 and ±0.5, average ultimate strength was dependent on site strength distribution with simulations with heterogeneous site strengths having higher mean ultimate strength. Width of site strength distribution had no effect peak strength when m_1 was in the range ±0.7.

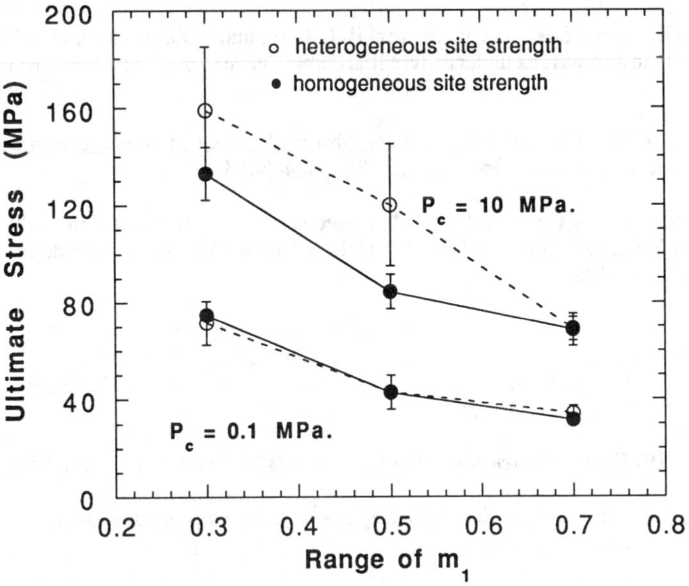

Figure 5: Average values of ultimate strength for simulations of uniaxial and confined compression tests. Results for both uniaxial and confined simulations show a strong inverse dependence on the geometric parameter m_1. Ultimate strength in uniaxial simulations is independent of disorder in site strength. Some simulations with more disorder in site strength had higher ultimate strength when confining pressure was applied.

This may be because increasing confining pressure reduces the effect of the crack interaction as illustrated in Equation 4. The trials where the heterogeneous site strength distribution gave higher ultimate strengths all had a significant number of sites break at relatively low levels of applied stress, and it is our preliminary interpretation that the effect of breaking these weak sites was to raise the mean site strength for the remaining sites, and it is these remaining sites that determine the ultimate strength in cases where crack interaction is suppressed.

4 SUMMARY AND CONCLUSION

A field theory model for evolution of fracture in brittle rock in compression has been used to explore the effect of heterogeneity for two grain scale parameters, site strength and grain geometry, on the macroscopic behavior in simulated compression tests. Results for these two dimensional simulations show that the model produces realistic stress strain curves. In addition, ultimate strength for uniaxial conditions had a strong inverse dependence on the geometric parameter but was independent of the width of the site strength distribution and may instead depend on the mean strength of the grains. Ultimate strength was higher for confined simulations, and was inversely dependent on the range of the geometric parameter.

However, when the geometric parameter was uniformly distributed over ±0.3 and ±0.5, dependence of ultimate strength on site strength distribution was observed.

REFERENCES

Myer, L.R., M.J. Kemeny, Z. Zheng, R. Suarez, R.T. Ewy, and N.G.W. Cook (1992). Extensile cracking in porous rock under differential compressive stress, *Appl. Mech. Rev.* 45 (8) 263–380.

Hadley, K. (1976). Comparison of calculated and observed crack densities and seismic velocities in Westerly granite, *J. Geophys. Res.* 81, 3484–3494.

Kemeny J.M., and N.G.W. Cook (1991). Micromechanics of deformation in rocks, *Toughening Mechanisms in Quasi-Brittle Materials,* SP Shaw (ed), Klewer Academic, The Netherlands, 155–188.

Blair, S.C., N.G.W. Cook and S.A. Daveler (1993). A Zeroth Order Statistical Model for Fracture in Rock in Compression, *Int. J Rock Mech. Min. Sci* 30, 7, Proceedings: 34th U. S. Symp. on Rock Mechanics, Madison, WI, June 27–30, 1993, and Lawrence Livermore National Laboratory, Livermore, CA, UCRL-JC-111977.

Jaeger and Cook, (1976). *Fundamentals of Rock Mechanics,* Wiley & Sons, New York.

Crouch, S.L., and A.M. Starfield (1990). *Boundary Element Methods in Solid Mechanics* (Unwin Hyman, London.

Stauffer, D. (1985). Introduction to Percolation Theory, Taylor and Francis, London & Philadelphia.

Zheng, Z. (1989). Compressive stress-induced microcracks in rocks and applications to seismic anisotropy and borehole stability, Ph.D. Thesis, University of California at Berkeley.

ACKNOWLEDGEMENTS

Work performed under the auspices of the U. S. Department of Energy by Lawrence Livermore National Laboratory under Contract W-7405-ENG-48 and supported specifically by the Office of Basic Energy Sciences.

Rock Mechanics, Nelson & Laubach (eds) © 1994 Balkema, Rotterdam, ISBN 90 5410 380 8

A new empirical peak shear strength criterion for rock joints

Pinnaduwa H.S.W. Kulatilake, Guohua Shou & Robert M. Morgan
University of Arizona, Tucson, Ariz., USA
Tsan-Hwei Huang
National Taiwan University, Taipei, Taiwan

ABSTRACT: The paper presents findings of an experimental as well as theoretical/analytical research initiated (a) to quantify joint surface roughness on a plane and then (b) to develop a new peak shear strength criterion for rock joints which has the capability of capturing the anisotropic behavior of peak shear strength due to the anisotropy of joint surface roughness. The results show clearly that at least two parameters are required to quantify joint roughness in one dimension. By limiting to only statistical parameters, an average slope of the profile (average I angle) is suggested to quantify the large scale undulations of the profile (non-stationary part of the profile) and the root mean square of the slope (z_2') is suggested to quantify the small scale roughness of the profile (stationary part of the profile). In order to capture the anistropic roughness, it was found necessary to calculate both averages I and z_2' in different directions on the joint surface. The shear strength results of a model material joint show that the suggested new equation, which is based on very limited experimental data, has a good capability of predicting the anisotropic peak shear strength of the joints.

1 INTRODUCTION

In the mining, civil, and petroleum engineering disciplines, the engineers confront problems associated with geotechnical systems which are in or on jointed rock. Due to the complex nature of natural joint patterns in rock, and the uncertainty involved in estimating geomechanical properties of joints and intact rock, the prediction of mass mechanical and hydraulic properties of rock at a large scale using available techniques presents a great challenge for the geo-engineering profession.

Strength and deformation behavior of joints and flow properties of joints depend very much on the surface roughness of joints. These effects arise from the fact that the surfaces composing a joint are rough and mismatched at some scale. The shape, size, number, and strength of contacts between the surfaces control the mechanical properties. The separation between the surfaces or the "aperture" determines the hydraulic properties. The "Joint Roughness Coefficient," JRC (Barton and Choubey, 1977), various statistical parameters and the fractal dimension, D, (Mandelbrot, 1983) have been suggested as parameters to quantify roughness of rock joint profiles along lines (or in one dimension).

Barton and Choubey (1977) have suggested two methods to choose an appropriate value for JRC. In the first method, the roughness profile of the natural joint is visually compared with ten standard profiles published by Barton and Choubey (1977). For these standard profiles JRC values have been assigned between 0 and 20 in steps of 2 starting from the smoothest to the roughest. This visual comparison method has been found to be subjective and unreliable by some investigators (Maerz et al 1990; Miller et al. 1990; among others). In the other method, tilt and pull tests should be performed on natural discontinuities to estimate JR by back analyzing the test results using the empirical equation suggeste for peak shear strength by Barton and Choubey (1977) along with the esti mations for normal stress and basic friction angle on the joint plane, and joint wall compressive strength. Some researchers (Miller et al., 1990, among others) feel that the second method has little practical merit because roughness preferably should be used to predict shear strength not vice-versa.

Researchers have tried statistical parameters such as centerline average value, mean square value, root mean square (RMS) value, mean square of the first derivative, (Z_2) RMS of the first derivative, RMS of the second derivative, probability density function, auto-correlation function, spectral density function, structure function (SF), roughness profile index (R_p) and micro-average i angle (A_i) [Wu and Ali, 1978; Tse an Cruden, 1979; Krahn and Morgenstern, 1979; Maerz et al. 1990] to quantif roughness. Some of these parameters are conceptually the same and hence they are dependent. At present, no firm conclusions are available in th literature evaluating how good these statistical parameters are in quantifying roughness. It is very important to note that only a single statistical parameter has been used to quantify rock joint roughness. It i highly questionable whether a single statistical parameter is adequate capturing the roughness arising from both stationary and non-stationary parts of a profile.

Some researchers (Brown & Scholz, 1985; Andrle and Abrahams, 1989; Chesters et al., 1989; Miller et al., 1990; Power et al., 1991; Huang et al., 1992) have investigated the possibility of using fractal dimension to provide a measure for roughness of rock joints. These investigation have led to controversial findings. The main reason for this is our limited understanding of the application of fractal theory in quantifyi the roughness of natural rock joints.

The brief literature review given above shows clearly that further research on roughness characterization of natural rock joints is badly needed. A part of the paper presents findings of a research initiated roughness quantification of natural rock joints.

The main contributors for the development of peak shear strength crit ria in the low normal effective stress range are Ladanyi and Archambaul (1969) and Barton and Choubey (1977). The joint peak shear strength shows anisotropic properties due to roughness variation with the sheari direction in direct shear tests (Huang and Doong, 1990). Even along on particular direction, the shear strength of a natural joint can be different between the forward and the backward directions. To capture the above observations, either the existing shear strength criteria should improved or a new peak shear strength criterion should be developed. A part of the paper presents findings of a research initiated to develop new empirical peak shear strength criterion.

Figure 1 Photographs illustrating the surfaces of (a) natural joint, (b) silicone rubber cast, and (c) model joint.

2 EXPERIMENTAL STUDY

Silicone rubber was used to obtain a pair of perfectly mated silicone rubber casts to represent the topographic features of a sandstone rock joint of about 100 mm diameter. These silicone rubber casts were then used to prepare about 65 model material joints (a mixture of plaster of paris, sand and water) having the same cross section and topographic features of the natural rock joint. The surfaces of the natural joint, silicone rubber cast and the model joint are shown in Fig. 1.

Since all the model material joints were made from the same silicone rubber cast, it was sufficient to measure the surface roughness of only one model material joint surface. A stylus profilograph was used to perform the roughness measurements. Roughness profiles were obtained in six directions (at every 30 degrees) on the model joint surface. In each direction, roughness profiles were obtained along nine parallel lines which are spaced at equal distance. Along each line the surface height was recorded at every 0.2 mm up to an accuracy of 0.01 mm.

Each pair of perfectly mated model material casts was used to prepare a model material shear sample to perform a direct shear test in a large Wykeham Farrance shear machine having a square cross section shear box of size 12 inches. Each shear sample was subjected to a selected normal stress in the range 0.5 - 10 kg/cm^2 and was sheared either in the forward or backward direction along one of the six directions used for roughness measurements. Using this testing scheme, a total of twelve direct shear test results were obtained for a selected normal stress. Five different normal stresses were used to obtain a total of 60 direct shear test results. All shear tests were performed until failure. The shear strength data obtained made it possible to study the anisotropic behavior of the peak shear strength of the joint resulting from anisotropic surface roughness. Four shear tests were performed under different normal stresses on smooth horizontal joint surfaces made out of the model material to estimate its basic friction angle. Two tests were performed to determine the joint compressive strength of the model material.

3 QUANTIFICATION OF JOINT ROUGHNESS

In this investigation, only the statistical parameters were considered in quantifying joint roughness. Future research will focus on investigating the possibility of using fractal theory in quantifying joint roughness.

All the statistical parameters mentioned in the Introduction were included in the investigation to select the best parameter(s) to quantify roughness. Out of these parameters, the statistical parameters associated with the slope of the profile (z_2, A_i, R_p) and SF were found to have strong correlations with JRC. Out of the slope parameters, (z_2) was found to be the best one because it changed significantly from the smoothest profile to the roughest of Barton's profiles and also showed the capability of reflecting the influence of anisotropic roughness on anisotropic peak shear strength. It is quite clear from the definitions for z_2, SF, A_i and R_p, that SF, A_i and R_p are all highly correlated to z_2. In addition, z_2 is unitless, and is better than SF, which has units. Therefore, out of the statistical parameters investigated, z_2 was found to be the best parameter.

If the joint surface is inclined (not horizontal) with respect to large-scale undulations, then the shear strength of the joint in the upward inclined direction will be different than the shear strength of the joint in the opposite direction. However, z_2 will be the same for both directions. Therefore, z_2 alone is not sufficient to quantify roughness. It seems that z_2 is a good parameter to quantify small-scale roughness. To quantify large-scale undulations, it is necessary to define another statistical parameter.

It may be important to note that all 10 of Barton and Choubey's surface profiles can be considered as stationary profiles. The stationary profiles satisfy the following second-order stationary properties: (a) The mean surface is horizontal with respect to spatial location; (b) The variance of the surface height around the mean surface is a constant with respect to spatial location; and (c) The covariance function of the surface height depends only on the lag distance irrespective of the spatial location.

However, many rock joint surface profiles can exhibit non-stationary properties. Neither JRC nor z_2 can capture the non-stationarity. At the simplest level, the non-stationarity of the joint surface can be represented by a linear function with a positive or a negative slope. This slope can be denoted by I and can be estimated through regression analysis. For most joint surfaces, this linear function will be sufficient to model the non-stationary portion of the joint surface profile. Therefore, in this investigation, to model the non-stationary part or the large-scale undulations of the surface profile, the parameter I was selected. To model the stationary part or the small-scale roughness of the surface profile, the statistical parameter z_2 was chosen.

To estimate the I of the surface in a certain direction, each I of the nine profiles in the selected direction was calculated; then, using the following formula, the weighted average value of I was estimated and used as one of the roughness parameters.

(1) $$\text{Average } I = \frac{1}{\sum_{j=1}^{9} l_j} \sum_{j=1}^{9} I_j l_j$$

where I_j = I angle estimated for *jth* surface height profile in the direction considered

l_j = length of the *jth* surface height profile in the direction considered.

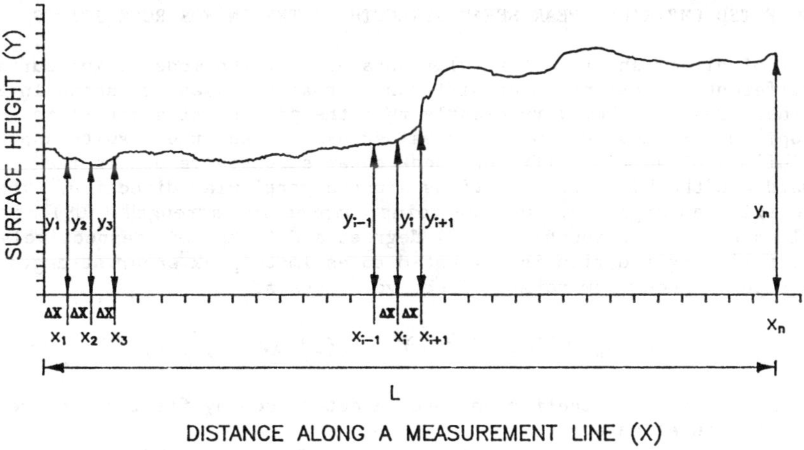

Figure 2 The diagram used to define statistical parameters for a joint surface profile.

For each profile, in a certain direction, a modified z_2 (z'_2) was calculated using the following equation:

$$z'_2 = \left[\frac{1}{n-1} \sum_{i=1}^{n-1} (((y_{i+1} - \bar{y}_{i+1}) - (y_i - \bar{y}_i))/\Delta x)^2 \right]^{1/2} \tag{2}$$

where y_{i+1} = measured surface height corresponding to x_{i+1} (See Fig. 2)
$\bar{y}_{i+1}$ = calculated surface height from linear trend equation at x_{i+1}
y_i = measured surface height corresponding to x_i (See Fig. 2)
$\bar{y}_i$ = calculated surface height from linear trend equation at x_i
$\Delta x = x_{i+1} - x_i$ for any i
n = number of y measurements made on the profile.

From equation (2) it is clear that the linear trend was removed from the measured values of surface height before calculating z'_2 values. Therefore, it provides a measure of small-scale roughness or roughness arising due to the stationary part of the surface profile. For each of the six directions considered, an average z'_2 value was computed according to the following equation:

$$\text{Average } z'_2 = \frac{1}{\sum_{j=1}^{9} l_j} \sum_{j=1}^{9} (z'_2)_j l_j \tag{3}$$

where $(z'_2)_j$ = calculated z'_2 value for jth profile in the direction considered
l_j = length of the jth profile in the direction considered.

Calculated average I and average z'_2 values clearly showed that the surface roughness of the natural joint considered is anisotropic.

4 SUGGESTED EMPIRICAL PEAK SHEAR STRENGTH CRITERION FOR ROCK JOINTS

A part of direct shear test results obtained for the model joint surface in different directions under different normal stresses are shown in Fig. 3. These results show very clearly that the peak shear strength is anisotropic and it depends on the normal stress. Also, the results show the possibility of getting different peak shear strength values in the forward and the backward directions along a particular direction. The basic friction angle (ϕ_b) and the joint compressive strength (JCS) of the model material were found to be 24 degrees and 40 kg/cm^2, respectively.

The following equation is suggested to estimate peak shear strength of a rock joint under low normal effective stresses:

$$(4) \qquad \tau = \sigma_n \tan(\phi_b + A(z'_2)^B \{\log(JCS/\sigma_n)\}^C + I)$$

where A, B and C are coefficients to be determined by fitting a curve to the experimental data.

The peak shear strength data for σ_n = 0.5, 3.2 and 10.0 kg/cm^2 were used along with the z'_2, I, ϕ_b and JCS values to estimate the coefficients A, B and C through multiple linear regression analysis. The following regression equation was obtained at a multiple R value of 0.9674.

$$(5) \qquad \tau = \sigma_n \tan(\phi_b + 25.28(z'_2)^{0.0646} \{\log(JCS/\sigma_n)\}^{0.9691} + I)$$

Equation (5) was then used to predict the peak shear strength at σ_n = 1.0 and 5.0 kg/cm^2 in different directions. These predicted peak shear strengths are compared with the measured peak shear strengths in Fig. 3. Three researchers who work in rock engineering were asked to estimate JRC values for all the roughness profiles considered in this research. The mean JRC value obtained for each profile from the three researchers were used in equation (6) to estimate an average JRC value for each of the six directions considered.

$$(6) \qquad \text{Average } JRC = \frac{1}{\sum_{j=1}^{9} l_j} \sum_{j=1}^{9} (JRC)_j l_j$$

where JRC = mean JRC value for jth profile in the direction considered
l_j = length of the jth surface height profile in the direction considered.

The average JRC value estimated for each direction was then used in Barton's shear strength equation to predict the peak shear strength of the model joint in the direction considered for σ_n = 1.0 and 5.0 kg/cm^2. These shear strengths were predicted in all the twelve directions and are shown in Fig. 3. It is clear from Fig. 3 that the shear strengths predicted by Barton's equation underestimates the actual peak shear strength significantly. Also, note that the Barton's equation is incapable of capturing the anisotropic shear strength occurring between the forward and the backward directions during shearing in any particular direction. Figure 3 shows clearly that the new equation has better capability for predicting joint peak shear strength compared to the Barton's equation.

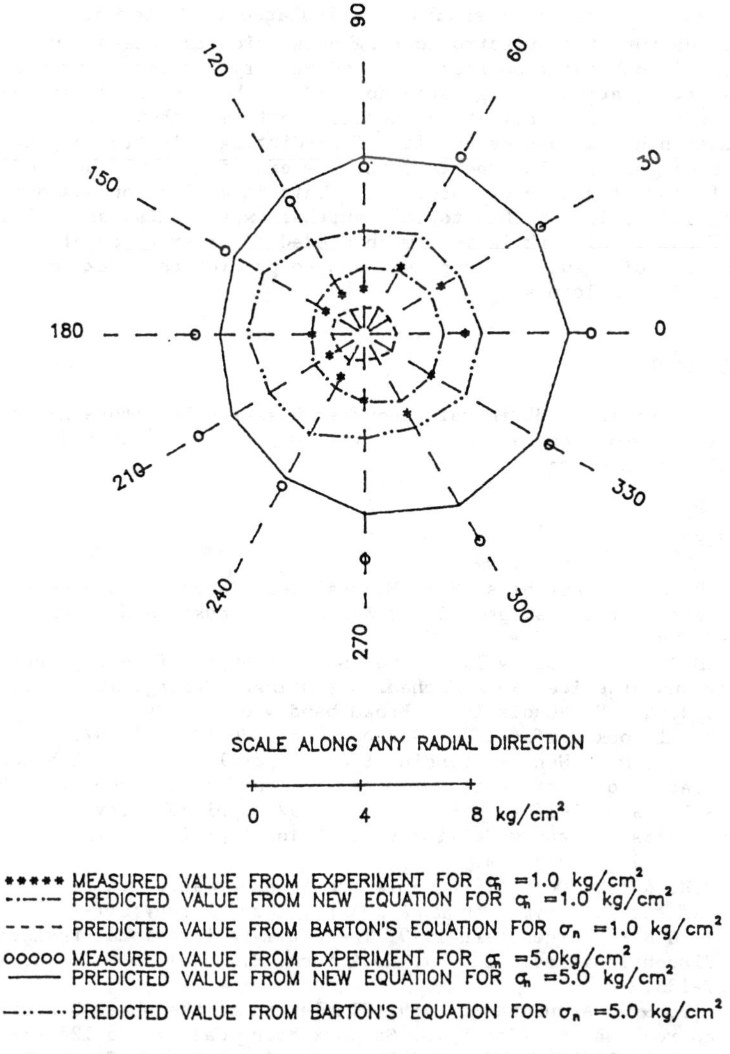

Figure 3 Comparison between measured and predicted peak shear strengths, based on the new equation and on Barton's equation.

Figure 3 also shows that the new equation has the capability of capturing the anisotropic behavior of the peak shear strength arising due to the anisotropic roughness.

5 CONCLUSIONS

The results of this investigation show clearly that at least two parameters are required to quantify joint surface roughness. By limiting only to statistical parameters, an average I angle is suggested to capture the large-scale undulations and z_2' is suggested to capture the small-scale roughness. Both z_2' and I should be calculated in different directions in order to capture the anisotropic roughness which is present in most of the natural rock joint surfaces. Based on very limited experimental data, a new equation is suggested to predict the peak shear strength of joint surfaces. The validation exercise performed shows clearly that the new equation has a good capability of predicting anisotropic peak shear strength of joints. In practice, the new equation should be used along with a factor of safety of about 1.5. This investigation was only a research initiation on this topic. Further experimental as well as theoretical/analytical work is very much needed in order to develop new tools to quantify joint surface roughness and to predict the peak shear strength of rock joints.

ACKNOWLEDGEMENT

The Arizona Mining and Mineral Resources Research Institute under Grant No. G1114104 provided partial financial support for this study. It is gratefully acknowledged.

REFERENCES

Andrle, R. & A.D. Abrahams 1989. Fractal techniques and the surface roughness of talus slopes. *Earth Surface Processes and Landforms* 14:197-209.

Barton, N.R. & V. Choubey 1977. The shear strength of rock joints in theory and practice. *Rock Mechanics (Springer-Verlag)* 10:1-54.

Brown, S.R. & C.H. Scholz 1985. Broad band width study of the topography of natural rock surfaces. *J. Geophys. Res.* 90:12575-12582.

Chesters, S., H.Y. Wen, M. Lundin, & G. Kasper 1989. Fractal based characterization of surface texture. *Applied Surface Science*, 40:185-192.

Huang, S.L., S.M. Oelfke, & R.C. Speck 1992. Applicability of fractal characterization and modelling to rock joint profiles, *Int. J. Rock Mech. Min. Sci.* 29:89-98.

Huang, T.H. & Y.S. Doong 1990, Anisotropic shear strength of rock joints. *Proc. of the Int. Symp. on Rock Joints, Loen, Norway*:211-218.

Krahn, J. & N.R. Morgenstern 1979. The ultimate frictional resistance of rock discontinuities. *Int. J. Rock Mech. Min. Sci. & Geomech. Abstr.* 16:127-133.

Ladanyi, B. & G. Archambault 1969. Simulation of the shear behavior of a jointed rock mass. *11th Symp. On Rock Mech. Cal.* 7:105-125.

Mandelbrot, B. 1983. *The fractal geometry of nature*. W. Freeman, San Francisco, 468 pp.

Maerz, N.H., J.A. Franklin, & C.P. Bennett 1990. Joint Roughness measurement using shadow profilometry. *Int. Jour. of Rock Mech. Min. Sci.*, 27/5:329-343.

Miller, S.M., P.C. McWilliams, & J.C. Kerkering 1990. Ambiguities in estimating fractal dimensions of rock fracture surfaces. *Proc. 31st U.S. Symp. on Rock Mech.* 471-478.

Power, W.L. & T.E. Tullis 1991. Euclidean and fractal models for the description of rock surface roughness. *J. Geophys. Res.* 96:415-424.

Tse, R. & D.M. Cruden 1979. Estimating joint roughness coefficients. *Int. J. Rock Mech. Min. Sci. & Geomech. Abstr.* 16:303-307.

Wu, T.H. & E.M. Ali 1978. Statistical representation of joint roughness. *Int. J. Rock. Mech. Min. Sci. & Geomech. Abstr.* 15:259-262.

10 Numerical modeling – Discontinuum analysis

Session developers:
Sitharam Thallak, The University of Texas at Austin
Jiannan Xiang, Bureau of Economic Geology

Rock Mechanics, Nelson & Laubach (eds) © 1994 Balkema, Rotterdam, ISBN 90 5410 380 8

Mechanical analysis of the accumulation of stress and slip on the San Andreas Fault at Parkfield

Deborah Hopkins
Energy and Environment Division, Lawrence Berkeley Laboratory, Berkeley, Calif., USA

William Foxall
Earth Sciences Division, Lawrence Berkeley Laboratory, Berkeley, Calif., USA

Rex Babcock
Chanute High School, Kans., USA

ABSTRACT: An analytical model, based on deformation of the half-spaces defining the fault, is used to analyze stress accumulation and slip on a segment of the San Andreas Fault near Parkfield. The study includes an analysis of the spatial and temporal relationship of 333 clustered earthquakes that occurred during the 32 month period from May, 1987, through December, 1989. The model predicts a progressive form of failure that results from mechanical interaction between contacting asperities. The results show that mechanical interaction can lead to clustering of events in time and space as is observed on the Parkfield segment. The spatial distribution of the asperities is shown to be an important factor in determining the progression of failure of the fault segment.

1 INTRODUCTION

The fault segment studied here is an approximately 420 square kilometer area on the central creeping section of the San Andreas fault, on which the 1966, magnitude 5.6, Parkfield Earthquake occurred. The creeping section of the fault lies between two locked segments that exhibit a very low level of microseismicity, and little, or no, measurable surface fault creep. In contrast, on the 150 km central creeping section, measured surface creep increases from less than one mm/yr at the ends, to more than 30 mm/yr at the center. The creep rate in the central zone is roughly equivalent to the long-term plate velocity assumed for the San Andreas Fault. The study area is located in the transition zone between the freely slipping central section, and the southern locked segment. For a complete description of the study area see Foxall (1992).

This segment of the fault is of particular interest because of a series of moderate earthquakes (magnitudes on the order of 5.5-6.0), occurring in 1857, 1881, 1901, 1922, 1934, and 1966, that suggest a periodic rupture process. Finding that these events were similar in epicenter, magnitude, seismic moment, and rupture area, Bakun and McEvilly (1984) described what they called the characteristic Parkfield earthquake. The concept of characteristic earthquakes has been used by other authors to describe fault segments that rupture repeatedly in similar earthquakes; e.g. see Shimazaki and Nakata (1980), and Schwartz and Coppersmith (1984).

Bakun and McEvilly (1984) applied both slip-prediction and time-prediction models to the data, concluding that with the exception of the 1934 earthquake, the events could be described as periodic in time. Ben-Zion et al. (1993) used three-dimensional finite element analysis to estimate recurrence intervals based on a model of the fault as an elastic upper crust over a Maxwell viscoelastic region. Assuming time dependent loading, they predict recurrence intervals that increase with time since the 1857 Fort Tejon earthquake.

In addition to the periodicity of the moderate level earthquakes, another important area of research is the study of the mechanism that triggers the events. Foxall (1992) has postulated that these repeating earthquakes result from the dynamic rupture of a major asperity which is evident in tomographic images as a high-velocity body in the lower, northwestern sector of the fault segment, and in inversions of geodetic data (see Harris and Segall, 1987).

Recent work indicates that many microseismic events on this segment also have common source parameters. Detailed analysis of these events is possible because of data recorded by the broad bandwidth, high resolution, seismic network in place at Parkfield. Approximately 50 percent of the earthquakes recorded during the period 1987-1991 have been identified by Nadeau et al. (1994) as cluster events based on their almost identical waveforms. These repeating events imply the presence of persistent asperities that may play an important role in the mechanical behavior of the fault.

The data recorded by the high resolution array at Parkfield provides a rare opportunity to study the temporal and spatial relationship of these clustered earthquakes. A model based on the deformation of the half-spaces defining the fault is used here to study the redistribution of stress, and the progression of failure, that results from the spatial distribution of the persistent asperities where multiple events have been observed. These are initial results from a study of the role of these microseismic events on the stability of the Parkfield fault segment.

2 DESCRIPTION OF THE CLUSTERED EARTHQUAKES

A total of 333 earthquakes were defined as cluster events during the 32 month study period, with magnitudes that ranged between -0.5 and 4.4. Figure 1 shows the location of these events on the fault plane for the period from May, 1987, through December, 1989, for four-month intervals. The earthquakes recorded during each interval are shown as black circles, and for ease of comparison, earthquakes occurring previously in the study period are indicated by open diamonds. The x-axis corresponds to the trace of the fault at the surface, and the y-axis measures the depth below the surface.

The analysis by Nadeau et al. identified the events plotted in Figure 1 as clustered based on their waveforms, without consideration of their time of occurrence. The plots in Figure 1 show that the events are clustered in both space and time. Out of the total of 333 events studied, 80 pairs of earthquakes occurred both sequentially in time, and at adjacent locations, suggesting that the redistribution of stress from failed asperities triggers failure at neighboring asperities. This result is consistent with the modeling results presented in Section 4.

3 DESCRIPTION OF THE MODEL

The fault is modeled as two rough half-spaces pressed together under the normal forces exerted by the bulk material on either side of the fault. Because the contacting surfaces are rough, contact occurs at discrete points, and likely varies with depth and rock type. The total contact area depends on the degree of roughness, and the total normal stress acting across the fault. The points of contact, and the normal and shear forces acting at each contact point, are calculated based on the deformation of the half-spaces defining the fault plane, and the deformation of the contacting asperities. Accounting for half-space deformation is important because it is the mechanism for mechanical interaction between contacting asperities, and describes deformation of the fault surfaces between asperities.

The asperities are modeled as disks that deform elastically. The normal deformation of the half-spaces is calculated from the Boussinesq solution for displacement below a loaded circle (Timoshenko and Goodier, 1970), and the shear displacement of the surfaces is calculated from modified solutions outlined by Johnson (1987). Assuming linear elasticity, the displacement at any point on the half-spaces is calculated as the sum of the displacements produced by the forces acting at all of the contact points. For specified far-field displacement boundary conditions, the forces acting at each asperity are calculated simultaneously as the solution that minimizes the strain energy in the system. An asperity is assumed to have failed when the shear stress is great enough to initiate sliding according to a failure criterion based on material properties and the normal stress acting on the asperity (see Jaeger and Cook, 1976). If one or more asperities fails, the minimization is repeated to recalculate the stresses because the redistribution of stress from failed asperities may cause additional failures. The process continues until either a stable state is reached (stress redistribution occurs without additional failures), or the entire plane fails. For a complete description of the model see Hopkins (1990), and for a discussion of the applicability of the model to fault planes, see Hopkins (1992).

4 RESULTS

To model the mechanical behavior of the persistent asperities, the fault segment was divided into a grid of squares with sides of length equal to 0.25 km. Persistent asperities were assumed to be located in squares where clustered events occurred during the period of study. Figure 2 shows these 154 locations, along with the number of events that occurred at each location. The difference in the number of cluster events at each location (Figure 2) indicates two important characteristics of the persistent asperities: some asperities are stronger than others, and after failure, asperities regain their ability to sustain shear stress. Neither of these features has been accounted for in this initial study of the effect of the spatial distribution on the progression of failure, but will be incorporated into future work.

The model described in Section 3 was used to calculate the normal stress acting at each asperity. Material constants were estimated from velocity profiles of the fault plane. To isolate the effect of the spatial distribution from other parameters, the asperities were assumed to be of equal dimension. The calculated normal stresses are plotted in Figure 3, where the area of the circles indicates the relative magnitude of the normal stress. Since the asperities are of equal dimension, the distribution of normal stress is determined by their spatial distribution. Because of mechanical interaction, asperities in the center of clusters have the lowest stresses, while isolated asperities have the highest stresses. Note that if mechanical interaction were not accounted for, all asperities would have equal normal stress, and would fail simultaneously (since they have equal dimensions and material constants).

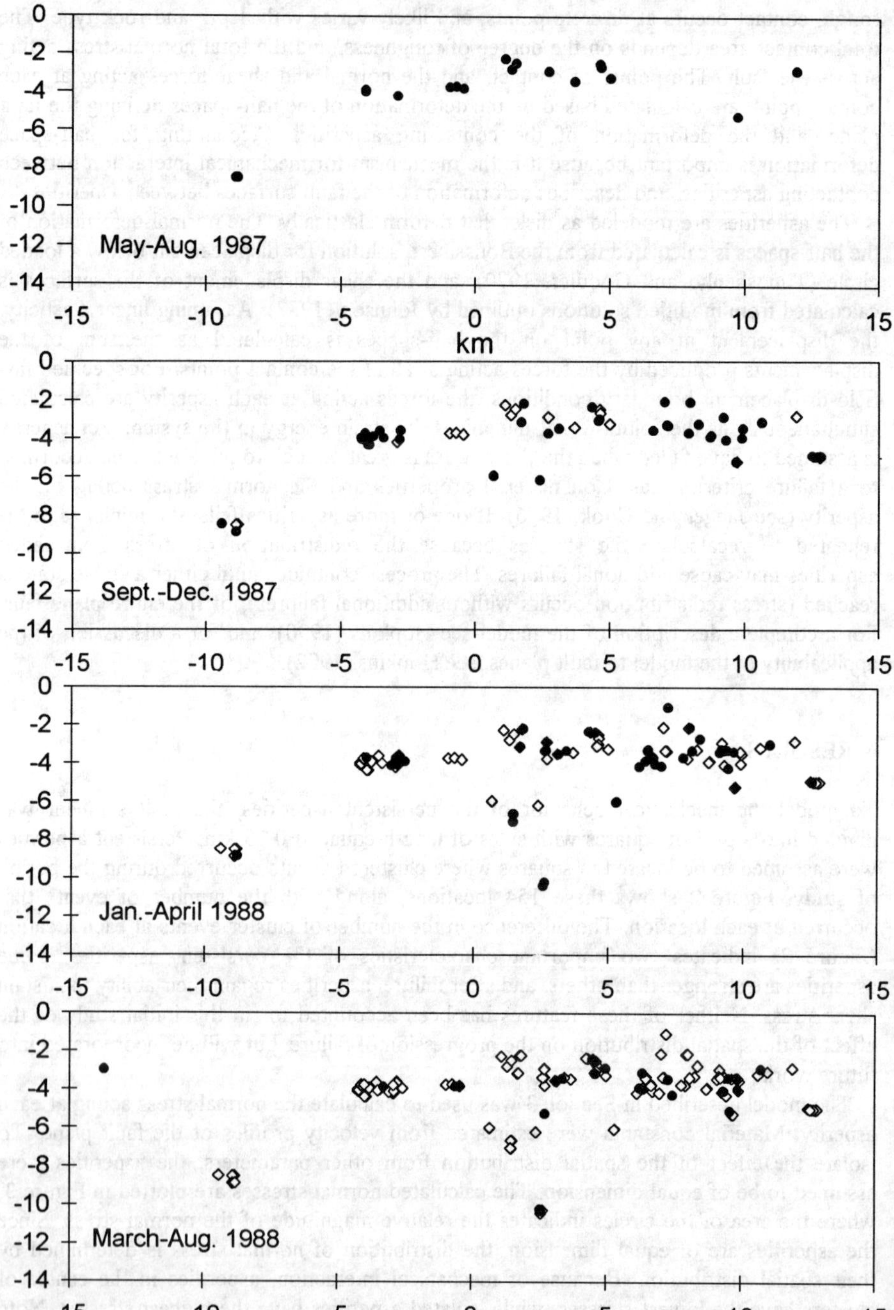

Figure 1. The location of 333 cluster events that occurred on the Parkfield fault segment during the 32 month period from May, 1987, through December, 1989, for four-month

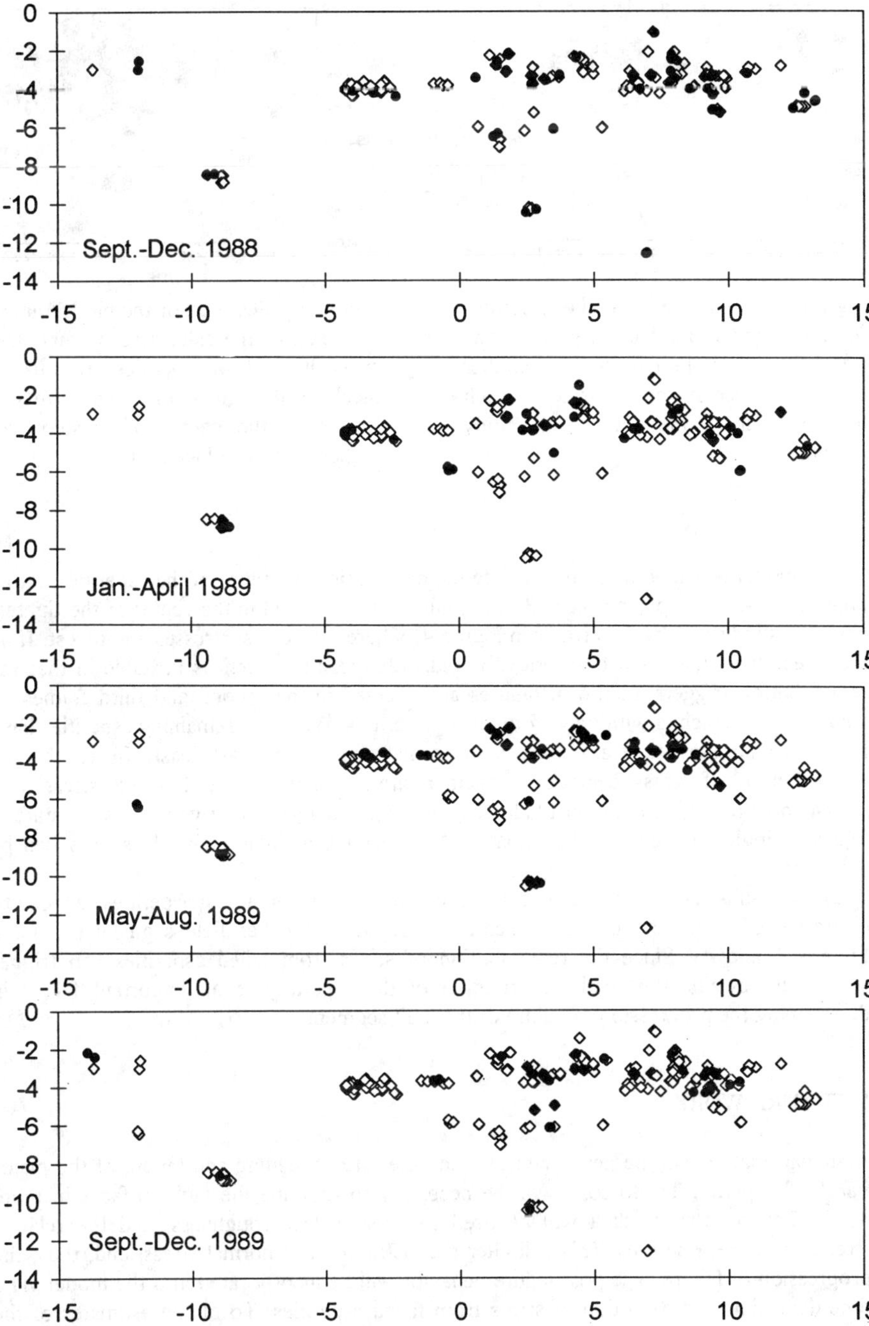

intervals. Eighty pairs of earthquakes occurred both sequentially in time, and at adjacent locations.

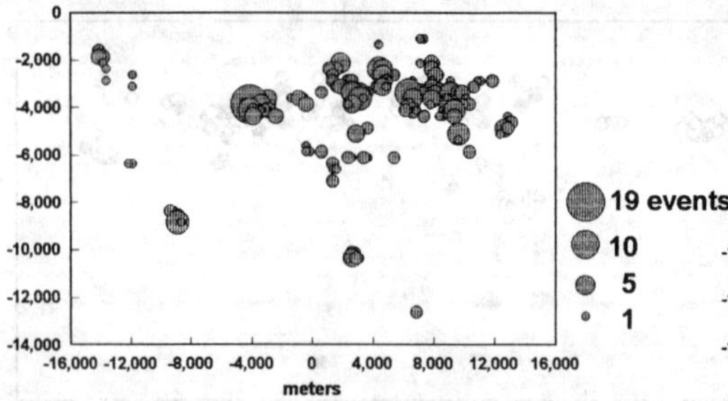

Figure 2. The location of the asperities used to model the fault segment. The area of each circle indicates the number of cluster events that occurred at each location during the 32 month study period.

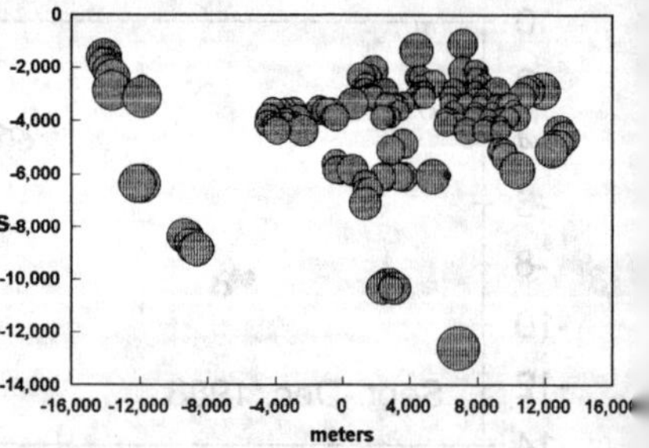

Figure 3. The area of the circles is proportional to the calculated normal stress at each of 154 asperities. Because of mechanical interaction, the stress is lowest in the center of clusters, and highest at isolated asperities.

The model was then used to calculate the progression of failure of the asperities. At an imposed far-field displacement of 35 mm, failure was initiated in the center of the tightest cluster (as indicated by the arrows in Figure 4) where the normal stresses are lowest. If all the stress from these failed asperities is redistributed, the system is unstable in that the redistribution triggers additional failures as indicated in the second and third frames. A stable state is reached with only 17 unfailed asperities. While the remaining asperities have high shear stresses, they also have high normal stresses, and are least affected by the redistribution of stress because of their relative isolation. The far-field stress was incremented to trigger additional failures (frame 5), and again to arrive at the final state in which a single unfailed asperity remains. As indicated, a single isolated asperity can be very strong.

These results show that mechanical interaction between contacting asperities can lead to clustering of events in time and space as observed on the Parkfield segment (compare Figures 1 and 4). Since the redistribution of stress from failed asperities can trigger subsequent events, the spatial distribution of the asperities is an important factor in determining the progression of failure of the fault segment.

5 FUTURE WORK

A similar analysis will be performed to study the rate of loading and failure of the major Parkfield asperity. To do so, it will be necessary to describe the fault surfaces in more detail. The occurrence data will be used to infer surface roughness, and the relative strengths of the asperities. This will alter the distribution of normal stress, and, thus, the progression of failure. The post failure behavior will be incorporated into the model via a time dependent redistribution of stress from failed asperities. To obtain estimates of the rate of loading of the major asperity, the background seismicity will have to be considered, and used along with tomographic data, to infer the location of all contact areas on the fault segment, as well as the location and size of the major asperity.

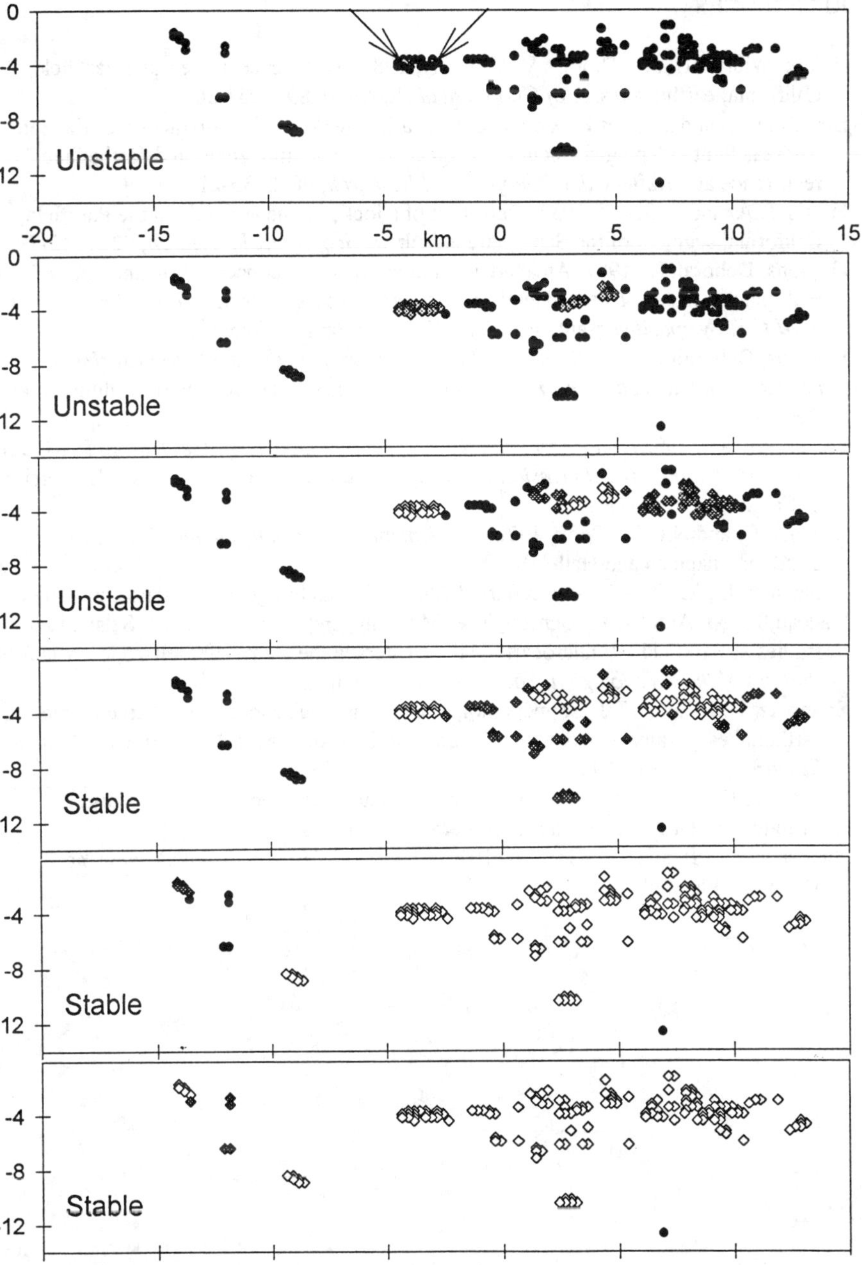

Figure 4. The progression of failure of the asperities as calculated by the deformation model. Failure is initiated in the center of the most dense cluster (indicated by the arrows) where the normal stress is the lowest. The redistribution of stress from these asperities triggers additional failures (frames 2 and 3). Isolated asperities are the last to fail because they have high normal stresses, and are least affected by the redistribution of stress.

REFERENCES

Bakun, William H. and Thomas V. McEvilly, 1984. Recurrence models and Parkfield, California, earthquakes. *J. of Geophysical Research*, 89: 3051-3058.

Ben-Zion, Yehuda, James R. Rice, and Renata Dmowska, 1993. Interaction of the San Andreas fault creeping segment with adjacent great rupture zones and earthquake recurrence at Parkfield. *J. of Geophysical Research*, 98: 2135-2144.

Harris, R.A. and P. Segall, 1987. Detection of a locked zone at depth on the Parkfield, California, segment of the San Andreas fault. *J. Geophysical Research*, 92: 7945-7962.

Hopkins, Deborah L., 1992. An analytical model for stress concentration and failure along fault planes. In J.R. Tillerson & W.R. Wawersik (eds.), *Rock Mechanics Proc. of the 33rd U.S. Symposium*, Santa Fe: 131-140. Rotterdam: Balkema.

Hopkins, Deborah Lynne, 1990. *The effect of surface roughness on joint stiffness, aperture, and acoustic wave propagation*. Ph.D. thesis, University of California at Berkeley.

Foxall, William, 1992. *Heterogeneous slip and rupture models of the San Andreas fault zone based upon three-dimensional earthquake tomography*. Ph.D. thesis, University of California at Berkeley.

Jaeger, J.C. and N.G.W. Cook, 1976. *Fundamentals of rock mechanics* (2nd ed.). London: Chapman and Hall.

Johnson, K.L., 1987. *Contact mechanics* (2nd ed.). Cambridge: Cambridge Univ. Press.

Nadeau, R., M. Antolik, P. Johnson, T.V. McEvilly, and W. Foxall, 1994. Seismological studies at Parkfield III: microearthquake clusters in the study of fault-zone dynamics. *Bulletin of the Seismological Society of America*, in press.

Schwartz, D. P. and K. J. Coppersmith, 1984. Fault behaviour and characteristic earthquakes: examples from the Wasatch and San Andreas faults. *J. of Geophysical Research*, 89: 5681-5698.

Shimazaki, K. and T. Nakata, 1980. Time-predictable recurrence model for large earthquakes. *Geophysical Research Letters*, 7: 279-282.

Timoshenko, S.P. and J.N. Goodier, 1970. *Theory of elasticity* (3rd ed.). New York: McGraw-Hill Book Co.

Rock Mechanics, Nelson & Laubach (eds) © 1994 Balkema, Rotterdam, ISBN 90 5410 380 8

Modeling fracturing of rock masses with the DDA method

B. Amadei, C.T. Lin & S. Sture
Department of Civil Engineering, University of Colorado, Boulder, Colo., USA

J. Jung
Sandia National Laboratories, Yucca Mountain Site Characterization Project, Albuquerque, N.Mex., USA

ABSTRACT: This paper presents two extensions to the Discontinuous Deformation Analysis (DDA) method originally proposed by Shi for modeling the response of blocky rock masses to mechanical loading. First, a sub-blocking capability is presented whereby blocks are discretized into sub-blocks. This sub-block capability can be used to obtain a better resolution of stress and strain within a block. An algorithm based on a three-parameter Mohr-Coulomb criterion that allows for shear or tensile fracturing of blocks into smaller blocks represents the second extension.

1 INTRODUCTION

Several numerical methods are used in rock mechanics to model the response of rock masses to mechanical loading. These methods include the Finite Element Method (FEM), Boundary Element Methods (BEM), and Discrete Element Methods (DEM). Although rock mass discontinuities can be modeled in a discrete manner using FEM and BEM, the description of discontinuities is usually difficult and there are often restrictions to the degree of deformation permitted. On the other hand, DEM methods are generally more tailored for problems in which there are many material discontinuities, placing special emphasis on how contacts are handled.

The Discontinuous Deformation Analysis (DDA) method is a recently developed technique that falls in the family of DEM methods. It was first proposed by Shi (1988) as part of his doctoral work. Computer programs based on the method were developed and applications were presented in Shi's thesis as well as in a more recent paper (Shi 1990). Over the past two years, we have explored a number of extensions to the DDA method to enhance its capability. The first extension presented in a recent paper (Lin et al. 1994), consisted of enforcing block-to-block contacts more precisely. This was done using an Augmented Lagrangian Method instead of the penalty method originally proposed by Shi (1988). The Augmented Lagrangian Method allows block contact forces to be calculated more accurately.

This paper first reviews some of the basic concepts of the DDA method. Then, two new extensions to the method are discussed. First, a sub-blocking capability is presented whereby blocks are discretized into sub-blocks. Contacts between sub-blocks are modeled using the Augmented Lagrangian Method. Then, a block fracturing algorithm is presented that allows intact blocks to be broken, creating smaller blocks. Fracturing is formulated using a three-parameter Mohr-Coulomb criterion.

2 MODELING OF JOINTED ROCK USING DDA

In the DDA method, a finite element type of problem is solved in which all the elements are physically isolated blocks bounded by pre-existing discontinuities. When blocks are in contact, Coulomb's law applies to the contact interfaces, and the simultaneous equilibrium equations are selected and solved at each loading or time increment. The large displacements and deformations are the accumulation of small displacements and deformations at each time step. Within each time step, the displacements of all points are small, hence the displacements can be reasonably represented by first order approximations. By adopting such approximations, the DDA method assumes that each block has constant stresses and strains throughout. The displacements (u,v) at any point (x,y) in a block, i, can be represented in two dimensions by six displacement variables

$$\mathbf{D_i} = (d_{1i}\ d_{2i}\ d_{3i}\ d_{4i}\ d_{5i}\ d_{6i})^T = (u_o\ v_o\ r_o\ \epsilon_x\ \epsilon_y\ \gamma_{xy})^T \quad (1)$$

where (u_0,v_0) is the rigid body translation at a specific point (x_0,y_0) within the block, r_0 is the rotation angle of the block with rotation center at (x_0,y_0) and ϵ_x, ϵ_y and γ_{xy} are the normal and shear strains in the block.

In the DDA method, individual blocks form a system of blocks through contacts among blocks and displacement constraints on single blocks. Assuming that n blocks are defined in the block system, Shi (1988) showed that the simultaneous equilibrium equations can be written in matrix form as follows

$$\begin{bmatrix} K_{11} & K_{12} & K_{13} & \cdots & K_{1n} \\ K_{21} & K_{22} & K_{23} & \cdots & K_{2n} \\ K_{31} & K_{32} & K_{33} & \cdots & K_{3n} \\ \cdot & \cdot & \cdot & \cdot & \cdot \\ \cdot & \cdot & \cdot & \cdot & \cdot \\ \cdot & \cdot & \cdot & \cdot & \cdot \\ K_{n1} & K_{n2} & K_{n3} & \cdots & K_{nn} \end{bmatrix} \begin{bmatrix} D_1 \\ D_2 \\ D_3 \\ \cdot \\ \cdot \\ \cdot \\ D_n \end{bmatrix} = \begin{bmatrix} F_1 \\ F_2 \\ F_3 \\ \cdot \\ \cdot \\ \cdot \\ F_n \end{bmatrix} \quad (2)$$

where each coefficient $\mathbf{K}_{ij}$ is defined by the contacts between blocks i and j. Since each block i has six degrees of freedom defined by the components of matrix $\mathbf{D}_i$ in equation (1), each $\mathbf{K}_{ij}$ in (2) is itself a 6 x 6 sub-matrix. Also, each $\mathbf{F}_i$ is a 6 x 1 sub-matrix that represents the loading on block i. Equation (2) can also be expressed in a more compact form as **K D** = **F** where **K** is a 6n x 6n stiffness matrix and **D** and **F** are 6n x 1 displacement and force matrices, respectively. The system of equations (2) is solved for the displacement variables. The solution is constrained by inequalities associated with block kinematics (e.g. no penetration and no tension between blocks) and Coulomb friction for sliding along block interfaces.

The penalty method was originally used in the DDA method to enforce contact constraints at block-to-block interfaces. The penalty method has three major limitations. First, the accuracy of the contact solution highly depends on the choice of the penalty number and the optimal number cannot be explicitly found beforehand. Secondly, the penalty approach satisfies the contact constraints only approximately. Finally, the contact forces must be calculated using auxiliary calculations. All three limitations can be overcome by using an Augmented Lagrangian Method as discussed by Lin et al. (1994). The essential concept behind the Augmented Lagrangian Method is to use, for each block-to-block contact both a penalty number (representing the stiffness of the contact spring) and a Lagrange multiplier (representing the contact force). To avoid increasing the number of governing equations, an iterative method is used to calculate the Lagrange multiplier until the distance of penetration of one block into the other is below a minimum specified tolerance.

3 SUB-BLOCK ANALYSIS

In the original DDA method, the first order displacement approximation implies that stresses and strains are constant within each block. Such an approach may not be well suited for some problems for which stress or displacement variations within each block are important. The accuracy of the DDA method and its ability to resolve stress variations can be increased by discretizing the blocks into finite cells or elements. Stresses within each block could be determined by coupling the DDA method to the finite element or finite difference method. As shown below, the same can also be done using a sub-block analysis where each block is divided into smaller sub-blocks. The continuity condition is preserved along the sub-block contacts and those contacts are themselves modeled using the Augmented Lagrangian Method.

Consider a large distinct block which is divided into many sub-blocks, as shown in Figure 1a. In order to satisfy the displacement compatibility inside the block, no relative displacement must be allowed along the contact interfaces between sub-blocks and all sub-blocks must move together. Consider a contact point Q at the interface between sub-blocks i and j (Figure 1b). The constrained conditions can be expressed in terms of relative displacements u and v and x, y coordinates of point Q in sub-blocks i and j as follows

$$\begin{aligned}(x_i+u_i)-(x_j+u_j) &= 0\\ (y_i+v_i)-(y_j+v_j) &= 0\end{aligned} \tag{3}$$

Using the Augmented Lagrangian Method used by Lin et al. (1994) for block-to-block contacts, two Lagrange multipliers, λ^*_x and λ^*_y and a penalty number, p, are introduced at the contact point Q in the x and y directions, respectively. At the k-th iteration, the strain energy, Π_c, resulting from the sub-block contact forces is equal to

$$\Pi_c=\Pi_{c\lambda}+\Pi_{cp}=[\lambda^*_{xk}\ \lambda^*_{yk}]\begin{bmatrix}(x_i+u_i-x_j-u_j)\\(y_i+v_i-y_j-v_j)\end{bmatrix}+\frac{p}{2}[(x_i+u_i-x_j-u_j)\ (y_i+v_i-y_j-v_j)]\begin{bmatrix}(x_i+u_i-x_j-u_j)\\(y_i+v_i-y_j-v_j)\end{bmatrix}. \tag{4}$$

In (4), $\Pi_{c\lambda}$, is the strain energy created by the iterative Lagrange multipliers λ^*_{xk} and λ^*_{yk} whereas Π_{cp}, is the strain energy created by the penalty constraint. The contribution of Π_{cp} to the equilibrium equations (2) is similar to Shi's formulation for fixing a point

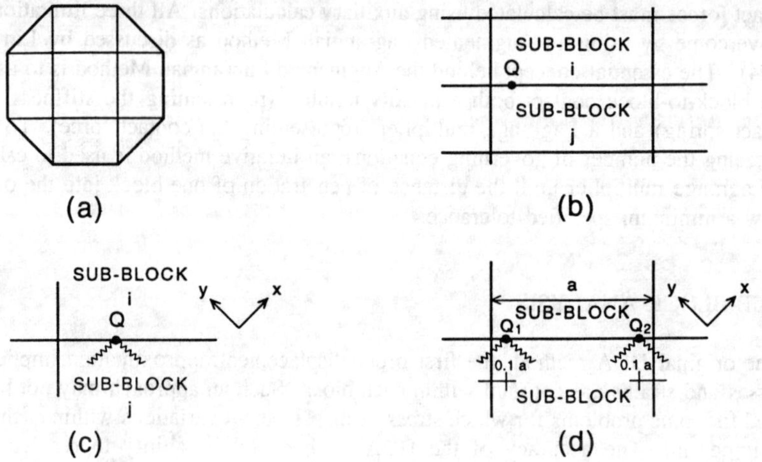

Fig.1. (a) Large block consisting of several sub-blocks, (b) Contact between sub-blocks i and j, (c) Two springs are required at contact point Q along the x and y directions, (d) A minimum of two points Q_1 and Q_2 is required along each sub-block contact.

of contact between two blocks using the penalty method (Shi, 1988). These penalty contributions are included in the left-hand side of equation (2) in the standard manner. The contribution of $\Pi_{c\lambda}$ to (2) is done in an iterative fashion by updating the **F** vector and is presented below.

As shown by Shi (1988), the displacements at point Q in blocks i and j are such that

$$\begin{bmatrix} u_i \\ v_i \end{bmatrix} = \boldsymbol{T_i}\, \boldsymbol{D_i} \quad ; \quad \begin{bmatrix} u_j \\ v_j \end{bmatrix} = \boldsymbol{T_j}\, \boldsymbol{D_j} \tag{5}$$

where $\mathbf{T_i}$ and $\mathbf{T_j}$ are two (2 x 6) matrices containing the coordinates of point Q in blocks i and j. Substituting (5) into the expression of $\Pi_{c\lambda}$ gives

$$\Pi_{c\lambda} = [\lambda^*_{xk} \quad \lambda^*_{yk}] \left\{ \boldsymbol{T_i D_i} - \boldsymbol{T_j D_j} + \begin{bmatrix} (x_i - x_j) \\ (y_i - y_j) \end{bmatrix} \right\} . \tag{6}$$

The derivatives of $\Pi_{c\lambda}$ with respect to d_{ri} and d_{rj} at 0 are defined as

$$f_{ri} = -\frac{\partial \Pi_{c\lambda}(0)}{\partial d_{ri}} = -\boldsymbol{T_i^T} \begin{bmatrix} \lambda^*_{xk} \\ \lambda^*_{yk} \end{bmatrix} \quad ; \quad f_{rj} = -\frac{\partial \Pi_{c\lambda}(0)}{\partial d_{rj}} = \boldsymbol{T_j^T} \begin{bmatrix} \lambda^*_{xk} \\ \lambda^*_{yk} \end{bmatrix} . \tag{7}$$

(with r = 1-6) which form two 6 x 1 matrices that are added to $\mathbf{F_i}$ and $\mathbf{F_j}$ in (2).

The combined effect of $\Pi_{c\lambda}$ and Π_{cp} to the equilibrium equations (2) can be summarized as follows:

$$p\ T_i^T\ T_i \rightarrow K_{ii} \tag{8}$$

is added to sub-matrix $\mathbf{K}_{ii}$ in (2),

$$-p\ T_i^T\ T_j \rightarrow K_{ij} \tag{9}$$

is added to sub-matrix $\mathbf{K}_{ij}$ in (2),

$$-p\ T_j^T\ T_i \rightarrow K_{ji} \tag{10}$$

is added to sub-matrix $\mathbf{K}_{ji}$ in (2),

$$p\ T_j^T\ T_j \rightarrow K_{jj} \tag{11}$$

is added to sub-matrix $\mathbf{K}_{jj}$ in (2),

$$-T_i^T \begin{bmatrix} \lambda_{xk}^{*}+p(x_i-x_j) \\ \lambda_{yk}^{*}+p(y_i-y_j) \end{bmatrix} \rightarrow F_i \tag{12}$$

is added to sub-matrix $\mathbf{F}_i$ in (2), and,

$$T_j^T \begin{bmatrix} \lambda_{xk}^{*}+p(x_i-x_j) \\ \lambda_{yk}^{*}+p(y_i-y_j) \end{bmatrix} \rightarrow F_j \tag{13}$$

is added to sub-matrix $\mathbf{F}_j$ in (2).

From a physical point of view, at the contact point Q between sub-blocks i and j, two springs with same stiffness p are needed along the x and y directions (Figure 1c). This is different from the contact between large blocks where only one spring normal to the contact is required.

Although all the sub-blocks inside a large distinct block can not have relative motion along their common sides, they can move and deform together without separating. All springs applied at the sub-block contacts are never removed during execution of the DDA program in order to preserve the displacement compatibility. Therefore, not only compressive but also tensile stresses can be transferred through sub-block contacts.

The sub-block methodology was implemented in the original DDA program of Shi (1988). In the new version of the program, each block is divided into sub-blocks whose number and shape are specified by the user. In order to prevent relative rotation of two adjacent sub-blocks, two contact points Q_1 and Q_2 are introduced along their interface and are arbitrarily located at 10 and 90% of the length of the contact (Figure 1d). At each contact point, two springs are applied along the x and y directions. The initial value of the spring stiffness, p, (initial penalty number) at the sub-block contacts is the same as that for the large block contacts.

As an illustrative example, Figure 2 shows an hexagonal block with a diameter of 120 ft (36.6 m) at the top of a 300 ft (90.0 m) high slope. The block contains nine sub-blocks. The intact material in the block has a Young's modulus of 2.88 x 10^8 psf (1.4 GPa), a Poisson's ratio of 0.25 and a unit weight of 130 pcf (0.02 MN/m^3). The ground consists of five blocks with a Young's modulus of 1.44 x 10^8 psf (0.7 GPa), a Poisson's ratio of 0.25 and a unit weight of 130 pcf (0.02 MN/m^3). A 150 ft (45.7 m) high rock fall barrier is placed at a distance of 350 ft (107 m) from the toe of the slope. The initial penalty number is equal to 10^9 lbs/ft (1.5 x 10^4 MN/m). Figure 2 shows the trajectory of the falling block after several time steps. The sub-block methodology maintained the continuity of the block along the sub-block interfaces during block movement.

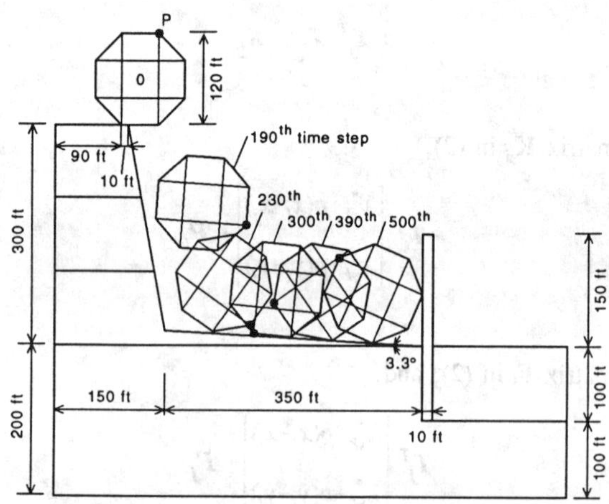

Fig.2. Rock fall example. Initial geometry and block trajectory after 190, 230, 300, 390 and 540 time steps. Each time step is equal to 0.03 seconds.

4 BLOCK FRACTURING

As a second extension of the DDA method, a fracturing capability that allows for shear or tensile fracturing of the intact block material was added to the original DDA program. This capability allows for intact blocks to be broken into smaller blocks.

The criterion selected here for block fracturing is a Mohr-Coulomb criterion with three parameters: s_o is the inherent shear strength of the block material, ϕ is its friction angle and T_o represents its tensile strength. Assuming that tensile normal stresses are positive, the three-parameter Mohr-Coulomb criterion is expressed in terms of the major and minor principal stresses σ_1 and σ_3 (with $\sigma_1 \geq \sigma_3$) as follows

$$\sigma_3 = -C_o + \sigma_1 \tan^2\left(\frac{\pi}{4} + \frac{\phi}{2}\right) \tag{14}$$

when $\sigma_3 < \sigma_{3c}$ corresponding to shear failure and

$$\sigma_1 = T_o \tag{15}$$

when $\sigma_3 \geq \sigma_{3c}$ corresponding to tensile failure. In (14), C_o is the unconfined compressive strength of the block material and is equal to $2s_o \tan(\pi/4 + \phi/2)$. The stress σ_{3c} is a transitional normal stress between the shear and tensile failure modes equal to

$$\sigma_{3c} = -C_o + T_o \tan^2(\frac{\pi}{4} + \frac{\phi}{2}) . \tag{16}$$

The three-parameter Mohr-Coulomb criterion was implemented in the original DDA program. For each block of the system, the major and minor in-plane principal stresses σ_{p1} and σ_{p3} are determined at the block's centroid. These two stress components are then compared to the third principal stress which is equal to 0 for a state of plane stress. The largest of the three principal stresses is taken as σ_1 and the smallest as σ_3.

In the program, the user divides the blocks into two groups: those that are allowed to break (breakable blocks) and those that are not (intact blocks). If condition (14) is satisfied within a breakable block, shear failure is assumed to occur with two failure planes passing through the block's centroid and inclined at $\pm$ $(\pi/4 - \phi/2)$ with respect to σ_3 and parallel to the intermediate principal stress, σ_2. Then, the block is divided into four blocks and the analysis is resumed with a new block configuration. If on the other hand, condition (15) is satisfied within a breakable block, tensile failure occurs with a failure plane passing through the block's centroid and oriented at right angle to σ_1. The block is divided into two blocks and the analysis is resumed. This process is repeated for all breakable blocks in the system. In this formulation, no energy dissipation is assumed to occur during shear or tensile failure. Upon breaking, the new blocks are assumed to have the same velocities as the original block.

As a second numerical example, the initial geometry of Figure 2 is considered again in Figure 3a. The falling block consists now of one element and is allowed to break on impact with the ground. The intact material in the block has a cohesion $s_o = 9 \times 10^4$ psf (4.3 MPa), a friction angle $\phi = 39.1°$ and a tensile strength $T_o = 6 \times 10^4$ psf (2.8 MPa). The blocks comprising the ground are not allowed to break. Figures 3b-3d show the block trajectory after several time steps.

5 CONCLUSIONS

This paper presents two extensions to the DDA method originally proposed by Shi (1988). A sub-blocking capability is used to determine stress variations in each block. This capability is based on the Augmented Lagrangian formulation. The block fracturing algorithm provides a valuable tool when looking at the mechanical fracturing of already fractured rock masses due to stress concentrations, impact or tool-rock interaction.

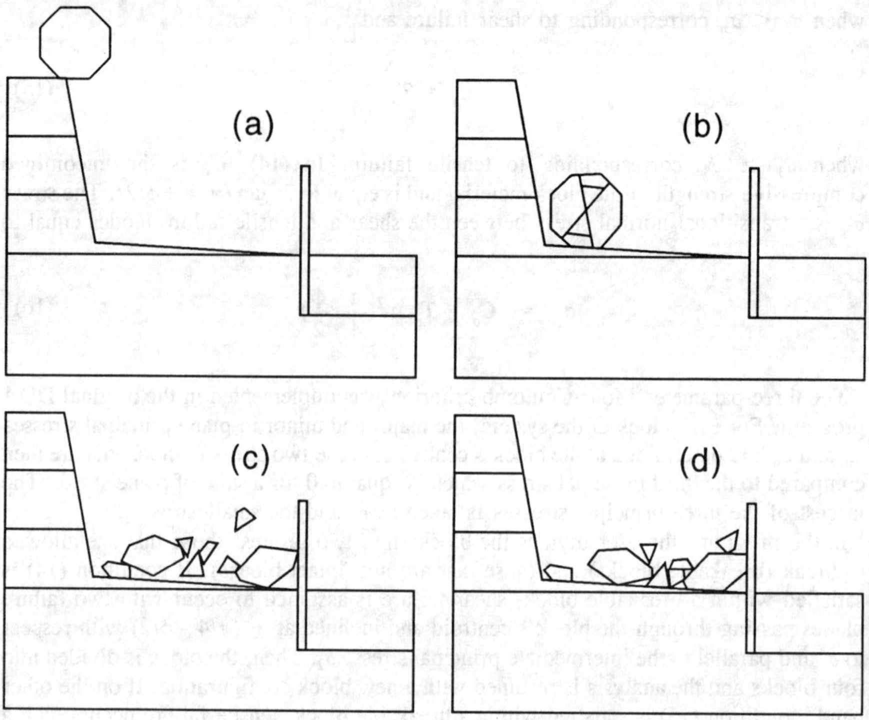

Fig.3. Rock fall example. Initial geometry in (a). Block trajectory and fracturing after 235, 355 and 400 time steps in (b)-(d). Each time step is equal to 0.03 seconds.

REFERENCES

Lin, C.T, Amadei, B., Sture, S. & Jung, J. 1994. Using an Augmented Lagrangian Method and block fracturing in the DDA method. *Proc. 8th. Int. Conf.Int. Ass. fr Computer Methods and Advances in Geomechanics*, Morgantown, WV.

Shi, G-H. 1988. Discontinuous deformation analysis. *PhD Thesis*, Univ. of California, Berkeley.

Shi, G-H. 1990. Forward and backward discontinuous deformation analyses of rock block systems. *Proc. Int. Conf. Rock Joints.* Loen (Norway): 731-743.

Acknowledgements: This work was performed under the auspices of the US Department of Energy, Office of Civilian Radioactive Waste Management, Yucca Mountain Site Characterization Project, under contract # DE-AC04-94ALB5000. The support of National Science Foundation under grant No. MS 9215397 is also greatly appreciated. The authors are grateful to Dr. Gen-Hua Shi for his comments and suggestions.

Rock Mechanics, Nelson & Laubach (eds) © 1994 Balkema, Rotterdam, ISBN 90 5410 380 8

Modeling dynamic rock failure by discontinuous deformation analysis with simplex integrations

Gen-hua Shi
Geotechnical Laboratory, US Army Engineer Waterways Experiment Station, Vicksburg, Miss., USA

ABSTRACT: The accuracy and efficiency of DDA algorithms to fulfill the basic physical laws depend upon mainly the integrations on complex rock shapes. The analytical formulae of the newly developed simplex integrations are derived and applied to DDA formulation. Simplex integrations are accurate solution on n-dimensional generally shaped domains. The integrand could be any multi-dimensional polynomials. The results depend on the coordinates of boundary vertices only. The DDA computations of three rock failure cases are presented.

1 PHYSICAL LAWS FOR ROCK FAILURE COMPUTATIONS

The failure process is a transition from continuous to discontinuous states. The discontinuous deformation analysis (DDA) has to fulfill physical laws of both continuous and discontinuous materials. When the computed prototype displacements and deformations are large enough to be visible, the mechanism of the failure and the final damage can be shown, and the ultimate strength of rock masses or structures can be intuitively estimated. The visible rock sliding and joint opening from the computation can demonstrate that the physical laws are satisfied.

The computations require equilibrium in both the discontinuous rock contacts and the continuous rock masses throughout the entire dynamic process. Following a real time sequence, the DDA uses a step by step approach. The displacements of each time step are so small that normal linear equations for small displacements can be adopted. At the end of each time step, the equilibrium in both discontinuous interfaces and continuous zones are reached.

As the step displacements are small, the kinematic relation and friction law are expressed as linear inequalities. Based on natural contact phenomena, an "entrance theory" was developed. The entrance lines are used to form linear inequality equations. The same linear inequalities are used to define the entrance distances, entrance points, entrance constraints and entrance criteria. The "open-close" iterations ensure that no tension and no penetration occur at all entrance positions. There are three entrance modes: open, sliding and locking. Coloumb's Law is also fulfilled at all entrance modes and all entrance positions.

The equilibrium of the DDA method is reached by minimizing the total potential energy. As the energy is computed by integrations, most of the DDA formulae are formed by the polynomial integrations over the generally shaped blocks. The simplex integration is developed and applied to DDA formulation. This new integration gives analytical solutions. The integrand can be multi-dimensional polynomials. In the two dimensional case, the integration domain can be any convex or concave polygons. In the three dimensional case, the integration domain can be any complex body with plane polygon boundary. The integration results are simply represented by the coordinates of boundary vertices. Based on the simplex integration, DDA algorithms are simple, efficient yet accurate. Most important, the accurate integral solution of mass matrix ensured the convergence of "open-close" iterations.

The simplex integrations can directly compute the volume and the center of gravity of a complex rock block without dividing it to simple blocks.

2 TRANSFER A NORMAL SIMPLEX TO THE COORDINATE SIMPLEX

A n dimensional simplex has $n+1$ vertices $P_0, P_1, P_2, \ldots, P_n$. The n dimensional coordinate simplex has $n+1$ vertices $U_0, U_1, U_2, \ldots, U_n$.

$$\begin{array}{llllll} P_0: & (x_{10}, & x_{20}, & x_{30}, & \ldots, & x_{n0}) \\ P_1: & (x_{11}, & x_{21}, & x_{31}, & \ldots, & x_{n1}) \\ P_2: & (x_{12}, & x_{22}, & x_{32}, & \ldots, & x_{n2}) \\ \vdots & \vdots & \vdots & \vdots & \vdots & \vdots \\ P_n: & (x_{1n}, & x_{2n}, & x_{3n}, & \ldots, & x_{nn}) \end{array} \quad \begin{array}{llllll} U_0: & (\ 0, & 0, & 0, & \ldots, & 0\ ) \\ U_1: & (\ 1, & 0, & 0, & \ldots, & 0\ ) \\ U_2: & (\ 0, & 1, & 0, & \ldots, & 0\ ) \\ \vdots & \vdots & \vdots & \vdots & \vdots & \vdots \\ U_n: & (\ 0, & 0, & 0, & \ldots, & 1\ ) \end{array} \tag{1}$$

The following coordinate transformation from $(u_1, u_2, u_3, \ldots, u_n)$ to $(x_1, x_2, x_3, \ldots, x_n)$ transfers coordinate simplex $U_0U_1U_2\ldots U_n$ to normal simplex $P_0P_1P_2\ldots P_n$.

$$\begin{array}{lcllllllll} x_1 & = & (1-\sum_1^n u_i) & x_{10} & +u_1 & x_{11} & +u_2 & x_{12} & +\ldots & +u_n \quad x_{1n} \\ x_2 & = & (1-\sum_1^n u_i) & x_{20} & +u_1 & x_{21} & +u_2 & x_{22} & +\ldots & +u_n \quad x_{2n} \\ x_3 & = & (1-\sum_1^n u_i) & x_{30} & +u_1 & x_{31} & +u_2 & x_{32} & +\ldots & +u_n \quad x_{3n} \\ \vdots & & & \vdots & & \vdots & & \vdots & & \vdots \\ x_n & = & (1-\sum_1^n u_i) & x_{n0} & +u_1 & x_{n1} & +u_2 & x_{n2} & +\ldots & +u_n \quad x_{nn} \end{array} \tag{2}$$

The Jacobi determinant is $J = \frac{D(x_1,\ x_2,\ x_3,\ \ldots,\ x_n)}{D(u_1,\ u_2,\ u_3,\ \ldots,\ u_n)} =$

$$\begin{vmatrix} x_{11}-x_{10} & x_{12}-x_{10} & x_{13}-x_{10} & \ldots & x_{1n}-x_{10} \\ x_{21}-x_{20} & x_{22}-x_{20} & x_{23}-x_{20} & \ldots & x_{2n}-x_{20} \\ x_{31}-x_{30} & x_{32}-x_{30} & x_{33}-x_{30} & \ldots & x_{3n}-x_{30} \\ \vdots & \vdots & \vdots & \vdots & \vdots \\ x_{n1}-x_{n0} & x_{n2}-x_{n0} & x_{n3}-x_{n0} & \ldots & x_{nn}-x_{n0} \end{vmatrix} = \begin{vmatrix} 1 & x_{10} & x_{20} & \ldots & x_{n0} \\ 1 & x_{11} & x_{21} & \ldots & x_{n1} \\ 1 & x_{12} & x_{22} & \ldots & x_{n2} \\ \vdots & \vdots & \vdots & \vdots & \vdots \\ 1 & x_{1n} & x_{2n} & \ldots & x_{nn} \end{vmatrix} \tag{3}$$

Simplex integration $V_{P_0P_1P_2\ldots P_n}(m_1, m_2, m_3, \ldots, m_n)$ on a n dimensional simplex $P_0P_1P_2\ldots P_n$ is defined as normal integration times the sign of determinant J.

$$\begin{aligned} & V_{P_0P_1P_2\ldots P_n}(m_1, m_2, m_3, \ldots, m_n) \\ &= sign(J)\int\int\int\cdots\int_{P_0P_1P_2\ldots P_n} x_1^{m_1}x_2^{m_2}x_3^{m_3}\cdots x_n^{m_n} \quad dx_1dx_2dx_3\ldots dx_n \\ &= sign(J)\int\int\int\cdots\int_{U_0U_1U_2\ldots U_n} x_1^{m_1}x_2^{m_2}x_3^{m_3}\cdots x_n^{m_n}|J|du_1du_2du_3\ldots du_n \\ &= J\int\int\int\cdots\int_{U_0U_1U_2\ldots U_n} x_1^{m_1}x_2^{m_2}x_3^{m_3}\cdots x_n^{m_n} \quad du_1du_2du_3\ldots du_n \end{aligned} \tag{4}$$

Denote $u_0 = 1-\sum_1^n u_i$, from the algebra, $x_l^{m_l} = (\sum_{k=o}^n u_k x_{lk})^{m_l}$

$$= \sum_{i_{l0},i_{l1},i_{l2},\ldots,i_{ln}\geq 0}^{i_{l0}+i_{l1}+i_{l2}+\ldots+i_{ln}=m_l} \frac{(i_{l0}+i_{l1}+i_{l2}+\ldots+i_{ln})!}{i_{l0}!i_{l1}!i_{l2}!\ldots!i_{ln}!} u_0^{i_{l0}}u_1^{i_{l1}}u_2^{i_{l2}}\cdots u_n^{i_{ln}}x_{l0}^{i_{l0}}x_{l1}^{i_{l1}}x_{l2}^{i_{l2}}\cdots x_{ln}^{i_{ln}} \tag{5}$$

Substituting (5) into (4), integral (6) is obtained.

$$\begin{aligned} & V_{P_0P_1P_2\ldots P_n}(m_1, m_2, m_3, \ldots, m_n) = J\int\int\int\cdots\int_{U_0U_1U_2\ldots U_n} \\ & \left(\sum_{i_{10},i_{11},i_{12},\ldots,i_{1n}\geq 0}^{i_{10}+i_{11}+i_{12}+\ldots+i_{1n}=m_1} \frac{(i_{10}+i_{11}+i_{12}+\ldots+i_{1n})!}{i_{10}!i_{11}!i_{12}!\ldots!i_{1n}!} u_0^{i_{10}}u_1^{i_{11}}u_2^{i_{12}}\cdots u_n^{i_{1n}} \quad x_{10}^{i_{10}}x_{11}^{i_{11}}x_{12}^{i_{12}}\cdots x_{1n}^{i_{1n}}\right) \\ & \left(\sum_{i_{20},i_{21},i_{22},\ldots,i_{2n}\geq 0}^{i_{20}+i_{21}+i_{22}+\ldots+i_{2n}=m_2} \frac{(i_{20}+i_{21}+i_{22}+\ldots+i_{2n})!}{i_{20}!i_{21}!i_{22}!\ldots!i_{2n}!} u_0^{i_{20}}u_1^{i_{21}}u_2^{i_{22}}\cdots u_n^{i_{2n}} \quad x_{20}^{i_{20}}x_{21}^{i_{21}}x_{22}^{i_{22}}\cdots x_{2n}^{i_{2n}}\right) \\ & \left(\sum_{i_{30},i_{31},i_{32},\ldots,i_{3n}\geq 0}^{i_{30}+i_{31}+i_{32}+\ldots+i_{3n}=m_3} \frac{(i_{30}+i_{31}+i_{32}+\ldots+i_{3n})!}{i_{30}!i_{31}!i_{32}!\ldots!i_{3n}!} u_0^{i_{30}}u_1^{i_{31}}u_2^{i_{32}}\cdots u_n^{i_{3n}} \quad x_{30}^{i_{30}}x_{31}^{i_{31}}x_{32}^{i_{32}}\cdots x_{3n}^{i_{3n}}\right) \\ & \ldots\ldots\ldots\ldots\ldots\ldots \\ & \left(\sum_{i_{n0},i_{n1},i_{n2},\ldots,i_{nn}\geq 0}^{i_{n0}+i_{n1}+i_{n2}+\ldots+i_{nn}=m_n} \frac{(i_{n0}+i_{n1}+i_{n2}+\ldots+i_{nn})!}{i_{n0}!i_{n1}!i_{n2}!\ldots!i_{nn}!} u_0^{i_{n0}}u_1^{i_{n1}}u_2^{i_{n2}}\cdots u_n^{i_{nn}} \quad x_{n0}^{i_{n0}}x_{n1}^{i_{n1}}x_{n2}^{i_{n2}}\cdots x_{nn}^{i_{nn}}\right) \\ & du_1du_2du_3\ldots du_n \end{aligned} \tag{6}$$

$$V_{P_0P_1P_2\ldots P_n}(m_1,m_2,m_3,\ldots,m_n)$$
$$= J\sum_{i_{10},i_{11},i_{12},\ldots,i_{1n}\geq 0}^{i_{10}+i_{11}+i_{12}+\ldots+i_{1n}=m_1}\quad \sum_{i_{20},i_{21},i_{22},\ldots,i_{2n}\geq 0}^{i_{20}+i_{21}+i_{22}+\ldots+i_{2n}=m_2}$$
$$\sum_{i_{30},i_{31},i_{32},\ldots,i_{3n}\geq 0}^{i_{30}+i_{31}+i_{32}+\ldots+i_{3n}=m_3}\cdots\sum_{i_{n0},i_{n1},i_{n2},\ldots,i_{nn}\geq 0}^{i_{n0}+i_{n1}+i_{n2}+\ldots+i_{nn}=m_n}$$
$$\frac{m_1!}{i_{10}!i_{11}!i_{12}!\ldots!i_{1n}!}\quad\frac{m_2!}{i_{20}!i_{21}!i_{22}!\ldots!i_{2n}!}\quad\frac{m_3!}{i_{30}!i_{31}!i_{32}!\ldots!i_{3n}!}\quad\cdots\quad\frac{m_n!}{i_{n0}!i_{n1}!i_{n2}!\ldots!i_{nn}!}$$
$$x_{10}^{i_{10}}x_{11}^{i_{11}}x_{12}^{i_{12}}\cdots x_{1n}^{i_{1n}}\quad x_{20}^{i_{20}}x_{21}^{i_{21}}x_{22}^{i_{22}}\cdots x_{2n}^{i_{2n}}\quad x_{30}^{i_{30}}x_{31}^{i_{31}}x_{32}^{i_{32}}\cdots x_{3n}^{i_{3n}}\quad\cdots\quad x_{n0}^{i_{n0}}x_{n1}^{i_{n1}}x_{n2}^{i_{n2}}\cdots x_{nn}^{i_{nn}}$$
$$\int\int\int\cdots\int_{U_0U_1U_2\ldots U_n} u_0^{i_0}u_1^{i_1}u_2^{i_2}\cdots u_n^{i_n}du_1du_2du_3\ldots du_n \tag{7}$$

where $i_l = i_{1l}+i_{2l}+i_{3l}+\cdots+i_{nl}, \qquad l=0,1,2,\ldots,n.$

3 MULTI-DIMENSIONAL SIMPLEX INTEGRATIONS

The coordinate simplex integration in (7) is another form of well known Dirichlet integration in classical analysis. The direct computation of the integration is given below.

$$V_1 = \int\int\int\cdots\int_{U_0U_1U_2\ldots U_n} u_0^{i_0}u_1^{i_1}u_2^{i_2}\cdots u_n^{i_n}du_1du_2du_3\ldots du_n$$
$$= \int\int\int\cdots\int_{u_0,u_1,u_2,\ldots,u_n\geq 0}^{u_0+u_1+u_2+\ldots+u_n=1} u_0^{i_0}u_1^{i_1}u_2^{i_2}\cdots u_n^{i_n}du_1du_2du_3\ldots du_n$$
$$= \int_0^1\int_0^{1-u_n}\int_0^{1-u_n-u_{n-1}}\cdots\int_0^{1-u_n-u_{n-1}-\cdots-u_2} u_n^{i_n}u_{n-1}^{i_{n-1}}u_{n-2}^{i_{n-2}}\cdots u_1^{i_1}u_0^{i_0}du_1du_2du_3\ldots du_n$$
$$= \int_0^1\int_0^{1-u_n}\int_0^{1-u_n-u_{n-1}}\cdots\int_0^{1-u_n-u_{n-1}-\cdots-u_3} u_n^{i_n}u_{n-1}^{i_{n-1}}u_{n-2}^{i_{n-2}}\cdots u_2^{i_2}$$
$$\left(\int_0^{1-u_n-u_{n-1}-\cdots-u_2} u_1^{i_1}\left(1-\sum_{k=1}^n u_k\right)^{i_0}du_1\right)du_2du_3\ldots du_n \tag{8}$$

After i_0 times of integration by parts, the inner integration of (8) can be computed.

$$\int_0^{1-u_n-u_{n-1}-\cdots-u_2} u_1^{i_1}\left(1-\sum_{k=1}^n u_k\right)^{i_0}du_1 = \frac{i_0!i_1!}{(i_0+i_1+1)!}\left(1-\sum_{k=2}^n u_k\right)^{i_0+i_1+1} \tag{9}$$

$$V_1 = \int_0^1\int_0^{1-u_n}\int_0^{1-u_n-u_{n-1}}\cdots\int_0^{1-u_n-u_{n-1}-\cdots-u_3} u_n^{i_n}u_{n-1}^{i_{n-1}}u_{n-2}^{i_{n-2}}\cdots u_2^{i_2}$$
$$\int_0^{1-u_n-u_{n-1}-\cdots-u_2} u_1^{i_1}\left(1-\sum_{k=1}^n u_k\right)^{i_0}du_1du_2du_3\ldots du_n$$
$$= \int_0^1\int_0^{1-u_n}\int_0^{1-u_n-u_{n-1}}\cdots\int_0^{1-u_n-u_{n-1}-\cdots-u_4} u_n^{i_n}u_{n-1}^{i_{n-1}}u_{n-2}^{i_{n-2}}\cdots u_3^{i_3}$$
$$\int_0^{1-u_n-u_{n-1}-\cdots-u_3} u_2^{i_2}(1-\sum_{k=2}^n u_k)^{i_0+i_1+1}du_2du_3du_4\ldots du_n\frac{i_1!i_0!}{(i_1+i_0+1)!}$$
$$= \int_0^1\int_0^{1-u_n}\int_0^{1-u_n-u_{n-1}}\cdots\int_0^{1-u_n-u_{n-1}-\cdots-u_4} u_n^{i_n}u_{n-1}^{i_{n-1}}u_{n-2}^{i_{n-2}}\cdots u_4^{i_4}$$
$$u_3^{i_3}(1-\sum_{k=3}^n u_k)^{i_0+i_1+i_2+2}du_3du_4du_5\ldots du_n\frac{i_0!i_1!}{(i_0+i_1+1)!}\frac{(i_0+i_1+1)!i_2!}{(i_0+i_1+i_2+2)!}$$
$$= \int_0^1\int_0^{1-u_n}\int_0^{1-u_n-u_{n-1}}\cdots\int_0^{1-u_n-u_{n-1}-\cdots-u_5} u_n^{i_n}u_{n-1}^{i_{n-1}}u_{n-2}^{i_{n-2}}\cdots u_4^{i_4}$$
$$\int_0^{1-u_n-u_{n-1}-\cdots-u_4} u_3^{i_3}(1-\sum_{k=3}^n u_k)^{i_0+i_1+i_2+2}du_3du_4du_5\ldots du_n\frac{i_0!i_1!i_2!}{(i_0+i_1+i_2+2)!}$$

$$= \int_0^1 \int_0^{1-u_n} u_n^{i_n} u_{n-1}^{i_{n-1}} (1 - u_{n-1} - u_n)^{i_0+i_1+i_2+\cdots+i_{n-2}+n-2} du_{n-1} du_n$$
$$\frac{i_0!i_1!i_2!\ldots!i_{n-2}!}{(i_0 + i_1 + i_2 + \ldots + i_{n-2} + n - 2)!}$$
$$= \int_0^1 u_n^{i_n} (1 - u_n)^{i_0+i_1+i_2+\cdots+i_{n-1}+n-1} du_n \frac{i_0!i_1!i_2!\ldots!i_{n-1}!}{(i_0 + i_1 + i_2 + \ldots + i_{n-1} + n - 1)!}$$
$$= \frac{i_0!i_1!i_2!\ldots!i_n!}{(i_0 + i_1 + i_2 + \ldots + i_n + n)!} \tag{10}$$

Substituting equation (10) into (7), the integrals of a n dimensional simplex $V_{P_0P_1P_2\ldots P_n}(m_1, m_2, m_3, \ldots, m_n)$ is obtained.

$$V_{P_0P_1P_2\ldots P_n}(m_1, m_2, m_3, \ldots, m_n)$$
$$= J \sum_{i_{10},i_{11},i_{12},\ldots,i_{1n}\geq 0}^{i_{10}+i_{11}+i_{12}+\ldots+i_{1n}=m_1} \sum_{i_{20},i_{21},i_{22},\ldots,i_{2n}\geq 0}^{i_{20}+i_{21}+i_{22}+\ldots+i_{2n}=m_2}$$
$$\sum_{i_{30},i_{31},i_{32},\ldots,i_{3n}\geq 0}^{i_{30}+i_{31}+i_{32}+\ldots+i_{3n}=m_3} \cdots \sum_{i_{n0},i_{n1},i_{n2},\ldots,i_{nn}\geq 0}^{i_{n0}+i_{n1}+i_{n2}+\ldots+i_{nn}=m_n}$$
$$\frac{m_1!}{i_{10}!i_{11}!i_{12}!\ldots!i_{1n}!} \quad \frac{m_2!}{i_{20}!i_{21}!i_{22}!\ldots!i_{2n}!} \quad \frac{m_3!}{i_{30}!i_{31}!i_{32}!\ldots!i_{3n}!} \quad \cdots \quad \frac{m_n!}{i_{n0}!i_{n1}!i_{n2}!\ldots!i_{nn}!}$$
$$x_{10}^{i_{10}} x_{11}^{i_{11}} x_{12}^{i_{12}} \cdots x_{1n}^{i_{1n}} \quad x_{20}^{i_{20}} x_{21}^{i_{21}} x_{22}^{i_{22}} \cdots x_{2n}^{i_{2n}} \quad x_{30}^{i_{30}} x_{31}^{i_{31}} x_{32}^{i_{32}} \cdots x_{3n}^{i_{3n}} \quad \cdots \quad x_{n0}^{i_{n0}} x_{n1}^{i_{n1}} x_{n2}^{i_{n2}} \cdots x_{nn}^{i_{nn}}$$
$$\frac{i_0!i_1!i_2!\ldots!i_n!}{(i_0 + i_1 + i_2 + \ldots + i_n + n)!} \tag{11}$$

where $i_l = i_{1l} + i_{2l} + i_{3l} + \cdots + i_{nl}, \qquad l = 0, 1, 2, \ldots, n.$

4 SIMPLEX INTEGRATIONS ON N-DIMENSIONAL POLYHEDRONS

Formula (11) directly gives many integrations on a simplex. One dimensional simplex P_0P_1 is a line segment. Two dimensional simplex $P_0P_1P_2$ is a triangle. Three dimensional simplex $P_0P_1P_2P_3$ is a tetrahedron. The Jacobi determinants J of a segment, a triangle and a tetrahedron are

$$J = \begin{vmatrix} 1 & x_{10} \\ 1 & x_{11} \end{vmatrix} \qquad J = \begin{vmatrix} 1 & x_{10} & x_{20} \\ 1 & x_{11} & x_{21} \\ 1 & x_{12} & x_{22} \end{vmatrix} \qquad J = \begin{vmatrix} 1 & x_{10} & x_{20} & x_{30} \\ 1 & x_{11} & x_{21} & x_{31} \\ 1 & x_{12} & x_{22} & x_{32} \\ 1 & x_{1n} & x_{2n} & x_{33} \end{vmatrix} \tag{12}$$

respectively.

$$V_{P_0P_1}(m) = sign(J) \int_{P_0P_1} x_1^m dx_1 = \frac{1}{m+1}(x_{11}^{m+1} - x_{10}^{m+1}) \tag{13}$$
$$V_{P_0P_1P_2}(0,0) = sign(J) \int\int_{P_0P_1P_2} dx_1 dx_2 = J \int\int_{U_0U_1U_2} du_1 du_2 = \frac{1}{2}J \tag{14}$$
$$V_{P_0P_1P_2}(1,0) = sign(J) \int\int_{P_0P_1P_2} x_1 dx_1 dx_2 = \frac{1}{6}J(x_{10} + x_{11} + x_{12}) \tag{15}$$
$$V_{P_0P_1P_2}(2,0) = sign(J) \int\int_{P_0P_1P_2} x_1^2 dx_1 dx_2$$
$$= \frac{1}{12}J(x_{10}^2 + x_{11}^2 + x_{12}^2 + x_{10}x_{11} + x_{10}x_{12} + x_{11}x_{12}) \tag{16}$$
$$V_{P_0P_1P_2}(1,1) = sign(J) \int\int_{P_0P_1P_2} x_1 x_2 dx_1 dx_2 = \frac{1}{24}J(2x_{10}x_{20} + 2x_{11}x_{21} + 2x_{12}x_{22}$$
$$+ x_{10}x_{21} + x_{10}x_{22} + x_{11}x_{20} + x_{11}x_{22} + x_{12}x_{20} + x_{12}x_{21}) \tag{17}$$
$$V_{P_0P_1P_2P_3}(0,0,0) = \frac{1}{6}J \tag{18}$$

One dimensional simplex integration (13) is same as conventional integration. Formulae (14)-(17) are 2-d simplex integrations over triangles. Since simplex integrations always have the Jacobian J as factor and J is an oriented area, the integrations on positive area and negative area can be neutralized. Denote $P_1P_2P_3\cdots P_n$, $P_{n+1}=P_1$ with $P_i=(x_{1i},x_{2i})$ as a polygon, the integrations (14)-(17) on $P_1P_2P_3\cdots P_n$ are computed by summations. (19)-(22) are the integrations of DDA method. Here P_0 can be any point, $P_0=(0,0)$ are chosen in order to have simpler formulae. Integration (18) is the volume of a tetrahedron. By summations of simplex integrations (18), the volume of any 3-d block can be computed. Assume $P_1^{[i]}P_2^{[i]}P_3^{[i]}\cdots P_{n(i)}^{[i]}$, $P_{n(i)+1}^{[i]}=P_1^{[i]}$ with $i=1,2,3,\ldots,s$ are all outward rotated polygons of a block, $P_j^{[i]}=(x_{1j}^{[i]},x_{2j}^{[i]},x_{2j}^{[i]})$ and $P_0=(0,0,0)$. The volume of this block is given by (23). Computed by simplex integrations, integrals (19)-(23) are represented by the coordinates of the boundary vertices only.

$$\iint_{(A)} dx_1dx_2 = \sum_{k=1}^{n} V_{P_0P_kP_{k+1}}(0,0) = \frac{1}{2}\sum_{k=1}^{n}\begin{vmatrix} x_{1\ k} & x_{2\ k} \\ x_{1\ k+1} & x_{2\ k+1}\end{vmatrix} \tag{19}$$

$$\begin{aligned}\iint_{(A)} x_1dx_1dx_2 &= \sum_{k=1}^{n} V_{P_0P_kP_{k+1}}(1,0) \\ &= \frac{1}{6}\sum_{k=1}^{n}(x_{1\ k}+x_{1\ k+1})\begin{vmatrix} x_{1\ k} & x_{2\ k} \\ x_{1\ k+1} & x_{2\ k+1}\end{vmatrix}\end{aligned} \tag{20}$$

$$\begin{aligned}\iint_{(A)} x_1^2dx_1dx_2 &= \sum_{k=1}^{n} V_{P_0P_kP_{k+1}}(2,0) = \frac{1}{12}\sum_{k=1}^{n} \\ &(x_{1\ k}^2+x_{1\ k+1}^2+x_{1\ k}x_{1\ k+1})\begin{vmatrix} x_{1\ k} & x_{2\ k} \\ x_{1\ k+1} & x_{2\ k+1}\end{vmatrix}\end{aligned} \tag{21}$$

$$\begin{aligned}\iint_{(A)} x_1x_2dx_1dx_2 &= \sum_{k=1}^{n} V_{P_0P_kP_{k+1}}(1,1) = \frac{1}{24}\sum_{k=1}^{n}(2x_{1\ k}x_{2\ k}+2x_{1\ k+1}x_{2\ k+1} \\ &+x_{1\ k}x_{2\ k+1}+x_{1\ k+1}x_{2\ k})\begin{vmatrix} x_{1\ k} & x_{2\ k} \\ x_{1\ k+1} & x_{2\ k+1}\end{vmatrix}\end{aligned} \tag{22}$$

$$\begin{aligned}\iint_{(V)} dx_1dx_2dx_3 &= \sum_{i=1}^{s}\sum_{k=1}^{n(i)} V_{P_0P_1^{[i]}P_k^{[i]}P_{k+1}^{[i]}}(0,0,0) \\ &= \frac{1}{6}\sum_{i=1}^{s}\sum_{k=1}^{n(i)}\begin{vmatrix} x_{1\ 1}^{[i]} & x_{2\ 1}^{[i]} & x_{3\ 1}^{[i]} \\ x_{1\ k}^{[i]} & x_{2\ k}^{[i]} & x_{3\ k}^{[i]} \\ x_{1\ k+1}^{[i]} & x_{2\ k+1}^{[i]} & x_{3\ k+1}^{[i]}\end{vmatrix}\end{aligned} \tag{23}$$

5 DYNAMIC ROCK FAILURE COMPUTATIONS BY DDA

Three cases of DDA computations are presented in the following. There are 1700 to 2300 rock blocks in each case. The dimensions of computed regions are about 40 meters. The numbers of time steps are from 300 to 600. The total elapsed times are from 0.6 to 1.0 second. The maximum total displacements are more than ten times the average block diameter.

ACKNOWLEDGEMENT The development of the theory presented herein is the result of research for Work Unit No. 31700. Program sponsored by the US Army Corps of Engineers (USACE). The author would like to thank Mr. Jerry Huie and Dr. Don Banks for their helps.

REFERENCES

Shi, Gen-hua 1993. Block system modeling by discontinuous deformation analysis. Southampton UK and Boston USA

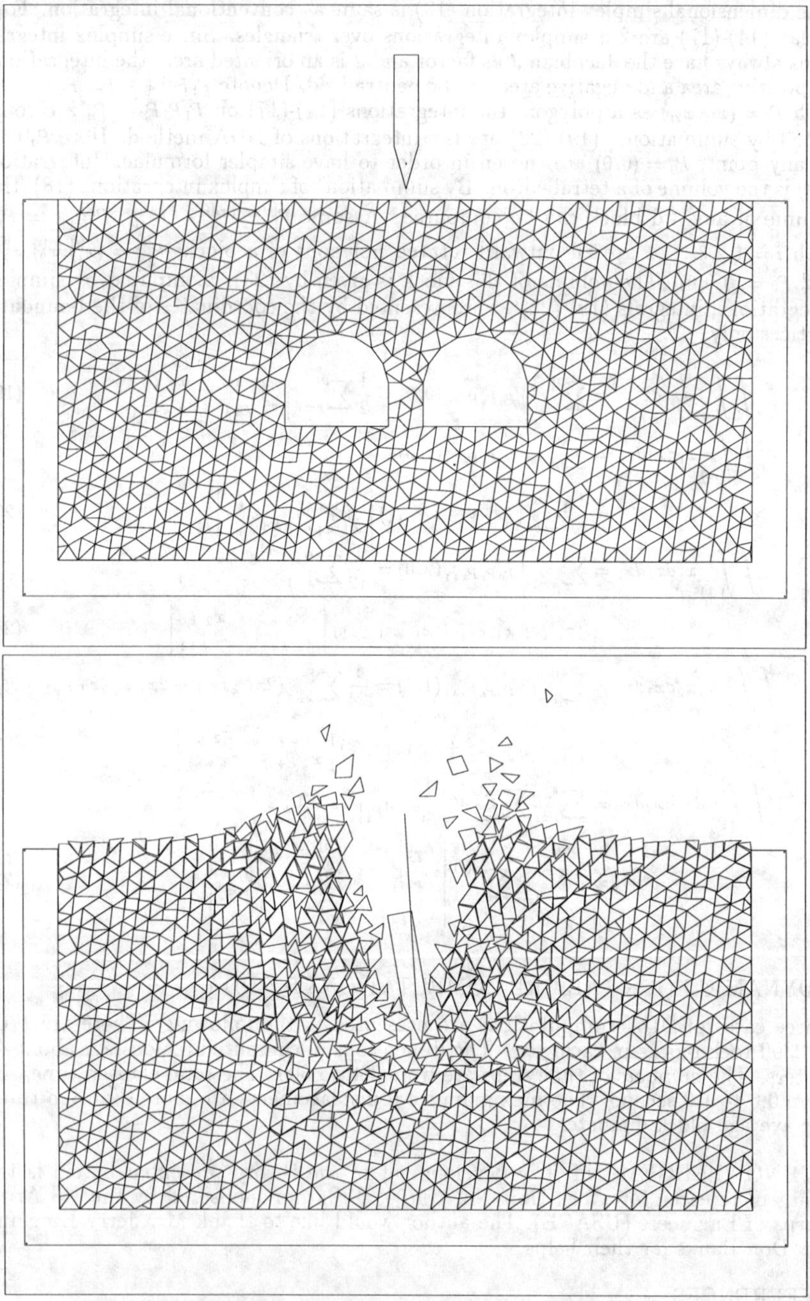

Figure 1. The penetration of a missile at a velocity of 300 meters per second into a blocky rock mass with two tunnels

Figure 2. A strong stress wave passing through three tunnels excavated in a blocky rock mass

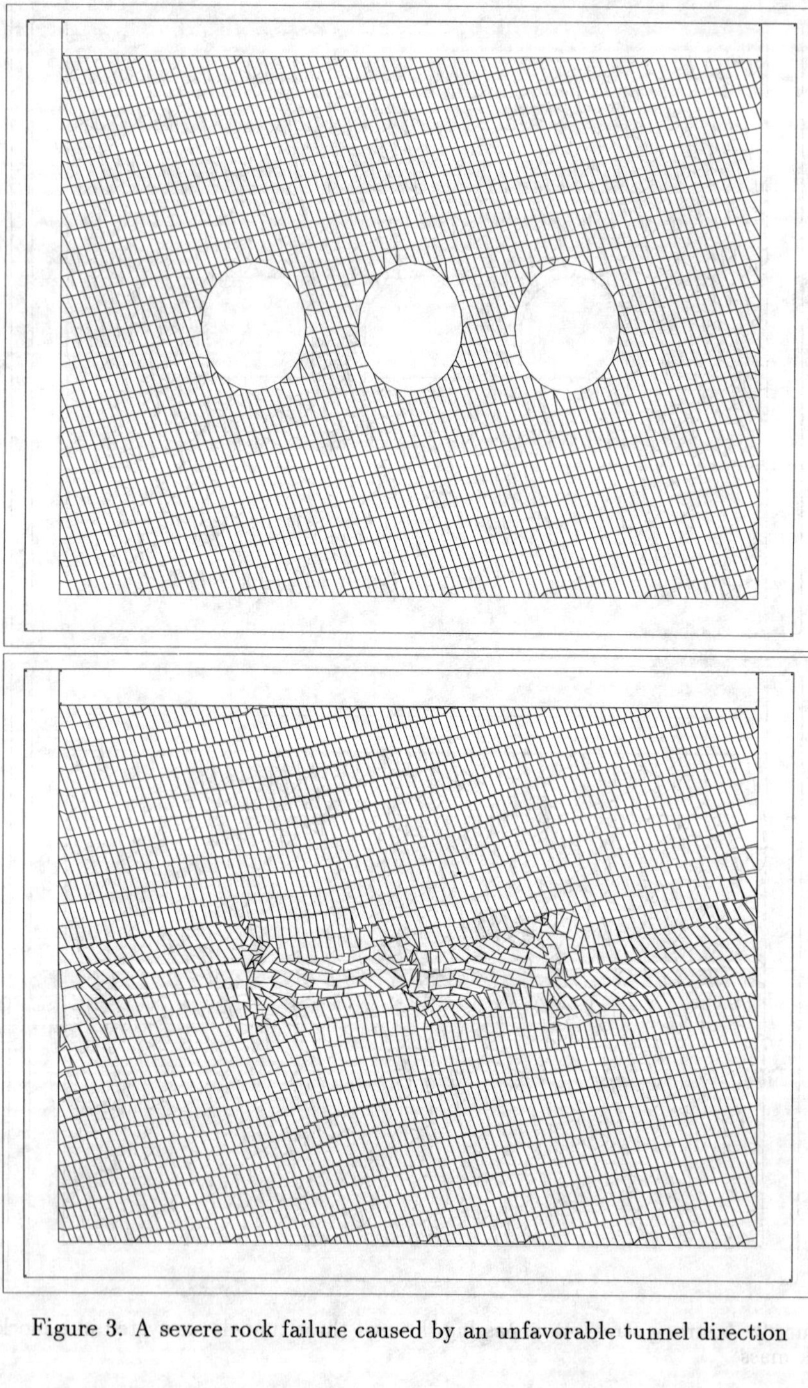

Figure 3. A severe rock failure caused by an unfavorable tunnel direction

Rock Mechanics, Nelson & Laubach (eds) © 1994 Balkema, Rotterdam, ISBN 90 5410 380 8

Discontinuous deformation analysis and 'the artificial joint concept'

T.-C. Ke & R. E. Goodman
Department of Civil Engineering, University of California, Berkeley, Calif., USA

ABSTRACT: This paper presents new capabilities of the discontinuous deformation analysis (DDA) when combined with "the artificial joint concept". These capabilities refine the determination of the refined stress distribution and permit simulating fracture propagation in intact blocks.

1 INTRODUCTION

Discontinuous deformation analysis (DDA), was introduced by Shi (1988) with constant stress within each block of a discrete block system. This paper describes an innovation for DDA modelling - "the artificial joint concept" that computes variable stress states within blocks, thereby allowing modelling of block problems with complex blocks containing holes, interior blocks, or cracks. The latter capability permits analysis of problems of fracture propagation across previously intact blocks. The method involves the addition of "artificial joints" which subdivide blocks. The artificial joints are assumed to be infinitely strong if the intra-block stress field is simply to be detailed within a block, but not perturbed. Assignment of finite strengths to the artificial joints allows perturbation of the block stresses as new fractures form and propagate.

2 DISCONTINUOUS DEFORMATION ANALYSIS

A system of simultaneous equations in DDA is formulated by minimizing the total potential energy of a block system. Inertia terms are included; the global stiffness matrix is positive definite and diagonally dominant. If a 'static' mode is chosen, the velocities of all the blocks are set to zero at the start of every time step and DDA resembles a pseudo-static analysis. If a 'dynamic' mode is specified, DDA yields a real-time response of a block system. DDA incorporates kinematics to fulfill constraints of no interpenetration of blocks and no tension between blocks[1]. The interactions between blocks are simulated by contact springs.

Shi's original DDA code can handle a system of linearly elastic blocks subjected to

[1] the 'no-tension' constraint can be relaxed for incipient joints which are assigned a tensile strength.

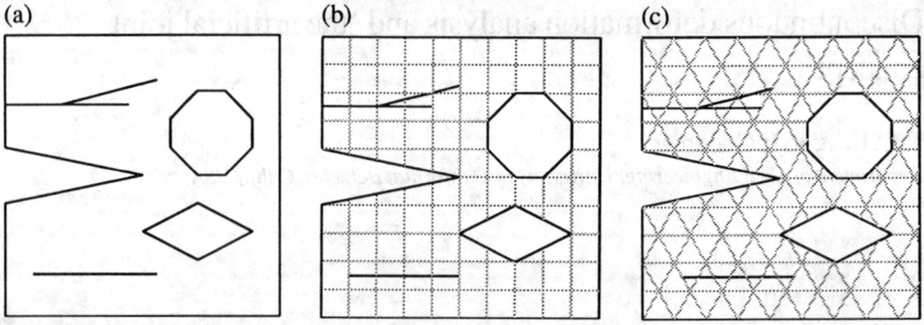

Figure 1 (a) A block with complex geometry and internal features; and refined geometry by: (b) Type A artificial joints; (c) Type C artificial joints

large displacements, rotations and deformations, and general loading conditions. However, it does not analyze fracture propagation because the modelled blocks are not allowed to degrade during the process of deformation. Further, since DDA uses a first-order displacement function, each block is modelled by a single constant stress state, no matter how complicated its geometry. Figure 1a shows a complex block with a notch at the left, one isolated crack and another with a branching crack, and two interior boundaries. (The spaces within the interior boundaries may be holes or other blocks.) Obviously it is unrealistic to assume this complex region has a single constant stress state. Under a certain loading condition, fracture may initiate from near or at the crack tips or the notch (Figure 1a).

3 THE ARTIFICIAL JOINT CONCEPT

The artificial joint concept includes two parts: *cutting* and *patching*. Cutting is a process of dividing a block into sub-blocks by artificial joints so that the block becomes an assemblage of sub-blocks created by imaginary boundaries. Patching is a process of assigning strengths to the artificial joints to *glue* sub-blocks together.

If the artificial joints are infinitely strong, the contacts between sub-blocks never fail, leading to continuous connection of two adjacent sub-block by contact springs. As the stiffness of contact springs increases, the contact sides of two adjacent sub-blocks will displace similarly. As a result, the assemblage of sub-blocks behaves as a continuous body like the original block but with refined stress distribution since DDA computes a constant stress state for each sub-block.

If the artificial joints are assigned finite strengths, they provide potential cracking paths wherever contact forces exceed limiting values. Therefore, fracture propagation of an intact block along a predestined path can be simulated. The failure criterion for (τ,σ) along the contacts between sub-blocks follows Coulomb's law with a tensile strength "cut-off". Ke (1993) added to the DDA code the procedures to deal with joints having cohesion and tensile strength.

Artificial joints intersecting the cracks, or boundaries of holes and interior blocks, allow these complex features to be modelled. In this case, after artificial joints have been added and a block system generated, DDA computes the contact behavior along real joints as well along artificial joints. Figures 1b and 1c show the resultant sub-block

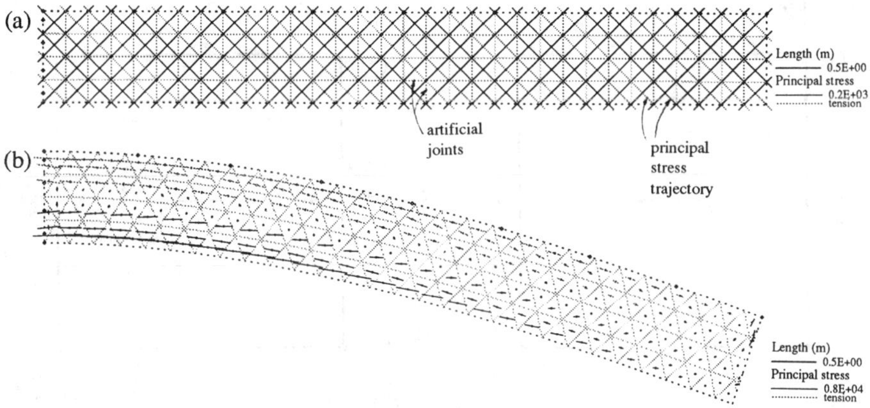

Figure 2 Principal stress field of: (a) Type A refined beam; (b) Type C refined beam

systems of the block in Figure 1a by adding artificial joint networks. Type A artificial joint network (Figure 1b) contains two perpendicular joint sets and generates rectangular sub-blocks; Type C artificial joint network (Figure 1c) contains three joint sets to produce equilateral triangular sub-blocks.

4 NUMERICAL EXAMPLES

The applicability of DDA combined with the artificial joint concept is illustrated by the following four examples. The 'static' analysis of DDA is used in the first three; while the 'dynamics' analysis of DDA is used in the last where a progressive failure is involved.

4.1 *Cantilever Beam*

This example examines the performance of Types A and C artificial joint networks for stress analysis in a cantilever beam. The beam, 8m long and 1m high, is fixed at its left end and is loaded at its right tip. The Young's modulus and Poisson's ratio of the beam are 100 MPa and 0.2, respectively; the tip load is 100 kN. Type A artificial joints divide the beam into 128 sub-blocks; while Type C artificial joints cut the beam into 228 sub-blocks. The added artificial joints have infinite strength.

Figure 2a shows the deformed geometry and principal stress field[2] of the Type A refined beam. The results are unsatisfactory; all sub-blocks have the same tensile principal stress at -45° and the same compressive stress at 45°. The whole beam responded in simple shear rather than bending, because the deformed shape of the rectangular sub-blocks generated by Type A artificial joint network is restrained by, the

[2] the principal stress field described in this paper means the DDA graphic output showing the principal stresses of each block at its centroid; compressive stresses are indicated by solid lines and tensile stresses by dotted lines.

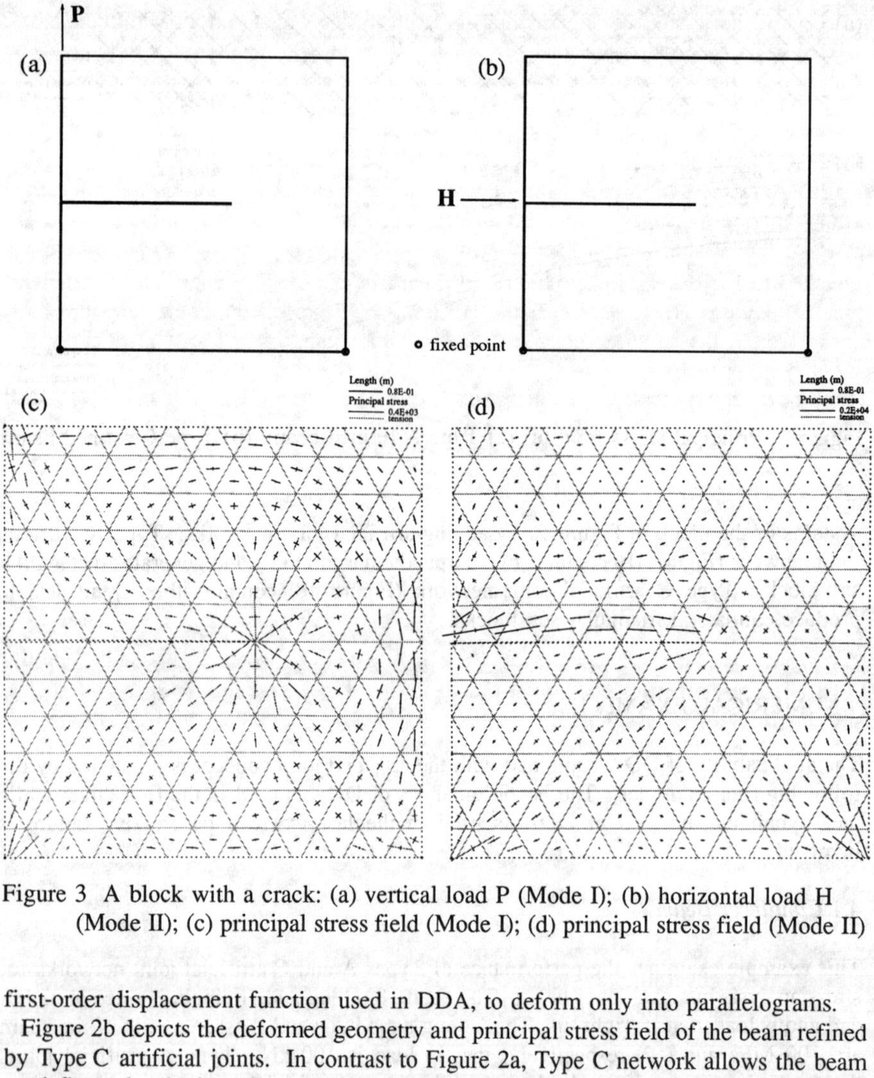

Figure 3 A block with a crack: (a) vertical load P (Mode I); (b) horizontal load H (Mode II); (c) principal stress field (Mode I); (d) principal stress field (Mode II)

first-order displacement function used in DDA, to deform only into parallelograms.

Figure 2b depicts the deformed geometry and principal stress field of the beam refined by Type C artificial joints. In contrast to Figure 2a, Type C network allows the beam to deflect, developing a neutral axis with tensile fiber stress in its upper part and compressive fiber stress in its lower part. This example shows the adequacy of Type C artificial joints to detail beam structures. Ke (1993) examined other cases with smaller deflection, in which Type C artificial joint network yielded results agreeing with those calculated using classical simple beam theory.

4.2 Stress Concentration Near a Crack Tip

Analytical solutions to two-dimensional crack problems for various loading conditions show that the stresses at a small distance r from the crack tip always vary as $r^{-1/2}$ (Rice, 1968). Two behavioral modes of a crack, Mode I (in-plane opening) and Mode II (in-plane shearing), are investigated here. Figures 3a or b show a block, 1m high and 1m wide, with a horizontal crack extending six tenths of the way across its width; the block is fixed at its two bottom corners. In Figure 3a, a vertical load *P* pulls the upper-left corner of the block upward, causing crack opening (Mode I); in Figure 3b, a horizontal load *H* at the left end of the top crack wall causes sliding (Mode II crack deformation). The block has the same Young's modulus and Poisson's ratio as in the cantilever beam example and the crack is assigned zero friction to signify the sliding behavior in Mode II. The cracked block was refined into 252 sub-blocks by Type C artificial joints with infinite strength.

Figure 3c shows the deformed geometry and principal stress field of the cracked block subjected to P=10 kN. In this figure, the crack is open, and higher values of tensile principal stresses are present near the crack tip in the radial direction. If more sub-blocks are used, the stress concentration near the crack tip in Mode I will be more precisely described.

Figure 3d depicts the deformed geometry and principal stress field of the cracked block subjected to H=100 kN. The adjacent crack walls are in contact at the tip and on the left end, but open between these two points. Higher compressive principal stresses run from the loading point, along the top crack wall, around the tip, then attenuate finally to merge toward the lower-right fixed corner; tensile stress is distributed below the crack wall and concentrated near the lower-left fixed corner.

4.3 Block-in-Matrix Material

A block-in-matrix material contains isolated (interior) blocks which are embedded within a matrix. The properties of the matrix and interior blocks are different; the interfaces between the matrix and interior blocks are potential discontinuities. The macro-mechanical behavior of a block-in-matrix material is a function of the proportion, orientation, arrangement, and size of interior blocks and, especially, the interface strengths.

Figure 4 shows two specimens from a particular case of block-in-matrix materials, *melange*, found in the subduction zones paralleling present and ancient convergent plate margins. Melange is distinguished by its oriented lenticular interior blocks. The specimen in Figure 4a contains 19 horizontally oriented blocks (shaded areas) which occupy 40% of the total "volume"; that of Figure 4b has 27 vertically oriented blocks (shaded areas) covering 58% of the "volume".

Young's moduli of the matrix and interior blocks are 250 and 420 MPa, respectively; both materials have Poisson's ratio equal to 0.4. The interfaces are assumed to be frictionless. A numerical uniaxial compression "test" was conducted on both specimens (Figure 4) by adding stiff loading plates on the top and bottom of the "specimen". The bottom plate is fixed; a vertical load, *P*, is applied to the top of the upper plate. The displacements of selected points around the specimens were used to compute their overall secant moduli and Poisson's ratios[3]. The matrix domain was refined by Type C artificial joints with infinite strength.

Table 1 lists the DDA results of two specimens, in which moduli are expressed in units of MPa. Both specimens behave non-linearly, indicated weakly by the computed secant moduli and more convincingly by Poisson's ratios[3]. The non-linear behavior comes from the large number of frictionless interfaces that undergo sliding in many

---

[3] ratios of average horizontal to vertical displacements revealed by the output.

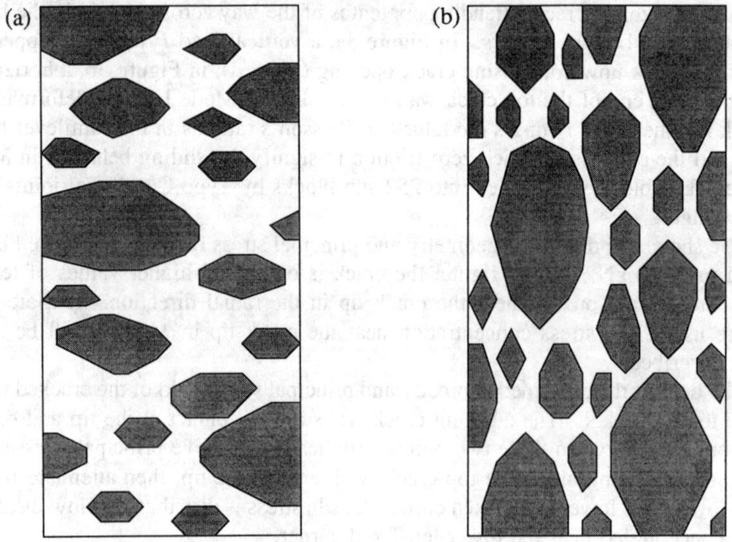

Figure 4 Melange specimens: (a) specimen 1 with 19 horizontally oriented blocks; (b) specimen 2 with 27 vertically oriented blocks

directions, leading to continuous reduction in stiffness with an increasing load. Some of the interfaces intersect the outer boundaries, especially on the free side walls of the specimens (Figure 4), further degrading the overall stiffness. Table 1 shows that although the block proportion of specimen 2 is higher than that of specimen 1 by 18%, the increase in modulus is not proportionally large. This is probably because specimen 2 loses some stiffness due to increased chance for sliding along longer frictionless interfaces.

Table 1 Computed secant moduli and Poisson's Ratios in Example 3

	P=1(kN)	P=2(kN)	P=5(kN)	P=10(kN)
Secant moduli: specimen 1	298.5	298.0	297.9	297.9
Secant moduli: specimen 2	312.2	310.3	309.9	309.9
Poisson's ratios: specimen 1	0.4819	0.5165	0.5262	0.5205
Poisson's ratios: specimen 2	0.5986	0.6595	0.8422	0.8755

4.4 An Arch Dam Subjected to Impact

As an example of fracture propagation capabilities, cracking and destruction of an arch dam subjected to enormous impact is simulated here. Figure 5a gives a plan view of an arch dam abutting on jointed rock and a huge block at t=0 approaching the dam in the indicated direction. Points 1 to 11 are fixed. Because an arch dam is always made

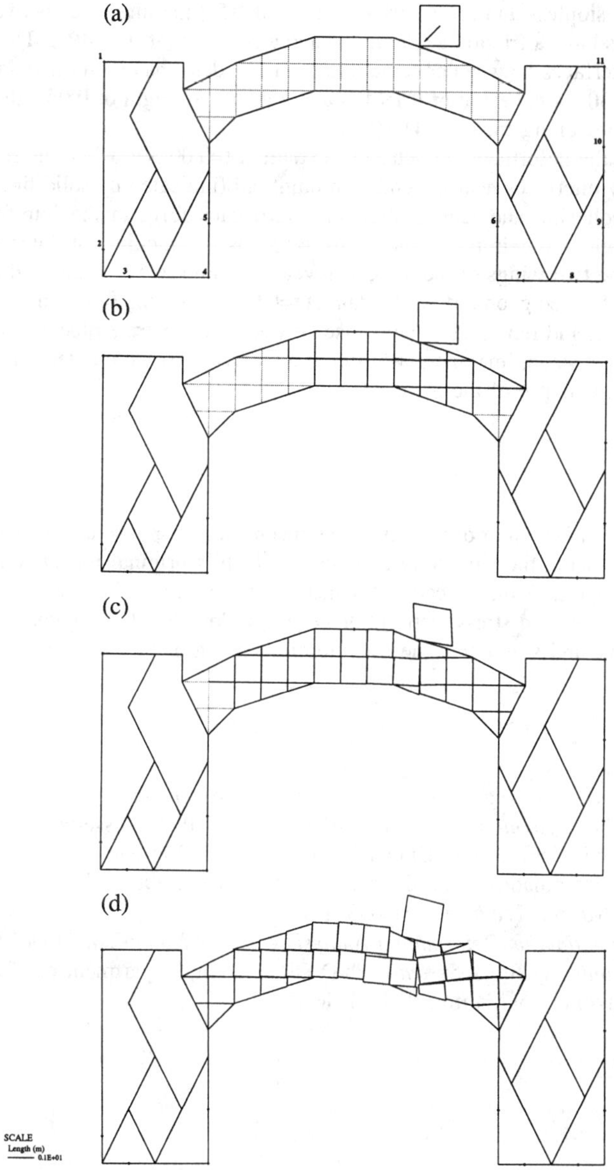

Figure 5 Cracking of an arch dam at t=: (a) 0; (b) 0.002; (c) 0.003; (d) 0.01sec

of rectangular concrete blocks, Type A artificial joints were used to refine the dam into 36 concrete sub-blocks, as depicted in Figure 5a. Each concrete sub-block was not further refined at the stage of this study.

Young's modulus and Poisson's ratio of the dam are 1000 MPa and 0.2, respectively, and those of the abutments are 200 MPa and 0.2, respectively. All joints in the abutments are cohesionless and have a friction angle of 15°; the interfaces between the dam and abutments have a friction angle of 30° and a cohesion of 0.1 MPa. The added artificial joints (interfaces between concrete blocks in the dam) have a limited strength: a friction angle of 30°, a cohesion of 0.1 MPa and a tensile strength of 0.03 MPa. The velocity of the approaching rock is 141.1(m/s).

Figure 5b shows the deformed geometry of the dam at t=0.002sec when the rock just hit the dam. Many interfaces near the collision point fail (indicated by solid lines), and lose their initial cohesion and tensile strength. Two gaps form in the interfaces of concrete blocks below the impact point. At t=0.003sec, cracking of interfaces is propagating into the two wings of the dam, as revealed in Figure 5c. Figure 5d depicts the significantly deformed geometry of the dam at t=0.01sec. In this figure, no interface survives at this time, and remarkable gaps arise between the concrete blocks below the hitting point. At this stage, imagine that water at a reservoir pressure enters these gap, leading to the total collapse of the dam.

5 CONCLUSION

DDA is a powerful numerical tool to analyze the mechanical response of discrete block systems. Without losing the simplicity and beauty of Shi's original formulation, "the artificial joint concept" was introduced to enhance its capabilities. The new capabilities include determining refined stress distributions even in blocks with complex geometry and internal defects, and simulating the fracture propagation of intact blocks.

6 REFERENCES

Ke, T. 1993. *Simulated Testing of Two-Dimensional Heterogeneous and Discontinuous Rock Masses Using Discontinuous Deformation Analysis*, Ph.D Dissertation, Department of Civil Engineering, University of California, Berkeley.

Rice, J. R. 1968. Mathematical analysis in the mechanics of fracture. Liebowitz (ed). *Fracture, and Advanced Treatise* (2):191-311.

Shi, G. 1988. *Discontinuous Deformation Analysis - A New Numerical Model for the Statics and Dynamics of Block Systems*, Ph.D Dissertation, Department of Civil Engineering, University of California, Berkeley.

Rock Mechanics, Nelson & Laubach (eds) © 1994 Balkema, Rotterdam, ISBN 90 5410 380 8

Application of the discontinuous deformation analysis to the evaluation of rock reinforcement for tunnel stabilization

M. Ronald Yeung & Stephen J. Klein
Woodward-Clyde Consultants, Oakland, Calif., USA
Max Y. Ma
Potsdam, N.Y., USA

ABSTRACT: The Discontinuous Deformation Analysis (DDA) is a numerical model for analyzing the mechanics of discontinuous, jointed rock systems. One application of the DDA method is to use the method to evaluate the behavior of tunnel openings in jointed rocks stabilized with rock reinforcement. This paper presents the results of DDA stability evaluations for a circular roof tunnel stabilized with rock reinforcement. DDA results include the effects of rock bolt length to spacing ratio, joint spacing, and loading condition. The rock bolt tensile forces required for stability obtained from the DDA evaluations are compared with those calculated using the Reinforced Rock Arch (RRA) method.

1 INTRODUCTION

Rock reinforcement is a cost effective method of stabilizing a tunnel excavation in jointed rock. A new approach to designing rock reinforcement involves the use of the Discontinuous Deformation Analysis (DDA) to model the behavior of the tunnel. DDA is a relatively new numerical modeling technique for evaluating the statics and dynamics of discontinuous rock block systems. In a two-dimensional DDA model of a rock block system, the blocks deform under applied driving forces and interact at their boundaries. The stability of a tunnel opening in such a rock block system can be evaluated, for an assumed rock reinforcement pattern, by modeling the rock bolts as elastic spring elements.

Another method of evaluating rock reinforcement requirement for stabilizing a tunnel opening in jointed rock is the Reinforced Rock Arch (RRA) design method that was developed by Lang et al. (1979). Rock reinforcement, consisting of a pattern of rock bolts installed around the tunnel perimeter is designed to develop the rock around the tunnel opening into a stable, competent structural element that is capable of maintaining the integrity of the opening. It was recognized that, for stability, interaction of adjacent rock bolts is essential and that this interaction is a function of the rock bolt length to spacing ratio, the rock bolt tension, the strength characteristics of the rock mass, and the loads resulting from deformation and relaxation of the rock mass surrounding the opening. The reinforced rock arch is formed by the zone of compression developed from the interaction of adjacent rock bolts (Figure 1). The RRA method has been used by the authors in the design of several tunnels, and the basic methodology has been reviewed and accepted by the U.S. Federal Highway Administration (Klein et al., 1989).

This paper presents the results of a parametric study in which DDA was used to study the effects of rock bolt length to spacing ratio, joint spacing, and loading condition on the stability of a tunnel with a circular roof stabilized with rock reinforcement. The rock bolt tensile forces required for stability are compared with those calculated using the RRA method.

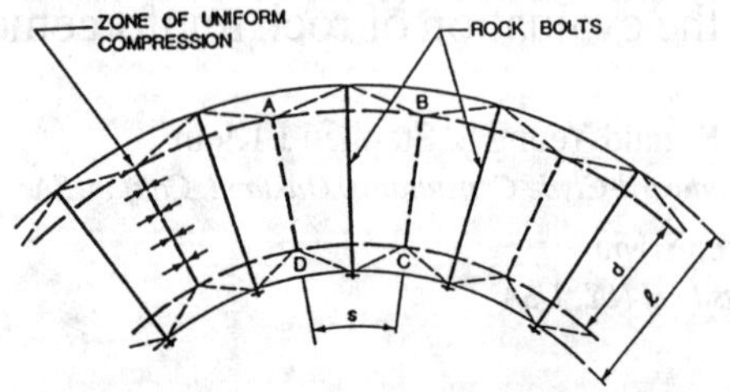

Figure 1. Reinforced rock arch concept.

2 DDA

DDA was first formulated by Shi and Goodman (1985) as an inverse method for calculating displacements and strains of an in-situ block model that best fit a set of experimentally measured displacements and deformations. Shi (1988) presented the complete theory of two-dimensional DDA, and Yeung (1993) applied it to study the fundamental behavior of a flat, rock bolted mine roof. DDA was developed to give a unique solution for large displacements and deformations of discontinuous block systems under static and dynamic loading conditions. It is a displacement method formulated using the principle of minimum total potential energy. Thus the equilibrium conditions for moments, forces and stresses are satisfied.

The major assumptions of DDA are: 1) it uses a first order approximation of the displacement of any point on a block; 2) the blocks are elastic; and 3) the contact forces obey the Mohr-Coulomb Law. The first assumption implies that the stresses and strains are constant within a block.

Displacements and deformations are accumulated in successive time steps, leading to large displacements and deformations. Within each time step, the equilibrium equations are solved implicitly and iteratively. Iteration is required to satisfy all contact conditions, which can be of four different types: opened, locked, sliding forward, and sliding backward. The technique of opening-closing iteration is used to achieve convergence: there is no change in contact conditions in successive iteration steps. To arrive at a set of consistent contact conditions, the equilibrium equations are updated and solved repeatedly. The opening-closing iteration is numerically stable, and guarantees a unique solution.

The Discontinuous Deformation Analysis Codes used to analyze the cases reported in this paper are written in BASIC. They were modified to provide the ability to model rock bolts. The function of a rock bolt as tensile reinforcement is simulated by modeling the bolt as an elastic spring connecting its end points. Input data required for the DDA programs include the block system geometry, the material properties of the blocks, the friction angle of block surfaces, and the stiffnesses and initial tensioning of the rock bolts. Output from the program include the deformed block system geometry, the principal stresses of each block, the contact forces between the blocks, and the tensile forces in the rock bolts.

3 APPLICATION OF DDA TO THE EVALUATION OF ROCK REINFORCEMENT

For a tunnel opening stabilized with rock reinforcement in a jointed rock mass, DDA was used to study the effects of rock bolt length to spacing ratio, joint spacing, and loading condition. The two different two-dimensional plain strain models used are called Model N1 and Model N2. Both models consist of a tunnel that has a circular roof with a span of 40 feet and vertical sidewalls 10 feet high. The tunnel is excavated in a jointed rock mass that is confined laterally and horizontally by two fixed, L-shaped blocks. Loads on the tunnel can be applied at 10 loading points located at the top surfaces of the models.

In Model N1 (Figure 2), there are two orthogonal joint sets in the rock mass; one joint set orients horizontally and the other vertically. Both joint sets have constant spacings of 10 feet.

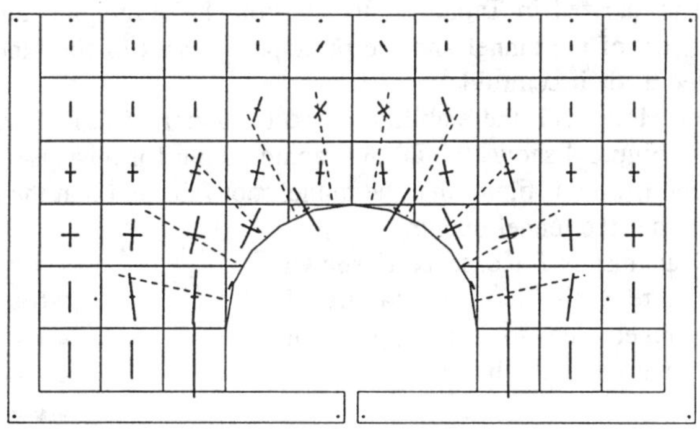

Figure 2. DDA graphic output for Case N1L.

Model N2 was used to study the effect of joint spacing. The two joint sets in Model N2 have the same orientations as those in Model N1, but the joint spacings are twice that in Model N1.

When rock reinforcement is included in the model, a uniform bolt pattern is used consisting of ten bolts spaced at 6 feet circumferentially around the tunnel arch and 5 feet longitudinally (perpendicular to the plane of analysis). The rock bolt length to spacing ratio is varied by using different bolt lengths while keeping the bolt spacings constant. A 5-foot thick slice perpendicular to the tunnel alignment was analyzed to be consistent with the longitudinal spacing of the bolts.

Six cases were analyzed using the two models and are summarized in Table 1. Typical DDA input data for the cases analyzed are given in Table 2. The material properties, such as the unit weight, are for a 5-foot longitudinal section of the tunnel. The material properties of the rock are that of a

Table 1. Summary of cases analyzed.

Case	Model	Bolt length to spacing ratio
N1U	Model N1	No bolts
N1L	Model N1	3.0
N1M	Model N1	2.0
N1S	Model N1	1.5
N2M	Model N2	2.0
N1MN	Model N1	2.0

typical hard sandstone, and a friction angle of 30° is assumed for sliding along the joint surfaces. In all cases except Case N1MN in which the loading is non-uniform, a vertical load of 100 kips (which is equivalent to a roof load of approximately 13 feet of rock) is applied at each of the loading points.

Table 2. Typical DDA input data.

Parameter	Value
Time step	0.5 second
Contact normal spring stiffness	$5X10^4$ kip/ft
Friction angle	30°
Unit weight	0.78 kip/ft^2/5 ft
Young's Modulus	1.4X10^6 kip/ft/5 ft
Poisson's Ratio	0.17

The downward displacement of the tunnel crown, D, is considered a measure of tunnel stability and is plotted against the number of time steps, N, for all cases in Figure 3. For each of the cases including rock reinforcement, the rock bolt tensile forces (for bolts numbered 1 through 10 from the left) required to stabilize the tunnel are summarized in Table 3. In addition, DDA graphic output shows the deformed geometry of the tunnel and the principal stresses of the blocks comprising the model plotted at their centroids.

For Model N1, the stability of the unsupported tunnel was first evaluated in Case N1U. Figure 4 shows the DDA graphic output for this case after 15 time steps. It can be seen from this figure that the tunnel roof fails and that the rock above the roof slides down into the tunnel opening. A plot of D against N for case N1U shown on Figure 3 also indicates failure of the tunnel due to the large roof deformations in the initial time steps.

3.1 *Effect of rock bolt length to spacing ratio*

Model N1 was analyzed with rock reinforcement used to stabilized the tunnel opening. Rock bolt length to spacing ratios of 3.0, 2.0, and 1.5 were used in Cases N1L, N1M, and N1S, respectively. DDA graphic output for Cases N1L, N1M, and N1S after 50 time steps are shown on Figures 2, 5, and 6, respectively. These figures show that arching develops within the reinforced zone around the tunnel arch, and the roof load is transferred by the arch to the tunnel sidewalls. The stress trajectories shown on these figures

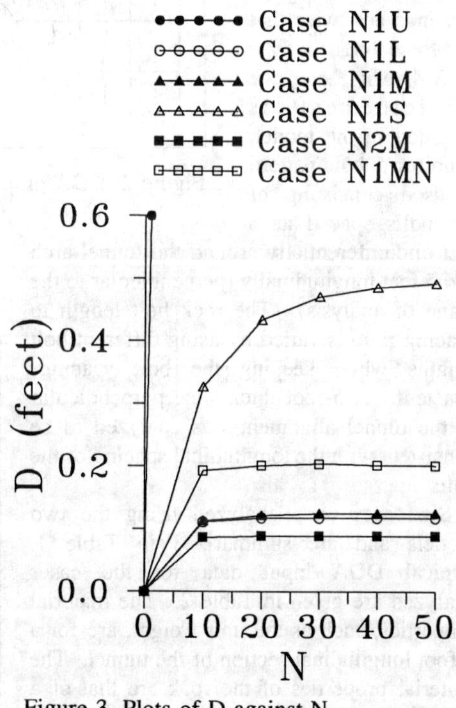

Figure 3. Plots of D against N.

Table 3. Summary of rock bolt tensile forces.

Case	Rock bolt tensile force (kips)										
	Bolt No.	1 Left	2	3	4	5	6	7	8	9	10 Right	From RRA
N1L		105	65	98	375	65	65	378	97	65	106	180
N1M		113	79	100	374	64	61	387	105	88	103	280
N1S		588	1074	53	380	80	79	377	54	1075	591	456
N2M		88	100	100	375	64	64	372	99	100	91	
N1MN		109	79	105	127	100	65	418	88	85	108	

therefore support the reinforced rock arch idealization shown on Figure 1.

Moreover, it can be seen that the thicknesses of the zones of compression in Cases N1L and N1M are about the same, while the thickness of the compression zone is considerably smaller in Case N1S. As a result, the rock bolt tensile forces required to stabilize the tunnel in Cases N1L and N1M are about the same and are smaller than the forces required in Case N1S. Therefore, the rock bolt tensile forces required for stability (Table 3) decrease significantly when the bolt length to spacing ratio is increased from 1.5 to 2.0 and remain virtually unchanged when the ratio is changed from 2.0 to 3.0. This variation of rock bolt tension with rock bolt length to spacing ratio is also consistent with the trend predicted by the RRA theory.

A study of the tensile forces in the 10 rock bolts (Table 3) revealed that the

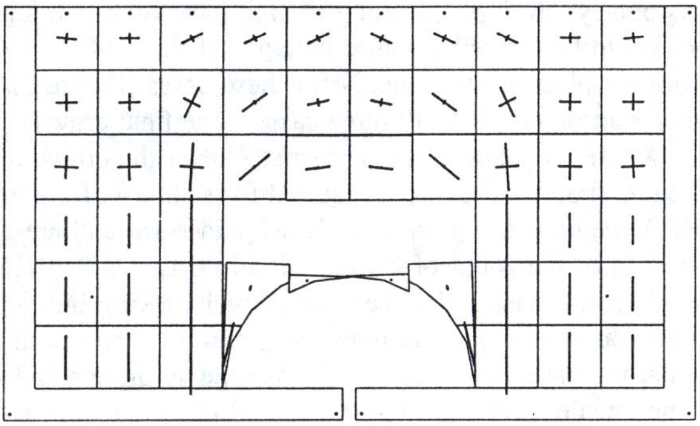

Figure 4. DDA graphic output for Case N1U.

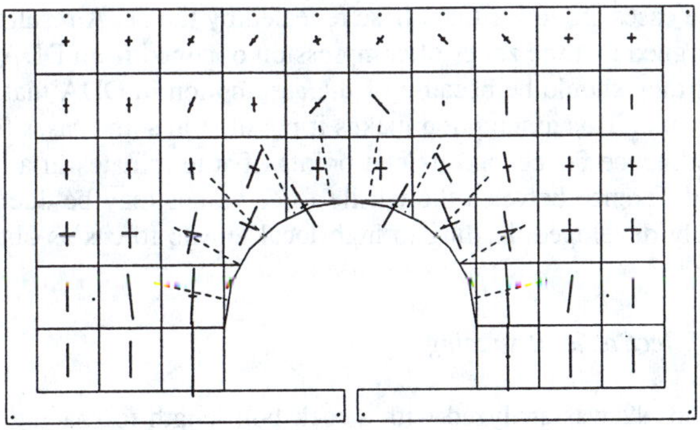

Figure 5. DDA graphic output for Case N1M.

distribution of tensile forces is basically symmetric about the model centerline but is non-uniform. In Cases N1L and N1M, the rock bolts have fairly uniform tensile forces except in bolt Nos. 4 and 7 which have higher tensile forces than the rest of the bolts; this indicates that the arch is fairly well-developed and is efficient in spreading and transferring the loads to the sidewalls. The distribution of bolt forces is less uniform in Case N1S, with relatively high tensile forces in bolt Nos. 1, 2, 4, 7, 9, and 10; this indicates that with the relatively short bolts in this case, the arch is poorly developed, leading to high local rock bolt tensile forces at places where hinges could be formed to cause collapse of the roof.

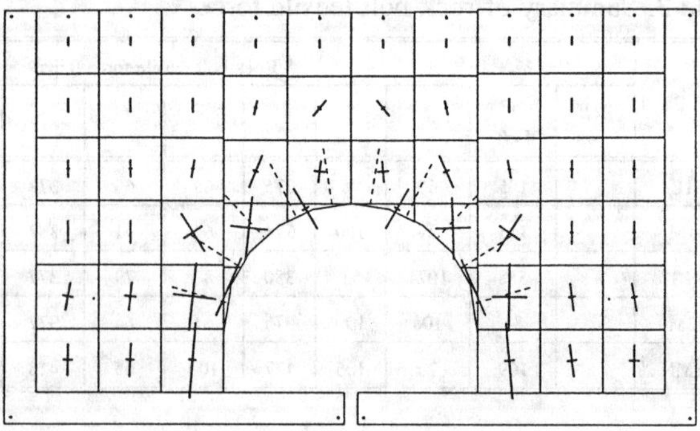

Figure 6. DDA graphic output for Case N1S.

From the plots of D against N for these cases (Figure 3), the tunnel is stabilized by the rock reinforcement in all three cases. The final crown displacements in Cases N1L and N1M are comparable and are smaller than the crown displacement in Case N1S.

The rock bolt tensile forces required for stability of the tunnel were calculated using the RRA method for Cases N1L, N1M, and N1S, and are given also in Table 3. The thicknesses of the zones of compression in Cases N1L, N1M, and N1S were assumed to be 13 feet, 7 feet, and 4 feet, respectively, taking into account the spreading of the bolt force at the anchor and tensioning points. The calculated rock bolt tensile force occurs at the base of the arch just above the springline and therefore can be compared with the tensile force in bolt No. 1 or 10 obtained from DDA. As can be seen from Table 3, the rock bolt tensile force calculated by the RRA method compares reasonably well with the tensile force in bolt No. 1 or 10 obtained from DDA. The difference between these tensile force values in Cases N1L and N1M may be explained by the fact that the thicknesses of the zones of compression assumed in using the RRA method for these cases are lower than those revealed by the DDA results (Figures 2 and 5). It may be argued that the zones of compression obtained from DDA for these cases are thicker than they should be because of the assumption in DDA that the stresses in a block are constant. This assumption makes it possible in some cases for a bolt to extend its zone of influence far beyond its end points, if it terminates in a large block. In Case N1S, the difference between the tensile force values may be due to the fact that the arch is poorly developed, leading to high local tensile forces as obtained from DDA.

3.2 *Effect of joint spacing*

Model N2 was analyzed with a rock bolt length to spacing ratio of 2.0 in Case N2M. Figure 7 shows the DDA graphic output for this case after 50 time steps. Comparing the plot of D against N in Figure 3 and the rock bolt tensile forces in Table 3 for this case with those for Case N1M, in which the joint spacing is half of that in Case N2M, it can be seen that the crown displacement is smaller in Case N2M than in Case N1M, while the rock bolt tensile forces are roughly the same in both cases.

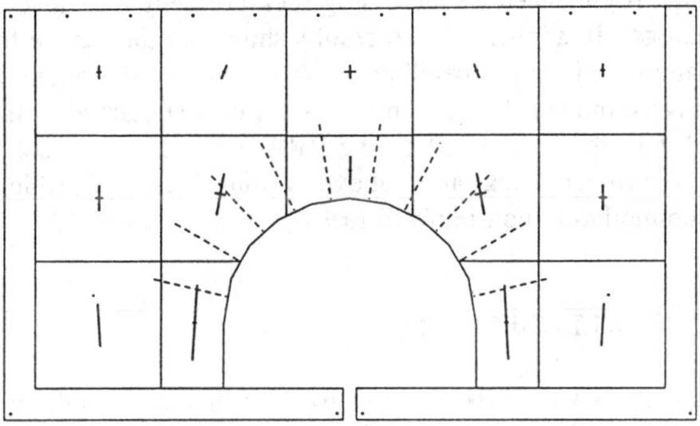

Figure 7. DDA graphic output for Case N2M.

3.3 *Effect of loading condition*

Model N1 was analyzed in Case N1MN with a rock bolt length to spacing ratio of 2.0 and non-uniform loading consisting of vertical loads of 50 kips at the five loading points left of the centerline of the model and 150 kips at the five loading points right of the centerline of the model. The total load is therefore 1,000 kips in both Cases N1MN and N1M. Figure 8 shows the DDA graphic output for Case N1MN after 50 time steps. It can be seen that Figure 8 is similar to Figure 5, the DDA graphic output for Case N1M in which the loading is uniform. This shows that the reinforced rock arch is a flexible structure capable of accommodating local non-uniform loads. Furthermore, the final crown displacement as shown in Figure 3 is higher in Case N1MN than in Case N1M, although the final state of stress surrounding the tunnel and the tensile forces in the rock bolts are roughly the same in both cases.

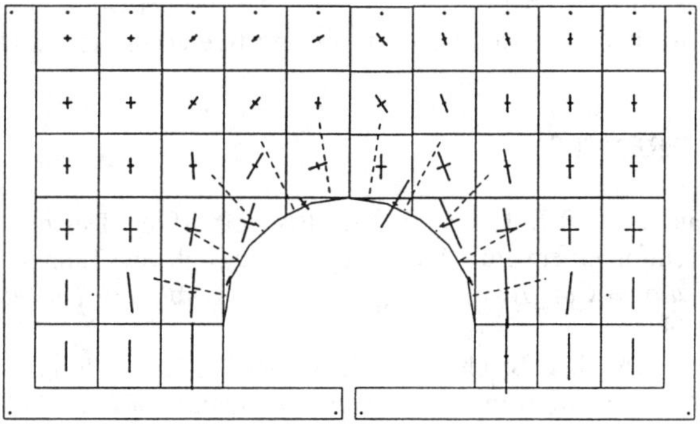

Figure 8. DDA graphic output for Case N1MN.

4 CONCLUSIONS

Results of analyses presented in this paper show that the DDA method can be useful in the evaluation of rock reinforcement for tunnel stabilization. They clearly demonstrate the arching phenomenon and the existence of a zone of compression created by rock reinforcement, which are fundamental assumptions in the RRA method. The load carrying capacity of the reinforced rock arch, and hence the stability of the tunnel, is enhanced by increasing rock bolt length to spacing ratio and increasing joint spacings. In addition, DDA results show that the tensile force of an individual bolt in a pattern can vary depending on its location in the tunnel arch. In Cases N1L, N1M, and N1S, the tensile force in bolt No. 1 or 10 obtained from DDA compares reasonably well with that calculated by the RRA method. The results of Case N1MN show that the reinforced rock arch above a tunnel is a flexible structure also capable of accommodating non-uniform loads.

ACKNOWLEDGEMENTS

The authors would like to acknowledge the significant contributions and the profound influence of Mr. Thomas A. Lang in the field of rock mechanics and, particularly, in the development of methods for designing rock reinforcement. Mr. Lang was 84, and he passed away on December 25, 1993. He initiated rock mechanics classes at the University of California, Berkeley and was an adjunct professor in the Civil Engineering Department from 1962 until 1970. The authors would also like to thank Dr. Gen-hua Shi for his guidance and for providing free access to his DDA programs.

REFERENCES

Klein, S.J., R.J. Essex & J.A. Bischoff 1989. Design and Construction of Rock Reinforcement for Three Large Span Highway Tunnels. *Proceedings of the International Congress on Progress and Innovation in Tunnelling*, Toronto, Canada. 321-329.

Lang, T.A., J.A. Bischoff & P.L. Wagner 1979. *A Program for Determining Optimum Roof Bolt Tension: Theory and Application of Rock Reinforcement Systems in Coal Mines, Volume 1*, Leeds, Hill and Jewett, Inc., San Francisco. Report prepared for the U.S. Bureau of Mines, Contract No. J0285006. 256p.

Shi, G. 1988. *Discontinuous Deformation Analysis: A New Numerical Model for the Statics and Dynamics of Block Systems.* Ph.D. Dissertation. Department of Civil Engineering, University of California, Berkeley. 378p.

Shi, G. & R.E. Goodman 1985. Two dimensional discontinuous deformation analysis. *International Journal for Numerical and Analytical Methods in Geomechanics.* 9:541-556.

Yeung, M.R. 1993. Analysis of a Mine Roof Using the DDA Method. *International Journal of Rock Mechanics and Mining Sciences.* 30:1411-1417.

Rock Mechanics, Nelson & Laubach (eds) © 1994 Balkema, Rotterdam, ISBN 90 5410 380 8

Environmentally motivated tracking of geologic layer movement during bench blasting using discrete element methods

Dale S. Preece
Sandia National Laboratories, Albuquerque, N. Mex., USA

D. Scott Scovira
ICI Australia Operations Pty Ltd, Singapore

ABSTRACT: The blast-induced movement and final location of geologic layers that may cause environmental problems can be predicted using discrete element methods. This prediction capability can be used by mine operators to locate the material in the muck pile during excavation which would allow encapsulation to prevent groundwater infiltration.

1 INTRODUCTION

Open-pit mining produces the majority of the coal marketed by the U. S., both domestically and abroad. An increasing emphasis on environmentally sound mining practices makes it necessary to be innovative in applying new technologies to open pit mining. An important challenge is ground water acidification attributed to geologic layers, typically a shale, that have a higher-than- average pyrite content. The acidification occurs as bacteria interact with the pyrite, producing hydrogen-sulfide which raises the PH of ground water flowing through the material. Groundwater acidification is accelerated significantly in open-pit mines as the relatively impermeable shale layers are rubblized by blasting, greatly increasing the hydraulic conductivity of the material. A viable solution for mine operators is to separate the pyritic material during muck-pile excavation and encapsulate it in some manner to prevent groundwater interaction. This solution depends on being able to locate the pyritic material in the post-blast muck-pile. The movement and final location of this material was analyzed by using a discrete element computer simulation.

2 DISCRETE ELEMENT MODELING

DMC is a two-dimensional spherical element discrete motion code that was originally developed by Taylor and Preece (1989 a&b and 1992) for modeling the motion associated with rock blasting. Spherical element bulking mechanisms have been added to allow spherical elements to behave more like rocks (Preece and Taylor, 1990). This program has been used for modeling a number of different blasting applications including: 1) oil shale retort blasting (Preece, 1990 a&b), and 2) bench blasting in rock quarries and coal mines (Preece and Knudsen, 1992b), (Preece,1993c). A concept called spherical element packing angle (Preece, 1993b) has been developed for more accurate modeling of bedding planes in sedimentary rocks. The current version of DMC has also been used to examine the influence of row delay timing on blasting

*. This work performed at Sandia National Laboratories supported by the U.S. Department of Energy under contract no. DE- AC04-94AL85000 and also supported by ICI Explosives USA.

induced rock motion (Preece, 1994a). DMC has been coupled with a gas flow computation capability so that the explosive loading is automatically treated when an explosive is specified along with its physical parameters and equation-of-state (Preece and Knudsen, 1992a) (Preece, 1993a). DMC has been used by industry to optimize blast designs for maximum movement of material. It compares favorably with other computer programs used to predict the results of a blast (Chung, McGill and Preece, 1994).

3 EXAMPLE PROBLEM DESCRIPTION

The example problem presented here is a coal mine bench blast at a mine located in the central Appalachian coal region. Actual field data from the mine including geometry, geology, blast design, and percent-cast were employed to guide the simulations presented here.

3.1 *Bench Blast Configuration*

Figure 1 illustrates a cross-section schematic of the bench blast where the second layer above the coal is a pyritic sandy shale. The mine operators load the blastwell with explosives above and

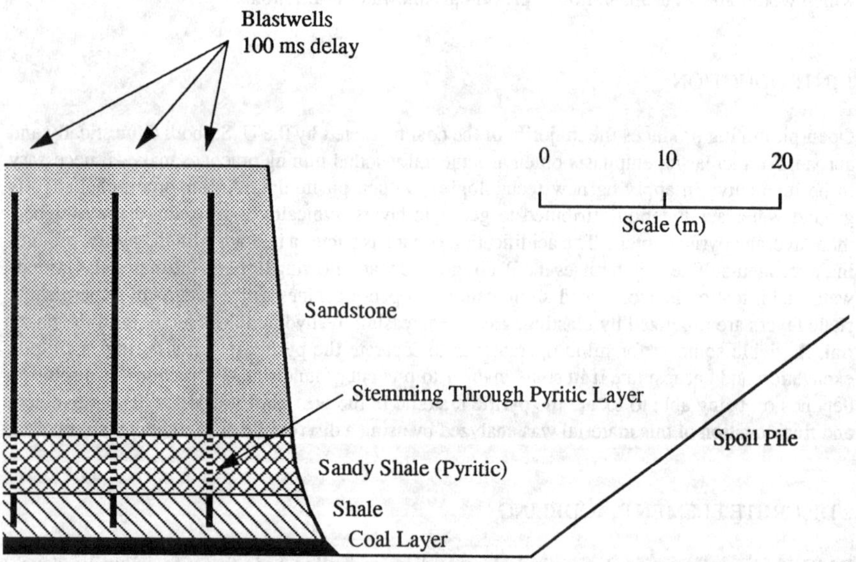

Figure 1: Schematic of Bench Blast Example Problem

below the pyritic layer. A deck of stemming material (gravel or drill cuttings) is placed through the pyritic layer to prevent it from being loaded directly by the explosives. That practice is modeled in this simulation. The induced motion in the pyritic layer results from frictional drag by the two layers above and below.

3.2 *Definition of Percent-Cast*

Figure 2 illustrates the definition of percent-cast, a concept commonly used to measure blast-induced rock motion. The cast-line is parallel to the angle of repose of the spoil pile and intersects the bottom of the bench face. Percent-cast refers to that portion of the overburden that is beyond the cast-line and thus resides in its final position, not requiring any rehandling. In this study the motion and final location of the pyritic layer is of greatest interest. Pyritic material that has moved very little will be easiest to recover during muck-pile excavation. Any of this material that has moved beyond the cast-line will be difficult to recover for two reasons: 1) excavation is normally not done beyond the cast-line, and 2) the material tends to scatter with increased motion which is indicated by a higher percent cast for the material. Thus, movement of pyritic material beyond the cast-line will be used as a measure of recoverability, and comparisons will be made between the overall percent-cast and that of the pyritic layer.

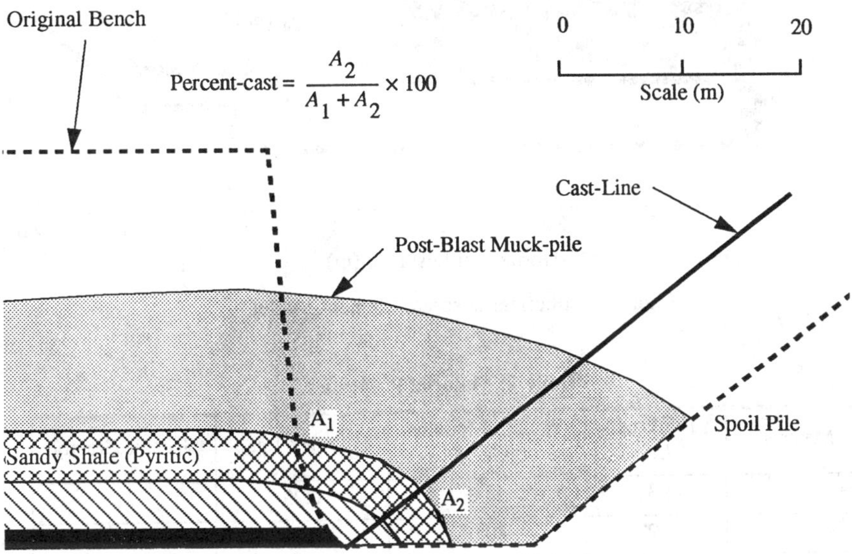

Figure 2: Definition of Percent-Cast Where A_2 Represents Material Beyond the cast-line, A_1 Represents Material in Front of the cast-line. Percent-Cast for Individual Materials are Determined in a Similar Manner.

4 ROCK MOTION SIMULATION

4.1 *Discrete Element Model*

The spherical discrete element model used in the rock motion simulation is illustrated in Figure 3. The coal is represented by the darkest layer of spheres just above the pit floor. The pyritic layer, which is of most interest in this calculation, is represented by the dark spheres constituting the second layer above the coal. The spherical elements beneath the coal and extending under part of the pit floor are included to move the hard bottom boundary away from the area where the action will occur during the simulation. Material parameters used in this calculation are given in Table 1, and explosive parameters are given in Table 2. In Table 2 the burden is the distance from the blastwell to either the free face (for the front row) or the blastwell row immediately in front of the current row, the spacing is the distance between blastwells parallel to the free face, and the powder factor is the mass of explosive per unit volume of rock. It is determined

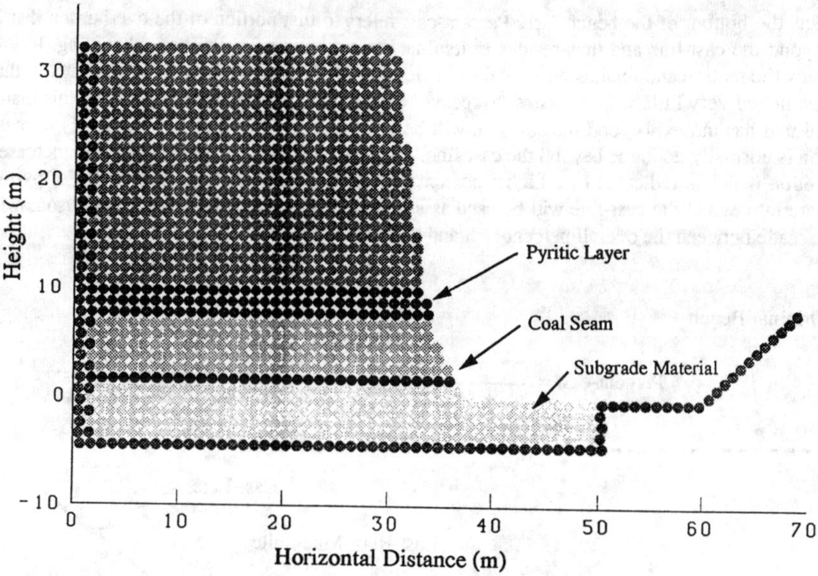

Figure 3: Spherical Discrete Element Model

Table 1: Material Properties

Material	Thickness (m)	Young's Modulus (GPA)	Poisson's Ratio	Specific Gravity
Coal	1.31	50.0	0.25	1.5
Shale	4.92	40.0	0.30	2.2
Sandy Shale	2.62	40.0	0.30	2.2
Sandstone	24.61	50.0	0.21	2.2

Table 2: Blast Design Parameters

Row No.	Burden (m)	Spacing (m)	Explosive	Powder Factor (Kg/m^3)	Hole Dia. (mm)	Row Delay (ms)
1	6.40	9.88	H-ANFO 60/40*	0.64	229	0
2	10.66	9.81	"	0.39	229	100
3	11.30	10.1	"	0.36	"	"

*Mixture ratio, emulsion to ANFO

from the blastwell geometry and explosive density along with the burden, spacing and face height. These values are typical for this type of mine. The powder factor is higher in the front row because the burden has been reduced compared with subsequent rows. This is a common practice to obtain better movement of the front row so that it doesn't choke the flow of subsequent rows.

4.2 Simulation Results

Two parameters were varied between simulations: 1) explosive loading to produce more or less rock motion and a different percent-cast for each simulation, and 2) the height of the pyritic layer above the pit floor. Simulations with different explosive loadings resulted in a relationship between percent-cast for the entire blast and percent-cast for the pyritic layer.This relationship is indicative of the percentage of pyritic material that moves beyond the cast-line and will likely not be recovered. Spherical element models with three different pyritic layer heights above the pit floor were modeled to determine the influence of vertical layer position on material scattering. Four snapshots in time, 100ms, 300ms, 1.0s and 7.0s, for the simulation with the pyritic layer closest to the pit floor and with a cast of 30% are illustrated in Figure 4.

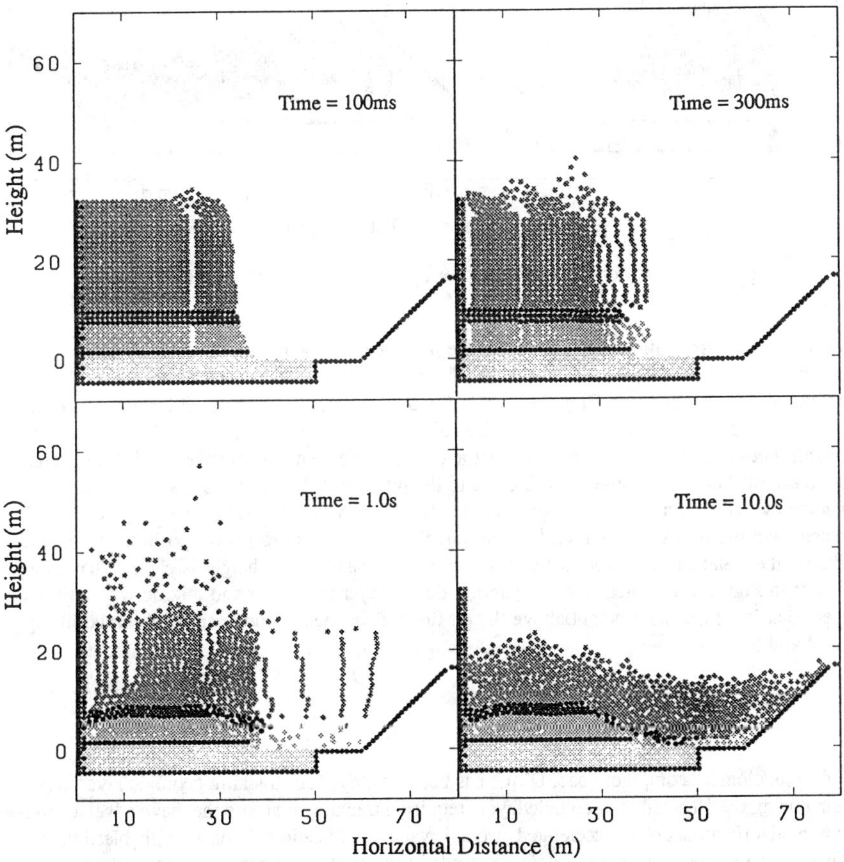

Figure 4: Rock Motion Simulation For Pyritic Layer Close to Pit Floor.

The 10.0s frame in Figure 4 shows the predicted final muck-pile shape and indicates that the material in the pyritic layer remains basically in place with only a small portion (14%) spilling into the pit, beyond the cast-line, where it will be difficult to recover. The simulation with the pyritic layer highest off the pit floor is illustrated in Figure 5 with two time snapshots, 100ms and 10.0s, Again the 10.0s frame presents the final muck-pile shape and shows considerable stretching of the pyritic layer and scattering amongst the surrounding material. It is evident from Figure 5 that this material would be almost impossible to locate in the post-blast muck-pile and points to vertical layer position as an important factor defining retrievability.

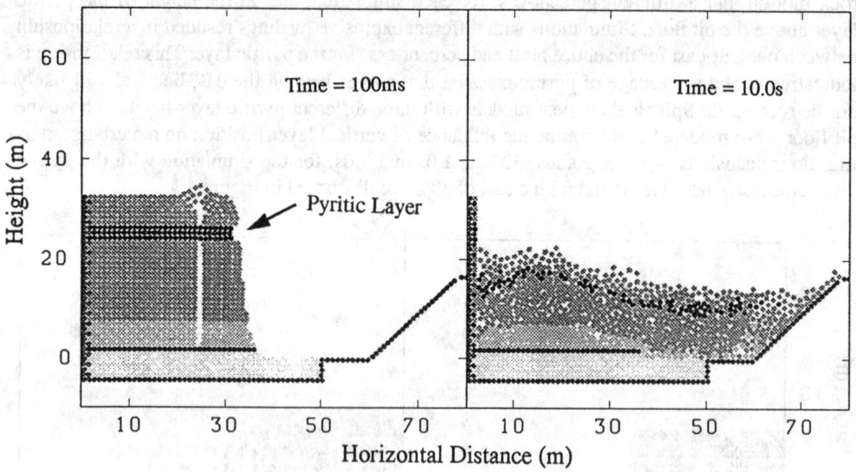

Figure 5: Rock Motion Simulation For Pyritic Layer High Off the Pit Floor

Results derived from all of the simulations are presented in Figure 6 in the form of pyritic layer percent-cast versus overall percent-cast for the three different layer vertical positions. Several conclusions can be deduced from the data in Figure 6. First, pyritic material loss due to a final position beyond the cast-line decreases with reduction of overall percent-cast. Second, the relationship between pyritic layer percent-cast and overall percent-cast changes with layer height above the pit floor. When the layer is close to the pit floor its' percent-cast is significantly less than the overall percent-cast. If the pyritic layer is relatively high in the bench its' percent-cast is greater than the overall percent-cast. Even though only three vertical layer positions were considered, the trend of more horizontal motion with increasing layer height is clear. What is not evident in Figure 6 is material irretrievability due to layer stretching and mixing with adjacent layers that increases with height above the pit floor. This phenomenon can be observed in Figures 4 and 5.

5 CONCLUSIONS

A discrete element computer code, DMC, has been employed for tracking geologic layer movement during bench blasting. Geologic layers requiring tracking are those that have adverse environmental effects such as accelerated ground water acidification. Even though blasting is a dynamic chaotic process, several definite trends in material movement have been identified. A number of DMC blast simulations have been used to demonstrate two important influences on the retrievability of material in the muck-pile, 1) overall percent-cast and 2) layer height above the pit floor. Pyritic layers that are relatively low in the geologic strata of the bench, close to the pit floor, are the most amenable to location prediction and recovery during muck-pile excavation. Layers high in the geologic strata of the bench will experience significant horizontal movement, stretching, scattering and intermixing with adjacent layers with a low probability of location and recovery during excavation. For all layer heights the scattering increases with overall percent-cast. The allowable layer percentage loss will dictate a blast design yielding a corresponding overall percent-cast.

Even though general trends have been shown in this paper, each blasting site is unique in terms of geometry, geology and blast design. Motion simulations that include specific site parameters would be necessary to attempt material location and retrieval at different sites than the one modeled here.

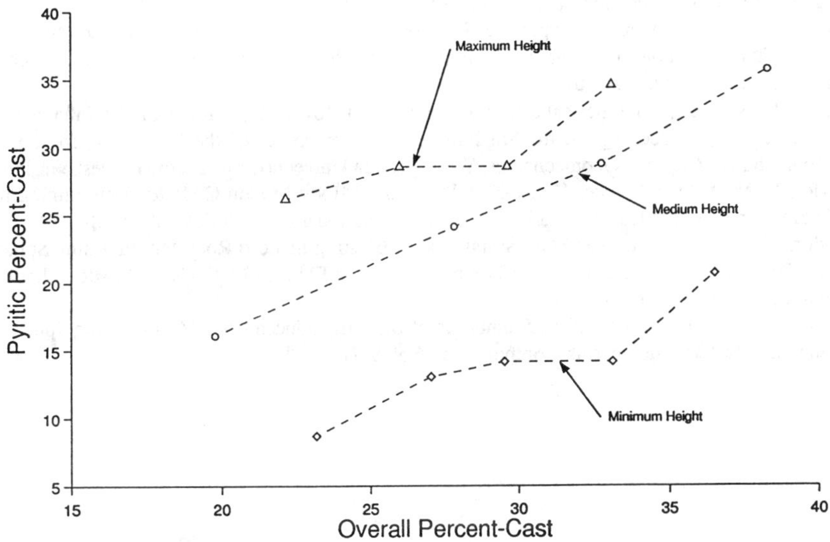

Figure 6: Pyrite Layer Percent-Cast Versus Overall Percent-Cast for Three Different Layer Heights Above the Pit Floor.

REFERENCES

Chung, S., McGill, M, and Preece, D. S., 1994, Computer Cast Blast Modeling, 20th Annual Conference on Explosives & Blasting Technique, International Society of Explosives Engineers, Austin, Texas.

Preece, D. S. & L. M. Taylor, 1990, Spherical Element Bulking Mechanisms for Modeling Blasting Induced Rock Motion, Third International Symposium on Rock Fragmentation by Blasting, Brisbane, QL, Australia.

Preece, D. S., 1990a, Rock Motion Simulation and Prediction of Porosity Distribution for a Two-Void-Level Retort, 23d Annual Oil Shale Symposium, Colorado School of Mines, Golden, Colorado.

Preece, D. S., 1990b, Rock Motion Simulation of Confined Volume Blasting, 31st U.S. Symposium on Rock Mechanics, Colorado School of Mines, Golden Colorado.

Preece, D. S., and S. D. Knudsen, 1992a, Coupled Rock Motion and Gas Flow Modeling in Blasting, Eighth Annual Symposium on Explosives and Blasting Research, International Society of Explosives Engineers, Orlando Florida.

Preece, D. S., and S. D. Knudsen, 1992b, Computer Modeling of Gas Flow and Gas Loading of Rock in a Bench Blasting Environment, 33d U. S. Symposium on Rock Mechanics, Santa Fe, New Mexico.

Preece, D. S., 1993a, Momentum Transfer From Flowing Explosive Gases to Spherical Particles During Computer Simulation of Blasting Induced Rock Motion, Ninth Annual Symposium on Explosives and Blasting Research, International Society of Explosives Engineers, San Diego, California.

Preece, D. S., 1993b, Variation of Spherical Element Packing Angle and Its Influence on Computer Simulations of Blasting Induced Rock Motion, 2nd International Conference on Discrete Element Methods, Massachusetts Institute of Technology.

Preece, D. S., 1993c, Coupled Explosive Gas Flow and Rock Motion With Comparison to Bench Blast Field Data, Fourth International Symposium on Rock Fragmentation by Blasting, Technical University, Vienna, Austria

Preece, D. S., 1994, A Numerical Study of Bench Blast Row Delay Timing and Its Influence on Percent Cast, Proceedings of the 8th International Conference of the Int. Assoc. for Comp. Meth. and Advances in Geomechanics, West Virginia University, Morgantown, West Virginia

Taylor, L. M. & D. S Preece, 1989a, DMC - A Rigid Body Motion Code for Determining the Interaction of Multiple Spherical Particles, Sandia National Laboratories, SAND88-3482.

Taylor, L. M. & D. S Preece, 1989b, Simulation of Blasting Induced Rock Motion Using Spherical Element Models, First U.S. Conference on Discrete Element Methods, Colorado School of Mines, Golden, Colorado.

Taylor, L. M. & D. S Preece, 1992, Simulation of Blasting Induced Rock Motion Using Spherical Element Models, Engineering Computations, Vol. 9, No. 2.

Rock Mechanics, Nelson & Laubach (eds) © 1994 Balkema, Rotterdam, ISBN 90 5410 380 8

A curved structural element model for rock surface linings in discontinuous rock analysis

B.Wang
Canada Centre for Mineral and Energy Technology (CANMET), Ottawa, Ont., Canada

ABSTRACT: This paper presents a curved structural element model for analyzing the supporting effect of the rock surface linings. The technique was developed for the Block-Spring Model (BSM), a numerical method for stress and deformation analysis of jointed rocks. BSM simulates the discontinuous rock masses by an assemblage of blocks divided by joints. The curved structural element acts as a curved beam. Each element is subjected to axial, transverse and flexural loads. The curved structural element does not cause the increase of the number of degrees of freedom of the block system. Stiffness of the structural elements can be assembled into the stiffness matrix of BSM with the size of the matrix remaining unchanged. The approach can be readily incorporated into other numerical models. An example of a circular tunnel is presented to demonstrate the proposed model in analyzing the supporting effect of surface lining on the surrounding rock. The results indicate that, with the increase of the lining stiffness, the rock deformation towards the tunnel decreases, the radial stress in the surrounding rock increases and the tangential stress decreases.

1 INTRODUCTION

Rocks around underground excavations are often supported by linings along the surface. The surface lining supports the rocks by imposing pressures on the rock surface upon rock deformation. To evaluate the supporting effect of the surface lining, special techniques must be employed.

Numerical approaches for rock surface linings in other numerical models, e.g., Finite Element Method (FEM), Boundary Element Method (BEM) and Distinct Element Method (DEM), can often be found in the literature (e.g., Gioda & Carini, 1985; Kumer & Singh, 1988; Lorig, 1984; Lorig & Brady, 1984). In these models, the rock surface linings are usually divided into elements interacting with the rock elements at nodal points through a kind of deformable elements, such as springs. A typical model is using beam elements to simulate the rock surface lining (e.g., Lorig, 1984; Lorig and Brady, 1984). The nodes of the lining element have certain degrees of freedom. As the rocks are described by a different type of elements, e.g., FEM, BEM or DEM, the size of the matrix can be significantly increased.

The present paper introduces a numerical technique for analyzing surface lining, e.g., shotcrete lining or steel set. It is proposed with the development of the Block-Spring

Model (BSM). BSM is a numerical model for analyzing jointed rock masses. A stiffness matrix is required in BSM to evaluate the displacements and stress of the rocks. The proposed numerical technique for surface linings uses curved structural elements to describe the surface lining materials, which do not cause the increase of the size of the global stiffness matrix in BSM.

2 THE BLOCK-SPRING MODEL

The Block-Spring Model was developed to analyze discontinuous rock masses (Wang, 1991; Wang and Garga, 1991, Wang and Garga, 1993). Various capabilities of the model have been demonstrated (Garga and Wang, 1993), which may include simulation of large movement of blocks along the joints; modelling of rockbolts and groundwater.

BSM simulates the jointed rocks by an assemblage of blocks divided by the discontinuities. The blocks are assumed in contact on surface which may separate and rejoin, or slide along the joint upon loading and unloading. Equilibrium conditions are applied to all the blocks. To each block, the equilibrium equations can be written as follows.

$$(1) \qquad \sum \mathbf{F} = \mathbf{0}$$

where F represents forces imposed on the block. The forces on each lock can be external loads, boundary restrictions, body forces, forces transformed from rockbolts, groundwater pressure and contact forces from other blocks. These forces can either be pre-determined as boundary conditions, or evaluated, based on certain assumptions, from the relative displacements between the blocks. Therefore, by assembling the equilibrium equations of the blocks, a set of global stiffness equation can be obtained as follows.

$$(2) \qquad [\mathbf{K}]\{\mathbf{U}\} = \{\mathbf{P}\}$$

where **[K]** denotes a global stiffness matrix with dimension 3N×3N for N blocks; **{U}** denotes an array of 3N displacements of the N blocks; **{P}** denotes boundary condition.

The displacements **{U}** of the blocks can be obtained after solving eq.(2). The contact forces between the blocks can be calculated thereafter from the relative displacements of the blocks.

An iteration scheme is used if any failure is detected along the rock discontinuity. The large displacements of the blocks can therefore be simulated incrementally.

3 PROPOSED STRUCTURAL ELEMENT IN BSM

Lining materials are often applied to rock surfaces around the excavations to impose a passive pressure on the rock, such as shotcrete lining and steel set. The passive pressure develops only when the rock deforms towards the opening. The pressure from the rock may cause the lining to bend. The deformation behaviour of the lining material may be very complicated. Ideally, extensive study may be necessary using numerical techniques, e.g., FEM, to discretize the lining into fine mesh if one expects to understand the precise behaviour of the lining material. However, the numerical technique proposed in the present paper is to study the structural effect of the rock lining on the surrounding rocks.

To analyze the supporting effect of the surface lining, the rocks can be modelled as an assemblage of rigid blocks as discussed earlier. The surface lining can be modelled by curved elements based on certain assumptions as discussed bellow.

3.1 Assumptions

In 2-D problem, the lining is discretized into curved structural elements determined by the size of the rock blocks. A curved element starts from the middle point of the covered surface of the block and ends at the middle point of the covered surface of the next block (Fig.1). Each element is considered as a curved beam (Fig.2). The curved beam may be subject to axial, transverse and flexural loads at its ends, which can be called nodes or nodal points. The structural elements interact with the rock blocks at the nodal points through rigid contact. The displacements of the nodal points can therefore be determined by those of the rigid blocks. The supporting effect of the surface lining can be evaluated by formulating the stiffness equations of the elements.

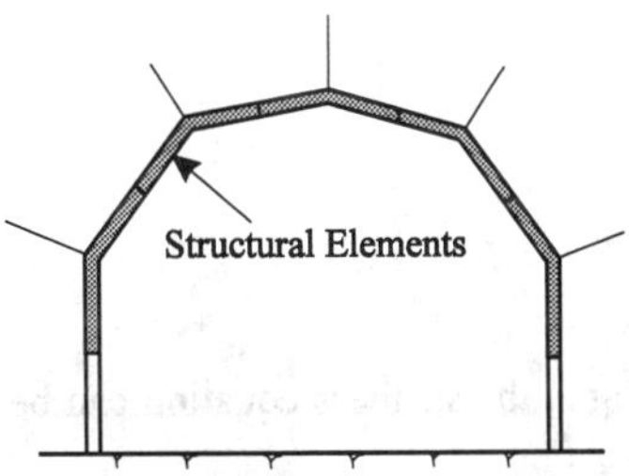

Fig.1. Determination of curved structural elements for surface lining.

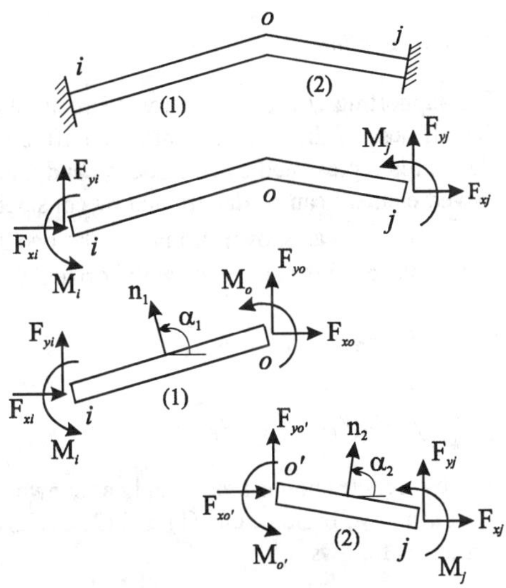

Fig.2. Description of a curved structural element.

It may be better to assume that the structural elements interact with the blocks through springs as commonly used in the conventional straight beam element as shown in Fig.3 (e.g., Lorig, 1984). The problem in such an assumption is that the size of the stiffness matrix can be greatly increased and the computation scheme as well as the procedure of coupling two different types of elements (rock blocks and structural elements) can be much more complicated. How much better the solution can be using such an assumption still remains to be investigated.

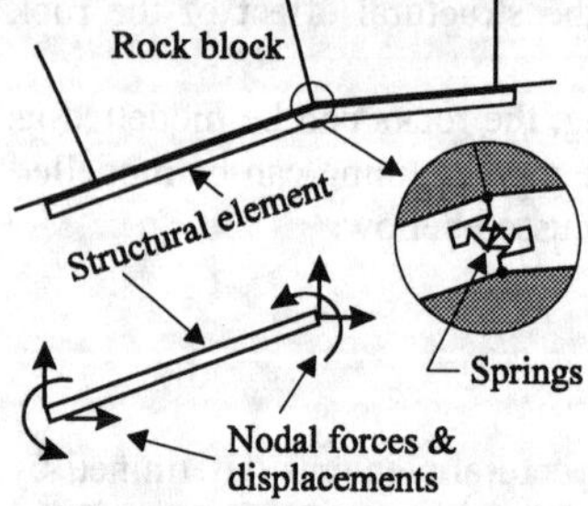

Fig.3. Conventional structural element

The proposed curved structural element significantly simplified the scheme of computation. Unlike the conventional model, the curved elements have no independent displacements at the nodal points. Introducing such elements into the block model will not increase the size of the stiffness matrix of the block system. It can be merged into the block model with very little effort.

3.2 Formulation

The supporting effect of the curved structural elements can be evaluated by assembling the stiffness of the elements into the stiffness matrix of the block system. The curved element is represented by two nodes *i* and *j*, and a turning point *o*. The stiffness of the curved element can be determined by considering two basic beam elements (1) and (2) (or $\overline{io}$ and $\overline{o'j}$) as shown in Fig.2. The two straight beams are subjected to the forces and displacements at the ends with horizontal, vertical and rotational components:

$$(3) \qquad \boldsymbol{F}_p = \{F_{px} \quad F_{py} \quad M_p\}^T$$

$$(4) \qquad \boldsymbol{u}_p = \{u_{px} \quad u_{py} \quad \theta_p\}^T$$

where *p* denotes points *i, o, o'* and *j* as shown in Fig. 2.

For the two basic beams (1) and (2) as shown in Fig.2, the stiffness equation can be written as follows:

$$(5) \qquad \begin{Bmatrix} \boldsymbol{F}_i \\ \boldsymbol{F}_o \end{Bmatrix} = \begin{bmatrix} \boldsymbol{K}_{ii} & \boldsymbol{K}_{io} \\ \boldsymbol{K}_{oi} & \boldsymbol{K}_{oo} \end{bmatrix} \begin{Bmatrix} \boldsymbol{u}_i \\ \boldsymbol{u}_o \end{Bmatrix} \qquad \text{Beam (1) } (\overline{io})$$

$$(6) \qquad \begin{Bmatrix} \boldsymbol{F}_{o'} \\ \boldsymbol{F}_j \end{Bmatrix} = \begin{bmatrix} \boldsymbol{K}_{o'o'} & \boldsymbol{K}_{o'j} \\ \boldsymbol{K}_{jo'} & \boldsymbol{K}_{jj} \end{bmatrix} \begin{Bmatrix} \boldsymbol{u}_{o'} \\ \boldsymbol{u}_j \end{Bmatrix} \qquad \text{Beam (2) } (\overline{o'j})$$

Each $\boldsymbol{K}_{pq}$ (*p, q=i, o, o', j*) in the above two equations represents a 3×3 stiffness matrix obtained from the corresponding beam, which is identified by the two subscripts. The matrix $\boldsymbol{K}_{pq}$ can be expressed in a general form:

$$(7)\ \boldsymbol{K}_{pq}=\begin{bmatrix} \delta_{pq}(\frac{EA}{L_m}\xi_m^2+\frac{12EI}{L_m^3}\eta_m^2) & \delta_{pq}(\frac{-EA}{L_m}\xi_m\eta_m+\frac{12EI}{L_m^3}\xi_m\eta_m) & \lambda_p\frac{6EI}{L_m^2}\eta_m \\ \delta_{pq}(\frac{-EA}{L_m}\xi_m\eta_m+\frac{12EI}{L_m^3}\xi_m\eta_m) & \delta_{pq}(\frac{EA}{L_m}\eta_m^2+\frac{12EI}{L_m^3}\xi_m^2) & \lambda_p\frac{6EI}{L_m^2}\xi_m \\ \lambda_q\frac{6EI}{L_m^2}\eta_m & \lambda_q\frac{6EI}{L_m^2}\xi_m & (1+\delta_{pq})\frac{2EI}{L_m} \end{bmatrix}$$

where $p, q = i, o, o'$ or j;

$$\delta_{pq}=\begin{cases}0 & if \quad p\neq q \\ 1 & if \quad p=q\end{cases};\qquad \lambda_p=\begin{cases}+1 & if \quad p=i \quad or \quad o' \\ -1 & if \quad p=o \quad or \quad j\end{cases};$$

$\xi_m=\sin\alpha_m$; $\eta_m=\cos\alpha_m$;

$m = 1$ or 2 represents beam (1) or (2) as shown in Fig.2;

α_m is an angle of normal $\mathbf{n}_m$ of beam m (=1, 2) from x axis (Fig.2);

L_m is the length of beam m;

E is the Young's modulus of the lining material;

I is the moment of inertia of the cross section of the lining.

From eqs.(5) and (6) and note that $\boldsymbol{F}_{o'}=-\boldsymbol{F}_{o'}$, the following expression of the nodal forces can be derived:

$$(8)\qquad \boldsymbol{F}_{ij}=\boldsymbol{D}_{ij}\,\boldsymbol{u}_{ij}$$

where

$$(9)\qquad D_{ij}=\begin{bmatrix}K_{ii} & 0 \\ 0 & K_{jj}\end{bmatrix}+\begin{bmatrix}K_{io} & 0 \\ 0 & K_{jo}\end{bmatrix}\left[K_{oo}+K_{o'o'}\right]^{-1}\left[-K_{oi},\ -K_{oj}\right]$$

$$(10)\qquad \boldsymbol{F}_{ij}=\{\boldsymbol{F}_i, \boldsymbol{F}_j\}^{\mathrm{T}}=\{F_{ix}, F_{iy}, M_i, F_{jx}, F_{jy}, M_j\}^{\mathrm{T}}$$

$$(11)\qquad \boldsymbol{u}_{ij}=\{\boldsymbol{u}_i, \boldsymbol{u}_j\}^{\mathrm{T}}=\{u_{ix}, u_{iy}, \theta_i, u_{jx}, u_{jy}, \theta_j\}^{\mathrm{T}}$$

The nodal displacement $\boldsymbol{u}_p$ ($p = i, j$) can be determined from the displacements of the blocks:

$$(12)\qquad \boldsymbol{u}_p=\boldsymbol{B}_p\ \boldsymbol{U}_p \quad (p=i,j)$$

where $\boldsymbol{U}_p=\{U_{px}, U_{py}, \Theta_p\}^{\mathrm{T}}$ is the displacement of the centroid of the block on which nodal point p (=i, j) is attached;

$$(13)\qquad B_p=\begin{bmatrix}1 & 0 & -\bar{y}_p \\ 0 & 1 & \bar{x}_p \\ 0 & 0 & 1\end{bmatrix}\qquad (p=i,j)$$

is a geometry matrix, in which $\bar{x}_p$ and $\bar{y}_p$ are the local coordinates of point p (= i, j) relative to the centroid c of the block.

From eqs.(8) and (12), the nodal forces $\boldsymbol{F}_{ij}$ can be expressed as follows:

$$(14) \quad \boldsymbol{F}_{ij} = \boldsymbol{D}_{ij} \boldsymbol{B}_{ij} \boldsymbol{U}_{ij}$$

where $B_{ij} = \begin{bmatrix} B_i & 0 \\ 0 & B_j \end{bmatrix}$; $\quad \boldsymbol{U}_{ij} = \{\boldsymbol{U}_i, \ \boldsymbol{U}_j\}^T$

The above forces are imposed on the nodal points of the structural elements from the rock blocks. The equivalent forces imposed on the blocks from the structural elements can be transformed to the centroids of the corresponding blocks as follows:

$$(15) \quad \boldsymbol{F}_{cp} = -[\boldsymbol{B}_p]^T \boldsymbol{F}_p \qquad (p = i, j)$$

where $\boldsymbol{F}_{cp} = \{F_{cx}, F_{cy}, M_c\}^T$ is the force with three components transformed to the centroid c of the block with node p ($p=i, j$) attached.

From eqs.14 and 15, the equivalent forces imposed on the centroids of the blocks can be determined as follows:

$$(16) \quad \boldsymbol{F}_{cij} = \overline{\boldsymbol{K}}_{ij} \boldsymbol{U}_{ij}$$

where $\boldsymbol{F}_{cij} = \{\boldsymbol{F}_{ci}, \boldsymbol{F}_{cj}\}^T$

$$(17) \quad \overline{\boldsymbol{K}}_{ij} = -[\boldsymbol{B}_{ij}]^T \boldsymbol{D}_{ij} \boldsymbol{B}_{ij}$$

The stiffness matrix $\overline{K}_{ij}$ in eq.(17) can be assembled into the global stiffness matrix **[K]** of the block system as described in eq.(2). The solution of eq.(2) will therefore include the supporting effect of the surface lining.

Mohr-Coulomb failure criterion may be used to determine the failure condition between the lining material and the rock blocks. Lining and rock block cease to be in contact at the point where the normal stress exceeds the tensile strength. Shear stress is reduced to the magnitude of shear strength at the point where shear failure occurs. An iteration scheme as applied in the Block-Spring Model (Wang and Garga, 1993) is used if any failure is detected.

4 AN ILLUSTRATIVE EXAMPLE

The following example illustrates the numerical analysis on the supporting effect of the surface lining around a circular tunnel. Fig.4 is a mesh of the Block-Spring Model with structural elements. The circular tunnel was analyzed in three different supporting conditions. The first case was without lining support. The second and the third case were with lining of 15 cm in thickness but with different Young's modulus. For demonstration purpose, the Young's modulus of the structural elements in the second case was assumed two times that of the rock material (i.e. $E_L = 2 \times E_R = 200$ GPa) and four times the Young's modulus of the rock in the third case. In all cases, the medium was assumed under hydro-static state of stress with the uniformly distributed pressure of

p_0 = 2.5 MPa along the boundary except the bottom part which was constrained by rollers (Fig.4). Two blocks at the centre point at the bottom were fixed in both directions.

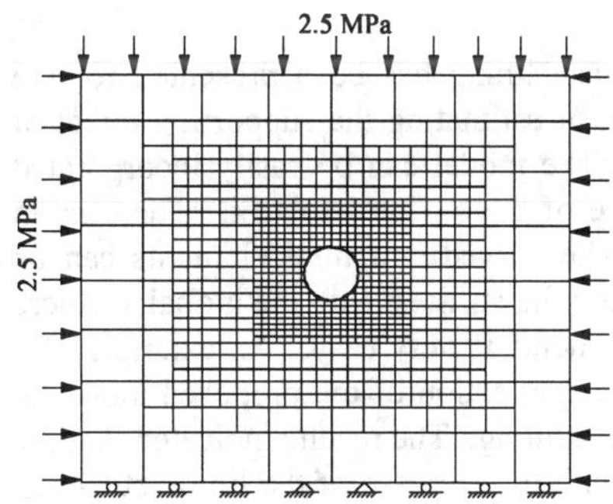

Fig.4. A BSM mesh for analysis of supporting effect of lining around a circular tunnel.

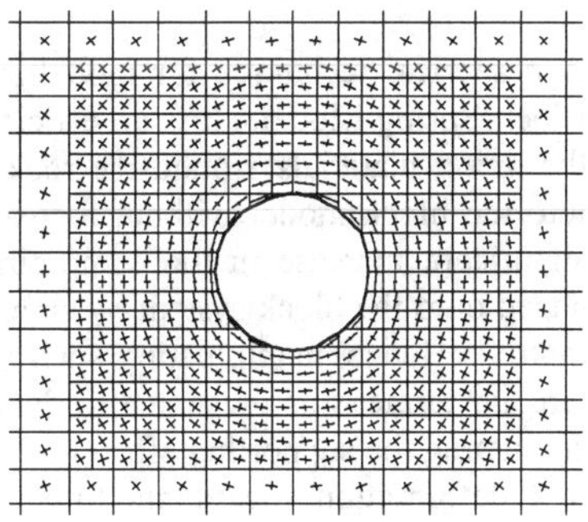

Fig.5. Principal stress distributions.

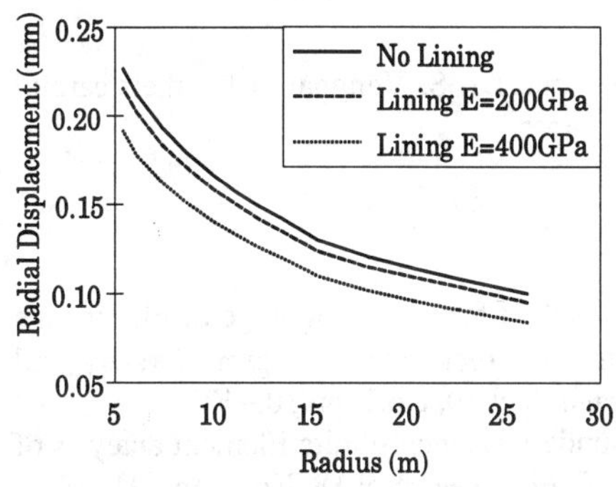

Fig.6. Vertical displacement along the radius above the crown of the tunnel.

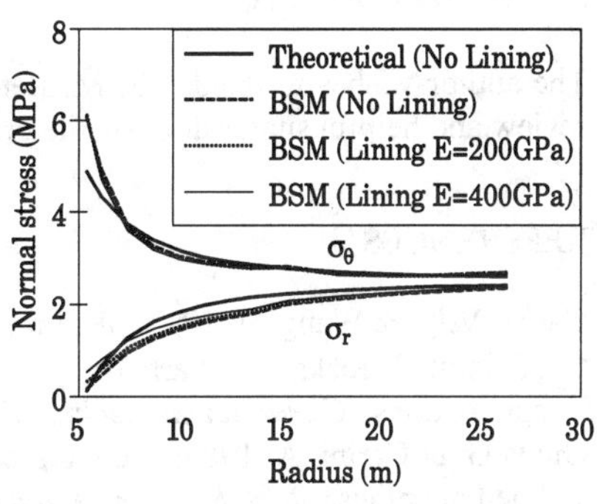

Fig.7. Radial and tangential stresses along the radius above the crown of the tunnel.

The distribution of the principal stresses around the tunnel in the second case is plotted in Fig.5. The BSM results of the three cases can be examined by comparing the displacement and stress distribution along the radius above the crown of the tunnel as shown in Fig.6 and Fig.7. It is obvious in Fig.6 that the displacement of the rock material towards the tunnel decreases with the presence of the surface lining and with the increase of the stiffness of the lining material. Such supporting effect of the surface lining can also be observed from the changes of the stress distribution. Due to the passive pressure imposed against the rock from the surface lining, the radial stress in the surrounding rock increases with the increase of the lining stiffness as shown in Fig.7. Such increase is more noticeable close to the tunnel surface. Slight decrease of the tangential stress with the increase of the lining stiffness can also be noted in Fig.7. A closed form solution of the stress distribution around the circular tunnel is also shown in Fig.7. Comparison of the numerical results with the closed form solution also indicates that the numerical solution well predicted the state of stresses.

5 CONCLUSIONS

A numerical technique for modelling rock surface lining has been presented for the Block-Spring Model. It is an efficient approach of estimating the supporting effect of the lining materials around the rock openings. The model can be easily incorporated into the block model without increasing the size of the stiffness matrix. There is no significant increase in the computing effort. The curved structural elements can be added onto the block system by simply assembling the stiffness into the global stiffness matrix. Hence, manipulating such structural elements is very easy. An example of a circular tunnel was presented to demonstrate the application of the proposed model in simulating the supporting effect of the rock surface lining. The results indicated that the rock deformation towards the tunnel decreases with the increase of the lining stiffness. It was also observed that, with the increase of the lining stiffness, the radial stress in the surrounding rock increases and the tangential stress decreases.

ACKNOWLEDGMENT

The author wishes to thank Mr. M. Bétournay and Dr. S. Vongpaisal for their careful review and helpful suggestions on the present paper.

REFERENCES

Garga, V.K. & Wang, B. 1993. A numerical method for modelling large displacements of jointed rocks — Part II: Modelling of rockbolts and groundwater, and applications. *Canadian Geotechnical Journal*, Vol.30, No.1, pp.109-123.

Gioda G. & Carini, A. 1985. A combined Boundary Element-Finite Element analysis of lined openings. *Rock Mechanics and Rock Engineering*, Vol.18, No.4, pp.293-302.

Kumer, P. & Singh, B. 1988. Pressure on lining due to initial stress field by Finite-Infinite-Interface Element Method. *Rock Mechanics and Rock Engineering*, Vol.21, No.3, pp.219-228.

Lorig, L. J. 1984. A hybrid computational model for excavation and support design in jointed media. Ph.D. thesis, University of Minnesota.

Lorig, L.J. & Brady, B.H.G. 1984. A hybrid computational scheme for excavation and support design in jointed rock media. *Design and Performance of Underground Excavations, Proc. ISRM Symp.*, Cambridge, U.K.; British Geotechnical Society, London. E.T.Brown and J.A. Hudson (eds); pp.105-112.

Wang, B. 1991. A Block-Spring Model for jointed rocks. CANMET Report, Energy, Mines & Resources Canada, Report No. MRL 92-023. 204p.

Wang, B. & Garga, V.K. 1993. A numerical method for modelling large displacements of jointed rocks — Part I: Fundamentals. *Canadian Geotechnical Journal*, Vol.30, No.1, pp.96-108.

Rock Mechanics, Nelson & Laubach (eds) © 1994 Balkema, Rotterdam, ISBN 90 5410 380 8

Numerical studies relating to micro fracture in granular materials

Zhong Liu & Neville G.W.Cook
Department of Materials Science & Mineral Engineering, University of California, Berkeley, Calif., USA

Larry R. Myer
Earth Science Division, Lawrence Berkeley Laboratory, Berkeley, Calif., USA

ABSTRACT: This paper describes a numerical analysis of the micro mechanical behavior of granular materials using a program MICRO-MECHANICS MODEL based on Boundary Element Method(BEM), elastic interaction, tensile fracture of grains and grain bonds. Some numerical tests have been done to investigate the effects of stochastic micro-structures, packing arrangement and confining stresses on the ultimate strength, fracture and macroscopic deformation.

1 INTRODUCTION

Geological materials such as sand, gravel and rock are composed of large numbers of individual particles packed in complex spatial arrangements. Such materials are commonly referred to as granular media. The most striking characteristic of a granular material is its dual nature that lies between a discrete material and a continuum. Although the individual grains of the granular material are solid, grains are only partly connected. This type of media transmits loading through discrete paths, which means that load transfer is primarily conducted through contact interactions between adjacent particles, and therefore the mechanical behavior of such materials under static and dynamic loading conditions is very complex and difficult to model. Experiments[1][2] have shown that the mechanical behavior of the granular materials is an integrated effect defined by the local micro structure, which includes the local geometrical arrangement of grains, grain shapes, grain sizes and pores, physics of contact, strength of the contact bonds and grains.

A lot of research on the mechanical behavior of granular materials has been reported in the literature. Early attempts at modeling granular material behavior used continuum mechanics, e.g. elasticity, plasticity or viscoelasticity. Photo elastic method has also been used to investigate the stress distributions inside particles, mechanical behavior and micro fracture under static loading[3], and wave propagation[[4]. However, it was gradually realized that the complex micro structural nature of these materials requires that Micro-mechanical models be used to more accurately predict the mechanical behavior. Micro mechanical models are used to study macroscopic properties of granular materials in terms of microscopic processes. Some of these models are based on a continuum containing micro cracks while others are based on a granular or discrete element approach[5][6][7].

In this paper we will use a numerical model called Micro mechanics Model based on Boundary Element Method(BE), the elastic interaction, tensile fracture of grains and grain bonds in order to simulate the effects of stochastic micro structures on macroscopic stress-strain relation and fracture under differential compressive stresses. We will restrict our attention to the effects of heterogeneity in grain size and loading on the mechanical behavior of granular materials.

2 TECHNICAL APPROACH

In this section, several major features of the Micro mechanics Model are described. The program, which is written in C language, provides a foundation for investigating the micro mechanical behavior of granular materials. The Micro mechanics Model program tracks the behavior of an assembly of two dimensional circular disc with arbitrarily radius.

2.1 Data structure

Data structures used in this program are designed according to the physical micro-structure of granular materials. At microscopic level, from a point of view of physics, granular materials consist of grains. Every grain has its own physical properties and geometric properties. When a grain is placed into a particle system, its contact information and neighborhood information also need to be described.

In the design of data structure for the Micro mechanics Model, a concept of layered structures is used. A granular material is represented as a global pointer at macroscopic level. This pointer is pointed to a GRAIN list. Every element of the GRAIN list represents a grain of a granular material. The sketch of the main data structures of the program is shown in figure 1. For every GRAIN, there are two parts. one part is information body describing all information for the grain. Another part is a pointer pointed to the next grain. The second layer includes data structures such as physical properties, geometric properties, contact information and neighbor information.

The sub data structure of physical properties of GRAIN is denoted as phy_ptr, which is a pointer pointed to one element of a list of constitutive models used in the particle system. The list of constitutive models is a database collecting all the constitutive models used in granular materials. The sub data structure of geometric properties of GRAIN is denoted as geo_ptr that is a pointer pointed to one element of a list of geometric properties, in which the shape, size and position in the granular system are defined. Contact information of grains is described in sub data structure con_ptr, which defines the properties of the contact point such as debonding strength, friction, sliding, contact forces and contact displacement etc.. The sub data structure nei_ptr defines all information for the neighborhood of a given grain. It specifies the detecting function to find out neighbor grains. Because the neighborhood situation may change due to movements of grains, nei_ptr also defines a rechecking criterion for recalculating the neighbor blocks of a given grain. By using this data structure, computing time can be reduced because only those neighbor grains on the link list are searched to find out if any new contact points are produced, not the whole grain system.

This layered data structure is very flexible. If a new feature needs to be added to the structure, there is no change for the existing program. All that is needed is to define a new sub data structure or a new function in a sub-layer. For example, if a new constitutive model needs to be added, a data structure for this model is designed and then put into the list of physical properties. By using the above data structure, the Micro mechanics Model is implemented in C language.

2.2 Numerical methodology

Local stress in individual grains are important in the overall response of granular materials and fracture propagation. In our research, a combined BEM(direct boundary integral method)-Energy Minimization Scheme is used to calculate the stress distribution within grains.

In the BEM, boundary stresses and displacements are chosen as variables. For every grain in the system, the boundary is divided into n segments, so there are 2n unknowns: σ_s^j, σ_n^j, u_s^j, u_n^j, with j = 1, 2, ..., n. σ_n^j, u_n^j are the normal stress and the normal

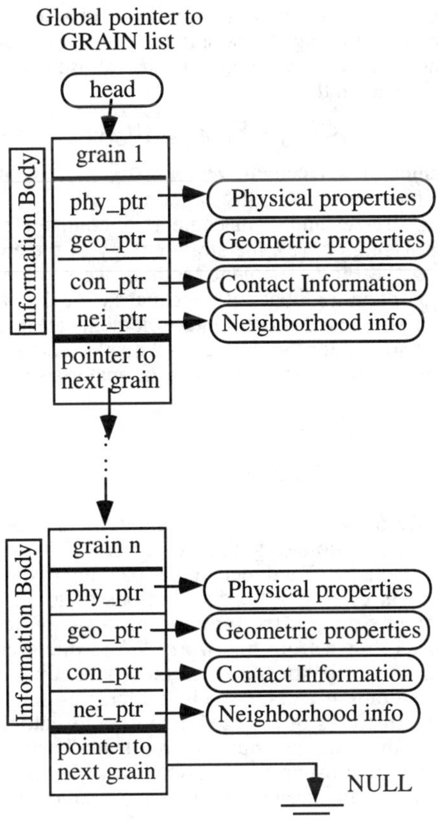

Figure 1 The sketch of data structures of granular materials.

displacement of segment j. σ_s^j, u_s^j are the tangential stress and displacement of segment j.

By using the integral representation:

$$u = \oint_C (G\sigma\hat{n} - \hat{n}\Sigma u)ds \qquad (1)$$

and choosing 2n test solutions[8], we can get an equation

$$B\{\sigma_B\} = A\{u_B\} \quad , \qquad (2)$$

which can be used to calculate the unknown stresses and displacements along the boundary. **B** and **A** in equation (2) are matrices of influence coefficients and dependent only on the geometry of the grain boundary. If two grains have the same shape with different boundary stresses and displacements, there is no change for either the **B** or the **A** matrix. $\{\sigma_B\}$ is a stress vector along the boundary and $\{u_B\}$ is a displacement vector along the boundary:

For granular materials, the detailed stress distributions within grains need to be computed to investigate the mechanisms of nonelastic behavior and micro fracture. Granular materials typically comprise a great number of grains. If every grain is divided into hundreds of segments, there are hundreds of thousands of variables in a system. We can not directly use the BEM to deal with this problem. The number of variables must be reduced. In the granular material system, external energy is input into grains through contacts. Stress distribution within grains can be described in terms of contact stresses and displacements along the boundary. Based on this, for every grain, only contact stresses and contact displacements are chosen as variables. After some derivation[9], it can be shown that

$$\{\sigma_{BC}\} = F_{stiffness}\{u_{BC}\} \quad , \qquad (3)$$

where $\{u_{BC}\}$ is a contact displacement vector, $\{\sigma_{BC}\}$ is a contact traction vector and $F_{stiffness}$ is a local stiffness matrix of a given grain.

In our simulation, stresses and displacements of contact elements can be chosen as the system variables. Using eq. (3), the strain energy of the entire system of grains is calculated. Following the Finite Element Method, we use the principle of minimum energy and eq. (3) to obtain a global stiffness matrix for the particle system, in which only displacements or stresses of those contact elements are variables. After the global system of equations is solved for the stresses and displacements of the contact elements, the stress distribution within the grains of granular materials can be calculated from eq. (1).

3. ANALYSIS OF MATERIAL BEHAVIOR

The software Micro-mechanics Model models the movement, deformation, interaction and stress distribution of grains which make up the granular materials. We have previously used hexagonal and rectangular packing of circular grains of the same size to study the effects of stochastic distributions of grain strength and confining stresses on the ultimate strength, fracture and macroscopic deformation[9]. Several conclusions have been drawn from this work. When micro cracks occur in granular materials they generate tensile stress concentrations in adjacent grains near the crack tip. These stress concentrations and the stochastic tensile strength distribution drive the propagation of macro fractures in hexagonal and cubic lattices. Results suggest that granular materials with homogeneous or small stochastic tensile grain strength distributions have only macroscopic strain-softening. Granular materials with large heterogeneous tensile grain strength distribution have both macroscopic strain-hardening and strain-softening.

In this paper, mechanical properties of granular materials with different grain sizes are described and analyzed. The sample of random structure is generated from a random generator based on local minimum potential energy. The particle systems used in our simulation consist of grains from 200 to 600 with the boundary of each grain divided into 100 elements. Within each grain, stresses at 400 points are calculated to find the maximum stress within the grain. With load increasing, the maximum stress point is searched through the whole system. When the maximum stress reaches the tensile strength of the grain, the grain is broken, and its Young's modulus is reduced 25% for the next step of the calculation. If contact stresses at interfaces are tensile and reach debonding strengths, these interfaces are broken and elements at these interfaces are traction free. If two grains contact again during subsequent loading, matrices A and B of these two grains must be recalculated due to the changed contact situation.

The stress distributions in granular materials with random structures are calculated by using our micro mechanics model. Fig. 2 is a higher magnitude of the stress distribution for a few selected grains from a random structure, which is very similar to those results gotten from photo elastic experiments. The stress distribution of the whole system is shown in Figure 3. From our calculation, it is found that stress concentrations arise at grain contacts in granular systems because external forces applied at the system boundary are transmitted from grain to grain over small contact areas. The intensity of a stress concentration at a contact depends primarily on micro structures in the region near that grain, and the orientation of the normal to that area relative to boundary loads. From Fig. 3, it is clear that some discrete paths have been established, along which major load transfer occurs. The local micro structure i.e. the local geometrical

Figure 2. Stress distribution of a few selected grains.

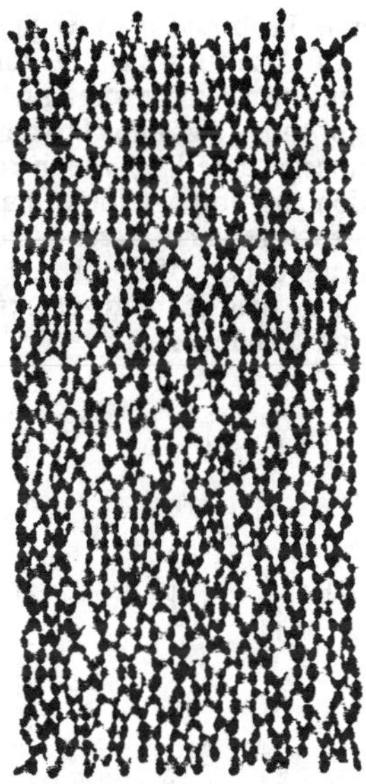

Figure 3. Stress distributions of a granular material with random structures.

arrangement of particles plays a dominant role in establishing this discrete load transfer phenomena. The discrete nature of the materials creates a structured network for the propagation of loading.

In the following, three cases have been chosen to study the mechanical behavior. Case #1 is a random structure consisting of grains with three radii: r1 = 1.9, r2 = 2.0, r3 = 2.1. Its porosity is 17.03%. Case #2 is a random structure consisting of grains with three radii: r1 = 1.9, r2 = 2.0, r3 = 2.1. Its porosity is 18.26%. Case #3 is a random structure consisting of grains with three radii over a border range: r1 = 1.8, r2 = 2.0, r3 = 2.2. Its porosity is 17.48%. For these random structures, the average radius is the same and all grains have a uniform tensile strength and debonding strength. Figure 4a, b, c show the micro structures of the above cases.

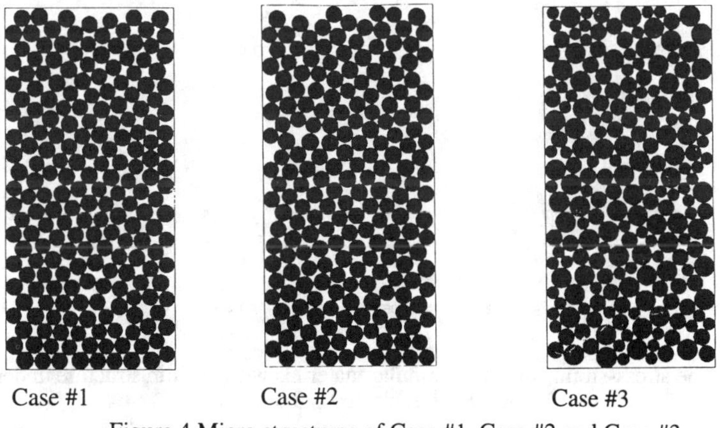

Case #1 Case #2 Case #3

Figure 4 Micro structures of Case #1, Case #2 and Case #3

Figure 5a, b and c show the micro-cracks produced and propagated in particle systems of Case #1, #2 and #3 under confining pressure 8.89. All grains in the systems have a uniform tensile strength. The length of these lines in the plots represents the number of times a grain is broken. When confining stresses reach certain values, micro-cracks merge together, which results in strain softening as shown in figure 6. The generation and propagation of micro-cracks are dependent on both the heterogeneity of micro-structures and stress concentration due to micro-cracks. Figure 6 shows the stress-strain curves of Cases 1, 2 and 3 under confining pressure 8.89. From Fig. 5 and 6, it is clear that, if micro-cracks in a particle system are more localized, more strain softening is produced.

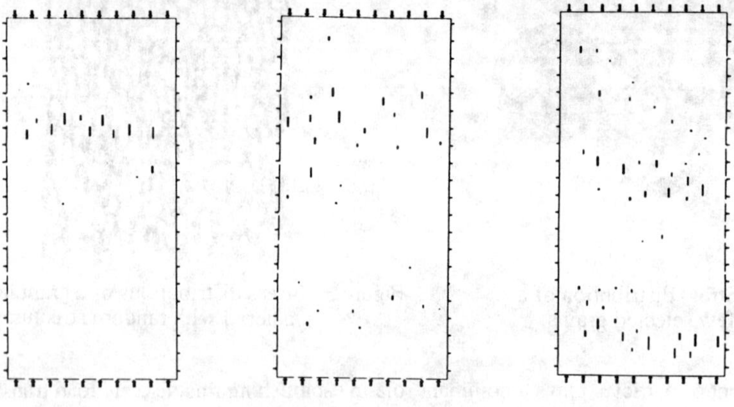

Figure 5. Distribution of micro-cracks of a granular material with random structure under confining pressure (p0 = 8.89).

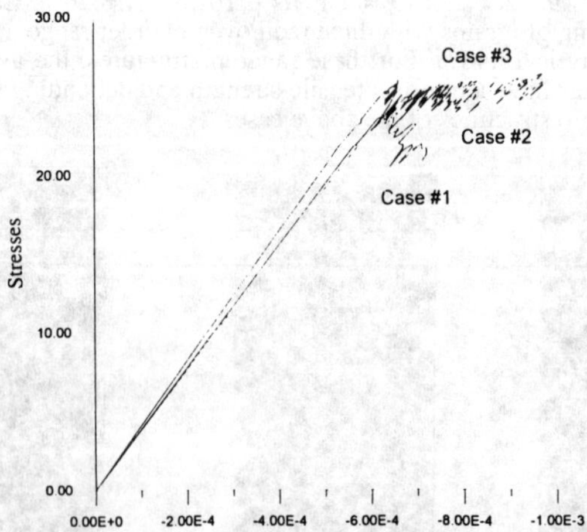

Figure 6. The stress-strain curves of granular materials with random structure under confining pressure (p0 = 8.89).

Figure 7a and b show the distribution of micro-cracks under confining stress 8.89 and 26.67 for Case #1. From Fig. 7, it is seen that, with 8.89 confining pressure, micro cracks merge together, which means under confining pressures, the stress concentrations have a bigger effect on determining which grain breaks next. When confining pressure reaches p = 26.67, a lot of micro-cracks are produced under hydrostatic condition, which results in strain-hardening as shown in Fig. 8.

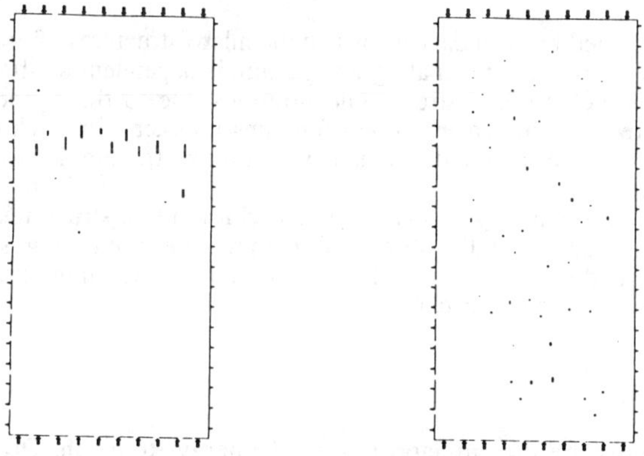

p=8.89 p=26.67

Figure 7. Distribution of micro-cracks of a granular material with random structure under confining pressure.

Figure 8 shows the stress-strain curves of Case #1 under different confining pressures. From these curves, it is clear that, under lower confining pressures, the system has little strain hardening and lots of strain softening. With confining pressures increasing, the yield strength and ultimate strength increase. When the confining pressure reaches certain values, strain hardening is a dominant behavior of the system.

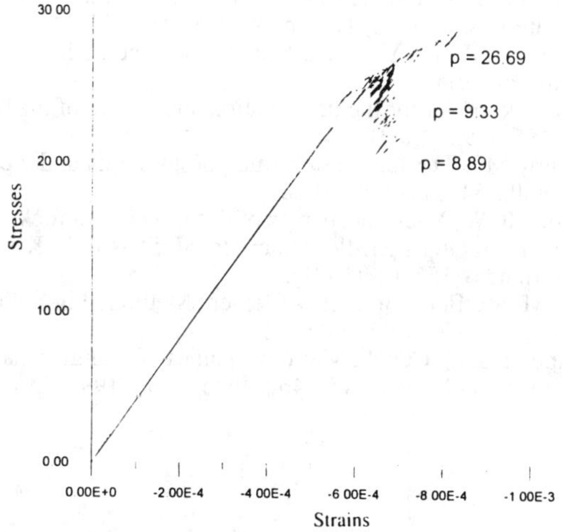

Figure 8. The stress-strain curves of granular materials with random micro structures under differential confining pressures(Case #1).

4 CONCLUSIONS

From the above discussion, the following conclusions can be drawn:

When micro cracks occur in granular materials they generate tensile stress concentrations in adjacent grains near the crack tip, as would occur in a continuum. These stress concentrations and the heterogeneity of random micro-structures determine the production of micro cracks and drive the propagation of macro fractures in granular materials.

Macro fractures are affected by boundary loads and the micro structures of granular materials. If the confining pressures are small, grain fracture is dependent on both the heterogeneity of micro-structures and stress concentrations due to these fractured grains. With confining pressures increasing, grain fractures interact with each other, stress concentrations due to these fractured grains have a strong effect in determining which grain is broken next.

Under lower confining pressures, granular materials with random structures have strain softening and low ultimate strength. With confining pressures increasing, so are the ultimate strength and yield strength of granular materials. When confining pressure reaches certain values, strain hardening occurs .

ACKNOWLEDGMENTS

This work was supported by the Director, Office of Energy Research, Office of Basic Energy Sciences, Engineering and Geosciences Division, of the US Department of Energy under Contract No DE-AC03-76SF00098.

REFERENCES

1. Zhang J, Wong TF, and Davis DM, Micromechanics of pressure-induced grain crushing in porous rocks, J. Geophys.Res., **95**, 341-352 (1990).
2. Myer LR, Kemeny JM, Zheng Z, Suarez R, Ewy RT, and Cook NGW, Extensile cracking in porous rock under differential compressive stress, Appl. Mech. Rev. **45**, n8, 263-280(1992).
3. Gallagher J.J., Friedman M., Handin J. and Sowers G.M., E, Experimental studies relating to microfracture in sandstone, Tectonophysics, 21, 203-247(1974).
4. Sadd M.H., Shukla A. and Zhu C.Y., The effect of voids and inclusions on wave propagation in granular materials,
5. Costin LS, A microcrack model for the deformation and failure of brittle rock, J. Geophys. Res., **88,** 9485-9492 (1983).
6. Sammis CG and Ashby MF, The failure of brittle porous solids under compressive stress states, Acta Metall., **34**, 511- 526 (1986).
7. Kemeny JM and Cook NGW, Micromechanics of deformation in rocks, in Toughening Mechanisms in Quasi-Brittle Materials, SP Shaw(ed), Klewer Academic, The Netherlands, 155-188(1991).
8. S. L. Crouch and A. M. Starfield, Boundary Element Methods] in Solid Mechanics. George Allen & Unwin, London(1983).
9. Z. Liu, L.R. Myer and N.G.W. COOk, Micromechanics of granular materials, 34th U.S. Symposium on rock mechanics, p459-462, June 27-30, 1993.

11 Numerical modeling – Continuum analysis

Session developers:
Ahmed Abou-Sayed, B.P. Exploration
Dan Schultz-Ela, Bureau of Economic Geology

Rock Mechanics, Nelson & Laubach (eds) © 1994 Balkema, Rotterdam, ISBN 90 5410 380 8

Evaluation of pressure and stresses due to fluid sources in a semi-infinite medium

M. Bai & J.-C. Roegiers
School of Petroleum and Geological Engineering, University of Oklahoma, Norman, Okla., USA

ABSTRACT: This paper presents a simplified analysis of fluid pressure and solid stresses induced by fluid injection or pumping in a semi-infinite medium. The coupling of volumetric strain with the change in fluid mass has been partially decoupled using a poroelastic potential. The solutions of pressure and stresses in a semi-infinite space are obtained via the method of superposition. The resulting nonphysical surface shear stresses are numerically eliminated. Fluid sources of various geometries can be constructed to adapt to complex three-dimensional settings of underground disturbances.

1 INTRODUCTION

In general, when a poroelastic medium is subjected to fluid injection or pumping stress redistributions may arise either due to nonuniform pore pressure distribution or as a result of external load, or combination of both causes. This phenomenon was first rationalized by Terzaghi (1923) in his effective stress law, and later generalized by Biot (1941) in a three-dimensional formulation for consolidation scenarios. Fundamental solutions for a homogeneous, isotropic, infinite poroelastic medium due to point fluid sources were given by Cleary (1977) based on the stress function method.

For semi-infinite media which resembles the spatial domain geometry for underground structures, various solutions are available based on the approximation of fully coupled poroelasticity. McNamee and Gibson (1960) examined the consolidation of clay in a semi-infinite stratum using two displacement functions. Apirathvorakij and Karasudhi (1980) presented solutions of a loaded elastic bar embedded in a poroelastic half space utilizing the concept of Biot's consolidation theory. Using Melan's solution for a central point dilatation, Segall (1985) solved the coupled flow-subsidence problem in a semi-infinite medium subjected to fluid withdrawal. Halpern and Christiano (1986) proposed the dynamic solutions for the media in a half space subjected to steady-state harmonic surface loads based on Biot's dynamic poroelasticity (Biot, 1962).

The complete analogy between poro- and thermoelasticity was identified by Biot (1955). As pointed out by Rice and Cleary (1976) the analogy held only when the coupling between the fluid pressure (or temperature) and stresses was rigorously maintained. In poroelasticity, complete decoupling may cause significant errors, especially in close vicinity of the loading source (Cryer, 1963). In an infinite medium, partial decoupling using displacement potentials appears to be a fairly accurate approximation even near a fluid source (Curran and Carvalho, 1987). As a result, a simple sequential solution procedure is permissible.

This paper presents solutions for pressure and stresses due to fluid sources in an unbounded three-dimensional space. The solutions are not completely analogous to those of decoupled thermoelasticity because the influence of volumetric strain on fluid pressure is implicitly coupled in the formulation. To extend the solutions for a semi-infinite space, the method of superposition as proposed by Nowacki (1962) for thermoelasticity is used. However, in order to eliminate the nonphysical surface shear stresses generated by using superposition, the technique proposed by St. John and Christianson (1980) is followed. This analysis offers a flexible tool to examine the magnitude and influence of pore pressure, solid stress and deformation within a three-dimensional domain.

2 FICTITIOUS DECOUPLING IN POROELASTICITY

For rigorous poroelasticity, coupling between pressure and displacements or stresses may be presented in a fictitious decoupled fashion for both displacement (Biot, 1941) and stress (Rice and Cleary, 1976) methods.

For the displacement method, fictitious decoupling between deformation and fluid pressure may be fulfilled through the solid deformation phase. In Biot's theory (Biot, 1941), the constitutive equation for the solid phase may be expressed as:

$$\epsilon_{ij} = \frac{1+\nu}{E}\sigma_{ij} - \frac{\nu}{E}\sigma_{kk}\delta_{ij} - \frac{1}{3H}p\delta_{ij} \tag{1}$$

where ϵ_{ij} and σ_{ij} are the strain and stress tensors, respectively; p is the fluid pressure, E is the elastic Young's modulus, ν is Poisson's ratio, H is Biot's constant, σ_{kk} is the total stress, and can be written as $\sigma_{kk} = \sigma_{11} + \sigma_{22} + \sigma_{33}$.

The volumetric strain, θ, can be derived from equation (1) as:

$$\theta = \epsilon_{ii} = \frac{(1-2\nu)}{E}\sigma_{kk} - \frac{1}{H}p = \frac{3(1-2\nu)}{E}(\sigma_c - \alpha p) \tag{2}$$

where α is the fluid pressure ratio which can be described as $\alpha = \frac{E}{3H(1-2\nu)}$; the mean stress σ_c may be expressed as $\sigma_c = \frac{\sigma_{kk}}{3}$. Assuming the effective mean stress $\sigma^e_{kk}/3$ to be equal to $\sigma_c - \alpha p$, then the effective total stress σ^e_{kk} is

$$\sigma^e_{kk} = \sigma_{kk} - 3\alpha p \tag{3}$$

As a result, the volumetric strain can be reformulated as $\theta = \frac{(1-2\nu)}{E}\sigma^e_{kk}$. If one assumes the effective stress tensor σ^e_{ij} as $\sigma_{ij} - \alpha p\delta_{ij}$, and substituting this relationship into equation (1), one obtains

$$\epsilon_{ij} = \frac{1+\nu}{E}\sigma^e_{ij} - \frac{3\nu}{E}\delta_{ij}\left(\frac{\sigma_{kk}}{3} - \alpha p\right) \tag{4}$$

Substituting equation (3) into (4), gives

$$\epsilon^e_{ij} = \frac{1+\nu}{E}\sigma^e_{ij} - \frac{\nu}{E}\sigma^e_{kk}\delta_{ij}. \tag{5}$$

Incorporating the strain-displacement relationship $\epsilon^e_{ij} = \frac{1}{2}(u^e_{i,j} + u^e_{j,i})$ and equation (5) into the equilibrium equation $\sigma^e_{ij,j} = 0$, yields

$$Gu^e_{i,jj} + (\lambda + G)u^e_{k,ki} = 0. \tag{6}$$

where λ is Lamé's constant, and G is the shear modulus.

Equation (6) represents a displacement based momentum conservation equation, which carries identical form as the elasticity theory except that the displacement is induced by the effective stress rather than by the total stress. Therefore, if the effective stress is known at the initial period and along the boundary in solving a poroelastic problem, it appears that only an elastic case lumping the effect of fluid pressure and solid displacement for porous media needs to be solved. In actuality, however, the total stress instead of the effective stress is known initially and at the boundary. As a result, a coupled scenario is still envisioned.

For the stress method, fictitious decoupling between deformation and fluid pressure may be similarly obtained, however through the fluid flow phase. Rice and Cleary (1976) proposed a stress-based method with the parameters distinguished by the loading stages. The change of fluid mass content per unit volume, $\Delta\zeta$, may be represented by

$$\Delta\zeta = A(\sigma_{kk} - \frac{3}{B}p) \tag{7}$$

where $A = \dfrac{3(\nu_u - \nu)}{BE(1+\nu_u)}$, ν_u and ν are the undrained and drained Poisson's ratios, respectively; B is Skempton's constant (Skempton, 1954).

Mass conservation can be satisfied by combining Darcy's velocity, q_i, with the continuity requirement to form $q_{i,i} - \Delta\dot{\zeta} = 0$; then substituting equation (7), yields:

$$-\kappa\nabla^2 p = A(\dot{\sigma}_{kk} - \frac{3}{B}\dot{p}) \tag{8}$$

where κ is the intrinsic permeability (ratio of permeability k to fluid dynamic viscosity μ). Combining the compatibility condition given by Rice and Cleary (1976) and equation (8), upon the consideration of equation (7), results in

$$c\nabla^2(\Delta\zeta) = \Delta\dot{\zeta} \tag{9}$$

where c is the diffusivity coefficient (Rice and Cleary, 1976).

Equation (9) shows that the change in fluid mass content per unit volume satisfies a homogeneous diffusion equation. It is interesting to note that the fictitious decoupling is completed through the fluid phase, as contrasted to the decoupling through the solid phase by the displacement method. Segall (1985) predicted stress change and subsidence in earthquake regions due to groundwater withdrawal assuming the fluid mass content per unit volume, ζ, as a primary unknown. In reality, however, it is difficult to quantify the initial and boundary conditions for ζ. As a result, the coupling effect is retained in the evaluation of the change in pressure and stresses using the stress method.

3 SOLUTIONS FOR PRESSURE AND STRESSES

The requirement of rigorous coupling in poroelasticity imposes significant restrictions on the flexibility of its mathematical modeling. When the pore pressure field is generated by applying fluid pressure or injection rate, and the changes of pressure and stresses at the reservoir boundary are negligible, the solid displacement may be expressed as a poroelastic potential (Curran and Carvalho, 1987). Consequently, a partial decoupling can be achieved.

Following the displacement method (Biot, 1941), the governing equations for solid deformation and fluid flow may be expressed as

$$\begin{cases} Gu_{i,jj} + (\lambda + G)u_{k,ki} + \alpha p_{,i} = 0 \\ \\ \dfrac{k}{\mu}p_{,kk} + \alpha\dot{\epsilon}_{kk} - \phi\dot{p} = 0 \end{cases} \tag{10}$$

where the fluid pressure ratio, α, and the relative compressibility, ϕ, may be alternatively evaluated by

$$\alpha = 1 - \frac{K}{K_s} = 1 - \frac{E}{3(1-2\nu)K_s}; \qquad \phi = \frac{1-n}{K_s} + \frac{n}{K_f} \tag{11}$$

where n is the porosity, K, K_s and K_f are the bulk moduli of rock skeleton, solid grains and fluid, respectively.

For an infinite medium, the poroelastic potential, Φ, may be expressed as $u_i = \Phi_{,i}$ (Goodier, 1936). Substituting this relationship into the first equation of (10), and integrating with respect to x_i, results in a Poisson equation

$$\Phi_{,kk} = \epsilon_{kk} = -mp \tag{12}$$

where $m = \frac{\alpha}{2G+\lambda}$. Substituting equation (12) into the second equation of (10), yields

$$c_k p_{,kk} = \dot{p} \tag{13}$$

where c_k is the modified diffusivity and may be given as

$$c_k = \frac{k}{\mu}(m\alpha + \phi)^{-1} = \frac{k}{\mu}\left[\frac{\alpha^2(1+\nu)(1-2\nu)}{E(1-\nu)} + \frac{1-n}{K_s} + \frac{n}{K_f}\right]^{-1} \tag{14}$$

c_k may also be written independent of the pressure ratio, α, as

$$c_k = \frac{k}{\mu}\left\{\frac{1+\nu}{1-\nu}\left[\frac{1-2\nu}{E} - \frac{2}{3K_s} + \frac{E}{9(1-2\nu)K_s^2}\right] + \frac{1-n}{K_s} + \frac{n}{K_f}\right\}^{-1} \tag{15}$$

Solutions of equation (13) subjected to an instantaneous fluid source can be described (Carslaw and Jaeger, 1959) by

$$p = \frac{Q}{8(\pi c_k t)^{3/2}} \exp\left[-\frac{R^2}{4c_k t}\right] \tag{16}$$

where the source strength Q may be represented by the product of the initial fluid pressure and the finite injected fluid volume at the source point (Carslaw and Jaeger, 1959); R is the distance from the source location (x_0, y_0, z_0), and $R^2 = (x-x_0)^2 + (y-y_0)^2 + (z-z_0)^2$. As $t \to 0$, p in equation (16) tends to zero at all points except at the source location where it becomes infinite.

Substituting equation (13) into (12), performing Laplace transform to the poroelastic potential Φ, and inverting the result, one has

$$\Phi = \frac{Qm}{4\pi R}\operatorname{erf}\left(\frac{R}{\sqrt{4c_k t}}\right) \tag{17}$$

Stresses can be determined through the following relationship

$$\begin{aligned}\sigma_{ij} &= 2G\epsilon_{ij} + \lambda\epsilon_{kk}\delta_{ij} + \alpha p\delta_{ij} \\ &= 2G\Phi_{,ij} + (\lambda - \frac{\alpha}{m})\Phi_{,kk}\delta_{ij} = 2G(\Phi_{,ij} - \Phi_{,kk}\delta_{ij})\end{aligned} \tag{18}$$

The previous solutions for the poroelastic potential, pressure and stresses are obtained for the fluid source applied instantaneously. For the medium subjected to the constant fluid source strength Q, the following expressions are applicable:

$$\Phi = \frac{Qm}{4\pi R}\int_0^t \operatorname{erf}\left[\frac{R}{\sqrt{4c_k(t-\tau)}}\right] d\tau$$

$$= \frac{QmR}{8c_k\pi}\left[1-\left(1+\frac{4c_kt}{2R^2}\right)\operatorname{erf}\left(\frac{R}{\sqrt{4c_kt}}\right)-\frac{1}{R}\sqrt{\frac{4c_kt}{\pi}}\exp\left(-\frac{R^2}{4c_kt}\right)\right] \tag{19}$$

$$p = \frac{Q}{4\pi {c_k}^{3/2}}\int_0^t \frac{1}{(t-\tau)^{3/2}}\exp\left[-\frac{R^2}{\sqrt{4c_k(t-\tau)}}\right] d\tau = \frac{Q}{4c_kR\pi}\operatorname{erfc}\left(\frac{R}{\sqrt{4c_kt}}\right) \tag{20}$$

where τ is an integral parameter. Equation (18) is still applicable for stresses, however, Φ from equation (19) should be used.

4 SOLUTIONS FOR SEMI-INFINITE MEDIA

The presented solutions are suitable for infinite media only. For the bounded media such as the one with a free surface in half space, an additional poroelastic potential satisfying the boundary conditions needs to be designed. The solutions of pressure and stresses are derived from the superposition of the solutions from an infinite medium as given previously and those due to the modification of the boundary conditions. The concept described by Nowacki (1962) in thermoelasticity for semi-infinite media, using the method of images, can be applied directly in poroelasticity. In a cartesian coordinate system, the superposition of fluid pressure and stresses due to a fluid source at $(x, y, z-a)$ and a sink at $(x, y, z+a)$ would result in a surface at (x, y, z) where the pressures and normal stresses vanish, and the shear stresses are doubled. To eliminate the nonphysical shear traction on the surface, another potential satisfying Love's function (Love, 1927) can be constructed. The summation of the pressure and stresses calculated from this potential with those obtained from an infinite medium, results in the poroelastic solutions in which a surface free from pressure and stresses is generated.

For a less rigorous approach, or if the underground disturbance is of limited scope, the nonphysical surface shear traction as a result of superposition can be artificially relieved by numerical means (St. John and Christianson, 1980). The basic concept is to design a surface shear relief area sufficiently large to cover the affected region due to the superposition of fluid sources and sinks as shown in Figure 1. The shear force equal magnitude but opposite sign to the shear due to the superposition is then applied on the designated surface area to create a shear-free surface. The point fluid sources and sinks create the surface point shear stresses, and can be subsequently eliminated in a piecewise fashion.

5 HYPOTHETICAL CASE STUDIES

The utilities of the presented solutions can be demonstrated by applying them in a few hypothetical case studies. The first case describes a constant point fluid injection situated at the depth 1000 m below the surface in the homogeneous semi-infinite poroelastic media. The selected parameters are listed in Table 1. Figure 2 illustrates that the different pressure decline profiles from the fluid source along the x direction under a constant y-z section subjected to the variation of permeabilities one hour after fluid injection. It is apparent that a lower formation permeability corresponds to a tighter reservoir, and subsequently leads to a higher generated pressure magnitude. For a same geometric configuration, Figure 3 describes the spatial pressure distribution for various elastic moduli of the formation at 1000 seconds after fluid injection. It appears that the suppression of the fluid pressure generation can be attributed to the increase in formation stiffness. Figure 4 depicts the temporal pressure changes at 0.1 m from the source center for various pressure ratios. The effect of coupling deformations with fluid pressure appears prominent, which is represented by the more than doubled pressure magnitude as α changes from 0.2 to 0.8.

The second case describes a series of fluid injection under constant rate along a borehole inclined 30 degree in the z-x section as shown in Figure 5. Using the parameters given in Table 1, Figure 6 provides the pressure distribution contouring in the vicinity of the fluid sources. Dramatic pressure change occurring in the close proximity of the sources can be readily observed. Although drastic, the pressure magnitudes are in general uniformly distributed along the source locations, which could be substantially modified if the source strength of variable intensities were applied. The displacement contours in the z direction are given in Figure 7. Dilatation of noticeable magnitude, as a result of fluid injection, can be found in the neighborhood of the sources. Higher magnitudes of displacement at the lower and upper part of the borehole may be attributed to the effect of structural inclination, which may become apparent in the analysis of shearing as shown in Figure 8.

6 CONCLUDING REMARKS

This paper presents the solutions of three-dimensional poroelasticity for a half-space with less restriction on coupling. The geometry of the semi-infinite medium is accommodated by adopting the technique of superposition frequently used in the hydraulics and heat transfer. The solutions in a three-dimensional configuration provide a flexible tool to evaluate the pressure and stresses due to the effect of underground disturbance which possesses a complicated spatial setting such as inclined boreholes, combination of vertical and horizontal wells, etc. Poroelastic analyses not only provide an assessment for potential fluid pressure control, but also give a stability appraisal for the affected facilities subjected to various loads. This preliminary analysis using the presented methodology indicates that the change in poroelastic parameters may have a significant influence on the variation of fluid pressure. The change in fluid pressure is also critically controlled by the fluid and mechanical properties of the reservoir. Depending on the intensities as well as geometries of the fluid sources, induced pressure and stresses may be nonuniformly distributed, and therefore may be subject to the influence of induced shear forces.

Table 1. Modeling parameters

Parameter	Value	Unit	Parameter	Value	Unit
Elastic modulus, E	0.5	GPa	Poisson ratio, ν	0.25	
Viscosity, μ	0.002	Pa s	Porosity, n	0.2	
Permeability, k	10^{-13}	m^2	Fluid modulus, K_f	1	GPa
Grain modulus, K_s	2	GPa	Pressure ratio, α	0.8	
Source strength, Q	1120	Pa m^3	Initial pressure, p_0	14	MPa

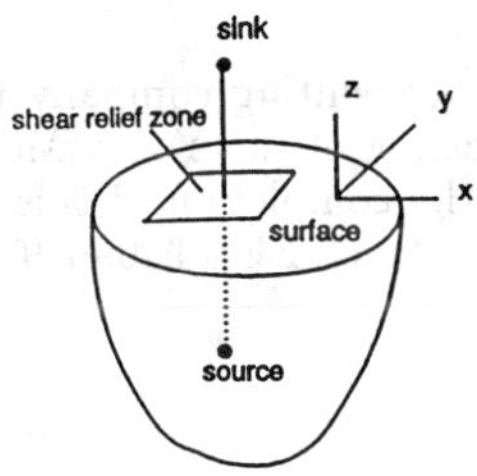

Figure 1. Schematic shear relief

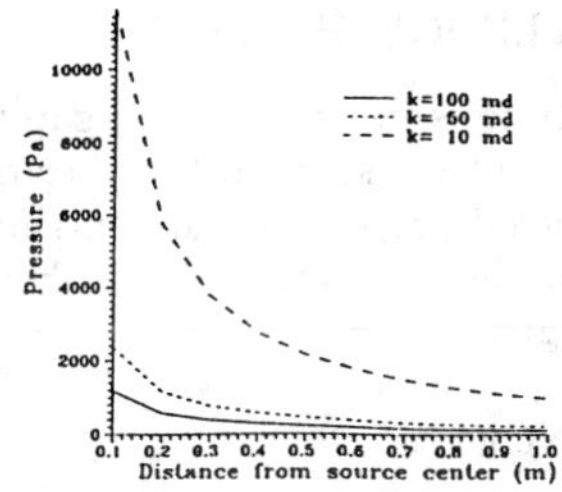

Figure 2. Pressure for permeabilities

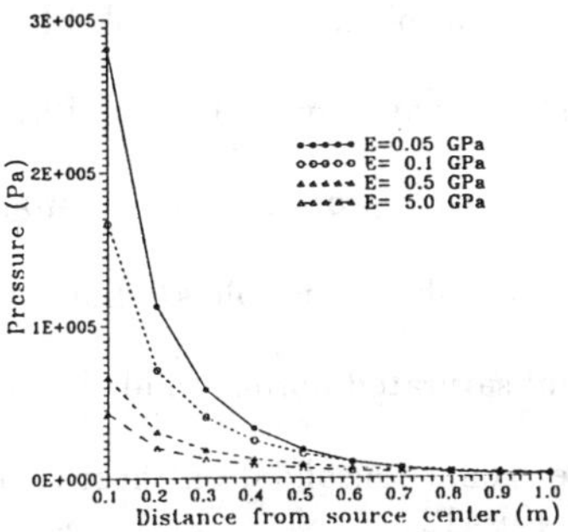

Figure 3. Pressure for moduli

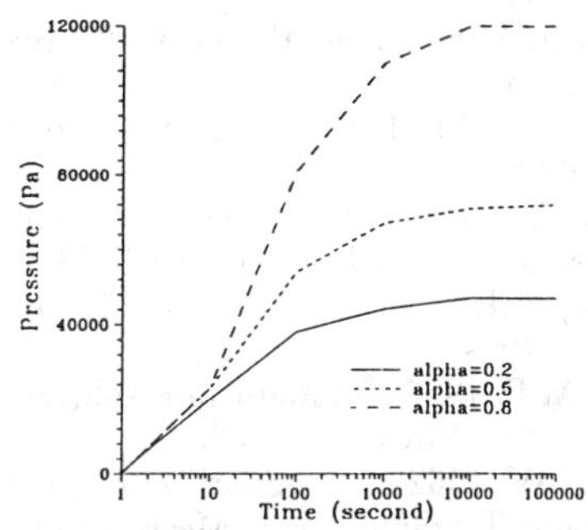

Figure 4. Pressure for ratios (α)

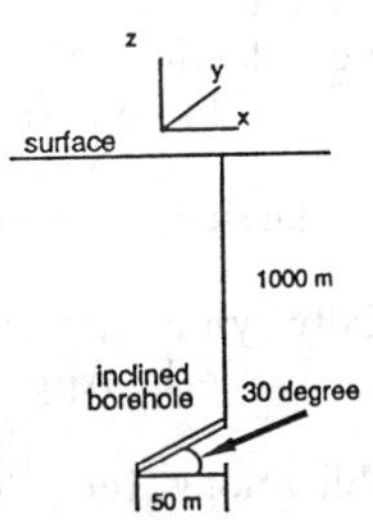

Figure 5. Inclined borehole

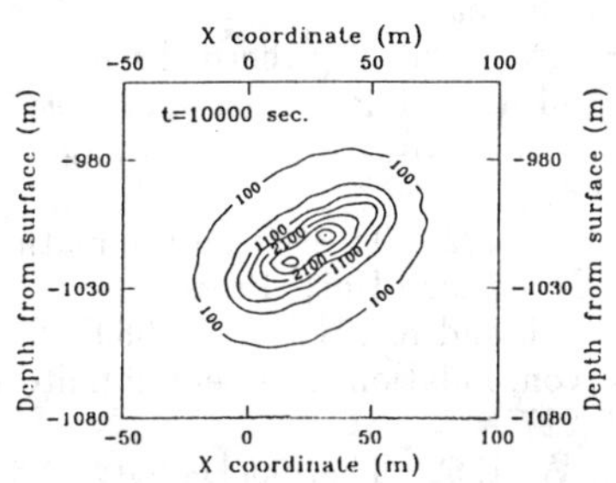

Figure 6. Pressure contour

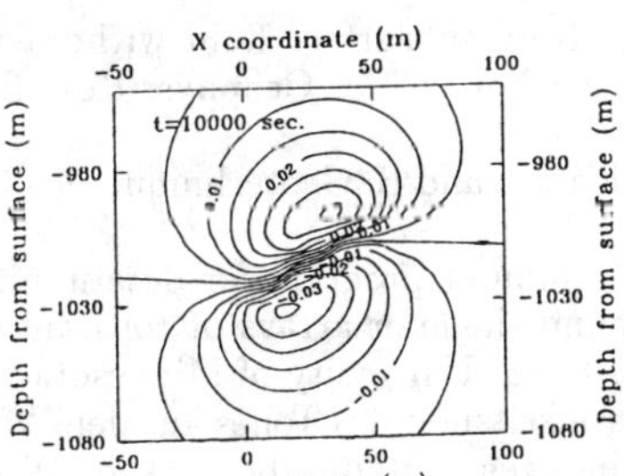

Figure 7. Displacement in z contour

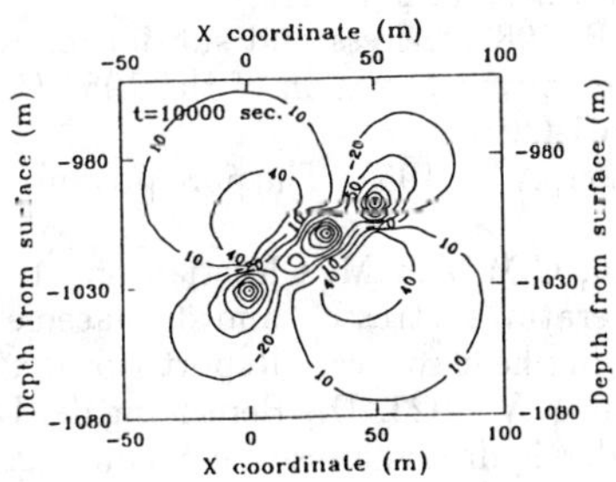

Figure 8. Shear stress in z-x contour

ACKNOWLEDGEMENTS

The gratitude is expressed to Agapito and Associates consulting company for the opportunity of working on the thermomechanical modeling of the Yucca Mountain waste disposal project which contributes partially to the completion of this paper. Support of the National Science Foundation under contract EEC-9209619 is also acknowledged.

REFERENCES

Apirathvorakij, V. and P. Karasudhi. 1980. Quasi-static bending of a cylindrical elastic bar partially in a saturated elastic half-space. Int. J. Solids Structures. 16:625-644.

Biot, M.A. 1941. General theory of three-dimensional consolidation. J. Appl. Phys. 12:155-164.

Biot, M.A. 1956. Thermoelasticity and irreversible thermodynamics. J. Appl. Phys. 27:240-253.

Biot, M.A. 1962. Mechanics of deformation and acoustic propagation in porous media. J. Appl. Phys. 33:1482-1498.

Carslaw, A.R. and C.J. Jaeger. 1959. Conduction of heats in solids. London: Oxford Press.

Cleary, M.P. 1977. Fundamental solutions for a fluid-saturated porous solid. Int. J. Solids Structures. 13:785-808.

Cryer, C.W. 1962. A comparison of the three-dimensional consolidation theories of Biot and Terzaghi. Q. J. Mech. Appl. Math. 16:401-412.

Curran, J. and J.L. Carvalho. 1987. A displacement discontinuity model for fluid-saturated porous media. Proc. 6th Int. Congress on Rock Mech. 73-78. Rotterdam: Balkema.

Goodier, J.N. 1936. The thermal stresses in a strip. Phys. 7:156.

Halpern, M.R. and P. Christiano. 1986. Response of poroelastic half space to steady-state harmonic surface tractions. Int. J. Num. Analy. Methods Geomech. 10:609-632.

Love, A.E.H. 1927. Treatise on the mathematical theory of elasticity. 4th ed. New York: Cambridge Univ. Press.

McNamee, J. and R.E. Gibson. 1960. Plane strain and axially symmetric problems of the consolidation of a semi-infinite clay stratum. Q. J. Mech. Appl. Math. 13:210-227.

Nowacki, W. 1962. Thermoelasticity. Addison-Wesley Publications, Inc. Reading: Mass.

Rice, J.R. and M.C. Cleary. 1976. Some basic stress diffusion solutions for fluid-saturated elastic porous media with compressible constituents. Rev. Geophys. Space Phys. 14:227-241.

Segall, P. 1985. Stress and subsidence resulting from subsurface fluid withdrawal in the epicentral region of the 1983 Coalinga earthquake. J. Geophys. Res. B8, 90:6801-6816.

Skempton, A.W. 1954. The pore pressure coefficients A and B. Geotechnique. 4:143-147.

St. John, C.M. and M. Christianson. 1980. A computer program for determining temperatures, stresses, and displacements around single or arrays of constant or decaying heat sources. Report No. RHO-BWI-C 78. University of Minnesota.

Terzaghi, K.V. 1923. Die Berechung der Durchassigkeitsiffer des Tones aus dem Verlauf der hydrodynamischen Spannungserscheinungen. Sitzungsber. Akad. Wiss. Wien Math Naturwiss. Kl. Abt. 2A, 132, 105.

Rock Mechanics, Nelson & Laubach (eds) © 1994 Balkema, Rotterdam, ISBN 90 5410 380 8

Dilation-induced permeability increase around caverns in salt

K. Fuenkajorn & S. Serata
Serata Geomechanics, Inc., Richmond, Calif., USA

ABSTRACT: A permeability model is developed from triaxial flow test data to predict the permeability increase of rock salt around storage caverns, as affected by excavation, confining and internal pressures, age and shape. The confining and internal pressures may increase the surrounding salt permeability from its in-situ value of 10^{-22} m^2 to as high as 10^{-16} m^2. The permeabilities increase with time and reach their ultimate value at within one year after excavation. Salt permeabilities around elliptical caverns are several orders of magnitude higher than those around the spherical cavern.

1 INTRODUCTION

A concern for utilization of the underground space in rock salt has been the increase in its permeability due to the inelastic dilation near the opening wall (e.g. Aubertin et al., 1993; Hunsche, 1993). The knowledge of the transport properties of salt becomes more important as a result of the rigorous requirements being imposed on the nuclear waste isolation projects in the U.S. and European countries (Stormont, 1990; Peach, 1991). Recent in-situ gas flow measurements by Sandia National Laboratories suggest that undisturbed bedded salt has permeabilities as low as 10^{-22} m^2, while due to excavation, salt near the openings can have a permeability greater than 10^{-18} m^2. For the nuclear waste isolation, the permeability increase zone around excavations near the repository could provide a preferential flow path for the radionuclides to bypass the seals or plugs. For oil, gas, brine or compressed-air storage caverns, the permeability increase zone induced by mining or by fluctuation of cavern pressures during operation may cause leakage to nearby caverns or to a high-permeability formation.

The objective of the present research is to predict the magnitude and extent of the permeability increase zone (PIZ) as a function of time, cavern shape, confining pressure (depth), and storage (internal) pressure. A permeability model is proposed to relate the mechanical behavior with the hydrological response of salt. The model parameters are calibrated with laboratory test results obtained by Stormont and Daemen (1992). The predictability is assessed by comparing the simulation results with the in-situ gas flow measurements around a rectangular opening at the WIPP repository. Spherical, elliptical and cylindrical cavern models are simulated under a variety of confining and internal pressures. Potential applications of the results to actual in-situ conditions are discussed.

2 DILATION-INDUCED PERMEABILITY INCREASE IN ROCK SALT

The changes of rock salt permeability discussed here are based on a series of measurements of gas permeability during quasi-static triaxial compression tests by Stormont and Daemen (1992). Their test results indicate that initially, under hydrostatic stress, salt specimens have no measurable gas permeability. As the deviatoric load is increased, the porosity and permeability increase as a result of sliding and dilation along grain boundaries. Continued loading results in the growth and coalescence of the secondary tensile cracks. The structure of the new pores induced in rock salt consists of narrow microcracks with large aspect (length-to-width) ratios. Derived from the equivalent channel flow equation, Stormont's flow model (Stormont and Daemen, 1992) defines the salt permeability (K) as

$$(1) \qquad K = A\,(\sigma_3^{*})^{\lambda}\,(\phi)^{s},$$

where A is permeability constant, σ_3^* is normalized minor principal stress (minimum stress divided by a unit stress, $\sigma_3^* = \sigma_3/1$), ϕ is porosity, λ is stress coefficient, and s is a constant related to flow path tortuosity. The triaxial flow test results obtained by Stormont and Daemen (1992) suggest that the inelastic volumetric strain (ϵ_v^{in}) is virtually identical to the gas-accessible porosity because the newly developed pore structure is largely interconnected. Since the porosity of the undisturbed salt is immeasurably small, they propose that the salt porosity ϕ equals to the volumetric dilation ϵ_v^{in}. We also found that a somewhat better back-fit to the laboratory data can be achieved by expressing s as a function of the minor principal stress (σ_3), plastic transition pressure (P), and an empirical constant (η). Equation 1 can be rewritten as

$$(2) \qquad K = A\,(\sigma_3^{*})^{\lambda}\,(\epsilon_v^{in})^{\{1-\exp(-\eta\sigma_3/P)\}}$$

The capability of the proposed permeability model is extended further by including the effect of the pronounced time-dependency of rock salt. Equation 2 is implemented into the finite element program GEO which is capable of describing elastic, viscoelastic, viscoplastic, strain-softening and dilation behavior of rock salt. The program carries out an explicit time-domain integration, and hence the permeability can be calculated using the current σ_3 and ϵ_v^{in} values. Since these values are updated for each time step, the change of salt permeability will be governed by the changes of stress state and creep deformation and dilation. When there is no dilation, the permeability is set to a default value which represents the undisturbed state. We use a value of 10^{-21} m^2 or lower as this represents a lower limit of the laboratory and field measurements of salt permeability (Stormont et al., 1991). The inelastic volumetric dilation is defined here as

$$(3) \qquad \epsilon_v^{in} = F\{\exp[C(\gamma_o-\gamma_c)/\gamma_c]-1\}\{\exp[-H\sigma_m/(P-\sigma_m)]\},$$

where F, C and H are coefficients related to strain-softening and confinement, γ_o and γ_c are induced and critical octahedral shear strains, and σ_m is the mean stress. The dilation is assumed to occur when $\gamma_o > \gamma_c$ and $\sigma_m < P$. The triaxial flow test data obtained by Stormont and Daemen (1992) are used to calibrate the parameters in equations 2 and 3, and the parameters related to the visco-elasticity and visco-plasticity in the GEO constitutive equations. Detailed derivations of the permeability model and the constitutive equations are given by Serata and Fuenkajorn (1991) and by Fuenkajorn and Serata (1992,1993). Figure 1 compares the permeability data with the best back-predicted curves from the GEO program. Deviation of the predictions from the test results is usually less than an order of magnitude, except for those tested under 7.6 MPa confining pressure. These curves yield the permeability parameters, A = 0.002 m^2, λ = 3.9, and η = 40, and the dilation parameters, F = 0.05, C = 0.2, H = 3.1, P = 41.4 MPa, and γ_c = 0.0085. The predictability of the model and its parameters has been assessed by simulating the increase of in-situ salt permeability around a rectangular opening at 600 m depth (room D of the WIPP Facility). Serata and Fuenkajorn (1991) give detailed boundary and loading conditions for the simulation. Figure 2 compares the in-situ permeability test data (Stormont, 1990) with the model prediction at 52 months after excavation. The permeabilities are presented as a function of horizontal distance from the opening boundary. The model agrees with the measurement on the extent (width) of the PIZ. Near the opening, the model overestimates the measured permeabilities by nearly four orders of magnitude. The discrepancy may be related to 1) insufficient laboratory data under a wide range of confining pressures, 2) inadequacy of the proposed model, and 3) uncertainties of the initial, boundary and loading conditions of the simulated opening. Since the proposed model overestimates the measured permeability, predictions of the PIZ around an actual storage cavern in rock salt would provide a conservative estimate with regard to its containment integrity.

3 SALT CAVERN MODEL SIMULATION

Using the proposed permeability model, an attempt is made at identifying the magnitude and extent of the permeability increase zone (PIZ) potentially developed around single (isolated) storage caverns in salt, as affected by cavern age, shape, confining pressure, and storage pressure. Spherical, elliptical and cylindrical cavern models (representing general shapes of cavern in bedded and domal salt formations) in a semi-infinite salt medium are simulated under isothermal condition with uniform confining pressures P_o ranging from 10 to 40 MPa and internal pressures P_i ranging from 0.1 P_o to 0.4 P_o. Analysis of these simplified models allows us to isolate the studied variables from other factors, such as irregular profile of cavern boundary, impurity, interference from nearby caverns, or from the overlying and underlying non-salt formations.

The simulation results indicate that for all cavern shapes the confining and internal pressure are the main factors governing the PIZ characteristics. Figure 3 shows the increase of salt permeability as a function of normalized distance (r/R) from the boundary of spherical model for P_o = 10, 20, and 30 MPa at 1 year after excavation. The permeabilities are largest near the cavern boundary, rapidly decrease with increasing distance from the cavern, and eventually approach the in-situ value of 10^{-22} m^2 (set as default value for this simulation). The maximum permeabilities decrease with the increase of P_o. At P_o = 10 MPa, the maximum permeabilities are up to 4 to 6 orders of magnitude above the in-situ value. This is probably because salt under low P_o is more brittle, and hence results in a high intensity of fracturing when it is subjected to shear stresses due to excavation. When P_o increases to 20 MPa, the mean stress is sufficiently large that the induced shear stress makes the salt deformed partly in plastic mode and to a much lesser degree in brittle failure. This results in a wider PIZ with a smaller

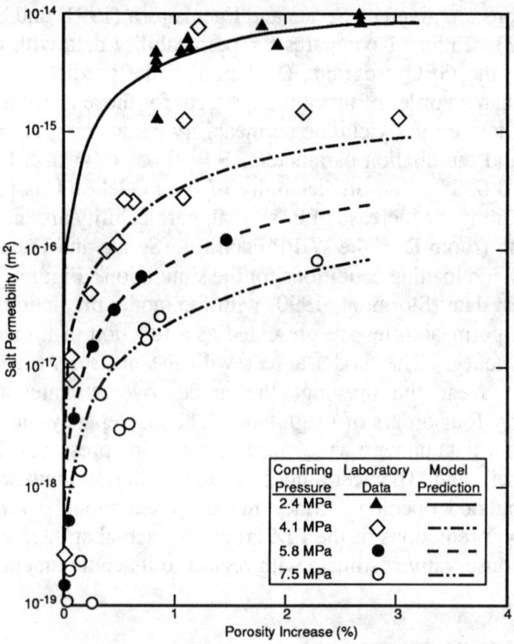

Figure 1. Permeabilities as a function of dilation, laboratory results versus prediction.

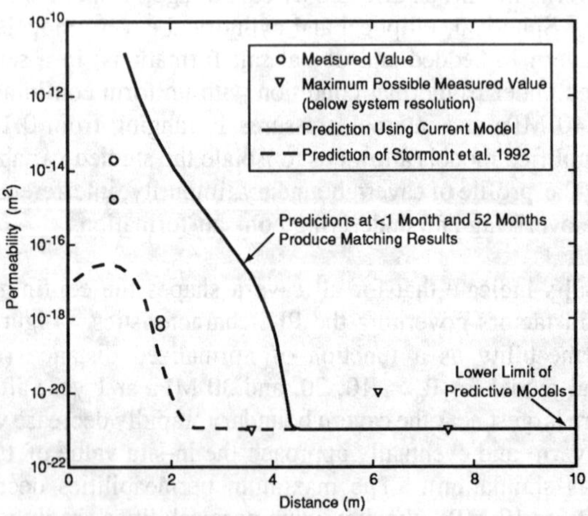

Figure 2. Comparison of permeability along a horizontal test borehole between model prediction and measurements.

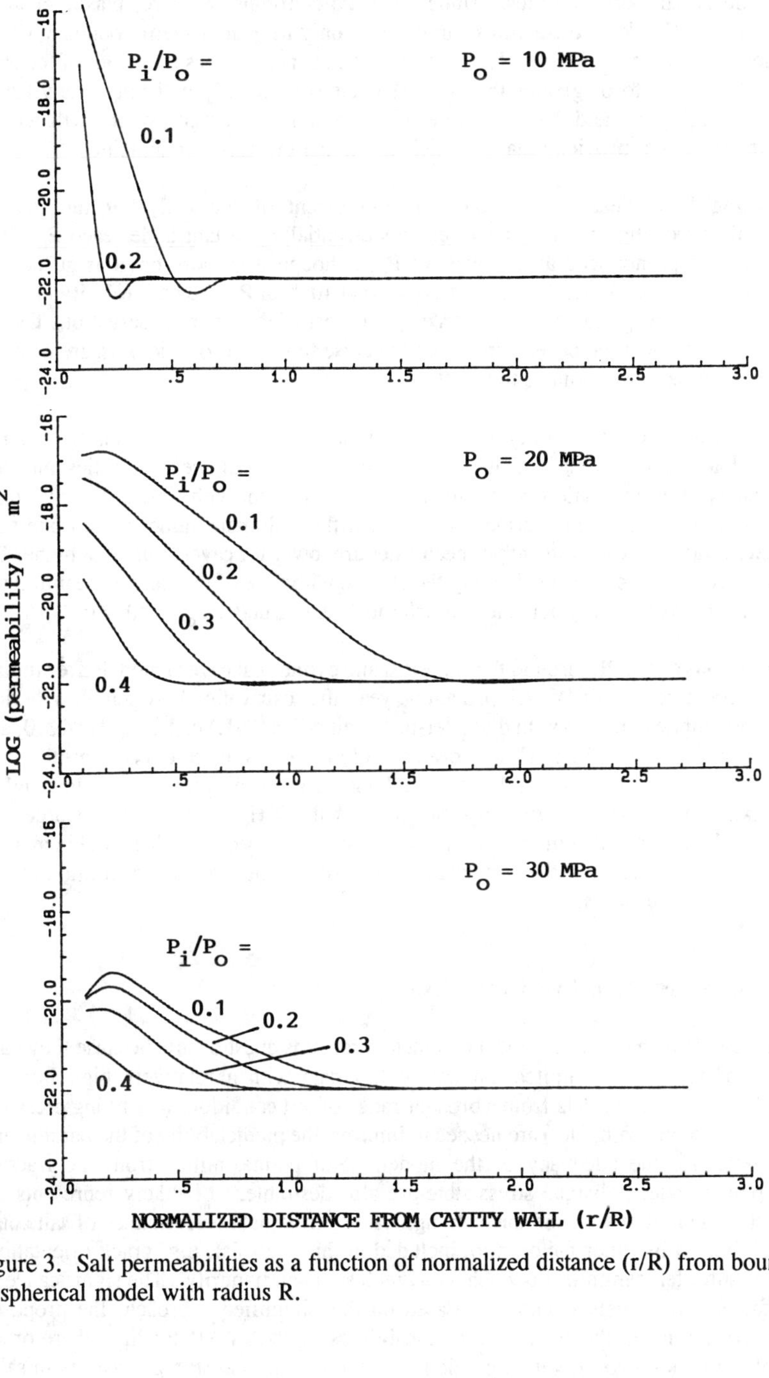

Figure 3. Salt permeabilities as a function of normalized distance (r/R) from boundary of spherical model with radius R.

magnitude of salt permeabilities. Under 30 MPa confining pressure, plastic creep is predominant. Cracks and fractures can develop only near the cavern boundary where σ_m is less than the plastic transition pressure P, and hence reduces the width of the PIZ. When P_o is equal to or greater than 40 MPa, there eventually will not be any brittle fractures developing, and the salt permeability around the storage caverns will remain at the in-situ value, provided that a sufficient internal pressure is maintained.

Increasing P_i decreases the magnitude and extent of the PIZ, but the rate of permeability reduction with distance remains essentially constant under each P_o. The effect of P_i is pronounced at all levels of P_o, although it is more obvious at the low confining pressures. When P_i is increased beyond 40 % of P_o, the permeability increase in salt becomes insignificant. This agrees with most of the actual observations that no leakage occurs when the cavern pressures are close to the hydrostatic pressure of brine where P_i approximately equals to 0.5 P_o.

The effect of cavern age (time) on the surrounding salt permeability is insignificant as compared to the confining and internal pressures. The salt permeabilities increase immediately after excavation, and remain virtually constant with time at their ultimate value. For most cases, the permeabilities reach their ultimate values within one year after excavation. These ultimate permeabilities are lower for caverns under a higher P_o. Figure 4 gives a typical result showing the change with time of the salt permeabilities at various locations from the boundary of spherical cavern under P_o = 20 MPa.

Distributions of the PIZ around the spherical and cylindrical cavern models are similar. Figure 5 compares the PIZ developed at one year after excavation between the spherical model and elliptical models with diameter-to-height ratios (D/H) of 1.5, 2.0 and 3.0, and under P_o = 20 MPa. The D/H ratio dramatically increases the salt permeabilities near the vertical axis (here assuming that cavern diameter is always greater than the height). The maximum permeability for elliptical cavern with D/H = 3.0 is up to 8 orders of magnitude higher than that of the spherical cavern. This is because when D/H increases, the mean stress in salt near the vertical axis decreases as the stress concentration moves further into the salt mass.

4 DISCUSSIONS AND CONCLUSIONS

The representativeness of the model parameters remains questionable because they have been calibrated from a limited amount of test data with a relatively high intrinsic variability. Laboratory data from a broader range of test conditions (confining pressure, stress rate, loading path, etc.) are needed to improve the predictability of the parameters, and to evaluate the adequacy of the model. Salt permeabilities from the fracture development under polyaxial stress state are also desirable, as it likely represents the actual stress conditions around the opening. The hydrological performance of salt could also be affected by other factors (not included in this analysis), e.g. crack orientation, pore pressure, temperature, flow behavior, inclusion and impurity. The significance of these factors are largely unknown. Based on this simplified approach, the proposed model overestimates the measured permeabilities at the WIPP facility, therefore it probably provides a conservative prediction of the PIZ around storage caverns in salt.

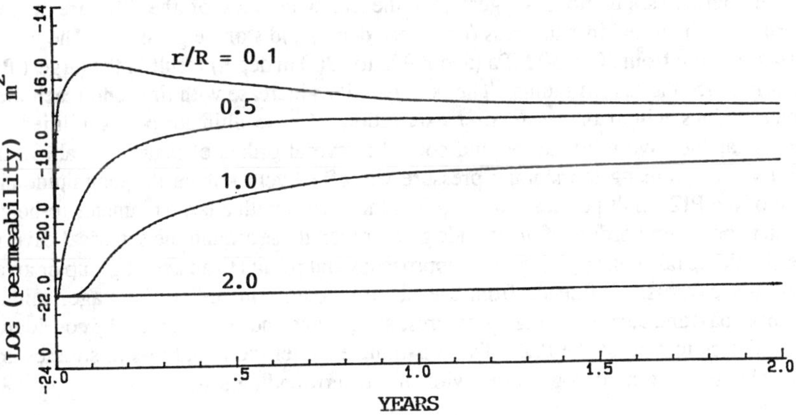

Figure 4. Permeabilities versus cavern age at various locations (r/R) from boundary of spherical model with radius R; P_o = 20 MPa, P_i = 0 MPa.

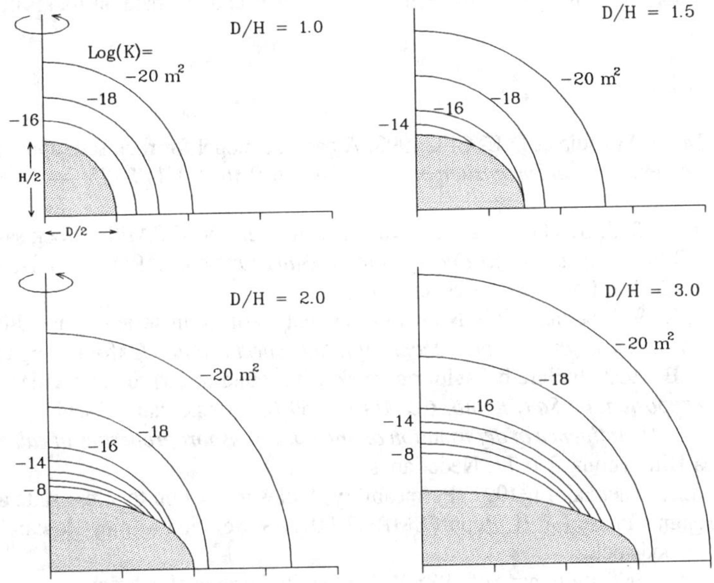

Figure 5. Permeability increase zone around spherical and elliptical models; D/H = 0, 1.5, 2.0 and 3.0; P_o = 20 MPa, P_i = 0 MPa.

Our numerical simulations suggest that the characteristics of the PIZ are mainly governed by the initial in-situ stress (or cavern depth) and storage pressure. The in-situ stresses ranging from 20 to 30 MPa (about 900 to 1300 m depth) result in the largest PIZ of nearly twice the cavern radius. The permeabilities increase with time and reach their ultimate values within one year after excavation. These ultimate permeabilities are largest near the cavern boundary, and could be several orders of magnitude above the in-situ value. Raising the storage pressure can effectively reduce the magnitude and extent of the PIZ. Salt permeabilities around the cavern with a large diameter-to-height ratio may be several orders of magnitude greater than those around the spherical cavern. Care should be taken in applying these approaches and results to an actual group or array of storage caverns. Influence from the nearby caverns including their age, mining sequence, past and current storage pressures, size, shape and spacing must be considered for the determination of the PIZ. These parameters alter the initial in-situ stresses and the mechanical and hydrological behavior of the surrounding salt.

ACKNOWLEDGEMENTS

The present research effort was funded by the Solution Mining Research Institute, under contract SMRI91-001A-S. We gratefully acknowledge Dr. John C. Stormont of Sandia National Laboratories for providing technical guidelines and test data on rock salt.

REFERENCES

Aubertin, M., J. Sgaoula & D.E. Gill. 1993. A damage model for rock salt: application to tertiary creep. *Seventh Symposium on Salt, Kyoto, 6-9 April.* 1:117-125. Amsterdam: Elsevier.

Fuenkajorn, K. & S. Serata. 1992. Geohydrological integrity of CAES in rock salt. *Proc. Int. Conf. Compressed-Air Energy Storage, San Francisco, 7-9 July.*:P4.1-P4.21. Palo Alto: Electric Power Research Institute.

Fuenkajorn, K. & S. Serata. 1993. Numerical simulation of strain-softening and dilation behavior of rock salt. *Int. J. Rock Mech. Min. Sci. Special Issue.* Oxford: Pergamon.

Hunsche, U.E. 1993. Failure behaviour of rock salt around underground cavities. *Seventh Symposium on Salt, Kyoto, 6-9 April.* 1:59-65. Amsterdam: Elsevier.

Peach, C.J. 1991. *Influence of deformation on the fluid transport properties of salt rock.* Geologica Ultraiectina. No.77. Nederlands.

Serata, S. & K. Fuenkajorn. 1991. Permeability studies in relation to stress state and cavern design - Phase I & II. Report SMRI91-001A-S, Solution Mining Research Institute, Woodstock.

Stormont, J.C. 1990. Summary of 1988 WIPP facility horizontal gas flow measurements. Report SAND89-2497, Sandia National Laboratories, Albuquerque.

Stormont, J.C. & J.J.K. Daemen. 1992. Laboratory study of gas permeability changes in rock salt during deformation. *Int. J. Rock Mech. Min. Sci. & Geomech. Abstr.* 29:325-342.

Stormont, J.C., C.L. Howard & J.J.K. Daemen. 1991. In-situ measurements of rock salt permeability changes due to nearby excavation. Report SAND90-3134, Sandia National Laboratories, Albuquerque.

Rock Mechanics, Nelson & Laubach (eds) © 1994 Balkema, Rotterdam, ISBN 90 5410 380 8

Time effect in rocks surrounding a horizontal tunnel

N. D. Cristescu
University of Florida, Gainesville, Fla., USA

ABSTRACT: Creep of the rocks surrounding a horizontal tunnel is considered. It is shown where the rock becomes dilatant, where compressible, if and where failure will occur and how much rock will be involved, where and when creep failure is to be expected, and how creep failure spreads into the mass of rock. The influence of an internal pressure on the above phenomena is also studied.

1. INTRODUCTION

The creep of rocks around horizontal tunnels is a phenomenon of great concern. It was studied by several authors for the case of both an empty tunnel and of a tunnel reinforced with lining (see literature mentioned in Cristescu 1989, Hudson 1993). Just due to the excavation, sometimes the rock becomes dilatant in some areas around the tunnel. This increases the permeability and also the danger of creep failure, which is to be expected after a certain time interval elapsed from the moment of excavation. In order to study this phenomenon both during transient and stationary creep, a general constitutive equation was developed by Cristescu (1993). This constitutive equation can describe irreversible dilatancy and/or compressibility during both transient and stationary creep, failure, work-hardening, evolutive damage and ultimate creep failure. The correctness of any solution strongly depends on the exact formulation of the initial data. That is why we investigate very throughout the state of the rock around a tunnel at the moment just following the excavation, when creep starts. This initial state can be obtained with analytical (exact) formulae and as such allows a parameter study. As a result we can determine quite accurately which parameter and how it influences a certain phenomenon. On the other hand, a mathematically correct initial solution ensures that the creep problem is correctly posed. Most of the conclusions obtained for the initial solutions are, in fact, independent of the constitutive equation itself, and depend only on some concepts which are obtained straightforward from the experimental data, as failure surface or compressibility/dilatancy boundary.

2. CONSTITUTIVE EQUATION

The constitutive equation describing instantaneous response of the rock, transient and stationary creep, is of the form (see Cristescu 1987,1989,1993,1994a)

$$\dot{\boldsymbol{\epsilon}} = \frac{\dot{\boldsymbol{\sigma}}}{2G} + \left(\frac{1}{3K} - \frac{1}{2G}\right)\dot{\sigma}\,\mathbf{1} + k_T\left\langle 1 - \frac{W_T(t)}{H(\boldsymbol{\sigma})}\right\rangle \frac{\partial F}{\partial \boldsymbol{\sigma}} + k_S \frac{\partial S}{\partial \boldsymbol{\sigma}} \tag{1}$$

where G and K are elastic parameters (generally not constant), H(σ) is the "yield" function so that

$$H(\boldsymbol{\sigma}) = W_T(t) \tag{2}$$

is the stabilization boundary for transient creep, with

$$W_T(T) = \int_0^T \sigma(t)\,\dot{\epsilon}_{VT}^I(t)\,dt + \int_0^T \boldsymbol{\sigma}'(t)\cdot\dot{\boldsymbol{\epsilon}}_T^{I'}(t)\,dt \tag{3}$$

the stress-work per unit volume, where "prime" stands for the "deviator". Further F(σ) is a viscoplastic potential for transient creep, S(σ) a potential for stationary creep, and k_T and k_S are two viscosity coefficients. Generally F≠H so that the constitutive equation is not associated. For irreversible volumetric rate of strain we have

$$\dot{\epsilon}_V^I = k_T\left\langle 1 - \frac{W_T(t)}{H(\boldsymbol{\sigma})}\right\rangle \frac{\partial F}{\partial \sigma} + k_S \frac{\partial S}{\partial \sigma} \tag{4}$$

with

$$\sigma = \frac{1}{3}(\sigma_1 + \sigma_2 + \sigma_3) \quad , \quad \tau = \frac{\sqrt{2}}{3}\bar{\sigma} = \sqrt{\frac{2}{3} II_{\sigma}'} \tag{5}$$

where σ is the mean stress, τ is the octahedral shear stress, $\bar{\sigma}$ is the equivalent stress and $II_{\sigma'}$ is the second invariant of the deviator. We assume that the constitutive functions depend on stress invariants σ and τ only. From (4) follows that the rock is dilatant if $\partial F/\partial\sigma<0$ and compressible if $\partial F/\partial\sigma>0$. Also during dilatancy $\partial S/\partial\sigma<0$, but during compressibility $\partial S/\partial\sigma=0$ (no stationary compressibility is possible).

3. FORMULATION OF THE PROBLEM

We make the standard assumption that the initial stresses before excavation, the "primary" stresses $\sigma^P(X)$, are known at the location X, where the tunnel is to be excavated. For the vertical primary component we use the overburden pressure at depth h. For the horizontal component we take $\sigma_h=n\sigma_v$, with n a constant in the range $0.3\leq n\leq 4$ (see Cristescu 1989 Chap.10, Cooling *et al.* 1988, Cunha 1990). Since the primary stress must satisfy the constitutive equation (1), we must have

$$H(\boldsymbol{\sigma}^P) = W^P = const. \quad and \quad \dot{\boldsymbol{\epsilon}}^P = k_S \frac{\partial S(\boldsymbol{\sigma}^P)}{\partial \boldsymbol{\sigma}} \tag{6}$$

where the first relation is the initial yield surface, and the second one may describe slow tectonic motions if τ^P has a significant value. Otherwise $\dot{\varepsilon}^P$ is negligible for time spans of interest for mining engineering.

Also we assume that the excavation is performed quite fast and the stresses just after excavation are the "secondary" stresses $\sigma^S(X,t)$. The deformation around the tunnel is produced by the relative components $\sigma^R(X,t)=\sigma^S(X,t)-\sigma^P(X)$.

The boundary conditions are formulated on the wall of the tunnel and at infinity. We assume that on the wall of the tunnel is prescribed a pressure p. The case when this pressure is due to a man-made lining is discussed by Cristescu (1989 Chap.13). Here we assume either p=0, or p=γh, or p=const. (pressure of a gas, or of a self-adjusting lining). For the conditions at infinity we assume $\sigma^R \to \mathbf{0}$ when we depart far enough from the excavation. Obviously we must also have for strains $\varepsilon^R \to \mathbf{0}$, and displacements $\mathbf{u}^R \to \mathbf{0}$ when departing very much from the excavation.

4. INITIAL DATA

A prerequisite of the study of wall closure of a tunnel is the correct formulation of the initial data. Since we have assumed a fast ("instantaneous") excavation, the stress state just after excavation is obtained from the elastic exact solution (see Cristescu 1985, 1989 Chap.12). For circular tunnels and plane strain we get for the secondary stress state

$$
\begin{aligned}
\sigma_{rr} &= p\frac{a^2}{r^2} + \frac{1}{2}(\sigma_h+\sigma_v)\left(1-\frac{a^2}{r^2}\right) + \frac{1}{2}(\sigma_h-\sigma_v)\left(1-\frac{4a^2}{r^2}+\frac{3a^4}{r^4}\right)\cos 2\theta \\
\sigma_{\theta\theta} &= -p\frac{a^2}{r^2} + \frac{1}{2}(\sigma_h+\sigma_v)\left(1+\frac{a^2}{r_2}\right) - \frac{1}{2}(\sigma_h-\sigma_v)\left(1+\frac{3a^4}{r^4}\right)\cos 2\theta \\
\sigma_{r\theta} &= \frac{1}{2}(\sigma_h-\sigma_v)\left(-1-\frac{2a^2}{r^2}+\frac{3a^4}{r^4}\right)\sin 2\theta \\
\sigma_{zz} &= \sigma_h - \nu(\sigma_h-\sigma_v)\frac{2a^2}{r^2}\cos 2\theta
\end{aligned}
\tag{7}
$$

where a is the initial radius of the tunnel. This stress state and the corresponding displacements are used as initial data for the problem of creep of rock around the tunnel. In order to visualize the initial data we plot two maps: the first one shows various domains existing in the stress space $\sigma\tau$ and maybe the stress distribution along the circumference of the tunnel, and another one shows the corresponding domains existing in the rock surrounding the tunnel. Fig.1 shows two such maps for a tunnel excavated in rock salt, depth 1 km, the assumed ratio of the far field stresses is $\sigma_h/\sigma_v=2$ and p=0. A significant failure is shown in both parts of the figure, the domain of dilatancy spreads quite far in the vertical direction (direction of minim component $\sigma^P_{min}=\sigma_v$ of the far field stress). In the horizontal direction (direction of σ^P_{max}) we get elastic (unloading) domains for the transient creep and two islands labelled "compressible" where stationary creep is present but producing incompressibility (for details concerning the constitutive equation and formulation of the problem see Cristescu 1994b). The interrupted line at the right figure is the initial yield surface with the rhomb showing the initial primary stress. Let us recall that all existing stationary creep laws are expressed as power functions of τ. That is why in the domains where τ has high values, one is expecting a much faster closure. For instance, in the case shown in Fig.1 the rate of strain for stationary creep would be of the order of magnitude 10^7 s^{-1} at the roof and floor (direction of σ^P_{min}), but only 10^4 s^{-1} in the horizontal direction. On the other hand, there where τ has high values, more damage is to be expected due to dilatancy, and maybe a creep failure as well.

If the tunnel is excavated at a much shallower depth, the instantaneous failure is no more present and the domain where the rock becomes dilatant shrinks significantly (Fig.2). Also the magnitude of τ diminishes everywhere so that creep failure may be expected after a very long time interval elapsed after excavation only, in the direction of σ^P_{min}. Also the closure is faster in the same direction. Fig.2 also shows that for the

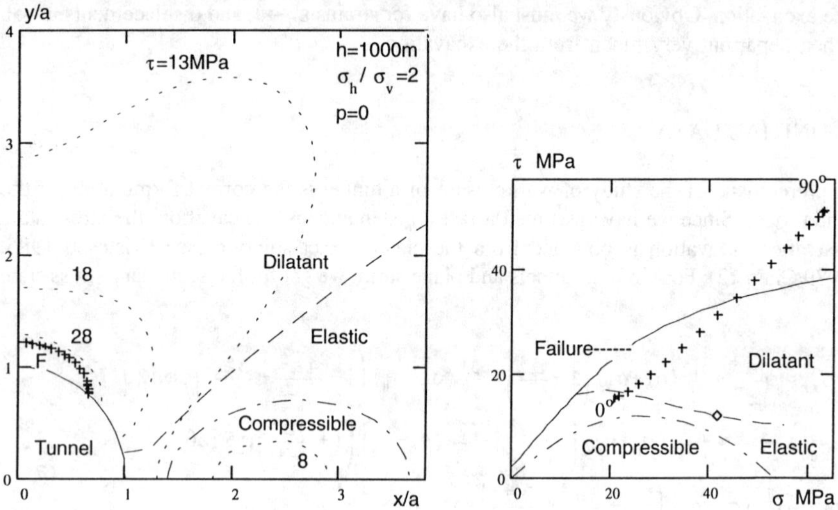

Fig.1 Domains of dilatancy, compressibility, elasticity and failure around a tunnel (left), and explanation in the stress space of the stress distribution along the contour (right)

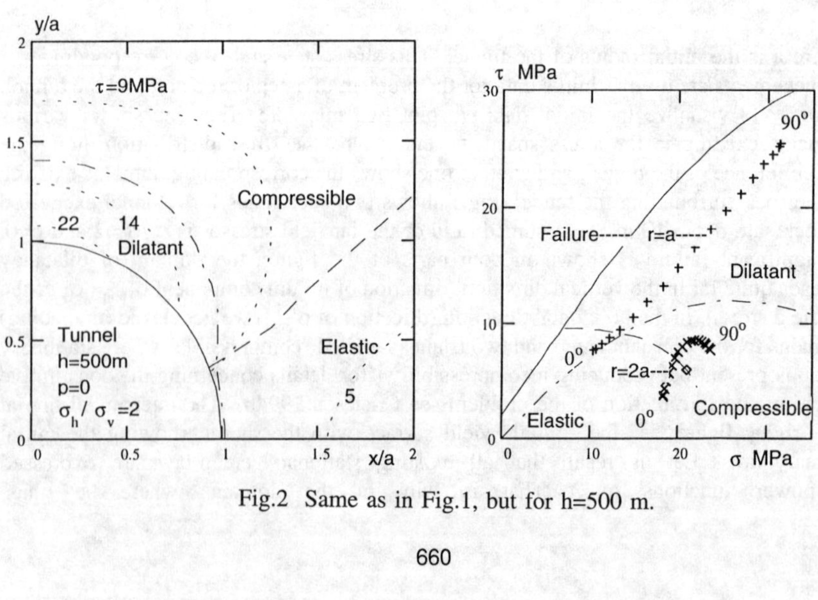

Fig.2 Same as in Fig.1, but for h=500 m.

contour r=2a the stresses are already close to the primary stress existing before excavation (rhomb).

Instantaneous failure may occur not only in the direction σ^P_{min}, but sometimes in the direction σ^P_{max} as well. That is shown, for instance, in Fig.3 (Cristescu 1994b) for the depth 400 m and σ_h/σ_v=4. Though instantaneous failure occurs in both vertical and horizontal direction, the two cases are not similar. In the vertical direction the instantaneous failure is followed quite soon by creep failure (a progressive phenomenon). Also the dilatant region extends very far from the tunnel contour. Furthermore, a significant creep closure will take place in the vertical direction (direction of σ^P_{min} where τ reaches very high values). Thus instantaneous failure occurs in both directions σ^P_{min} and σ^P_{max}, but creep failure, main damage and faster closure are in the

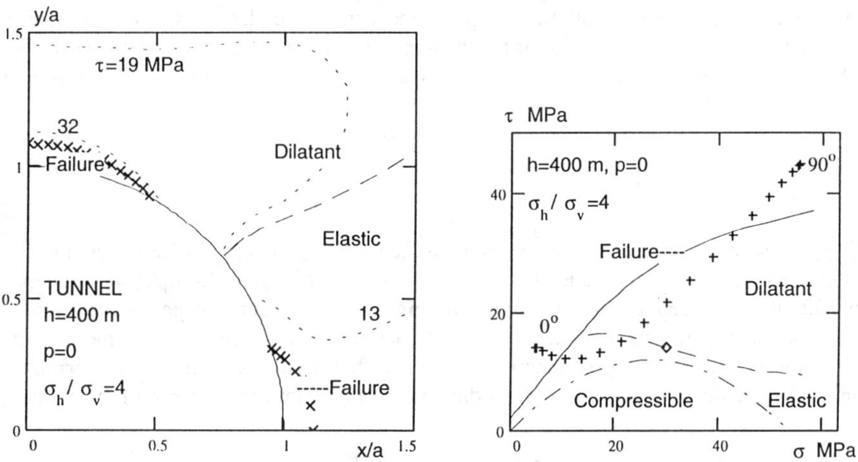

Fig.3 Same as in Fig.1 but for h=400 m and σ_h/σ_v=4.

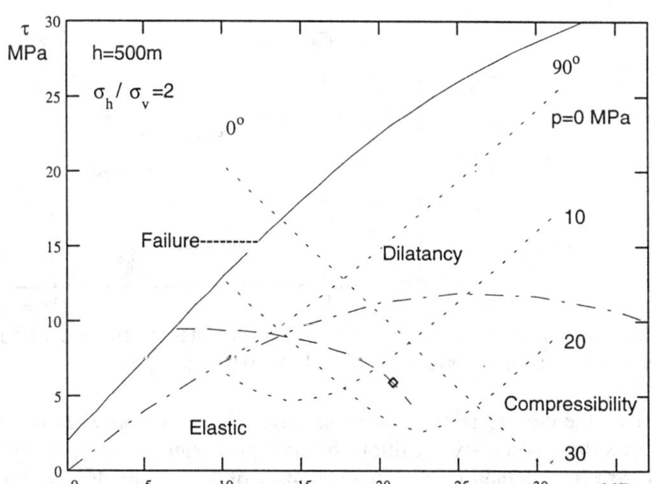

Fig.4 Stress distribution on the contour of a tunnel for various internal pressures.

direction of σ^P_{min}. In the domain labelled "elastic" there is no transient creep, but the stationary creep is still there producing dilatancy and change in shape. $\dot{\varepsilon}^I_S$ is of the order of magnitude 10^7 s^{-1} in the vertical direction, and 10^4 s^{-1} in the horizontal direction.

The presence of an internal pressure changes the stress distribution around the tunnel. For instance, if we consider again the case shown in Fig.2, but this time with various internal pressures, we obtain the stress distribution shown in Fig.4 by several dotted lines. For p=0 the closure by creep is faster in the direction of σ^P_{min}, while for p=30 MPa in the direction of σ^P_{max}, there where an instantaneous failure also occurs, and where creep failure is also to be expected. An intermediate pressure (p=10 MPa,say) is very favorable in the sense that instantaneous failure is no more possible and also because everywhere in the rock surrounding the tunnel the value of τ is significantly reduced, and so is the closure by creep, while creep failure is no more to be expected. Thus for various depths and ratios σ_h/σ_v, one can try to determine the value of p (pressure of a gas, or of a liquid, or maybe a pressure exerted by a self-adjusting lining) which would optimize the behavior of the rock around the tunnel (avoiding failure and creep failure, and slowing down the creep closure).

5. CREEP CLOSURE

Creep closure of a tunnel can be described quite easy with a very simple formula in the case σ_h/σ_v=1, or when this ratio is close to this value. For instance for σ_h/σ_v=1 and depth h=700 m, the creep closure is shown in Fig.5. The "present shape of the tunnel" corresponds to the dimensionless time $k_S(t-t_o)= 1.1\times10^7$, where t_o is the moment of excavation. The spreading of creep failure up to that time is shown by an interrupted line. Also two dotted lines show how dilatancy spreads into the creeping rock mass.

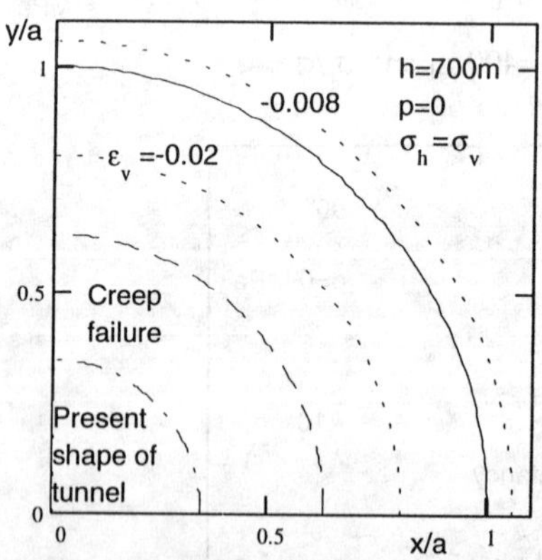

Fig.5 Creep closure of a tunnel showing also creep failure and dilatancy spreading.

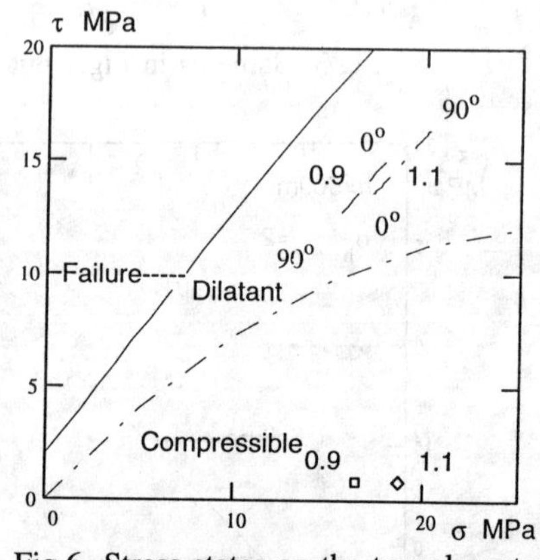

Fig.6 Stress states on the tunnel contour for σ_h/σ_v=0.9 and σ_h/σ_v=1.1

For $\sigma_h/\sigma_v\neq1$ the closure is no more symmetric. Figures 6 and 7 show two such cases. Fig.7 shows three successive positions of the tunnel wall during convergence for σ_h/σ_v=1.1 and σ_h/σ_v=0.9 . In the first case, the boundary of the domain damaged by creep failure up to time t_3 is shown by crosses. No creep failure has been obtained for the case σ_h/σ_v=0.9 up to the same time. It is to be mentioned again that the closure is faster in the direction σ^P_{min}, and in the same direction a creep failure may be expected sooner or later. The time when it may occur can be theoretically predicted. The reason is that in the direction σ^P_{min} the value of τ is greater. That is clearly shown in Fig.6. For the case σ_h/σ_v=0.9 the highest value of τ is in the horizontal direction, while for the case σ_h/σ_v=1.1 the maximum τ is in the vertical direction. These directions are just the direction of σ^P_{min} for the two cases. Since the contribution to the closure of the transient creep is quite small, and since the constitutive equation for stationary creep involves power functions of τ, there where τ has higher values these components are with several order of magnitude greater than in some other domains around the tunnel. In Fig.6 the square marks the primary stress for the case σ_h/σ_v=0.9, while the rhomb for the case 1.1. If we examine now the problem in what direction is the orifice larger during creep, one cannot give a general answer. From Fig.7 left, it follows that the orifice is larger in the σ^P_{max} direction due to the creep, but at the same time, creep failure enlarges the orifice in the σ^P_{min} direction; thus, depending on the ratio σ_h/σ_v, on depth and on the mechanical properties of the rock during creep, we may find the orifice larger either in the σ^P_{max} direction or in the σ^P_{min} direction.

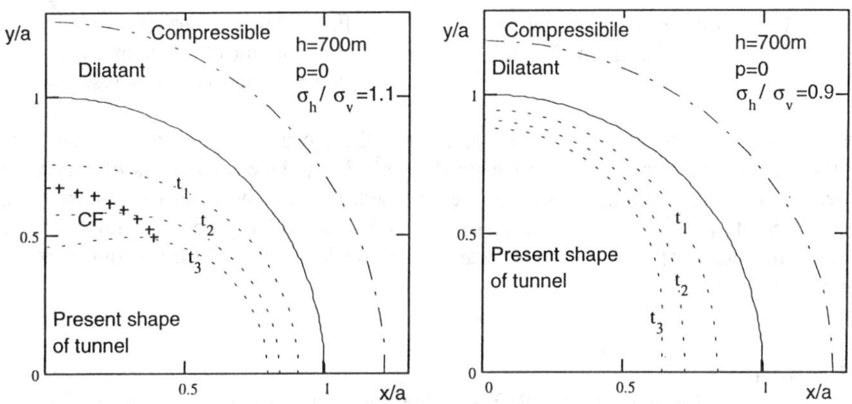

Fig.7 Creep closure for $\sigma_h \neq \sigma_v$. CF means "creep failure"

An internal pressure may significantly reduce the creep rate. Fig.8 shows the influence of the internal pressure on the creep closure of the rock surrounding a tunnel; the conditions are the same as for the Fig.7 left. A small internal pressure reduces only little the creep closure rate, and it is only a pressure of the order of 10 MPa which may practically stop the creep closure. A pressure exerted by a tunnel lining is generally too small to stop the creep closure but it can do it at a ceetain advanced stage of closure. The pressure exerted by a column of brine can stop the creep closure at an early stage of closure. A very high pressure will expand the orifice. Since for high pressures the maximum value of τ is in the direction of σ^P_{max} (see Fig.9), the expansion of the orifice is faster in the horizontal direction. It is also in the same direction that a sudden failure is to be expected if the pressure is increased very much. That is shown very clearly in Fig.9, where the various segments shown represent the stress variation on the contour of the tunnel, from the horizontal direction 0° up to the vertical direction 90°. The rhomb is again the primary stress state. For p=0, the maximum τ and faster closure are in the direction of σ^P_{min} (i.e. 90°), but for very high pressures the maximum values for τ and faster expansion are in the direction of σ^P_{max} (i.e. 0°). For intermediate values (for the

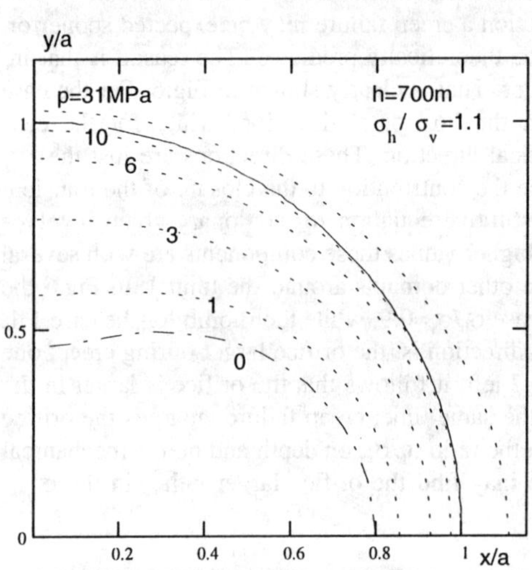

Fig.8 Influence of an internal pressure on the creep closure.

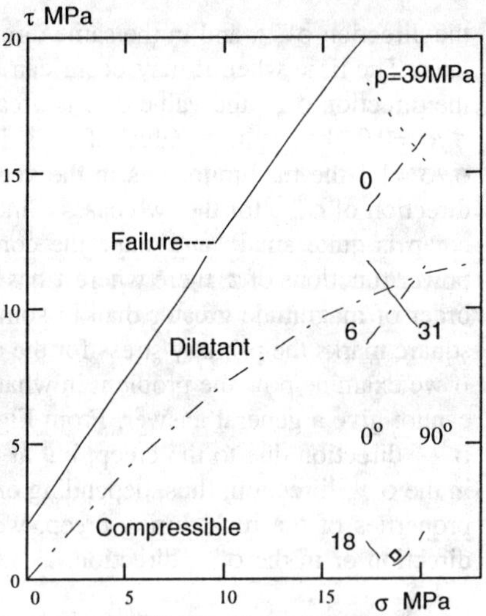

Fig.9 Stress distribution along the contour of the tunnel for various internal pressures.

present example 6 MPa ≤ p ≤ 20 MPa, say), the creep of the walls is small. The optimum pressure can be obtained from $\partial\tau/\partial p=0$. No unique solution is obtained for $\sigma_h/\sigma_v\neq1$, but an approximated one can be obtained. For the particular case considered, a pressure of 18 MPa will allow very small creep rates, which produce a small closure around 90°, but a very small expansion around 0°, while around 45° the contour does not change over time.

REFERENCES

Cooling, C.M., J.A.Hudson & L.W.Tunbridge 1988. In situ Rock Stresses and Their Measurements in the U.K. Part II. Site Experiments and Stress Field Interpretation. *Int.J.Rock Mech.Min.Sci.& Geomech.Abstr.*,**25**,6, 371-382.

Cristescu, N. 1985. Viscoplastic Creep of Rocks around Horizontal Tunnels. *Int.J.Rock Mech.Min.Sci.& Geomech.Abstr.*,**22**,6, 453-459.

Cristescu, N. 1987. Elastic/Viscoplastic Constitutive Equations for Rock. *Int.J.Rock Mech.Min.Sci.& Geomech.Abstr.*,**24**,5, 271-282.

Cristescu, N. 1989. *Rock Rheology.* Dordrecht: Kluwer Academic Publishers.

Cristescu, N. 1993. A general constitutive equation for transient and stationary creep of rock salt. *Int.J.Rock Mech.Min.Sci.& Geomech.Abstr.*, **30**, 2,125-140.

Cristescu, N. 1994a. A procedure to determine nonassociated constitutive equations for geomaterials. *Int.J.Plasticity,* **10**,2 (in press).

Cristescu, N.D. 1994b. Viscoplasticity of Geomaterials.In N.Cristescu and G.Gioda (eds), *Time-dependent Behaviour of geomaterials,* Udine Oct.11-15 1993, Wien: Springer (in press).

Cunha, A.P. 1990. Scale effects in rock mechanics. In A.Pinto da Cunha (ed), *Scale Effects in Rock Masses,* 3-27, *Rotterdam: Balkema.*

Hudson, J.A. (ed.). 1993. *Comprehensive Rock Engineering.* Oxford: Pergamon Press.

Rock Mechanics, Nelson & Laubach (eds) © 1994 Balkema, Rotterdam, ISBN 90 5410 380 8

Complex tunnel intersections of the SSC: The inter-relationship of analysis, design and construction

Fwu-Ding (Kim) Lin & Steven H. Baldauf
The PB/MK Team, Dallas, Tex., USA
M. Amanda Elioff
Parsons Brinckerhoff/EMC, Los Angeles, Calif., USA

ABSTRACT: Two SSC tunnels with a 46 foot vertical separation are connected by transfer tunnels which crisscross in a vertical plane in "bow-tie" fashion. Finite difference ground-structure interaction analyses are described, as well as the application of analysis results to design, and the close coordination with construction sequences and procedures.

1 INTRODUCTION AND PROJECT OVERVIEW

The Superconducting Super Collider (SSC) tunnel forms an underground oval around the city of Waxahachie, located 30 miles south of Dallas, Texas. The accelerator and its particle detectors comprise two strings of superconducting magnets housed in an underground system of tunnels, interaction halls, galleries, niches and shafts.

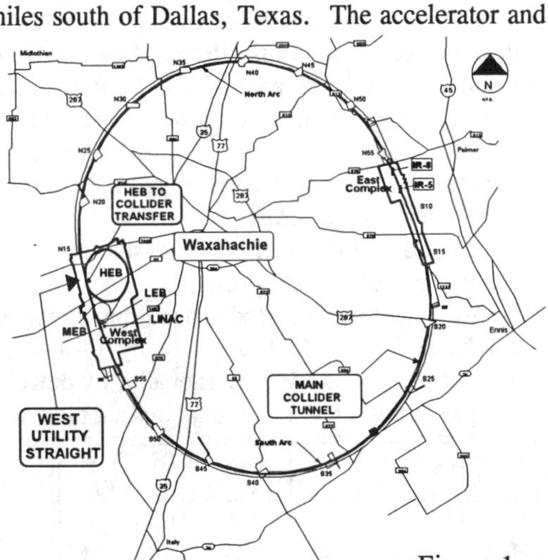

Figure 1

This paper is concerned with a portion of the West Complex, known as the West Utility Straight, which is a portion of the main collider tunnel and its connection to the last stage of the particle beam injector, the High Energy Booster (HEB). In this straight portion of the tunnel, the large kicker magnets necessary for directing the beam into the collider accelerator, as well as all of the radio frequency (RF) equipment required to accelerate the beam are located. The typical collider tunnel diameter is 14 ft. nominal; in order to accommodate the larger magnets, the tunnel diameter in the Straight was increased to 17 ft. minimum inside dimension. The design contract included a portion of the HEB tunnel in the transfer area, 2.4 miles of collider tunnel, seven shafts, transfer tunnels between the HEB and collider, and six galleries to house additional equipment.

2 GEOLOGY

The Texas site was selected for the SSC in part because geologic conditions are excellent for tunneling. Over 60 percent of the main collider tunnels are in the Austin Chalk Formation (AC), a soft to medium-hard limestone. Above and below the AC are Clay-shale formations, the Taylor Marl (TM) and the Eagle Ford Shale (EFS), respectively. All geologic formations are from the Cretaceous age. The formations dip one to two degrees to the southeast, such that a portion of the collider tunnel on the west side encounters the EFS. The West Utility Straight section of the collider tunnel will be constructed in this area, within the EFS, just below the contact with the AC. The HEB is expected to lie entirely within the AC formation, and the crisscrossing transfer tunnels between the two will encounter the contact.

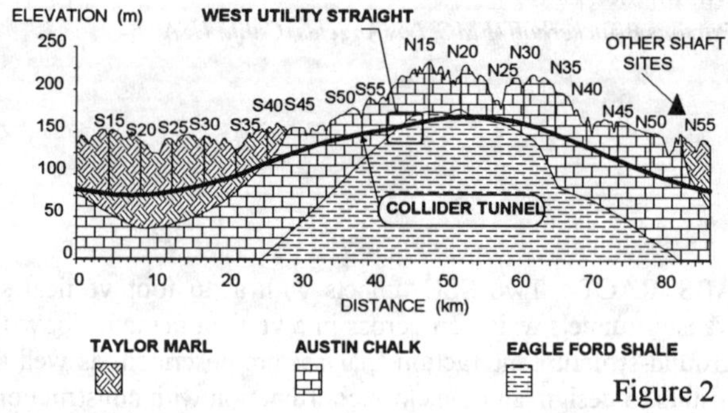

Figure 2

The EFS is a relatively weak shale with a high montmorillonite content and a compressive strength averaging 280 psi. The shale is a highly moisture-sensitive material which deteriorates when exposed to air and swells with exposure to water or with reduction in confining stress. Some overstressing of the EFS will occur at the tunnel depths, typically around 200 ft. Previous experience in other SSC contracts has shown that support and protection of the material must be installed within hours of excavation. Standard tunnel designs in the EFS included precast concrete segmental linings to be erected in the tail shield of the tunnel boring machine (TBM). In adits and chambers, larger cross sections were designed for sequential excavation with top headings and multiple drifts. Extensive use of spiling and short unsupported lengths of excavations were also specified.

The AC is not overstressed around tunnel openings, as average uniaxial compressive strengths are over 2000 psi. Many of the collider tunnels are unlined and are supported by dowels and, occasionally, by dowels with a few inches of shotcrete.

The properties of the ground, as adopted for computer modeling, are shown in Figure 3. These properties are based on the most typical values for the soil-like rock in the vicinity of the West Utility Straight, with modifications developed to account for variations in strength found at the interface. These variations were based on experience obtained from tests performed on samples taken from an "exploratory" shaft drilled not far from this project.

Property	AC	Wea. AC	EFS	Wea. EFS
E (ksi)	500	400	100	80
Unit Wt.(pcf)	147	147	137	137
Poisson's Rat.	0.3	0.3	0.3	0.3
phi (deg)	30	24	25	20
Cohesion (psi)	635	500	50	40
K_o	2.0	2.0	1.5	1.5
Tens Str. (psi)	250	200	30	24

Figure 3

3 GROUND SUPPORT SYSTEMS

For scheduling and contractual reasons, the construction of the project was divided into "basic" and "finish" contracts. This was a form of "fast-track" scheduling which allowed excavation to begin while the details of the technical systems were still being worked out by the scientists. The separation of contracts also kept the cost of each phase within the range that could be bonded by contractors. Basic contracts included tunnel and shaft excavation, initial ground support and surface infrastructure work. Finish contracts included final linings where required, invert concrete, and mechanical and electrical systems. For the West Straight, to the greatest extent possible, concrete support elements were to be constructed in the basic contract because of the complex geometry and potential cost savings in design and construction. The unusual geometry gave preference to shotcrete and dowel-supported sections. The initial ground support systems consisted of combinations of the following:

1. Shotcrete systems: Steel fiber-reinforced shotcrete (f'c = 5000 psi), and unreinforced shotcrete (f'c = 7000 psi).
2. Rock Dowels: Rock dowels are used (a) as specified pattern bolting systems and (b) to reinforce local rock weakness. These dowels are used in conjunction with the shotcrete in the EFS, or by themselves in the AC.
3. Precast Concrete Segmental Liners of 8000 psi concrete, 9 inches thick, are used in EFS where the tunnel is excavated by TBM.
4. Ribs and Boards: Structural-grade lumber and A36 steel ribs and pipes are used as temporary supports, when further enlargement of the tunnel section or a break-through for connecting transfer tunnels is required after the TBM operation.
5. Lattice Girders/Wire Mesh: Used as reinforcement for shotcrete liners where required in the EFS, usually in large caverns excavated in multiple drifts.

In the running tunnel, the precast segmental concrete liner also served as the final liner after it was grouted in place. In adits, galleries and transfer tunnels, either cast-in-place reinforced concrete or shotcrete is specified for final liners. A drainage system was designed to be placed between the initial and final liners to control any anticipated groundwater inflow.

4 CONSTRUCTION METHODS

The design team worked closely with construction managers to ensure that construction methods and sequences of excavation were practical and the most economical. At the "bow-tie" transfer region, at least two shafts would be excavated first, to allow access for a Tunnel Boring Machine to excavate the collider tunnel and a roadheader to mine the HEB tunnel and transfers. A precast concrete segmental lining or steel ribs with wood lagging (ribs and boards) at selected locations would be installed in the collider tunnel in EFS. Rock dowels with shotcrete support would be used in the AC in the HEB tunnel. The anticipated construction sequence was driven in part by muck removal considerations, and is illustrated in Figure 4:

1. Excavate the HEB tunnel in AC using either roadheader or TBM methods; install initial ground support. (Stage 1 excavation.)

2. Excavate the main collider tunnel in EFS using a TBM to place the liner segments. (Stage 2 excavation.)

3. Excavation and support of the enlarged transfer portion of the HEB tunnel by roadheader. (Stage 3 excavation.)

4. Place concrete slabs on the tunnel invert to serve as future HEB floor slabs as well as providing crown support for the initial section of transfer tunnels.

5. Continue the excavation toward one transfer tunnel by roadheader below floor slabs until a minimum distance between the transfer and collider tunnel is reached. (Stage 4 excavation.) Repeat Stage 4 excavation for the opposite transfer tunnel.

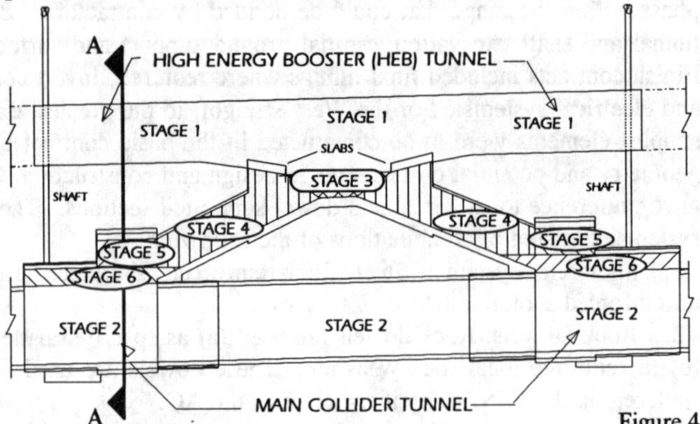

Figure 4
(NTS)

6. During the above stages, temporary support would be added inside the collider tunnel, just below the transfer area.

7. Excavate and support the transfer tunnel directly above the pre-supported collider tunnel. (Stage 5 excavation.)

8. Pre-support the ground above the collider tunnel with reinforcement such as fiberglass dowels to facilitate mining. Remove the temporary ribs and boards liner and enlarge the tunnel by roadheader, and install additional initial ground support. (Stage 6 excavation.)

9. Provide additional support for breakouts to niches and galleries from the precast concrete segmental linings, then complete excavation and ground support for niches, adits and galleries.

5 NUMERICAL MODELING

The analysis views the excavation and lining operation as a time-dependent, three-dimensional phenomenon. To simulate the transient behavior of the ground interaction with the excavation and lining process, a three-dimensional non-linear finite element analysis which implements an elastic/plastic material model to approximate the Sequential Excavation Method (SEM) is most desirable. However, experience on the project has shown that the ground deformation and relaxation at critical sections can be reasonably estimated in these geological formations using two-dimensional models. Critical sections within the transfer zone were analyzed with two goals in mind. The first was to examine the stability of the excavation, and to predict the stresses and deformations in the surrounding ground, as influenced by sequential liner installation. The second was to determine the forces in the supporting elements, as well as the magnitudes of the loads in the temporary linings utilized in the various stages of construction described above.

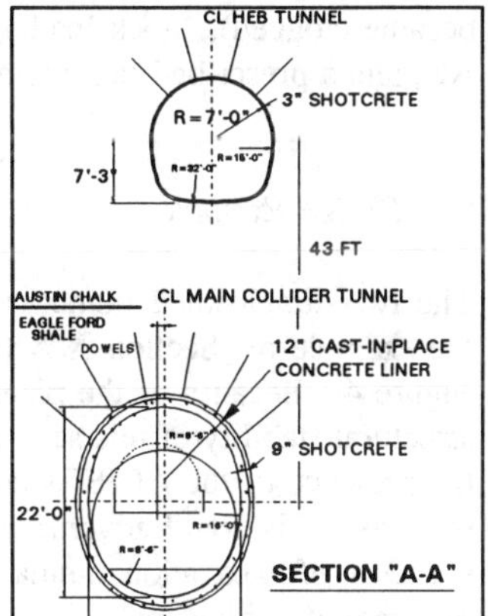

Figure 5

5.1 *Methodology*

Due to the soil-like nature of the clay-rich EFS, with its low but measurable porosity and permeability (10^{-8} cm/sec) characteristics, it was determined that pore pressures and consolidation of the shale would be considered in the analysis. A program which implements an explicit finite difference solution, known as FLAC, was used to model not only the ground's mechanical stress/strain response to the SEM, but also to account for the change in pore water pressure resulting from excavation and flow over time. This program solves the dynamic equations of motion at each grid point by using small time increments, and thus simulating the time-step nature of the excavation and lining sequence.

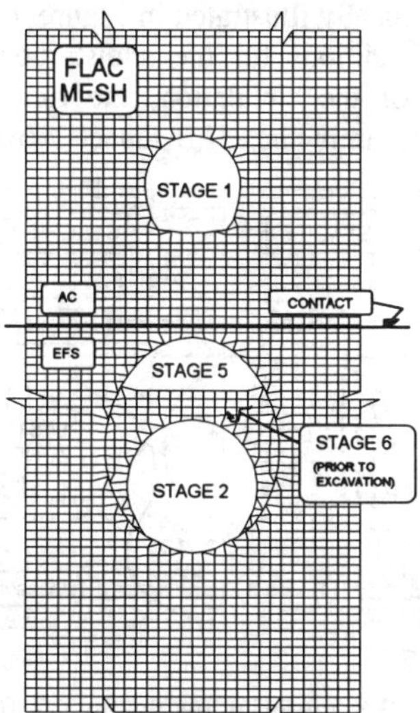

Figure 6

The initial shotcrete liner was subjected to a mechanical analysis coupled with pore water pressure simulating the sequential excavation and liner installation. Using a Mohr-Coulomb elastic/plastic material model, a "no flow" model was developed to approximate short-term behavior under effective stress conditions within the shale. This approach is valid for the sequentially installed initial shotcrete liners in that it assumes that there is insufficient time for significant flow to occur, and thus there is no appreciable drawdown of the ground water. This type of coupled analysis accounts for the ground deformation resulting from excavation, as well as the change in pore water pressure in response to changes in stress distributions around the opening. The effect of confining pressure exerted on the ground by the liner is thus accounted for at the various stages of excavation.

Some of the galleries and tunnels have a final liner in conjunction with a drainage system installed at some time after the placement of the initial shotcrete liner. In order to provide an "envelope" solution for the response of the liner to the anticipated ground conditions, a number of independent, non-additive, analytical cases were investigated. In these cases, the final liner is assumed to carry all the equipment and other long-term loads, including any build-up of hydrostatic pressure and stress changes in the ground due to consolidation or expansion resulting from long-term changes in pore water pressure. In addition to the above, the final liners, whether shotcrete or reinforced concrete, were subjected to loads resulting from a potential build-up of hydrostatic pressure behind the lining (if the drainage system became clogged), block loadings applied to liners in faulted or fractures zones in the AC, and a prescribed minimum long-term consolidation/swell load in the EFS.

5.2 *Design Example*

The two-dimensional numerical modeling and design processes are demonstrated by the example of Section A-A (Figure 5). The location of Section A-A is identified in Figure 4. Because of the planned construction sequence, it is necessary to verify the structural stability of this section during construction. The problem is compounded by the presence of the AC/EFS contact 5 feet above the combined Transfer/Main Tunnel sections. This could adversely affect the inherent arching strength of the rock. The sequence of excavation simulated in the model corresponds directly to the construction sequence described previously. A portion of the mesh used in the FLAC model is shown in Figure 6. Except for the natural ground surface, the boundaries of the model are located at a minimum of 8 tunnel diameters away from the center of the excavation. Since the tunnels are symmetric about the centerline, only half of the mesh was included in the model.

The four phases of the step-by-step simulation is graphically illustrated in Figure 7. The analysis was begun by establishing equilibrium conditions for the undisturbed ground at in-situ stress levels. For the first phase of the simulation, the finite difference mesh elements representing the AC material within the HEB tunnel were removed as indicated by the Stage 1 excavation. The model was then allowed to reach equilibrium after this ground disturbance. As expected, the idealized AC was found to be stable, without need for further support. For the second phase of the simulation, the elements representing the EFS material inside the 17 ft diameter collider tunnel were removed, as indicated by the Stage 2 excavation, and temporary tunnel liner elements (ribs and boards) were introduced after approximately 70% of the total deformation had occurred. The simulation was continued to permit full deformation for this stage of the excavation, and the model was again allowed to reach static equilibrium. Some yielding of the material around the opening was observed, prior to the installation of the liner elements.

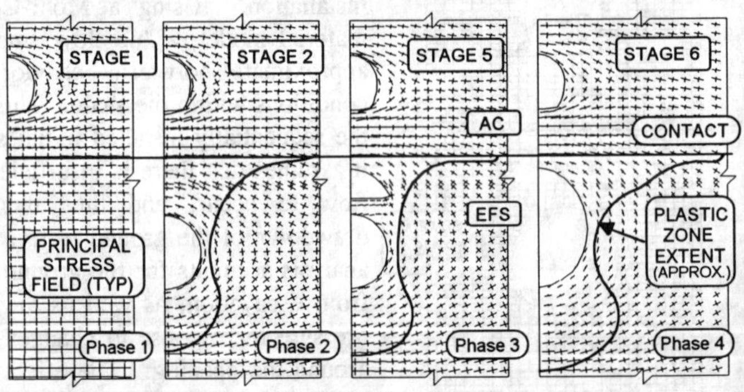

Figure 7

In the third phase of the analysis, the material just above the collider tunnel (Stage 5 excavation) was removed. Since it was anticipated that this portion of the transfer tunnel was to be excavated by a roadheader, a uniformly distributed construction load (500 psf), with a 30% allowance for machine impact, was applied to the invert of the Stage 5 drift.

The fourth and final phase of the simulation was executed by removing the bracing (described below), the temporary ribs and boards liner and the Stage 6 excavation, by slashing the roof down from the collider tunnel below. Also, at each stage of the simulation, protective layers of shotcrete were applied to the EFS to maintain confining pressure in the model (and to prevent deterioration in the field). In the final configuration, which was a 22 ft. high by 18 ft. wide tunnel, a maximum initial shotcrete thickness of 9 inches (placed in 3" lifts) was utilized.

The development and growth of the plastic zone around the openings is also illustrated in Figure 7. It is noted that the rock is overstressed in the Eagle Ford Shale right along the interface, and extends to approximately one diameter below the opening in the Eagle Ford Shale at the end of each stage of excavation. No tension failure is observed in the rock with the proposed opening size and the construction sequence. The principal stress plots (Figure 7), clearly show the forming and shifting of the ground arching around the openings with the progress of the excavation. Finally, as a displacement vs. time (model time, not real time), Fig. 8 displays the ground displacement history at several monitor points of Sect. A-A.

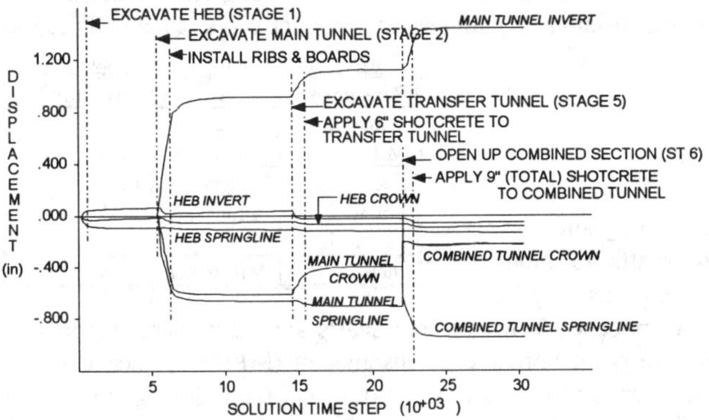

Figure 8

During the third phase of the analysis, after redistribution of the ground stresses and equilibration of the model, it was apparent that the temporary tunnel liner was overstressed under the combined action of ground loading plus overhead construction activity. Accordingly, internal pipe strut bracing was placed inside the collider tunnel at 4'-0" centers to match the spacing of the steel ribs. In addition, the Stage 5 drift was lined with a 6" thickness of shotcrete. A minimum of 5 foot rock thickness was maintained between the Main and Transfer tunnels at the end of Stage 5 excavation. This rock mass served as a strut which allowed the rock arching effect to envelope both openings. Figure 9 depicts the exaggerated ground movement during this phase. The cross-shaped pipe supports installed prior to the phase 3 Transfer Tunnel excavation (i.e., Stage 5 excavation) were instrumental in reducing the upheave tendency of the Main Tunnel

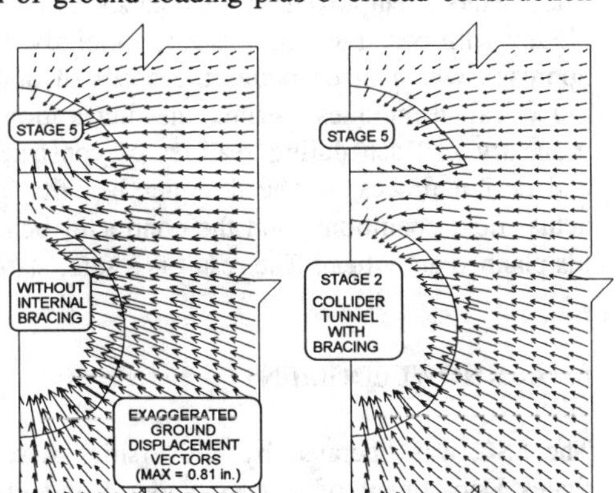

Figure 9

when the rock above was removed. The bracing is critical for the stability of both the Main Tunnel and the Transfer Tunnel during this stage of construction. Figure 9 illustrates the upward ground movement of the same model with and without the temporary supports. With this construction bracing, the liner and opening were stable throughout this intermediate construction phase.

In addition to the above, four elastic/plastic analyses utilizing both FLAC and the STAAD III finite element computer program were performed to evaluate the forces on, and determine the need for, additional long-term support in the form of a final liner. The results indicated that at the design depth of 220 ft. below ground surface, the controlling long-term load was the prescribed 25% hydrostatic loading. A 12" thick reinforced concrete liner was anticipated to handle long-term loads at these locations, and to support the machine, equipment and magnet support loads throughout the life of the Collider.

Conversion Factors from U.S. Customary to SI

To Convert	To	Multiply by
Pound mass	Kilogram	0.454
Mile	Kilometer	1.61
Feet	Meters	0.3048
Pound force per sq. in.	Kilopascal	6.89
Inch	Millimeter	25.4

6. INSTRUMENTATION

An instrumentation program was implemented at nearby shafts to monitor excavations, to verify design assumptions, and to calibrate analytical models. Two adjacent shafts and adits were instrumented with inclinometers, multiple point borehole extensometers (MPBX's), heave gauges, convergence gauges and piezometers. Deformations observed during construction confirmed the validity of two- and three-dimensional computer models for the EFS and AC geologies.

7. CONCLUSION

The methods employed in this paper demonstrate that tunnel designers can work closely with construction managers to jointly develop a design/excavation/construction approach which incorporates the needs of both the designer and builder. Available numerical techniques permit the integration of the constraints of the physical, monetary and scheduling realities of construction into the ground-structure analysis and design process. The designer benefits by being able to match the analysis to actual field conditions, and the contractor benefits in that the design is a response to his planned activities. The result is a safe, stable and economical project.

8. ACKNOWLEDGMENT

The SSC was operated by Universities Research Association, Inc. for the United States Department of Energy under Contract DE-AC35-89ER40486. The PB/MK Team was responsible for the design and construction management of SSC conventional facilities. The material in this paper represents the efforts of many talented individuals, and the procedures described have evolved over the course of the project.

Rock Mechanics, Nelson & Laubach (eds) © 1994 Balkema, Rotterdam, ISBN 90 5410 380 8

Recovery of a double width stope at Golden Giant Mine: A successful application of numerical modelling and conventional methods for stability analysis

P. Rocque
Hemlo Gold Mines Inc., Golden Giant Mine, Marathon, Ont., Canada

W.F. Bawden
Department of Mining Engineering, Queen's University, Kingston, Ont., Canada

ABSTRACT: Poor ground conditions in one area of the Golden Giant mine made the recovery of the stopes very difficult. In order to minimize ground rehabilitation and the dilution coming from the hanging wall a modified stope geometry was proposed. It was decided to extract these stopes as double width (28 m along strike rather than 15 m). The design was based on results from a detailed rock mechanics study which included the Modified Mathew's Stability Analysis, an instrumentation programme and a numerical model. This methodology has been effective in predicting potential ground conditions problems and has been proven as a valuable tool for the successful recovery of the first double width stope.

1.0 INTRODUCTION

Hemlo Gold Mines Inc. is operating the Golden Giant Mine located about 700 km northwest of Toronto and 1 000 km east of Winnipeg, Canada. The Golden Giant is one of three mines which have been producing from the Hemlo camp since 1985. The mine's active areas are located in the central portion of the ore body and they are bounded on the East and West side by different mining operations. The Hemlo deposit is hosted in a quartz-feldspar-granofels rock type. It strikes at about 295° and dips at an average of 65° North. The primary form of mineralization is disseminated gold and molybdenite within potassium-feldspar rich rock. Occasionally coarser grains of gold, stibnite and realgar have been observed within deformed quartz pods and veinlets (Kusins *et al.* 1991). Current mining is taking place at depths varying from 200 m to about 900 m and the main mining method consist of open stoping with delayed cemented rockfill. The average stope size is about 15 m along strike by 20 m thick and 100 m high. The current production rate is set at 3 000 t/d, which allows Golden Giant Mine to pour over 400 000 ounces of gold annually.

The stopes located at the bottom East end of Golden Giant Mine's property are named "*H stopes*" (see Figure 1). These stopes carry a significant grade of gold (up to 35 g/t with an average of about 20 g/t) but the dilution coming from the sericitic hanging wall rock can be as high as 20 % and it often delays production activities because of additional ground rehabilitation. Stress conditions prevailing in the western abutment of the H stopes coupled to existing development configuration further complicated the recovery of these stopes. As shown in Figure 2, a drift (labelled "4490") was developed about 5 to 10 m above the 4466 Sub-Level. The proximity of that drift induced stability problems in the back and the hanging wall. Buckling failure was observed in the hanging/wall for about 45 to 50 m from the H6 backfilled stope on the 4466 Sub-Level.

In order to extract these highly valuable stopes with minimal dilution, the stability of different stope geometries was assessed using the Modified Mathew's Method (Bawden 1993). A numerical model was set up to verify the influence of a stope length increase on the adjacent area. In order to correlate visual observations with the model predictions, an instrumentation programme was set up.

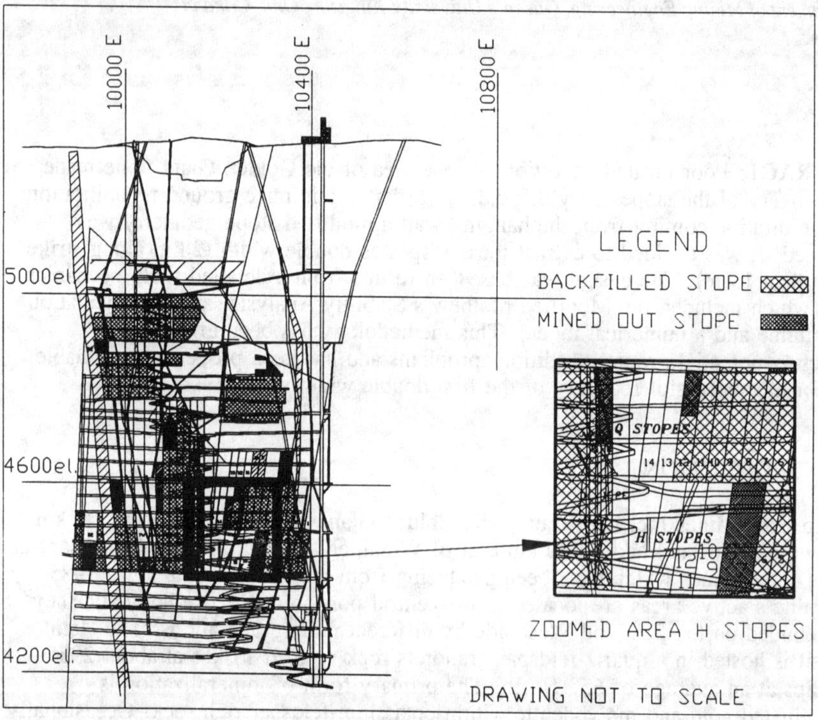

Figure 1: Location of the H Stopes.

2.0 MATHEW'S STABILITY ANALYSIS

The semi-empirical Modified Mathew's Method is used to assess the stability of a surface (back, hanging wall, etc.). This method takes into account the geometry of the wall, the induced stresses generated from the mining activities and the effect of the orientation of structural features and stope walls. Because of its empirical nature, it needs to be calibrated at each mine site. This method has been used consistently for the past five years for stope design at the Golden Giant mine and as a result, it is well calibrated.

Geomechanical parameters used for the Mathew's analysis are shown in Table 1 and the results are displayed on Figure 3 for the back, end wall and hanging wall (Bawden 1992, Rocque 1992). The parameters used are defined as follows:

- "A" represents the rock stress factor (UCS/induced compressive stress);
- "B" represents the joint orientation adjustment factor;
- "C" represents the gravity adjustment factor for gravity fall and sliding;
- "N'" represents the Modified Stability Number and it is the result of the multiplication of Q'*A*B*C;
- "S" represents the Hydraulic Radius (area/perimeter).

Results from the stability analysis indicated that a stope length of 28 m by 15 m width and 100 m vertical height would not compromise the extraction of the ore provided that additional hanging wall support is installed.

Field testing at the Golden Giant mine of a novel type of cable bolt in Canada - the nutcage cable - had been very promising (Hyett *et al.* 1993). This type of cable was consequently selected over traditional types of cables. It was decided to provide pattern cable bolt hanging wall support from special hanging wall drives (refer to Figure 4).

Table 1: Stability Analysis Parameters.

Parameter	Q'	A	B	C	N'	$S_{maximum}$	S_{actual}
Back	3.3-6.7	0.15	0.3	2.0	0.3-0.6	7.5-8.0	4.9
Hanging Wall	1.4-2.9	1.00	0.3	5.7	2.4-5.0	9.5-11.5	11.2
End Wall	3.3-6.7	0.15	0.3	7.5	1.1-2.3	8.5-9.5	6.6

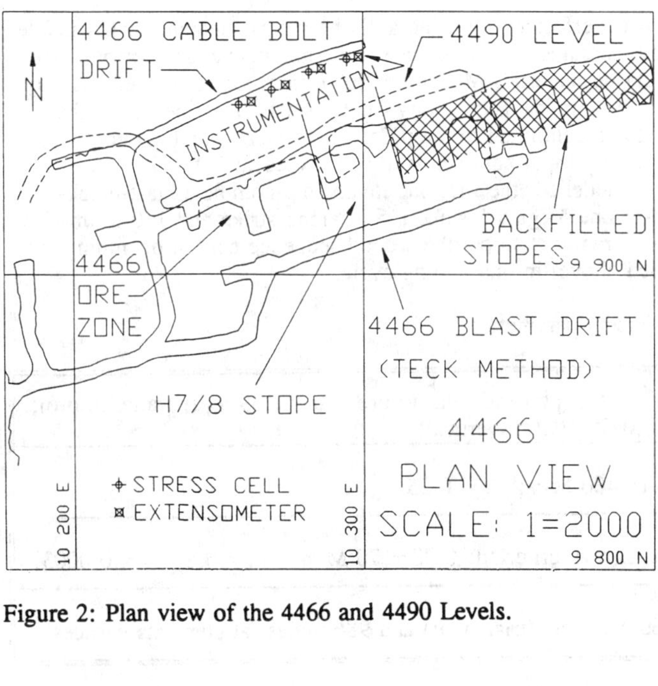

Figure 2: Plan view of the 4466 and 4490 Levels.

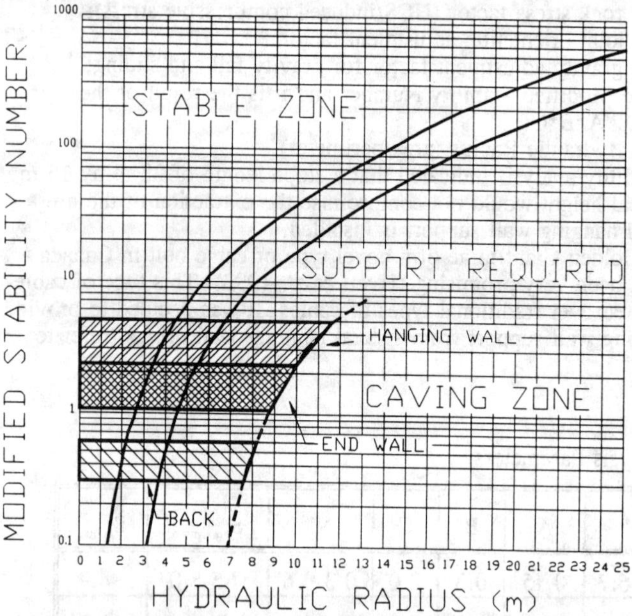

Figure 3: Modified Mathew's Stability Graph for the H stopes.

3.0 NUMERICAL MODELLING

Numerical modelling was performed by using a 3D boundary element package called *EXAMINE^{3D}*. This model assumes the rock mass to be homogeneous, isotropic and to behave as a linear elastic material. Modelling has been accomplished extensively for the Hemlo deposit by Noranda Technology Centre staff. Over the years, the model has been refined and calibrated to the point where there was high confidence in the validity of the results. Input parameters are presented in Table 2.

Two sequences were modelled: the first one involved a complete blast-muck-backfill mining cycle for one "regular" stope (15 m along strike and 100 m in height) then for another "regular" stope; the second sequence consist of mining the two stopes (as described above) in one mining cycle.

Table 2: *EXAMINE^{3D}* model parameters.

Stresses	σ_1=2.04*σ_3 (perpendicular to ore); σ_2=1.39*σ_3 (parallel to ore); σ_3=0.027 MPa/m (vertical)
Elastic Constants	E= 16 400 MPa; ν= 0.25
Failure Criterion	Hoek-Brown criterion: UCS= 70 MPa; m= 0.8; s= 0.0003
Mesh Parameters	1 960 elements (triangular) and 986 nodes (at elements vertices)

4.0 INSTRUMENTATION

In addition to the extra support, an instrumentation programme was also put in place to monitor stress changes and deformation that occurred in the hanging wall. The analysis of the data gathered from the H7/8 stope blast are expected to provide a better understanding of the rock mass response to a double width stope extraction and ultimately to improve the stope design for the remaining H stopes. Four pairs of stress cells and four extensometers were placed in the ground at a distance of about 1 m from the ore/waste hanging wall contact. Figure 4 shows a typical cross-section.

The extensometers measured about 22 m long and came with five anchors installed approximately at 22, 20, 18, 16 and 10 m. They were grouted in a 56 mm downhole which was drilled perpendicular to the ore/waste contact; this hole orientation was preferred because it minimized the measured displacements which occurred other than along the extensometer axial direction. Stress change was monitored using vibrating wire cells manufactured by CANMET. These "Herget cells" were mounted in pairs in order to allow measurement of both minor and intermediate stress changes. The interest for measuring the minor principal stress change lies in the variation of cable bolt radial confinement which can significantly reduce the loading capacity of the cable bolts (Hyett *et al.* 1994).

The extensometers and stress cells were installed from the 4466 Sub-Level since it was expected that movement and stress changes would be more significant near the mid-height of the stope. The four extensometers were installed within four cable bolt rings spaced 15 m apart. Since the stress cells needed a different hole size diameter, four NQ downholes were drilled within 1 m of the extensometers holes. The cells were installed at about the same depth as the deepest anchor for each extensometer in order to correlate measured movements to calculated stress changes.

The installation of the extensometers was fairly straightforward but problems were encountered with the stress cells. These cells were wedged in the hole by applying a preset force with the installation tool. Due to problems with the installation tool the force applied by the tool did not achieve the required values. As a result three cells out of eight were not properly installed and their results were discarded.

Data was recorded by a datalogger located about 75 m away from the furthest extensometer. Automatic reading was set at a 12 hour interval which allowed for sufficient data collection prior to the blast. On the blasting day, the datalogger was set to read every 15 min in order to provide a more accurate observation of the initial rock mass response to the blast. Costs of the instrumentation programme are summarized in Table 3.

Table 3: Instrumentation costs.

STRESS CELLS	
4 pairs @ 2 140 $/pair	8 560 $
Drilling NQ size: 96 m @ 125 $/m	12 000 $
EXTENSOMETERS	
4 units @ 2 560 $/unit	10 240 $
MISCELLANEOUS (bolts, etc.)	200 $
TOTAL:	31 000 $

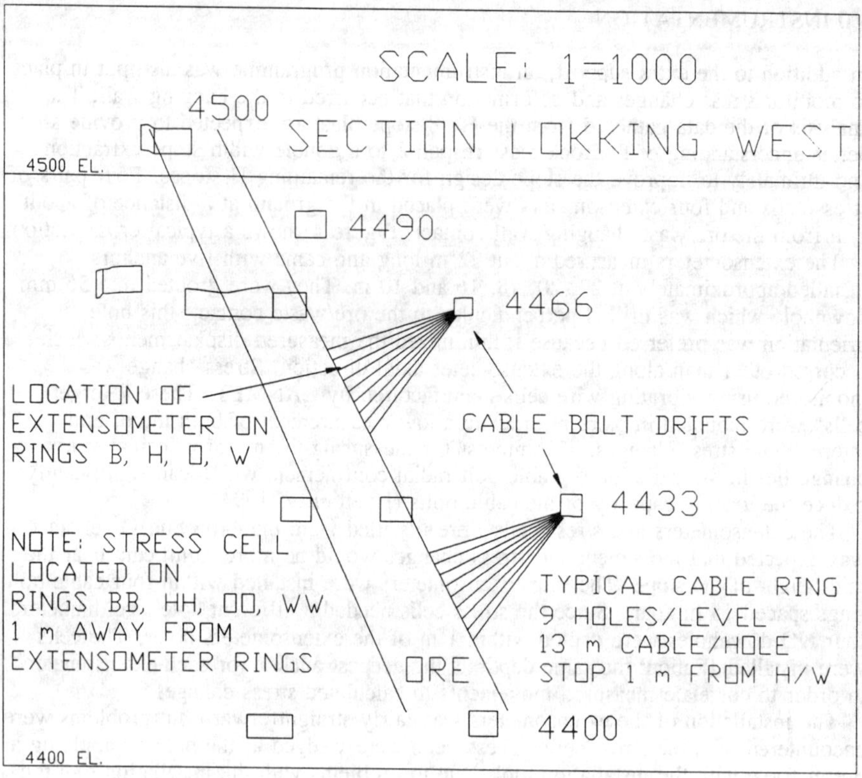

Figure 4: Typical Cross-Section Through Double Width H Stopes.

5.0 INTERPRETATION AND DISCUSSION OF PRE-BLAST RESULTS

5.1 Numerical modelling

The modelling results did not show any significant differences between the induced stresses predicted for the first and the second sequence; however, there are some differences worth noting:

- there is more extensive yield of the sill (between 4466 and 4500 Sub-Levels) predicted for a double width stope than for a single width (assuming a regular cycle of blast-and-muck one lift at the time), and this could lead to blasthole instability (shear failure) and limited stand up time;
- the distribution of σ_3 across the stope hanging wall implies that destressing is more important in the case of a double width stope and would affect mainly the East portion of the mined out stope. Hence a loss of confinement that could influence the cable bolt performance in the East stope (H7) but not in the West stope (H8) would be predicted.

5.2 Extensometers and stress cells data

Preliminary results from the extensometers are indicating that the delamination process which took place between the 4466 and 4433 Sub-Levels began before the first lift was blasted and displacements increased at a steady rate to about 4 mm in the H8 stope and about 0.5 mm in the H9 stope. The stress cell located in the ore zone on 4466 Sub-Level indicated a sudden change in the N-S stress in the range of about 5 to 10 MPa. This result confirmed the visual observations made in that drift. Since the remaining stress cells located in the hanging wall did not show any significant stress changes prior to the final blast it is believed that the severe degradation observed on the 4466 Sub-Level is the result of stress concentration in the top portion of the sill (4466 to 4490 Sub-Levels) and excessive stand up time resulting from the delay between the first lift and the final blasts.

5.3 Pre-blast visual observations

The first lift from 4400 to 4433 Sub-Levels was blasted in September of 1993. The unsupported back portion of the H7 Stope on 4433 Sub-Level failed (as predicted) and arched 5 to 7 m above this Sub-level.

The final mass blast of the remaining lifts (4433 to 4500 Sub-levels) was deferred until the beginning of February '94 due to a change in mining priorities. During this time the exposed hanging wall (4400 to 4433 Sub-levels) has remained stable, showing the effectiveness of the cable bolt design.

At the same time on 4466 Sub-level ground conditions began to deteriorate on the back and walls. New cracks emerged, cleaning of the blastholes became difficult and rebar heads snapped off. These events indicated a stress increase over time, again as predicted.

6.0 CONCLUSIONS

Due to difficult ground conditions a decision to alter the geometry of the stopes was made in order to minimize the dilution coming from the hanging wall. This decision was based on conventional techniques used in rock mechanics and on a numerical modelling method which both have been successfully implemented at the Golden Giant mine over the years.

An instrumentation programme was set up in order to correlate the ground conditions forecasted by both stability analysis and modelling. So far, the predictions have been verified in numerous ways: stress increase was measured and observed in the top portion of the sill pillar around the 4466 Sub-Level and the hanging wall between 4433 and 4400 remained stable after the H7/8 first lift blast.

This methodology of combining stability analysis, numerical modelling techniques and an instrumentation programme has been effective in predicting and monitoring potential ground conditions problems and has been proven as a valuable tool for the successful recovery of the H7/8 double width stope. It is expected that the final results will be valuable in the design of the extraction process of future double width stopes.

ACKNOWLEDGMENTS

The authors thank the Golden Giant Mine management for the permission to publish this paper. Acknowledgment is extended to Alain Gendron, Richard Hamilton and Simon Nickson from NTC. Miners from the cable bolting crew and colleagues from the Engineering and Geology Department at Golden Giant also provided useful assistance.

REFERENCES

Bawden, W.F. 1992. Proposed Extraction of Stopes H6 and H7 as Single Stope, a Stability Analysis. Report to Golden Giant Mine.

Bawden, W.F. 1993. The use of Rock Mechanics principles in Canadian Underground Hard Rock Mine Design. Comprehensive Rock Engineering, Vol. 5, Surface and Underground Project Case Histories, John A. Hudson Ed., Pergamon Press.

Hyett, A.J., Bawden, W.F., Powers, R. & Rocque, P. 1993. The Nutcase Cable Bolt. Proceeding from the Innovative Mine Design for the 21st Century, Bawden and Archibald Ed., Balkema, pp. 409-419.

Hyett, A.J., Bawden, W.F., MacSporran, G.R., Moosadi, M. 1994. Constitutive Law for Bond Failure of Fully Grouted Cable Bolts Using a Modified Hoek Cell. Submitted to the IJRM.

Rocque, P. 1992. Stability of Modified Geometry H Stopes. Internal report.

Kusins, R., Chong, A., Johnston, P., McIlveen, D. & McNena, K. 1991. Geology of the Golden Giant Mine and Golden Sceptre Orebody. Geology and Gold Deposits of the Hemlo Area, T.L. Muir, B.R. Schnieders & M.C. Smyk Editors, pp. 16-33.

Rock Mechanics, Nelson & Laubach (eds) © 1994 Balkema, Rotterdam, ISBN 90 5410 380 8

Constitutive modeling of rock masses containing fracturing joints and analysis of large-scale excavation

H. Horii & H. Yoshida
Department of Civil Engineering, University of Tokyo, Japan

ABSTRACT: The mechanical behaviors of a jointed rock mass are strongly affected by the property and geometry of joints. In the present study, a constitutive model of rock masses with multiple fracturing joints is proposed. It reflects the effect of orientation and density of joints. Then, the proposed constitutive model is implemented into a finite element analysis code. As an example, a problem of excavation of a power station cavern is analyzed. Characteristic features of displacement distribution, the opening and sliding of joints, and the change in the tensile forces in prestressed anchorage bars during excavation process are captured in the present analysis.

1 INTRODUCTION

Rock masses are characterized by distributed joints. It is well known that the mechanical behaviors of jointed rock masses are strongly affected by not only the mechanical property but also geometry of joints. The number of joints in the rock masses, however, is so enormous that it is almost impossible to deal with each joint individually. Thus, it is necessary to replace the jointed rock mass with an equivalent continuum body for analysis. Cai and Horii [1) 2)] proposed a constitutive model which presents the effects of density, orientation and connectivity of joints as well as the property of the joint itself. This micromechanics-based model is for the analysis of rock deformation before the failure of joints with small displacement.

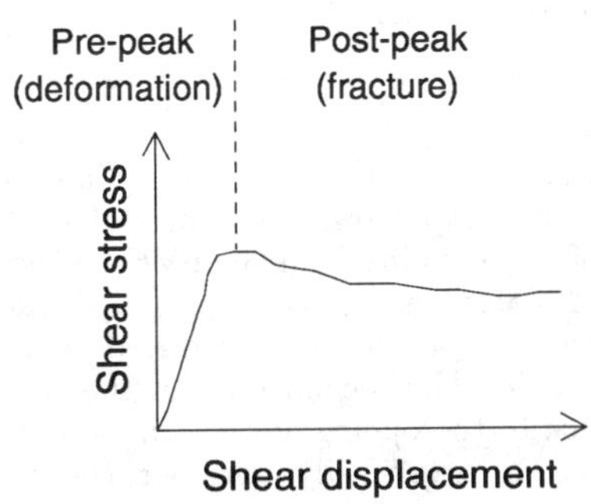

Fig.1 Shear stress - shear displacement relation

In the case of underground excavation for, for instance, an underground power station, joints initially closed by earth pressure may slide and open due to stress relaxation during excavation. Joint behaviors in unloading, not in loading, are important in excavation. The amount of sliding and opening is very large, and the joint behavior corresponds to "fracture", post-peak behavior in shear tests of joints rather than "deformation" before the peak (see, Fig.1). Though the joints are flat locally, they may have more or less undulation in a large scale. In the case of excavation, such large-scale undulation of joints is considered to play an important role. Sliding along a part of a joint leads to opening along the other part. A large compressive stress is left in the hoop direction around a cavern after excavation, causing the joint to open widely. A constitutive theory reflecting these features is necessary in order to analyze problems of the excavation of a large-scale underground cavern. In this article, a constitutive equation of jointed rock masses is derived from the relationship between average stress and average strain over a representative volume element (RVE) which contains a number of joints. The behavior of each joint in the RVE is estimated on the basis of micromechanics. As is already mentioned, Cai and Horii [1) 2)] proposed a constitutive model of jointed rock masses where the pre-peak behavior of joints is considered. In the present study, their theory is modified and applied to rock masses with 'fracturing joints'. Then, the proposed constitutive model is implemented into a finite element analysis code to analyze a large-scale underground excavation.

2 CONSTITUTIVE MODEL OF ROCK MASSES CONTAINING JOINTS

2.1 *Average stress - average strain relation*

Consider an RVE of volume V and external surface ∂V that contains a lot of joints of surface Ω. The average stress ($\overline{\sigma}$) and average strain ($\overline{\varepsilon}$) in the RVE are defined as

$$\overline{\sigma}_{ij} = \frac{1}{V}\int_V \sigma_{ij} dV, \quad \overline{\varepsilon}_{ij} = \frac{1}{V}\int_V \varepsilon_{ij} dV. \tag{1}$$

It is noted that the average stress of the RVE ($\overline{\sigma}$) is the same as that of intact rock ($\overline{\sigma}^R$), while the average strain of the RVE ($\overline{\varepsilon}$) is different from that of intact rock ($\overline{\varepsilon}^R$). The average strain over a body containing discontinuities is expressed in the following form [3)].

$$\overline{\varepsilon}_{ij} = C^R_{ijkl}\overline{\sigma}_{kl} + \frac{1}{2V}\int_\Omega ([u_i]n_j + [u_j]n_i) dS, \tag{2}$$

where C^R_{ijkl} is the elastic compliance tensor of the intact rock and $[u_i]$ is the displacement jump across the joint surface. The integration is taken over the upper surface of all joints inside of the RVE. In the derivation, the relationship between strain and displacement, $\varepsilon_{ij} = \frac{1}{2}(u_{i,j} + u_{j,i})$ and the divergence theorem are employed and no approximations are introduced. The intact rock surrounding joints is assumed to be elastic. The second term of the right hand side in Eqn.(2) is the additional strain due to the opeing and sliding joints. Since the mechanical response of jointed rock masses is generally non-liner, the incremental expression is employed, i.e.

$$\Delta\overline{\varepsilon}_{ij} = C^M_{ijkl}\Delta\overline{\sigma}_{kl} + \frac{1}{2V}\sum_\alpha \int_{\Omega^\alpha} (\Delta[u_i]n_j + \Delta[u_j]n_i)_\alpha dS, \tag{3}$$

where Ω^α is the surface of joint α. Thus, if the incremental displacement jump across the joint surface, $\Delta[u_i]$, is given as a function of the incremental average stress, $\Delta\overline{\sigma}_{ij}$, the constitutive equation of rock masses containing joints is obtained from Eqn.(3) as

$$\Delta\overline{\varepsilon}_{ij} = \overline{C}_{ijkl}\Delta\overline{\sigma}_{kl}, \tag{4}$$

where $\overline{C}_{ijkl}$ is the tangential compliance of jointed rock masses.

2.2 *Behavior of joints in rock masses*

To express the incremental displacement jump across a particular joint as a function of the incremental average stress, let us consider the decomposition of the problem as shown in Fig.2; the original problem (a) is decomposed into a homogeneous problem (b), an open slit with non-zero traction problem (c), and a cut-out joint problem (d). The relationship between the average incremental displacement jump across the slit and the average incremental stress acting on the slit is given as follows [1];

$$\Delta\overline{\sigma}_n - \Delta\overline{\sigma}^J_n = \overline{K}_n\Delta[\overline{u}^S_n], \quad \Delta\overline{\sigma}_s - \Delta\overline{\sigma}^J_s = \overline{K}_s\Delta[\overline{u}^S_s], \tag{5}$$

where $\Delta[\overline{u}^S_n]$ and $\Delta[\overline{u}^S_s]$ are normal and shear components of the average incremental normal and shear displacement jump across the slit on which the incremental normal stress $\Delta\overline{\sigma}_n - \Delta\overline{\sigma}^J_n$ and shear stress $\Delta\overline{\sigma}_s - \Delta\overline{\sigma}^J_s$ are acting, respectively. $\overline{K}_n$ and $\overline{K}_s$ are normal and shear system stiffness, which are assumed for a joint in the rock mass surrounding the joint and are of form,

$$\overline{K}_n = \frac{\overline{E}}{\lambda_n L^J}, \quad \overline{K}_s = \frac{\overline{G}}{\lambda_s L^J}, \tag{6}$$

where $\overline{E}$ is the tangential effective Young modulus in a direction normal to the joint and $\overline{G}$ is the tangential effective modulus for the shear deformation. λ_n and λ_s are the dimensionless factors to consider the connectivity of joints.
The compatibility condition of displacement jump between the joint and the slit leads to

$$\Delta[\overline{u}^S_n] = \Delta[\overline{u}^J_n], \quad \Delta[\overline{u}^S_s] = \Delta[\overline{u}^J_s], \tag{7}$$

where $\Delta[\overline{u}^J_n]$ and $\Delta[\overline{u}^J_s]$ are the average incremental displacement jump across the joint surface, respectively.

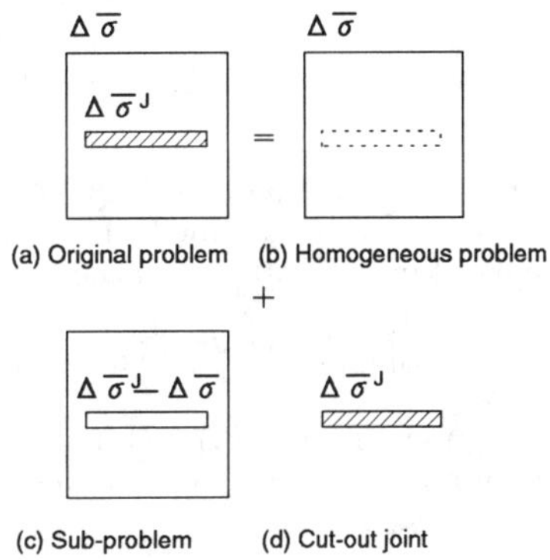

Fig.2 Decomposition of joint problem

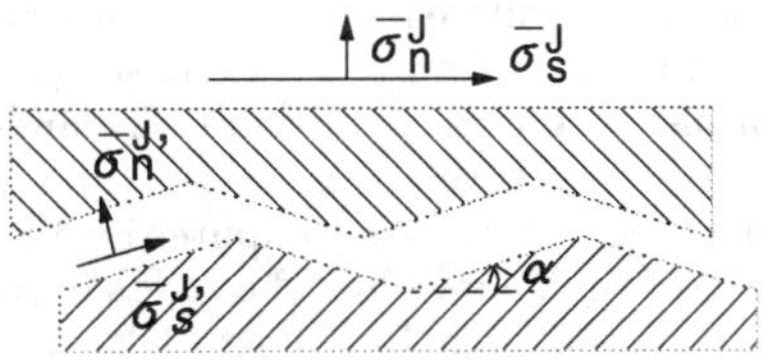

Fig.3 Joint with undulation

The saw-teeth shape shown in Fig.3 is considered to model the large-scale undulation of joints. It is assumed that deformation of the joint is caused by sliding along one part of its surfaces and the associated opening of the other part.

The Columb fracture criterion is employed, i.e, the deformation of a joint is initiated when the following condition is satisfied;

$$|\overline{\sigma}^{J'}_s| = -\overline{\sigma}^{J'}_n \tan\phi, \quad \overline{\sigma}^{J'}_n < 0, \tag{8}$$

where ϕ is the frictional angle on the joint surface, $\overline{\sigma}^{J'}_n$ and $\overline{\sigma}^{J'}_s$ are the average normal and shear stresses on the sliding joint surface, respectively. Only the case when the sliding of one part leads to the opening of the other part is considered. At this moment, the opening part of the joint surface is assumed to be traction free.

From the equilibrium condition, the relationship between the nominal joint stresses on the cut-out joint plane and those on the sliding joint with undulation (see, Fig.3) is given by

$$\Delta\overline{\sigma}^J_n = \frac{1}{2}(\Delta\overline{\sigma}^{J'}_s \tan\alpha + \Delta\overline{\sigma}^{J'}_n), \quad \Delta\overline{\sigma}^J_s = \frac{1}{2}(\Delta\overline{\sigma}^{J'}_s - \Delta\overline{\sigma}^{J'}_n \tan\alpha), \tag{9}$$

where α is the undulation angle of the joint.

It is assumed that the fracture criterion, Eqn.(8), is maintained during sliding of the joint, and the average incremental stresses on the sliding joint surface satisfy the following condition;

$$\Delta\overline{\sigma}^{J'}_s = \begin{cases} -\Delta\overline{\sigma}^{J'}_n \tan\phi & \alpha > 0 \\ \Delta\overline{\sigma}^{J'}_n \tan\phi & \alpha < 0 \end{cases}. \tag{10}$$

Along the sliding joint surface, no normal component of displacement jump across the joint surface is assumed. Then, the relationship between global and local components of incremental displacement jump across the joint surface is give as

$$\Delta[\overline{u}^J_n] = \Delta[\overline{u}^{J'}_s]\sin\alpha, \quad \Delta[\overline{u}^J_s] = \Delta[\overline{u}^{J'}_s]\cos\alpha, \quad \Delta[\overline{u}^{J'}_n] = 0. \tag{11}$$

By combining Eqns.(5), (7), (9), (10) and (11), the average incremental displacement jump across the joint surface is obtained as a function of the average incremental stress.

3 RESULT OF CONSTITUTIVE MODEL

As an example of the response of the proposed constitutive model, we consider a case in which a rock mass with one set of joints normal to the x_1-axis is initially under 10 MPa compressive stresses from three directions (see, Fig.4). We compute the displacement of the joints while the stress in x_1-axis is decreased from 10 MPa to 0 MPa. This reproduces the phenomenon of joints sliding and opening due to stress relaxation during excavation. The material properties are set as follows; the Young modulus E=5000 MPa, the Poissons ratio ν=0.25, joint frictional angle ϕ=30°, joint undulation angle α=20°, joint characteristic length L^J=10 m, average distance of the joints =1 m, and the normal and shear coefficient of joint connectivity λ_n=1.592, λ_s=12.0, respectively.

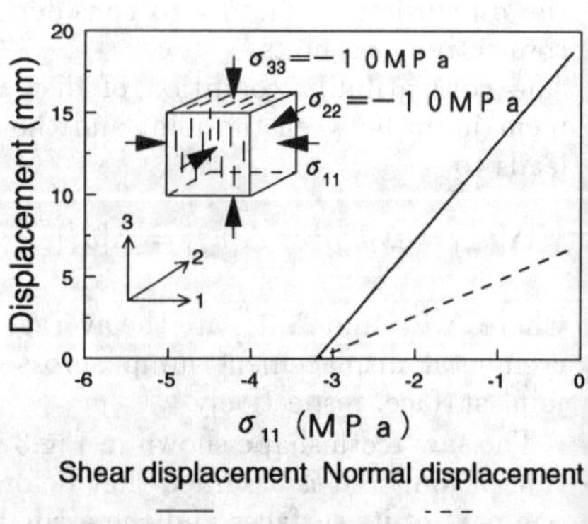

Fig.4 Displacement jump of joint during stress relaxation

The relationship between the displacement jump across joints and the stress in x_1-direction is shown in Fig.4. The sliding and associated opening of joints starts at around 3 MPa.

4 FINITE ELEMENT ANALYSIS

The proposed model is implemented into a finite element analysis code to analyze the underground structure under arbitrary loading and boundary conditions. As an example, the problem of excavation of a power station cavern is analyzed. Two dimensional problem under a plane strain condition is solved although the analysis code is for general three dimensional problems. The distribution of the displacement jump across joints and the deformation of rock masses, the tensile force of prestressed anchorage bar, are obtained.

The input data are as follows: two sets of joint whose dip directions are 30° clockwise and 60° counterclockwise, respectively. Their strike is parallel to the axis of power station. The joint frictional angle of both joints is ϕ=25°. The size of power station and the initial principal compressive stresses are shown in Fig.5. This stress condition corresponds to the pressure at 200 m depth. The access tunnels for carrying the manufacture of machinery are located on the right and left hand sides of the power station cavern. The FEM mesh is prepared to take care of the effect of a fault on the right hand side of the cavern. However, the same material properties is assumed in the fault portion for the analysis presented in this paper. The other input data are the same as those in the previous section.

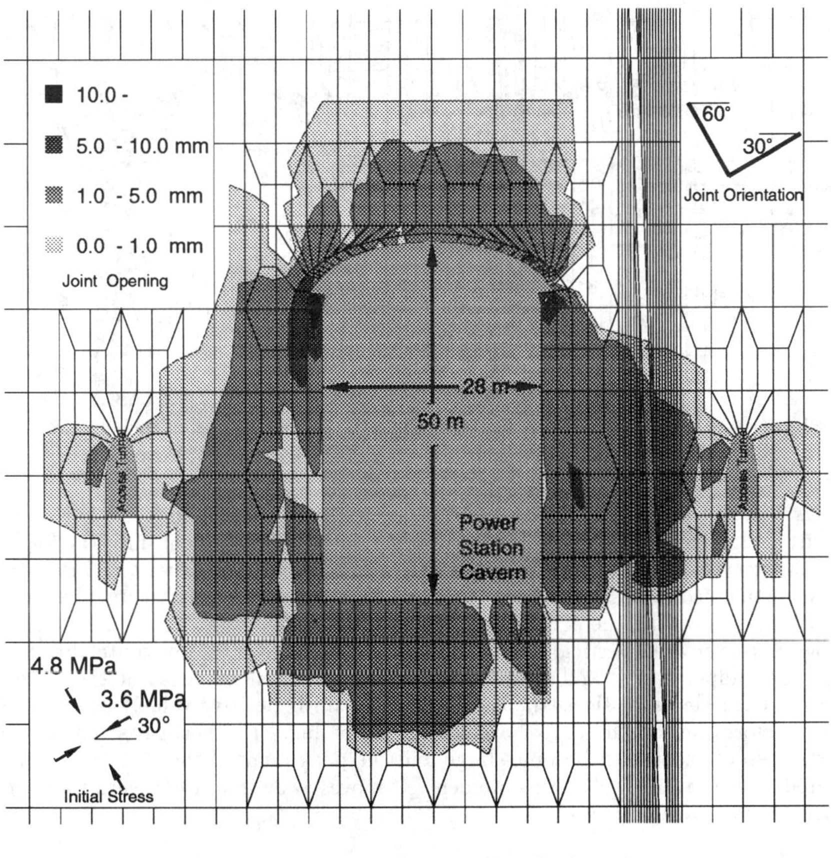

Fig.5 Distribution of joint opening displacement

In this analysis, we have eight steps of excavation. Between the first and second steps of excavation, an arch concrete lining is constructed at the top part of the cavern. The elastic modulus of the concrete is assumed to be 20000 MPa. After the second step of the excavation, the prestressed anchorage bars are cast at both sides of cavern wall. The anchorage bar of length 20 m connects the nodal points in the finite element mesh. Initial force is applied just before the two nodal points are connected by a bar element. The Young modulus of the bar is assumed to be 200000 MPa. The amount of initial forces and area of the bar element are calculated for the anchorage density of $\frac{1}{3}$ bars per 1 m^2 with 96 t and 13.9 cm^2 for one anchorage bar which consists of four twisted steel bars.

The distribution of joint opening displacement after the last step of excavation is shown in Fig.5. Regions with large opening displacement are formed around side walls. The maximum opening displacement, in this analysis, is about 10 mm.

Figs.6 show the absolute displacement of rock masses at the first step and the last step of excavation. The prestressed anchorage bars are cast at the points shown in Fig.6(a); from A to H. According to Fig.6(b), the wall displacement takes its largest value around the middle point of left hand side of the cavern, which is about 37 mm.

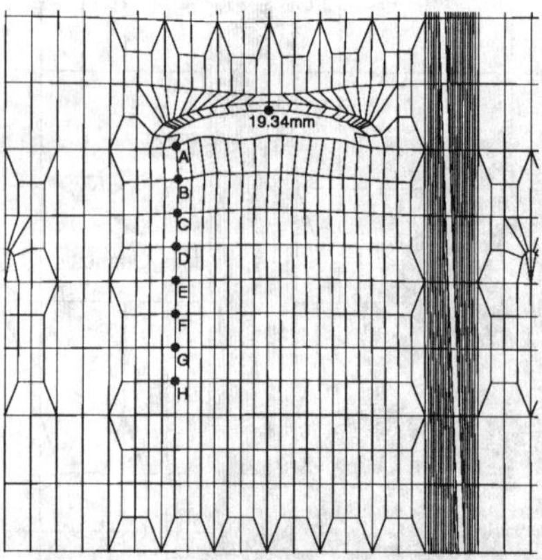

Fig.6(a) Deformation of rock masses (First step of excavation)

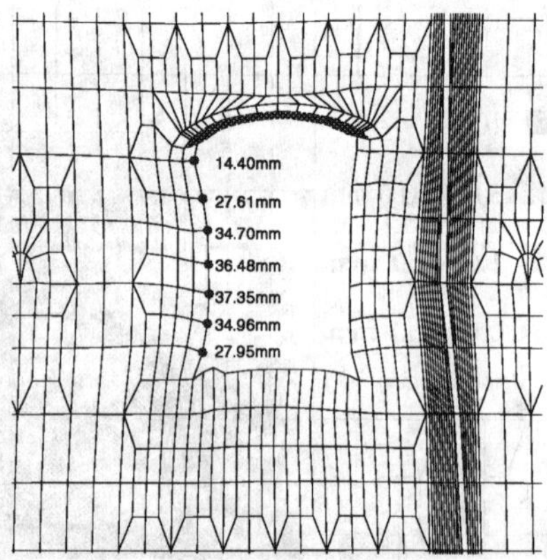

Fig.6(b) Deformation of rock masses (Last step of excavation)

Figs.7 show the relative displacement of jointed rock masses along the horizontal line from the point D shown in Fig.6(a). The relative displacement until the fourth step of excavation shown in Fig.7(a) is very small, while those after the forth step of excavation shown in Fig.7(b) are lager, which is quite similar to the real behavior of rock masses during the excavation of such scale.

Figs.8 also show the vertical strain of jointed rock masses in the horizontal direction from the midpoint of C and D in Fig.6(a). The strain is computed at the gravity point of each element. Here, the compressive strain is positive value. The vertical strain before the fourth step of excavation is not big. The vertical strain at the fourth step of excavation becomes large and the peak point of the vertical strain is located at the cavern wall at this moment. It moves away from the cavern wall with proceeding excavation steps.

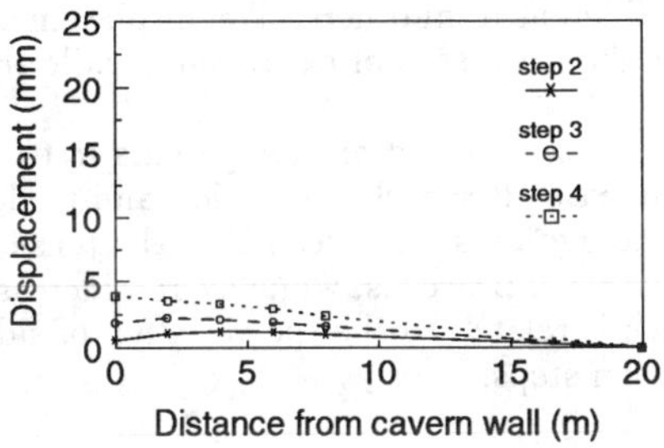

Fig.7(a) Displacement of rock masses (before forth step of excavation

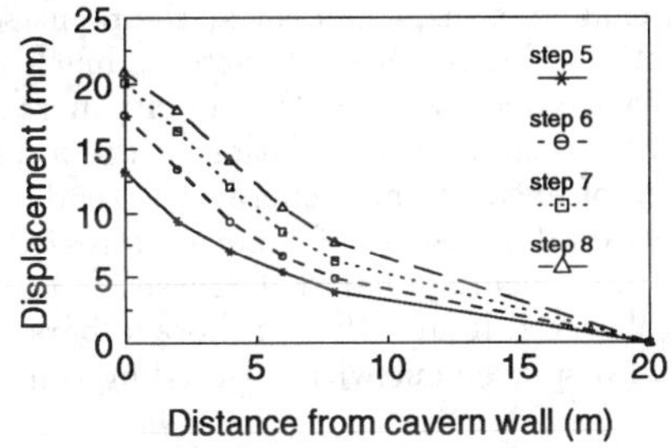

Fig.7(b) Displacement of rock masses (after forth step of excavation

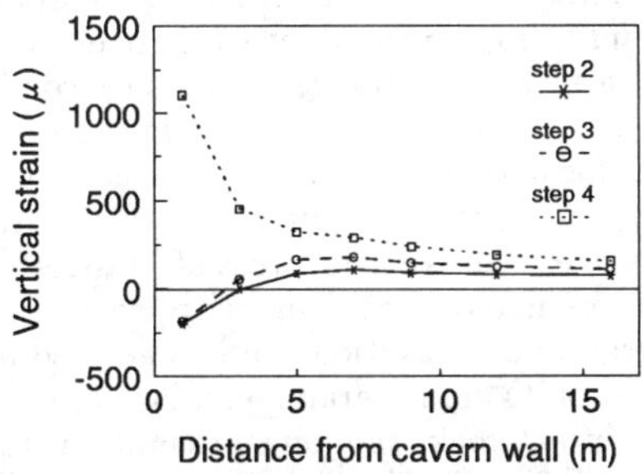

Fig.8(a) Vertical strain (before forth step of excavation

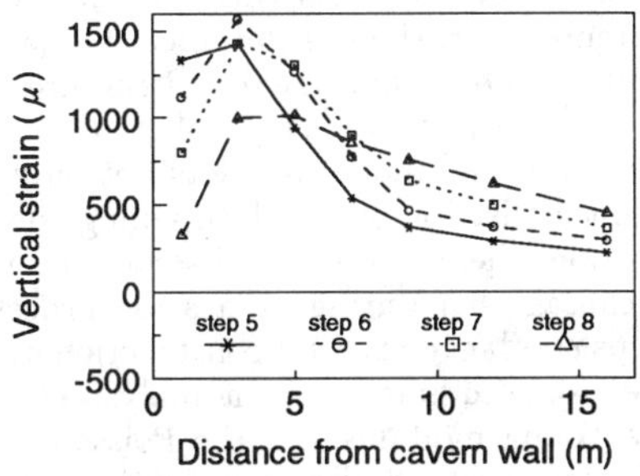

Fig.8(b) Vertical strain (after forth step of excavation

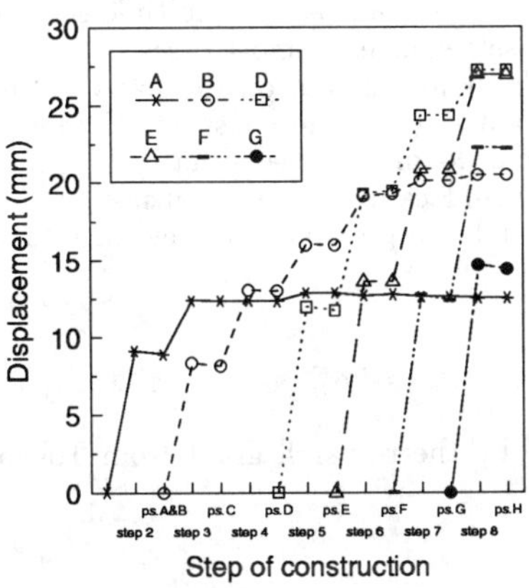

Fig.9 Displacement of rock masses of cavern wall

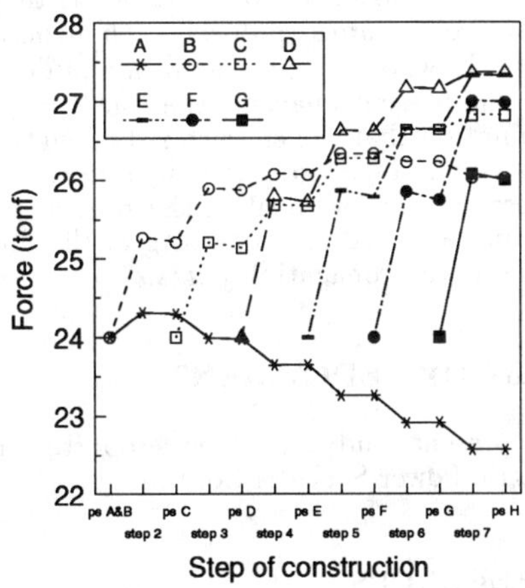

Fig.10 Force of prestressed anchorage bar

Figs.9 show the variation of side wall displacement with proceeding construction steps. A-G also correspond to points shown in Fig.6(a). The displacement after the cast of the prestressed anchorage bar reduces due to the effect of the prestressed anchorage bar, while the displacement after each excavation increases. The tendency of change of displacement at the point A is different from others. The displacement at the point A does not increase much after the third step of excavation, while the displacements at the other points increase.

Fig.10 shows the variation of the force in the prestressed anchorage bars in each step of excavation. Generally, it tends to increase after each excavation and to decrease after casting of other prestressed anchorage bars. It is seen that the tension in the anchorage bar-A decreases as the excavation proceeds, while overall increase is observed in all other anchorage bars, which is related to the the variation of side wall displacement with proceeding construction steps.

5 CONCLUSION

In this study, a constitutive model of rock masses containing fracturing joints is formulated on the basis of micromechanics, and is implemented into a finite element analysis code to analyze rock structures under arbitrary loading and boundary conditions. As an example, a problem of excavation of a power station cavern is analyzed. The proposed model predicts not only the deformation of jointed rock masses but also the distribution of the opening and sliding displacements jump across joints. In addtion, the distribution of strain and the change in the tensile forces of prestressed anchorage bar during excavation process are predicted. Problems with excavation steps similar to the real construction of underground excavation can be analyzed by the proposed finite element analysis code. Most of its input parameters, for example, the Young modulus and the Poisson ratio of intact rock, the joint orientation, the joint density and the joint frictional angle can be determined easily, for example, from the Schmidt net, observed average distance of joints and the room tests on core samples. It is, however, necessary that the other parameters such as the joint undulation angle, the joint characteristic length and the coefficient for joint connectivity are presumed properly. The constitutive model can handle multiple sets of joints. However, the consideration of dominant sets suffices in most cases.

In the present analysis, the constitutive model is formulated on the basis of the assumption that the characteristic length of a joint is given and constant. However, it may be changed gradually by stress relaxation or loading in real situations. Thus, a possible improvement of the present model would capture the expansion of the opening and sliding parts or growth of joints and the possible breakage of intact rock due to propagating joints.

ACKNOWLEDGEMENT

The present study has been supported in part by the research grant from Tokyo Electric Power Services Co.,Ltd.

REFERENCES

1) Cai M. and Horii H. A Constitutive Model of Highly Jointed Rock Masses, *Mechanics of Materials*, 13, pp.217-246 (1992)

2) Cai M. and Horii H. A Constitutive Model and FEM Analysis of Jointed Rock Masses, Int. J. Rock Mech. Min. Sci. & Geomech. Abstr Vol.30, No.4, pp.351-359 (1993)

3) Horii H. and Nemat-Nasser S. Over all Moduli of Solids with Microcracks: Load-Induced Anisotropy, *J. Mech. Phys. Solids*, 31 ,pp.155-171 (1983)

Rock Mechanics, Nelson & Laubach (eds) © 1994 Balkema, Rotterdam, ISBN 90 5410 380 8

The edge function method: An efficient new numerical technique in rock mechanics

J.F. Dwyer & B. Amadei
Department of Civil Engineering, University of Colorado, Boulder, Colo., USA

ABSTRACT: The edge function method for anisotropic elasticity is presented. Analytical solutions model field behavior in critical parts of the domain of interest. Boundary conditions are matched using a variational principle. Accurate results are obtained for a range of rock mechanics problems involving gravity forces, cavities, cracks and singular loadings.

1 INTRODUCTION

The most popular numerical methods used for the analysis of problems in rock mechanics are the finite element method and the boundary element method. A major disadvantage of using finite elements is the exceedingly high number of degrees of freedom required to model any kind of singular behavior. A complicated mesh generating algorithm is usually required. In general, finite element approaches require discretization of the entire region under consideration while boundary element approaches require discretization of the boundary only. The edge function method (EFM), which is described in this paper, is a more direct approach and does not require these discretizations.

The edge function technique involves the use of analytical solutions to model the field behavior in various critical parts of the domain under investigation. The essence of the edge function approach is the approximation of the solution of a boundary-value problem by a linear combination of solutions of the field equations. A set of solutions is generated which can model arbitrary effects on each boundary (edge functions) and which exhibit rapid decay away from that boundary. Singular solutions corresponding to vertices, cracks, holes and concentrated loads are also included to provide a rapidly convergent analytical solution. The edge functions and singular solutions are based on the complex variable formulation of Lekhnitskii (1963). The unknowns in the linear combination are obtained from a system of equations which follows from the approximation of the boundary conditions by a boundary-Galerkin energy method.

Some interesting features of the edge function approach are noted :

(a) The method is characterized by the extremely accurate results obtained (Dwyer 1986) and the small number of degrees of freedom required.

(b) The analytic functions involved are derived from solutions for infinite regions which make them immediately applicable to many rock mechanics problems.

(c) The model can handle isotropic data as a special case simply by perturbing the material constants by a very small amount and the solution has been shown to be highly stable for such perturbations.

(d) The method can be coupled with other numerical approaches wherein edge functions techniques are used to model singularities while non-singular field behavior could be modelled by another method such as finite elements.

In this paper, the edge function method for two dimensional problems in anisotropic elasticity is briefly described. Then, several examples are included to show the overall efficiency of the method and to exhibit its applicability for various rock mechanics problems involving gravity forces, cavities, cracks and singular loadings. The final section discusses both the advantages and limitations of the new approach to these problems.

2 THE EDGE FUNCTION METHOD

2.1 *Edge and vertex functions*

Finite polygonal regions and infinite regions are considered (Figure 1). Regions may contain cracks and circular or elliptical cavities.

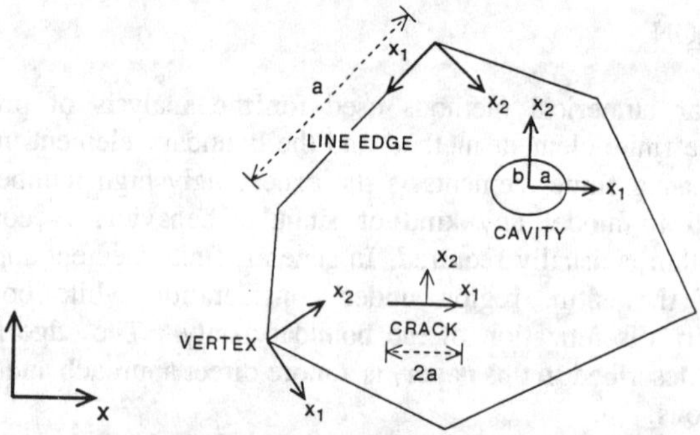

Fig. 1. Polygonal region deforming in plane strain.

Edge function potentials are required which can model arbitrary displacements or tractions on each straight line edge or elliptical edge. System matrix stability considerations require that each edge function displacement and traction field decays away from the boundary with which it is associated. Each line edge function is identified by its harmonic number (the order of the trigonometric polynomial to which it reduces on its characteristic edge).

Cavity edge functions are based on a conformal mapping of the exterior of the ellipse onto that of the unit circle. Crack edge functions can be obtained from the elliptical cavity edge functions by setting the length of the minor semi-axis to zero. The displacements and tractions decay away from the crack and the tractions reduce to polynomials on the crack faces.

Singular solutions which satisfy homogeneous boundary conditions in the neighbourhood of a vertex are included in the representation of the approximate solution in order to accelerate convergence. Such solutions are termed homogeneous vertex functions. Singular loadings are modeled by inhomogeneous vertex functions.

2.2 *Gravitational loading*

The field behavior for each of the cases discussed above is represented by a complex variable function which satisfies the homogeneous form of the equilibrium equations of elasticity. A particular solution for the non-homogeneous case, corresponding to a gravity body force term, has also been developed (Dwyer and Amadei 1994a). This solution is subtracted from the original problem to give a modified boundary value problem which satisfies the homogeneous form of the equilibrium equations.

2.3 *Boundary Galerkin method*

The solution is approximated by a linear combination of functions of the above types. Since the equilibrium equations are satisfied a-priori it remains only to satisfy the boundary conditions approximately. In this work an energy method (boundary Galerkin) is used. In the case of traction or displacement problems the method is equivalent to the minimization of the strain energy error. The system equations then have the form $A\underline{c} = \underline{b}$ where $\underline{c}$ is the vector of coefficients of the potentials and $\underline{b}$ is determined by the boundary conditions. The matrix A is symmetric and positive definite for displacement or traction problems (Dwyer 1986).

Boundary-values and errors of the approximate solution are computed at a number of equidistant points on each boundary. The errors are computed in terms of the differences (residuals) between the prescribed and computed quantities. Displacements and tractions on specified lines and curves are also computed as well as cicumferential stresses on the surface of each cavity and stress intensity factors for each crack.

The relationship between the number of potential functions in the representation of the solution and the total number of degrees of freedom in the system matrix is explained in detail in Dwyer and Amadei (1994a). In general there are four degrees of freedom associated with each harmonic level of each potential function on each critical part of the region.

3 NUMERICAL EXAMPLES

Several numerical examples are presented below to illustrate the applicability of the edge function method in determining stresses in various rock mechanics problems. In these examples, the material parameters are defined in terms of Young's moduli, shear moduli and Poisson's ratios in a global (x, y, z) coordinate system such that the x and z-axes are horizontal and the y-axis is vertical upwards. The rock is assumed to deform in plane strain in the (x, y) plane and to be orthotropic or transversely isotropic in the (x, y, z) coordinate system. The standard notation (Lekhnitskii 1963) is adopted in the specification of the elastic properties of the following five sets of rock materials which are considered in the examples :

.Rock A is transversely isotropic in the vertical (y, z) plane and is such that $E_1 = 34.47$ GPa, $E_2 = E_3 = 17.23$ GPa, $G_{12} = G_{13} = 8.62$ GPa, $G_{23} = 6.89$ GPa, $\nu_{21} = \nu_{31} = 0.15$ and $\nu_{32} = 0.25$ (Jaeger and Cook 1969).

.Rock B is similar to Rock A except that the larger modulus is now in the vertical y

direction with E_1 = 17.23 GPa, $E_2 = E_3$ = 34.47 GPa, $G_{12} = G_{13}$ = 8.62 GPa, G_{23} = 13.78 GPa, $\nu_{21} = \nu_{31}$ = 0.30 and ν_{32} = 0.25.
.Rock C is an orthotropic granite whose properties are E_1 = 42.4 GPa, E_2 = 44.5 GPa, E_3 = 30.3 GPa, G_{12} = 19.6 GPa, G_{13} = 16.9 GPa, G_{23} = 17.2 GPa, ν_{21} = 0.176, ν_{32} = 0.142 and ν_{31} = 0.145 (Gerrard 1975).
.Rock D is transversely isotropic in the horizontal (x, z) plane and is such that $E_1 = E_3$ = 34.47 GPa, E_2 = 17.23 GPa, $G_{12} = G_{23}$ = 8.62 GPa, G_{13} = 13.78 GPa, $\nu_{21} = \nu_{23}$ = 0.15 and ν_{13} = 0.25.
.Rock I is isotropic with $E_1 = E_2 = E_3$ = 34.47 GPa, $G_{12} = G_{13} = G_{23}$ = 13.78 GPa and $\nu_{21} = \nu_{31} = \nu_{32}$ = 0.25.

Throughout the following examples, only finite regions are considered. However, the size of the outer boundaries of those regions is made sufficiently large compared to their internal boundaries to assume that the regions represent, for all practical purposes, infinite rock masses. Also, all stress distributions are presented in dimensionless form by introducing characteristic stresses. Because of the theory of linear elasticity, the dimensionless stresses depend only on ratios of elastic constants and ratios of geometric dimensions. Therefore, the geometric units used in the following examples are immaterial.

Example 1:

This first example is included as a simple comparison with a standard problem in the literature and serves as a benchmark to test the edge function method. It corresponds to the case of an isotropic square region of side length, L, with a hole of radius, a, subject to a uniaxial stress σ_o in the y direction. In this example, L = 2.0 cm and a = 0.1 cm (L/a = 20). The distribution of the dimensionless vertical stress σ_y/σ_o along the x-axis, determined with the edge function method, is shown in Figure 2. It is compared with the analytical solution for the stresses around a hole in an infinite plate given by Timoshenko and Goodier (1970). The edge function solution was obtained using only seven degrees of freedom consisting of three polar functions and four cavity edge functions. For comparison, Becker (1992) achieved results similar to those shown in Figure 2 by using a boundary element program with 33 elements.

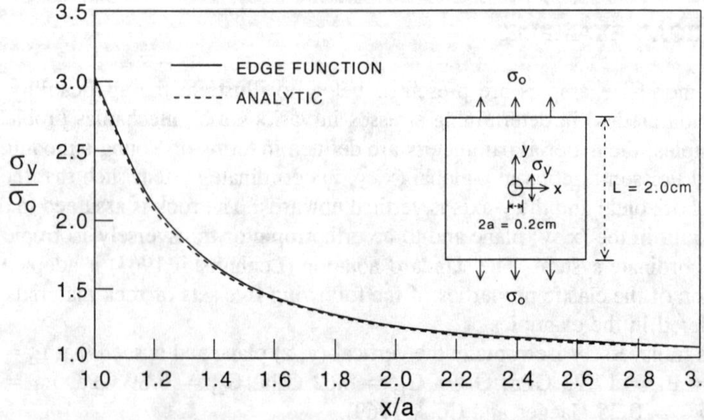

Fig. 2. Dimensionless vertical stress along x-axis in isotropic region with a hole.

The accurate stress concentration factor here was also obtained using the ALGOR finite element package (Algor 1990). In this case the quarter symmetry of the problem was utilized. Nevertheless 400 degrees of freedom were required to achieve a similar accuracy. The elements used were four-noded quadrilateral plane strain elements. The problem was also examined using the COSMOS finite element package (Cosmos/M 1992) and utilizing plane strain triangular elements. The computed stress concentration factor was in error by 15% while using 398 degrees of freedom.

Example 2:

This example is also used as a benchmark for the edge function method. It corresponds to the case of a uniform compressive pressure, P, on a strip area of width B = 1 m acting over a finite region of length L = 100 m and height H = 100 m (Figure 3). The region is assumed to consist of an isotropic medium, is fixed at its base and is free to deform laterally in the x direction. Figure 3 shows the variation of σ_x/P along the strip load center line (x/L = 0.5) calculated with the edge function method and that of Poulos and Davis (1974) for a strip load in an infinite medium. A good agreement is found between the two solutions.

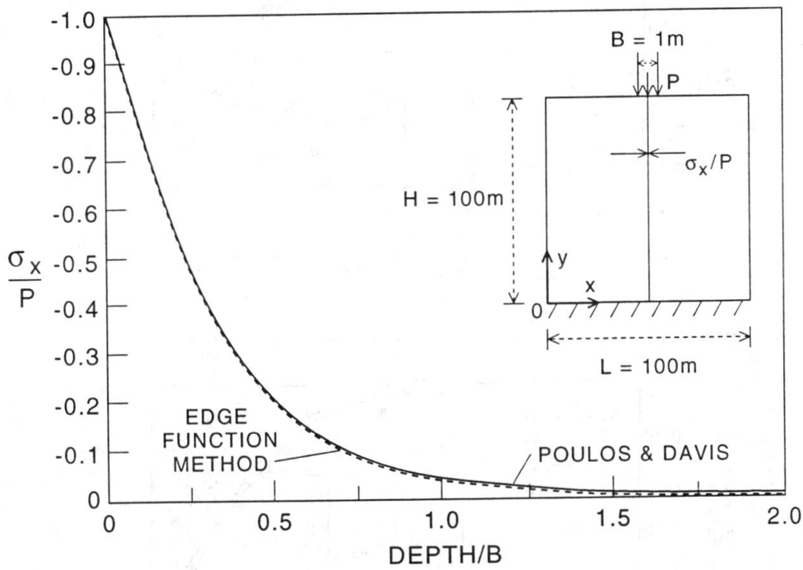

Fig. 3. Dimensionless horizontal stress on line beneath center of strip load.

A total of 97 degrees of freedom, resulting from using 4 harmonic levels of line edge functions, 4 levels of polar functions and 18 vertex functions, were used here. An extremely accurate solution was obtained, the highest root mean square residual being less than 1% of the applied pressure P. An attempt to solve the same problem using the ALGOR finite element package (Algor 1990) failed to produce comparable accuracy even though half symmetry was utilized and 400 four noded quadrilateral plane strain elements were used to model the region. This resulted in approximately 400 degrees of freedom but the results were in error by 50%.

Example 3:

The effects of a gravity body force term are considered in this example which presents the stress distribution in a finite region representing a rock mass of length L = 1000 m and height H = 1000 m under the influence of gravity only. The region is assumed to be fixed at its base and is free to deform laterally in the x direction. Figures 4a and 4b show, respectively, the variations of the vertical stress $\sigma_y/\rho gH$ along the horizontal line y = H/2 and the horizontal stress $\sigma_x/\rho gH$ along the vertical line x = L/2 for rocks B, C and D and I. This latter is interesting in that it shows a region of tension near the top of the rock mass, which may be significant in the possible formation of cracks. Similarly, the area of compression lower down may inhibit crack propagation.

A total of 93 degrees of freedom were used and maximum root mean square residuals of 1% of ρgH indicate that an accurate solution has been obtained at this level. The

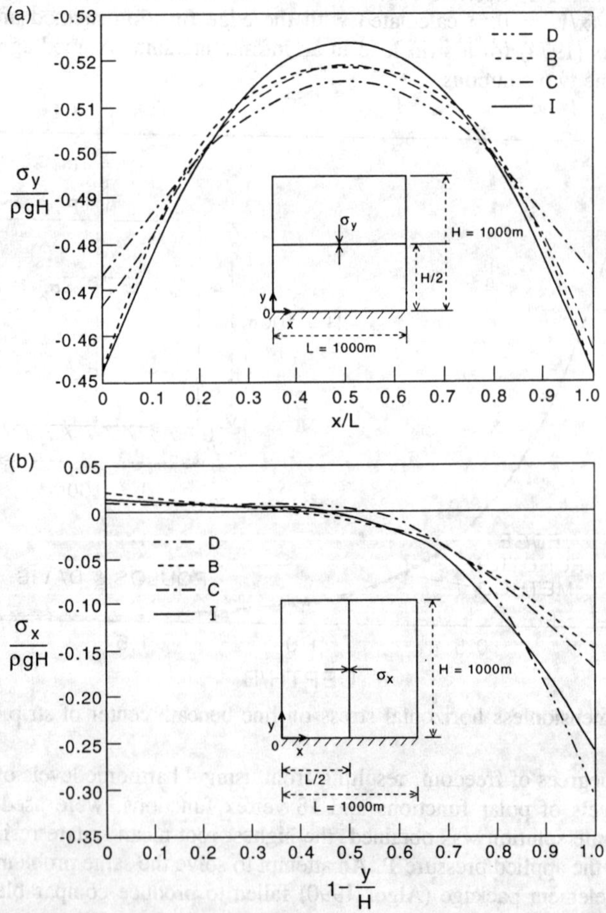

Fig.4. Dimensionless (a) vertical and (b) horizontal stresses along center lines in rectangular rock mass under gravity.

symmetry in the computed stresses (Figure 4a) also provides confidence in the accuracy of the solution. Furthermore, an increase in the region's length to width ratio, L/H, led, as expected, to a stress distribution, over much of the domain, which approached that of an infinite rock mass (Dwyer and Amadei 1994a).

Dwyer and Amadei (1994a) also considered the same finite region, consisting of rocks A, B, C or D, subject to gravity only, with a circular excavation, of radius 10 m, included at its center (x = 500m, y = 500m). In this case, a very accurate solution was obtained using 109 degrees of freedom. This is quite significant as it shows that the effect of the hole is not detrimental to the convergence and that the latter is of the same order as above. This confirms that high stress concentrations do not prove to be an obstacle to obtaining accurate solutions once the gravity term is successfully modeled.

Example 4:

The effect of the combination of gravity and an underground fracture is addressed in this example. A finite crack of length 2a is contained in a finite region of length L = 100 m and height H = 100 m (L/H = 1). The crack is such that 2a/L = 0.2 and is inclined at an angle θ with respect to the x-axis. Its center has coordinates $x_1 = y_1 =$ 50 m (= L/2 = H/2). The region is assumed to consist of rock B, D or I, is fixed at its base and is free to deform laterally in the x direction. Gravity is the only active force. Figure 5 gives the variation, with the angle θ, of the two dimensionless stress intensity factors $|K_I|\sqrt{a}/\rho gH$ and $|K_{II}|\sqrt{a}/\rho gH$ calculated at the right crack tip for each rock type. The rock anisotropy and the orientation of the planes of transverse isotropy do not have an effect on the values of the two stress intensity factors. As in the previous example, an accurate solution was obtained using about 100 degrees of freedom. This shows that the singularities are well modeled by the edge function method. This was confirmed by a further example which studied the effects of two cracks in a rock mass and where accurate solutions were obtained with 125 degrees of freedom (Dwyer and Amadei 1994b).

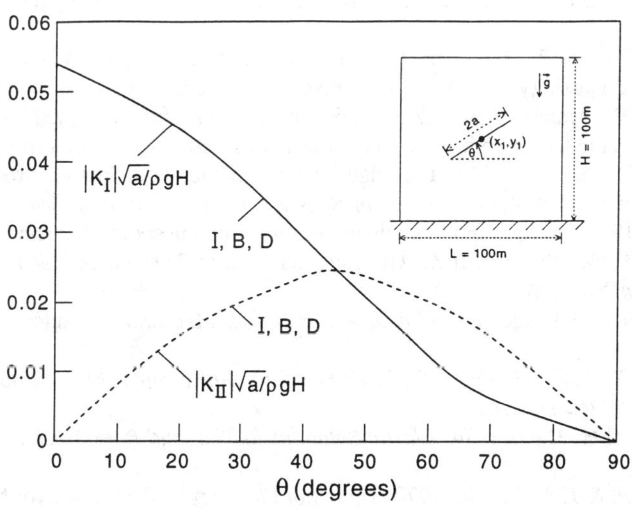

Fig. 5. Dimensionless stress intensity factors for finite crack in rectangular rock mass under gravity.

4 CONCLUSIONS

The edge function method has been successfully applied to a range of problems in rock mechanics. Some simple test cases were used to exhibit the accuracy of the method and results compared favorably to those in the literature. Singular problems involving cracks and discontinuous loads were also modeled and the interactions of the singularities with gravity forces did not unduly affect the accuracy.

The main advantage of using the edge function technique is the accuracy obtained with a small number of degrees of freedom. This is in contrast to the finite element methods currently in use, as was clearly shown in the test cases discussed above. Another advantage is the elimination of time consuming mesh generation algorithms since the edge function method simply uses a boundary matching principle to directly generate a set of equations.

The greatest potential for the edge function method, at present, is to incorporate it into more general finite element, boundary element or discrete element codes. The problems of crack propagation could also be examined and the edge function method has a particular advantage for this kind of analysis since only new crack edge functions have to be generated for a new crack position while the remainder of the potential function representation does not have to be recomputed.

Acknowledgments: The assistance of Neil Wahl and Tom Ottenheimer of Montana State University in constructing finite element models is greatly appreciated. The support of the National Science Foundation under grant MS 9215397 is also greatly appreciated.

REFERENCES

Algor User's Manual. 1990. Pittsburgh: Algor Inc.

Becker A.A. 1992. *The Boundary Element Method in Engineering: a Complete Course.* New York: McGraw-Hill.

Cosmos/M User's Manual. 1992. Santa Monica: Structural Research and Analysis Corp.

Dwyer J.F. 1986. Application of the edge function method to singular plane problems in anisotropic elasticity. *Ph.D Thesis*. University College Cork.

Dwyer J.F & B. Amadei 1994a. Application of the edge function method to rock mechanics problems. Submitted to *Rock Mechanics and Rock Engineering*.

Dwyer J.F & B. Amadei 1994b. The edge function method and singular problems in rock mechanics. Submitted to *Int. J. of Rock Mech., Min. Sc. and Geom. Abstr.*

Gerrard C.M. 1975. Background to mathematical modeling in geomechanics: The roles of fabric and stress history. In G. Gudehus (ed), *Finite Elements in Geomechanics*, 33-120. New York: Wiley.

Jaeger J.C. & N.G.W Cook 1969. *Fundamentals of Rock Mechanics*. London: Methuen & Co. Ltd.

Lekhnitskii S.G. 1963. *Theory of Elasticity of an Anisotropic Elastic Body.* San Francisco: Holden-Day Inc.

Poulos H.G. & E.H. Davis 1974. *Elastic Solutions for Soil and Rock Mechanics*. New York: Wiley.

Timoshenko S.P. & J.N. Goodier 1970. *Theory of Elasticity*. 3rd edn. Tokyo: Mcgraw-Hill.

Rock Mechanics, Nelson & Laubach (eds) © 1994 Balkema, Rotterdam, ISBN 90 5410 380 8

Numerical simulation of stress wave interaction with cracks in layered rock

H.P.Rossmanith, K.Groschupf & R.E.Knasmillner
Institute of Mechanics, Technical University Vienna, Austria
K.Uenishi
Department of Aeronautical Engineering, University of Tokyo, Japan

ABSTRACT: The interaction of impact generated elastic stress waves with discontinuities in solids is considered. In this contribution, the evolution in time of damage at a bimaterial interface is described by an exponential growth law of Kachanov-type and a finite difference method has been adapted for the investigation of progressive deterioration of an interface. The particular case of a shear wave impacted stick-slip-interface is presented.

1 INTRODUCTION

In recent years, the response of joints and faults in fractured rock to wave interaction induced by seismic, mining or percussion drilling activities has become of great importance in geotechnical engineering, mining and rock mechanics.

When waves of seismic or other origin are propagated through fractured rock mass not only will the reflection and transmission coefficients depend on the acoustic impedance mismatch and the nature of the interface layer, but also the physico-mechanical behaviour of the interface will change as a function of time. Degradation and aging processes of the interface material, which may quantitatively be formulated by means of continuum damage mechanics principles, may be made responsible in part.

Elastic wave behaviour across imperfectly bonded linear slip interfaces separating two elastic media is investigated in Schoenberg (1980), where the displacement discontinuity, or slip, is taken to be linearly related to the stress traction which is continuous across the interface for both the normal and tangential direction. Applying the theoretical results of Schoenberg (1980) to seismic wave propagation model experiments in a jointed columnar basaltic rock formation, acceptable agreement between observed and predicted results has been found by Myer et al (1990).

A proper constitutive model for rock joints must be able to predict the deformation behaviour involving aging effects on a long time scale as well as deterioration effects due to repeated loading and unloading. In percussion drilling, the interface is dynamically loaded at high intensity and the interface characteristics will change with each consecutive percussive blow of the tool. Improved knowledge about the impact-induced time-variation of the mechanical behaviour of the rock joints appears to be instrumental and vital for the improvement of drill efficiency and the development of *smart mining machinery.*

This paper discusses the dynamic behaviour of single rock joints subjected to single

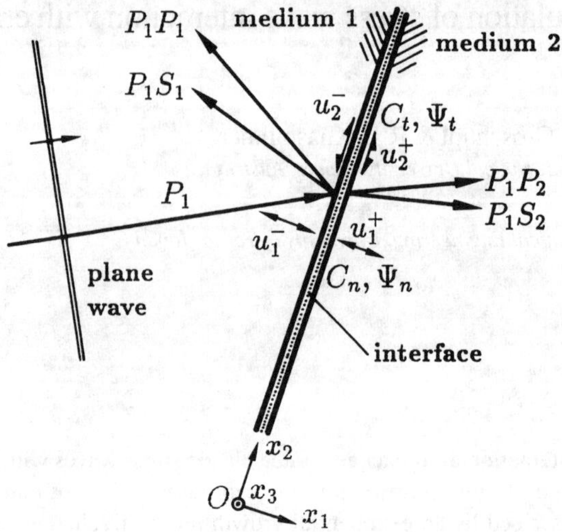

Figure 1: Interaction of incident P-waves with a damaged interface between two elastic half-spaces

pulse loading and to medium frequency harmonic wave loading. The theory of elastic wave propagation across a displacement discontinuity at the boundary between two half-spaces is combined with continuum damage mechanics (Kachanov 1986, Lemaitre and Chaboche 1985). This theory predicts changes of the reflection and transmission coefficients on the basis of damage-controlled time-varying rock joint characteristics. The evolution laws for normal and shear damage were allowed to be dissimilar but coupling effects have been neglected. In addition, plane interfaces with linear normal and slip conditions have been employed up to complete degradation, i.e. complete separation, of the joint surfaces.

2 LINEAR INTERFACE CONDITIONS WITH DAMAGE

Consider a nominally plane interface between two dissimilar linear elastic solid half-spaces as shown in Figure 1. Imperfect bonding, which will increase as a function of time or as a function of the number of blasts or impacts, is assumed to prevail. Tractions are assumed to be continuous across the interface but the small displacement field is not. The condition of continuous displacement is replaced by a so-called *linear slip condition* (Schoenberg 1980).

A linear damage-controlled viscoelastic slip interface constitutive law of the form

$$P_j(\Delta u_j) = Q_j\left(\frac{\sigma_{nj}}{1 - D_j}\right) \tag{1}$$

with

$$P_j() = \alpha_{0j} + \alpha_{1j}\frac{\partial}{\partial t} + \cdots + \alpha_{mj}\frac{\partial^m}{\partial t^m}, \quad Q_j() = \gamma_{0j} + \gamma_{1j}\frac{\partial}{\partial t} + \cdots + \gamma_{lj}\frac{\partial^l}{\partial t^l} \tag{2}$$

and $j = n, t$ (n = longitudinal wave, t = shear wave) will be employed. P and Q are polynomial operators with respect to time and $\underset{\sim}{D}$ represents the damage tensor. For mutually independent damage behaviour in the normal and shear directions, the damage tensor diagonalizes and the traction-displacement discontinuity relations decouple.

At the interface normal and tangential damage is assumed to develop independently according to Kachanov-type evolution laws:

$$\frac{d\Psi_j}{dt} = -A_j \left(\frac{\Delta u_j}{\Psi_j} \right)^{m_j} \tag{3}$$

where $\Psi_j = 1 - D_j$, A_j and m_j are constitutive parameters and Δu_j are the normal and shear displacement jumps responsible for the damage.

3 REFLECTION AND REFRACTION OF PLANE P- AND SV-WAVES

Consider two dissimilar isotropic, homogeneous, linear elastic half spaces in contact along a plane (possibly predamaged) interface along $x_1 = 0$.

Generally, a harmonic plane incident P- or SV-wave gives rise to two reflected and two transmitted elastic waves with associated coefficients R_P, R_S, T_P, and T_S and these are present in the general wave solution. Time-varying normal and tangential damage has been introduced by Rossmanith et al. (1993) where the combination of Hooke's law (valid for the elastic halfspaces), the interface constitutive equations, the damage laws and the four interface conditions determine the four reflection and transmission functions R_P, R_S, T_P, and T_S, respectively (Rossmanith et al. 1993). Figure 2 shows the variations of the reflection (R_P and R_S) and transmission coefficients (T_P and T_S) as a function of progressive damage for obliquely impinging P- and SV-waves for various angles of wave incidence, N_{max} is the number of impacts needed for totally destroying the interface.

As Figure 2 pertains to a situation where normal and tangential damage develop simultaneously, wave pulses impacting at 45° (Figure 2 c) yield similar damage evolutions. Normal damage will progressively exceed tangential damage for pulses impinging at small angles and vice versa for large angles of incidence. As a result of the decoupled normal and shear damage no shear damage is induced in the case of normal P-wave incidence (Figure 2 a).

Notice, that the scale in Figures 2 a,b,c,d is not the same. A much larger number of impacts is required for larger angles of incidence.

In practice, the current state of damage by harmonic wave pulses is assumed to be concentrated at the interfaces. The interfacial damage imparted by the impacting stress waves leads to a decreasing stiffness and likewise to increasing reflection coefficients as a function of the damage.

When sensing a nearby interface and trying to determine its characteristics, sensing pulses will intermittently be radiated from a source and the response signal from the interface will be recorded for further processing and evaluation.

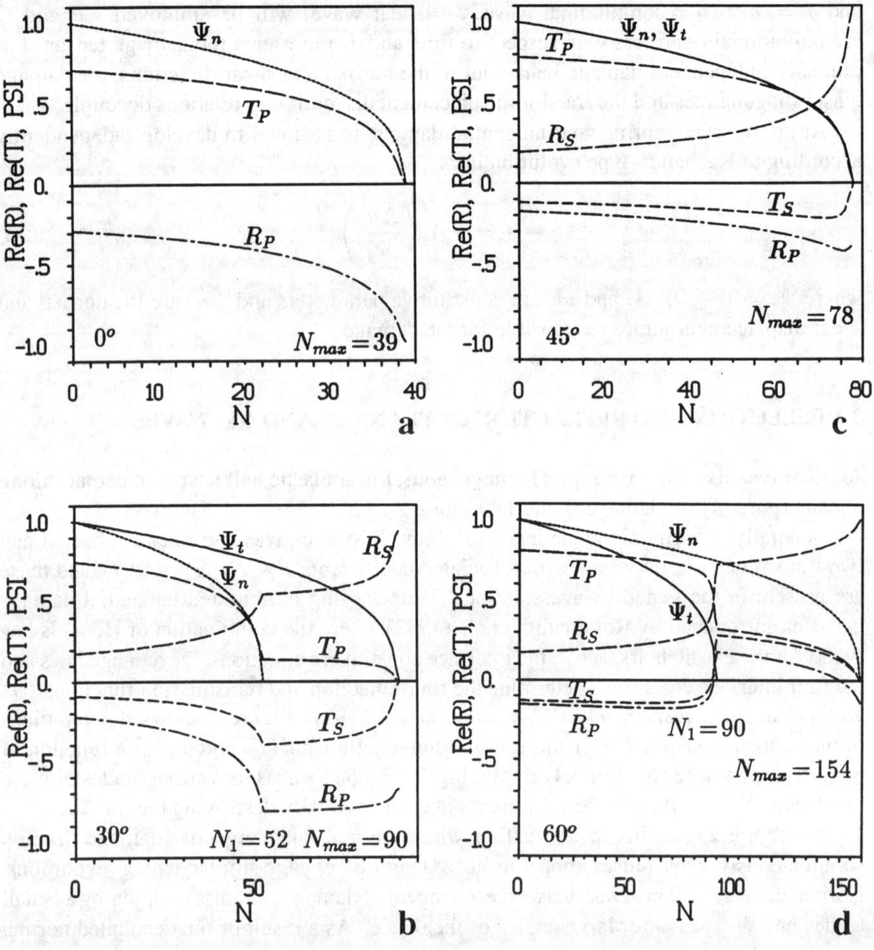

Figure 2: Variation of reflection and transmission functions as a function of similar damages Ψ_N and Ψ_T ($A_n = A_t$) of a viscoelastically reacting interface between two similar layers for an obliquely impinging P-wave.

4 INTERACTION OF P- AND S-WAVES WITH INTERFACE CRACKS

The effect of sequential impact of waves on a pre-damaged interface is to increase the level of deterioration with the possible appearance of a crack. Pre-damage may be the result of various preceding natural history of geological structure or man-made activities (mining, blasting etc.) (Rossmanith et al. 1990, Barton and Stephansson 1990, Atkinson 1987). Mechanically, the strength of an interface may range from practically zero (open,

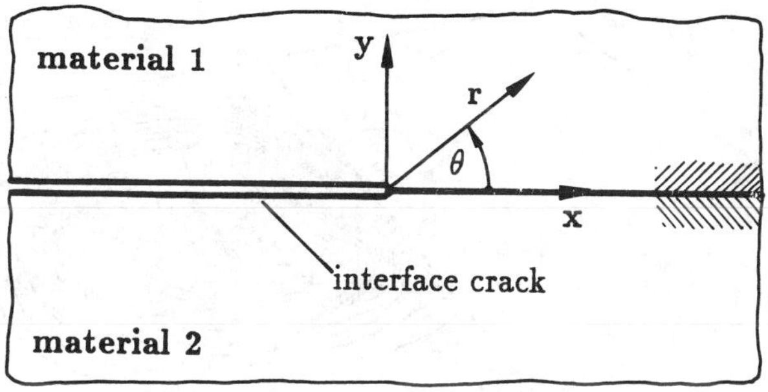

Figure 3: Interface crack in dissimilar media

fluid-filled, mud-filled) to extremely strong (quartzite veins, etc).

The general interface crack problem (see Figure 3) is governed by the near crack tip stress field

$$\sigma_{ij} = \frac{\mathbf{Re}[Kr^{i\varepsilon}]}{\sqrt{2\pi r}}\tilde{\sigma}^1_{ij}(\theta,\varepsilon) + \frac{\mathbf{Im}[Kr^{i\varepsilon}]}{\sqrt{2\pi r}}\tilde{\sigma}^2_{ij}(\theta,\varepsilon) + \frac{K_3}{\sqrt{2\pi r}}\tilde{\sigma}^3_{ij}(\theta,\varepsilon) + \ldots \tag{4}$$

where K_1, K_2 and K_3 are the stress intensity factors K = [stress][length]$^{1/2-i\varepsilon}$; K_{III} = [stress][length]$^{1/2}$ and ε is the bimaterial constant (Williams 1959). For $\varepsilon = 0$, the classical standard equations are obtained.

The displacement jumps δ at a distance r behind the crack tip are given by (Shih et al 1991):

$$(\delta_y + i\delta_x)_{\theta=\pi} = \frac{1}{(1+2i\varepsilon)\cosh\pi\varepsilon}\frac{4Kr^{i\varepsilon}}{E^*}\sqrt{\frac{2r}{\pi}}; \qquad (\delta_z)_{\theta=\pi} = \frac{2K_{III}}{\mu^*}\sqrt{\frac{2r}{\pi}} \tag{5}$$

where E^* and μ^* are the effective moduli.

5 SHEAR-WAVE IMPACT AT A STICK-SLIP BIMATERIAL INTERFACE

A central finite difference program was developed for the determination of dynamic stick and slip behavior for the problem of elastic wave interaction with a weak interface. Along the interface cohesive threshold values for tensile fracture and shear slip can be activated. Figure 4 shows the distribution of isochromatic fringes (lines of constant maximum shear stress) around the stick-slip section of the bimaterial interface interacting with an elastic shear wave. In the upper area, a leak of energy across the contacting boundaries can be detected, whereas in the lower half-plane an open crack due to tensile stresses prevents energy transmission into the adjacent layer.

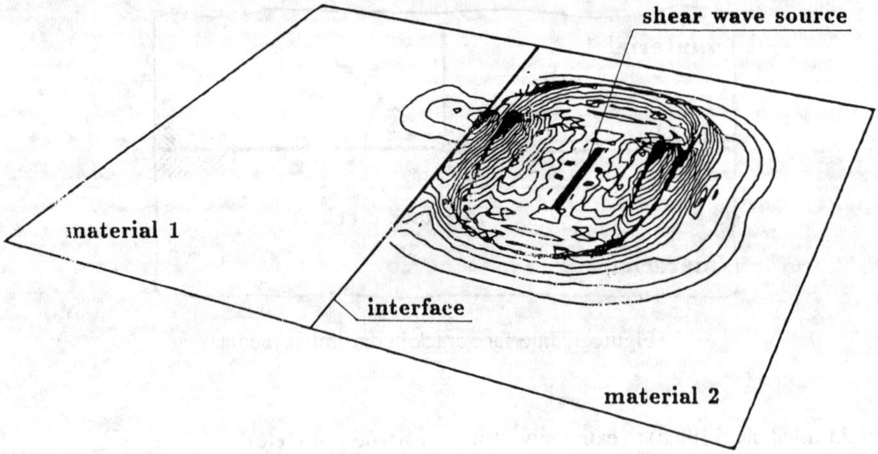

Figure 4: Shear wave impact at a stick-slip interface

6 CONCLUSIONS

Progressive evolution of damage at interfaces between rock layers is modeled by combining the fundamental law of continuum damage evolution and the constiutive equations for a viscoelastic interface materials.

A new finite difference method allows for the study of wave-interaction processes in mining engineering where a very wide variability in terms of crack geometry, material distribution, boundary conditions and form and shapes of excitation can be taken into account. Only very limited analytical results are available for this kind of practical problem.

One result of the finite difference calculations is shown in Figure 4. For qualitative assessment the numerically generated isochromatic fringe patterns allows the comparison with natural isochromatic fringe patterns from high-speed recordings of model tests in the laboratory.

7 ACKNOWLEDGEMENT

The authors would like to acknowledge financial support of this work by the *Austrian National Science Foundation FWF* under project number *P*6989.

REFERENCES

Atkison, B.K. (1987) Fracture Mechanics of Rock. Academic Press, London.

Barton, N. and Stephansson, O. 1990. Proc. Int.Conf. on Rock Joints, Loen, Norway, 1990. A.A. Balkema, Rotterdam.

Kachanov, L.M. (1986) Introduction to continuum damage mechanics. Martinus Nijhoff Publishers.

Lemaitre, J. and Chaboche, J.L. (1985) Mécanique des matériaux solides. Dunod, Paris.

Myer, L.R., Pyrak-Nolte, L.J. and Cook, N.G.W. (1990) Effects of single fractures on seismic wave propagation. Proc.ISRM Symp.on Rock Joints, Loen, Norway (eds Barton & Stephansson), Balkema, Rotterdam, 467-473.

Rossmanith, H.P. and Knasmillner, R.E. (1990) Application of photomechanics to studies of fracture in jointed rock. Proc. Int.Conf. on Mechanics of Jointed and Faulted Rock, Vienna, Austria, 1990. A.A. Balkema, Rotterdam, 285-293.

Rossmanith, H.P., Xu Chongfeng, Knasmillner, R.E. and Wegner, J.L. (1993) Effect of stress waves on pre-damaged interfaces in rock mass. Proc. 4th Int.Symp.on Rock Fragmentation by Blasting - FRAGBLAST-4, Vienna, (ed Rossmanith H.P.), A.A. Balkema, Rotterdam, 201-210.

Schoenberg, M. (1980) Elastic wave behavior across linear slip interfaces. J.Acoust. Soc.Am., 68(5), 1516-1521.

Shih, C.F. et al. (1991) Cracks on bimaterial interfaces: Elasticity and plasticity effects. Materials Science and Engineering.

Williams, M.L. (1959) Bulletin of the Seismological Society of America, 49, 199-204.

Kanninen, M.F. [illegible] Introduction to computational [illegible] mechanics. [illegible]

[illegible]

[illegible] (19[illegible]) [illegible] In Proc. [illegible] (ed. [illegible]), [illegible] 172–173.

[illegible] (19[illegible]) [illegible] Appl. Mech. [illegible]

[illegible] (19[illegible]) [illegible] Part [illegible] Conference [illegible]

Schroeder, [illegible] (1930) [illegible]

[illegible] (19[illegible]) [illegible] Materials [illegible] and [illegible].

Williams, M.L. (19[illegible]) [illegible] Bulletin of the Seismological Society of America, [illegible] 199–204.

12 Mechanics of weak rock – Creep/Rheology

Session developers:
Michel Aubertin, École Polytechnique, Université de Montréal
Moses Karakouzian, University of Nevada Las Vegas

Rock Mechanics, Nelson & Laubach (eds) © 1994 Balkema, Rotterdam, ISBN 90 5410 380 8

The influence of geological burial on the deformation characteristics of bonded carbonate rocks

M.E.Jones & S.Willis
Department of Geological Sciences, University College London, UK
N.Kågeson Loe & R.N.Mortimore
Department of Civil Engineering, University of Brighton, UK
M.J.Leddra
School of the Environment, University of Sunderland, UK

ABSTRACT: Mechanical behaviour of bonded carbonate sediments is dominated by a transition from an intact-elastic deformation regime at small strains to a ductile compression regime at larger strains. Characteristics of the transition depend upon porosity, with both the elastic limit and the transitional stress interval increasing as porosity decreases. The transitional behaviour is demonstrated using uniaxial compression experiments that define two uniaxial stress (k_o) ratios. These represent the stress conditions prevailing during one dimensional compression in the intact-elastic and ductile-compression regimes. When the sediment achieves a volume strain of about 1%, it leaves the intact-elastic regime and makes an excursion to the ductile-compression regime where further deformation occurs. This excursion is irreversible in mechanical terms, but intergranular bonding is re-established during diagenesis causing a return to the intact-elastic regime. During its geological history, a sediment may make repeated excursions between the two states: in each excursion pore volume is lost.

1 INTRODUCTION

The presence of intergranular bonding in porous sediments has a pronounced and recognised influence on the manner in which they deform both during compaction and shear deformation as has been shown for chalk by Kågeson-Loe et al (1993, 1992) and carbonate sands (e.g. Coop & Atkinson 1993). In detail however, the deformation characteristics of chalks, which have been subject to appreciable burial depths (several km's) and large *in situ* effective stresses during their cementation (diagenesis) differ from those of naturally and artificially cemented materials which became cemented at, or close to the surface of the original sediment substrate in the presence of low effective stresses.

Previous experimental studies on chalk (Kågeson-Loe 1993, Kågeson-Loe et al 1993, 1992, Leddra et al 1990a, 1990b) have collectively demonstrated that by using conventional soil mechanics, the mechanical behaviour can be described by a revised Mohr-Coulomb type model. Similarly Coop & Atkinson (1993) have shown that the deformation characterisitcs of cemented carbonate sands can be described using an equivalent model. The Mohr-Coulomb model needs to be revised for both materials as deformation is complicated by the presence of high stiffness at small strains due to intergranular cement bonding and/or grain interlocking. The loss of this stiffness (referred to as 'yield' in this text), during compression and/or shear appears to be related to the material's geological history. Porous rock samples deformed under conditons of uniaxial (one-dimensional) compression show this transition in material compressibility most clearly and are therefore used to illustrate the discussion.

2 DESCRIPTION OF EXPERIMENTAL METHODOLOGY

The chalk experiments reported in this text were conducted in a standard high pressure hydraulic triaxial cell using a high accuracy closed loop servo control system to regulate confining pressures. The deviatoric component of stress is applied directly by a load frame via a balanced ram mechanism. Changes in pore volume were monitored using a low pressure 'Imperial College' volume gauge operating at a 1MPa back pressure. Axial and radial strains were measured during the tests using sensitive axial and radial displacement transducers. Similar experiments on carbonate sands reported by Coop & Atkinson (1993) were conducted in a computer controlled high pressure stress path triaxial cell.

For uniaxial (one-dimensional) compression experiments, the rock sample is deformed in such a way that radial deformation is prevented (i.e. radial strain equals zero). This is acheived by continuously adjusting confining pressure in response to minor variations in sample diameter caused by the applied deformation, thereby maintianing the original sample diameter. The experiment is conducted under drained conditions: pore fluids displaced from the sample as it deforms are free to drain at a constant back pressure. Under these deformation conditions, the ratio of effective confining pressure (σ'_3) to effective axial stress (σ'_1) is given by the term, k_o, where $k_o=\sigma'_3/\sigma'_1$.

The stress and strain variables used to illustrate the mechanical behaviour during deformation follow standard geotechnical useage: deviatoric stress ($q=\sigma'_1-\sigma'_3$); mean effective stress ($p'=(\sigma'_1+2\sigma'_3)/3$); and volumetric strain, represented as void ratio e, where $e=n/(1-n)$ and n is the sample porosity.

3 MECHANICAL PROPERTIES

The mechanical behaviour of the chalks studied during this work, and of carbonate sands (Coop and Atkinson 1993) is illustrated in Figures 1 and 2. These chalks are characterised by a wide range of strengths which depend primarily on porosity but in uniaxial (k_o) compression, each effective stress/void ratio curve has a similar form (Figure 1). Each curve has a distinct change in the slope at about 1% volume strain (Figure 1) which corresponds to a change in material state from intact elastic to disaggregated ductile material (Kågeson-Loe et al 1993, 1992, Leddra et al 1990a, 1990b). This change of state takes place over a distinct interval of stress and strain which increases in magnitude with increasing rock strength. This is termed the yield transition. The stress paths recorded during uniaxial (k_o) compression (Figure 2) also show consistant changes of slope, although in this case the curves each show three distinct regions. Figure 2 illustrates that for both the chalks and bonded carbonate sand, irrespective of yield strength, there exists two limiting envelopes for possible stress states that accompany k_o compression. These correspond to stress ratio values of $k_o\approx$ 0.25 and $k_o\approx$ 0.65 and relate to the intact (elastic) and post-yield (ductile) material state. The space between these lines represents the yield transition when k_o has a nominal value of 1.0. It is seen that the yield transition occupies progressively larger ranges of mean effective stress, the greater the strength of the sample.

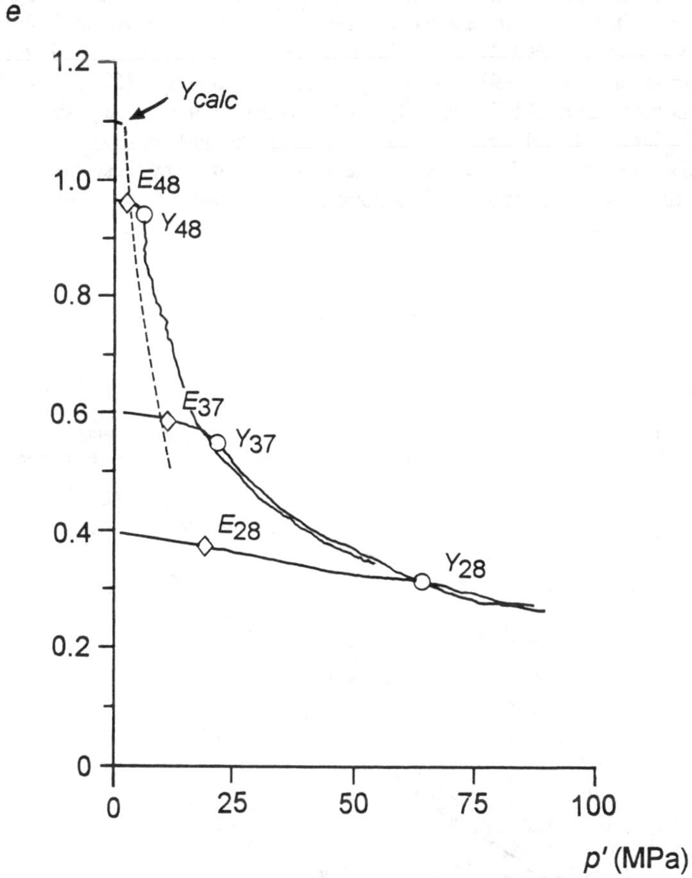

Figure 1. The relation between void ratio and mean effective stress for chalks of different porosity and bonded carbonate sand (dashed line). E=elastic limit (number is sample porosity in %pore space); Y=yield point (where number is again sample porosity); Ycalc is the yield point for the carbonate sand. Each curve shows two distinct sections, which represent the intact-elastic regime and the ductile compression regime. The former has a gentle slope and lies between the origin and the elastic limit, the latter a steeper slope and lies beyond the yield point. The section of the curve between the elastic limit and the yield point is the transitional regime.

4). DISCUSSION

The experimental results illustrated in Figures 1 and 2 can be described using a modified Mohr-Coulomb model which evokes a two stage yield transition from an intact state to a fully disaggregated state (Kågeson-Loe 1993, 1992). This transition begins at the elastic limit, (which is the stress-strain condition pertaining at the end of recoverable elastic deformation) and ends at the yield point (where large strain ductile deformation begins) (Figure 1). The yield transition therefore delineates a progressive loss of material stiffness with increasing compression. This is similar to the double-yield concept of Vaughan et al (1988) for bonded soils and is also seen in bonded carbonate sands (Coop and Atkinson 1993). In the highest-porosity chalks (>45% porosity), and bonded carbonate sands (>50% porosity) yield occurs as a relatively abrupt event confined to a small interval of stress and/or strain (Kågeson-Loe 1993, Leddra et al 1990a, Addis 1987). This is because at the low strengths of these high porosity carbonates, the elastic limit and the yield point occur in close proximity under similar conditions of stress and strain.

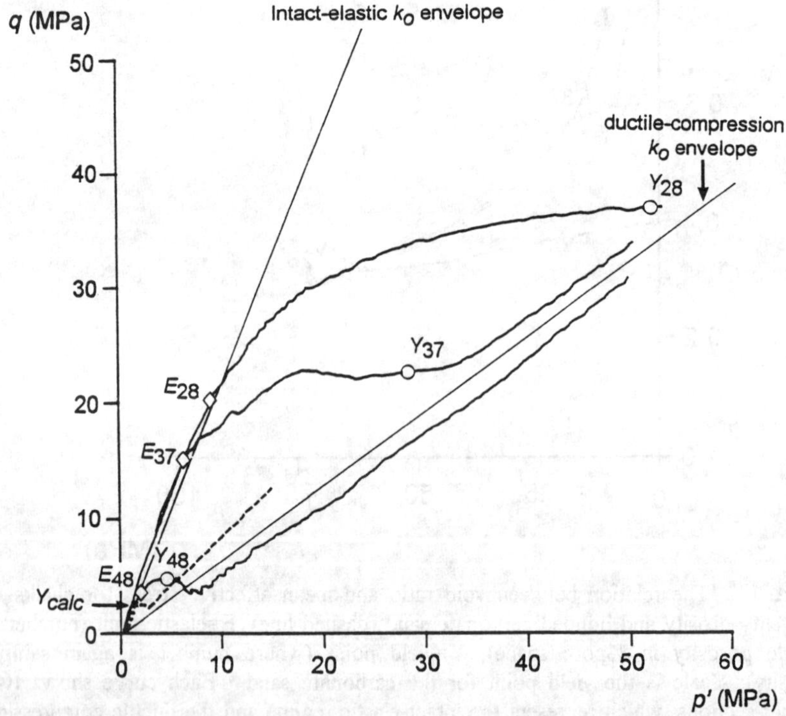

Figure 2. Stress paths for uniaxial compression of chalks of different porosities and bonded carbonate sand (dashed). (For explanation of symbols see Figure 1.) This diagram shows the k_o envelope for chalk in the intact-elastic regime and the envelop for the ductile-compression regime. Only three chalks are illustrated, but the two envelopes were determined using the results of about 150 uniaxial compression experiments on different chalks. Each experimental stress path shows the excursion that is made between the two regime when the intact structure begins to fail at about 1% volume strain. This change is mechanically irreversible, but the system can be reset to a intact bonded state by diagenetic processes.

The double yield model for one of the chalks tested is illustrated schematically in Figure 3, together with a single stress path from a k_o compression test on the same material. The elastic failure envelope, yield surface and critical state line have all be defined by drained triaxial deformation experiments (Kågeson-Loe 1993, Kågeson-Loe et al 1993, 1992, Leddra et al 1990a, 1990b) on companion samples. It can be seen that the first inflexion on the k_o stress path corresponds with its intercept with the brittle-elastic failure envelope. At this point in the experiment brittle deformation begins in the sample, and if the state of lateral deformation was not constrained to be zero, a shear plane would eventually develop and the sample would fail. In order to prevent lateral deformation, the rate of increase in the confining stress (σ'_3) increases substantially and becomes equal with the rate of increase in axial stress (σ'_1): the k_o ratio ($=\sigma'_3/\sigma'_1$) changes from k_o=0.25 to k_o=1.0. As lateral strain and the associated development of a shear plane is prevented a progressive and (eventually) pervasive microfracture population develops allowing sliding of grains to begin. This microfracture population and the proportion of grains free to move increases with the stress until the yield envelop is reached. At this condition, the proportion of free grains in the sample exceeds a critical condition and pervasive ductility with grains sliding into pore spaces becomes the dominant deformation mechanism. This is the yield point and is characterised by the onset of large compressive strains. The mechanism of yield is pervasive non-dilatant cataclastic flow.

It is evident from Figures 2 and 3 that samples which have experienced more natural densification (infilling of pore space by cements, or significant compression prior to cementation) enter the yield transition at high effective stress states and experience greater changes in effective stress whilst in the transition. Thus increasing densification inhibits both the onset of yield and the transition to ductile compression. Geologically, this could, in the most simple cases, be equated to increasing burial depth and diagenesis. However, the stress and strain states described in Figures 1 and 2 pertain to pore fluid expulsion (they are fully drained experiments) and the associated strains (Figure 1) include volume change (compression). Thus position within the stress path diagram is not simply a consequence of geological burial. Sediments cemented at an early stage and subsequently subject to excess pore fluid pressures may retain low yield strengths. Examples of this are the high porosity chalks in the Valhall and Ekofisk oilfields. At the outset of production these chalks at a nominal depth of 3000m were subject to vertical effective stresses of only 5 and 14MPa respectively. Interestingly, these stresses correspond to the yield strength of the highest porosity chalk present in either reservoir.

Both the k_o=0.25 and the k_o=0.6 stress paths are paths that are characterised by increasing compression. On the former (intact) path, increase in stress is accompanied by limited volume strain (the elastic limit is reached at about 1% volume strain) before the change of state to ductile compression begins. On the k_o=0.6 (ductile compression) stress path, permanent compressive strains accumulate in the largely particulate material and are limited only by the state of compression of the material. The transition from intact elastic to ductile compression regime is therefore a change in state that is strain controlled and, in mechanical terms, is irreversible. Because of this, unloading of the material in the ductile compression regime does not lead to a re-establishment of the original elastic stiffness and elastic limit (Figure 3), but gives rise to overconsolidated states.

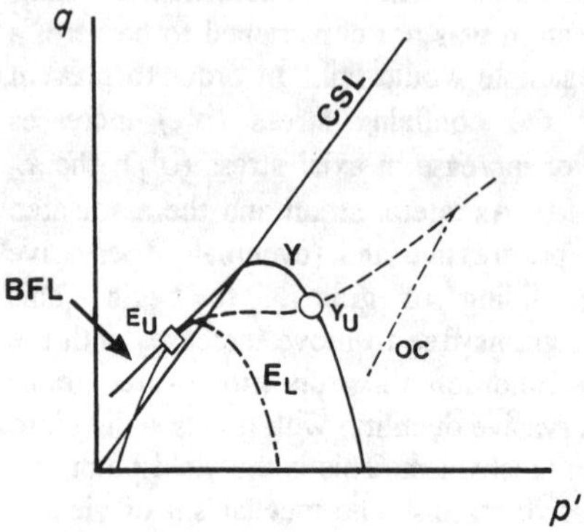

Figure 3. The relationship between the ko stress path for a chalk and its yield envelop (as defined by a family of conventional undrained triaxial deformation experiments on companion samples of the same chalk. The deformation in the intact-elastic regime ends when the path encounters the elastic-brittle failure envelope (BFL) at the point labelled Eu. The material is in the transitional regime until the yield envelope (Y) is reached where the yield point (Yu) is encountered. It is beyond this point that large compressional strain accumulate. (EL = the locus of elastic-limits encountered on other stress paths; CSL = critical state line; OC = the unloading/reloading path (over-consolidation) in the ductile-compression regime.

The transition from any state of compression in the ductile compression regime back to the intact elastic regime can only be achieved by imposing the reverse change in state on the material. This requires the introduction of a mineral cement, or the development of chemical bonding between grains such that grain motion becomes restricted by the elastic strength of these bonds and intergranular friction ceases to control rock strength. Further deformation will now occur in the the intact elastic regime and strain energy must be gained by the rock to once again promote bond (or grain) breakage to allow a further excursion to the ductile compression regime. All such yield excursions are accompanied by irreversible loses of pore volume and stiffness. This requires that the more densified sediment must accumulate greater elastic strain energy and sustain greater degrees of bond and/or grain failure in each successive excursion. The zone occupied by the transition regime therefore widens in the stress path diagram. Eventually, when virtually all porosity has been lost, or all grains are thoroughly interlocked in a close packed structure, excursions to the ductile compression regime become impossible and other deformation mechanisms must become operative.

The difference in behaviour that accompanies the change in state, from intact-elastic to ductile-compression regime, is pronounced and potentially important to engineering in carbonate environments. Thus the continuing large-scale land subsidence above the Ekofisk oilfield in the North Sea (Jones & Mathiesen 1993), the failure of piles driven into carbonates cemented at shallow depths (King & Lodge 1988, Carter et al 1988), production of chalk solids from oil wells completed in high porosity chalks (Ruddy et al 1988) and the sudden, often catastrophic failure of high-porosity chalk rock masses during tunnel excavation (Lake 1990) are all consequences of this change of state. Conversely, chalks cemented in the presence of large effective stresses, those highly cemented whilst still highly porous and those which, during diagenetic history, have experienced repeated excursions between the two regimes, do not exhibit these specific, unfavorable engineering characteristics, and are strong brittle rocks.

5 REFERENCES

Addis, M.A. 1987. *Mechanisms of sediment compaction responsible for oil field subsidence*. University of London: PhD Thesis.

Carter, J.P., I.W. Johnston, M. Fahey, G.A. Chapman, E.A. Novello & W.S. Kaggwa 1988. Triaxial testing of North Rankine calcarenite. *Proc. Int. Conf. Calcareous Sediments, Perth*: 515-530. Rotterdam:Balkema.

Coop, M.R. & J.H. Atkinson 1993. The mechanics of cemented carbonate sands. *Geotechnique* 43: 53-67.

Jones, M.E. & E. Mathiesen 1993. Pore pressure, compaction drive and reservoir compaction in the North Sea chalk hydrocarbon reservoirs. *Int. J. Rock Mech. Min. Sci. & Geomech. Abstr.* 30:

Kågeson-Loe, N.M. 1993. *The strain behaviour of chalk.* University of London: PhD Thesis.

Kågeson-Loe, N.M., M.E. Jones, D.N. Petley & M.J. Leddra 1993. Fabric evolution during the deformation of chalk. *Int. J. Rock Mech. Min. Sci. & Geomech. Abstr.* 30: 739-745.

Kågeson-Loe N.M., M.J. Leddra & M.E. Jones 1993. Strain states during stress path testing of the chalk. *Proc. 33rd U.S. Symp. Rock Mech., Santa Fe*: 927-936. Rotterdam:Balkema.

King, R. & M. Lodge 1988. North west shelf development - the foundation enginerring challenge. *Proc. Int. Conf. Calcareous Sediments, Perth*: 333-342. Rotterdam:Balkema.

Lake, L.M. 1990. Underground excavations in chalk. *Proc. Int. Chalk Conf., Brighton:* 461-467. London: Thomas Telford.

Leddra, M.J., M.E. Jones & A.S. Goldsmith 1990a. Laboratory investigation of the compaction of chalk under conditions of increasing effective stress. *Proc. 3rd North Sea Chalk Symp. Copenhagen.*

Leddra, M.J., M.E. Jones & A.S. Goldsmith 1990b. Flow of chalk under conditions of large shearing stress and excess pore fluid pressure. *Proc. 3rd North Sea Chalk Symp. Copenhagen.*

Ruddy, I., M.A. Andersen, P.O. Pattillo, M. Bishiawi & N. Foged 1988. Rock compressibility, compaction and subsidence in a high porosity reservoir: A case study of Valhall Field. *Proc. Soc. Pet. Eng. 63rd Tech. Conf., Houston.* SPE: 18278, 179-186.

Vaughan, P.R., M. Maccarini & S.M. Mokhtar 1988. Indexing the engineering properties of residual soil. *Qtly. J. Engin. Geology* 21: 69-84.

Rock Mechanics, Nelson & Laubach (eds) © 1994 Balkema, Rotterdam, ISBN 90 5410 380 8

Probability distributions for parameters of the Munson-Dawson salt creep model[1]

A. F. Fossum, T.W. Pfeifle & K.D. Mellegard
RE/SPEC Inc., Rapid City, S.Dak., USA

D.E. Munson
Sandia National Laboratories[2], Albuquerque, N.Mex., USA

ABSTRACT: Stress-related probability distribution functions are determined for the random variable material model parameters of the Munson-Dawson multi-mechanism deformation creep model for salt. These functions are obtained indirectly from experimental creep data for clean salt. The parameter distribution functions will form the basis for numerical calculations to generate an appropriate distribution function for room closure. Also included is a table that gives the values of the parameters for individual specimens of clean salt under different stresses.

1 INTRODUCTION

The Environmental Protection Agency (EPA) has issued Standards against which the performance of the Waste Isolation Pilot Plant (WIPP) will be evaluated for compliance. Since probabilistic measures will often be used in the regulatory framework, it follows that uncertainties in WIPP performance assessment calculations must be quantified in terms of probabilistic measures. In these calculations, random selection of the appropriate model parameters from their distributions are used with the model to determine, through repetitive realizations, the distribution of the calculated response.

An important model that describes the creep behavior of salt in performance assessment calculations is the multi-mechanism deformation creep model (M-D model) of Munson and Dawson (1984), which is used to make creep closure calculations of the repository rooms. Details of the creep response of the rooms provide the underlying bases for the time to reconsolidation of room backfill and encapsulation of the waste, the final porosity of the room contents, and the response to potential gas generation. In addition, the creep response is a critical component of the design and assessment of the engineered seals of the repository.

The probability distributions presented in this paper are based on laboratory data and analysis. The paper evaluates the parameters in the M-D model that contain the material uncertainty. Estimates of these random parameters were obtained from triaxial compression creep data generated from tests on WIPP salt. Different probability

---

[1]Work supported by U.S. Department of Energy (DOE), Contract No. DE-AC04-94AL85000
[2]A U.S. DOE facility

distribution functions for each of the parameters are fitted to the estimates and their relative goodness-of-fit tested using the Kolmogorov-Smirnov (*K-S*) statistic (Ang and Tang, 1975). The best-fit-probability distribution, i.e., the one with the smallest *K-S* statistic, is identified for each of the random creep parameters. The work described in this paper extends earlier preliminary work, Pfeifle et al., (1992), in which probability distributions were determined from experimental data for steady-state creep strain rate, the transient strain limit, and the workhardening parameter at constant values of temperature and applied stress differential. However, it is necessary to frame these parameters into stress dependent functions to give a representation relevant to the numerical calculation of repository room closure.

2 CONSTITUTIVE MODEL

The modified Munson-Dawson constitutive model (Munson et al., 1989) describing the time-dependent deformation behavior of salt is based on micromechanistic concepts as determined using the deformation mechanism map for salt. The mechanism map defines regions of stress and temperature in which a unique deformation mechanism controls or dominates the steady-state creep characteristics of the material. Expected repository conditions indicate that only three mechanisms operate: (1) dislocation climb at high temperature, (2) an undefined mechanism at low stress and temperature, and (3) dislocation slip at high stress. Although one mechanism dominates the steady-state deformation in a specific region of the map, the mechanisms are parallel processes and are summed to give a total steady-state strain rate, $\dot{\varepsilon}_s$, that is expressed as

$$\dot{\varepsilon}_s = \sum_{i=1}^{3} \dot{\varepsilon}_{s_i} \tag{1}$$

where $\dot{\varepsilon}_{s_i}$ is the steady-state strain rate of the ith mechanism. The individual steady-state strain rates for the dislocation climb, undefined, and dislocation slip mechanisms are, respectively:

$$\dot{\varepsilon}_{s_1} = A_1 \exp(\frac{-Q_1}{RT})(\frac{\sigma}{\mu})^{n_1} \tag{2}$$

$$\dot{\varepsilon}_{s_2} = A_2 \exp(\frac{-Q_2}{RT})(\frac{\sigma}{\mu})^{n_2} \tag{3}$$

$$\dot{\varepsilon}_{s_3} = |H|[B_1 \exp(\frac{-Q_1}{RT}) + B_2 \exp(\frac{-Q_2}{RT})] \times \sinh[\frac{q(\sigma-\sigma_0)}{\mu}] \tag{4}$$

where the A_i's and B_i's are constants, Q_i's are activation energies, T is the absolute temperature, R is the universal gas constant, the n_i's are the stress exponents, q is the stress constant, σ is the equivalent stress measure, μ is the shear modulus, σ_0 is the stress limit of the dislocation slip mechanism, and $|H|$ is the Heaviside step function with argument $\sigma - \sigma_0$.

Transient creep in the M-D model is incorporated through the function F such that

$$\dot{\varepsilon} = F\,\dot{\varepsilon}_s \tag{5}$$

where $\dot{\varepsilon}$ is the total creep strain rate. F, is a transient function with three branches:

$$F = \begin{cases} \exp[\Delta(1-\frac{\zeta}{\varepsilon_t^*})^2] \ , & \zeta \le \varepsilon_t^* \\ 1 & , \zeta = \varepsilon_t^* \\ \exp[-\delta(1-\frac{\zeta}{\varepsilon_t^*})^2], & \zeta \ge \varepsilon_t^* \end{cases} \tag{6}$$

The three branches represent a workhardening branch, an equilibrium branch, and a recovery branch, respectively. The parameters, Δ and δ, are the workhardening and recovery parameters, respectively, and ε_t^* is the transient strain limit. The variable, ζ, is an internal variable whose rate of change is governed by

$$\dot{\zeta} = (F-1)(\dot{\varepsilon}_s) \tag{7}$$

The transient strain limit, ε_t^*, can be expressed as

$$\varepsilon_t^* = K_0 e^{cT} (\frac{\sigma}{\mu})^m \tag{8}$$

where K_o, c, and m are material constants. In this work, which deals primarily with stress loading, only the stress dependent workhardening expression

$$\Delta = \alpha + \beta \log(\frac{\sigma}{\mu}) \tag{9}$$

is relevant. Here α and β are constants.

3 EXPERIMENTAL DATABASE

The database used in the evaluation of the M-D model parameters comprises all 29 triaxial compression creep tests performed on 100-mm-diameter by 200-mm-long, cylindrical specimens of clean salt obtained from the WIPP. Each specimen was prepared from cores recovered either from the ERDA 9 borehole or from the WIPP workings. Test conditions for each creep test (including temperature, stress difference, and confining pressure) are given in Table 1. During each test, axial and lateral strain were measured as a function of time, while the temperature, stress difference and confining pressure were held constant at the values shown in Table 1. Stress difference was held constant by continuously monitoring lateral deformation and adjusting the axial load to account for changes in specimen diameter.

The results of a typical creep test are shown in Figure 1a. The creep strains can be separated into those that result from the steady-state creep, ε_s, and those that result from the workhardening transient creep, ε_t. The workhardening transient accumulates to a maximum value which is the transient strain limit, ε_t^*, and then remains constant. Thereafter, further strain accumulates only through the steady-state creep process. In

Table 1 Test conditions and parameter estimates

	Test Conditions				Measured			Calculated
Test I.D.	T (K)	σ_1-σ_3 (MPa)	σ_3 (MPa)	$\dot{\varepsilon}_s$ (10^{-10}s^{-1})	ε_t^* (10^{-2})	Δ	ζ_a (10^{-2})	ζ_0 (10^{-2})
ERDA9/88/2124-0/1	295	11.7	20.7	1.22	.633	9.50	0.228	0.690
ERDA9/88/2127-0/1	298	10	15	.306	.321	9.00	0.166	0.634
DX19-5/7-1-2	298	10	15	1.31	.194	7.50	0.093	0.561
DX16-2/7-1-2	298	10	15	.778	.161	8.00	0.074	0.542
C1X01-04/1-3/1-2	298	10	15	.994	.367	8.50	0.151	0.614
C1X01-1/3-3/7-1	298	11.5	15	.947	.428	7.50	0.096	0.554
DX19-5/5-1-2	298	13	15	2.69	.718	7.00	0.229	0.681
ERDA9/88/2126-0/1	298	15	15	5.02	1.68	7.50	0.438	0.890
C1X01-1/3-3/1-1	298	15	15	7.10	1.32	6.00	0.164	0.616
C1X01-1/3-2/6-1	298	15	15	17.5	.551	8.00	0.159	0.611
C1X01-04/1-3/2-2	298	15	15	6.68	1.256	10.0	0.399	0.608
DX16-2/1-1-2	298	15	15	11.9	.832	8.00	0.156	0.851
C1X01-02/1-3/1-2	298	15	15	3.51	1.24	9.00	0.216	0.668
C1X01-1/3-3/4-1	298	17	15	21.4	2.01	7.50	0.316	0.761
DX16-2/4-1-2	298	17	15	22.8	1.49	6.00	0.344	0.789
DX16-2/8-1/2	298	19	15	34.5	3.50	8.00	0.408	0.847
DX19-5/6-1-2	298	19	15	40.5	2.25	8.00	0.366	0.805
DX16-2/6-1-2	298	21	15	57.9	3.43	6.00	0.615	1.047
C1X01-1/3-3/2-1	298	21	15	60.2	4.07	7.00	0.471	0.903
DX19-5/1-1-2	298	25	15	199.	5.40	6.50	0.890	1.309
DX19-5/2-1-2	298	30	15	588.	7.87	6.00	1.390	0.793
DX19-5/3-1-2	298	30	15	588.	6.86	5.0	1.366	1.769
DX16-2/3-1-2	373	3.5	15	1.48	.059	Not	0.025	0.739
C1X01-1/3-3/6-1	373	3.5	15	.558	.163	used	0.028	0.769
C1X01-1/3-2/7-1	373	5	15	2.89	.200	---	0.075	1.234
DX19-4/4-1-2	373	7	15	41.2	1.02	---	0.059	1.067
MCE36-1/1-1/2-2/2	373	7	15	25.2	.394	---	0.059	1.067
DX19-5/4-1-2	373	17	15	3100.	5.85	---	1.000	1.445
DX16-2/2-1-2	473	2	15	4.4	.026	---	0.005	0.994

addition to the creep strains, there are strains associated with the elastic, ζ_e, and plastic, ζ_a, strains accumulated during stress application, load-up of the creep test specimen, and strains associated with the in situ or excavation strain, ζ_x, contained in the material from which the test specimens are fabricated. An excavation strain value of $\zeta_x = 0.005$ produces good agreement between model and experiment. These strains can be represented in the test results by $\zeta_0 = \zeta_a + \zeta_x - \zeta_e$ where $\zeta_e = \sigma/E$ (E = 31.0 GPa). In this study, ζ_0 is added to the creep strain of the creep test.

The results of each creep test were used to estimate four quantities shown in Equations 5 and 6, i.e., $\dot{\varepsilon}_s$, ε_t^*, Δ, and ζ_a. Table 1 gives the estimates of these four quantities determined from each of the 29 creep tests. Two parameters, $\dot{\varepsilon}_s$ and ε_t^*, were determined by successively fitting a straight line to the data within a restricted window

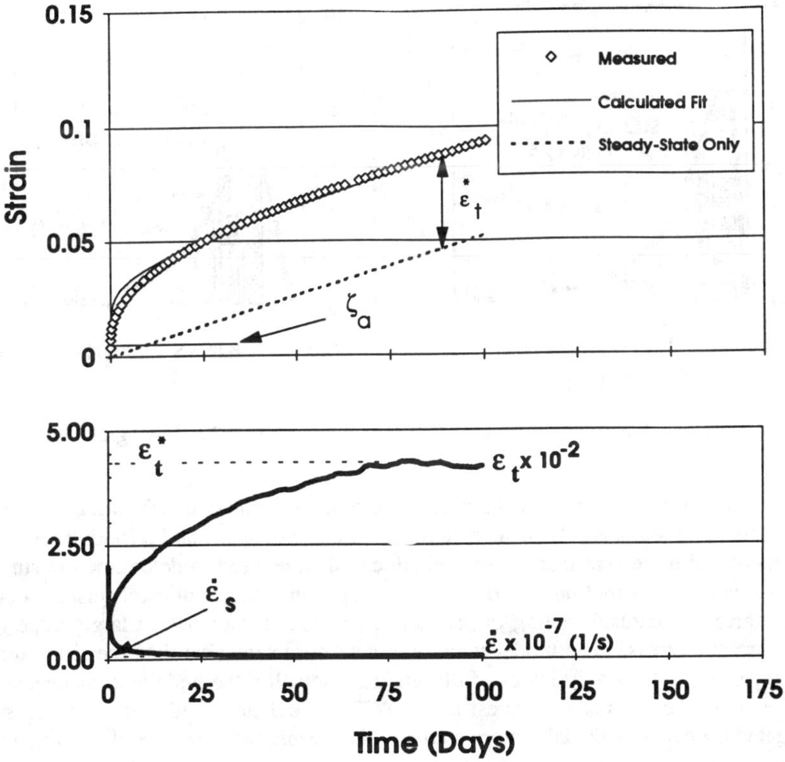

Figure 1 Analysis of typical creep test results and model calculation

of three data points that slides along the creep data in time. The slopes determined in the fitting process represent instantaneous values of the strain rate, $\dot{\varepsilon}$, while the intercepts represent instantaneous values of ε_t. With sufficient time, both of these quantities reach their asymptotic values of $\dot{\varepsilon}_s$ and ε_t^*, respectively, as shown in Figure 1b. The parameter, Δ, is determined as the natural logarithm of the ratio of the initial strain rate to the steady-state strain rate, $\dot{\varepsilon}_s$. This ratio was determined by extrapolation to zero time from a plot of $ln(\dot{\varepsilon}/\dot{\varepsilon}_s)$ versus time. The application strain ζ_a is a measured quantity of each test.

4 PARAMETER ESTIMATION AND DISTRIBUTION FUNCTIONS

All of the clean salt data have now been reevaluated through the process just discussed and illustrated in Figure 1 to determine the experimental values of the steady-state rate, $\dot{\varepsilon}_s$, the transient strain limit, ε^*_t, and the workhardening parameter, Δ. However, these experimental values cannot be used directly in finite element calculations, but must be framed in terms of the appropriate parameters of the stress dependent functions. In this way, the uncertainty in the *M-D* model is quantified by

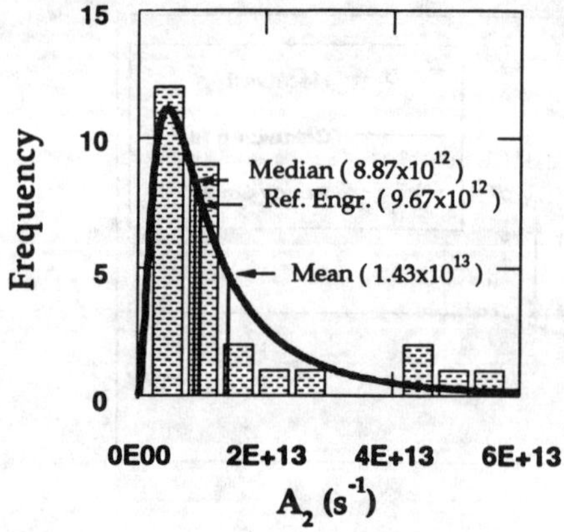

Figure 2a Histogram for A_2

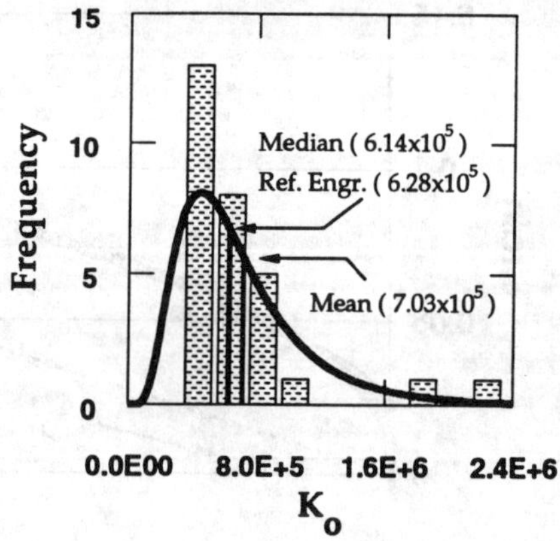

Figure 2b Histogram for K_0

assuming that four parameters of the model are random (probabilistic) variables. The four random parameters are A_2 (steady state), K_0 (transient), and α and β (both transient) of the undefined or second mechanism, which can then be used to determine the stress related parameter distribution functions. As is apparent, the multimechanism model contains three mechanisms acting in parallel to define behavior over a large range of temperature and stress. The model range is much more extensive than the laboratory data base or the field conditions can fully utilize. Actually, the undefined mechanism dominates over the stresses of interest in the WIPP calculations and is the mechanism investigated primarily in the laboratory studies. As a result, the parameter determination can center on the undefined mechanism of the three mechanism model. However, even though the other mechanisms are thought to be minor contributors at the extremes of the test conditions, the uncertainty in the parameters of these mechanisms is treated indirectly in the analysis of the undefined mechanism by using Equations 2-4 to calculate A_2, taking A_1, B_1, B_2, Q_1, Q_2, n_1, and q as constants. In this manner, errors in the assumed constant values of the first and third mechanisms are contained in the uncertainty of the second mechanism parameters. The only remaining second

Table 2 Summary of Distribution Fitting for Random Creep Parameters

Creep Parameter	Distribution Function	Distribution Parameters	K-S Statistic, D
K_0	Lognormal	λ = 13.31, ζ = 0.557	0.138
A_2	Lognormal	λ = 29.91, ζ = 0.873	0.178
α	Normal	μ = -8.26, σ = 4.41	(a)
β	Normal	μ = -5.45, σ = 1.52	(a)

(a) Random parameters assumed to be normal.

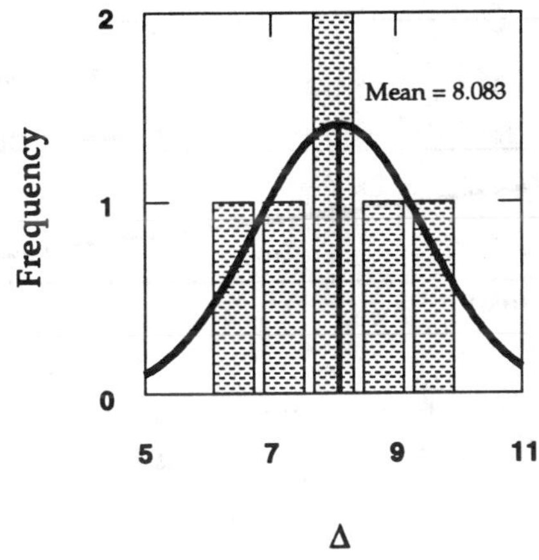

Figure 3 Histogram for Δ

mechanism parameter, n_2, is directly coupled to A_2 and must be held constant. In the determination of K_0, both c and m, which is a theoretical value, are constant. All values assumed constant are given by Munson et al. (1989).

Histograms of A_2 and K_o are shown in Figure 2. Six different probability distribution functions were fitted to the experimental histograms to identify those distributions that best represent the variability in these two random parameters. The Kolmogorov-Smirnov (K-S) goodness-of-fit test (Ang and Tang, 1975) was applied to the fit of each distribution function to discriminate among the distributions. Based on this procedure, the best fitting distribution functions for A_2 and K_o were both lognormal. These distributions are superimposed on the histograms in Figure 2 and their parameters and the K-S statistic, D value, are summarized in Table 2. Also shown in Figure 2 are the mean, median, and the reference engineering value given by Munson et al. (1989).

The quantity Δ is related to the parameters, α and β, as given in Equation 9 and was determined experimentally for each creep test with the values summarized in Table 1. Because the temperature dependence of Δ is not yet known, only the tests conducted at room temperature (298°K) were used. Several repeat tests were performed at a stress difference of 15 MPa. A histogram of the values of Δ at this test condition indicated that Δ is normally distributed as shown in Figure 3. Since the linear combination of normally distributed random variables produce another normally distributed random variable, it was assumed that α and β were both normally distributed as shown in Figure

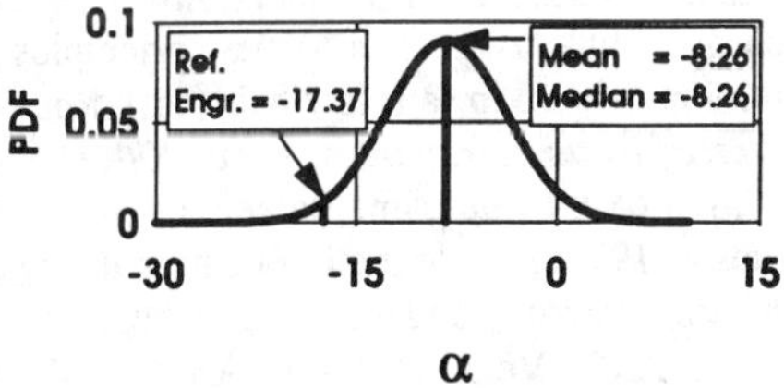

Figure 4a Probability distribution for α

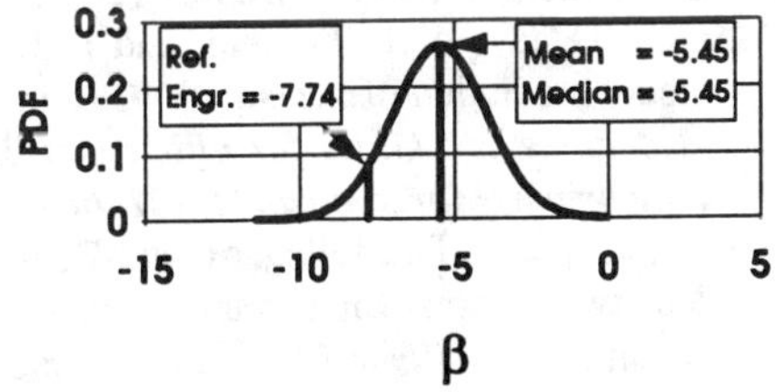

Figure 4b Probability distribution for β

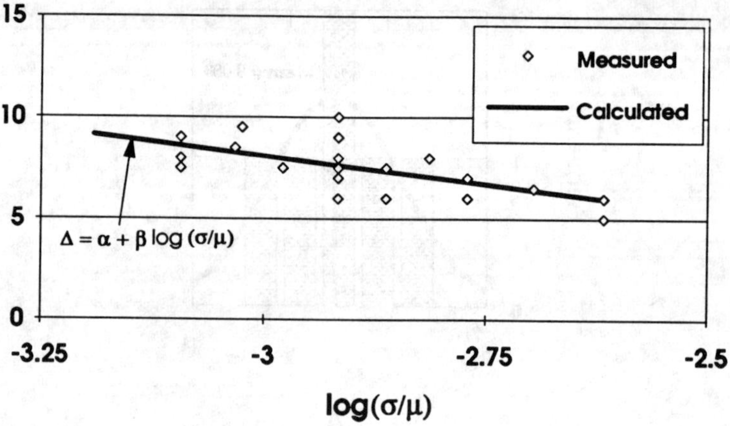

Figure 5 Least squares fit to Δ

4. We recognize that this database is quite sparse and therefore a normal distribution seems appropriate in this circumstance. The means and standard deviations of α and β were determined from a linear least-squares fit of Equation 9 to the data shown in Table 1. The data and the fitted line are shown in Figure 5. Table 2 also summarizes the distribution information for α and β.

5 SUMMARY AND CONCLUSIONS

Probability distribution functions were determined from the WIPP creep database for the material parameters of the Munson-Dawson salt creep model. Parameters analyzed were those thought to contain material uncertainty. Quantification of uncertainty is necessary for probabilistic structural calculations that will be used in WIPP performance assessment calculations. The distribution functions will be used as input in numerical calculations to determine the distribution function for room closure.

REFERENCES

Ang, A. H.-S., & W.H. Tang 1975. *Probability concepts in engineering planning and design, vol 1, basic principles*. John Wiley & Sons.

Munson, D.E. and P.R. Dawson 1984. Salt constitutive modeling using mechanism maps. in H.R. Hardy, Jr., & M.L. Langer, (eds.), *Proc. First Int. Conf. on the Mechanical Behavior of Salt:*717-737. Clausthal: Trans. Tech. Publications.

Munson, D. E. , A. F. Fossum, and P. E. Senseny 1989. Approach to first principles model prediction of measured WIPP in situ room closure in salt. In Khair (ed.), *Rock Mechanics as a Guide for Efficient Utilization of Natura Resources, Proc. 30th U.S. Symposium, Morgantown, 19-22 June 1989*:673-680. Rotterdam: Balkema.

Pfeifle, T. W., K. D. Mellegard, and D. E. Munson 1992. Determination of probability density functions for parameters in the Munson-Dawson model for creep behavior of salt. *Probabilistic Method in Geomechanics*, AMD-Vol. 134:41-56. ASME: NY.

Rock Mechanics, Nelson & Laubach (eds) © 1994 Balkema, Rotterdam, ISBN 90 5410 380 8

Experimental identification of kinematic and isotropic hardening in rocksalt

M.Aubertin, S.Servant & D.E.Gill
École Polytechnique, Montréal, Que., Canada

ABSTRACT: In recent years, internal state variable models have gained popularity for describing the behavior of rocksalt. Among the various advantages of such models, the possibility to distinguish between kinematic and isotropic hardening phenomena is of the outmost importance, as physical (microscopic) evidences clearly show that both aspects must be considered in properly formulated constitutive equations. However, if the mixed nature of hardening is fairly well established from the micromechanical point of view, it is usually not clear how one should proceed to separate the components of hardening from macroscopic test results. In this paper, the authors present a general approach aimed at obtaining experimentally these informations from laboratory tests on rocksalt samples. The proposed procedure is based on the use of triaxial compression tests. Using schematic and experimental results on rocksalt, it is shown how such tests can be used to isolated the kinematic and isotropic components. A previously developed model is used to illustrate the procedure.

1 INTRODUCTION

The design and monitoring of underground excavations in rocksalt requires time-dependent inelastic analysis of the stress, strain and displacement fields around the openings. For that purpose, it is now often recognized that, in order to adequately reproduce the complex behavior of rocksalt, fairly elaborate models are needed. In that regard, internal state variable models formulated in the context of unified theories of inelastic flow with mixed hardening have received increased attention in recent years, because of the enormous possibilities they offer (e.g. Aubertin et al. 1991, 1993a, Pudewills and Hornberger 1993).

A variety of constitutive equations have been produced in the last two decades to allow more accurate modelling of engineering materials behavior. The large majority of these equations have been built on the basis of one-dimensional data, and these usually give a good description of inelastic behavior under proportional loading. However, because most practical situations involve multiaxial non-proportional loading, often with some rotation of the principal stress directions increments, it is necessary to properly generalize the one-dimensional formulation to the three-dimensional case. In this instance, one must consider that hardening associated to ductile flow may involve isotropic (non-directional) effects as well as kinematic (directional) effects. Unfortunately, most traditional experimental procedures do not allow the distinction between these two components of the hardening process.

In this paper, the authors first present a general viscoplastic modelling framework for mixed (isotropic and kinematic) hardening, and then describe how one could establish experimentally from triaxial compression tests on rocksalt the relative contribution of each of these components. Mixed hardening has important implications for the analysis of underground excavations in rocksalt, as stress redistribution creates both unloading and rotation phenomena in the material around the openings.

2 EQUATIONS FOR HARDENING OF ROCKSALT

In the ductile regime, the behavior of rocksalt is quite similar to that of metals (e.g. Senseny et al. 1992). The inelastic flow of such material usually results in hardening. At the microscopic level, one can relate hardening with the raising dislocations density and with their increased interaction with various types of obstacles, that result in the activation of dynamic recovery mechanisms and in the formation of heterogeneous dislocations substructures, inducing both isotropic and kinematic hardening phenomena (Aubertin 1993).

At the macroscopic level, hardening translates into an apparent yield strength increase during constant strain rate (CSR) tests and in the ensuing enhanced strain resistance. The decreasing strain rate in a constant stress test during the transient (primary) creep phase is also a macroscopic manifestation of hardening of the material.

The cumulated inelastic strain is often considered as a natural hardening variable. In fact, in classical creep models, the hypothesis of strain hardening (SH) is usually favoured over that of time hardening (TH) as experimental observations during incremental creep tests have shown that the latter generally underestimate inelastic strain. Such results have been demonstrated in many occasions for rocksalt (e.g. Durup and Xu 1993), as well as for other crystalline materials (e.g. Chaboche 1983). These observations have led to the hypothesis of a state equation that, when written in a one-dimensional form, becomes:

$$F(\sigma, \epsilon^i, \dot{\epsilon}^i) = 0 \tag{1}$$

where σ is the stress, ϵ^i is the inelastic strain, and $\dot{\epsilon}^i$ is the inelastic strain rate. Although various specific formulations have been used to describe F, a product of power functions has been adopted by many, as it often provides good results over a fairly wide range of conditions for transient creep, as well as for other simple loading conditions such as uniaxial relaxation tests for instance. This simple formulation can be written as:

$$\sigma = A_0 (\epsilon^i)^{1/m} (\dot{\epsilon}^i)^{1/n} \tag{2}$$

where A_o, m and n are considered material constants for a given temperature. This simple equation is very appealing for engineering purposes as only three constants have to be determined experimentally. Furthermore, equation 2 can be easily generalized to a three-dimensional form using von Mises equivalent stress and strain, if one accepts the isotropic hardening hypothesis.

Unfortunately, it has been shown that, for various materials, such equation is only valid in the case of nearly proportional monotonous loading, that is when all tensors (including stress σ_{ij} and inelastic strain rate $\dot{\epsilon}^i_{ij}$) are codirectional. In fact, experimental results on crystalline materials have shown that a rotation of principal stress directions in multiaxial loading conditions induces a transient behavior and a temporary non-colinearity between the stress and inelastic strain rate tensors (e.g. Murakami et al. 1986). Such evidence implies that inelastic flow of materials cannot be considered as an isotropic phenomena, but that anisotropic effects have to be introduced in the hardening function. None of the classical theories of creep can adequately describe such phenomena (Chaboche 1983). And although it is possible to modify such theories to account for these effects in the inelastic strain space, it is often considered more practical to describe these by using a viscoplastic theory expressed in stress space. In this case, internal stresses should be introduced in the constitutive equations.

The use of an internal stress σ_i in constitutive laws is far from new. For instance, it was introduced by Orowan (1947) in Norton's steady-state creep law to obtain a stress exponent which was closer to physical theories. In this case, the uniaxial power law becomes:

$$\dot{\epsilon}^i = A\left(\frac{\sigma - \sigma_i}{K}\right)^N \tag{3}$$

where A and K are material parameters, and N is a constant. If σ_i is considered to evolve during transient flow, then equation 3 becomes the basis for a viscoplastic model with internal state variables (Aubertin et al. 1991).

In a three-dimensional formulation, the internal stress may be a scalar or a tensor depending on the types of interaction created during inelastic flow. If one considers both types of internal stress, one could then write, using a von Mises type formulation:

$$X_{ae} = R + \sigma_a \tag{4}$$

$$\text{with} \quad \sigma_a = K\left(\frac{\dot{\epsilon}_e^i}{A}\right)^{1/N} \tag{5}$$

$$X_{ae} = \left[\frac{3}{2}\,(S_{ij} - B_{ij})\,(S_{ij} - B_{ij})\right]^{1/2} \tag{6}$$

where S_{ij} is the deviatoric stress tensor, and $\dot{\epsilon}_e^i$ is the equivalent inelastic strain rate. For the one-dimensional case, equation 4 becomes:

$$\sigma - \mathrm{B} = R + \sigma_a \tag{7}$$

In these equations, there are three different types of internal state variables. B_{ij} (or $\pm$ B in the uniaxial case) is a second-order tensor, often called the backstress, which is associated to directional (kinematic) hardening due to non-homogeneous stress distribution within the crystalline structure. Scalar R is an isotropic internal stress serving as a threshold that defines the elastic field. Its value increases according to dislocations density and configuration; its initial value, R_o, represents the initial yield strength of the material which, in the case of rocksalt, is often considered to be very low (close to zero). Finally, σ_a is the active (or viscous) stress which, in this case, has been expressed as a power function; it includes the effect of K, the dragstress, which act to slowdown the motion of dislocations (Aubertin 1993). Each of the internal state variables evolves during transient flow according to its own growth law.

Using equations 4 and 5, developed in the proper manner, one can obtain the following kinetic law (Aubertin et al. 1991):

$$\dot{\epsilon}_{ij}^i = A\left\langle\frac{X_{ae} - R}{K}\right\rangle^N n_{ij} \tag{8}$$

where < > are the MacAuley brackets and n_{ij} is the unit directional tensor. Equation 8 represents a fairly general viscoplastic formulation producing mixed hardening. When the effect of the internal state variables are neglected (B_{ij}, R, and K fixed to their nominal values), it becomes a standard steady-state power creep law. On the other hand, when the evolution laws of the internal state variables are properly formulated, this equation provides a very similar response to that of the SH transient creep model (equation 2) for idealized uniaxial incremental creep tests (Aubertin et al. 1991).

3 UNIFIED THEORY OF INELASTIC FLOW

Equation 8 presented above offers much more flexibility than conventional creep laws based on the isotropic hypothesis. For instance, it allows a proper description of decremental creep tests (Aubertin et al. 1993b) which, as it is now well established, cannot be correctly described by a solely isotropic hardening model (Munson et al. 1993). Furthermore, it provides an explicit description of the so-called creep-plasticity interaction. In that regard, it represents a unified view of inelastic flow. According to this unified approach, inelastic behavior can be expressed entirely from the present state of the material, which depends on the present values of both the observable variables (such as stress, strain and temperature) and a set of internal state variables that, if properly selected and formulated, should reflect thermomechanical history effects. According to this increasingly popular view, which is based on sound physical basis, results of different types of tests, such as constant strain rate (CSR, which includes relaxation for $\dot{\epsilon} = 0$) and constant rate of stress (CRS, which includes creep for $\dot{\sigma} = 0$), are particular response for specific loading conditions obtained from the same set of equations comprising the kinetic law and the growth laws. And although a given type of test can enhance a particular component of the material behavior, all aspects are always present in any type of tests. Accordingly, various types of tests can be used to study inelastic behavior. In practice however, it is more convenient to rely on specific tests to investigate specific aspects, such as "short term" CSR tests for dynamic recovery and "long term" creep tests for static recovery (Aubertin et al. 1993c).

It should also be remembered that, according to this unifying view, the steady-state flow is considered to exist for all loading conditions, and that its characteristics are generally independent of the thermomechanical history of the material. Because of that, a single equation can give the steady-state relationship between stress, inelastic strain rate and temperature. This expression can be based on external (observable) variables only, as is the case with most stationary creep laws.

4 EXPERIMENTAL IDENTIFICATION

4.1 *Laboratory procedure*

The basic experimental procedure proposed here relies on the use of the triaxial cell for cylindrical samples, such as the well known Hoek-Franklin device that allows an independent control of the radial and axial compressive stresses. The test should be performed at sufficiently high minor principal stress to suppress possible microfracturing in the rocksalt sample; at room temperature and for strain rate below 10^{-4} s^{-1}, a stress of 15 to 20 MPa should be sufficient at low strain ($\epsilon < 2$ - 5%).

The test is performed according to a standard procedure for conventional triaxial compression (CTC). The sample is first submitted to isotropic loading at a given loading rate (typically 0.01 to 0.1 MPa/s), and then to a deviatoric stress, at a given confining pressure, under CSR condition. During the CSR test, partial unloading and reloading cycles must be performed at the same strain rate. After substracting the elastic strain (ϵ^e) of the material response, which in this case is simply given by $\Delta\sigma_{ax}/E$ (were σ_{ax} is the axial stress and E, the elastic modulus), one should obtain a curve of the type shown schematically with a full line in Figure 1. Performing multiple unloading and reloading cycles leads to an estimate of the

σ_a value (equation 4) throughout the test; for that, one simply has to locate the end of the linear portion on the reloading stress-strain curves, as shown in Figure 1. One should remember that in the (re)loading phase, the inelastic strain appears whenever the deviatoric stress leaves the elastic field given by radius R. Also shown in Figure 1 are curves representing the evolution of the internal state variables during inelastic straining.

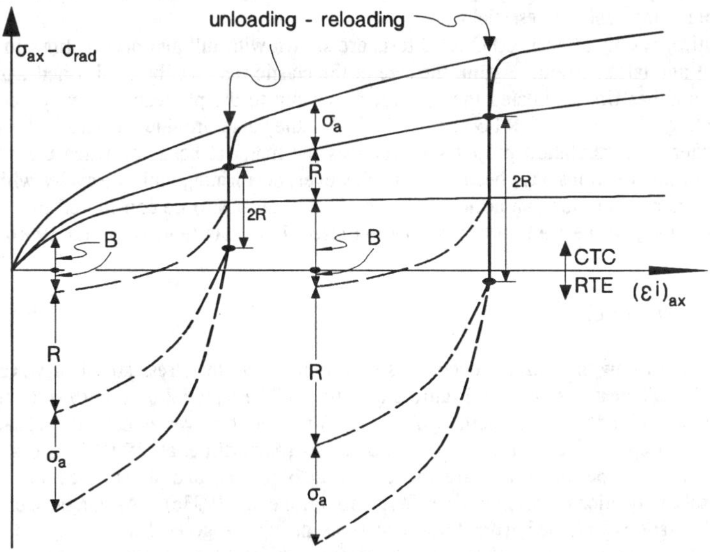

Figure 1 - Schematical representation of CTC-RTE tests results, at constant strain rate, with the evolution of internal state variables.

It should be recognized however that the precise identification of the elastic field limit on the reloading curve is not always an easy task. Nevertheless, repeating the process with careful measurements should give a good estimate of σ_a. For rocksalt submitted to a CSR-CTC test, the early value of σ_a should be fairly low, often less than 10% of the applied deviatoric stress, and it should reach its saturation value, which is about 20 to 50% of the stress, when steady-state is reached. Performing such tests at various strain rates gives information on the effect of $\dot{\varepsilon}^i$ on the evolution of σ_a and K.

The unloading portion of the stress-strain curve should be closely monitored as one does not want to leave too much the elastic field because of possible hysteresis in the reloading cycle. A good control of unloading provides enough information so that the size of the elastic field can be determined. On Figure 1, it is given by the linear segment on the unloading and reloading curve. The location of the center of the elastic field, considering it to be symmetric relative to the axe perpendicular to the deviatoric stress, can then be easily determined; it corresponds to the value of B (or B_{ij}), the directional component of the internal stress. In most cases, it should be observed that B initially evolves faster than R and K as indicated by physical observations on the generation of internal stresses in crystalline materials (Aubertin 1993).

At large strain, exceeding a few percents, it is possible that the size of the elastic field may not be entirely defined by even a complete unloading, as shown by the second unloading cycle in Figure 1. In this case, one has to alternate CTC loading with reduced triaxial extension

(RTE) loading. Under loading states that render the inelastic flow of rocksalt independent of the mean stress σ_m, CTC-RTE testing is equivalent to producing a Bauschinger effect, a well known phenomena in metals. In the latter case however, this type of behavior is studied with uniaxial tensile and compressive stresses. Because this cannot be done with rocksalt, which is much too brittle when loaded in tension, CTC-RTE tests at high radial stress provide a mean to study directional hardening without actually imposing a tensile stress on the sample. In fact, a RTE test is equivalent to applying a uniaxial tensile stress on a sample initially submitted to an isotropic compressive stress field.

Schematical results of such CTC-RTE tests are shown with full and pecked lines in Figure 1 for small and larger strain. Again, the size of the elastic field can be established from such results, if one has first estimated the σ_a value according to the procedure already described. An interesting aspect of CTC-RTE test results is that they also provide a mean to check if the mixed hardening formulation properly reproduces the material behavior when the principal stress and strain directions have been rotated. However, performing multiple cycles with CTC-RTE loadings may produce an accumulation of strain that has to be considered in a specific manner, similarly to the well known ratchetting effect in metals (e.g. Nouilhas 1990).

4.2 *Experimental results*

In order to show how the above procedure is used in practice, the stress-strain curve resulting from a CTC-RTE test is shown on Figure 2 (full line). The material used here is an artificial salt which, when loaded in the ductile (fully plastic) regime, behaves as natural rocksalt, both from the microscopic and macroscopic point of views (Aubertin et al. 1991). Also shown in Figure 2, with the pecked lines, are the evolution of σ_a, B, and R as calculated from a complete set of equations presented elsewhere (Aubertin et al. 1993c). As can be seen on this Figure, simulations obtained from these equations provide a good description of the CTC portion and a good prediction of the RTE portion of the test. Such test result provides the starting point in the process of constants value determination (Aubertin et al. 1993c). In this

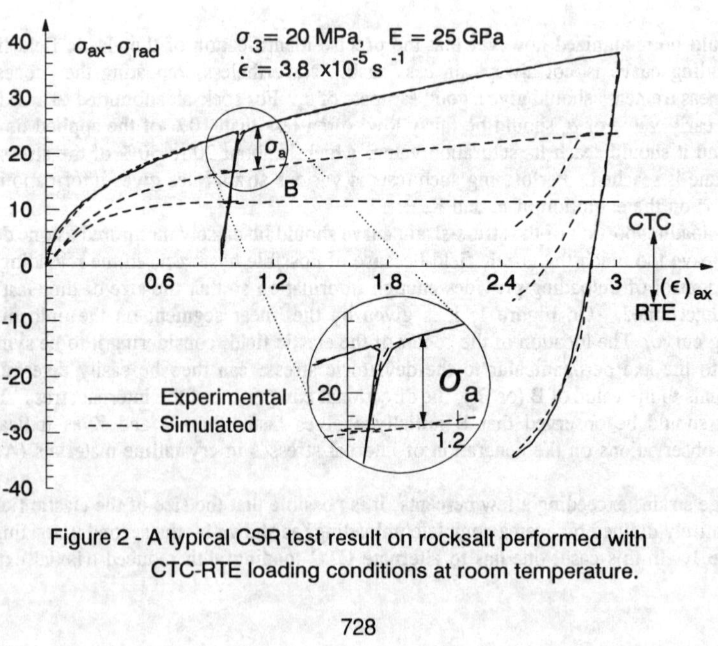

Figure 2 - A typical CSR test result on rocksalt performed with CTC-RTE loading conditions at room temperature.

case, the constants determined during CTC loading and unloading are used to predict the behavior during RTE loading. As on can see on Figure 2, the results are very encouraging.

These results only serve to illustrate, in a simplistic manner, the proposed procedure. An extensive testing campain is however needed to perperly ascertain all the material constants for a unified model.

5 DISCUSSION

The above mentioned procedure is one that has been known for many years for metals, using uniaxial testing. It is inspired largely by typical hardening theories in elastoplasticity extended thereafter for viscoplastic models (Aubertin and Gill 1993). Although the contribution of kinematic hardening has long been recognized in traditional elastoplastic models, it took the original contribution of Malinin and Khadjinsky (1972) to introduce such effect into creep (and viscoplasticity) theories. Accordingly, creep tests could also be used to distinguish between the various hardening components. Classical dip tests, occasionally performed on rocksalt, give some information on the hardening components. It is however preferable to study the stress-strain curve upon unloading and reloading after a creep period to estimate, using an approach similar to the one given above, the size of the elastic field (2R) and the location of its center (B). In fact, alternating CTC and RTE loading could also be performed for creep conditions. And although the authors are not aware of anyone having performed creep tests under alternative CTC-RTE loading on rocksalt, tests on metals have clearly shown the mixed nature of hardening during creep, where the kinematic component appears to produce a temporary softening of the material after a stress reversal (Murakami et al. 1986).

Finally, the various components of hardening could also be investigated with tests on thin wall cylinders, which allow different loading paths (e.g. Mellegard et al. 1992). The size, location, and shape of the elastic field could then be defined more precisely, so that one would be able to follow more closely the internal state variables evolution.

6 CONCLUSIONS

In this paper, the authors have presented a general framework to unified modelling of mixed (isotropic and kinematic) hardening of rocksalt using internal state variables. Then, after discussing the necessity of using such mixed hardening formulation, it is shown how to properly isolate the isotropic component from the kinematic component, by performing triaxial compression tests with unloading and reloading cycles. Using a previously developed model and a typical CTC-RTE test result on rocksalt, the procedure is illustrated. Other alternatives to identify the various components of hardening are also briefly discussed.

ACKNOWLEDGEMENTS

This research work was supported by financial contributions from NSERC (OGP0089749 OGP0003412).

REFERENCE:

Aubertin, M. 1993. On the physical origin and modelling of kinematic and isotropic hardening of salt. Preprints, *3rd Conf. Mechanical Behavior of Salt*, Palaiseau, 1-15.

Aubertin, M. and Gill, D.E. 1993. Elastoplastic modeling for the ductile behavior of

polycrystalline sodium chloride with SUVIC. *Int. J. Plasticity* 9:479-509.
Aubertin, M., Gill, D.E. and Ladanyi, B. 1991. A unified viscoplastic model for the inelastic flow of alkali halides. *Mechanics of Materials* 11:63-82.
Aubertin, M., Gill, D.E. and Ladanyi, B. 1993a. Modeling the transient inelastic flow of rocksalt, *7th Symp. on Salt*, Kyoto, I:93-104, Elsevier Science Pub.
Aubertin, M., Gill, D.E. and Ladanyi, B. 1993b. Some phenomelogical attributes of a new viscoplastic model for rocksalt. *Proc. 7th Int. Cong. on Rock Mech.*, Aachen, 3:1831-1835, Balkema.
Aubertin, M., Gill, D.E. and Servant, S. 1993c. Preliminary detemination of constants for an updated version of the SUVIC model. Preprints, *3rd Conf. Mechanical Behavior of Salt*, Palaiseau, 17-29.
Chaboche, J.L. 1983. On the constitutive equations of materials under monotonic or cyclic loadings. *Rech. Aérosp.* 5:31-43.
Durup, G. and Xu, J. 1993. Comparative study of certain constitutive laws used to describe the rheological deformation of salt. Preprints, *3rd Conf. Mechanical Behavior of Salt*, Palaiseau, 75-83.
Mellegard, K.D., Callahan, G.D. and Senseny, P.E. 1992. Multiaxial Creep of Natural Rock Salt. SAND 91-7083, Sandia National Laboratories, Albuquerque.
Malinin, N.N. and Khadjinsky, G.M. 1972. Theory of creep with anisotropic hardening. *Int. J. Mech. Sci.* 14:235-246.
Munson, D.E., De Vries, K.L., Fossum, A.F. and Callahan, G.D. 1993. Extension of the M-D model for treating stress drops in salt. Preprints, *3rd Conf. Mechanical Behavior of Salt*, Palaiseau, 31-44.
Murakami, S., Ohno, N. and Tagami, H. 1986. Experimental evaluation of creep constitutive equations for type 304 strainless steel under non-steady multiaxial states of stress. *J. Engng Materials and Technology*, ASME, April '86, 108:119-126.
Nouilhas, D. 1990. Prediction of ratchetting and creep effects at room temperature on 304 stainless steel. *Res Mechanica* 30:223-243.
Orowan, E. 1947. The creep of metals, *J. West. Scot. Iron St. Inst.* 54:45-68.
Pudewills, A. and Hornberger, K. 1993. A unified viscoplastic model for rock salt. Preprints, *3rd Conf. Mechanical Behavior of Salt*, Palaiseau, 45-52.
Senseny, P.E., Hansen, F.D., Russell, J.E., Carter, N.L. and Handin, J.W. 1992. Mechanical behaviour of rock salt: phenomenology and micromechanisms. *Int. J. rock Mech. Min. Sci. & Geomech. Abstr.* 29:363-378.

Rock Mechanics, Nelson & Laubach (eds) © 1994 Balkema, Rotterdam, ISBN 90 5410 380 8

Thermomechanical damage recovery parameters for rocksalt from the Waste Isolation Pilot Plant[1]

Nancy S. Brodsky
RE/SPEC Inc., Rapid City, S. Dak., USA
Darrell E. Munson
Sandia National Laboratories[2], Albuquerque, N. Mex., USA

ABSTRACT: Laboratory healing tests were conducted on specimens of intact WIPP salt that were predamaged by straining under constant strain-rate triaxial stress conditions at low confining pressure (0.5 MPa) and near room temperature (25°C). Introduction of dilatant damage caused specimen volumes to increase and ultrasonic compressional wave velocities and amplitudes to decrease. Specimens were then loaded hydrostatically to 15 MPa at temperatures up to 70°C. Under these conditions, damage was partially recovered and specimen volume and compressional wave characteristics approached the values of the undamaged state. From these data, time constants and an activation energy were determined for the damage recovery process. The results have significant implications for sealing systems because they demonstrate that the healing process is quite rapid compared to the creep closure process.

1.0 INTRODUCTION

The Waste Isolation Pilot Plant (WIPP) is a research and development facility sited in a bedded salt formation in southeastern New Mexico. The purpose of the facility is to aid in the development of the technology necessary for the safe disposal of radioactive transuranic waste from the defense programs of the U.S. Excavation of rooms and shafts at the WIPP will induce deviatoric stresses in the surrounding rock, resulting in inward creep deformation which may also produce a microcracked zone (Holcomb, 1988) referred to as the Damaged Rock Zone (DRZ). Formation of a DRZ has the potential to increase the permeability of the salt adjacent to the excavations and to degrade the effectiveness of an emplaced seal system. Although initial deformation is expected to form the DRZ, it is anticipated that emplaced seals will take up the overburden load so that further creep damage will stop and the subsequent repressurization of the salt surrounding the seal will close void spaces in the DRZ, lowering its permeability and limiting the potential migration of radionuclides. In evaluating seal design and performance it is necessary to predict the behavior of the DRZ in the seal system and the kinetics of the healing of the repressurized fractures initially formed in the DRZ. Although the mechanics of this damage recovery process have not been studied in detail, the work described here is a preliminary attempt to define some of the thermomechanical quantities that control recovery.

---

[1]This work supported by U.S. Department of Energy (DOE) Contract DE-AC04-94AL85000
[2]A U.S. Department of Energy facility

Very few studies have documented the process of void closure or crack healing. Costin and Wawersik (1980) measured fracture toughness by inducing damage in short rod specimens of salt, pieced the specimens back together, and remeasured fracture toughness after subjecting the specimens to hydrostatic pressures, at two temperatures, for various lengths of time. Typically, specimens pressurized at 10 to 35 MPa regained 70 to 80 percent of their original fracture toughness. Ultrasonic wave velocities and attenuation have been used extensively to indicate crack opening and closure in geologic materials (Birch, 1960; Soga et al., 1978; Sobolev et al., 1978; Granryd et al., 1983). This technique was applied qualitatively to the study of crack closure and healing in salt by Brodsky (1990). In that study, compressional wave ultrasonic data were used to assess the extent of crack closure during hydrostatic compression of damaged WIPP salt specimens at 20°C. The two test variables studied were damage level and the hydrostatic pressure used for damage recovery. It was found that the recovery of ultrasonic velocities depended on pressure. As expected, the higher the pressure, the greater the velocity recovery during crack closure and healing. Recovery of ultrasonic velocity and amplitude also depended on damage level. Recovery was more complete in specimens with the least damage and it was concluded that recovery is slower when damage is sufficient to cause changes in the geometry of the crack walls.

This report extends the earlier work of Brodsky (1990) to the elevated temperature regime. The laboratory experiment measured damage recovery quantitatively in the form of crack closure and healing. Specimens were predamaged by straining under constant strain-rate triaxial conditions at low (0.5 MPa) confining pressure and near room temperature (25°C) until 1.5 percent axial strain was reached. Damage recovery, as indicated by the recovery of volume strains and of ultrasonic compressional wave velocities and amplitudes, was measured as a function of time at a hydrostatic pressure of 15 MPa and at temperatures of 20°C, 46°C, or 70°C. Introduction of dilatant damage caused specimen volume to increase and compressional wave velocities and amplitudes to decrease. During damage recovery under confining pressure, these trends reversed and specimen volume and compressional wave characteristics approached the values of the initial undamaged state. The data from each test were fitted using a first order kinetics law to determine characteristic time constants for damage recovery at each temperature. The rates of damage recovery determined at different temperatures were fitted to an Arrhenius plot to determine an activation energy for the recovery process.

2.0 SPECIMENS

Specimens were prepared from two separate specimen stocks (MCE and C1X01) of clean salt, both obtained from the same stratigraphic layer in the WIPP underground facility. Most specimens were taken from the east rib of Room M (MCE), and have test numbers beginning with LDR (Loading and Damage Recovery). Data from one test, reported previously by Brodsky (1990), used a specimen taken from the rib of experimental Room C (C1X01) and has a test number beginning with LCH (Loading and Crack Healing). The locations from which specimen stocks were obtained were less than 38 m apart.

3.0 EXPERIMENTAL TECHNIQUE AND DATA REDUCTION

Tests were performed in a computer-controlled, servohydraulic test system shown in Figure 1. A controller was used for the actuator servoloop, and a servocontroller was used for maintaining confining pressure. A microprocessor provided data acquisition and programmable control.

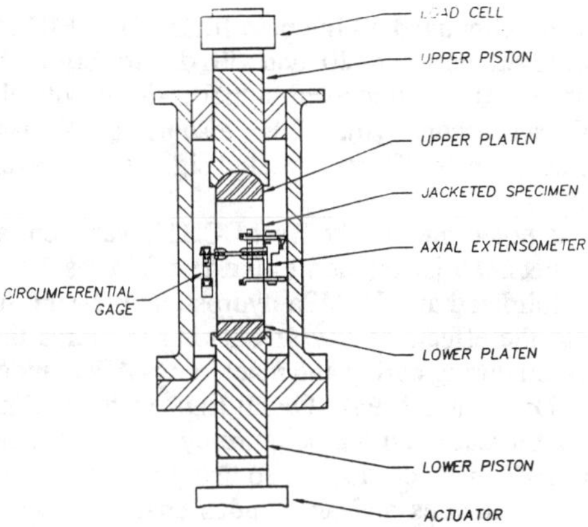

Figure 1. Loading column and pressure vessel.

Specimens were instrumented with direct contact axial and roller-link circumferential extensometers as illustrated in Figure 1. During testing, the axial extensometer and the chain of the circumferential extensometer were centered with respect to the specimen midheight. The influence of confining pressure and temperature on the outputs of the strain extensometers was measured before testing by calibration, as described by Senseny (1987). Thin 0.127 mm (0.005 in) teflon sheets that adhered to the platens were placed at both ends of the specimen to minimize end effects. Uniformity of deformation was checked by comparing post-test diameter measurements in two perpendicular directions made at equally spaced intervals along the specimen length. For all tests the post-test diameter measurements were uniform to within 0.6 percent of one another.

The ultrasonic velocity measurement system is shown schematically in Figure 2 and includes two pairs of compressional wave velocity transducers. One set of velocity transducers was mounted in the platens above and below the specimen for measurements parallel to the specimen axis, and a second set of transducers was held by springs against the sides of the specimen for measurements perpendicular to the axis. The second set of transducers was located just above the specimen midheight to allow room for the circumferential extensometer. All transducers had a 500 kHz resonant frequency.

Viton jackets were used to shield the specimen from the silicon oil confining fluid. Holes, or ports, were cut out of the Viton jackets at the velocity transducer locations and the locations required for the circumferential extensometer. In the holes, 0.051-mm-thick (0.002 in.) steel shims were bonded to the Viton and the assembly was held tightly against the specimen by the confining pressure.

During velocity measurements, a 250-V pulse generated by the pulser/receiver was sent through a switching box to one of the pulsing transducers. Simultaneously, the pulser/receiver sent a 1-V signal to the external trigger of the oscilloscope. The main pulse traversed the specimen, and the received signal was preamplified before it was

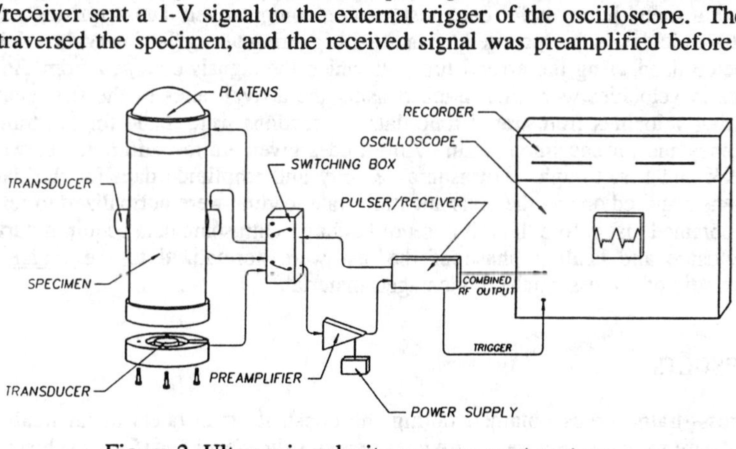

Figure 2. Ultrasonic velocity measurement system.

sent to the pulser/receiver. The preamplifier contained a high pass filter of 0.3 MHz to eliminate noise. All signals were amplified between 20 and 50 dB to bring the maximum expected amplitudes to within ± 1.0 V. Received signals, along with the trigger signal were recorded on a digital oscilloscope with a 20ns resolution. Voltage resolution was ± 0.4 percent of full scale or ± 4-mV. Thirty repeated signals were averaged to increase the signal-to-noise ratio.

The test matrix is given in Table 1. As noted previously, Test LCH15 was from an earlier study (Brodsky, 1990). Leaking specimen jackets led to the test failures.

Before testing, all specimens were consolidated at a 20 MPa hydrostatic pressure for 2 weeks at room temperature to eliminate the effects of any preexisting fractures that formed during coring and machining (Allemandou and Dusseault, 1993). On all LDR tests, preliminary pressure and temperature cycles to 15 MPa and 70°C were carried out so that initial ultrasonic wave velocities and amplitudes characteristic of the essentially undamaged material at particular pressures and temperatures could be measured and used for normalization of data obtained in the damaged state. The same preliminary sequence of temperature and pressure cycles was applied to all specimens, except the LCH test which was cycled to a pressure of 15 MPa at 20°C.

Table 1. Test Matrix

Test	Closure/Healing Condition	
	Temperature (°C)	Duration
LCH15	20	2.4 days
LDR1	70	45 min
LDR2	70	12 days
LDR3	46	failed
LDR4	46	failed
LDR5	46	7 days
LDR6	25	failed

After completion of preliminary cycling, a constant strain-rate triaxial compression test was conducted at 25°C and 0.5 MPa confining pressure to introduce damage into the specimen. Load was applied at an axial strain rate of approximately $10^{-6}s^{-1}$ until the axial strain reached 1.5 percent. Specimens were unloaded quickly at constant stress rate to a stress difference of 0.25 MPa, which was maintained to provide good axial coupling. Specimens were then slowly heated to the selected temperature for the specific individual test. Once the temperature had stabilized, they were loaded hydrostatically to 15 MPa.

All force, pressure, and displacement transducers were initially calibrated in their operating orientations, and outputs were observed at normal data collection points. Calibrations were verified at the conclusion of the testing program.

Time, axial force, axial strain, circumferential strain, temperature, and pressure were recorded at timed intervals for each test using a computerized data acquisition system. The ultrasonic data were recorded by manual triggering at regular intervals. Velocities were determined using the arrival times at which the signals emerged from the noise. Changes in velocities were then tracked using the arrival times of the first peaks. In calculating velocities from travel time data, corrections were made for the transducer delay times and for specimen strain. Amplitudes given are the difference between the first peak and first trough. Ultrasonic velocity and amplitude data for the damaged specimens acquired during the constant strain-rate loading were normalized to reference values obtained just before the initiation of loading. Ultrasonic data acquired during the crack closure and healing phase of the test were normalized to the initial values characteristic of the essentially undamaged material.

4.0 RESULTS

The stress-strain curves obtained during the constant strain-rate triaxial loading test segments are shown in Figure 3. As noted previously, Test LCH15 was subjected to a

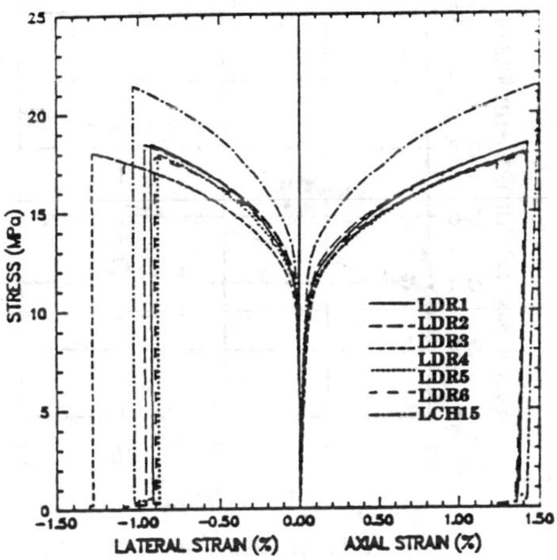

Fig. 3. Stress-versus-strains for all tests.

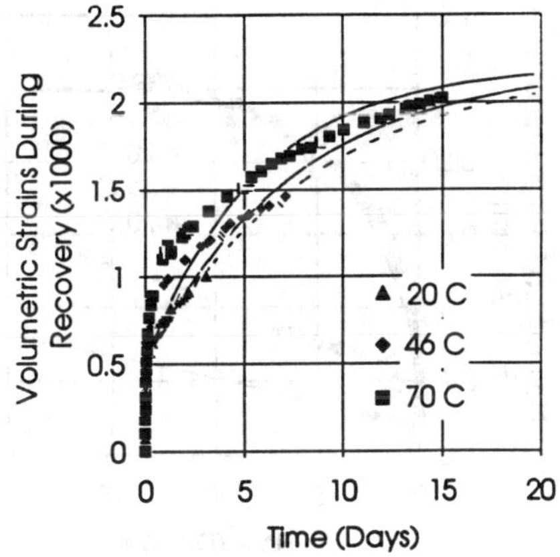

Fig. 4. Volume strains during recovery. Symbols: data; Lines: fits to data.

much simpler preliminary cycling sequence than the rest of the specimens. The LDR tests show very good reproducibility and higher strains at a given stress level than the LCH test.

Changes in volumetric strain occurring during damage recovery are plotted versus time in Figure 4. All curves show decrease of volumetric strain or closure of void volume. Volumetric strain rates increase with increased temperature.

Normalized amplitude data for the crack closure and healing phase are given in Figures 5 and 6 for waves traveling perpendicular, $P_{\perp}$, and parallel, $P_{||}$, to the specimen axis, respectively. Amplitudes are normalized to values obtained from the undamaged specimen during preliminary cycling at recovery temperature and pressure conditions. These data show that recovery proceeds faster for tests at higher temperatures. The absence of $P_{||}$ data at 70°C is due to a mechanical problem during testing. Unload/reload cycles are evident for tests at 46°C and 70°C. It was hoped that the low pressure data obtained during cycling could be used to differentiate between the closure of cracks under load and the permanent rejoining of surfaces. However, because jacket leaks terminated many of these tests prematurely, insufficient data were obtained to be used for this purpose.

Normalized velocity data for damage recovery are given in Figures 7 and 8 for waves traveling perpendicular and parallel to the specimen axis, respectively. Velocities were also normalized using data from the preliminary cycles. Velocity data provide a less well defined correlation between recovery rates and temperature.

5.0 DISCUSSION AND DATA ANALYSIS

Time constants for the recovery process were calculated using a first-order kinetics law given in differential form as

$$\frac{dV}{dt} = -\frac{1}{\tau} * (V_{\infty} - V) \tag{1}$$

where *dV/dt* is the rate of recovery of a measured value (e.g., volumetric strain, ultrasonic velocities, or amplitudes), V_{∞} is the parameter value at infinite time, V is the

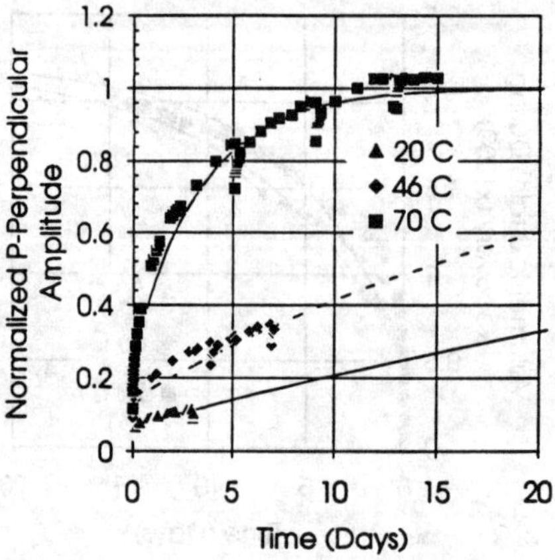

Fig. 5. Amplitudes during recovery. Symbols: data; Lines: fits to data.

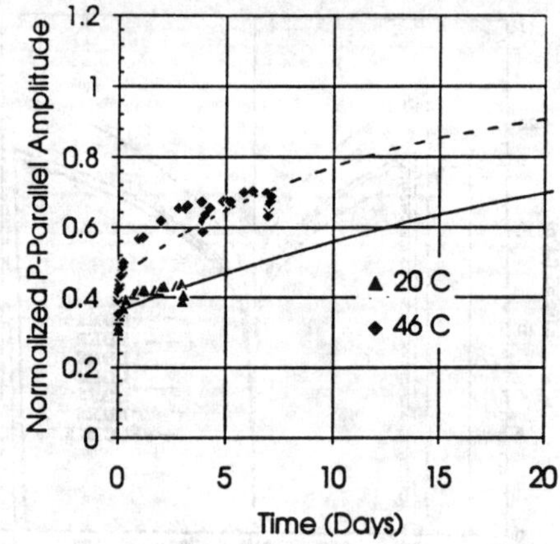

Fig. 6. Amplitudes during recovery. Symbols: data; Lines: fits to data.

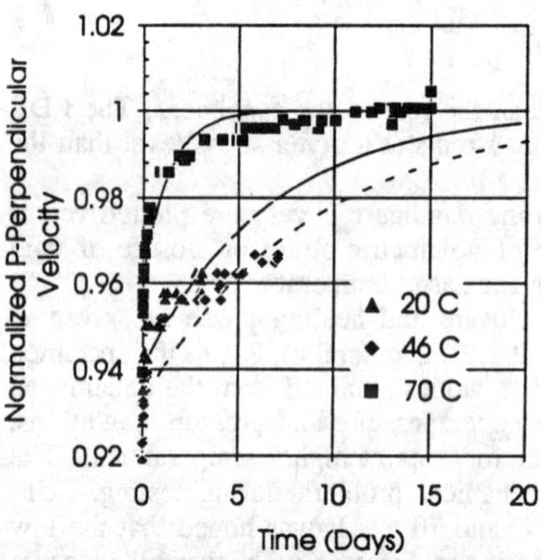

Fig. 7. Velocities during recovery. Symbols: data; Lines: fits to data.

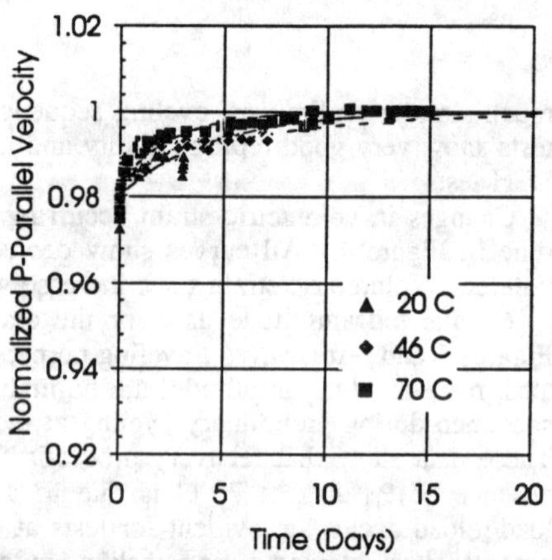

Fig. 8. Velocities during recovery. Symbols: data; Lines: fits to data.

value at time t, and τ is the time constant for the recovery process. This equation describes an exponential decay of the difference between the measured value and its post-recovery value. This equation was integrated as follows

$$V=V_{\infty}-(V_{\infty}-V_0)e^{(-t/\tau)} \tag{2}$$

where V_0 is the value at the start of recovery. Equation (2) was fitted separately for each test and for each of five experimentally determined values: volumetric strain, ultrasonic compressional wave amplitudes perpendicular and parallel to the specimen axis, and compressional wave velocities perpendicular and parallel to the specimen axis. For the ultrasonic data, the fitting was performed under the constraint that post-damage-recovery experimental values (V_{∞}) asymptotically approached those of the initial values of the essentially undamaged state. For the volumetric strains, the data provided very

Table 2. Calculated Time Constants

Test	Time Constant (Days)		
	From Velocity Data	From Amplitude Data	From Volumetric Strain Data
LCH15 (20°C)	5.4[a]	43[a]	8.3
LDR5 (46°C)	8.1[a]	19[a]	7.5
LDR2 (70°C)	2.0[a]	3.4	5.6

(a) Mean Value

little constraint on the fits, and so the fits were forced to the same starting point (a strain of 0.0005) and asymptote (a strain of 0.0022). The fits to the data are shown in Figures 4 to 8 and the fitted time constants are summarized in Table 2. Time constants calculated from amplitude and strain data decrease as temperature increases, as expected for a thermally activated process; however, time constants determined from velocity data show no systematic change with temperature. All time constants are very short; i.e., on the order of days.

The volumetric strain data for each temperature were fitted to the following simple model so that an activation energy for the recovery process could be approximated:

$$\dot{\varepsilon}_v = A e^{\left(-\frac{Q}{RT}\right)} \tag{3}$$

where $\dot{\varepsilon}_v$ is the initial time rate of change of volumetric strain, A is a constant, Q is the activation energy, R is the gas constant, and T is temperature. Initial volumetric strain rates were used because structures were most equivalent at the start of damage recovery. Linear least squares fitting on a ln($\dot{\varepsilon}_v$)-versus inverse temperature plot (Arrhenius plot) provided an activation energy of 10.5 kJ · mole^{-1} (see Figure 9). As would be expected, this is considerably less than the range of creep values: 105 kJ · mole^{-1} for dislocation climb and 41.8 kJ · mole^{-1} for the undefined mechanism of creep, respectively (Munson and Dawson, 1979). Although thermally activated healing is possibly related to a surface process, no specific process could be established at this time.

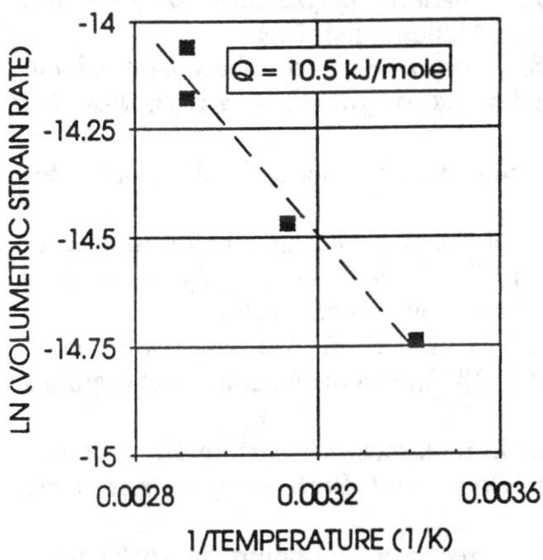

Fig. 9. Arrhenius plot for calculation of activation energy.

6.0 SUMMARY AND CONCLUSIONS

Laboratory experiments were conducted in which introduction of damage and damage recovery were observed in intact WIPP salt specimens. Damage was introduced during constant strain-rate tests at low confining pressure and damage recovery occurred during subsequent hydrostatic pressurization at elevated temperature. Throughout each test, axial and lateral strains were monitored as were ultrasonic compressional waves traveling both perpendicular and parallel to the direction of maximum principal stress.

Data were modeled using a first-order kinetics law and applying fitting constraints. Ultrasonic data obtained at 70°C recovered to values characteristic of the undamaged state within days. Based on the modeling results, time constants associated with the recovery process at all temperatures are short, implying that the rate of crack closure of the DRZ under confining pressure will be rapid. Rates of recovery obtained at different temperatures were used to determine an activation energy for the recovery process of 10.5 kJ $\cdot$ mole^{-1}.

The implications of the rapid recovery of the damaged specimens under confining pressure are significant to seal performance evaluation. Clearly these results indicate that once the seal system is repressurized by the creeping salt, the DRZ will reclose and heal quite rapidly; i.e., on a short time frame compared to the creep closure process.

7.0 REFERENCES

Allemandou, X. & M.B. Dusseault, 1993. Procedures for cyclic creep testing of salt rock, results and discussion. *Proc. 3rd Conf. on Mech. Behav. of Salt, Palaiseau*:193-204.

Birch, F. 1960. The velocity of compressional waves in rocks to 10 kilobars, part 1. *J. of Geophys. Res.* 65:1083-1102.

Brodsky, N.S. 1990. *Crack closure and healing studies in WIPP salt using compressional wave velocity and attenuation measurements: test methods and results,* SAND90-7076, Sandia National Laboratories, Albuquerque, NM.

Costin, L.S., and W.R. Wawersik 1980. *Crack healing of fractures in rock salt.* SAND80-0392, Sandia National Laboratories, Albuquerque, NM.

Granryd, L., I.C. Getting, and H. Spetzler 1983. Path dependence of acoustic velocity and attenuation in experimentally deformed westerly granite. *Geophys. Res. Let.* 10:71-74.

Hadley, K. 1976. The effect of cyclic stress on dilatancy; another look. J. Geophy.Res. 81:471-2474.

Holcomb, D.J. 1988. Cross-hole measurements of velocity and attenuation to detect a disturbed zone in salt at the Waste Isolation Pilot Plant. *29th U.S. Symp. on Rock Mech., Minneapolis, 13-15 June 1988:* 633-640. Rotterdam:Balkema.

Munson, D.E. and P.R. Dawson 1979. *Constitutive model for the low temperature creep of salt (with application to WIPP)*, SAND79-1853, Sandia National Laboratories, Albuquerque, NM.

Senseny, P.E. 1987. Deformation measurements in rock mechanics using direct-contact extensometers. *Proc., 4th Annual Hostile Envir. and High Temp. Meas. Con., Windsor Locks:*94-99.

Sobolev, G., H. Spetzler, and B. Salov. 1978. Precursors to failure in rocks while undergoing anelastic deformations. *J. Geophys. Res.* 83:1775-1784.

Soga, N., H. Mizutani, H. Spetzler, and R.J. Martin III. 1978. The effect of dilatancy on velocity anisotropy in westerly granite. *J. Geophys. Res.* 83:4451-4458.

13 Intact rock testing – Influence of pores and fluid transport on properties

Session developers:
Michael Batzle, ARCO Exploration and Production Technology
Jon Holder, The University of Texas at Austin

Rock Mechanics, Nelson & Laubach (eds) © 1994 Balkema, Rotterdam, ISBN 90 5410 380 8

Laboratory testing of Lägerdorf Chalk

Panayiotis Chryssanthakis & Lloyd Tunbridge
Norwegian Geotechnical Institute, Oslo, Norway
Geir Helge Landa
Phillips Petroleum Company Norway, Tananger, Norway

ABSTRACT: The mechanical behaviour and conductivity of intact and fractured chalk have been studied in laboratory tests. The objectives of the tests were: to investigate the effects of fractures on deformation due to pore collapse; and to investigate the effects of shearing and normal load on fracture conductivity and matrix permeability. Cylindrical specimens of Lägerdorf Chalk, approximately 100 mm diameter by 200 mm length, were made, some of which included a natural fracture. The specimens were tested in a triaxial cell with the capability of measuring fracture conductivity, matrix (bulk) permeability, vertical and horizontal deformation.

The results indicated that: changes in the conductivity of a fracture are dependant on the normal stress during fracture shear displacement; the pore collapse mechanism may begin at lower stresses in fractured chalk due to stress concentrations caused by the fractures; and that fracture conductivity may be significantly reduced under high normal stresses and may not recover when stresses are relaxed.

1. INTRODUCTION - OBJECTIVE OF THE STUDY AND TESTS

The laboratory testing conducted at NGI aimed at giving a better understanding of the coupled mechanical and hydraulic behaviour of natural fractures in a typically fractured chalk reservoir. The changes in the state of stress in a reservoir while, for example depleting or injecting, contribute to the complexity of the behaviour. It is of particular interest though to examine what happens to a fracture at stresses sufficient to initiate the pore collapse mechanism. Is the fracture hydraulically closed at these stress levels? What happens when the stresses are reduced during water injection? Is the fracture still hydraulically closed? Another problem that the laboratory testing is trying to address, is the behaviour of the fractured chalk material when undergoing pore collapse. Is there a difference in properties between the fractured material and the intact material?

The laboratory testing was intended to determine the feasibility of performing comparisons of the behaviour between fractured and intact specimens, as will be explained later in the paper. The objectives of the tests were:

- To measure and compare the permeability of the matrices of the fractured and intact chalk specimens.
- To measure and compare the fracture conductivities for the specimens containing natural fractures, under various conditions of normal stress and shear displacement.
- To determine the mechanical properties of fractured and intact specimens (K_0 value, deformation modulus, etc.).
- Determine changes in chalk porosity in the above tests.

A total of ten specimens of Lägerdorf Chalk were tested. A description of this type of chalk and index test results follow in next chapter.

2 INDEX TESTS ON INTACT LÄGERDORF CHALK

A series of index tests have been performed in order to determine the properties of Lägerdorf chalk. A total of ten tests were performed in NGI's helium porosimeter apparatus. The porosity measurements varied between 44.19% and 44.49% with a mean value of 44.35%. The matrix permeability was measured in three of these specimens with values ranging from 5.03 mD to 5.11 mD and mean value of 5.07 mD. The mean uniaxial compressive strength was found to be 2.08 MPa with a mean saturatedwater content of 21.8%.

The mean of 2.1 MPa for uniaxial compressive strength (σ_c) shows that this material is significantly weaker than typical values from a fractured chalk reservoir. The corresponding values for a typically fractured chalk reservoir are between 4 and 40 MPa.

The porosity value of 44.3% (mean) and permeability value of 5.1 mD (mean) indicate that Lägerdorf Chalk is more porous and more permeable than a typically fractured chalk reservoir.

3 TRIAXIAL TESTING EQUIPMENT DESCRIPTION

All tests were performed in NGI's high pressure triaxial cell apparatus which has a specially designed base and top cap for the permeability and fracture conductivity measurements. The cell is capable of testing specimens up to 102mm diameter with a length of diameter ratio of 2.5:1 and the maximum confining pressure that the system can operate now is 70 MPa. The specimens were cylindrical and measured approximately 200mm in length and 100mm in diameter. All tests were performed at ambient temperature (approx. 22°C). Details of the different parts of the triaxial cell can be seen on Figure 1.

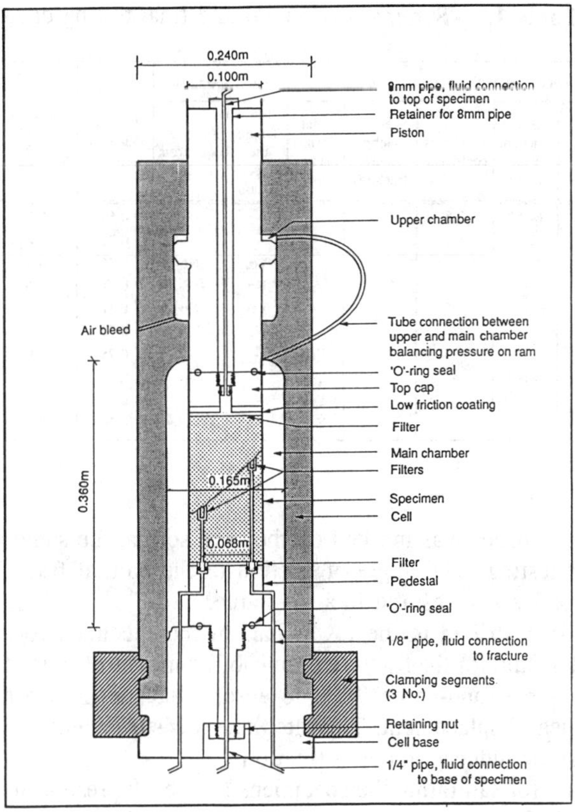

Figure 1. Schematic section of the NGI's triaxial cell.

Internal LVDT's were used for measuring the vertical deformation and strain gauge cantilevers used for the horizontal deformation relative to the top and base pedestals. Circuits and calibrations were checked before each test.

4 SAMPLE PREPARATION

Three types of tests were performed during this testing programme. Two tests were designed to investigate the effect of normal stress and shear displacement on the fracture conductivity of natural fractures; four tests were designed to investigate the effect of fractures on deformation due to pore collapse; and four tests were designed to investigate the effect of normal stress on fracture conductivity. A summary of all tests performed is shown at table 1.

The specimens were drilled out using a thin wall diamond impregnated drill with waterflush. The specimen fracture surface was examined and dip angles over the area were measured, as was the average dip of the fracture. Roughness profiles were made in both the shearing direction and at right angles to this. The

Table 1. Summary of initial and final testing data.

Specimen	Specimen with natural fracture					Intact specimen		$\bar{B}$-value saturation	Initial permability (mD)	Final water content %	Final permability (mD)
	Natural fracture angle dip angle degree	Tilt test friction angle degree	Joint roughness coefficient JRC	Shearing along natural fractures	Joint Normal loading only	Brittle failure after consolidation	K_o conditions from start				
P1	47.5°	64.5°	7.8	Yes	-	-	-	0.60	7.7	25.0	4.4
P2	47.5°	52.5°	5.1	Yes	-	-	-	0.71	3.6	24.8	3.5
P3	nf	-	-	-	-	Yes	No	0.76	4.3	24.4	2.3
P4	nf	-	-	-	-	No	Yes	0.74	4.6	22.4	1.6
P5	nf	-	-	-	-	Yes	No	0.82	3.3	21.1	1.4
P6	nf	-	-	-	-	No	Yes	0.72	3.0	23.2	1.2
P7	44°	-	-	-	Yes	-	-	0.75	7.0	24.8	1.8
P8	46°	>66°	-	-	Yes	-	-	0.76	7.8	27.8	4.6
P9	37°	-	-	-	Yes	-	-	0.70	8.5	-	-
P10	41°	-	-	-	Yes	-	-	0.74	5.5	24.9	3.0

(nf = non-fractured)

location of the profiles was marked on the core so that the same profiles could be taken after testing. The angle of friction of the natural fracture under near zero normal load was measured in a tilt frame.

Holes were then drilled in the lower half of the specimen for installation of pipes to transfer fluid for the fracture permeability measurements. The holes were filled with a coarse sand-epoxy filter to ensure that the fluid pathway remains open during shear displacement. The filter size was sufficient to ensure free fluid flow despite shear induced gouge formation.

The fluid used for saturating the specimens (at a back presure of 0.5 MPa) was a synthetic pore fluid comprised of distilled water and calcium carbonate. The B-value obtained measured at the end of this saturation process varied between 0.6 and 0.82. The tests were thereafter run under drained conditions (Bishop & Henkel 1964).

5 TESTING PROCEDURE - RESULTS

5.1 TEST PROCEDURE PART 1 - RESULTS

Two triaxial tests, designated P1 and P2, were performed on chalk specimens with natural fractures inclined at about 47° degrees. The objective of these tests, was to study the fracture conductivity behaviour while the upper part of the joint was sheared against the lower part (Barton & Choubey 1977).

The tests were performed with constant rate of vertical deformation, the stress path being controlled by changing the horizontal (confining) stress.

The first of the specimens (test P1) was loaded such that the normal stress across the fracture was higher than σ_c (uniaxial compressive strength). The second specimen (test P2) was loaded such that the normal stress was less than σ_c.

The total deformation for P1 specimen measured vertically was 9.5mm, while sliding along the fracture appeared to be about 7mm after the test. Matrix permeability was decreasing when vertical deformation increased in test P1 while it kept constant during test P2. Fracture conductivity was normalized by dividing by the pressure head to give units in mℓ/s/MPa. These are plotted against vertical deformation and presented in Figure 2. Generally matrix permeability tests were carried out by halting deformation for a period and measuring flow rates through the matrix. Fracture conductivity measurements were obtained continuously during shearing.

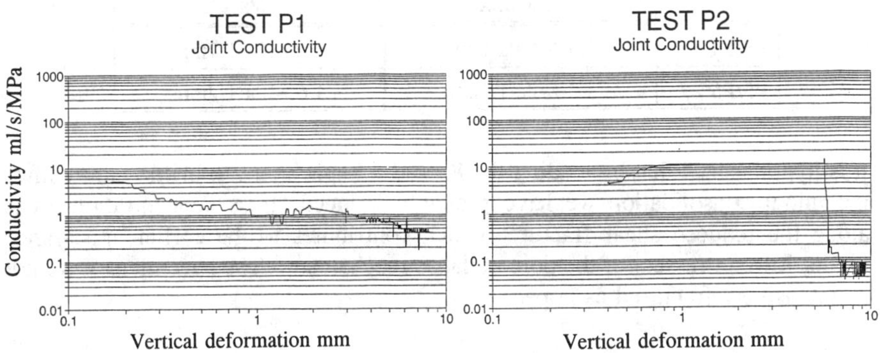

Figure 2. Fracture flow for tests P1 and P2 (conductivity (ml/sec/MPa) versus vertical deformation.

5.2 TEST PROCEDURE PART 2 - RESULTS

The testing of intact specimens comprised two twin pairs of specimens named P3/P4 and P5/P6, where the two specimens in each pair were drilled out side by side in the same block of Lägerdorf Chalk.

Four triaxial tests were performed on *intact* rock samples with dimensions of 200mm length, 100mm diameter. Two of these specimens (P3 and P5) were loaded such that brittle failure occurred and several failure planes were formed. The stresses were then increased such that vertical deformation occurred (about 5 mm) while horizontal (radial deformation) was held constant (uniaxial strain path). The specimens were loaded until the specimens reached pore collapse conditions (plastic failure region). Two other specimens P4 and P6 were loaded after consolidation in such a way that uniaxial strain conditions prevailed throughout the test. The specimens were also loaded until they reached pore collapse conditions (plastic failure region).

Thus, ideally, the difference between each specimen in a pair was the fracturing and some deformation in specimen P3 and P5 before the uniaxial strain path was followed.

Another interesting point in the behaviour of the Lägerdorf Chalk is the point where the pore collapse mechanism is initiated. In Table 2 below, it is observed that the pore collapse mechanism started under lower stresses in the fractured specimen than in the intact specimens. The total vertical deformation at the point where the pore collapse mechanism was initiated, is also less in the fractured specimens than in the intact specimen.

Table 2 The confining pressure and vertical deformation level when the pore collapse process was initiated

Specimen No.	Horizontal stress (MPa)/ Vertical deform. (mm) Fractured	Horizontal stress (MPA)/ Vertical deform. (mm) Intact
P3/P4	4.4 MPa/1.0mm	5.6 MPa/1.7mm
P5/P6	3.3 MPa/1.1mm	4.6 MPa/1.2mm

It is hypothesized that by loading the P3 and P5 specimens to brittle failure after the initial consolidation we have probably induced stress concentration areas along the induced shear fracture, which contributes to the earlier appearance (under lower stresses) of the pore collapse mechanism. The specimens were not so intensively fractured as hoped.

5.3 TEST PROCEDURE PART 3 - RESULTS

Four triaxial tests were performed on chalk specimens with natural fractures inclined between 37° and 46° degrees. The objective of the tests, designated P7, P8, P9 and P10, was to investigate changes in conductivity of natural fractures under cyclic loading. The stress normal to the fracture loading during this testing varied between 0.7 and 10 MPa. These specimens were not subject to shear stress along the fracture. A typical test is shown on Figure 3.

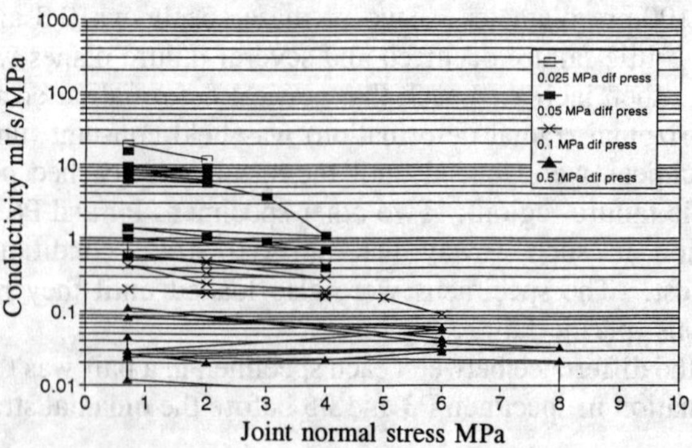

Figure 3. Joint conductivity plot for test P10 (conductivity (ml/sec/MPa) versus joint normal stress.

A typical test specimen was isotropically consolidated at 0.7 MPa, at which point fracture conductivity and matrix permeability were measured.
The specimen was then cyclically loaded. At the first loading interval (2.0 MPa) an elastic recovery of conductivity was typically observed. At the next interval (4 MPa) the elastic recovery of the conductivity was reduced. At the 6 MPa, 8 MPa and 10 MPa intervals the conductivity of the joint was typically dramatically reduced. These stresses were in the plastic region of the failure envelope. There was little recovery of conductivity when stresses were relieved.

An attempt was made to hydraulically fracture the P10 specimen on the last unloading cycle by increasing the injection pressure to greater than the confining stress. No significant increase in conductivity was observed. The fracture surface of this specimen was rather planar, slightly stained, slickensided and possibly remineralized with calcite.

6. CONCLUSIONS

- The behaviour of naturally fractured Lägerdorf Chalk specimens during shear displacement, is strongly dependent on the imposed stress level. The test results indicated that if the normal stress level on the joint exceeded the uniaxial compressive strength (σ_c), both the fracture conductivity and the matrix permeability would be reduced throughout the test indicating joint contraction and fracture of intact material. If the normal stress level on the joint did not exceed the uniaxial compressive strength on the joint, then the fracture conductivity increased, indicating dilation of the joint while shearing. The matrix permeability remained constant during the test.
- The behaviour of the intact chalk specimens showed very little vertical deformation in the elastic region. The pore collapse mechanism was initiated under lower stresses in the fractured specimens than the intact specimens. The fractured specimen exhibited greater total vertical deformation than the intact specimens. This variation was between 32% to 95% of total deformation or about 3.5% vertical strain.
- The tests on the fractured rock specimens (test P3 and P5) generally confirm the results from the numerical UDEC-BB analyses performed also by NGI at an earlier stage where it was seen that fractures caused stress concentrations and redistribution of stresses. The numerical models with greater numbers of fractures and the same boundary conditions, also experienced greater total deformation.
- In the horizontal stress interval between 6 MPa and 8 MPa, the fractured samples were deforming vertically less than the intact specimens. This is believed to be due to the pore collapse mechanism starting under lower stresses on the fractured specimens than on the intact specimens. The K_0 values for zero lateral deformation conditions varied between 0.48 and 0.65. There was not a consistent difference between fractured and intact specimens.
- A relatively good elastic recovery of the conductivity for cyclic loading

that did not exceed the uniaxial compressive strength (0.7 MPa to 2.0 MPa) was observed for the normal closure behaviour of the chalk fractures. When stresses reached the plastic deformation region of the failure envelope, the fracture conductivity was significantly reduced and recovered poorly when stresses were removed. Test P7, though, showed an elastic response of the conductivities up to 8 MPa normal stress on the fracture surface. When the normal stress level reached 10 MPa, no signs of fracture conductivity recovery could be observed when the stress was reduced.

7 IMPLICATIONS FOR A FRACTURED CHALK RESERVOIR

If a fractured chalk reservoir material behaves similarly to Lägerdorf and if there are consistent differences in behaviour between fractured and intact material as indicated in these tests, then the implications are that:

- There is significantly greater initial deformation for fractured material than predicted by laboratory tests on intact material.
- The greater deformation occurs at lower stresses and the increase in deformation for a given increase in stress becomes much less for fractured material than for intact material at higher stresses.

The results may nevertheless have a significant potential in predicting that present compaction in a fractured chalk reservoir is more advanced than predicted due to the presence of numerous fractures in the reservoir and none in the laboratory specimens that have been used as a basis for predicting compaction. Numerical modelling performed by NGI in an earlier stage for the same project confirms the "catalytic effect" of fractures in stimulating the onset of pore collapse and early compaction followed by strain hardening. Less final compaction is suggested through the presence of joints.

8 ACKNOWLEDGMENTS

The authors acknowledge permission to publish the above paper from Phillips Petroleum Company Norway and co-ventures including Fina Exploration Norway Inc., Norsk Agip A/S, Elf Aquitaine Norge A/S, Norsk Hydro a.s, Den Norske Stats Oljeselskap a.s, Total Norge A/S, Elf Rep Norge A/S, Elf Rex Norge A/S and Norminol. The opinions expressed here are the opinions of the authors.

9 REFERENCES

1. Bishop A.D. Henkel (1964). *The measurement of soil properties in the triaxial cell*. Published by Edw. Arnold Ltd. London.
2. Barton N. and V. Choubey (1977). *The shear strength of rock joints, in theory and practice*. Rock Mechanics, Springer, Vienna No. 1/2, pp. 1-54. Also NGI publication No. 119, 1978.

Rock Mechanics, Nelson & Laubach (eds) © 1994 Balkema, Rotterdam, ISBN 90 5410 380 8

A study of the coupling between mechanical loading and flow properties in tuffaceous rock

Runqi Wang & John M. Kemeny
Department of Mining and Geological Engineering, University of Arizona, Tucson, Ariz., USA

ABSTRACT: Coupled mechanical/flow tests were performed on a number of samples of Apache Leap tuff. Both the axial and lateral permeabilities were measured while samples were subjected to uniaxial loading. Measurements were made in the linear portion of the stress-strain curve, during unloading, in the strain-hardening region near the peak stress, and in the strain-softening region after macroscopic fractures had formed. The axial and lateral permeability tests show very interesting relationships between mechanical loading and hydraulic conductivity. The axial permeability (permeability in the direction of uniaxial loading) decreases dramatically with mechanical loading in the linear-elastic portion of the stress-strain curve, due to the axial closing of cracks and pores. On unloading the permeability returns to near the original level. In the strain-hardening region, the axial permeability increases due to the formation of microcracks, and in the strain-softening region, the axial permeability of post-failure specimens is 30 times higher than that of the original unstressed samples. The lateral permeability (permeability perpendicular to the direction of uniaxial loading) shows some significant differences with the axial flow results. In the linear-elastic portion of the stress-strain curve, the lateral permeability increases due to the propping open of cracks and pores perpendicular to the direction of maximum compression. Also, the lateral permeability is noticeably higher than the axial permeability in the strain-hardening and strain softening regions.

1. INTRODUCTION

Yucca Mountain, Nevada is the proposed site for the underground storage of high-level civilian nuclear waste. The site is located in Topopah Spring tuff at a depth of approximately 350 meters. The host rock surrounding the underground repository will be subjected to mechanical loads from the in-situ and excavation-induced stresses, and thermal stresses from the emplacement of the nuclear waste. Based on the design as given in the SCP (DOE, 1988), the waste canisters will heat up the surrounding rock to a maximum temperature of 206°C twenty years after emplacement, followed by a rock temperature above 100°C for over 1000 years. This thermal load will result in a thermal stress in the rock, which, in conjunction with the mechanical loading, could change the near-field and far-field flow properties of the rock.

Rock is a very heterogeneous geological material, containing natural discontinuities and weaknesses at all scales. These include grain boundaries, pores, and pre-existing cracks on the small scale, and joints, faults, and bedding planes on the large scale. When rock is subjected to mechanical loading, these pre-existing discontinuities and weaknesses can close, open, slide, or induce crack growth, which can in turn change the structure of the rock and alter the fluid flow properties of the rock. In particular, apertures of pre-existing fractures and pores may be reduced due to an increased normal stress perpendicular to the fractures, apertures along pre-existing fractures and pores may be increased due to slip along the fractures or a reduced normal stress perpendicular to fractures, and cracks may grow and interact with pre-existing fractures to form high aperture fracture networks. In the regions where pre-existing cracks and pores close, the hydraulic conductivity will be reduced, and in the areas where crack and pore apertures are increased or fracture networks are created, the hydraulic conductivity will be increased. Also, these effects will be directional and result in anisotropic flow properties.

These stress-induced changes in flow properties could impact the underground storage of nuclear waste in many ways. First of all, in the rock immediately around the waste canisters, it has been shown that crack initiation and growth are possible, potentially creating a highly permeable "damage zone" around the canisters (Kemeny and Cook, 1991). At the same time, compressive tangential stresses in the rock around the canisters may cause the permeability of pre-existing fractures to be reduced. On the large scale, the combination of thermal and mechanical stresses around the emplacement drifts may cause certain flow paths to be shut down and others to be enhanced, resulting in significant changes in the far-field flow field.

Laboratory tests have been performed to measure the axial and lateral permeabilities of rock specimens subjected to differential compressive stresses. These include measurements in the initial linear portion of the stress-strain curve, during unloading and reloading cycles, and most importantly, measurements in the strain-hardening and strain-softening portions of the stress-strain curve where crack propagation and fracture formation are occurring. Samples of unwelded Apache Leap tuff from Superior, Arizona have been used, which is considered to be an analog rock to the host rock at Yucca Mountain. Special experimental facilities have been developed for the coupled mechanical/flow tests. The experimental facility for the axial permeability measurements utilize cylindrical specimens and the permeability is measured through specially-designed porous metal disks. The experimental facility for the lateral permeability tests use specially prepared rectangular rock samples. Plexiglass plates are glued to two sides of the specimen, and the lateral permeability is measured through tubes connected to the plexiglass plates. Both the axial and lateral permeabilities can be measured while specimens are subjected to both axial and confining stresses. Only the results of axial and lateral permeability measurements under uniaxial loading are presented in this päper. Coupled mechanical/flow tests under triaxial conditions will be carried out in the near future and reported at a future date.

The experimental results show very interesting relationships between mechanical loading and the flow properties along various portions of the complete stress-strain curve. The axial permeability (permeability in the direction of the uniaxial load) decreases with mechanical loading in the linear portion of the stress-strain curve, due to the axial closing of pre-existing cracks and pores. On unloading, the permeability returns to near the original level. In the strain hardening region, the axial permeability increases due to the formation and propagation of microcracks. In the stain-softening region, the axial permeability increases dramatically due to the formation of macroscopic fractures. In contrast, the lateral permeability (permeability perpendicular to the direction of maximum compression) increases with mechanical loading in the linear-elastic portion of the stress-strain curve due to the lateral propping open of pre-existing cracks or pores. Also, the lateral permeability becomes higher than the axial permeability in the strain-hardening and strain-softening regions. Details of these results will be presented in the next two sections.

2. EXPERIMENTAL TECHNIQUES

2.1. Rock Type and Specimen Preparation

The rock type used in this study was Apache Leap tuff from Superior, Arizona. Tuff is a kind of volcanic ash that has been compressed under high temperature during deposition. The mechanical properties are very different depending on the degree of welding, and based on degree of welding, Apache Leap tuff is divided into five different units (from top to bottom): white, gray, brown, vitrophyre and basal tuff. Samples of both the brown and white units have been collected for the coupled mechanical/flow experiments. Based on its mineralogy and low porosity and permeability, the densely welded brown unit of Apache Leap tuff is considered a reasonable analog to Topopah Springs tuff, which is the proposed host rock for underground nuclear waste storage at Yucca Mountain (Fuenkajorn and Daemen, 1991; Wang and Kemeny, 1992). The physical and mechanical properties of the welded brown unit Apache Leap tuffs have been reported by Wang and Kemeny (1992). The unwelded white unit Apache Leap tuff has been collected from the same location. It has a similar mineral composition but higher porosity and lower strength than the brown unit tuff. The dry uniaxial compressive strength of the white unit is 41 MPa compared with 147 MPa for the welded brown unit. In this paper, all specimens are tested fully saturated. The fully saturated wet uniaxial compressive strength of unwelded white unit tuff is 12 - 24 MPa (12 MPa tested from rectangular specimens, 24 MPa from standard cylindrical specimens) which is roughly 30 - 60% of the standard dry strength.

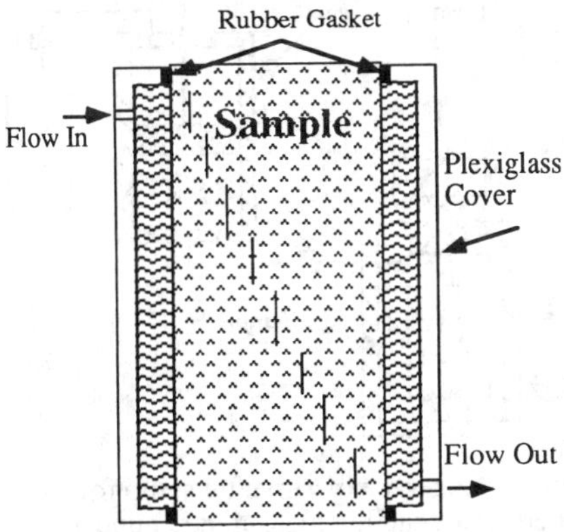

Fig. 1. Specimen for lateral permeability test.

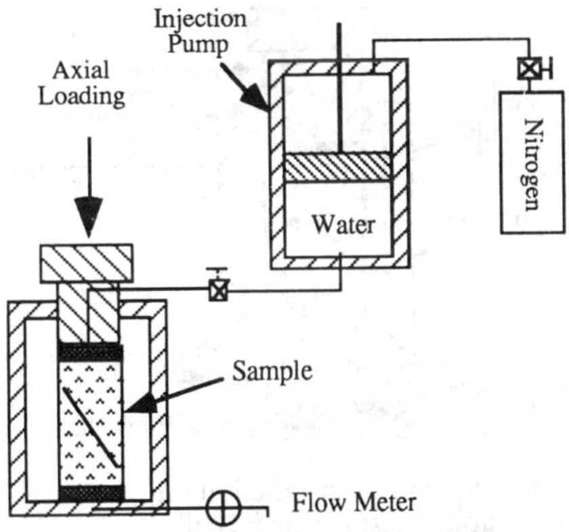

Fig. 2. Experimental setup for axial permeability testing under uniaxial loading.

Two kinds of tuff specimens were prepared. For the axial permeability tests, cylindrical specimens 50 mm in diameter and 125 mm in length were prepared (length to diameter ratio 2.5), similar to specimens for standard triaxial tests. For the lateral permeability tests, special rectangular specimens with dimensions 25×40×125 mm^3 were prepared. Two 1/8 inch thick 40×125 mm^2 plexiglass plates were prepared for each rectangular specimen along with two 1/8 inch thick rubber gaskets, as shown in Figure 1. Tubes are connected to both of the plexiglass plates and two water reservoirs are therefore formed on both sides of the specimen. By applying water pressure to one side of the reservoir, water penetrates though the specimen to the other reservoir. By measuring the input and output flow rate from one side of the specimen to the other, the lateral permeability is calculated. Care is taken in applying the correct water pressure. High pressures will induce tension between the rock and rubber gasket, which could cause leaking during the test. For the lateral permeability tests under uniaxial loading, the maximum water pressure used was 5 psi. Even though not reported in this paper, the specimens can also be tested under triaxial stress conditions. The lateral confinement under triaxial stress conditions is expected to reduce the problem of leakage between the rock and rubber gasket.

2.2 Experimental Facilities and Test Procedure

The experimental system for the axial permeability measurement is shown in Figure 2. The system includes a triaxial cell and a water injection pump. The detailed structure of the triaxial cell and specimen installation is illustrated in Figure 3. The cell has a piston that is hollow to allow internal pressure to be applied and controlled independently. The hollow piston allows water under pressure to axially flow through the specimen while subjected to both axial loading and a confining pressure. Shrinkage tubing is used to seal the sample. Two specially designed porous metal disks are placed at both ends of the rock specimen to allow the water to be evenly distributed across the sample. The water injection pump utilizes pressured nitrogen to provide the water head, and the water flow from the injection pump has the same pressure with the applied gas. In order to avoid leakage between the specimen and shrinkage tube, the water pressure is carefully controlled. For the uniaxial test results presented in this paper, the applied pressure is less than 30 psi. A constant room temperature is maintained throughout the experiment.

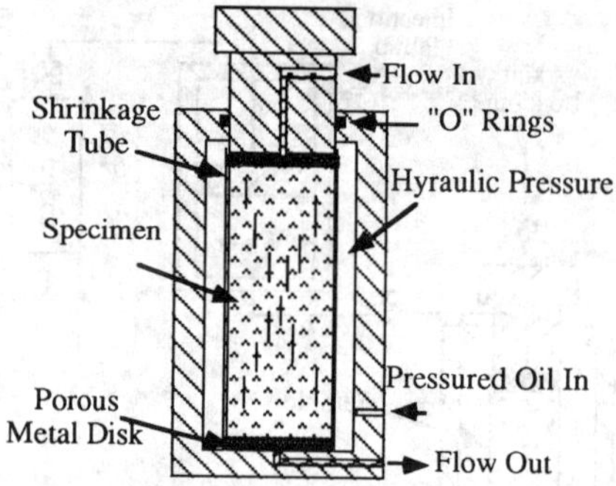

Fig. 3. Specimen installation for axial permeability test in triaxial cell.

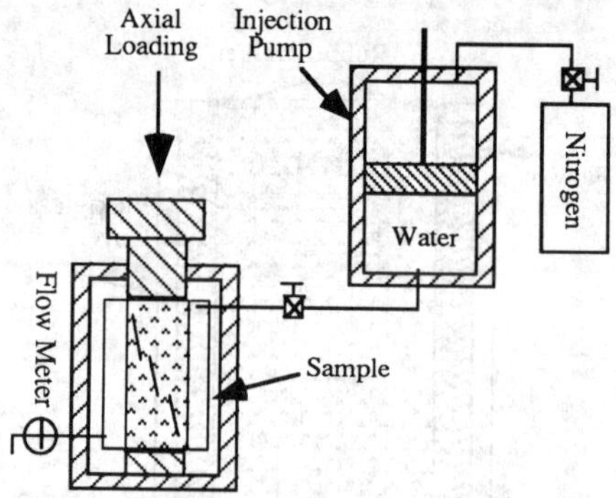

Fig. 4. General experimental setup for lateral permeability measurement under axial loading.

The procedure for conducting the coupled mechanical/flow test is similar to the procedure for a normal triaxial compression test. The specimen with shrink tubing is placed in the testing frame. A small amount of axial load is applied before applying water pressure to the specimen. Axial load is applied with an SBEL CT-500 loading frame controlled by an SBEL 447 servo controller. A load cell and a linear vertical displacement transducer (LVDT) are used to record the displacement-load curve during each test. A constant water pressure is applied with an injection pump. The axial loading rate is set to about 10% of the rate normally used in a standard uniaxial compression test. Under the lower loading rate, additional time is available for the sample to reach hydraulic equilibrium. Several load-unload cycles are conducted to measure the permeabilities at different magnitudes of load. When axial load is applied, flow input and output rates from the specimen can be measured by using the flow meter and the scale in the injection pump. The permeability at different loads are then calculated using the flow rate and the cross sectional area of specimen. A test is completed when the sample fails in the strain-softening portion of the stress-strain curve, and this can be detected from either the rapid drop in the load-displacement curve or a dramatic increase in the flow meter reading.

The lateral permeability test system is shown in Figure 4. The major facilities are similar to that of the axial system, including the triaxial cell (optional) and injection pump. The system can be used to measure the lateral permeability under triaxial compressive stresses, even though only the results from uniaxial loading are reported here. Prior to conducting the lateral permeability test the sample with plexiglass plates is checked for leaks. A special plexiglass pressure chamber was designed for specimen leakage checking. The specimen is first placed in the clear plexiglass chamber containing clear water. The plexiglass chamber contains valves that allow a flow test with colored water to be conducted through the sample with plexiglass plates. A leak can easily be detected through the clear water and is then fixed before conducting the coupled mechanical/flow test. The lateral permeability test is conducted as follows. First, the specimen is placed in the loading frame and a uniform displacement rate is applied axially to the specimen. A constant water flow head is applied during the measurement. The pressure head depends on the quality of the sealing, and a maximum pressure of 5 psi is used in all the tests reported in this paper. Cyclic loading is then applied to the specimen until its failure. By knowing the flow rate and cross sectional area, the lateral permeability corresponding to the particular load can be determined along various portions of the stress-strain curve. Lateral permeability measurements are made in the linear-elastic portion of the stress-strain curve, during unloading cycles, in the strain-hardening portion of the stress-strain curve, and in the strain-softening region after specimen failure. The rubber gasket between the sample and plexiglass plates fails at the time that the specimen fails. The permeability of the failed sample is measured by repairing the gasket and making a permeability measurement of the failed specimen with zero axial load.

3. EXPERIMENTAL RESULTS AND DISCUSSIONS

3.1. Axial Permeability Results

Axial permeability tests were performed on cylindrical unwelded Apache Leap tuff specimens according to the procedures described in section II. For each test, a complete stress-strain curve was recorded to help in interpreting the experimental results. All specimens were fully saturated before the tests. The original axial permeability was measured as a reference before applying any load to specimen.

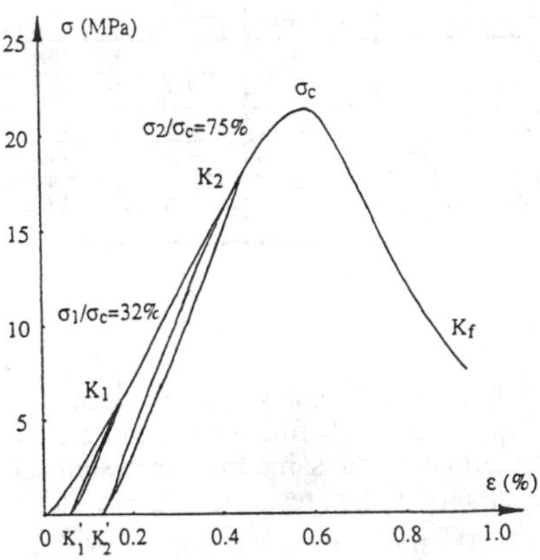

Fig. 5. Complete stress-strain curve of specimen AP-A-1 under uniaxial cyclic loading.

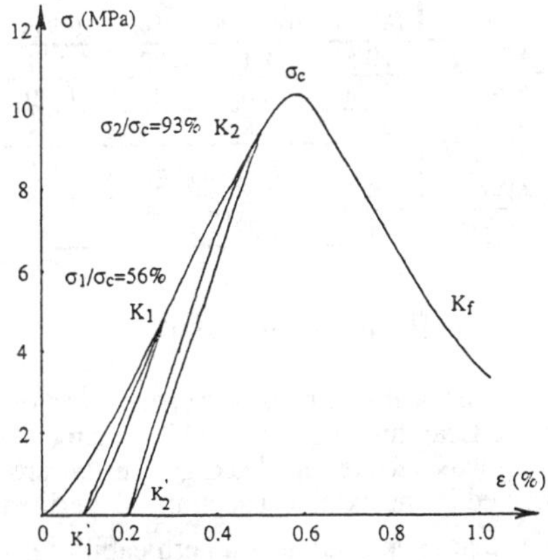

Fig. 6. Complete stress-strain curve of specimen AP-L-3 under uniaxial cyclic loading.

Figure 5 shows a complete stress-strain curve of specimen AP-A-1 under uniaxial cyclic loading. The original permeability K_0 for this sample is 4.1×10^{-6} m/s. A uniform displacement rate was applied in the axial direction, and a permeability K_1 of 0.11×10^{-6} m/s was measured at a load of 32% of the uniaxial compressive strength. This permeability is only 3% of the original value. This dramatic decrease in the axial permeability is due to the axial compression of pores and pre-existing fractures. The specimen was then unloaded, and a permeability K_1' of 2.4×10^{-6} m/s was measured, which is 60% of the original unstressed permeability. The K_1 measurement is still in the linear portion of the stress-strain curve for the sample, and new fractures are unlikely to have formed at this stage. The specimen was then reloaded to 75% of the uniaxial compressive strength. The permeability K_2 measured at this load was 5.6×10^{-6} m/s, which is about 130% of the original value. New fractures have probably been created at this value of load, and this accounts for the increased permeability. After unloading, the permeability K_2' is 6.8×10^{-6} m/s, which is larger than the value under the loading. This is probably due to the reopening of fractures that were closed at the high value of load. The specimen was finally loaded to failure. A post-failure permeability K_f for specimen AP-A-1 of 89×10^{-6} m/s was measured, which is about 20 times the permeability of the original measurement before loading.

Five axial permeability tests were performed, and the results are presented in Table 1. The average initial permeability K_0 is 4.32×10^{-6} m/s, and the average post-failure K_f of 126×10^{-6} m/s is about 30 times higher. Other test results corresponding to various portions of the stress-strain curve are also listed in Table 1. The results indicate a strong coupling between mechanical loading and axial flow (flow in the direction of the maximum compression) in Apache Leap tuff. The permeability is found to decrease in the elastic region (applied load less than 50% of the uniaxial compressive strength), increase slightly in the strain hardening region (> 70% of σ_c), and then increase significantly in the strain-softening region. This correlates with the axial compression of pores and cracks in the linear-elastic region, the growth of isolated fractures in the strain-hardening region, and the formation of connected fracture networks and macroscopic fractures in the strain-softening region.

Table 1 Axial Permeability Test Results

Specimen No.	σ_c (MPa)	K_0 10^{-6} m/s	σ_1/σ_c (%)	K_1 10^{-6} m/s	K_1' 10^{-6} m/s	σ_1/σ_c (%)	K_2 10^{-6} m/s	K_2' 10^{-6} m/s	K_f 10^{-6} m/s
AP-A-1	20.5	4.1	32	0.11	2.4	66	4.0	6.8	89
AP-A-2	17.8	2.8	35	0.09	2.2	77	3.2	7.7	103
AP-A-3	27.4	3.8	44	0.28	3.6	80	4.3	6.3	190
AP-A-4	24.6	6.1	29	0.6	5.4	69	4.5	5.6	113
AP-A-5	29.7	4.8	31	0.08	4.2	81	5.8	9.1	134
Average	24.0	4.32							126

3.2. Lateral Permeability Results

Lateral permeability tests were performed on specially prepared rectangular unwelded Apache Leap tuff specimens. All specimens were fully saturated before testing and gasket sealing was checked for leakage. Tests were performed under the same loading system as described in the axial test section. Experimental results for all five tests are listed in Table 2. The average initial lateral permeability K_0 is 4.34×10^{-6} m/s, which is similar to the initial permeability for the axial measurements. This suggests isotropy for the unstressed hydraulic conductivity for Apache Leap tuff. The procedure for the lateral permeability test follows the one described in section II.

As an example of the lateral permeability results, consider sample AP-L-3. Figure 6 shows the complete stress-strain curve for this specimen under cyclic uniaxial loading. The original permeability K_0 is 3.7×10^{-6} m/s, which is slightly less than the average value. During the first loading cycle, the specimen was loaded to 56% of the peak stress. The permeability K_1 at that value of load is 6.1×10^{-6} m/s, which is 150% of the initial value. This result differs from the axial permeability measurement, and indicates that some lateral opening of pre-existing pores or cracks has occurred. This agrees with theoretical results that predict that cracks can be propped open in the direction perpendicular to the direction of maximum compression, and also with experimental results showing lateral dilatation. When the load is removed, the permeability K_1' returns to the original value (4.1×10^{-6} m/s). A second cyclic loading was then applied to 93% of the peak stress. The permeability

K_2 is 10.3×10^{-6} m/s, which is 3 times the initial value. The stress in the strain hardening region is expected to propagate cracks in the direction of the maximum compression and prop them open in the lateral direction. After unloading, the permeability K_2' decreased to 8.3×10^{-6} m/s which is still 2 times the original value. This indicates that permanent damage was created in the specimen, and this damage contributed to the higher hydraulic conductivity. The lower permeability values after both unloading cycles again supports the idea that cracks are propped open perpendicular to the direction of maximum compression, and relax when the load is removed. The test was terminated at the point of specimen failure. The gasket sealing was then repaired, and a post-failure permeability K_f of 196×10^{-6} m/s was measured, which is 50 times the initial value. The reason for the dramatic increase in the post-failure lateral permeability is illustrated in Figure 7. Two macroscopic fracture zones have been formed, which provide pathways for significant water flow.

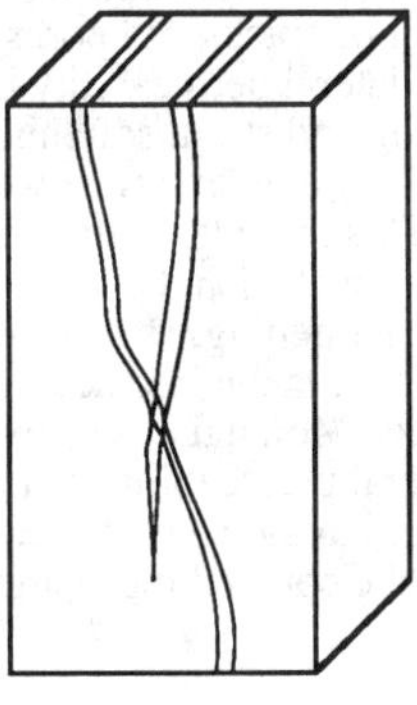

Fig. 7. Post-failure specimen AP-L-3

Table 2 Lateral Permeability Test Results

Specimen No.	σ_c (MPa)	K_0 10^{-6} m/s	σ_1/σ_c (%)	K_1 10^{-6} m/s	K_1' 10^{-6} m/s	σ_1/σ_c (%)	K_2 10^{-6} m/s	K_2' 10^{-6} m/s	K_f 10^{-6} m/s
AP-L-1	12.1	3.7	78	4.1	3.8	91	8.9	6.8	224
AP-L-2	12.4	4.2	61	4.0	4.3	76	7.8	7.7	284
AP-L-3	11.8	3.9	56	6.1	4.1	93	10.3	8.3	196
AP-L-4	13.1	5.1	73	5.6	5.4	80	12.8	8.6	208
AP-L-5	14.2	4.8	69	5.2	4.9	81	15.8	12.1	235
Average	12.72	4.34							229.2

The lateral permeability tests indicate that the permeability perpendicular to the direction of maximum compression can be enhanced in all portions of the stress-strain curve. In the linear-elastic region of the stress-strain curve, it is found that the applied uniaxial stress opens pores or pre-existing cracks in the direction perpendicular to the load. In the strain hardening region, new fractures are formed and the lateral permeabilities are greatly enhanced. The new fractures permanently damage the sample, so that the permeability is enhanced even when the load is removed. All specimens failed by axial splitting similar to the illustration in Figure 7. This mode of failure is expected to give significant increases in permeability since these are macroscopic tensile fractures. Under triaxial loading, shear fractures will most likely be formed, and the increase in permeability may be less pronounced.

4. CONCLUSIONS AND FURTHER WORK

Coupled mechanical/flow tests were performed on a number of samples of Apache Leap tuff. Both the axial and lateral permeabilities were measured while samples were subjected to uniaxial loading. Measurements were made in the linear portion of the stress-strain curve, during unloading, in the strain-hardening region near the peak stress, and in the strain-softening region after macroscopic fractures had formed. The axial and lateral permeability tests show very interesting relationships between mechanical loading and hydraulic conductivity. The axial permeability (permeability in the direction of uniaxial loading) decreases dramatically with mechanical loading in the linear-elastic portion of the stress-strain curve, due to the axial closing of cracks and pores. On unloading the permeability returns to near the original level. In the strain-hardening region, the axial permeability increases due to the formation of microcracks, and in the strain-softening region, the axial permeability of post-failure specimens is 30 times higher than that of the original unstressed samples. The lateral permeability (permeability perpendicular to the direction of uniaxial loading) shows some significant differences with the axial flow results. In the linear-elastic portion of the stress-strain curve, the lateral permeability increases due to the propping open of cracks and pores perpendicular to the direction of maximum compression. Also, the lateral permeability is noticeably higher than the axial permeability in the strain-hardening and strain softening regions.

In summary, a significant coupling exists between mechanical loading and flow properties in Apache Leap tuff. This includes stress-induced anisotropy in the flow properties, and regions of both diminished and enhanced flow depending on whether the rock is in the linear-elastic, strain-hardening, or strain-softening portions of the stress-strain curve. This coupling should be taken into account in the design of the proposed underground waste facility at Yucca Mountain. The experimental system has been designed to measure axial and lateral permeabilities under triaxial stress conditions, and such triaxial tests will be carried out in the near future. Also, micromechanical models are now being developed to simulate the coupled mechanical/flow response in heterogeneous rocks.

ACKNOWLEDGMENT

This work was supported by the Nuclear Regulatory Commission under Grant No. NRC 0493088, and by the Natural Science Foundation Solid and Geomechanics Program under Grant No. MSS 9022381.

REFERENCES

DOE, 1988, Overview, Yucca Mountain Site, Nevada Research and Development Area, Nevada in Site Characterization Plan, U.S. Department of Energy Report DOE/RW-0198.

Fuenkajorn, K. and Daemen, J.J.K., 1991, Mechanical characterization of the densely welded Apache Leap Tuff, Technical Report, U.S. Nuclear Regulatory Commission.

Kemeny, J.M. and Cook, N.G.W., 1991, Micromechanics of deformation in rocks, In Toughing Mechanism in Quasi-Brittle Materials,(Edited by S.P. Shaw), Kluwer Academic Publishers, Dordrecht, Netherlands.

Wang, R. and Kemeny, J.M. 1992. Fracturing Mechanisms in Apache Leap Tuff under Compressive Stress, Fractured and Jointed Rock Masses Conference, Lake Tahoe, CA.

Wang, R. and Kemeny, J.M. 1993a. Micromechanics of deformation in Topopah Spring tuff, Yucca Mountain, Nevada, Proceedings of the International High Level Radioactive Waste Management Conference, Las Vegas, NV.

Wang, R. and Kemeny, J.M. 1993b. Micromechanical modeling of tuffaceous rock for application in underground nuclear waste storage, 34th US Rock Mechanics Symposium, Madison, WI.

Rock Mechanics, Nelson & Laubach (eds) © 1994 Balkema, Rotterdam, ISBN 90 5410 380 8

Deformation of Wilcox shale: Undrained strengths and effects of strain rate

Ohmyoung Kwon & Andreas K. Kronenberg
Department of Geology and Geophysics, Center for Tectonophysics, Texas A&M University, College Station, Tex., USA

ABSTRACT: Mechanical and fluid transport properties of brine-saturated illite shale from the Wilcox Formation have been investigated to examine the effects of strain rate on fracture strength and to determine the conditions over which rates of deformation and internal fluid flow compete. Permeability (k) measurements predict time scales for fluid pressure equilibration that are long relative to those of deformation experiments; values of k are small even at relatively low effective pressures P_e (assumed $\cong P_c - P_f$) and they drop below 10^{-20} m^2 at pressures of ≥ 5 MPa. Failure strengths (σ_1-σ_3) of Wilcox shale specimens saturated (as nearly as can be achieved) with brine and shortened at strain rates of $2.8\times10^{-7} \leq \dot{\varepsilon} \leq 8.2\times10^{-3}$ s^{-1} fit an exponential law $\dot{\varepsilon} = A \exp\{\alpha(\sigma_1-\sigma_3)\}$ of the same form as determined for Wilcox shale specimens deformed at much higher pressures in the absence of pore fluids. The positive slope and comparable value of α (= 0.29 MPa^{-1}) to that determined for specimens without pore fluids ($\alpha = 0.3$ MPa^{-1}) suggest that Wilcox shale exhibits undrained behavior at strain rates > 10^{-7} s^{-1} and that strain rate effects associated with pore pressure perturbations and pore fluid transport occur at much lower strain rates.

1 INTRODUCTION

Clay-rich shales deform readily in a variety of geologic settings in response to tectonic and depositional loads that are relatively small, yet laboratory experiments performed on clay-bearing shales in the absence of pore fluids reveal significant strengths that are comparable to those measured for higher grade rocks such as slates and mica-bearing schists (Shea and Kronenberg 1992, Ibanez and Kronenberg 1993). Low strengths are required of shales that make up basal detachments of thin-skinned thrust sheets (Chapple 1978) as well as shale units that develop into diapir structures (Bishop 1977). Fluid-saturated shales have low strengths primarily due to elevated pore pressures and correspondingly low effective pressures that may be maintained over extended periods of time. Experimental studies on shales indicate that shear strengths of fluid-saturated shales are lower than those of dry shales and undrained compressive strengths are lower than drained strengths (Mesri and Gibala 1972, Van Eeckhout 1976, Bauer 1979, Cook 1989, Steiger and Leung 1990, Nuesch 1991).

Shales may exhibit undrained mechanical response over a wide range of conditions due to permeabilities that are very low. Changes in stress state during deformation experiments can give rise to pore pressure perturbations within specimen interiors that are not reflected in fluid pressures measured at specimen extremities and failure will be governed by deformation in regions of elevated pore pressure. As strain rates are reduced and times available for fluid transport are increased, rates of deformation and fluid flow may compete giving rise to strengths that depend upon strain rate (Swan et al. 1989), and at very low strain rates, shales may exhibit drained response.

A variety of deformation mechanisms may exhibit critical thresholds for their activation that depend on the rate of deformation and previous experiments performed on Wilcox shale specimens at high pressures in the absence of pore fluids reveal an intrinsic effect of strain rate on inelastic strength (Ibanez and Kronenberg 1993). Compressive differential stresses (σ_1-σ_3) measured at confining pressures (P_c) of 250 MPa both at the onset of yielding and at failure increase with increasing strain rate ($\dot{\varepsilon}$) and fit an exponential law $\dot{\varepsilon} = A(T) \exp\{\alpha(\sigma_1-\sigma_3)\}$ where A(T) depends on temperature and α is a material parameter.

Similarly, increases in failure strengths of fluid-saturated Kimmeridge Bay shale specimens were observed with increasing strain rate (Swan et al. 1989) at strain rates ($\dot{\varepsilon} > 10^{-4}$ s^{-1}) sufficiently rapid to preclude loss of any pore fluids during deformation. However, Swan et al. (1989) also reported a transition in mechanical response at lower strain rates with an inverse relationship between strength and strain rate that was inferred to reflect internal fluid flow and the decay of local pore pressure perturbations at longer time scales. Characteristic times over which fluid pressures equilibrate may differ amongst different shales depending upon their permeabilities and the length scales over which fluid flow must occur. The strain rates at which other shales exhibit the transition in mechanical behavior reported for Kimmeridge Bay shale therefore depend upon the complex distributions of interconnected pores that govern permeability and interactions with microcracks that are generated during deformation.

In this paper, we present experimental determinations of fluid flow and fracture strength over a wide range in strain rate for brine-saturated Wilcox shale in order to determine the conditions required for drained, undrained, and transitional mechanical responses. The results suggest that strain rate effects associated with transient pore pressure perturbations occur at very low strain rates (< 10^{-7} s^{-1}), well below those investigated in this study. The observed dependence of failure strength on strain rate appears to reflect the undrained mechanical behavior of Wilcox shale. Remarkably, the undrained strength-strain rate relationship parallels the intrinsic material response of Wilcox shale observed at high confining pressures in the absence of pore fluids.

2 EXPERIMENTAL METHODS

2.1 Starting material

Shale specimens were prepared from cores of the Wilcox Formation recovered from depths of approximately 3,960 m from West Baton Rough Parish, Louisiana. Wilcox shale is rich in illite and chlorite, and clay minerals make up 55% of this shale. Porosities range from 2.5% in the clay-rich sections to 15% in the silty and quartz-rich sections; however, most of the specimens tested in this study are rich in clay and low in porosity. X-ray pole figure goniometry and SEM observations indicate that clay minerals are aligned parallel to the bedding plane with an illite preferred orientation that is comparable to those of many other deeply-buried shales but weak relative to layer silicate preferred orientations of slates and schists. This material has been characterized by Ibanez and Kronenberg (1993) and its mechanical properties in the absence of pore fluids were determined at strain rates of 2.3×10^{-7} to 1.0×10^{-3} s^{-1} at high confining pressures ($P_c \geq 200$ MPa).

2.2 Experimental techniques

Right cylindrical specimens of Wilcox shale were cored parallel to bedding (12.5 mm diameter) and ends ground parallel (specimen length = 25.4 mm). Measurements of permeability and compressive strength in the presence of brine were made at pressure using the same apparatus (Fig. 1). Measurements of permeability along the axes of specimens (parallel to bedding) were made using a simple transient pulse technique (Brace et al. 1968, Sutherland and Cave 1980). Fluid pressures were measured at both

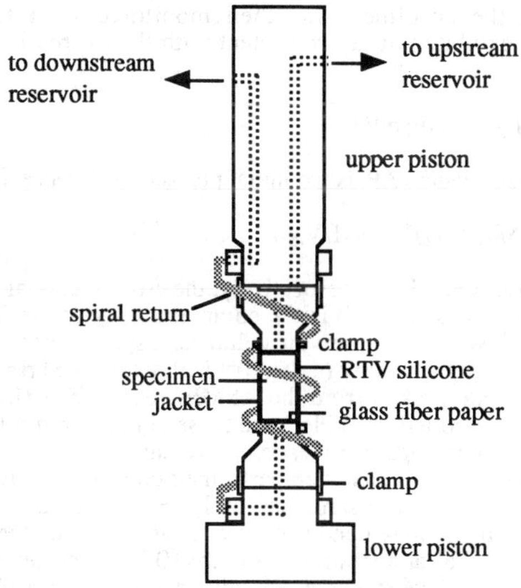

Figure 1. Schematic configuration of specimen assembly used in triaxial deformation apparatus, both for permeability measurements and for deformation experiments.

ends of each specimen by means of two access ports within the top piston, a spiral return line (Fig. 1), and high sensitivity Dynisco PT130 transducers. A small stepwise change in fluid pressure was applied to one end of the specimen and decay in the difference in fluid pressures across the specimen was measured as a function of time. Failure strengths were measured for specimens that were confined laterally and loaded axially (parallel to bedding) with known fluid pressures at specimen ends.

Permeabilities were determined using 1M NaCl solutions that filled the volume of the apparatus fluid pressure system and were introduced into specimen pores by multiple vacuum impregnation procedures performed over several days. A thin coat of RTV silicone was applied on the lateral surfaces of specimens before inserting them into heat shrinkable polyolefin jackets in order to prevent fluid flow between the jacket and specimen. Layers of glass fiber paper were placed on both ends of the specimen to ensure uniform access of pore fluid to the specimen ends. Tests performed on an aluminum specimen using these methods indicate that flow between the lateral specimen surface and jacket does not occur for confining pressures (P_c) that are greater than fluid pressure (P_f) by ≥ 0.5 MPa. Effective volumes of upstream (15 ml) and downstream (24 ml) reservoirs of the fluid pressure system were determined after purging the system of air and these values were used for permeability determinations.

After the specimen assembly was placed within the pressure vessel, the specimen and that portion of the pore pressure system that required disassembly between experiments were placed under a vacuum for 1 hour to remove air trapped in the system. The system was then filled with NaCl solution and bled repeatedly to ensure complete filling by the fluid. After filling the system with NaCl solution, confining pressure (P_c) and fluid pressure (P_f) were increased stepwise over an extended period of time to the desired experimental conditions, taking care that the difference between confining and fluid pressures did not exceed the pressure difference (P_c - P_f) at which permeability was to be measured. Several (2 to 4) days were generally required for fluid pressures to reach equilibrium, depending on the values of P_c and P_f desired. Once fluid pressures achieved equilibrium, the upstream reservoir was isolated from the downstream reservoir and the pressure in the upstream reservoir was increased stepwise at time $t = 0$ by a small amount (ΔP_i). The decay of the fluid pressure difference ΔP (= P_{up} - P_{dn} where P_{up} and P_{dn} are fluid pressures in the upstream and downstream reservoirs, respectively) across the specimen was then monitored as a function of time. Measurements of ΔP exhibited decay consistent with the expression (Sutherland and Cave 1980):

$$(1) \qquad (P_{up} - P_{dn}) = \Delta P_i \, e^{-\alpha t}$$

within experimental error, where ΔP_i is the initial pressure difference at time $t = 0$ and

$$(2) \qquad \alpha = (kA/\mu\beta L)(1/V_{up} + 1/V_{dn})$$

where k is permeability, A and L are, respectively, the cross-sectional area and length of the specimen, μ is the viscosity and β is the compressibility of the fluid, and V_{up} and V_{dn} are upstream and downstream reservoir volumes, respectively. Values of viscosity ($9.9\text{x}10^{-10}$ MPa·s) and compressibility ($4.56\text{x}10^{-4}$ MPa^{-1}) for the brine were assumed to be similar to those of water, taken from the ASME steam tables (Meyer et al. 1979). The fluid pressures (P_f) reported in Table 1 are those at equilibrium with the specimen before applying the step in the upstream reservoir pressure.

Triaxial compression experiments were performed on shale specimens impregnated with 1M NaCl solution, using the same assembly (Fig. 1) as used for permeability measurements. Experiments were conducted over a range of initial effective pressures, P_e ($\cong P_c - P_f = 3$ to 10 MPa) at a strain rate of $1.5\text{x}10^{-5}$ s^{-1} and over a range of strain rates (from $2.8\text{x}10^{-7}$ to $8.2\text{x}10^{-3}$ s^{-1}) at an effective pressure of 5.5 MPa (with $P_c = 10$, 13, and 21.5 MPa, and externally determined $P_f = 4.5$, 7.5, and 16 MPa, respectively). Because pressures were allowed to equilibrate over extended times prior to the deformation experiments, external measures of fluid pressures may provide true values of pore pressures within specimens at the onset of deformation. However, we cannot be certain that internal pore pressures were equivalent to externally measured fluid pressures during the deformation experiments once differential stresses were applied. Specimen dimensions and procedures to fill the pore pressure system were the same as those used for permeability measurements. However, the upstream and downstream fluid reservoirs were not isolated during deformation experiments. Fluid pressure was applied at both specimen ends and fluids were allowed to drain at both ends of each specimen. Fluid pressures were not adjusted during deformation experiments; instead, external fluid pressures were monitored during deformation. All permeability measurements and deformation experiments were performed at room temperature.

3 RESULTS AND DISCUSSION

Representative results used to determine permeabilities are illustrated in Figure 2A and permeability measurements are summarized in Table 1 and Figure 2B. Results of deformation experiments are shown in Figures 3 and 4 and listed in Tables 2 and 3.

3.1 Permeability measurements

Wilcox shale exhibits permeabilities (k) to brine that are very small and long times were required for each measurement of k. The decay in fluid pressure P_{up} and growth of P_{dn} (Fig. 2A) were monitored continuously for each determination of k until the difference between the two dropped below 50% of the initial pressure difference ΔP_i imposed. Daily fluctuations in fluid pressures due to fluctuations in room temperature (±2°) were minimized by insulation that was wrapped around the fluid reservoirs and plumbing. However, the difference between the two pressures ($P_{up} - P_{dn}$) was much less sensitive to temperature fluctuations and values of k were determined from the decline in $\log(P_{up} - P_{dn})$ with time (Fig. 2A) using eqns. (1) and (2).

Permeabilities of rocks are generally observed to vary with effective pressure and

Table 1. Permeability results for Wilcox shale.

Specimen Number	P_c (MPa)	P_f (MPa)	$P_c - P_f$ (MPa)	ΔP_i (MPa)	Permeability ($x10^{-21}$ m^2)
WFS 2-7A	13	10	3	2	141 (±2.5)
WFS 6-1A	15	12	3	1.5	16.8 (±0.2)
WFS 7-6	16.5	13.5	3	1.5	26.9 (±0.4)
WFS 6-1B	16	12	4	1.5	12.9 (±0.1)
WFS 6-1C	17	12	5	1.5	9.0 (±0.3)
WFS 5-9	13	7.5	5.5	1.5	8.9 (±0.2)
WFS 7-2	25.5	20	5.5	2	4.4 (±0.3)
WFS 2-7B	18	10	8	2	2.4 (±0.1)
WFS 6-1D	20	12	8	1.5	6.6 (±0.1)

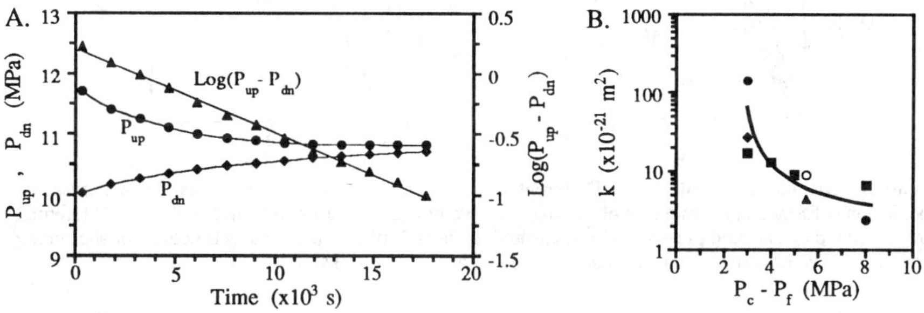

Figure 2. Permeability results. (A) Pressure changes in upstream (P_{up}, shown as circles), and downstream (P_{dn}, as diamonds) reservoirs with time, as well as $\log(P_{up} - P_{dn})$ with time (shown as triangles) for Wilcox shale (WFS2-7A) at $P_c - P_f = 3$ MPa where P_c and P_f are confining and fluid pressures, respectively, before imposing ΔP_i. A permeability $k = 1.41x10^{-19}$ m^2 is given by the slope in $\log(P_{up} - P_{dn})$ vs. t and application of eqns. (1) and (2). (B) Permeability (k) as a function of ($P_c - P_f$). Results shown for experiments performed at fluid pressures of 7.5 (open circle), 10 (filled circle), 12 (square), 13.5 (diamond), and 20 MPa (triangle).

measurements of k for transport in Wilcox shale parallel to bedding (Fig. 2B) show a strong dependence on ($P_c - P_f$), the difference between confining pressure and fluid pressure. The pressure difference ($P_c - P_f$) may not be a true measure of effective pressure for transport properties and values of k measured at the same ($P_c - P_f$) but different P_c show some scatter. Further experiments are needed to explore the dependence of k on P_c and P_f; nevertheless, a general trend is clearly defined and permeabilities drop below 10^{-20} m^2 at ($P_c - P_f$) ≥ 5 MPa (Fig 2B).

Depending upon the scale at which fluid flow occurs, characteristic times for pore pressure equilibration can be determined from values of k (Rice and Cleary 1976, Roeloffs 1988) and used to evaluate whether perturbations in pore pressure associated with deformation are likely to be generated. For pore pressure equilibrium at the specimen scale, time intervals greater than $2x10^5$ s (56 hours) are required (using $k = 8.9x10^{-21}$ m^2 at $P_c - P_f = 5.5$ MPa, specimen WFS5-9, assuming that flow is one-dimensional, eqn. (23) of Roeloffs, 1988, and that fluid pressure perturbations decay by ≥ 90%). Fluid flow at the specimen scale is negligible at times less than $1.5x10^3$ s (25 minutes) and external measures of fluid pressure over time intervals less than this cannot be expected to track pore pressures within the specimen interior. Alternatively, fluid flow may be important at much finer scales associated with equilibration of fluid pressures between nearby pores and dilatant microcracks. For fluid flow over length scales of ~0.5 mm, near-equilibrium can be achieved in times of > 370 s (6 minutes) while fluid pressures may be spatially very heterogeneous at times < 3 s.

Table 2. Mechanical results for Wilcox shale at $\dot{\varepsilon} = 1.5x10^{-5}\ s^{-1}$.

Specimen Number	P_c (MPa)	P_f (MPa)	$P_c - P_f$ (MPa)	$\sigma_1 - \sigma_3$ (MPa)
WFS 2-4	10	7	3	8.0
WFS 7-6	16.5	13.5	3	32.0
WFS 7-5	18	15	3	19.3
WFS 7-1	21.5	16	5.5	37.0
WFS 2-7	20	10	10	47.0

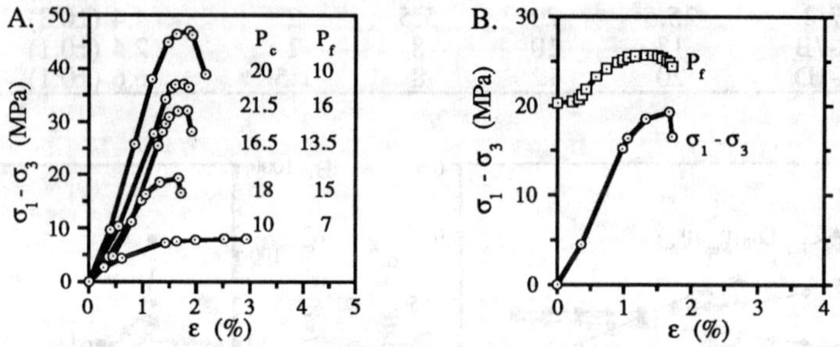

Figure 3. Mechanical results. (A) Differential stress-strain (σ_1-σ_3 vs. ε) curves for Wilcox shale specimens deformed at a strain rate of $1.5x10^{-5}\ s^{-1}$ (values of P_c and P_f given in MPa). (B) Differential stresses (σ_1-σ_3) and fluid pressures (P_f) measured at the ends of the specimen as functions of shortening strain (ε) for Wilcox shale (WFS7-5) at P_c = 18 and initial P_f = 15 MPa.

3.2 Triaxial deformation

Triaxial deformation experiments conducted at a strain rate of $1.5x10^{-5}\ s^{-1}$ over a range of confining pressures and initial fluid pressures reveal a general trend of increasing strength with increasing effective pressure (Fig. 3A) associated with the development of sharply defined shear fractures, similar to results obtained at low confining pressures for Wilcox shale without pore fluids (Ibanez and Kronenberg 1993). Confining and fluid pressures were applied over times sufficiently long prior to the deformation experiments to achieve equilibrium pore pressures at their outset. However, time scales for fluid flow at the specimen scale predicted by permeability measurements exceed the time intervals over which the stress state changes at $\dot{\varepsilon} = 10^{-5}\ s^{-1}$, and the results for three specimens deformed at initial effective pressures of 3 MPa (Fig. 3A) but at different confining pressures show wide variations in strength. Fluid pressures monitored during deformation increased as differential stresses increased and began to decrease near failure (Fig. 3B) indicating that some fraction of the pore fluid was initially expelled and later drawn into specimens during deformation. However, the variations in failure strength at given values of initial effective pressure (P_c - P_f) suggest that specimens deformed at $\dot{\varepsilon} = 10^{-5}\ s^{-1}$ exhibit essentially undrained behavior.

Given that small fluctuations in fluid pressures at specimen extremities were detected at $\dot{\varepsilon} = 10^{-5}\ s^{-1}$, deformation experiments on Wilcox shale were conducted at strain rates from $2.8x10^{-7}$ to $8.2x10^{-3}\ s^{-1}$ to test for effects of transient pore pressures at the same confining and fluid pressures (P_c = 13 MPa and P_f = 7.5 MPa, respectively) as were applied during permeability measurements. The mechanical results of two additional specimens (WFS7-1 and WFS5-7) deformed at an initial effective pressure (P_e) of 5.5 MPa were also used to examine effects of strain rate, although the confining pressures applied during these experiments differed from those of the other experiments (P_c = 21.5 MPa and P_c = 10 MPa, respectively). Combined, these experiments reveal an increase in failure strength with increasing strain rate (Fig. 4A), a result that differs from anticipated strain rate effects associated with pore fluid transport and is more readily explained if specimens remain undrained during deformation over the entire

Table 3. Mechanical results for Wilcox shale at $P_c - P_f = 5.5$ MPa.

Specimen Number	P_c (MPa)	P_f (MPa)	$P_c - P_f$ (MPa)	$\dot{\varepsilon}$ (s^{-1})	$\sigma_1 - \sigma_3$ (MPa)
WFS 7-8	13	7.5	5.5	2.8×10^{-7}	20.5
WFS 5-7	10	4.5	5.5	3.0×10^{-7}	31.1
WFS 5-1	13	7.5	5.5	8.1×10^{-7}	29.5
WFS 7-1	21.5	16	5.5	1.5×10^{-5}	37.0
WFS 5-2	13	7.5	5.5	8.2×10^{-5}	40.2
WFS 5-4	13	7.5	5.5	2.3×10^{-3}	57.9
WFS 5-8	13	7.5	5.5	8.2×10^{-3}	60.8

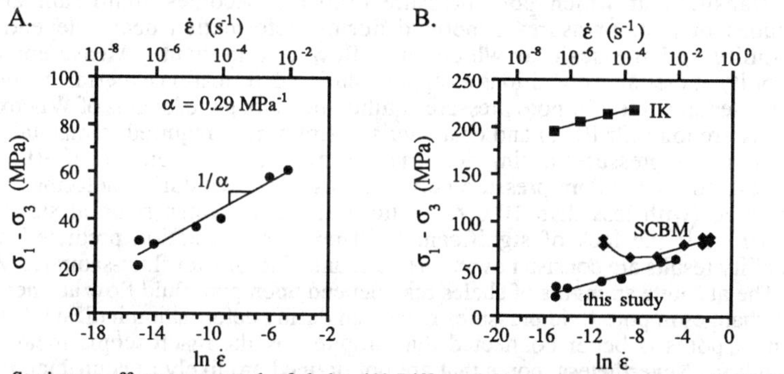

Figure 4. Strain rate effects on strength of shale. (A) Differential stress (σ_1-σ_3) measured at failure vs. ln $\dot{\varepsilon}$ for experiments performed on Wilcox shale specimens at an initial effective pressure (given by externally measured P_f) of 5.5 MPa. Assuming eqn. (3), the slope is given by $1/\alpha$ with $\alpha = 0.29$ MPa^{-1}. (B) Comparison of the results of this study with those for dry Wilcox shale at high confining pressures ($\alpha = 0.3$ MPa^{-1}, indicated by IK, Ibanez and Kronenberg 1993) and for fluid-saturated Kimmeridge Bay shale ($\alpha = 0.27$ MPa^{-1} at high strain rates, indicated by SCBM, Swan et al. 1989).

range in strain rate tested.

Previous investigation of Wilcox shale deformed at high confining pressures without pore fluids (Ibanez and Kronenberg 1993) revealed a relationship between ultimate strength and strain rate given by an exponential glide law of the form:

(3) $$\dot{\varepsilon} = A(T) \exp\{\alpha(\sigma_1 - \sigma_3)\}$$

where A(T) depends on temperature and α is a material parameter. Although our experiments were conducted at relatively low effective pressures, the results over the range of strain rates investigated can similarly be fit by eqn. (3). The results displayed as (σ_1-σ_3) vs. ln $\dot{\varepsilon}$ define a slope (= $1/\alpha$, Fig. 4A) with a value of $\alpha = 0.29$ MPa^{-1} (determined by least-squares methods) that is surprisingly similar to that ($\alpha = 0.3$ MPa^{-1}) determined for Wilcox shale (Ibanez and Kronenberg 1993) without pore fluids present (Fig. 4B). Strengths of Wilcox shale at elevated pressures are much greater than the failure strengths measured in this study at low effective pressures and brittle mechanisms of deformation play a much larger role in the deformation experiments reported here; yet the relationship between strength and rate of deformation is very similar. The results presented in this study reflect the undrained mechanical response of Wilcox shale and comparison with the results of Ibanez and Kronenberg (1993) suggests that strengths of shales, both for drained and undrained conditions, exhibit similar intrinsic dependencies upon the rate of deformation. Swan et al. (1989) reported an intrinsic effect of strain rate on the undrained strength of fluid-saturated Kimmeridge Bay shale for strain rates of $1.7 \times 10^{-5} \leq \dot{\varepsilon} \leq 1.7 \times 10^{-1}$ s^{-1}. Recasting their results as (σ_1-σ_3) vs. ln $\dot{\varepsilon}$ (Fig. 4B), the increases in strength they observed with increasing strain rate (at $\dot{\varepsilon} > 10^{-4}$ s^{-1}) yield a value of $\alpha = 0.27$ MPa^{-1}, consistent with the results of this study and the relationship between strength and strain rate reported for high confining pressures (Ibanez and Kronenberg 1993).

The results of this study differ with those of Swan et al. (1989) at $\dot{\varepsilon} \leq 10^{-4}$ s^{-1}. However, the conditions for fluid exchange between internal pores of specimens and reservoirs of the fluid pressure systems differ as well. The rate-dependent deformation of brine-saturated Wilcox shale is described by eqn. (3) over the entire range of strain rate ($2.8 \times 10^{-7} \leq \dot{\varepsilon} \leq 8.2 \times 10^{-3}$ s^{-1}) investigated, whereas experiments performed on Kimmeridge Bay shale (Swan et al. 1989) show a transition in mechanical response at a strain rate of $\sim 10^{-4}$ s^{-1} (Fig. 4B) with an inverse relationship between strength and strain rate at low strain rates attributed to partial drainage and relaxation of pore pressures.

The transition at which pore pressure drainage becomes significant and the magnitudes of pore pressures generated during deformation decay depend upon permeability and the scale at which fluid flow is important. Measurements of permeability at the same conditions as applied during deformation experiments indicate that characteristic times for pore pressure equilibration along the lengths of Wilcox shale specimens are long (2×10^5 s) and characteristic strain rates required to maintain near-equilibrium pore pressures during deformation (over strain increments of ~0.1%) are low ($\dot{\varepsilon} < 4 \times 10^{-9}$ s^{-1}). Pore pressure perturbations generated during deformation may remain large (with less than 10% relaxation) at the specimen scale at strain rates $\dot{\varepsilon} > 6 \times 10^{-7}$ s^{-1}; the lack of significant fluid pressure relaxation predicted by our permeability results are consistent with the mechanical results for fluid-saturated Wilcox shale. The absolute strengths of shales may depend upon pore fluid flow at finer scales and the changes in pore fluid pressures measured during deformation indicate that some fraction of pores is better connected than implied by the macroscopic measures of permeability. Nevertheless, pores that are not drained are likely to contribute most to weakening specimens and the results for Wilcox shale at strain rates as low as 2.8×10^{-7} s^{-1} appear to reflect this.

The permeability of Kimmeridge Bay shale ($k = 8 \times 10^{-21}$ m^2 at an effective pressure of 10 MPa, Swan et al. 1989) is essentially the same as determined for Wilcox shale (at $P_c - P_f = 5.5$ MPa). However, the experiments of Swan et al. (1989) allowed for radial exchange of pore fluids between specimens and fluid reservoirs of their apparatus. The shortened length scale for radial fluid transport shifts the transitional mechanical behavior between fully drained and undrained responses to higher strain rates and helps explain the differences in mechanical results reported for these two shales.

The results for Wilcox and Kimmeridge Bay shales combined indicate that the intrinsic mechanical properties of clay-bearing shales depend upon rates of deformation with similar stress-strain rate relationships for drained and undrained conditions. An inverse relationship between strength and strain rate, as reported by Swan et al. (1989), may be expected over an interval in strain rates; however, the conditions for this transitional behavior depend upon permeabilities and length scales of fluid flow that result in partial relaxation of pore pressure variations during deformation.

4 CONCLUSIONS

Based on the results of permeability measurements and deformation experiments on Wilcox shale, the following conclusions can be drawn:

1. The permeability of illite-rich shale depends upon effective pressure; measured values of k decrease with increasing effective pressure and are very low ($\sim 10^{-21}$ m^2) at $P_c - P_f \geq 5$ MPa.

2. Permeability measurements for Wilcox shale indicate that equilibrium pore pressures cannot be maintained during deformation experiments at $\dot{\varepsilon} > \sim 10^{-9}$ s^{-1}, consistent with the apparent undrained response of specimens at $\dot{\varepsilon} > \sim 10^{-7}$ s^{-1}. Mechanical results over the range in strain rates investigated depart from a simple effective pressure law and appear to represent the response of illite-bearing shale without

significant relaxation of pore pressure perturbations generated during deformation.

3. Fracture strengths (σ_1-σ_3) of Wilcox shale specimens saturated (as nearly as can be achieved) with pore fluids increase with increasing strain rates $\dot{\varepsilon}$ over the range $2.8\text{x}10^{-7} \leq \dot{\varepsilon} \leq 8.2\text{x}10^{-3}$ s^{-1} and fit an exponential relationship $\dot{\varepsilon} = A \exp\{\alpha(\sigma_1\text{-}\sigma_3)\}$ with a value of $\alpha = 0.29$ MPa^{-1} nearly identical to that measured at high confining pressures ($\alpha = 0.3$ MPa^{-1}) without pore fluids.

5 ACKNOWLEDGMENTS

Special thanks go to Dr. J. Holder for many helpful discussions during the course of this study and for technical suggestions that made these measurements possible. We also thank Dr. R. Berg for discussions regarding the geologic roles of shales and for providing the shale core for study. This study was supported in part by the U.S. Department of Energy (Grant No. DE-FG05-87ER13711) and by a cooperative Industrial Associates Research Program with member companies including Conoco Inc., Shell Development Co., Union Pacific Resources, and Unocal Corporation; their contributions to this study are gratefully acknowledged.

REFERENCES

Bauer, R.A. 1979. The loss of natural moisture content of Pennsylvanian shale and effect on physical properties. *Proc. Ill. Mining Inst., 87th Annual Meeting*: 13-28.

Brace, W.F., J.B. Walsh & W.T. Frangos 1968. Permeability of granite under high pressure. *J. Geophys. Res.* 73: 2225-2236.

Bishop, R.S. 1977. Shale diapir emplacement in South Texas - LaWard and Sherriff examples. *Gulf Coast Geol. Soc. Trans.* 27: 20-28.

Chapple, W.M. 1978. Mechanics of thin-skinned fold-and-thrust belts. *Geol. Soc. Amer. Bull.* 89: 1189-1198.

Cook, J.M. & M. Thiercelin 1989. Indentation resistance of shale: The effects of stress state and strain rate. In *Rock mechanics as a guide for efficient utilization of natural resources, Proc. 30th U.S. Symp. Rock Mech.*, ed A.W. Khair, 757-764.

Ibanez, W.D. & A.K. Kronenberg 1993. Experimental deformation of shale: Mechanical properties and microstructural indicators of mechanisms. *Int. J. Rock Mech. Min. Sci. & Geomech. Abstr.* 30: 723-734.

Mesri, G. & R. Gibala 1972. Engineering properties of a Pennsylvanian shale. In *Stability of rock slopes, Proc. 13th U.S. Symp. Rock Mech.*, 57-75.

Meyer, C.A., R.B. McClintock, G.J. Silvestri & R.C. Spence, Jr. 1979. *ASME Steam table, Thermodynamic and transport properties of steam*, The American Society of Mechanical Engineers.

Nuesch, R. 1991. Das mechanische *Verhalten von Opalinuston*. Ph.D. thesis, Eidgenossischen Technischen Hochschule, 244 pp.

Rice, J.R. & M.P. Cleary 1976. Some basic stress diffusion solutions for fluid-saturated elastic porous media with compressible constituents. *Rev. Geophys. Space Phys.* 14: 227-241.

Roeloffs, E.A. 1988. Fault stability changes induced beneath a reservoir with cyclic variations in water level. *J. Geophys. Res.* 93: 2107-124.

Shea, W.T. & A.K. Kronenberg 1992. Rheology and deformation mechanisms of an isotropic mica schist. *J. Geophys. Res.* 97: 15201-15237.

Steiger, R.P. & P.K. Leung 1990. Predictions of wellbore stability in shale formations at great depth. In *Rock at great depth*, eds. V. Maury & D. Foumaintraux, 1209-1218.

Sutherland, H.J. & S.P. Cave 1980. Argon gas permeability of New Mexico rock salt under hydrostatic compression. *Int. J. Rock Mech. Min. Sci. & Geomech. Abstr.* 17: 281 288.

Swan, G., J. Cook, S. Bruce & R. Meehan 1989. Strain rate effects in Kimmeridge Bay shale. *Int. J. Rock Mech. Min. Sci. & Geomech. Abstr.* 26: 135-149.

Van Eeckhout, E.M. 1976. The mechanisms of strength reduction due to moisture in coal mine shales. *Int. J. Rock Mech. Min. Sci. & Geomech. Abstr.* 13: 61-67.

Rock Mechanics, Nelson & Laubach (eds) © 1994 Balkema, Rotterdam, ISBN 90 5410 380 8

Pore collapse mechanisms in Cordoba Cream limestone

S. Mowar & M. Zaman
Civil Engineering & Environmental Science, The University of Oklahoma, Norman, Okla., USA
D.W. Stearns
Geology & Geophysics, The University of Oklahoma, Norman, Okla., USA
J.-C. Roegiers
Petroleum & Geological Engineering, The University of Oklahoma, Norman, Okla., USA

ABSTRACT: Controlled experiments were conducted in an attempt to better understand pore collapse during reservoir compaction resulting from an increase in an effective confining pressure. Micromechanism changes that accompany stages of pore collapse were recorded for the high porosity, test material (Cordoba Cream Limestone). Though cataclastic flow and crystal plasticity are both operative throughout pore collapse, the contribution of a single mechanism to the total deformation causing pore collapse varies during specific collapse stages.

1 INTRODUCTION AND OBJECTIVES

Reservoir compaction due to pore collapse in weak, porous rocks as effective stress increases due to pressure drawdown during production can have detrimental effects on both producing facilities and recoverable reserves (Addis, 1987; Smits et al., 1988). A large decrease in permeability due to pore collapse can lead to a significant reduction in recovery or the pore collapse may induce local subsidence which, in turn, can lead to drilling and production equipment damage. In addition, pore collapse can also trigger sand production, lead to wellbore instability, and casing damage.

The underlying assumption that motivated this study is that a more detailed understanding of the exact deformation mechanisms that cause pore collapse might lead to a resolution of some of the problems caused by pore collapse. It was further assumed that the macroscopic effects of pore collapse are ultimately caused by the cumulative effects of deformations at the microscopic scale. Therefore, natural pore collapse was simulated in controlled laboratory experiments so that these microscopic mechanisms could be delineated for, and correlated to, various stages of macroscopic reservoir compaction. More specifically, answers were sought to three basic questions: (i) what micromechanisms are responsible for pore collapse? (ii) which of the different mechanisms dominate during particular stages of the collapse? and, (iii) does the microscopic response correlate with distinct stages of the macroscopic response as identified from hydrostatic pressure-strain curves. Questions (i) and (ii) are the main focus of this paper. For data concerning question (iii), see Mowar (1993).

2 LABORATORY TESTS

The test material is a high porosity (≅ 25%), low permeability (≅ 15 md) fossiliferous limestone (commercially known as Cordoba Cream), collected from a quarry near Austin, Texas. The major mineral composition (obtained via an electron microprobe) is calcite (85%), clay (13.6%) and quartz (0.9%) and the rock fabric is primarily one of foraminifera, or fragments, cemented together with fine-to-medium grained calcite.

Cubic samples of the limestone were subjected to hydrostatic confining pressure in a High Capacity Cubical Device. Sample volume reduction was recorded as the hydrostatic confining pressure increased to 25 ksi (172.5 MPa). In addition, tests were interrupted at selected confining pressures in order to delineate microscopic features at various stages of pore collapse.

Data were reasonably reproducible and three distinct regions having different slopes along the hydrostatic confining pressure-strain curves were delineated, indicating three distinct phases of deformation: elastic, pore collapse and post-pore collapse (Figure 1). In this use of Cordoba Cream limestone, pore collapse begins at about 6 ksi (41.4 MPa) hydrostatic confining pressure and continues to a confining pressure of about 15 ksi (103 MPa).

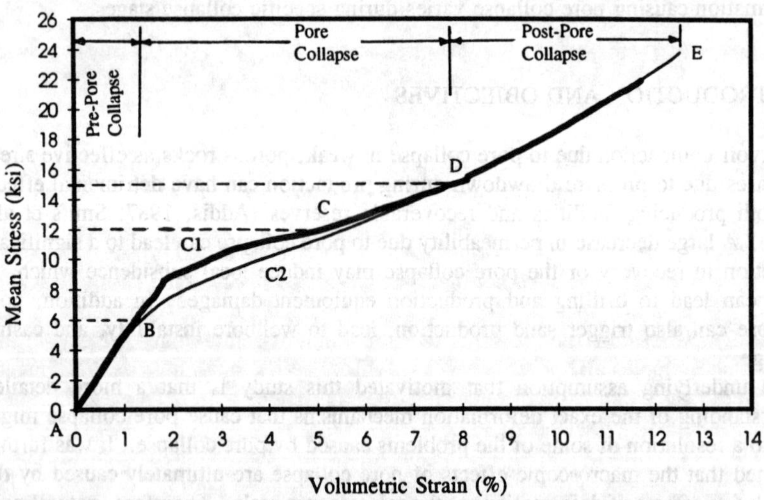

Figure 1 Comparison of volumetric response between specimens C1 and C2 for Hydrostatic Compression tests. The three different phases of pore collapse and their beginnings and ends are marked.

In order to determine any change in the micromechanisms producing pore collapse, three arrested tests were run. One was terminated at the beginning of pore collapse (6 ksi hydrostatic confining pressure), another high in the pore collapse phase (12 ksi hydrostatic confining pressure) and the third carried into the post-pore collapse phase

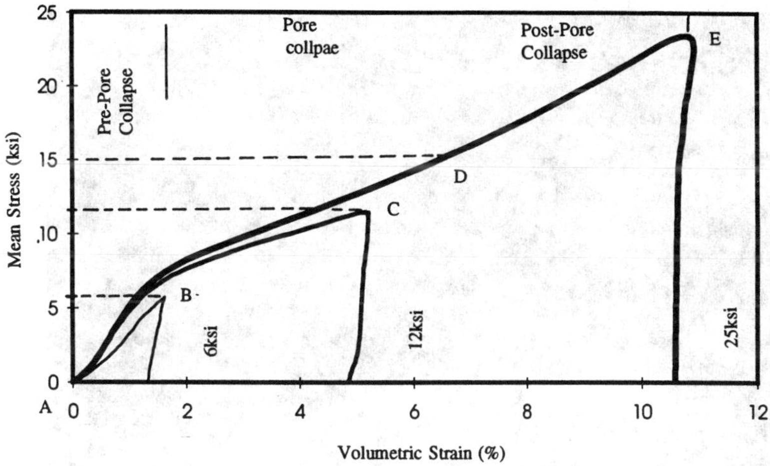

Figure 2 Volumetric responses of three Hydrostatic Compression tests.

(25 ksi). Volumetric changes were also recorded for each of the three arrested tests (Figure 2).

3 MICROSCOPIC OBSERVATIONS

Epoxy impregnated thin sections were made of the test samples from each of the three arrested tests, as well as from the undeformed rock, in order to study, and compare, the operative microprocesses.

Pore distribution in the undeformed rock seems to be of two types; (i) scattered small pores close enough together that they are interconnected and (ii) more widely spaced, larger pores that are not interconnected, at least in two dimensions (see Figure 3). Few, if any, calcite twins or microfractures are present in the undeformed samples and pores are clear of disconnected small host rock fragments. The overall appearance of the untested rock is that of a loosely packed, lightly cemented skeletal framework.

At 6 ksi (41.4 MPa) hydrostatic confining pressure some pores have just begun to collapse while others are still intact (Figure 4). Collapse within the smaller pores is almost entirely due to grain crushing at the periphery of the pores (cataclastic flow). The rock fragments spall off from the host and collect in the open small pores, thereby reducing the pore throat aperture size. The larger pores seem to change shape by flow within the matrix calcite grains as evidenced by an increase in twinning in the matrix grains near the larger pores. Because the boundary stress condition is hydrostatic it could not cause the observed twinning. The stress difference required for the twinning must originate at the pore boundaries where the stress state is non-hydrostatic.

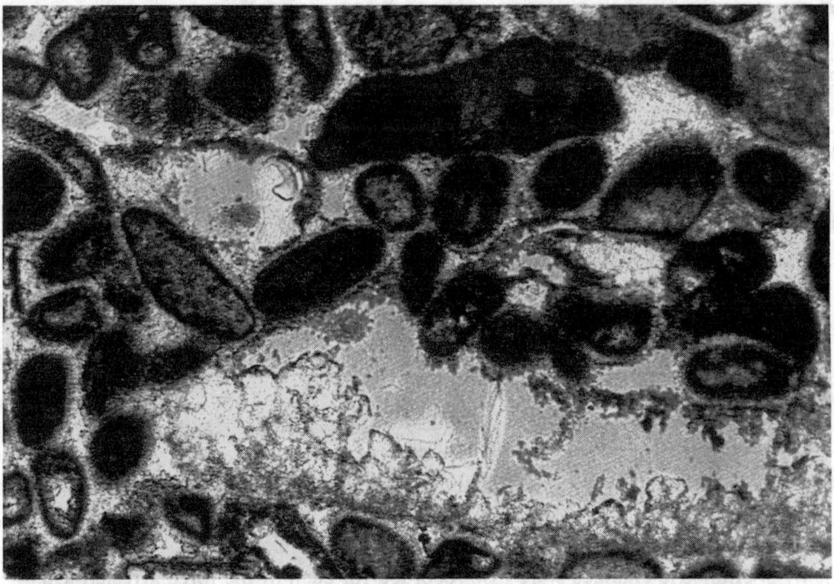

Figure 3 Photomicrograph of an untested sample in plain polarized light showing pores, calcite grains and fossil fragments.

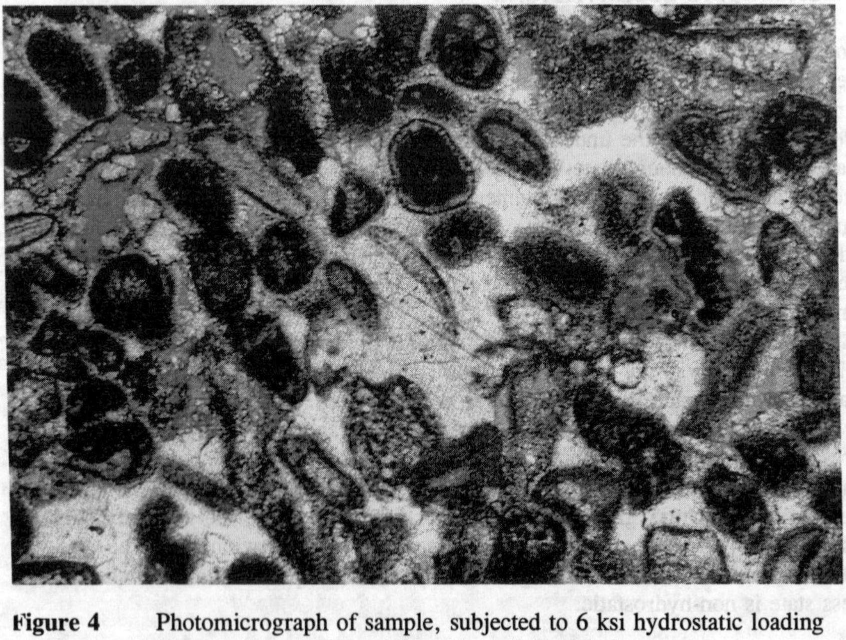

Figure 4 Photomicrograph of sample, subjected to 6 ksi hydrostatic loading (onset of pore collapse) in plain polarized light.

At 12 ksi (82.8 MPa), well into the pore collapse region, there is a considerable change in the rock fabric (Figure 5). Here virtually all pores show signs of collapse mostly as the result of continued grain crushing into pores of all sizes. In addition, there is a distinct increase in the calcite twinning with most grains being moderately to heavily twinned. In other words, at this stage both cataclastic flow and crystal plasticity are active mechanisms. Through-going microcracks are, for the first time, observed. Due to grain crushing and changes in pore geometry through twinning, very small, even capillary sized, pores are being created along the boundaries of old pores. The compaction at this stage has greatly reduced the average pore throat size.

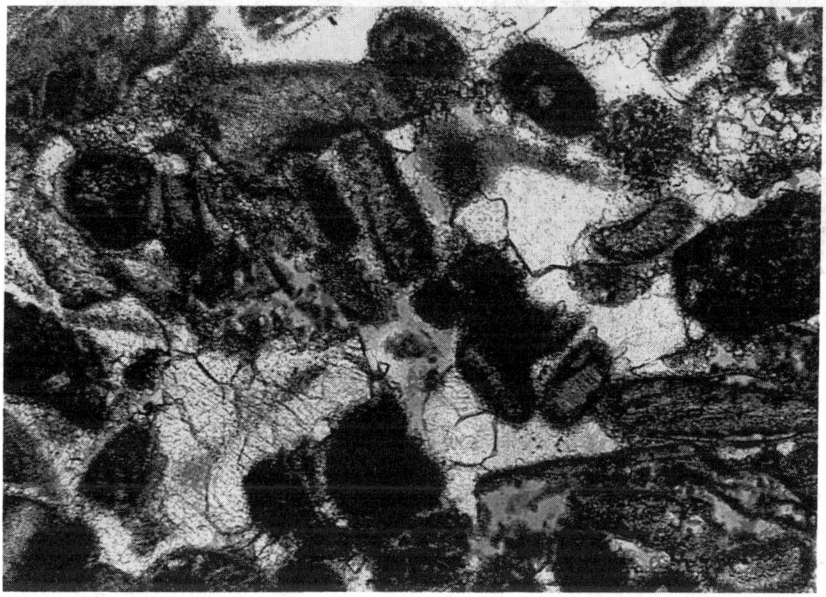

Figure 5 Photomicrograph of sample, subjected to 12 ksi hydrostatic loading, in plain polarized light.

At 25 ksi (172.5 MPa) hydrostatic confining pressure (post-pore collapse phase) pore collapse is nearly complete (Figure 6). Thin sections show evidence of both increased cataclastic flow and increased crystal plasticity (twinning). The remaining pores are so small that they are stable and the matrix is carrying most of the load. Therefore, further deformation will require a non-hydrostatic stress state and continued strain will involve shape changes along the boundaries of the sample as opposed to the simple volumetric strains that dominated up to the post-pore collapse stage.

The untested rock has strong preferred orientation in which elongate fossil fragments are nearly parallel to bedding planes (Mowar, 1993). Pore size distribution in the untested rock is fairly uniform for both the larger and smaller undamaged pores. Though progressive pore collapse did not totally destroy the bedding parallel fabric of the fossil fragments it did weaken it considerably. This

Figure 6 Photomicrograph of sample, subjected to 25 ksi hydrostatic loading, in plain polarized light.

observation argues that the cataclastic flow accompanying pore collapse occurred not only at the scale of spalling into the pores but, also, at a larger scale by the rigid body rotation and rupturing of fossil fragments. The more or less homogeneous fabric of pores and pore size in the untested rock was nearly obliterated during pore collapse.

4 CONCLUSIONS

In the Cordoba Cream limestone pore collapse due to hydrostatic confining pressure begins after an initial purely elastic response and ceases at a confining pressure beyond which it is no longer operative. Cataclastic flow at the grain crushing scale and crystal plasticity (calcite twinning) are the two major micromechanisms causing pore collapse that can be documented in this study. There is some evidence for rigid body rotation at a scale larger than discussed here and there is undoubtedly some grain boundary slip (not documented here) involved in the total collapse but the two mechanisms of cataclastic flow and crystal plasticity dominate the process of pore collapse. Pore collapse initiates by grain spalling into small pores but as the process continues crystal plasticity and larger scale cataclaysis become more important.

ACKNOWLEDGEMENT

The authors are grateful to The University of Oklahoma Rock Mechanics Consortium members, the Oklahoma Center for Advancement of Science and Technology (OCAST) and the National Science Foundation (Rock Mechanics Research Center) for financial support.

REFERENCES

Addis, M.A. 1987. "Material Metastability in Weakly Cemented Sedimentary Rocks," Memoir of the Geological Society of China, pp. 495-512.

Mowar, S. 1993. "Experimental Study of Pore Collapse in Limestone: Macroscopic and Microscopic Aspects," M.S. Thesis, University of Oklahoma, Norman.

Smits, R.M.M., J.A. de Waal & J.F.C. van Kooten 1988. "Prediction of Abrupt Reservoir Compaction and Surface Subsidence Caused by Pore Collapse in Carbonates," SPE Formation Evaluation (June), pp. 340-346.

Rock Mechanics, Nelson & Laubach (eds) © 1994 Balkema, Rotterdam, ISBN 90 5410 380 8

Acoustic velocity anisotropies in Cordoba Cream limestone during different deformational stress paths

M. Azeemuddin, T. E. Scott, Jr & J.-C. Roegiers
Petroleum & Geological Engineering, The University of Oklahoma, Norman, Okla., USA
M. M. Zaman
Civil Engineering & Environmental Science, The University of Oklahoma, Norman, Okla., USA

ABSTRACT: Ultrasonic compressional (P) and shear (S) waves were transmitted through the axial and lateral directions of cylindrical samples of a high porosity rock formation subjected to triaxial, hydrostatic and uniaxial strain deformational paths. Triaxial tests which terminate in a brittle shear fracture generate a P-wave anisotropy and a lateral shear wave birefringence in both the elastic and dilatant phases of loading. Post-failure sliding along the shear fracture does not result in changes in anisotropy. Triaxial tests at high confining pressure generate ductile yielding of the limestone. In the initial elastic phase no anisotropies are evident. At yielding significant P-wave and S-wave anisotropies develop which continue to increase until termination of the tests. Uniaxial strain (k_o) tests develop initial acoustic anisotropies at the point of ductile yielding. These anisotropies continue to increase after yielding, for about 1 to 2% volumetric strain, at which point the acoustic velocities begin to increase and the velocity anisotropies remain constant for the duration of the test. Hydrostatic compaction tests evidence small acoustic anisotropies which develop during the initial (elastic) phase but begin to decrease after hydrostatic yielding.

1 INTRODUCTION

High-porosity formations subjected to increased effective stresses, resulting from the production of oil and gas, are found to undergo sudden and drastic volume reductions through a micro-level mechanism called *pore collapse*. An analysis of pore collapse using correlated observations from the macroscopic deformational behavior (e.g. strain-stress relationships), the microscopic deformational mechanisms, and the acoustic properties will ultimately lead to an understanding and prevention of the damaging effects of pore collapse. Some of the consequences resulting from this pore collapse are loss of reserves due to decreased permeability and damage to equipment incurred by reservoir compaction and subsidence.

Ultrasonic velocity anisotropy measurements have been made in the past on Westerly granite (Lockner et al., 1977) and sandstones (Lo et al., 1986; Sammonds et al., 1989; Sayers et al., 1990; Scott et al., 1993). The only study on limestone (Gupta, 1973) used prismatic samples, loaded biaxially, with one P- and one S-wave velocity recorded in each of the three directions.

Scott et al. (1993) studied the stress-induced P- and S- wave anisotropies through the use of one P- and two mutually-orthogonal S-waves in both axial and lateral directions in Berea sandstone. They showed, from a series of experiments using triaxial loading, that significant P-wave anisotropy and a lateral S-wave birefringence developed during the elastic phases of both brittle shear failure and shear-enhanced compaction. However, it is known that sections of reservoir rock subjected to an increase in effective stress due to the extraction of oil and gas tend to deform uniaxially; hence, a similar loading path ought to be simulated (Johnson et al., 1989).

The objective of this study is to investigate the acoustic velocity anisotropy changes in a limestone along different deformational stress paths with the aim of providing useful data which could be used to identify or predict the occurrence of pore collapse in the field.

2 EXPERIMENTAL PROGRAM

2.1 *Material used*

Cordoba Cream limestone was used in this study for its high porosity (23%) and weakly cemented nature. This type of rock undergoes pore collapse, marked by a sudden increase in compressibility and volumetric strain, under hydrostatic compression, uniaxial strain conditions, and in triaxial compression experiments at high confining pressures (Azeemuddin et al., 1992). The behavior is similar to that observed in many other high porosity rocks (e.g., Addis, 1987).

2.2 *Experimental procedure*

Ten cylindrical samples of Cordoba Cream limestone were prepared from the same lithologic interval, with their axes oriented normal to the bedding plane fabric, following the ISRM suggested methods. The finished samples were 2.125 inches (5.4 cm) in diameter and 4.25 inches (10.8 cm) in length. Crystals for acoustic pulse transmission in the axial direction were housed inside the steel loading platens, their coupling being accomplished with thin lead sheets, 0.8 mm thick, inserted between the platens and the sample. The crystals for lateral pulse transmission were glued on the opposite sides of the sample using a silver conductive epoxy. One P- and two mutually-orthogonal S-waves were transmitted using PZT-5a piezoelectric crystals with a nominal frequency of 600 kHz. The sample assembly and experimental setup was similar to that used by Scott et al. (1993).

The compressional and shear wave velocities are denoted by V_P and V_S, respectively.[1] A total of six waves (V_{P11}, V_{S12}, V_{S13}, V_{P22}, V_{S21} and V_{S23}) were transmitted through the sample.

Triaxial compression tests were performed on separate samples at initial confining pressures of 0.2, 1, 2, 3, 5, 7, and 11 ksi (1.38, 6.9, 13.8, 20.7, 34.5, 48.2 and 75.8 MPa) at a constant displacement rate of .0011 in./min (.028 mm/min) (4.3x10^{-6}/sec strain rate). The uniaxial strain (k_0-) tests were conducted under circumferential strain control mode, starting from different initial hydrostatic stress states simulating

[1]The velocity V_{S23}, for example, represents the shear wave velocity propagated in the 2-direction and polarized in the 2-3 plane (parallel to bedding).

different depths of burial. The hydrostatic compression tests were carried up to a maximum mean stress of 20 ksi (138 MPa). The loading rate in k_0- as well as hydrostatic compression tests was equal to 200 psi/min (1.38 MPa/min).

3 RESULTS AND DISCUSSION

Figure 1 shows the stress paths followed during the hydrostatic, triaxial and uniaxial strain tests along with the yield cap obtained from the stress-strain curves.

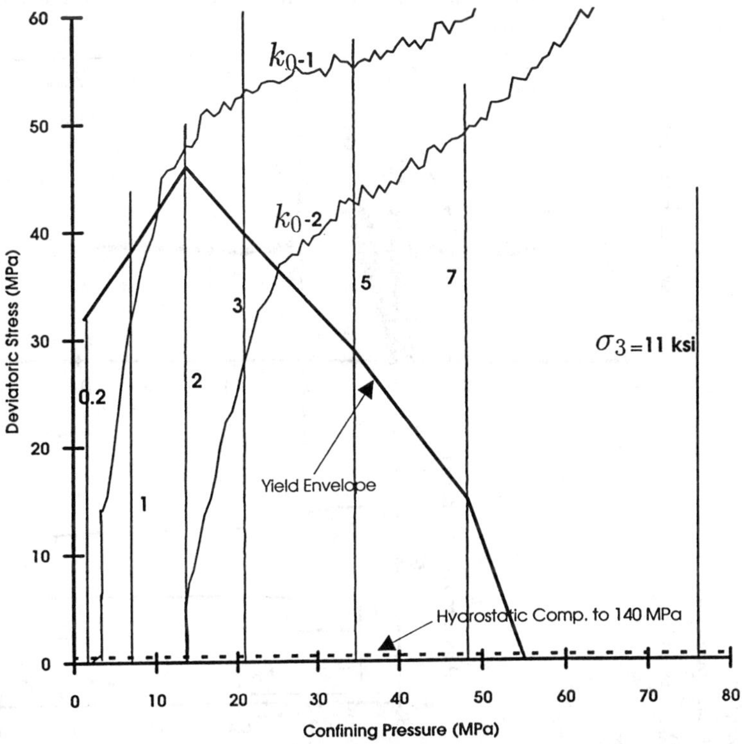

Figure 1. Yield envelope and stress paths for various tests.

3.1 *Triaxial compression stress paths*

Figure 2 shows the variation of compressional and shear wave velocities for a triaxial test which ended in brittle failure (at an initial confining pressure of 0.2 ksi or 1.38 MPa) at an axial strain of about 0.22%. The topmost graph shows the three strains (axial, radial and volumetric) as a function of the deviatoric stress, $\sigma_1 - \sigma_3$. The middle and the lowermost figures show the variation of P- and S-wave velocities normalized with respect to the corresponding velocities at the initiation of shear loading versus the axial strain (volumetric strain in case of hydrostatic compression test).

As seen from Figure 2, the axial P- and the two axial S-wave velocities increase in the early stage of loading, an effect attributed to the closure of microcracks oriented perpendicular to the direction of loading. The velocities in the lateral direction remain initially unaffected. On further loading, both the axial and lateral P- and S-wave velocities decrease due to the opening of microcracks oriented parallel to the axis (dilatancy). Acoustic anisotropies in these plots are evidenced by a divergence in the normalized axial and lateral velocity curves. The acoustic anisotropies develop at the initiation of axial loading and continue to increase throughout the test up to the point of brittle shear failure. After the formation of shear fracture, the velocities do not change (and therefore anisotropies are held constant) during sliding along the shear fracture. These observations are consistent with those for Indiana limestone by Gupta (1973), for Darley Dale sandstone by Sammonds et al. (1989) and for Berea sandstone by Scott et al. (1993).

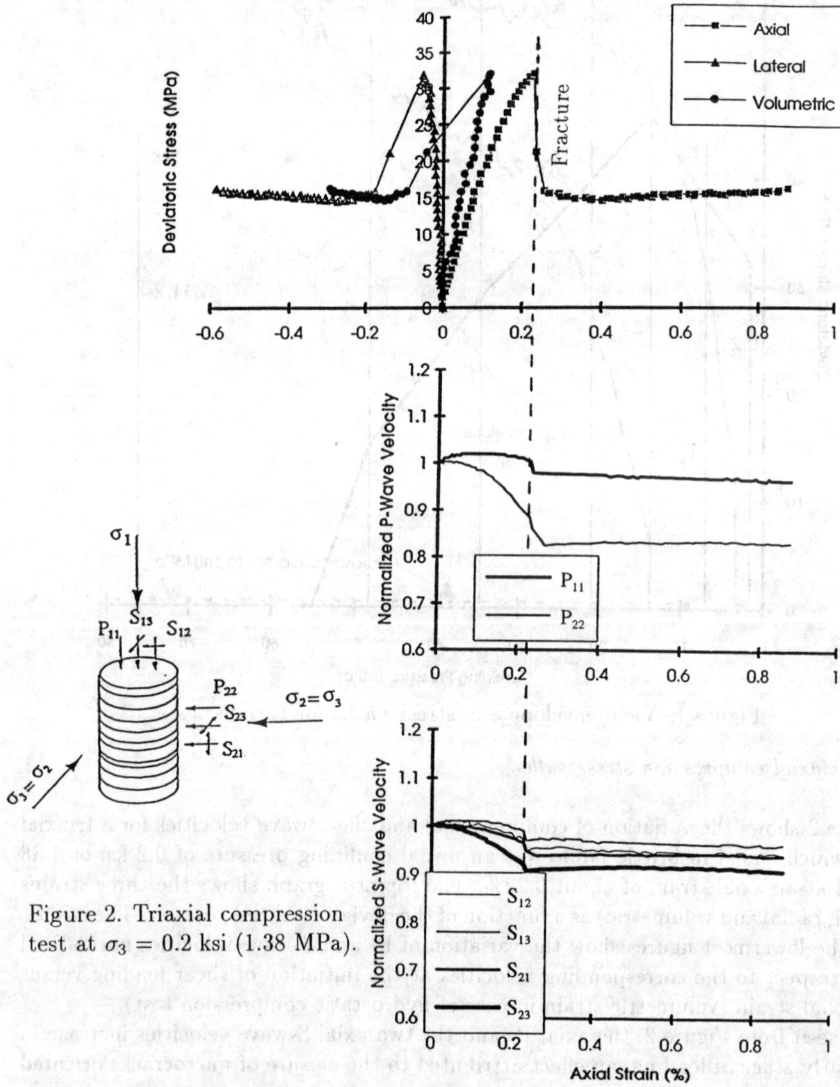

Figure 2. Triaxial compression test at $\sigma_3 = 0.2$ ksi (1.38 MPa).

In triaxial compression tests at higher confining pressures, the operating micromechanisms cause compaction and pore collapse rather than dilatancy and brittle shear failure. At a confining pressure of 3 ksi (20.7 MPa) there were no anisotropies observed during the initial stages of axial loading (Figure 3). Acoustic velocities also decrease during this stage. The possible micromechanisms thought to be active during this deformational stage are those of grain boundary slip, grain rotation and calcite twinning, based on observations by Mowar (1993) in other types of tests (hydrostatic compression and uniaxial strain paths).

At yielding (0.34% axial strain) there is a marked development of a compressional wave anisotropy and a shear wave birefringence which continues to increase for the

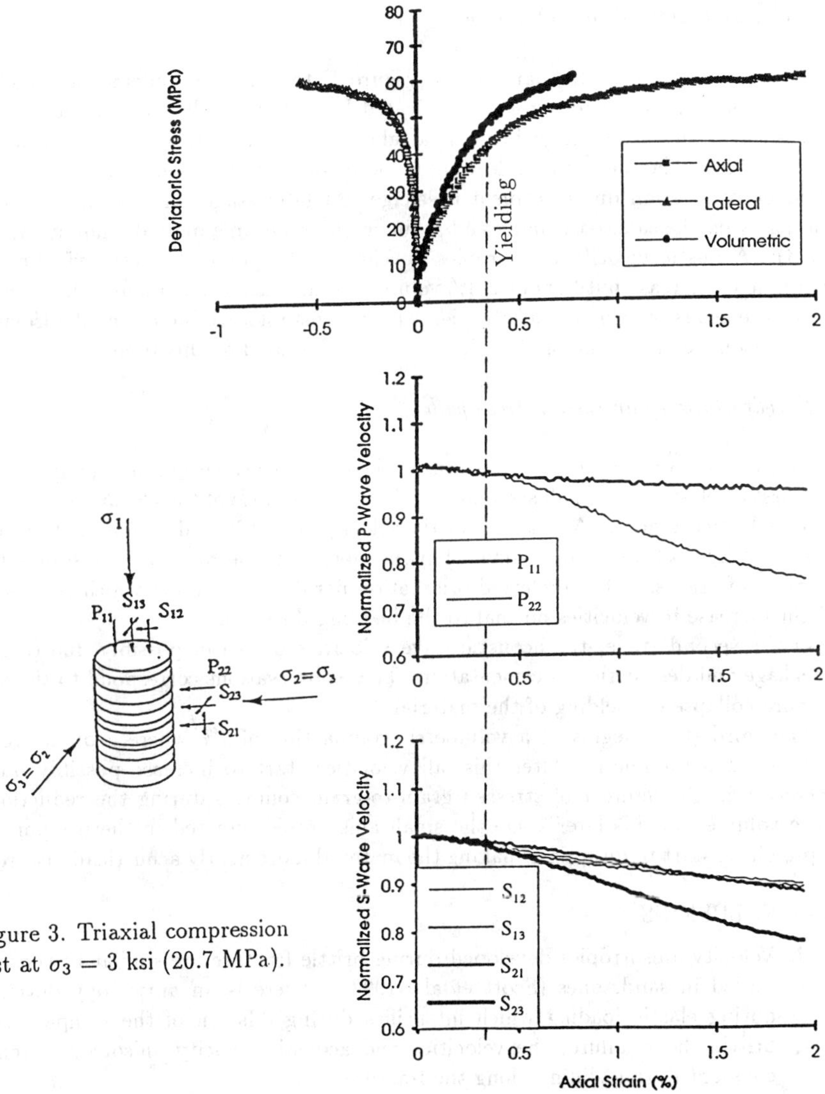

Figure 3. Triaxial compression test at $\sigma_3 = 3$ ksi (20.7 MPa).

remainder of the test. The shear wave birefringence generated during ductile deformation always results in a slower shear wave polarized in the horizontal plane (V_{S23}) than those polarized in the vertical plane (i.e., V_{S12}, V_{S21}, and V_{S13}).

In the later stages there is a decrease in the rate of change of the acoustic velocities possibly due to the effects of two competing processes: first, the destruction of cementation, the development of calcite twinning, and grain crushing which are thought to result in decreases in acoustic velocities; and second, the increase of highly stressed grain-to-grain contacts resulting from the grain crushing and pore collapse which are speculated to generate increases in velocities. The latter effect is assumed to become more important in the later stages of deformation.

3.2 *Uniaxial strain deformational path*

For the uniaxial strain (k_0-path) tests (Figure 4) the samples undergo pore collapse after an initial elastic response, as indicated by the sharp decrease in slope of the stress-strain curves at the yield point at about 0.2% axial strain and 40 MPa deviatoric stress. The mechanisms leading to pore collapse are grain boundary slip and rigid body rotation due to cement breakage. At later stages of pore collapse other mechanisms like calcite twinning and grain crushing become more dominant (Mowar, 1993). Acoustic velocity anisotropies develop at the point of ductile yielding and continue to increase until about 2-3% volume strain is achieved (Figure 4). After this point the acoustic velocity anisotropies remain constant with increasing deformation. The reasons for this type of behavior are not understood at this time.

3.3 *Hydrostatic compression stress path*

This test shows three distinct patterns in the acoustic velocity profiles (Figure 5). In the first or elastic region, all six velocities increase because of the closure of randomly oriented microcracks. A small acoustic velocity anisotropy develops in the initial stages of hydrostatic compression. This is thought to be due to closure of a large number of cracks with a preferred orientation parallel to bedding which results in a slight increase in velocities normal to the bedding direction.

In the second stage, the acoustic wave velocities decrease, possibly due to grain breakage and destruction of cementation. These observations correspond to the onset of pore collapse or yielding of the material.

The third stage begins at a volumetric compaction of 2% where acoustic velocities are at a minimum. After this, all velocities start to increase possibly due to increases in the number of stressed grain-to-grain contacts during the reduction in pore volume. In this later stage the small anisotropies created in the pre-pore collapse phase start to disappear making the material more nearly acoustically isotropic.

4 CONCLUSIONS

1. Velocity anisotropies developed during brittle fracture are similar to those observed in sandstones (Scott et al. 1993). There is an anisotropy developed during elastic loading which intensifies during dilation of the sample. After brittle shear failure, the velocities and acoustic velocity anisotropies remain constant during sliding along the fracture.

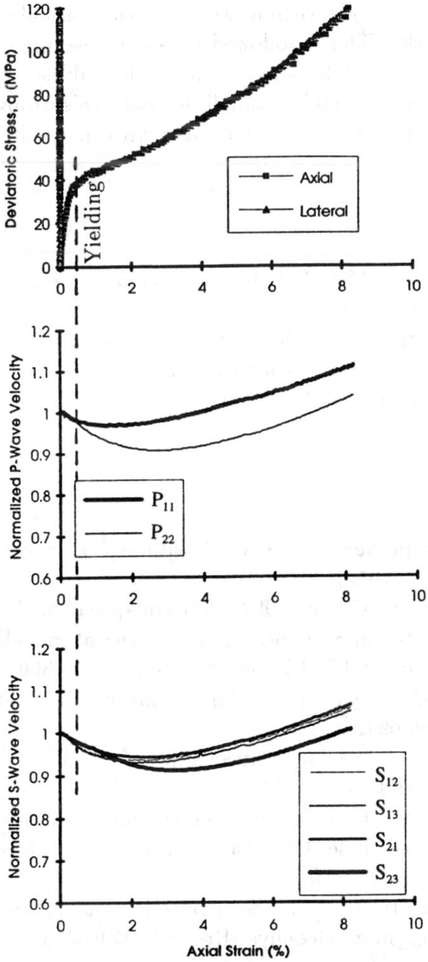

Figure 4. Uniaxial strain (k_0-) test.

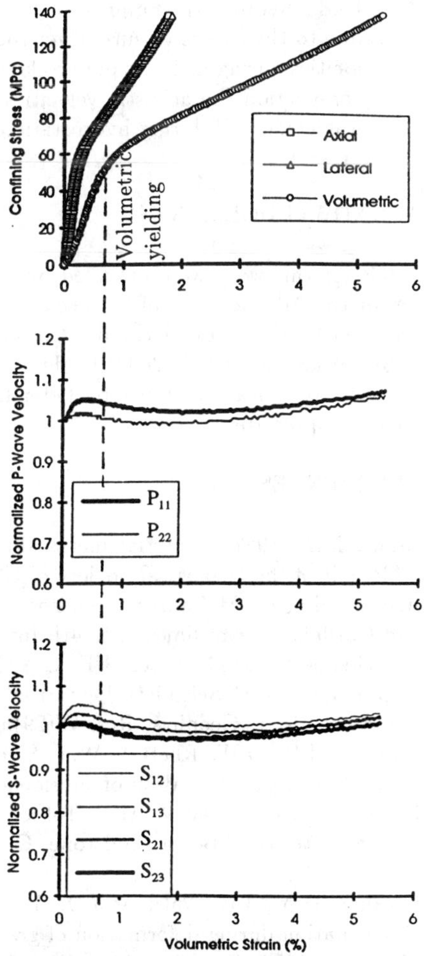

Figure 5. Hydrostatic compression test.

2. P-wave anisotropy and S-wave birefringence development are associated with yielding or pore collapse in ductile triaxial tests and uniaxial strain tests. In ductile triaxial tests the acoustic anisotropies continue to increase with increasing deformation. In uniaxial strain (k_0-) tests the anisotropies developed in early stages of loading are arrested after yielding and then are constantly maintained with increasing deformation even though acoustic compressional and shear wave velocities are increasing.

3. In cases where a strong shear wave birefringence is generated, such as during the later stages in uniaxial strain tests and ductile deformation in triaxial tests, the shear wave velocity polarized in the horizontal plane (V_{S23}) is slower than shear waves polarized in the vertical plane (i.e., V_{S12}, V_{S21}, and V_{S13}).

4. Under hydrostatic compression, the initial increase in wave velocities is probably due to the elastic closure of microcracks. This is followed by a decrease in velocities during inelastic pore collapse. In the later stages of inelastic hydrostatic compaction the acoustic velocities increase. Only a small acoustic anisotropy was evidenced during hydrostatic compaction and it decreases with increasing pressure.

ACKNOWLEDGEMENTS

Funds for this work were provided by a grant (ARO-114) from the Oklahoma Center for the Advancement of Science and Technology (OCAST). Matching funds were provided by the Rock Mechanics Consortium and the National Science Foundation (NSF Award no. EEC-9209619). Funding was also provided by Saga Petroleum of Norway. Technical assistance by Peter Lemmon and Tony Haas in the laboratory is greatly appreciated.

REFERENCES

Addis, M.A. 1987. Material metastability in weakly cemented sedimentary rocks. Memoir of the Geological Society of China. 495-512.

Azeemuddin, M., M. Zaman, J.-C. Roegiers & S. Mowar. 1992. Pore collapse behavior of Cordoba Cream limestone. 4th International Symposium on Numerical models in Geomechanics, Swansea, UK. G.N. Pande and S. Pietruszczak (eds.), 291-300

Gupta, I.N. 1973. Seismic velocities in rock subjected to axial loading up to shear fracture. Jour. Geophys. Res. 78(29):6936-6942.

Johnson, J.P., D.W. Rhett & W.T. Siemers. 1989. Rock mechanics of the Ekofisk reservoir in the evaluation of subsidence. JPT (7):717-722.

Lo, T.-w., K.B. Coyner & M.N. Toksöz. 1986. Experimental determination of elastic anisotropy of Berea sandstone, Chicopee shale, Chelmsford granite. Geophys. 51(1):164-171.

Lockner, D.A., J.B. Walsh & J.D. Byerlee. 1977. Changes in seismic velocity and attenuation during deformation of granite. Jour. Geophys. Res. 82:5374-5378.

Mowar, S. 1993. Experimental study of pore collapse in limestone: macroscopic and microscopic aspects. M.S. thesis, The University of Oklahoma, Norman, OK.

Sammonds, P.R., M.R. Ayling, P.G. Meredith, S.A.F. Murrell & C. Jones. 1989. A laboratory investigation of acoustic emission and elastic wave velocity changes during rock failure under triaxial stresses. In Rock at Great Depth, Maury and Fourmaintraux, eds., A.A. Balkema, Rotterdam, 233-240.

Sayers, C.M., J.G. van Munster & M.S. King. 1990. Stress-induced ultrasonic anisotropy in Berea sandstone. Int. J. Rock Mech, Min. Sci. & Geomech. Abstr., 27(5)429-436.

Scott, T.E., Q. Ma & J.-C. Roegiers. 1993. Acoustic velocity changes during shear enhanced compaction of sandstone. Proc. 34th U.S. Symp. on Rock Mechanics, U. of Wisconsin, Madison. 1:29-32.

Rock Mechanics, Nelson & Laubach (eds) © 1994 Balkema, Rotterdam, ISBN 90 5410 380 8

The effect of clay-fluid interactions on ultrasonic seismic wave propagation

S.L.Ita & N.G.W.Cook
Department of Materials Science & Mineral Engineering, University of California, Berkeley, Calif., USA

L.R.Myer
Earth Sciences Division, Lawrence Berkeley Laboratory, Berkeley, Calif., USA

ABSTRACT: This laboratory study found that small changes in the fluid type, fluid content, and amount of a reactive clay film can have significant effects on seismic wave propagation in unconsolidated geomaterials.

1 INTRODUCTION

Many researchers have found that clay content affects seismic wave propagation in granular rocks (Han et al, 1986; Klimentos & McCann, 1990). Anstey (1991) discussed the large effects contact properties can have on seismic velocities. Kowallis et al (1984) noted that framework cementation in a sandstone, partially due to authigenic clay, had an important influence on observed seismic velocities. Suárez-Rivera (1992) showed that S-wave attenuation across a single, clay-coated planar interface changed in response to different fluid types and amounts. Attenuation increased as the water content increased, but was less in the presence of non-polar fluids. These studies suggest that clay and fluids at contacts in granular materials may have notable effects on seismic wave propagation.

This experimental work investigated the physicochemical effects of clay-fluid interactions at interfaces on seismic wave propagation. Fluids can change the properties of the clay, leading to rheologic changes at the grain-to-grain contacts. Both polar and non-polar fluids were used. Polar fluids were expected to have a greater effect due to their ability to cause clay to swell. Measurements were conducted at ultrasonic wavelengths and are interpreted for mechanisms. Understanding these mechanisms at ultrasonic frequencies provides a basis for determining the effects at lower frequencies. This work is part of a group effort to assess the use of seismic techniques for the imaging and characterization of rock and soil properties and processes.

2 EXPERIMENTAL DESIGN

Precise control was required over sample preparation to isolate the effects of clay-fluid interactions from other factors that can affect seismic wave propagation. Spherical soda-lime glass beads were chosen as the matrix material. This insured that the grain-to-grain contacts were uniform, Hertzian (point-to-point) contacts. The use of glass beads also allowed for grain size selection. 50 to 100 μ diameter beads were used to correspond to the lower range of sand-size particles. The uniform composition of the glass beads resulted in beads with identical density. Because of these attributes, the matrix of each specimen was uniform and repeatable. Any changes in seismic wave propagation could thus be attributed to the effects of clay-fluid interactions.

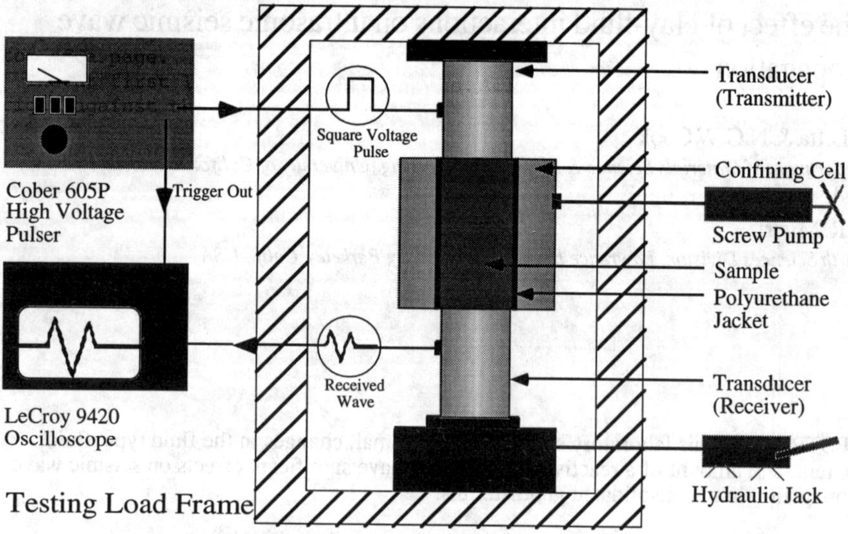

Figure 1: Schematic diagram of laboratory testing apparatus. Transducers include both P and S-wave piezoelectric crystals.

Montmorillonite clay was deposited onto the glass beads to provide an active clay surface at grain-to-grain contacts. Repeatable clay films were produced by stirring 100 grams of glass beads into ~20 ml of $4x10^{-3}$ (gm-clay/gm-water) montmorillonite-distilled water suspension. The clay was dried onto the surface using both vacuum and desiccant drying. Measurements after removal from the drying chamber indicated a clay content of 0.8% (wt-clay/wt-specimen) and an average clay water content of 5% (wt-water/wt-clay). One test was conducted with clay dried in a 100° C oven to determine the effects of dry clay.

Water, ethyl alcohol or hexadecane were added to the specimen by vapor deposition in a sealed evaporation box. The beads were mechanically separated before insertion into the box to facilitate fluid adsorption by the clay. The fluid content of the clay was controlled by regulating the vapor pressure and exposure time.

Prepared specimens were tested in a load frame, as shown in Figure 1. Hydrostatic stress was maintained to minimize breakage of the glass beads. To consolidate the specimen, it was stress-cycled three times to the highest testing stress, 14.7 MPa. Ultrasonic wave propagation tests were conducted on the fourth stress cycle. The parameters for each specimen tested are given in Table 1.

Table 1: Parameters for Tested Glass Bead Specimens

Test Condition	Fluid	Clay Fluid Content (wt-water/wt-clay)	Specimen Fluid Content (wt.-water/wt-specimen)	Length
Dry, Clean Beads	none	0 %	0 %	.0254 m
Oven-Dry, Clay Beads	none	0 %	0 %	.0254 m
Clay-coated Beads	water	50 %	0.40 %	.0254 m
Clay-coated Beads	water	80 %	0.64 %	.0256 m
Clay-coated Beads	water	140 %	1.12 %	.0277 m
Clay-coated Beads	water	200 %	1.60 %	.0269 m
Clay-coated beads	water	300 %	2.40 %	.0274 m
Clay-coated beads	ethyl alcohol	200 %	1.60 %	.0236 m
Clay-coated beads	hexadecane	50 %	0.40 %	.0261 m

3 TEST RESULTS

The data from the seismic propagation tests were analyzed in a number of different ways. The first-break arrival times were measured and used to calculate group velocities. In addition, the peak-to-peak voltage of the first wave cycle was measured as a first order estimate of attenuation. For more detailed analysis, the wave train was fast fourier transformed to obtain the spectral amplitude versus frequency information. Measured peak frequencies and calculated quality factors at this frequency were used to compare the seismic wave attenuation between the specimens.

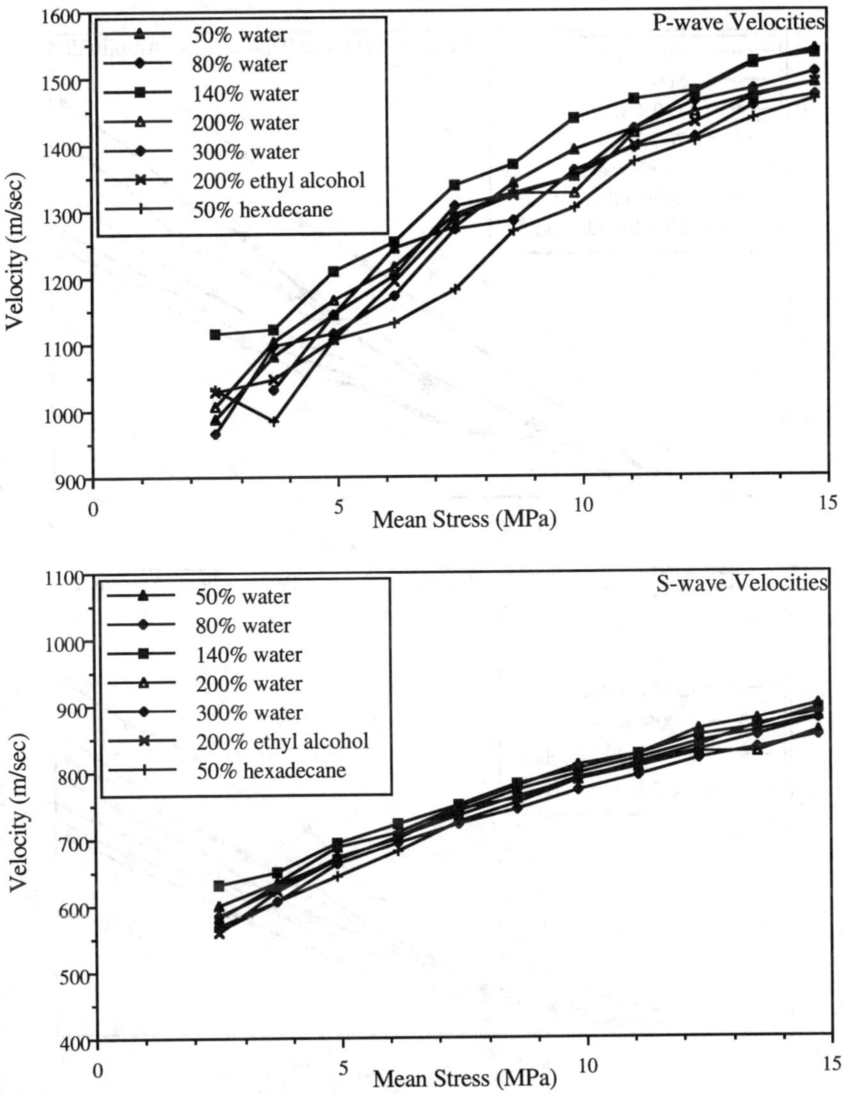

Figure 2: (a) P-wave and (b) S-wave velocities in clay-coated specimens that contain different amounts and types of fluid. The scales are the same, but the ranges differ.

The velocities from specimens with clay and fluid are shown in Figure 2. The P-wave velocities exhibit systematic but small differences between the specimens. The 140% water content and 50% hexadecane content P-wave velocities are slightly higher and lower, respectively, than average. The higher velocity for 140% water content indicates an increased contact stiffness. This could result from a larger contact area due to clay swelling while maintaining a high rheologic viscosity. The lower velocity for 50% hexadecane content indicates a decreased contact stiffness. The non-polar nature of hexadecane may allow unbound fluid in the macropores to form a thin fluid film at the contact, reducing contact stiffness. The S-wave velocities are similar between the specimens, reflecting their lower sensitivity to contact stiffness variations.

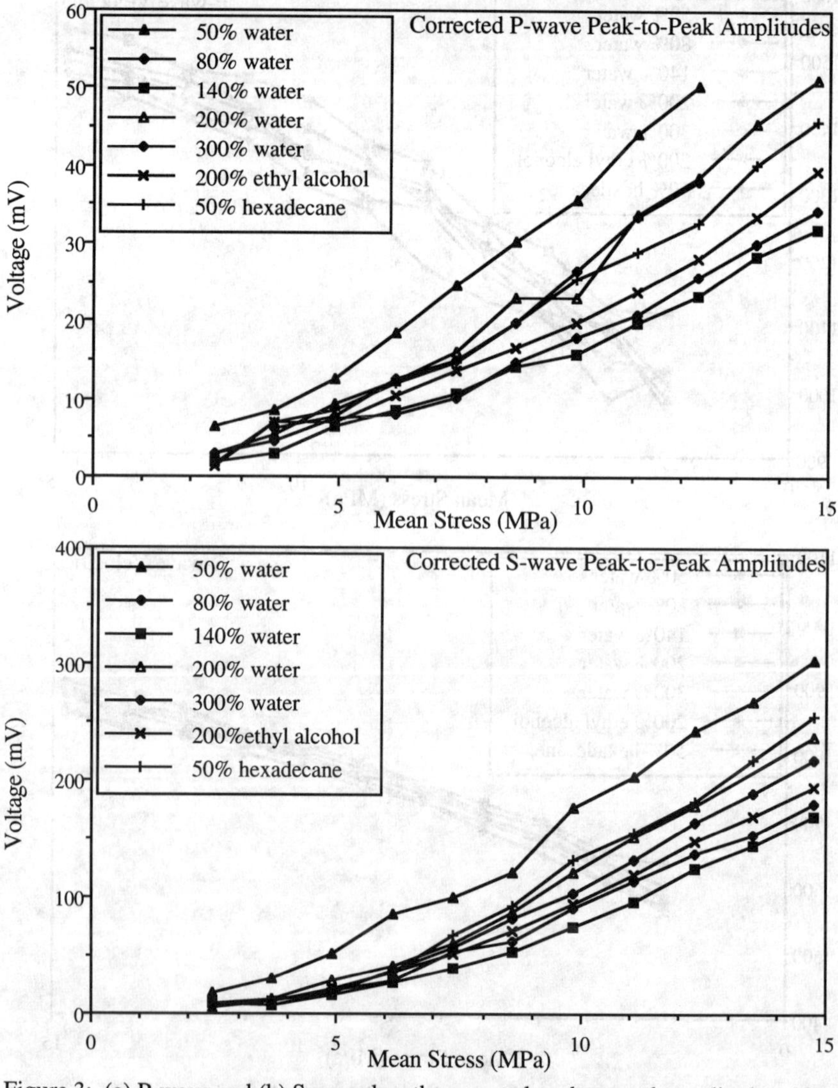

Figure 3: (a) P-wave and (b) S-wave length-corrected peak-to-peak amplitudes. Note the different scales. These are used as a first-order indication of attenuation.

The important trend in the velocity data is that both P and S-wave velocities increase with increasing stress. This indicates that the contact stiffness increases with stress. This is due to an increase in the number of contacts and in the size of the contact areas.

The measured peak-to-peak amplitudes required correction for variations in sample length before comparisons could be made between specimens. The correction was calculated using:

$$U_s(L_s) = U_m(L_m)e^{-\frac{\pi f \Delta L}{QC}} \quad (1)$$

where U_S is the corrected amplitude, U_m the measured amplitude, f the frequency, $\Delta L = L_S - L_m$ the sample length correction, Q the quality factor, and C the seismic velocity. The quality factor corresponding to each specimen was used (Table 2). This correction is valid because the peak spectral frequency was equivalent to the dominant frequency observed in the time domain waveform.

The length-corrected peak-to-peak amplitudes are shown in Figure 3. Both P and S-wave corrected amplitudes exhibit the same trend with changing clay conditions. The corrected amplitudes are highest for 50% water content, then decrease as water content increases to 140%. The 200% and 300% water content corrected amplitudes lie between the 50% and 80 % water content data. In general, the 50% hexadecane corrected amplitudes were similar to 200% water content. The 200% ethyl alcohol content corrected amplitudes were typically lower than 200% water content and 50% hexadecane content, but higher than 80% water content data. The seismic transmission implied by these results will be analyzed in more detail after spectral analysis.

Although the P and S-wave corrected amplitudes have the same trends, the loss mechanisms are distinct. P-wave losses are probably due to pumping, where the clay film is pushed into then retreats from the surrounding open pore space. This type of movement is not possible at large planar interfaces. S-wave losses are probably due to viscous dissipation, where the clay-water particles shift within the clay layer.

As with the velocity data, the peak-to-peak corrected amplitudes increase with increasing stress. However, the general ranking for the different fluid and fluid contents remains unchanged. This important result indicates that seismic energy transmission is universally sensitive to contact stiffness. It is therefore useful to compare the seismic behavior of the different specimens at the same contact stresses and thus mean stress.

Spectral amplitudes were calculated for each specimen at 8.6 MPa hydrostatic mean stress. The spectra for 50%, 80%, and 200% water content are shown in Figure 4. For comparison, the peak frequencies are ~800 and ~600 kHz for P and S-waves through a 0.0254 m aluminum standard. The spectral amplitudes are ~8 and ~45 mv/1024 points, respectively, at these peak frequencies. These are an order of magnitude larger than the spectral amplitudes in glass bead specimens. Above 200 kHz, the 200% water content spectral amplitudes are between the 50% and 80% water content values. Also, the 80% water content peak frequencies are slightly lower for both P and S-wave tests than for the other specimens. This trend agrees with the peak-to-peak corrected amplitudes.

The effects of clay-fluid interactions are best seen in the peak frequency-quality factor data presented in Table 2. The quality factor, Q, is a measure of the energy loss per cycle. It was calculated at the peak spectral amplitude frequency using a spectral ratio comparison to an aluminum standard as follows:

$$\alpha_S = \alpha_{AL}(f) + \frac{1}{L}\ln\left(\frac{|U_{AL}(f)|}{|U_S(f)|}\right) \quad (2)$$

$$Q_S = \frac{\pi f_S}{\alpha_S C_S} \quad (3)$$

where α is the attenuation coefficient, L the length of the sample, U the spectral amplitude, C the velocity, and f the frequency. Subscript S refers to the specimen; subscript AL to the aluminum standard. Geometrical spreading factors were computed, but found to be negligible and so were not included.

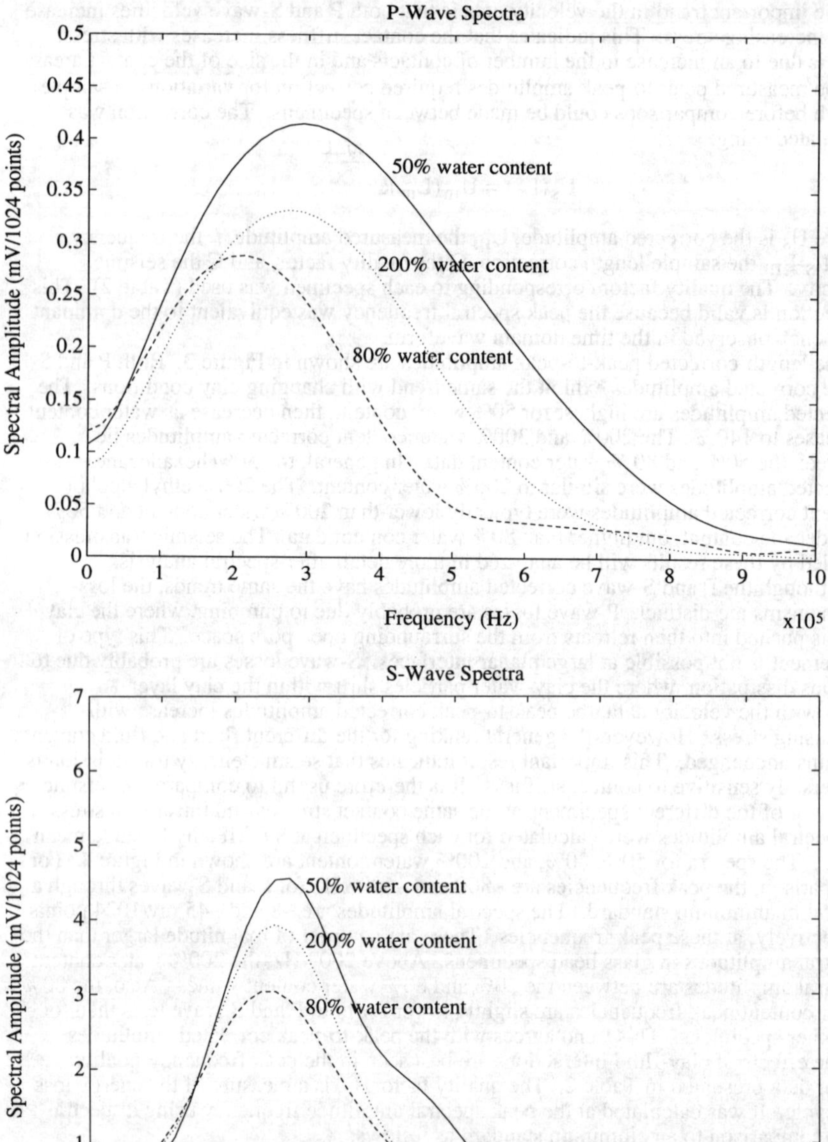

Figure 4: (a) P-wave and (b) S-wave frequency spectra from fourier transforms of transmitted wave for three different water content specimens. Note the different amplitude scales.

Table 2: Peak Frequency and Quality Factor Data for Tested Specimens

Test	Peak Frequency P-wave (kHz)	Quality Factor P-wave	Peak Frequency S-wave (kHz)	Quality Factor S-wave
Dry, Clean	322	10.424	293	15.478
Oven-dry	313	11.835	254	20.495
50% water	293	10.169	273	16.347
80% water	215	8.634	234	13.378
140% water	176	7.274	234	11.811
200% water	273	9.549	254	16.066
300% water	254	9.275	254	13.324
200% ethyl alcohol	234	8.190	254	13.842
50% hexadecane	234	9.947	234	15.016

Adding a dry clay film has a significant effect on seismic wave propagation as seen in the dry clean and oven-dry data. The peak frequencies decrease as clay is added; the quality factors, however, increase. The changes are more significant for S-waves, with a 13% decrease in peak frequency and a 32% increase in quality factor. This is consistent with clay being a visco-elastic medium. The introduction of clay increases the viscous losses in the material, thus decreasing the peak frequency. At the same time, the clay stiffens and/or increases the contact area, leading to an increase in transmitted energy and thus decrease in attenuation.

For clay with fluids, the peak frequency and quality factor data exhibit the same trend for both P and S-waves. As the clay water content increases from 0 to 140%, the peak frequencies shift to lower frequencies. The P-wave frequency shifts are much more significant than the S-wave shifts. The quality factors also decrease: P-waves by 39% amd S-waves by 42%. This is consistent with the polar water interacting with the charged clay particles and causing swelling, e.g. an increase in the clay volume due to the absorption of fluid. The swelling of the clay decreases its viscosity, leading to a more dissipative clay layer between the grain-to-grain contacts.

The 200% and 300% water content data are larger than the 140% water content data, with a 55% increase in P-wave peak frequency and a 34% average increase in quality factor. The values lie between the 50 and 80% water content data. This can be attributed to a change in clay structure near 200% water content. Scratch tests showed that up to 200% water content, swelling clay retains its structural cohesion (Suárez-Rivera, p.145). At 200% water content, however, clay loses some of its cohesion and can flow under stress (ibid., p.147). Therefore, at water contents at and above 200%, the clay mixture may be forced from the grain-to-grain contacts. This would result in a toroidal clay mass around the contact. A much thinner clay layer may still exist at the contact. This increases the peak frequency and quality factor over specimens with lower water contents,. The slightly lower values for 300% as opposed to 200% water content may be due to either a less viscous swelled clay or the presence of free water, e.g. not bound to the clay.

The 200% ethyl alcohol specimen peak frequencies and quality factors are generally 14% less than for the 200% water specimen. (The S-wave peak frequencies are similar.) Given their polar nature, both fluids swell clays. The lower seismic transmission in ethyl alcohol may be due to its larger molecular size. This may keep the ethyl alcohol molecules in the macropores instead of the micropores. At the same fluid content, ethyl alcohol will thus swell the clay less but also will be present as free fluid. The presence of unbound fluid will cause more losses at the grain-to-grain contacts and lead to lower peak frequencies and quality factors as seen.

The 50% hexadecane peak frequencies are significantly lower than for the 50% water content specimen, 20% and 14% for P and S-waves respectively. The quality factors are only slightly lower, 2% and 8% respectively. The non-polar nature of hexadecane will not cause swelling of the clay. However, hexadecane molecules may be absorbed into the macropores of the clay and be present as free fluid. This would lead to increased viscous losses. While the peak frequencies of the 50% hexadecane specimen are similar to the 300% water and 200% ethyl alcohol content specimens, the quality factors are higher. This can be attributed to hexadecane's higher liquid viscosity (3.34 cp) as opposed to water and ethyl alcohol (1.00 and 1.20 respectively).

4 CONCLUSIONS

Both P and S-wave transmissions exhibited the same trend with fluid type and content changes. This is in contrast to previous study with a single clay-coated quartz interface. For that single interface, only the S-wave transmission changed in response to fluid changes. The difference appears to be due to a point-to-point contact geometry and the presence of open pore space in this multi-grain study. This allows for pumping loses during P-wave propagation that are not possible at large planar interfaces.

Seismic velocities and peak-to-peak amplitudes both increased with increasing stress. This suggests that seismic wave propagation is sensitive to contact stiffness. Likewise, contact stiffness changes due to clay film rheologic changes at and near grain contacts can have significant effects on seismic wave propagation. Changes were most significant for attenuation. An increase from 0 to 140% water content in the clay film, producing a change from 0 to 1.12% in the overall specimen water content, decreased the P and S-wave quality factors by 40%.

The rheologic viscosity changes are the result of clay-fluid interactions. Polar fluids swell reactive clay. This decreases the viscosity of the clay film and initially increases seismic attenuation. At the point where the swelling clay loses structural coherence, the low viscosity of the clay film allows it to be pushed from the grain-to-grain contact into the surrounding open pore space. The smaller distance between grains and the larger contact area result in decreased seismic attenuation. Polar fluids at high fluid contents and non-polar fluids at measurable fluid contents may be present as unbound fluid. Seismic attenuation will be higher with unbound fluid than with incorporated fluid and will depend on the unbound fluid's viscosity. In general, higher viscosity fluids will have lower attenuation loses at ultrasonic frequencies.

This experimental study shows that small changes in fluid content, fluid type, and amount of a reactive clay film can have significant effects on seismic wave propagation in unconsolidated materials. The data suggest that in principle attenuation measurements may be used to detect clay and fluid contents in the subsurface.

ACKNOWLEDGEMENTS
This work was supported by the Gas Research Institute under Contract 5086-260-1830 and by the Earth Sciences Division, Lawrence Berekeley Laboratory.

REFERENCES

Anstey, N.A., 1991. "Velocity in Thin Section", *First Break*, v.9, n.10, p.449-457.

Han, D.H., A. Nur, and D. Morgan, 1986. "Effects of Porosity and Clay Content on Wave Velocities in Sandstones", *Geophysics*, v.51, n.12, p.2093-2107.

Klimentos, T. and C. McCann, 1990. "Relationships among Compressional Wave Attenuation, Porosity, Clay Content, and Permeability in Sandstones", *Geophysics*, v.55, n.8, p.998-1014.

Kowallis, B.J., L.E.A. Jones, and H.F. Wang, 1984. "Velocity-Porosity-Clay Content Systematics of Poorly Consolidated Sandstones", *JGR*, v.89, n.B12, p.776-792.

Suárez-Rivera, F.R., 1992. "The Influence of Thin Clay Layers Containing Liquids on the Propagation of Shear Waves", Ph.D. Thesis, University of California, Berkeley, California.

14 Natural fracture systems

Session developers:
Mel Friedman, Texas A&M University
John Lorenz, Sandia National Laboratories
Neil Hurley, Marathon Oil Company

Rock Mechanics, Nelson & Laubach (eds) © 1994 Balkema, Rotterdam, ISBN 90 5410 380 8

Fractal point patterns and fractal fracture traces

Robert L. Kranz
US Bureau of Mines, Denver, Colo., USA

ABSTRACT: Fracture traces and seismic event location data were examined and numerical experiments performed to test whether scale-invariant fracture trace patterns can be represented by point patterns generated at discrete trace sites. Point pattern data censored in different ways were examined with respect to the effects on fractality. A sparse fractal point data set remains fractal, but the scale range of fractality decreases as sparseness increases. Thus, a seismic event location point pattern may not adequately represent the full scale over which the parent faults are fractal. An example of the inverse problem of determining the fractal nature of unknown fault traces from seismicity is given.

1 INTRODUCTION

Given information about the organization of cracks, joints or faults within a rock mass, and knowledge of the fracture and frictional strengths, one might be able to predict the locations where seismic events will occur for a set of imposed boundary stresses. The inverse problem is more common, however. Given a pattern of points that represent acoustic emission (AE) or earthquake (EQ) source locations, and knowledge of the stress state, what can be inferred about the organization of cracks and faults on which the seismic events occur? This problem of correlating unknown structure elements with seismicity is made particularly difficult because of the non-unique geometric relationship between a few points and the lines and planes that may contain them. If correlations could be made between seismic event point patterns and geologic structure, mining into unexplored rock mass could potentially be less dangerous. Furthermore, seismologists might have a stronger basis for evaluating earthquake probabilities in areas where faults are not exposed to view. If the point and line patterns have common characteristics that persist over a range of scales, extrapolations (predictions) might be made from scales where only seismicity or only geologic structure data exist to different scales of the other.

The study of fractal geometry has become established as a way to quantify spatial relationships and extrapolate between different scales of various geologic data (Korvin 1992). Evidence of fractality in both seismicity and related structure speaks to common, persistent physical processes as the source of each. Spatial point patterns representing AE and EQ source locations have been shown to be fractal, or self-similar, over various scales (Kagan and Knopoff 1980; Sadovskiy et al. 1984; Hirata et al. 1987; Kusunose et al. 1991; Hirata 1989; De Rubeis et al. 1993). Likewise, distributions of crack, joint and fault traces have been shown to be fractal within specific, bounded scales, from hand samples to tectonic regimes (Barton and Larsen 1985; Chiles 1988; Hirata 1989; Barton and Hsieh 1989; Merceron and Velde 1991. If we confine ourselves to the scale limits in which fractality holds, what relationship can be inferred between the spatial organization of cracks and faults and the AE or EQ source location patterns generated? Specifically, in this paper we consider the representability of fractal line patterns by point patterns.

2 NUMERICAL EXPERIMENTS

Barton and Larsen (1985) mapped tuff fracture traces intersecting flat pavements on Yucca Mountain, Nevada. All traces greater than 20 cm in length were mapped. (Additional data and results are given in Barton and Hsieh, 1989). These traces were shown to be fractal in that over a specified scale range, they obeyed the power law relation

$$N = r^{-D}, \tag{1}$$

where N is the number of square boxes of length r containing portions of one or more fracture traces, and D is the capacity or box dimension of the distribution, obtained by fitting log N as a function of log(1/r). D was constant for any particular pavement over the scale of one half to several tens of meters. D values ranged from 1.4 to 1.7 for different pavements. Chris Barton generously provided the digitized trace data for these pavements upon which numerical experiments were performed.

For the first experiment, we ask: given a set of fractal line traces, can they be represented by a point set, is that point set fractal, and if so, what is the relationship between the fractal dimensions of the sets? Initially, two sets of points for each of the pavement fracture trace data sets were created. One set contained trace line midpoints and the other trace line endpoints. Figure 1 shows an example of mapped fracture traces for a single pavement and the derived trace line midpoint pattern. Both the endpoint and the midpoint data sets were analyzed to see if they, also, were spatially fractal. This was done in two ways. First, equation 1 was employed to calculate a box dimension, where N is the number of cells of size r containing at least one point. Second, a correlation dimension was calculated using the correlation integral

$$C(R) = \frac{2}{n(n-1)} N(r<R) \tag{2}$$

where n is the total number of points in the data set, and N is the cumulative number of location point pairs (X_i, X_j) having an inter-point distance r less than some scale length R. This means that in any data set of X_n event location points, *all* possible inter-event distances are calculated and used to determine the distribution. The log of C(R) is then fit as a function of log R, using traditional least-squares methods, and the slope of the line gives the correlation dimension.

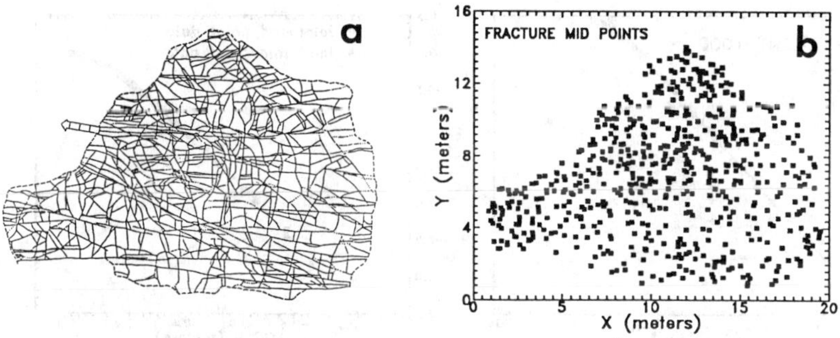

Figure 1. a) Fracture pavement 1000 (Barton and Hsieh 1989) and b) corresponding fracture midpoint map.

To distinguish these two different fractal dimensions, D_B and D_C are used to represent the box and correlation dimensions, respectively. Examples of applying equations 1 and 2 are shown in figure 2. Table 1, which shows the results (D_B and D_C) of this experiment for different pavements, indicates that values of the capacity or box dimension D_B, determined from the end point patterns, are in reasonably good agreement with the D_B calculated by Barton and coworkers. The scale range over which the log-log plots are linear, and therefore over which the pattern is

TABLE 1: ANALYSES BASED ON BARTON'S DATA

DATA SET	# PTS	D_C	D_B	RANGE OF FRACTAL BEHAVIOR (m)	BARTON'S D_B
P100.END	414	1.64	1.38	1.3 - 12	1.52
P100.MPT	208	1.69	1.26	1.3 - 12	
P200.END	196	1.43	1.23	1.6 - 14	1.38
P200.MPT	99	1.89	1.19	2 - 14	
P300.END	478	1.62	1.47	1.1 - 12	1.5
P300.MPT	240	1.67	1.34	1 - 10	
P400.END	924	1.71	1.52	2 - 20	1.61
P400.MPT	463	1.87	1.46	3 - 20	
P500.END	606	1.71	1.61	1.1 - 7	1.59
P500.MPT	304	1.97	1.55	1.1 - 7	
P600.END	642	1.63	1.55	1.4 - 20	1.54
P600.MPT	321	1.83	1.48	1.6 - 20	
P1000.END	1226	1.77	1.73	0.6 - 5	1.7
P1000.MPT	614	1.79	1.69	0.6 - 5	

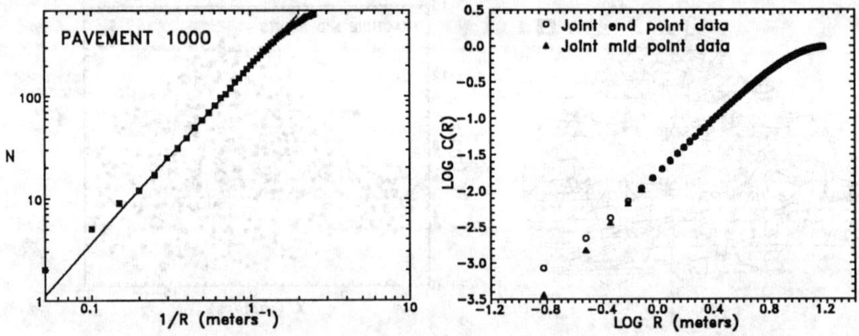

Figure 2. Examples of determination of box dimension (left) and correlation dimension (right) using equations 1 and 2 respectively. Best fitting slopes give the dimensions D_B and D_C listed in Table 1.

represented by a single fractal dimension, is typically one order of magnitude. One can expect the lower end of the fractal range to be slightly different for the midpoint data than for the end point data because the mean nearest-neighbor distance is larger for midpoint representations. Note the faster roll-off at the low end for the midpoint data in figure 2. The linear range here is almost the same for the box and correlation dimension analyses.

Next one needs to consider the effect of censoring the point patterns. All seismic event location data sets are censored to some extent by instrumentation, computational or temporal limitations. Accordingly, 50%, 75% and 90% of the midpoint data shown in figure 1b were randomly removed. These censored data sets were then analyzed using equation 2. Results are shown in figure 3, where it can be seen that as the data sets get sparser, the range of scale invariance diminishes. For example, the data set containing only 10% of the original points appears to match the slope of 1.75 only within the small range of $1.5 < R < 4$ meters. Many additional straight lines with larger slopes (higher fractal dimensions) might be fitted to the sparse data for short ranges, yet any connection to real geologic structure at those ranges would be fortuitous. Thus, while censored point data may reflect the underlying structure's fractality, the entire range of structural scale-invariance may not be fully represented by sparse point sets.

Some fractures will be aseismic and some multiseismic. In any rock mass subjected to differential stresses, pre-existing cracks, joints or faults oriented within about 30 degrees of the maximum principal stress are those most likely to fail, and therefore are those most likely to emit seismic energy. New cracks also tend to form at small angles to the maximum principal stress. Accordingly, next analyzed were only the subset of fractures in any pavement having orientations within 5, 15, or 30 degrees of the X axis, assumed for this experiment to be the maximum principal stress direction. Figure 4a shows the midpoints of such fractures extracted from the set displayed in figure 1a, and figure 4b is the result of applying the correlation integral equation 2 to each subset. One can see that the point sets representing the oriented fractures retain their fractal character. The same analysis on oriented subsets from the other pavements (Table 1) gave similar results. Thus a

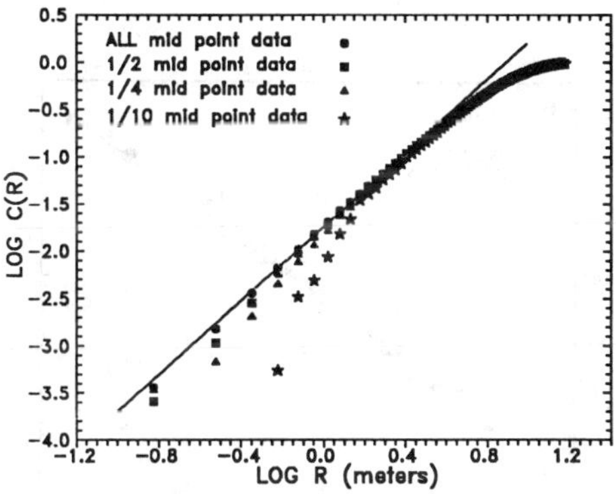

Figure 3. Effect of random data censoring on the determination of correlation dimension. Slope of the line gives $D_C = 1.7$.

"seismically active" subset of fractal fractures is similarly fractal.

The three preceding experiments give one confidence that fractal fracture distributions can be represented by corresponding seismic event point locations. The sufficiency of the representation will depend on the number of such points and the scale of area or volume in which they occur. Other, similar experiments have shown that non-fractal trace line patterns do not give rise to fractal point patterns, except in certain pathological cases. An example problem of inferring unknown structure from a seismic event location point set is now presented.

Figure 5 shows a portion of the State of Utah, approximately 25 km by 30 km in extent, that contains a number of active underground coal mines. Although most of the mapped, nearly vertical, normal faults are in the northeast quadrant of this area, most of the seismicity is associated with coal mines in the southern half of this area (Wong 1985; Wong et al. 1989; Williams and Arabasz 1989). Event locations were obtained from the University of Utah Seismograph Stations catalog. Only the 334 events with magnitudes larger than 2.0 were analyzed. Although depths were not well-constrained, the vast majority of these have assigned depths greater than 3 km, far below the mines in the area. Smaller events are typically shallower and more likely to be associated with a mine itself. Focal mechanisms of these probably deep events are consistent with the prevailing extensional tectonic regime (Wong et al. 1989). However, very few faults of a size sufficient to generate M > 2 events are known to intersect the mine workings. Is there a connection between these larger, deep events and a fault system below the mines that is similar to the known fault system further north?

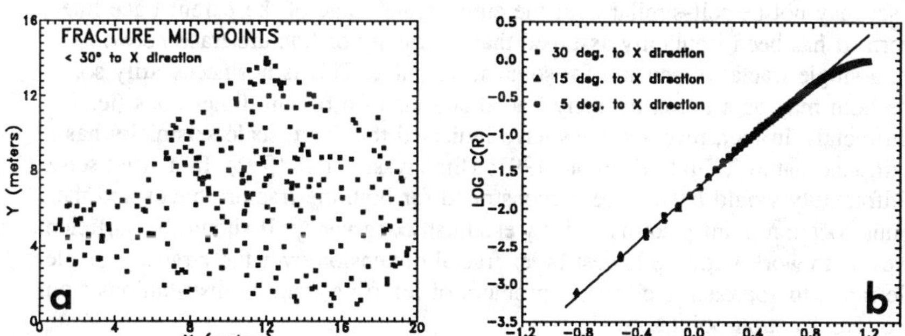

Figure 4. Analysis of points representing fracture traces censored according to orientation with respect to a hypothetical maximum stress in the X direction.

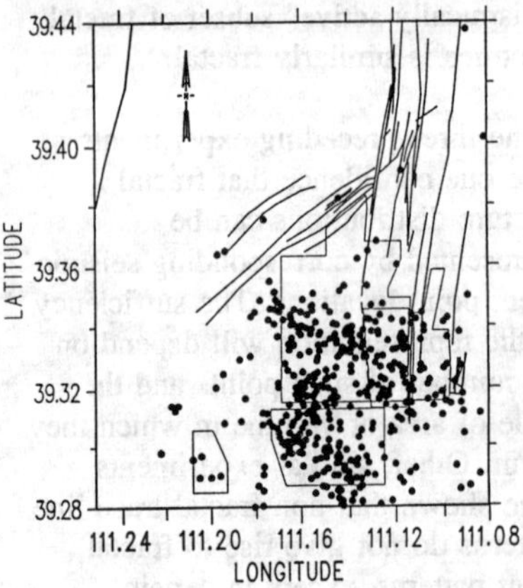

Figure 5. Map of a portion of the State of Utah, U.S.A., showing surface-breaking faults, seismicity (1962-1992, M > 2.0) and approximate areas of underground coal mines.

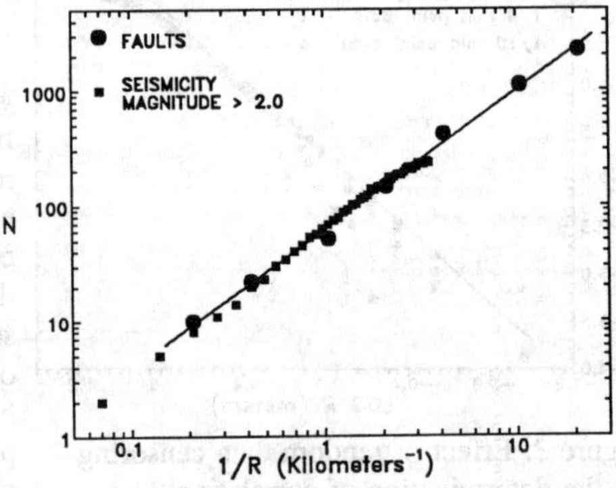

Figure 6. Box dimension determination of surface visible fault traces and seismicity shown in Fig. 5. Slope of line gives $D_C = 1.21$.

The M > 2 seismicity and the fault traces in figure 5 were each analyzed using equation 1. Both the seismicity and the fault traces were found to be fractal with box dimensions D_B equal to 1.30 and 1.21 respectively. The data are shown in figure 6 where it can be seen that the scales of the data sets overlap, extending together over more than two orders of magnitude. Although the fractal dimensions of the two data sets are not identical, they are close enough to suggest that the deep events occurred on a fault system similarly organized as the one to the north. This similarity applies only to the scale ranges investigated.

3 DISCUSSION AND CONCLUSIONS

It has been demonstrated that, if fracture traces are fractal over some scale range, they can be represented by a set of points on a plane with the same fractality, but the set may not be self-similar over the entire scale range of the parent trace line pattern. It has been implicitly assumed that seismicity or fracture trace networks have a single fractal dimension for some scale range. This is not necessarily so, since both may be a union of many fractal subsets of different dimensions (ie, multifractal). In fact, recent studies have indicated that large scale seismicity has a multifractal nature (Hirata & Imoto 1991; Hirabayashi et al 1992). In a strict sense, multifractality would have to be demonstrated for both the fracture traces and the seismic location point patterns to fully establish congruency. It should be sufficient, however, to work with the lowest order fractal dimension over the largest possible scale range to approach the inverse problem of inferring fracture distributions from AE or EQ location point sets.

For a rock mass, all planar slices through the fracture network may not yield the same fractal dimension. In this paper, we worked exclusively with horizontal planes. However, the volumetric organization of the fractures can be studied with a representative three-dimensional point set by sequentially selecting those points lying within thin tabular regions that cut the network in different orientations. A minimum of three such point subsets from orthogonal tabular regions would be necessary to determine if the fracture network is a homogeneous or self-affine fractal. To obtain information on *individual* fracture planes within the rock mass, an algorithm that finds points associated with that plane would be needed (eg. Fehler and Johnson 1989).

In addition to providing information about the spatial organization of the underlying fracture network, the fractal dimension of the point pattern may provide information about the proportion of the fracture population that is seismically active. La Pointe (1988) investigated the relationship between fracture numbers, "blockiness" created by intersecting fracture boundaries and the associated fractal dimensions. He found that the lower the fracture intensity (sparser trace lines), the more variable was the resulting block density and the lower was the fractal dimension of the network. From the works initially cited above, we know that the greater the degree of clustering of points, the lower the fractal dimension. Therefore, if the event location points represent seismically active fractures, lower point pattern fractal dimensions suggest fewer active fractures and less uniform deformation. Higher fractal dimensions would suggest greater numbers of fractures or a larger proportion of the fracture network participating in the regional deformation. This idea is consistent with laboratory rock mechanics experiments that have shown fairly uniform, sample-wide deformation, diffuse AE locations and high fractal dimensions initially, but less uniform deformation, more clustered AE locations and lower fractal dimensions as failure is approached.

By examining seismicity in a mine, use may be made of these results to infer changes in seismically active fracture networks before and after blasting, or to infer differences in active fracture networks in different loaded pillars or seams. If seismic event locations are fractal, it is likely that the parent fractures are also. Any maps of fracture traces that intersect mine openings could then be extrapolated at least up to the scale of the seismicity in the rock mass to infer otherwise non-obtainable density or spacing information.

REFERENCES

Barton, C.C. & E. Larsen 1985. Fractal geometry of two-dimensional fracture networks at Yucca Mountain, southwest Nevada, in: O. Stephannson (ed), *Fundamentals of rock joints, Proc. of the International Symp.*, Bjorkliden, Sweden. 77-84. Rotterdam: Balkema.

Barton, C.C. & Hsieh 1989. Physical and hydrologic-flow properties of fractures, in: *Field Trip Guidebook T385, 28th International Geological Congress*, Washington, D.C.

Chiles, J.P. 1988. Fractal and geostatistical methods for modeling of a fracture network, *Math. Geol.* 20: 631-654.

De Rubeis, V., P. Dimitriu, E. Papadimitriou & P. Tosi 1993. Recurrent patterns in the spatial behavior of Italian seismicity revealed by the fractal approach, *Geophys. Res. Lett.* 20: 1911-1914.

Fehler, M. and P. Johnson, 1989. Determination of fault planes at Coalinga, California by analysis of patterns in aftershock locations, *J. Geophys. Res.*, 94: 7496-7506.

Hirabayashi, T., K. Ito & T. Toshii 1992. Multifractal analysis of earthquakes, *Pure and Appl. Geophys.*, 138: 591-610.

Hirata, T. 1989. Fractal dimension of fault systems in Japan: fractal structure in rock fracture geometry at various scales, *Pure Appl. Geophys.* 131: 157-170.

Hirata, T. & M. Imoto 1991. Multifractal analysis of spatial distribution of microearthquakes in the Kanto region, *Geophys. J. Int.*, 107: 155-162.

Hirata, T., T. Satoh & K. Ito 1987. Fractal structure of spatial distribution of microfracturing in rock, *Geophys. J. R. Astron. Soc.*, 90: 369-374.

Kagan, Y.Y. & L. Knopoff 1980. Spatial distribution of earthquakes: the two-point correlation function, *Geophys. J. R. Astron. Soc.*, 62: 303-320.

Korvin, G. 1992. *Fractal models in the earth sciences*, 396 pp, N.Y.: Elsevier.

Kusunose, K., X. Lei, O. Nishizawa & T. Satoh 1991. Effect of grain size on fractal structure of acoustic emission hypocenter distribution in granitic rock, *Phys. Earth & Plan. Int.*, 67: 194-199.

LaPointe, P.R. 1988. A method to characterize fracture density and connectivity through fractal geometry, *Int. J. Rock Mech. Min. Sci.*, 25: 421-429.

Merceron, T. & B. Velde 1991. Application of Cantor's method for fractal analysis of fractures in the Toyoha mine, Hokkaido, Japan, *J. Geophys. Res.*, 96: 16641-16650.

Sadovskiy, M. A., T.V. Golubeva, V.F. Pisarenko & M.G. Shnirman 1984. Characteristic dimensions of rock and hierarchical properties of seismicity, *Izvestiya, Earth Phys.*, 20: 87-96.

Williams, D.J. and W.J. Arabasz 1989. Mining-related and tectonic seismicity in the East Mountain area Wasatch Plateau, Utah, U.S.A.,*Pure & Appl. Geophys.* 129: 345-368.

Wong, I.G., 1985. Mining-induced earthquakes in the Book Cliffs and Eastern Wasatch Plateau, Utah, U.S.A., *Int. J. Rock Mech. Min. Sci.*, 22: 263-270.

Wong, I.G., J.R. Humphrey, J.A. Adams and W.J. Silva 1989. Observations of mine seismicity in the Eastern Wasatch Plateau, Utah, U.S.A.: a possible case of implosional failure, *Pure & Appl. Geophys.* 129: 369-405.

Rock Mechanics, Nelson & Laubach (eds) © 1994 Balkema, Rotterdam, ISBN 90 5410 380 8

On the paradox of systematic, contemporaneous, orthogonal opening-mode fractures

Stephen J. Martel
Department of Geology and Geophysics, University of Hawaii, Honolulu, Hawaii, USA

ABSTRACT: Simple two-dimensional, plane strain, linear elastic analyses of the stress field around a mode I fracture indicate that opening and closing of fractures could contribute to the formation of contemporaneous orthogonal joints. However, either nonlinear elastic processes or non-elastic processes apparently must operate if a well-developed orthogonal network of joints is to form at depth during a single period of deformation.

1 INTRODUCTION

The engineering behavior of rock masses is strongly influenced by natural macroscopic fracture patterns. For example, it is well known that a preferred orientation of fractures can cause a rock mass to be mechanically and hydrologically anisotropic, and that a highly connected network of fractures can substantially weaken a rock mass and increase the overall hydraulic conductivity. Correspondingly, engineering operations involving fractured rock are strongly influenced by theoretical arguments as to how natural fractures might or might not form. Two examples illustrate this point well. First, in many, if not most, engineering situations we can observe directly only a small portion of a fracture network. Theory provides the rationale for extrapolating the available direct data to predict the macroscopic fracture pattern in a large volume of rock. Second, theory influences where we search to find or avoid fractures in the subsurface. As a result, macroscopic fracture patterns and our theoretical considerations regarding those patterns impact the planning and execution of many engineering operations.

This paper addresses patterns of joints (i.e., opening-mode fractures) that form in a single episode of fracturing. The particular focus is on vertical joints, which are perhaps the most common type of fracture in horizontal sedimentary rocks. Such joints open perpendicular to the local least compressive horizontal stress. They can form systematic and unsystematic patterns (Pollard & Aydin 1988). Systematic joints (e.g., Figure 1A) are planar, and in plan view their traces form an orderly arrangement, whereas unsystematic joints (Figure 1B) have irregular, non-parallel traces. Olson & Pollard (1989) showed that if the difference between the regional horizontal principal stresses is large relative to the stress driving the joints open, then a single systematic set of parallel joints can develop; strong anisotropy in the regional horizontal stress field results in a strong ordering of the joints. Unsystematic joints tend to develop where the horizontal principal stresses are nearly equal; here the isotropy of the regional horizontal stress field can lead to an irregular disordered fracture pattern. The two

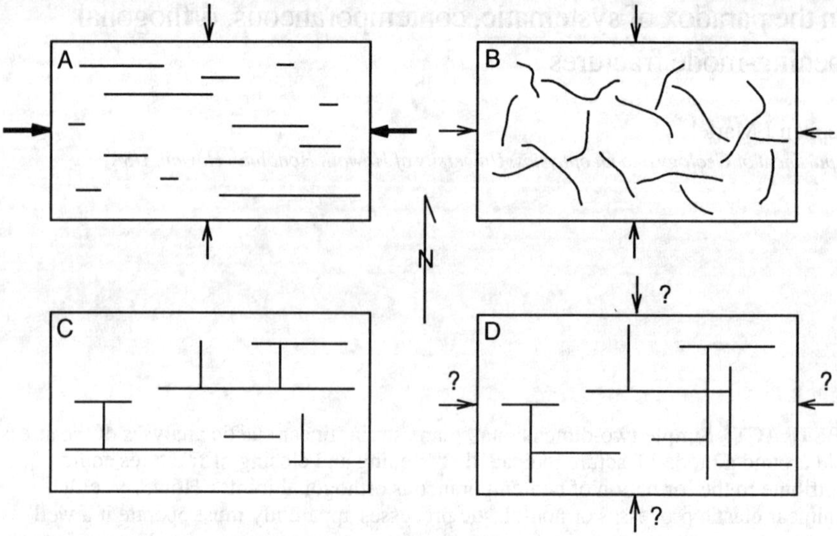

Figure 1. Plan view diagrams of the traces of vertical joints. Arrows denote orientations of principal stresses where fractures form contemporaneously. A) A single set of parallel systematic joints. B) Unsystematic joints. C) Two sets of orthogonal joints, with the east-west set being older. D) Contemporaneous orthogonal joints. In (D), east-west joints abut against north-south joints with the same frequency as north-south joints abut against east-west joints.

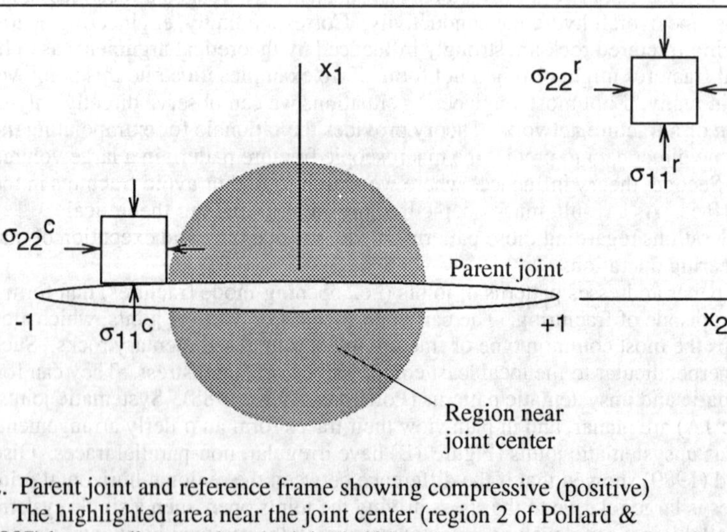

Figure 2. Parent joint and reference frame showing compressive (positive) stresses. The highlighted area near the joint center (region 2 of Pollard & Segall 1987) has a radius equal to the joint half-length.

styles of fracture patterns will have significantly different effects on the mechanical and hydrologic behavior of a rock mass.

The particular focus here is on contemporaneous orthogonal joints (e.g., Turner & Hancock 1989, Dunne & North 1990). A highly interconnected network of orthogonal joints is one of the most common fracture patterns. In some cases one set of joints can be identified as being older than the other (Figure 1C). Multiple deformations at depth clearly could lead to such a pattern. Some investigators also have concluded that an orthogonal network of fractures could form due to a single deformation at depth followed by a relief of stresses during uplift and erosion of the overburden (Nickelsen & Hough 1967, Engelder 1987). Recent experimental work on fractures in plastics supports the idea that orthogonal fractures can form in a relatively simple loading and unloading process (Rives & Petit 1990). However, the orthogonal fracture pattern shown in Figure 1D has posed a mechanical paradox. Systematic joints that form in one episode of deformation are generally expected to be of a single orientation, yet the abutting relationships of the joints in Figure 1D indicate that those joints formed in a single episode of deformation. An explanation of the pattern in Figure 1D seems to require that the least compressive stress be in two different orientations during one episode of deformation. If the two regional horizontal principal stresses were equal, then one might expect an unsystematic pattern to develop (Olson & Pollard 1989). One alternative explanation, that contemporaneous orthogonal joints arise by repeated rotations of the regional principal stresses by precisely 90°, is patently implausible. Another alternative is that the regional principal stresses are approximately equal, and that minor oscillations in one or both of them could be cause the local principal stresses to alternate by 90°. An additional possibility is that, under constant regional stresses, the fracturing process itself causes local 90° reorientations of the principal stresses, and that is examined here.

2 MECHANICAL CONTINUUM ANALYSES

We now examine a few different scenarios for how the growth of a "parent" joint might promote the formation of a nearby orthogonal "daughter" joint. In all analyses the rock mass is treated as an isotropic, isothermal, homogeneous, linear elastic material. A large body of work shows that processes other than linear elastic ones play a key role in the formation of fractures (e.g., Costin, 1987); such processes are for the most part unaccounted for in the analyses described here. However, as noted by Biot & Willis (1957), the elastic effects described here should be applicable to at least incremental changes in a nonlinear elastic medium.

All the analyses deal with a single joint under two-dimensional, plane-strain loading conditions. Following standard fracture mechanics treatments (e.g., Pollard & Segall 1987), the parent joint is treated as a narrow crack (Figure 2). The joint is normal to the x_1 axis. The x_3 dimension of the joint is taken as large relative to its x_2 dimension and is vertical. The half length of the joint along the x_2 axis is set to unity. Just two perpendicular normal stresses are considered, σ_{11}, which acts normal to the joint, and σ_{22}, which acts parallel to the joint. Far from the joint, the least compressive stress is σ_{11}^r and the most compressive stress is σ_{22}^r. Compressive stresses are treated here as positive, so $\sigma_{22}^r > \sigma_{11}^r$. The normal stress acting on the joint walls is σ_{11}^c and is equal to the fluid pressure (if any) in the joint (P^c). Pure mode I loading conditions are applied, so no shear stress is resolved on the walls of the joint (i.e., $\sigma_{12}^r = \sigma_{12}^c = 0$). The parent joint will be taken to be open slightly, with the stress state in the surrounding rock having relaxed back to the ambient level. The aperture of the parent joint thus will be able to increase or decrease.

The discussion here does not treat mechanical effects of a fluid pressure P^r in the rock matrix explicitly. However, a uniform pore pressure in the matrix does act to decrease the effective normal compressive stresses (σ') felt by the mineral framework of a rock mass. Total normal stresses (σ) and effective normal stresses are related as follows:

(1) $\sigma' = \sigma - \alpha P^r$,

where α is a constant, under a given set of environmental conditions, between zero and one (Nur and Byerlee 1971). A positive matrix pore pressure can cause effective tensile stresses to exist where the total stresses are compressive, and in that sense it contributes to the opening of new fractures. However, because the total stresses and the corresponding effective normal stresses differ by a constant, a uniform matrix pore pressure will not cause a reorientation of the principal stresses at any point. As a result, unless σ_{22} is locally less than σ_{11}, a uniform pore pressure will not promote the formation of a daughter joint orthogonal to the parent. In the stress analyses that follow, the changes in the mean normal stress in the rock matrix are for the most part small, and hence variations in pore pressure in the rock matrix are also likely to be small (Rice & Cleary 1976). Changes in the total normal stresses are thus the key issue.

In order for the fracturing process to cause a daughter joint to open orthogonal to a parent joint, two criteria must be met: 1) σ_{22} must locally become sufficiently tensile to fracture rock, 2) σ_{22} must become more tensile than σ_{11}. The fracturing process also should be unlikely to produce significant numbers of daughter fractures oblique to the parent joint.

2.1 *Effect of opening of joints*

Following Secor (1965), the parent joint will open if either (a) the fluid pressure in it (P^c) is positive and exceeds the least compressive far-field stress normal to the fracture, or (b) the regional principal stress normal to the fracture is tensile. For a uniform pressure in the joint, both these conditions can be stated as follows:

(2) $\sigma_{11}{}^r - P^c < 0$.

The difference ($\sigma_{11}{}^r - P^c$) is known as the driving stress $\Delta\sigma_I$. If compressive stresses are treated as positive, then the joint will open if the driving stress is negative, that is

(3) $\Delta\sigma_I < 0$.

Once the parent joint opens, the stress state in the rock around it is perturbed. At distances more than a few fracture half-lengths from the parent the perturbing effects become small. We focus now on what might cause σ_{22} to become be tensile (i.e., negative) and cause an orthogonal joint to nucleate either along the walls of the parent joint or in the rock matrix a small distance from the center of the joint. The walls of the parent are free of shear stress, so the principal stress trajectories, which are potential paths for opening-mode fractures, are locally parallel and perpendicular to the walls of the parent joint.

The normal stresses along the walls of the parent joint are (Pollard & Segall 1987):

(4) $\sigma_{22}{}^c = \sigma_{22}{}^r - \Delta\sigma_I = \sigma_{22}{}^r - \sigma_{11}{}^r + P^c$,

(5) $\sigma_{11}{}^c = P^c$,

If the driving stress is negative and the joint opens, equation (4) shows that the compression along the walls of the parent will *increase* above the ambient level $\sigma_{22}{}^r$ along the entire length of the parent. This holds even if $\sigma_{22}{}^r$, the most compressive regional effective stress, is tensile. Opening of a parent joint thus works against the nucleation of an orthogonal daughter joint everywhere along the walls of the parent. This holds regardless of whether opening is due to a regional tensile stress perpendicular to the joint or to a high fluid pressure in the joint. On the other hand, if the driving stress is positive, the joint will close, and the compression along the entire length of the parent will decrease. The issue of how the parent joint might close given that it is perpendicular to the most tensile regional effective stress will be discussed below.

We now turn to a broader region near the center of the joint (Figure 2). Approximate values for the effective normal stresses there are (Pollard & Segall 1987):

(6) $\sigma_{22} = (\sigma_{22}{}^{r} - \Delta\sigma_{I}) + \{\Delta\sigma_{I}\,(2|x_1|)\} = \sigma_{22}{}^{c} + \{\Delta\sigma_{I}\,(2|x_1|)\}$

(7) $\sigma_{11} = (\sigma_{11}{}^{r} - \Delta\sigma_{I}) = \sigma_{11}{}^{r} - (\sigma_{11}{}^{r} - P^{c}) = P^{c}$;

Equation (6) reveals that if the driving pressure is negative and the joint opens, then the effective compressive stress parallel to the joint walls initially will decrease with distance from the joint. This stress reaches a minimum at $x_1 = \sqrt{2}$:

(8) $\sigma_{22}\,(x_1=\sqrt{2}, x_2=0) \approx \sigma_{22}{}^{r} + [0.1]\,\Delta\sigma_{I}$.

(9) $\sigma_{11}\,(x_1=\sqrt{2}, x_2=0) \approx \sigma_{11}{}^{r} - [0.5]\,\Delta\sigma_{I} .= (\sigma_{11}{}^{r} + P^{c})/2$.

Under a restricted set of conditions (e.g., $\sigma_{22}{}^{r} = 0$, and $\sigma_{11}{}^{r} = -P^{c}$), orthogonal daughter joints could form off to the sides of a parent joint in response to opening of the parent (Figure 3). A key shortcoming with this scenario is that stresses become even more tensile in other locations around the joint (Figure 3), and daughter joints oblique to the parent should be able to open in a variety of locations. Opening of an isolated parent joint seems unlikely to lead to planar orthogonal daughter fractures.

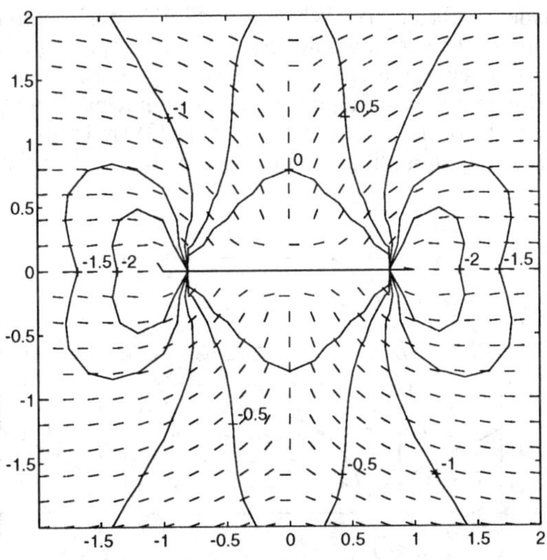

Figure 3. Stress state for a joint with a unit pressure under a remote uniaxial tension of unit magnitude. The contour map of the most tensile principal stress is superposed upon a plot of stress trajectories parallel to the local most compressive stress. The parent joint is centered in the plot. Daughter joints would be able to open everywhere except between the contours labeled zero and the parent joint. Daughter joints would be oriented along the stress trajectory tick marks. Daughter joints orthogonal to the parent could open along the x_1 axis where $0.9 < |x_1| < 1.5$. Daughter joints oblique to the parent could open at many places and are most likely to form near the tips of the parent.

2.2 *Effect of closing of joints*

Equation (4) establishes that if the driving stress were positive and the aperture of the parent joint decreased, then the compressive stress parallel to the walls of the parent would be diminished. Equations (1) and (4) together indicate that if the fluid pressure in the joint decreased by an amount greater than the difference between the regional principal stresses, then a tensile stress will develop along the walls of the joint. Such a pressure drop could conceivably occur during propagation of the parent joint. A condition that allows fracturing at the joint walls is thus:

(10) $|\Delta P^c| > \sigma_{22}{}^r - \sigma_{11}{}^r$,

with the sign of ΔP^c being negative. Equation (10) reveals that nearly any sharp drop in the fluid pressure in the parent joint would lead to tensile stresses along the walls of the joint provided that the difference between the regional principal stresses is very small. Given that $\sigma_{22}{}^c$ would be constant along the walls of the parent, orthogonal "daughter" joints could nucleate anywhere along the walls of the parent joint. Also note that the tensile effective stress at the joint walls is independent of the length of the parent joint. This means that short joints as well as long joints could serve equally well as parents. As the fluid pressure builds back up in the parent the fluid pressure would also build in the any daughter joint that extended from the parent walls. The daughter joint could then extend away from the parent at right angles.

Although this mechanism is in some ways appealing, it does not provide a completely satisfactory explanation for orthogonal fractures. A comparison of equations (4) and (5) shows that $\sigma_{11}{}^c$ and $\sigma_{22}{}^c$ will both change by the same amount, so even though $\sigma_{22}{}^c$ becomes more tensile, so does $\sigma_{11}{}^c$. Whereas $\sigma_{22}{}^c$ was more compressive than $\sigma_{11}{}^c$ to begin with, closing of the parent would appear to favor opening of daughter cracks parallel to the parent rather than perpendicular to it (Figure 4). The difficulty is only partly eased by noting that flaws of equal length are somewhat more likely to propagate if they intersect the walls of the parent (Tada et al. 1973), and pre-existing flaws normal to the parent are obviously more likely to intersect the parent than flaws parallel to it.

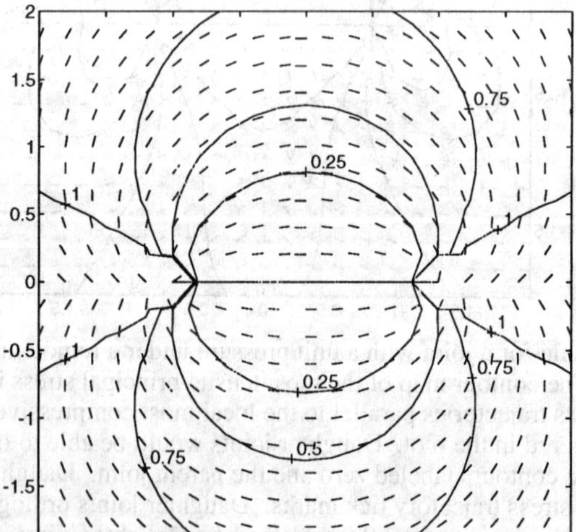

Figure 4. Stress field about a closing joint. Contours of the least compressive principal stress are superposed upon stress trajectories parallel to the local most compressive stress. The compressive far-field stresses are of unit magnitude. The joint, centered in the diagram, has a pressure of zero. Daughters near the parent would tend to parallel it.

2.3 *One possible effect of joint interaction*

Joints of special configurations can interact to yield conditions that would favor the develop of orthogonal daughter joints. Figure 5 shows such an example, where the upper joint has closed and the parallel lower joint has opened. The bull's-eye pattern of contours mark the regions where opening of daughter joints is especially favored. One of the areas where daughter joints are most likely to form is along the center of the upper joint, and the daughters there would be orthogonal to the parent. Daughters could also form near the ends of the lower parent joint; they would be roughly parallel to that joint. Clearly not all configurations would yield similar results.

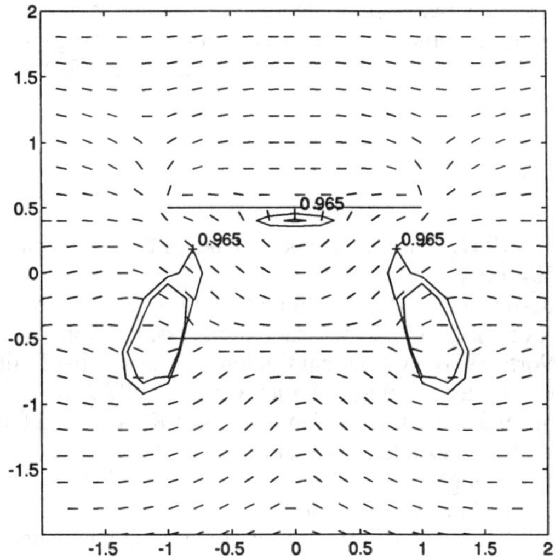

Figure 5. Stress state near an opening joint and a neighboring closing joint. Contours are of the least compressive stress (normalized with respect to the mean normal remote stress) superposed upon a plot of stress trajectories parallel to the local most compressive stress. The relevant boundary conditions are: $\sigma_{11}{}^{r} = 0.99$, $\sigma_{22}{}^{r} = 1.01$, $\sigma_{11}{}^{c,\ upper} = 0.98$, $\sigma_{11}{}^{c,\ lower} = 1.02$. The pairs of closed contours are in the regions where the least compressive stress is smallest. The compressive stresses decrease towards the centers of the enclosed regions, with the unlabeled contours denote stress magnitudes of 0.96. Stress calculations were done using a boundary element technique.

3 DISCUSSION AND CONCLUSIONS

The elastic effects associated with opening and closing of joints can contribute to the formation of contemporaneous orthogonal joints, but based on the analyses described here they do not seem sufficient by themselves to fully account for such a pattern. The elastic effects could contribute significantly only if the difference between the regional principal stresses is small relative to the driving pressure changes in the joints. In the absence of other effects, the elastic effects described here seem unlikely to prevent daughter joints from forming parallel to the most compressive regional stress, even if the difference between the regional principal stresses is small. Factors not examined here might enhance the effectiveness of the proposed elastic mechanisms; grain-scale anisotropy might play a particularly important role. Joints of different geometry than those considered here might react differently also.

Certain results of the elastic analyses could be tested by careful fieldwork. The elastic mechanisms discussed seem most likely to cause orthogonal daughter joints to nucleate at the walls of a parent. Detailed examinations of surface textures on joint walls could indicate where daughter fractures do nucleate relative to parents and whether daughters grow towards the parents or away from them. The analyses also suggest that in a set of contemporaneous orthogonal joints, joints of one orientation should be distinctly more common than joints of the other if the rock is isotropic.

There has been a continuing debate in the earth sciences community regarding how natural fractures form in the earth. The simple analyses presented here indicate that although an orthogonal joint network could develop to a limited extent at depth during a single episode of deformation, either multiple deformations or a single deformation with a pronounced non-elastic component would be much more effective in producing sets of systematic orthogonal joints.

REFERENCES

Biot, M.A. & D.G. Willis 1957. The elastic coefficients of the theory of consolidation. J. Appl. Mech.. 24:594-601.

Costin, L.S. 1987. Time-dependent deformation and failure. In B.K. Atkinson (ed.), Fracture mechanics of rock, p. 167-215. Academic Press, London

Dunne, W.M. & C.P. North 1990. Orthogonal fracture systems at the limits of thrusting: an example from southwestern Wales: J. Struct. Geol. 12:207-215.

Engelder, T. 1987. Joints and shear fractures in rock. In B.K. Atkinson (ed.), Fracture mechanics of rock, p. 27-69. London: Academic Press.

Nickelsen, R.P. & V.N.D. Hough 1967. Jointing in the Appalachian Plateau of Pennsylvania. Geol. Soc. Am. Bull. 78: 609-630.

Nur, A. & J.D. Byerlee 1971. An exact effective stress law for elastic deformation of rock with fluids: J. Geophys. Res. 76: 6414-6419.

Olson, J. & D.D. Pollard 1989. Inferring paleostresses from natural fracture patterns: a new method: Geology. 17: 345-348.

Pollard, D.D. & P. Segall 1987. Theoretical displacements and stresses near fractures in rock: with applications to faults, joints, veins, dikes, and solution surfaces. In B.K. Atkinson (ed.), Fracture mechanics of rock, p. 277-349. London: Academic Press.

Pollard, D.D. & A.A. Aydin 1988. Progress in understanding jointing over the past century: Geol. Soc. Am. Bull. 100: 1181-1204.

Rice, J.R., & M.P. Cleary 1976. Some basic stress diffusion solutions for fluid-saturated elastic porous media with compressible constituents: Rev. Geophy. & Space. Phys. 14: 227-241.

Rives, T. & Petit, J-P 1990. Experimental study of jointing during cylindrical and non-cylindrical folding. In H.P. Rossmanith (ed.), Mechanics of Jointed and Faulted Rock, p. 205-211. Rotterdam: Balkema.

Secor, D.T. 1965. Role of fluid pressure in jointing. Am. J. Sci. 263: 633-646.

Tada, H., P.C. Paris, & G.R. Irwin 1973. The stress analysis of cracks handbook. Hellertown, Pa: Del Research Corporation.

Turner, J.P. & P.L. Hancock 1989. Relationships between thrusting and joint systems in the Jaca thrust-top basin, Spanish Pyrenees. J. Struct. Geol. 12: 217-226.

Rock Mechanics, Nelson & Laubach (eds) © 1994 Balkema, Rotterdam, ISBN 90 5410 380 8

Fracturing and stress history of the Devonian Antrim Shale, Michigan Basin

Ted Apotria
Shell Development Company, Bellaire Research Center, Houston, Tex., USA (Presently: Exxon Production Research Company, Houston, Tex., USA)

Charles J. Kaiser & Bruce A. Cain
Shell Western Exploration and Production Inc., Houston, Tex., USA

ABSTRACT: The Antrim Shale is a fractured Devonian gas reservoir in northern Michigan with multiple natural fracture sets. At least three regional fracture sets are observed in the Antrim Shale which developed independent of local structure. Vitrinite reflectance data constrain the maximum burial of the Antrim Shale in the productive trend to about 6000 ft (1800 m) near the end of the Paleozoic era. Stress history analysis supports a model in which NW-striking fractures formed by natural hydraulic fracturing (NHF) as hydrocarbons were generated near the end of the Alleghenian orogeny. NE-striking fractures formed as a result of cooling and unloading during nearly 4000 ft (1220 m) of basin uplift since the Permian, and are aligned with the contemporary NE-directed maximum horizontal stress in the mid-continent region. Continued uplift and local reversal of the minimum and intermediate principal stress resulted in a second set of NW-striking cross fractures, orthogonal to NE-striking fractures.

Elastic property data indicate that the organic-rich, black shale has the lowest Young's modulus, Poisson's ratio, and fracture toughness in a sequence containing gray shales and limestones. It is also the lithology which is preferentially fractured, in contrast to other reservoirs where the stiffest lithology is most fractured. Fracture preference and containment in layered sedimentary rocks is not only a function of lithology and rock properties, but depends on the mechanism of fracture. The mechanism of NHF favors lithologies with low Poisson's ratio, low Biot poroelastic parameter, low fracture toughness and high fluid pressures. Fracturing under conditions where strain (bending, layer-parallel stretching) or uplift (thermoelastic contraction) mechanisms are dominant favors fracturing in stiff, brittle lithologies with high Young's moduli.

1 INTRODUCTION

The delineation of joint and fracture systems in sedimentary basins remains a critical problem in the exploration and production of oil and gas in low permeability reservoirs. The Antrim Shale in the northern Michigan Basin is a fractured Devonian gas reservoir with an estimated 2.3 TCF of known recoverable reserves. The Antrim Shale represents a desorption-type reservoir that requires natural fractures to provide the essential reservoir permeability (McGuire and Oliver 1988; Kuuskra et al. 1992). We sought a better understanding of the natural fracture systems in the Antrim Shale during exploration of this unconventional gas resource.

Several studies have addressed the origin and distribution of fractures across the Michigan Basin. Holst and Foote (1981) documented four systematic fracture sets (NE, NW, EW, and NS-striking) in outcrops dominated by carbonates of the Devonian Traverse Group. The two Antrim Shale outcrops in their study showed fewer than four orientations. Holst and Foote

(1981) suggested that the NE-striking set are extension fractures related to the contemporary stress field, while the NW-striking set may be genetically related to NW-trending folds in the Michigan Basin. Versical (1991) noted that the NW-striking fractures are consistent in orientation with the principal shortening direction measured in deformed calcite twins from Paleozoic carbonates throughout the mid-continent (Craddock et al. 1993) and probably are related to the Alleghenian orogenic event. In the Devonian strata of the Appalachian Basin, cross fold fractures, which are commonly NW-striking, were found to be the oldest and deepest in the Devonian Catskill Delta, whereas NE-striking fractures were younger, shallower and formed during unloading in the contemporary stress field (Evans 1980, Engelder 1985).

We inferred from the work of Evans (1980), Engelder (1985), and Versical (1991), that the timing and mechanisms of fracture formation in the Michigan Basin could be quite analogous to that in the Appalachian Basin, and that similar stress regimes may have existed throughout the mid-continent region. In order to test this hypothesis, we used outcrop and subsurface fracture data in the Michigan Basin, together with burial and thermal history data from the Antrim Shale. These data provided the input to a stress history analysis which forms the basis of a geologic model for fracturing in the Antrim Shale.

2 FRACTURE DATA

The Antrim Shale interval consists of interbedded organic-rich, black shales, gray-green shales, and gray limy shales, and is underlain by carbonates and shales of the middle Devonian Traverse Group. In northern Michigan, outcrop and well data suggest the Antrim Shale has at least three systematic fracture sets. The two most common are NW- and NE-striking with subvertical dips (Figure 1). Locally, a third fracture set (~EW-striking) supplements or replaces one of the other sets. One group of NW-striking fractures is the oldest as evidenced by continuity and abutting relationships in outcrop. Shorter and less systematic NW-striking cross fractures which abut against NE-striking fractures also occur, suggesting two episodes of fracturing with a NW strike orientation. Other workers have described a fracture set which strikes ~NS. This set is rarely seen in Antrim outcrops, and is best developed in the underlying Traverse Limestone (Holst and Foote 1981, Versical 1991).

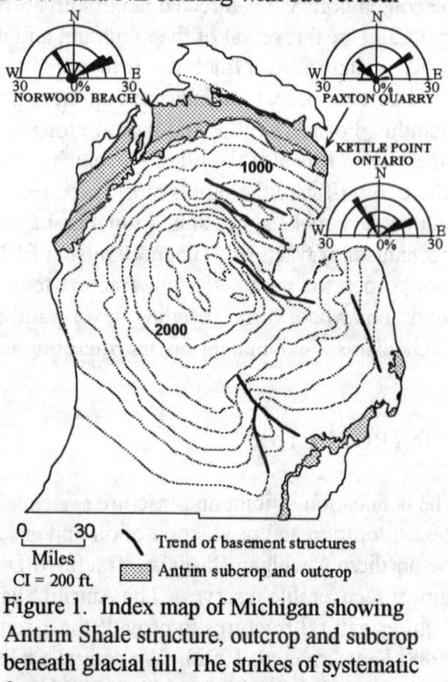

Figure 1. Index map of Michigan showing Antrim Shale structure, outcrop and subcrop beneath glacial till. The strikes of systematic fracture sets at three Antrim outcrops are indicated by rose diagrams.

Fractures in the Antrim Shale have the characteristics of regional fractures, in that they are widely distributed independent of structure, have spacings which are lithology and bedding thickness dependent, and often occur in orthogonal sets. The far-field stresses that cause regional fracture systems may be tectonic, yet the fractures are not restricted to local structures. In contrast, tectonic fractures develop locally during folding or faulting, and have orientations and spacing which are influenced by structural position.

Core and borehole imaging data from Antrim gas wells, and outcrop observations from the Paxton shale quarry near Alpena, Michigan, indicate that the fractures are best developed and more closely spaced in organic-rich black shales and are less well developed in gray shales and limestones.

3 ROCK PROPERTY DATA

The mechanical properties of various lithologies within the Antrim Shale interval and the underlying Traverse Group are critical to the issue of fracture mechanism and containment in an Antrim gas reservoir. Kim (1978) and Kim and Mubeen (1980) reported standard elastic properties for cored specimens from the Antrim interval. Inspection of their data reveals two mechanical end-members representing the black shale and limestone lithologies. Typical elastic properties for organic-rich black shale ($E=2.2x10^6$psi, $\nu=.20$, $\rho=2.10$, $K_{IC}=990$ psi $in^{1/2}$) and limestone ($E=7.8x10^6$psi, $\nu=.25$, $\rho=2.66$, $K_{IC}=1350$ psi $in^{1/2}$) were used as input into stress history models, where E is Young's modulus, ν is Poisson's ratio, ρ is bulk density, K_{IC} and is fracture toughness.

In comparing the rock property data to outcrop and well observations, we note that the organic-rich black shales which have relatively low Young's modulus and low Poisson's ratio, are preferentially fractured. This is in contrast to many fractured reservoirs where the stiffest lithologies (i.e. those with the highest Young's moduli) are preferentially fractured (e.g. Stearns and Friedman 1972). We address this apparent contradiction below.

4 BURIAL AND THERMAL HISTORY

In order to constrain the timing, depth and stress history conditions which contribute to fracturing, we modeled the burial and thermal history of the Antrim Shale. Well information used in the model comes primarily from the Shell Oil Co. Fisher & State Frederic 1-16 well in Crawford County. The model is based on the following data: (a) stratigraphic thicknesses, absolute ages, and lithologies, (b) vitrinite reflectance profiles in the Fisher & State Frederic well and other wells from northern Michigan, (c) the present-day surface and subsurface temperatures. Other variables such as the thickness of eroded sections and the rates of uplift and erosion for each unconformity are much less constrained, and make a unique solution impossible. However, of the range of solutions tested, the following represents the best approximation to the burial and thermal history.

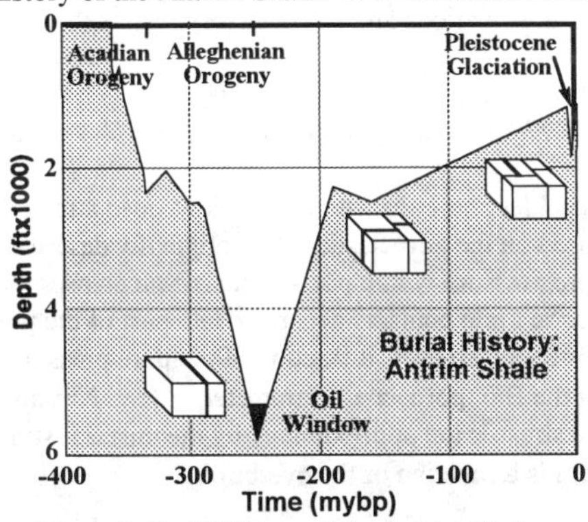

Figure 2. Burial history of the Antrim Shale from maturation modeling. The approximate depth the Antrim entered the oil window is indicated. Block insets indicate the relative timing of primary fracture sets (long axis is NW-directed).

In northern part of the Michigan Basin, maximum burial of the Antrim Shale to about 5800 ft. (1170 m) occurred at about 250 to 240 mybp, near the end of the Alleghenian orogeny in the Appalachians (Figure 2). Nearly 4000 ft (1220 m) of uplift since the Permian has brought the Antrim Shale to a present depth of about 1950 ft. This includes an estimate of 1000 ft (305 m) of glacial loading and unloading in the Pleistocene. The Antrim Shale reached its maximum temperature at about 250 to 240 mybp.

5 STRESS HISTORY MODELING

Stress history models have been used for fracture analysis in several sedimentary basins (e.g. Price 1974, Engelder 1985, Warpinski 1989). During burial, lithifying sediments experience variations in overburden, pore pressure, and temperature as rock properties change during diagenesis. During uplift, decreases in overburden and temperature occur without significant changes in rock properties. As a result, the horizontal stress at a given depth may be quite different during uplift relative to burial.

In the simplest stress history models, rocks are assumed to be elastic, homogeneous and isotropic. Rock properties are assumed to vary linearly during burial, attain their final (measured) values at maximum burial, and remain constant during uplift. We initially assume the horizontal stresses are governed by the uniaxial strain model, which assumes zero horizontal strain. A simplified version of the burial history described above is input into the stress history models, which smoothes the episodes of second order uplift and burial, and glacial loading during the Pleistocene. We suggest that this simplification does not alter the first order effects of burial and uplift on the modeled state of stress.

Clearly, some of these assumptions are not strictly met for a number of reasons. Rocks do not behave homogeneously and elastically during burial due to compaction, diagenesis, and stress concentrations around flaws. In addition, strains induced by Alleghenian compression or by the contemporary stress field are difficult to quantify. The use of one stress history model to analyze several episodes of fracturing also is troublesome, because a simple elastic analysis becomes invalid after the first failure criterion is met. To avoid this problem we have run several models independently, extracting the salient results of each model. Despite these limitations, the models provide a first-order sensitivity analysis and lead us to a geologic model which is consistent with our observations and stress history results.

5.1 *The Poroelastic Effect on Natural Hydraulic Fracturing*

Engelder and Lacazette (1990) developed a model for natural hydraulic fracturing (NHF) in rocks based on the poroelastic effect originally described by Biot (1941). In general, an increase in pore pressure within poroelastic rock increases the total horizontal compressive stress. However, elastic grain contacts accommodate part of the pore pressure increase, so that the net increase in horizontal stress is a fraction of the pore pressure increase. This fractional increase is indicated by the Biot poroelastic parameter, α (e.g. Nur and Byerlee 1971). In a tectonically relaxed basin, the stresses are approximated by the uniaxial strain model in which the total horizontal stress (σ_h) is a function of the overburden:

(1) $$\sigma_h = \frac{\nu}{1-\nu}\left(\sigma_v - \alpha P_f\right) + \alpha P_f$$

where ν is Poisson's ratio, σ_v is the vertical stress (= ρgz), and P_f is fluid pressure. Assuming α equal to 1.0, total horizontal stress is always greater than the fluid pressure, until P_f and σ_h converge at fluid pressures equal to the overburden. However, for cases where $\alpha < 1.0$, P_f may exceed σ_h, leading to conditions favoring NHF. In addition to the total horizontal stress (Eqn. 1), the fluid pressure necessary to propagate a fracture is also a function of the fracture toughness, initial crack (flaw) length, and crack shape (Engelder and Lacazette 1990). Therefore, the fluid pressure must not only exceed the in situ horizontal stress, but also the fracture toughness or "strength" term. For the purpose of this discussion, this term is not included in the stress history models, although we note that fracture toughness is 50% greater for the limestone.

For unconsolidated sediments, α is typically 1.0. However, as sediments become compacted, lithified, and cemented, α is expected to decrease because only a fraction of the rock stress is supported by fluid pressure. Part of the load is supported by elastic grain contacts. Laurent et al. (1993) found that α in limestones decreases from 0.83 to 0.34 as porosities decrease from 23 to 4.5%. Warpinski and Teuful (1992) measured a range for α in tight sandstones between 0.65 and 0.95 and about 0.9 in chalk, with some variation as a function of axial stress and pore pressure.

Because α of the Antrim Shale has not been measured, we test a range of plausible values in the stress history model. The equation for total horizontal stress (Eqn. 1) is integrated throughout the burial history (Figure 3a). This model assumes hydrostatic fluid pressures such that $\lambda = P_f/\sigma_v$ is about 0.45. An initial Poisson's ratio of 0.40 and final value of 0.20 is also assumed (representing a typical value for the Antrim black shale). The average density of overburden is assumed to increase from 2.0 at deposition to 2.4 g/cc at maximum burial and remains constant during uplift. Hence the total vertical stress is greater during uplift than at the corresponding depth during burial. As expected from Equation (1), σ_h increases with increasing depth and fluid pressure. Note that for decreasing values of α, the fluid pressure may exceed σ_h, leading to a condition favorable for NHF. For example, fluid pressure equals σ_h at maximum burial for α=0.6, and at shallower depths for lower values of α.

In the next model, fluid pressures are increased from a hydrostatic gradient (λ~0.45) to a modest gradient (λ~0.65) below a depth of 5000 ft (1525 m) (Figure 3b). This fluid pressure scenario simulates the effects of thermal maturation and hydrocarbon generation near maximum burial ("oil window"). The poroelastic parameter decreases from 1.0 at deposition to the final value of 0.7 at maximum burial. Increased fluid pressures promote NHF at shallower depths than for hydrostatic fluid pressure, and allow rocks with a relatively higher α and ν to fracture. Fluid pressure exceeds the total σ_h at depths between 4200 ft (1280m) and 5000 ft (1525 m) for α between 0.4 and 0.7. Other models (not shown) vary Poisson's ratio only and indicate that relatively low ν also promotes NHF (as expected from Eqn. 1). Also, it is likely that the organic-rich shales preferentially achieved higher fluid pressures during thermal maturation, also making them favored for NHF.

We conclude that the NHF mechanism is favored by rocks with low Poisson's ratio, low poroelastic parameter, low fracture toughness, and high fluid pressure. The models also suggest that conditions leading to NHF (fluid pressures which exceed σ_h) occurred between 4500 ft and maximum burial (assuming an estimate of α between 0.5 and 0.8). At that time, the maximum horizontal compression was NW-directed, controlled by the Alleghenian orogeny in the Appalachians.

5.2 *The Effects of Unloading and Cooling*

In the following models, we test the effects of cooling *in addition to* unloading by including the thermal stress term in the expression for effective horizontal stress. Assuming the uniaxial strain model:

(2) $$\sigma_h' = \frac{\nu}{1-\nu}\left(\sigma_v - \alpha P_f\right) + \frac{E n \Delta T}{1-\nu}$$

where σ_h' is the effective horizontal stress, *n* is the coefficient of thermal expansion, and T is temperature.

We tested two models of this type. In the first, elastic stresses were maintained during burial and uplift. However, this case overestimates the magnitude of the effective horizontal stress during burial due to the large thermal term, such that it may exceed the total vertical stress. In reality, inelastic processes relax the stresses somewhat during burial so that large thermal compressive stresses do not develop. In the second model, the σ_h' is "hardwired" to equal 0.8 of the vertical effective stress during burial, consistent with observations in many sedimentary basins (e.g. Breckels and van Eekelen 1981). During the uplift phase, rock properties are assumed to remain constant, and the elastic stresses including the thermal term (Eqn. 2) are calculated (Figure 3c). Various rock property and fluid pressure scenarios incorporated into the stress history calculation consistently demonstrate that uplift may reduce the effective horizontal stress to the tensile field. We noted previously that a low Poisson's ratio contributes to lower stresses. However, in this suite of models, uplift *and* cooling generate effective tensile stresses. The slope of the minimum effective stress curve during uplift is dominated by Young's modulus in the thermal stress term. The depth where effective tension is reached varies significantly with Young's modulus, which is a factor of three greater in limestone that in black shale. If cooling were the only fracture mechanism operative during uplift, fracture in limestone would be favored over black shale, and could occur at greater depths.

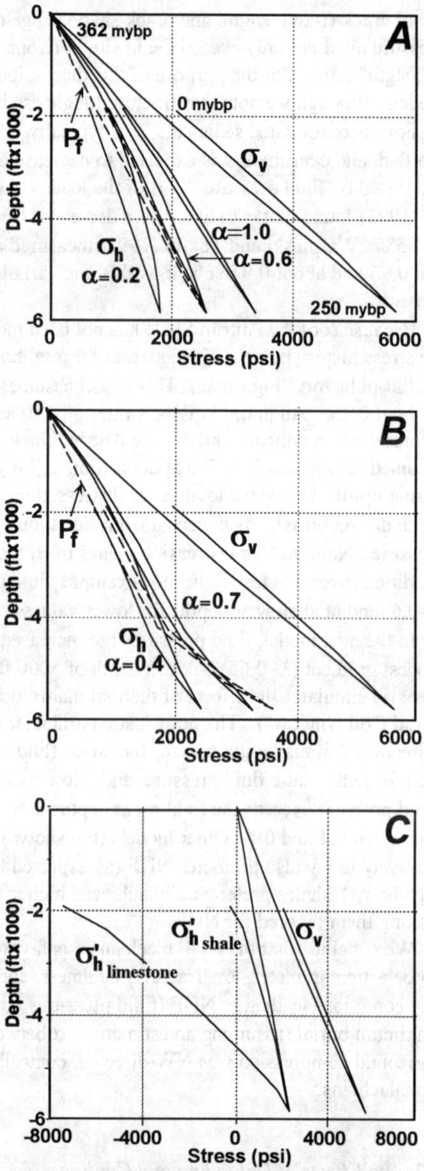

Figure 3. Stress history models of the Antrim Shale with (A) hydrostatic fluid pressure and total horizontal stress for different values of α, (B) increased fluid pressures below a depth of 5000 ft. (C) thermal stresses included during uplift. Rocks presumably fracture before large tensile stresses develop.

6 GEOLOGIC MODEL FOR THE ORIGIN OF FRACTURES

Combining our observations from outcrop and well data with the results from stress history models, we propose the following model for the origin of the primary fracture sets found in the Antrim Shale of the Michigan basin. The oldest near-vertical fractures in the basin are NW-striking and formed as natural hydraulic fractures (Figure 2). Fracturing occurred during burial and was assisted by thermal maturation and hydrocarbon generation in the late Paleozoic coincident with the Alleghenian orogeny in the Appalachians. During this time, the regional maximum horizontal stress direction was oriented NW. Fracturing occurred preferentially in organic-rich black shales which have lower Poisson's ratios and probably had higher fluid pressures due to hydrocarbon maturation relative to carbonates. That NW-striking fractures in Michigan are Alleghenian in age is also consistent with the results of Craddock et al. (1993) who suggested that Alleghenian stresses affected the mid-continent region based on the analysis of twinned calcite from Paleozoic carbonates.

NE- and EW-striking fractures formed during basin uplift and consequent unloading and cooling. The onset of uplift marked a transition from a NW-directed maximum horizontal stress (Alleghenian orogeny) to the presently observed ENE-directed maximum horizontal compression (contemporary stress field) (Zoback and Zoback 1991). The precise depth above which NE- and EW-striking fractures form depends on lithology and elastic properties, and the extent to which other mechanisms contribute to fracturing during uplift (e.g. NHF or extensional strain). The fact that many of these fractures do not abut older fractures suggest that appreciable traction existed across NW-striking fractures when the NE-striking set formed. Such conditions would exist at greater depths than the near-surface. Uplift also provides a mechanism for the formation of a second generation of NW-striking cross fractures which are less systematic (younger than NE-striking fractures). As the NE-striking fracture set forms, the local minimum effective horizontal stress is reduced to zero, and may become larger (less tensile) than the intermediate effective stress. With continued uplift, the new minimum effective horizontal stress is further reduced, and may result in fracture in a direction perpendicular to the previous set. Such cross fractures are probably limited to the near-surface region. This principal stress-reversal mechanism may occur more than once during the uplift history depending on the orientation of the minimum stress and the magnitude of compressive normal stress on pre-existing fractures.

7 SUMMARY

The models presented above highlight several important points. First, fractures in the Michigan Basin are regional in extent. The presence of several interconnected fracture sets in outcrop and in the subsurface does not require high curvature due to folding.

Second, fracture preference and containment in layered sedimentary rocks is not only a function of lithology and rock properties, but depends on the mechanism of fracture. The NHF mechanism favors lithologies with low Poisson's ratio, low Biot poroelastic parameter, low fracture toughness and high fluid pressures. The poroelastic effect predicts lower fluid pressures required for failure as α is reduced (Engelder and Lacazette 1990). Fracturing under conditions where strain (bending, layer-parallel stretching) or uplift (thermoelastic contraction) mechanisms are dominant favors fracturing in stiff, brittle lithologies with high Young's moduli.

Third, stress history models predict NHF near maximum burial during the Alleghenian orogeny. This is consistent with our observation that NW-striking fractures are the oldest in the basin. It also supports the contention that Alleghenian stresses affected the mid-continent region as evidenced by calcite twinning strains in Paleozoic limestones (Craddock et al., 1993).

Fourth, the Antrim Shale has experienced 4000 ft (1220 m) of uplift since the Permian, in which unloading and cooling provide a viable mechanism for fracture. NE-striking fractures which formed during uplift are consistent in orientation with the contemporary stress field.

8 ACKNOWLEDGMENTS

We wish to acknowledge our Shell colleagues who offered their critiques of our work, especially Steven Naruk who generated the maturation models of the Antrim Shale. We also thank Shell Western E&P Inc. and Shell Development Company for permission to publish this paper.

9 REFERENCES

Biot, M.A., 1941, General theory of three-dimensional consolidation, Journal of Applied Physics, v. 12, p. 155-164.

Brekels, I.M. and van Eekelen, H.A.M., 1981, Relationship between horizontal stress and depth in sedimentary basins, in, 56th annual Fall Technical Conference, Society of Petroleum Engineers of AIME, October 5-7, 1981, Paper 10336, 7 p.

Craddock, J.P., Jackson, M., van der Pluijm, B.A., and Versical, R.T., 1993, Regional shortening fabrics in eastern North America: a far-field stress transmission from the Appalachian-Ouachita orogenic belt, Tectonics, v. 12, n. 1, p. 257-264.

Engelder, T., 1985, Loading paths to joint propagation during a tectonic cycle: an example from the Appalachian Plateau, U.S.A., Jour. Struct. Geol., v. 7, p. 459-476.

Engelder, T., and Lacazette, A., 1990, Natural hydraulic fracturing, in Rock Joints, Barton, N., and Stephansson, O., (eds), Balkema Press, p. 35-44.

Evans, M.A., 1980, Fractures in oriented Devonian shale cores from the Appalachian Basin, M.S. Thesis, W. Va. Univ., 276 p.

Holst, T.B., and Foote, G.R., 1981, Joint orientation in Devonian rocks in the northern portion of the lower peninsula of Michigan, Geol. Soc. Amer. Bull., v. 82, p. 85-93.

Kim, K., 1978, Mechanical characteristics of the Antrim Shale, DOE/FE/2346-24, 85 p.

Kim, K., and Mubeen, A., 1980, Fracture toughness of the Antrim Shale, DOE/FE/2346-68, 18 p.

Kuuskra, V.A., Wicks, D.E., and Thurber, J.L., 1992, Geologic and reservoir mechanisms controlling gas recovery from the Antrim Shale, Society of Petroleum Engineers, SPE-24883, p. 209-224.

Laurent, J., Bouteca, M.J., Sarda, J.P., and Bary, D., 1993, Pore pressure influence in the poroelastic behavior of rocks: experimental studies and results, SPE Formation Evaluation, June 1993, p. 117-122.

McGuire, D.R., and Oliver, S.J.P., 1988, Devonian shale gas production in the Michigan Basin, Eastern Devonian Shale Technology Review, v. 5, n. 2&3, p. 28-40.

Nur, A., and Byerlee, J.D., 1971, An effective stress law for elastic deformation of rocks with fluids, J. Geoph. Res, v. 76, p. 6414-19.

Price, N. J., 1974, The development of stress systems and fracture patterns in undeformed sediments, Proc. 3rd Rock Mech. Congr. ISRM, v. 1, p. 487-496.

Stearns, D.W., and Friedman, M., 1972, Reservoirs in fractured rock, In Kling, R.E. (ed.), Stratigraphic Oil and Gas Fields, Classification, Exploration Methods, and Case Histories, AAPG Bulletin, Memoir 16, p. 82-106.

Versical, R.T., 1991, Basement Control on the Development of Selected Michigan Basin Oil Fields as Constrained by Fabric Elements in Paleozoic Limestones, M.S. Thesis, Western Michigan University.

Warpinski, N.R., and Teufel, L.W., 1992, Determination of the effective stress law for permeability and deformation in low-permeability rocks, SPE Formation Evaluation, June 1992, p. 123-131.

Warpinski, N.R., 1989, Elastic and viscoelastic calculations of stresses in sedimentary basins, SPE Formation Evaluation, v. 4, p. 522-530.

Zoback, M.D. and Zoback, M.L., 1991, Tectonic stress field of North America and relative plate motions, in Slemmons et al. (eds.), Neotectonics of North America, Geol. Soc. Amer., Decade Map Volume 1, p. 339-366.

Rock Mechanics, Nelson & Laubach (eds) © 1994 Balkema, Rotterdam, ISBN 90 5410 380 8

Fracture mechanics analysis of limestone cantilevers subjected to very long-term tensile stress in natural caves

Thomas M.Tharp
Department of Earth and Atmospheric Sciences, Purdue University, West Lafayette, Ind., USA
Thomas J.Holdrege
Holdrege and Kull, Nevada City, Calif., USA

ABSTRACT: Many long-lived (10^3 - 10^6 years) natural cantilevers in five limestone caves are characterized by maximum tensile stress of 1 MPa. This is interpreted as the stress necessary to cause subcritical propagation of cracks having a wide variety of lengths. We propose that in these rocks, crack propagation velocity $v = c\ (K_I/K_Q)^n$, where c and n are constants and K_Q (Ingraffea and Schmidt 1978) is mode I critical stress intensity, which varies with crack length for short cracks.

1 INTRODUCTION

Natural cantilever beams in caves present a unique opportunity to study the effect of a known tensile load acting over time spans many orders of magnitude longer than accessible in laboratory investigations. We measured natural cantilever beams, both intact (still attached) and failed (detached), in five limestone caves in south-central Indiana (Tables 1 and 2). Length is maximum length perpendicular to the line of attachment and thickness is measured where the cantilever is (or was) attached to the wall. Tensile stress is outer fiber stress at the wall, calculated by elementary beam theory for the bending moment calculated from measured dimensions.

Age of the cantilevers may be estimated from the age of the caves and from the morphology of the cave passages. The geomorphic history and stratigraphic position of the caves is essentially that of such caves as Wyandotte, in Indiana, and the Mammoth-Flint Ridge system, nearby in Kentucky. Ages of the Mammoth-Flint Ridge system determined by U-Th radiometric methods and paleomagnetism date the youngest passages at 141,000 years (Hess and Harmon 1981) and the oldest, highest level passages at a probable 2 million years (Schmidt 1982). This range is consistent with geomorphic inferences for Indiana caves. Palmer and Palmer (1975) attribute upper passages of many prominent caves such as Wyandotte, in the geomorphic province of the caves studied here, to the late Tertiary. Powell (1968) gives probable late Tertiary and early Pleistocene ages for the two levels of Wyandotte cave. Wilson (1985) studied all caves of the present study but Sullivan. His work suggests that most passages in which cantilevers were measured are probably early Pleistocene.

In some cases, cantilevers overlie or lie upon floors showing no clear sign of strata that may have fallen to free the cantilevers. Because this previous breakdown has been obscured or removed by the slow processes of sedimentation, dripstone deposition or dissolution, such cantilevers probably remained attached for thousands and possibly more than one million years. Where breakdown is abundant on the floor, age of the

Table 1. Intact cantilerers

Number	Cave	Rock type	Length (cm)	Thickness (cm)	Tensile Stress (MPa)
1	Buckner	Sparite	129.5	13.06	0.999
2	Buckner	Sparite	91.4	5.08	1.280
3	Salamander	Micrite	116.8	13.97	0.356
4	Buckner	Fine sparite	94.0	12.95	0.530
5	Buckner	Fine sparite	170.0	20.02	0.106
6	Buckner	Sparite	96.0	12.95	0.556
7	Buckner	Fine sparite	169.9	14.99	0.950
8	Grotto	Micrite	135.9	13.34	1.080
9	Grotto	Micrite	163.8	10.74	1.930
10	Grotto	Micrite	198.1	30.48	1.000
11	Grotto	Micrite	109.2	10.79	0.901
12	Brinegar	Micrite	259.2	30.48	1.710
13	Brinegar	Micrite	97.8	12.70	0.586
14	Sullivan	Fine sparite	170.2	10.16	2.220
15	Sullivan	Fine sparite	119.4	8.89	1.160
16	Sullivan	Fine sparite	134.6	45.72	0.308
17	Sullivan	Fine sparite	189.2	25.40	1.100
18	Sullivan	Fine sparite	124.5	17.78	0.678
Mean			142.7	17.20	0.925

Table 2. Failed cantilevers

Number	Cave	Rock type	Length (cm)	Thickness (cm)	Tensile Stress (MPa)
1	Salamander	Micrite	845.80	188.00	2.960
2	Buckner	Sparite	83.82	11.43	0.481
3	Buckner	Sparite	60.96	10.16	0.235
4	Salamander	Sparite	68.58	20.32	0.180
5	Salamander	Sparite	83.82	18.41	0.099
6	Salamander	Micrite	438.10	340.40	0.439
7	Salamander	Oolite	576.60	274.30	0.750
8	Salamander	Micrite	226.10	31.24	0.592
9	Salamander	Sparite	68.58	19.05	0.192
10	Buckner	Sparite	120.00	8.13	1.378
11	Sullivan	Fine sparite	210.80	40.64	0.850
12	Sullivan	Fine sparite	373.40	134.60	0.805
13	Sullivan	Fine sparite	142.20	35.56	0.442
14	Sullivan	Fine sparite	121.90	38.10	0.303
15	Sullivan	Fine sparite	297.20	50.80	1.352
16	Sullivan	Fine sparite	165.10	12.70	1.669
Mean			242.69	77.12	0.795

associated cantilevers is even less certain. However, in most cases roofs with breakdown lie at nearly the same elevation as adjacent roof with original solution features (Wilson 1985). This indicates that progressive roof breakdown is a very slow process and that a given cantilever may remain intact for a long period. The intact cantilevers are known to act as cantilevers because they are separated from the roof

Table 3. Observed and calculated deflection for intact cantilevers

Number	Length (cm)	Thickness (cm)	Measured Deflection (cm)	Elastic Deflection (cm)	Deflection, one crack[1] (cm)	Deflection, multiple cracks[2] (cm)
1	129.5	13.06	14.99	0.015	0.054	0.134
2	91.4	5.08	5.72	0.024	0.076	0.286
3	116.8	13.97	12.70	0.009	0.049	0.112
4	94.0	12.95	4.06	0.004	0.045	0.098
5	170.0	20.02	11.99	0.019	0.049	0.096
6	96.0	12.95	8.99	0.005	0.046	0.098
7	169.9	14.99	6.10	0.033	0.064	0.133
8	135.9	13.34	3.81	0.017	0.054	0.129
9	163.8	10.79	14.60	0.056	0.087	0.188
10	198.1	30.48	5.08	0.015	0.041	0.075
11	109.2	10.79	8.89	0.011	0.054	0.142
12	259.1	30.48	7.62	0.043	0.058	0.085
13	97.8	12.70	12.70	0.005	0.046	0.110
14	170.2	10.16	4.45	0.072	0.101	0.210
15	119.4	8.89	8.26	0.023	0.065	0.180
16	134.6	45.72	8.89	0.001	0.032	0.042
17	189.2	25.40	12.06	0.018	0.045	0.088
18	124.5	17.78	6.35	0.007	0.045	0.094

[1]One crack at fixed end with length such that $K_I = K_{IC}$.
[2]Crack spacing equal to cantilever thickness, starting at fixed end. All cracks have the same length, determined by the condition that $K_I = K_{IC}$ for the crack at the fixed-end.

above by an open, horizontal joint resulting from deflection that exceeds elastic deflection by orders of magnitude. Table 3 shows observed deflections and calculated elastic deflections for uncracked beams. It also shows deflections that would occur if there were a single crack of critical length ($K_I = K_{IC}$) at the point where the cantilever is attached to the wall. Also shown are cantilever deflections given cracks of this same length with spacing equal to cantilever thickness. Deflections with the hypothesized cracks are calculated using a cracked beam-column finite element (Tharp 1987). Clearly no sparse array of cracks can account for the large observed deflections.

The cantilevers studied are generally developed in the Levias limestone member of the Ste. Genevieve formation (Holdrege 1986). Three-point bending tests were performed on 3 cm thick beams of micrite and sparite from the Levias limestone to characterize short-term tensile strength in bending. All speciments were saturated with water. Tensile strength (modulus of rupture) is 27.2 MPa for the micrite and 25.1 MPa for the sparite.

Thickness and length of both intact and failed cantilevers vary widely (Tables 1 and 2) and volumes of the cantilevers span almost two orders of magnitude for intact and almost four orders of magnitude for failed, yet these cantilevers are in one respect surprisingly similar. Mean maximum bending stress is 0.93 MPa for the intact and 0.80 MPa for the failed cantilevers. For the intact cantilevers 72% are within a factor of two of the mean stress. The ratio of short-term, three-point bending strength to long-term cantilever strength is approximately 30.

Short-term failure of a rock cantilever is generally a weakest link failure, which places it in the restricted class of structures to which a Weibull analysis is appropriate (Bazant *et al.* 1991). A Weibull analysis for a self-loaded cantilever (Tharp 1983) was performed using scatter of the three-point bending results to determine size effect.

The size effect model for sparite was used for cantilevers of sparite and oolite and the micrite model for those of micrite. The Weibull model predicts that the six failed cantilevers less than 20 cm thick should have strength ranging from 10.6 to 13.6 MPa, whereas strength of the four largest cantilevers should be 2.9 to 6.2 MPa. Predicted strengths for the thinner cantilevers are more than an order of magnitude greater than observed strengths. Observed strengths vary little with size (Table 2) rather than decreasing as the Weibull model predicts. Subcritical crack propagation will be shown to offer a better explanation for the observed behavior.

2 SUBCRITICAL CRACK PROPAGATION

The relative consistency and smallness of bending stress in the cantilevers are best explained as the result of subcritical propagation of initially short cracks. For a corrosive environment and small K_I/K_{IC}, velocity of crack propagation v is given by

$$v = c\,(K_I/K_{IC})^n \tag{1}$$

where c is $V_0 \exp(-H/RT)$, H is activation enthalpy, V_0 and n are empirical constants, T is absolute temperature, and R is the gas constant. We calculated different values of c and n for Carrara marble (Atkinson 1984), Falerans micrite and St. Pons marble (Henry *et al.* 1977) in water. All were calculated for linear portions of the log (velocity) vs log (K_I) curves interpreted by Atkinson (1984) to correspond to region 1, i.e. control by reaction rate at the crack tip. Because this region corresponds to the lowest stress intensities, it may be extrapolated to lower stress intensities than have been investigated experimentally. Atkinson (1984) reports H of 63-147 KJ mol^{-1} measured by Henry (1978) for Falerans micrite. Given the uncertainty of this parameter, it is fortunate that the 13 °C mean annual temperature in the region of the caves (Wilson 1985) is close to the 20 °C temperature of the laboratory crack propagation experiments from which the other parameters are calculated. For 13 °C and the reported range of H, crack propagation rate is 23 to 53% of that at 20 °C, and 38%, an average of those values, is used in calculations.

For an assumed cantilever load and initial crack length, crack propagation may be integrated until K_I reaches K_{IC} and the beam fails. Stress intensity for any given crack length is calculated according to formulas for edge-cracked beams in pure bending (see Tharp 1987). The effect of K_{II} is small. Two models for subcritical crack propagation will be considered. The first, a generally accepted model for long cracks and small K_I/K_{IC}, is stated in eq. 1, where K_{IC} is constant. The second model follows.

3 K_Q MODEL FOR SUBCRITICAL CRACK PROPAGATION

Effective K_{IC} for short cracks (called K_Q by Ingraffea and Schmidt (1978)) is a function of crack length for lengths less than some value. K_Q decreases with decreasing crack length because for very short cracks the process zone must be small and the energy dissipated in crack propagation is correspondingly less. For Indiana limestone, Ingraffea and Schmidt (1978) find that K_Q decreases from $K_Q = K_{IC} = 0.97$ MPa $\cdot$ m$^{1/2}$ for crack length $a = 5$ cm to $K_Q = 0.5$ MPa $\cdot$ m$^{1/2}$ for $a = 0.5$ cm. It is useful to extend their curve to smaller crack lengths. For very short cracks, the finite width of the crack tip becomes important and the zone of stress concentration at the crack tip shrinks. When it is too small to be intersected by a significant number of dislocations or other defects, no mechanism exists for formation of a process zone. This crack will be approximated by a thin ellipse. Radius of curvature at its tip is approximated by d, half the 2×10^{-9} m cleavage rhomb unit cell dimension for calcite. Tensile strength approaches the theoretical strength of $E/10$, where E is Young's modulus. Equating the externally applied failure stress for the ellipse with that for a crack obeying standard fracture mechanics assumptions will give K_Q for very short cracks.

For an ellipse with long and short axes a and b respectively, at the ends of the ellipse tangential stress $\sigma = P(1 + 2\,a/b)$, where P is the remotely applied tension perpendicular to the long axis. Radius of curvature at the ends of the ellipse is b^2/a. If σ equals theoretical strength $E/10$ and radius of curvature equals d, the stress equation above may be written as $P = E/(10(1 + 2\,a/(ad)^{1/2}))$ where P becomes the stress required for tensile failure at the ends of the ellipse. This is substituted into the equation for stress intensity of a crack of half-length a, $K_I = P(\pi a)^{1/2}$ to yield $K_Q = E(\pi a)^{1/2} / 10\,(1 + 2\,a^{1/2}/d^{1/2})$. For $d = 10^{-9}$ m and $a > 10^{-7}$ m, the second term in the sum in the denominator is dominant and $K_Q \cong E(\pi d)^{1/2}/20 = 0.168\ \mathrm{MPa \cdot m^{1/2}}$. As expected, this is close to K_{IC} for cleavage of a single crystal (Atkinson 1984).

A further hypothetical extension of existing theory helps to explain the cave cantilevers. If K_Q replaces K_{IC} in eq. 1, the resulting equation is:

$$v = c\,(K_I/K_Q)^n \qquad (2)$$

For this relationship propagation behavior varies substantially with imposed stress. As seen in Fig. 1, for high stress the curve of K_I vs. crack length a is much steeper than K_Q vs. a. The ratio K_I/K_Q reaches a high value at short crack length (Fig. 2) with the result that for a particular cantilever to deflect inelastically yet be preserved for a long time the initial crack must fall within a narrow length range. For low stress, K_I remains below the K_I/K_Q range of 0.1 to 0.3 at which meaningful propagation first becomes possible until a is quite large. Again the range of crack lengths for long-lived, inelastically deflected cantilevers is narrow, and in this case it is a large fraction of the thickness of the cantilever. Between these stress extremes lies a small range (0.5 to 1.0 MPa) in which K_I/K_Q is nearly constant within the range 0.1 to 0.3 that allows substantial propagation in 10^3 to 10^6 years. Significantly, the associated wide range of crack lengths extends into the range of grain size and intergranular pores.

Cantilever behavior in these three stress regimes is illustrated in Fig. 3. In the low stress regime, for a cantilever preserved for 10^3 to 10^6 years, the initial crack must lie within a length range varying by only a factor of 1.9 and it must nearly equal the critical length ($K_I = K_{IC}$). For high stress, the possible length of the initial crack is similarly narrow, but it lies within the range of pore and grain size at which defects are common. In a small, intermediate stress range, cantilever life of 10^3 to 10^6 years is achieved over a very broad range of initial crack lengths, with its lower end in the pore and grain size range. This small stress range was centered about 1 MPa for the subcritical propagation parameters for Falerans micrite (Fig. 3) and for two of three curves for St. Pons marble. The stress is 1.25 MPa for the third St. Pons marble curve and 0.5 MPa for Carrara marble. These stresses are nearly insensitive to cantilever thickness over the wide range studied in the caves. The c and n values for Falerans micrite are used for propagation calculations in Figs. 3-6.

Simulation of crack propagation with any of these models in the intermediate stress range (all close to calculated stress in the cave cantilevers) produced time to failure of 10^3 to 10^6 years for a 70-fold range of crack sizes for cantilever thicknesses spanning the range observed in the caves. The effective long-term tensile strength of 1 MPa is analogous to the higher stress for which $K_I \cong K_Q$ over a wide range of crack lengths, which was shown by Ingraffea and Schmidt (1978) to approximate short-term tensile strength in Indiana limestone.

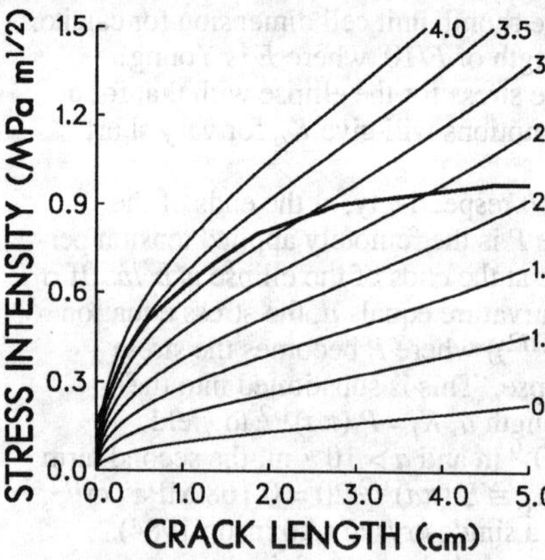

Fig. 1. Bold curve is K_Q, other curves are stress intensity for beam 15 cm thick in pure bending. Stress for each curve (MPa) is maximum for an uncracked beam.

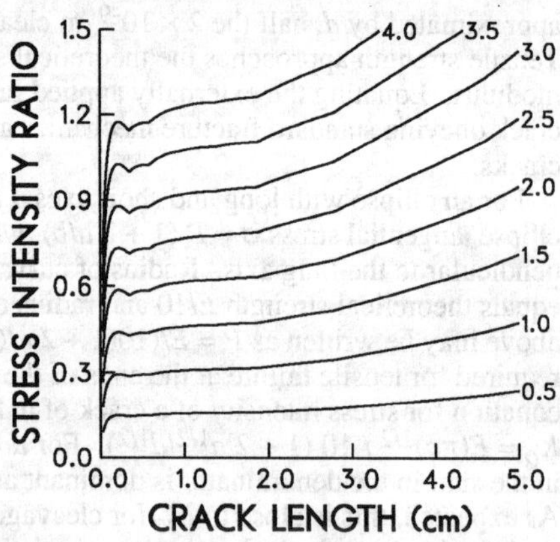

Fig. 2. Stress intensity divided by K_Q for stress intensity distributions in Fig. 1.

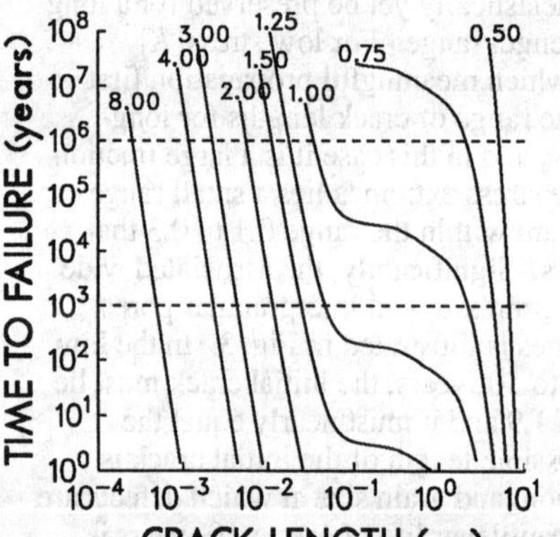

Fig. 3. Time to failure for 15 cm thick beam under different bending loads indicated by maximum stress for an uncracked beam. Variable K_Q model (eq. 2).

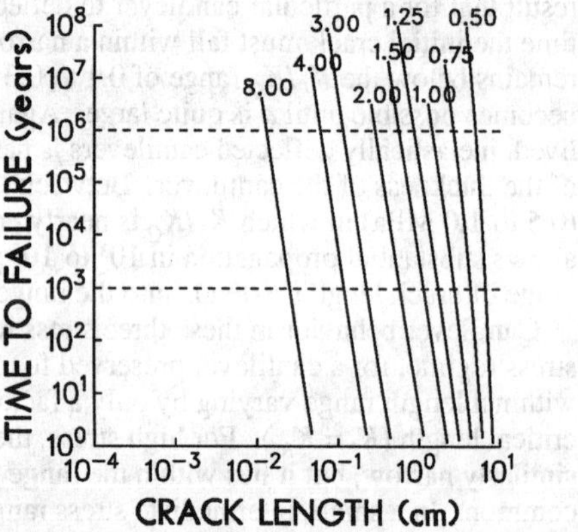

Fig. 4. Same as Fig. 3 but with the standard model (eq. 1).

For the standard subcritical crack propagation model with constant critical stress intensity (eq. 1) initial crack length (depending on c and n) must exceed 1 cm for all but five of the 34 cantilevers measured (Fig. 6). This is far above the size of intrinsic defects defined by pore and grain size, and for such relatively long cracks wide scatter in length is expected. The small range of crack lengths that allow a cantilever under a particular load to deflect significantly away from the roof and yet be preserved for 10^3 to 10^6 years (Figs. 4 and 6) would make intact cantilevers rare and their maximum

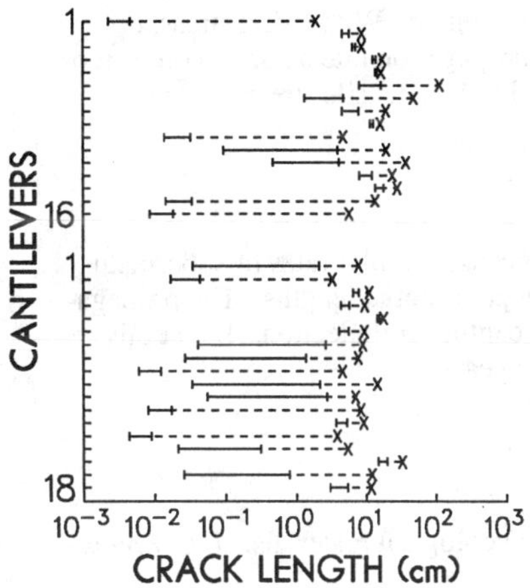

Fig. 5. Reference crack lengths calculated for failed (upper group) and intact (lower group) cantilevers. Solid line with brackets shows initial crack lengths for life of 10^3 - 10^6 years. Dashed line is crack length interval of subcritical propagation (eq. 2), ending with beam failure at the x.

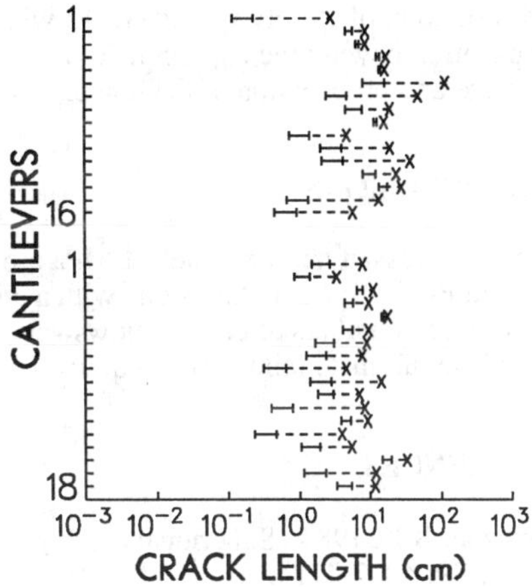

Fig. 6. Same as Fig. 5 but with standard model (eq. 1).

bending stress far more variable than observed. As shown in Fig. 5, the variable K_Q model allows the observed long-lived cantilevers for short initial cracks which should be abundant, and in many cases for a wide range of crack lengths. Only in those cantilevers under very small load would crack propagation require an initial crack near the critical length ($K_I = K_{IC}$). The standard model requires such a crack in most cases (Fig. 6).

One strong argument for the variable K_Q model is that the very large deflections observed in intact cantilevers can be accommodated by cracking only if many cracks propagate to a considerable length (Table 3). This is probably possible only if propagation velocities are similar over a wide range of crack lengths. Of course, calcite cementation of propagated cracks may help to accommodate large deflections without failure, but its role is uncertain. Another confounding variable is that although humidity was always high, some of the cantilevers were dry when visited. Because measured crack propagation rates in calcite rocks in humid air range from much lower than to very similar to those for water saturation (Atkinson 1984), this effect is difficult to evaluate.

Our hypothesis for cantilever behavior in limestone may apply to other rock types and geologic environments subject to long-term effective tension (normal stress - pore pressure). With rising effective tensile stress, a rock mass would pass through three regimes. At low stress, crack propagation would occur only in rare, large flaws. If stress passes into the intermediate range there is a large expansion of the range of cracks subject to propagation. Because K_I / K_Q is nearly constant over that range, cracks with very different initial lengths will propagate at the same velocity. Thus, while a long crack lengthens by 10%, a crack initially an order of magnitude shorter will double in length. This will result in an apparent homogenization of crack lengths and formation of closely spaced cracks with similar length. With rising stress, or propagation beyond the K_Q regime of a few cm, propagation rate again becomes more selective and propagation of the longer cracks will relieve tensile stress.

4 CONCLUSIONS

A tensile stress of approximately 1 MPa probably causes similar rates of subcritical propagation in cracks in limestone with a wide range of initial lengths. The propagation of large numbers of cracks allows very large cantilever deflections, but at this stress level ultimate failure may require 10^3 to 10^6 years.

REFERENCES

Atkinson, B.K. 1984. Subcritical crack growth in geological materials. *J. Geophys. Res.* 89: 4077-4114.

Bazant, Z.P., Y. Xi, and S.G. Reid 1991. Statistical size effect in quasi-brittle structures: I. Is Weibull theory applicable? *J. Engr. Mech.* 117: 2609-2622.

Henry, J.P. 1978. Mecanique lineaire de la rupture applique a l'etude de la fissuration et de la fracture de roches calcaires. Ph.D. Thesis, Univ. des Sci. et Tech. de Lille, Lille, France.

Henry, J.P., J. Paquet & J.P. Tancrez 1977. Experimental study of crack propagation in calcite rocks. *Int. J. Rock Mech. Min. Sci. & Geomech. Abstr.* 14: 85-91.

Hess, J.W. & R.S. Harmon 1981. Geochronology of speleothems from the Flint Ridge-Mammoth cave system, Kentucky, U.S.A. *Proc. 8th Int. Congress of Speleology, Bowling Green, Kentucky:* 433-436.

Holdrege, T.J. 1986. Failure of limestone roof-beams in south-central Indiana caverns. M.S. Thesis, Purdue University.

Ingraffea, A.R. & R.A. Schmidt 1978. Experimental verification of a fracture mechanics model for tensile strength prediction of Indiana limestone. *Proc. 19th U.S. Symp. on Rock Mechanics*: 247-253, University of Nevada-Reno, Reno.

Palmer, M.V. & A.N. Palmer 1975. Landform development in the Mitchell Plain of southern Indiana: origin of a partially karsted plain. *Zeits. fur Geomorph.* 19: 1-39.

Powell, R.L. 1968. Geology and geomorphology of Wyandotte Cave, Crawford County, Indiana. *Indiana Acad. of Science Proc.* 77: 236-244.

Schmidt, V.A. 1982. Magnetostratigraphy of sediments in Mammoth Cave, Kentucky. *Science* 217: 827-829.

Tharp, T.M. 1983. Mechanics of failure for rock masses subjected to long-term tensile loading-analysis of large naturally occurring cantilevers. *Proc. 14th U.S. Symp. on Rock Mechanics*: 309-318, Association of Engineering Geologists.

Tharp, T.M. 1987. A finite element for edge-cracked beam columns. *Int. J. Num. Meth. Engr.* 24: 1941-1950.

Wilson, W.L. 1985. Lithologic and base level controls of ground water flow paths in the Garrison Chapel area, Indiana. M.S. Thesis, Indiana State University, Terre Haute, pp. 211.

Rock Mechanics, Nelson & Laubach (eds) © 1994 Balkema, Rotterdam, ISBN 90 5410 380 8

The widespread occurrence of healed microfractures in siliciclastic rocks: Evidence from scanned cathodoluminescence imaging

K. L. Milliken
Department of Geological Sciences, The University of Texas, Austin, Tex., USA

ABSTRACT: Photomultiplier-based imaging of electron beam-induced luminescence, applied to a variety of sandstones of different compositions, ages, and burial histories, reveals that brittle deformation of detrital quartz is more prominent and widespread than previously recognized. Healing of microfractures by precipitation of authigenic quartz within detrital grains masks much of this fracturing from observation using standard petrographic methods.

In addition to its role in compactional grain rearrangement, brittle deformation of detrital quartz is linked to pressure dissolution. Simple concavo-convex grain boundaries, sutured grain boundaries, and stylolites are all accompanied by brittle extension resulting in physical redistribution of material on a microscopic scale.

The crack-seal mechanism of fracture and vein formation is also dramatically confirmed in scanned cathodoluminescence images. Repeated brittle extension along close-spaced (micron-scale) subparallel microfractures, that are subsequently filled by multiple generations of authigenic quartz, attests to a complex history in the development of veins and open fractures.

The abundant grain fractures observed with scanned cathodoluminescence suggest that grain fracturing plays a more important role in sandstone consolidation than has been previously recognized. Sandstones may have hydrodynamic properties greatly different from systems characterized solely by intergranular porosity, and thus these observations have implications for understanding the hydrologic characteristics of siliciclastic rocks in the subsurface.

1 OBSERVING NATURAL MICROFRACTURES

Natural microfractures in rocks are difficult to observe. Distinguishing naturally-formed open fractures from defects induced during sample preparation requires specially impregnated and polished rock sections and considerable expertise (e.g. Simmons and Richter, 1976). Healed natural microfractures in rocks can also be observed and provide a wealth of useful information on the kinematic history of rocks (e.g. Laubach, 1988a,b; Onash, 1990; Wu and Groshong, 1991). Detecting healed microfractures however requires examination of micron-scale trails and arrays of liquid- and/or gas-filled inclusions or equally minute zones of mineral precipitates that are relatively obscure using standard light microscopy. Filled microfractures are most inconspicuous in the case of authigenic precipitates within deformed crystals of similar mineralogy (Elliot, 1973, p. 2658), common examples being authigenic quartz filling fractures in detrital quartz (e.g. Morad et al., 1991) and authigenic albite filling fractures in detrital Ca-plagioclase (e.g. Milliken, 1989; Ramseyer et al., 1992).

Scanned cathodoluminescence (CL) imaging permits clear discrimination between detrital quartz and subsequently precipitated authigenic quartz in microfractures at the micron scale (Dickinson and Milliken, 1993; Dunn, 1993). This paper summarizes observations made on 12 sandstone formations, ranging in age from Oligocene to Pennsylvanian, as well as a few sandstone-associated macroscopic veins. These observations suggest that brittle deformation of detrital quartz, manifested as healed intragranular and transgranular microfractures, plays a key role in porosity evolution and fabric development in quartz-rich sandstones. This method has promise for documenting the characteristics and history of natural microfractures in quartz- and feldspar-bearing rock types in a variety of geological contexts.

1.1 *Causes of cathodoluminescence in quartz*

The causes of CL in quartz are not fully understood. Likely controls on the wavelength and intensity of photon emission from quartz include trace impurities and crystal defects (Marshall, 1988; Walker and Burley, 1991). Regardless of the primary controls, it is well established that there is a general tendency for quartz precipitated from aqueous solution at relatively low temperatures to have less intense CL than does quartz crystallized at high temperatures (e.g. Sipple, 1968; Zinkernagel, 1978) and that this CL heterogeneity in quartz persists through a wide range of burial diagenetic conditions before it is ultimately homogenized during low grade metamorphism (Sprunt et al. 1978; Owen, 1991).

1.2 *Instrumentation and operating conditions*

Images were produced using an Oxford Instruments photomultiplier-based CL detector installed on a JEOL T330A scanning electron microscope. Light collection is performed with a parabolic mirror inserted approximately 1 mm above the surface of a carbon-coated, epoxy-impregnated polished thin section. An accelerating voltage of 10 KV produced adequate photon emission for examining luminescence variations in both detrital and authigenic quartz and yielded sharper images than higher voltages. The T330A does not provide a numerical readout of sample current; sample currents near 90 percent of maximum for the instrument were used.

Several factors control the intensity of a CL signal recorded with this type of imaging. Some of these variables (e.g. carbon coat thickness and sample heating) are difficult to ascertain (Hogg et al., 1992, p. 424). Scanned CL images are very useful with regard to textural features, but do not provide quantitative information about either CL intensity or color.

Microscope-based CL observations prompted several workers to emphasize the presence of brittle deformation in sandstones (e.g. Sipple, 1968; Onash, 1990; Morad et al., 1991). Scanned CL imaging, however, has several advantages over this and other common petrographic methods (including standard light microscopy (plane light and x-polar observation), secondary electron imaging, backscattered electron imaging, and X-ray mapping) for the documentation of healed microfractures in quartz (and, excepting backscatter imaging, in feldspar).

Scanned CL:

1. allows a clear distinction to be made between detrital quartz and the authigenic quartz which fills healed microfractures;
2. can be applied at the relatively high magnifications (~50x to 5000x) characteristic of the SEM;
3. benefits from the well-controlled observing conditions associated with high vacuum operation; and,
4. is easy to photograph compared to microscope-based CL which typically requires exposure times too long for metering.

2 SIMPLE INTRA- AND TRANS-GRANULAR MICROFRACTURES

Scanned CL-images show that intra- and trans-granular microfractures (Figure 1) are present in all of the sandstones examined. In every case, a substantial proportion of the microfractures revealed by scanned CL-imaging are not apparent in other modes of observation. In most samples inherited microfractures coexist with microfractures formed *in situ*.. The post-depositional (*in situ*) versus inherited nature of the fracturing and subsequent quartz precipitation can be deciphered in many cases by the relationship of fractures to grain contacts and by the cross-cutting relationships between cement within the fractures versus that on grain surfaces. Microfractures cross grains from points of contact with adjacent grains. Some grains contain arrays of close-spaced microfractures. Microfractures which apparently formed in situ display a wide range of apparent dips.

Figure 1. Simple intra- and trans-granular fractures (arrows), healed with authigenic quartz, in a porous sandstone. Quartz cement (q) on detrital quartz grains (g) fills a portion of the primary porosity (p). Inhomogeneous grains are siliceous rock fragments (Rf).

In plane light some microfractures appear as zones of micron-scale gas- and liquid-filled inclusions. In cross-polar observation many fractured quartz grains appear polycrystalline, with no clear evidence of the origin of the polycrystallinity.

2.1 *Discussion*

The morphology of simple intra- and trans-granular microfractures in sandstones resembles closely the microfractures observed following uniaxial compaction of wet sand packs (Gallagher et al., 1974; Schutjens, 1991; Zhang et al., 1990; Elias and Hjash, 1992). The variable orientations are consistent with the heterogeneity of fracture-producing loads transmitted by stacked grains of irregular shapes. All the fractures appear to be extensional. In a recent study using scanned CL, Dunn (1993) describes increases in apparent volumes of individual grains of up to 30 percent.

3 MICROFRACTURES ASSOCIATED WITH PRESSURE SOLUTION

3.1 *Concavo-convex grain-to-grain contacts*

Several samples contain examples of grains having indented contacts of the type which are conventionally interpreted to represent grain-to-grain interpenetration by pressure solution. The most striking of these examples is the Mesozoic Etjo sandstone (Karoo Group), a porous eolian unit exposed at the surface in northern Namibia (Dickinson and Milliken, 1993). The scanned luminescence images of Etjo sandstones reveal that a large proportion of the material redistribution around arcuate contacts has arisen from fracturing and extension of the indented grain rather than volume loss (Figure 2).

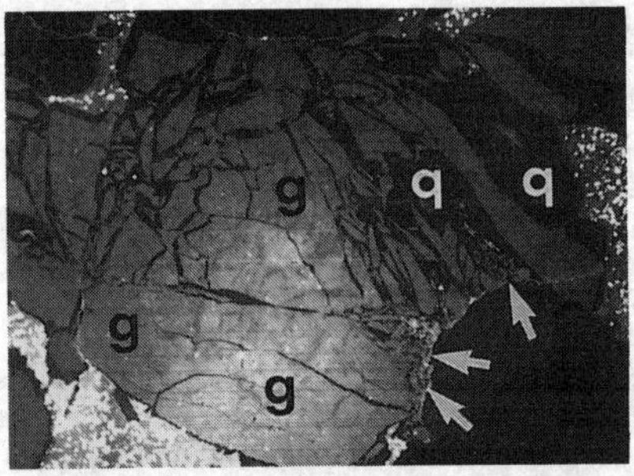

Figure 2. Opening-mode microfractures, healed with authigenic quartz (q), in a quartz sand grain (g) from the Etjo Formation.. In more conventional types of imaging the arcuate contact between the grains would be interpreted to represent pressure solution and volume loss from the indented grain. Arrows indicate regions of finely comminuted (<5 μm) grain debris near the grain contact.

Comparison of scanned CL images and cross-polar photomicrographs reveals that most polycrystalline grains correspond to the grains that are indented and fractured.

Many grains show a pronounced increase in intensity of brecciation immediately adjacent to the region of maximum curvature of the grain contact. In these areas, intense crushing of the indented grain produces a finely comminuted aggregate of <5 to 10 μm particles that are near the limit of the instrument's spatial resolution (Figure 2, arrow). Minor suturing along grain contacts (Figure 2) appears to represent actual pressure solution in these samples, but the volume loss represented is vastly less than that estimated by grain indentation as interpreted from grain indentation. Unfortunately, the burial history of the Karoo Group is not well constrained and it is not possible to determine the depth of burial at which deformation and re-cementation of the grains took place.

3.2 *Sutured grain-to-grain contacts*

Pervasive microfractures are also observed in grains with more complex, sutured boundaries (Figure 3). In these cases, both grains may manifest substantial fracturing with the regions of most intense brecciation corresponding to areas within the indented grains, proximal to regions of maximum curvature. Microfracture termini are generally localized on the most steeply curving portions of the sutures, the microfractures propagating into grains on the convex sides of steep projections (arrows, Figure 3).

Figure 3. Sutured quartz grains with quartz-filled extension fractures localized (arrows) within the indented grain along regions of maximum curvature of the sinuous contact.

3.3 *Stylolites*

Prominent microfracturing also occurs in the vicinity of stylolites (Figure 4). Stylolite-

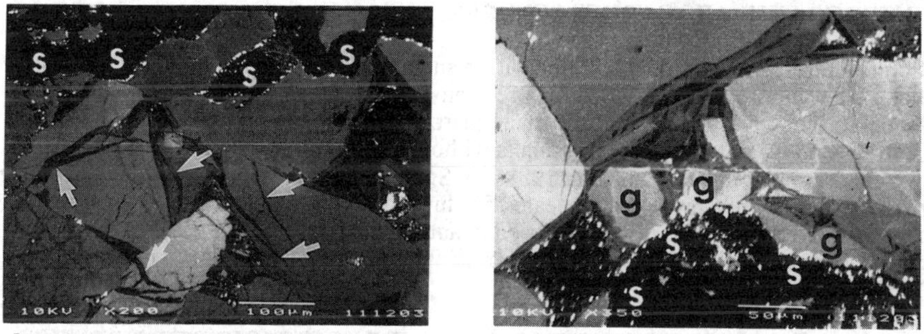

A B

Figure 4. Microfractures proximal to a stylolite seam.
A. Quartz-filled fractures (arrows) near a stylolite seam (s). Apart from the stylolite, this field of view appears as uniform quartz in back-scattered electron images.
B. Brecciated grain debris (g) cemented by zoned authigenic quartz (q) near the top of a prominent stylolite column (s).

associated microfracturing includes both simple intra- and trans-granular fracturing of grains adjacent to the stylolite (Figure 4A) and the formation of complex, granulated breccias at the tops of stylolite columns (Figure 4B).

3.4 *Discussion*

A loose genetic association between brittle extension and pressure solution has been suggested previously based upon theoretical models of stress fields (Fletcher and Pollard, 1981; Wood, 1981), relationships between stylolites and macroscopic extension veins in outcrop and core (Choukroune, 1969; Nelson, 1981; Heydari and Moore, 1993), and associated stylolites and microfractures at the thin section scale (Onash, 1990; Raynaud and Carrio-Schaffhauser, 1992). In general, brittle extension fractures are oriented normal to stylolites and are assumed to at least partially compensate the volume loss associated with stylolite development.

Scanned CL images show that fracturing which accompanies the pressure solution process is substantial. Clearly, brittle deformation is a prominent control on the morphology of all three types of grain contacts described above, in each case leading to volume loss estimates less than would interpreted for pressure solution alone. Furthermore, the localization of microfractures on grain projections and the finely crushed grain debris concentrated along pressure-solved contacts suggest that brittle fracture may be integral to the actual pressure solution mechanism. More detailed observations are needed to decipher the exact process by which brittle deformation interacts with dissolution to produce the various morphological features categorized as pressure solution.

It is plausible, however, that enhanced solubility of detrital particles may result at least in part from the extreme reduction of particle size taking place in what might be termed a "brittle process zone" at grain contacts (e.g., see comments by Green, 1984). In this view, fracturing is the primary process active in the pressure solution mechanism, followed by the secondary process of dissolution. If further substantiated, the role of fracturing in pressure solution will require a substantial reevaluation of some models for subsurface porosity evolution.

4 INTERGRANULAR MICROFRACTURES

Microfractures are prominent features within and around some open fractures and quartz-filled veins (Figure 5) A common feature is an array of subparallel microfractures parallel to the long dimension of the main quartz-lined fracture or vein. These types of fractures tend to cross grains and cements without regard to the localizing effects of individual grain contacts (Laubach, 1988a,b). Some individual sand grains have been greatly extended by a process of repeated fracturing and authigenic mineral precipitation. Crosscutting relationships of CL-zonation in authigenic quartz reveals that early-formed authigenic quartz is itself subject to fracturing and rehealing (Figure 5B, arrows).

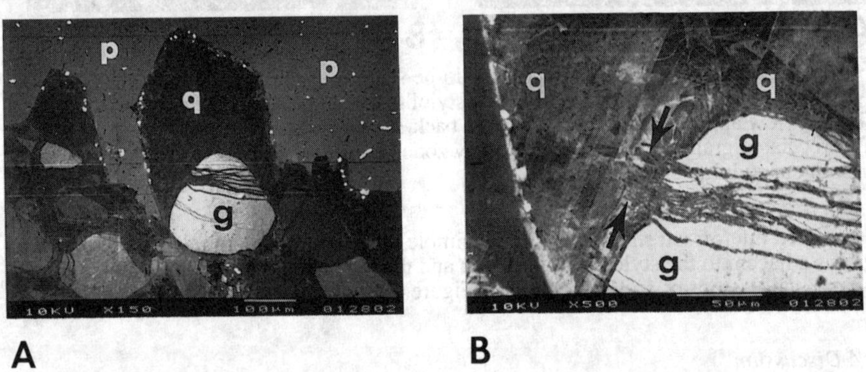

Figure 5. An array of healed microfractures with a crack-seal history of development.
A. The large euhedral quartz crystal (q) is localized as an overgrowth on a quartz grain (g) which projects into an open fracture (p). The long dimension of the fracture is subparallel to the long dimension of the CL-photomicrograph. The grain has been repeatedly fractured and extended.
B. Close-up view of the left edge of the grain shown in A. Arrows indicate filled microfractures which crosscut earlier-formed zones of authigenic quartz. Such textural relationships can be used to decipher the interrelated chemical and kinematic histories of the sandstone.

4.1 *Discussion*

Scanned CL images of microfractures associated with partially open fractures and quartz-filled veins constitute vivid confirmation of the crack-seal mechanism of fracture propagation (Ramsay, 1980). Textural relationships between multiply fractured and highly extended grains and zoned, fractured, and rehealed authigenic quartz attest to an episodic history of repeated fracturing and cementation. Extension as measured on veins formed in this manner must be accounted by excluding that portion of the vein comprising detrital quartz, a procedure that is readily accomplished on CL-images (Onash, 1990). Likewise, documenting the amount of fractured and highly extended detrital material within veins provides critical information for oxygen isotopic and fluid inclusion studies of vein quartz.

5 CONCLUSIONS AND IMPLICATIONS

It is clear that in some sandstones a substantial, even major, portion of the total authigenic quartz may be sequestered within brittlely extended detrital grains. This intragranular authigenic quartz constitutes an increase of the apparent detrital volume (as conventionally observed), a gain accommodated by a decrease the intergranular volume (IGV, cements + primary porosity). This observation contradicts conventional assumptions of either constancy of detrital quartz volume in the case of simple compaction or of detrital volume loss in the case of pressure solution. Assessing true volume losses and gains for a variety of geological problems (e.g. chemical mass balance) requires the detailed textural information available from scanned CL-imaging. Many opportunities exist for using scanned CL-images to obtain quantitative determinations of authigenic quartz content, brittle grain extension, and microfracture orientation. Relationships between authigenic crystal zonations and microfracture development may also provide data to relate chemical and kinematic history in sandstones.

6 ACKNOWLEDGMENTS

During the course of this study the luminescence detector at the Department of Geological Sciences, University of Texas at Austin was on loan from Oxford Instruments. Paul Wright and Glenn Kinnear of Oxford Instruments kindly arranged for the installation of the detector. My views on brittle deformation and pressure solution benefited greatly from discussions with Lynton Land, Sharon Mosher, Earle McBride, and Warren Dickinson. Sandstone samples examined for this study were provided by Warren Dickinson, Shirley Dutton, Nina Harun, Dave Hicks, Steve Johansen, Ralph Kugler, Steve Laubach, Alaa Salem, Sally Sutton, and Suzanne Weedman. I thank Steve Laubach for encouraging me to write this paper and for his helpful comments on an early draft.

REFERENCES

Choukroune, P. 1969. An example of mesocopic analysis of concentric folding in limestone series. *Tectonophysics*. v. 7, p. 57-70.

Dickinson, W.W. and K. L. Milliken. 1993. Sandstone compaction by interrelated brittle deformation and pressure solution, Etjo Sandstone, Namibia. *Geol. Soc. Amer. Abstr.* v. 25, no. 6, p.65.

Dunn, T. L. 1993. Early quartz grain fracturing and annealing: an important brittle deformation mechanism of early compaction in first cycle orogenic sandstones, revealed by scanning electron cathode luminescence microscopy. *Geol. Soc. Amer. Abstr.* v. 25, no. 6, p. 335.

Elias, B. P. and A. Hjash. 1992. Changes in quartz solubility and porosity due to effective stress: An experimental study. *Geology*. v. 20, p. 451-454.

Elliot, D. 1973. Diffusion flow laws in metamorphic rocks. *Geol. Soc. Amer. Bull.* v. 84, p. 2645-2664.

Fletcher, R. C. and D. D. Pollard. 1981. Anticrack model for pressure solution surfaces. *Geology*. v. 9, p. 419-424.

Gallagher, J. J., M. Friedman, J. Handin, and G. M. Sowers. 1974. Experimental studies relating to microfractures in sandstone. *Tectonophysics*. v. 21, p. 203-247.

Green, H. W., II 1984. "Pressure solution" creep: Some causes and mechanisms. *J. Geophys. Res.* v. 89, B6, p. 4313-4318.

Heydari, E. and C. H. Moore. 1993. Zonation and geochemical patterns of burial calcite cements: Upper Smackover Formation, Clarke County, Mississippi. *J. Sedimentary Petrology*. v. 63, p. 44-60.

Hogg, A. J. C., E. Sellier, and A. J. Jordan. 1992. Cathodoluminescence of quartz cements in Brent Group sandstones, Alwyn South, UK North Sea. pp. 421-440, in A. C. Morton, R. S. Haszeldine, M. R. Giles, and S. Brown, eds., *Geology of the Brent Group, Geol. Soc. Special Pub.* No. 61.

Laubach, S. E. 1988a. Fractures generated during folding of the Palmerton Sandstone,

eastern Pennsylvania. *J. Geol.*. v. 96, p. 495-503.
Laubach, S. E. 1988b. Subsurface fractures and their relationship to stress history in East Texas Basin sandstones. *Tectonophysics*. v. 156, p. 37-49.
Marshall, D. J. 1988. *Cathodoluminescence of Geological Materials*. Allen and Unwin, Inc., Winchester, Massachusetts, 146 p.
Milliken, K. L. 1989. Petrography and composition of authigenic feldspars, Oligocene Frio Formation, South Texas. *J. Sedimentary Petrology*. v. 59, p. 361-374.
Morad, S. A. Bhattachayya, I. S. Al-Aasm, and K. Ramseyer 1991. Diagenesis of quartz in the Upper Proterozoic Kaimur Sandstones, Son Valley, central India. *Sedimentary Geol.* v. 73, p. 209-225.
Nelson, R. A., 1981. Significance of fracture sets associated with stylolite zones. *Amer. Assoc. Petroleum Geologists Bull.* v. 65, p. 2417-2425.
Onash, C. M. 1990. Microfractures and their role in deformation of a quartz arenite from the central Appalachian foreland. *J. Structural Geol.* v. 12, p. 883-894.
Owen, M. R. 1991. Application of cathodoluminescence to sandstone provenance. pp. 67-75, in in C. E. Barker and O. C. Kopp, eds., *Luminescence microscopy: Quantitative and Qualitative Aspects*. SEPM, Tulsa, Oklahoma, 196 p.
Ramsay, J. G., 1980. The crack-seal mechanism of rock deformation. *Nature*. v. 284, p. 135-139.
Ramseyer, K, J. R. Boles, and P. C. Lichtner. 1992. Mechanism of plagioclase albitization. *J. Sedimentary Petrology*. v. 62, p. 349-356.
Raynaud, S. and R. Carrio-Schaffhauser. 1992. Rock matrix structures in a zone influenced by a stylolite. *J. Structural Geol.* v. 14, p. 973-980.
Schutjens, P. 1991. Experimental compaction of quartz sand at low effective stress and temperature conditions. *J. Geol. Soc. London*. v. 140, p. 725-740.
Simmons, G. and D. Richter. 1976. Microcracks in rocks. pp. 105-137, in R. G. J. Strens, ed., *The Physics and Chemistry of Minerals and Rocks*, John Wiley and Sons, New York, 697 p.
Sipple, R. F. 1968. Sandstone petrology: evidence from luminescence petrography. *J. Sedimentary Petrology*. v. 38, p. 530-554.
Sprunt, E. S., L. A. Dengler, and D.Sloan. 1978. Effect of metamorphism on quartz cathodoluminescence. *Geology*. v. 6, p. 305-308.
Walker, G. and S. Burley .1991. Luminescence petrography and spectroscopic studies of diagenetic minerals. pp. 83-96, in C. E. Barker and O. C. Kopp, eds., *Luminescence microscopy. Quantitative and Qualitative Aspects*. SEPM, Tulsa, Oklahoma, 196 p.
Wood, J. R. 1981. Fracture pressure solution: Experiment and theory. *Geol. Assoc. Canada, Mineralogical Assoc. Canada, Abstr. with program*.. v. 6, p. A63.
Wu, S. and R. H. Groshong. 1991. Low temperature deformation of sandstones, southern Appalachian fold-thrust belt. *Geol. Soc. Amer. Bull.* v. 103, p. 861-875.
Zhang, J., T.-F. Wong, and T. M.Davis. 1990. Micromechanisms of pressure-induced grain crushing in porous rocks. *J. Geophys. Res.* v. 95, B1, p. 341-352.
Zinkernagel, U. 1978. Cathodoluminescence of quartz and its application to sandstone petrology. *Contrib. Sedimentology* 8. 69 p.

Rock Mechanics, Nelson & Laubach (eds) © 1994 Balkema, Rotterdam, ISBN 90 5410 380 8

Containment of natural fractures in brittle beds of the Austin Chalk

M. Friedman & O. Kwon
Department of Geology and Center for Tectonophysics, Texas A&M University, College Station, Tex., USA

V. L. French
School of Geology and Geophysics, University of Oklahoma, Norman, Okla., USA

ABSTRACT: The vertical continuity of natural fractures in a formation such as the Austin Chalk is dependent upon whether or not the fractures tend to be throughgoing or are contained within brittle chalk layers separated by shale/marl layers. Holifield (1992) emphasizes the economic importance of recognizing a horizontally stratified reservoir when planning the trajectories of vertical, slant and horizontal boreholes. Field observations of the Atco Member of the chalk at the Leigh Quarry in Waco indicate that fractures tend to propagate upward from brittle layers; fan and die out in the shale/marl members, and exhibit a spacing about equal to brittle-bed thickness. Laboratory experiments involving the propagation of a mode 1 fracture in a brittle layer toward a clay layer at 10 and 50 MPa confining pressures indicate that fracture containment in the brittle layer is enhanced by increasing thickness of the clay layer, decreasing effective confining pressure, using wet versus room dry clay, reducing the displacement rate and changing the brittle member from Berea Sandstone to Austin Chalk.

1 INTRODUCTION

Field observations focused on the vertical continuity of fractures in the Austin Chalk indicate that many of the fractures are contained within brittle chalk members when the latter are bounded by shales or marls that exceed 10-cm in thickness (Corbett et al. 1987, 1991a, 1991b, Wagner 1991). There are exceptions to this generality, however, and some fractures are observed to terminate where the shale/marl is much thinner and other exceptions where the fractures propagate through several shale/marl breaks. Clearly a better understanding of the factors that control fracture-bed containment is needed, especially in light of the importance of this topic to the vertical connectivity within naturally fractured reservoirs. For example, Holifield (1992) emphasizes the importance of recognizing whether or not the fractures tend to be bed-contained when planning the trajectories of vertical, slant and horizontal boreholes.

A fracture contained within a brittle layer, the so-called bed-contained fracture, dies out within the bounding shale/marl member and contributes to the porosity and permeability only of the brittle member. Such bed-contained fractures are particularly conspicuous in the Lehigh Quarry, Waco, Tx (Figure 1a,b) where the basal Atco Member of the Chalk is exposed. The stratigraphic architecture at Waco is thought to be similar to that extant in the Giddings Field, 100 miles to the SE. In contrast, the Atco Member in the Old Alamo Quarry, San Antonio has few shale breaks and the fractures are in good vertical communication, particularly within fault zones (Corbett et al. 1991a, 1991b).

Accordingly, presented herein are (a) a series of laboratory experiments designed to identify the conditions that promote the containment of fractures within brittle layers; (b) data on the spacing of bed-contained fractures relative to the thickness of the brittle layers;

Figure 1. Photographs illustrate aspects of the Atco Member of the Austin Chalk at the Lehigh Quarry, Waco, Tx. (a) Overview shows brittle chalk layers separated by conspicuous shale/marl layers (dark) in a minor graben. Note vehicle for scale. (b) Well developed fractures in brittle chalk terminate at upper shale/marl (dark layer at top). Note scale in photo; black short bars on left are each 1-cm long. (c) Fractures in brittle chalk fan and terminate in overlying shale/marl bed (middle of photo) thereby showing fractrure containment and an upward direction (arrow) of fracture propagation. Distance between vertical lines of arrow is 1 cm.

and (c) observations of a polar fabric element calibrated in the laboratory to indicate the direction of fracture propagation.

2 STRENGTH AND DUCTILITY OF THE AUSTIN CHALK

The mechanical response of outcrop samples of the Austin Chalk deformed at 10-MPa confining pressure, dry, varies over a wide range from relatively strong and brittle to relatively weak and ductile (Figure 2). The specimens from the seven locations included in Figure 2 are all from chalk layers (as opposed to shale/marl breaks) and show in the field various manifestations of brittle fracture. Even the weakest and most ductile of these specimens (Figure 2, Waco) exhibits conspicuous fracturing. This spectrum of behavior probably occurs at each of the seven field locations -- clearly one specimen does not describe the behavior of all the rocks at each location. These specimens contain no clay but do show a porosity range of 10 to 30 percent. There is an excellent correlation between porosity and ultimate strength with strength increasing with decreasing porosity (R = 0.95, plot not shown).

3 RESULTS OF EXPERIMENTS WITH CLAY LAYERS

To gain some insight into the factors that control fracture-containment within brittle layers, laboratory experiments were made inwhich a mode 1 fracture initiates from a notch in a brittle rock and propagates toward a "bedding plane" (Figures 3 and 4, Table 1). The bedding plane defines the interface between the brittle host rock and a shale bed (bentonitic clay), the thickness of which is varied up to 15 mm. This clay, precompacted at 30 MPa, is extremely ductile when deformed at a confining pressure of 10 MPa (Figure 2) and has a maximum strength of only 21 MPa at 18% shortening. Whether or not the fracture crosses through the clay member or is contained within the brittle member as it terminates within the clay (Figure 4) is investigated as functions of confining pressure (10 and 50 MPa), displacement rate (10^{-2} and 10^{-3} in/sec), thickness of the clay layer, the wet versus room-dry nature of the clay layer and for a change in the host rock from Berea Sandstone to two types of Austin Chalk . Data in Table 1 indicate that conditions favoring containment of fractures within the brittle layer (i.e., termination within the clay layer) are (1) increasing thickness of the clay, (2) low effective confining pressure, (3) wet versus dry condition of the clay, (4) slower fracture propagation, and (5) when the fracture propagates from the notch in Austin Chalk rather than Berea Sandstone.

In addition, the laboratory specimens indicate that when the fractures do not propagate across the clay they fan-out and terminate within the clay layer (Figure 4). The direction of fanning in the clay therefore defines a polar fabric element that shows the direction of propagation of the fractures in the brittle member. This same fabric element is conspicuous in the exposures at the Lehigh Quarry, Waco as described below.

4 SPACING OF BED-CONTAINED FRACTURES

Fracture spacing versus bedding thickness was studied in exposures of the lower Austin Chalk (Atco Member) at the Lehigh Quarry, Waco Texas. Fractures striking from N30°W to N40°E were measured in nine traceable chalk layers along an 80-m long vertical face trending N82°W . The fractures are genetically related to the regional Balcones Fault trend and to cross faults producing grabens within the Balcones trend (Corbett et al. 1991a, 1991b). The data for the "regional" fractures (i.e., those not immediately adjacent to a local fault) show there is a strong correlation (R = 0.9) between bed thickness and fracture spacing with the spacing being about equal to the bedding thickness (Figure 5). In marked contrast, there is no such relationship for superposed tectonic fractures in the hangwall of and within 10 m of a listric normal fault exposed along the line of section (Figure 5). The brittle-behaving Atco chalks in this quarry are conspicuously separated by shales/marls such that most of the fractures are bed-contained. This same stratigraphic architecture characterizes the producing horizons in the greater Giddings field 100 miles to the SE.(see Item 6, below).

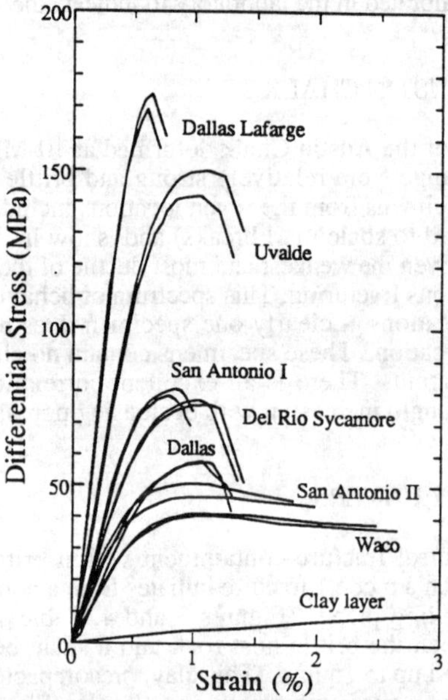

Figure 2. Stress/strain curves for seven specimens of Austin Chalk show a relatively large spread in strength and ductility when deformed room dry, at 10 MPa confining pressure and at 24°C and $10^{-4}s^{-1}$. Two curves for each specimen show good reproducibility. Also shown is the mechanical behavior of the clay layer deformed under similar conditions. (see text [Sec 3], Figures 3 and 4, Table 1).

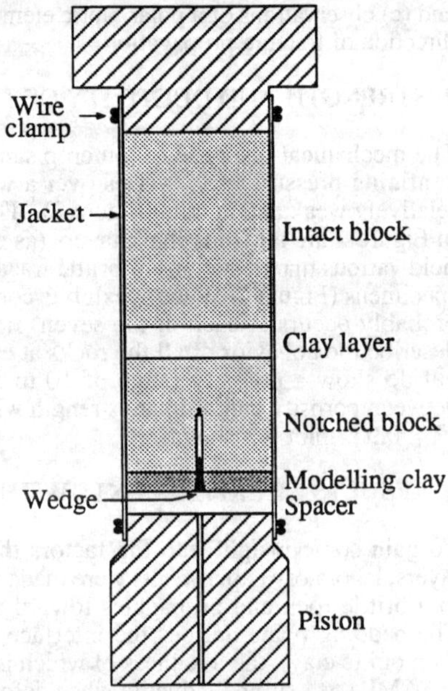

Figure 3. Schematic drawing showing design of experiment to evaluate factors that control fracture bed-containmen (see text, Figure 4 and Table 1).

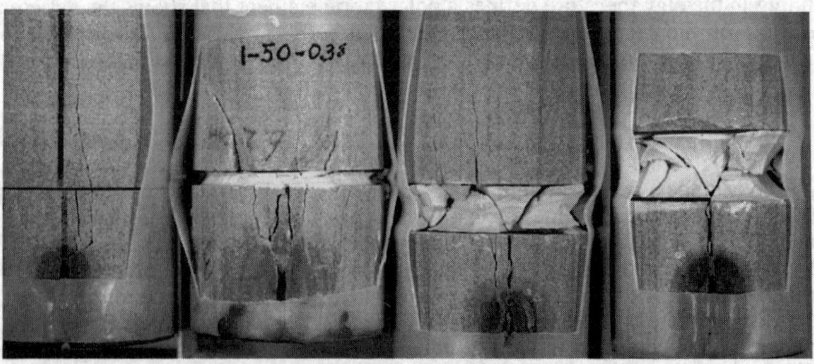

Figure 4. Photographs of specimens showing fractures propagating upward from notch (base of each specimen) across or into clay layer thichness of which varies from 0 to 15 mm (from left to right), (see Table 1). Fractures either cross interface or fan-out and terminate within clay layer thereby creating the polar fabric element also found in outcrop (see Figure 1c).

TABLE 1 - EXPERIMENTS ON FRACTURE CONTAINMENT

EFFECTIVE Pc* (MPa)	CLAY THICKNESS** (mm)	COMMENTS Fracture:
Displacement rate: 10^{-3} in/sec:		
Berea Sandstone***		
10	0.0	propagates across interface
50	0.0	propagates across interface
50 (wet clay)****	5.0	contained
50 (wet clay)****	5.0	contained
10	5.1	contained
10	10.9	contained
50	3.8	propagates across interface
50	4.6	contained
50	6.8	contained
50	10.0	contained
50	10.0	contained
50	10.6	propagates across interface
50	15.2	contained
Austin Chalk****		
10 (weak chalk)	9.5	contained
50 (weak chalk)	9.2	contained
50 (strong chalk)	9.5	contained
50 (strong chalk)	9.8	contained
Displacement rate: 10^{-2} in/sec		
Berea Sandstone		
10	10.5	propagates across interface
10	10.6	propagates across interface
10	14.7	contained

*Pc, the confining pressure, also is the magnitude of the normal stress across the interface. ** All clay layers are precompacted to Pc of 30 MPa. *** The Austin Chalk and Berea Sandstone at these low effective confining pressures are equally brittle, however at Pc = 10 MPa, the sandstone has an ultimate strength of 125 MPa while the two chalks have ultimate strengths of 170 and 50 MPa, respectively. The Mohr envelope for the sandstone is $\tau = 20$ (MPa) + $.55\sigma_n$ (Handin et al. 1963); that for the chalk averages $\tau = 9$ (MPa) + $.64\sigma_n$ (Corbett et al. 1987). ****Wet clay is 50:50, H_2O to bentonitic clay, by weight.

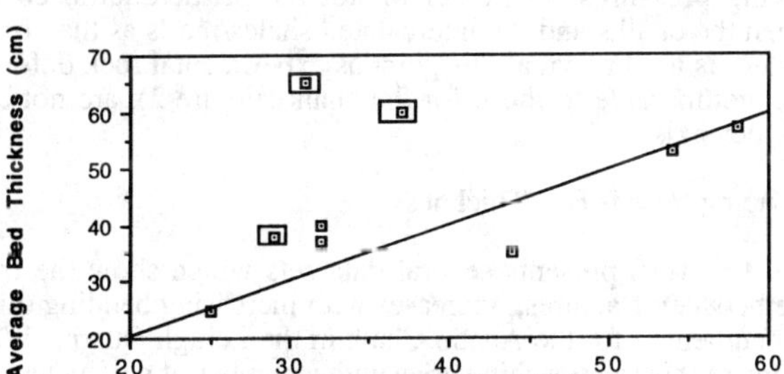

Figure 5. Plot of fracture spacing vs bed thickness data from Austin Chalk at Lehigh Quarry, Waco Tx. Line shows slope when spacing equals bed-thickness. Three data points in squares are for tectonic fractures related to local faults.

5 DIRECTION OF FRACTURE PROPAGATION

The observation that fractures tend to fan-out and terminate in the clay layer of experimentally deformed specimens relative to the known direction of fracture propagation gives rise to a polar fabric element, the polarity of which indicates the direction of fracture propagation. Tens of similar structures were observed in the Lehigh Quarry, Waco (Figure 1c) and all show that the fractures propagated vertically from an underlying brittle chalk member upward into an overlying shale/marl.

6 DISCUSSION AND SIGNIFICANCE

6.1 Fracture Bed Containment

If the stratigraphic architecture of a given section involves the intercalation of relatively brittle and ductile layers then fracture containment is possible and with it, reduced fluid transmissivity normal to bedding. It is a fundamental point in the approach to the development of such a stratified reservoir to plan borehole trajectories with fracture containment in mind (Holifield 1992, Friedman and Wiltschko 1992). While at least five factors control fracture containment (see section 3, above) the thickness of the shale/marl breaks and the existance of low effective confining pressure at the time of fracturing are perhaps most important and tractable. Cores and logs can be used to ascertain the thicknesses of the shale/marl layers at depth and along with outcrop studies calibrate the thickness of shale/marl breaks across which fractures do not propagate. The thicker the shale/marl layer the greater the tendency for the containment of fractures in the adjacent brittle layer. As cited earlier, outcrop observations suggest that rarely do fractures transect shale/marl layers in the Austin Chalk that are > 10-cm thick (e.g., Corbett et al. 1987).

The existance of low effective confining pressures at the time of fracturing can often be inferred. For examples: (a) fracturing early in the burial history of the Austin Chalk, as suggested by faulting prior to the deposition of the Pecan Gap formation (B. R. Holifield, Personal Communication, 1992), would be at relative shallow depth of burial where the effective pressures would be low even with normal pore-fluid pressure; or (b) fracturing at greater depths of burial but at the time of pore-fluid pressure increase associated with oil maturation and/or late dewatering of clay minerals, particularly in the Eagleford Shale directly underlying the Austin Chalk would create abnormally high pore pressures and give rise to low effective pressures. Note that a low effective pressure must also have existed when rocks at the present outcrop fractured as the cover of sediments along the present outcrop belt could have been no more than about 1 km. Accordingly, one could rely on extrapolation of certain fracture characteristics from the outcrop to the subsurface (such as bed-containment) since both the outcrop and the rocks at depth fractured under similar low effective pressures.

In general, the strength and ductility of rocks are decreased when deformed at low effective confining pressures. Whether or not the brittle/ductile contrast would be enhanced between the chalks and the intercalated shales/marls as the effective confining pressure is reduced is not known at this point as experimental rock deformation tests for the shales/marls, comparable to those for the chalk (Figure 2), are not available to this writer, if indeed they exist.

6.2 Fracture Spacing Versus Bed Thickness

Nelson (1985, p. 157-160) presents several data sets which show the fracture spacing, i.e., the distance between fractures, increases with increasing bedding thickness. This is same trend as documented for the Austin Chalk in the Lehigh Quarry, Waco, Tx (Figure 5). The exact slope of this relationship varies with a number of parameters, but it is always most reliable for regional fractures (or those associated with small differential stresses and strains) as opposed to tectonic ones (R. A. Nelson, Personal Communication, 1993). A point also documented herein (see section 4, above).

As mentioned in section 4, the fracture spacing data in Figure 5 indicate that the fractures spacing is about equal to bedding thickness. What are the ramifications of this statement? First, it needs to be clarified that these data are only 2-dimensional since only fractures oriented at >45° to the trend of the outcrop would have been counted without geometric bias. Assume, however, that orthogonal sets of vertical fractures exist (as is the case within this quarry [Corbett et al. 1991b, p. 41]) and that the thinner beds are broken into smaller rectangular blocks than are the thicker beds (Figure 5). Secondly, assume that longer fractures have wider aperture than shorter ones, as predicted by linear elastic fracture mechanics (e.g., Pollard and Aydin 1988). It follows, therefore, that even though thin brittle beds, would contain more fractures per unit volume of rock mass than thicker units, the latter would support the longer and wider fractures. It follows that since fracture permeability is dependent upon the cube of the fracture aperture (e.g., Jones, 1975, Nelson 1985, p. 58-59) the thicker units would exhibit greater fracture permeability than smaller ones. Moreover, Jones (1975) shows that fracture porosity is proportional to the cube root of the fracture permeability. Accordingly, the thicker -bedded units thus might be the better reservoirs because their permeability and fracture porosity are greater than those of the thinner-bedded units even though the latter are "more highly fractured".

A second ramification of Figure 5 is that it serves, along with 3-dimensional data on fracture spacing, to access the distribution of intact blocks sizes (each block is bounded laterally by nearly vertical fractures and vertically by nearly horizontal bedding planes). For applications such as CO_2 water flooding, the corresponding fracture surface area in contact with the CO_2-laced waters is critical in calculating the potential success of the treatment. Again, relative to bed thickness, the blocks in the thinner beds would have more fracture surface area per unit volume than those in the thicker units such that oil migration from matrix to fracture network (natural or induced by the CO_2 treatment) would be better in the thinner units. Accordingly, in units like the Austin Chalk that may be self-sourcing (Hinds and Berg, 1990) it is possible that CO_2 treatments would be more productive in the naturally fractured thinner-bedded units.

6.3 Direction of Fracture Propagation

Mode 1 fractures propagating into and terminating within the clay layers of the labortory experiments (Figure 4) provides a new polar fabric element for the determination of fracture propagation directions in naturally deformed rocks (Figure 1c). It thus joins hackle marks, arrest lines, slickensides and en passant configurations etc., as similarly useful fabric elements. Observations of this polar fabric element in cores, not only would enable the observer to determine if fractures were propagating upward or downward into intercalated shales/marls, but also would provide evidence that fractures were terminating within the shale/marl breaks and thus direct evidence of fracture containment or otherwise.

7 CONCLUSIONS

The main conclusions warranted from this study are:

1. Laboratory experiments indicate that the containment of fractures within brittle layers and their termination in adjacent clay layers is favored by increasing thickness of the clay layer, decreasing effective confining pressure, using wet versus room dry clay, decreasing the fracture propagation rate, and using Austin Chalk rather than Berea Sandstone as the brittle member.
2. When not immediately and genetically related to local faults, bed-contained fractures in the Atco Member of the Austin Chalk exposed in the Lehigh Quarry, Waco, Tx., exhibit a well ordered spacing such that the average spacing is equal to the thickness of the brittle layer.
3. Observations in the Lehigh Quarry of the polarity of the fabric element that in laboratory experiments defines the direction of fracture propagation from brittle chalk

layers into shale/marl layers, indicate that the bed-contained fractures propagated upward into the bounding shale/marl breaks, without exception.

8 ACKNOWLEDGMENTS

This work was sponsored by the U. S. Department of Energy, Grant No. DE-FG22-89BC14444, the State of Texas through Center for Energy and Mineral Resources, Texas A&M University, and discretionary funds from the Mollie B. and Richard A. Williford Professorship in Petroleum Geology. We also wish to thank Mr. Charles Moore, manager of the Lehigh Quarry, Waco, TX for his thoughtful cooperation enabling us access to the quarry.

REFERENCES

Corbett, Kevin, Melvin Friedman, & John Spang 1987. Fracture development and mechanical stratigraphy of the Austin Chalk, Texas. *Am. Assoc. Petroleum Geol. Bull.* 71: 17-28.

Corbett, K.P., M.Friedman, D.V.Wiltschko, & J.H. Hung 1991a. Characteristics of natural fractures in the Austin chalk outcrop trend. p. 3 - 11*in* Stewart Chuber,Ed.,*Proc. South Texas Geological Society Symposium on the Austin Chalk,* Febr. 25-26, San Antonio, Tx. 131p.

Corbett, K.P., M.Friedman, D. V.Wiltschko, & J.H. Hung 1991b. Controls on fracture development, spacing, and geometry in the Austin Chalk formation, Central Texas: Considerations for Exploration and Production. *Dallas Geological Soc. Field Trip No. 4,* AAPG Ann. Conv., Dallas, Tx, April 1991. 49p.

Friedman, M. & D.V. Wiltschko 1992. Exploration for naturally fractured reservoirs-a generic approach with examples from the Austin Chalk. in *Geological Studies Relevant to Horizontal Drilling: Examples from Western North America.* J. W. Schmoker, E. B. Coalson, C. A. Brown, eds., 89-99.

Handin,John, Rex V.Hager, Melvin Friedman &, J.N. Feather 1963. Experimental deformation of sedimentary rocks under confining pressure: Pore pressure tests. *Am. Assoc. Petroleum Geol., Bull,* 47:717-755.

Hinds, G.S. & R. R. Berg 1990. Estimating organic maturity from well logs, Upper Cretaceous Austin Chalk, Texas Gulf Coast. *Gulf Coast Association of Geological Societies Trans.* 40:295-300.

Holifield, B. R. 1992. Experiences in the Austin Chalk. *in* S. W. Poston, ed.*Symp.on the Austin Chalk Producing Trend,* Texas A&M University, June 3, 13p

Jones, F. O. Jr. 1975. A laboratory study of the effects of confining pressure on fracture flow and storage capacity in carbonate rocks. *Jour. Petroleum Technology,* 21-27, January.

Nelson, R.A. 1985.*Geological Analysis of Naturally Fractured Reservoirs.* Gulf Pub. Co, Houston, Tx. 320p.

Pollard, D. D. & A. Aydin 1988. Progress in understanding jointing over the past century. *Geol. Soc. Amer. Bull.* 100:1181-1204.

Wagner, D.T. 1991. Controls on fracture distribution in the Giddings Austin Chalk (Abstract).p. 125 *in* Stewart Chuber, Ed., *Proc. South Texas Geological Society Symposium on the Austin Chalk,* Febr. 25,26, San Antonio, Tx. 131 p.

15 Mechanics of poorly consolidated weak rock

Session developers:
Andreas Kronenberg, Texas A&M University
John Kemeny, University of Arizona

Rock Mechanics, Nelson & Laubach (eds) © 1994 Balkema, Rotterdam, ISBN 90 5410 380 8

Strength and deformation properties of a physical model melange

E.S. Lindquist & R.E. Goodman
Department of Civil Engineering, University of California, Berkeley, Calif., USA

ABSTRACT: Physical model melanges made up of stronger stiffer blocks in a weaker softer matrix were created to study the effect block proportion and orientation have on melange mass strength and deformation properties. Test results indicate that increasing the block proportion generally results in a decrease in cohesion, an increase in angle of internal friction and an increase in modulus of deformation. The highest block proportion specimens have a cohesion as low as half of that of the matrix and an angle of internal friction as much as 16.5° higher than that of the matrix. Block orientation also affects the strength, most notably the cohesion, and the modulus of deformation. The specimens with the most adversely oriented blocks (30°) have the lowest cohesion, and the increase in modulus with block proportion is most pronounced for the models with vertically (0°) oriented blocks and least pronounced for those with horizontally (90°) oriented blocks.

1 INTRODUCTION

Melange is the French word for "mixture" and is defined by Bates and Jackson (1984) as, "a mappable body of rock that includes blocks of all sizes, both exotic and native, embedded in a fragmented and generally sheared matrix." In essence, therefore, a melange is a rock body made up of stronger blocks in a weaker matrix (i.e. a block-in-matrix rock or bimrock). Other examples of bimrocks are conglomerate and sheared serpentinite.

Due to their complex nature, melanges are difficult materials with which to work. D'Elia, et al. (1984) state the following regarding the "Argille Scagliose" (an Italian melange unit), "Given the lithologic and structural characters of the "Argille Scagliose" it was not possible to carry out laboratory or in-situ tests capable of providing data on the mechanical properties of the mass formation." This fact has been realized by many engineers and geologists working with melanges all over the world. The reasons for this difficulty include:

1. Obtaining "undisturbed" samples of a mixture of harder blocks in weaker matrix is virtually impossible by coring. The drilling resistance of the harder and softer materials are significantly different. This difference in drilling resistance can cause the harder materials to gouge into the weaker materials resulting in significant sample disturbance or even a complete loss of the weaker material.

2. Even if one were able to recover an undisturbed sample, it is a virtual impossibility that the sample would be representative of the melange mass of interest. The testing of large in-situ samples is perhaps the most promising approach to finding a representative specimen, but even this possibility seems remote.

Common engineering practice has been to design for the strength and deformation properties of the weak matrix. In some cases this might prove to be overconservative, though. For these reasons, a different approach for determining the strength (cohesion and angle of internal friction) and stiffness (modulus of deformation) of a melange needs to be developed. It has been found that it is possible to recover and test samples of pure matrix and pure block. Given the mechanical properties of these two components and some other properties of the melange mass (namely the volumetric proportions of the two components and the orientation of the melange fabric), it may be possible to estimate a melange's mass properties. For example, Volpe, et al. (1991) suggest that the strength of a melange mass can be represented by the weighted average of the strengths of the weaker matrix and the stronger blocks based on their volumetric proportions. Unfortunately, this approach has no theoretical basis. This research has used physical model melanges to study the effect of block proportion and fabric orientation on the strength and stiffness of some model melanges.

2 PHYSICAL MODEL MELANGES

Melanges tend to have a foliation due to both block alignment and prevalent shearing around the blocks. This is an important structural characteristic because, as both Jaeger (1960) and Donath (1964) discuss, anisotropic rock has marked anisotropic strength characteristics. The physical models therefore incorporate oriented blocks and mock shearing. The model block shape and size distribution are based on personal observations of melange made along California's north coast and various photographs and descriptions of melanges from around the world.

Cemented soils were chosen as the materials to model both the block and matrix components. A sand-portland cement-fly ash mixture was used for the blocks and a bentonite-portland cement mixture for the matrix. Thin layers of wax coated with talcum powder were used to model matrix shearing.

Six-inch diameter cylindrical specimens of model melange have been fabricated and tested triaxially to determine their Mohr-Coulomb strength parameters and stress-strain behavior. Models with four different block orientations, each with three block proportions, were created. Figure 1 schematically shows the different model types. The arrows in the diagram indicate the axial loading direction, and the angle indicated (0°, 30°, 60° or 90°) is simply the angle between the axial loading direction and the orientation in which the blocks are aligned.

Five specimens of each type, along with 7 6-inch diameter pure matrix and 10 2-inch diameter pure block specimens were created. This means a total of 67 6-inch diameter specimens and 10 2-inch diameter specimens were tested for this study. These numbers do not include all of the trial mixes required to finalize the model materials and specimen preparation methods. In total, over 100 6-inch diameter and 80 2-inch diameter specimens were tested.

3 TEST RESULTS

3.1 Effect of block proportion and orientation on strength

The Mohr-Coulomb strength parameters, cohesion (c) and angle of internal friction (ϕ), are used to represent the physical models' strengths. Table 1 presents the results for all specimen types.

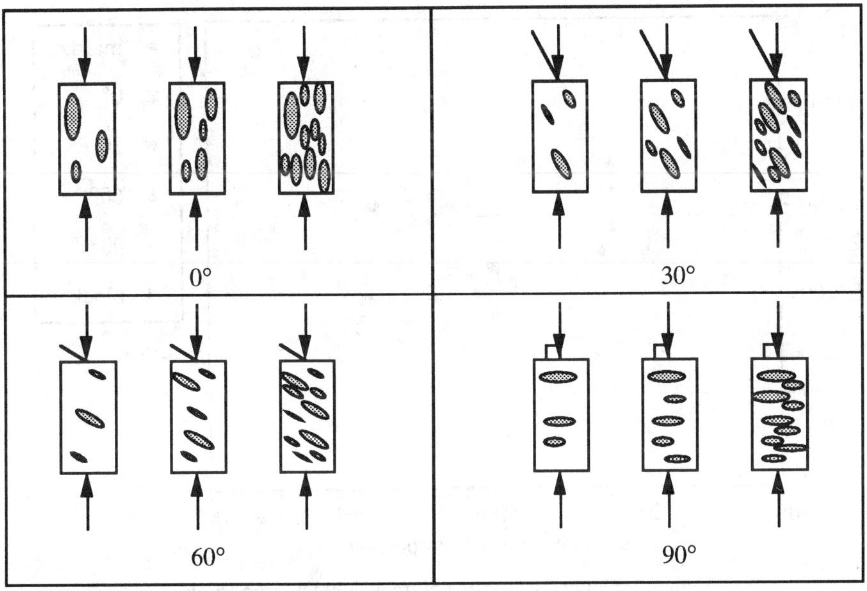

Figure 1. Specimen Types

Table 1. Values of cohesion and angle of internal friction

Block orientation	Average block proportion	c (psi)	ϕ (°)
N/A	0%	330	24.7
N/A	100%	445	38.3
0°	29%	345	26.2
30°	31%	250	32.9
60°	33%	377	23.4
90°	29%	276	32.8
0°	50%	233	33.5
30°	53%	206	32.3
60°	54%	229	33.3
90°	57%	231	37.6
0°	72%	199	39.6
30°	74%	163	38.2
60°	73%	180	41.2
90°	71%	302	34.0

The effect of block proportion on cohesion is summarized by Figure 2. This plot shows that as the block proportion increases, the cohesion generally decreases. In fact, the specimens with high block proportions (approximately 70 percent) have a cohesion only about half that of the matrix alone.

As discussed previously, Jaeger (1960) and Donath (1964) found that fabric

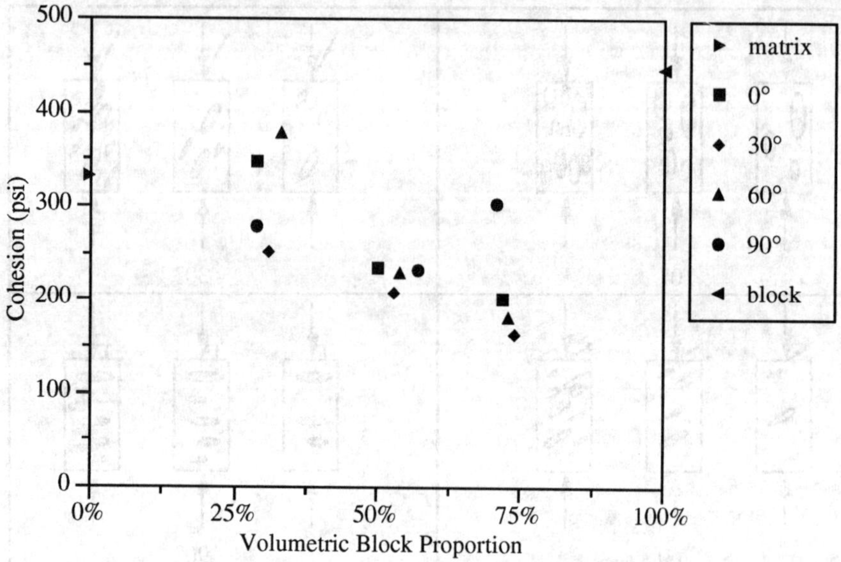

Figure 2. Cohesion versus volumetric block proportion

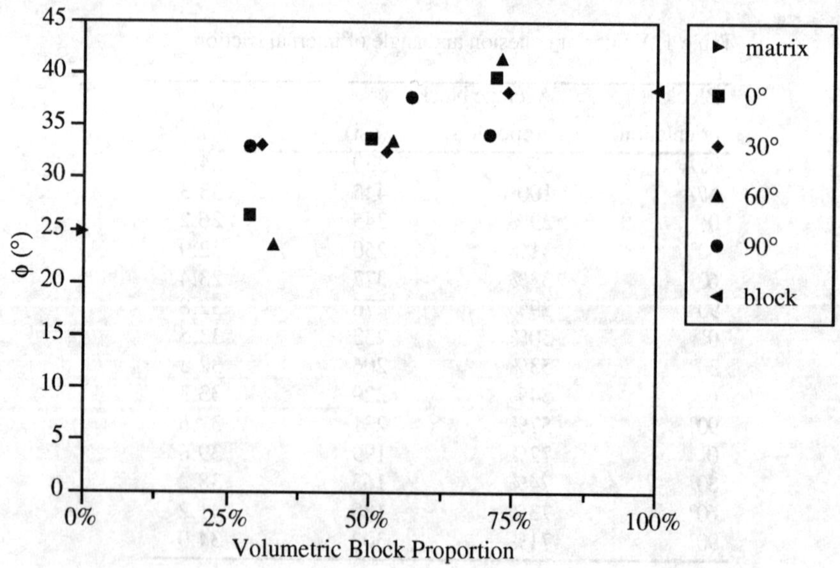

Figure 3. Angle of internal friction versus volumetric block proportion

anisotropy creates strength anisotropy in rocks. When planes of weakness (e.g. cleavage or shears) are oriented "adversely" with respect to the loading the result is a significantly lower strength. The block-in-matrix models used in this study are anisotropic due to the orientation of the larger blocks in each specimen. Although the "foliation" created by orienting the blocks is not nearly as strong as that formed by, for example, slaty cleavage, it still results in at least minimal strength anisotropy. Note that the specimens with blocks oriented at 30° to the axial loading direction (the most "adverse" orientation tested) have the lowest cohesion when compared to the models with similar block proportions and different block orientations.

Figure 3 clearly shows that the angle of internal friction increases with increasing block proportion. This increase is as much as 16.5° for the 60° orientation high block proportion specimens over the pure matrix specimens.

The values of angle of internal friction for the medium and high block proportion specimens do not vary very significantly with block orientation. This corroborates the assertion by Goodman (1989) that, "the variation of the friction angle with direction proves generally less severe than the variation of the shear strength intercept," in anisotropic rock.

Another observable result is that the low block proportion specimens show evidence of a threshold block proportion below which the presence of blocks has little effect on strength. Note that the cohesion and angle of internal friction for the 0° and 60° specimens are very close to those of the matrix material, while the 30° and 90° specimens have much lower cohesions and higher angles of internal friction. These data suggest that this threshold may be close to 30%. Evidence of a similar threshold has been found for heterogeneous soils. Heterogeneous soils are mixtures of coarser particles in a finer grained matrix, much like a melange. In fact, based on a review of the heterogeneous soil literature and some of their own laboratory work Irfan and Tang (1993) propose a threshold volumetric block proportion for heterogeneous soils of 25%.

Looking at the failed specimens provides an explanation of how the block proportion affects the cohesion and angle of internal friction. The failures tend to form along the block-matrix contacts rather than through the blocks even though the blocks are only about twice as strong as the matrix in unconfined compression. The block-matrix contacts are surfaces of weakness just as they are in real melange. The specimens with a higher block proportion therefore have a higher density of weakness surfaces resulting in a lower cohesion. On the other hand, the angle of internal friction increases because the failure surface becomes more tortuous (it has to fail around more blocks) at higher block proportions. An increase in tortuousity of the failure surface due to the presence of blocks in bimrocks has been hypothesized but never demonstrated in the past by D'Elia, et al. (1988) for melange and Savely (1990) for conglomerate.

Another interesting discovery that can be made by studying the failed specimens is that the failure surface only rarely passes through the blocks. For this reason it does not appear that the strength (cohesion and angle of internal friction) of the blocks plays a role in the strength of the mass so long as the blocks are stronger than the matrix. The strength contrast between the blocks and matrix required to prevent failure through the blocks is not known, but, as was stated previously, in this case the blocks were only about twice as strong as the matrix in unconfined compression which indicates the strength difference certainly does not have to be immense.

3.2 Effect of block proportion and orientation on the modulus of deformation

The modulus of deformation is defined in this study as the slope of a line drawn from the origin on the stress-strain plot to the point on the virgin loading curve corresponding to 40 percent of the maximum stress difference ($(\sigma_1-\sigma_3)_{max}$). This measure is typically called the secant modulus. The term modulus of deformation is used rather than the modulus of elasticity because during virgin loading the specimen undergoes both recoverable (elastic) and nonrecoverable (plastic) deformations.

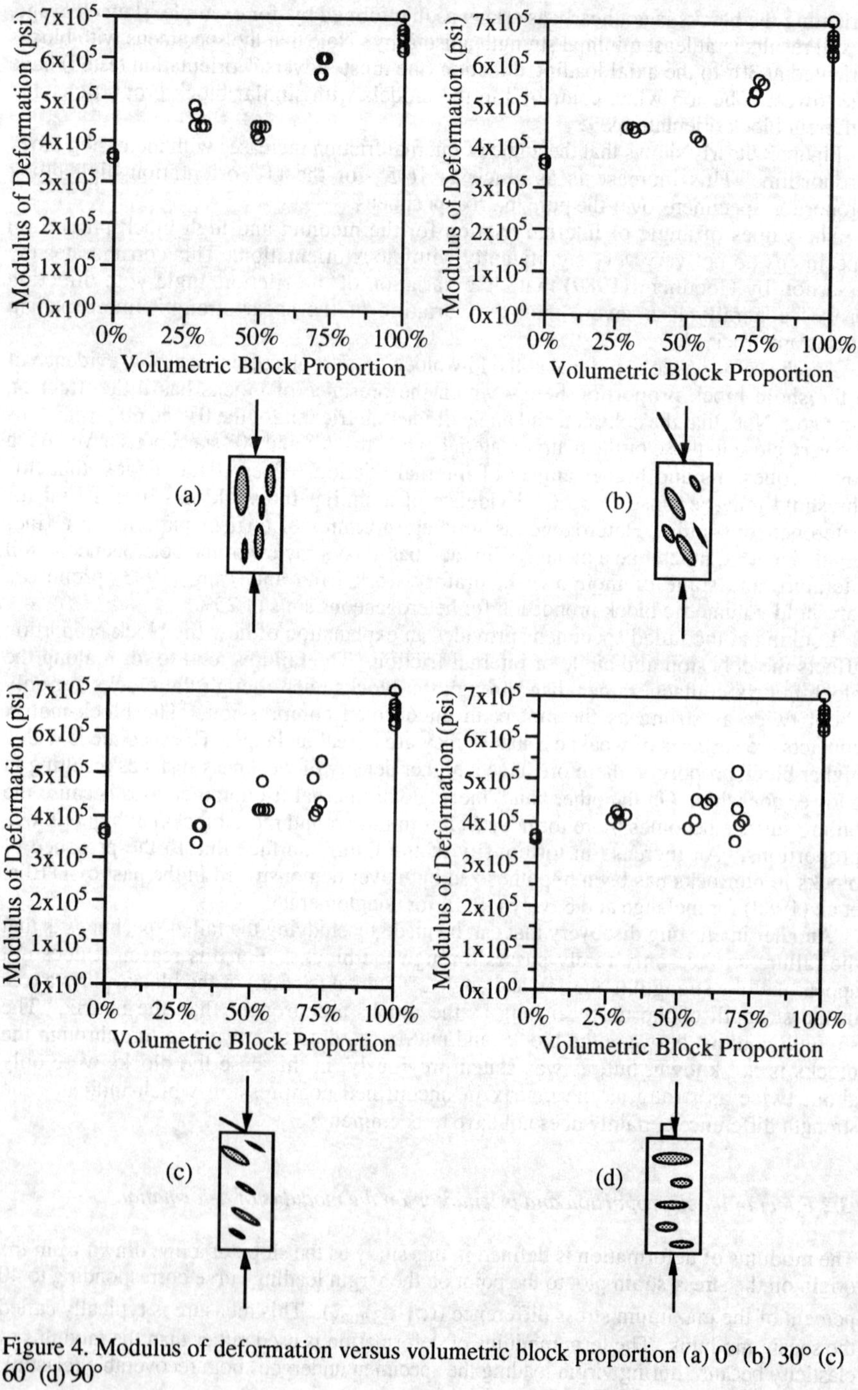

Figure 4. Modulus of deformation versus volumetric block proportion (a) 0° (b) 30° (c) 60° (d) 90°

Figure 4 shows that increasing the block proportion generally increases the modulus of deformation of the models. Each plot is for a different block orientation as indicated by the schematic specimen sketch below each plot. Note that the medium (approximately 50 percent) block proportion specimens with 0° and 30° block orientations fall below the trend set by the low and high proportion specimens. It has yet to be determined why this behavior was exhibited.

This increase in modulus is expected because as the block proportion is increased, the volume of stiffer inclusions in the softer matrix is increased, resulting in a stiffer model. Normal strength concrete can be viewed as a sort of melange (i.e. it is made up of stronger aggregate in a weaker matrix), and similar increases in modulus have been reported in the concrete literature for many years.

Another finding evident in Figure 4 is that the rate of increase in modulus with block proportion decreases as the block orientation is rotated from vertical to horizontal. This result is also not unexpected. Simple models such as the one proposed by Hirsch (1962) for concrete can be used to show that when the blocks are vertically inclined they attract more of the axial stress. The result is a stiffer model. A second reason why the increase in modulus is smaller for the specimens with horizontally oriented blocks is wax layer alignment. Aligning the blocks tends to align the wax "shears" resulting in subparallel layers of wax in the specimen. As these layers are rotated closer to horizontal they significantly reduce the stiffness of the model.

4 SUMMARY AND CONCLUSIONS

Triaxial tests on physical models indicate that the strength and deformation properties of a melange are significantly affected by block proportion and fabric orientation.

The test results indicate that the cohesion decreases and the angle of internal friction increases with increasing block proportion, and the strength, particularly the cohesion, decreases as the block orientation becomes more adverse. It is postulated that the strength parameters are affected in these ways because the failures form along the block-matrix contacts.

The test results also indicate that the modulus of deformation increases with increasing block proportion and this increase is greatest for specimens with vertically (0°) aligned blocks and smallest for specimens with horizontally aligned blocks (90°).

Given this information, it is clear that the block proportion is an important value to determine in the field. Some ideas on how this value can be determined are given by Medley and Goodman (1994). It must be remembered, though, that melange in the field can be significantly more complex than the models used in this study, and many factors beyond the block proportion and fabric orientation may be important. For example, one may find major zones of sheared material that will play an important role in the field behavior. Even so, it is believed that the basic behavior exhibited by these models is indicative of that which will be found in actual melange.

5 ACKNOWLEDGMENTS

Lab testing was out with the aid of research assistants Dimitrios Protogirou and Jason Choi. Funding for this research was provided by the Pacific Gas and Electric Company, San Francisco. Their support is gratefully acknowledged.

REFERENCES

Bates, R.L. & J.A. Jackson 1984. *Dictionary of geological terms.* New York: Doubleday.

D'Elia, B., D. Distefano, F. Esu & G. Federico 1988. Deformation and stability of high cuts in a structurally complex formation: analysis and prediction. *Proc. 5th Int. Symp. on Landslides:* 599-604. Rotterdam:Balkema.

D'Elia, B., D. Distefano, G. Federico & S. Oliva 1984. Full-scale study of a high cut in a structurally complex formation. *Proc. 4th Int. Symp. on Landslides:* 2:57-62. Ontario:Canadian Geotechnical Society.

Donath, F.A. 1964. Strength variation and deformational behavior in anisotropic rock. In W.R. Judd (Ed.), *State of stress in the Earth's crust,* p.281-298. New York:Elsevier.

Goodman, R.E. 1989. *Introduction to rock mechanics.* New York:Wiley.

Hirsch, T.J. 1962. Modulus of elasticity of concrete affected by elastic moduli of cement paste matrix and aggregate. *ACI Journal.* 59:427-447.

Irfan, T.Y. & K.Y. Tang 1993. *Effect of the coarse fractions on the shear strength of colluvium,* GEO Report #23. Hong Kong:Geotechnical Engineering Office.

Jaeger, J.C. 1960. Shear failure of anisotropic rocks. *Geol. Mag.* 97:65-72.

Medley E.W. & R.E. Goodman 1994. Estimating the volumetric block proportions of melanges and similar block-in-matrix rocks (bimrocks). *Proc. 1st NARMS.* Rotterdam:Balkema.

Savely, J.P. 1990. Determination of shear strength of conglomerates using a Caterpillar D9 ripper and comparison with alternative methods. *Int. J. Min. & Geol. Eng.* 8:203-225.

Volpe, R.L., C.S. Ahlgren, C.S. & R.E. Goodman 1991. Selection of engineering properties for geologically variable foundations. *Proc. 17th Int. Cong. on Large Dams:* 1087-1101. Paris:ICOLD.

Rock Mechanics, Nelson & Laubach (eds) © 1994 Balkema, Rotterdam, ISBN 90 5410 380 8

Estimating the block volumetric proportions of melanges and similar block-in-matrix rocks (bimrocks)

Edmund Medley & Richard E.Goodman
Department of Civil Engineering, University of California, Berkeley, Calif., USA

1 INTRODUCTION

Although the chaotic fabric of melange (from the French, *mélange*, or mixture) is characteristic of rocks formed within the accretion prisms of subduction zones, geologists disagree about the details of melange formation (Raymond and Terranova, 1984), as reflected in the many aliases of melange (e.g., chaotic formations, wildflysch, mega-breccia, argille scagliose and friction carpets). For engineering purposes, Medley (1994) considers melange to be a *bimrock* (block-in-matrix rock) which he defines as "a mixture of relatively large, competent blocks within a bonded matrix of finer and weaker texture", a definition that ignores rock genesis. Bimrocks similar to melange that are formed from cataclasis and fragmentation include breccias, coarse pyroclastics, lahars and tillites. Other bimrocks form from weathering (decomposed granite) or sedimentation (boulder conglomerates). Analagous soils, such as colluvium and till are termed *bimsoils*.

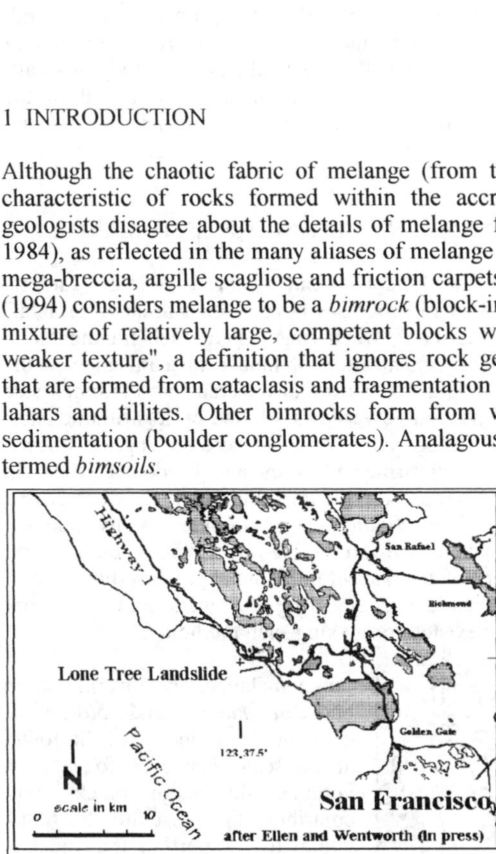

Figure 1: Regional -scale blocks in the Franciscan north of San Francisco, California

Melange bodies have been identified in the mountains of over 60 countries (Medley, 1994) and are exemplified by the Franciscan Assemblage (the Franciscan) a regional-scale jumble in northern California. Typically, the Franciscan contains chaotic zones of relatively strong blocks of greywacke sandstone, chert, basalts, limestone and exotic metamorphic rocks embedded within a matrix of pervasively sheared shale and argillite. The blocks range between sand particles and mountain masses, are irregularly shaped, and generally trend NNW-SSE (Figure 1). Shears are confined to the weak shale matrix, which flows around the closely jointed blocks. The sheared shale, and the sheared serpentinite bodies common in the Franciscan, are both responsible for the myriad earth-flow landslides of the Fransciscan. Also, encounters with unpredictably distributed blocks in melanges cause expensive surprises during earthwork excavations and foundation preparations.

When working with bimrocks or bimsoils, it is universal geotechnical engineering practice to design using the strength of the weaker matrix. But extensive laboratory testing of physical model melanges (Lindquist and Goodman, 1994) showed that the strength of a physical model melange increased with the volumetric proportion of blocks when the proportion exceeded approximately 30 percent as a result of the increased tortuosity of failure surfaces. Relationships between block content and overall strength have been observed qualitatively for melanges in California (Bedrossian, 1980) and Italy (D'Elia et al, 1986)

But the practical use of a relationship between block volumetric proportion and overall bimrock strength requires the determination of the in-situ volumetric proportion of blocks. In this paper, we first show that much can be learned from the study of small-scale images of melanges; secondly, that stereological methods are useful in the estimate of volumetric proportions and block size distributions; and finally, we present the results of a field application of our approach.

2 GRAPHIC MODELS BASED ON APPARENT MELANGE SELF SIMILARITY

Melanges are commonly characterized as "chaotic". But recent work (e.g., Turcotte (1992)) shows that many chaotic geological processes have a self-similar or "fractal" nature. A structure possesses self similarity if, when fragmented, each fragment is a replica of the parent whole. The fractal organization of blocks in a melange at the outcrop scale was demonstrated by Lindquist[1] when he discovered that the block size frequency distribution of the melange obeyed power laws. Plotted on logarithmic axes, the data were organized linearily: such plots are referred to here as "log-log linear". Lindquist's preliminary work has been confirmed (Medley and Lindquist, 1994; in preparation) for melanges of many scales (compare Figure 1 and Figure 2, which have a scale ratio of greater than a million). These findings have a fundamentally practical implication because, given sufficient data to define a log-log linear plot, predictions can be made about the sizes and numbers of blocks within a melange. Any predictions, though simple, are useful to geotechnical engineering designers and earthwork contractors. Up to now, no method has existed for making such predictions.

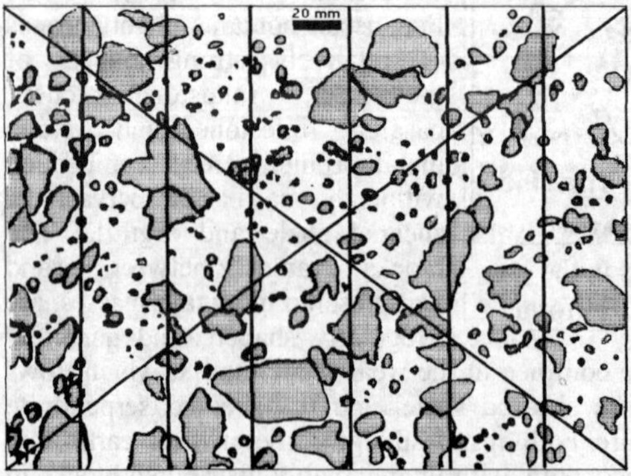

Figure 2: Graphic model of melange with scanlines. Scale bar is 20 mm long

In a melange, the discrimination between *matrix*, and *blocks*, is difficult since blocks will be found at any scale. However, for a given volume, the large blocks will contribute the most to the block volumetric proportion, thus one can select an arbitrary threshold block size of 1 percent of the maximum observed dimension of the largest block in the population

[1] E. Lindquist, 1991; "Fractals - Fractures and Franciscan"; Term Paper for CE 280, Rock Mechanics; Dept. Civil Eng., University of California, Berkeley, California

being measured. Assuming spherical blocks and a fractal block size distribution, blocks with characteristic dimensions less than a 1 percent threshold will contribute less than 1 percent of the volume and less than 5 percent of the total surface area of the blocks in the rockmass (Medley, 1994).

Because of the apparent self-similarity of melanges, drawings and photographs at the outcrop scale are *graphic models* of larger scale melanges. Graphic models allowed us to develop characterization methods useful in estimating the block volumetric proportion and block size distributions of melanges at engineering scales. For example, Figure 2 is the hand-tracing of a photograph of melange from Caspar Headlands, near Mendocino, California. The model cross-section is 180 mm in "depth".

We used manual and computer-assisted image analysis methods to determine the areal proportions of the blocks and the block size distributions of graphic models. Block size is characterized by the length of the maximum observeable dimension of the block. The pictures were scanned and the scanned image digitized into an array of *pixels* (picture elements), each with a value of between 0 (black) and 255 (white) representing some shade of gray. Image analysis software measured the areas, perimeters and axial dimensions of the individual blocks in the images. In the case of the image of Figure 2, the sum of individual block areas yields a block areal proportion of 35.6 percent. The frequency distribution of the block sizes (maximum observeable dimensions) plots as a log-log linear curve (power law fit), as shown on Figure 4

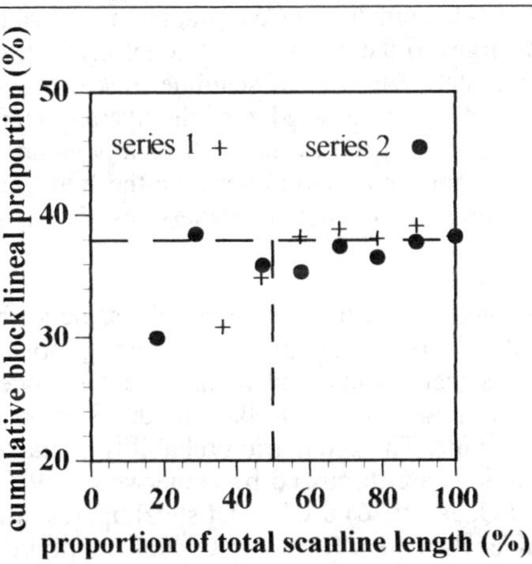

Figure 3: Lineal block proportion for scanlines shown in Figure 2. The cumulative block lineal proportion is the ratio of the sum of scanline intercepts through blocks, to the cumulative length of the scanlines. Note that the convergence of the block lineal proportion to approximately 38% which is close to the measured block areal proportion of 36%. Convergence is effectively achieved at approximately 50% of the total scanline length of 1684 mm (840mm).

In practice one rarely has the good exposure of melange modelled in Figure 2, and outcrop mapping and drill core sampling must be performed in an attempt to characterize the rockmass. The continuous detailed geological log of a borehole or the core itself, if available, are sampling *scanlines*. In the case of the graphic models, scanlines (modelled boreholes) were drawn over the graphic model of Figure 2, and measurements made of the lengths of chord intercepts through each block intersected by the scanline. The block lineal proportion for each scanline is the total length of chord intercepts along the scanline divided by the length of the scanline. If an image is sampled by a sufficient total length of scanlines, the areal block proportion is estimated. The equivalence of lineal, areal and volumeteric proportions is a fundamental law of *stereology*, using which structures are characterized by zero-, one-, and two-dimensional measurements (Underwood, 1970). The practice of making linear ("Rosiwal") traverses across microscopic images to estimate the areal and volumetric proportion of particulates within a matrix was a standard laboratory technique for petrographers, biologists and metallographers before the advent of computer-assisted image analysis.

The areal proportion of blocks in the image of Figure 2 was estimated by accumulating the block linear proportions of eight scanlines totalling 1684mm in length. The individual block lineal proportions varied between 53 percent and 30 percent. To test the influence of varying the order of scanlines, the incremental cumulative proportion (running average) and incremental proportion of total scanline length were computed. separately for two series scanlines in arbitrarily chosen sequences. Figure 3 shows that the cumulative block lineal proportion converges to 38 percent, close to the block areal proportion of 35.6 percent. Convergence effectively occurs at 40 percent of the total scanline length, or some 840mm of scanline measurements.

Holmes (1921) reported a guideline to selecting a sufficient length of linear traverse for estimating the areal proportions of mineral grains in microscope images: a total traverse length of 100 times the average grain size in the area traversed was recommended. If this simple criterion is adopted for the scanline measurements of the graphic model of Figure 2 (average block size of 8.25mm), then a cumulative length of scanlines of 825 mm would be sufficient to estimate the block areal proportion. The results shown in Figure 3 validate Holmes' (1921) guideline for the graphic model since the cumulative block lineal proportion converges to the known block areal proportion, for both series of scanlines, within approximately 840 mm of scanline measurement. However, the guideline is of little use without prior knowledge of the average block diameter. Moreover, if the length of scanlines is reduced to more efficiently estimate the block areal proportion, then there will be poorer correlation between the frequency distributions of block intercept lengths (chords) and block diameters, as discussed below.

The lengths of chords across blocks intercepted by scanlines are rarely the same as the block characteristic dimensions (or "diameters"). But to simplify a complex problem, we assume that the distribution of block cross-sections in the plane of a graphic model, or the distribution of chord lengths from scanlines, is related to the true distribution of the three dimensional blocks in the parent volume. The geometric probability character of this problem and some stereological solutions are discussed by Underwood (1970). From the geotechnical viewpoint, Tang and Quek (1986) used a statistical approach to estimate the frequency distribution of boulder diameters from a frequency distribution of borehole chord intercepts through boulders. Savely (1990) used Monte Carlo simulations to determine correction factors between the frequency distribution of boulder dimensions in a cemented boulder conglomerate, and the frequency distribution of drilled core intercepts through the boulders. Correction factors range between 1.0 and 1.6 (David Nicholas, Call and Nicholas, Inc; pers. comm). Interestingly, the eight scanlines across the Figure 2 graphic model resulted in 86 chord intercepts, with an average length of 7.29 mm, but the actual average block diameter (in two-dimensions) was 8.25 mm, larger by a factor of 1.13. However, for melanges, the notion of *average* dimensions is misleading, since average block sizes are greatly skewed toward the small sizes due to the fractal distribution of blocks. Hence, chord to diameter conversions may not be as simple for fractally distributed and irregular shaped melange blocks as it appears to be for normally distributed and rounded boulders in conglomerates.

The frequency distributions of chord intercepts for graphic models and field scanlines generally plot as power-law curves (log-log linear) like that of Figure 4, which also shows the distribution of block diameters for the graphic model (mini-melange) of Figure 2. The plots of Figure 4 suggest that the distributions are related since the slopes of the plots are similar, so that an empirical and general adjustment between the frequency distributions of chord intercepts and block maximum dimensions for melanges may yet be discovered. The plots of Figure 4 show that for any given chord intercept length, the number of equivalent block diameters is many times greater; and for any given frequency, the estimates of true block dimensions will be severely underestimated if one assumes the chord dimensions to be diameters.

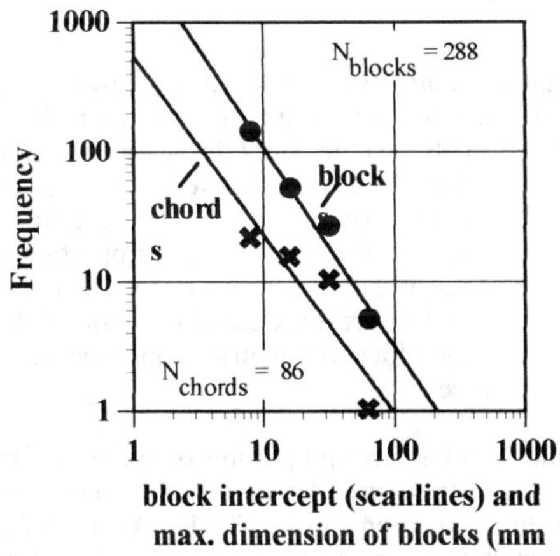

Figure 4: Plots of the block size distribution (from image analysis measurements) and the chord length distribution (from scanline measurements) of blocks shown in the graphic model of Figure 2.

3 CASE HISTORY: LONE TREE LANDSLIDE, CALIFORNIA

Shortly after the Loma Prieta earthquake in October, 1989, California Highway 1 was closed by the Lone Tree Landslide (Figure 1). The California Department of Transportation (CALTRANS) restored access by excavating 765,000 m^3 (10^6 yd^3) of Franciscan melange to an average depth of 37m(Van Velsor and Walkinshaw, 1993). Relatively intact rock was excavated from behind the interpreted landslide surface to provide stable cut slopes (Figure 5). Blocks up to 25m were blasted during the excavation work (Michael Hobbs, Ford Construction, Inc.; pers. comm).

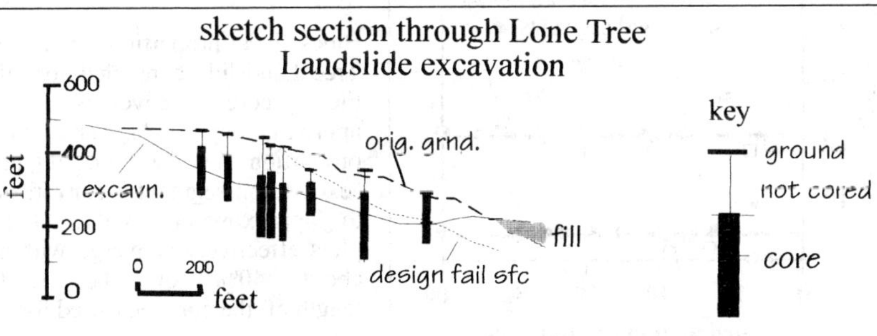

Figure 5

courtesy of CALTRANS

Prior to construction, nine exploratory borings (Figure 5) were drilled to between 37m and 82m deep to investigate the stability of the melange that would be exposed in the cut slopes. Some 375m of HQ size core (61mm diameter) was recovered, and eventually given to us for our research. Coring started some distance below the ground surface and only 82m of the core was from those borehole segments located above the future excavated surface. The core from these segments was assumed to represent the melange excavated, referred to here as the "excavation segments", and averaged 10m in length. The average length of all the core was 42m. Since the largest blocks in the vicinity of the slide are some 30m in size (maximum observed dimension), we chose the matrix/block threshold to be 0.3m using an arbitrary 1 percent criterion (as described in section 2), but nevertheless measured block intercepts in the core to as small as 5 cm in length.

Some 40 percent of the core (approximately 150m) was sufficient to show that the block lineal proportion of all the core effectively converged to approximately 21 percent (Figure 6). The longest block intercept in all the core was 7.9m, with an average length of 0.43m. The intercepts from the excavation segments core were much shorter, ranging to 9.3cm long, with an average length of 3.1cm. Some 38 percent of the excavation segments core was sufficient to estimate the block lineal proportion of approximately 4.5 percent (Figure 6). For both sets of data, the ratios of total length of lineal traverse (core) to the average intercept lengths, greatly exceed the ratio of 100 recommended by Holmes (1921). We estimate the block volumetric proportion of the unexcavated melange to be approximately 28 percent.

Medley, (1994) tested the estimate of block volumetric proportion of the excavated material as measured from the excavation segments core (4.5 percent), by measuring the areal proportion of blocks exposed in the excavation slopes. The blocks are relicts of larger blocks that had been blasted or ripped during excavation. The site (37,780 m^2, or 8.1 acres) was divided into 21 sub-areas and the maximum observed dimension of blocks greater than 1m in each sub-area was visually estimated. The block areas were estimated using the assumption that the maximum observed dimension was a circular diameter. The total areal proportion of blocks exposed on the slopes of the excavation is estimated as 4.2 percent. This proportion is remarkably close to the 4.5 percent lineal proportion estimated from measuring the core from the excavation segments. It appears that the material removed from the excavation was relatively deficient in blocks, and that much of it was landslide material, as indicated by the sheared shale in the core of the excavation segments.

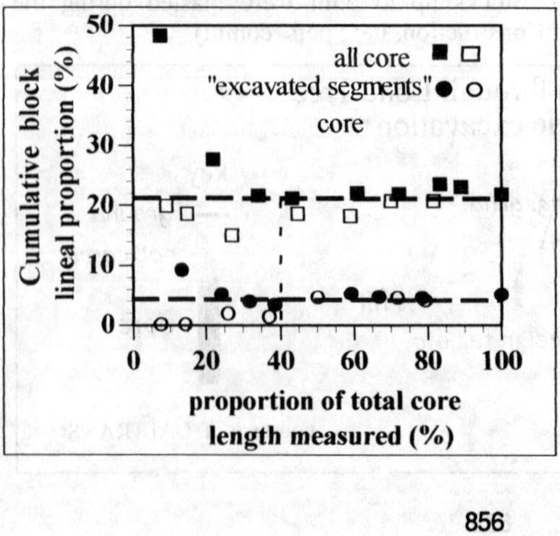

Figure 6:

Block lineal proportion for Lone Tree Landslide core: that for all the core coverges to approximately 21%, and the proportion for the core of the "excavation segments" converges to approximately 4.5%. Both plots effectively converge within about 40% of the total length of the core measured for each exercise.

Figure 7 shows the frequency distribution of the chord intercepts of the blocks measured for all the core and also the frequency distribution of the maximum observed dimensions of the blocks estimated in the field. There is reasonable correlation between the slopes of the plots of the fieldwork and core measurements (of all the core), but considerable difference between the maximum size of blocks predicted: the fieldwork predicts a block of 100m, and the core a block of 15m. Based on our work with graphic models, it is likely that the true distribution of block characteristic dimensions ("diameters") has been severely underestimated. The plot based on the measurements from the core from the excavated segments is poorly correlated with the other two, although it too plots log-log linear. The steep slope of the plot may be related to the sparse block population of the material excavated, but in a manner not yet clear to us.

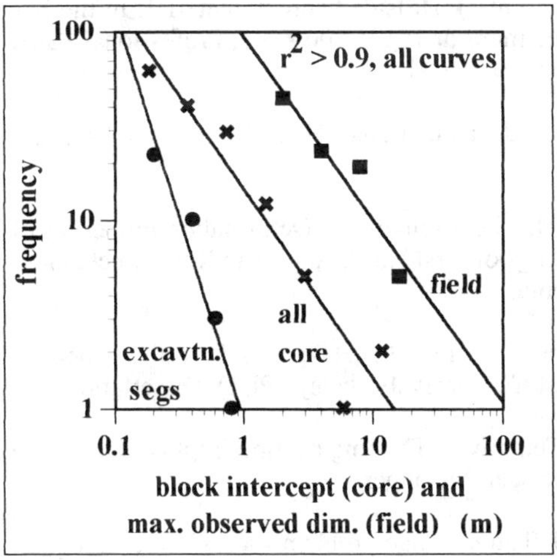

Figure 7:

Compilation of the distributions of block lengths in the "excavation segments" of core and all the core, together with the distribution of maximum observed dimensions of blocks, as mapped in the field

4 SUMMARY AND CONCLUSIONS

When working with a bimrock (block-in-matrix rock), the application of a block volumetric proportion/strength relationship requires that the volumetric proportion be estimated. The apparent self-similarity of the block distribution in melanges allowed the use of small-scale graphical models to develop a measurement technique useful for estimating the volumetric proportion of blocks by measuring the lineal proportion of chord intercepts recovered in drill core. Field mapping and measurement of drill core recovered from the Lone Tree Landslide, California, indicate that the technique may predict the in-situ proportion of blocks in a melange. Yet more work needs to be performed to allow predictions to be made about the block-size distribution of in-situ melange from chord intercepts measured from drill core.

5 ACKNOWLEDGEMENTS

We are grateful to the Pacific Gas and Electric Corporation, San Francisco, for supporting our research; and to the Geotechnical Engineering Department, CALTRANS District 4, Oakland, for allowing us access to their Lone Tree Landslide records and core.

REFERENCES

Bedrossian, T.L., 1980, Geology and Slope Stability in Selected Parts of the Geysers Geothermal Resources Area, Special Reports No. 142; California Division of Mines and Geology, 66 p.

D'Elia, B., Distefano, D., Esu, F., and Federico, G., 1986, Slope Movements in Structurally Complex Formations, in Tan Tjong Kie, Li Chengxiang, and Yang Ling, eds., Engineering in Complex Rock Formations; Proc. of the Int. Symp; Beijing, China.

Ellen, S.L and Wentworth, C.M, 1994 (in press), Hillside bedrock materials in the San Francisco Bay Region, California; three maps at 1: 125,000 scale, Professional Paper No.??, United States Geological Survey.

Holmes, A., 1921, Petrographic Methods and Calculations:, Thos. Murray and Co., London, England.

Lindquist, E.S., and Goodman, R.E., 1994, The Strength and Deformation Properties of a Physical Model Melange, in Proceedings of First North American Rock Mechanics Symposium, Austin, Texas; A.T.Balkema.

Medley, E., 1994, The Engineering Characterization of Melanges and Similar Block-in-Matrix Rocks (Bimrocks): University of California, Berkeley, Ph.D. Dissertation.

Medley, E., and Lindquist, E.S., 1994, (Tentative): The Engineering Implications of the Apparent Self Similarity of Some Franciscan Melanges.

Raymond, L.A., and Terranova, T., 1984, The melange problem - a review. (Prologue), in Raymond, L.A., ed., Melanges: their nature, origin and significance; Geological Society of America, Boulder, CO, p. 1-5.

Savely, J.P., 1990, Determination of shear strength of conglomerates using a Caterpillar D9 ripper and comparison with alternative methods: *International Journal of Mining and Geological Engineering*, v. 8, p. 203-225.

Tang, W., and Quek, S.T., 1986, Statistical model of boulder size and fraction: *Journal Of Geotechnical Engineering;* ASCE, v. 112, p. 79-90.

Turcotte, D.L., 1992, Fractals and Chaos in Geology and Geophysics: New York, NY, Cambridge University Press, 221 p.

Underwood, E.E., 1970, Quantitative Stereology, Addison-Wesley Publ. Co, Reading, MA, 273 p.

Van Velsor, J.E., and Walkinshaw, J.L., 1992, Accelerated movement of a large coastal of Annual Meeting of the Transp Res. Board, Jan 1992, Washington, DC.

Rock Mechanics, Nelson & Laubach (eds) © 1994 Balkema, Rotterdam, ISBN 90 5410 380 8

Translational slope instabilities in weak rock

R. Berardi, R. Passalacqua & C. Scavia
University of Genova, Italy
F. Forlati
Regione Piemonte, Torino, Italy

ABSTRACT: This paper examines the problem of translational slope instabilities occurring in complex geological formations of weak rock as are typical of the Langhe region in Italy. Unstable rock volumes slide along slightly dipping planar surfaces that are formed where adjacent layers of marl and sand are in contact. Four sites were identified and examined. Special attention was devoted to the development of a laboratory method able to simulate the behaviour of the thin layer of marl which is deemed to play an active role in these phenomena in conjunction with the sand layers providing a preferential path for groundwater circulation. Direct shear tests were carried out and the results obtained show that the fundamental role played in instability phenomena by the de-structuring effect of water on marl cannot be taken into account through the customary laboratory techniques.

1 INTRODUCTION

The Langhe territory (in North-west Italy, see fig. 1) consists of a vast hilly region comprising several small towns and roads. The area is characterised by the presence of translational slope instabilities occurring in complex geological formations of weak rock (alternate layers of marl and sandstone being prevalent). Unstable rock volumes (up to 1 million m^3) slide along planar surfaces dipping from 8 to 14 degrees (fig. 2).

Several authors have studied the problem systematically and have pointed out the following geometric and kinematic characteristics (Govi and Sorzana, 1982):

- sliding mostly occurs in slopes showing clear signs of earlier gravitational phenomena of the same type as the current ones (where the geometry of landslide bodies is greatly affected by pre-existing discontinuities, such as ancient scarps and closed or open fractures);

- the movements generally affect the rock mass to an average depth of about 10 m, with planimetric movements of up to a few hundred meters;

- the main direction of the movements is seen to be appreciably parallel and often

Figure 1. Geographical position of the Langhe region.

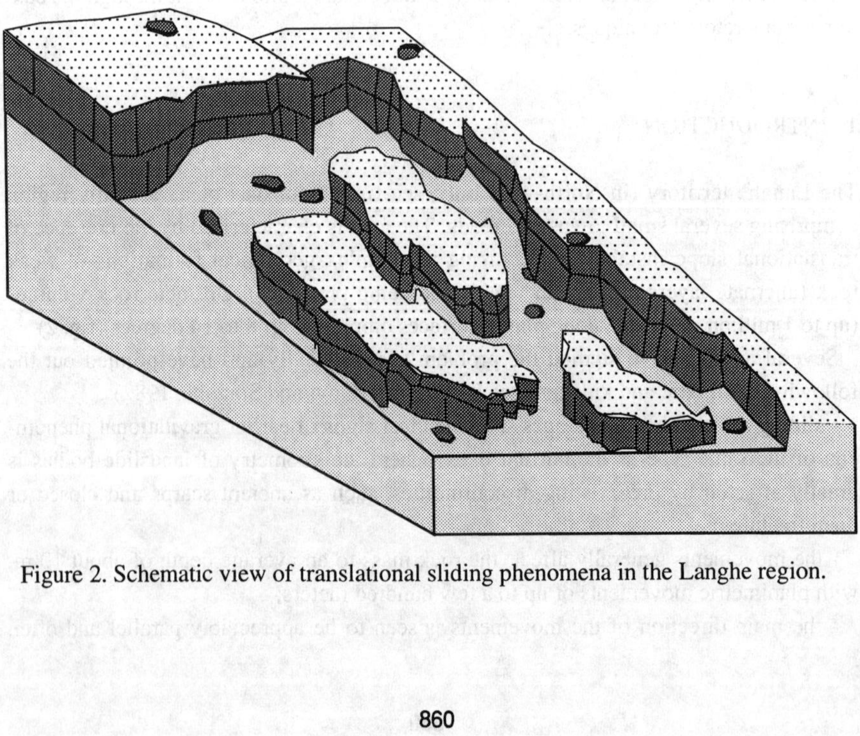

Figure 2. Schematic view of translational sliding phenomena in the Langhe region.

coinciding with the dip direction of the strata, even in those situations in which the general inclination of the slope is altogether different;

- the layer of material primarily involved in the sliding phenomena is from 10 to 40 mm deep.

Several important aspects require further investigations, and namely:

- the role of main discontinuity systems (orientation, persistence, spacing, strength characteristics) on the instability phenomena;
- rock mass characterisation;
- the identification and characterisation of the sliding surfaces;
- the analysis of the mechanical behaviour of the material directly involved in the failure phenomenon;
- groundwater circulation and the effects of water pressure.

In this connection, four sites were identified as being representative of the entire area and were examined through a methodology combining field surveys and laboratory tests, specially developed to take into account the special needs of this problem.

2 GEOLOGICAL BACKGROUND

Without going into too much detail, it can be pointed out that the main rock types of this area belong to the Oligocene-Miocene sequence. They exhibit a wide monocline with North-West dip direction and average dips of between 7 and 15 degrees. The layers run parallel and exhibit down slope dipping. Marl, silty-marl, sandstone and sand prevail, in rhythmical sequences. The depth of these layers generally ranges from a few cm to 1-2 m, most of them being from 5 to 50 cm deep.

The layers and sandy levels overlying marl, clayey marl and, secondarily, clay, form aquifer layers which are often exploited for agricultural purposes.

The lithological units include 2 - 3 systems of subvertical discontinuities, of high to medium persistence, and are arranged in the predominantly NW-SE direction.

The morphology of the area is clearly affected by the structural situation, giving rise to long slopes dipping in the NW direction with down slope stratification, and to short, steep slopes dipping in the SE direction with counter slope stratification (fig. 3).

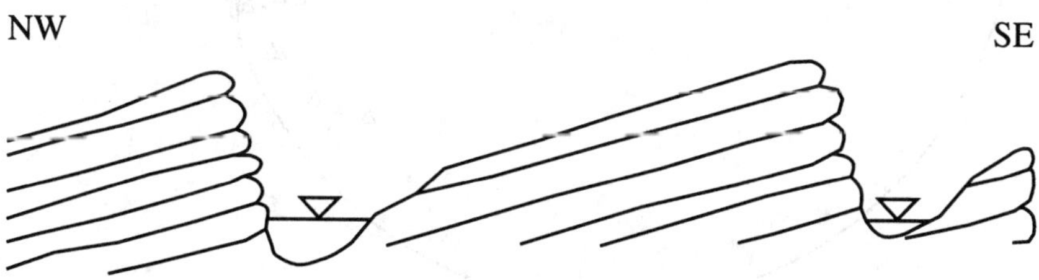

Figure 3. Morphology of the area.

3 GEOMECHANICAL MODEL

The four sites being considered all exhibit clear signs of the occurrence of earlier planar instability phenomena. The following studies were carried out for each of them:

1) historical research on the evolution of past sliding phenomena;

2) aerial photogrammetry analyses;

3) field surveys, aimed at determining:

- the main rock types and their characteristics (pocket's penetrometer, carbonate content, Barton's joint compressive strength index),
- the main discontinuity sets and their geometric characteristics (orientation, spacing, persistence),
- location and characteristics of the sliding surfaces,
- water percolation patterns.

The results obtained in the four sites examined may be briefly summarised as follows:

1) Four discontinuity sets are present (see fig. 4): Set K1 which coincides with the bedding planes, K2 and K3 delimiting the unstable blocks on both sides, and K4 on the back. The blocks are exposed at their foot by the erosion of water courses.

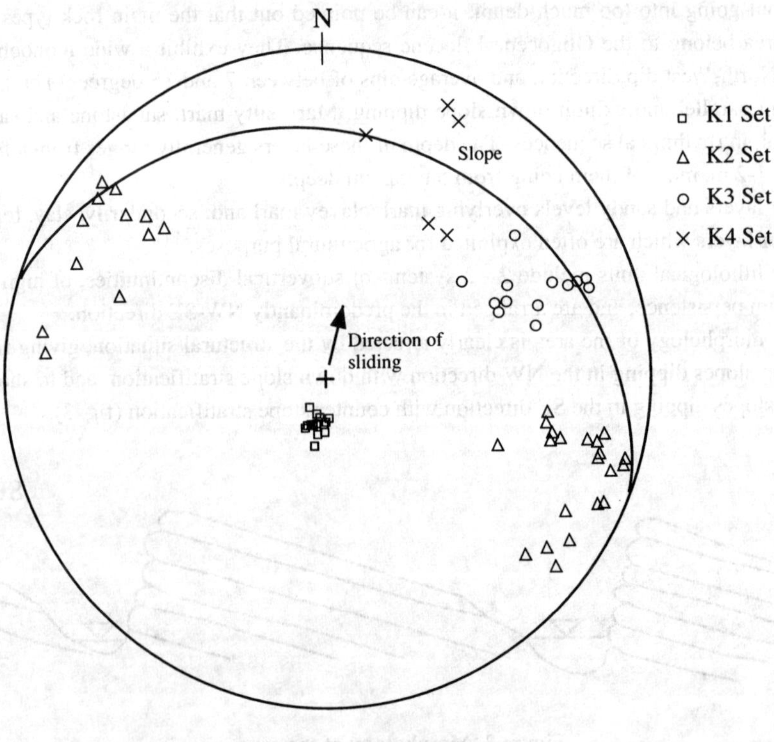

Figure 4. Discontinuity sets.

2) Sliding occurs on one of the planes belonging to K1, with dip ranging from 8 to 14 degrees. The sliding surface is produced either at interface between a marl layer and a sand layer or along a bedding plane between two layers of marls, at a depth varying from 6 and 8 m.

3) The marl layer involved in the sliding phenomena (from 1 to 4 cm deep) is destructured by the action of the water seeping through the discontinuity sets to a considerable depth and circulating at the interface of the adjacent marl layers or in the sandy level. Carbonate content is always lower (between 8 and 20%, approx.) than the content observed in the marl not directly involved in the failure phenomena (ca 30 - 40%).

A simple slope stability back analysis was performed on the earlier instability phenomena mentioned above through the Limit Equilibrium Method. The results show that the friction angle (assuming cohesion to be nil) at failure varies between 13 and 20 degrees, depending on the assumptions made on the distribution of water pressures. Shear resistance appears to be much lower than the values obtained through standard laboratory testing and generally given in the literature.

On the basis of these observations, a series of laboratory tests was carried out on samples of marl taken from the area.

4 LABORATORY TESTS

A series of oedometric compression and direct shear tests was performed on samples taken from two of the four chosen sites (Gottasecca and Prato Noero) with the aim of determining their characteristics and mechanical properties, grain size distribution (fine content, CF), water (W) and calcium carbonate ($CaCO_3$) content, and Atterberg's limits and indices (liquid limit, LL; plastic limit, PL; plastic index, PI).

It proved almost impossible to obtain undisturbed core samples because of the presence of a sequence of relatively stiff marl strata just above and under the slip plane, which sometimes also included thin layers of sand. Direct shear and confined compression tests were therefore performed on samples produced in the laboratory from blocks taken directly from the sites. Before being subjected to the tests, the Gottasecca samples were enclosed in restraining ring and kept for 15 days under a static water head. Some of the tests concerned mechanically remolded samples with a water content close to their Liquid Limit.

By means of a special device fitted out with a free inner sampler, an undisturbed sample was drilled out from a block obtained from the Prato Noero site. This sample, which included a portion of the slip surface, was subjected to a shear test.

Some of the results obtained are summarised in table 1.

Direct shear tests were performed through repeated cycles on cylindrical samples (diameter: 6.4 cm, height: 3.0 cm) with maximum horizontal displacement of 1.2. The results of these tests, which enabled the failure enveloped to be determined, are summarised in table 2 .

Table 1 Results of laboratory tests

SITE	CF(%)	W*(%)	LL(%)	PL(%)	PI(%)	$CaCO_3$(%)
Gottasecca	<20	17/21	34-39	26-28	8-11	40-43
Gottasecca remolded	<20	35/25	35	27	8	43
Prato Noero slip surface	-	35	52	28	24	25

* measured before/after consolidation.

Table 2 Friction angle, ϕ', and cohesion, c', values

SITE	ϕ'(°)	residual ϕ'(°)	c' (kPa)	residual c' (kPa)
Gottasecca	29	29	20	17
Gottasecca Remolded	30	30	10	10

From these values it is evident that the rock's geotechnical properties alone cannot account for the instability phenomena observed in the Langhe region. Friction angle values, however, agree with the values given by Lupini et al. (1981) and Skempton (1985) for materials with CF ≤ 20% and PI = ca. 10%.

We can make the following observations:

a) non-remolded samples were disturbed to some extent during field drilling and laboratory cutting;

b) after repeated direct shear cycle testing, they displayed a rough failure surface. This fact, due to the schistose texture of the material, explains some of the irregularities observed in the shear - displacement ($\tau-\delta$) curve obtained experimentally:

c) in remolded samples, the orientation of clay particles was arbitrary and the schistose texture observed in situ along the true slip surface was lost;

d) the values given as 'residual' in table 2 should actually be considered as 'conventional residual', in that δ_{max} = 1.2 cm;

e) however, following Skempton (1985) and since CF ≤ 20%, we get:

$$(\tau/\sigma'n)_{peak} \geq (\tau/\sigma'n)_{CS} \approx (\tau/\sigma'n)_{residual}$$

where CS stands for incipient critical state;

f) it is therefore possible to state that the orientation of clay particles has limited influence on the strength of these materials in relation to full scale instabilities.

5 CONCLUSIONS AND FURTHER DEVELOPMENTS

Field surveys and laboratory test results seem to suggest that the alteration of the marl strata brought about by groundwater is the main cause of failure. The steady seepage of water down through the dipping surfaces and inside preferential strata (sand or fissured marl), in fact, certainly caused swelling and was responsible for a weathering process resulting in the progressive deterioration of marl either by reducing its cementation bonds or by dissolving its calcium-carbonate content.

This de-structuring process was probably responsible for the considerable decrease in shear strength (Chandler, 1969).

Further developments of this research will include:

- additional laboratory tests (direct shear, triaxial tests) on both undisturbed and remolded samples;
- tests on slip surfaces;
- direct shear tests on an artificially weathered slip surface, to be obtained by forcing a steady flow of water, under strictly controlled conditions, over the top of a cored or remolded sample: this will make it possible to reproduce the causes and effects of the alteration (or weathering) and to examine them in a systematic way.

Through a better understanding of these instability it may prove possible to develop more effective means of protection for inhabited areas.

ACKNOWLEDGEMENTS

The authors wish to express their gratitude to the Servizio Prevenzione Rischio Geologico Meteorologico e Sismico of the Regione Piemonte for the assistance provided. They also wish to thank F. Campagnoni for his valuable contributions to the laboratory testing program.

REFERENCES

Chandler, R.J. 1969. The effect of weathering on the shear strength properties of Keuper marl, *Geotechnique*, 19, No. 3: 321-334.

Govi, M. and Sorzana, P.F. 1982. Frane di scivolamento delle Langhe cuneesi, febbraio-marzo 1972, febbraio 1974. *Bollettino Associazione Mineraria Subalpina*, XIX, No. 1-2.

Lupini, J.F., Skinner, A.E.and Vaughan, P.R. 1981. The drained residual strength of cohesive soils. *Geotechnique*, 31, No. 2: 181-213.

Skempton, A.W. 1985. Residual strength of clays in landslides, folded strata and the laboratory. *Geotechnique*, 35, No. 1:. 3-18.

Rock Mechanics, Nelson & Laubach (eds) © 1994 Balkema, Rotterdam, ISBN 90 5410 380 8

The impact of tropical weathering on the mechanical behaviour of mudrocks

C.-H. Fan & M. E. Jones
Department of Geological Sciences, University College London, UK

ABSTRACT: Scaly clay is an argillaceous sediment, with a pervasive fabric of lenticular, highly polished, curved, slickensided surfaces and is commonly associated with mélange formations. In regions subject to tropical weathering regimes outcrops of this material erode rapidly giving rise to extensive badlands topography. Islands such as Barbados and Taiwan lose considerable areas of agricultural land to badlands generation each year and associated landslides disrupt settlements and communications.

A programme of laboratory studies has been completed to characterise the physical, mineralogical and mechanical properties of undisturbed samples of these materials. The samples, collected from different depths within weathering profiles, between ground surface and fresh materials with no sign of weathering at depth were each subject to similar experiments. These data illustrate important changes during weathering, including destruction of the original sedimentary structure, a large increase in pore volume and variations in strength.

1 INTRODUCTION

In tropical environments which are tectonically active, such as Barbados and Taiwan, weathering is at its most severe due to the high temperatures, high humidity and high uplift and denudation rates. Clay rich mélange materials in such environments often form steep, unvegetated slopes which develop a special landscape known as badlands topography (Fan, *et al.* 1994, Figure 1). On these steep slopes landslides and mass movements are frequent, allowing the badlands to increase in extent as the highly weathered near-surface material is continuously washed away during the torrential rainstorms that characterise tropical environments. Large amounts of agriculture land are lost each year to this cause. The widespread development of shallow mass movements in these terrains indicates that the mechanical characteristics of mélange clays must change rapidly as the materials come towards the ground surface.

Scaly clay is a naturally structured argillaceous rock characterised by a pervasive fabric of highly polished, slickensided, curved surfaces. The scaly fabrics are the result of local preferred alignment of platy clay particles caused by previous tectonic deformation when the clays were resident at depth. The natural deformation processes responsible for this fabric means that scaly clays at shallow depths are both bonded and very highly overconsolidated. The mechanical behaviour of such highly structured sediments cannot be simply interpreted in terms of their void ratio/ effective stress history (Leroueil & Vaughan, 1990; Aversa, *et al.*, 1993). The scaly fabrics of the material, may therefore, play an important role in localising the mechanical weakening associated with weathering to a relatively thin surface layer.

Figure 1 Typical badlands topography in mélange area, photo taken from Joe's River Formation scaly clay, Barbados.

2. STUDY AREAS

Undisturbed samples were extracted from selected weathering profiles within the Joe's River Formation scaly clay, Barbados and the Lichi Mélange, Taiwan. Both locations are tectonically active and are characterised by oceanic tropical climates.

The island of Barbados is a part of the Lesser Antilles Forearc complex. It is a tectonically active area where the Atlantic Plate is being subducted beneath the Caribbean Plate at an average rate of 2cm y^{-1} (Pudsey & Reading, 1981). Presence of mud diapirs associated with underconsolidated sediments at depth has been shown by seismic surveys and sonar images of the sea floor to the east of Barbados. It is these sediments that are believed to form the diapiric mélange of which the Joe's River Formation is a part (Brown & Westbrook, 1987). The Joe's River Formation outcrops in the Scotland District of Barbados (Figure 2) and is thought to represent a diapir mélange deposit, as opposed to an olistostrome mélange (Pudsey & Reading, 1981). The Formation includes a number of large olistoliths embedded in a dark green to brown, silty clay matrix, which appears to be highly inhomogeneous, containing clay rich and silt rich areas. It also contains wisps of uncemented, fine grained sand and streaks of solid hydrocarbon The most distinctive features of the mélange clay are its pervasive fracture pattern and the badlands topography it forms (Enriquez-Reyes *et al.*, 1990).

The island of Taiwan is located on the collision zone between the continental Eurasian and oceanic Philippine plates, where an island arc on the Philippine plate has been obducted on to the Eurasian margin. At present the rate of plate convergence is about 7cm y^{-1} in a northwest-southeast direction (Seno, 1977). The Lichi mélange underlies a narrow 3-5km wide belt of low rolling hill country along the western foot of the southern Coastal Range for a distance of about 65km (Figure 3). It consists of disturbed scaly mudstone matrix and chaotically distrbuted blocks and fragments, largely of sandstones and ophilitic rocks. There has been an argument concerning the origin of the Lichi mélange (eg. Page, 1978; Teng & Wang, 1981; Chen, 1991) between olistostrophic and tectonic causes. However, it is generally accepted that the Lichi Formation is somehow, directly or indirectly, related to the tectonic collision between the two plates.

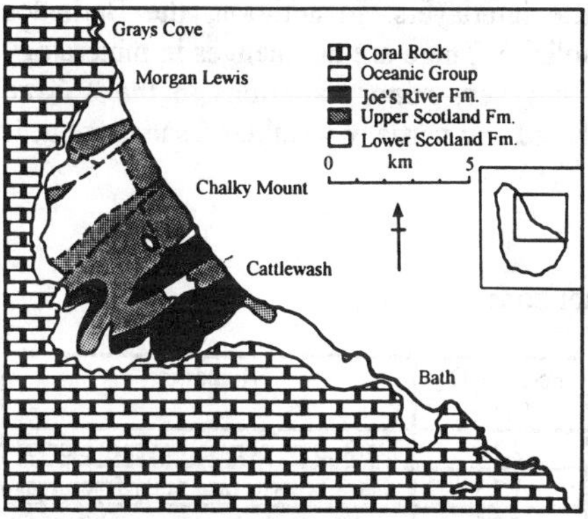

Figure 2 Geological map of the study area at Barbados.

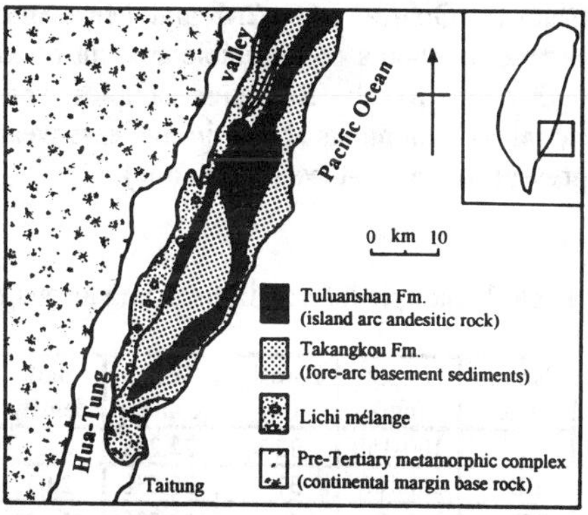

Figure 3 Geological map of the study area at Taiwan.

3. EXPERIMENTAL PROGRAM

The experimental program involved physical characterisation of the test materials, measurement of their degree of oxidation and determination of their mineralogy as well as a series of mechanical tests. The identification of different degrees of weathering was initially based on field observation such as changes in colour and structure, and the depth of burial. Subsequently, an examination of the Ferric/Ferrous iron ratio was undertaken to give a more quantitative measure of weathering.

Mechanical behaviour was investigated using a standard soil-mechanics traxial cell, in which sample specimens were subject to standard undrained triaxial experiments following consolidation to different isotropic effective stress states. During shearing cell pressure was held constant as was strain rate. Due to its extremely low permeability the Taiwan material was tested using small right cylindrical specimens 25.4mm diameter and 51mm long, to shorten the consolidation time. The more permeable Barbados material was tested using samples 38mm in diameter and 76mm in length. The axial deformation rates used were 0.05 mm/min for Barbados samples and 0.02 mm/min for the Taiwan material.

At least three experiments at different consolidation pressures were conducted on materials from each weathering horizon. Samples were isotropically consolidated to the required pressures and then loaded axially under undrained conditions. The consolidation pressures used were 100, 200 and 400kPa. Each test was terminated after the specimen failed in undrained shear. Failure was deemed to have occured when either deformation became locallised on a discrete shear surface or continued at constant efeective stress state.

4. RESULTS

The physical parameters, oxidation index and brief field descriptions of the tests materials are summarised in Table 1. A gradual change is seen in porosity, moisture content, bulk density and oxidation of the material with decreasing sample depth. X-ray diffraction data (Table 2) show that both materials have very similar clay mineralogies, namely Chlorite, Quartz, Mica-2M and Mica-Smetite interlayers. In addition, the Barbados samples contain a considerable amount of Kaolinite. There are no changes in mineralogy with depth for the Barbados samples, and only very minor variations in the Taiwan samples (Kaolinite is absent in the least-weathered and partially weathered samples but is present in the most-weathered one).

Table 1 Sample informations from various horizons

Sample No.	Depth (cm)	Porosity %	Water content	Bulk density	FeO/Fe_2O_3 Ratio	Remarks
BDL	100-150	32.8	22.2%	2.10	2.14	Dark grey, pervasive scaly fabrics
BDP	50-100	37.2	29.6%	2.04	1.18	Dark brown, rare scaly fractures
BDM	10-50	42.2	32.3%	1.91	0.36	Brown with organic matter
TWL	100-150	20.3	9.3%	2.36	2.94	Dark grey, pervasive scaly fabrics
TWP	50-100	22.9	10.5%	2.28	2.20	Dark grey, less scaly fabrics
TWM	10-50	25.4	12.9%	2.19	1.50	Light grey, scaly fabrics absent

BD: Barbados; **TW**: Taiwan; **L**: Least-weathered; **P**: Partially-weathered; **M**: Most-weathered

Stress paths and stress-strain data recorded during the experiments are presented in Figures 4-7. Figures 4 and 5 show the deviatoric stresses at specific axial strains as a fuction of initial effective confining pressure and Figure 6 and 7 are the stress paths. All the tests except BDL-200 were taken to large axial strains (ie. above 15%). BDL-200 was terminated at an axial strain of about 7%. This was because the specimen developed a shear failure plane at an early stage which, by 7% axial strain, threatened to puncture the membrane.

Table 2. X-ray diffraction data of the materials

Sample No.	Chlorite	Quartz	Mica-2M	Mi-Sm intl.	Kaolinite
BDL	*	*	*	*	*
BDP	*	*	*	*	*
BDM	*	*	*	*	*
TWL	*	*	*	*	
TWP	*	*	*	*	
TWM	*	*	*	*	*

Mi-Sm intl.: Mica-Smectite interlayers.

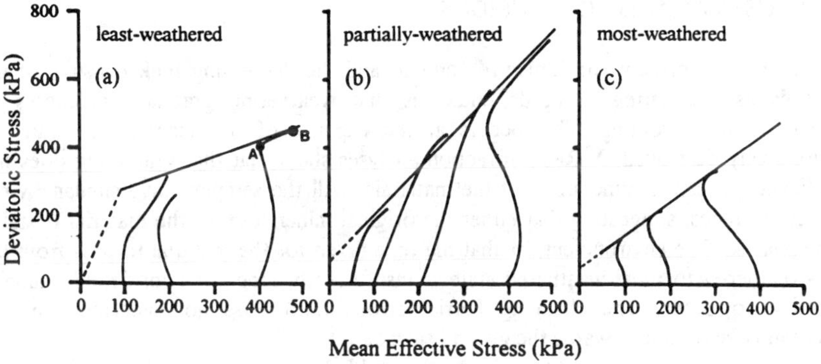

Figure 4 Stress paths of the BD experiments. (a) least-weathered samples. (b) partially-weathered samples. (c) most-weathered samples.

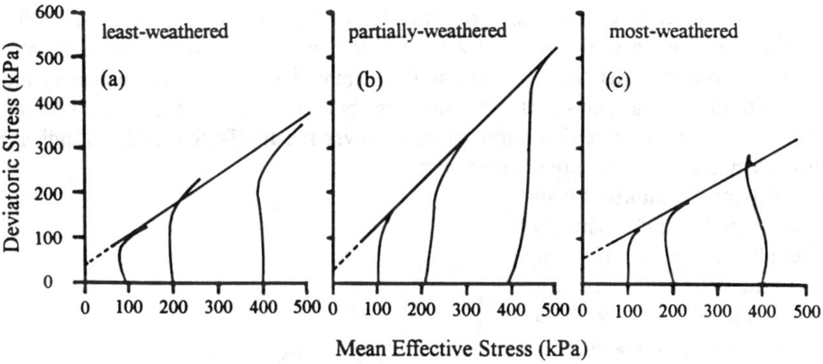

Figure 5 Stress paths of the TW experiments. (a) least-weathered samples. (b) partially-weathered samples. (c) most-weathered samples.

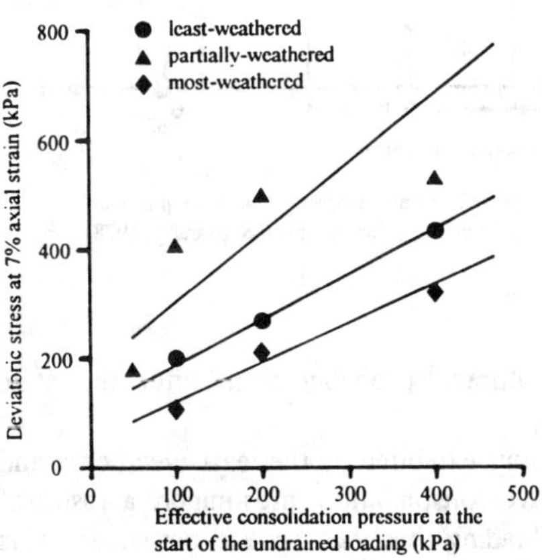

Figure 6 Strength-consolidation pressure relationship of the BD experiments.

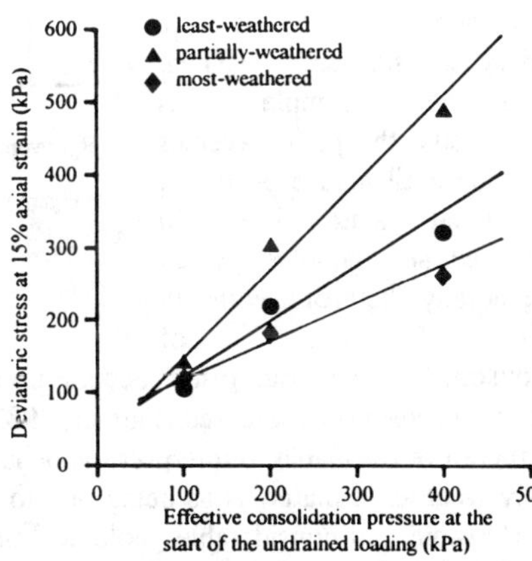

Figure 7 Strength-consolidation pressure relationship of the TW experiments.

5. DISCUSSION AND CONCLUSIONS

The increasing porosity and moisture content, and the decreasing bulk density of the materials as weathering proceeds, indicates that weathering gradually promotes an increase in pore volume. This occurs at the expense of the scaly fabric which is progressively destroyed. X-Ray Diffraction analyses show that the weathering does not significantly alter the mineralogy of the materials. All the samples have similar X-ray diffraction traces, suggesting that either the original mineralogy of the material is stable in the near surface environment, or that the time taken for the material to pass from the least-weathered to most-weathered state is insufficiently long for considerable mineral alteration to occur. The changing Ferric/Ferrous ratio does however indicate that oxidation is increasing towards the ground surface.

The triaxial test results show that the mechanical behaviour of the scaly clay from both locations is strongly influenced by the impact of weathering. Inspection of the two families of stress path data (Figures 4 and 5) reveals that although the samples were loaded under undrained conditions, pore pressure development is restricted, especially in the partially-weathered samples. The least-weathered samples show steeply rising stress paths that curve towards higher stress at the failure line. This effect becomes more pronounced in the partially-weathered samples, but in contrast, the most-weathered material shows stress paths that curve towards lower mean effective stresses indicating more appreciable pore pressure development.

The behaviour exhibited by the least-weathered and partially-weathered materials is very similar to that of an overconsolidated clay as described by Atkinson and Bransby (1978). Undrained test paths for overconsolidated materials (Figure 8) start from the q=0 plane of the yield figure and rise almost vertically to the state boundary surface. Then, providing the sample is homogeneous, the path traverses the surface until failure occurs at the critical state line. In undisturbed soil samples (which are generally heterogeneous in detail), failure of overconsolidated material often occurs prematurely, probably soon after the state boundary surface is encountered (Farmer, 1983).

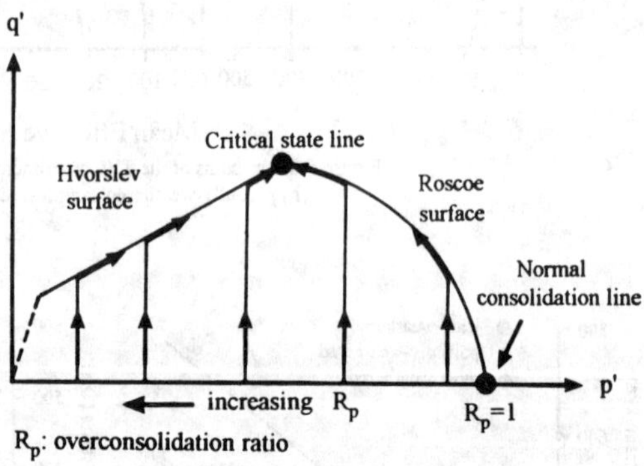

Figure 8 Expected undrained stress paths of samples with different R_p. (after Atkinson & Bransby, 1978)

It is therefore resonable to interpret the behaviour exhibited by the least-weathered and partially-weathered materials as being due to overconsolidation, presumably a result of the fairly rapid and considerable geological unloading that they have experienced. This cannot however be the full story, because the behaviour of the most-weathered samples is far closer to that of a normally consolidated soil.

The overconsolidation behaviour exhibited by the partially-weathered samples is controlled primarily by dilation. The samples deformed under the highest confining pressures exhibit the greatest pore pressure response. A similar effect is not seen in the least-weathered material. All of the most-weathered samples show pore pressure generation, but in these experiments the effect is most pronounced in the samples deformed under the largest confining pressures.

The overconsolidated behaviour exhibited by the least-weathered and partially weathered samples is also reflected in their strengths, irrespective of the confining pressure (Figures 6 and 7). These data show that in both cases the partially weathered clay is also by far the strongest. Thus the onset of weathering appears to emphasise the overconsolidated nature of these scaly clays and also enables them to mobilise a greater strength. These strengthening effects, seen in the deeper parts of the weathering horizons are not present in the behaviour of the most-weathered samples. These materials appear to be returning to a normally consolidated state and this change in characteristic is accompanied by an appreciable reduction in strength, both with respect to the partially-weathered and the least-weathered material.

The onset of weathering in scaly clay materials is thus unusual in causing a general strengthening of the clay as it weathers. This effect appears to be related to the presence of the scaly fabric which exists as an array of interconnected surfaces of low shear strength. As these surfaces become exposed to that physical and chemical changes accompanying weathering, the clays along them become restructured and reorientated leading initially to an increase in shearing resistance. With continuing weathering, restructuring and reorientation of the clay fabric continues and the scaly fabric is largely eradicated from the material. These changes are akin to a partial remoulding of the clays and result in a behaviour close to that of the normally consolidated material, and the reduced strength.

Thus weathering of scaly clays leads to the localisation of physically-weakened and restructured material in a relatively thin layer near to the exposed surface of the clay. The largely destructured nature of this material combined with its low strength accounts for the ease with which it is removed from slopes. This results in poor soil layer and vegetation development accounting for the widespread development of badlands topography in regions of scaly clay outcrop.

ACKNOWLEDGEMENTS

Use of the facilities of the Sediment Deformation Research, and that of the Research School of Geological & Geophysical Sciences, Birkbeck College & University College London are warmly acknowledged. Thanks also go to Mr. A. Osborne for the preparation of the X-Ray Diffraction samples and Mr. T. Stiles for processing the photograph.

REFERENCES

Atkinson, J.H. and Bransby, P.L. 1978. The mechanics of soils: An introduction to critical state soil mechanics. McGraw-Hill, London.

Aversa, S.; Evangelista, A.; Leroueil, S. and Picarelli, L. 1993. Some aspects of the mechanical behaviour of 'structured' soils and soft rocks. Geotechnical Engineering of hard soils-soft rocks, Anagnostopoulos et al. (eds) Balkema, Rotterdam.

Brown, K. and Westbrook, G. 1987. The tectonic fabric of the Barbados Ridge Accretionary Complex. Marine and Petroleum Geology Vol.4, p. 71-81.

Chen, W.S. 1991. Origin of the Lichi Mélange in the Coastal Range, eastern Taiwan. Spec. Pub. of Central Geol. Serv. , Taiwan. No.5 pp.257-266.

Enriquez-Reyes, M.P., Allison, R.J. and Jones, M.E. 1990. Slope stability in scaly clay terrains. In: Cripps, J.C. and Moon, C.F. (Eds), The Engineering Geology of Weak Rock, Geological Society Engineering Group, London, p. 422-432.

Fan, C; Allison, R.J. & Jones, M.E. 1994. Effects of tropical weathering on the physical characteristics of argillaceous rocks. in Rocks, rock weathering and landform evolution. (Eds.) Robinson, D.A. & Williams, R.B.G. John Wiley & Sons, Chichester.

Farmer, I. 1983. Engineering behaviour of rocks, 2nd ed. Chapman and Hall Ltd., London.

Leroueil, S. and Vaughan, P.R. 1990. The general and congruent effects of the structure in natural soils and weak rocks. Géotechnique 40(3): 467-488.

Page, B.M. 1978. Franciscan mélange compared with olistostromes of Taiwan and Italy. Tectonophysics No.47 pp.223-246.

Pudsey, C.J. and Reading, H.G. 1981. Sedimentology of the Scotland Group, Barbados. Geol. Soc. London Sp. Pub. 10:291-308.

Seno, T. 1977. The instantaneous rotation vector of the Philippine Sea Plate relative to the Eurasian Plate. Tectonophysics No.42 pp.209-226.

Teng, L.S. and Wang, Y. 1981. Island arc system of the Coastal Range, eastern Taiwan. Proc. Geol. Soc. China. No.24 pp.99-112.

Wood, D.M. 1990. Soil behaviour and critical state soil mechanics. Cambridge University Press, Cambridge.

16 Deep mine design in South Africa and Canada

Session developers:
Roger Tyler, Bureau of Economic Geology
Will Bawden, Queen's University

Rock Mechanics, Nelson & Laubach (eds) © 1994 Balkema, Rotterdam, ISBN 90 5410 380 8

Integrated seismic-stress-geomechanical analysis of a cable bolted back failure, Mines Gaspe, Canada

W.F. Bawden
Queen's University, Kingston, Ont., Canada

ABSTRACT: Mining at Mine Gaspe is carried out via longitudinal blasthole open stoping with delayed backfill. Increasing seismic activity was observed as mining progressed into the second panel. An integrated study of mine induced seismicity occurring over roughly eight months following system installation, plus seismic velocity, stress and geomechanical data, indicated that the majority of microseismic activity was related to the development of a large stress arch over the mined out panels. During extraction of the 400 panel a back failure in previously cable bolt supported ground occurred in the 400-10 stope. Analysis suggests a significant mine induced stress decrease occurred with extraction of the 400-10 stope. It is believed that this resulted in the observed back failure. The overall integrated analysis was critical to ultimately understanding the interaction of mine induced stresses, mine structure and cable bolt capacity. This technique shows promise as a tool for enhanced design of underground mine openings.

1. SITE LOCATION

Mines Gaspe is located in the town of Murdochville, Quebec, in Eastern Canada. The E-32 deposit is being mined using a bulk longitudinal blasthole open stoping method with delayed cemented rockfill, and is planned for 100% extraction (i.e. no permanent pillars). Longitudinal stoping blocks are mined in approximately 35 meter stopes which are 15 meters thick (north - south) by the full orebody height (up to 60 meters maximum), starting with the 450 block at the lowest point of the orebody and moving up and to the south, to the 150 block; there are a total of six longitudinal blocks planned for the E-32.

2.REGIONAL GEOLOGY OF THE GASPE COPPER DEPOSITS

The geological setting of the Murdochville region consists of lower Devonian impure limestones and calcareous siltstones and shales that were moderately folded, faulted and underwent hydrothermal alteration from mineral bearing fluids associated with nearby plutonic activity during the Acadian Orogeny (Kloeren, 1986, Procyshyn et al, 1989). The E-32 ore zone, located below the Murdochville townsite, originally contained mining reserves of 4.5 Mtonnes grading 2.8% Cu. It occurs as a bedded replacement deposit hosted in a 90 meter sequence of calcareous siltstones and an argillaceous carbonate (impure limestone), metasomatically altered into porcellanite and marble respectively.

3.SUMMARY OF GEOMECHANICAL DATA FOR THE E-32 ORE ZONE

Structural Mapping in the E-32 ore zone was carried out in 1991. Approximately 1200 data points were recorded, including strike and dip, quantity of features and average spacing, feature type (i.e. joint, bedding, fault), surface condition (planarity and roughness), and presence of any infilling material (usually calcite or chlorite). In addition to surface classification, roughness profiles were recorded using a 100-pin profile comb, and joint surface strength measurements were done using a Schmidt Hammer.

Table 1.0: **M**ean Orientations of Principle Joint Sets. E-32 Ore Zone, Mines Gaspe

STRUCTURAL FEATURE	Strike (deg)	Dip (deg)	Spacing (meters)	Freq. (j/m)
Stratigraphic Bedding	271	28	0.6235	1.59
North South Set	158	80	0.5142	1.95
East West Set	085	86	0.4050	2.49
Inverse Bedding	122	56	0.6630	1.56

Table 1 gives the mean orientations for the four (4) principle joint sets identified plus the mean joint spacing and frequency (in meters and joints/meter respectively). Tables 2 and 3 summarize input for numerical stress analyses as determined from geomechanical data collection and rock mass classification.

Table 2: Rock Mass Classification Information with Hoek-Brown Failure Criterion parameters (After Connors, 1992).(*Q': Modified NGI rockmass classification after Bawden 1993)

	Q'	RMR	m	s	E_{RM} GPa
Lower Bound	5.1	58.6	1.590	0.010	17.6
Mean	5.9	60.2	1.678	0.012	20.6
Upper Bound	7.4	62.3	1.815	0.015	24.6

Table 3: Summary of In-situ stress field used in numerical analysis

Stress Component	MPa	Azimuth (degrees)	Dip (degrees)
Major Principle Stress	33.5	153	14
Intermediate Principle Stress	22.0	051	39
Minor Principle Stress	10.7	262	46

4. SEISMIC ANALYSES AT MINE GASPE

Seismic methods deal with the measurement and examination of seismic waves traveling through a rock mass and the characterization of stress induced failure processes occurring within the medium. Two seismic methods were used in this study: (1) tomography, which characterizes the variation in seismic wave velocity throughout the rock mass; (2) microseismic monitoring, which monitors the failure of the rock mass caused by induced stress change, providing information regarding source location, failure mode, energy release and other physical parameters (Young, 1992).

A 32 channel, full wave-form microseismic monitoring system was installed around the E-32 deposit in 1990. A three dimensional seismic velocity survey was conducted June 1991. All seismic installations and the velocity survey were done by the Queen's University Rock Physics Group.

4.1 Integrated Seismic-Stress-Geomechanical Analysis

Connors (1993) conducted an integrated seismic-geomechanical-stress analysis using a subsection of the Mine Gaspe microseismic database in order to investigate factors controlling observed mine induced seismic activity and to test this approach as a possible means to constrain and validate selected numerical modeling techniques. In this case, initial stress analysis was restricted to linear elastic, using a 3D boundary element method (Examine 3D), which was also used to display and interpret the seismic data.

Connors et al (1993) reported key findings from the above integrated analysis. A comparison of the average nodal plane orientations determined from fault plane solutions (done on appropriately filtered pure shear seismic events) with the structural mapping results suggest a strong correlation between the Stratigraphic Bedding (270/30 N) and fault plane A (256/36 N) in Figure 1. No correlation was found between any mapped structure and fault plane B.

To validate input parameters used in the numerical modeling, principal stress orientations calculated at the seismic event locations were compared to the fault-plane solution average P, B, and T axes. The contoured principal stresses (Figure 2) effectively give a 1:1 correlation to the contoured P, B, and T axes (Figure 3), suggesting that the model predicted stress regime is similar to that obtained independently from the seismic analysis.

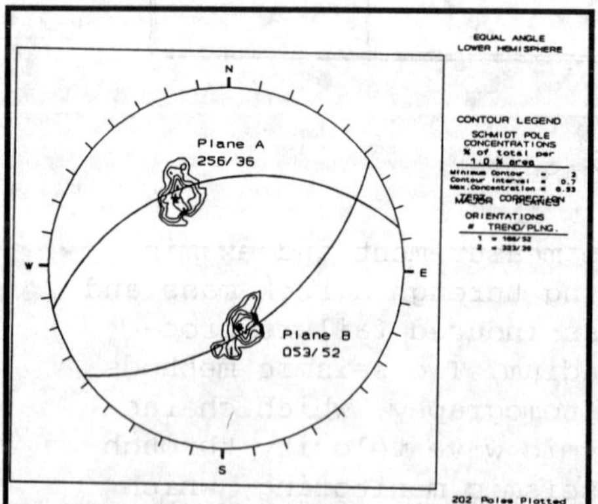

Figure 1: Fault Plane solution for Data Set #1

In order to provide an

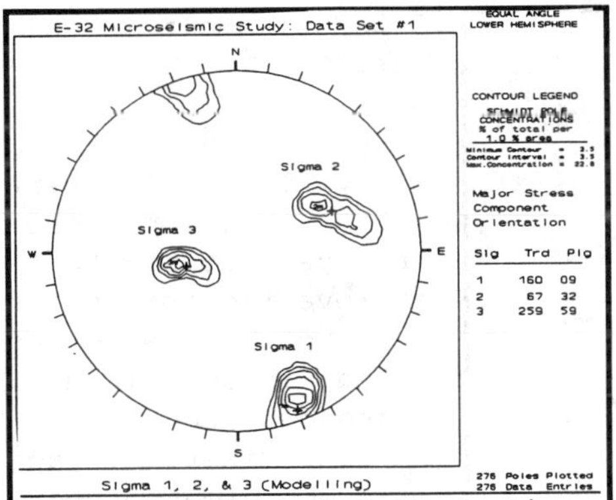

Figure 2: Major pole concentrations of major, intermediate and minor principal stresses: Data set #1. (after Connors, 1993)

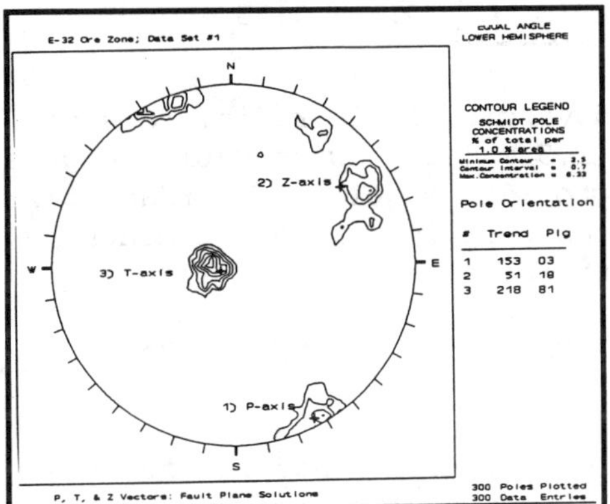

Figure 3: Major pole concentrations of the compressional, dilational and null axes: Data set #1. (After Connors, 1993)

integrated analysis between the stress, geomechanical and seismic data, a ubiquitous joint analysis was conducted. This analysis also strongly suggested that failure was associated with slip along Stratigraphic Bedding, as had been indicated by mine induced seismicity, further supporting the use of elastic modeling in this case (Connors et al 1993).

Figure 4 summarizes the overall integrated interpretation from the Mine Gaspe data. This suggests that mine induced microseismicity at Gaspe is predominantly related to a stress arch process mobilizing shear on bedding structure over the mined out stopes. Mine induced seismicity concentrates in the velocity gradient and high(er) velocity zone above the stopes, while the low(er) velocity zone in the immediate stope back is predominantly aseismic. Interestingly, the cable support failure described later in this paper occurred in this aseismic, low velocity region.

5 MINE GASPE STOPE 400-10 CABLE BOLT FAILURE

5.1 Description of Failure

The 400-10 stope cable bolt failure occurred on the north side (back/hangingwall) of the stope. Cables were observed hanging from the back stripped of grout and rock. Some appeared to show "pig-tailing", indicating they may have taken considerable load, (15t to 25t), while others appeared to have simply unraveled, indicating minimal loads (5t to 15t), (MacSporran, 1992).

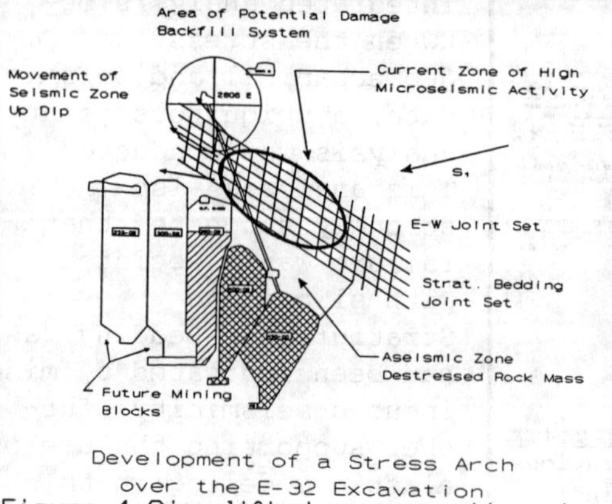

Figure 4: Simplified cross section view of E-32 excavation showing seismic & asesimic zones (After Connors, 1993)

5.2 Cable Bolt Installation

Gaspe uses a high-early (Type 30) cement with a friction reducing agent to try to improve flowability of the grout paste. Grout mixing is done in a homemade batch shearer-mixer. Average measured grout properties are: density = 2.07; UCS = 89 MPa; E = 16 GPa; w:c ratio = 0.3 - 0.35.

Cable bolts were drilled vertically in radial fans of 5 to 8 cables. Row spacing is 1.5 m (5ft.) with cable lengths ranging from 12.5 m (41 ft) to 18.6 m (61 ft). Surface fixtures were not used and cables were installed one month prior to the excavation of the stope.

5.3 Calculation of Effective Embedment Length and Radial Stiffness

Using the measured orientation and normal spacing for the various joint sets, (Table 1.0), the effective embedment length (i.e. average fracture spacing along the cable) was calculated for each cable bolt hole using the relation developed by Priest and Hudson (1976). Embedment lengths between 0.2m and 0.4m were predicted.

Using techniques suggested by Hyett et al, (1992), field classification values were used to obtain the destressed field rock mass modulus (E_Q') and radial stiffness (K_Q'). Intact rock properties from core tests were used to derive initial radial stiffness approximations for highly stressed (confined) conditions, (Table 4).

6.0 NUMERICAL MODELING

Elastic stress analysis was performed using the EXAMINE 3D software discussed earlier. Specific locations ("field points"), corresponding to the mid and end points of se-

Table 4: Radial Stiffness ($K_{Q'}$ & K_{Intact}) for a 2.5 inch cable bolt hole for destressed ($E_{Q'}$) and stressed (E_{Intact}) environments respectively (After MacSporran, 1992)					
	Q'	$E_{Q'}$ (GPa)	$K_{Q'}$ (MPa/mm)	E_{Intact} (GPa)	K_{Intact} (MPa/mm)
Lower Bound	5.1	17.6	466	47.6	1259
Mean	5.9	20.6	545	50.9	1347
Upper Bound	7.4	24.6	651	52.8	1397

lected cable bolt locations, were entered into the program and the relative stress changes were calculated at these locations for various extraction periods leading up to cable bolt failure.

Eight different rings were selected from the cable bolt pattern, with every other cable being used for the analysis. A few rings were selected in areas where stable conditions occurred in order to compare against stresses in the failed area. Numerical modeling results indicated a substantial decrease in the minimum principal stress over the 400-10 stope cable mid sections, whereas the end sections only underwent a modest decrease. In all locations within the 400-10 failure area σ_1 increased (MacSporran, 1992).

7.0 CABLE BOLT FAILURE ANALYSIS

Installed grout quality was good and within design specifications. The effect of radial confinement and embedment length were then considered probable controlling factors contributing to cable bolt failure.

The intial rock mass modulus assumed a highly confined environment (K_r = 1347 MPa/mm; Line A in Figure 5). Predicted cable capacity is then 12t to 20t. The amount of decrease in s3 was substantial at the cable mid length. Assuming that the rock mass modulus following destressing can be calculated using EQ', radial stiffness reduces to approximately 545 MPa/mm, (Line B in Figure 5), and predicted cable capacity reduces to 7t to 13t,(cable capacity reduction of 35% to 40%).

8.0 DISCUSSION

The integrated seismic-stress-geomechanical analysis suggested that simple 3D linear elastic stress analysis was appropriate for conditions at Mine Gaspe. This is believed to be the case because of a combination of factors including moderate depth (about 600m), relatively strong rock and 100% extraction, retreat mining. The analysis indicates that the destressing has occurred in an area typified by aseismic

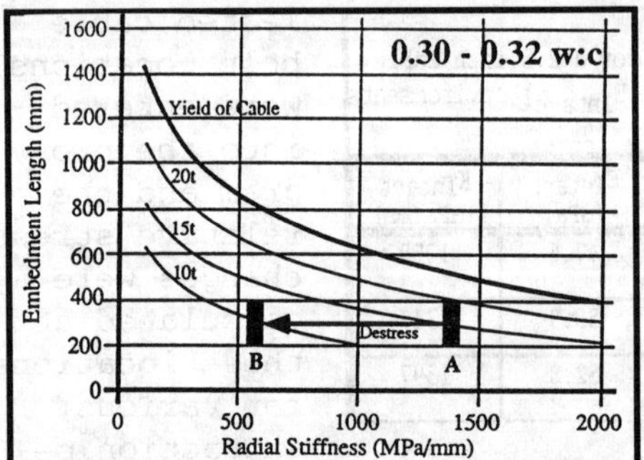

Figure 5: Performance assessment chart, 400-10 stope (After MacSporran, 1993)

behaviour and showing lower than average velocity. Seismic data tended to corroborate the simple 3D elastic model results.

Failure of the 400-10 stope back/hangingwall appears to have resulted from a combination of factors including: 1) large stress change (decreased radial confinement); 2) short embedment length; and, 3) cable bolt pattern. The preceding analysis suggests that factor (1) was dominant in this case.

To improve cable performance under such conditions an improved understanding of the effect of stress decrease is essential. With conventional cables, application of surface fixtures, (Plates or straps), is recommended. Adverse effects of large stress change, (decrease), and short embedment length, could probably be better accommodated through the use of modified geometry cables (birdcage, nutcase or bulbed cables). These geometry's are less sensitive to stress change and provide a stiffer support system than conventional strand, (Hyett, et al 1993). A tighter cable bolt pattern might assist, but this is very costly and the outcome is uncertain. The only time a tighter pattern can be fully justified is when broken cables are observed.

The Mine Gaspe case study suggests that mine induced seismicity can be used as a tool to help validate numerical modeling techniques for specific geomechanical domains. The models can then be used, in combination with simple empirical cable bolt design techniques, to predict cable capacity and potentially unsatisfactory cable bolt support behaviour related to mine induced stress change caused by destressing and/or rock mass failure.

9.0 ACKNOWLEDGMENTS

The author wishes to acknowledge Mr. C.F. Connors and Mr. G.R. MacSporran for their excellent work on the analyses discussed in this paper. Financial support from Mine Gaspe, the Natural Sciences and Engineering Research Council of Canada and the Mining Research Directorate of Ontario is gratefully acknowledged.

10.REFERENCES

Bawden, W.F. 1993. The use of rock mechanics principles in Canadian underground hard rock mine design. Chapter 11, V. 5, Comprehensive Rock Engineering. Pergamon Press. p. 247 - 291

Connors, C. F. 1993. A study of microseismic event mechanisms and their significance to open pit mining of the E-32 ore zone, Noranda Minerals, Mines Gaspe Division. M. Sc. Thesis. Queen's University, Department of Mining Engineering. Unpublished.

Connors, C. F., Urbancic, T. I. Bawden, W. F. & Young, R. P. 1993. Constraining numerical models with geomechanical data and microseismic fault-plane solutions. In Proc. 34th U.S. Rock Mechanics Symposium, Madison, Wisconsin. To be published in I.J.R.M. Journal.

Hyett, A. J.,Bawden, W. F., Powers, R. & Rocque, P. 1993. The nutcase cable bolt. In Proc. International Congress on Mine Design. A. A. Balkema. Kingston, Ontario, Canada. p. 409-419

Hyett, A.J., Bawden, W. F. and Reichert, R. D. 1992. The effect of rock mass confinement on the bond strength of fully grouted cable bolts: Int. J. Rock Mech. Min. Sci. & Geomech. Abstr. Vol. 29, No. 5, pp 503 -524.

Kloeren, C. J. 1986. Stratigraphy and horizon-markers at mines Gaspe, Quebec. 88th Annual General Meeting of the CIM, Montreal, Canada.

MacSporran, G. R. 1993. An empirical investigation into the effects of mine induced stress change on standard cable bolt capacity. M. Sc. Thesis. Queen's University, Department of Mining Engineering. Unpublished.

Priest, S. D. and Hudson. 1976. Discontinuity spacings in rock. Int. J. Rock Mech. Min. Sci. & Geomech. Abstr. V. 13, p.135-148.

Procyshyn, E. L. & Bernard, P. 1989. The copper ore deposits at mines Gaspe, Murdochville, Quebec. Geological Association of Canada Guidebook.

Young, R. P. 1992. Invited paper: correlations between seismic velocity and induced seismicity in underground mines. Proc. 33rd U. S. Rock Mechanics Symposium. A. A. Balkema. p. 1113 - 1123

Rock Mechanics, Nelson & Laubach (eds) © 1994 Balkema, Rotterdam, ISBN 90 5410 380 8

An integrated rock mechanics program at the David Bell Mine

C.W. Pelley, K. McKirdy & E. De Souza
Department of Mining Engineering, Queen's University, Kingston, Ont., Canada

ABSTRACT: This paper describes an integrated program of rock mass classification, instrumentation, and numerical modelling used at the David Bell Mine. This program was used as a planning tool so that the extraction of a critical sill pillar in the mine could be successfully achieved.

1.0 INTRODUCTION

Mining levels at the David Bell Mine, one of the three gold operations in the Hemlo area of Ontario, are spaced 100 m vertically from the number 10 level, approximately 1000 m below surface, up to the number 6 level. The initial bulk mining operation began on the 8 and 7 levels shown in figure 1, proceeding upwards from both on intermediate levels spaced 25 meters vertically. The resultant sill pillar between the 8C and 7 levels was forseen as a problem mining area.

Successful extraction of this sill pillar was critical as it contained high grade ore. In addition, it was critical to document the rock mass behaviour as the sill was extracted to help in planning the extraction of a pillar to be created as the number 9 and 10 levels are mined upwards beneath the number 8 level. The increased depth of that pillar will increase stress related problems, thus making detailed extraction planning critical. An integrated program of rock mass classification, instrumentation and numerical modelling was begun to assist in planning both pillar extractions.

2.0 ROCK MASS CLASSIFICATION

Extensive collection and assessment of rock mass data was performed in 1991 (1). Joint line mapping was carried out in eight mine areas and the joints were analyzed for spacing, orientation, and structure as required for the NGI classification system. Diamond drill core analysis was also implemented by the geology department.

2.1 Rock Quality Designator

A summary of the rock quality designator (RQD) for various rock types as given by Leduc (1) is provided in Table 1.

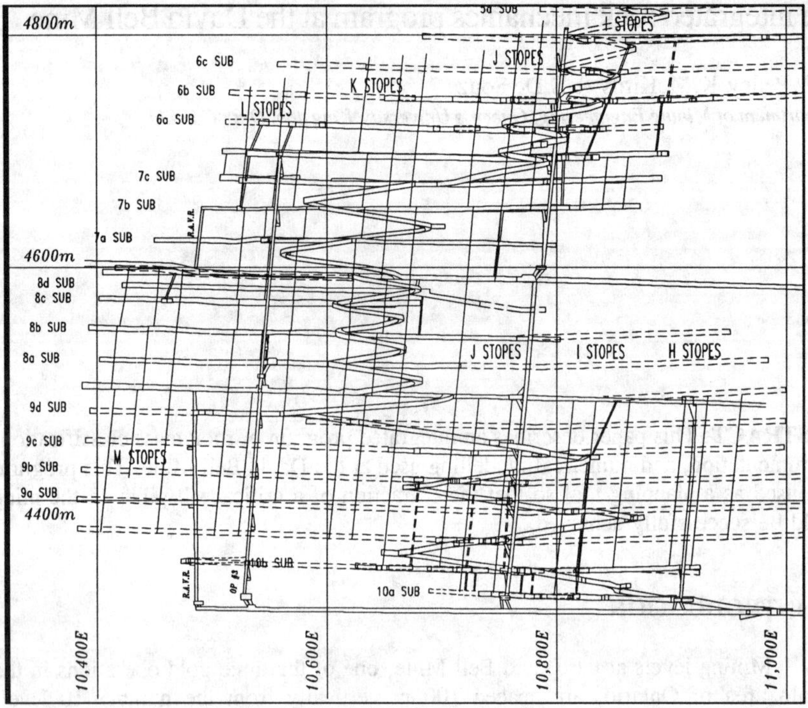

Figure 1 A Longitudinal Section of the Mining Area

Table 1. RQD OF THE DIFFERENT ROCK MASSES

HOLE	RQD F/W	RQD ORE	RQD H/W
6A-1	63.9	87.0	35.2
6A-2	---	80.1	54.3
6A-4	---	---	69.9
6-19	---	60.6	79.3
9D-14	90.2	70.2	---
9D-15	89.5	70.9	94.1
7B-23 (C-ZONE)	84.7	---	---

2.2 Joint Set Description

In total, 1335 joints were mapped in 8 different traverse locations and the results as analyzed (2) are shown in Table 2. Three main joint sets were identified. Set A is roughly parallel to the orebody and lithological contacts with the northerly dip increasing from 57° on the fifth level to 65° on the 10A sublevel. The B set is a steeply dipping joint set with a strike of 340°. This set varies over the mine and tends to dilate and grow longer if there is an open stope immediately adjacent to the mapping area. The C set is a flat-lying set with a general 090° strike and 20° dip. The spacing on the C set is generally greater than the other two but it can be quite persistent. The presence of the C set can cause wedge type failures in the backs of both the undercuts and the haulage drifts.

Table 2. JOINT SET ANALYSIS

LEVEL	# OF SETS	RQD	LOCATION
5 WEST	2 + RANDOM	59.5	F/W
6 EAST	3 + RANDOM	82.7	H/W
7C WEST	3	66.0	H/W
7B WEST	3	63.7	H/W
7 WEST	3	60.4	H/W
7 EAST	2	35.4	H/W
9D CENTRE	2 + RANDOM	70.3	ORE
10A CENTRE	3	63.4	H/W

The three main joint sets provide almost ideal conditions for block failures. In development openings and ore undercuts, the failure would be a wedge failure along the C set joints abutting the B set joints and bounded along strike by the A set joints. In the hangingwall, failure can occur along the A set joints, which are parallel to the ore planes, topped by the C joint sets and bounded East and West by the B joints, which are roughly perpendicular to the ore planes.

2.3 Rock Strength Analysis

Determination of the strength characteristics of the various rock types to be used in numerical modelling studies, was carried out at Queen's University. The initial core, tested early in 1988, came mainly from footwall drilling of the orebody and this initial data was used for the bulk of the preliminary modelling work.

No hangingwall core had been tested as little surface drilling had been carried out into the main area of the orebody. In 1990, a limited test was carried out on core from the hangingwall provided by drilling to install an extensometer on 7 level. In 1991, 10 additional samples were tested with core provided by exploration drilling. Table 3 gives the results of both the 1991 tests and the previous 1990 tests.

3.0 INSTRUMENTATION PROGRAM

As part of an instrumentation program initiated in 1988, the mine has used Geocon vibrating wire type stress cells for stress monitoring. Additional cells were purchased and installed during 1991 and early 1992. Generally, there are approximately 15 active stress cells and data exists on more than 40 inactive cells

Table 3. ROCK STRENGTH ANALYSIS

SAMPLE #	UNCONFINED STRENGTH (MPa)	YOUNG'S MODULUS (MPa)	POISSON'S RATIO	ROCK TYPE	COMMENTS ON TEST ROCKS
1-1	54.17	13079	0.07	7D	Good shear, high biotite
2-1	91.70	12776	0.17	7F	Good shear,medium biotite
2-2	116.84	16729	0.16	7F	Good shear, medium biotite
3-1	70.04	12708	0.16	7C	Good shear, very high course sericite
4-1	98.50	15098	0.11	7C/F	Good shear,fine biotite
5-1	112.06	18000	***	7C/E	Good shear,silicified
5-2	162.19	19251	0.25	7C/E	Good shear,silicified
6-1	19.17	4084	0.26	7C	Good shear, very high medium sericite
6-2	10.15	4673	0.07	7C	Schistosity shear, very high medium sericite
7-1	36.55	9549	0.03	5D	Schistosity shear, high moly along shear
1	91.5	20525	0.15	H/W	1991 SAMPLE
2	43.8	13840	0.19	H/W	1991 SAMPLE
3	40.8	9493	0.14	H/W	1991 SAMPLE
4	43.0	10239	0.14	H/W	1991 SAMPLE

from mined out areas. This data aids in the analysis of the mining sequence and to help validate numerical models.

The stress cells are installed in the back of the ore undercut or in the back of the haulage ways. The cells are placed in 5 metre diamond drill holes drilled up dip into the back and are aligned to detect changes in stress perpendicular to the orebody. Some haulage cells in the footwall have been aligned to detect changes along strike.

The cells in the ore show increases in stress as mining progresses towards them but the final stress results are often not recorded as the cells or access to them are lost as the stope immediately below is blasted. Cells in the footwall haulage show small stress increases as mining advances towards them but show stress reductions when the stope immediately above is blasted. These cells, because of the orientation of the rock strata, are immediately down structure from the overlying stope.

A thorough analysis of the stress data was carried out during the summers of 1991 and 1992. The stress results were compared to the sequence of stope removal using a spread sheet. Figure 2 shows an example from a stress cell installed in the 8D undercut in the L84D stope, central to the sill pillar, with the stress increasing until

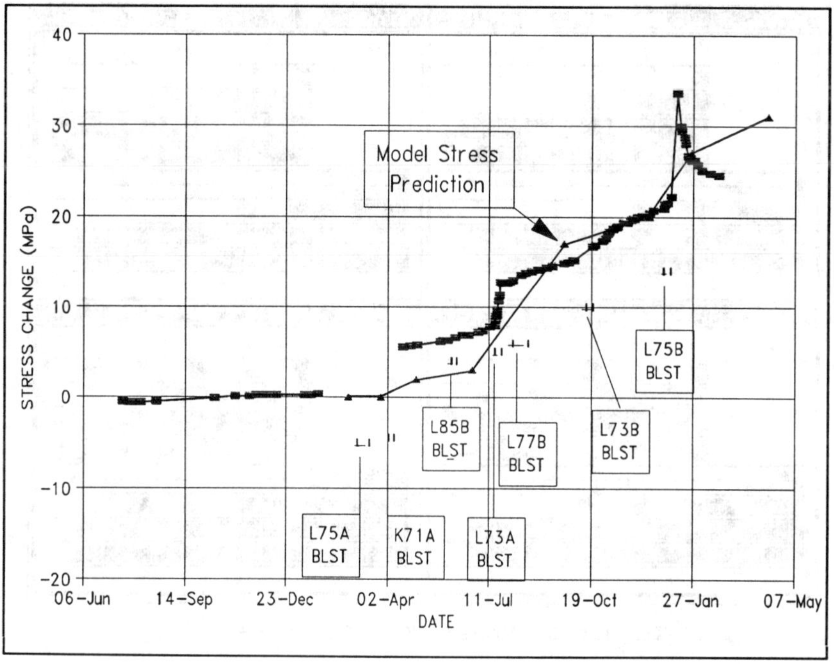

Figure 2 Stress Cell: L84D, 8D Undercut

pillar failure occurred. Understanding that failure had occurred was critical in developing a mining method and sequence for the pillar.

Initially extensometers were installed to monitor hangingwall to footwall closure but results were contradictory. An extensometer was planned for the 7 level to measure closure as the sill pillar was removed, but delays in its installation and the early failure of the sill pillar prevented continuation of this phase of the program.

4.0 THE NUMERICAL MODELLING PROGRAM

Numerical modelling is an accepted method of evaluating mining extraction sequences and is required under the Ontario mining regulations (3). Numerical modelling requires the magnitude and direction of the stress field surrounding the orebody and models have to be calibrated with underground observations as mining progresses, rock failure occurrences for example, or with instrumentation results.

The majority of the numerical modelling was carried out using MINTAB (4). The initial modelling of the five year plan was done in 1988 as the first stope was being mined. As mining progressed, the models were updated to the actual extraction sequence and stope geometry changes. Figure 3 shows four steps of the fourth model which was run after extraction of the first fifty stopes and shows the sill pillar below the 7 level. Models were run of differing extraction sequences and stope sizes as a guide the the best method to recover the pillar while minimizing the risk of ore loss.

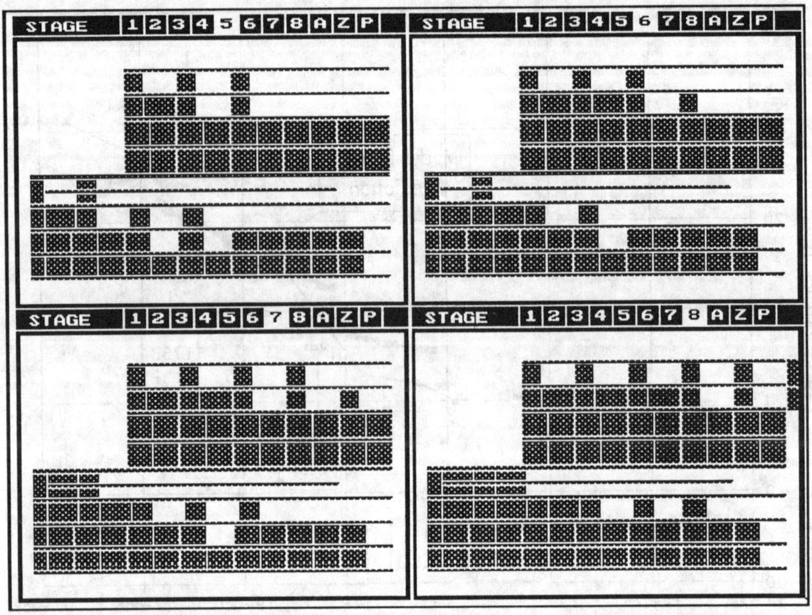

Figure 3 Model TC4, Steps 5-8

4.1 Measurement of in-situ Stress

The initial numerical modelling was carried out using stress data from adjoining operations as summarized by Kazakidis (5). Two in-situ tests were conducted in the first week of June 1992. These tests were carried out in the 10B west drift at the mine and the results along with the data from the core material testing are given in Table 5. The results are similar to the data from the adjoining mines and correspond to the general theory used in all the models that σ_1 is generally perpendicular to the orebody with a ratio of 1.8 to σ_3, which is steeply dipping and comparable to what would be expected from gravitational loading.

4.2 Calibration of Numerical Models

Calibration of the numerical models was carried out on site using underground observations and data from the stress cells. The stress prediction of the L84D stope area above the 8D undercut is plotted as the thinner line in figure 2. As can be seen, very close agreement exists between the modelling predictions and the stress cell data until the premature failure of the 7 level sill occurs. After sill failure, the elastic models were not valid for quantitative results but qualitative comparisons of various extraction sequences proved useful throughout the sill removal.

This project has provided a very comprehensive data base on the rock mass and the stress levels encountered during the mining sequence and this information can be used to validate models as they continue to be developed and improved in future.

Table 4. IN-SITU STRESS FROM THE DAVID BELL MINE
(Depth approximately 1000 m.)

Test #	10B West 1(a) 1(b)	10B West 2(a) 2(b)
U.C.S.	220.9 244.0 Avg. - 232.5	193.7 200.8 Avg.- 197.3
Young's Modulus	55.8 61.0 Avg.- 58.4	67.0 51.9 Avg.- 59.3
Poisson's Ratio	0.26 0.24 Avg.- 0.25	0.28 0.20 Avg.- 0.24
Principal Stress Magnitude	σ_1 = 44.6 σ_2 = 25.6 σ_3 = 23.5	σ_1 = 34.7 σ_2 = 25.5 σ_3 = 18.5
Vector Azimuth	3.5° 99.8° 218.3°	359.1° 267.4° 137.4°
Vector Dip	23.4° 14.1° 62.2°	10.4° 9.0° 76.2°

5.0 CORRELATION OF GROUND FAILURE MODES

A problem mineral, barite, is found in bands up to 1 metre wide which contain up to 75% barite. These bands are generally found in the wider western part of the orebody and have been associated with falls of ground in the ore undercuts. The failure of these bands occurs slowly upwards with small rocks falling out. As the failure advances higher into the back of the undercut, a brow is created in the adjacent rock which can then fail into the undercut.

To better correlate this failure mechanism with the ground failures which had occurred, a barite contour map was produced (6). This contour plot, shown in figure 4, shows high correlation between areas of high barite content and the ground failure occurrences. In addition, numerical modelling carried out at Queen's university using the FLAC (7) modelling code showed that failure of these baritic zones could be expected under a variety of confinement situations. This has proven to be an important discovery to aid in future mine planning, in fact a failure on a lower level was predicted by mine personnel based on this understanding of the problem.

6.0 CONCLUSIONS

The extraction of the 8 level sill was completed in early 1993. The program of rock mass classification, instrumentation and numerical modelling were used by the mine as tools to plan the successful extraction of this valuable economic resource. The

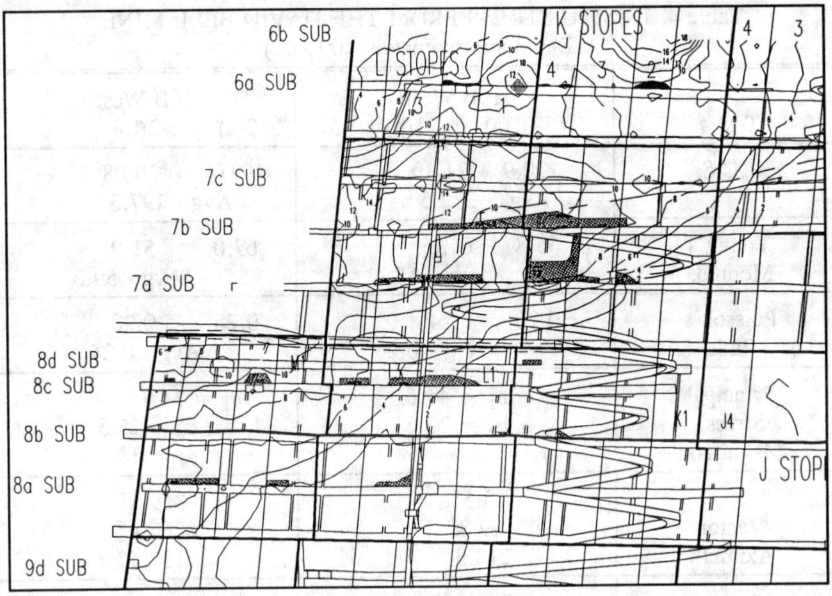

Figure 4 Barite Content Contour

data base developed as a result of these projects is one of the most complete data sets available in Canadian mining and will be used in future for various research purposes.

7.0 ACKNOWLEDGEMENTS

The authors would like to thank the Teck-Corona Operation for funding and support during this project. The financial support for the project provided by the University Research Incentive Fund of the Ontario Ministry of Colleges and Universities is also gratefully acknowledged.

8.0 REFERENCES

1. Leduc, M., **Rock Mass Behaviour at the David Bell Mine, Hemlo, Ontario,** B.Sc. Thesis, Queen's University at Kingston, 1992.
2. DIPS User's Manual, Rock Engineering Group, University of Toronto.
3. Section 5a, Ontario Regulation 694 under the Occupational Health and Safety Act, R.S.O. 1980, chapter 321, as amended.
4. Yu, Y.S. et al, **MINTAB User's Guide**, CANMET Division Report MRP/MRL 83-25, March 1983.
5. Kazakidis, V.N., **An Integrated Study of Fractured Rock Mass Behaviour Using Structural Mapping, Microseismic Monitoring and Numerical Modelling**, Master's Thesis, Queen's University at Kingston, July 1990.
6. Bos, E., **Barite and its Effect on Stability at the David Bell Mine,** B.Sc. Thesis, Queen's University at Kingston, 1993.
7. FLAC User's Manual, Itasca Consulting Group, Inc., Minneapolis, Minnesota.

Rock Mechanics, Nelson & Laubach (eds) © 1994 Balkema, Rotterdam, ISBN 90 5410 380 8

The nature and possible source mechanisms of seismic events from a shallow pillar supported, platinum mine in South Africa

K. Aref, A. L. Jager & S. M. Spottiswoode
Division of Mining Technology, CSIR, Johannesburg, South Africa

ABSTRACT: Seismic data recorded by near surface geophones over a period of 20 months at a 900 m deep platinum mine in the Bushveld Igneous Complex (BIC) have been analysed. The mine is supported by a system of reef pillars and the objective of the analysis was to distinguish between pillar failure related seismicity and that caused by shearing on geological discontinuities. An area of relatively high seismic activity was highlighted for study. The evaluation of several parameters, which, together can be considered to indicate seismic risk, correlated well with the increased seismic activity. A newly developed data analysis technique was applied to investigate source mechanisms. Three possible mechanisms which fit the data were postulated. Two of these have been observed underground, the most common being pillar punching.

1 INTRODUCTION

The main economic platinum bearing horizon is the Merensky Reef within the Bushveld Igneous Complex (BIC). It is a tabular orebody which dips at approximately 10° towards the centre of the Complex and is mined at a stoping width of 1 m. Mining operations on the Wildebeestfontein North mine extend down to a depth of 1 000 m below surface. Reef pillars provide the main support to the mine workings. The pillar system comprises a series of dip oriented barrier pillars 10 to 20 m wide to provide regional stability; and yield pillars, 6 m x 3 m, aligned on strike and spaced 33 m apart on dip, to support the immediate hangingwall to a height of about 40 m. (SANGORM Meeting: Kotze, 1993, Lougher & Mellowship, 1993.) An additional function of the pillars is to reduce stress concentrations at the stope faces and stope closure, and hence the incidence of seismicity. Nevertheless some major seismic events have occurred causing damage to workings. The circumstances leading to these rockbursts were not understood. A preliminary study was therefore carried out to investigate the nature of the seismicity and failure mechanisms on mines in the BIC. During the investigations, an area, known as the 5S block, experienced a relatively high level of seismicity accompanied by occasional pillar bursts and footwall damage in stopes. The results of the study to analyze seismicity in this area showed that if real time analysis of the seismic data had taken place the increased seismicity could have been anticipated. The study also suggests that changes to the design of the pillars could reduce seismicity.

2 SEISMIC NETWORK DESCRIPTION

A set of three orthogonal geophones was grouted in a 50 m vertical borehole drilled from surface. Seismograms recorded at, or near the earth's surface show clear P waves but are complicated by surface reflections. At incident angles greater than 30°, vertically polarized S waves (SV) are converted into Rayleigh waves that travel along the surface. At incident greater than 40°, Rayleigh waves have large amplitudes and dominate the seismograms. Therefore, the source parameters are more accurate for those events located within a 40° incident angle area. The seismic analyses discussed in this paper, are restricted to the events located within the incident angle area (Figure 1). Seismic events recorded from January 1, 1992 to September 31, 1993, and having local magnitudes $M_L > 0$ were considered in this investigation.

3 GENERAL ASSESSMENT OF SEISMIC DATA

The location and range of magnitude of seismic events are shown in Figure 1. A total number of 230 events were located. 70% of the events are concentrated in the 5S block on the W15/67 raise line and on the pillars close to this raise. This area was relatively active in the first and second quarters of 1993 and several pillars in the area spalled violently during this period, often accompanied by damage to the footwall caused by punching of the pillars. In the remainder of the analysis, only the 5S block is considered. The seismicity rate was reasonably constant up to December 1992, after which an increased rate was recorded. Figure 2.

The data (Table 1) show that seismic energy was generated at a variable rate with the highest in the first quarter of 1993 and in general, 1993 being more active than 1992.

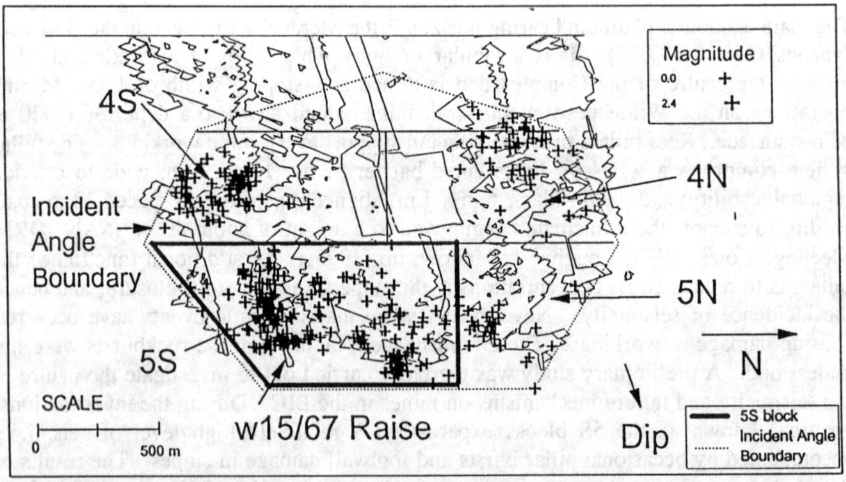

Fig 1: Epicentres of seismic events(M_L>0) within the incident angle area on the Wildebeestfontein North Mine - recorded between Jan 1992 - Sep 1993
The boundary of the area within which the incident angle between the hyocentres of seismic events located near reef and the geophone is less then 40 degrees is indicated and also the limits of 5S and surrounding blocks.

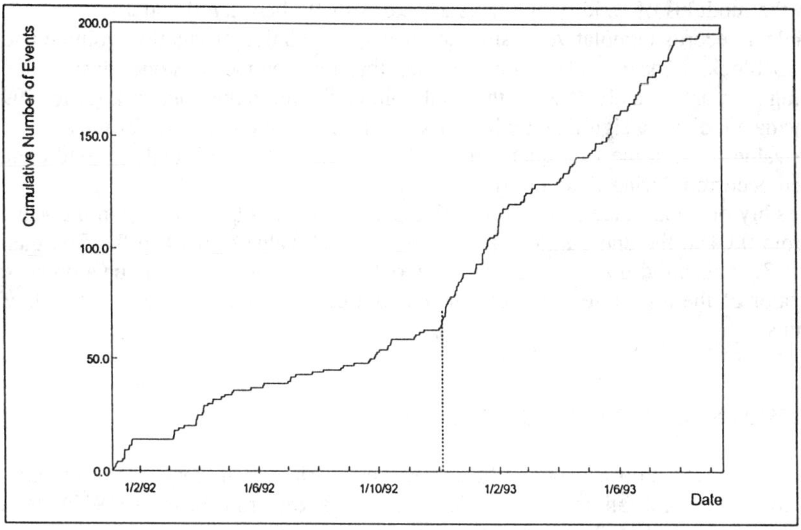

Fig 2: Cumulative number of events (M >0) in 5S block.
Note the elevated sesimicity rate from December 1992 onwards

4 THE VOLUME CHANGE METHOD

McGarr (1976) presented a method of equating the change in volume due to mining ΔV, with the seismic deformation produced by the volume change. The relationship is expressed as:

$$\sum Mo = \gamma_c . G . \Delta V \quad or \qquad \gamma_c = \frac{\sum Mo}{G . \Delta V}$$

Where $\sum Mo$ is the sum of the seismic moments of the population of seismic events,

Table 1. Seismic energy release rates for the 5S block

Period	Total Energy (kJ)	Area Mined (m^2)	Energy (kJ/m^2)	$\sum Mo$ (N-m)	γ_c
1st quarter - 92	2.70E+5	16 775	16.7	7.4E+12	0.9
2nd quarter - 92	1.65E+5	22 680	7.3	4.8E+12	0.04
3rd quarter - 92	0.91E+5	17 050	5.3	1.8E+12	0.12
4th quarter - 92	1.76E+5	20 141	8.7	1.8E+13	0.19
1st quarter - 93	3.75E+5	19 442	19.3	4.1E+13	0.44
2nd quarter - 93	1.94E+5	20 392	9.5	1.4E+13	0.14
3rd quarter - 93	2.25E+5	19 169	11.7	1.7E+13	0.18

G is the modulus of rigidity and γ_c is a factor usually between 0 and 1.

Table 1 records cumulative seismic moments, cumulative production volume and γ_c for 5S block. The γ_c calculation assumes that the volume of stope closure due to mining is approximately 15% of the total volume for the mined out area (Kriel 1993).

The data indicate a significantly higher seismic deformation in 1993 for the area. The high value of γ_c in the first quarter of 1993 was associated with underground damage which occurred during this period.

Possibly of importance from a predictive point of view is the build-up in the value of γ_c from 0.04 in the 2nd quarter of 1992, to the peak value of 0.44 in the first quarter of 1993. This build-up of γ_c from the 2nd to 4th quarter of 1992 could be a precursory indicator of the higher seismic deformation and concomitant damage in the following months.

5 ASSESSMENTS OF THE STATIC STRESS DROP

The static stress drop can be applied to express variations in rock condition and the potential for violent failure. To calculate the static stress drop, Brune's (1970) method, employing moment (M_o) and source radius (r_o) was used:

$$\Delta\sigma = \frac{7M_o}{16\, r^3_o}$$

In order to compare regions of consistently high or low stress drops, seismic events were categorized into stress drop intervals of 0 to 2 MPa, 2 to 8 MPa, 8 to 16 MPa and greater than 16 MPa and are plotted in Figure 3 for the period October 1992 to September 1993. There are only a few events with stress drops less than 2 MPa, the majority lie between 2 MPa and 8 MPa but a significant number have stress drops up to 16 MPa or more. They mainly cluster around the W15/67 raise line. For the first three quarters of 1992 the situation is reversed - 50% of events having stress drops less than 2 MPa. Table 2. The increase in the value of stress drop coincides with the period in which damaging events occurred.

The change in stress drop with time is shown in Table 2. Considering stress drops of >2 MPa it is seen that there is a steady increase throughout 1992 peaking in the first quarter of 1993.

The pattern in stress drop with time is similar to that for γ_c, both these parameters show a steady increase in value through 1992 to peak in the third quarter of 1993. The question is raised that if two independent parameters show the same trend do they together have a stronger predictive reliability than the parameters individually?

6 SEISMIC EVENT MECHANISMS

Mine seismologists usually assume that mine seismic events involve pure shear slip within an infinite homogeneous rockmass and seismograms are interpreted accordingly. There is, however, evidence that the mining stopes play an important role. Recorded seismograms (Spottiswoode & Churcher, 1988) are strongly affected by the close vicinity of open stopes. It has been proposed (Anon, 1988) that seismic events can be classified as "crush-type" when failure occurs in the immediate vicinity of the face, or

Table 2. Total number and percentage of events ($M_L > 0$) with stress drops falling within fixed intervals for the 5S block

Period	$0 < \Delta\sigma \leq 2$ (MPa)	$2 < \Delta\sigma \leq 8$ (MPa)	$8 < \Delta\sigma \leq 16$ (MPa)	$\Delta\sigma \geq 16$ (MPa)
1st quarter - 92	2 (40%)	3 (60%)	0 (0%)	0 (0%)
2nd quarter - 92	3 (50%)	2 (33%)	0 (0%)	1 (17%)
3rd quarter - 92	5 (56%)	2 (22%)	0 (0%)	2 (22%)
4th quarter - 92	9 (22%)	16 (44%)	10 (28%)	2 (5%)
1st quarter - 93	10 (25%)	17 (43%)	9 (23%)	4 (10%)
2nd quarter - 93	10 (24%)	24 (56%)	7 (25%)	2 (5%)
3rd quarter - 93	4 (16%)	15 (58%)	6 (23%)	1 (4%)

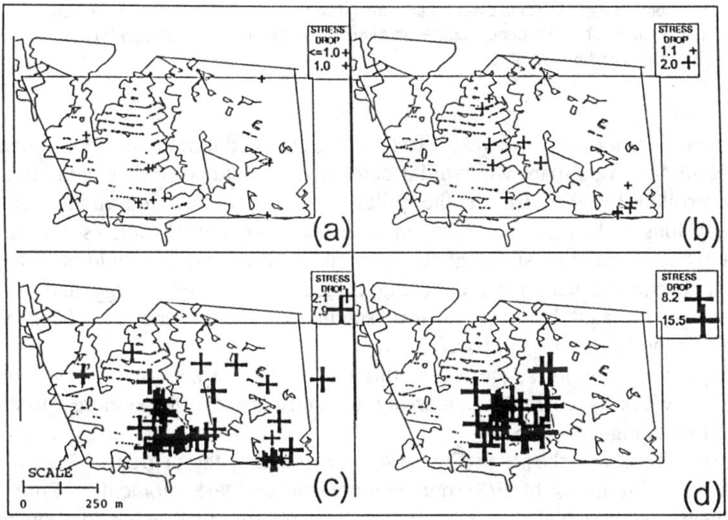

Fig3 : Plots of events ($M_L > 0$) with stress drop (in MPa) falling within various fixed intervals for 5S block Oct 1992 - Sep 1993.

when pillars fail. "Shear-type" events involve slip on a shear plane and are the idealised event usually assumed. Let us now consider the implication of "crush-type" or pillar failure on seismic radiation. Figure 4 is a sketch of shear and crush mechanisms. We envisaged that the seismic radiation of the crush event is enhanced in P waves, compared to shear events, in the following manner:

1. Increased contribution to P-wave seismic moment. Stope closure can exceed shear slip by an order of magnitude, and will appear as enhanced seismic moment as measured from the P waves.

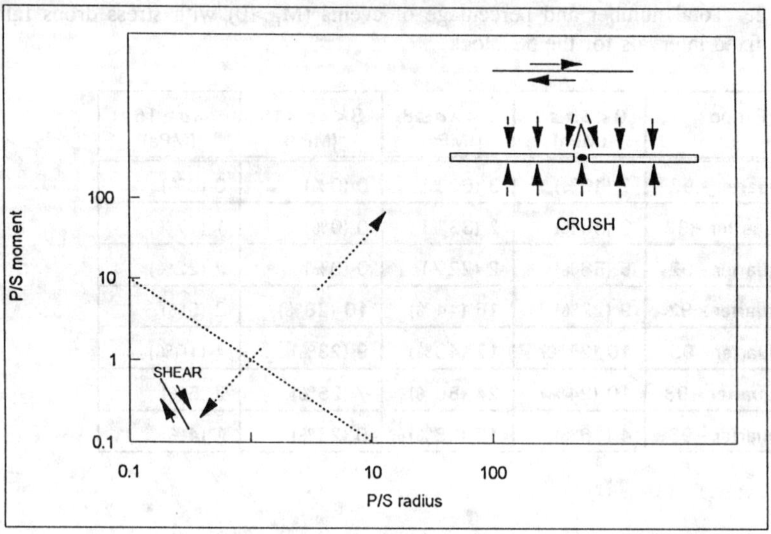

Fig 4 : P/S moment vs P/S radius diagram indicating conceptual regions for shear and crush mechanisms. The horisontal shear is an alternative interpreteation of high P content if it occured in the footwall.

2. Longer duration of P waves. The characteristic duration of stope closure is determined by the stope span, while pillar deformation could be complete within a time period controlled by the size of the pillar. Again, P-wave durations could be significantly longer than S-wave duration, and therefore source radii estimated from corner frequencies could be significantly enhanced by P waves. It should be noted that these P-wave enhancement have opposing effects on radiated energy and that the seismic energy radiated by stope closure should be smaller than the shear energy radiated from the failure region itself.

These effects are simply suggested in this paper and the analysis is based on this suggestion. Work is currently underway to study these effects more closely by numerical modelling.

Both these effects are shown in Figure 4 by expressing the degree of "crush-type" failure in terms of ratios of P/S source dimension and P/S moments. Pure shear seismic events are enriched in S wave component relative to P wave component and would plot in the area close to the (1,1) in Figure 4 whereas "crush-type" events would plot along a line pointing away from the origin. The P/S moment and radii ratios for events from the 5S block are plotted in Figure 5. Very few events plot in the region below line 10-10, describing pure shear behaviour. The dominant mechanism is thus crush-type failure. Note that the ratio of radii is limited to 10, which is the span/width ratio identification. The data for the mining area indicates that a range of different failure modes may exist. This concept of event mechanism derived from analysis of seismograms was then compared to underground observations.

Seismic events with high magnitudes and those which caused damage have been documented by mine personnel. It was found that in the majority of cases, the footwall of the stope was damaged by the seismic events (Kriel, 1993). Examination of damage and photographs of pillars have led to the hypothesis that punching of a solid

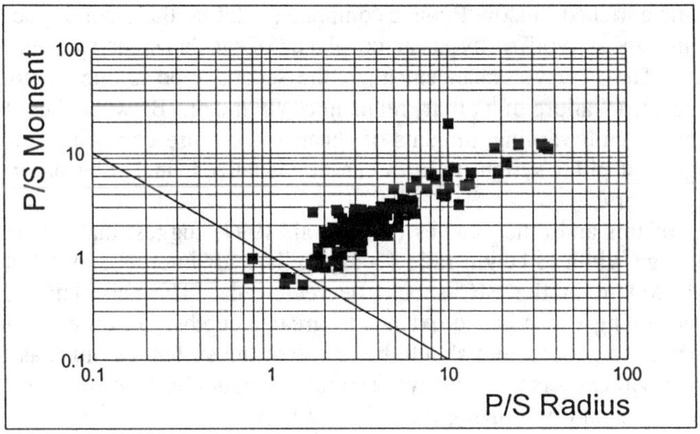

Fig5: Plots of events for 5S block on P/S moment vs. P/S radius diagram showing relative propensity for various mechanisms.

pillar core along two or more converging planes into the footwall causes the damage to excavations in the proximity of pillars. Other incidents involved violent pillar fracturing, spalling and blasting. The observations suggest that most of the damaging seismic deformation at Wildebeestfontein North Mine has been the result of pillar punching which is in accordance with the interpretation of seismic data.

However, because the geophones are located in a 50 m borehole drilled from surface the seismograms obtained are considerably more difficult to analyze and interpret than data recorded from an underground seismic network. Therefore, the proposed seismic event mechanisms should not be regarded as conclusive. Further research coupled with field observations are required to support the concept.

7 CONCLUSIONS

The combination of seismic assessment and underground observations at 5S block has resulted in an improved understanding of the seismic activity on the mine.

1. The increased seismicity is probably due to the build up of high load as mined out area expanded as indicated by the increasing values of stress drop encountered at greater depth with failure mainly being related to the pillar/hangingwall/footwall interaction.

2. Values of the parameter, γ_c, were found to vary systematically, with the highest values calculated for the first quarter in 1993. This period corresponds with elevated seismic activity and damage in the 5S area. The gradual increase in the value of γ_c with time could possibly be used as an indicator of future more severe seismicity.

3. The stress drops for recorded seismic events vary in a similar way to γ_c. The highest stress drops were recorded during the period when damaging seismic activity took place. Thus the increase in average stress drop with time could also possibly be used to predict future higher levels of seismicity.

4. Based upon source parameters, two types of seismic event mechanism have been proposed. Seismic events with a large S-wave component characterize the first type,

whereas events enriched in slow P-wave components define the second type. The first type of mechanism is typified by shear on planes of weakness or rockburst fracture, shear events. The P wave enhancement of the second type is due to stope closure consequent on shear failure of a pillar, remnant or abutment. Between these two limits, failure can occur with varying amounts of shear and closure components for which a range or degree of pillar system failures can be visualized, including pillar punching, as frequently observed.

The results of this and other studies (Aref et al, 1994) suggest that pillar design for mines using large width to height ratio (>3:1) pillars for internal support at depths of more than 1 000 m in the BIC should be reviewed. The seismicity is likely to deteriorate as mining expands and extends to greater depths. In order to manage the seismic activity, the general aim should be to soften the system as much as possible in order to reduce violent seismic activity. The mine layout could be softened by making use of crush pillars and optimizing the effective span.

ACKNOWLEDGEMENTS

The work described in this report forms part of the SIMRAC project SA9305. The co-operation of staff at Wildebeestfontein North Mine is greatly appreciated. The authors would also like to thank Impala Mines for permission to publish this paper.

REFERENCES

Anon. 1988. *An industry guide to methods of ameliorating the hazards of rockfalls and rockbursts*. Chamber of Mines Research Organization.

Aref, K., Spottiswoode, S.M. & Jager, A.L. 1994. A comparison of seismicity from two mines in the Bushveld Igneous Complex using different pillar systems. *The 1994 ISRM International Symposium and VI South American Congress on Rock Mechanics*. Santiago, Chile.

Brune, J.N. 1970. Tectonic stress and the spectra of shear waves from earthquakes. *Journal of Geophysical Research* 75, 4997-5009.

Kriel, G., & Spencer, D. 1993. Personal communications.

McGarr, A. 1976. Seismic moments and volume changes. *Journal of Geophysical Research*, Volume 81, 1487-1494.

Mendecki, A.J. 1993. Real time quantitative seismology in mines. *3rd International Symposium on Rockbursts and Seismicity in Mines*, Kingston, Canada. 287-297.

SANGORM Meeting. 1993. Kotze, T.J. Lougher, D.R., & Mellowship. P., Spencer, D.A. *Symposium on Rock Engineering Problems Related to Hard Rock Mining at Shallow to Intermediate Depth*. Rustenburg.

Spottiswoode, S.M., and Churcher, J.M. 1988. The effect of backfill on the transmission of seismic energy, Backfill in South African Mines, *SAIMM special publication series SP2*, 203-221.

Rock Mechanics, Nelson & Laubach (eds) © 1994 Balkema, Rotterdam, ISBN 90 5410 380 8

A conceptual model of a hard rock, deep level, tabular ore body that incorporates the potential for face bursting as a natural product of mining

N. Lightfoot & D. H. Kullmann
Division of Mining Technology, CSIR, Johannesburg, South Africa

A. R. Leach
Anglo American Corporation of South Africa, West Rand Region, South Africa

ABSTRACT: A discontinuum model has been developed to simulate the highly fractured rock mass around a deep level gold mine stope. The model has been used to investigate mechanisms of face bursting particularly when mining towards natural joints. Numerical simulations have been compared to a field study of face bursting and it has been shown that natural joint orientations have an effect on the potential for face bursting in deep level mines.

1. INTRODUCTION

Mining of tabular ore bodies causes the redistribution of the super-incumbent load carried by the in situ rock mass. At depth the stresses on the mine abutments are large and result in fracturing of the previously intact rock around the stope. The intensity, and extent of the mining induced fracture zone is a function of depth, mining, rock strength and geology. On deep level gold mines, where this fracture zone is particularly well developed, seismicity has increased when mining with faces parallel to dominant fractures. For example, in scattered mining, remnants are frequently created which are generally elongate in the dip direction. Fractures develop parallel to the long axis of the remnant, and the optimum mining method is to advance up-dip, with the mining facc lying perpendicular to the fracturing (Berrange, 1956). In addition to the mining induced fractures natural jointing also affects ground conditions. The incidence of seismic events has been observed to increase when mining with faces parallel to major joint sets.

Western Deep Levels South Mine (WDLS) has successfully applied a strategy of up-dip mining on problem longwalls for some years now (Hobday and Leach, 1991). Recently, detailed joint mapping has been carried out on one particular longwall at WDLS, where seismicity caused considerable problems during mining in 1991. The aim of this paper is to quantify the influence of jointing and face shape on seismicity through analyses of the geology and mining history and the use of numerical modelling.

2. DEVELOPMENT OF A DISCONTINUUM MODEL

During the period 1980-1987, Chamber of Mines Research Organisation (now CSIR Miningtek) personnel carried out detailed mapping of mining induced fracturing (Brummer, 1987) which was used to establish a model to describe the geometry and intensity of fracturing around deep narrow stopes (Figure 1). An important consideratioi

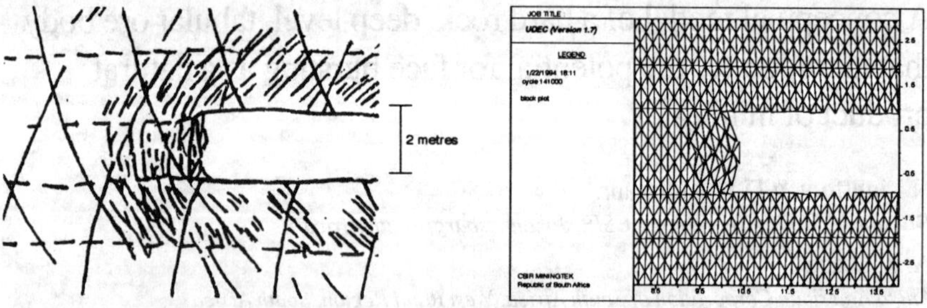

Figure 1. The conceptual model of the fracture zone around a deep level gold mine proposed by previous researchers (left) represented as a blocky rock mass for simulation with the Distinct Element Code UDEC (right).

in this model is that by the time the rock is exposed in the stope, it has already failed; that is the fractures already exist.

The model recognised the highly discontinuous nature of the rock mass and, by implication, also recognised that the response of the rock mass around the mined out stope was controlled not by the characteristics of the intact rock, but by shear on, and extension of, the discontinuities themselves. Representation of this fracture zone for numerical analyses is usually as an inelastic continuum that is allowed to fail according to a constitutive law governed by a shear failure envelope and tensile strength (e.g. Kirsten and Howell, 1993). In some cases, preferential failure orientations may be allowed by means of an ubiquitous joint constitutive model. A limited number of discontinuities may be represented as crack type constructs (e.g. de Villiers, 1989).

The Distinct Element Method (DEM) (Cundall, 1971) which is capable of representing the rock mass as a true discontinuum offers a number of advantages over the more common continuum and discontinuity analyses such as large strains and displacements, block rotations and global rock mass behaviour that is dependent upon the discontinuities that exist in the rock. The DEM requires prior definition of discontinuities but this may be viewed as equivalent to the fractures existing before their exposure at the mining face.

Using the DEM a conceptual model of the rock mass surrounding a deep level stope has been developed based upon the field observations reported in Brummer (1987) and those of the authors. The model assumes the rock mass immediately surrounding the stope is composed of an assemblage of tightly packed, triangular blocks. The blocks interact with each other through their boundaries according to a joint constitutive law, defined in terms of a friction angle, cohesion and peak tensile strength. In the implementation described here four fracture orientations exist, horizontal, vertical and a conjugate set aligned 30° about the vertical. The choice of these fracture sets and their orientations are based on underground observations. The symmetric conjugate fracture set introduces a potential for bifurcation when subjected to a coaxial oriented biaxial stress field. Which of the conjugate surfaces slides is indeterminate and in simulations is dependent, initially, on numerical round-off error. This is equivalent to having minute perturbations in the rock or joint strength in the physical rock mass.

In order to investigate the behaviour of the blocky rock mass a stope model was subjected to simulated incremental mining using the DEM code UDEC (Itasca, 1992). The fractures in the model were cohesionless Mohr-Coulomb surfaces with a friction angle of 30°. The discontinuum was surrounded by an elastic continuum to which the boundary conditions were applied. Axes of symmetry were applied to the vertical boundaries of the mesh. Five mining steps were incrementally advanced from right to left into the discontinuum. The resultant stress distribution ahead of the face is complex (Figure 2) but the trend is towards a destressed face with stresses increasing through the fracture zone to some peak and then decaying to the primitive stress level. However, stress concentrations exist at points close to the mining face. These concentrations are caused by 'lock-up' at block corners that prevent slip. Such 'lock-up' is an inherent instability and may form the initiation mechanism for a face burst.

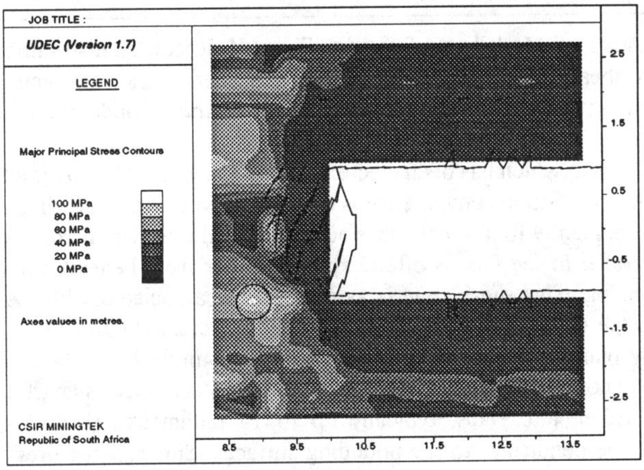

Figure 2. UDEC simulation of the stope advance after 5 mining steps showing contours of the major principal stress. Notice the stress concentration (circled) at the footwall contact horizon approximately 1 metre ahead of the current face position.

The blocky rock mass model of the fracture zone has been used to investigate mechanisms for preconditioning of stope faces to reduce rockbursts (Lightfoot, 1993, Adams, et al., 1993), and was able to demonstrate a mechanism for dissipating energy from the rock fracture zone as a result of the application of transient explosive gas pressures within the fracture zone.

More refined simulations indicate that blasting close to the stope face in the fracture zone should have the most beneficial effects on stress redistribution by preconditioning. Blasting in more highly confined rock close to the stope face, is generally ineffective, and could result in increasing the stress on the face (Figure 3). Seismic monitoring of a preconditioning site has shown shifts in seismicity both towards and away from the face (Figure 4) indicating stress redistributions similar to those simulated by the modelling.

3. FIELD OBSERVATIONS OF ROCKBURST DAMAGE AND IN SITU JOINTING

WDLS is mining the Ventersdorp Contact Reef, a gold bearing conglomerate band typically up to 1 m thick, lying on an unconformity, underlain by quartzites and overlain by lavas. Dip is 21° to the Southeast.

When faces have been mined on strike, parallel to jointing, and seismic events in the magnitude range 1 to 2 have occurred, rockburst damage at the stope face occurs as a

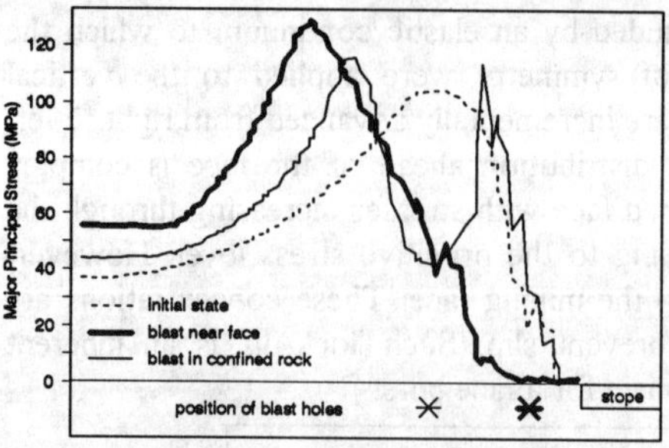

Figure 3. Results of Modelling of stress redistribution ahead of the stope due to preconditioning at different positions.

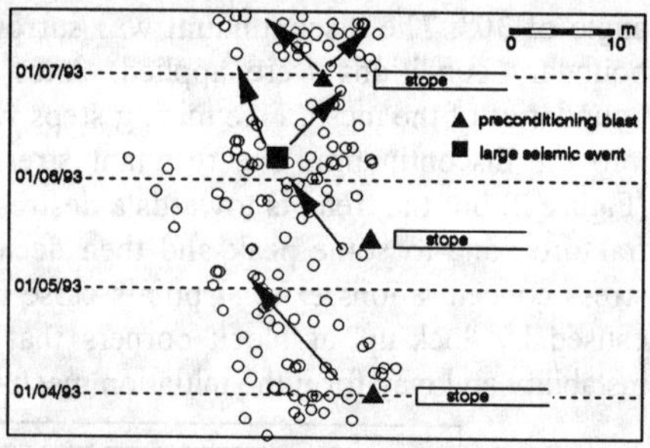

Figure 4. Recorded distribution of seismic events ahead of an advancing stope face subjected to preconditioning.

pile of reef material, which has been ejected from the face, overlain by joint bound blocks of lava hangingwall. Stope convergence is rarely observed. After clearing the debris from the face, it is frequently found that it has advanced well over 1.5 metres. A steep dipping shear zone parallel to the face is often exposed, either from beneath the debris or after subsequent mining. This shear zone is assumed to be associated with the seismic event which caused the damage and may be complex in structure (Figure 5). In places, it may be completely planar, apparently following a pre-existing joint surface. Elsewhere, the shear zone may comprise two surfaces spaced 10 to 15 centimetres apart, between which the rock is broken into slabs, typically up to 15 millimetres thickness, by fractures running nearly perpendicular to the bounding surfaces. Similar structures were noted by Gay and Ortlepp, 1979 and Ortlepp, 1978 while mapping a mining induced fault zone at

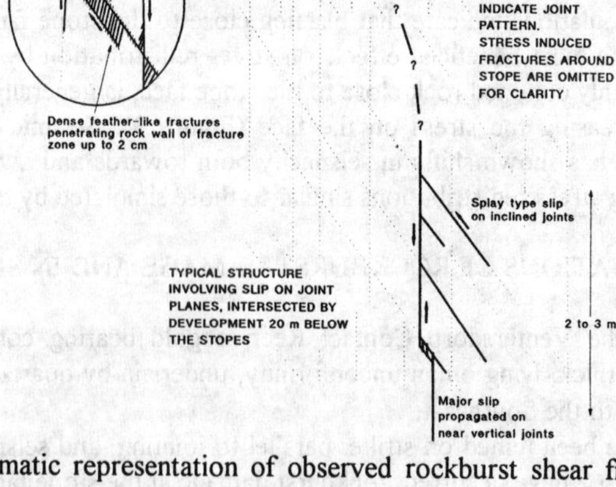

Figure 5. Schematic representation of observed rockburst shear fractures. Note: both senses of shear have been observed ahead of the face.

ERPM. Similar structures have been noted on various scales in faulting (Segall and Pollard, 1983) and earthquake ruptures where they have been referred to as jogs (e.g. Sibson, 1985). In some instances at WDLS the bounding surfaces appear to have been closely spaced joint planes, less frequently they seem to have developed through rupture of intact rock, as described by Gay and Ortlepp at ERPM. These zones have been observed to extend continuously over several metres in the direction of slip (usually vertical), disappearing into the stope hangingwall and footwall. Horizontally they may be exposed for up to 10 metres along a face before terminating abruptly and changing to a single rupture plane or a parallel zone of shattered rock some short distance away. Characteristically, major fractures within a shear zone show several millimetres of white powdered gouge infilling, and the rock on either side of the principal slip surfaces may be intensely fractured to a depth of approximately 1 centimetre by fine parallel feather-like fractures inclined at 45° to 60° to the main slip surface. Such fractures have been observed in natural (Segall and Pollard, 1983) and mining-induced faulting (Gay and Ortlepp, 1979, McGarr et al, 1979).

The interaction between the shear zones and the extension fractures which develop ahead of a mining face is highly variable. In places the shear fractures displace the extension fractures indicating that shear occurred within the pre-existing fracture zone. Crushed rock at the corners of blocks adjacent to such shears is consistent with the notion of violent release of 'lock-up' on pre-existing fracture surfaces. Shear zones have been seen to move around corners, particularly where one mining panel leads the next. Short sections of shear zone are joined by jumps on cross jointing. For shear zones to be affected by such local changes in panel geometry they must form close to the mining face , perhaps on the edge of the mining induced fracture zone, no more than 15 metres ahead of the face.

Off-reef extensions of shear zones have been intersected by follow-behind footwall tunnels which were developed 20 metres below the longwall. The shears occur as single gouge-filled planes, all following near vertical joint surfaces. Splay type movement may be seen along other joint sets which obliquely intersect the vertical set. The structures away from the reef plane, have much in common with mining-induced seismicity resulting from slip on pre-existing fault planes, and are distinctly different from the mining-induced shear mapped by Gay and Ortlepp (1979), which was interpreted as having originated by rupture of intact rock.

Joint mapping was carried out in roadways and portions of footwall haulages in the 84-49 West longwall at WDLS, where past seismicity may have been joint related .The contoured stereographic projection of the poles of measured joints (Figure 6) shows four principal sets, three nearly vertical, the other dipping steeply to the East.

Joint set	Strike	Strike Variation	Dip	Dip Variation
1	138	120-161	86SW	67SW-82NE
2	95	86-105	87N	86S-80N
3	160	145-187	63E	40-75E
4	15	2-27	88E	81E-Vertical

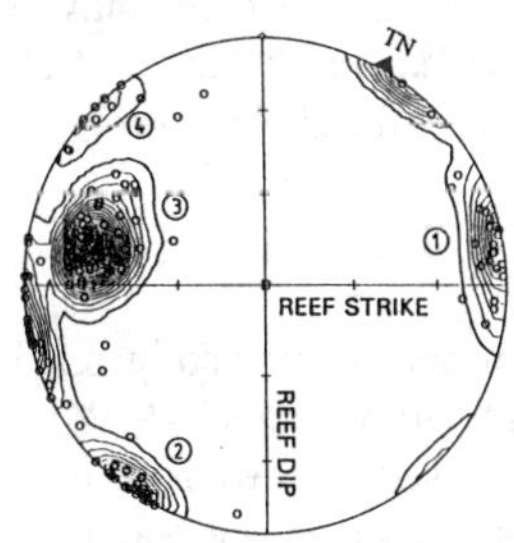

Figure 6. Joint set orientations mapped in the WDLS 84-49 West longwall roadways and haulages (the stereonet is a lower hemisphere projection).

On reef, the joint survey was conducted along roadways which were frequently over 3 metres high, hence few readings were taken from joint surfaces in the lava hangingwall. In the quartzite footwall, the orientation of joint set 3 made it almost indistinguishable from the mining induced fractures. However, this set was clearly detected through mapping in the footwall haulages which were developed in overstoped, de-stressed ground. The four sets correspond well with those mapped previously in the hangingwall lavas 1500 metres Southeast of the 84-49 West longwall (Cambitzi, 1987), indicating that the same sets developed throughout the region, irrespective of the host rock.

4. SIMULATIONS OF STOPING TOWARDS A JOINT

The discontinuum model was used to investigate the influence of geological joint sets and the mining induced fracture zone on the incidence of face bursting. Incremental mining was simulated in five mining steps in geologically undisturbed ground (Figure 7). At mining step five a small face burst was initiated as a result of strain energy stored in the fracture zone. Ejection velocities were around 0.2 ms^{-1}. The simulation was repeated with a footwall joint ahead of the advancing face (Figure 8). As the fracture zone intersected the joint the joint slipped and a face burst occurred with ejection velocities more than double those in the unjointed case and more of the face was mobilised.

Figure 7. UDEC simulation of stope advance in unjointed ground. The figure to the left shows the blocky rock mass and vectors of face ejection velocities and that to the right shows the shear displacement on the joints (magnitude is proportional to line thickness).

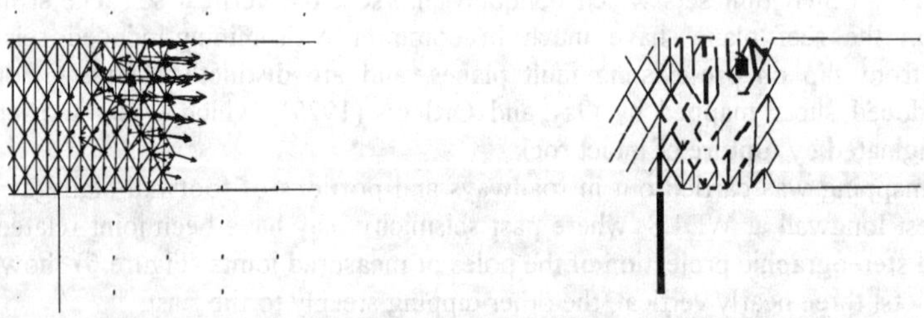

Figure 8. UDEC simulation of stope advance towards a face parallel joint in the footwall. This results in significantly more ejection at higher velocities (left plot) and greater shear displacements (right plot) close to the hangingwall than in the previous simulation (see Figure 7 for explanation).

5. CASE STUDY

High seismicity and ground control problems attributable to joint orientations were identified on the 84-49 West longwall at WDLS in early 1991. At that time the longwall comprised a series of 30 metre panels advancing West on strike with a slightly underhand longwall shape (i.e. each panel lags behind the adjacent up-dip panel). Frequently, face bursts occurred and it was necessary to advance a roadway heading and re-establish a panel by mining a 10 metre wide up-dip face following a face burst. Face bursts tended to affect the upper panels in the longwall, resulting in a higher rate of advance in the lower panels. By January 1992 the bottom four panels were in-line with each other, and parallel to joint sets 1 and 3. Early in 1992, the seismicity on the longwall reached a peak with several large magnitude events (Figure 9) when all faces in the longwall came in line parallel to jointing. By mid 1992 a decision was made to radically change the longwall shape such that faces were no longer advancing parallel to jointing. All panels stopped mining while a heading was advanced at the bottom of the longwall and a series of six up dip faces, each typically 11 metres wide, was created. During this period the seismicity drastically reduced because of reduced production. Full production on the longwall was regained in 1993, and the rockburst problem was virtually eliminated.

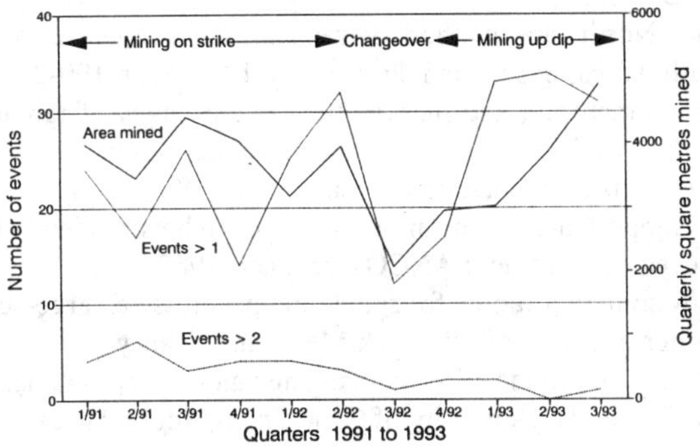

Figure 9. Quarterly seismicity and production on the 84-49 West longwall.

During 1993 production with the up-dip method has progressively increased, and seismicity has reached a plateau of nearly 50 events of magnitude 0.8 and greater per month, where it has remained, despite further increases in the area being mined. In terms of number of events this level of seismicity matches that of early 1992. However, the seismic network has been improved gradually during this period and some events may have been missed in 1992. The frequency of events larger than magnitude 2 has significantly lower in 1993 than in 1991 and early 1992, and, in general the magnitude of events has decreased. WDLS intend to continue mining up-dip panels with faces perpendicular to dominant jointing to help reduce the rockburst hazard associated with mining at depth.

6. CONCLUSIONS

The model demonstrates a mechanism for increasing the face burst hazard when mining towards face parallel joints. Experience on WDLS confirm that mining breast panels with faces parallel to major joint sets results in frequent face bursting. The frequency of face bursting was reduced when the mining geometry was changed to ensure that faces ran perpendicular to the main joint sets.

It is important to take cognisance of the orientation of major joint sets during mine planning on deep level mines to ensure that faces are not aligned parallel to them. Face preconditioning by means of the introduction of transient gas pressures through blasting in the fracture zone could further reduce the hazard of face bursting under the conditions outlined in this paper.

7. ACKNOWLEDGEMENTS

Much of the work presented in this paper forms a part of the SIMRAC project GAP030 (Rockburst Control) being undertaken by the Division of Mining Technology CSIR. The authors would like to thank the Management of Western Deep Levels Limited (South) for permission to publish some of the details included in this paper.

8. REFERENCES

Adams D.J., Gay N.C. and Cross M. 1993. Preconditioning - A technique for controlling rockbursts. Proc. 3rd Int. Symp. On Rockbursts And Seismicity In Mines, Kingston, Ontario, Canada. Balkema.

Berrange C.L. 1956. North face method of remnant extraction. Association of Mine Managers of South Africa, papers and discussions, 1956-57, p 199-215.

Brummer R.K. 1987. Fracturing and deformation at the edges of tabular gold mining excavations and the development of a numerical model describing such phenomena. Ph.D. thesis, Rand Afrikaans University, Johannesburg, South Africa.

Cambitzi C. 1987. Support design of large winder chambers at depth on No.1 shaft, Western Deep Levels Gold Mine. SANCOT seminar, 1987.

Cundall P.A. 1971. A computer model for simulating progressive, large-scale movements in blocky rock systems. Symp. of Int. Rock Mechanics, Nancy.

de Villiers N.A. 1989. Beam models for the hangingwall of deep, tabular excavations in stratified rock. M.Sc. thesis, University of Cape Town, South Africa.

Gay N.C. and Ortlepp W.D. 1978. The anatomy of a mining-induced fault zone. Bulletin Geol. Soc. Am., part 1, vol 90, p 47-58.

Hobday H.M.D. and Leach A.R. 1991. Some case studies in rock engineering application on Western Deep Levels South mine. Sangorm symposium: impact of rock engineering on mining and tunnelling economics, September 1991.

Itasca Consulting Group. 1992. Universal Distinct Element Code. Version 1.8. User's Manual.

Kirsten H.A.D. and Howell G.C. 1993. Effect of depth, span, stope support, and discontinuity strength on the stress and stiffness characteristics of the fractured rock around a tabular mining excavation. J. S. Afr. Inst. Min. Metall., vol. 93, no. 10. Oct. 1993 pp. 253-266.

Lightfoot N. 1993. The use of numerical modelling in rockburst control. Proc. 3rd Int. Symp. On Rockbursts And Seismicity In Mines, Kingston, Ontario, Canada. Balkema.

McGarr A., Pollard D.D., Gay N.C., and Ortlepp W.D., 1979. Observations and analysis of structures in exhumed mine induced faults, Analysis of Actual Zones in Bedrock, U.S. Geological Survey, Open File Report, 79-1239, 101-120.

Ortlepp, W.D. 1978. The mechanism of a rockburst. 19th U.S. rock mechanics symposium, Stateline, Nevada, May 1978.

Segall P. and Pollard D.D. 1983. Nucleation and Growth of Strike slip faults in granite. Journal of Geophysical Research, vol 88, p 555-568.

Sibson R.H. 1985. Stopping of earthquake ruptures at dilational fault jogs. Nature, vol 316, p 248-251, July 1985.

Rock Mechanics, Nelson & Laubach (eds) © 1994 Balkema, Rotterdam, ISBN 90 5410 380 8

Numerical modelling of the behaviour of concrete stabilising pillars compared to reef pillars in deep level narrow reef mining

G. York & G. L. Smith
CSIR Miningtek, Johannesburg, South Africa

ABSTRACT

Reef and concrete stabilising pillars are modelled in the two dimensional plane strain code FLAC. The rock mechanics criteria used for comparison are hangingwall stability, pillar stress distribution and stope closure profile, as well as the punching effect of the pillar into the foundation. Foundation failure is qualitatively the same as that shown by Wagner and Schümann (1971). Results show that hangingwall stability may be promoted, and the punching effect much reduced, by the use of a concrete pillar, thus making the use of a concrete pillar justified.

1 INTRODUCTION

In the deep level South African gold mining industry, some 15 to 20 per cent of gold bearing reef is left behind in the form of stabilising pillars. Concrete pillars can be used as a means of obtaining a 100 per cent extraction ratio. Smith and York (1993) discussed the economic details as well as the system requirements for the placing of concrete pillars.

The aim of this paper is to compare the rock mechanics behaviour of concrete stabilising pillars to that of reef stabilising pillars by the use of the numerical modelling code FLAC. FLAC, a two dimensional plane strain code, is ideal for modelling strip pillars of infinite length. The factors to be considered are hangingwall conditions adjacent to the pillar, the overall pillar stress profile and the overall closure profile. The punching effect of the two materials into the hangingwall will also be compared. This will be done by modelling the inelastic strain softening of the material.

The modelling of the failure mode of the pillars is based upon observations and analyses of punch tests on laboratory rock specimens by Wagner and Schümann (1971). Wagner and Schümann showed that a circular punch does not affect the ultimate failure in compression and

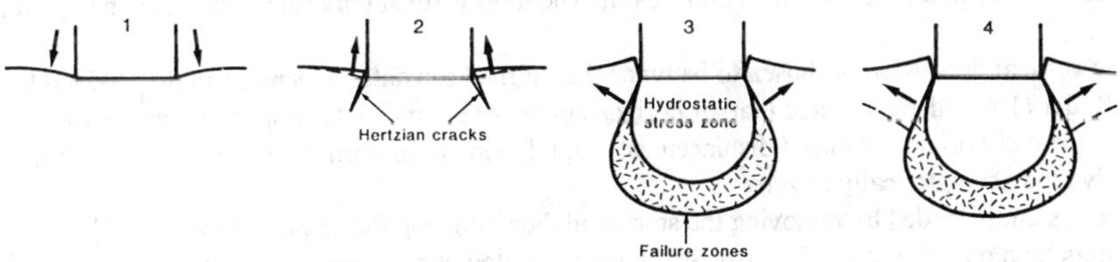

FIGURE 1 : Failure of laboratory rock specimen under punch loading (after Wagner and Schümann, 1971)

shear, when the failure lobes meet under the punch (Figure 1). The circular punch test can therefore be used to give an indication of the failure mode of strip pillars.

Elastic analyses are insufficient to study and model the failure of the rockmass. As the rockmass undergoes plastic strain, the values of cohesion, friction angle and dilation angle change. The material in the hangingwall was thus allowed to strain soften/harden, while the pillar was modelled as a Mohr-Coulomb material.

2 PILLAR STABILITY AND FOUNDATION FAILURE

The commonly assumed mechanism for foundation failure is that of fracture lobes meeting under the foundation as shown in Figure 1 (Wagner and Schümann, 1971). In a series of stamp load tests, the failure load tended to about four times the uniaxial compressive strength (UCS) of the material as the stamp diameter increased. A figure of three times the UCS of the foundation material has been adopted by the South African gold mining industry as a criterion for the design of strike stabilising pillars. More recently, subsequent to further analysis on pillar related rockbursts, a figure of 2,5 times the UCS has been recommended (ANON, 1988).

Elastic theory shows that pillars need to be 20 m wide at a dip spacing of 140 m to avoid stresses higher than three times the UCS at a depth of 3000 m (Cook *et al*, 1973). However, it has been found in practice that such narrow pillars become fractured throughout their width (Hagan, 1987), thus reducing their load carrying capacity and increasing the risk of failure. Other workers have shown that fracturing seven metres in from the edge of abutments can reasonably be expected (Brummer, 1984). Due to these inelastic effects, most strike stabilising pillars in South African gold mines have a width to height ratio of approximately 40:1, with stoping widths of 1 to 1,5 m.

Hanging- and footwall stability adjacent to reef pillars has been a problem in the use of stabilising pillars, particularly with relatively narrow pillars of width to height ratios of less than or equal to 20:1 (Hagan, 1987). Hard reef pillars have the effect of punching or shearing into the footwall and/or hangingwall (Lenhardt and Hagan, 1990). Fracture growth in the footwall can make gullies unstable and fracture growth further into the hangingwall or footwall can cause damaging seismic events (Ozbay and Ryder, 1989).

3 MODEL GEOMETRY AND BOUNDARY CONDITIONS

The infinitely repeating strike stabilising pillar geometry as modelled in two dimensions is assumed to be in plane strain. The geometry lends itself to vertical lines of symmetry at the stope and pillar centre lines. Dip has been ignored, but will be included in later studies. The lack of dip allows horizontal symmetry through the centre line of the pillar. These lines of symmetry allow one quarter of the full repeating unit to be modelled (Figure 2). The left, right and bottom boundaries are axes of symmetry (i.e. no normal displacement is allowed), while the top boundary is a vertical stress boundary of 80 MPa.

A stoping width of 1,2 m is assumed and the pillar is 48 m wide (40:1 ratio). An extraction ratio of 85 per cent was chosen, yielding a stope span of 272 m. Thus in the model the half span is 160 m, with the pillar half width equal to 24 m. The total vertical dimension into the hangingwall is 320 m.

The vertical dimension is chosen to be twice that of the horizontal dimension in the model. Ozbay and Ryder (1990) demonstrated that, at horizons greater than twice the stope half span above or below the reef horizon , induced displacements rapidly tend to uniformity while induced stresses rapidly tend asymptotically to zero.

The system is loaded by removing the stope and then allowing the stope to converge. This transfers load onto the pillar. The stresses in the stope do not disappear in one step, but are gradually redistributed (by means of FLAC FISH routines) into the surrounding rockmass so as not to shock the system and lead to spurious results.

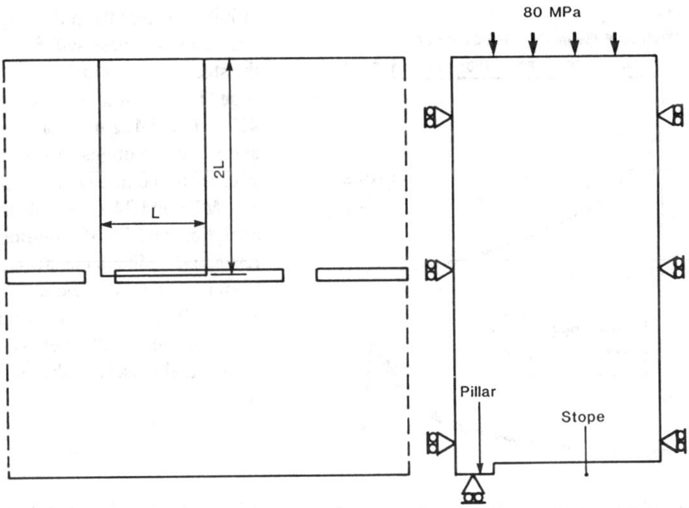

FIGURE 2: Repeating pillar layout (cross section) and model boundary conditions

4 MATERIAL PARAMETERS

Strain softening curve fitting was performed on a set of Witwatersrand quartzite laboratory test results. The Mohr-Coulomb and strain softening parameters are shown in the Addendum. There is some debate as to what degree the softening parameters should be downgraded to more closely mirror the real *in situ* values. The parameters as they stand are deemed sufficiently representative for the purposes of a comparison between the performance of concrete and reef pillars.

At this stage, full triaxial cell curves are unavailable for the particular concrete modelled, so that strain softening parameters could not be obtained for concrete. The concrete parameters are those used by Smith and York (1993), where the aggregate was obtained from the waste product of a particular mine. Further parameters required for the Mohr-Coulomb model have been obtained by means of a literature search.

The concrete pillar is modelled as a Mohr-Coulomb material. For comparison purposes, the reef pillar is also modelled as a Mohr-Coulomb material, even though strain softening parameters are available for the quartzite. In both cases, however, the hangingwall is modelled as a strain softening material.

5 RESULTS

5.1 Elastic Results

The thickness and orientation of shear bands as simulated in a code such as FLAC are to some extent mesh dependent. The elastic FLAC solutions for pillar stress and stope closure were compared to analytic elastic solutions based on Salamon's equations (Salamon, 1968). An important factor to be considered is that the analytic results predict that the pillar stresses tend to

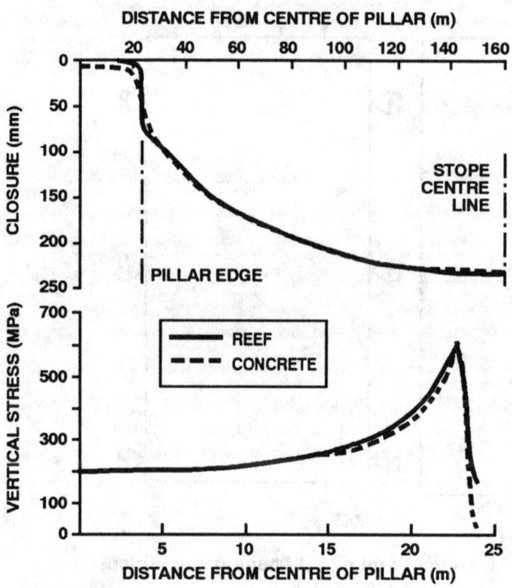

FIGURE 3: Stope closure profiles and pillar stress distributions for reef and concrete pillars

infinity towards the pillar edge. A coarse FLAC mesh will fail to reflect the steep stress gradient near the pillar edge. The maximum stope closure was 427 mm and 422 mm for the analytical and FLAC solutions respectively. The stresses in the middle of the pillar were 342 MPa and 340 MPa for the analytical and FLAC solutions respectively, while the stresses 0,45 m from the pillar edge were 1746 and 1773 MPa respectively. The FLAC elastic model results therefore correspond closely to the analytic results.

5.2 Strain Softening Results

The material in the hangingwall is strain softening, while the reef and concrete pillars are Mohr-Coulomb. The model was run until foundation failure occurred, that is, until the failure lobes extended under the pillar to the vertical centre line of the pillar. The mode of failure is the same as that of laboratory specimens under punch loading (Figure 1) where failure lobes have met under the foundation.

5.2.1 Pillar Stress Distribution

The concrete pillar needs to withstand the same pillar loads that the reef pillar withstands, without causing greater stope closure and hangingwall damage. The stress distributions in the concrete and reef pillars are similar, except at the edge of the pillar (Figure 3). A stress drop off at the edge of both pillars can be seen, signifying crushing of the material at the pillar edge. The result of this is that the peak load is moved away from the edge of the pillar. As can be expected, the concrete sustains less load at the edge of the pillar. The average pillar stress (APS) of the reef pillar is 273 MPa, while the APS of the concrete pillar is 263 MPa. This small difference indicates that the concrete pillar will reach this stage slightly earlier in the mining cycle than will the reef pillar, and that the concrete pillar will reach similar stress levels at greater strain.

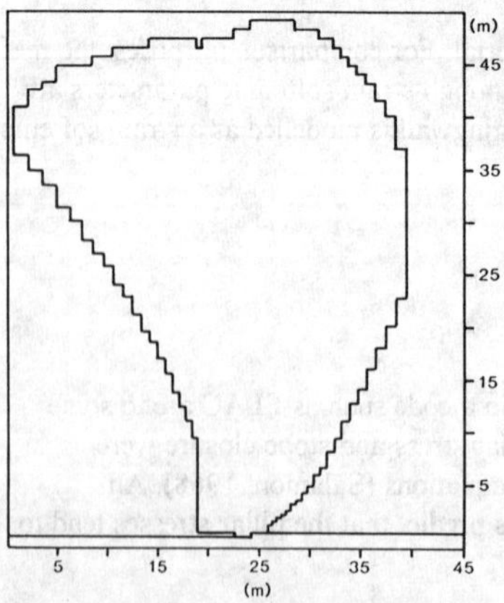

FIGURE 4: Failure lobes in reef pillar foundation

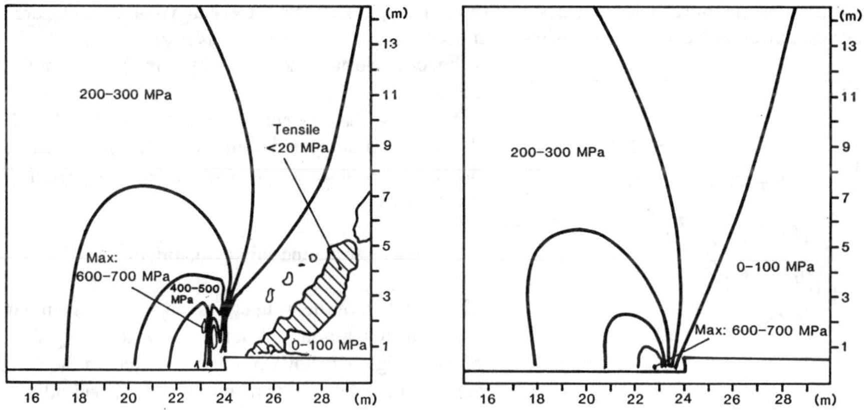

FIGURE 5: Vertical stresses at the edge of the pillar for reef (left) and concrete (right) pillars

5.2.2 Stope Closure Profile

Evidence exists that the overall amount of mine seismicity is related to the amount of elastic volumetric closure (McGarr and Wiebols, 1977). Therefore, the closure allowed by the concrete pillar should not be significantly greater than that for the reef pillar. The concrete pillar itself deforms much more in the vertical direction than the reef pillar due to its smaller Young's modulus. Despite this, the closure at mid span is almost the same for both pillars. The volumetric closure is 25,75 m^3 for reef pillars and 26,06 m^3 for concrete pillars, a difference of 0,31 m^3. In this case, due to the two dimensional nature of the model, cubic metres refer to the volume per unit metre in the third (strike) direction. So as mining advances 100 m, the difference would be 31 m^3. These figures include displacement in the pillar, thus including the much larger vertical deformation of the concrete pillar shown in Figure 3. If only the stope closure is considered, then the volumetric closures are 25,81 and 25,72 m^3 for the reef and concrete pillars respectively, a difference of 0.09 m^3. For the full stope span, this would be 0,18 m^3.

This difference is small compared to the volume mined (less than 0,3 per cent). Bearing in mind

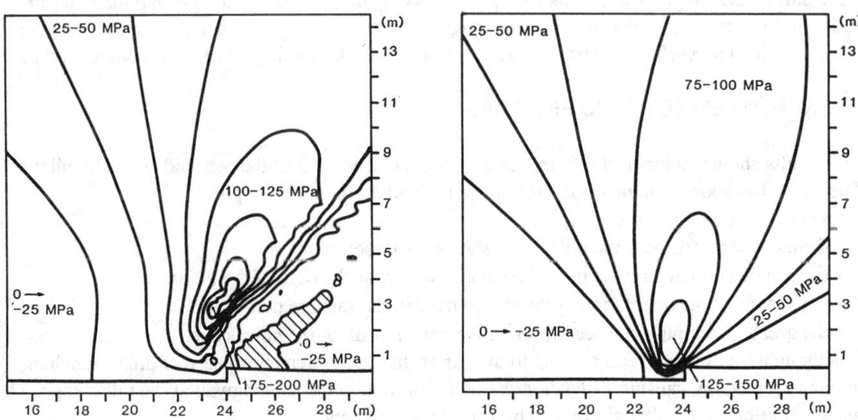

FIGURE 6: Shear stresses at the pillar edge for the reef (left) and concrete (right) pillars

that this closure figure includes inelastic closure, it can be seen that the elastic volumetric closure is very similar in both cases. This indicates that, to the extent that seismicity is related to elastic volumetric closure, the seismicity is the same in both cases.

There is a sharper increase in closure at the edge of the reef pillar compared to the concrete pillar, because the reef pillar undergoes less vertical deformation near the pillar edge.

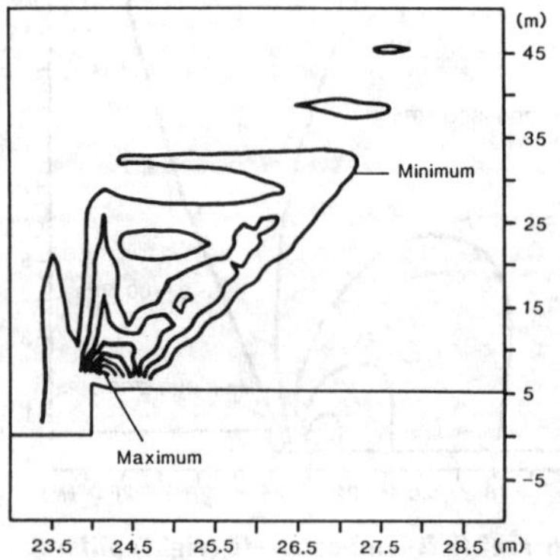

FIGURE 7: Cumulative maximum shear strain in the hangingwall at the edge of the reef pillar.

5.2.3 Hangingwall and Pillar Conditions

The condition of the hangingwall adjacent to the pillar is extremely important from the practical mining point of view. Figure 4 shows plots of the extent of the failed zones in the pillar and hangingwall for the reef pillar. The failure pattern is very similar for the concrete pillar. The zones have failed according to the Mohr-Coulomb shear criterion, with material strain softening as plastic strain increases. The model shows failure over the pillar and in the hangingwall adjacent to the pillar, indicating fracturing over the hangingwall and back into the pillar.

For a better assessment of the hangingwall stability in the two cases it is necessary to examine the conditions at the edge of the pillar more closely. The vertical stresses in the pillars are comparable except at the edges of the pillars (Figure 5). The area of high stress (400-700 MPa) is larger in the case of the reef pillar. This implies a greater danger of punching failure in the case of the reef pillar (Lenhardt and Hagan, 1990).

The shear stress in the reef pillar hangingwall changes sign in the hatched zone shown in Figure 6. This indicates a change in the shear sense in that region of the hangingwall, resulting in the hatched part of the hangingwall attempting to break off in the direction of the stope and the pillar. This zone is completely absent in the case of the concrete pillar (both in the vertical and shear stress field).

The tensile zone in the hangingwall of the reef pillar (Figure 5), together with the positive shear zone in the same part of the hangingwall, indicates an initial fracture or failure direction. The plot of the cumulative maximum shear strain for the reef pillar, which shows a concentration in the same part of the hangingwall, further supports this (Figure 7). The cumulative maximum shear strain in this area is an order of magnitude greater than the corresponding values in the concrete pillar, in which no such concentration exists. Hence the lack of a diagram for the concrete pillar.

6 PUNCHING EFFECT OF REEF PILLAR

The results show a number of differences between the behaviour of the reef and concrete pillars. The reef pillar shows a number of phenomena not seen in the concrete pillar.
These are :

- the closure profile near the pillar edge shows a sharper increase
- the tensile vertical stress zone in the hangingwall near the edge of the pillar
- the change in the orientation of the shear stress in the same region
- the greater accumulated shear strain in the hangingwall at the pillar edge.

It is thought that these effects are due to the stiffer, harder, extremely squat reef pillar punching into the foundation, causing a higher stress concentration in the hangingwall at the pillar edge. The much softer concrete pillar shows an absence of these factors.

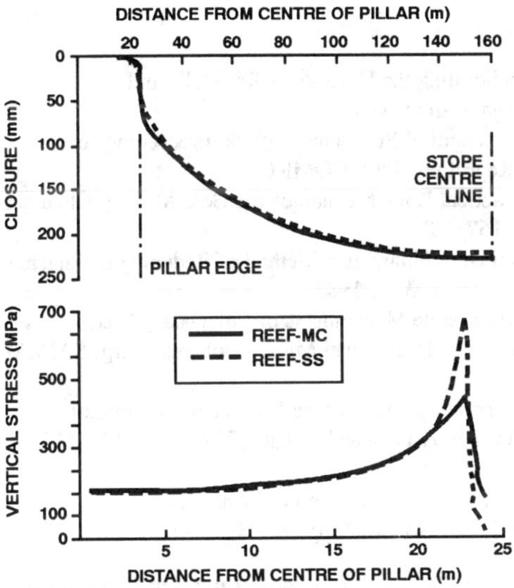

FIGURE 8 : A comparison between Mohr-Coulomb and strain softening reef pillars.

7 VALIDITY OF USING MOHR-COULOMB PILLARS

A FLAC run was performed using the reef strain softening parameters for the pillar. The boundary conditions and model geometry were the same as already discussed in every respect. The results are compared in Figure 8.

In the closure profile, it can be seen that the punching effect (as seen by the sharp increase in closure) at the edge of the pillar is less marked, but still present. It may be expected that the stronger Mohr-Coulomb material would allow greater closure before foundation failure occurred, but the overall closures are similar. Thus the Mohr-Coulomb pillar inhibits closure to the same degree as does the strain softened pillar.

As expected, the strain softening material softens dramatically at the pillar edges, transferring load to the interior of the pillar. This load transfer accounts for the much higher stress peak. An examination of the hangingwall stability, according to the same criteria as discussed previously for the Mohr-Coulomb reef pillar, shows that the strain softened pillar produces the same results qualitatively. The APS values are 273 MPa and 267 MPa for the Mohr-Coulomb and strain softened pillars respectively. The Mohr-Coulomb pillar thus allows foundation failure at the same stress level, rather than at higher stress levels as might be expected for a stronger material.

These results show that a Mohr-Coulomb reef pillar represents the overall pillar behaviour and its effects on the surrounding rockmass well compared to the strain softened reef pillar. Thus it is expected that a similar result will be obtained for a strain softened concrete pillar, except that there will be more failed material at the edge of the pillar, thus pushing the stress peak more inwards towards the centre of the pillar. This is more advantageous for hangingwall stability, as the zone of highest stress in the hangingwall is more confined as it moves away from the stope area.

8 CONCLUSIONS

Pillar foundation failure has been simulated in qualitative agreement with laboratory results. The nature of the failure is the same for the reef and concrete pillars. The average pillar stresses and closures at which foundation failure occurs are similar in both cases, implying a similar regional stability performance. This also implies that the concrete pillar is withstanding the same loads as the reef pillar, despite its much softer properties which cause the pillar to deform more vertically. This signifies that the controlling parameter is probably the pillar width to height ratio, rather than material properties.

The modelled punching effect is much reduced in the case of the concrete pillars. From the local (as opposed to regional) rock mechanics point of view, then, the use of concrete pillars to replace reef stabilising pillars is justified, and may even lead to an improvement in the hangingwall conditions adjacent to pillars.

REFERENCES

ANON, An Industry Guide to Methods of Ameliorating the Hazards of Rockfalls and Rockbursts. Chamber of Mines Research Organisation, 1988

Brummer R.K., Fracturing, Deformation and Strength of Stabilising Pillars. Proceedings of a Seminar Organised by COMRO, Research Report no. 8/87, COMRO

Cook N.G.W., Klokow J.W., White A.J.A., Practical Rock Mechanics for Gold Mining. Chamber of Mines of South Africa, P.R.D. Series no. 167, 1973

Hagan T.O., An Evaluation of Systematic Stabilising Pillars as a Method of Reducing the Seismic Hazard in Deep and Ultra-deep Mines. Ph.D Thesis, Wits, 1987

Lenhardt W.A., Hagan T.O., Observations and Possible Mechanisms of Pillar-associated Seismicity at Great Depth. Technical Challenges in Deep Level Mining, Johannesburg, SAIMM, 1990

McGarr A., Wiebols G.A., Influence of Mine Geometry and Closure Volume on Seismicity in a Deep-level Mine. Int. J. Rock Mech. Min. Sci. & Geomech. Abstr., Vol 14, pp 139-145, 1977

Ozbay M.U., Ryder J.A., The Effect of Foundation Damage on the Performance of Stabilising Pillars. J. S. Afr. Inst. Min. Metall., Vol 90, no. 2, pp. 29-35, 1990

Ryder J.A. Personal communication

Salamon M.D.G., Two-dimensional Treatment of Problems Arising from Mining Tabular Deposits in Isotropic or Transversely Isotropic Ground. Int. J. Rock Mech. Min. Sci. & Geomech. Abstr., Vol 5, pp 159-155, 1968

Smith G.L., York G., Concrete Strike Stabilising Pillars in Narrow-reef, Deep-level Gold Mining. Minefill 93, Johannesburg, SAIMM, 1993

Wagner H., Schümann E.H.R., The Stamp Load Bearing Strength of Rock - An Experimental and Theoretical Investigation. Rock Mechanics, Vol 3/4, pp 185-207, 1971

ADDENDUM : MATERIAL PROPERTIES

Mohr Coulomb properties

Quartzite rock properties :

density=2700 kg/m^3 ; shear modulus=32,19 GPa ; bulk modulus=44,77 GPa
friction angle=45,3° ; cohesion=37,9 MPa ; tensile strength=37,5 MPa
dilation angle=10°

Concrete properties :

density=2400 kg/m^3 ; shear modulus=4,17 GPa ; bulk modulus=5,56 GPa
friction angle=37° ; cohesion=5 MPa ; tensile strength=2 MPa
dilation angle=5°

Strain Softening parameters

COHESION		FRICTION ANGLE		DILATION ANGLE	
Plastic shear strain mm/m	Value MPa	Plastic shear strain mm/m	Value Degrees	Plastic shear strain mm/m	Value Degrees
0	44	0	16.4	0	0
0.005	12	0.0005	37.6	0.002	9
0.06	7.2	0.001	44.8	0.004	17
0.0065	5.2	0.002	46.8	0.006	22
0.008	2.8	0.004	46	0.0075	22
0.01	2	0.0075	40.4	0.0085	21
0.0125	1.2	0.01	39.2	0.01	0

Rock Mechanics, Nelson & Laubach (eds) © 1994 Balkema, Rotterdam, ISBN 90 5410 380 8

The development of a rationale for the design of stope and gully support in tabular South African mines

A.P.Squelch, M.K.C.Roberts & P.N.Taggart
Mining Technology Division, CSIR, Johannesburg, South Africa

ABSTRACT: A stope and gully support rationale is being developed for application in tabular stopes in South African mines. At present there exists no design rationale for optimum support. Present support systems differ widely in both type and unit spacings, even under similar mining conditions, and different support systems may be used in adjacent stopes, both with apparently similar success. Some of these support systems provide no or little back area support, while others provide substantial support resistance. Many of the systems have evolved empirically as a result of trial and error, with little attempt to quantify their effectiveness or contrast the effectiveness of different support systems. This highlights the need for establishing a sound engineering design rationale for support systems for given mining conditions in terms of maximizing safety and if possible, reduced cost.

1 INTRODUCTION

Rockbursts and rockfalls currently account for more than half of the injuries and fatalities in the South African gold mines (Roberts and Jager, 1991) and platinum mines. Most of these accidents occur in the stope, in particular at the stope face. Figure 1 shows the typical distribution of fatal accident locations in one gold mining area. These accidents lead to severe production losses and low morale, apart from the tragedy of fatally or severely injured workers. In order to address this problem, the Safety in Mines Research Advisory Committee is funding the development of a rationale for the design of stope and gully support.

At present there is no design rationale for optimum gully support and permanent stope support systems. Existing support systems are empirically based, as a result of trial and error. Moreover, they often differ widely in both type and unit spacings, even though under similar mining and loading conditions they have the same apparent success. There are estimated to be 139 variations of support systems in use in the gold mining industry, which highlights the need for a rationalisation of support systems as well as a design rationale. Some stope support systems provide little or no back area support, while others provide substantial support resistance in the back area. Similarly, in the case of gullies, there are a wide variety of stope gully geometries and support systems used in similar mining conditions. Thus optimum gully geometries and support resistance criteria need to be determined for different mining and loading (i.e. static and dynamic) conditions.

Gold and platinum mining is generally carried out in open tabular stopes with stope widths of one or two metres, at depths below surface varying from hundreds of metres to three thousand six hundred metres. Stopes are either of the longwall (Figure 2a) or scattered

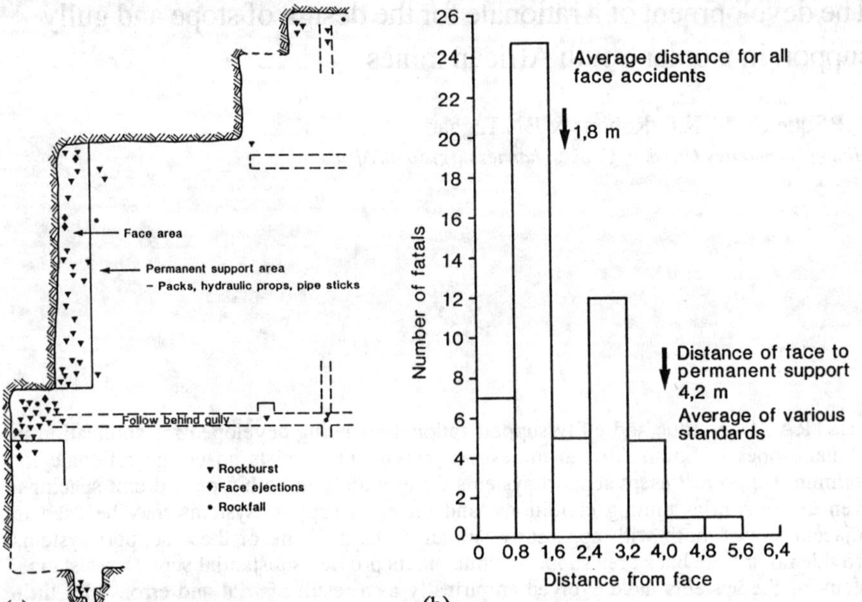

(a) (b)

Figure 1. Location of stope fatalities; (a) in the stope and (b) in relation to the stope face for one mining district (after Roberts and Jager, 1991).

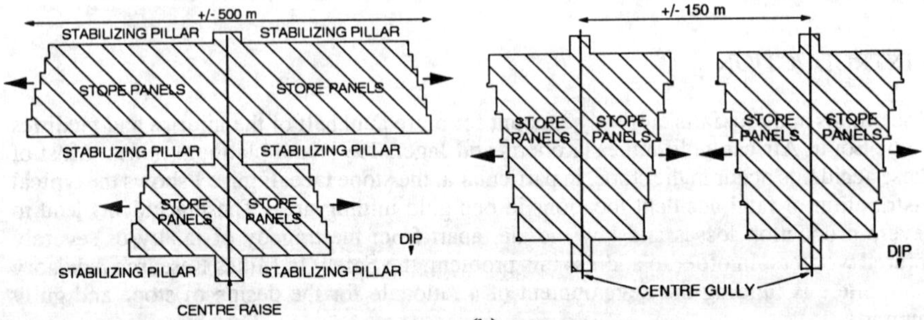

(a) (b)

Figure 2. Schematic of typical stope layouts; (a) longwall and (b) scattered

(Figure 2b) type and can extend on strike for many hundreds, even thousands (in the case of a longwall layout) of metres and on dip for tens or hundreds of metres. Scattered mining layouts usually involve short dip and strike spans and are dictated by the presence of faulting and dykes.

Figure 3 shows a typical stope and stope support layout in a longwall stope panel. Variations in permanent support are mainly to do with using different types of packs, e.g. timber mats, skeleton, composite (timber and concrete blocks), end-grain, or grout packs and timber elongates (e.g. profile props, wedge props or pipe-sticks) or backfill. Temporary face area support also varies from rapid yield hydraulic props, through mechanical props to timber sprags. The dip and strike spacings of support units also vary from one system to another.

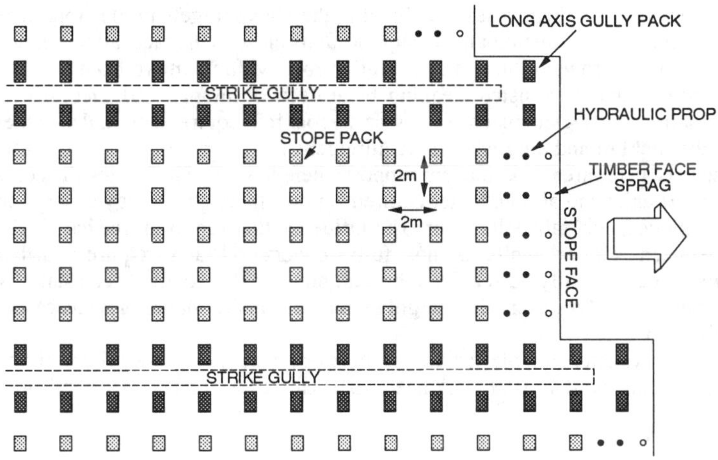

Figure 3. Schematic of a typical stope and support layout

2 STOPE SUPPORT

Research (Gürtunca *et al*, 1989) indicates that support systems do not affect the gross deformation of the rockmass around the stope until a support resistance of 1 000 kN/m^2 (1 MPa) is exceeded. This support resistance, which is able to reduce the inelastic stope closure, can only be achieved by backfill. This implies that conventional stope support systems are affecting only the stability of the skin of the excavation. The function of conventional support systems is therefore to stabilize the skin of the stope hangingwall and in addition to absorb energy during a rockburst. The evaluation of current support practice, and an investigation into rockmass-support interaction can therefore contribute to the development of a design rationale for stope support.

2.1 Current stope support systems

Numerous stope support systems are in use in the mining industry and an evaluation of their effectiveness is necessary. This evaluation, and a record of the dimensions of the rockfall associated with each accident, is being made via an analysis of the Government Mining Engineer's inquiries into rock related fatalities. This analysis has shown, that out of all fatal accidents in the gold and platinum mines for the period 1988-1992: 39 per cent occurred at the stope face, 6 per cent occurred in the strike gullies and 4 per cent in the centre gully/tip areas. In terms of stope accidents, 80 per cent occurred at the stope face, 12 per cent in strike gullies and 8 per cent in the centre gully/tip areas. By far the majority of these stope accidents are rockfall or rockburst related. These figures indicate that current support systems used at the stope face (stope working area) are not effective.

From underground investigations into falls of ground it has been determined that under quasi-static conditions a 50 kN/m^2 support resistance is required from the stope face support system to stabilize the stope face hangingwall. Under rockburst conditions the stope face support system is required to absorb 60 kJ of energy per square metre. These values provide a basis for the design of stope face support systems and probably apply also to areas further back in the stope.

There are, however, some difficulties in achieving the above targets in the stope face area. For practical reasons it is difficult to get support units close to the face, although two systems that use multiple rows of blast-on hydraulic props within a metre from the face in conjunction with water-jet assisted cleaning have been developed and successfully implemented on some mines. One system addresses the rockfall requirements and the other addresses both the rockfall and the rockburst requirements.

Problems also exist in the area of permanent support where fall-out of hangingwall occurs between the support units and although the required support force and energy absorption criteria can be achieved, rockfalls still occur. The influence that support force has on the hangingwall between support units is not fully understood and requires further investigation which can partially be achieved through numerical modelling. The stability of the support units themselves as mining height increases is also a matter of concern and requires investigation.

Currently support systems suitable for use in the stope face area, under rockfall and rockburst conditions, are available but work needs to be carried out on back area and gully support.

2.2 Static and dynamic behaviour of support units

Support units have been tested to establish their behaviour under static and dynamic conditions, in order to provide the necessary input for the comprehensive design of appropriate support systems and the numerical modelling investigation. The dynamic load rate behaviour of timber support units (e.g. packs and elongates) determines the units' ability to absorb energy during seismic events. Tests on packs carried out at various load rates show marked increases in pack force with increasing loading rates (Figure 4). This has important implications for stope support and gully support, particularly with respect to the absorption of energy during rockbursts where it is believed that closure rates in excess of 3 m/s can occur.

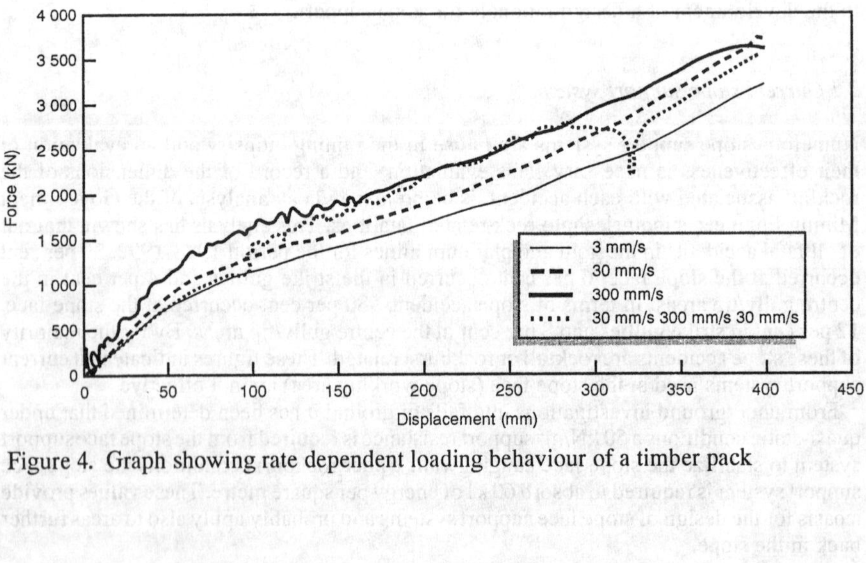

Figure 4. Graph showing rate dependent loading behaviour of a timber pack

2.3 Stope support and hangingwall interaction

An understanding of the interaction between support units and the immediate rock mass, and the effect of support systems on the overall stability of the stope hangingwall is required to ensure the design of safe and economical support systems. Numerical modelling will be used to obtain part of this understanding. Below are some of the rock mass-support interaction variables that will be investigated:

- effect of spacing of support units on the propensity of blocks of hangingwall between the units to fall out, under both static and dynamic conditions;
- effect of varying the support unit force on the stability of hangingwall blocks between the units, under both static and dynamic conditions;
- effect of various support systems (point loaded elongates or packs) on the stability of the fractured rock making up the immediate stope hangingwall.
- effect of increasing or decreasing the support resistance on the stability of the skin of the excavation.
- effect of area coverage on the stability of the skin of the excavation.

3 GULLY SUPPORT

A high incidence of rock related casualties occur in strike and dip gullies in stopes. However, a support resistance criterion for stope gullies under static and dynamic loading has not yet been determined. A wide variety of stope gully geometries and support systems are used in similar mining conditions and in addition various gully geometries are used. Thus, a priority need is to determine optimum gully geometries for different mining conditions, as well as to determine gully support resistance criteria for static and dynamic loading under different mining conditions. Together these developments will do much to improve the stability of stope gullies and reduce casualties.

The development of improved gully support systems will depend primarily on the determination of gully support resistance criteria for static and dynamic loads, by evaluating current practice and using numerical models to simulate gully response during mining activities.

A major problem is that current gully packs damage the gully sidewalls on which they stand and that these packs and their foundations would not survive rockbursts. As input into the design of better gully support, the performance of current gully support is being evaluated *in situ* including the determination of forces at which damage occurs. Numerical modelling is also being carried out to supplement the underground measurements.

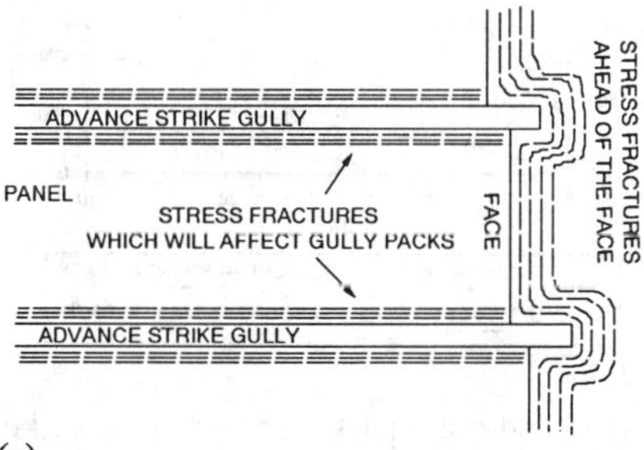

(a)

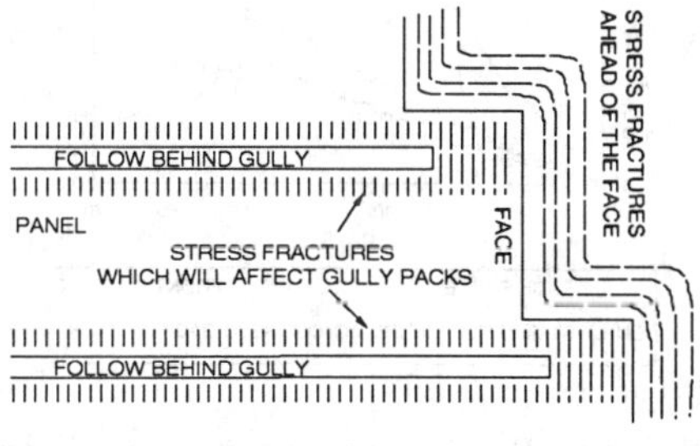

(b)

Figure 5. Schematic of fracture patterns affecting (a) advance strike gullies and (b) follow behind gullies

Gully geometries in the tabular gold and platinum mines can be separated into two basic categories, either; advance strike gully (ASG) (Figure 5a) or follow behind gully (Figure 5b). Advance strike gullies by their nature are developed ahead of the stope face and therefore induce stress fracturing similar to a tunnel i.e. parallel to the gully's sidewalls and arching over the gully. The portion of these fractures in the footwall of the reef will underlie the gully packs installed after the stope face has been driven forward. These stress induced fractures, because of their orientation parallel to the gully, are a major cause of gully pack foundation failure.

Follow behind gullies are developed in the footwall of already stoped reef, and therefore do not induce stress fractures. The fractures affecting these gullies and their sidewalls are therefore those induced by and parallel to the stope face. They thus are perpendicular to the line of the gully and are less likely to affect the stability of the packs' foundations. The depth of a gully will also affect the stability of the gully's sidewall and hence the stability of the packs' foundations and is a factor to be considered in the design of a gully support system.

3.1 In situ monitoring of gully packs

Gully packs have been monitored at several underground sites by means of pressure load cells installed within packs and measuring stations to record stope closure and gully sidewall closure. Initial instrumentation of gully packs under static conditions indicates that forces of the order of 1 000 to 2 000 kN are reached at which point damage to the gully sidewall occurs and further load build-up is prevented (Figure 6). This damage is usually confined to the mobilization of pre-existing (pre-pack installation) fractures rather than the creation of new fractures. A visual record was also made of the fracturing in the gully sidewall beneath the packs so as to correlate pack load and gully sidewall damage; this confirmed the anticipated fracture patterns described above.

Conclusions at this stage are tentative, however the majority of the packs, of which the results in Figure 6 are only a sample, exhibited flat force-deformation curves at peak loads near or below 2 000 kN and in some cases below 1 500 kN. This was accompanied by an inability to generate increased load due to gully sidewall movement beneath the packs.

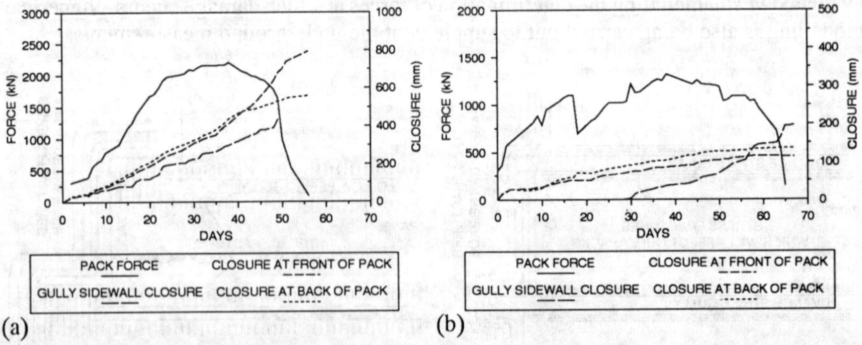

Figure 6. Results from *in situ* monitoring of (a) an end-grain pack on an advance strike gully and (b) a mat pack on a follow behind gully

This implies that expensive high force-capability packs are not necessary for gully support and could cause additional damage during seismic events. Damage will occur to the gully sidewall as a result of the increased force generated in the packs during a seismic event, as found from the dynamic testing described above. If the *in situ* findings are confirmed by further pack instrumentation then gully packs need to be re-designed for static and dynamic conditions such that the peak force on the pack does not exceed that which causes gully sidewall damage. The results of dynamic testing of timber units done to date show that timber is not suitable for rockburst gully support because the increase in load during rapid loading will accelerate damage to the gully sidewall and render the pack useless.

3.2 Numerical modelling of gully packs

Numerical modelling has been carried out to investigate gully parameters not readily amenable to *in situ* investigation and to allow subsequent design work to be carried out. As a first step, simulation of some measured behaviour was attempted. UDEC (Universal Distinct Element Code) was selected for this purpose because the fractured nature of the gully sidewall requires a non-linear discrete block modelling approach. The geometry of the model depicted a dip section through a gully with packs on each side and with an unfavourably orientated fracture cutting through the footwall beneath one of the gully packs (Figure 7). This geometry was chosen as a starting point because it suits the 2D nature of UDEC and because similar examples were observed in relation to some of the monitored packs. The graphs in Figure 8 show a relatively good similarity between modelled and measured results, although closure and force magnitudes are different.

Further development of this modelling will enable some different gully and fracture geometries to be simulated and investigated for their suitability and consequences.

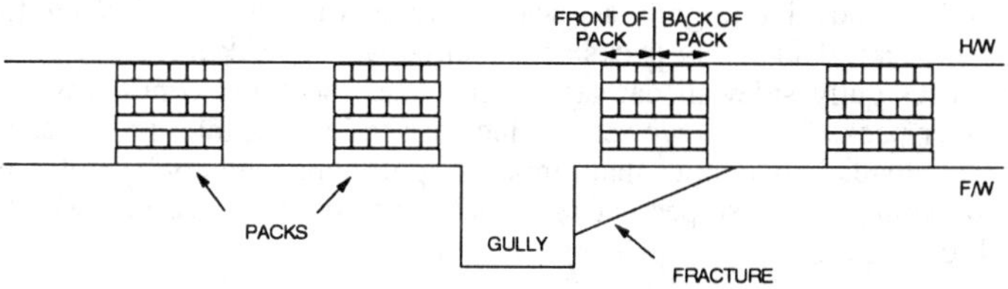

Figure 7. Gully and fracture geometry modelled with UDEC

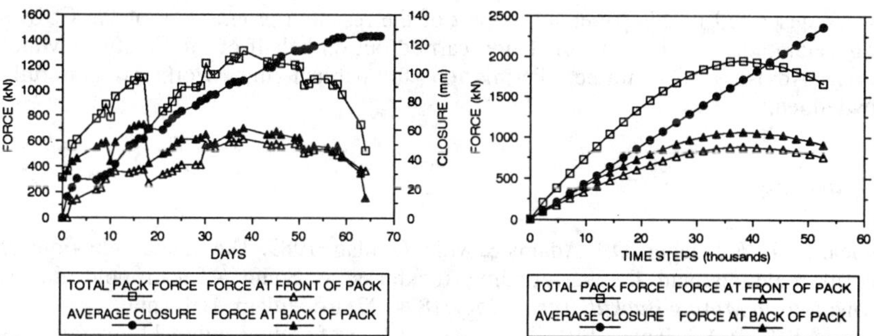

(a) (b)

Figure 8. *In situ* pack force results (a) compared to modelling results (b)

3.3 *Support resistance in gullies*

A further consideration that has to be made in the design of gully support is that of the support resistance required to maintain stability of the gully hangingwall between support units, especially across the gully. One way of estimating this requirement is to analyse the records of the Government Mining Engineer's accident enquiries for the dimensions of falls of ground associated with fatalities in gullies. An initial analysis of the records for 1990 to 1992 shows that 90 per cent of the rockfalls in gullies could have been prevented by a support resistance of less than 50 kN/m^2.

4 CONCLUSIONS

Two effective face area support systems, incorporating hydraulic props, have been developed and are available for implementation. One system meets the rockfall support requirement of 50 kN/m^2 and the other meets both the rockfall requirement and the rockburst energy absorption requirement of 60 kJ per square metre for that region of a stope. The use of these support systems should reduce the high incidence of rock related accidents and fatalities that occur in the stope face area.

An understanding has been gained of the different ability of timber support units to absorb energy depending on the loading rate they are subjected to. This knowledge will assist in the design of efficient energy absorbing support systems.

The interaction of support units with the stope hangingwall is currently being studied with the use of numerical models with the objective of gaining an insight to the design parameters required to devise efficient energy absorbing support systems which are suited to the prevailing rock conditions.

It has been determined that at the majority of instrumented sites gully packs exhibit flat force-deformation curves near or below 2 000 kN and that this is accompanied by an inability to sustain higher loads due to gully sidewall movement resulting in damage to the sidewall below the packs. Thus gully packs are required whose peak force does not exceed that which causes gully sidewall damage, even under rockburst conditions. In addition, tests on timber packs have shown that an increase in loading rate produces a marked increase in pack load. This implies that timber support units may not be suitable as gully support in rockburst prone stopes because peak load could exceed that which causes gully sidewall damage.

ACKNOWLEDGEMENTS

The work described in this paper forms part of the research programme of the CSIR's Mining Technology Division and is being carried out on behalf of the Safety in Mines Research Advisory Committee. Permission to publish this work is gratefully acknowledged.

REFERENCES

Gürtunca, R.G., A.J. Jager, D.J. Adams & M.D. Gonlag. 1989. The in situ behaviour of backfill materials and the surrounding rockmass in South African gold mines. *Innovations in Mining Backfill Technology:*187-197. Rotterdam: Balkema.

Roberts, M.K.C. & A.J. Jager. 1991. An analysis of falls of ground and rockburst fatalities in three mining districts. *Proceedings of Mine Safety and Health Congress:*181-200. Johannesburg: Chamber of Mines of South Africa.

Rock Mechanics, Nelson & Laubach (eds) © 1994 Balkema, Rotterdam, ISBN 90 5410 380 8

Spotlight on new developments in the support of tunnels in South African mines under high stress conditions

L.Wojno & A.J.Jager
Mining Technology Division, CSIR, Johannesburg, South Africa

ABSTRACT: For some years it has become increasingly apparent that control of the kinetic energy imparted to the rock by the seismic ground motion is the crucial consideration for the design of rockburst support. Another essential requirement is increased area coverage provided by a support system, which not only maintains the integrity of a zone of intensely fractured rock surrounding deep mine tunnels, but also provides a large area coverage support right at the development end where falls of ground are common.

One of the strategies to overcome these problems and to achieve the associated benefits is to develop support units and systems which are better able to control the fractured rock wall, particularly when rockburst conditions occur and the rock walls can be subjected to velocities of 6 m/s or more.

Accordingly, the Chamber of Mines Research Organization, now the CSIR Mining Technology Division, developed and successfully tested two new support systems for tunnels: a new generation of grouted yielding tendons and a flexible water based, spray-on temporary support. These developments address two important needs: improved seismic energy dissipation; and improved area coverage by support systems.

1 INTRODUCTION

At current mining depths of 2,5 - 3,5 km the stability of tunnels developed under high stresses is of major concern. Field stresses are as high as 150 MPa to 200 MPa and stress increases of 40 MPa or more, due to stoping in the vicinity are common. Moreover, after overstoping, stress relaxation occurs, which further complicates the situation .Another stress change condition occurs in longwall mines where follow-on development is carried out in overstoped, stress relieved ground but later is subjected to increasing stress as closure of the stope occurs.

In general the support methods developed for tunnels have been effective for most situations. However, where large quasi-static deformations or rockbursting occurs damage often results due to lack of yieldability of tendons and insufficient area coverage. Recent surveys indicate that approximately 85 km of tunnels have to be comprehensively rehabilitated, redeveloped or abandoned each year. An estimate of the direct cost of rehabilitation is R45 million annually, and that of tunnel deterioration due to inadequate support for the prevailing stress conditions, R100 million per annum.

In this paper new solutions for the support of tunnels in high stress conditions are discussed: they address the important aspects of safety and cost saving benefits, by improving overall stability of tunnels at depth.

2 ROCK CONDITIONS

2.1 *Stress Fracturing*

The sidewalls of tunnels fracture when the maximum field stress is greater than about 30 % of the uniaxial compressive strength of the rock (UCS). The density of fracturing is related to the rock type and the magnitude of the prevailing stresses. Strong, brittle orthoquartzites tend to fracture more intensely than do argillaceous rocks, forming near vertical sheets of several square metres in extent and 15 mm or less in thickness. This fracturing is usually parallel to the tunnel walls (Figure 1). Extension or cleavage fractures predominate but shear fractures also occur.

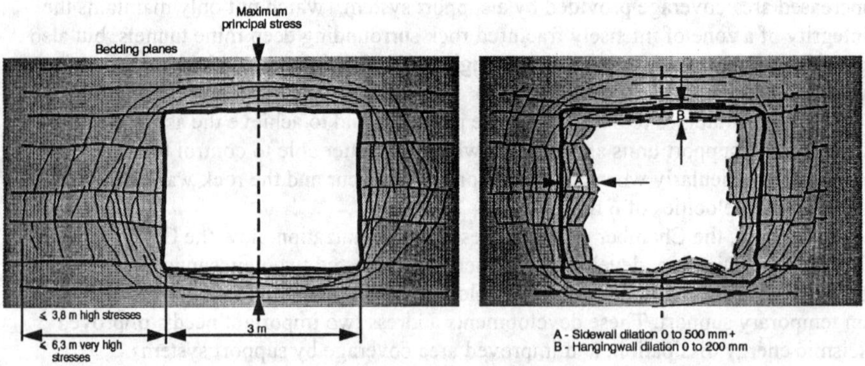

Figure 1. Extent and deformation of fracture zone around tunnels

The depth of fracturing into the wall rock is dependent on the field stresses. For moderately high stress conditions, 70 MPa, the fracturing can extend into the sidewall to a distance equivalent to the height of the tunnel. For stresses in excess of 100 MPa the depth of fracturing may be twice this distance. The extent of fracturing into the hangingwall and footwall is usually half or less of that into the sidewall. The sidewalls in particular are broken into numerous slabs which require total area support such as mesh and lacing to maintain their integrity. Tendon support alone is totally inadequate.

2.2 *Dilationary Deformations and Forces under Quasi-Static Loading*

The process of fracturing around a newly exposed rockwall commences immediately after the blast and continues until the tunnel has advanced several metres. The amount of dilation during this period is likely to be relatively small because of the restraint provided by close proximity of the face. Large deformations occur only when the highly stressed tunnels are subjected to significant stress increases induced by stoping.

Deformations of over 500 mm per sidewall can develop during this stage when the tunnel is subjected to high abutment stresses. The dilation causes a general slow loosening of the fracture zone but on occasions can be rapid and catastrophic when the rockmass is subjected to rapid seismic ground motion.

The forces driving the dilation were measured using load cells installed on four 450 kN cables against large base plates covering an area of one square metre.

The total force generated at a depth of 2 metres was in excess of 500 kN at a dilation of only 25 mm. This force, is far greater than the support resistance developed by the most dense support systems in use on the mines. Thus for tendons to remain effective when installed in rock which will dilate more than the failure strain of rigidly anchored or full column grouted tendons, the integrated support system as a whole, must be able to yield sufficiently to accommodate the dilation.

2.3 *Deformation under Rapid Loading*

During seismic loading the wall rocks can be accelerated to peak velocities of up to 4,4 m/s in the source region of a seismic event (McGarr and Bicknell 1988). Preliminary estimates indicate that about 80 % of rockbursts are caused by seismic deformations where maximum velocities are up to 3 m/s (Jager, 1992). However, calculations of the velocities required to propel large blocks of rock a known distance from the tunnel sidewall indicate that velocities up to 6 m/s have occurred (Turner, 1989).
Any solutions to improve the stability of tunnels in highly stressed conditions must therefore be designed to cater not only for extensive fracturing and large quasi-static deformations but also for rapid deformations caused by seismic events.

3 PERFORMANCE REQUIREMENTS FOR YIELDING ROCK TENDONS IN HIGH STRESS TUNNELS.

The main requirement for ensuring the stability of highly stressed tunnels is to design a support system that yields and spreads the restraining forces as evenly as possible, and can absorb the kinetic energy imparted to the rock by seismic ground motion. A basic element required for such a tunnel support is a yielding tendon, recognized by Ortlepp (1970). In 1986 COMRO initiated programme to design and develop a yielding tendon that would perform satisfactorily both under large quasi-static rock deformations and rapid loading conditions. The initial task was to set performance specifications for yielding tendons. Four fundamental parameters: the work done during yielding, the static and dynamic yield forces and displacement, were specified and formed the basis for the specifications put forward in 1987 as guidelines for the design of yielding rock tendons. The main requirement was to absorb sufficient kinetic energy imparted to the rock during a rockburst to maintain the stability f the tunnel walls (Wojno et al 1987). A minimum figure of 25 kJ per tendon was prescribed, sufficient to arrest a 2 m thick slab in the sidewall accelerated to a velocity of 3 m/s, assuming a support density of 1 tendon per metre square. Two other parameters were static and dynamic yield loads of 100 kN and 50 kN, respectively. These yield loads were based on the most generally accepted value for the load carrying capacity of tendon support, and the average support resistance required to prevent rapid deterioration of tunnel rockwall conditions.

These minimum requirements were tempered by known difficulties experienced in achieving sufficiently high combined values of yield force and yield length. In the light of current knowledge, a figure of 50 kJ/m^2 would appear to be a suitable specification. Figure 2 illustrates work requirements in terms of yield force and yield distance which satisfy the criteria of 25 kJ and 50 kJ. Also shown is the work done by a conventional tendon W_1 and, for comparison, what is achieved by the cone bolt W_2.

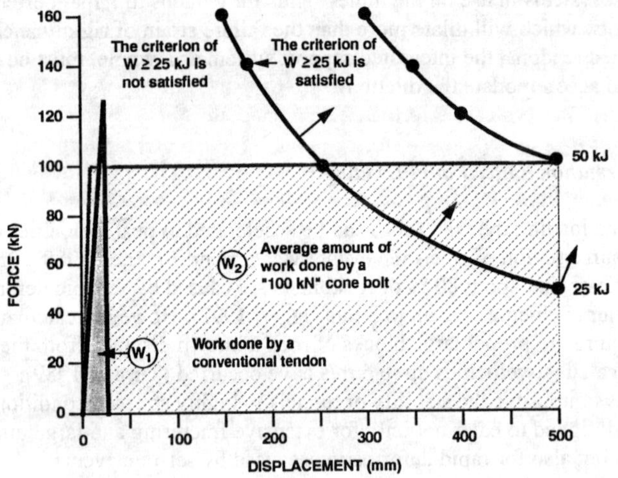

Figure 2. Force-Displacement combinations which satisfy the minimum work criterion of 25 kJ and 50 kJ for yielding tendons

4 THE CONE BOLT

The cone bolt is a grouted yielding tendon for the provision of effective support in regions prone to seismic events or high stress changes and is capable of controlling ground movements in excess of 500 mm and of reducing damage resulting from rockbursts and large quasi static rock deformations.

4.1 *Principle of Operation*

The displacement of the dilating rock in which a cone bolt is installed is transferred to the tendon via the bearing plate, thus pulling the conical end piece through the grout. In this way work is done and energy is absorbed from the rock. This is made possible by coating the yielding tendon to prevent the formation of a bond between the bar and the grout. Among the most important characteristics of cone bolts are;

* the ability to yield by 0,5 m or more at displacement rates ranging from slow to over 12 m/s, while maintaining the designed support forces of up to 200 kN,
* easy integration with other elements of the support system thus requiring minimum training of personnel,

* as easily installed in boreholes drilled in highly fractured rock as conventional tendons, and
* not significantly sensitive to grout strength, but performs most effectively for grouts with uniaxial compressive strengths ranging from 25 MPa to 60 MPa.

4.2 *Performance Characteristics of Cone Bolt*

a) Laboratory tests

Pull tests were carried out on conventional rockbolts, grouted tendons and two geometries of cone bolts. The loading rate for quasi-static tests was kept at 0,001 m/s. Under rapid loading, two rates were used; 1,6 m/s and 3 m/s.

Figure 3 illustrates typical performance characteristics for conventional tendons and cone bolts. For example the performance of a 22 mm bar cone bolt indicates that it is capable of absorbing 100 kJ of energy. Such a tendon can bring to rest, within 0,5 m, a 2 m thick block in the hangingwall accelerated to 3 m/s or a 4 m thick block in the sidewall accelerated to over 4 m/s.

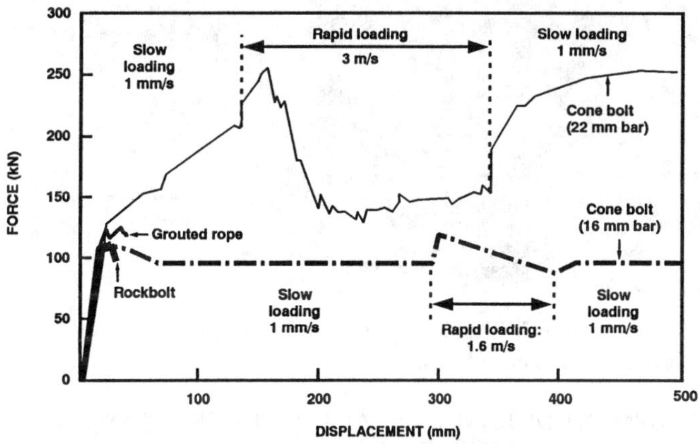

Figure 3. Comparison of typical performance for conventional tendons and cone bolts.

b) Field tests

To be able to test rock tendons under conditions, that would compare with even higher loading rates observed underground, a collaborative experiment by Steffen, Robertson & Kirsten, Anglo American Corporation West Rand Region, Strata Control Systems and Miningtek was carried out. The test facility was constructed on a floor of a granite quarry. The concept forming the basis of the test procedure was to use explosives to provide sufficient, rapidly imparted energy to a set of rock tendons, achieve the ejection velocities inferred from observations underground, and simulate the destruction of tendons that frequently result from a rockburst. The explosive force was applied across a horizontal plane to elevate a concrete block anchored to the rock floor in a quarry.

Since each concrete block had the same mass and gravity is constant, the energy "balance" of the block is easily determined once the block has been torn free from its anchoring.

In the tests, cone bolts loaded by explosives have survived rapid loading in excess of 12 m/s. At the same time, they arrested the 3,8 ton block after only 500 mm displacement.

Under the same loading conditions, conventional 25 mm diameter rebars grouted over the same length of 1,1 m as the cone bolts, either snapped or pulled out, allowing the block to be thrown more than 4,6 metres into the air, as illustrated in the Figure 4. In the process of yielding 0,5 m , the cone bolts suffered no damage whatsoever and absorbed 20 times as much energy as would cause complete failure of a conventional, 25 mm diameter rebar.

Figure 4. A 3,8 ton block flies through the air during an experiment using rebars, as opposed to cone bolts.

5 DEVELOPMENT OF LARGE AREAL COVERAGE SPRAY-ON COATING FOR HANGINGWALL SUPPORT

Rock fall related accidents in the South African gold mining industry account for more than 50 % of the total fatalities and about 30 % of the total injuries.

It appears that under poor hanging wall conditions, current temporary face and gully support methods are not able to adequately prevent accidents due either to rockbursts or rockfalls, despite the fact that much attention has been given to areal coverage of support, its spacing and resistance, which are the most important parameters which affect both stope stability and worker safety. Many falls of ground have been investigated resulting in a reasonable understanding of what causes them. For example most large falls of ground initiate as a result of the loosening and collapse of relatively small key blocks. Once dislodged, the fall of ground can extend as the lateral forces confining adjacent blocks are removed, permitting more blocks to collapse. The philosophy behind the new technique is that if the relatively small, initial falls can be prevented timeously then the majority of larger falls will not occur. It is envisaged that a support system that could prevent or significantly reduce fall out of the initial key blocks, would go a long way to reduce the occurrence of larger falls of ground and could possibly also reduce rockburst damage.

Against this background, a concept of a thin , flexible, spray-on coating has been developed. In the process, a number of coating materials have been eliminated as being either too expensive, environmentally hazardous or difficult to apply in the extremely demanding mining environment where the ambient temperature can be in excess of 32 ^{0}C, the relative humidity reaches 100 %, rock face temperature exceeds 40 ^{0}C and the air velocity is very slow.

Intensive investigations carried out by the CSIR Mining Technology Division in collaboration with a manufacturer have resulted in the development of an acrylic based, fibre reinforced coating material that appears to have the potential for providing temporary support for mine excavations in poor rock conditions. The material is easily sprayable and adheres strongly to the rock surface producing a uniform coating even if sprayed on irregular surfaces.

The basic properties of the coating material have been quantified as follows:

* Tensile strength: ranging from 3 MPa to 5 MPa depending on the fibre content
* Elongation: >100 %
* Water based; easily washable before curing
* Non-toxic vapors during application, and if exposed to open flame
* Optimal thickness of the coating: 1,5 to 2 mm.

At present, preliminary investigations are being carried out on the mines to evaluate the potential use of coating, with the main objective being to quantify the environmental

Figure 5. A coating is sprayed on a single layer of bricks (packed in a square). After the coating has cured, the bonded slab is inverted and supported only at its corners. The flexible coating is able to support almost 200 kg pillar.

conditions so as to determine whether acrylic based coating can be used effectively. Figure 5 illustrates an example of the support potential of the acrylic coating where not only loose and laterally unconfined blocks are kept together by a thin coating but in addition, almost 200 kg load is carried by such a "structure".

6 CONCLUSIONS

Field evaluation of cone bolts, together with promising preliminary data on the potential use of an acrylic based, flexible, spray-on temporary support systems, has shown that these support developments have considerable potential to make working conditions in South African mines safer. In particular, if utilized effectively they should result in reductions in falls of ground and cost of tunnel rehabilitation, as well as provide better control of rockburst damage and improved hanging wall support, especially in wide stopes.

7 ACKNOWLEDGMENT

The work discussed in this paper forms part of the research programme carried out by CSIR Mining Technology Division on behalf of the South African mining industry and partly funded by the Safety in Mines Research Advisory Committee, Department of Mineral and Energy Affairs.

8 REFERENCES

Jager, A.J. 1992. Two new support units for the control of rockburst damage. In Peter K. Kaiser & Dougal R. McCreath (eds), *Rock Support in Mining and Underground Construction - Proc. Int. Symposium on Rock Support, Sudbury, Ontario, Canada, 16-19 June 1992*: 621-632. Rotterdam:Balkema

Jager, A.J. and Wojno, L. Z. 1991. Damage to tunnels in gold mines - are there cost effective solutions? *Proc. SANGORM Symposium, Impact of Rock Engineering on Mining and Tunneling Economics, 7 8 October 1991*. Welkom, R.S.A. *Technical Challenges in Deep Level Mining.* Johannesburg:1155-1171.

McGarr, A. and Bicknell, J. 1988. Estimation of the near-fault ground motion of mining-induced tremors from locally recorded seismograms in South Africa. *Second International Symposium on Rockbursts and Seismicity in Mines. University of Minnesota, June,* Minneapolis, U.S.A.

Ortlepp, W.D. 1970. An empirical determination of the effectiveness of rockbolt support under impulse loading. Proc. *International Symposium on Large Permanent Underground Openings.* Oslo: 197-205.

Turner, P.A. 1989. Personal Communication

Wojno, L.Z., Jager, A.J. and Roberts, M.K.C. 1987. Recommended performance requirements for yielding rock tendons. *SANGORM Symposium - Design of Rock Reinforcing: Components and Systems,* Johannesburg: 71-74

17 Deep mine design/bursting/pillar design

Session developers:
Will Bawden, Queen's University
Roger Tyler, Bureau of Economic Geology

Rock Mechanics, Nelson & Laubach (eds) © 1994 Balkema, Rotterdam, ISBN 90 5410 380 8

Analysis of induced seismicity in a hard-rock mine in the Sudbury Basin, Ontario, Canada

S.Talebi, M.Ge, P.Rochon & P.Mottahed
Mining Research Laboratories, CANMET, Sudbury, Ont., Canada

ABSTRACT: Full-waveform signals of mine-induced seismic events recorded in a hard-rock mine were processed in order to study attenuation effects, source parameters and scaling laws for such events. Attenuation along the source-sensor path had a significant effect on the recorded signals. Quality factor of S waves (Q_S) varied in the range 20-500, the peak being observed in the 50-150 range. The relationships between seismic moment and source radius on one hand and peak velocity parameter and seismic moment on the other were compatible with those expected from theoretical shear failure models. This strongly indicates a self-similar behavior of seismic sources.

1 INTRODUCTION

Induced Seismicity is a general term defining seismicity caused by human activity, i.e. not directly originating from natural processes. This phenomenon is caused in a number of ways, including removal of part of a rock mass (mining, tunneling), perturbation of the natural hydraulic regime (fluid injection or extraction, dam construction) and perturbation of rock temperature (thermal cracking). In the first case, excavation of large volumes of rock at depth and the resulting stress redistribution can give rise to crack propagation and movement along preexisting fracture planes.

Two scales can be considered for mine-induced seismic events: small events due to stress concentration and sudden failure of rock close to mining faces and large seismic events due to the interaction of tectonic and mine-induced stresses on pre-existing fault planes. The primary concern from the point of view of the mining industry is to deal with large events which could cause fatalities and severe damage to mine openings and equipment. Although most daily recorded events are small and the level of activity is, in general, an indication of rockburst potential, the relationship between large and small events is complex. For example a major rockburst with a magnitude of 3.4 occurred in Creighton mine, studied in this paper, in 1989. The level of microseismic activity began to increase three days before the occurrence of the main event, but subsided the day before the rockburst. Most of these events were close to the hanging wall, some 150 m away from the rockburst located near a drift in the footwall (Ge and Kaiser, 1991).

In this paper, we study source parameters and scaling relations for mine-induced events in the magnitude range 0-3 in order to examine any fundamental differences between the physical properties and characteristics of events of different magnitudes.

2 DESCRIPTION OF THE EXPERIMENT

2.1 *The mine site*

Creighton mine is one of the most seismically active mines in the Sudbury Basin where some of the strongest mine-induced events in Canada have been observed. The orebodies at Creighton mine extend to a depth in excess of 2200 m and high in-situ stresses are present. The mine has a history of rockbursting dating back to the early 1930's. On average, a rockburst of at least a magnitude of 2 occurs about once a month, and a magnitude of at least 3 occurs once a year (Hedley, 1992).

2.2 *The full-waveform data acquisition system*

The data acquisition system installed at Creighton Mine uses five triaxial sensors located around the active mining area. The differential signals are amplified and transmitted to an underground multiplexer unit and then sent to surface via a single fiber optic link. A demultiplexer unit on surface reconstructs the signals and sends them to the signal conditioning and recording systems. The anti-aliasing filter is set at 1500 Hz at 72 dB/octave. The sampling rate is 4500 points per second and per channel. The sensor amplifiers are dual gain having both a low and a high output for each axis. This allows for a larger system dynamic range to record both small and large seismic events. The selected outputs are then routed to both a triggering detection controller and a data acquisition board located in a PC- compatible host computer. System calibration consisted of generating sine waves of known amplitude and frequency within the recording window at the input of underground preamplifiers and recording them. The recorded signals were then compared to the input signals and calibration curve were calculated and used for further signal processing.

3 SOURCE STUDIES OF MINE-INDUCED SEISMIC EVENTS

Studies of mine-induced events observed around mining districts have shown that most of these events are caused by shear failure along fault planes or preexisting weakness planes in the rock mass (e.g. Spottiswoode and McGarr, 1975; McGarr et al, 1979). Most seismic models predict a flat low-frequency trend for the displacement spectra and a descending trend at high frequencies, inversely proportional to some power of frequency (Brune, 1970 and 1971; Madariaga, 1976). The region where the two trends intersect is dominated by frequencies with wavelengths related to the duration of rupture on the fault. For a constant rupture speed, this reveals to be simply the fault dimension. The high-frequency descending trend of the spectra has a slope of -2 or lower. This slope is affected by attenuation properties of the medium the waves propagate through. Several studies have shown that path effects cannot be neglected in source parameter determinations (e.g. Talebi and Cornet, 1987; Spottiswoode, 1993).

The parameters characterizing the source of induced seismic events, commonly called source parameters, are calculated in time and frequency domains and measure three types of parameters (see figure 1): the strength of an event (seismic moment, magnitude, seismic energy), source dimensions (source and asperity radii) and estimates of stress release (static, dynamic or rms stress drop, apparent stress) and their various

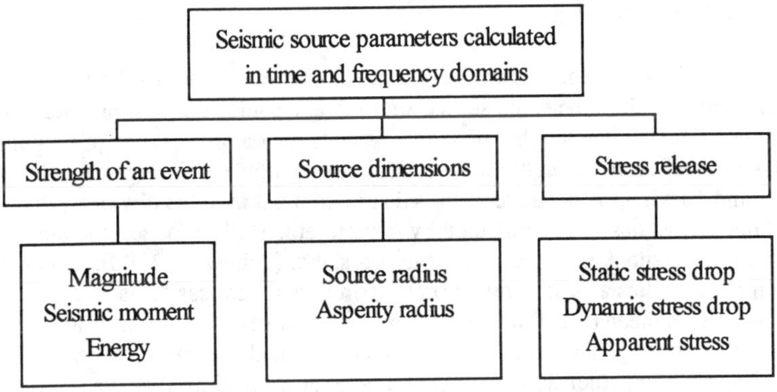

Figure 1 - Three types of seismic source parameters.

combinations (Gibowicz, 1990).

The most popular parameter used to describe the strength of seismic events is magnitude. Seismic moment, however, is a more reliable model-independent parameter used to describe source strength. Seismic moment is usually calculated from the low-frequency horizontal trend of displacement spectra of P or S waves as following:

(1) $$M_0 = 4 \pi \rho C^3 R \; \Omega_0 / F_C$$

where ρ = rock density,
C = P- or S-wave velocity,
R = source-sensor distance,
Ω_0 = plateau level of the spectrum,
F_C = P- or S-wave radiation coefficient,
Average values of 0.52 and 0.63 are used for F_C if the focal mechanism of the event is not known.

It is generally accepted that the dimensions of the failure area depend inversely on the "corner frequency" which corresponds to the intersection of the high- and low-frequency trends of the displacement spectra. For example in the Brune's (1970) model, the corner frequency (f_0) is interpreted as being inversely proportional to the rupture duration. Taking a circular source and a constant rupture velocity, source radius is calculated from the displacement spectra of S waves as following:

(2) $$r_0 = 2.34 \; V_S / 2 \pi f_0$$

where V_S it the velocity of S waves.

There are many different measurements of stress release because of a seismic event. The most used estimate of stress drop representing the most accurately stress change along the fault area is the "static stress drop" defined as the average difference between the initial and final stress levels over the fault plane. This parameter can be calculated from the estimations of seismic moment (M_0) and source radius (r_0) according to Brune (1970, 1971) as following:

(3) $\Delta\sigma = 7 M_o / 16 r_o^3$

Scaling relations of seismic sources are the relationship between the two first types of source parameters, i.e. those between source strength and source dimensions. Studies of source parameters for a large number of seismic events over a broad range of magnitude has shown that stress drop is roughly independent of magnitude and ranges mostly between 0.1 and 10 MPa. A dependence of seismic moment on r_o^3 is obviously observed. This pattern has been confirmed by innumerable studies for seismic and volcanic events, aftershock sequences, mine tremors, etc. (Gibowicz, 1990). A constant stress drop model implies a "self-similar" rupture process regardless of the scale of the seismic events. This means, in simple terms, that earthquakes, mine tremors and microseismic events are generated in a similar manner, but along different-size failure areas. There is growing evidence, however, of a breakdown in similarity for very small events. This effect, observed at the lower end of the spectrum, consists of seismic moment decreasing much faster than r_o^3, suggesting that there might be a lower limit to source dimensions (see e.g. Gibowicz et al, 1991). This phenomenon can be explained by any phenomena at the source, in the rock or the recording system reducing the influence of high frequencies in the recorded signals.

4 PROCESSING OF THE FULL-WAVEFORM DATA

More than 150 events recorded over a period of 8 months were analyzed. Spectral analysis was performed using interactive computer graphics allowing the choice of signal and noise windows on any seismic channel. The time series were tapered using a cosine function prior to the calculation of the spectra using an FFT (Fast Fourier Transform) algorithm. The spectra were corrected for instrumental response prior to the calculation of spectral parameters (figure 2).

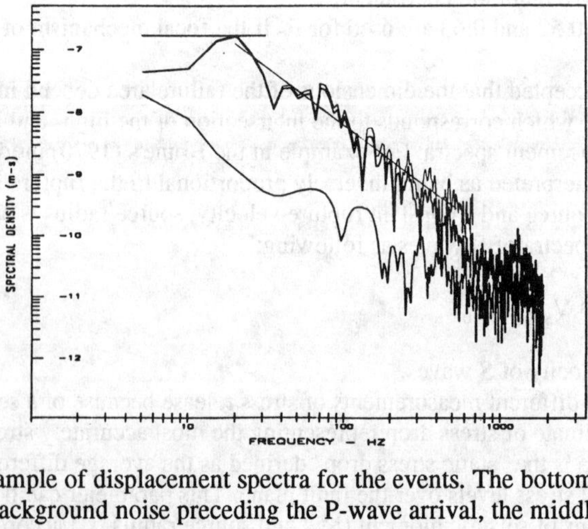

Figure 2 - Example of displacement spectra for the events. The bottom trace is the spectrum of background noise preceding the P-wave arrival, the middle trace is the spectrum of the recorded S-wave signal and the top trace is the spectrum of the S-wave signal corrected for attenuation.

4.1 *Attenuation correction*

Attenuation is a fundamental aspect of wave propagation and originates from the loss of energy observed in seismic waves due to their propagation through the earth. To correct for attenuation along the path, the spectra should be multiplied by the exponential term $\exp(\pi f R / C Q)$ where f is the frequency and Q is the quality factor of the body wave. As mentioned earlier, the high-frequency descending trend of the displacement spectrum has a fall-off slope of -2 according to Brune's model. The difference between the observed slope and -2 is usually attributed to attenuation and a number of authors have used this property for attenuation measurements (e.g. Spottiswoode, 1993).

The procedure adopted in this study consisted, in the first instance, of measuring the slope of the high-frequency descending trend of the S-wave displacement spectra in the range from a few Hertz to a few hundred Hertz. Q_S was then measured from the slope of the best-fit line to the high-frequency descending trend of the spectrum. Figure 3 shows a histogram of the results. The lower and upper bounds of Q_S were clearly 20 and 500. The data are rather scattered due to the fact that ray paths pass through areas of fractured rock, backfill, etc. and are heavily affected by them (Talebi, 1993).

4.2 *Source parameter determinations*

Two independent parameters were calculated directly from the displacement spectra of S waves corrected for attenuation: the amplitude of the low-frequency constant trend of the displacement spectra (Ω_o) and the energy flux (J_C), the integral of the square of ground velocity for the S-wave window. The method used is that of Snoke (1987). Corner frequencies were then calculated from the following relationship:

(4) $$f_0 = (J_C / 2\pi^3 \Omega_o^2)^{1/3}$$

Determinations of corner frequencies for different events were made for each sensor where the estimates of Ω_o and J_C were available. The low-frequency spectral levels were calculated for each sensor as the vector sum of the determinations for the three components of recording on that particular sensor. The energy flux for each sensor was the sum of the measurements for the 3 components. The parameters characterizing the source could then be calculated using the methodology described in the previous section. The following is the range of source parameters:

Seismic moment: 3.6 to 7470 GNm
Source radius: 8.8 to 153 m
Stress drop: 0.03 to 10.6 MPa

4.3 - *Scaling relations*

Figure 4 shows peak velocity parameter Rv plotted versus seismic moment for the present results. McGarr (1984) has analyzed a large number of earthquakes and mine-induced events recorded over a large range of magnitude and obtained a slope of 0.44 for the best-fit line. However, most models lead to a slope of 0.33 for this scaling, making this a model-independent observation. McGarr (1984) shows in his study that indeed Rv scales as $M_0^{1/3}$, as expected, and attributes the discrepancies to several factors such as

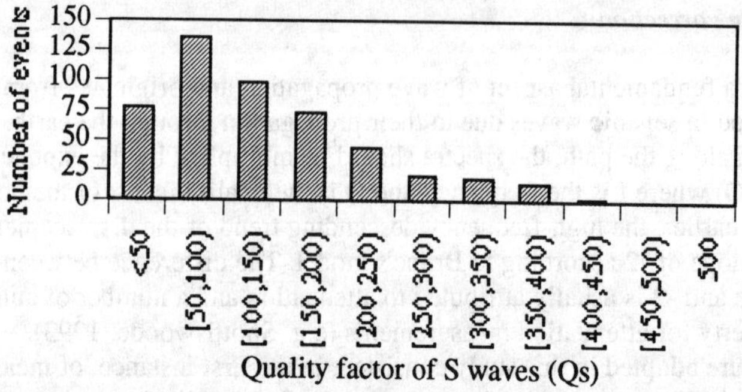

Figure 3 - Histogram showing the results of (Q_S) measurements.

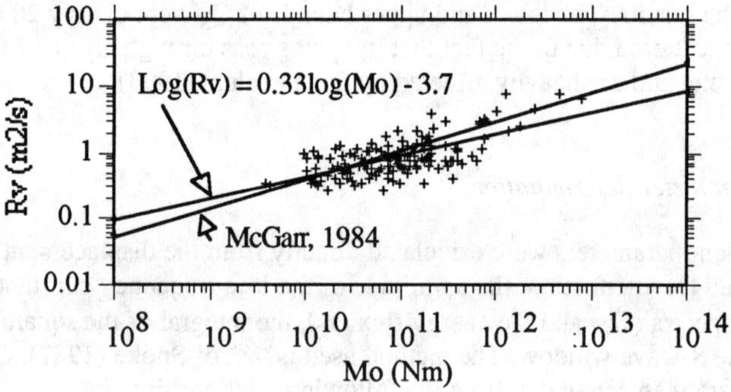

Figure 4 - Peak velocity parameter (Rv) as a function of seismic moment (M_0) for the present case (see the text for more details).

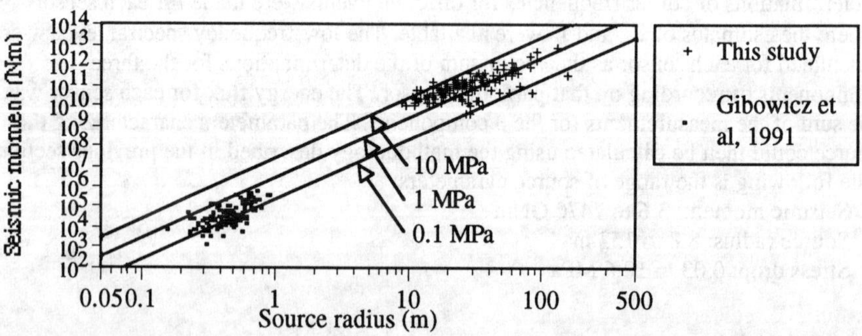

Figure 5 - Seismic moment versus source radius for the present case and the modified results of Gibowicz et al (1991). Straight lines show constant values of stress drop.

crustal stress and focal depth or local site conditions. Interestingly, the best-fit line for the present data-set shows a slope of 0.33, compatible with a self-similar rupture process.

Figure 5 shows seismic moment versus source radius both for the results obtained in this study and those of Gibowicz et al (1991). The latter results were modified to allow comparison with those of the present study, mainly by using Brune's model instead of Madariaga's model. Although stress drops between the two cases are different by a significant factor of about 10, as the clustering of the points on this figure obviously demonstrates, the large majority of the results show a stress drop within the 0.1-10 MPa, as reported in the literature and described in the previous section, and therefore confirm previous observations of self-similarity of seismic sources.

5 CONCLUSION

- Attenuation along the source-sensor path plays a major effect on recorded signals and estimates of source parameters of mine-induced events. Such estimates should be made after correction of spectra for system response and attenuation effects.
- S-wave attenuation factor Q_S varied in the range 20 - 500, the peak being observed between 50 and 150. The amount of scatter observed is due to the presence of areas of fractured rock, backfill, etc., in the mine.
- The large majority of the results show stress drops within the 0.1-10 MPa range. A typical dependence of source parameters expected from theoretical shear models was observed, between seismic moment and source radius on one hand and between peak velocity parameter and seismic moment on the other. These observations strongly indicate a self-similar source behavior.

ACKNOWLEDGMENTS

This work was supported by the Canadian Rockburst Research Program (CRRP) and the Mining Research Directorate (MRD). Their support in making this research possible is gratefully acknowledged. The authors would like to thank their colleagues at CANMET for their help in different aspects of data acquisition and processing. The contribution of Jean-Charles Condo is particularly acknowledged.

REFERENCES

Brune, J.N. 1970. Tectonic stress and the spectra of seismic shear waves from earthquakes. *J. Geophys. Res.* 75: 4997-5009.

Brune, J.N. 1971. Correction. *J. Geophys. Res.* 76: 5002.

Ge, M. and P.K. Kaiser. 1991. Microseismic source location study at Creighton and North mines. Report to INCO Limited. Copper Cliff, Ontario.

Gibowicz, S. J. 1990. Keynote Lecture: The mechanism of seismic events induced by mining: a review. *Proc. 2nd Int. Symp. Rockbursts and Seismicity in Mines*: 3-27, Rotterdam: Balkema.

Gibowicz, S. J., R.P. Young, S. Talebi and D.J. Rawlence 1991. Source parameters of seismic events at the Underground Research Laboratory in Manitoba, Canada: Scaling relations for events with moment magnitude smaller than -2. *Bull. Seism. Soc. Am.* 81: 1157-1182.

Hedley, D.G.F. 1992. Rockburst handbook for Ontario hard-rock mines. CANMET special report SP92-1E.

Madariaga, R. 1976. Dynamics of an expanding circular fault. *Bull. Seism. Soc. Am.* 66: 639-666.
McGarr, A., S.M. Spottiswoode, N.C. Gay and W.D. Ortlepp 1979. Observations relevant to seismic driving stress, stress drop and efficiency. *J. Geophys. Res.* 84: 2251-2261.
McGarr, A. 1984. Scaling of ground motion parameters, state of stress, and focal depth. *J. Geophys. Res.* 89: 6969-6979.
Spottiswoode, S.M. 1993. Seismic attenuation in deep-level mines. *Proc. 3rd. Int. Symp. Rockbursts and Seismicity in Mines*: 409-414. Rotterdam: Balkema.
Spottiswoode, S.M. and A. McGarr 1975. Source parameters of tremors in a deep-level gold mine. *Bull. Seism. Soc. Am.* 65: 93-112.
Snoke, J.A. 1987. Stable determination of (Brune) stress drop. *Bull. Seism. Soc. Am.* 77: 530-538.
Talebi, S. and F.H. Cornet 1987. Analysis of the microseismicity induced by a fluid injection in a granitic rock mass. *Geophys. Res. Lett.* 14: 227-230.
Talebi, S. 1993. Source studies of mine-induced seismic events over a broad magnitude range (- 4 < M < 4). CANMET divisional report # MRL 93-046(CL) .

Rock Mechanics, Nelson & Laubach (eds) © 1994 Balkema, Rotterdam, ISBN 90 5410 380 8

On the use of the brittleness index modified (BIM) to estimate the post-peak behavior of rocks

Michel Aubertin, Denis E.Gill & Richard Simon
École Polytechnique, Montréal, Que., Canada

ABSTRACT: Various types of brittleness measures have been proposed in the past twenty years or so. These are usually aimed at describing, in a relative manner, the potential for a rock to collapse suddenly when loaded up to its ultimate strength. One of these measures is the Brittleness Index Modified (BIM), which was developed by the authors as a variant to the well known W_{et} index. In this paper, the authors first introduce some of the basic ideas behind the BIM. An empirical relationship is then developed between the BIM value and the ratio of the pre-peak modulus over the post-peak modulus, based on a large number of uniaxial compression test results on igneous rocks. Also, using an extended data base on over 50 different types of rocks of various origins, the authors propose some critical BIM values that can be used to evaluate, in a preliminary manner, the bursting liabilities of highly stress rock around underground excavations.

1 INTRODUCTION

The design process of rock structures imply a proper knowledge of rock mass strength properties, as well as the evolving stress field resulting from the excavations. Many rock engineering problems give rise to situations where the induced stresses exceed the peak strength of the rock mass. In such cases, fracturing inevitably occurs in the rock mass surrounding the excavation. Because some portion of the surrounding rock can be brought to the "brittle" state (Wagner, 1989), one must then consider its post-peak behavior, as defined by the descending portion of the stress-strain relationship. It is well known, however, that the post-peak response of brittle rocks depends on the stiffness of the loading system. In that regard, it has been shown that when a rock specimen is submitted to monotonous straining under uniaxial compression, the post-peak behavior can only be followed as long as the absolute value of the post-peak modulus of the specimen is smaller than that of the testing machine (Salamon, 1974). Otherwise, a violent failure occurs. As this phenomenon has been proposed by some authors as a model for specific types of rockburst, the evaluation of the descending portion of the stress-strain curve becomes important when it is expected, in the design process, that the induced stress could exceed the peak strength.

The post-peak behavior of brittle rocks is difficult to obtain from conventional testing facilities, and specially designed stiff or servo-controlled loading systems have to be used (e.g., Hudson et al., 1972; Hakami, 1988). It would nevertheless be practical to be able to make use of test results conducted with "soft" testing machine to estimate the post-peak response of such brittle rocks. For that purpose, the authors are proposing in this paper a brittleness measure called BIM (Brittleness Index Modified) which is obtained from the pre-peak portion of the stress-strain curve of a rock sample submitted to uniaxial compression, and which can be related to the post-peak behavior. It is also shown that this index could be used as a bursting liability index for underground openings.

2 BEHAVIOR OF BRITTLE ROCKS UNDER UNIAXIAL COMPRESSION

2.1 *Load-displacement curve*

If one is concerned that violent failure might take place around an underground excavation, then one must consider the stiffness of the surrounding rock mass and that of the post-failure behavior of the rock, which together determine wether or not a sudden rock failure is likely to occur (Wagner, 1989; Gill et al., 1993).

Since the late fifties, there has been a great deal of effort devoted to the study of the strength properties of brittle rocks using various testing techniques. These studies have focused on defining the peak (ultimate, failure) strength, establishing the strain resistance in both the pre-peak and post-peak stages, and investigating the physical processes that controls the inelastic behavior of these materials (e.g., Bieniawski, 1967; Paterson, 1978; Li and Nordlund, 1993). And although various types of tests have been employed in these numerous laboratory studies, the uniaxial compression test is still the most widely used for engineering applications (Pells, 1993).

One can see on Figure 1 an idealized load-displacement curve for a brittle rock sample submitted to a uniaxial compression test with monotonous straining, together with the mechanical response of the loading system as defined by slope k_{sc}. If k'_r represents the varying post-peak slope of the load-displacement curve, it can be shown that the system is in a stable equilibrium only when the absolute value of k'_r is smaller then that of k_{sc} (Salamon, 1974). The system is stable up to the failure point, but after that, stability depends on the relative values of the two stiffness k_{sc} and k'_r. On Figure 1, the post-peak equilibrium is stable until k_{sc} becomes tangential to the load-displacement curve of the rock sample at point C. At this point, a violent failure occurs. So, in order to follow the post-peak curve of brittle samples, very stiff testing machines have to be used. But even for "infinitely" stiff loading systems, some believe that it may not be possible to follow the behavior of certain very brittle rocks, which appear to show a positive gradient in the post-failure region (e.g., Wawersik and Fairhurst, 1970; He et al., 1990). This type of behavior, known as class II, will not be considered in the following.

2.2 *Physical processes*

Figure 2 shows schematically a typical stress-strain relationship for a rock specimen submitted to a uniaxial compression test. Some phase boundaries are also added. The first phase, curved upward, is associated to the reversible closure of microcracks; in dense rocks with very low porosity, this first phase is almost non-existent. Then follows a linear phase due to the elastic response of the rock, which extends up to the microfracturing threshold where stable crack propagation starts. The onset of microcrack growth, which greatly proceeds the peak strength, commonly begins above about 50% of the ultimate load, as shown by studies on volumetric measurements, acoustic emissions, wave velocity, etc. (e.g., Paterson, 1978; Hakami, 1988; Cox and Meredith, 1993). When approaching the peak strength, the size and density of cracks increase, and cracks interaction becomes more important, so that some unstable crack propagation can be initiated.

Damage accumulation during crack propagation leads to a rapid increase in dilation and eventually to strain localization. It is known that, for brittle materials such as rocks, localization associated to a loss of homogeneity of the strain field, usually occurs in the vicinity of the peak load (e.g., Wawersik et al., 1990). In the post-peak phase, localization phenomena become more important, and usually produce a gradual reduction of the sample cohesion with increasing inelastic strain. This causes a pronounced softening of the material, that is a progressive decrease of strength as strain accumulates.

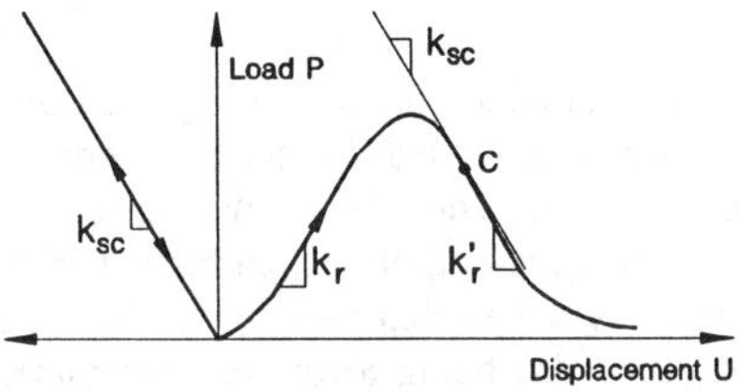

Figure 1 - Schematical load-displacement relationship of a rock sample submitted to a uniaxial compression test (after Salamon, 1974).

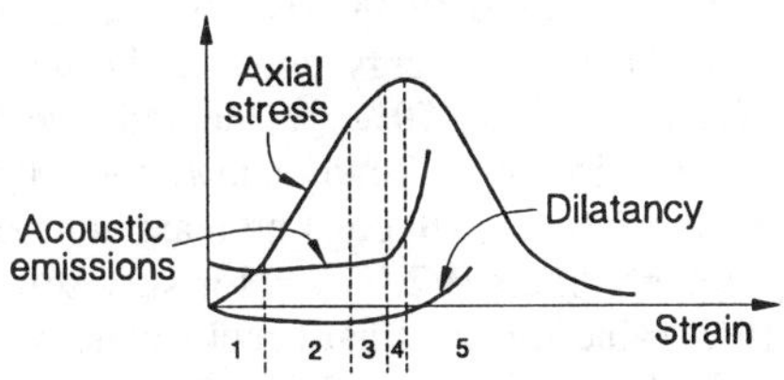

Figure 2 - Schematical stress-strain curve of low porosity brittle rocks (after Bieniawski, 1967, and Paterson, 1978).

3 THE BRITTLENESS INDEX MODIFIED

3.1 *Origin and definition*

It is difficult to define the appropriate measure for the brittleness of a rock sample. One such measure is to establish the rate of softening during post-peak phase (Sheorey and Singh, 1990). The most brittle rocks then show the largest (in absolute value) post-peak modulus (Homand et al., 1990). However, because the post-peak behavior of rock samples is often difficult to follow during compression tests, especially for hard rocks loaded under low confining pressure, other measures of relative brittleness have been developed. Some of these measures, sometimes called "bursting liability indices" or "burst proneness indicators", have been reviewed by Kidybinski (1981), Singh (1989) and Homand et al. (1990). Among the various measures that have been used for that purpose, the W_{et} index proposed by Polish workers (Neyman et al., 1972) is probably the best known and most widely used. This relatively simple index is obtained from the ratio of the recovered (elastic) energy stored in the sample over the dissipated energy. These are determined from hysteresis loop parameters in a uniaxial compression loading and unloading test, as shown in Figure 3. The higher the W_{et} index, the more likely the sample could fail violently if loaded up to its peak strength.

The main problem for the application of the W_{et} index is that, according to the defined procedure, one is expected to load the sample up to about 80 to 90% of the peak strength C_o before unloading it. Unfortunately, the C_o value, which has a stochastic nature, is not known precisely a priori, so that one cannot be sure of the level reached in the loading and unloading cycle. As expected, experimental results have shown that the value of the W_{et} index is sensitive to this stress level (Homand et al., 1990), because the size of the hysteresis loop due to damage accumulation caused by fracturing evolves progressively in the crack propagation phase (Hueckel, 1987).

In order to eliminate this problem, the authors have proposed a modified version of this brittleness measure. It is based on the analysis of a uniaxial compression curve up to failure, considering that the elastic energy can be calculated from the tangential modulus in the elastic phase. Using a typical stress-strain curve, shown schematically on Figure 4, one can then write for the "Brittleness Index Modified", or BIM (Aubertin and Gill, 1988):

$$BIM = A_2/A_1 \tag{1}$$

where A_2 represents the energy given by the total area below the stress-stain curve, and A_1 is the elastic energy stored in the sample, which is obtained here by using the slope of the curve at 50% of the ultimate strength (as often used in elastic modulus computation; e.g., Brady and Brown, 1985; Pells, 1993). The minimum value of the BIM is 1.0 and, although no maximum exists in theory, the authors have observed that it seldom exceeds 2.0 even for very soft rocks such as shale, chalk and rocksalt. The BIM value for most hard rocks is between 1.05 and 1.35, being lower for ultramafic and other very dense rocks.

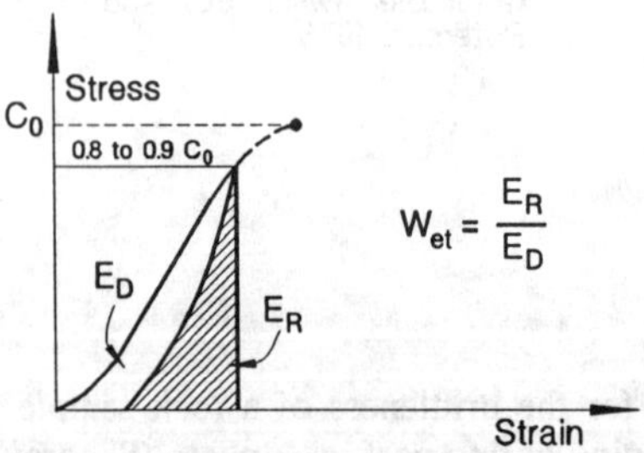

Figure 3 - Calculation of the W_{et} index from a uniaxial compression stress-strain curve (after Kidybinski, 1981).

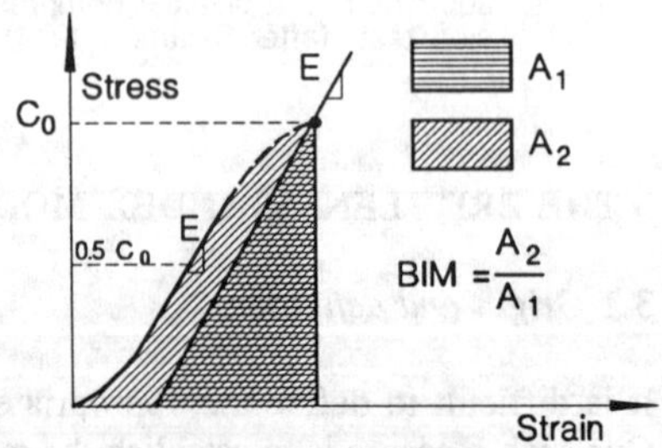

Figure 4 - A schematical representation of the BIM value determination from a uniaxial compression test result (after Aubertin and Gill, 1988).

3.2 *Correlation with the post-peak stiffness*

In the past, some authors have tried to establish correlations between brittleness measures and other rock properties (e.g., Singh, 1989). In order to investigate the possibility that some basic mechanical properties could be related to the proposed index, the authors have compared BIM values with the elastic modulus E, the uniaxial compressive strength C_o and the absolute value of the post-peak modulus $|E'|$ (when available). In this study, the different modulus values were usually obtained numerically after digitalizing stress-strain curves of various rock types, most of which taken from the literature; all the test results considered used specimen with a slenderness ratio of 2.0 or above (Aubertin et al., 1994). The $|E'|$ values were obtained from the maximum negative slope of the contour curve given by the stress-strain relationship. Only negative gradient post-peak behavior have been considered.

This exercise has first shown that no correlation could be established between BIM and E, and although one could perceive a tendency for the BIM value to decrease when C_o and $|E'|$ were increased, results do not seem to indicate statistically significant relationships for any rock types or rock groups (Aubertin et al., 1994). This may simply signify that none of these parameters, taken alone, reflects the brittleness of a rock.

On the other hand, an interesting correlation was established between the BIM value and the $|E/E'|$ ratio for igneous rocks of various origins. This ratio has been considered by some as a measure of the relative brittleness of rocks (Homand et al., 1990). This correlation, which is shown on Figure 5, can be written as follows:

(2) $$BIM = 0.42\ |E/E'|\ + 0{,}98$$

The statistical significance of this correlation is relatively weak (correlation coefficient of 0.53) but, as seen on Figure 5, there is a clear trend for the $|E/E'|$ ratio to increase when BIM increases.

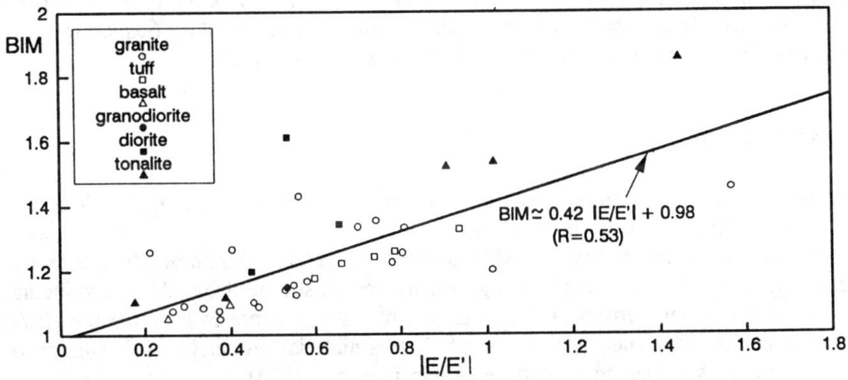

Figure 5 - An empirical linear correlation between the BIM and I E/E' I values for igneous rocks (details given in Aubertin et al., 1994).

This correlation could be used to estimate, at least for preliminary calculations, the value of E' for the intact rock, which is sometimes used in stress analysis calculations. It could also be used in the establishment of the relative stiffness of the failed rock versus that of the surrounding rock mass (e.g., Gill et al., 1993). This correlation does not however take into account slenderness effects for equivalent width to height ratio above 0.5; it could be modified according to existing relationships developed for that purpose, such as the one presented in Brady and Brown (1985) for instance.

3.3 *BIM and bursting liabilities*

A brittleness measure such as the BIM represents the ease with which a rock sample can dissipate energy by damage accumulation due to crack propagation before failure. This expenditure of energy means that less elastic strain energy is stored in the sample, thus reducing the amount that can be released in the post-peak phase. Considering that a higher BIM value corresponds to a greater amount of energy expended before failure and to lesser energy available to be released after peak, then one can expect that a higher BIM value leads to a smaller chance of having a violent failure in situ. As others have done with the W_{et} index, the authors have used BIM values to establish, in a preliminary manner, the bursting liabilities of rock masses submitted to high stress states. Using more than 300 tests results on over 50 different types of rocks, the authors have selected indicative values of the BIM in relation with in situ bursting liabilities based on case studies. These indicative values are presented in Table 1.

Table 1 - Indicative values of BIM as related to bursting liabilities.

BIM	Bursting liabilities
between 1.00 and 1.20	high
between 1.20 and 1.50	moderate
above 1.50	low

One should mention here that almost every cases where rockbursts have been identified showed BIM values between 1.05 and 1.20. This was the case for igneous rocks (such as tuff, andesite and porphyry), as well as for sedimentary (jasperoid, coal) and metamorphic (quartzite, schist) rocks (Aubertin et al., 1994).

4 DISCUSSION

Although the above mentioned results can be appealing for practical applications, one must be careful in the way these are used.

One is first reminded that the BIM is based on various hypotheses that, although frequently adopted in rock engineering practice, could still be debated. These include the use of the linear portion of the uniaxial stress-strain curve to estimate the elastic modulus of the rock, and the adoption of C_o as a material property. A discussion on these elements is presented elsewhere (Aubertin et al., 1994).

The authors have also used E', the post-peak modulus, as an intrinsic material characteristic in establishing the correlation given by equation 2. However, various studies have shown that the post-peak behavior of brittle materials is rather a structural response which depends on various factors, including the global stiffness of the loading machine and the performance of the (servo-) controlled system, the size and shape of the sample, the rate of loading, and the boundary conditions on the sample surface (e.g., Hudson et al., 1972; Hakami, 1988; Labuz and Biolzi, 1991). Because the authors have used experimental results from different sources, many of which were taken from the literature, some of these factors were probably not uniform. Thus, the results were certainly affected by these differences. Nevertheless, care was taken to select results which appeared consistent with the above mentioned restrictions. More details on this aspect are also given in Aubertin et al. (1994); in this same report, the authors also discuss the reasons for selecting only class I stress-strain curves.

It should be further emphasized that the use of the BIM to establish possible bursting liabilities should only serve as a preliminary indication. Only a few real situations encountered underground can be directly linked to the uniaxial behavior of rocks, the loading of pillars being probably the best example. There are many other possible causes for rockbursts, such as those induced by slip along weak planes for instance, and these are related to loading conditions that are far from that of a uniaxial compression test. A possible practical use of the BIM in that regard would be to evaluate, in a relative manner, the bursting liability of different zones in a mine or portion of a mine. It could also be combined with other parameters, such as the well-known Energy Release Rate (ERR) or the Excess Shear Stress (ESS), to establish more precisely rockbursting liabilities.

It is finally interesting to note that, when comparing the BIM and W_{et} index, one can show that a BIM value of 1.20 corresponds to a W_{et} index value of 5 for the case where the complete loading curve is considered (up to failure), while a BIM value of 1.50 corresponds to a W_{et} index value of 2. For the latter index, these values have been associated to high and non-existing bursting liabilities respectively (Kidybinski, 1981).

5 CONCLUSION

In this paper, the authors have introduced the brittleness index modified (BIM) which is a variant of the W_{et} index previously developed by Polish workers. The BIM is based on the use of the complete stress-strain curve of rock samples submitted to uniaxial compression up to failure. Using a large number of test results on various types of rocks, an interesting correlation is established between the BIM and the post-peak modulus of rock samples. This correlation could be used, for preliminary calculations, in order to establish the relative stiffness of failed rock and of the surrounding rock masses around an underground opening, as is required when one wants to evaluate the rockburst potential following a methodology such as the one proposed by Gill et al. (1993). Then, the computed BIM values are used to establish various classes of bursting liabilities according to observed in situ behavior of rock mass around underground excavations in highly stress conditions.

ACKNOWLEDGEMENTS

This project was financed by grants from IRSST and NSERC (OGP0089749 and OGP0003412).

REFERENCES

Aubertin, M. and Gill, D.E. 1988. Une méthodologie d'évaluation du potentiel de coups de terrain dans les mines d'Abitibi. Proc. Colloque sur le Contrôle de Terrain (AMMQ), Val d'Or, 47-77.

Aubertin, M., Gill, D.E. and Simon, R. 1994. Une étude sur l'utilisation de l'indice de fragilité modifié (BIM). Report EPM/RT-94/xx, École Polytechnique.

Bieniawski, Z.T. 1967. Mechanism of brittle fracture of rock, Part I, II and III. Int. J. Rock Mech. Min. Sci., 4:395-430.

Brady, B.H.G. and Brown, E.T. 1985. Rock Mechanics for Underground Mines, George Allen and Unwin.

Cox, S.J.D. and Meredith, P.G. 1993. Microcrack formation and material softening in rock measured by monitoring acoustic emissions. Int. J. Rock Mech., Min. Sci. and Geomech. Abstr., 30:11-24.

Gill, D.E., Aubertin, M. and Simon, R. 1993. A practical engineering approach to the evaluation of rockburst potential. Proc. 3rd Int. Symp. on Rockbursts and Seismicity in Mines, Kingston:63-68, Balkema.

Hakami, H. 1988. Post-failure Behaviour of Brittle Rock. Doctoral Thesis, Luleå University of Technology, Sweden.

He, C., Okubo, S. and Nishimatsu, Y. 1990. A study of class II behaviour of rock. Rock Mechanics and Rock Engineering, 23:261-273.

Homand, F., Piguet, J.P., Revalor, R. and Josien, J.P. 1990. Dynamic phenomena in mines and characteristics of rocks. Proc. 2nd Int. Symp. on Rockbursts and Seismicity in Mines, Minneapolis, 139-142, Balkema.

Hudson, J.A., Crouch, S.L. and Fairhurst, C. 1972. Soft, stiff and servo-controlled testing machines: a review with reference to rock failure. Engineering Geology, 6:155-189.

Hueckel, T. 1987. Hysteresis Evolution and Failure in Cyclic Loading of Dilatant Rocks. Constitutive Laws for Engng. Materials: Theory and Applications. 565-572, Elsevier Science Pub.

Kidybinski, A. 1981. Bursting liability indices in coal. Int. J. Rock Mech. Min. Sci. Geomech Abstr., 18:295-304.

Labuz, J.F. and Biolzi, L. 1991. Class I vs class II stability: A demonstration of size effect. Int. J. Rock Mech. Min. Sci. and Geomech. Abstr., 28:199-205.

Li, C. and Norlund, E. 1993. Deformation of brittle rocks under compression - with particular reference to microcracks. Mechanics of Materials. 15:223-239.

Neyman, B.Z., Szecowska, Z. and Zuberek, W. 1972. Effective methods for fighting rockbursts in Polish collieries. Proc. 5th Int. Strata Control Conf., London, 23.1-23.9.

Paterson, M.S. 1978. Experimental Rock Deformation - The Brittle Field. Springer-Verlag.

Pells, P.J.N. 1993. Uniaxial strength testing. Comprehensive Rock Engineering, Principles, Practice and Projects. 3:67-85, Pergamon Press.

Salamon, M.D.G. 1974. Rock mechanics of underground excavations. Advances in Rock Mechanics, Proc. 3rd Cong. Rock. Mech.,Denver, 1-B:951-1099, National Academy of Science.

Sheorey, P.R. and Singh, B. 1990. Studies on bump proneness of Dishergarh seam at Barmondia Colliery. Rockbursts - Global Experience. 131-143, Balkema.

Singh, S.P. 1989. Classification of mine workings according to their rockburst proneness. Mining Science and Technology, 8:253-262.

Wagner, H. 1989. Underground openings in overstressed rock. Proc. 6th Cong. Rock Mech., Montreal, 3:1443-1456, Balkema.

Wawersik, W.R. and Fairhurst, C. 1970. A study of brittle rock fracture in laboratory compression experiments. Int. J.Rock Mech. Min. Sci., 7:561-575.

Wawersik, W.R., Rudnicki, J.W., Olsson, W.A., Holcomb, D.J. and Chau, K.T. 1990. Localization of deformation in brittle rock: theoretical and laboratory investigations. Micromechanics of Failure of Quasi-Brittle Materials, 115-124, Elsevier Applied Science.

Rock Mechanics, Nelson & Laubach (eds) © 1994 Balkema, Rotterdam, ISBN 90 5410 380 8

The (in?)significance of creep in a prestressed polyester resin grouted rock anchor

T.S. Avery
Woodward-Clyde Consultants, Oakland, Calif., USA

J.J.K. Daemen
Department of Mining Engineering, Mackay School of Mines, University of Nevada, Reno, Nev., USA

ABSTRACT: One of the major concerns regarding polyester resin grouted rock anchors is whether the resin creeps when loaded. The most extreme case would be that the resin behaves in a visco-elastic manner, exhibiting continuous movement or load decay. Examination of pull test results, however, indicates that polyester resin does not behave visco-elastically. The exact rheological behavior is unknown, however, time-dependent behavior is load-sensitive. When either the grout or rock-grout interface strength is exceeded, the excess load will transfer along the bolt. Should the anchor length "saturate" with load, the excess load will be shed onto the free length of the rock anchor, loading it in compression. The load distribution in the free length will require proportionately less transfer length than the equivalent load in the anchor zone due to the compressive loading of the free length. The added design length due to safety factors should prevent anchor failure by creep if the bond strength is adequately determined.

INTRODUCTION

Concern has existed over whether or not polyester resin grouted rock bolts or anchors creep when loaded. The fact that polyester resin is a plastic material (in the chemical sense) with an inert filler (ground limestone and other minerals) adds credibility to the notion that resin creeps. There has yet to be a definitive study to determine the exact rheological nature of polyester resin. Consequently, questions often arise regarding the stability of a loaded anchor when grouted with polyester resin.

Some of the earliest studies on the time-dependent nature of polyester resin was conducted by Whittaker, Breckels and Daws (1977). They conducted several tests on short anchorages in sandstone. Their findings indicated that in sedimentary rock, the standard resin design formula of:

$$L = 25P + 50 \qquad (1)$$

where

L = Length of anchor (mm)
P = Load in tonnes

adequately prevents the occurrence of creep. Whittaker and Grant (1980) state that this equation also works well for igneous and sedimentary rocks. One obvious aspect of the equation is that it does not take into account material constants of the rock, polyester resin, or the anchor bar. While this equation prevents the occurrence of creep in weaker rocks, it results in an overconservative design in strong rocks.

A later report by Miller (1980) investigated the load decay of a polyester resin grouted rockbolt. A 1-inch diameter, grade 60 length of rebar was grouted into 1 1/2-inch diameter hole in concrete to achieve a 12-inch bonded length. A 19 kip load was applied to the bolt by tightening a nut. The load was monitored over time. The load decayed to 16.38 kips after 22 hours. The nut was tightened on the bolt to bring the load back up to 19 kips, and the load was monitored for 24 hours after which the load decayed to 18.2 kips. The pattern of loading, observing the load decay and retightening the nut continued for several days. After the seventh day, the 19 kip load was reapplied, and the bolt monitored for 21 months. The load decayed to approximately 17.6 kips. A final 40 month load decay test was conducted which resulted in an approximate 17.4 kip load. At the end of 61 months of testing, a pull-test was conducted. The bar failed at a load of 51 kips.

A recent report published by the Army Corps of Engineers (Best and McDonald, 1990) discussed the data collected from 6 months of creep testing. Lengths of 3/4-inch diameter, grade 60 rebar were grouted into a 1 1/4-inch diameter hole to achieve a 15-inch bond length. The bolts were subjected to a constant load of 16,360 pounds. The creep tests on the dry installed polyester resin grouted bolts exhibited primary creep characteristics with 95% of the displacement occurring during the first month of testing. The tests were not conducted long enough to determine if secondary creep characteristics were occurring. One of their conclusions was that "creep data should be considered in the selection of an anchor grout where the frictional resistance and bond strength between the two masses to be anchored together are important."

The above creep studies have yielded specific information on narrow aspects of polyester resin grouted anchoring. Studies have yet to address general questions regarding polyester resin creep in various rock types and at various loads outside the general empirical equation presented by Whittaker, et al (1977). The intent of this paper, as the title implies, is to discuss the significance, or perhaps insignificance, of the occurrence of creep in a prestressed, two stage polyester resin grouted anchor. The results of this investigation can also be applied to a polyester resin grouted rock bolt.

Several factors must be investigated to determine those conditions which lead to load redistribution or time-dependent behavior. The first factor is the polyester resin itself, and whether it is inherently subject to creep as a visco-elastic substance, or is there a threshold stress beyond which creep occurs. Second, if creep is dependent on a threshold stress, the load or stress distribution along the anchor will indicate the degree to which the resin is overloaded. And third, if the strength of the grout-rock interface is less than a creep threshold stress, it will govern the performance of the anchor. The relationships between the resin's time-dependent behavior, the load transfer mechanics and the rock-grout strength should lead to an understanding of the potential for creep to occur in a specific rock anchor system. Should creep occur, its effect can be assessed on the load capacity of that support to carry out its design function.

CREEP

Creep is a general term covering time-dependent behavior of a material. The time-dependent behavior has been classified into primary, secondary, and tertiary behavior. An object exhibiting primary creep characteristics displaces over time at an exponentially decaying rate under a constant load. A linear increase in displacement over time is secondary creep. Tertiary creep is exponentially increasing displacement with time, which leads rapidly to failure.

Rheological models to simulate a material's time-dependant behavior have been developed that permit predictions to be made on the long-term behavior of that material under load. Of greatest concern is whether the material is a visco-elastic substance, which would exhibit continuous linear deflection under a constant load, or an exponential load decay trending to zero in time. The time at which the load reaches 1/e times the initial load is called the relaxation time (Jaeger and Cook, 1979). The equation governing visco-elastic load decay is given by:

$$\sigma = k\varepsilon_{(0)}e^{-t/t(1)} \quad (2)$$

where

σ = stress at time, t

k = spring constant of the elastic portion of the visco-elastic model

$\varepsilon_{(0)}$ = instantaneous strain induced at load application

$t_{(1)}$ = relaxation time.

Using Miller's data at various times should yield a consistent value for $t_{(1)}$ if the material behaves according to the visco-elastic model. Calculated relaxation times range from 9,000 min for the 22 hour load decay test to 30.5×10^6 minutes for the 40 month load decay test. The data available appears to preclude the assumption that polyester resin behaves in a visco-elastic manner.

A model that might more accurately model the resin's behavior is a Zener solid (Farmer, 1983). If the material behaves according to this model, the stress should decay asymptotically to a constant proportional to the internal spring constants of the model. Fitting the behavior to an actual model should yield important results. However, the data available on the above test is insufficient to match the characteristics of the resin to a specific rheological model.

The assumption that polyester resin behaves according to a Zener solid is based on another assumption that the system behaves in a linear fashion. Non-linear models incorporating a St. Venant, or frictional substance take into account displacements induced at higher loads. The incorporation of a frictional material in the investigation may accurately model the behavior of a grouted bolt inside a borehole. An examination of the grout-rock interface, usually the weak link in a resin grouted anchor or bolt, may yield the key to understanding the time-dependent behavior of polyester resin grouted rock anchors.

LOAD DISTRIBUTION

The load along a polyester resin grouted bolt has been shown both theoretically and experimentally to be transferred into the rock over a relatively short length. The tensile load is transferred into the rock as shear stress through the anchor grout. A comparison of the theoretical and measured stress distribution along a resin grouted anchor was made by Farmer (1980). As long as the system remains elastic, the load transfer from

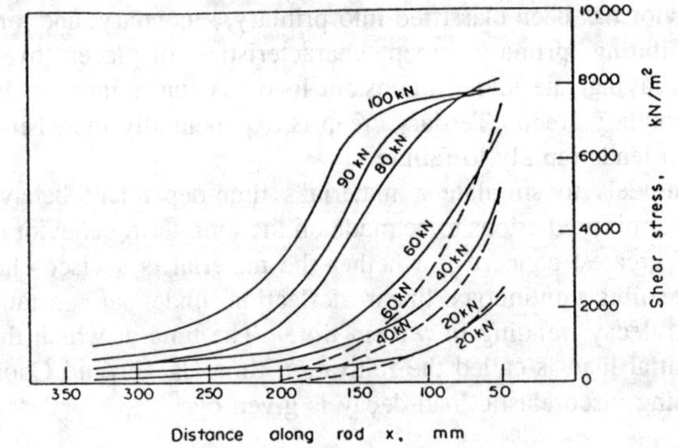

Figure 1. Solid line represents computed shear stress distribution from strain gauge measurements on 350 mm long anchors grouted into concrete. Dashed line is theoretical shear stress distribution (from Farmer, 1975).

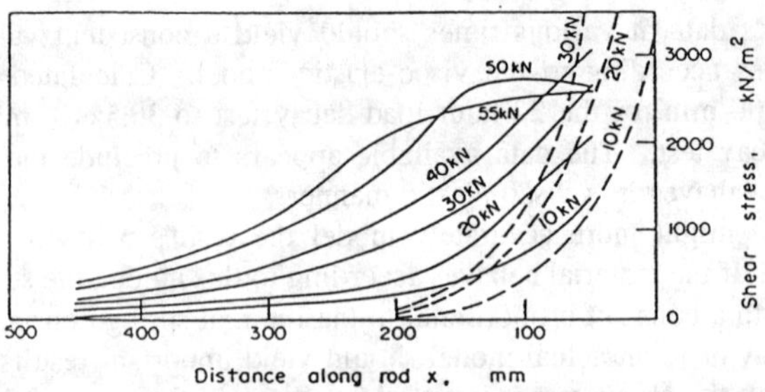

Figure 2. Shear stress distribution curves for 500 mm long anchors grouted into limestone (from Farmer, 1975).

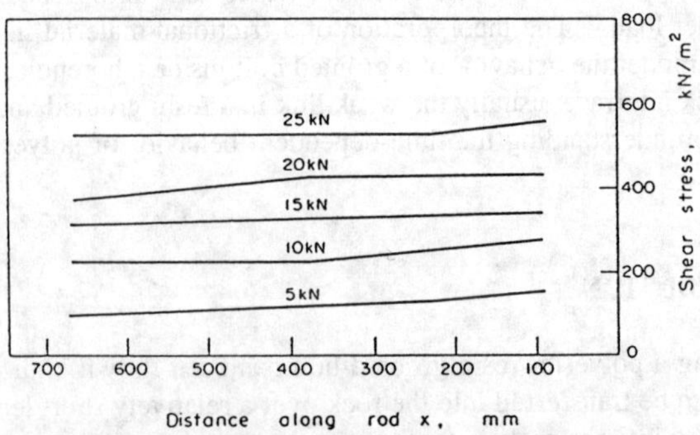

Figure 3. Shear stress distribution curves for 700 mm long anchors grouted into chalk (from Farmer, 1975).

bolt to the rock can be represented by the following closed-form decay formula:

$$\sigma_{(x)} = \sigma_{(0)}e^{-\alpha x} \tag{3}$$

where

$\sigma_{(x)}$ = axial normal stress in bolt at a distance y from the loaded end, psi,
x = distance along bar from the applied load

Alpha is a decay coefficient which is a function of the system's stiffness, and can be expressed as:

$$\alpha^2 = E_{(g)} /[E_{(a)}a^2(\ln(R/a))], \tag{4}$$

where

$E_{(g)}$ = grout modulus,
$E_{(a)}$ = the deflection at the load applied end of the bolt.
a = anchor radius
R = Borehole radius

A significant simplification of the governing differential equation (refer to Farmer (1975) for a complete discussion on the development of these equations) assumes L is much larger than $1/\alpha$, thus the decay becomes:

$$\xi_x = \sigma_{(0)}/(\alpha E_{(a)}) \bullet e^{-\alpha * x} \tag{5}$$

where

ξ_x = displacement of the bolt at distance x from the loaded end

Farmer compared the theoretical stress distribution with finite element modelling and actual bolt pull-out tests. The comparison between the theoretical stress distribution and pull tests showed that the two correlated at low anchor loads (approximately 3,000 pounds) in moderate strength concrete blocks. At higher anchor loads, the theoretical and measured stress levels diverged, indicating that the anchor was "debonding", mobilizing the residual strength of the rock/resin interface. Figure 1 shows the pull test results of 350-mm long strain-gaged rock bolts installed in a concrete block. In softer limestone and chalk test blocks, debonding occurred at lower loads, as shown on Figures 2 and 3, respectively.

The alpha term in the load distribution equation (5) is of great significance. If x is set to zero, the displacement measured at the end of the anchor becomes:

$$\xi_x = \sigma_{(0)}/\alpha E_{(a)} \tag{6}$$

If an anchor behaves in a linearly elastic manner, alpha should remain constant as $\sigma_{(0)}$ increases. Plotting alpha as a function of load will result in a straight line with a horizontal slope. Divergence from the line indicates non-linear behavior of the load distribution. The point at which alpha diverges may represent the shear strength of the rock-resin grout interface. From a practical standpoint, careful measurement of the displacement while loading or stressing a bolt will allow the determination of alpha.

Pull test data obtained in Avery and Friant (1993) was used to develop the plot of alpha versus load shown on Figure 4. Two interesting observations can be made from the alpha plot. The first is the variation in average alpha values for the five bolts,

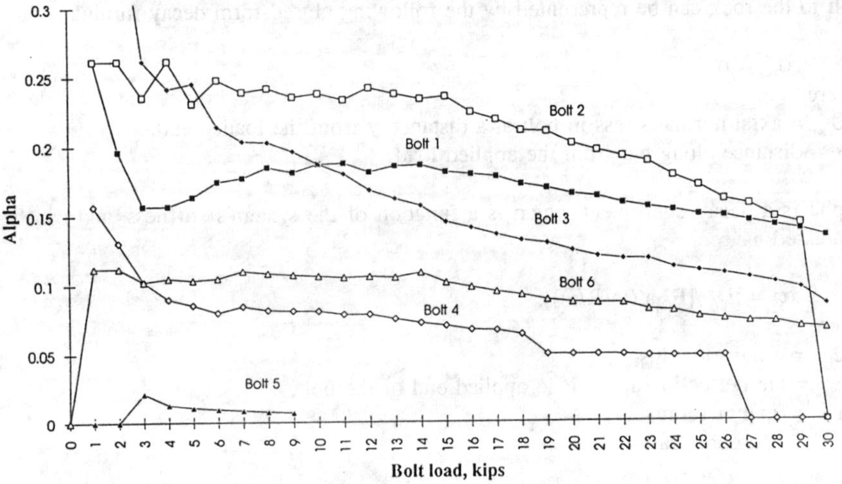

Figure 4. Alpha versus load for bolts grouted into concrete. Bolt 1 - 24 inch grouted length installed in a dry hole; Bolt 2 - 24 inch bolt in damp hole; Bolt 3 - 24 inch bolt in wet hole; Bolt 4 - 24 inch bolt in submerged hole; Bolt 5 - 12 inch bolt in submerged hole; Bolt 6 - 36 inch bolt in submerged hole.

ranging from a high of approximately 0.235 to a low of 0.01. These differences reflect various installation conditions and bolt lengths. The installation conditions ranged from dry to damp to submerged boreholes; bolt lengths ranged from 17 to 36 inches. The highest alpha constant was for a 26 inch bolt installed in a dry borehole, the lowest was for a 17 inch bolt installed in a submerged borehole. The quality of the installation can be assessed by the alpha value (in the elastic region).

The second observation is the point at which the curve deviates from a constant alpha value. This indicates debonding or mobilization of the skin friction of the proximal portion of the bolt. Bolts 1, 2 and 6 show a departure from a constant alpha value at loads between 15 and 17 kips. If 15 kips is the load at which debonding or mobilization of skin friction occurs, the shear stress at that point equals the shear strength. The shear stress distribution along the bolt-grout interface from Farmer (1975), is governed by the equation:

$$\tau_{(x)} = 1/2 \cdot a\alpha\sigma_{(0)}e^{-\alpha x} \tag{7}$$

Therefore, at a 15 kip load (33,953 psi for the 3/4 inch bolt), the shear stress at the bolt-grout interface is 1,423 psi, which is 1,067 psi at the grout-concrete interface (one inch diameter borehole). Farmer's strain gaged bolts showed a shear stress of approximately 8,000 kN/m^2 at the bolt-grout interface, which is 1,160 psi at the bolt-grout interface and 828 psi at the grout-concrete interface. If the material is not visco-elastic, the load should be stable below the shear strength of the grout-rock interface.

PERFORMANCE

The consequences of load redistribution can only be assessed in light of the design use of that particular support element. Two primary uses of polyester resin could be classified as rock bolting and rock anchoring. One main distinction between the two is that a rock bolt is untensioned or provided with only a slight alignment load and a rock anchor has a specified prestress load. In rock bolting, the bolts act as passive supports requiring rock movement to mobilize rock bolt strength. When used in anchoring, a prestress is applied which prevents movement of the structure, thus requiring a constant load to be maintained.

ROCK BOLTING

The first case, rock bolting, requires rock movement to mobilize the rock bolt. An example of this is a pattern rock bolt support scheme in a tunnel. The rock bolts are installed before the rock mass completely relaxes. The load that develops in a particular rock bolt due to joint dilation will be governed by (5), where ξ_x represents the displacement induced across the joint. The load distribution along the rock bolt will develop according to (3). If the bolt is long enough to distribute the applied load, the rock load will stabilize and movement of the rock arch will cease.

ROCK ANCHOR PERFORMANCE

The load developed on a two stage resin grouted anchor is developed during prestressing. Generally, the prestress load will equal or exceed the load required in the structure being anchored. The main requirement of the anchor is that it transmit the design load into the ground. Ignoring ground-induced loadings, changes in the prestress load should only occur prior to the set of the slow set resin. A slight decrease in tension may occur while the slow set resin is still fluid due to load redistribution in the anchor zone. The load decay will be minimized by stored tension in the free length of the anchor.

Should the load decay be greater than anticipated, ie, the strength of the grout-rock interface be less than anticipated, load will be transferred into the free length of the anchor. The load transfer will occur according to equation (5). Alpha, which is a function of the grout modulus, will increase by the square root of the ratio of grout's compressive to tensile modulus. The load transfer length decreases linearly with an increase in alpha. Therefore, if the modulus increases by an order of magnitude, the load transfer distance decreases (for a given load) by 66%.

As stated earlier, the theoretical load distribution along a bolt yields to the shear strength of the grout-rock interface. While the increased modulus equates to a shorter load transfer length, the actual grout-rock strength will govern the transfer length. DuPont introduced a bolt for coal mines that capitalized on loading the resin in compression. The bolt, although not a commercial success, in some cases tripled the strength of a resin grouted bolt in weak rock. The tripled grout-rock strength would result in a load transfer distance of approximately one third than that of a grout column loaded in tension. The load transfer from the anchor length to the free length will only affect the structure it reinforces if the transfer in the free length crosses into the stabilized wedge. The prestress load across the two blocks would decrease by the value of the load at the interface. As is the convention with cement grouted permanent ground anchorages, an additional length of one-fifth the vertical slope height is added to the free length of the anchor. This length provides an additional factor of safety against anchor pullout or loss of prestress.

CONCLUSION

It has been tentatively shown that polyester resin is not a visco-elastic substance. Verification of this would show that long term behavior under load would be governed by the lower of either the shear strength of the resin or of the rock - resin interface. The ability of polyester resin to redistribute loads exceeding the governing strength then requires a sufficient anchor length to effectively transfer the load. Should the anchor length be overstressed, the excess load will be transferred to the free length, loading it in compression. The compressive loading will distribute the load over a shorter length than if loaded by tension. Prestress loading across a given plane will not decrease until the load transfer along the bolt crosses that plane

REFERENCES

Avery, T.S. and Friant, J.E. 1993. Applications and Testing of Resin Grouted Rockbolts.Technical Report REMR- GT-17, US Army Engineer Waterways Experiment Station, Vicksburg, MS.

Best, J.F., and McDonald, J.E. 1990. Evaluation of Polyester Resin, Epoxy and Cement Grouts for Embedding Reinforcing Steel Bars in Hardened Concrete. Technical Report REMR-CS-23, US Army Engineer Waterways Experiment Station, Vicksburg, MS.

Farmer, I. W. Stress Distribution Along a Resin Grouted Rock Anchor. Int. J. Rock. Mech. Min.Sci. and Geomech., v. 12, 1975, pp. 347-351.

Farmer, I. W. 1983. Engineering Behaviour of Rocks, Chapman and Hall, London

Jaeger, J.C. and Cook, N.G.W. 1979. Fundamentals of Rock Mechanics, Chapman and Hall, London.

Whittaker,B.N., Breckels, I.M., & Daws, G. Performance of Resin Grouted Reinforcing Bars in Carboniferous Rocks with Special Reference to Time- Dependent Behaviour. Proceeding of a Conference on Rock Engineering, Newcastle upon Tyne, U.K. April, 1977.

Whittaker, B.N., Grant, A.S. Time-dependence, load relaxation and performance aspects of polyester resin reinforcing bars in sedimentary rocks, International Conference on Structural Foundations on Rock, Sydney, May, 1980.

Rock Mechanics, Nelson & Laubach (eds) © 1994 Balkema, Rotterdam, ISBN 90 5410 380 8

Rockbolt behaviour under dynamic loading: Field trials and numerical simulation

R.K.Brummer, X.Yi, D.D.Tannant & G.M.McDowell
Geomechanics Research Centre, Laurentian University, Sudbury, Ont., Canada

ABSTRACT: As a part of the Canadian Rockburst Research Programme (aimed at investigating the performance of rock support under rockbursting conditions) in-situ dynamic monitoring of various types of rockbolts loaded by blasts was performed. The aim of the monitoring was to investigate and explain the interaction of the dynamic load with the rock-rockbolt system. The measured strain waves on the bolt and the dynamic movement of the rock surface were successfully simulated numerically using a one dimensional finite difference representation of the mechanical system involving the wave source, the rockmass, the mechanical rockbolt, and the anchor and head connections. The model was calibrated using field data from the blasts, and parametric studies of the influence of various system parameters on the dynamic behaviour of the rockbolt were carried out. It was found that the wave source, rockmass modulus, tensile strength, connections at the rockbolt ends, and rockbolt pre-tension all play important roles in determining the dynamic stress levels in rockbolts.

1. Introduction

The Canadian Rockburst Research Programme has as one of its objectives the development of support systems which are capable of withstanding the damaging effects of rockbursts. As part of this overall programme, the Geomechanics Research Centre has conducted several field investigations into the behaviour of rockbolts under various forms of dynamic loading. During 1993, field trials at the Bousquet Mine and at the CANMET Experimental Mine in Quebec were carried out. These trials involved dynamically loading various different support systems (rockbolts, screen and shotcrete in different combinations) using explosives, and monitoring the damage caused, as well as the dynamic motions and strains in the support components. A numerical model was developed to assist in the interpretation of the dynamic records from the rockbolt measurements.

2. Field Trials

The field trials described in this paper were carried out at the CANMET

Experimental Mine near Val d'Or, Quebec. This mine no longer produces ore and is operated by CANMET as an experimental underground facility. The trials described here were carried out in two parallel drifts at shallow depth in the mine. The rockmass at the test site is a good quality porphyritic granodiorite with a compressive strength of 190 MPa and a dynamic Young's modulus of 60 GPa. Figure 1 is a plan of the drift in which the blasting took place and the parallel observation drift where three instrumented rockbolts and six geophones were installed. The wall monitored in the observation drift was located 6.4 m from the centreline of the blasted drift.

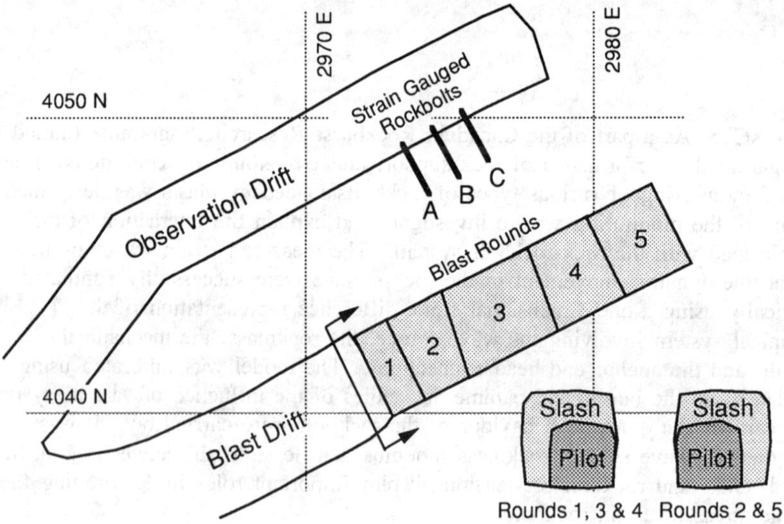

Figure 1 Plan view of observation and blast drift showing location of monitored rockbolts and the five slash rounds

The particle velocities on the rock surface were measured immediately adjacent to three strain-gauged rockbolts. This allowed comparisons between particle velocities in the rock and dynamic strains or stresses in the rockbolts. The blasted drift was excavated in five stages. Each stage consisted of a slash into a previously excavated pilot opening. Figure 1 also shows the location of the slash with respect to the pilot round for each excavation stage. For the slash blasts, blasthole diameters were 32, 45, and 57 mm. The blastholes were charged with either pneumatically loaded ANFO at an in-hole density of 0.9 g/cm³ or small diameter cartridge MineRite at a nominal density of 1.2 g/cm³. To prevent excessive acquisition recording times, short period delays were used so that the total blast initiation period was limited to 450 milliseconds. All five slashes were monitored during the course of the experiment.

Several tubular end-anchored rockbolts were installed in the observation drift and these were instrumented with pairs of electrical resistance strain gauges mounted longitudinally along the shank of the bolts. Strain gauge pairs were attached on opposite sides of the shank at distances of 0.3 m and 0.8 m from the heads of the rockbolts. The tension in the bolts (45 to 70 MPa) created by tightening the nuts was sufficient to cause a slight dishing of the 150 mm square, 6.4 mm thick flat steel plates. Two uniaxial, long travel geophones (McSeis-101 LT 8 Hz) were installed on the rock surface immediately adjacent to each rockbolt to measure the rock vibrations. The strain and velocity waveforms were recorded at sampling rates of 10 kHz using a portable seismic system.

3. Experimental Results

It was found that the end-anchored rockbolts were loosened by the blasts. The loss in tension in the rockbolts was at least 10 to 12 MPa during the course of the experiment. The closest blast (the "S3" blast - a pattern of 7 57 mm by 2.4 m long blastholes charged with ANFO) caused slight to moderate spalling damage to the wall of the observation drift. For this blast, peak particle velocities (ppv) at the surface of the observation drift of 1.0 to 1.2 m/s were measured and the highest dynamic stresses induced in the rockbolts were on the order of 200 MPa. Results from the second delay of the S3 blast and strains measured on rockbolt B are discussed further in this paper since this rockbolt was oriented perpendicular to the blast's dynamic compression wavefront. Figures 2(a), 2(c) & 2(e) show the velocity measured on the free surface of the rock, the axial dynamic strain in the bolt at 0.8 m from the bolt head and the Fourier spectra at 0.3 m and 0.8 m from the head. Further details on the field experiment are described by Tannant *et al* (1994).

Note that the strain waveform in Figure 2(c) contains only the dynamic axial strains - the pre-existing static strain at the beginning of the dynamic loading was removed from the plot, and bending effects were minimised by averaging the gauge readings from opposite sides of the bolt. The measured peak dynamic stresses (obtained by multiplying the dynamic strain by the Young's Modulus for the bolt) were within the elastic range. The strain records showed very clearly that for some delays there was a permanent reduction in the bolt tension presumably caused by factors such as anchor slip or rock damage. This bolt loosening can be seen clearly in Figure 2(c) where a 10 millistrain reduction occurs between the beginning and end of the waveform.

4. The Numerical Model

The rheological components and discretization of the 1-D model are shown in Figure 3. The rock face is treated as a planar free surface. The end-anchored tensioned mechanical bolt is connected to the surrounding rockmass through the shell anchor and to the free surface through the face plate. The rockmass is treated as a series-connected array of masses and springs. Dashpots are also incorporated as shown, to allow for damping. In order to allow for the fracturing or blast damage which usually occurs near the free surface, a "fractured zone" was also incorporated. This region is treated as an elastic material whose stiffness in tension is small compared to its stiffness in compression.

The bolt shank and the adjacent rockmass are discretized into equal-length elements, with each element represented by a mass, a spring, and a dashpot. These elements are employed to model the inertial, elastic and viscous-damping properties

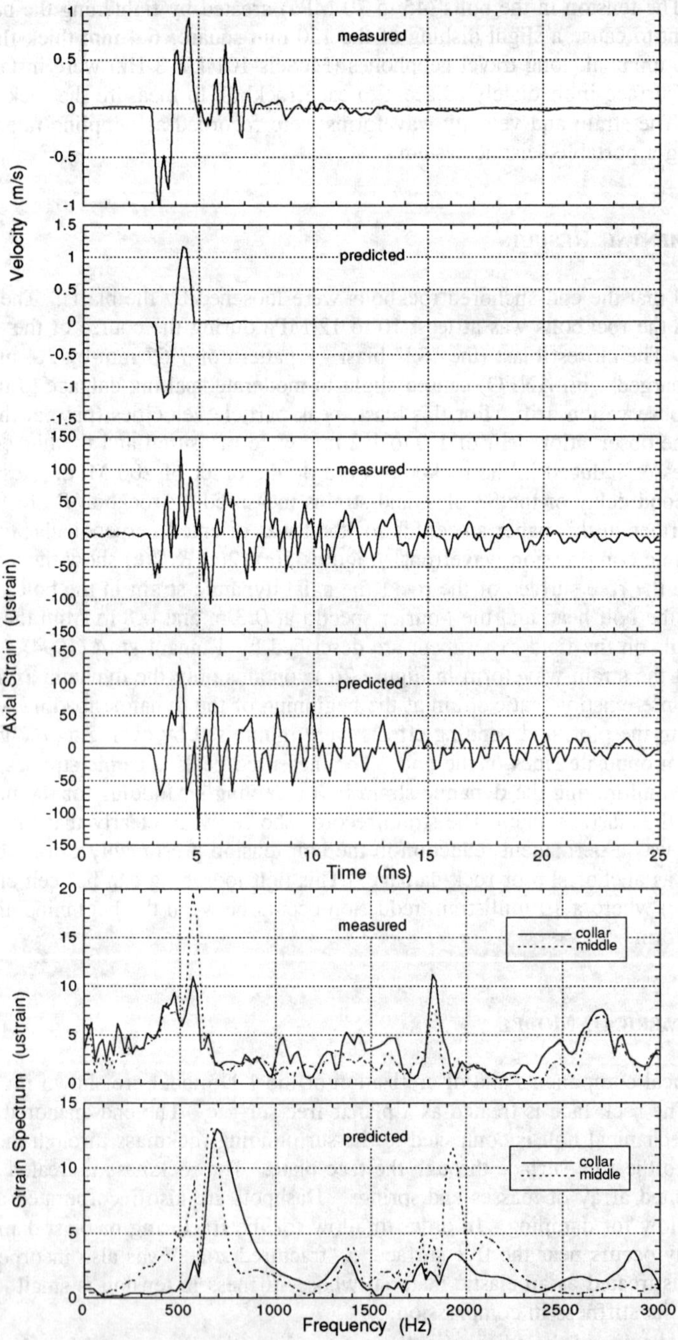

Figure 2 Measured and model surface velocities (a) & (b), axial strains (c) & (d) and dynamic strain spectra (e) & (f)

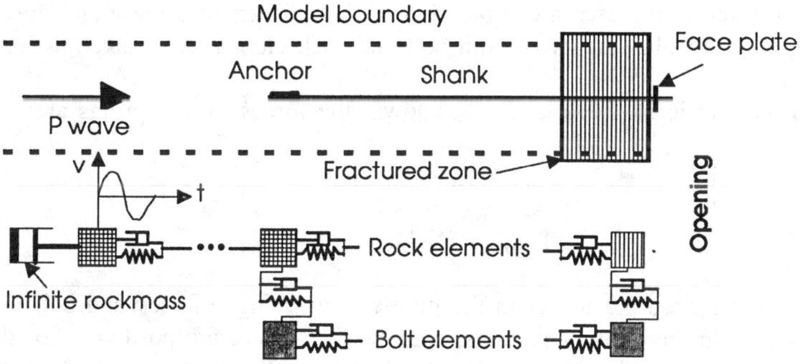

Figure 3 Rheological components of the one-dimensional rock/bolt model

of the rockmass and steel. The anchor and the face plate are modelled as spring-dashpot combinations with constant yield loads to allow simulation of slip which may occur between the rock and the anchor. In order to allow for the modelling of fully or partially grouted bolts, springs and dashpots between the rock elements and the bolt elements were also incorporated into the model, although this feature was not used in the series of analyses described here since the bolts were end-anchored. The input wave (from the blast) was simulated by applying a single cycle of a 600 Hz sinusoidal velocity-time history (ppv = 0.6 m/s) to the rock element furthest from the free surface.

The model is an explicit finite difference time-marching scheme, where we solve for displacements, forces, accelerations, velocities, and then displacements once again, repeating the cycle as often as necessary. The displacements of the jth rock and bolt elements, D_{rj} and D_{bj} at a particular time, can be expressed respectively as:

$$D_{rj} = D'_{rj} + v'_{rj}\,dt$$
$$D_{bj} = D'_{bj} + v'_{bj}\,dt$$
$$j = 1, 2, \ldots, (N) \qquad \ldots\ldots\ldots \quad (1)$$

where "r" denotes rock element and "b" bolt element, j is the element number, N the number of elements, *dt* the time interval between two consecutive iteration steps, and D' and v' are the displacement and velocity of the an element at the previous time step. Displacements and velocities towards the free surface are defined as positive. The magnitude of *dt* is chosen to be a small fraction of the Courant Limit (the time required for the wave to propagate from one element midpoint to the next), i.e.,

$$dt = \alpha\, l \sqrt{(E/\gamma)} \qquad \ldots\ldots\ldots \quad (2)$$

where α is a coefficient much less than 1.0, and l, γ, and E are element length, density, and elastic modulus of either steel or rock. This choice of a very short timestep was necessary since the wave propagation velocities in the rock and the steel are different, so there is no unique efficient timestep for the problem. It was found that acceptably accurate results and reasonable run times (300 elements on a 20 MHz 80386 PC) were obtained when the factor α was less than 0.05. The rockbolt pre-tension causes a compression in the adjacent rockmass and the resulting static displacements of both bolt and rock elements are taken as initial conditions.

With the elastic stiffness k_r and k_b known, the forces in the springs are:

$$F_{rj} = -k_r [D_{r(j+1)} - D_{rj}]$$
$$F_{bj} = -k_b [D_{b(j+1)} - D_{bj}]$$
$$j = 1, 2, ..., (N-1) \qquad .... \quad .... \quad .... \quad (3)$$

where F_{rj} and F_{bj} are the forces in the jth element springs of rock and bolt respectively. In the above equations, compressive forces are positive. For the fractured rock zone (Fig. 3), k_r is decreased to a small fraction of its original value when the rockmass is in tension. The springs for the anchor and head connections are non-linear, i.e., the stiffnesses increase with increasing elongations. The forces in the anchor and head plate connections can be expressed as:

$$F_{anchor} = k_{anchor}[1 + A\,|(D_r - D_b)/D_{anchor}|]\,(D_r - D_b)$$
$$F_{head} = k_{head}[1 + H\,|(D_r - D_b/D_{head}|]\,(D_r - D_b) \qquad .... \quad .... \quad .... \quad (4)$$

where k_{anchor} and k_{head} are the respectively initial stiffnesses of the anchor and head connection springs before dynamic displacements take place, D_{anchor} and D_{head} are deformations of the springs due to pre-tension which are used here as normalizing denominators, D_r and D_b are the displacements of the rock and bolt elements in connection with the springs, and A and H are non-linearity factors for the connection springs. k_{head} is set to zero if the load in the head connection is tensile, since the plate is in contact with the rockmass only under compression. The viscous damping forces in the rock and bolt dashpots are expressed as follows:

$$VF_{rj} = -h_r [v_{r(j+1)} - v_{rj}]$$
$$VF_{bj} = -h_b [v_{b(j+1)} - v_{bj}]$$
$$j = 1, 2, ..., (N-1) \qquad .... \quad .... \quad .... \quad (5)$$

where h_r and h_b are the viscous damping coefficients for rock and bolt, which can be expressed as percentages of the corresponding critical damping coefficients. The critical damping coefficient is 2(mk), where m and k are the mass and stiffness of an element (Den Hartog 1985). The damping forces in the connections are equal to $h_c (v_r - v_b)$, where h_c is the damping coefficient for the connections. The velocities of elements can then be obtained as follows:

$$v_{rj} = v'_{rj} + a_{rj}\,dt$$
$$v_{bj} = v'_{bj} + a_{bj}\,dt$$
$$j = 1, 2, ..., N \qquad .... \quad .... \quad .... \quad (6)$$

where the rock element acceleration is, $a_{rj} = [(F_{r(j-1)} + VF_{r(j-1)}) - (F_{rj} + VF_{rj})] / m_r$, m_r being the element mass. For bolt elements, r is replaced with b. The velocities obtained from this equation must be substituted into equation (1) to perform the iterations until the pre-determined number of time steps is reached.

A dash-pot is used to simulate the infinite rockmass (Fig. 3). The normal stress in rockmass at the dashpot is given as follows (White, 1983):

$$s = b\, v_{r(j=0)} \qquad \text{.... ....} \quad (7)$$

where the constant b equals approximately 1 for non-reflection at the model boundary.

5. Comparison Between Numerical Model and Field Measurements

Simulations for several of the monitored blasts were done but only the results for the second delay, bolt B, blast S3, are presented as examples here. In this case, the bolt shank and the rock around it were discretized into 50 elements and the rock between the bolt anchor and the blast holes was discretized into 100 elements. The bolt diameter was 15 mm, pretension was 10 kN, the initial stiffnesses of the anchor and head connections were 22 MN/m and 0.5 MN/m respectively and the yield load of the anchor and head connections was 120 kN. Figure 2(b), (d) & (f) show the results produced by the model for the velocity on the free surface of the rock, the axial dynamic strain at 0.8 m from the head and the Fourier spectra at 0.3 m and 0.8 m from the head of the bolt. The computed dynamic strain in the bolt agrees with the measured strain in amplitude, shape, and frequency content, as can be seen by comparing Figures 2(c) & (d) and Figures 2(e) & (f). The computed particle velocity on the rock surface agrees qualitatively with the measured velocity record, but the amplitude is 1.2 m/s (twice the amplitude of the incoming wave) whereas the measured ppv was just under 1 m/s. A high frequency velocity reversal near the positive and negative velocity peaks was measured in the field (Figure 2a), but we were unable to reproduce this in the model. Also, the free rock face in reality continues to oscillate for some time after passage of the wave front, whereas the model does not permit this to occur. This is likely to be due to the inherent inhomogeneity of, and presence of discontinuities in the real rock. Each such discontinuity will result in a partial reflection which will prolong the duration and reduce the maximum value of the propagating wave. The field situation is more complex than the model which does not account for these inhomogeneities.

6. Conclusions

The field trials revealed that significant unloading of end-anchored bolts can occur during dynamic loading. This unloading is probably due to slip of the anchor, since the maximum dynamic stresses induced in the bolt were less than 200 MPa, insufficient to cause yielding in the bolt. It was also found that spalling of the rock surface took place at measured peak particle velocities of 1.0 to 1.2 m/s.

The simple one-dimensional numerical model proved to be a useful tool for understanding the behaviour of rockbolts under dynamic loading. The model was able to simulate the measured dynamic strain in the bolt. There was good agreement in amplitude, shape, and frequency content between the model and the measured dynamic strains. The model overestimated the peak particle velocity on the rock surface, but this can be explained by considering the effect of the many reflections which probably occur within the discontinuous rockmass. These discontinuities appear to prolong the duration of the wave and reduce the peak particle velocity. In the model, the vibrational behaviour of the bolt was controlled mainly by the stiffness of the head and anchor connections rather than by the stiffness of the bolt itself. These connections behave like soft springs which support the bolt. Through parametric analyses it was found that the initial stiffnesses of these connections must be relatively low in order to reproduce the measured field behaviour.

7. Acknowledgements

This work was funded by the Mining Research Directorate as part of the Canadian Rockburst Research Programme. The assistance and cooperation of CANMET's Val d'Or experimental mine and GRC staff is gratefully acknowledged.

8. References

Den Hartog J. P. (1985). *Mechanical Vibrations*. New York: Dover Publications Inc., 427p.

Tannant, D.D., Brummer, R.K., and Kaiser, P.K. 1994. Response of rockbolts to nearby blasts. Integral Approach to Applied Rock Mechanics, *IV South American Congress on Rock Mechanics*, Santiago, 1994.

White J.E. (1983). *Underground Sound - Application of Seismic Waves*. Elsevier: Amsterdam-Oxford-New York.

Rock Mechanics, Nelson & Laubach (eds) © 1994 Balkema, Rotterdam, ISBN 90 5410 380 8

Geomechanics of sill pillar mining

M. R. Hudyma, *Brunswick Mining, Bathurst, N.B., Canada*

Y. Potvin & D. R. Grant, *Centre de Technologie Noranda, Pointe-Claire, Que., Canada*

D. Milne, *University of British Columbia, Vancouver, B.C., Canada*

R. K. Brummer, *Geomechanics Research Centre, Laurentian University, Sudbury, Ont., Canada*

M. Board, *ITASCA, Minneapolis, Minn., USA*

ABSTRACT: A detailed geomechanics study was carried out during the mining of two sill pillar stopes at Brunswick Mining. Microseismics, conventional ground control instruments, numerical modelling and visual observations were used to understand the observed failure mechanisms. Severe rock mass failure was initiated in several locations. The complex nature of the failures was unexpected. Only through the use of all the available tools, was it possible to understand some of the rock mass problems.

1 INTRODUCTION

This paper is a technical summary of a geomechanics research project carried out at Brunswick Mining from the summer of 1992 to the spring of 1993. The project was sponsored by Brunswick Mining, the Noranda Technology Centre, and the Mining Research Directorate.

Brunswick Mining is a 10,500 tonne per day lead-zinc-copper-silver producer in northern New Brunswick, Canada. The mining methods used have created four major sill pillars at depths varying from 600 metres to 1000 metres underground. More than 20 million tonnes of ore reserves are held in these sill pillars. Significant mining has already been carried out in the mine's sill pillars. This mining was subject to severe stress and macroseismic problems.

In 1992, a research project was initiated to study the geomechanics problems associated with the sill pillar mining. Two adjacent sill pillar stopes in the 725 metre level sill pillar (approximately 600 metres below surface) were chosen as test sites. The 16N and 13N sill pillar stopes contained about 225,000 tonnes of ore. Previous mining in nearby stopes in this sill pillar had generated rock bursts and stress related rock mass failure.

A final report on all aspects of the research project (Hudyma, Milne and Brummer, 1994) will be completed by the Spring of 1994.

1.1 *Study preparation*

The test stopes chosen had a series of exceedingly difficult geomechanics problems, including:

- nearby geological structures (faulting and a porphyry dyke inclusion) with a history of seismic problems,
- a poor quality rock mass in the hanging wall which was prone to failure,
- and a high extraction ratio resulting in elevated local stress concentrations.

To investigate each of these problems, a detailed suite of geomechanical tools were used, including 11 multipoint borehole extensometers, 10 one dimensional stress cells, and a 32 channel full waveform microseismic system. Additionally, detailed structural mapping of drift accesses and of several boreholes (using a borehole camera) was carried out. Visual observations and photographic logging were an integral aspect of the data collection.

The stress cells were installed in 100 millimetre diamond drill holes. Before the stress cells were installed, a campaign of stress measurements was carried out in some of the diamond drill holes using a borehole slotter (see Kanduth et. al. 1991, for details about the borehole slotter). A total of 94 slotting tests were completed in three non-coplanar holes in the abutment of the sill pillar. The results were comparable to past measurements determined using triaxial overcoring.

1.2 *Geomechanics problems during mining*

Mining of the 16N and 13N sill pillar stopes was started in July 1992 and concluded in February 1993. A total of twenty blasts were used to extract the two stopes. During the blasting of the two stopes more than 20,000 microseismic events were located with the full waveform microseismic system. These events ranged from about Magnitude -3 to Magnitude 2.6 (Nuttli).

High stress related rock mass failure occurred in a number of locations near the mining. Falls of ground and severe rock mass deterioration occurred in mine development near the two sill pillar stopes. Two of these failures almost prevented further mining in the sill pillar stopes.

The poor quality hanging wall had excessive overbreak after about 25 metres of exposure. Uncontrollable dilution ultimately stopped mining in the 16N stope and a rib pillar to support the hanging wall had to be left. A hanging wall cable bolt pattern was only temporarily successful at minimizing waste rock overbreak.

2 RESPONSE OF THE GEOMECHANICS INSTRUMENTATION

The following section details the most significant conclusions and results interpreted from the seismic system, conventional ground control instruments, and numerical analyses.

2.1 *Microseismic system*

Over a period of nine months, more than 20,000 microseismic events were associated with the mining of the 16N and 13N sill pillar stopes. Source location error on the events located was estimated to be less than 5 metres for more than 80% of the events. The microseismicity had several general tendencies:

1. Approximately 60% to 70% of the microseismic events were located on known geological discontinuities. There were eight major structural seismic sources that were active throughout the study period.
2. Approximately 30% to 40% of the microseismic events occurred close to a newly blasted stope face.
3. Approximately 5% of the microseismic events occurred along unexpected, or previously undetected geological discontinuities.
4. Microseismic event magnitude and frequency was directly correlated to locations of underground rock mass deterioration and failure.

A local magnitude scale had been developed to provide a quantitative measure of the size of seismic events in the mine. This scale was defined by a triaxial accelerometer, a near-field triaxial geophone, and a qualitative assessment based on the uniaxial accelerometers of the microseismic system. Frequency-magnitude 'b' factors were found to be well related to seismic source mechanisms. Microseismic events within 15 to 20 metres of the open stopes had a 'b' factor of about 1.2. This is indicative of relatively uniformly distributed volumetric fracturing. Microseismic events occurring along known structural features had a frequency-magnitude 'b' factor of about 0.86. Seismic events along one active but remote feature had a 'b' factor of 0.65. A clear relationship was observed between seismic source mechanism and the frequency-magnitude characteristic of the emitted seismic events.

Three hundred seismic events in the month prior to the start of blasting were analyzed for location. This seismicity was located on seven of the eight most microseismically active geological features in the study region (see Figure 1). All of these microseismic events had a magnitude of Magnitude -2 or smaller. In effect, seven of the eight most commonly active features could be identified before stope blasting started. This type of interpretation is only possible with high accuracy source locations for events Magnitude -2 and -3 in size.

A disappointing aspect of the seismic study was the inconsistency of the first motion source mechanism analyses. There were 26 uniaxial accelerometers in the region of the sill pillar stopes. First motion analyses were attempted for events ranging from Magnitude -1 to 2.6. As a general rule, the larger the event, the more difficult it was to find a good double-couple solution. Many of the Magnitude -1 events gave excellent solutions compared to the orientation of known geological features. However, for the largest events, it was not possible to determine solutions. We believe that this should be an expected conclusion. The large seismic events are not point sources compared to the size of the microseismic array, however, events smaller than Magnitude 0 were effectively point sources compared to the size of the microseismic array.

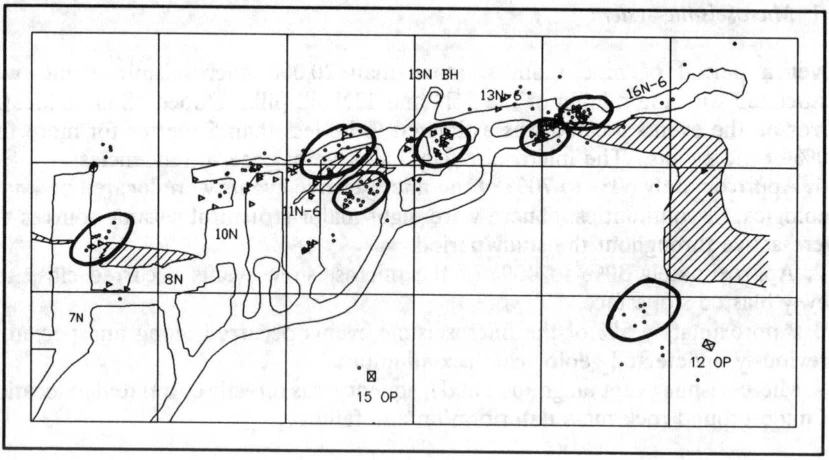

Figure 1. The location of about 300 seismic events during the month prior to the start of mining highlighted seven of the eight most geological features that were most active during mining. The location of the features (circled areas) and the precursory events are shown on the mine plan.

2.2 *Conventional ground control instrumentation*

The conventional instrumentation consisted of 11 multipoint extensometers, 10 CANMET strain cells, closure stations and movement monitors. They were installed to answer specific questions about the expected rock mass response to mining. In more than one instance, the instruments did not respond in the expected fashion. The preconceived notions of the ground response were found to be far too simplistic and often did not match the underground observations.

A key component of the conventional instruments was the connection of the extensometers and strain cells to an automatic datalogger. The instruments were read hourly for a period of eight months. The instruments were read every minute after the large mine blasts. This gave some interesting information. In all cases, between 50% and 100% of the stress change due to blasting occurred within one minute of the blast. Figure 2 shows the strain response to a blast more than 50 metres from the strain cell location.

All of the strain cells were installed in the massive sulphide orebody at various distances from the mining. The results were compared to three dimensional linear elastic numerical modelling. In general, most of the cells indicated greater stress increases than shown by the modelling. However, one pair of cells showed a lower stress increase than predicted by numerical methods. These cells exhibited repeated non-linear behaviour, showing a sharp increase in strain directly after blasts, followed immediately by a significant strain decrease. An example of this behaviour is shown in Figure 3. The strain cells served to highlight the non homogeneous nature of the rock mass in terms of stress as well as strength properties.

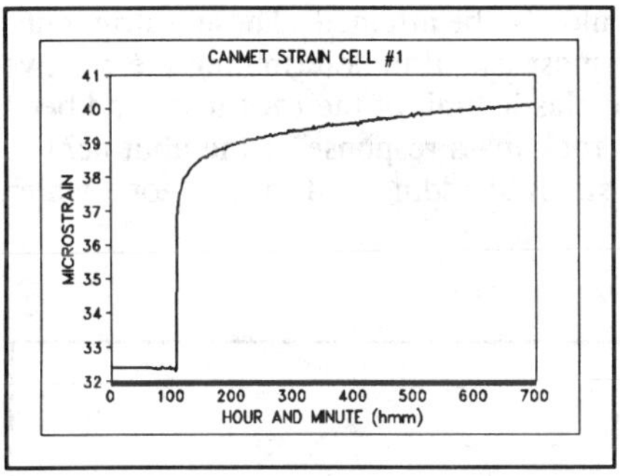

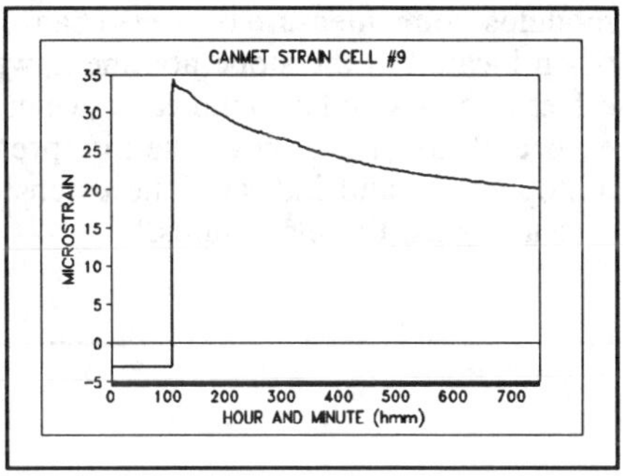

Figures 2 and 3. The response of the strain cells to a mine blast. The strain cells are being every minute. In figure 2 (left), a strain cell shows more than 50% of the strain change within the first minute of a mine blast more than 50 metres away. In figure 3 (right), a strain cell is showing non-linear stress shedding. One hundred percent of the strain increase at the cell location occurs in the first minute, followed by a gradual strain decrease.

The extensometers had two basic functions. The first function was to measure hanging wall dilation as the hanging wall span increased. The second function was to measure strain increases in the orebody that could be related to increased stress.

The extensometers installed to measure hanging wall dilation provided one of the more unexpected results in the study. It was found that tensile zones in both the ore and the hanging wall coincided with slip surfaces which were shearing and dilating. The slip surfaces were generated by shear stresses induced around bends in the orebody. Three slip surfaces were delineated by the extensometers and followed projections of the ore / hanging wall contact after the contact bent to a different plane. These slip surfaces generated planes of weakness which the hanging wall tended to fail to and gave an indication an appropriate depth for artificial rock mass support. This shear movement may also explain some of the difficulties in correlating the strain cell results with the numerical results.

Several of the extensometers extended from the stope hanging wall into the orebody to obtain strain between anchors and an estimated stress increase with mining. It was found that instead of indicating a general stress increase in the abutment of the approaching mining face, the extensometers identified alternating zones of compression and tension associated with the three shear planes. These alternating compression and tension zones were also measured in the hanging wall abutment and were highlighted by significant quantities of small microseismic events (mostly magnitude -2 and smaller). The movement in the shearing zones masked the expected compression due to stress increase.

Simple three point closure stations were installed in several locations in the study area. Measuring points were installed in the back and in each wall and three measurements were taken. A simple program was developed to convert these measurements to a two dimensional principal strain change, and with an assumed modulus, some measure of stress change could also be attained. These instruments, when located in the stope abutment, were most useful in determining the relative effect of blasts and also indicated when the elastic limit of the rock mass had been exceeded. Figure 4 shows the interpreted rock mass response of the abutment to a stope blast and indicates the extensive stress shedding and failure zone which extended past the blast limits.

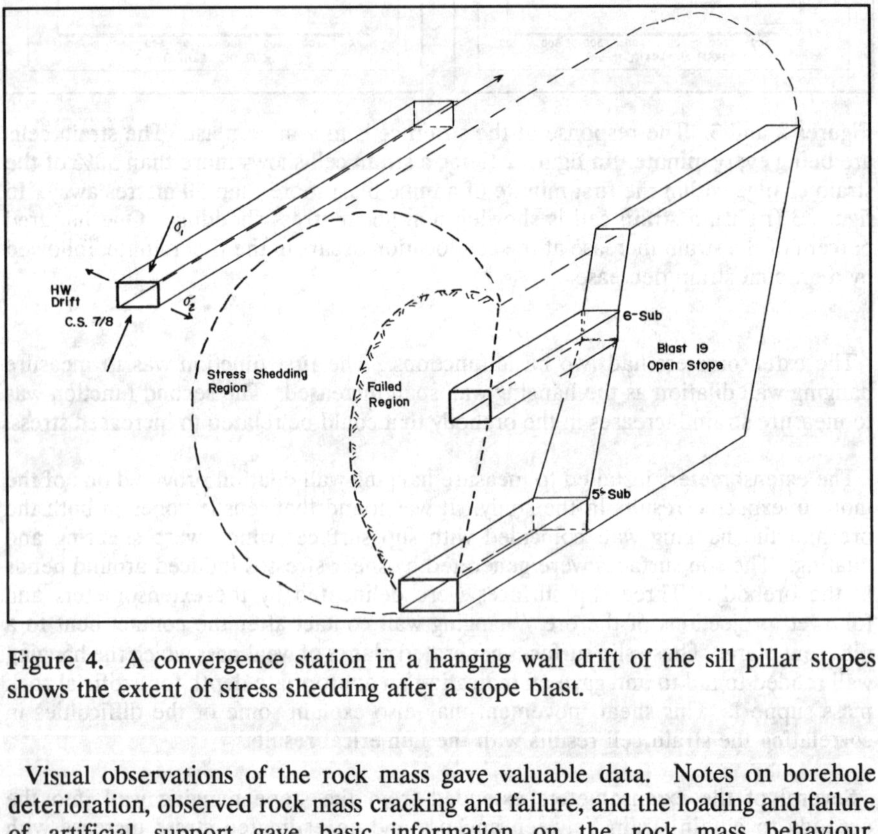

Figure 4. A convergence station in a hanging wall drift of the sill pillar stopes shows the extent of stress shedding after a stope blast.

Visual observations of the rock mass gave valuable data. Notes on borehole deterioration, observed rock mass cracking and failure, and the loading and failure of artificial support gave basic information on the rock mass behaviour. Interpretations of the more sophisticated instruments were often made with the aid of knowledge gained from observations.

2.3 *Numerical analyses*

Numerical analyses of the mining of the two sill pillar stopes had two major goals. The first objective was to estimate the degree of linear elastic behaviour occurring in the sill pillar by comparing three dimensional boundary element modelling with measured stress change, microseismics, and underground observations. The second objective was to analyze the stability of the sill pillar using a distinct element technique.

The linear elastic model used was BEAP (acronym for Boundary Element Analysis Program) developed by J.A.C. Diering (1987). Linear elastic simulations of the sill pillar mining were performed for each of the major stope blasts. Grant (1993) found that the microseismicity near the newly blasted face correlated well with the theoretical change in maximum principal stress. These seismic events were generally Magnitude -2 or smaller. There was a poor correlation between theoretical stress change and the large seismic events, or the seismicity located on known geological structures.

There was modest agreement between the stress change predicted by BEAP and the results of the strain cells. Some cells gave relatively consistent results throughout the mining of the sill pillar stopes, while others gave very erratic and inconsistent results. Based on the overall observations and measurements of the rock mass conditions in the sill pillar, there were many instances of inelastic rock mass behaviour that could not be simulated using only a linear elastic numerical technique.

ITASCA was contracted to analyze the mining and overall stability of the 725 level sill pillar test site (Sjoberg and Board, 1993). The goal of this work was to develop hypotheses for why the largest seismic events associated with the sill pillar mining were occurring at distances of up to 200 metres from the stope blasting. The numerical code 3DEC was used to simulate the sill pillar rock mass. 3DEC is a three dimensional distinct element program in which the rock mass is represented as blocks separated by discontinuities. Linear-elastic or elastic-plastic material properties can be attributed to the blocks.

Conclusions of the 3DEC analyses were as follows:

1. The fault structures identified in the 725 sill pillar are unstable and have the potential for significant movement and the creation of large seismic events.
2. The location, orientation and properties of the discontinuities have a fundamental influence on the predicted stability of the sill pillar. The location and timing of large seismic events was highly dependent upon local properties of the geological features.
3. Geological discontinuities which are persistent along the strike of the sill pillar tend to control the instability in the sill pillar. Major cross-cutting structures help define the blocks which may move but do not have a fundamental influence on the rock mass dynamics.
4. Some of the major geological features in the sill pillar had no influence on the stability. Traditionally, a cross-cutting inclusion of quartz-porphyry dyke was believed responsible for seismic problems in the 725 sill pillar. Numerical analyses showed the influence of this structure to be relatively benign.
5. The wider regions of the orebody had much greater potential for block movement and the creation of large seismic events. Fault-slip events were more likely to occur in wide regions of the orebody and could potentially occur quite remote from active mining. Strain burst seismic events were more likely to occur close to active mining faces.

3 CONCLUSIONS

Many of the failure mechanisms observed during the sill pillar mining were well understood. Microseismics, conventional ground control instrumentation, numerical modelling and underground observations all provided valuable information about the behaviour of the sill pillar rock mass. However, the actual failure mechanisms were much more complicated than originally anticipated. In many cases, all of the sources of rock mechanics information were needed to interpret the ground reactions.

ACKNOWLEDGMENTS

Many individuals aided in the design and execution of this study. The results and conclusions would not have been possible without the aid of: Terry Mac Donald and Ashley Jagoe of Brunswick Mining, and Alain Gendron, Paul Germain, Denis Jacob and Harald Kanduth of the Noranda Technology Centre. Brunswick Mining, the Noranda Technology Centre and the Mining Research Directorate are acknowledged for their financial support of the sill pillar mining project.

REFERENCES

Diering, J.A.C. 1987. Advanced elastic analysis of complex mine excavations using three-dimensional boundary element techniques. PhD Thesis, Pretoria University, South Africa, 201 pp.

Grant, D. 1993. BEAP modelling results of the Brunswick Mines 725 sill pillar mining of the 13N and 16N stopes. Internal Noranda Technology Centre report.

Hudyma, M.R., Milne, D., & Brummer, R.K., 1994. Geomechanics of sill pillar mining in rock burst prone conditions. Phase I final report to the Mining Research Directorate.

Kanduth, H.H., Corthesy, R., & Gill, D.E. 1991. Validation of borehole slotting as a method for in situ stress measurements. Proceedings 7th International Congress on Rock Mechanics, Aachen, Germany. Rotterdam:Balkema.

Sjoberg, J., & Board, M. 1993. Three-dimensional numerical analysis of sill pillar mining at the Brunswick Mine. ITASCA report prepared for the Centre de Technologie Noranda, 59 pp.

Rock Mechanics, Nelson & Laubach (eds) © 1994 Balkema, Rotterdam, ISBN 90 5410 380 8

Composite source mechanism solutions for rockbursts at the Lucky Friday Mine, Mullan, Idaho

C.J.Wideman
Spokane Research Center, US Bureau of Mines, Spokane, Wash. & Montana College of Mineral Science and Technology, Butte, Mont., USA

J.M.Girard, T.J.Williams & T.L.Nichols
Spokane Research Center, US Bureau of Mines, Spokane, Wash., USA

ABSTRACT: U.S. Bureau of Mines researchers have analyzed full-waveform digital seismic records from moderate-to-large rockburst events at the Lucky Friday Mine, Mullan, ID. The goal of this work is to characterize source mechanisms for rockburst events. Because the seismometer array used for the recordings usually consists of fewer than 10 instruments, a method of obtaining composite source solutions has been developed. These solutions are determined by selecting events with similar arrival patterns at the seismometer array and with hypocentral locations in the same general region of the mine. Some events have all-dilatational first arrivals, while others display arrival patterns with both compressional and dilatational arrivals. There are at least three types of events in the mine. Some of the more complex arrival patterns seem to be associated with right-lateral, strike-slip motion, while others seem to be associated with left-lateral or dip-slip movement.

The type of event seems to be independent of source location, although geologic conditions in different regions vary. Even though there is a fundamental difference in the geometry of the mining environment, the three types of events are still recognizable.

This work is continuing, and data from all large events (Richter magnitude, $m_L > 2.0$) at the mine are being analyzed and added to a database.

1 INTRODUCTION

The complex geologic setting of the Lucky Friday Mine, Mullan, ID, and the interaction of mine-induced changes in the regional stress field present a unique opportunity for rockburst research. The mine is located in a region of intense folding and faulting, and there is some evidence that tectonic deformation continues at the present time (Sprenke et al. 1991).

Full waveform recording of seismic events and the use of three-component seismometers were initiated in 1989 and were described by Jenkins et al. (1990). The original data were obtained by five seismometers placed at strategic positions in the mine. The original array, although minimal, demonstrated the usefulness and desirability of a more complete seismic data collection system. The receiver array was modified in January 1991 and now includes a mixed array of three-component and single-component seismometers. The number of receivers was increased to 11, and this new array has operated successfully ever since.

Because of the limited number of geophones available for analyzing individual rock-burst events, U.S. Bureau of Mines (USBM) researchers adapted a method used in studies of earthquake seismology—composite source solutions. Composite solutions are used to analyze events occurring in the same general region of the mine and having the same general first-motion arrival patterns. Composite solutions do not represent a particular source mechanism, but instead represent classes of source mechanisms occurring throughout the mine. This paper outlines a method for creating composite source solutions for digitally recorded events from the Lucky Friday Mine.

1.1 *Geologic Setting*

Figure 1 shows a plan view of the 5100 level of the Lucky Friday Mine and three major regions in the mine (labeled A, B, and C in Figure 1). In general, the trend of the mine workings is northeast-southwest; however, Figure 1 shows that some mine workings trend north-south or east-west.

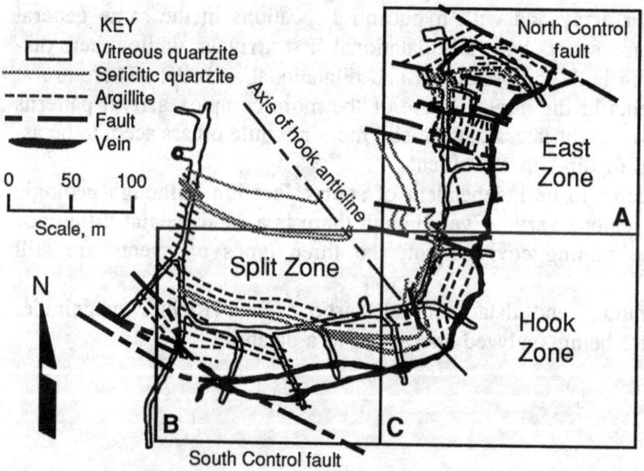

Figure 1. Plan view of 5100 level, Lucky Friday Mine.

There are also abrupt major changes in orientations of geologic formations throughout the mine. In the northeast portion of the mine, bedding trends north-south and dips steeply to the east. However, close to the North Control Fault (NCF), strikes and dips change rapidly. The NCF is a major tectonic feature that strikes northwest and dips at a near-vertical angle. Offset on the fault is believed to be left-lateral (Blake 1989), and near the fault the strike of the bedding can change by as much as 90°.

In the southwest portion of the mine (region B in Figure 1), the general strike of the bedding is approximately east-west and dips are to the south. Another major tectonic feature, the South Control Fault (SCF), is located in this region of the mine. The intersection between bedding and faults in region B is not as complex as that in region A. Because trends of bedding and faulting are similar and because some tectonic separation is thought to have occurred through slip along bedding planes, the identification of faulting in the southwest portion of the mine is somewhat difficult (Scott et al. 1993).

1.2 *Seismic Monitoring*

Data from three types of seismic monitoring systems have been collected from the Lucky Friday Mine. A seismometer on the surface near the mine workings was used to determine event magnitudes. Approximate average distance between events and the surface seismometer was 1.5 km, and event magnitude (M_L) was estimated by measuring millimeters of displacement recorded on a drum. An event that displaces 50 mm on the seismometer is approximately an $M_L = 1$ event.

Event locations are determined by a mine-wide microseismic monitoring system consisting of single-component seismometers and accelerometers. In this system, approximately 40 detectors are placed about 60 m apart. The microseismic network is an event detection-and-location system that records only first-arrival information.

Event locations are also calculated by the USBM macroseismic network and a least-squares inversion program implemented by Girard (1990b). The distinction between the microseismic and macroseismic systems is the ability of the latter to record full waveform data. With the macroseismic array, the distance between event and detector can range from 50 m to as much as 2 km.

Events detected by the macroseismic system are recorded by an IBM-compatible computer[1] modified by the installation of an Atlab analog-to-digital board. The system is controlled by the XDETECT software developed by the International Association of Seismology and Physics of the Earth's Interior (IASPEI), with sample rates for each detector set as high as 1,290 samples per second. For data analysis, a batch file entitled GOFPS, developed by Girard (1990a), is used. GOFPS calls two USBM programs, RBURST and RPEAK (Nichols 1990) that allow P-wave and S-wave picking and spectral analysis. Nichols patterned his programs after the IASPEI program PCEQ, but the former programs differ from PCEQ primarily in their ability to output arrival time information to 0.1-ms accuracy rather than the 0.01-s limitation of PCEQ. The RBURST program is used for analysis of single-component data and the RPEAK program is used for analysis of three-component data. Figure 2 is an example of the computer screen display for RPEAK.

After the arrival times of the phases are picked, the GOFPS program puts arrival times and event location data in the correct format to be input to the FPFIT and FPPLOT programs described by Lee et al. (1988) and modified by Wong (1990). The resulting output is a double-couple source mechanism solution that best fits the data. Oppenheimer (1991) has suggested that some minimal number of seismometers is required for good source mechanism determinations and that the number may be as high as 32. Because the macroseismic system was originally very limited, the GOFPS

---

[1]Reference to specific equipment or tradenames does not imply endorsement by the U.S. Bureau of Mines.

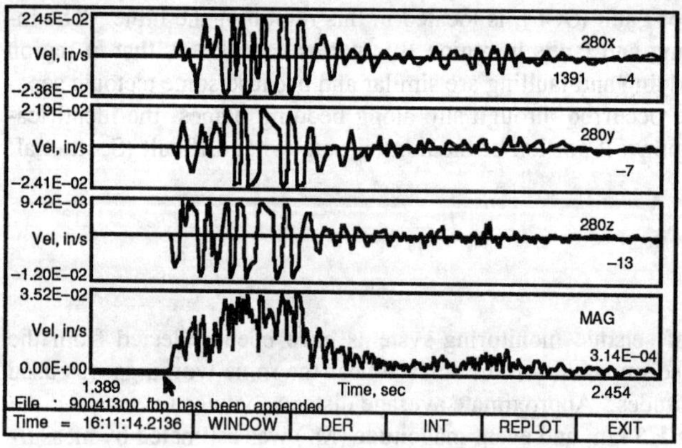

Figure 2. Computer screen display of program RPEAK.

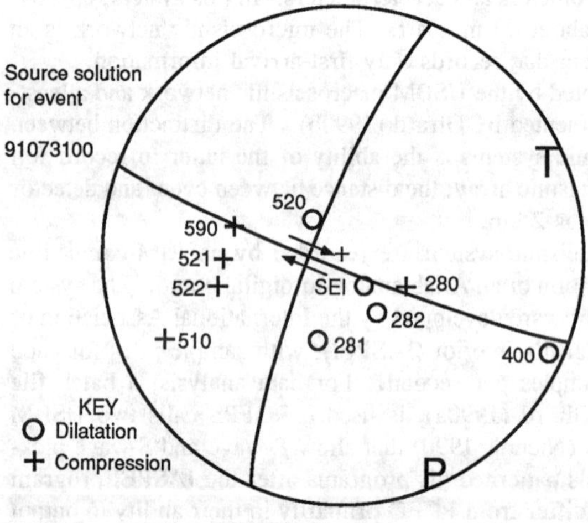

Figure 3. Source solution for a single event showing possible right-lateral, strike-slip motion on a northwest-trending fault.

program allows a researcher to use as few as six receivers before being warned of possible problems. Figure 3 is a typical source mechanism solution for a single event.

2 DATA ANALYSIS

Several hundred events were detected and recorded during the first 2 years of system operation. Individual events were analyzed for location and source mechanism. Several operating characteristics were established during this time: (1) The array geometry was such that some detectors were in the "near field" and some were in the "far field." (2) Because some of the seismometers were near the same level as rock-burst events, placement of the hypocentral location in the vertical direction controlled the source mechanism solution. Small changes in source location estimates often had correspondingly large changes in source solutions. (3) Only a few compressional events have been detected by the surface seismometer. (4) Some source mechanism solutions were compatible with right-lateral, strike-slip motion; some were compatible with left-lateral, strike-slip motion, and others were compatible with dip-slip motion.

Before the macroseismic system was installed, there was a tendency to lump events from the same general area of the mine and assume that these events had similar source mechanisms. However, after computing several source mechanism, it became apparent that each region of the mine had a variety of source mechanisms. Because of the mine's geologic complexity, this finding was not unexpected, but this was the first time instruments had verified the complexities of the source mechanisms.

Because the small number of detectors placed severe limitations on determining source mechanisms, a method of making composite solutions was adapted. Of particular concern was that the layout of the seismometer array had caused certain events to be recorded as all-dilatational events having arrivals from the same stereonet quadrant. Such events have source solutions that would not be uniquely determined.

As source solutions were obtained, they were inspected to determine the general patterns of first-motion arrivals. If, for instance, the geophones of the array all plotted within the southwest quadrant of the focal sphere, and all the arrivals were dilatational, the focal mechanism solution was obviously poorly constrained. This was one of the problems encountered during the data analysis and probably represents a worst-case scenario. Usually the receiver array, although constrained by mine geometry, was deployed so that arrivals would be plotted in at least three quadrants, if not all four, of the focal sphere. Events from region A that had similar allowable source solutions were grouped together as a single event, and the composite solution, as shown in Figure 4, was obtained.

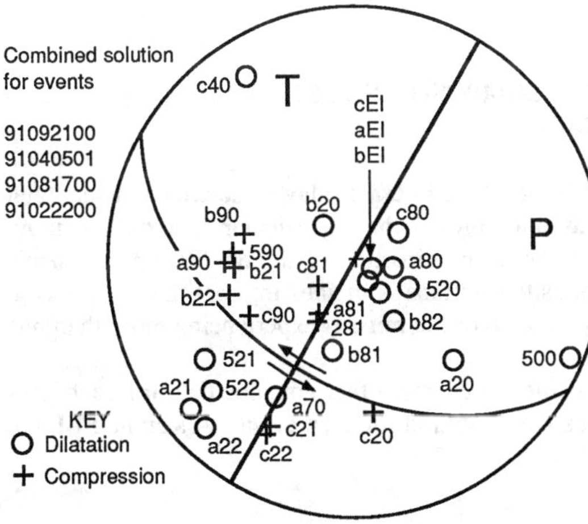

Figure 4. Composite source solution showing possible left-lateral, strike-slip motion on a northwest-trending fault.

As a result of grouping events into composite solutions, at least three distinct classes of events were determined. Figures 4 and 5 illustrate the types of source mechanism solutions obtained by combining the recordings for selected events. Each of the events selected originated in the northeast part of the mine. The actual events used for the composite source mechanisms are indicated on the figures. Because the major faults of the area trend northwest-southeast, the source mechanism classification was related to the sense of motion that could have occurred on these faults. Some source mechanisms were right lateral, some were left lateral, and some were dip slip.

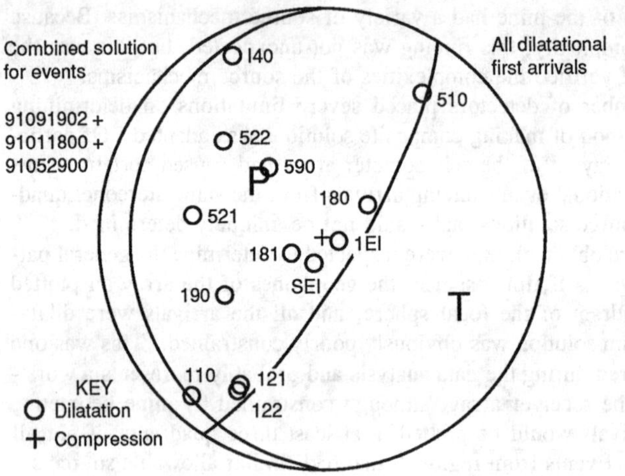

Figure 5. Composite source solution generated from events having all-dilatational first arrivals.

3 METHODS OF DETERMINING COMPOSITE SOURCE SOLUTIONS FOR THE LUCKY FRIDAY MINE

Composite solutions developed for events at the Lucky Friday Mine are not actual fault plane solutions and should not be used to "prove" that a particular rock mass is moving along a fault or geologic formation in the direction of the double-dipole seismic solutions. The correct use of composite solutions is in showing that there are recognizable types of events and that each region of the mine is experiencing more than one type of event.

As events are analyzed, a *.FBP file is generated that contains data for each geophone used in the computation of a source solution for each event. A sample FBP file is shown as follows:

Example FBP file

```
281Z  1817   1:18:54.5220    -6
520X  1632   1:18:54.3791    -1
520Y  1632   1:18:54.3791     1
520Z  1632   1:18:54.3791    -3
400Z  1719   1:18:54.4463     4
440Z  1679   1:18:54.4154    -3
590Z  1655   1:18:54.3969    -5
521X  1611   1:18:54.3629   -23
521Y  1611   1:18:54.3629    34
521Z  1611   1:18:54.3629    19
522X  1607   1:18:54.3598   -10
522Y  1607   1:18:54.3598    -5
522Z  1607   1:18:54.3598   -27
SEIS  2010   1:18:54.6709   -18
570H  1659   1:18:54.4000    95
```

The first column is a list of geophone names and components. For example, geophone 520 has three components: the vertical Z component (+ indicates upward motion); the east-west X component (+ indicates eastward motion); and the north-south Y component (+ indicates northward motion). The next column consists of sample numbers from the waveform file for the P-wave phase pick. The third column shows the time corresponding to the sample number in column 2. The format of column 3 is hours: minutes:seconds, and times are recorded to tenths of milliseconds. Column 4 is a list of magnitudes and polarities of motions at the time the sample was recorded.

Data from the first and last columns of the file, along with event location information, are used to make source mechanism estimates by incorporating this information into *.DAT files to be input to the FPFIT program. An example DAT file is illustrated below:

EXAMPLE DAT FILE

```
1
  DATE      ORIGIN     LAT N      LONG W     DEPTH      MAG
 92062703        .00 47-26.10 115-43.09 -19.24        .30

  STN  DIST AZM AIN PRMK
  281   2.8  28 150 IPD
  521    .3 332 102 IPC
  590   1.0 312  51 IPC
  440   1.3 104 131 IPD
  520    .5  33  91 IPD
  522    .2 298 100 IPD
  SEI   5.3 352 167 IPD
  570    .9 224  40 IPD
  400   1.8 313 132 IPC
```

The information contained in the first three rows of the DAT file is required to run the source mechanism program, but includes some dummy variables. The event name (in this case, 92062703) and the event depth (in this case, -19.24 hundred feet) are real. However, LAT/LONG refers to general mine location, and the ORIGIN and MAG fields are filled only to satisfy software format requirements and are not used in the analysis. The remaining information is determined by comparing the previously calculated event hypocenter with geophone (STN) locations. The distance (DIST) between the event and the geophone is indicated, as well as the azimuth (AZM) and takeoff angle (INC) for the seismic wave leaving the source and arriving at the receiver. The last column contains information about the seismic wave: IPC indicates an *I*mpulsive *P*-wave with *C*ompressional motion, while IPD indicates an *I*mpulsive *P*-wave with *D*ilational motion.

After a DAT file is created, the information is in the correct format to be input to the U.S. Geological Survey programs FPFIT and FPPLOT. These programs are used to generate fault plane solutions and to view them on lower hemisphere stereonet projections. Each solution is analyzed to see if the pattern of geophone arrivals is unique or if the pattern can be incorporated in composite solutions.

If the fault plane solution is similar to several other solutions for events in the same region of the mine, a composite solution is made. Composite solution files are obtained by concatenating FBP files for similar events. A new DAT file is computed for the concatenated data and is then used for the FPFIT and FPPLOT programs.

4 CONCLUSIONS

Conditions for inclusion of data into a composite solution is somewhat arbitrary. Composite solutions for events at the Lucky Friday Mine are obtained as follows:

1. Events are separated by region.
2. Events are chosen from similar depths and the same general area of the mine.
3. Arrival patterns are examined as they are recorded by the receivers. Groups of compressional and dilatational arrivals must be similar.

The composite solutions are not used as actual double-dipole seismic solutions, but instead represent classes of solutions. Preliminary characterization of source mechanisms has shown there are at least three types of events occurring in the mine and arrival patterns seem to be associated with right-lateral, left-lateral, or dip-slip motions. This work is continuing, and large events are being analyzed and added to the current database.

ACKNOWLEDGMENTS

The authors would like to thank the personnel of Hecla Mining Co. for long-standing cooperation in field experiments at the Lucky Friday Mine. The authors also thank Doug Scott, geologist, Spokane Research Center, and Mike Friedel, geophysicist, Twin Cities Research Center, for their helpful reviews of this manuscript.

REFERENCES

Blake, W. 1989. Personal communication.

Girard, J.M. 1990a. GOFPS.BAT. Software program available from J. Girard, Spokane Research Center, U.S. Bureau of Mines, Spokane, WA.

Girard, J.M. 1990b. INV.FOR. Software program available from J. Girard, Spokane Research Center, U.S. Bureau of Mines, Spokane, WA.

Jenkins, F.M., T.J. Williams, and C.J. Wideman. 1990. Rockburst mechanism studies at the Lucky Friday Mine. In *Rock Mechanics Contributions and Challenges: Proc., 31st U.S. Symposium*, CO Sch. of Mines, Golden, CO, June 18-20. Balkema:Rotterdam:955-962.

Lee, W.H.K., D.M. Tottingham, and O.J. Ellis. 1988. A PC-based seismic data acquisition and processing system. U.S. Geol. Surv. Open File Report 88-751, 31 pp.

Nichols, T. 1990. RBURST.BAS and RPEAK.BAS. Software programs available from T. Nichols, Spokane Research Center, U.S. Bureau of Mines, Spokane, WA.

Oppenheimer, D.H. 1991. Personal communication.

Scott, D.F., B.G. White, and T.J. Williams. 1993. Host structures for slip-induced seismicity at the Lucky Friday Mine. In *Rockbursts and Seismicity in Mines 93. Proc., 3rd International Symp. on Rockbursts and Seismicity in Mines*, Kingston, ON, Aug. 16-18. Balkema:Rotterdam:245-248.

Sprenke, K.F., M.C. Stickney, D.A. Dodge, and W.R. Hammond. 1991. Seismicity and tectonic stress in the Coeur d'Alene Mining District. Bull. Seismo. Soc. Amer. 81(4):1145-1156.

Wong, I. 1990. Personal communication.

Rock Mechanics, Nelson & Laubach (eds) © 1994 Balkema, Rotterdam, ISBN 90 5410 380 8

Chimneying disintegration failure mechanisms of hard rock mines

M.C. Bétournay
Canada Centre for Mineral and Energy Technology (CANMET), Ottawa, Ont., Canada
H.S. Mitri & F. Hassani
Mining Engineering, McGill University, Montréal, Que., Canada

ABSTRACT: Chimneying disintegration, one of several shallow stope failure mechanisms of hard rock mines, develops rapidly to surface in weak rock masses from depths as much as 275 m. A limit equilibrium equation is presented, which is based on the inability of the rock mass to support its own weight and fails by shearing along lines of active pressure. The material's ability to breakdown to small sizes allows an effective shear rupture to develop. Four cases have been studied to apply and evaluate the analytical equation, numerical modelling and a dedicated shallow stope empirical method. The equation has returned factors of safety representative of the failed and unfailed cases. A numerical correlation between critical span and rock mass cohesion is presented, and the span and material strength limits are discussed for chimneying disintegration to occur. Numerical modelling can represent weak rock mass behaviour and indicate where failures start but not the outline or continuity of failure to surface. The dedicated empirical method has limitations in predicting the peripheral location and stope depth from which failure can occur.

INTRODUCTION

Canadian hard rock mine extraction has commonly created shallow stopes, 12 % of which have caved from gravity-induced failure mechanisms: plug failures, block ravelling, strata failures, chimneying disintegration and block caving.

Research into the subject of shallow stopes of hard rock mines (Bétournay 1994) established that few numerical or empirical methods existed to represent these failure mechanisms and that stability analysis used generic rock mechanics methods. The research led to the development of failure specific analytical equations, addressing the mechanics of the failure process and incorporating the capability to arrive at the ultimate failure outline and potential for reaching surface.

Chimneying disintegration is a common failure creating isolated "chimneys" in weak rock often reaching surface, either following a vertical path in a homogeneous mass, Figure 1, or up-dip in a weak unit bounded by competent rock, Figure 2. In the former, failure occurs by progressive collapse of a disintegrating rock mass which leaves behind intact, steep walls. The progression of this degradation can develop

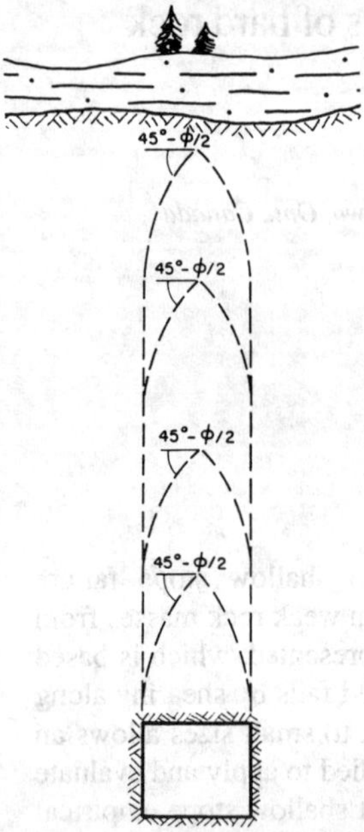

Figure 1. Progression of a vertical chimneying disintegration in homogeneous material (Bétournay 1994).

quickly: over a 4.3 m x 8.5 m opening cutting across a relatively incompetent graphitic slate dipping at 60°, working vertically 275 m through to surface in about one year (Rice 1934). Supplementary cases have been reported in severely altered rock, Figure 2, (Strata Engineering 1987), above coal measures in weak shale (Jones 1993), and in sericitic or gouge-filled schists (Picciaccia 1988) (Commission of Enquiry Belmoral Mine Accident 1981). The Belmoral mine failure (Val d'Or, Quebec), involved the development of a chimneying disintegration in a chloritic gouge-filled schist reported to have insufficient "cohesion" to support its own weight resulting in the inflow of 100,000 tons of wet overburden underground, fatalities and the cessation of operations. In these failures the width of the chimney can be as low as 2 to 4 m. Though instances of chimneying disintegration have choked themselves, known failures to surface occurred as a result of failed material removed or fallen into large stopes. However, such failed materials do not have high bulking factors.

ANALYTICAL EQUATION

Plug failure, piping in caved material and chimneying disintegration are generically classed as chimney caving. Chimneying disintegration is causative with distinct mechanics of failure incorporated in the following limit equilibrium equation, which compares the influences of loads against the capacity of the rock mass to resist movement. Although a limit equilibrium approach simplifies the problem to one of strength relationship without considering deformation, it is usually adopted as a norm in rock mechanics (Brady and Brown 1985).

The approach considers failure as the inability of the rock mass to support its own

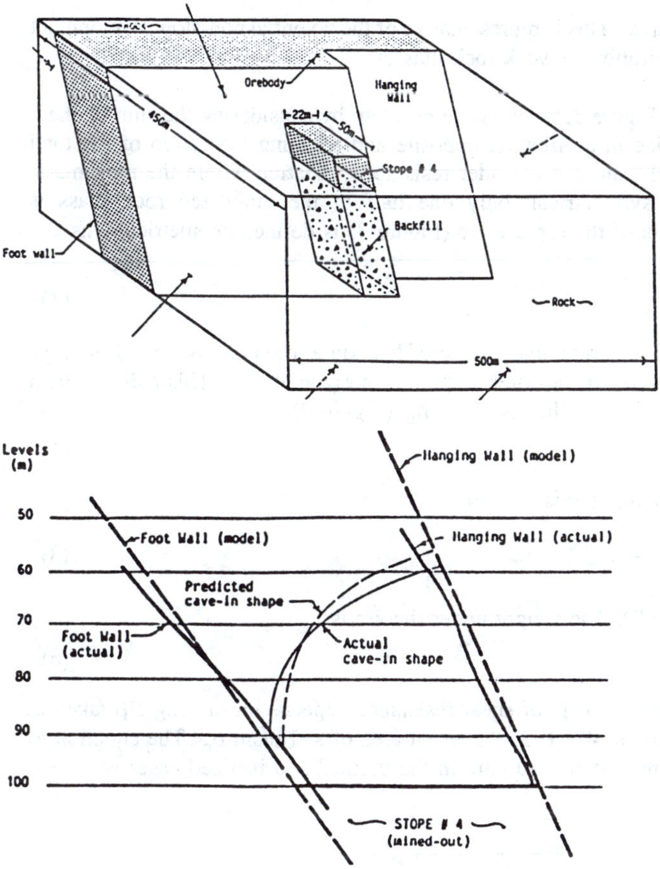

Figure 2. Development of Selbaie chimneying disintegration failure, numerical modelling and actual behaviour (Strata Engineering 1987).

weight and ruptures along lines of active pressure, exceeding the material's cohesion, Figure 1. A simple approximation of a 3-D rupture line is to use a 2-D circular slip surface, used in active pressure shear failures in slopes of cohesive soils, weak rock material and rock rubble (Hoek and Bray 1977). Because materials known to host such failures (e.g. schists) can inherently break down to small sizes, they are able to develop approximate circular failure shapes. Discontinuities are not the most important feature to control failure. The failure surface is free to find the line of least resistance through the material. Rock mass cohesion is low, allowing the ease with which shearing can occur. Each rupture line is then composed of symmetric circular arcs intersecting at an apex, a point of passive earth pressure. In the vertical case further ruptures develop on this basis resulting in parallel lines to the first and shearing vertically rock between arc segments. This progression in effect creates vertical walls as the failure continues. This would explain the shearing of the material noted. The failure is self progressing because there will always be sufficient weight to mobilize other active shear failures. This is representative of the potential non-stop development of chimneying disintegration in weak rock masses.

A method of slices, Figure 3, resolves the problem by considering the sum of shear resistance for each slice in a direction opposite and resisting the action of the total weight. This approach can also consider resistance variation within the rock mass. Since the rupture is symmetrical, only one half of the mobilized rock mass is considered. The radius of the rupture arc (Figure 3) is defined geometrically as

$$r = \frac{b}{[1 - \cos\ (\alpha/2)]} \tag{1}$$

where $\alpha/2 = 45° + \phi/2$, ϕ = rock mass angle of friction at $\sigma_n = 0$, b = 0.5L, L = stope span. The center of the radius lies on a stope roof horizontal line. This radius is then used to calculate the height of the apex line h_{1h} (Figure 4)

$$h_{1h} = \sqrt{b\ (2r\ - b)} \tag{2}$$

The area subtended by this arc is

$$A_T = \frac{(45° + \phi/2)}{360}\ \pi r^2 - \frac{hN}{2} \tag{3}$$

where N = r-b (Figure 3). The weight under the arc is

$$W_T = A_T\ \gamma_r \tag{4}$$

The vertical component $c_m \cos\ \beta_i$ of shear resistance depends on the angle β taken at the slice mid-point (Figure 4). It acts over an area of s_i•1m/sin β_i. The equation to evaluate stability against the first rupture in the vertical and inclined cases is

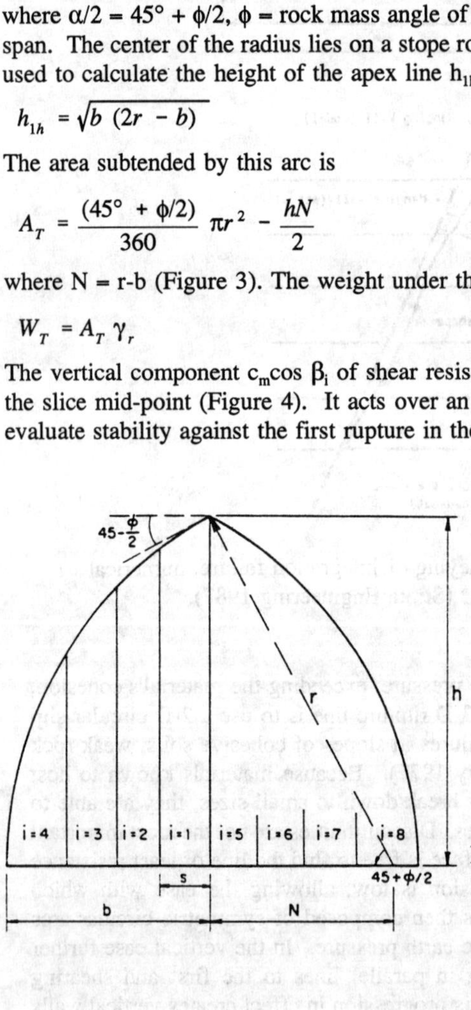

Figure 3. Chimneying disintegration outline with arc and slice definitions (Bétournay 1994)

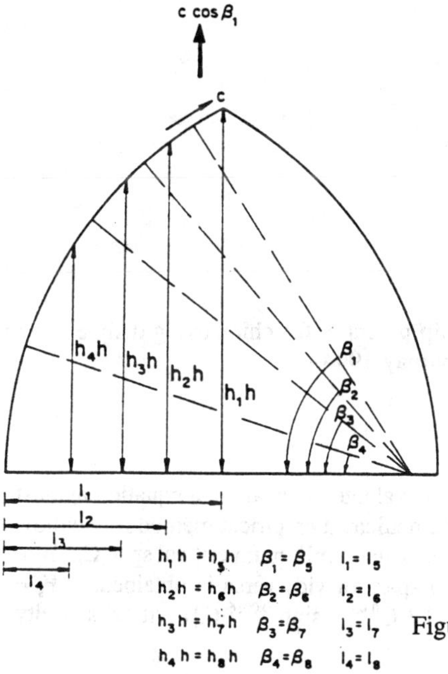

Figure 4. Chimneying disintegration rupture resistance per slice (Bétournay 1994)

$$(F_s)\,\gamma_r \left[\frac{(45^\circ + \phi/2)\,\pi r^2}{360} - \frac{h_{1h}\,N}{2}\right] = \sum_{i=1}^{n} \frac{s}{\sin\left[\tan^{-1}\left(\dfrac{(h_{ih} + h_{(i+1)h})\,0.5}{N + (i-1/2)s}\right)\right]} \bullet$$

$$c_m \cos\left[\tan^{-1}\left(\frac{(h_{ih} + h_{(i+1)h})\,0.5}{N + (i-1/2)s}\right)\right] \tag{5}$$

where $\beta_i = \tan^{-1}\left[\dfrac{(h_{ih} + h_{(i+1)h})\,0.5}{N + (i - \frac{1}{2})s}\right]$, $h_{ih} = \sqrt{l_i\,(2r - l_i)}$

and l_i=0.5L-(i-1)s, i=slice number, s=width of each slice, γ_r = rock unit weight.

Bounded by competent hangingwall, chimneying disintegration proceeds up-dip with one arc, Figure 5. As for the vertical case, sufficient weight can be mobilized for the second and subsequent failures. The limitation to the up-dip process will occur for the second rupture arc when the dip of the weak zone is sufficiently shallow for the normal component of the weight on the footwall plus the resistance against tensile rupture are sufficient to resist the driving force. This limit condition is

$$W_T \sin\psi = \Sigma\, T_i + W_T \cos\psi \tan\phi \tag{6}$$

where Σ T_i = tensile resistance force of slices along an arc parallel to the first rupture.

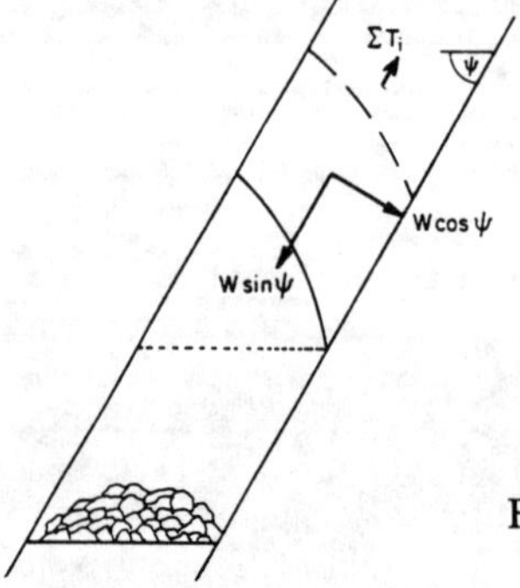

Figure 5. Up-dip potential for chimneying disintegration (Bétournay 1994).

CASE STUDIES

Four case studies have been used to apply and evaluate the analytical equation against the performance of numerical modelling and a dedicated empirical method for shallow stopes. The latter (Golder Associates 1990), provides a minimum critical span C_s, from which a desired pillar thickness t for a given span, or vice versa is obtained, at F_s = 1, using Figure 6: the material quality, the 3.3 $Q^{0.43}$ x sinh $^{0.0016}$ (Q) critical stability line and

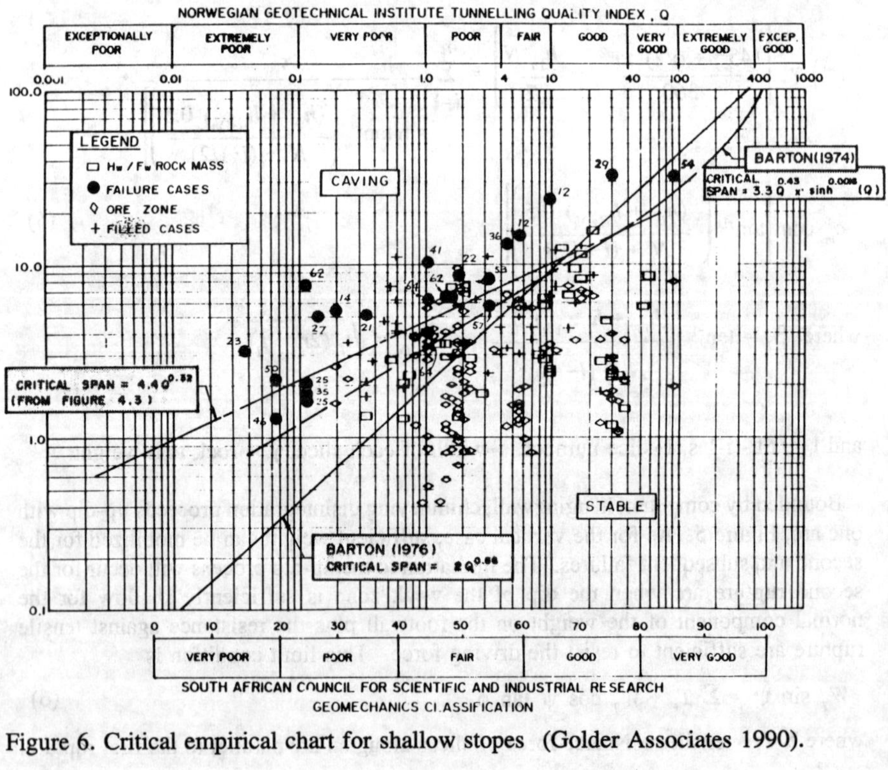

Figure 6. Critical empirical chart for shallow stopes (Golder Associates 1990).

$$C_s = S\left[\frac{\gamma_r}{\left[t\,(1+S/L)\ \ (1-0.4\cos\psi)\right]}\right]^{0.5} \tag{7}$$

where ψ is the controlling discontinuity dip, S the stope span, L the strike length, in meters, γ_r the unit weight in T/m^3.

The cases studied and the data used are found in Table 1. In cases where in situ characterization and lab tests were performed to quantify the Hoek and Brown (1988) field m and s values and calculate rock mass strength envelope parameters, the factor of safety returned by the analytical equation is representative of failed and unfailed conditions encountered. Owing to having only one qualitative source of reference (Rice 1934) to define the geomechanical parameters, the selection of m and s field parameters from approximate values (Hoek and Brown 1988) yielded a representative value for the Brier Hill Mine failure.

Table 1. Summary of results from the application of analytical, numerical and empirical methods to analyzing for chimneying disintegration

Case Study	Material	Rock Mass Parameters (σ_n=0)					Analytical Equation F_s	Stope Width, Depth	NGI Q
		m	s	c_m (MPa)	ϕ degrees	γ_r MN/m^3		m	
Dumagami*	talc covered foliated volcanics	0.40	0.0001	0.20	57.2	0.027	5.1	4.0, 20.0	0.71
Belmoral*+	chloritic gougy schist	0.35	0.00008	0.04	71.8	0.027	0.98	3.8, 25.0	0.01
Brier Hill**+	graphitic slate	0.90	0.0006	0.14	69.9	0.029	1.4	4.2, 275.0	0.32
Selbaie*+	severely altered rock	0.20	0.000005	0.04	34.0	0.023	0.22	22.0, 30.0	0.004

* from field and lab work ** from Hoek and Brown (1988) approximate m & s values + failed case(s) at the site

Figure 7 indicates that based on calculations with equation 5 a numerical correlation exists between the span L of the stope and rock mass cohesion c_m at $F_s = 1$

$$L_{critical} = 100\ c_m \tag{8}$$

The limit for chimneying to develop may occur when shearing of the rock mass at low confinement is precluded, such as the case for blocky ground where a failure surface implies shearing of large interlocking asperities, as was the case for the back analysis
of the Athens Mine plug failure (Bétournay 1994) which yielded $F_s = 1.04$ for chimneying disintegration over a span of 80 m and $c_m = 0.8$ MPa Therefore the application of the analytical equation is limited to cases where the largest rock mass particle size will allow such shearing for the scale, stope span, considered. However, a large span may force the development of block caving instead of chimneying disintegration, as was the case at the Brier Hill Mine (Rice 1934).

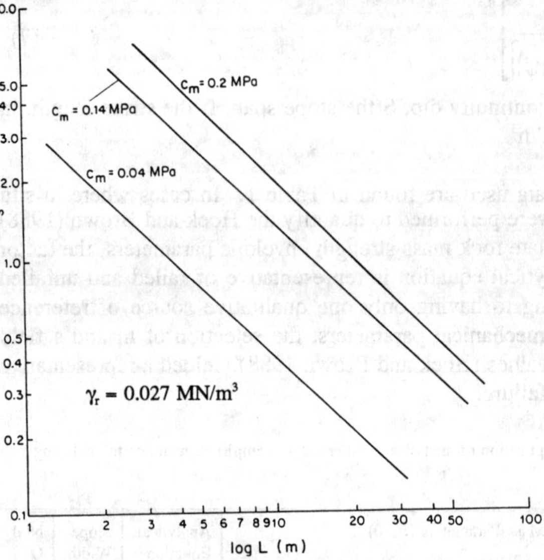

Figure 7. Relationships between stope span L, rock mass cohesion c_m and factor of safety F_s, chimneying disintegration stability analysis (Bétournay 1994).

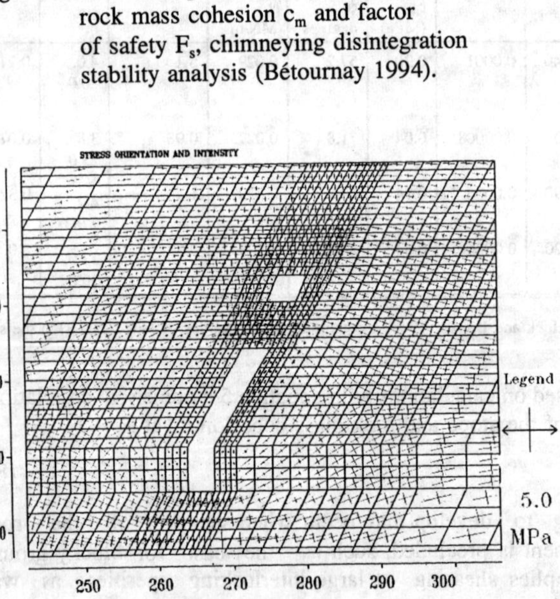

Figure 8. Stress orientation and intensity using numerical modelling, Belmoral Mine failure, dimensions in meters (Bétournay 1994).

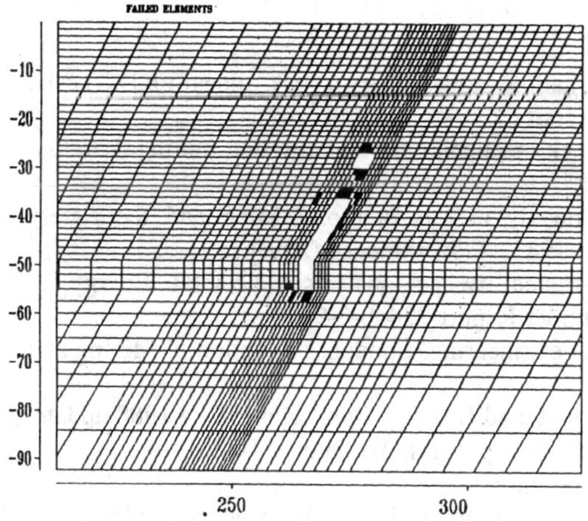

Figure 9. Failed portions of the weak rock mass using numerical modelling, Belmoral Mine failure, dimensions in meters (Bétournay 1994).

Numerical modelling was used and evaluated as a stability analysis tool and as an indication of stress distribution. To simulate the longitudinal continuity of geology and mine openings a 2-D linear elastic finite element code (Mitri 1988) was used for the Brier Hill case. A 2-D elasto plastic finite element code (Yu and Toews 1988) was used for the schist/foliated rock case studies (Belmoral, Dumagami Mines). A large strain finite element code, ADINA, was used for the altered rock of the Selbaie case study (Strata Engineering 1987). Numerical modelling predicts failures in weak rock masses where chimneying disintegration has historically started. However, failed elements must be removed, by using the code's birth/death option or using sequential mining steps to delineate the potential of failure to surface. These failures occur from compressive redistributed stresses, Figures 8, 9, helping gravity action to begin.

The dedicated empirical method underestimated the minimum surface crown pillar thickness (depth) 13.1 m at Brier Hill where the failure started at a depth of 275 m, 226 m at Belmoral where failures continue to occur from depths of 265 m and could not locate the failure at the stope periphery. Small variations in defining the quality of such poor rock masses yields large differences in the C_s parameter.

CONCLUSIONS

Chimneying disintegration commonly affects shallow stopes of hard rock mines. Weak rock masses (cohesion 0.04 - 0.14 MPa) host this type of failure which can be as narrow as 2 m. Mobilization of crown or hangingwall material into the stope occurs along lines of active shear. This may be helped by compressive redistributed ground stresses high enough to fail the material. A new limit equilibrium analytical equation has successfully back-analyzed known case studies. Failures are not expected where the rock mass is incapable of shearing under its own weight. Numerical modelling, can indicate where failures can start but will not predict the nor the chimneying outline or continuity to surface. A shallow stope empirical stability analysis method is limited in predicting such failures. Proper definition of the rock mass cohesion and angle of friction is essential. Rapid application of ground support such as cable bolting will prevent mobilization of the weak rock mass, and the beginning of the failure process.

REFERENCES

Bétournay, M.C. 1994. Stability of shallow stopes of hard rock mines. Ph.D Thesis, Mining Engineering, McGill University.

Brady, B.H. & Brown, E.T. 1985. Rock mechanics for underground mining. London: George Allen and Unwin.

Commission of Enquiry on the Tragedy at the Belmoral Mine and the Safety Conditions in Underground Mines, 1981. Government of Quebec.

Golder Associates 1990. Crown pillar stability back-analysis, CANMET, Energy, Mines and Resources Canada, Contract Report 23440-8-9074.

Hoek, E. & Bray, J. 1977. Rock slope engineering. London: Institution of Mining and Metallurgy.

Hoek, E. & Brown, E.T. 1988. The Hoek and Brown failure criterion - a 1988 update. Proc.15th Canadian Rock Mechanics Symposium, Toronto: 31-38.

Jones, D. 1993. Personal Communication.

Mitri, H.S. 1988. Finite element applications in mining engineering using MSAP2D. Professional seminar text, McGill University.

Picciaccia, L. 1988. Personal communication.

Rice, G.S. 1934. Ground movement from mining in Brier Hill Mine, Norway, Michigan, Proc. Am. Inst. Min. & Met. Eng. 109: 191-206.

Strata Engineering 1987. Weak rock mass model for support of surface crown pillars at Les Mines Selbaie. CANMET, Energy, Mines and Resources, Canada, Contract Report 23440-5-9017.

Yu, Y.S. & Toews, N.A. 1988. PCEPFE user's guide - a 2-D elastic-plastic finite element stress analysis package using a personal computer; CANMET, Energy, Mines and Resources, Canada; Report MRL 88-95(TR).

18 Coal and surface mine design

Session developers:
Robert Bauer, Illinois State Geological Survey
Will Bawden, Queen's University

Rock Mechanics, Nelson & Laubach (eds) © 1994 Balkema, Rotterdam, ISBN 90 5410 380 8

Three-dimensional elasto-plastic analysis of coal pillars for highwall mining

M.S.Craig, M.E.Duncan Fama & R.Trueman
Division of Exploration and Mining, CSIRO, Qld, Australia

ABSTRACT: Highwall mining is being introduced into central Queensland open cut mines so that coal can be mined from existing highwalls, where the overburden depth is too great for conventional surface methods to operate economically. With this mining method, remote controlled continuous miners extract coal from a series of parallel unsupported roadways driven from the base of the highwall, perpendicular to the highwall face, for a distance of up to 300m. The optimal design of the width of the pillars left between the roadways is one of the factors that is crucial to achieving maximal recovery whilst maintaining the integrity of workings

This paper describes a three-dimensional elasto-plastic model of a highwall and the driving of roads into a dipping seam. The model is used to estimate a factor of safety for a particular combination of pillar and roadway widths, and depth of mining. This analysis method was applied to formulate a preliminary pillar design for a central Queensland coal mine.

1 INTRODUCTION

The mining of coal by open cut methods is not economically viable when the ratio of overburden thickness to coal seam thickness beecomes excessive. For coal seams with a thickness of between three and four metres the limiting overburden thickness for typical Australian conditions is in the range 50-60m. In virtually all cases the seams are dipping at between 3 and 15 degrees and the recovery of coal from parts of a seam at greater depths using surface methods is a major focus of the industry because of the large investment in equipment and the high cost of developing an underground mine.

Highwall mining offers a lower cost alternative to underground mines. When the economic depth limit is reached for open cut mines, coal may be extracted from a series of parallel unsupported roadways driven into the dipping seam from the base of the highwall by a remote controlled continuous miner.

The width of the pillar separating the roadways is of crucial importance to the success and viability of highwall mining. The width must be large enough to ensure the pillar does not collapse, but leave the minimum of coal unmined. Coal pillars in underground mines are usually designed using empirical formulae to determine their strength. These formulae are based on a significant body of data and have had a measure of success for underground mines. There are no significant data on pillar stability for highwall mining and the extrapolation of the empirical design rules to this mining method cannot be justified, as the geometry, loading and initial damage of the pillars are quite different.

A combination of numerical models was used to quantify the effect of these factors in Duncan Fama et al. (1994). Two-dimensional plane strain models were used to study vertical slices perpendicular to the roadway and the effects of many parameters can be examined economically. However, the effect of the dipping seam, the transfer of stress from one section to another and the regions near the highwall and the roadway end, where plane strain assumptions are not appropriate, require a much more computationally intensive three-dimensional model and this latter modelling is the subject of this paper

A three-dimensional model of a highwall and the driving of roadways from the base of the highwall is described in Section 2. The coal properties used in the model are based on a back analysis of square plan coal pillars in underground mines to fit a modified Hoek-Brown yield criterion with strain softening. Parameters are chosen to fit existing empirical strength formulae for such pillars and are thus appropriate for the global average strength for coal.

Estimates of the average pillar vertical stress and a means of estimating a factor of safety for a seam dip of 6°, a pillar width of 2.5m, a roadway width of 3.4m, a seam height of 3.5m and a penetration depth of 300m, are given in Section 3. This method was used for various pillar widths and heights, in combination with plane strain studies, to formulate a preliminary pillar design for a central Queensland mine, as reported in Trueman et al. (1994),

2 THE NUMERICAL MODEL OF HIGHWALL MINING

A three-dimensional numerical model of parallel roadways driven into a sloping coal seam from the base of a highwall was constructed. Details of the model are described in this section.

2.1 *Geometry of Highwall Mining*

The geometry of the model was based on a central Queensland coal mine. A simplified section perpendicular to the highwall is shown in Figure 1.

The vertical height of the highwall was taken as 52m. The roadways were 300m long and 3.4m wide. The maximum depth at the end of the roadways was 83m.

2.2 *The Properties of the Coal*

The coal seam was modelled as an elasto-plastic material with a modified Hoek-Brown yield criterion:

(1) $$\sigma_1 = \sigma_3 + (A\sigma_3 + B)^a,$$

where σ_1 and σ_3 are respectively the major and minor principal stresses in the coal and A, B and a are parameters which need to be determined. A and B may be expressed in terms of the standard Hoek-Brown parameters, m_b and s, and the unconfined compressive strength, σ_c, by

(2) $$A = m_b\sigma_c^{1/a-1} \text{ and } B = s\sigma_c^{1/a}.$$

Strain softening is introduced by reducing A as the plastic strain increases according to:

$$A = A_2 + (A_1 - A_2)\, e^{-\alpha\gamma_p}. \tag{3}$$

Here the plastic strain, $\gamma_p = \varepsilon_1 - \varepsilon_3$ where ε_1 and ε_3 are the major and minor principal plastic strains and A_1, A_2 and α are constants. A_1 is the pre-failure value of A (i.e. its value when $\gamma_p = 0$) and A_2 is the residual value of A (i.e. its value when γ_p is large).

At the stage the initial pillar design was required, no in situ mass strength data for coal were available for fitting to this model. Also, scale effects make experimental results (based on small, disturbed samples) unreliable. It was decided to use the empirical rules developed by Salamon and Munroe (1967), Bieniawski (1987) and Obert and Duvall (1967) for square plan pillars and to back analyse for parameters in the coal model.

Details of this back analysis, which used the same three-dimensional program, are reported in Duncan Fama et al. (1994) and the selected parameters (as well as the Young's modulus E and Poisson's ratio ν) are given in Table 1. For an unconfined compressive strength, σ_c, of 30MPa, the pre-failure value of m_b is 2.83 and of s is 0.0004. These parameters fall within the range of accepted Hoek-Brown parameters for coal, but are not unique in giving a reasonable fit to the empirical curves.

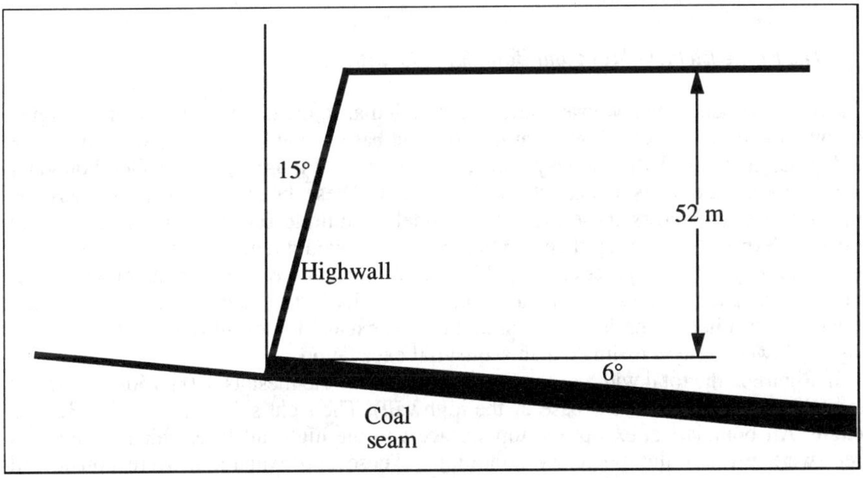

Figure 1 Schematic of highwall mining geometry

Table 1 Parameters used for coal properties

E	ν	a	B	A_1	A_2	α
2.5 GPa	0.24	0.5	0.368 (MPa)^2	85 MPa	21.25 MPa	100.

2.3 *The Properties of the Roof and Floor*

The geological structure above and below the coal seam was simplified and the roof and floor of the roadways were modelled as elastic materials with Young's modulus 3.4 GPa and Poisson's ratio 0.2, except for the damaged regions near the highwall. Two-dimensional plane strain studies, Duncan Fama (1993), indicated that for this model there was little dependence of pillar strength on the precise geology of the roof and floor, assuming that they remained intact. However, the pillar strength was influenced markedly by the properties of the interfaces between the coal and the roof and floor; this effect was discussed by Duncan Fama et al. (1994). Near the highwall, damaged zones were observed by Leisemann and Follington (1993). No other information was available to give good estimates of the properties of the damaged rock and a set of properties was assumed that were consistent with the observations. The damaged rock was modelled as an elasto-plastic material with Young's modulus lower than the intact rock and failure parameters of the roof the same as the coal, given in Table 1. The values of reduced Young's moduli that were used at different distances from the highwall face are given in Table 2.

Table 2 Young's modulus in the damaged region

Distance from face (m)	0-3	3-6	6-10	10-20	20-30	30-40
Young's modulus (GPa)	0.05	0.1	0.2	0.5	1.5	2.5

2.4 *The Finite Element Mesh and Boundary Conditions*

The finite element mesh, whose section perpendicular to the highwall is shown in Figure 2, contains the coal seam dipping at 6° from the base of the highwall. The mesh is one half pillar and one half roadway thick, as shown in Figure 3, with roller boundary conditions on the sides indicated by thick lines. These boundary conditions assume mirror symmetry across these faces and model an infinite set of roadways excavated together. Explicit modelling of sequential excavation would require a much larger mesh, which is beyond the practical capabilities of current computer technology. Two-dimensional studies, Duncan Fama et al. (1994), indicate some effect of excavation sequence on pillar strength and the simultaneous excavation of all roadways leads to a slightly lower strength estimate than sequential excavation.

In Figure 2, the total width of the lower boundary of the mesh is 550m (50m to the left and 500m to the right of the base of the highwall). The right side boundary is 132m in length. All boundaries, except the top surface and the highwall face which were stress free, were given roller boundary conditions. These approximations were considered adequate as they gave reasonable results for the stresses near the boundaries.

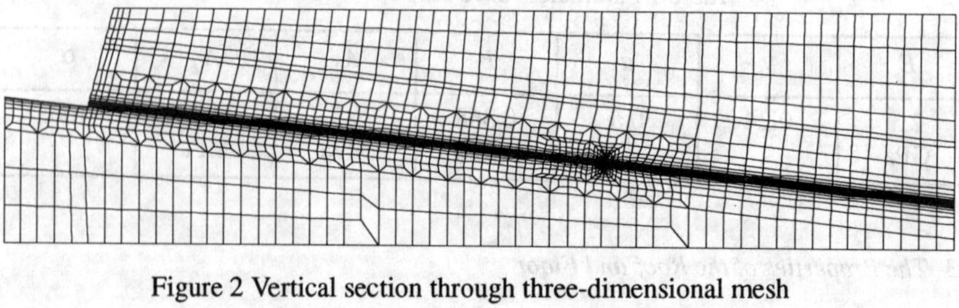

Figure 2 Vertical section through three-dimensional mesh

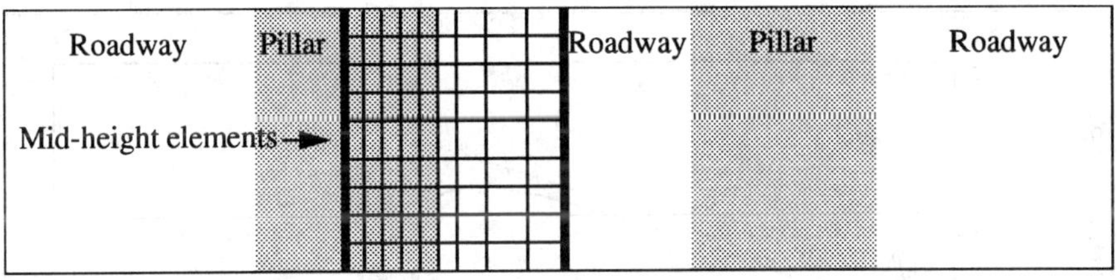

Figure 3 Vertical section through the coal seam parallel to the highwall

2.5 *Initial Stress Field*

The stress field, before the excavation of the highwall, was assumed to be hydrostatic, with each principal stress equal to ρgy, where ρ is the rock density (2602 kg/m^3), g is the acceleration due to gravity and y is the (negative) distance below the surface. Horizontal stresses measured by Enever and Follington (1993) in the vicinity of the highwall were found to be approximately equal to or a little less than vertical stresses.

2.6 *Details of the Analysis*

The three-dimensional analysis model used in this study is 3STRESS, developed by the CSIRO Division of Exploration and Mining. It is an explicit finite element code for the analysis of non-linear problems in geomechanics.

Using a mesh whose cross-section is shown in Figure 2 and one element thick to simulate plane strain, the stress field following the excavation of the highwall was estimated by letting the mesh (with the highwall excavated, the damaged region near the highwall and the initial stress discussed in Section 2.5) reach equilibrium.

The displacements, stresses and plastic strains from the excavation of the highwall are then used as initial conditions for the driving of the infinite array of roadways using the mesh described in Section 2.4. The entire length (300m) and height (3.4m) of the roadway in the computer model are removed in four width stages to avoid excessive impulse unloading and the resultant overestimate of plastic strain. The pillar width selected for this paper is 2.5m and the pillar height is 3.5m.

3 THE AVERAGE PILLAR VERTICAL STRESS

The average vertical stress along the pillar and in the coal seam up to a distance of 500m from the highwall is shown in Figure 4. The average stress is defined as the mean stress in the five mid-height elements in the pillar—these are the five larger pillar elements shown in Figure 3. For comparison purposes, the initial vertical cover stress is plotted for distances greater than 300m (i.e. beyond the end of the roadway) and the initial cover stress, multiplied by (3.4+2.5)/2.5 (i.e. the factor to account for the removal of the roadway elements) is plotted for distances less than 300m. These modified cover stresses are the vertical stresses that would be expected without any redistribution of stress along the pillar.

Between 60m and 250m from the highwall, the vertical stress is close to the modified cover stress, as would be expected if the pillar is able to support the increased load. The material properties in the damaged region near the highwall were rough estimates only and the details of the stress field here are somewhat uncertain. The peak in the stress near the damaged region is displaced slightly in from the highwall and the peak is slightly

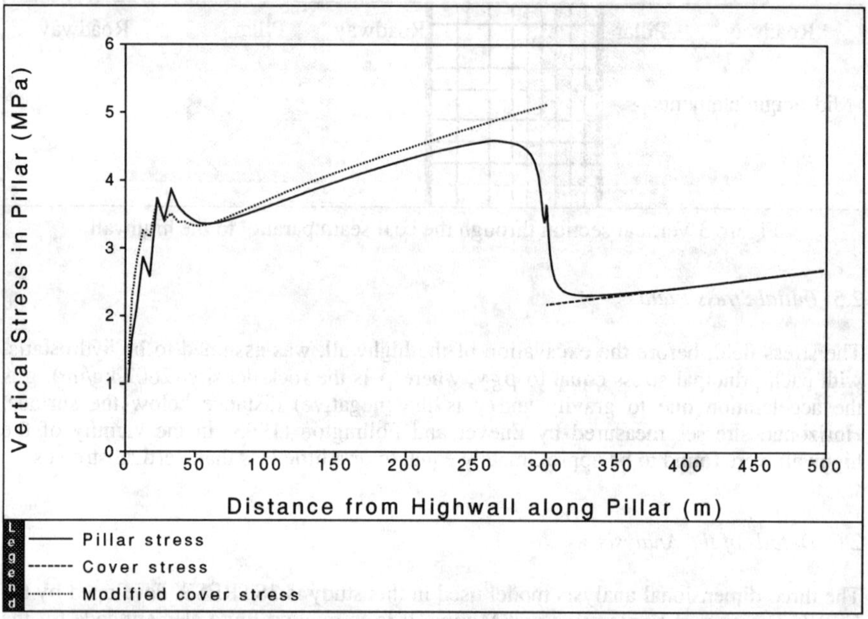

Figure 4 Average vertical stress carried by the pillar and coal seam

higher than in the modified cover stress. Some of the pillar vertical stress is redistributed to be carried by the coal seam beyond the roadway. The excess stress (over the cover stress) in the seam beyond the end of the roadway is carried by the full width (pillar plus roadway), not just the pillar. The extra load carried by the intact seam is sufficient to equilibrate the load deficit of the pillar, compared with the cover load.

To assist the engineer in selecting a suitable pillar width, a factor of safety may be defined as the factor by which the initial stresses may be increased before equilibrium cannot be achieved and pillar collapse occurs. The factor of safety provides a measure of the robustness of the pillar width design to errors in estimation of material parameters and stress conditions.

Figure 5 shows the average equilibrium vertical stress in the pillar and coal seam if the pre-roadway stresses and the gravitational body forces are increased by a factor of 1.2. Equilibrium is achieved and the pillar is stable.

If these stresses and body forces are increased by a factor of 1.3, equilibrium is not achieved and the modelled pillar fails. This last condition is illustrated in Figure 6, which shows the average vertical stress in the pillar and coal seam at some time after mining of the roadway. The pillar stresses are unable to support the pillar loads, indicated by the dotted line. The modelled failure begins near the end of the roadway and propagates back towards the highwall, due to stress redistribution and strain softening. At this time, the damaged zone near the highwall is still stable, but the modelled collapse will eventually propagate to the full pillar length. The detailed consequences of failure of part of a pillar may be different with sequential roadway excavation. The coal seam just beyond the end of the roadway carries a high stress, but this is insufficient to stop the collapse. The factor of safety for the pillar width of 2.5m is, therefore, between 1.2 and 1.3.

In two-dimensional plane strain studies of vertical slices perpendicular to the roadway, using methods described in Duncan Fama et al. (1994), the authors estimated factors of safety between 0.95 (250m from the highwall) and 1.3 (50m from the

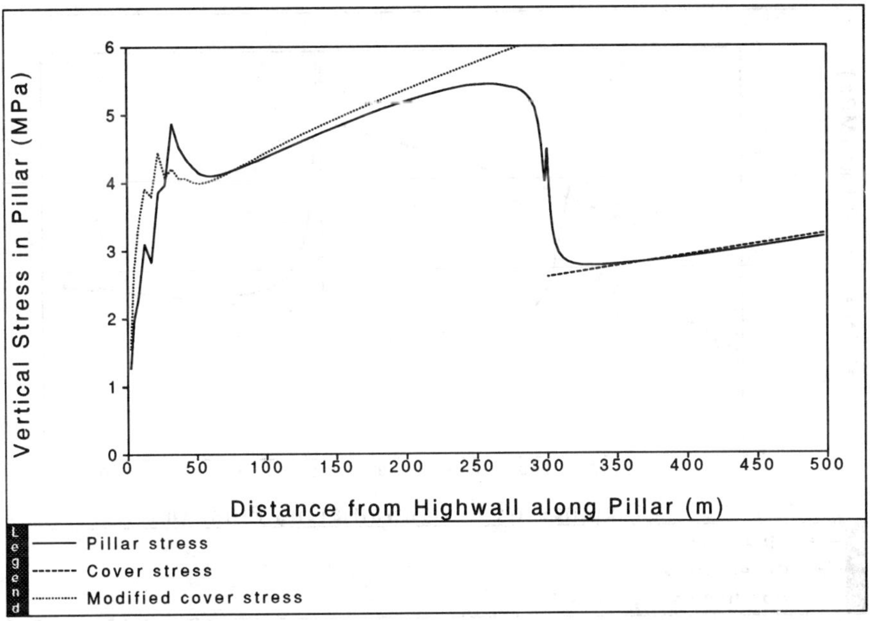

Figure 5 Average vertical stress carried by the pillar and coal seam with 1.2 times pre-excavation stress

highwall) at different points along the roadway. The three-dimensional model's ability to redistribute stresses along the length of the pillar and the roadway end accounts for its apparently higher estimate of factor of safety for the entire pillar.

4 CONCLUSIONS

A three-dimensional model of highwall mining was constructed to complement two-dimensional plane strain studies. Computer resource constraints necessitated the assumption of an infinite row of roadways, excavated simultaneously, rather than the more realistic sequential excavation of roadways. The roof and floor materials are assumed to remain intact. The model was used to estimate the factor of safety of pillars with a particular mining geometry. Use of this technique for various geometries, in combination with two-dimensional models, enabled a preliminary pillar design for a central Queensland coal mine to be formulated, as described in Trueman et al. (1994).

REFERENCES

Bieniawski, Z.T. 1987. *Strata Control in Mineral Engineering*. AA. Balkema, Rotterdam, 59-90.

Duncan Fama, M.E. 1993. A parametric study of the factors affecting the use of numerical modelling for pillar design. *CSIRO Division of Geomechanics Internal Report (New Series) No. 181.*

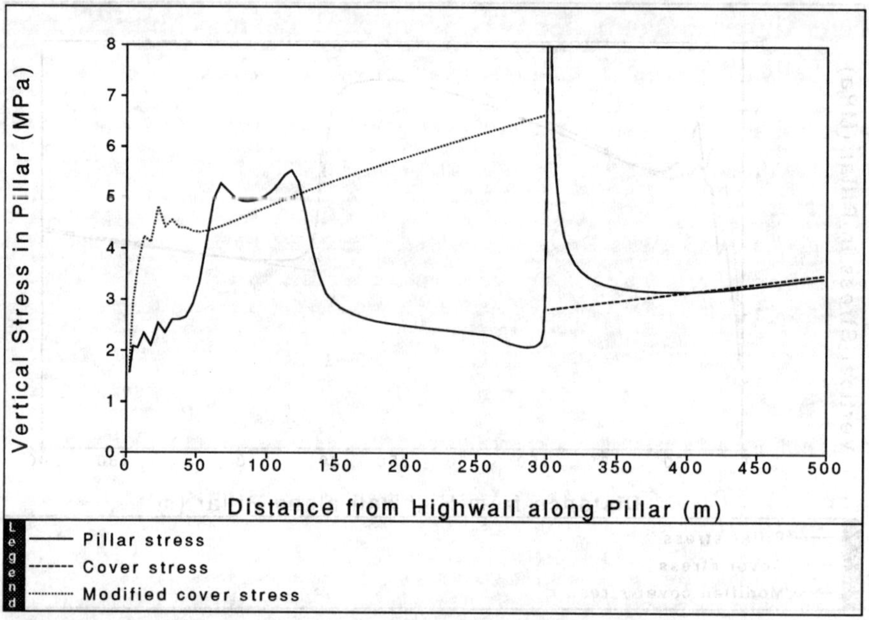

Figure 6 Average vertical stress carried by the pillar and coal seam with 1.3 times pre-excavation stress—equilibrium is not achieved and the modelled pillar fails

Duncan Fama, M.E., R. Trueman & M.S. Craig 1994. Two and three dimensional elasto-plastic analysis for coal pillar design. Submitted to *Int. J.Rock Mech. Min. Sci*

Enever, J.R. & I.L. Follington 1993. Stress measurements in the highwall of pit 17DU South Moura mine. *CSIRO Division of Geomechanics Internal Report (New Series) No 180*.

Leisemann, B.E. & I.L. Follington 1993. Structure, strength and relaxation of interburden for input into highwall mining design in the Baralba coal measures of Queensland, Australia. *Twelfth Conference on Ground Control in Mining*, Morgantown, WV, August 3-5, 1993, 301-306.

Obert, L. & W.I. Duvall 1967. *Rock Mechanics and the Design of Structures in Rock*. Wiley, New York, 545-547.

Salamon, M.D.G. & A.H. Munroe 1967. A study of the strength of coal pillars. *J. S. Afr. Inst. Min. Metall.*, 55-67.

Trueman, R., M.E. Duncan Fama & M.S. Craig 1994. Pillar design method for highwall mining. Submitted to *Integral Approach to Applied Rock Mechanics*, ISRM International Symposium, Santiago, Chile, May, 1994.

Rock Mechanics, Nelson & Laubach (eds) © 1994 Balkema, Rotterdam, ISBN 90 5410 380 8

Three dimensional modeling of a four-way coal mine intersection for analysis of roof-pillar-floor interaction

K.Chandrashekhar
Indiana Department of Natural Resources-Reclamation, Jasonville, Ind., USA
Y.P.Chugh
Department of Mining Engineering, Southern Illinois University, Carbondale, Ill., USA

ABSTRACT: Roof and pillar interaction with weak floor strata in a partial extraction room-and-pillar mining system (such as those found in Illinois) have lead to problems of a) pillar settlement, b) floor heave and squeeze, c) instability due to differential settlement of pillars and d) surface subsidence. A three dimensional (3D) finite element model (FEM) of an underground intersection was used to study the effect of varying weak floor thickness and differential loading conditions on the mechanisms underlying the above field observations. A linear elastic analysis with scaled material properties compared favorably with actual field observations.

INTRODUCTION AND SPECIFIC OBJECTIVES

The design of underground room-and-pillar coal mines in the Illinois Coal Basin, particularly those at shallow depths below 400 ft depth, appear to be governed by the weak floor strata (Chugh, et. al., 1989). Common related problems include a) excessive pillar settlement and instability, b) floor heave and squeeze, c) roof falls due to differential pillar settlements, and d) subsidence movements that may adversely affect surface land use. Problems related to the roof strata and coal pillars have been extensively studied in the past in Illinois Basin mines (Bauer, 1984; Hanna et. al., 1986 among others). Limited work has also been done by several researchers to investigate specific effects of weak floor strata on mining. Rockaway and Stephenson (1979) used moisture content to estimate cohesion of underclay to modify Vesic's bearing capacity equation for weak floor conditions. Chugh et. al. (1986a, 1986b) characterized the weak floor strata for several mines in Illinois and reported correlations between laboratory and field measured parameters. Bandopadhay (1982) and Caudle et. al. (1988) studied the interaction problem in entries with a 2D model. However, these studies were not adequate to address the effects of interaction near an intersection. To the best of the authors' knowledge, very limited studies have been done to study such interaction with a 3D perspective. This paper therefore discusses the results of such a study using a 3D FEM to study stability at 4-way intersections for comparisons with field observations to develop a predictive capability.

The current study was done as an extension to previous 2D research to analyze the stability and roof-pillar-floor interaction at 4-way intersections associated with weak floor conditions using a 3D model. A detailed study was planned to include laboratory and field measurement together with finite element modeling. The results of extensive laboratory and field studies have been reported separately (Chugh et. al., 1988a; Caudle et. al., 1988; Chandrashekhar, 1990). The goals of the modeling portion of the overall study were to

(1) better understand the effects of roof-pillar-floor interaction on mine stability under weak floor conditions through sensitivity analysis and (2) develop analytical tools to predict these interactions and their effects on mine stability.

REVIEW OF PERTINENT LITERATURE

Gercek (1982) conducted a parametric 3D finite element study of underground coal mine intersections without weak floor strata and concluded that the disturbed zone around an intersection is limited to about one entry width around it. Bandopadhay (1982) reported an increase in shear stress at the roof-coal and coal-underclay interface, pillar settlement, and convergence with larger underclay thickness using a 2D model for a Western Kentucky mine. Wang, et al. (1985) conducted 2D and 3D finite element analyses and found that it was enough to model 1.5 times the width of the opening because the stress distribution beyond this height was relatively uniform. Prasad (1985) showed close correlation between measured and model predicted values using a 2D FEM in analyzing a southern Illinois coal mine. Caudle et. al. (1988) identified potential failure zones around an opening due to weak floor strata using a 2D elastic FEM (Figure 1). The review indicated that only 2D FEMs have been used with field studies to study the effect of weak floor strata on structural instability. There existed a need for a 3D analysis of the effects of weak floor strata around an intersection.

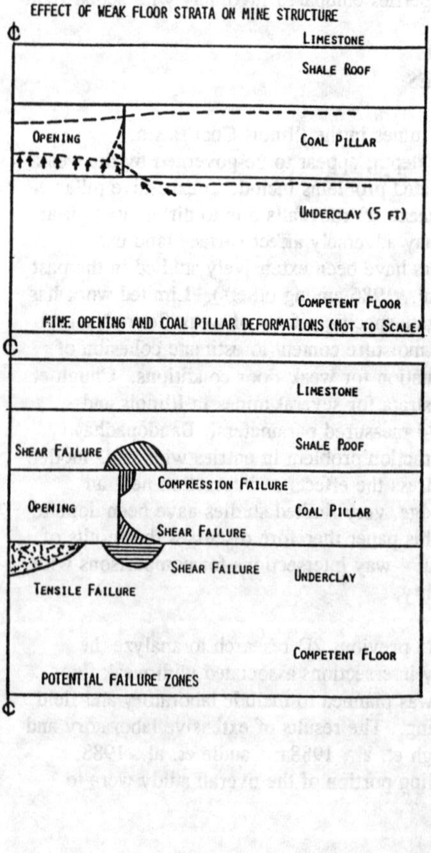

Figure 1 Deformation around opening & predicted failure zones due to weak floor strata (Caudle et. al., 1988)

MODEL DESCRIPTION

A 4-way intersection in a room-and-pillar mine near the center of a panel was modeled (Figure 2) to minimize the effects of the barrier pillar and the working face. This facilitated modeling the effect of roof, pillar, and floor interaction due to lateral variability in floor thickness and differential settlement of adjacent pillars. Fourteen element groups capable of being assigned different material properties allowed the flexibility of simulating varied geological settings such as changing strata thickness and rock characteristics. The position of the nodes and element sizes were planned to get model information equivalent to that obtained during field measurement and enhance both qualitative and quantitative comparison.

The coal seam is located at a depth of approximately 280 ft. A vertical stress of 310 psi corresponding to this depth (τ=160 pcf) was applied to the model. An in-situ lateral stress of about 1500 psi was applied based on previous stress measurements. The laboratory values of material properties were scaled by a factor of 2.5 based on past studies (Caudle et. al., 1988). Table 1 summarizes the pertinent material properties data that were used in the model. The thickness of strata modeled (67 ft) above and below the coal seam was adequate because the stress concentration factors around the opening do not change significantly when larger thicknesses are considered (Wang, et al., 1985, Caudle et. al.,1988). The boundary conditions were assigned so that the assumed symmetry conditions were fulfilled.

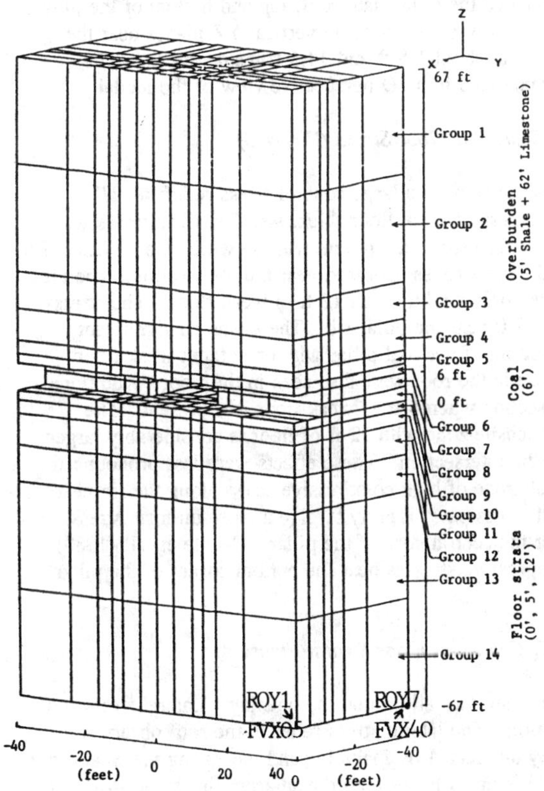

Figure 2 Perspective view of the 3D finite element mesh

Table 1 Material properties used in the model

Material Type	Youngs Modulus (psi)	Poisson's Ratio	Compressive Strength (psi)	Tensile Strength (psi)
Limestone	1.00×10^6	0.15	8000	500
Shale	0.35×10^6	0.15	3000	300
Coal	0.175×10^6	0.35	2500	100
Underclay	0.050×10^6	0.40	600	0

RESULTS - EFFECT OF UNDERCLAY THICKNESS

The stratified model with isotropic linear elastic properties for each layer was analyzed for the effect of varying weak floor thickness (0, 5, and 12 ft). Contours of vertical stress concentration factors (VSCF), horizontal stress concentration factors (HSCF), safety factors (SF) based on shear stress, and vertical displacements were drawn for different sections strategically located in the model. Data analysis consisted of evaluating these parameters in the roof and floor from the model and comparing them with field measurement. The horizontal (XY) planes of interest included the immediate roof, top and bottom of the pillar, and the immediate floor. Similar analysis was also done on vertical YZ planes near the center of the entry, at the pillar edge, and about 1.5 ft and 14.5 ft inside the rib. The observations have been generally summarized in a 3D perspective view of the model.

Vertical Stresses - Immediate Roof, Coal, and Floor Strata (Figure 3)

The central half of the entry and crosscut showed a large zone of stress relief for all stratified cases irrespective of the thickness of weak floor thickness. Tensile stresses were seen in the roof along the crosscut and adjacent to the pillars with no weak floor. When 5 ft of weak floor was present, additional tensile zones appeared near the intersection. The area and depth of this tensile zone increased only slightly with 12 ft of weak floor. The stress concentration rapidly increased to 1.5-2.0 near the pillar rib. There was no significant change in stresses near the top and bottom of the coal pillar with or without weak floor strata. Tensile stresses similar to those in the roof were also seen in the floor. Additional tensile zones appeared near the intersection when 5 ft of weak floor was present. The increase in the area and depth of this tensile zone with 12 ft of floor is considerably larger in the floor strata especially close to the intersection. These effects were less pronounced away from the intersection. The small zone of high compressive stress along the rib near the top and bottom corners of the pillar dissipated rapidly to give a more uniform stress distribution in the roof and floor over the central core of the pillar. Caudle et. al. (1988) also reported a slight increase in compressive stresses near the bottom corner of the pillar with their 2-D analysis.

Horizontal Stresses - Immediate Roof, Coal, and Floor Strata (Figure 4)

In general, there is a decrease in lateral stresses all around the opening. Lateral stress relief is more pronounced near the intersection. The lateral stress relief in the roof observed without weak floor strata is practically unaffected by including and increasing the weak floor thickness because the stiffness of shale is much lower than the adjacent limestone strata. A similar behavior was seen near the top of the coal pillar. Near the bottom of the coal pillar,

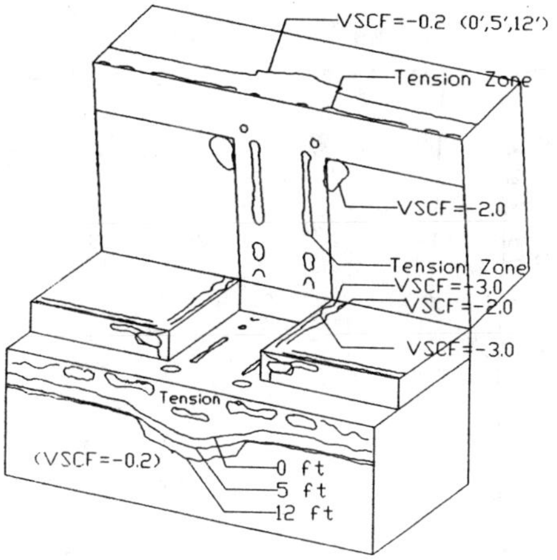

Figure 3 Vertical stress concentration factors in the roof, pillar, & floor

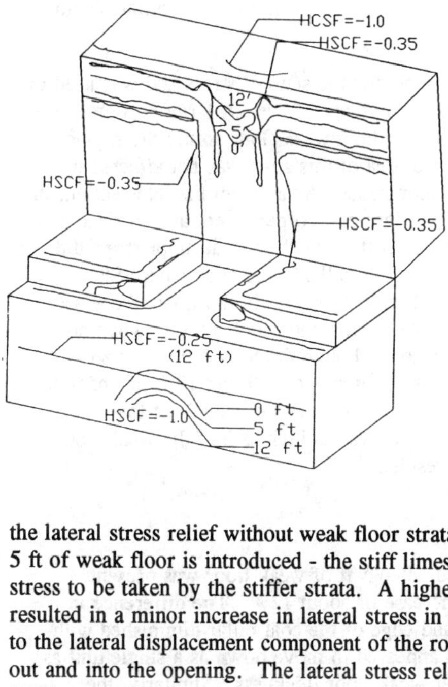

Figure 4 Horizontal stress concentration factors in the roof,pillar, & floor

the lateral stress relief without weak floor strata is further reduced by almost 25-50% when 5 ft of weak floor is introduced - the stiff limestone strata is always close enough for the stress to be taken by the stiffer strata. A higher thickness of weak floor strata (12 ft) resulted in a minor increase in lateral stress in the upper and lower portions of the pillar due to the lateral displacement component of the roof sag and weak floor material being pushed out and into the opening. The lateral stress relief in the floor was reduced considerably (75-85%) with 5 ft of weak floor; there was no change when weak floor thickness was increased to 12 ft. In all stratified cases, the horizontal stresses rapidly increased to 75% of the applied stress within a few feet of the stiff limestone strata. The entire weak floor strata was always a part of larger lateral stress relief.

Safety Factor Analysis

The safety factor indicated by Pariseau and Sorensen (1979) derived from the

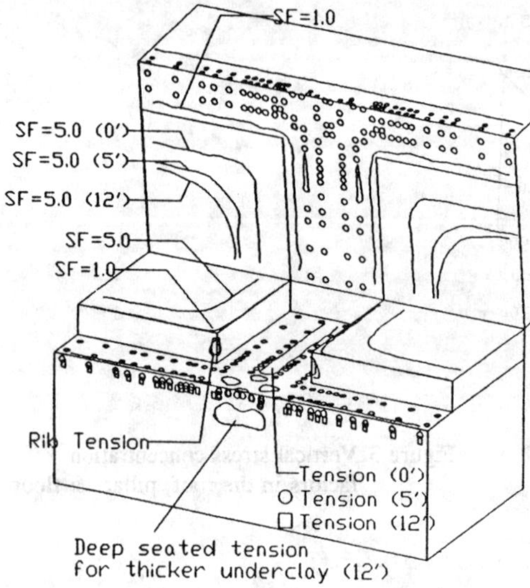

Figure 5 Distribution of safety factors in the roof, pillar, & floor

isotropic-extended von Mises yield criterion accounts for the state of stress and was used to evaluate stability. Figure 5 summarizes the discussion that follows. A large portion of the roof over the entry is likely to fail (SF < 1.0) even without weak floor strata; the region extended into the limestone strata. The height and area of this zone was not affected by including and increasing the thickness of weak floor strata. Away from the intersection, the failure zone still intercepted the limestone. Pillar edges and corners were also critical zones of failure even without weak floor strata. Including 5 ft or 12 ft of weak floor strata did not affect the safety factor near the rib and inner core of the pillar. The zone where SF < 1.0 near the rib related well with the zone of vertical tensile stresses. In the floor, there was a small zone of tensile principal stress along the center of the entry and crosscut when no weak floor strata was present that may fail in tension although the SF > 1.0. This zone increased to cover the entire opening width with the presence of 5 ft and 12 ft of underclay. The floor failure was deeper with increasing weak floor thickness especially at the intersection. Field measurement of the affected depth was as high as half the weak floor thickness and it compared well with the model results.

Vertical Displacements

The maximum roof convergence almost doubled when 5 ft of weak floor was present (Figure 6); 12 ft of weak floor resulted in an increase of about 12%. The difference in downward roof movement between the center and edge of the coal pillar diminished with increasing weak floor thickness. The roof thus appeared to move down as a single unit as the pillars tend to punch into the floor for larger weak floor thickness. Similarly, the maximum floor heave was four times larger when 5 ft of weak floor was present (Figure 7); 12 ft of weak floor increased this an additional three times. The pillar settlement (center of pillar) doubled with 5 ft of underclay and then doubled again for 12 ft of weak floor. For a given increase in weak floor thickness, floor heave increased more rapidly than pillar settlement. The floor therefore was more likely to be the zone of instability due to differential displacements under weak floor conditions. At the intersection, higher strains were observed due to differential convergence near the top and bottom corners of the pillars. Caudle et. al. (1988) reported similar findings with 2-D analysis of a similar problem

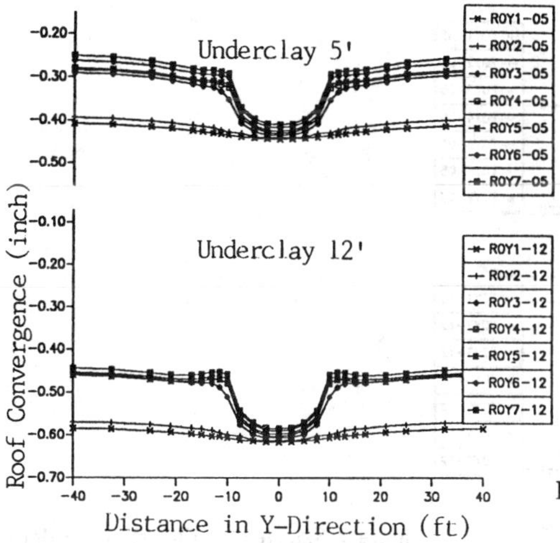

Figure 6 Selected profiles showing roof convergence along the entry

(Figure 1). These observations also compared favorably with field measurement of relatively uniform pillar settlement across the panel and floor heave on the order of several inches. The center of the pillar was in compression as reported by Chugh et. al. (1988b) from field measurement. Sub-floor movement were relatively smaller as seen in the field measurement using the sondex setup.

Horizontal Displacement in the Weak Floor Strata (Figure 8)

Lateral floor movement increased considerably when 5 ft of weak floor is introduced and almost doubled for 12 ft of weak floor. In general, movement across an opening was larger compared to those along its axis and mostly limited to the weak floor with lower elastic moduli. Shear zones developed at the interface due to rapidly increasing modulus into limestone. This compared well with lateral movement down to 20 ft below the coal seam where only 5 ft of underclay was present; existence of shear zones was also verified.

Field behavior led the authors to believe that underclay may be flowing outward from beneath the pillars into the opening and may tend to distort the pillar in an unstable manner. Although field instrumentation was inadequate to verify this, the model did lend itself for such analysis. Interestingly, lateral displacement increased and reached a maximum below the edge of the pillar and then decreased towards the center of the pillar. The middle portion of the weak floor always moved the most resulting in floor heave in the center of the opening. The zones of predicted instability due to such underclay movement compared well with those predicted by Caudle et. al. (1988). However, the 3D model predicted these effects near top and bottom corners of the pillar at the intersection (Figure 9).

RESULTS - EFFECT OF DIFFERENTIAL LOADING OF ADJACENT PILLARS

Differential loading conditions can arise in the field due to a roof fall or non-uniform pillar settlement. This was modeled with 5 ft of weak floor strata and a trapezoidal loading (110 psi at one edge to 510 psi at the other) of the model resulting in an average stress of 310 psi. Overall, there was good interaction between adjacent pillars. A larger height of vertical stress relief zone and tension along the entry was seen in the case of non-uniform loading. Vertical displacement in the roof was asymmetric and larger on the side with

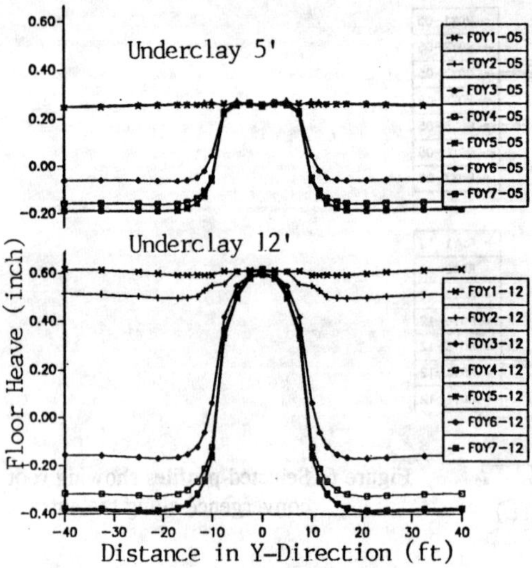

Figure 7 Selected profiles showing floor heave along the entry

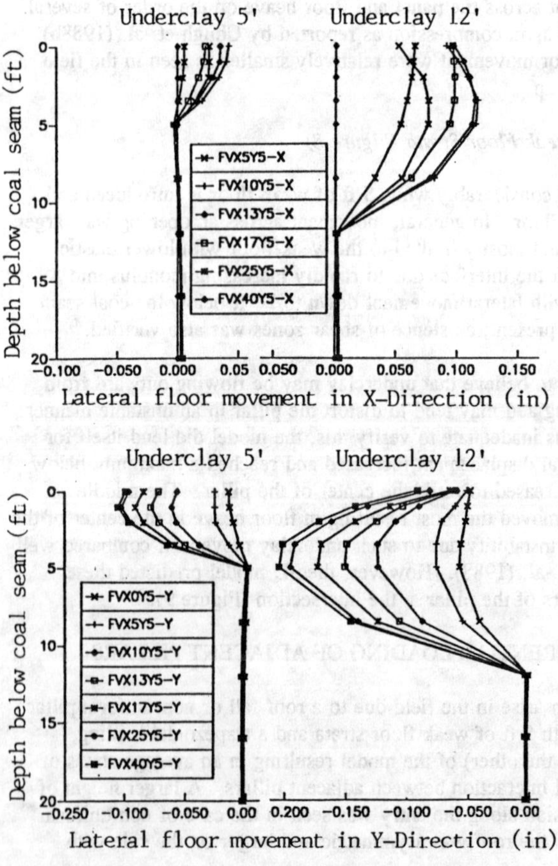

Figure 8 Lateral sub-floor movement near the pillar rib

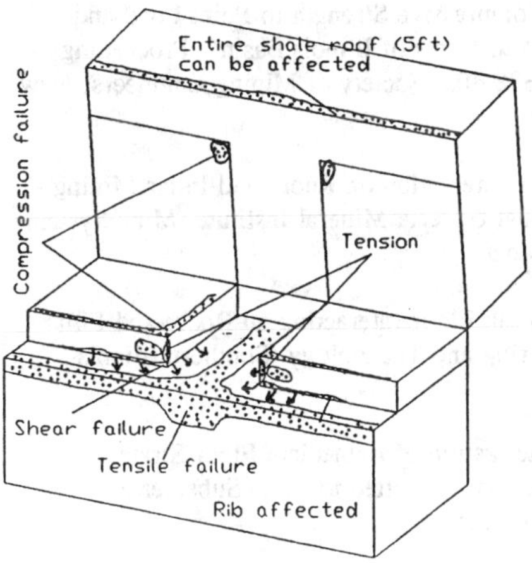

Figure 9 Perspective view of anticipated failure zone due to weak floor

higher loading. The vertical stresses near the top of pillar showed similar asymmetric behavior but was more uniform near the floor. A slightly larger floor heave on the side of higher loading dissipated the vertical stresses making the effect of differential loading more prominent in the weak floor strata. These effects were most visible near the intersection. The zone of vertical stress relief and reduced safety factor was larger for non-uniform loading. An additional zone of tension developed in the floor along the rib. There was a general reduction in lateral stresses in the coal pillar. The change in pillar safety factors was insignificant.

CONCLUSIONS AND RECOMMENDATIONS

Model behavior was in very close agreement with observed field behavior in entries and around an intersection using a linear elastic analysis and scaled laboratory values for material properties as derived from earlier 2D analysis by Caudle, et.al (1988). Field behavior of relatively uniform pillar settlement and interaction between adjacent pillars due to differential settlement was verified. The model also indicated that the immediate floor strata were more likely to be the zone of instability in the event of asymmetric displacement or loading conditions unless the roof strata was weak. Significant floor movements occurred through the entire weak floor strata although detectable lateral movements could extend deeper based on sub-floor properties; most lateral movement was across the opening. Significant rib sloughing was caused in the lower half of the pillar especially near the intersections as predicted by Caudle et. al. (1988). Vertical tensile stresses acting immediately below the opening could cause bed separations as observed in the field. Figure 9 summarizes the observations regarding pillar and floor deformation and expected failure zones around an intersection. Although a good agreement was seen with previous studies and observed field behavior, non-linear material behavior of immediate floor strata should be investigated to model larger floor movement seen in the field. Effective quantitative estimation and prediction of floor heave, roof sag, pillar settlement, subsidence, and zone of instability in and around mine openings may require scaling factors adjustments and time-dependent analytical modeling.

REFERENCES

Bandopadhay, C., 1982, "Analysis of Soft Floor Interaction on Underground Mining at a Western Kentucky Mine", M.S. Thesis, Southern Illinois University at Carbondale.

Bauer, R. A., 1984, "Relationship of Uniaxial Compressive Strength to Point Load and Moisture Content Indices of Highly Anisotropic Sediments of Illinois Basin", Proceedings - 25th U.S. Symposium on Rock Mechanics, pp. 398-405, Society of Mining Engineers, New York, N.Y.

Caudle, R. D. et al., 1988, "Effects of Soft Floor Interaction on Room-and-Pillar Mining - A Progress Report", Paper Published in the Annual Generic Mineral Institute (Mine Systems and Ground Control) Meeting, Tuscaloosa, Alabama.

Chandrashekhar, K., August 1990, 'Effects of Weak Floor Interaction on Room-and-Pillar Mining', Ph.D. Dissertation, College of Engineering and Technology, Southern Illinois University, Carbondale, Illinois.

Chugh, Y. P. et al., 1986a, "Laboratory Characterization of Immediate Floor Strata Associated with Coal Seams in Illinois", Final Report Submitted to Mine Subsidence Research Program, ISGS, Champaign, Urbana, Illinois.

Chugh, Y. P., 1986b, "In-Situ Strength Characteristics of Coal Mine Strata in Illinois", Final Report to the U.S. Bureau of Mines, Contract J0256002, 154 pp.

Chugh Y. P., A. Atri, and J. Dougherty, 1988a, "Laboratory and Field Studies of Immediate Floor Strata Associated with Coal Mining in Illinois", Final Report, Illinois State Geological Survey, Champaign, Illinois, 123pp.

Chugh, Y. P., K. Chandrashekhar, R. D. Caudle, 1988b, "A Field Geotechnical Study of the Effects of Weak Floor Strata on Underground Coal Mining in Illinois", Proceedings - 29th U.S. Symposium on Rock Mechanics, Minneapolis, Minnesota, pp. 681-690.

Chugh, Y. P., O. Pula, W. M. Pytel, 1989, "Design of Coal Pillars for Weak Floor Strata Conditions - A Rational Approach", Proceedings - Seventh Annual Workshop, Generic Mineral Technology Center: Mine Systems, Design and Ground Control, October 22-23.

Gercek, H., 1982, "Stability of Intersections in Room-and- Pillar Coal Mining", Ph.D. Dissertation, Department of Mineral Engineering, The Pennsylvania State University, PA.

Hanna, K. et al., 1986, "Structural Stability of Coal Mine Entry Intersections-Case Studies", Proceedings - 27th U.S. Symposium on Rock Mechanics, Tuscaloosa, AL, pp. 512-519.

Pariseau, W. G., and W. K. Sorensen, 1979, "3-D Mine Pillar Design Information from 2D FEM Analysis", Int. Journal for Numerical and Analytical Methods in Geomechanics, Vol. 3, No. 2, April-June 1979, pp. 145-157.

Prasad, K. V. K., 1985, "An Analytical Study of a Retreat Mining System in Illinois", M.S. Thesis, Southern Illinois University at Carbondale.

Rockaway, J. D. and R. W. Stephenson, 1979, "Investigation of Effects of Weak Floor Conditions on the Stability of Coal Pillars", Report No. BUMINES-ofr-12-81.

Wang, S. T., et al., 1985, "2-D and 3-D Finite Element Analyses of Room-Pillar Mining Systems with Flat and Rolling Coal Seams", Proceedings - 26th U.S. Symposium on Rock Mechanics, Rapid City, South Dakota. pp. 231-238.

Rock Mechanics, Nelson & Laubach (eds) © 1994 Balkema, Rotterdam, ISBN 90 5410 380 8

The role of local mine stiffness in pillar bump prediction

Y. Pen & K. Barron
University of Alberta, Edmonton, Alb., Canada

ABSTRACT: The local mine stiffness concept has been investigated by a numerical analysis to provide guideline for longwall pillar design in bump-prone coal beds. The results show that local mine stiffness can be used in evaluation of longwall mine design and in control of pillar bumps.

1 INTRODUCTION

Longwall mining has proved to be a highly productive and safe mining method in North America. In this method, a series of longwall panels are developed in parallel and are mined in succession. The panels are separated by panel entries and chain pillars on either side of each panel. When the longwall face advances, the roof behind the face is allowed to cave into the gob.

Analysis indicated (Hsiung 1984) that during advance of the longwall face, pressure builds up on the coal at the face and at the two sides of the panel. The side abutment pressure is supported by the chain pillars along the panel sides.

1.1 Chain pillar design

The design requirements for the chain pillars separating the panels are primarily that they should maintain the stability of the head and tail entries throughout all the mining phases necessary for complete panel recovery. At the same time there are economic incentives to minimize the size of these chain pillars to ensure the maximum resource recovery. Hence, the dimensioning of the chain pillars is a major factor in longwall design.

For longwall mining, more and more mines are using a combination of the stiff and yield pillars to obtain safer and more economical design. Usually, there are rows of abutment pillars sided with the rows of the yield pillars. The yield pillars are designed to yield to the development load so as to relieve the stress on the nearby entries, and also to act as buffer zones between entries and the stiff abutment pillars, if violent failures of the abutment pillars occur.

1.2 Pillar bumps in longwall mining

The chain pillars at the T-junction of tail entry and face are especially vulnerable to the severe abutment pressure because of the interaction between front abutment pressure and side abutment pressure. For the bump-prone coal beds, which are identified in North America as the coal beds with strong and brittle coal, overburden greater than 500 ft, and strong roof and floor (Peng 1986 and Iannacchione 1990), very often the chain pillars at or near the T-junction burst violently into the entry. This violent pillar failure is called a pillar bump. The pillar bumps occurring at the tail entries may lead to the collapse of the entry roof and eventually close the entries.

A pillar bump is defined as an instantaneous release of high stressed pillar rib material which is propelled into a mine opening (Haramy 1988). When a pillar bumps, its load-bearing capacity is totally or partially lost and the load is immediately transferred to adjacent pillars. This may lead to the bumping of the adjacent pillar and to the so-called 'Domino bump failure" in which the collapse of the whole mining section occurs.

1.3 Pillar bump mechanism

The mechanism of pillar bumps was first explained by Cook in his historical paper (Cook 1965) and was later theoretically summarized by Salamon (1968). According to the theory, the structure will be stable if the equation below is always satisfied, no matter how much displacement the pillar has experienced.

(1) $$k_{li} + \lambda_i > 0,$$

where k_{li} is the local mine stiffness with respect to the i-th pillar and λ_i is the stiffness of the i-th pillar. In the elastic range, the pillar stiffness λ_i is positive. Hence the equation (1) is always satisfied. In the post-failure range, λ_i becomes negative and instability will occur whenever $|\lambda_i|$ is greater than k_{li}. If Equation (1) is satisfied by all pillars in the mine, the structure is stable. Therefore, if the post-failure pillar stiffness and local mine stiffness can be obtained and the stress level in the pillars can be estimated, the stability of these pillars can be evaluated based upon the above theory.

1.4 Evaluation of pillar bumps

Clearly, stability evaluation should be included in the design process of the chain pillar design. But in practice, there are still problems in the evaluation. The stress level can be estimated by analytical or numerical methods, but the determination of the post-failure pillar stiffness and the local mine stiffness in a mining layout is difficult because the stiffness not only depends on the material's physical properties, but also depends on the other factors such as pillar configuration and material properties.

Usually, local mine stiffness is determined by the so-called "perturbation method" proposed by Starfield and Wawersik (Starfield 1968). In this method a pillar is substituted by an imaginary hydraulic jack under equilibrium. A small uniformly distributed load increment is imposed at the interface of the roof and the jack which are initially in the equilibrium, and the accompany displacement on the jack is determined analytically or numerically. The ratio of the force increment over the resulting displacement is defined as the local mine stiffness for that particular pillar position. For tabular ore bodies like coal beds, this can be best achieved by using computer programs in boundary element analysis (Brady 1980, Ozbay 1989 and Zipf 1992).

The most reliable method to determine the pillar stiffness would be to directly load the pillar to failure in the field and obtain the load-deformation curve for the pillar. Such measurements have been made by Bieniawski (1968), Cook (1971), Wagner (1974) and Heerden (1975), but these direct measurements were only be made on small size of pillars (width of 1-4 m with width/height ratio no more than 3) because of the restrictions of equipment and cost. The limited results revealed that the post-failure stiffness is not a material constant but decreases with the increase of the pillar width/height ratio. Most data on the post-failure stiffness comes from the laboratory testing on the small cylinder or block specimens (Crouch 1973, Das 1986 and Seedsman 1991). Their results are significant because the width/height ratio reached to 10 and showed a similar trend to the field data.

1.5 Problem summary

Currently, chain pillar design is mainly based on the ultimate strength consideration and the pillar bump problem is not considered or is neglected in the design. This may either lead to a oversized pillar which is a waste of the resources, or a yield pillar which is bump-prone. Most analyses have been restricted to the two dimensional room-and-pillar cases and longwall face advance has not been considered in these cases. Therefore, the application of the results is limited. Although the local mine stiffness theory has been well established for a long time, there are only limited applications of this theory to practical mining problems, especially in three dimensional longwall cases. There is also controversy in the application of the local mine stiffness concept to solve the pillar bump problem (Zipf 1992). Further study and clarification is clearly needed.

2 MODELING APPROACH TO DESIGN

MULSIM boundary element program developed by US Bureau of Mines (Zipf 1992) was modified to calculate local mine stiffness. MULSIM is a three dimensional displacement discontinuity boundary element computer program which is specifically suitable for analyzing the stress and displacement of the tabular, multiple ore seams in elastic and homogeneous surrounding rock.

2.1 Local mine stiffness evaluation

Starfield's perturbation method was integrated into MULSIM boundary program by modifying the boundary conditions of the specific fine elements. Instead of the single element disturbing approach used by Zipf (1992), the all elements of a pillar were disturbed by giving them a unit displacement. The local mine stiffness is then defined as the average load difference over the unit displacement.

The modified program was first tested against the analytical solution of a simple case provided by Salamon (1970) and then was further tested by application to a simplified elastic longwall model (Pen 1993). The results indicated that the local mine stiffness decreases with the advance of longwall face. The greatest changes of the local stiffness coincide with the position where the face is close to or in the line with the test pillar location during the mining of the first and the second panels. This indicates that the critical condition for pillars to bump occurs when face is very close or in the line with the pillar. Because the model remains elastic through the mining simulation, the change in local mine stiffness represents the affect of geometry change, i.e., longwall face advance.

A more detailed study has been carried out to investigate how other factors, such as the ratio of the surrounding rock modulus to the coal modulus, the non-linear behavior of the coal and the configuration of the chain pillars, will affect the changes of the local mine stiffness.

Model with four entries and three columns of chain pillars was setup as shown in figure 1.

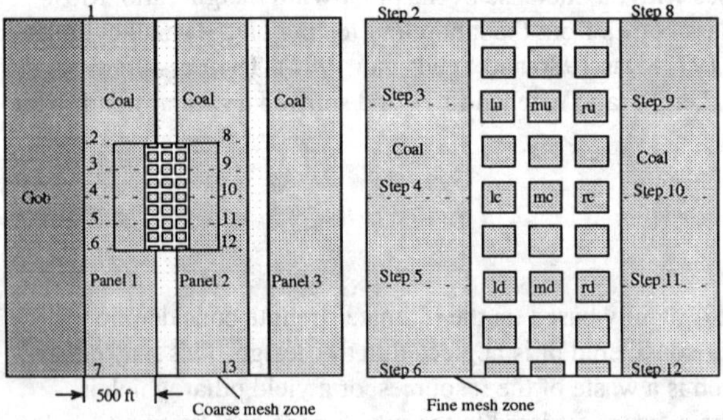

Figure 1. Simulated longwall panels (pillar size 80×80 ft).

The rock and gob remain elastic with Young's modulus 145000 psi and 2900 psi and Poison's ratio 0.25 and 0.3 respectively. However, the coal was specified as strain-softening with peak stress 2900 psi, peak strain 0.01 and residual strain 0.015. The residual stress was varied to simulate different post-failure coal stiffness. Mining was simulated in 13 consecutive mining steps. The stress and closure changes with the face advance clearly showed that the stress initially builds up around the boundary of the pillars and gradually migrated into the core part of the coal until the pillars have totally yielded.

It is shown by figure 2 that the locations along the entry direction did not affect the behavior of the pillars, because all rows of the three pillars show the same strain, stress and local mine stiffness change pattern.

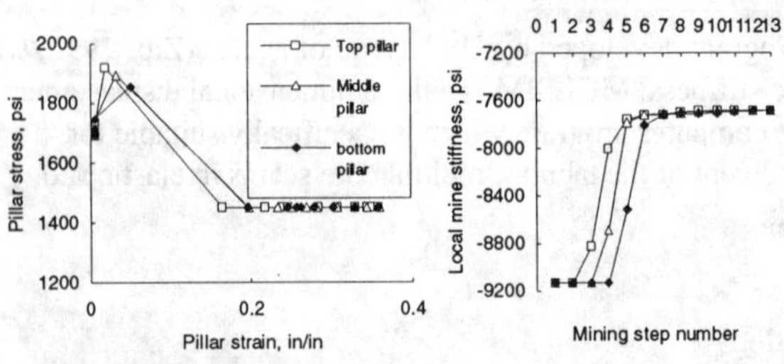

Figure 2. Affect of pillar location along the entries on the local mine stiffness (top pillar - pillar lu; middle pillar - pillar lc; bottom pillar - pillar ld, refer to the pillars in figure 1).

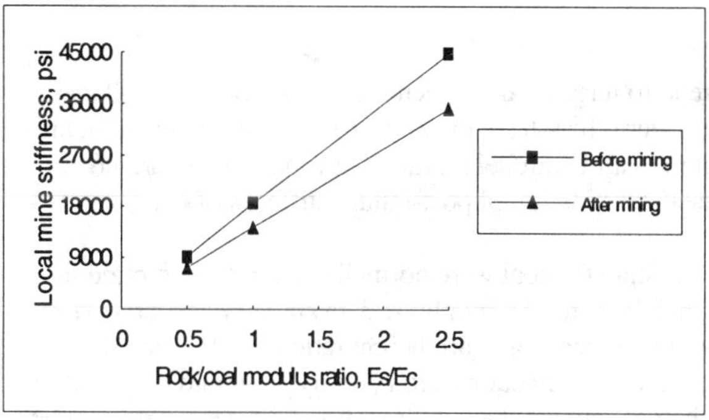

Figure 3. Relation of local mine stiffness and rock/coal elastic modulus ratio.

This is significant because it eliminates the need to check every row of pillars in the panel to evaluate the stability of the pillars. It is also because of this, that the following discussion only deals with one row of the three pillars in the middle of the layout.

Figure 3 shows that the calculated values of the local mine stiffness was linearly proportional to the rock/coal elastic modulus ratio, which indicates that the rock elastic modulus was the dominant factor for the local mine stiffness. However, the value of the local mine stiffness was significantly less than the value of the rock elastic modulus. Figure 3 also shows that after the two panels were extracted, the local mine stiffness dropped 20 percent with respect to the three different rock/coal elastic modulus ratios. This indicated that the longwall face advance did have a significant affect on the local mine stiffness.

The affect of the strain-softening of coal on the local mine stiffness was also examined by varying the residual stress (see figure 4) so as to vary the slope of the strain-stress curve of the coal model . Figure 4 shows that when coal gets stiffer (with lower residual stress), the changes of the local mine stiffness become more abrupt; this means that the possible bump area is closer to the face for this coal, because of the close relationship between the local mine stiffness change and the passage of the longwall face.

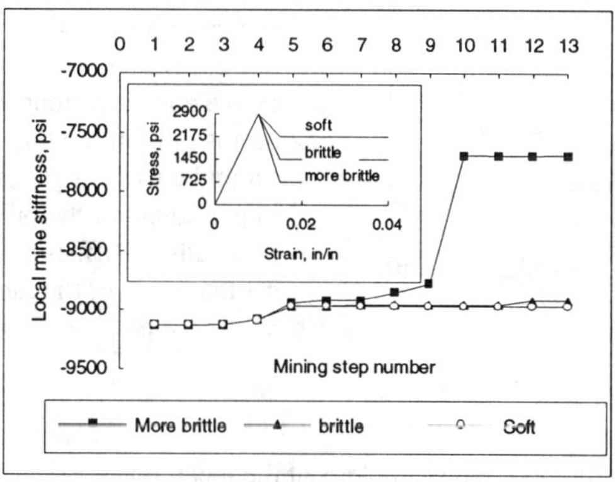

Figure 4. Affect of strain-softening behavior on the local mine stiffness. The diagram in the middle shows the strain-stress curves of three strain-softening material models.

All the results showed that for a pillar in the panel, the sudden drop of the local mine stiffness always occurs when face is very close to the pillar site. The drop of the local mine stiffness also corresponded to the abrupt change of the stress and strain of the pillar. Therefore, the change of the local mine stiffness might be used as an index to indicate the critical state of the pillar behavior under the specific roof and floor conditions.

2.2 Pillar stiffness determination

In longwall mining, there is a trend to increase the dimension of the abutment pillars to avoid pillar bumps, because the wider pillars have lower post-failure stiffness. A literature review indicated that for the pillars with width/height ratios beyond 9, there are no reliable data available. Hence the determination of post-failure stiffness of this size of pillars is still difficult.

The available field and laboratory data for coal were normalized and were plotted in figure 5 against the pillar width/height ratio. Although the data are very scattered, the trend of stiffness decreasing with the increasing width/height ratio is still clear. Attempts were made to fit this type of the data into a linear relation (Brady 1980 and Ozbay 1989). Based on the linear relation, Ozbay concluded that a pillar with width-height ratio greater than 5 could not bump because from the relation, the post-failure stiffness would be zero. However, it has been reported that even pillars with a width-height ratio greater than 8 experienced bump failures (Babcock 1984 and Campoli 1990). The laboratory data from Das (1989) and Crouch(1972) indicated that the post-failure stiffness cannot be considered to be zero even after the width/height ratio has passed 8. The post-failure stiffness data estimated by the field measurements (Wang 1976 and Iannacchione 1989) also showed that even for the pillars with 13.4 width/height ratio, the post-failure stiffness is still close to the original elastic modulus of the pillar. Therefore, it appears that any kind of interpolation of the data may be misleading. Instead of trying to fit the data with a specific relation, a polynomial upper bound of the scattered data can be identified. Obviously, this polynomial relation will provide the maximum possible post failure stiffness, which can only lead to a conservative pillar design. Based on the above considerations, it seems that Figure 5 might be used to give an upper bound of the pillar post-failure stiffness estimation over the range of the data.

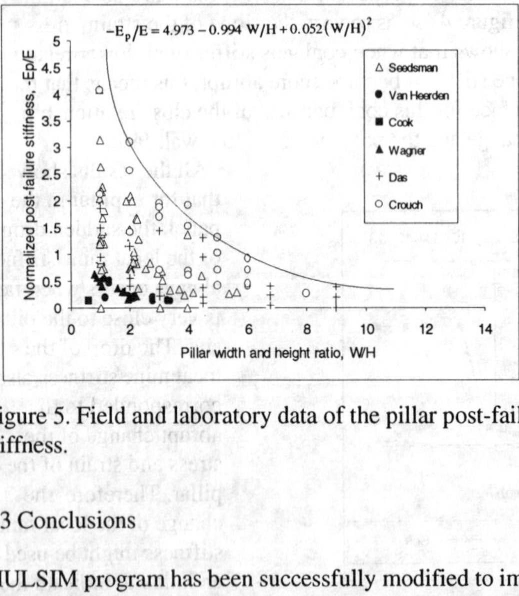

Figure 5. Field and laboratory data of the pillar post-failure stiffness.

2.3 Conclusions

MULSIM program has been successfully modified to implement the local mine stiffness calculation during the regular numerical simulation of the longwall mining. The local mine stiffness is defined as the ratio of the unit perturbation displacement over the average load difference. The parametric study of the local mine stiffness has been performed. The parameters studied includes longwall face advance, elastic modulus of the strata and the seam, softening behavior of coal and location in the panel. The following tentative conclusions are drawn from the investigation:

(1) The modified program works consistently and the local mine stiffness can be easily obtained;
(2) Local mine stiffness decreases with the advance of longwall face;
(3) The greatest drop of the local mine stiffness occurs when face is close to the pillar location or when the pillar starts yielding;
(4) Brittle coal has greater drop of the local mine stiffness value than the soft coal during passage of the face;
(5) The modulus of the surrounding strata dominates the magnitude of the local mine stiffness but the longwall face advance also has significantly effect;
(6) The pillars along the panel entry behave similarly, as far as strain, stress and local mine stiffness are concerned.

3 PILLAR DESIGN METHODOLOGY

Based on the results presented above, a chain pillar design approach can be constructed which should be especially suitable for the deep and bump-prone coal mines.

The core of the design approach is to include the stability evaluation into longwall mining design. Therefore, the numerical model is first built according to the initial design and the results of site investigation and testing. The model is fine-tuned by fitting the calculated stress and deformation to the measured stress and deformation. The calculated stress and deformation should agree with the measured stress and deformation throughout the panel extraction, or agree with the estimated results from the experience if the measured results are not available.

Then, the local mine stiffness at the sites of interest are calculated through all the mining steps. Because the study showed that the critical stages occur when the longwall face is close or in the line with the pillars, therefore, the interested sites should be the pillar positions near the face. It is also seen that each pillar in a column parallel to the face advance direction behaves in a similar manner, whereas each pillar in the row perpendicular to the face advance direction behaves differently. Therefore, only one row of pillars across the section of the entries needs to be considered.

Finally, the calculated local mine stiffness of a specific pillar is compared to the estimated post-failure stiffness of the pillar. To avoid pillar bumps, the local stiffness should be higher than the estimated post-failure stiffness through the panel excavation. Dimensions of the pillars and configurations of the pillars in the panel can be varied to control the time when local mine stiffness approaches the post-failure stiffness, thus the time and place of the pillar bumps can be controlled in case the bumps cannot be avoided. It is also clear that, with the longwall face advancing, a design in which the changes of the local mine stiffness are slow and smooth is better than a design in which the changes are fast and abrupt.

REFERENCES

Babcock, C.O. a. B., D.L. 1984. Constraint - the missing variable in the coal burst

problem. *Proceedings of the 25th U.S. Symposium on Rock Mechanics,* 639-647.
Bieniawski, Z.T. 1968. In situ strength and deformation characteristics of coal, *Engineering Geology* , 2(5), 325-340.
Brady, B.H.G., & Brown, E.T. 1980. Energy changes and stability in underground mining: design applications of boundary element methods. *Trans. Instn Min. Metall.*, 90(Section A), A61-68.
Campoli, A.A., et al. 1990. Mitigating destructive longwall bumps through conventional gate entry design, *RI 9325*, U.S. Bureau of Mines.
Cook, N.G.W. 1965. A note on rockburst considered as a problem of stability. *Journal of the South African Institute of Mining and Metallurgy*, 65(March), 437-446.
Cook, N.G.W., et al. 1971. A 100-MN jacking system for testing coal pillars underground. *Journal of the South African Institute of Mining and Metallurgy*, (June), 215-224.
Crouch, S.L. and Fairhurst, C. 1973. The mechanics of coal mine bumps and the interaction between coal pillars mine roof and floor, *U.S.B.M Contract No. H0101778.*
Das, M.D. 1986. Influence of width/height ratio on post-failure behaviour, *International Journal of Mining and Geological Engineering*, N0. 4, 79-87
Heerden, W.L.V. 1975. In situ complete stress-strain characteristics of large coal specimens. *Journal of the South African Institute of Mining and Metallurgy*, (March), 207-217.
Holland, C.T. 1955. Rock burst or bumps in coal mines. *Colliery Engineering*, (April), 145-153.
Hsiung, S.M., & Peng, S.S. 1984. Chain pillar design for U.S. longwall panels. *SME-AIME Fall meeting* (pp. 1-17). Denver, Colorado.
Iannacchione, A.T., & DeMarco, M.J. 1991. Optimum mine designs to minimize coal bumps: A review of past and present U.S. practices. *Proceedings of the Conference on Deposit Exploitation in Natural Hazard Conditions,* Cracow, Poland.
Ozbay, M.U. 1989. The stability and design of yield pillars located at shallow and moderate depths. *J. S. Afr. Int. Min. Metall.*, 89(3), 73-79.
Pen, Y. and Barron, K., A Model Approach to Chain Pillar Design in Longwall Mining, paper for CANMET/University of Alberta Mining Seminar, CANMET Mining Research Laboratories, Sydney, Nova Scotia, pp. 1-28, Aug. 19, 1993.
Salamon, M.D.G. 1968. Two-dimensional treatment of problems arising from mining tabular deposits in isotropic or transversely isotropic ground. *Int. J. Rock Mech. Min. Sci.*, (5), 159-185.
Salamon, M.D.G. 1970. Stability, instability and design of pillar workings. *Int. J. Rock Mech. Min. Sci.*, 7, 613-631.
Seedsman, R.W. and Hornby, P. 1991. Controlled and uncontrolled pillar collapse, *Nerddc End of Grant Report*, 1-75.
Starfield, A.M., & Wawersik, W.R. 1968. Pillars as structural components in room-and-pillar mine design. *The 10th U.S. Symposium on Rock Mechanics,* 793-809.
Wagner, H. 1974. Determination of the complete load-deformation characteristics of coal pillars. *Proceedings of the 3rd ISRM Congress,* 1076-1081, Denver.
Zipf, R.K., Jr. 1992. Analysis of stable and unstable pillar failure using local mine stiffness method. *Proceedings of the Workshop on Coal Pillar Mechanics and Design,* 128-143, Santa, Fe, New Mexico, U.S. Bureau of Mines.

Rock Mechanics, Nelson & Laubach (eds) © 1994 Balkema, Rotterdam, ISBN 90 5410 380 8

Overburden failure due to longwall mining

D.K.Ingram, M.A.Trevits & P.W.Jeran
US Bureau of Mines, Pittsburgh Research Center, Pa., USA

ABSTRACT: The objective of this U.S. Bureau of Mines investigation was to characterize the change in the overburden due to mining of a 270 m wide longwall panel. The overburden thickness ranged from 95 to 105 m. Five-anchor, multipoint borehole extensometers (MPBX) were installed in each of five coreholes to measure subsurface displacement. Survey data of an array coincident with the boreholes show that surface movement began when the longwall face was about 45 m in front of any point. Over 90% of the total subsidence was achieved when the face was one overburden thickness past a point. Overburden failure was detected by the MPBX units when a borehole was undermined. Overburden failure occurred first in the center of the panel and then migrated outward toward the panel margins. The overburden appeared to collapse vertically as a unit and was laterally restricted to the mined-out area of the panel.

INTRODUCTION

A majority of the previous work in subsidence has focused on surface deformations through correlation of mining data with surface ground movement data. The role of geology at a specific site or across coal regions is usually generalized or overlooked in subsidence research because these data are typically unavailable in sufficient detail or very expensive to obtain. The goal of this Bureau research effort was to identify and characterize the dynamic response of the overburden to longwall mining. Studies of this type provide key information essential to the development of an understanding of the magnitude, duration, and impact of longwall mining on the overburden rock mass. Ultimately, these data can be used in longwall support design, studies of hydrologic response, and overburden modeling.

SITE DESCRIPTION

The study area is situated over a longwall panel in southeastern Ohio. The panel measures 270 m by 2,940 m and was flanked on each side by a 69 m wide, four-entry three-pillar gate entry system. The extraction thickness varied between 1.7 and 1.8 m. Overburden thickness is shallow, ranging from 96 to 105 m. The strata have a regional dip of about 1 degree towards the southeast. There are no major geologic structures and the topography has a maximum relief of approximately 26 m.

To delineate the geology of the overburden, five NX size (88.9 mm) coreholes were drilled. Analysis of the recovered core samples shows that the overburden consists of approximately 78% sandstone, 11% shale, 10% claystone, and less than 1% each of limestone and coal (figure 1). Sandstone units are predominately massive with unconfined compressive strengths ranging from 36,500 to 117,200 kPa. A limestone unit, situated about 0.9 m above the mined coalbed, averages 2.1 m thick and has an average unconfined compressive strength of about 155,100 kPa. Values of unconfined compressive strength were not determined for the shale, claystone, and coal units because representative samples could not be obtained.

MONITORING PROGRAM

Overburden Failure

The five coreholes drilled for geologic information were utilized as overburden deformation monitoring boreholes. Three of the holes were drilled along the centerline of the panel and were positioned 45 m apart. The other two boreholes were drilled along a profile line perpendicular to the middle centerline borehole (figure 2). Boreholes 1, 2, and 3 were located in the center of the panel where maximum amount of subsidence was expected. Borehole 4 was located 33 m from the ribline, over the panel at a position where the inflection point was expected to occur (V. Adamek, P. W. Jeran, and M. A. Trevits). Borehole 5 was situated 27.6 m from the ribline over the gate entries to observe the lateral extent of overburden deformation.

The configuration of the MPBX units was designed to identify the temporal, three-dimensional development of overburden failure. Each hole was outfitted with a five-anchor, MPBX system. During the undermining, all MPBX anchor displacement data

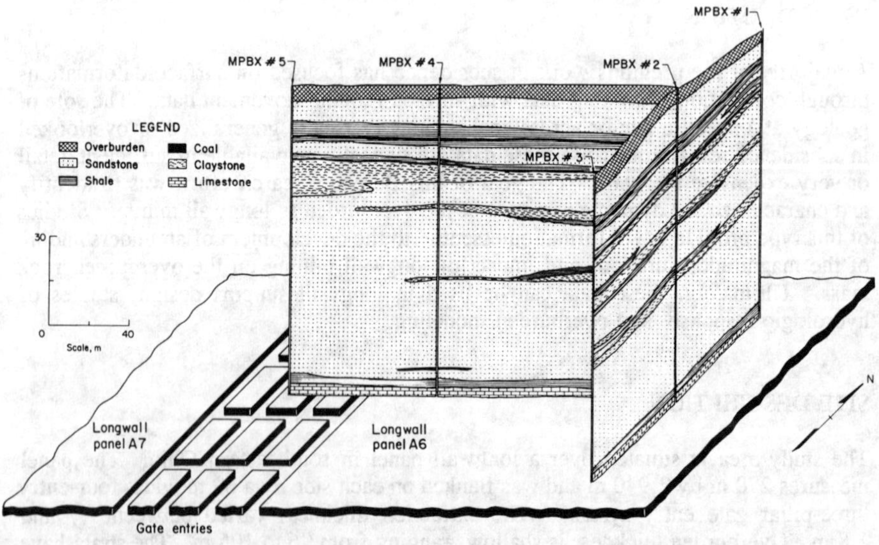

Figure 1. Cross-section of the lithology with MPBX borehole locations.

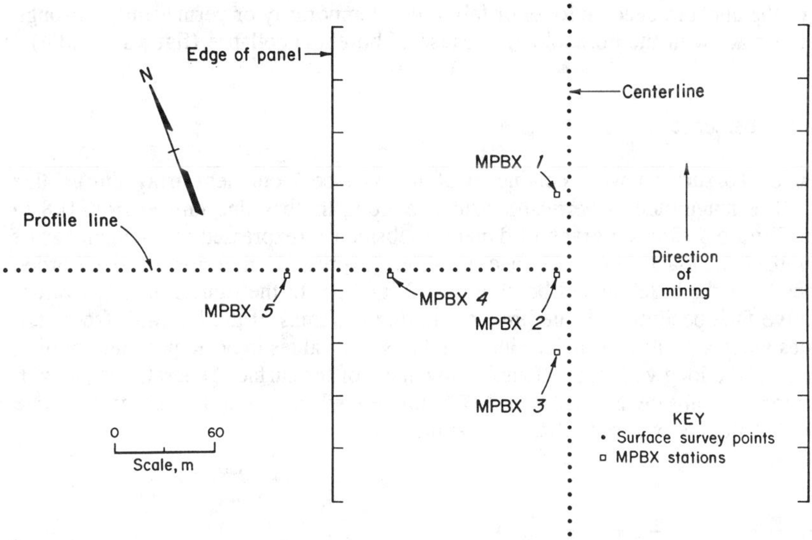

Figure 2. Site map identifying the MPBX borehole and surface monitoring array.

were recorded every minute by a data logger. The MPBX anchors were selectively positioned throughout the boreholes in an attempt to identify the theoretical zones of overburden deformation (S. S. Peng and H. S. Chaing). The anchors are numbered 1 to 5, with anchor 1 being the deepest and 5 being closest to the surface.

Surface Subsidence Monitoring

Survey monuments were positioned 7.5 m apart and were installed as an array over the study area to monitor the surface deformation during longwall mining. The array consisted of a 300 m long baseline (along the centerline) and a 330 m long profile line (figure 2). Periodic surveying of the array was necessary to identify the dynamic characteristics of subsidence, the final subsidence profile, and to provide surface reference data for the MPBX units. Surveys were conducted before, during, and after mining.

FIELD DATA RESULTS

MPBX Boreholes

Theoretically, the first strata to collapse and show the greatest displacement are located immediately above the void created by the extraction. The magnitude of deformation is progressively reduced moving upward and outward through the overburden from the center of the panel towards the surface (Ingram and Trevits). These same deformational characteristics should be recorded by the MPBX units provided the anchors remain intact during the entire event. The deepest anchor (anchor 1) should displace first and show the largest movement. In turn, the overlying anchors should show progressively less and less displacement. Inevitably some of the anchors become loose or fail, either temporarily or permanently, through loss of contact with the borehole or because of borehole collapse (figures 3 and 4).

Surface Subsidence

A total of 16 surveys were conducted of the surface monument array during this study. The magnitude of vertical subsidence along the baseline ranged from 0.8 to 1.1 m. Figure 5 shows a graph of dynamic subsidence (expressed as a percentage of total subsidence) and longwall face position (expressed in terms of overburden thickness) for the baseline and boreholes 1, 2, and 3. In the figure, negative values of relative face position indicate preundermining positions of the longwall face; zero indicates when a point was undermined; and positive values indicate postundermining positions of the longwall face. Notable movement of the surface generally began with undermining. Subsidence was over 90% complete when mining had progressed one overburden thickness past any surface point.

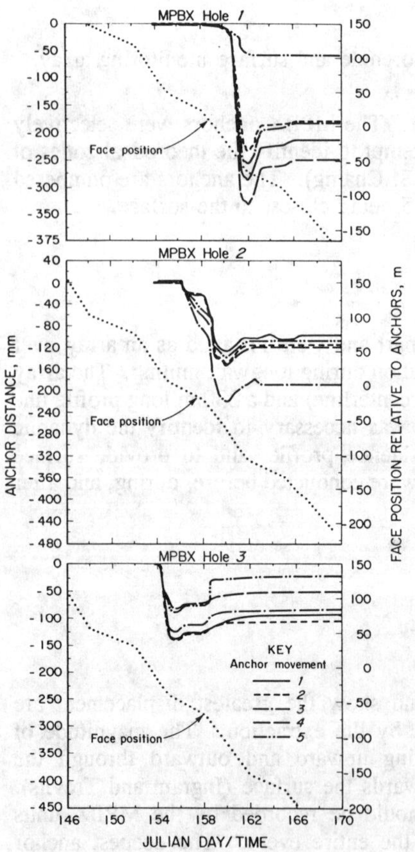

Figure 3. Anchor displacement for MPBX boreholes 1, 2, and 3.

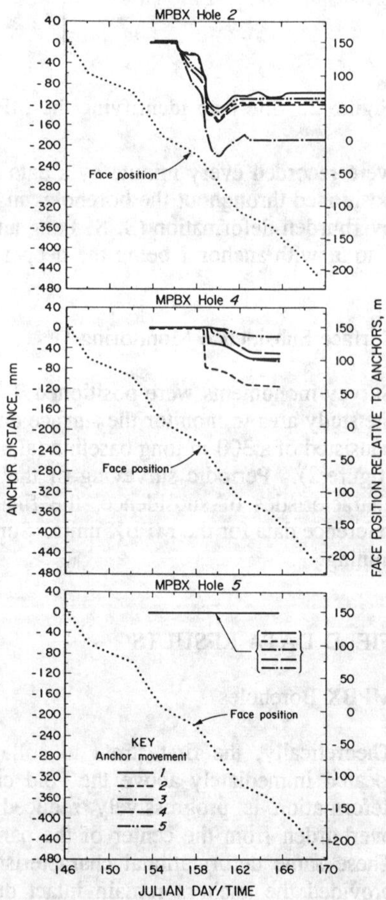

Figure 4. Anchor displacement for MPBX boreholes 2, 4, and 5.

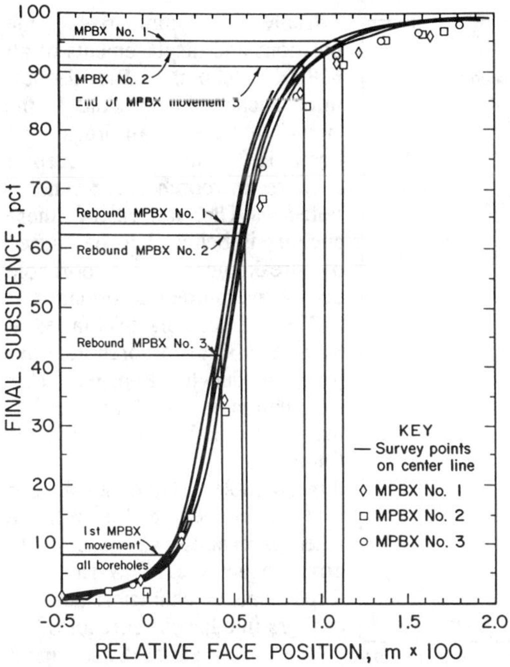

Figure 5. Final surface subsidence baseline data.

Figure 6 shows the final subsidence along the profile line. The magnitude of subsidence over the gate entries was less than 0.1 m. The maximum subsidence (S_{max}) is in the center of the panel and the inflection point (1/2 S_{max}) is located 51 m inside the panel ribline.

DISCUSSION AND INTERPRETATION OF OVERBURDEN RESPONSE

MPBX Data

The development of overburden deformation was first detected at borehole 3 and then progressed to boreholes 2, 4, and 1. At borehole 3, all of the anchors began to displace when the longwall face was approximately 12 m beyond the borehole. The positive displacement of the anchors in borehole 3, indicate that the anchors were beginning to stabilize while the surface continued to move downward (figure 3). This holds true for all anchors in the other boreholes that show a positive displacement. Finally, at borehole 3, differential movement between the anchors and the surface ceased when the longwall face had advanced 90 m past the borehole.

The next borehole to show deformation was borehole 2. Similar to borehole 3, all anchors in borehole 2 started to displace at the same time. This occurred when the longwall face had advanced 13.5 m past the borehole. Differential movement in borehole 2 stabilized when the face was 114 m beyond the borehole.

Borehole 4 then detected the strata movement (figure 4). The deepest, reliable anchor (anchor 2 located 25.8 m above the coalbed) began displacing when the face had progressed 24 m beyond the borehole. Movement at the position of anchor 3 in borehole 4 did not occur until the longwall face had progressed 48 m beyond the borehole. It was not until the face advanced to 81 m beyond borehole 4 that the upper two anchors (4 and 5) in the borehole started to displace. Stabilization of the anchors in borehole 4 was observed when the face advanced 141 m past the borehole.

When the face advanced 12 m past borehole 1, all of the anchors started to show displacement. Differential movement in borehole 1 stabilized when the face was 102 m beyond the borehole. No displacement was observed in the anchors in borehole 5 which was positioned over the gate entries.

In summary, the chronological development of overburden deformation was as follows. Movement was first detected in borehole 3, next, all of the anchors in borehole 2 started to displace, next the lower anchors in borehole 4 moved followed by all of the anchors in borehole 1, and finally the upper anchors in borehole 4 moved.

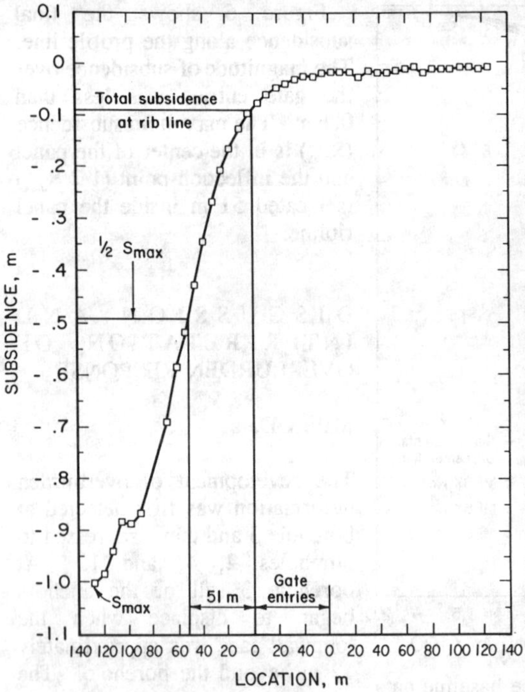

Figure 6. Final surface subsidence profile.

Table 1 summarizes the positions and displacements of all MPBX anchors. The deepest three anchors in borehole 1, the two deepest anchors in borehole 2, and the deepest anchors in boreholes 3 and 4 failed. The failure of these anchors is believed to be caused by breakdown of the borehole during overburden deformation.

The two anchors that failed at 81 m and 57 m in borehole 3 are probably due to slippage. This explanation is based on the fact that two reliable anchors were found below the 57 m point in this borehole. Note, an anchor was determined to be reliable when its response was observed to correlate well with the displacement characteristics of anchors in adjacent boreholes.

The magnitude of displacement (expressed in terms of the percentage of extraction thickness) for the reliable anchors indicates a uniform decrease in magnitude moving upward toward the surface in all the boreholes. The deepest, reliable anchor in this study is anchor 2 in borehole 3 (located 27 m above the coalbed). This anchor displaced 62% of the extraction thickness. All other reliable anchors in boreholes 1, 2, and 3 correlate with one another with respect to their position in the overburden and the magnitude of displacement.

Table 1. MPBX anchor location and displacement.

Borehole	1		2		3		4		5	
	H[1]	D[2]	H	D	H	D	H	D	H	D
Ground Surface	96	51	105	57	102	57	105	2	104	0
Anchor 5	83	55	80	63	81	F	82	2	82	0
Anchor 4	56	66	57	63	57	F	56	5	55	0
Anchor 3	39	F[3]	39	67	39	61	38	5	37	0
Anchor 2	25	F	25	F	27	62	26	8	27	0
Anchor 1	7	F	7	F	8	F	8	F	9	0

[1]H = Height above coalbed rounded to the nearest meter.
[2]D = Total displacement expressed as a percentage of extracted thickness.
[3]F = Failed anchors.

The difference between total displacement for boreholes 1, 2, and 3 (as shown in table 1) is relatively small. These similar displacement magnitudes indicate that the overburden responded as a unit. The small values of displacement in borehole 4 as compared to the centerline boreholes is attributed to the position of borehole 4 relative to the longwall panel.

The overburden can be divided into a lower and upper zone of deformation by using the failed anchor locations and the total magnitude of displacement of the reliable anchors. The lower zone, having characteristics of a caved zone, begins at the top of the coalbed and ranges upwards 20 to 50 m in the center of the panel (height of the failed anchors) and is approximately 20 m along the margin of the panel. The smaller thickness of the lower zone at the margin of the panel can most likely be attributed to the overburden strata being partially supported by pillars in the gate entries.

The upper zone extends from the top of the lower zone to the surface. The difference between the lower and upper zone can be attributed to bulking characteristics. Given the small variance in the magnitude of displacement of the reliable anchors in all the boreholes, it appears that the upper zone collapsed as a unit with minimal bulking expansion due to deformation. This implies that the majority of bulking occurred in the lower zone.

The vertical and horizontal distribution of the zone of deformation was confined to the area outlined by the extracted panel. Evidence from borehole 5 and surface subsidence monitoring indicates that there was no deformation over the gate entries.

Surface Subsidence

Several observations can be made when the surface subsidence data from the baseline array is compared to the displacement of the MPBX surface assemblies for boreholes 1, 2, and 3. Figure 5 illustrates the dynamic surface subsidence for the baseline. Plotting surface subsidence from boreholes 1, 2, and 3 on this curve (as indicated in the figure by triangles, squares, and circles) shows the validity of using the composite curve to describe the dynamic movement of the surface locations for boreholes 1, 2, and 3.

When differential movement between the anchors and the surface began in boreholes 1, 2, and 3 (anchors started to displace), the longwall face was about 10% of the overburden thickness beyond each borehole location. Surveys indicate that the surface had already subsided approximately 8% of the final subsidence at this time. When the anchors in borehole 3 started to rebound or when the surface began converging towards the anchors, the longwall face was approximately 40% of the overburden thickness past the borehole. When the face was at that position, the surface had already subsided approximately 43% of the final subsidence. For boreholes 1 and 2, the rebounding effect occurred when the face was approximately 55% of the overburden thickness beyond each borehole. At that face position, the surface at boreholes 1 and 2 had subsided about 64% of the final subsidence. When the anchors stabilized or the differential movement between the anchors and the surface ceased in borehole 3, the longwall face was about 90% of the overburden thickness past the borehole. At boreholes 1 and 2, the longwall face was about 100% and 110% (respectively) of the overburden thickness past each borehole. At all three boreholes, subsidence was over 90% complete at this point in the study.

Based on these observations, it is apparent that the overburden was subsiding as a whole shortly before and after the MPBX anchors detected movement. It is also probable that the overburden was subsiding as a whole even during the differential movement between the anchors and the surface.

SUMMARY AND CONCLUSIONS

1. The development of overburden deformation occurred in such a manner that the center of the panel collapsed as a unit, whereas the margin of the panel collapsed as segments starting above the coalbed and propagating upward towards the surface. From a three-dimensional perspective, the deformation in the center of the panel developed in advance of the margin of the panel. An additional 27 m of longwall face advance occurred after the center of the panel stabilized before the margin of the panel stabilized.

2. At this site, the overburden response can be divided into a lower and upper zone of failure. In the center of the panel, the lower zone begins just above the extracted area and extends upward 20 to 50 m. Along the margin of the panel this lower zone extends upward from the extraction zone approximately 20 m. The upper zone extends from the top of the lower zone to the surface. Magnitudes of displacement, as shown by the MPBX units, indicate that the upper zone collapsed as a unit with minimal deformation.

3. There was no movement in the overburden over the gate entries during the mining of the panel.

4. Initial, differential movement between the centerline MPBX units and the surface began when the longwall face advanced about 10% of the overburden thickness beyond each borehole. Differential movement between the MPBX units and the surface stabilized when the longwall face advanced between 90 to 110% of the overburden thickness beyond each borehole. Less than 10% of the final subsidence occurred before the onset of differential movement between the MPBX anchors and the surface and an equal amount of subsidence occurred after stabilization.

REFERENCES

Adamek, V., P. W. Jeran, and M. A. Trevits, Prediction of Surface Deformations Over Longwall Panels in the Northern Appalachian Coalfield, BuMines RI 9142, 1987, 19 p.

Ingram, D. K. and M. A. Trevits, Characteristics of Overburden Deformation Due to Longwall Mining, Proceedings of the Third Subsidence Workshop Due to Underground Mining, Morgantown, West Virginia, June 1-4, 1992, pp. 280-289.

Peng, S. S. and H. S. Chaing, Longwall Mining, Chapter 2, Strata Mechanics, John Wiley & Sons, NY, 1984, pp. 17-72.

Rock Mechanics, Nelson & Laubach (eds) © 1994 Balkema, Rotterdam, ISBN 90 5410 380 8

High amplitude stress wave generation and damage induced by roof caving in mines

B.T. Brady & K.Y. Haramy
US Bureau of Mines, Denver Research Center, Colo., USA

ABSTRACT: There is controversy as to the cause of catastrophic failures occurring when mining coal under strong roof. The U.S. Bureau of Mines presents a hypothesis that catastrophic mine roof failure is due to the formation of a macrocrack (dilatant zone) in the rock mass above a pillar or longwall panel immediately before roof collapse. The macrocrack closes upon itself during roof caving. From appropriate energy calculations, sufficient energy is available from this model to produce tremors whose energy release is equivalent to local earthquake magnitudes between $M_L = 2.0 \rightarrow 3.5$. This energy release is applied to a model of damage criterion for propagation of high-amplitude compressive stress waves through the mine workings. Using a particle velocity damage criterion of 0.30 m/s as needed to induce tensile cracking at free surfaces in coal, the area of damage from a $M_L = 3.6$ mine tremor (NEIS) in a Utah coal mine is calculated to be about 10^5 m^2, corresponding closely to the area (7.6 x 10^4 m^2) of observed damage.

1 INTRODUCTION

Coal mine tremors (sometimes called coal bumps), and the damage associated with some of these tremors, are major hazards to underground coal mining throughout the world. These events have inflicted severe injury to mine personnel. Invariably, production is disrupted and, sometimes, mines close. Although these events have been observed in mining for years and have been extensively studied, there is little quantitative understanding of their origin. Experience shows most coal mines that experience dynamic failures all have in common strong sandstone roof or floor members near the mine workings, which is usually strong coal. In retreat mining practices when the main roof does not readily cave, considerable stress can be transferred to the face that may undergo catastrophic failure. Depending on the physical properties and thickness of the coal and surrounding rock mass, the roof may cave in large blocks creating air blasts and damage to the workings. In addition, the coal seam can be constrained in these situations by frictional stresses induced by elastic mismatch between the competent roof and the weaker coal seam. Coal located ahead of the face near to the peak abutment zone will be confined and highly stressed and may fail catastrophically if confinement is suddenly reduced. The dynamics of these failure mechanisms, whose physics is not yet clearly understood, may cause considerable damage to the gate roads and shields by the rapid (dynamic) loading.

Studies of large South African mine tremors confirm that they are caused by shear fractures on fault planes in a rock mass and that they can be modeled mathematically by a double-couple point source (McGarr, 1984; Spottiswoode and McGarr, 1975; Gibowicz, 1990). Studies in South Africa and elsewhere also show mine tremors are similar to earthquakes with regard to stress drops, seismic moment and magnitude, source dimensions, and the like. However, some studies associated with mine tremors suggest alternate mechanisms other than double-couple mechanisms are needed to explain observational data. In most cases of large shallow earthquakes, the deviation of moment tensor solutions from a double-couple model may be considerable, although there are reasonable explanations for these deviations other than physical processes at the source (Gibowicz 1990). For example, a combined shear-tensional and shear-implosional mechanism has been proposed by Teisseyre (1980) to describe the nondouble-couple behavior observed from earthquakes and mine tremors. Although it has been well documented that mine tremors can and often do involve the fracturing of intact rock, seismic data show that shear failure dominates the focal process.

Questions often asked after the occurrence of a mine tremor are how large was it, what caused it, did damage occur, and if damage occurred, what was the physical process responsible for the damage? With regard to the first question, studies have shown that the local magnitude scale M_L, originally devised by Richter for earthquakes in Southern California, is applicable to mine tremors in South Africa (McGarr 1984). M_L provides a rudimentary but useful measure of the size of a mine tremor. This scale is crude in that it is simply proportional to the logarithm of some quantity, such as trace amplitude, that can be read off a seismogram. The local magnitude has been used successfully in South Africa in conjunction with the Gutenberg-Richter relationship to obtain a rough measure of the energy released in mine tremors (McGarr 1984). Satisfactory answers to the physics of and the damage associated with a mine tremor are not yet available.

Several implications of recent studies on the dynamics of brittle fracture conducted by the U.S. Bureau of Mines are reviewed, and the relevance of these studies to questions as to the causes of mine tremors and associated damage are presented.

2 GROUND DISPLACEMENT VELOCITIES AND SCALING LAWS FOR DAMAGE ASSOCIATED WITH MINE TREMORS

Field and laboratory evidence suggest that dynamic fracture of brittle materials under applied compression may occur by tensile stresses generated locally at the source (nucleation zone) during the dynamic phase of fracture (Gay and Ortlepp 1979; Ortlepp 1992; Brady and Rowell 1986; Brady and Rowell 1994). For example, Ortlepp (1992) discovered several mine tremors in a South African mine exhibited features that support tensile failure along the fault produced by these events even though the focal mechanisms of the tremors displayed typical double-couple focal mechanisms. Results from laboratory studies also support these field observations and together have been used to construct a hypothesis that the mechanism of brittle fracture is an explosive process (Brady and Rowell 1994). This conjecture is adopted as a working hypothesis to explain damage associated with tremors in mines.

Dimensional analysis is a powerful technique by which it is possible to learn a great deal about very complicated situations if one can identify the essential features of the problem. In the subject area of explosive processes, a well-known example of the applicability of dimensional analysis is G.I. Taylor's deduction of the yield of the first nuclear explosion from a series of photographs of an expanding fireball. Taylor recognized he was seeing a strong shock expanding into an undisturbed medium. The photographs gave him the radius as a function of time, r(t). He reasoned that the only important parameters to determine r(t) are the initial energy release, E_o, and the density of the undisturbed medium, ρ (air = 0.001 gm/cm^3). The only possible combination is

(1) $$r(t) = a[\frac{E_o}{\rho}t^2]^{\frac{1}{5}},$$

where "a" is a constant of proportionality. If a process is explosive, a log-log plot of r versus t will give a slope of 0.40.

Equation 1 has been used to successfully describe the motion of fragmented pieces of a brittle rock (Brady and Rowell 1986; 1994) stressed uniaxially to failure in the laboratory. With spectrographic evidence, Equation 1 provides evidence that rock fracture is an explosive process. These laboratory studies and field observations from rock bursts in mines suggest fracture of brittle materials under applied compression (shear failure) occurs by tensile stresses generated locally in the source or nucleation zone during the dynamic phase of fracture (Brady and Rowell 1994; Gay and Ortlepp 1979; Ortlepp 1992). A working hypothesis of how this process occurs is as follows: a high aspect ratio narrow macrocrack forms in the nucleation zone due to coalescence of microcracks. This is followed by macrocrack closure and its growth during the dynamic phase of (fault) failure. Shear displacement will occur along the closed fault surface following closure. This hypothesis is consistent with observed double-couple solutions that dominate the focal mechanism solutions of mine tremors. The closure or implosion phase would be detectable only in the near field.

Equation 1 can be differentiated and rewritten

(2) $$v(R) = \frac{2}{5}a^{\frac{5}{2}}\sqrt{\frac{E_o}{\rho_s}}R^{-\frac{3}{2}} = v_o[\frac{r_o}{R}]^{\frac{3}{2}}, R>0$$

where the particle velocity v(r) (= dr/dt) is evaluated at R, the distance from the sensor location to the source, ρ_s is the medium density, v_o, evaluated at r_o is the particle velocity at the source (~ macrocrack closure velocity), r_o is the source radius, and E_o is the energy radiated from the source(s) within the rock mass. E_o is assumed to be corrected for geometric spreading, attenuation, and surface sensor response. Constant "a" is defined to be the *transmission coefficient* for the medium and must be evaluated for each site. Applied to laboratory fracture data of Salida granite (Brady and Rowell 1986), equation 2 gives (v and v_o on the order of 20 m/s and 1 km/s (shear velocity), R = 40 mm) a source length on the order of 3.0 mm. This length is on the order of the mean grain size for Salida granite.

Assume for calculational purposes that the Gutenburg/Richter relationship (log E_o = 11.8 + 1.5 M_L), where M_L is the local magnitude (E_o is in ergs) holds for the medium. This relationship, using M_L, as opposed to the 20-second surface wave magnitude M_S, has been used successfully in South African mines to calibrate energy release and M_L (McGarr 1984; Lenhardt and Hagan 1992). In addition, we assume Hedley's (1990) proposed relationship between weight of charge W and M_L, $\log W = M_L + \alpha$, ($\alpha = 0.87$ when W is expressed in kg of charge) holds. Equation 2 can be written in the form $v(R) = \text{const.}\ (R/\sqrt{W})^{-3/2}$. This proposed law for scaling particle velocity is identical in form to that observed empirically to predict particle velocities for a diverse assemblage of quarries examined by the U.S. Bureau of Mines (Nicholls et al. 1971). Equation 2 can also be written as

$$\log R^{\frac{3}{2}} v = 5.90 - 0.50 \log \rho_s + 2.5 \log a + 0.75 M_L, \tag{3}$$

where particle velocity is in cm/s, source distance to sensor R measured in cm, and ρ_s is in gm/cm^3. Bureau of Mines blasting data gives typical values for "a" of order 0.81 ($\rho_s \approx 2.0$ gm/cm^3, (Nicholls et al. 1971); 0.80 ($\rho_s \approx 2.6$ gm/cm^3) for Ontario mines (Hedley 1990); South African mines, 1.82 ($\rho_s \approx 2.65$ gm/cm^3) (McGarr 1984); and 2.86 ($\rho_s = 2.65$ gm/cm^3) for the 1971 San Fernando earthquake recorded at the Pacoima dam site (McGarr 1984). In the laboratory, the calculated "a" is 1.18 ($\rho_s \approx 2.65$ gm/cm^3) for Salida granite (Brady and Rowell 1994) with the assumption that the source is located at the center of the specimen. Naismith (1984) examined vibration levels and blasting parameters at an underground coal mine in South Africa. His data give "a" to be approximately 0.77, using a medium density ρ_s of 1.5 gm/cm^3. These "transmission coefficient" values of "a" and their variations are reasonable considering the wide range of rock masses and soils considered. If v_{min} denotes the minimum particle velocity to cause damage and R_{max} is the distance from the source to the point of no detectable damage, we conjecture that equations 2 or 3 can be used to estimate damage associated with an underground tremor once E_o (or M_L) and "a" are known. Note we propose a $R^{3/2}v$ scaling law for damage as opposed to the conventional Rv law commonly used by the seismology community.

3 HIGH AMPLITUDE STRESS WAVE GENERATION FROM COAL MINE ROOF FAILURE

A hypothesis is presented that damage associated with and the physical cause of the tremor due to catastrophic roof failure in an underground coal mine can be explained by the preceding model. For simplicity, assume that the coal seam is strong, competent, and elastic. Figure 1 is an idealized model for a strong and competent roof overhang (plate) in an underground coal mine. Assume for the following argument that the competent roof is in immediate contact with the coal seam and that the roof mass that will eventually collapse due to mining is initially firmly attached to the above rock mass. Let "h" denote the surface to seam (overburden) distance and L, the total overhang length. BB' denotes the position of the hinge point for the beam, the point of no deflection resulting from overhanging roof. The hinge point is located a distance δL from the face position at the time of main roof collapse. AB and A'B' denote the original interface positions immediately before main roof separation. ABC denotes the dilatant region (the crack opening between the roof and the layers above) above the seam following bed separation, and AÇ depicts the fracture (thickness of fracture plus deflection resulting from freeing the roof) when the roof collapses. Roof separation relaxes the normal and shear stresses along AB and causes these stresses to redistribute into the abutment zone. As the crack begins to grow along the roof, tensile stress in the detaching beam develops near A as the crack propagates toward the hinge point B. Because the tensile crack is propagating rapidly before caving, the preexisting shear and normal stresses are dynamically redistributed into the abutment zone beyond BB'. After roof detachment, the dilatant zone ABC begins to close, beginning first at B and propagating toward A. Portions of the upper coal surface will likely be inelastically deformed during the final loading process, that is, δL, defined above to be the length of the coal seam deforming elastically, will decrease. However, it can be shown that the hinge position BB' will move outby the face during bed separation, thereby increasing δL. The effects of inelastic deformation will be considered elsewhere in a future publication.

For calculation purposes, assume there is no inelastic deformation due to crushing of the seam along A'B'. During and following roof detachment, the normal and shear stresses along AB then approach their pre-fracture (equilibrium) configuration. If Δ denotes the mean crack opening just before failure, w, the width of the roof slab, and ρ_s, the mean density of the overburden rock, an estimate of the energy imparted along AB during contact is

$$\Psi = k\rho_s g h w \delta L \Delta , \tag{4}$$

where the average compressive force acting along the roof boundary during contact is $F_t = k\sigma_v w\delta L$. The average normal stress along the boundary is $k\rho_s gh$, where k is a stress concentration factor ($\succ$ 1.0) along AB. For ease of calculation, assume k = 1.0. Plausible values for typical roof falls based on experience for w, δL, Δ, ρ_s, and h are 50 m, 3 m, 50 mm, 2.6 gm/cm^3, and 300 m. These parameters give an estimated total energy of impact equal to be 5.8 x 10^{14} ergs (58 MJ). This energy would correspond to an earthquake of equivalent local magnitude M_L = 2.0. The model for the focal mechanism associated with this process will be extremely complex but grossly will correspond to dilatational first motion corresponding to the crack ABC opening along the bed at its late stages, near B in Figure 1. This is followed by roof convergence

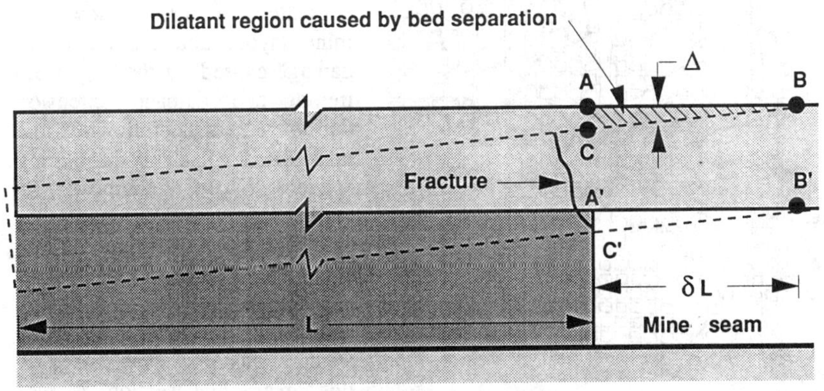

Figure 1. Idealized model of roof over-hang illustrating dilatant region along bed separation plane.

(crack ABC closing) and compressional motion within the rock mass as the compressive and shear stresses are dynamically redistributed following roof caving. Roof collapse and caving and damage associated with the tremor will also be contained within the seismic signature.

The initial amplitude of the compressive stress pulse generated by this mechanism will be of order σ_v. Reflection of the compression pulse from free surfaces generates tensile waves that may induce tensile fractures along the face and openings leading to an outburst. Equation 3 can be used to estimate the distance that damage will occur along the face and gate roads. Using a typical value of v to be 0.30 m/s (Rupert and Clark 1977; Langefors and Kihlstrom 1963) as needed to induce tensile fracture in coal and the above parameters for an event releasing 58 MJ give R ≈ 25 m, respectively, using a value of "a" = 0.77 (Naismith 1984). An event releasing 15,800 MJ, corresponding to an earthquake of equivalent local magnitude of 3.6, induced by caving results in R ≈ 286 m, respectively, as the distance from the source where damage can be expected. This example illustrates that damage associated with moderate outbursts in mines should be confined to a relatively narrow zone surrounding the source. In longwall coal mines, this damage would occur primarily in the tailgate region where mining-induced stresses are the largest.

4 DYNAMIC MINE FAILURE: AN EXAMPLE

A seismic event registering M_L = 3.6 (NEIS) occurred in the Book Cliffs area of Utah with the epicenter coinciding with the location of an underground coal mine. The mine uses the room-and-pillar extraction method to mine a 2.7-m-thick coal seam. The overburden thickness varies between 450 m and 900 m, with approximately 600 m in the immediate damaged area. The immediate roof consists of a competent 12-m-thick sandstone. The uniformity and competency of the sandstone makes it difficult to cave regularly during retreat mining.

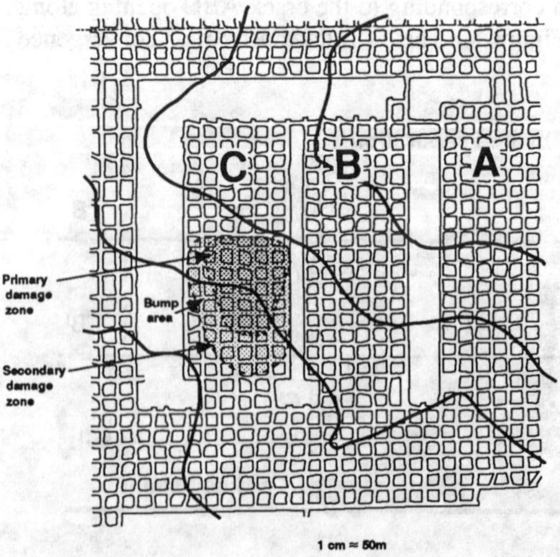

Figure 2. Schematic of Utah coal mine showing primary damage from M_L = 3.6 mainshock.

Due to unfavorable conditions, the mine had a history of tremor occurrences with severe damage occurring during full pillar extraction. Figure 2 shows the mine layout and the extent of damage caused by the M_L = 3.6 tremor and damage extension during the days following the main event. Each of the mining sections (A,B,C) composed of seven entries was developed using 18.2-m-square pillars on 24.3-m centers. A 45.5-m-wide barrier pillar exists between sections A and B. Damage due to small tremors was first experienced in this area while section A was retreated. Full pillar extraction started at the fifth pillar line and continued outby until the end of the section. Several minor tremors occurred and became more frequent after retreating six rows of pillars. Mining in section B also began at the fifth pillar line. With the completion of fully extracting six rows of pillars, a major mine tremor with no outbursting occurred. The company was forced to abandon three rows of pillars for support. Mining commenced with only partial pillar extraction.

An 18.2-m-wide barrier pillar was left between sections B and C as shown in figure 2. Realizing that the barrier pillar was insufficient to withstand the abutment pressures, a row of pillars in panels B and C were sacrificed for extra support to the main roof. Full pillar extraction started at the fourth pillar line of panel C.

Following the extraction of four rows of pillars, stresses became extremely high as evidenced by the $M_L = 3.6$ and resulted in section failure. Major structural damage occurred in the section extending seven rows of pillars outby. Aftershocks were later recorded and extended the damage to two or three rows of pillars. The severity of the damage increased from east to west and from south to north. The barrier pillar on the east side was severely damaged. In some areas the coal was almost pulverized against the roof. Many entries were filled with intensely fragmented coal while the roof remained intact with large gaps observed between the roof and coal.

The above model is applied to explain some of the damage at this site. Here the width of the caved region is approximately w = 276 m, the overburden depth is h = 600 m, and values for δL and Δ are approximated at 6 m and 0.2 m, respectively, giving a predicted energy release of approximately 1.05×10^{17} ergs or 1.05×10^4 MJ. The overhang is assumed to be hinged along the line of outer undamaged pillars. If this event were a fault-slip event, the equivalent magnitude is calculated to be approximately $M_L = 3.5$. The deformation Δ is only approximate and corresponds to the known modulus of the coal, the overburden, and the extraction ratio. Expected damage could extend to approximately 300 m using a value 0.30 m/s for v_{min}, well in the range of observed damage. Precise determinations of v_{min}, "a", Δ, and δL are, of course, needed to provide accurate estimates of the damage extent. This subject will be taken up in later studies.

5 CONCLUSIONS

A testable hypothesis was presented describing catastrophic roof failure in mines to be the result of the formation of a macrocrack or dilatant zone above a pillar or longwall panel immediately prior to roof collapse. This dilatant region closes upon itself as the weight of the overhanging roof on the panel or pillar(s) is released during the caving process. The energy released during caving generates a high-amplitude stress pulse, whose magnitude is on the order of the stress field existing prior to roof detachment. The result was applied to a particle velocity damage criterion to predict expected damage from this roof caving model.

REFERENCES

Gibowicz, S. J. 1990. The Mechanism of Seismic Events Induced by Mining: A Review, in Rockbursts and Seismicity in Mines, Balkema, pp. 2-27.

Ortlepp, W. D. 1992. Note on Fault-Slip Motion Inferred from a Study of Micro-Cataclastic Particles from an Underground Shear Rupture, Pure and Appl. Geophys., v. 139, pp. 677-695.

Brady, B. T., and G. A. Rowell 1986. Laboratory Investigation of the Electrodynamics of Rock Fracture, Nature, v. 321, pp. 488-492.

McGarr, A. 1984. Some Applications of Seismic Source Mechanism Studies to Assessing Underground Hazard, Rockbursts and Seismicity in Mines, Balkama, pp. 199-208.

Spottiswoode, S. M., and A. McGarr 1975. Source Parameters of Tremors in a Deep-Level Gold Mine, Bull. Seismo. Soc. Amer., pp. 93-112.

Gay, N. C., and W. D. Ortlepp 1979. Anatomy of a Mining Induced Fault Zone, Bull. Geo. Soc. Amer., pp. 47-58.

Wong, I. V., and A. McGarr 1990. Implosional Failure in Mining-Induced Seismicity: A Critical Review, in Rockbursts and Seismicity in Mines, Balkema, pp. 45-51.

Tesisseyre, R. 1980. Some Remarks on the Source Mechanism of Rock Bursts in Mines, Acta Mont., v. 55, pp. 7-14.

Hedley, D. G. F. 1990. Peak Particle Velocity for Rockbursts in Some Ontario Mines, in Rockbursts and Seismicity in Mines, Balkema, pp. 345-349.

Nicholls, H. R., Johnson, C. F., and W. I. Duvall 1971. Blasting Vibrations and Their Effects on Structures, USBM Bulletin 656, 105 pp.

Naismith, W. A. 1984. The Influence of Large-scale Surface Blasting on the Stability of Underground Coal Mine Workings, in Rockbursts and Seismicity in Mines, Balkema, pp. 183-191.

Rupert, G. B., and C. B. Clark 1977. Criteria for the Proximity of Surface Blasting to Underground Coal Mines, Proc. 18th U.S. Rock Mechanics Symposium, pp. 3C310-3C331.

Langefors, H., and B. Kihlstrom 1963. Rock Blasting, John Wiley and Sons, Inc., New York, 425 pp.

Brady, B. T., and G. A. Rowell 1994. A Laboratory Investigation of the Precursors of Rock Failure, in Theory of Earthquake Precursors and Fracture Processes, ed. R. Tesisseyre (in press).

Lenhardt, W. A., and T. O. Hagan 1992. Observations and Possible Mechanisms of Pillar-Associated Seismicity at Greath Depth, in Induced Seismicity, ed. P.Knoll, Balkema, pp. 149-162.

Rock Mechanics, Nelson & Laubach (eds) © 1994 Balkema, Rotterdam, ISBN 90 5410 380 8

Roof fall monitoring and prediction at the Waste Isolation Pilot Plant

Christopher T. Francke & Stanley J. Patchet
Waste Isolation Division, Westinghouse Electric Corporation, Carlsbad, N. Mex., USA
Saeid Saeb
IT Corporation, Albuquerque, N. Mex., USA

ABSTRACT: Through analysis of geomechanical mechanisms controlling deformation and fracture development around excavations in bedded salt, analysis of ten years of geomechanical instrument data, and direct monitoring of a large roof fall, a basis for roof fall monitoring and prediction has been developed for excavations at the Waste Isolation Pilot Plant.

1 INTRODUCTION

The Waste Isolation Pilot Plant is located in southeastern New Mexico about fifty kilometers east of Carlsbad. The WIPP was authorized by Congress in 1979 (Public Law 96-164) to provide "...a research and development facility to demonstrate the safe disposal of radioactive wastes resulting from the defense activities and programs of the United States exempted from regulation by the Nuclear Regulatory Commission." The WIPP is intended to receive, handle, and permanently dispose of transuranic waste. To fulfill this mission, the U.S. Department of Energy is constructing a full scale facility to demonstrate both technical and operational principles of the permanent isolation of transuranic waste.

The underground facilities are excavated approximately 655 meters beneath the surface in the Permian Salado Formation, a thick, predominantly halite evaporite sequence. Shaft sinking at the WIPP began in 1980, and most of the excavation was completed by 1984. The operating life required of the underground facility is approximately twenty-five years after first waste receipt which is currently scheduled for 1998. Therefore, many of the access drifts must remain usable for at least forty years.

The creep behavior of halite makes long term performance difficult to predict (Baar 1977). Fracture development that starts soon after excavation further complicates estimation of a stand up time for WIPP excavations. The ability to assess the stability of the excavations at the WIPP is necessary to protect the safety of site workers, the environment, and the integrity of barriers that confine the transuranic waste. Although roof falls are not expected to occur in maintained excavations at the WIPP, roof falls have been allowed to occur in unused, barricaded rooms so that the pre-collapse behavior of the rooms could be studied. This study concentrates on estimating the time a roof fall would occur in an existing excavation.

2 DEFORMATION MECHANISMS

Potential deformation mechanisms affecting the facility excavations have been proposed. The mechanisms described here are simplifications of the very complex forces acting on complex materials. They attempt to explain the causes of the deformation and fracture patterns that are observed at the WIPP. The stratigraphic details around the excavations are very important to the deformation and are discussed first.

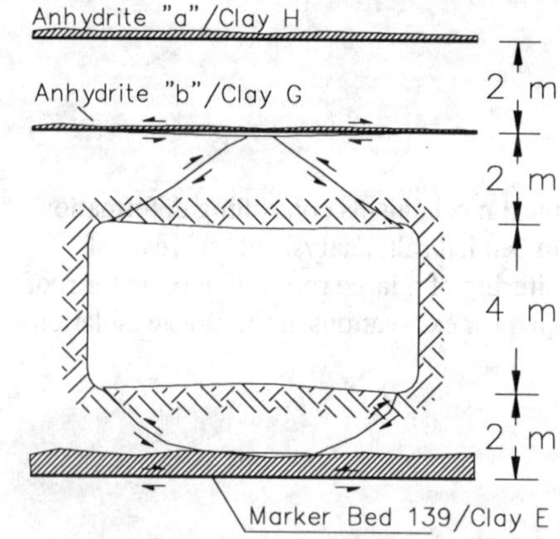

Figure 1. Typical WIPP stratigraphy and fracture pattern.

2.1. Geology

The facility horizon lies within a twelve-meter thick evaporite sequence consisting of halite, argillaceous halite, and polyhalite (Figure 1). These horizontal beds are laterally consistent throughout the site. MB139, a persistent 50 to 80 centimeter thick bed of anhydrite and polyhalite lies about 1.5 meters below the repository floor. MB139 is underlain by clay seam E. Anhydrite "a," about twenty centimeters thick, is about four meters above the repository roof and is underlain by clay H. Anhydrite "b," about seven centimeters thick, is about two meters above the roof and is underlain by clay G. The clay seams are all less than twenty-five millimeters thick.

2.2 Description of Deformation Mechanisms Around WIPP Excavations

The state of stress and associated displacement and strain fields around an opening in a nonlinear and time dependent environment, such as salt, are quite complicated and can only be analyzed through comprehensive and detailed computer modeling. However, computer modeling has its own limitations. General constitutive models that simulate the behavior of the material address limited aspects of the material performance that are considered to be the governing behavior. Furthermore, there is no material model that can simulate the fracturing and failure process in salt. Progress has been made (Chan 1992) towards development of "smeared" fracture models. It appears, however, that discrete, macroscopic fractures must be numerically simulated to produce model results optimal for stability analysis. To simulate the formation of discrete fractures, the geomechanical processes that cause their formation must be understood.

As the rooms are excavated, the in situ stresses normal to the perimeter of the excavation are relieved, and far-field stresses tend to push the salt mass toward the opening. Because the subhorizontal clay seams do not provide much, if any, shear resistance, the salt between clay G in the roof and clay E in the floor moves horizontally toward the excavation faster than the adjacent layers. This movement concentrates the

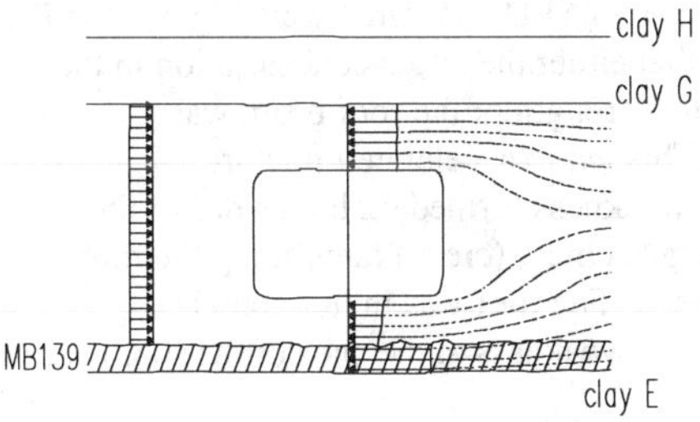

Figure 2. Concentration of Lateral Stress in the Roof Beam.

stresses on the portion between the roof and clay G, referred to as the roof beam, and the portion between the floor and clay E, referred to as the floor beam (Figure 2).

The concentration of horizontal stress and lack of vertical confinement in the roof beam cause large strains to develop with time in the roof beam. This causes strain softening due to dislocation along crystal boundaries which in turn may form shear bands and macroscopic fractures.

The dominant, excavation induced fractures in the roof beam are fairly consistent throughout most WIPP excavations, regardless of opening size or geometry (see Figure 1). The size of the excavation only appears to effect the timing of the formation of the fractures (Francke, et al., 1993). The location of the dominant, diagonal fractures in the roof beam appears to be controlled by a combination of several mechanisms.

There are no vertical planes of weakness in the roof beam. In the midroof, the salt mass can displace into the excavation with much less resistance than it can near the ribs where the pillars provide some support. This may induce large shear strains in the roof close to the ribs which in turn weaken the rock and contribute to a weakened zone.

Geometric eccentricity in the axially loaded roof beam can generate significant additional bending, which affects the strength of the rock. The stress concentration at the corners of rectangular excavations increases the potential for fracture initiation near the corners. Finally, the excavation sequence of the excavation and of nearby drifts affects the stress distribution around the excavation and, therefore, the fracturing.

3 FIELD DATA ANALYSIS

Observation of the geomechanical information collected from an instrumented drift that eventually suffered a large roof fall has been used to relate the deformation mechanisms presented earlier to field data. A 700 ton roof fall occurred in an experimental WIPP excavation (SPDV Room 1) on February 4, 1991. The section that fell was approximately 33 feet wide by seven feet high by 150 feet long, had a triangular cross-section, and was located primarily in the north end of the room. The time and shape of the roof fall were predicted using methods that will be discussed later. Furthermore, the potential for a roof fall in SPDV Room 1 was identified and the room was cleared and barricaded two years before the roof fell. This room had been instrumented shortly after its excavation in March of 1983 and was closely monitored until the roof fall. This is the only large (room-scale) fall which has occurred at the WIPP. No ground support had been installed in the room.

3.1 Assessment of SPDV Room 1 Behavior

From the beginning, displacement rates in the roof of SPDV Room 1 were higher than in similar rooms (US DOE, 1993). This implies that either the stress concentration in the roof of SPDV Room 1 was unusually high or the strength of the roof beam was unusually low, compared to the other rooms. This led to accelerated fracture development in the roof of Room 1. Bands of weakness formed on both sides of the roof near the ribs due to differential vertical displacement (creep) rates along the roof and to the concentration of stresses in the corners. The roof was in horizontal compression from the beginning, and the zones of weakness at the ends of the roof beam continued to grow as strain accumulated.

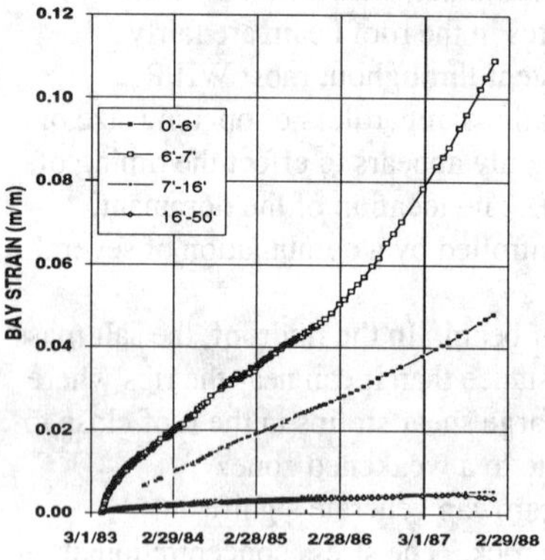

Figure 3. Typical Vertical Strain Distribution in SPDV Room 1 Roof

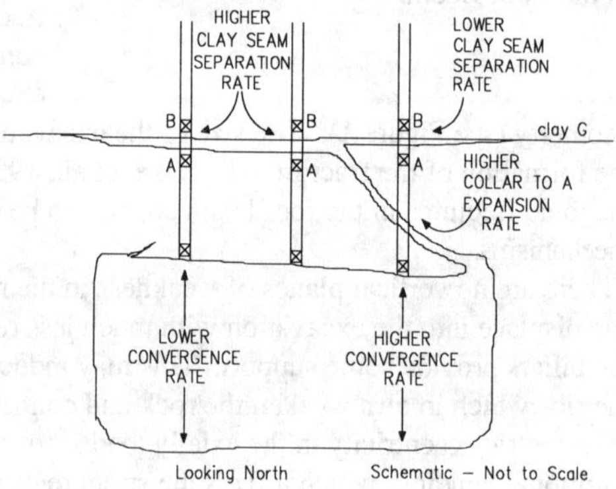

Figure 4. Fracturing in SPDV Room 1 Roof

In early 1986, the clay G bay strain (separation) rates of the roof extensometers in Room 1 increased over a three or four month period (Figure 3), indicating a significant change in loading or strength of the roof beam at that time. It is postulated that the increase resulted from the formation of a fairly consistent fracture or band of fractures on the east side of the roof beam (Figure 4). This effectively turned the roof beam into a cantilever relatively fixed on the west side and relatively free on the fractured east side. Exactly why the east side fractured first is not apparent. It does not appear to be related to the excavation sequence of the room or of the panel of rooms.

Other evidence for the formation of a cantilevered beam is found in the distribution of the clay seam bay strains, the roof-to-floor convergence rates, and the borehole fracture data. The clay seam bay strains (the strain between two borehole extensometer anchors straddling the clay seam) are higher at the center and the west side, which is consistent with the cantilevered section bending down into the excavation. The convergence rates (Figure 5) are higher on the east side of the cross-section which is consistent with the free end of the cantilever moving down faster than the middle or the fixed end. Finally, the borehole horizontal offset data show the center of the roof beam moving in the same direction as the fixed end of the cantilever, but in the opposite direction of the free end (Figure 6). This is consistent with the cantilevered part of the roof beam moving as one piece to the east and the stub on the east side of the roof beam moving to the west.

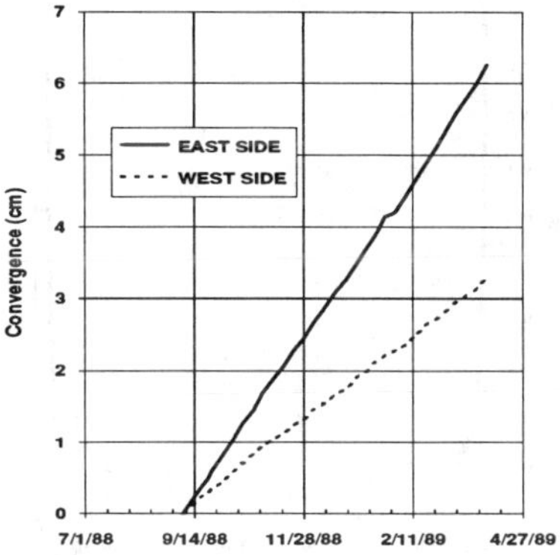

Figure 5. SPDV Room 1 Convergence Data Showing Asymmetry

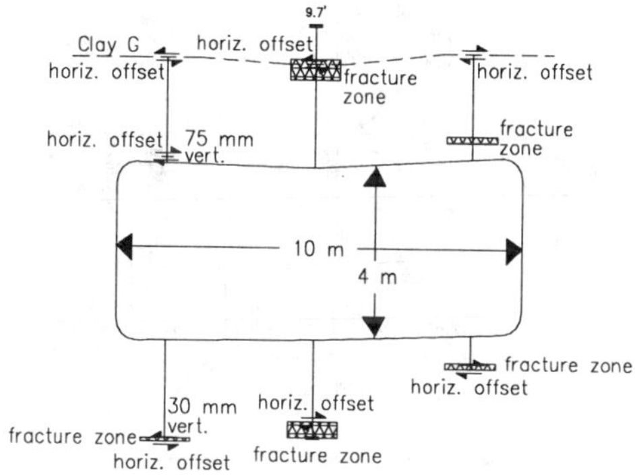

Figure 6. Fracture and Clay Seam Offset Data

This situation continued with fairly steady clay seam bay strain rates and only slightly increasing closure rates until early 1988, when the bay strain rates and closure rates again increased. It is postulated that this increase was caused by the development of another set of fractures on the west side of the roof beam (Figure 7). This may be due to a combination of tensile stresses and the pre-existing band of weakness. The tensile stresses could be induced in the fixed end in several ways such as by deflection of the cantilever due to gravity or by the free end of the cantilever being forced downward along the older shear fracture on the east side by horizontal movement. It is likely that both sides of the roof beam formed bands of weakness well before the first major fracturing occurred. The tension induced in the roof beam by the deflecting cantilever probably combined with previously formed fractures and weaknesses. The fractures thus formed on both sides of the roof beam produced a wedge shaped slab primarily supported at the north and south ends.

Beginning with the formation of the wedge, and ending with the roof fall, convergence rates accelerated. It is likely that the wedge first formed in a small area or in several small unconnected areas. As the roof continued to deform, the length of the wedge (along the length of the room) grew, and the independent wedges connected. As the total length of the wedge grew, so did its weight, which put an increasing burden on what little support remained. Both the strain undergone by the supporting rock as well as its reduced confinement undoubtedly lowered the roof strength. This led to ever-increasing convergence rates (Figure 8). After the wedge formed, the deformation and continued fracturing of the wedge was due almost entirely to dead-weight gravity loading. It was then essentially acting independently from the stress field around the excavation as a whole. This is consistent with the lack of any measurable changes in creep rates at locations outside SPDV Room 1. For example, the extensometer and convergence points at the north intersection of Room 1 and N1420 show no signs of the roof fall that occurred about forty feet away. This is solid evidence that gravity forces on the semi-detached wedge drove the final failure, and that properly installed ground support would prevent or significantly delay a similar roof fall.

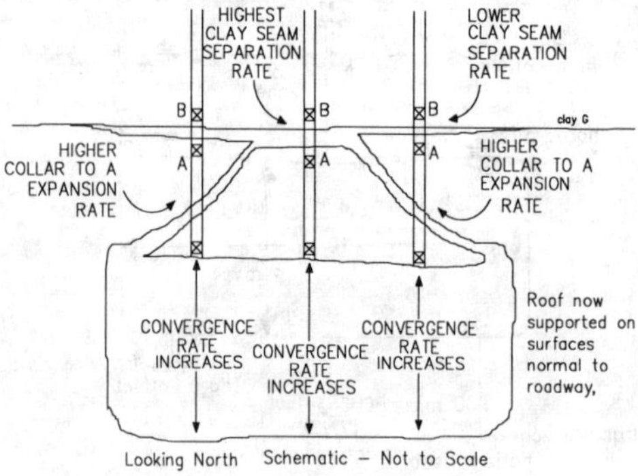

Figure 7. Advanced Fracturing in SPDV Room 1

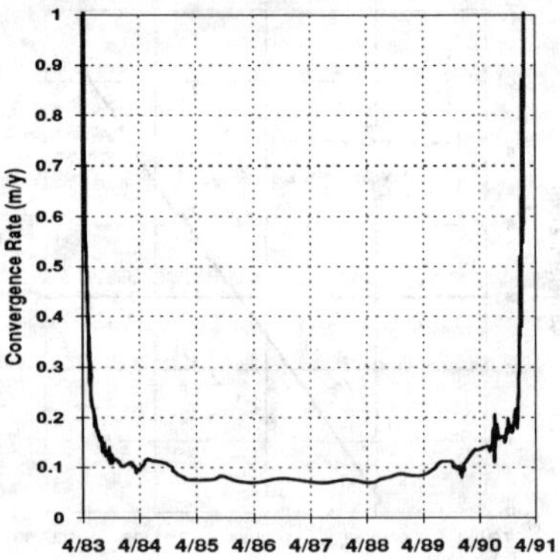

Figure 8. SPDV Room 1 Vertical Convergence Data

4 PREDICTION FOR SPDV ROOM 2

SPDV Room 1 is located in a four room panel in the northwest section of the WIPP underground facility (Figure 9). The four SPDV Rooms were all excavated within a two month period. SPDV Room 2, located next to SPDV Room 1, is the only monitored excavation at WIPP that is currently displaying signs of instability. New data from Room 2 are currently limited to vertical convergence gauges. Visual observations are limited because the room is barricaded. Because no additional data can be obtained from Room 2, prediction of the remaining stand-up time must be based on the vertical convergence rate alone.

A curve fitting approach has been developed to predict the time of failure of the roof of SPDV Room 2. Statistical analyses of the convergence data from both SPDV Rooms indicate that the convergence rates in SPDV Room 2 are changing in a manner very similar to those of SPDV Room 1 in the months before the roof fall. Although SPDV Room 2 has deformed and developed fracturing slower than SPDV Room 1, it is reasonable to expect the two rooms to deform at nearly the same rate as they approach roof collapse, when fracture formation and separation are the primary contributors to convergence. The fracture formation is primarily caused by the deformation of the roof beam under its own weight. Therefore, two equal sized drifts could be expected to behave similarly after the roof is well fractured (as in SPDV Rooms 1 and 2), regardless of the past performance of the excavations.

The curve fitting makes the following assumptions:

1. the deterioration of the slab in SPDV Room 2 will be similar to that of SPDV Room 1;

2. both convergence meters were located in parts of their respective slabs with similar relative levels of deterioration.

The curve fitting was repeated on the data from each of the four SPDV Room 1 convergence meters. Equations were derived of the form:

$$\log(t) = \begin{cases} m_1\dot{C} + b_1 & \text{for } \dot{C} > \dot{C}_c \\ m_2\dot{C} + b_2 & \text{for } \dot{C} < \dot{C}_c \end{cases}$$

Where t is the number of days between the time of the reading and the roof fall, $\dot{C}$ is the convergence rate in inches per year, and m, b, and $\dot{C}_c$ are curve fitting parameters. The curve fitting results for each of the SPDV Room 1 convergence meters are:

Instrument	m_1	b_1	m_2	b_2	$\dot{C}_c$
51X CW-0001 (CW1)	-0.01224	1.484	-0.2238	3.631	10.15
51X CW-0002 (CW2)	-0.01475	1.638	-0.2050	3.639	10.52
51X CW-0003 (CW3)	-0.02044	1.488	-0.3392	4.174	8.43
51X CW-0004 (CW4)	N/A	N/A	-0.2958	3.599	N/A

There are no values for m_1, b_1, and $\dot{C}_c$ for CW4 because the closure rate did not accelerate until shortly before it fell.

At the time of writing (November 15,1993), the highest monthly convergence rate in SPDV Room 2 was 7.258 inches per year. Solving the above equation using the parameters from each SPDV Room 1 instrument analysis and 7.258 inches per year convergence rate gives from 102, 142, 52, and 28 days until the roof fall based on CW1, CW2, CW3, and CW4 data, respectively. Figure 10 shows this graphically. In the plot, actual convergence rate data from SPDV Room 2 up to November 15, 1993, is shown with predicted convergence rates based on the four curve fitting equations appended on after November 15. Data from SPDV Room 1 is also superimposed on the plot for comparison. Confidence levels are based first on the degree of fracturing at the SPDV Room 1 instrument locations and second on the convergence rate. The degree of fracturing around CW1 was greater than the other SPDV Room 1 instruments. CW2 had the highest convergence rate, but CW1 was very close. Therefore, confidence is highest for CW1 followed closely by CW2.

The range of likely dates for a SPDV Room 2 roof fall is December 13, 1993, to April 5, 1994, with February 24, 1994, the most likely date for the roof fall. Confidence in this prediction is high as long as the assumption holds that SPDV Room 2 convergence rates are changing on a timetable similar to that observed in SPDV Room 1. This continues to be the case.

Changes take place slowly and, as the time of the roof fall approaches, the prediction quality will improve and the range of dates will narrow. A major roof fall in SPDV Room 2 or in any other excavation in the facility level at WIPP will not occur abruptly. There will be months or years of warning from the geotechnical instrument data which are watched carefully for signs of instability or other signs such as fracture formation or bolt failures. Instrumentation and visual observations indicate that the access drifts to the SPDV panel are quite stable. As was shown earlier, the instrumentation immediately adjacent to SPDV Room 1 showed no effects of the roof fall. This indicates that, unless indications of instability develop in the access drifts themselves, there is no danger of roof collapse in an access drift due to instability in an adjacent room.

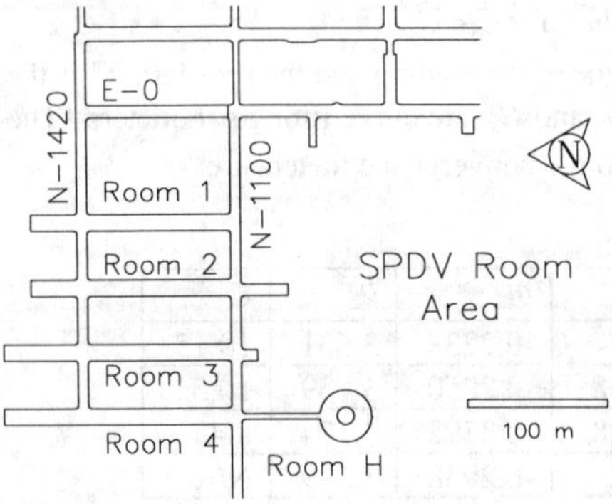

Figure 9. SPDV Room Panel Layout

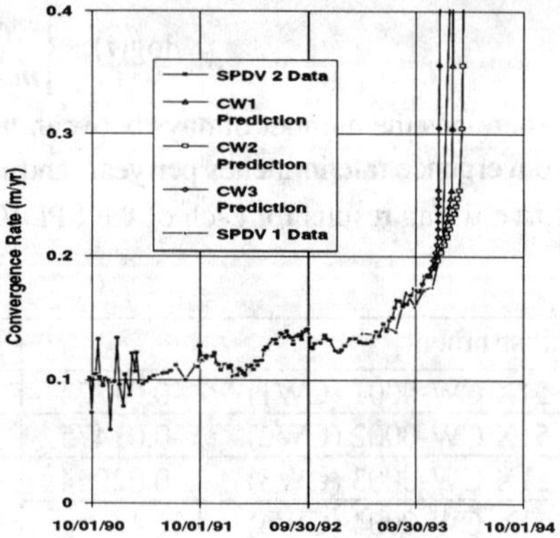

Figure 10. SPDV Room 2 Convergence and Prediction

5 CONCLUSIONS

Geomechanical data from an excavation that collapsed and another that is believed to be nearing collapse has been used to characterize the behavior of unstable excavations at WIPP. The excavations were carefully monitored and the precursors to the roof fall were identified. These precursors were used to develop an empirical method for predicting the timing of a roof fall in the second excavation. However, this prediction is based on data from just two drifts. The roof falls at WIPP have only occurred in unsupported, barricaded openings. In every case, the roof of these excavations could have been reinforced or otherwise remediated to prevent the roof fall. Additional data will be obtained as other monitored excavations become unstable and are not remediated. Future work will concentrate on refining the conceptual model and applying the results to numerical models.

Work sponsored by the U.S. Department of Energy under DOE contract No. DE-ACO4-86AL31950.

REFERENCES

Baar, C.A., 1977, Applied Salt Rock Mechanics I. New York: Elsevier.

Chan, K.S., S.R. Bodner, A.F. Fossom, and D.E. Munson, 1992. "A constitutive model for inelastic flow and damage evolution in solids under triaxial compression," Mechanics of Materials, Vol. 14, Elsevier.

Francke, C.T., and L.J. Terrill, 1993. "The Excavation Effects Program at the Waste Isolation Pilot Plant," Innovative Mine Design for the 21st Century, proceedings of the International Congress on Mine Design, Balkema.

U.S. Department of Energy, 1993, Geotechnical Field Data And Analysis Report, DOE/WIPP 93-019, Waste Isolation Pilot Plant, Carlsbad, New Mexico.

19 Rock mechanics in dam and hydro-project construction

Session developers:
Gregg Scott, US Bureau of Reclamation
Alex Grenoble, Stone and Webster, Inc.

Rock Mechanics, Nelson & Laubach (eds) © 1994 Balkema, Rotterdam, ISBN 90 5410 380 8

Rock mechanics dam safety studies for Seminoe arch dam

R.E.Goodman
Department of Civil Engineering, University of California, Berkeley, Calif., USA
G.A.Scott
U.S. Bureau of Reclamation, Denver, Colo., USA

ABSTRACT: The Bureau of Reclamation is performing dam safety studies for all of its existing dams, including approximately 60 concrete storage dams. Seminoe Dam, located on the North Platte River in Wyoming, is one of these. It is used here to illustrate the types of evaluations that are being performed. An overview of rock mechanics studies related to the safety of the dam are presented, including those addressing rock mass deformability, static and dynamic abutment sliding stability, overtopping erosion, spillway tunnel stability, and reservoir landslides.

1 INTRODUCTION

Seminoe Dam (Figure 1), an existing concrete arch dam located on the North Platte River about 120 km southwest of Casper, Wyoming, retains a reservoir volume of over 1.2 X 10^9 m^3. Construction for the dam began in 1936 and was completed in 1939. The dam has a structural height of 90 m and a crest length of 162 m, with a maximum base width of 26.8 m and a crest thickness of 5.2 m. The spillway is a gated inclined shaft that connects with the 9.14-m diameter original diversion tunnel on the right abutment. The surface powerplant, capablc of producing 32.4 MW, is located in the center of the river channel at the base of the dam. Seminoe Dam was constructed on Pre-

Figure 1. Photo of Seminoe Dam (north is almost directly downstream).

Cambrian granite and gneissic granite, intruded by diorite and lamprophyre dikes. The foundation rock is cut by persistent joints of one set and less persistent joints of at least two other sets, as well as by a number of small shears. Two fault zones also cross the foundation. Foundation treatment included excavation and cleaning to sound granite, extensive grouting, installation of a drainage curtain, and excavation and concrete backfilling of the fault zones.

Historically, about 70 percent of all concrete dam failures have been attributed to foundation or geotechnical deficiencies (ICOLD, 1974). These statistics warrant careful consideration in performing dam safety studies for concrete dams. The studies presented below for Seminoe Dam illustrate the types of rock mechanics studies currently being performed for Bureau of Reclamation (Reclamation) concrete dams. They attempt to account for pertinent rock mechanics properties and potential modes of instability to ensure the integrity of the dam, its foundation, and the appurtenant structures critical to its safety.

2 FOUNDATION MODULUS

The deformation properties of the foundation rock must be considered in modeling the dam-foundation interaction. Despite what intuition may tell us, assuming a low modulus value may not be conservative. An arch dam is a complex three-dimensional structure, and deformation of the foundation affects the load distribution within the structure. In addition to geometric irregularities and variable deformation properties, a very stiff foundation often attracts load and causes a more adverse stress distribution than a more flexible foundation. Therefore, it is important to develop reasonable estimates for foundation modulus values.

For engineering purposes, the foundation rock at Seminoe Dam is reasonably uniform. Although an exploratory hole had been drilled, water tested, and logged on each abutment, no in situ or laboratory rock compression tests were performed. Rock Quality Designation (RQD) and surface mapping data suggested close fracture spacings. These values were not thought to be indicative of the more open joints that would contribute to foundation deformability below the zone of foundation excavation. Therefore, instead of evaluating the joint data directly, an approach that used data from the water tests was pursued based on the work of Snow (1970).

The fractures or joints that conduct water must be open to a certain extent. It is these fractures that would be expected to deform under application of normal stress from the dam, and therefore lower the effective foundation modulus. The proportion of borehole packer tests that yielded no flow into the rock were assumed to represent sections of the test hole that intersected no open fractures. A Poisson distribution for the occurrence of fractures in a short length of borehole was used to calculate the average spacing of conducting fractures. These spacings proved to be much greater than the spacings calculated from RQD or determined from the surface mapping. From the permeability test values and average spacings calculated for the conducting joints, it was possible to compute the average hydraulic aperture. Based on the spacing and aperture, it was estimated that the in situ modulus should be approximately 60 per cent of the intact rock modulus. Using a modulus value for the intact rock of about 34.5 GPa, based on experience from similar rock, results in an in situ modulus of about 20.7 GPa.

A check was performed based on the Rock Mass Rating (RMR) system. The RMR for the rock developed from judged parameters was found to be about 62. The correlation offered by Bieniawski (1989) gives an in situ modulus value of 24 GPa, which compares favorably with that determined by the other approach. Interestingly, values used for joint spacing were closer to those obtained from surface mapping than from the packer testing.

3 FAULT TREATMENT

Two significant fault zones were encountered in the channel area of the foundation during construction, termed the "red seam fault" and the "black seam fault". The red seam fault strikes parallel to the axis of the dam and dips downstream at about 50°. The fault zone has a thickness of about 2.1 m, consisting of crushed rock and soft clay gouge. The black seam fault strikes upstream to downstream, dipping 70° to the west at foundation level in the channel, and flattening to 40° at the crest elevation on the right abutment. The thickness of the chloritic gouge and breccia-filled zone varies from 1.8 to 3.7 m in the lower foundation, thinning to less than 0.3 m on the right abutment slopes. Although shown to be seismically inactive during seismotectonic investigations, faults in the foundation raise other issues including: (1) compressibility of the fault zone under loads transmitted by the dam, (2) stability of a block that might be created by intersections of the fault with other discontinuities in the foundation, and (3) blowout or piping of the fault zones due to seepage emerging at high gradient downstream (applicable for black seam fault only since it is the only fault oriented transverse to the dam axis).

No obvious problems have been observed at the dam related to faults. This is thought to be largely due to the extensive fault treatment installed at the time of construction. The faults were treated by mining from a succession of shafts and backfilling with concrete to depths of 3 to 26 m for the black seam fault and 9 to 11 m for the red seam fault. Guidelines published by Reclamation (1977) indicate the minimum treatment depth should be 3 m to ensure smooth deformation across the zones. Examination of possible intersections of the red and/or black seam faults with other discontinuities in the central part of the dam foundation revealed no potentially unstable foundation blocks. The exit gradient calculated for the black seam fault using the method reported by Scott (1982) is about 2.3. The critical exit gradient for the material in the fault zone is not known. However, small amounts of seepage exit shears with similar filling materials without piping and without treatment. The black seam fault has been treated, and the moderate gradient with no observable problems suggest the treatment is adequate.

4 DYNAMIC STABILITY OF THE DAM FOUNDATION

The foundation of the dam has performed in a satisfactory manner for many years under static loading conditions. However, the extreme maximum credible earthquake (MCE) loading may be sufficient to cause instability under dynamic conditions. To examine this possibility, shear and joint sets, and their orientations were defined for each area of the foundation. This was done primarily from high quality color photographs onto which the geologists had mapped the prominent discontinuities and their orientations. The discontinuity sets were analyzed using block theory (Goodman and Shi, 1983) to determine the critical combinations of three joint sets that form removable blocks with the abutment free surfaces. Those combinations that could form reasonable blocks within the dam foundation were studied further in a deterministic manner. This was accomplished by carefully searching the photographs for evidence of the blocks on the well exposed rock surfaces downstream of the dam. No information was available concerning discontinuities in the dam foundation or upstream of the dam. Analyses were conducted for those blocks that could be identified as projecting under the dam.

Only one block was identified for further analysis. This occurred in the right abutment of the dam. The kinematically removable block is seen to belong to joint pyramid (JP) 101, as shown in Figure 2. Continuous planes of sets B and C could

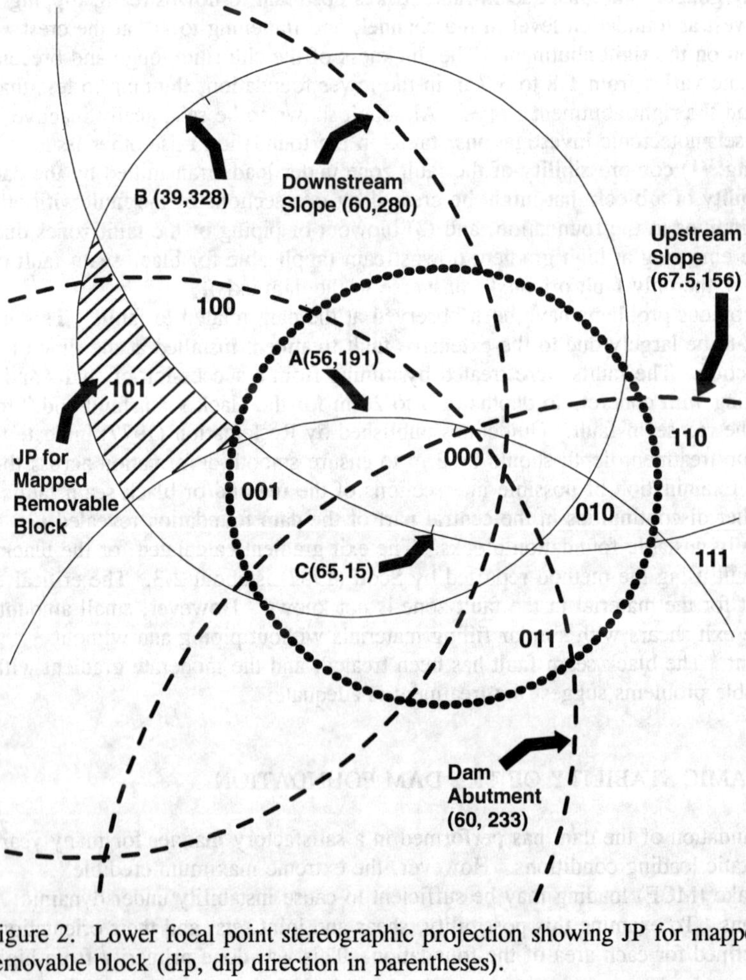

Figure 2. Lower focal point stereographic projection showing JP for mapped removable block (dip, dip direction in parentheses).

be located on a photograph of the right abutment, such that the block lies above joint B and below C, as shown in Figure 3. The planes were assumed to be continuous where obscured by the powerplant and the dam. The block was assumed to be truncated by an A joint upstream of the dam. The A joints are the most numerous and prominent joints at the site.

Figure 3. Photo of right abutment showing mapped removable block.

Static and dynamic forces applied to the block were calculated from three-dimensional finite element structural analyses. Water forces acting on the block were based on assuming full reservoir driving force, since little information was available concerning the pressure distribution within the abutments. Time-varying earthquake structural and inertia forces were applied to the block in combination with the static loads corresponding to normal reservoir water elevation and winter temperatures. Figure 4 shows calculated permanent displacements of the block, using a "Newmark" approach, for various friction angles. It was assumed that the load is not redistributed by the arch (which in reality would occur). A friction angle of 34° reduces the displacement to zero. The applicable friction angle is believed to be considerably higher than 34° due to the roughness of the joints. Thus, this block is not considered to be a problem under MCE loading.

5 TUNNEL SPILLWAY AND INTAKE STABILITY

The tunnel spillway was not designed to be pressurized. However, during high tailwater conditions associated with the probable maximum flood (PMF), the tunnel becomes pressurized. The calculated hydraulic pressure head was compared to the rock cover to determine the possibility of spillway blowout, and subsequent undermining of the dam foundation. The criteria developed empirically from Norwegian experience, as reported by Brekke and Ripley (1989), were used. The required minimum rock cover, in the direction normal to the nearest free surface, was found to be consistently much smaller than the available minimum rock cover. In addition, the tunnel is lined with reinforced concrete which would increase its pressure capacity. However, a more serious condition could result from partial blockage of the tunnel whereby the pressure in the tunnel could approach full reservoir head. Under such a condition, it is possible that the tunnel could burst out into the rock cliff, precipitating a landslide that could rapidly extend headward in the

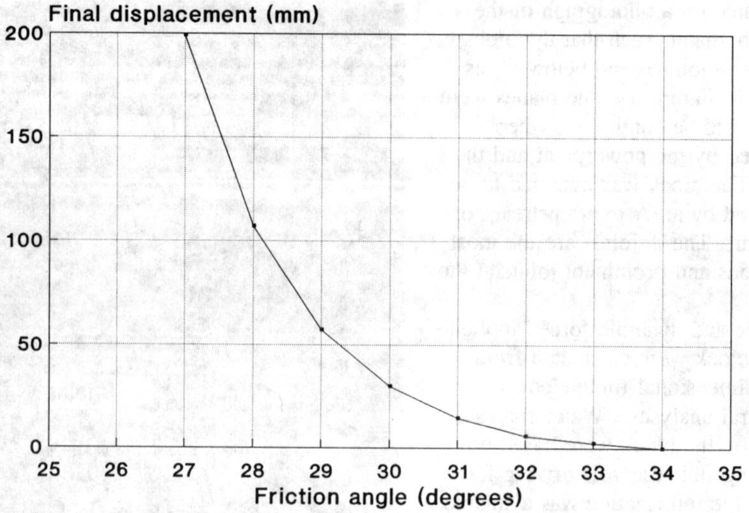

Figure 4. Cumulative permanent displacements of block for dynamic analyses.

jointed rock downstream of the right abutment. It was judged that a massive rock slide at the intake or downstream section of the tunnel could potentially form such a blockage. Examination of conditions upstream and downstream of the tunnel indicate that the most unstable conditions probably occur at the intake. The intake was excavated into a dip slope, undercutting the parallel persistent joints and minor faults of set A, as shown in Figure 5. A large slide would likely involve large blocks that could become lodged in the tunnel forming an obstruction. In order to protect the spillway gates and tunnel from the consequences of a slide, which might be triggered by heavy rains, rock reinforcement was recommended for this slope.

6 OVERTOPPING PROTECTION

The spillway capacity is inadequate to pass the large flows associated with the PMF. Therefore, the dam would overtop during this unusual loading condition, and the consequences of this must be considered. The ability of the abutments to resist overtopping flows was evaluated by evaluating the likelihood that rock blocks would be removed by the flows. The approach used the critical key block concept as described by Goodman and Hatzor (1993). The relative block failure likelihood is found for each possible three joint combination in the rock mass. This is the product of the following three parameters: (1) the relative probability of occurrence of each joint combination (density parameter), (2) the removability or kinematical freedom based on block theory (shape factor), and (3) the stability of the blocks based on limit equilibrium analysis (instability parameter). Based on this analysis, the blocks most likely to be removed were determined. A search for the blocks on the geologic photos revealed that indeed these were present on both abutments. An analysis of the blocks based strictly on their own weight and hydrostatic water forces in the joint planes showed the blocks are not stable. Under the additional hydrodynamic loading resulting from the overtopping flows, it was judged that hydraulic mining of the abutments would occur. Overtopping protection in the form of a thin concrete paving on the abutments, secured with anchor bars and rock bolts, and containing water stops in the contraction joints, was proposed to keep flows from adversely impacting the abutments.

Figure 5. Photo of dip slope at spillway intake.

7 LANDSLIDES

Large reservoir landslides of the type related to the Vaiont Dam disaster are of course a concern for any dam safety evaluation. Aerial photos were reviewed and suspicious areas were examined on the ground. Although one area about 250 m upstream of the dam was identified, field reconnaissance indicated that it was formed by erosional features rather than by ground movement. Failure of a large block of rock above the black seam fault on the right abutment was deemed probable. Although this block does not lie within the foundation of the dam, and its failure would not compromise the dam, it could have an adverse impact on the powerhouse. Treatment by excavation or rock reinforcement was recommended.

8 GEOTECHNICAL INSTRUMENTATION AND MONITORING

It is beyond the scope of this paper to discuss the instrumentation and monitoring aspects of Seminoe Dam in detail. Monitoring is necessary to evaluate the long term performance of the parameters critical to the dam's performance. This would include seepage rates, foundation pressures, and movements. At the same time, with increasing demands for the available funding, the program must be as cost effective as possible. A detailed evaluation of the monitoring program and schedule, including necessary instrumentation and visual monitoring is required for dam safety studies. This should include parameters for assessing whether the performance is within expected limits.

9 CONCLUSIONS

Dam safety evaluations have been completed for Seminoe Dam. Geotechnical studies indicated that the foundation and abutments were stable under static and dynamic loading conditions. However, overtopping flows during large flood events could result in severe erosion through plucking of identified abutment blocks, possibly leading to undermining of the dam abutments. Therefore, overtopping protection is warranted.

Large floods and the corresponding high tailwater elevations will result in spillway tunnel pressures in excess of design values. This condition will not result in distress to the tunnel. However, blockage of the tunnel due to rock instability could increase pressures to the point of tunnel instability. The probable mechanism for such a blockage would be sliding of large blocks on a dip slope at the spillway intake. The studies indicate that this slope should be stabilized.

Treatment of the larger foundation faults was sufficient to facilitate load transfer across the zones and prevent weak material from piping. No large reservoir landslides, capable of generating waves that would endanger the dam were identified, although a potentially unstable mass requiring treatment was identified above the powerhouse on the right abutment of the dam. Long term monitoring of the dam must be considered during dam safety studies. Future work will concentrate on corrective action designs to mitigate the identified deficiencies.

REFERENCES

Bieniawski, Z.T. 1989. *Engineering rock mass classifications*. New York: Wiley

Brekke, T.L. & B.D. Ripley 1989. Geotechnical engineering challenges in the design of pressure shafts and tunnels. *Symposium Honoring Ralph Peck*: 521-535 Englewood Cliffs: Prentice Hall.

Bureau of Reclamation 1977. *Design of arch dams*. Denver: U.S.G.P.O.

Goodman, R.E. & Y. Hatzor 1993. Determination of the "design block" for tunnel supports in highly jointed rock. *Comprehensive Rock Engineering*. 2: 263-292. Oxford: Pergamon.

Goodman, R.E. & G.H. Shi 1983. *Block theory*. Berkeley: University of California.

International Commission on Large Dams (ICOLD) 1974. *Lessons from dam incidents*, complete edition. Paris: ICOLD.

Snow, D.T. 1970. The frequency of apertures of fractures in rock. *Int. J. Rock Mech. & Min Sci*. 7:23-40.

Scott, G.A. 1982. Piping potential of weak zones under concrete dams. *J. Geotech. Engr. Div., Proc. ASCE.* 108 GT3: 488-493.

Rock Mechanics, Nelson & Laubach (eds) © 1994 Balkema, Rotterdam, ISBN 90 5410 380 8

Stability of a concrete gravity structure

Carlos A.Jaramillo
Harza Engineering Co., Chicago, Ill., USA

ABSTRACT: A large block slide precipitated by blasting occurred during the foundation excavation for the tailrace channel of the Main Dam Spillway at the Rocky Mountain Project. The block slid along a previously unidentified layer of laminated calcareous sandstone and shale which continued upstream under the already constructed spillway. This paper reports on the studies performed for characterization of the sandstone/shale layer, and the stability studies performed to evaluate the safety of the spillway.

1 PROJECT SETTING

The Rocky Mountain Project is a 850 Mw pumped storage development currently under construction in NW Georgia. The Lower Reservoir is formed by the waters of Heath Creek contained by the Main Dam, a 37 m high structure (Figs. 1 & 2) formed by an embankment section and a concrete gravity section, founded on a sequence of shale, sandstone and limestone.

The foundation rock of the gravity section, striking nearly parallel to the river (NE), and dipping approximately 16° SE, can be divided in an upper unit composed of limestone with calcareous shale, calcareous limestone, and laminated calcareous sandstone and shale, and a lower unit composed of sandy limestone and calcareous sandstone. The laminated calcareous sandstone and shale layer contains evidence of shear parallel to bedding, including polished and slickensided bedding planes. The slickensided bedding planes indicate updip movement of the overlying rock mass with respect to the underlying rock. This direction of relative displacement is consistent with the general deformation that produced the Rocky Mountain syncline.

Three main sets of moderately to widely spaced joints are present in the foundation area: a northeast set (N55°E to N65°E) with dip of 80° north; a near vertical northwest set (N50°W to N75°W); and an easterly set (N85°E) dipping 80° N. Three persistent, weathered joints of the northeasterly set cross the Main Dam spillway foundation nearly perpendicular to the dam centerline (Fig. 3). All three joints are undulating and consist of a zone of weathered and fractured rock 5 cm to 20 cm wide.

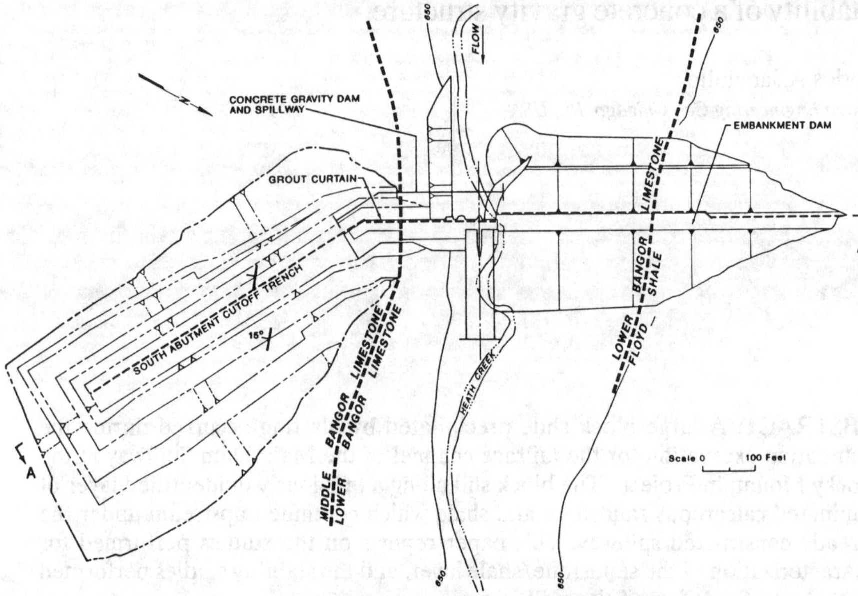

FIGURE 1. MAIN DAM – PLAN

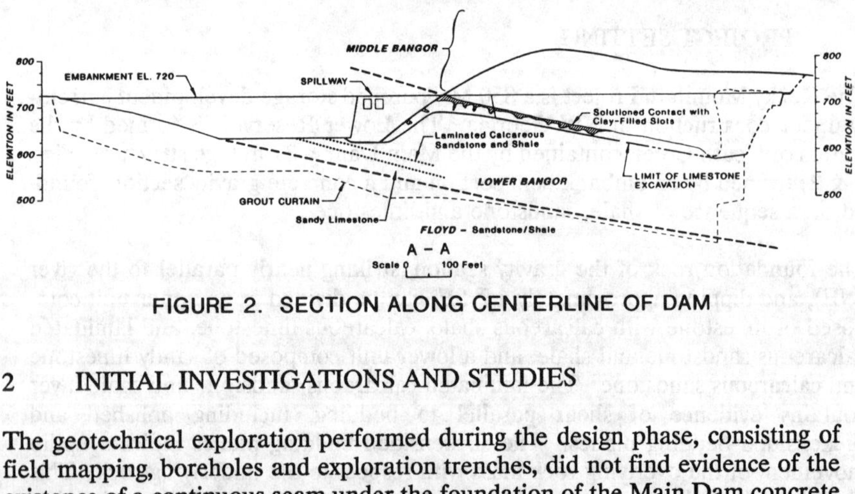

FIGURE 2. SECTION ALONG CENTERLINE OF DAM

2 INITIAL INVESTIGATIONS AND STUDIES

The geotechnical exploration performed during the design phase, consisting of field mapping, boreholes and exploration trenches, did not find evidence of the existence of a continuous seam under the foundation of the Main Dam concrete gravity structures. The dimensions of the structures were selected based on findings of this exploration and on the initial pre-construction geological assessment, and laboratory testing of samples recovered during the exploration. Construction proceeded following plans developed accordingly. The existence of upstream-downstream fractures (N55°E to N65°E) was not indicated by the field exploration nor by observations made during construction of the South Abutment cutoff trench, which was excavated and backfilled before starting construction of the Main Dam. These features were first exposed during excavation for the spillway foundation.

The stability of the spillway concrete gravity section was evaluated following the classic two dimensional limit equilibrium approach, assuming that the most critical failure surface was a bedding plane at the base of the structure. The procedure was considered satisfactory and conservative because the postulated failure plane was horizontal, while bedding dipped gently towards the reservoir, a favorable direction for stability purposes.

The shear strength parameters used for the initial stability analysis ($\phi=36°$ and c=0.35 MPa) were based on an evaluation of the peak strength, the ultimate strength, and the residual strength (Nicholson 1983, 1987) obtained from direct shear tests of samples recovered during the subsurface exploration, and the Hoek-Brown failure criterion (Hoek 1990, Hoek & Brown 1988). The shear strength evaluation considered the possibility of slickensided planes along the assumed failure surface. Uplift was assumed to vary from reservoir head upstream, to tailwater plus 50% of the difference between headwater and tailwater, at the drain line, to tailwater at the toe.

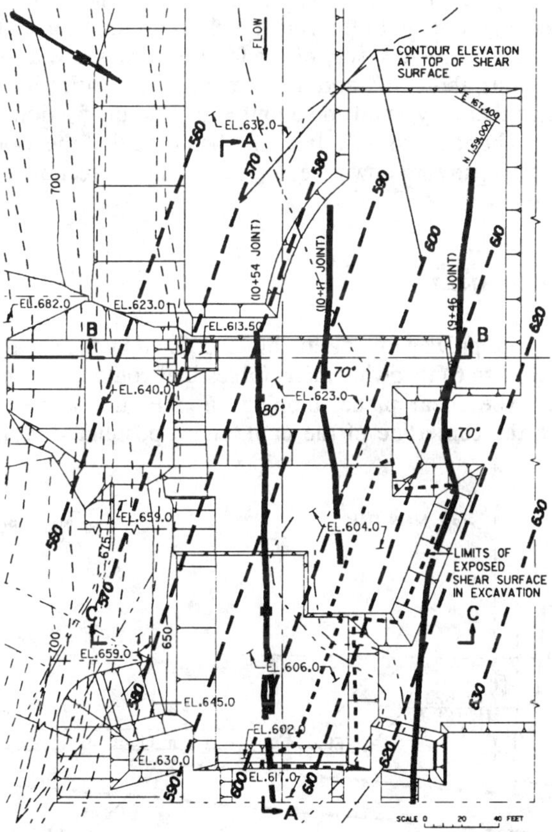

FIGURE 3. PROMINENT NORTHEAST JOINTS MAIN DAM SPILLWAY EXCAVATION

3 CHARACTERIZATION OF SLIDING SURFACE

Detailed mapping of the block slide area after its removal identified a thin (1 to 8 mm) gouge seam near the contact of the laminated calcareous shale and sandstone unit and the overlying calcareous mudstone. The gouge seam consisted of dark gray clay and silt with flat angular rock chips. In general the seam was a planar feature, but locally it branched across bedding.

Previous residual direct shear testing of slickensided surfaces of samples recovered from the foundation indicated a friction angle of 22°. Liquid and plastic limits testing of the seam material showed LL=29, PI=11 pointing to a residual friction angle of 29° (Mesri & Cepeda-Díaz 1986).

The failure of the rock block in the left side of the stilling basin excavation was used to back calculate the shear strength of the shear surface. Based on information from the blasting (loadings, loading pattern, delays, and recorded particle velocities at the constructed structures) that forced the displacement of the block, and using a pseudo-static approach, it was found that the excavation blast provided enough energy to the rock block to account for additional 3° to 6° on the friction angle. Field survey of the failure plane showed that the dip of the shear surface was about 16°, and a survey of its asperities in the down-dip direction indicated that they could add about 3° to 5° to the friction angle (ISRM 1978), reaching to about 22° to 27°. It should be noted that the asperities of the shear plane in the upstream-downstream (across-dip) direction contribute an even larger value.

4 STABILITY ANALYSIS

The geometry of the spillway foundation block defined by the shear surface is 3-dimensional (Figs. 4 to 6) involving many forces that could not be considered in a 2-D analysis. The shear plane, dipping 16° NE, the subvertical features perpendicular to the centerline of the dam, and a postulated vertical fracture

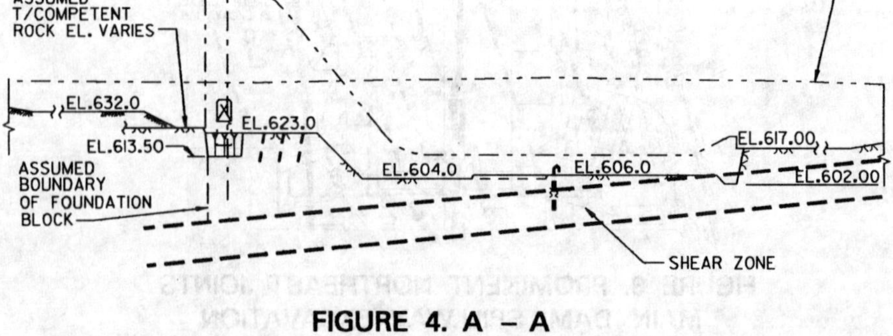

FIGURE 4. A – A

assumed to be located at the heel of the dam and parallel to the its centerline were assumed to be the boundaries of the foundation block. The spillway - stilling basin - foundation block was considered monolithic for this analysis, a departure from established practice, but justified by the geometry of the problem, as both structures are anchored to the foundation rock, and displacement (i.e. failure) of the spillway block could not occur without simultaneous displacement of the stilling basin block.

Only sliding on the upstream-downstream direction was considered, and although there was a component of the resultant acting in the downdip direction, normal to the right boundary of the foundation block, it was assumed that there was no contribution of friction on this right plane, and only a nominal cohesion of 0.35 MPa. No resistance from the stilling basin closure walls, the right concrete gravity dam section, or the backfill was included.

Uplift on the failure surface was assumed to vary linearly from reservoir level upstream , to tailwater plus 50% of the difference between reservoir level and tailwater level at the intersection between the plane formed by the drainholes and the shear surface, to tailwater at the upstream end of the stilling basin. The same variation of uplift was assumed for the vertical bounding plane at the right of the structure.

5 RESULTS

The stability analysis was performed following a parametric approach. The friction angle of the shear zone was assumed to vary from a minimum of 16°, the dip of the failure surface of the failed rock block, to a maximum of 30°, the peak shear strength of slickensided samples tested in the laboratory. The calculated factors of safety for the normal loading condition varied between 1.7 and 3.4, if

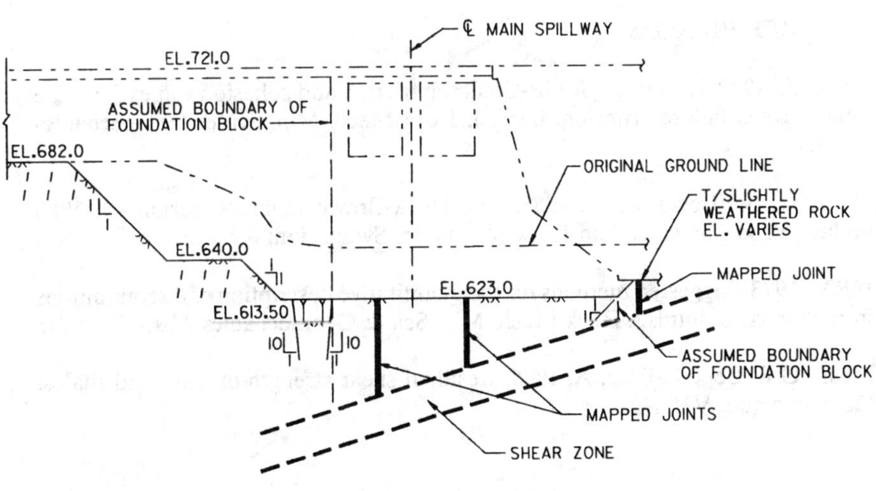

FIGURE 5. B – B (℄)

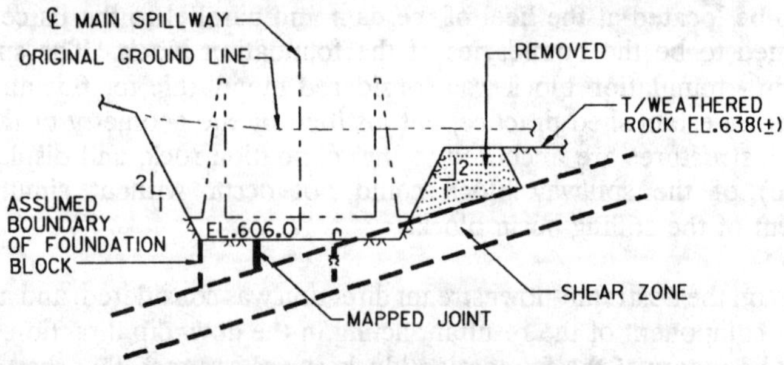

FIGURE 6. C – C (152')

the lateral resistance due to the right boundary plane was not considered, or between 3.2 and 4.8, otherwise. The spillway - stilling basin - foundation block was found to be acceptably stable (FS>1.5) under the assumptions made.

Piezometers were installed in the shear zone and above it. Reservoir filling started September 1992 (First Stage), and continued in November 1993 (Second Stage). It is expected that the data will confirm the uplift assumptions, arguably the most important factor for the stability of the structure.

The three dimensional stability analysis performed allowed a more realistic representation of the geometry of the structure being analyzed, the forces acting on it, and, given the geological characteristics of a site foundation, provide a more realistic assessment of the sliding stability of a gravity structure. It was considered the right approach for the structure discussed above, but its general use depends on a careful evaluation of the geometry of the structure being studied.

6 REFERENCES

Hoek, E., 1990, Estimating Mohr-Coulomb friction and cohesion values from the Hoek-Brown failure criterion, Intnl. J. Rock Mech. Min. Sci. & Geomechanics Abstr.

Hoek, E., & Brown, E. T., 1988, The Hoek-Brown failure criterion - a 1988 update, Proc. 15th Canadian Rock Mechanics Symposium

ISRM, 1978, Suggested methods for the quantitative description of discontinuities in rock masses, Intnl. J. Rock Mech. Min. Sci. & Geomechanics Abstr., V15 N6

Mesri, G & Cepeda-Díaz, A, 1986, Residual shear strength of clays and shales, Géotechnique, V36 N2

Nicholson, G. A., 1983, Design of gravity dams on rock foundations: Sliding stability assessment by limit equilibrium and selection of shear strength parameters, Technical Report GL-83-13, Waterways Experiment Station.

Nicholson, G. A., 1987, Methodology for selecting shear strength parameters of rock, REMR Technical Note GT-RE-1.2, Waterways Experiment Station.

Rock Mechanics, Nelson & Laubach (eds) © 1994 Balkema, Rotterdam, ISBN 90 5410 380 8

Seepage investigations and 3-D finite element analysis

Wendell O. Miller & John B. Palmerton
Geotechnical Laboratory, US Army Engineer Waterways Experiment Station, Vicksburg, Miss., USA

ABSTRACT: A series of pressure injection tests to determine the rock permeability were conducted prior to the construction of Cerrillos Dam. Upon the first filling of the reservoir, foundation seepage exceeded the design estimate by a significant amount. An extensive field investigation to better identify the causes of the excessive seepage was initiated and a 3-D Finite Element seepage analysis was conducted. The ability to include cross-valley components of the water bearing rock units into the 3-D analysis provided a more accurate estimate of the seepage rates.

1 PROJECT DESCRIPTION

Cerrillos Dam is a 323 foot high earth and rockfill dam located northeast of Ponce, Puerto Rico. The dam was designed (Ref. 1) and constructed by the U. S. Army Corps of Engineers, Jacksonville District. A digitized graphical representation of the dam and the vicinity is shown in Figure 1. The rock at the damsite is divided into geologic units that include limestones, sandstones, siltstones, tuffs, intrusive andesites, and some claystones. These geologic units are shown in Figure 2. To a varying degree essentially all the rock is tuffaceous and most of the rock contains at least some form of calcium carbonate. The principal joint systems strike northeast and northwest, with both systems generally dipping between 60 and 90 degrees. The bedding planes of the rock units cross the dam centerline at about a 45 degree angle (to the northeast) and dip downstream at about 45 degrees. The foundation along the dam axis consists primarily of steeply dipping limestones, siltstones, and tuffs that are exposed in the reservoir wall and pass beneath the dam axis, being exposed again in the valley, downstream of the dam. The Ridge Forming Limestone unit, which is considered to be the primary source of seepage beneath the dam, crosses the core trench from the base of the left abutment to halfway up the left abutment.

2 PRESSURE INJECTION TESTS

A series of rock mass permeability tests were conducted in 1981 at the dam site prior to construction. These tests were conducted during excavation of the core trench for the purpose of developing design estimates for foundation seepage. The tests were conducted using a subsurface hydraulic injection test system developed at the U. S. Army Waterways Experiment Station (WES). The system uses a downhole control unit that maintains a preset differential pressure between the packer(s) and the test section, allows changes from single packer to double packer configurations with control commands from ground surface, and provides redundancy in downhole pressure transducers located above the upper packer, within the test section, and below the lower packer. The system is shown schematically in Figure 3.

Site characterization with respect to permeability concentrated primarily upon the limestone units within the foundation. Tests were conducted in 3 borings located along the axis of the dam. These borings averaged approximately 200 feet deep, and each penetrated the steeply dipping limestones,

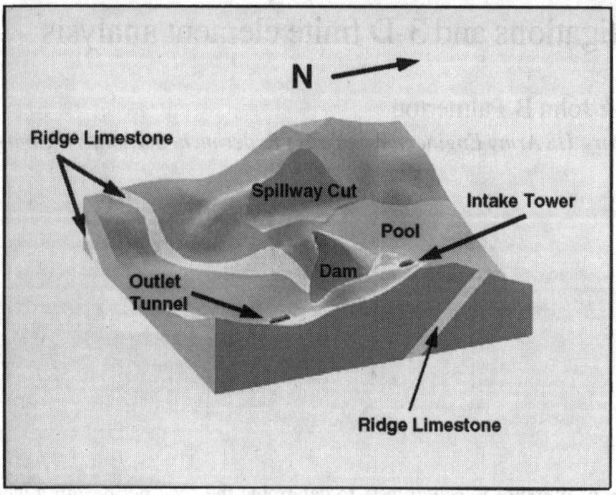

Figure 1. Cerrillos Dam and vicinity

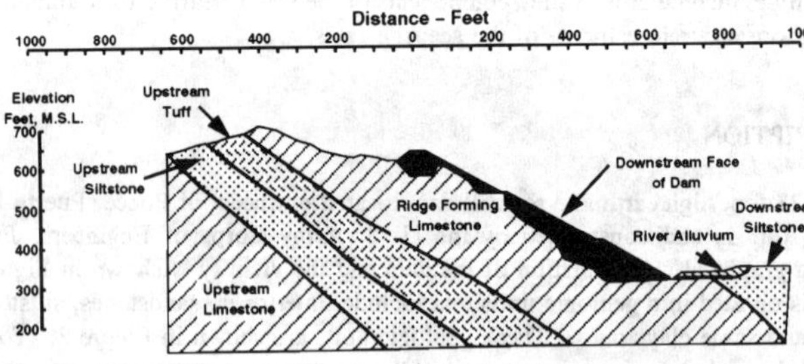

Selected dam cross section near top of left abutment.

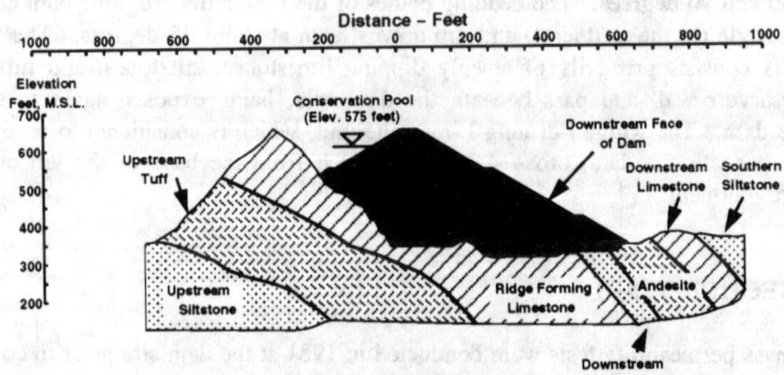

Selected dam cross section near base of left abutment.

Figure 2. Selected dam cross sections

siltstones, and tuffs units. Double packer tests were conducted in each boring, starting at the bottom and incrementally progressing to the top. Equivalent coefficients of permeability were calculated for each test section interval assuming radial flow, isotropy, and homogeneity with respect to permeability using the following equation; $K_e = (Q/2\pi LH_o)(\ln R/r_o)$; where Q is the volume flow rate, L is the test

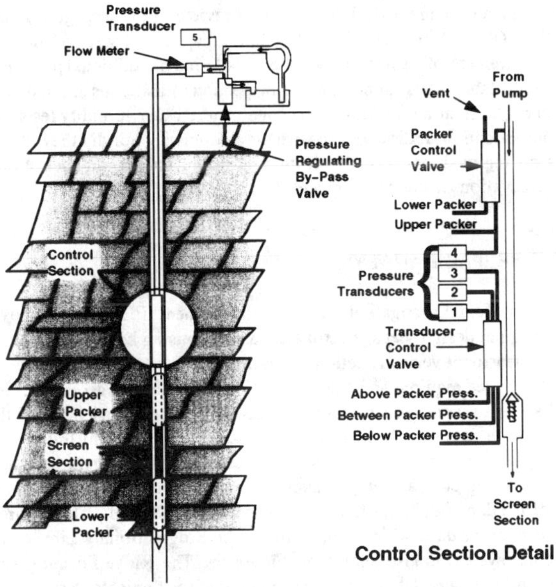

Figure 3. Downhole control unit schematic

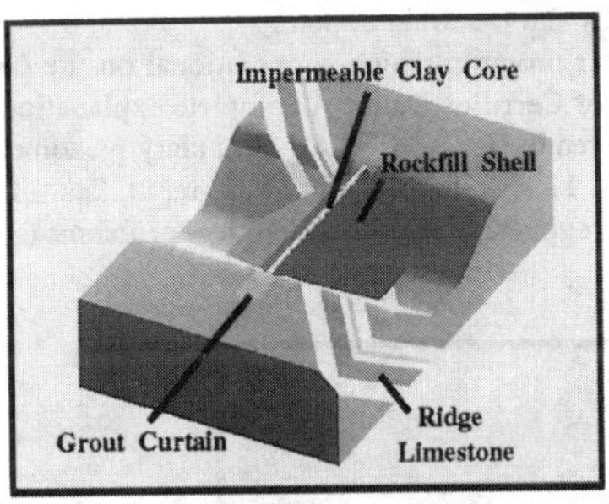

Figure 4. Simplified 3-D model dam, grout curtain and Ridge Limestone

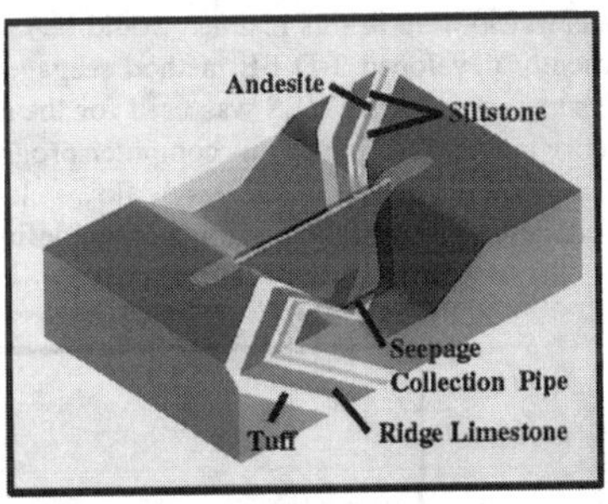

Figure 5. View showing geology (Rockfill shells shown as being transparent)

section length, H_o is the excess pressure (over hydrostatic) at the center of the test section, R is the radius of influence (assumed to be equal to L/2, and r_o is the radius of the boring.

3 SEEPAGE FIELD INVESTIGATIONS

After construction, foundation seepage in excess of design estimates occurred following the early stages of the first filling of the reservoir (fall of 1992). Seepage through the dam is controlled by a clay core located within the center of the dam and by a two line grout curtain that extends 200 feet into the foundation rock. A drainage collection system located within the downstream toe of the rockfill shell collects and monitors the seepage discharge. Since the measured seepage discharges were approximately six times greater than the design estimate, a review of the seepage regime was undertaken. These investigations (Ref. 2) included the installation of additional seepage monitoring devices to provide continuous measurement of seepage flows, the installation of additional piezometers to measure water pressures within the dam and its foundation, an inspection of materials and moisture content from the impervious clay core, thermal monitoring of seepage water, water turbidity tests and dye injection tests. None of these investigations indicated a construction deficiency nor did they identify any isolated flow path. There is no regional evidence of solutioning in the foundation rock units and none was observed during the excavation of the spillway and dam foundation.

4 FINITE ELEMENT SEEPAGE ANALYSIS

Concurrent with the recent field investigations, a 3-D Finite Element (FE) seepage analysis (Ref. 3) was also conducted. The original design seepage estimate was made using hand drawn flow net procedures for an idealized two-dimensional vertical section taken through the dam.

The 3-D FE method analyses required some simplifications of the actual geometry. In particular, the valley floor and walls were assumed to be uniform and straight in the vicinity of the dam and the cross-valley curvature of the dam was eliminated. These assumptions are illustrated on Figure 4 which depicts the relationship of the dam, grout curtain and the Ridge Forming Limestone. Figure 5, in which the rockfill shell is depicted as being transparent, shows the exposure of the Ridge Forming Limestone upstream of the dam and beneath the downstream rockfill shell. This figure also shows the location of the tuff and siltstone units which are parallel to the Ridge Forming Limestone. The location of the seepage collection system is also shown on Figure 5. The Ridge Forming Limestone unit is positioned such that water may percolate from the pool into that rock unit along the left valley wall (just upstream of the dam and within the dam's permeable upstream rockfill shell) and be carried in a relatively unimpeded fashion (compared to the surrounding rock) under the grout curtain (which also intercepts a portion of the Ridge Forming Limestone) and then emerge from beneath the downstream rockfill shell and from the lower elevations of the downstream face of the left abutment. Seepage water which is transported in this manner would be collected by the toe drain system.

A recently developed 3-D FE method seepage computer program which is operational on the Cray Y-MP supercomputer at WES was used for the analyses of Cerrillos Dam. A complete explanation of the theory and application of this computer program is given by Tracy (Ref. 4). Boundary piezometric heads and boundary flow quantities or flow velocities can be specified as part of the input data. This FE formulation can simulate confined or unconfined flow regimes. For unconfined flow problems (such

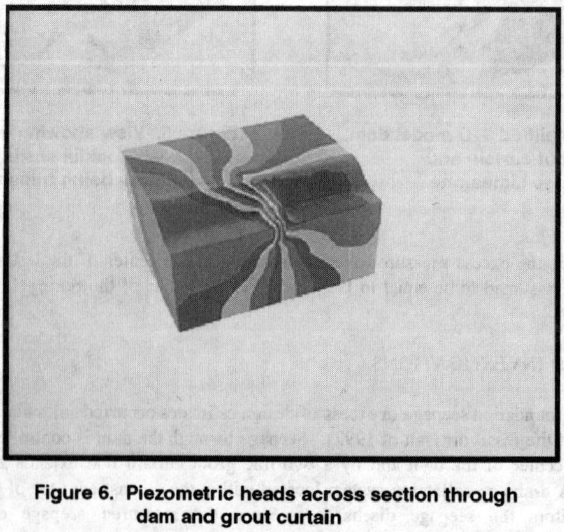

Figure 6. Piezometric heads across section through dam and grout curtain

as Cerrillos Dam) the location of the phreatic surface is established by an iterative process. The output of the program consists of the total seepage discharge quantity, the piezometric head at each node point, and the flow velocities. Three dimensional contouring programs may be used to produce 3-D plots of piezometric heads as shown in Figure 6.

The Water pressure injection tests conducted at the site indicated permeability values ranging from 5.1×10^{-3} cm/sec in the Ridge Forming Limestone to 2.5×10^{-5} cm/sec at greater depths. A decrease in permeability with depth is observed at most sites. This decrease is due to overburden induced stresses causing cracks and rock joints to be more tightly closed at depth. The 2-D case shown on Figure 7 yielded a design seepage foundation estimate of 1.0 cfs with the water level at the elevation of the conservation pool. For this analysis, the typical vertical section as shown assumed Darcy permeability coefficients which decreased with depth. The seepage through the abutments not specifically included, nor was the continuity of the more permeable valley crossing Ridge Limestone unit considered. This 2-D case was simulated with the 3-D FE program by assigning the permeability coefficients as listed in Figure 7 and setting the permeability coefficients beyond the base width of the valley floor to a very low value (1×10^{-12} cm/sec). The seepage discharge was then computed by multiplying the 3-D result by the ratio of the length of the hand calculated section (1555 ft) to the width of the valley floor in the 3-D model (428 ft). As desired, a result very near to the seepage discharge of 1 cfs was computed by the 3-D FE program. The simulation of the actual 3-D situation was accomplished by computing various trial computer simulations in which various combinations of k values for the Ridge Forming Limestone, the surrounding rock formations, and the permeability of the grout curtain were assumed. After an examination and discussion of these simulations with involved engineers and geologists, the set of permeability coefficients listed in Table 1 was selected. The three-dimensional delineation of the zones is shown on Figure 8. The combination of permeability coefficients listed in Table 1 yielded an excellent estimate for the observed seepage discharges at the project for a pool depth of about 200 ft (which was the depth at the time of the study). The predicted seepage discharge versus pool elevation is shown on Figure 9. As the pool elevation continues its rise, a seepage discharge of approximately 6.5 cfs (2920 gpm) at the maximum conservation pool is predicted.

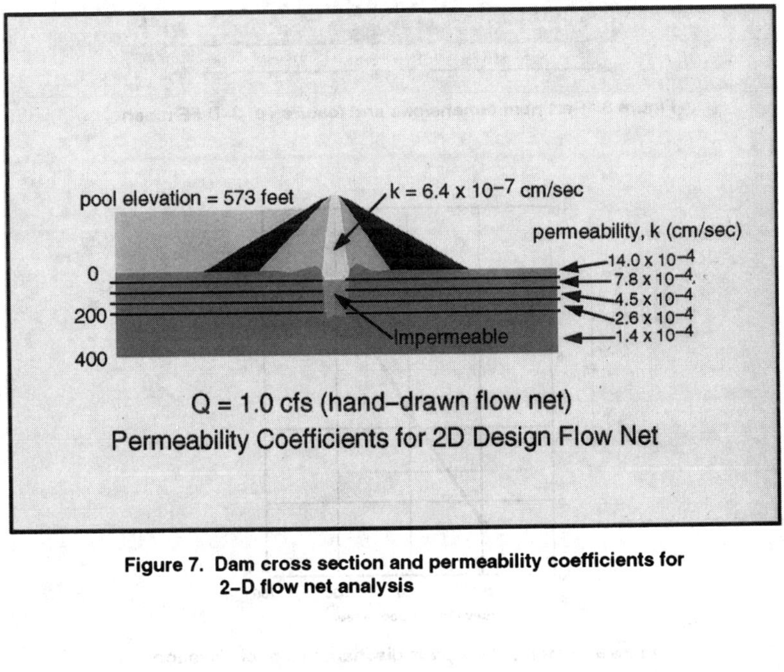

Figure 7. Dam cross section and permeability coefficients for 2–D flow net analysis

Table 1

Grout Curtain (200 foot depth below clay core)	1.0×10^{-6} cm/sec
Impermeable clay core of dam	6.4×10^{-7} cm/sec
Ridge Forming Limestone unit	5.0×10^{-3} cm/sec
All other foundation rock units above el -105 ft*	1.0×10^{-4} cm/sec
All foundation rock units below el -105 ft	Impermeable

* el -105 is 400 ft below the valley floor which is at el 295

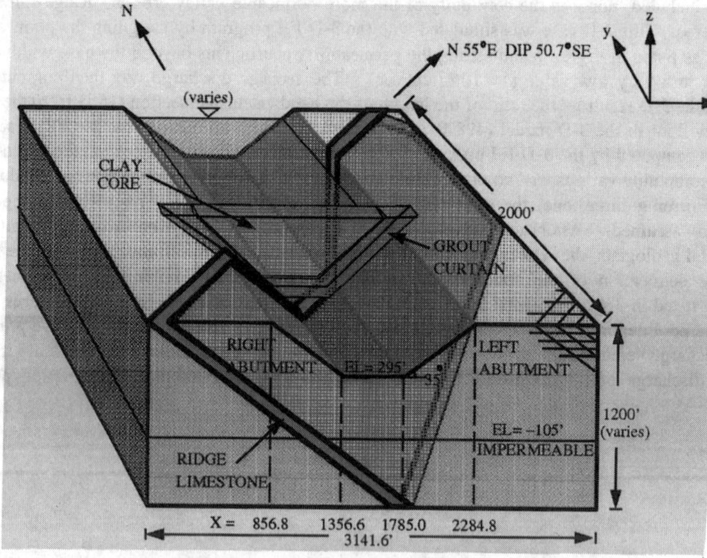

Figure 8. Pertinent dimensions and features of 3-D FE mesh

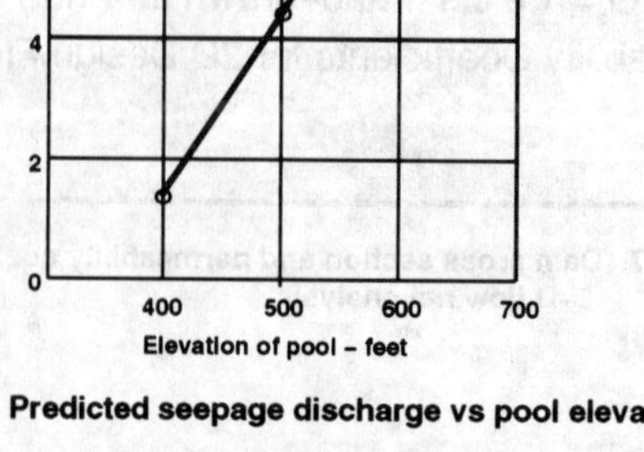

Figure 9. Predicted seepage discharge vs pool elevation

5. CONCLUSIONS

An important usefulness of the 3-D FE analysis has been the demonstration that the observed higher than estimated seepage discharges at Cerrillos Dam have been replicated in a rational manner that did not require any major modification of the understanding of the rock permeability properties which existed during the design phase. Further, it is doubtful that 2-D (whether FE of flow net) analysis could ever have rationally predicted the higher seepage discharges. The complicated geologic conditions necessitate a 3-D analysis. Had a 3-D FE seepage analysis of Cerrillos Dam been attempted during the design phases of the project, it is questionable if the now measured seepage discharges would have been exactly predicted. However, since the Ridge Forming Limestone had been identified as a very (relatively) permeable geologic unit, any reasonable 3-D analysis in which different permeability combinations were varied would certainly have served to alert the designers to a potential problem. The foundation permeability model presented in Table 1 shows the Ridge Forming Limestone to be 50 times more permeable that the other foundation rock. Results of the FE modeling bear out the validity of the field permeability tests and the permeability coefficient values determined during the design phase.

Because of the availability of 3-D FE (or similar) computer programs, it is recommended that analyses of these types be conducted for all projects which present the possibility of a three-dimensional effect on design computations. Even for cases involving seepage that may adequately be formulated in two-dimensional form, the use of numerical models (e.g., 2-D FE models) should receive serious consideration for use. These numerical models generally do not place restrictions on the manner in which different permeability zones are conceptually assimilated. Conceptual restrictions generally do apply for manually constructing flow nets; especially if transformed sections must be considered. That is to say that flow net construction techniques may require unnecessary oversimplification of the conceptual permeability model. For situations where multiple zones and anisotropic permeability must be included, a numerical model is, for all practical purposes, mandated.

ACKNOWLEDGEMENTS

The investigations described herein for Cerrillos Dam were conducted for the U. S. Army Engineer District, Jacksonville by personnel of that District, the Huntington and Wilmington Districts and the U. S. Army Engineer Waterways Experiment Station. The Chief of Engineers granted permission to publish this information. However, this study is ongoing and all opinions, findings, and conclusions expressed are those of the authors and do not necessarily reflect the official views of the U.S. Army Corps of Engineers.

REFERENCES

1. _, 1983, *Feature Design Memorandum No. 17, Cerrillos Dam and Spillway, Volume 1,* U.S. Army Engineer District, Jacksonville, Jacksonville, FL.

2. __, 1993, *Cerrillos Dam - Underseepage and Outlet Works Leakage Investigation,* U.S. Army Engineer District, Jacksonville, Jacksonville, FL.

3. Palmerton, John B., 1993, *Cerrillos Dam - 3-D Seepage Analysis,* U.S. Army Engineer Waterways Experiment Station, Miscellaneous Paper GL-93-25, Vicksburg, MS.

4. Tracy, Fred T., 1991, *Application of Finite Element, Grid Generation, and Scientific Visualization Techniques to 2-D and 3-D Seepage and Groundwater Modeling,* U.S. Army Engineer Waterways Experiment Station, Technical Report ITL-891-3, Vicksburg, MS.

Rock Mechanics, Nelson & Laubach (eds) © 1994 Balkema, Rotterdam, ISBN 90 5410 380 8

Long-term stability of the Santa Rosa Dam left abutment

Francisco Gonzalez-Valencia & Carlos Montoya-Beltran
Federal Commission of Electricity, Mexico City, Mexico

ABSTRACT: The Santa Rosa Dam left bank stability problems date back to the construction stage in 1963 and continued until new geological exploration, geotechnical analyses and stabilization works where carried out in 1982. The results obtained from the instrumentation shows that from 1984 to 1993 there was a satisfactory performance.

1 GEOLOGY

The site is located within the physiographic province of the East-West volcanic alignment across Mexico called "Eje Neovolcanico". Tertiary and Quaternary volcanoes are typical and their lava flows of basalt, andesite and rhyolite interbedded with pyroclastics cover the region. Differential erosion of the Santiago River formed canyons and deep valleys. The river's course follows a main regional fault which crosses the damsite at the right abutment and at this point the river was diverted by ancient slides, and eroded the igneous formations.

The dam is founded upon rhyolitic flows split by joints which dip towards the Northwest with inclinations varying from 20 to 45 degrees and covering a layer of dense tuffs (fig 1). The original canyon walls were almost vertical and about 100 m deep whereas the left abutment slope was considerably flattened due to weathering and large rock falls along the joints of the rhyolitic mass. Four sets of faults of tectonic origin were detected, having strikes and dips of N38°W-70°SW, N45°E-80°NW, NS-90° and EW-90°, respectively. Among the fault systems found in the left bank two discontinuities are conspicuous: Fault A with a N20°W strike and 80°SW dip, filled with cemented brecciated material and having a width of about 10 cm; and fault B of the N38°W system dipping 70° towards the SW wich crosses the outlet of the tailrace tunnel and the left abutment at the dam crest zone (fig 2).

The abundant apparent discontinuities associated to the flow lines of the rhyolitic mass which at the left bank dip towards the river, serve as a boundary of rock blocks. Their movements have been monitored since the first filling of the reservoir (1963).

2. LEFT BANK BEHAVIOUR

In order to secure the surface blocks, at the construction stage 442 tension bolts of 1 1/2" ϕ high strength steel bars having a design load of 42 ton and a length of 28 m each, were placed in five zones of the left bank below the contour at elevation 750. The anchors that did not fail were subsequently retensioned to their original design load in 1967, 1970, 1974, 1980 and 1982. The percentage of failed anchors in each zone are shown in figure 3, where it can be noticed that the most active areas have been zones 1,3 and 4.

The spillway structure is located on the right bank and in front of the entrance to the Tunnel One, former access to the powerhouse. The maximum discharge flow is 6700 m^3/s with a drop height of about 80 m. The original riverbed was at elevation 655 and the rock at elevation 641. During the flood seasons of 1966 and 1967 the rate of discharge reached up to 3000 m^3/s, and the impact of the water eroded a pool at the riverbed down to elevation 616. In 1973 another flood with a peak discharge of 5800 m^3/s scoured the bottom of the pool down to elevation 613, and has remained

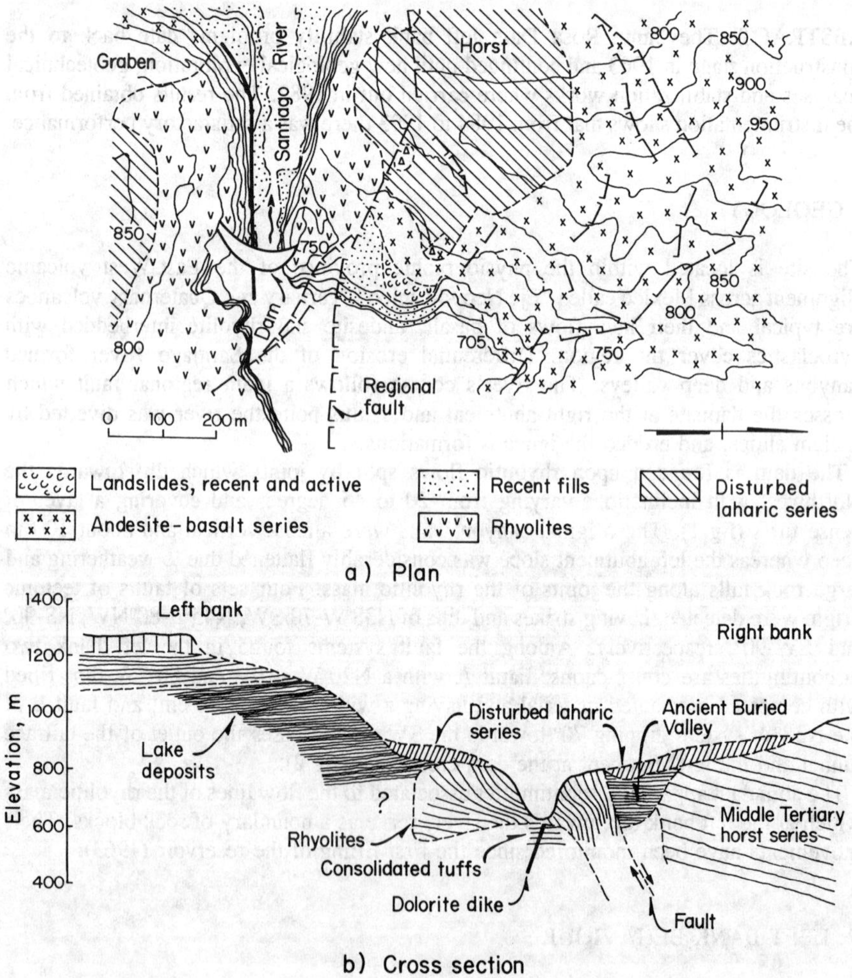

Fig 1 Geology of Santa Rosa damsite

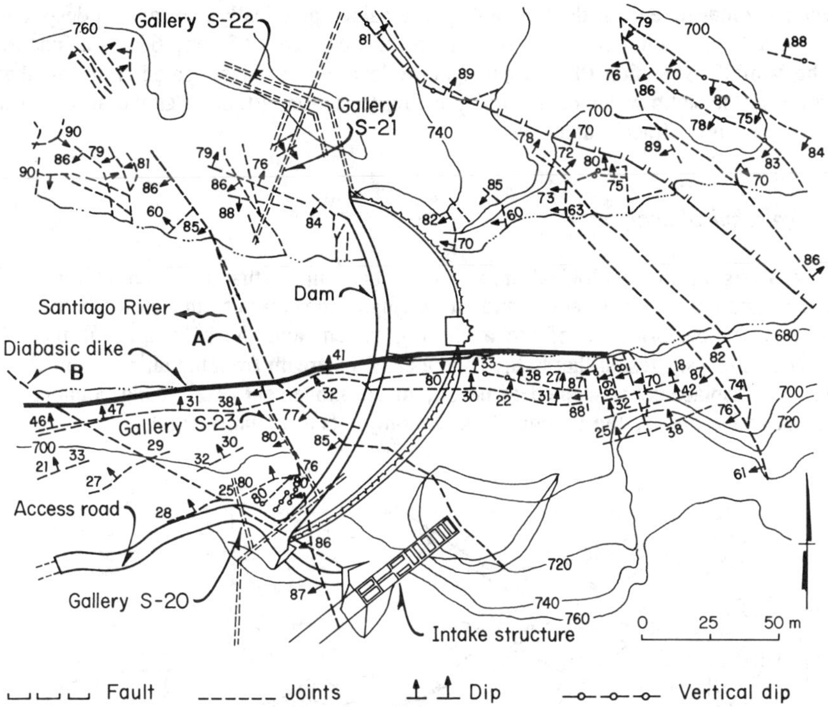

Fig 2. Principal structural weaknesses of the rock

practically unchanged since then. It was necessary to drive an auxiliary access tunnel to the powerhouse, because the access road to Tunnel One was eroded by the spillway discharges, and rock falls of the "peel off" type that has been taking place along a 50 m zone from the Tunnel One entrance to downstream.

2.1 *Geotechnical Instrumentation*

First instrumentation was designed to permit analysis of the overall performance of the dam and verify the behaviour of their foundation.

In 1982 there were installed additional plumb lines, rock extensometers and joint meters in order to improve the knowlegde of the left bank behaviour and mainly find out the extension of the unstable mass. The location of some of the instrumentation is shown in figure 3, and a complete description and the resulting measurements can be found in Dawson et al (1976) and Alberro et al (1985).

The results of the measurements taken from 1963 to 1982 at the left abutment plumb line, indicate a slow but continuous displacement towards the river and downstream (N15°W resultant direction) between galleries S-20 (elev 712) and S-23 (elev 676). The movement reached 11 mm in 1982 but it was observed that the displacement along the X direction slowed down the years when the anchor retensioning works were carried out, but soon afterwards it increased once again. At the plumb line installed in 1982 at the shaft of the powerhouse, no movements could be detected.

The piezometric level al the left bank has remained practically constant, in despite of the fluctuations in the pool level, but varied between elevs 658 and 676 as a function of the rainfall. At station PD, the piezometric level was always located at no less than 5 m above the static water table corresponding to stations PB and PC and at least 7 m higher than the tailrace water level.

2.2 *Geotechnical analyses*

The analysis of the geological information, particulary that obtained in the 1982 detailed geological exploration and surveyings, the former and new instrument readings, the performance of the anchoring system and the behaviour of the left abutment itself, suggested that in the zone located upstream from the tailrace tunnel the measured displacements seemed to be due to the movement of a wedge bounded by faults A and B and by the pseudo-stratification joints at its base (fig 4).

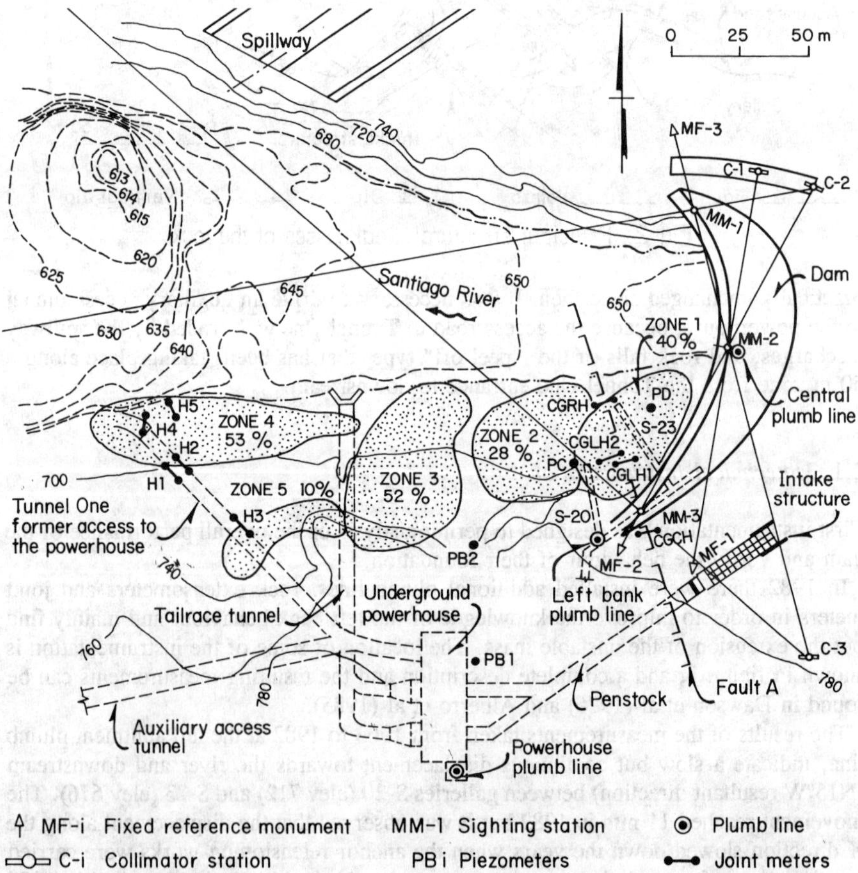

Fig 3 Layout, anchor zones, instrumentation of the left bank and riverbed erosion

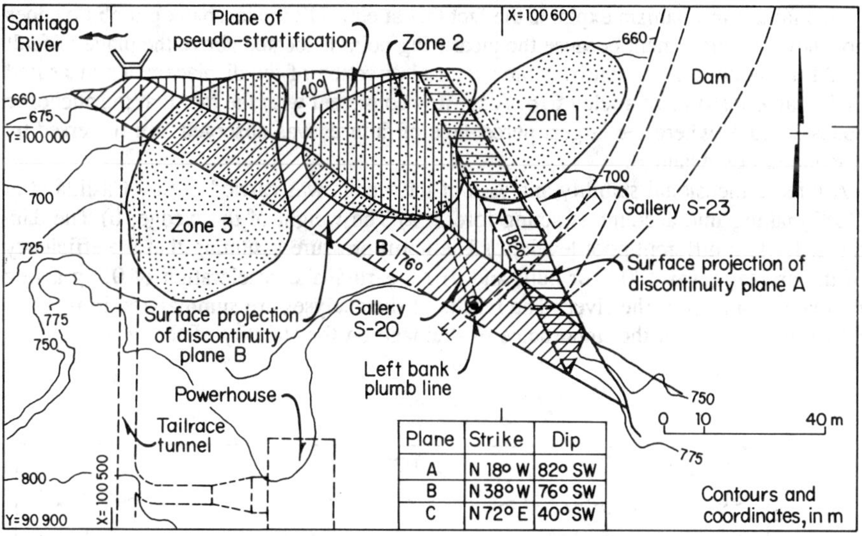

Fig 4 Unstable wedge analysed

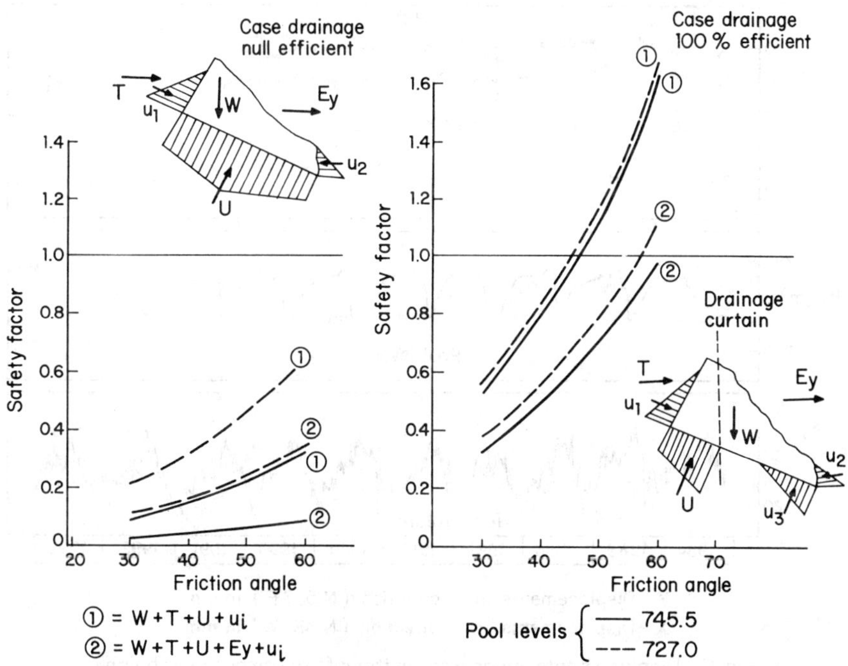

Fig 5. Safety factors of the wedge for different load conditions and drainage curtain efficiency and pool level

This failure mechanism explains the fact that at elev 712 the left bank plumb line does not show any movement because the measuring point is located above the plane of fault B. At the same time, it confirms the vectorial direction of the displacements measured in the same instrument at elev 676. Also, it is consistent with the fact that the anchored zones 1 and 3, where the largest percentage of failures was observed, are traversed by discontinuities A and B.

A three-dimensional stability analyses of the wedge was carried out by Castilla et al (1982) taking into account as acting forces: a) The weight of the wedge; b) The dam thrust for two different pool levels; c) The uplift pressure assuming that the efficiency of the drainage was 100% or null; and d) An earthquake acceleration of 0.2 g acting normal or parallel to the river. The results of the analyses are summarized in figure 5 where it can be seen the importance of drainage on the stability of the wedge.

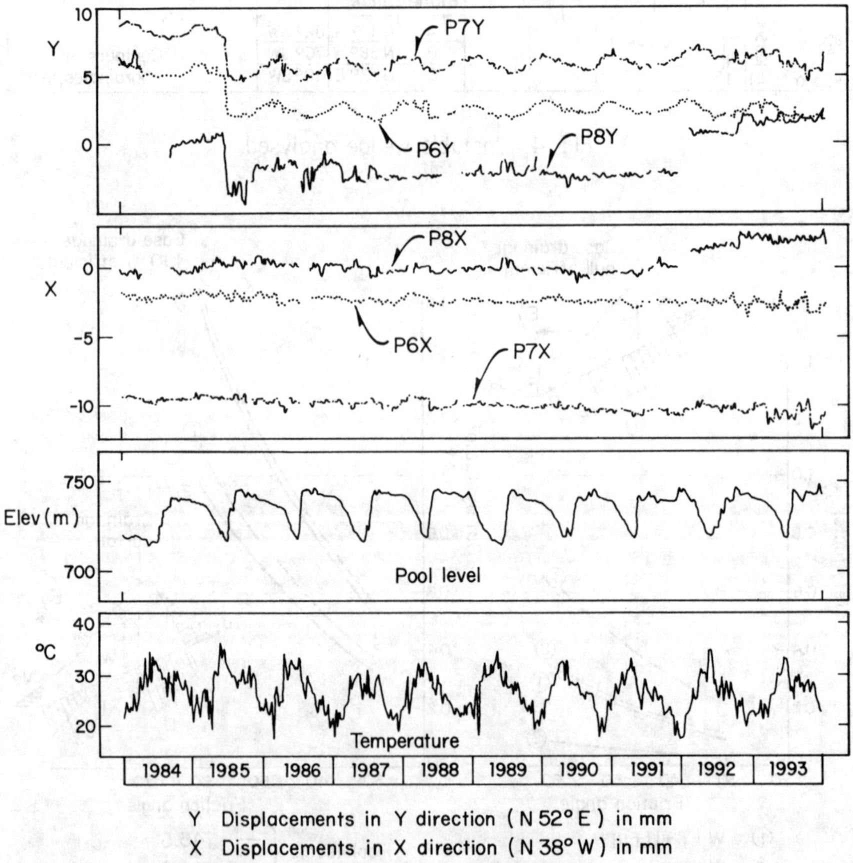

Fig 6. Displacements measured on the left abutment plumb line

3. STABILITY WORKS AND LONG TERM BEHAVIOUR

To improve the stability conditions of this wedge the boreholes of the original drainage curtain were redrilled and supplemented with new ones; both drain holes 20 m long drilled from the cliff in a square pattern of 15 x 15 m, and drain holes 30 m long drilled in a fanned-type pattern from the diversion tunnel. Also, the left abutment was covered with shotcrete to avoid the percolation of rain or spillway spray water. Finally, in 1984 finished the anchorage support complement with 333 new friction anchors 1 1/2" ϕ and 18 m long made of AE 60 steel rebars.

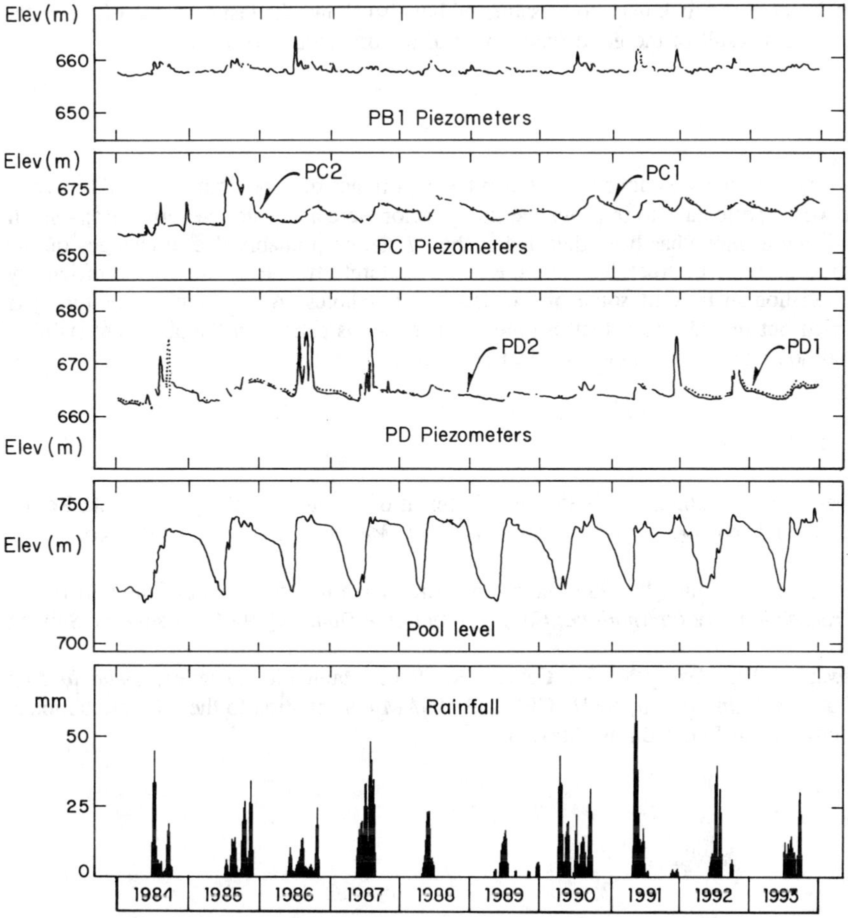

Fig 7. Water piezometers level measured on the left abutment

The left abutment is monitored continuously and the data obtained from the plumb line is presented in figure 6. As it can be seen, the movement of the wedge has been stopped although a very small tendency to move of about 0.125 mm per year can be observed in station P7 in the X direction. From the end of 1992 the change of the measuring instrument and personnel cause noticeable variations on the readings, but the mean remains the same. In the Y direction a calibration work done in 1985 permitted to adjust the readings and it can be observed the seasonal cyclic influence of the dam thrust. Also in 1992 the inverted portion of the plumb line (P8) was affected by external factors that conducted to unreliable readings.

The piezometric level has remained almost constant with some slight fluctuations in the rainy seasons (fig 7). The peaks observed in piezometers PD show the sustained capability of the rock mass to be saturated but also shows the fast decrease of the water levels as a result of the good functioning of the drainage boreholes.

4. CONCLUSION

The stability analyses done have shown the importance of good drainage conditions and the works performed to improve the safety factor were adequate. Nevertheless, a small tendency to move has been detected at the P7 station probably due to changes on the permeability of the rock by the increase on shotcrete fracturing and or the efficiency degradation in time of some of the drainage boreholes. A continuous monitoring is carried out in order to detect in time any anomalous change in the abutment stability condition that leads to new stabilization works.

REFERENCES

Alberro, J. & Colina, P. 1985. Left abutment of Manuel M. Dieguez (Santa Rosa) Dam, Jalisco. *Behaviour of Dams Built in Mexico Vol 2.* CFE. México.

Castilla, J. & Colina, P. 1985. Santa Rosa Arch Dam Reinforcement of the Left Bank. *Transactions XV International Congress on Large Dams*. Q59 R53. Lausanne Suisse.

Dawson A.W., Garay N. and Cabrera R. 1976. Manuel M. Dieguez. *Behaviour of Dams Built in Mexico*, SRH, CFE and UNAM contribution to the XII International Congress on Large Dams, México.

Rock Mechanics, Nelson & Laubach (eds) © 1994 Balkema, Rotterdam, ISBN 90 5410 380 8

Durability issues in the production of rock for erosion control

D.A. Lienhart
US Army Corps of Engineers, Cincinnati, Ohio, USA

ABSTRACT: Durability is the one factor that has proven to be of utmost concern in the cost of rock for erosion control. Three areas of knowledge having the greatest effect on selecting a durable product if properly utilized are (1) intrinsic rock properties, (2) quarrying techniques, and (3) weathering processes. A correlation between rock properties and durability may improve the results of the exploration efforts. Employment of proper quarrying techniques may result in production of a more durable stone. Knowledge of the relationship between weathering processes and both rock properties and quarrying techniques will lead to a more durable stone product.

1 INTRODUCTION

Rock for erosion control may take many shapes and sizes. It may be eight-inch diameter cobbles or rock fragments for gabions. It may be 30 to 500 pound graded riprap for slope protection on a dam. It may be one to three ton cut or blasted stone used as armor stone for shoreline protection or it may be three to 20-ton stone used as breakwater stone.

Each year millions of dollars are spent on rock for erosion control. The one problem that most affects the cost of these rock products is durability. Durability is the ability of rock to withstand, to a satisfactory degree, the effects of production, transportation, placement and service conditions to which it will be subjected, such as wear, chemical action and weathering. Just based on this definition, it is quite evident that the factors affecting durability are numerous. This study focuses on three factors which have the greatest effect on resulting durability of rock for erosion control:

1. Intrinsic rock properties;
2. Production techniques and time of year of production;
3. Weathering processes.

2 INTRINSIC ROCK PROPERTIES

Through measurement of a few index properties, a set of curves may be developed for each rock type such that non-durable rock may be distinguished from durable rock. Although the development of the curves may be time-consuming, the long-term payback is substantial since the need for more expensive and lengthy accelerated weathering tests is negated. This type of information will also provide a good background for the exploration and materials investigation phase of the project.

A set of four curves was developed by Lienhart and others (1993). These curves were based on three intrinsic properties (specific gravity, absorption and adsorption). The data used for development of these curves were taken from test data of 125 limestones and dolomites in the north central U.S. Figures 1 through 4 show the set of curves developed for quick durability determinations.

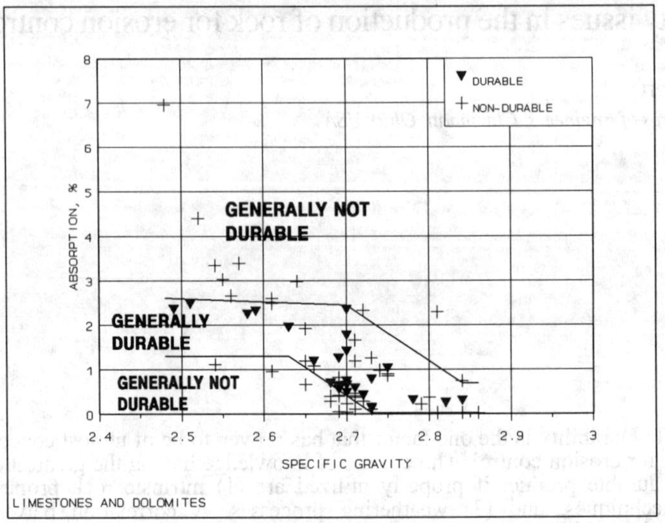

Figure 1 - Durability versus absorption and specific gravity (after Lienhart, Fisher and Robinson, 1993).

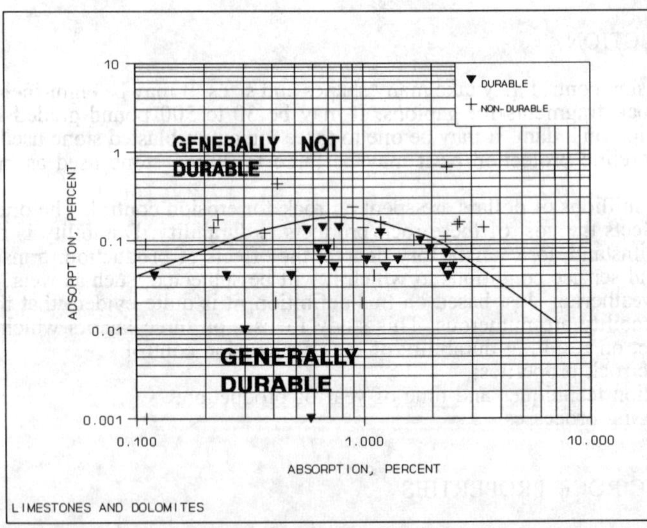

Figure 2 - Durability versus adsorption and absorption (after Lienhart, Fisher and Robinson, 1993).

It should be noted that the figures contain areas labeled as "Generally Not Durable" and "Generally Durable." The lines separating the "Generally Not Durable" rock from the "Generally Durable" rock were arbitrarily drawn to separate most of the data points for non-durable rock from the data points of the rock that performed well in the accelerated weathering tests. The position of these lines of separation defines the ranges of values for index properties that determine which rock materials are non-durable. This

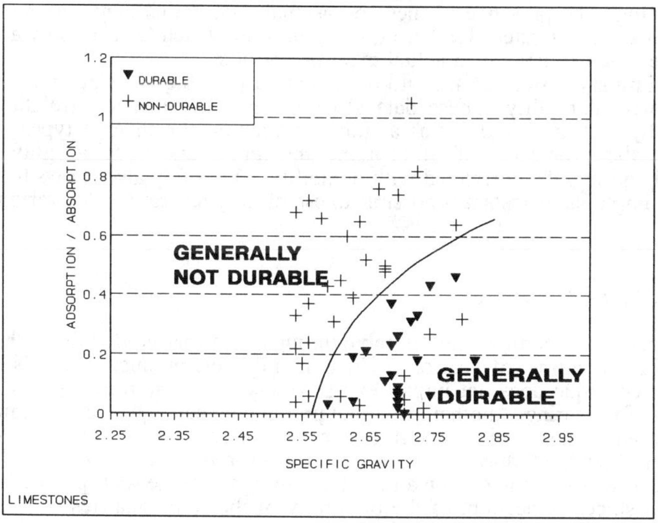

Figure 3 - Durability versus adsorption/absorption ratio and specific gravity for limestones (after Lienhart, Fisher and Robinson, 1993).

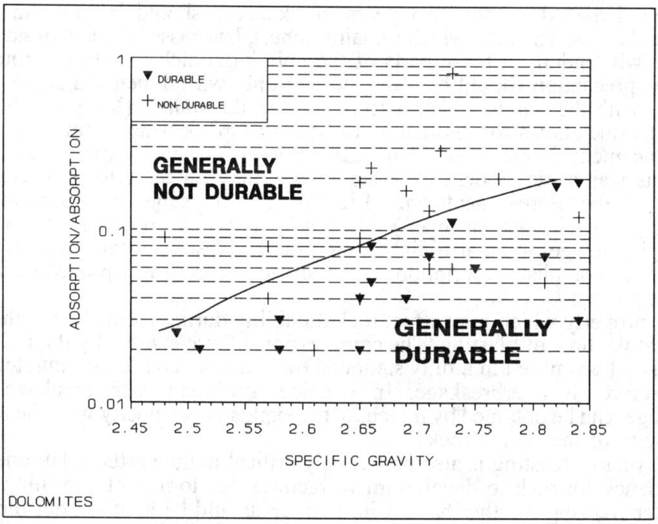

Figure 4 - Durability versus adsorption / absorption ration and specific gravity for dolomites (after Lienhart, Fisher and Robinson, 1993).

is the case for Figures 1 through 4. The areas of these figures labeled as "Generally Not Durable" contain few or no data points for stone that performed with no change in the accelerated weathering tests (durable stone). The areas labeled as "Generally Durable" however, contain occasional data points for stone that suffered some minor changes during the accelerated weathering tests. These could be non-durable rocks but are believed to be generally durable with minor losses due to small spalls which originated during blasting. The presence of these occasional data points may also be due to the harshness of the accelerated weathering test procedure which leads to only a 70 percent success rate when compared to actual exposure results.

The preliminary results of this study indicate a possible correlation of index rock properties with durability. These data also indicate that such a correlation must be performed by rock type and not as a single correlation for all rock types. This study shows that the properties of adsorption, absorption and specific gravity are all interrelated and may be correlated with durability. This study also shows that by using a series of simple index tests it is possible to establish specifications for various qualities of riprap.

3 PRODUCTION TECHNIQUES

A durable rock may be made non-durable through incorrect production techniques. By observing limits on time-of-year of production, a durable product results as long as the production technique does not impart excessive energy to the rock which could result in stress relief fracturing. Another important production technique was recognized by the ancient Roman architect, Vitruvius. It is generally believed that Vitruvius wrote his "Ten Books of Architecture" during the reign of Emperor Augustus but even at this early stage in the history of construction a need for "curing" of stone was specified as follows.

> "But since, on account of the proximity of the stone quarries of Grotta Rossa, Palla, and the others that are nearest to the city, necessity drives us to make use of their products, we must proceed as follows, if we wish our work to be finished without flaws. Let the stone be taken from the quarry two years before building is to begin, and not in the winter but in the summer. Then let it lie exposed in an open place. Such stone as has been damaged by the two years of exposure should be used in the foundations. The rest, which remains unhurt, has passed the test of nature and will endure in those parts of the building which are above ground. This precaution should be observed, not only with dimension stone, but also with the rubble which is to be used in the walls" (Morgan, 1960).

This technique allows for dissipation of "quarry sap" or water from the rock product prior to placement. It has been found that through "curing," a more durable product results. This was made evident by a laboratory study on the effects of curing on the durability of the Berea sandstone. Lienhart (1975) subjected "head-size" pieces (averaging 25 to 30 kg in mass) of freshly quarried, uncured Berea sandstone (Mississippian in age) to an artificial curing test. The results of this study are presented in Figure 5. A complete description of the artificial curing test procedure is given by Lienhart and Stransky (1981).

Another property which may affect rock durability during production is the degree of saturation at the time of blasting. The compressional force exerted by the P and S waves is more easily transmitted in a fully saturated rock than in a partially saturated rock. The result therefore, is overbreakage. In practice, sufficient rock breakage, but little overbreakage, can be obtained by matching the explosive's velocity with the longitudinal wave velocity of the source rock.

Time of year of blasting is also particularly critical in the northern latitudes. Because of the tendency for rock to develop microfractures due to loss of confining pressure if blasted after freezing weather has set in, no rock should be blasted after the first frost. Any rock produced after the first frost should be stockpiled over winter, examined the following spring for defects and microfractures and set aside as waste if any such edfects are found.

4 PHYSICAL WEATHERING PROCESSES

There is generally an incomplete understanding of the relationship of weathering processes to the intrinsic rock properties and to the quarrying techniques. The effects of the weathering processes may be enhanced by certain rock properties which, in turn, may be altered by production efforts and quarrying techniques, resulting in an inferior

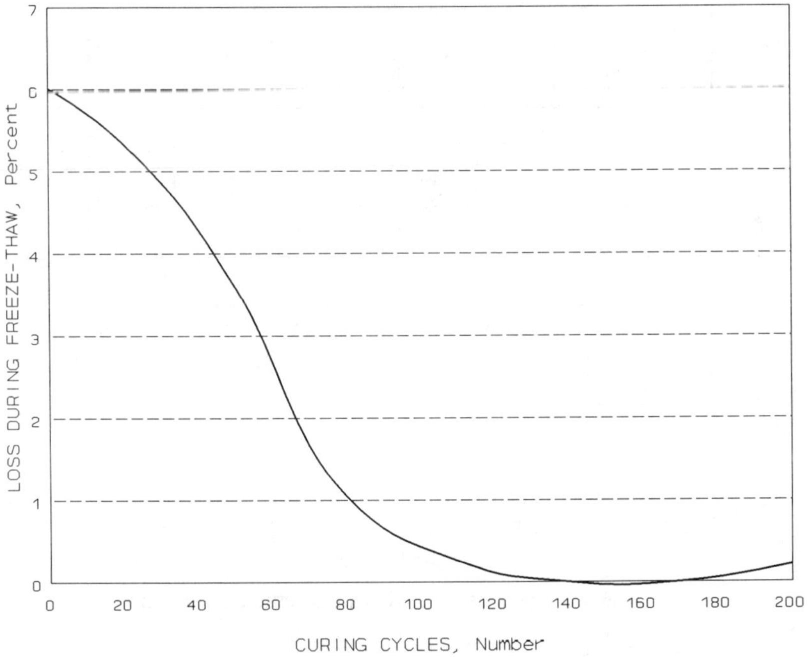

Figure 5 - Freeze-thaw durability versus curing cycles (after Lienhart, 1975).

rock product. The physical processes of concern are basically of two types:

1. Those which are due to stress applied from internal conditions (such as stress-relief due to unloading and expansion of swelling clays contained as a component of the rock matrix). (Because of the vast amount of literature on swelling clays, this subject is not discussed in this paper.)
2. Those which are due to stress applied as a result of external conditions (such as frost weathering and salt-burst weathering).

4.1 *Stress-relief*

Unloading effects are of particular concern in areas covered by the last continental glaciers and in areas of high tectonic stress. For rock that is completely or nearly saturated cold weather production may have a two-fold effect:

1. Fracture development due to unloading of residual stress;
2. Sudden volumetric expansion due freezing of water contained in the pores of the rock (much like the opening of a frozen can of soda).

4.2 *Frost weathering or congelifraction*

Frost weathering or congelifraction is related to three processes that result in freezing expansion (volumetric expansion, adsorptive suction, and expansion of non-freezable or adsorbed water). These processes are controlled by four environmental factors (freezing intensity, freezing rate, freezing duration, and available moisture) and by three mechanical or physical property factors (moisture content, total porosity and pore size, and Young's Modulus in tension).

Volumetric expansion is the dominant process involved in congelifraction. The 9%

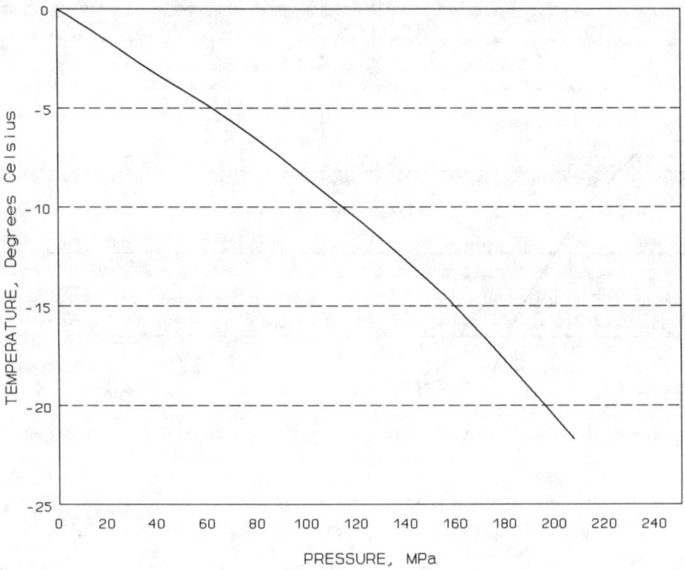

Figure 6 - Stresses associated with volumetric expansion.

volume change which occurs when ice forms produces a shattering force related to Young's Modulus in tension. When the shattering force becomes greater than the tensile strength of the rock, failure occurs. Due to the fatigue caused by cycling between freezing and thawing, failure may occur without the shattering force exceeding the tensile strength of the rock (Mellor, 1970). The pressures involved with this process may best be illustrated by Figure 6. This figure shows the development of pressure with decreasing temperature in a closed system of porosity under fully saturated conditions.

Although the adsorptive suction process or water migration theory has been referenced in the literature as far back as 1930 (Black and Hardenberg, 1991), it was not until 1983 that it was actually observed (Fukuda, 1983). The process involves the movement of pore water toward the freezing front (the boundary between the frozen and unfrozen portions of the rock) by capillary action and by suction due to negative pore pressure development just ahead of the freezing front.

Non-freezable water is the connate water trapped in the smallest of rock pores. This water can never be totally replaced if the rock particle is allowed to dry for a long period of time prior to being exposed once again to high moisture conditions. This process is one of the principle causes of rock deterioration. It is also the reason that freshly quarried rock was termed "green" and was set aside to "cure" by old-time quarry workers. Recent investigations in Germany have found that this water expands in volume by 0.6 percent as the temperature decreases from +4° C to -10° C (Winkler, 1992). The pressures associated with this expansion are due to the hydration or "ordering" of the capillary or adsorbed moisture. As with volumetric expansion, the theoretical pressures associated with hydration of adsorbed water are approximately 200 MPa. This implies that even if the rock particle does not contain sufficient moisture for volumetric expansion to take place, or if pore sizes are not of critical size for adsorptive suction to work, expansion of the non-freezable water can still result in disruption.

Additional ramifications of this particular process are as follows:

1. Rock quarried during temperatures of less than -2° C stand a greater chance of suffering some freezing damage or disintegration.

2. Rock quarried within a short period of time prior to the onset of temperatures of less than -2° C stand a chance of suffering some freezing damage or disintegration.

3. Some, but not all, rock needs to be "cured" prior to use.

Freezing intensity (the critical temperature for development of sufficient freezing strain to cause crack development) is one of the controlling factors. Walder and Hallet (1985) suggest that the range of -4° to -15° is most critical for crack propagation but Mellor (1970) has shown that the freezing strain increases rapidly just below freezing but slows with decreasing temperatures. Therefore, critical freezing intensity is taken to be in the range of -2° C to -5° C.

Because pore ice reacts plastically over a period of time, low freezing rate may result in low freezing strain and therefore, may result in less disruption to the rock particles. This factor may be useful in preventing disruption of rock particles in a vital portion of a structure if a way can be found to slow down the freezing rate at that particular point.

Freezing duration and moisture availability are mutually interdependent. Walder and Hallet (1985) found that the duration of freezing is a critical factor in those circumstances where the rock is in intimate contact with a source of moisture (e.g., shoreline exposure). It has been shown that freezing strain will continue to increase for the duration of the freezing cycle, thereby creating significant damage in only one cycle.

Critical saturation is not only dependent on the available moisture factor but is intimately dependent on total porosity and pore size. The volumetric expansion and adsorptive suction processes of rock disintegration depend on a critical degree of saturation. Without sufficient moisture within the rock particle or available moisture to supply moisture through water migration to the freezing front, sufficient pressures for fracture propagation may not result. Dunn and Hudec (1965) found that a critical degree of saturation was 75 percent or greater for total porosities of 25 percent or less and for pore size distributions that would allow for an adsorption value of less than 45 percent at 100 percent relative humidity. For pore size distributions that allow for adsorption values greater than 45 percent, the criteria become more complex.

It is quite obvious that fracture development in frozen rock is directly related to both the freezing strain and the tensile strength of the rock in question. It follows then, that theoretically, Young's Modulus in tension is another factor that should provide some indication of the ability of a rock at greater than critical saturation to withstand freezing at a fast rate for a long period of time at a critical temperature.

4.3 *Salt-burst weathering*

When salts change from one state of hydration to another the hydration pressures associated with these changes are not as severe as those associated with volumetric expansion of water below the freezing point but they are substantial. For example, the pressures approach 25.5 MPa for magnesium sulfate and 48.25 MPa for the sodium sulfate at 20°C and 100 percent relative humidity. These hydration pressures are very sensitive to both temperature and relative humidity. These pressures can vary by as much as 1.65 MPa per degree change in temperature and one percent change in relative humidity for sodium sulfate. For magnesium sulfate, the pressure will vary by only 0.027 MPa per degree change in temperature but will vary by over 1.38 MPa for every one percent change in relative humidity (Winkler and Wilhelm, 1970). Salt-burst weathering, therefore, can be almost as severe as frost weathering, particularly in areas of high temperature and humidity variability.

5. CONCLUSIONS

Durability of rock for erosion control is controlled by intrinsic rock properties and production techniques. Intrinsic properties may be altered, to the detriment of the durability of the rock, through improper production techniques. Production techniques and proper testing and selection of quality rock may be enhanced through a thorough understanding of the weathering processes and their controlling factors.

6. REFERENCES

Black, P. B. and Hardenberg, M. J., Eds. 1991. Historical Perspectives in Frost Heave Research: The Early Works of S. Taber and G. Beskow. *U.S. Army Cold Regions Research and Engineering Laboratory Special Report 91-23*. Hanover, NH.

Dunn, J.R. and Hudec, P.P. 1965. The influence of clays on water and ice in rock pores: *Physical Research Report RR 65-5*. State of New York Dept. of Public Works.

Fukuda, M. 1983. The pore water pressure profile in porous rocks during freezing: *Proceedings, Fourth International Conference on Permafrost*. Washington, D.C.: National Academy Press.

Lienhart, D.A. 1975. Special Study on the Effect of "Curing" on the Durability of the Berea Sandstone. *Ohio river Division Laboratory Open-File Report #103/75.618B*. Cincinnati, OH: U.S. Army Corps of Engineers.

Lienhart, D.A. and Stransky, T.E. 1981. Evaluation of potential sources of riprap and armor stone - methods and considerations. *Bulletin of the Association of Engineering Geologists*. 18 (3): 323-332.

Lienhart, D.A. 1988. The geographic distribution of intensity and frequency of freeze-thaw cycles. *Bulletin of the Association of Engineering Geologists*. 25 (4): 465-469.

Lienhart, D.A., Fisher, H.H. and Robinson, E.F. 1993. The correlation of index tests with rock durability: *Presentation at the International Riprap Workshop, 12-16 July 1993, Fort Collins, CO.*

Mellor, M. 1970. Phase composition of pore water in cold rocks: *Research Report 294*. Hanover, NH: U.S. Army Engineers Cold Regions Research and Engineering Laboratory.

Morgan, M.H. (translator) 1960. *Vitruvius. The ten books on architecture*. New York: Dover.

Walder, J. and Hallet, B. 1985. A theoretical model of the fracture of rock during freezing: *Geological Society of America Bulletin*. 96 (3): 336-346.

Winkler, E.M. and Wilhelm, E.J. 1970. Salt burst by hydration pressures in architectural stone in urban atmosphere: *Geological Society of America Bulletin*. 81 (2): 567-572.

Winkler, E. M., Professor Emeritus, University of Notre Dame, South Bend, IN, personal communication, 1992.

20 Rock mechanics in the design of rock slopes

Session developers:
Luis Vieitez Uteza, Centro de Investigacion Sismica
Gregory Matthews, Malcolm Pirnie, Inc.

Rock Mechanics, Nelson & Laubach (eds) © 1994 Balkema, Rotterdam, ISBN 90 5410 380 8

Rock mechanics at the Fiesta Texas Theme Park

Gary W. Hall
Project Control of Texas, San Antonio, Tex., USA
Mark R. Funkhouser
Golder Associates Inc., Houston, Tex., USA
Richard W. Humphries & Wayne L. Warburton
Golder Associates Inc., Atlanta, Ga., USA

ABSTRACT: This paper describes the application of rock mechanics in the design and construction of the Fiesta Texas Theme Park in San Antonio, Texas. The park was constructed in a former limestone quarry and covers approximately 80 hectares. As a result of quarry operations, the quarry walls had extensive blast damage. Rock mechanics was involved because the park developers and theme park architects were interested in maximizing the available space in the quarry and bringing park visitors as close to the quarry walls as safely as possible. The quarry walls were also planned to be the basis for several of the major attractions. Work completed as part of park development that involved rock mechanics included cliff stabilization, tunnel construction and rockfall hazard mitigation. The visual impact of these works was a primary concern of the park architects and was an important factor in design. Design and construction of the park involved a diverse team of professionals including owners, landscape architects, ride manufacturers, theme park architects, rock mechanics engineers, park developers, and contractors.

1 INTRODUCTION

The Fiesta Texas Theme Park is located approximately 16 km northwest of San Antonio, Texas, within what used to be a limestone quarry. The park and associated facilities cover approximately 80 hectares and have a perimeter length of 8 km The perimeter of the park consists almost entirely of vertical limestone cliffs up to 37 m high, with only a few hundred meters of at-grade length where the park entrance and exit are located.

The park was developed by the USAA and Opryland USA organizations. Project management for development of the theme park was provided by Project Control of Texas. Rock mechanics engineering services were provided by Golder Associates, Inc.

2 GEOLOGY

The Fiesta Texas Theme Park is constructed on the fall line just northwest of San Antonio, Texas in an area comprised of Cretaceous Limestones. Structurally, the area is characterized by the Balcones Fault Zone, which trends northeast - southwest, and separates Texas Hill Country from the Coastal Plain. The park is dissected by one major and several minor faults related to the Balcones Fault Zone which result in the southeastern half of the park being surrounded by cliffs of the Edwards Group while the northwestern half of the park is surrounded by the older Glen Rose Formation.

The Edwards Group (in the park) is generally a thickly bedded, very fine grained, gray, horizontally bedded limestone. Locally, some beds have clay filled vugs and solution cavities. The Glen Rose Limestone is tan, very thickly bedded, with beds on the order of 6 m thick. It is generally less solutioned and more arenaceous than the Edwards Group.

The rock in the vicinity of the major fault which separates these two rock types in the park is much more jointed and solutioned than in the other areas of the park. This caused some difficulties during construction of the roller coaster tunnel which was excavated through this rock. There was also somewhat more moisture encountered in this area as the clay layers and clay filled solution features allowed build-up of some perched groundwater.

The major structural features in the rock mass are the horizontal bedding and near vertical jointing. Locally, solution cavities as well as nearly vertical fault planes are important structural features. The bedding thickness and strength of the limestone units vary significantly and some of the beds have high clay contents. The bedding planes are relatively smooth and undulating. The joints are generally tight (except where disturbed by the blasting) and the joint surfaces are rough, wavy and irregular. Where the limestone is relatively fresh, the unconfined compressive strength of rock is generally high and was assessed to vary from about 75 MPa to 100 MPa.

Because there are extensive exposures of the rock conditions in the quarry faces, the geotechnical investigations for the stability analyses could be limited to mapping the quarry walls. With the horizontal bedding, the near vertical joint sets and faults, and relatively strong rock mass, stability analyses showed that the vertical cliff faces have adequate safety factors, except in the limited areas of the faulting and the larger solution features.

3 CLIFF STABILIZATION AND ROCKFALL MITIGATION

The site had been used for a limestone quarry from the early 1900's through to the 1980's. The quarry operators were interested in maximizing blast fragmentation, and placed little value on the resulting stability of the quarry cliffs. Consequently, when the quarry was finally abandoned, the quarry cliffs had an, irregular and rugged appearance. Close examination of the cliffs revealed extensive blast damage with wide, open cracks behind the crests and loose slabs, columns and solution cavities in the faces, which are typically shown in Figure 1.

With much emphasis being given to incorporating the cliffs as a feature of the park, it was essential to increase the stability of the cliffs and reduce the rockfall hazard. In areas where park features were planned close to the cliffs, rock support measures were required, but where adequate space was available at the base of the cliff, a "safety zone" was provided for rockfall and rockfall runout. Due to the varying rock conditions throughout the park and the occasional park design changes, a rock mechanics engineer was on site full time during the cliff stabilization and construction of the rockfall mitigations to adjust the designs of these measures as required.

The first step to enhance cliff stability involved an extensive trim blasting operation intended to remove much of the loose, blast damaged rock that remained following the quarrying operation. Prior to trim blasting, backbreak cracks up to 2 m wide and 3 to 9 m deep were visible at many locations around the crest of the cliffs. The trim blasting was performed by lightly loading 150 mm diameter vertical drill holes spaced at 2.7 m They were loaded with Ireco Iresplit E prepackaged emulsion with 50 grain detonating cord for the full depth. The burden varied from hole to hole and from shot to shot, but was typically in the range of 3 m to 9.5 m. The trim blast loading was optimized through experimentation.

This initial trim blasting program removed nearly all of the unstable slabs and columns around the perimeter of the park. However, some new, smaller backbreak cracks developed during the trim blasting operation, and some loose rocks remained at the cliff crest and on the cliff face. To further reduce the rockfall potential, the cliffs were scaled using a Caterpillar EL 240

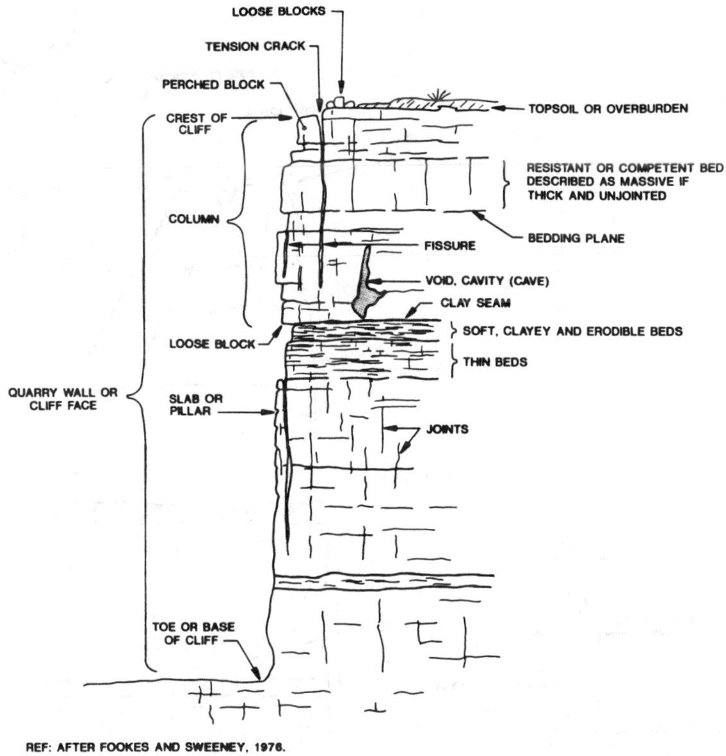

Figure 1: Typical Quarry Cliff Conditions Before Trim Blasting

B backhoe with a slurry trench arm from both above and below the cliff. The extended arm had a reach of approximately 18 m, so almost all of the cliff faces could be scaled with this equipment. When working from above the cliff, a spotter directed the backhoe operator to the loose pieces of rock on the cliff face.

In areas where there are no facilities close to the cliffs, no other stability measures were taken and a safety zone was provided at the base of the cliff, as shown in Figure 2. The width of the safety zone and the size of the boulder barrier were designed by a combination of rockfall trials (rolling rocks off the cliff and measuring impact location and runout) and the use of the CRSP (Colorado Rockfall Simulation Program) and the Golder Associates Rockfall Simulation Program. The trials proved invaluable in calibrating the CRSP and Golder rockfall simulation models and emphasized the large effect that even a small, inclined ramp has at the toe of vertical cliffs. It also became apparent that for the near vertical quarry walls, it was not necessary for rocks to "roll" down the cliff face to develop rotational energy. Significant rotational energy was developed by the impact of irregular (or angular shaped) rocks on the ground surface. As shown in Figure 2, the safety zone has a crushed stone base for energy absorption and a boulder barrier to limit rock runout and to exclude park visitors.

Additional support measures were required in areas where facilities are within the rockfall and rock runout zone. Grouted rock dowels were used extensively around the park perimeter to enhance cliff stability, as shown in Figure 2. The dowels used were typically Number 8, Grade 60, epoxy coated, deformed reinforcing steel. Lengths varied depending on the location and intended use, and ranged from 3 m to 9 m in length. They were installed in 76

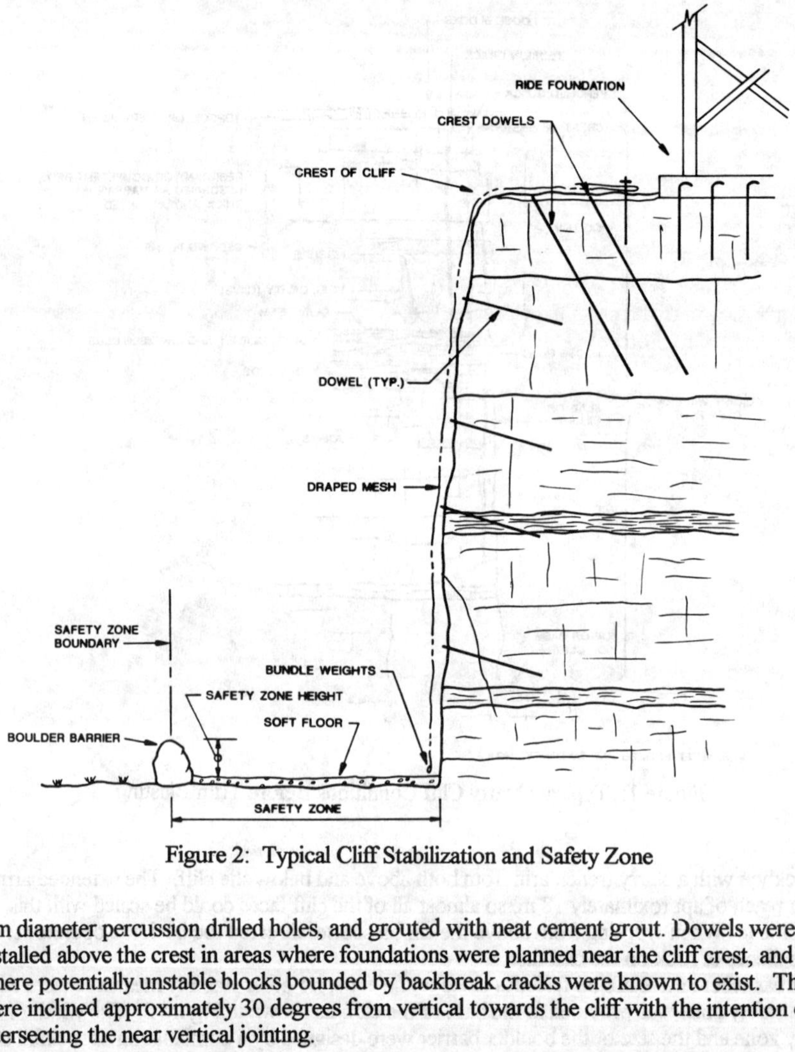

Figure 2: Typical Cliff Stabilization and Safety Zone

mm diameter percussion drilled holes, and grouted with neat cement grout. Dowels were installed above the crest in areas where foundations were planned near the cliff crest, and where potentially unstable blocks bounded by backbreak cracks were known to exist. These were inclined approximately 30 degrees from vertical towards the cliff with the intention of intersecting the near vertical jointing.

Dowels installed in the cliff face were very similar to those installed in the crests. However, the installation method was significantly different. A crane with a specially fabricated platform was used to lift a track-mounted pneumatic drill and a two man drill crew into position to drill the dowel holes and to grout and install the dowels. This installation method proved to be quite effective after the initial learning curve. However, drilling rates using the platform were lower than usual because drill pressure was limited by the reaction of the weight of the platform.

In addition to rock dowels, draped mesh was installed in some areas of the park. The use of draped mesh was limited to areas where facilities or public access were within the rockfall zone. Initially, there was concern that the mesh would be aesthetically objectionable, but after installation, consensus was that the mesh was hardly noticeable. The mesh used was double twisted, 9 gauge, galvanized wire mesh supplied by Maccaferri Gabions, Inc.. The mesh was anchored behind the crest of the cliff with dowels grouted in vertical holes and wire cable extending parallel to the cliff crest behind the dowels.

4 PARK RIDES AND FACILITIES

In addition to trim blasting and scaling, dowels and draped mesh were required at the following locations around the park:

- At the roller coaster, where the track is close to the cliff;
- At the spillwater ride, where the upper part of the ride is notched into the cliff and the riders are within 3 m of the cliff face;
- Along the portals of both tunnels; and
- At one of the parking areas, where the paved surface is close to the cliff and the cliff face is closely jointed.

The roller coaster tracks go over the crest of the cliff six times, so there are numerous trestle foundations close to the crest and close to the base of the cliff. Cliff stabilization measures had to be developed to protect wooden bents that support the coaster from rockfalls and to provide adequate safety for the riders. Stabilization measures include installation of 3 m long fully grouted No. 8 dowels on a 2.5 m by 2.5 m pattern over the entire 22 m high face adjacent to the roller coaster. Additionally, 6 m long dowels were installed along the crest of the cliff. Draped mesh was placed over the face in critical areas. In addition, a roof (shown in Figure 3) was constructed over the coaster in areas where small rock fragments could be a

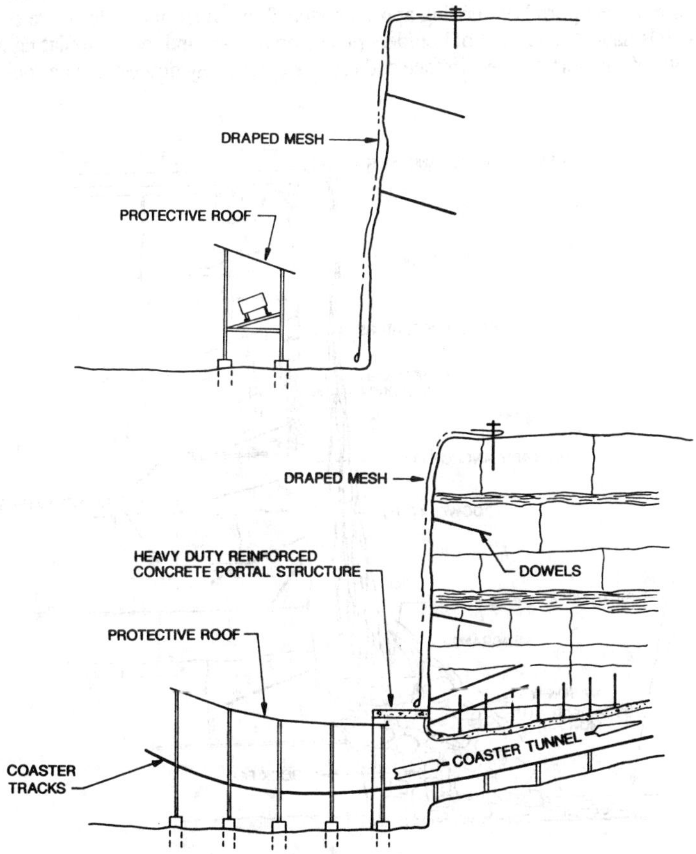

Figure 3: Typical Rockfall Protection for Roller Coasters

hazard to the riders. The roller coaster travels through a tunnel that was excavated in the quarry walls. Each portal has a heavy duty reinforced concrete structure extending from the cliff face which is designed to withstand a rockfall from the face, as shown in Figure 3.

The artificial waterfall is a prominent feature of the Park. It has flow capacity of up to 1.6m^3 per second and spills over the 18 m to 31 m high quarry walls. Of primary concern in the waterfall area was the erosion of soft and vuggy limestone on the quarry face, and potential instability resulting from this erosion. It was originally proposed to protect these areas with shotcrete and dowels. However, the owners and landscape architects were strongly opposed to placing shotcrete on the natural rock faces. Subsequently, a plan was devised to use dowels and local limestone masonry facing to protect these areas. The area was scaled thoroughly and the soft, weak zones were excavated using high pressure water jets and handheld pneumatic tools. These zones were then replaced with reinforced limestone masonry. Dowels were also installed in all potentially unstable blocks and beds of rock. All of this work was conducted from a platform suspended by a crane. Cliff stabilization measures used for the waterfall are shown in Figure 4. This method of protection has worked remarkably well, and the masonry blends in very well with the cliff face.

The park railroad traverses the majority of the park and goes through a tunnel which was excavated in the quarry wall. Extensive rockfall analysis was necessary during the design phase of the railroad tunnel portal structures. At the tunnel portals, the cliff is on the order of 37 m high and is composed of some of the more blocky rock in the park. Ultimately, reinforced concrete portals extending approximately 9 m out from the cliff were constructed. These portals have a "cushion" of boulders placed on and around them simulating a talus pile. Above both of the portals, the cliff face and crest is extensively doweled. The rock dowels are

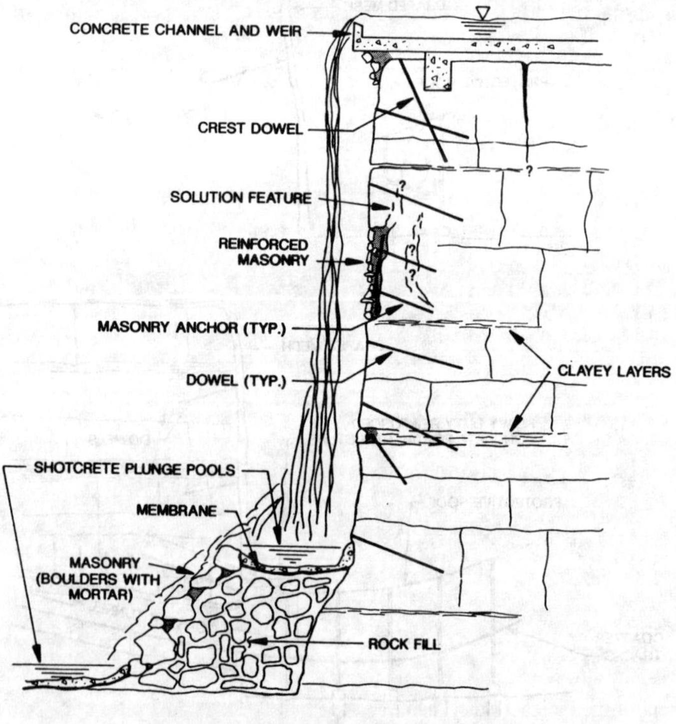

Figure 4: Cliff Reinforcement at Waterfall

intended to improve the stability of the cliff, and to limit the size of any individual rock blocks that could dislodge and fall on the tunnel portals. The design rockfall for the heavy duty reinforced concrete portal structure is a 1 cubic meter rock falling from the crest of the cliff.

5 TUNNELS

As described above, two tunnels were constructed for the Park rides; the roller coaster tunnel and the park railroad tunnel. The tunnels are 3.7 m and 5.5 m wide and 55 m and 52 m long, respectively. They are inverted U-shaped and sized to provide sufficient clearance while providing the maximum thrill for the riders. A typical section of the coaster tunnel is shown in Figure 5.

Because the quarry cliffs provided excellent exposures of the limestone beds in the area in which the tunnels would pass, no subsurface investigations were conducted. The rock cover over the tunnels is relatively low (15 m to 31 m) and the rock strength is moderate, so the rock mass behavior is controlled by the rock structure. The observational approach was used for placing the rock support, and ensuring that active rock support was placed close to the tunnel heading. Generally, tensioned bolts at approximately 1.2 m by 1.2 m in the crown and a 100 mm thick fiber reinforced shotcrete layer on the walls and crown were sufficient support, although additional strategically placed rock bolts and additional shotcrete were required locally, particularly when solutioned zones were encountered and at the portals.

The tunnels were excavated by the drill-and-blast method. A 3 boom jumbo was used for drilling and rock bolting and conventional front-end loaders were used for mucking and hauling. The observational method for placing rock support required the full time presence of a qualified rock engineer. This proved effective and economical and is considered to be particularly appropriate for short tunnels where there is little time for a learning curve.

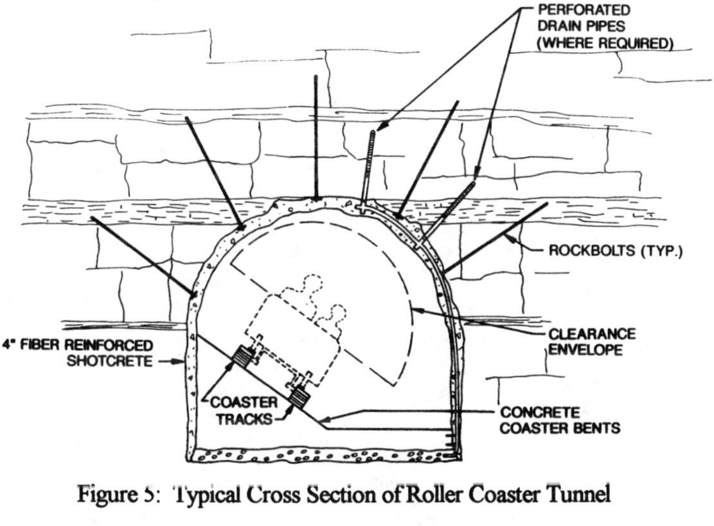

Figure 5: Typical Cross Section of Roller Coaster Tunnel

6 CONCLUSIONS

Development of the Fiesta Texas Theme Park required close cooperation of a diverse group of professionals. Roller coasters, rapid river rides and artificial waterfalls are not the usual realm of rock mechanics, but by applying modern rock engineering practice, it was possible to construct a safe, innovative and exciting theme park.

REFERENCES

Fookes, P.G. & Sweeney, M. 1976. Stabilization and Control of Local Rockfalls and Degrading Rock Slopes. Q. Jl Engineering Geology. Vol. 9, p.37-55.

Ritchie, A. 1963. Evaluation of Rockfall and its Control. Highways Research Record, 17, 1963.

Rock Mechanics, Nelson & Laubach (eds) © 1994 Balkema, Rotterdam, ISBN 90 5410 380 8

Application of the adaptive filtering technique for rock mass displacement monitoring by the Global Positioning System (GPS)

N. Shimizu & K. Nakagawa
Department of Civil Engineering, Yamaguchi University, Japan

S. Sakurai
Department of Civil Engineering, Kobe University, Japan

ABSTRACT: In this paper, the authors propose a method for improving the accuracy of GPS displacement measurements. Adaptive filtering is adopted to formulate this method. The proposed method is one of the back analysis methods used for assessing field measurement results. In order to demonstrate the applicability of the method to practical problems, a numerical simulation is demonstrated.

1 INTRODUCTION

The Global Positioning System (GPS) is an innovative surveying system using artificial satellites. This system has the potential to measure the rock mass deformation of an extensive area in various rock engineering projects, such as high cut slopes, large open pit mines, landslides and subsidences due to underground excavation or oil production. Actually, some applications of GPS to ground displacement measurements have already been studied (Chrzanowski et al. 1988, Murria & Abi Saab 1988, Strange 1989, Blodgett 1990).

The authors have applied GPS to the rock mass displacement measurements of a high cut slope and it was found that GPS is very useful for monitoring the stability of cut slopes (Shimizu & Sakurai 1992, Sakurai, Shimizu & Iriyama 1992, Sakurai & Shimizu 1992). The accuracy of GPS displacement measurements was also found within ten millimeters for the long baselines through practical applications. This accuracy may sometimes be enough to monitor rock mass deformation. But if more precise rock mass monitoring is required, higher accuracy is needed for GPS displacement measurements.

In this paper, the authors propose a method for improving the accuracy of results of GPS displacement measurements. Adaptive filtering is adopted to formulate this method. The proposed method is one of the back analysis methods (Gioda & Sakurai 1987) used for assessing field measurement results. In order to demonstrate the applicability of the method to practical problems, an example of numerical simulations is illustrated.

2 OUTLINE OF ROCK MASS DISPLACEMENT MONITORING BY GPS

GPS has been developed to support navigation and surveying. It provides 24-hour all weather navigation and surveying capability. GPS offers several advantages over conventional surveying methods. For example, three dimensional baseline vectors can be determined with high accuracy, intervisibility between observation points is unnecessary, measurements can be taken night or day, under varying weather conditions and with portable equipment, etc. (Wells 1986, Leick 1990). Any person who has a GPS receiver to record the signal transmitted from GPS satellites can use the system. GPS cannot be applied, however, at places where the signal cannot be received, i.e. underground or in buildings.

Photo 1 GPS receiver and antenna, and slope monitored by GPS

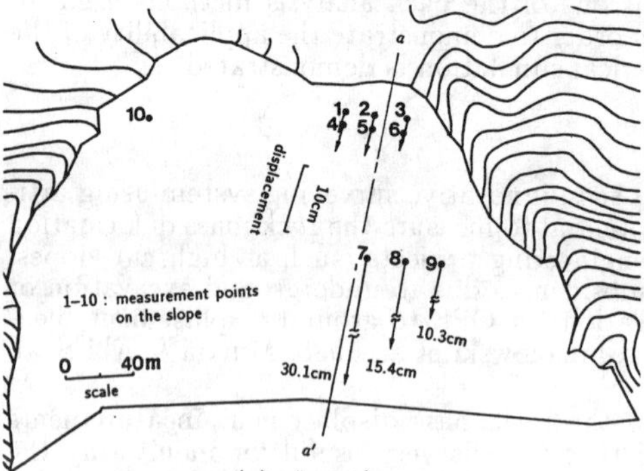

(a) plan view

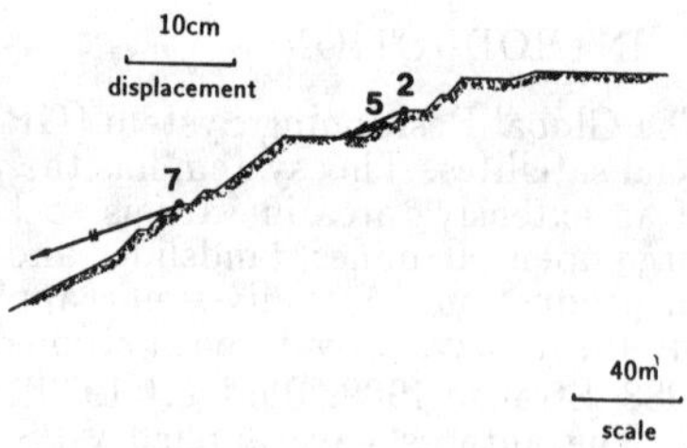

(b) vertical section $(a - a')$

Figure 1 Displacement vectors on the slope measured by GPS (Sakurai & Shimizu 1992)

To survey with GPS, set up a GPS antenna over a bench mark and another antenna over a measuring point. Data from satellites are then recorded by receivers connected to those antennas. Three dimensional relative coordinates of the measuring point from a bench mark are obtained by computing recorded data. This method of GPS surveying is called *relative positioning.*

The authors applied the GPS relative positioning method to displacement measurements while cutting a high slope (Sakurai & Shimizu 1992). The height and width of the slope were about 120 m and 200 m, respectively. Two bench marks were set up about 600 m away from the slope and ten measuring points were set on the slope. Photo 1 shows the GPS receiver and antenna on the tripod set over a bench mark.

The observation time of the relative positioning is 90 - 180 minutes for each baseline from bench mark to measuring point in one session. Figure 1 shows vectors of measured displacements at the measuring points set on the slope. In order to assess the stability of slope, a back analysis of displacements measured by GPS was conducted (Sakurai, Shimizu & Iriyama 1992).

On the other hand, total station surveying was also conducted to verify the accuracy of GPS displacement measurements. The results comparing displacements measured by both GPS and total station revealed that accuracy of GPS displacement measurements falls within 10 mm in the horizontal direction and 20 mm in the vertical one.

3 FORMULATION OF THE PROPOSED METHOD

This chapter presents the formulation and solution procedure of the proposed method.

3.1 *Governing equation of displacement behavior – the state equation*

In this study, rock mass deformation is assumed to show steady-state creep behavior. That is, each component of displacement vector $u_i(t)$ ($i = 1, 2, 3$: component index, t: time) is supposed to be independently governed by the following equation for a linear viscoelastic model:

$$a_{i0}u_i(t) + a_{i1}\frac{du_i(t)}{dt} + a_{i2}\frac{d^2u_i(t)}{dt^2} + \ldots + a_{in}\frac{d^nu_i(t)}{dt^n} = p_i(t) \qquad (1)$$

where a_{ij} is denoted by coefficient of creep behavior and $p_i(t)$ represents a function related to loading/excavation procedure.

In discrete time Δt, Eq. (1) becomes

$$\alpha_{i0}u_{i.t} + \alpha_{i1}u_{i.t-1} + \ldots + \alpha_{in}u_{i.t-n} = p_{i.t} \qquad (2)$$

where $u_{i.t-j} = u_i(t - j\Delta t)$, $p_{i.t} = p_i(t)$ and

$$\alpha_{ij} = (-1)^j \sum_{k=1}^{n} \begin{pmatrix} k \\ j \end{pmatrix} \frac{a_{ik}}{k\Delta t}, \quad \begin{pmatrix} k \\ j \end{pmatrix} : \text{combination symbol.}$$

Equation (2) is similar form to a equation of consolidation derived by Asaoka & Matsuo (1980).

From Eqs. (2) and $u_{i.t-j} = u_{i.t-j}$ ($j = 1, 2, \ldots, n$) the following equation can be deduced:

$$\boldsymbol{u}_{i.t} = \boldsymbol{A}_i\boldsymbol{u}_{i.t-1} + \boldsymbol{b}_i p_{i.t} \qquad (3)$$

where $\boldsymbol{u}_{i.t} = \{u_{i.t}\ u_{i.t-1}\ \ldots u_{i.t-n+1}\}^T$, $\boldsymbol{b}_i = \{\frac{1}{\alpha_{i0}}\ 0\ \ldots\ 0\ 0\}^T$ and

$$\boldsymbol{A}_i = \begin{pmatrix} -\frac{\alpha_{i1}}{\alpha_{i0}} & -\frac{\alpha_{i2}}{\alpha_{i0}} & \cdots & -\frac{\alpha_{i\,n-1}}{\alpha_{i0}} & -\frac{\alpha_{in}}{\alpha_{i0}} \\ 1 & 0 & \cdots & 0 & 0 \\ \cdots & \cdots & \cdots & \cdots & \cdots \\ 0 & 0 & \cdots & 1 & 0 \end{pmatrix}.$$

Setting $\boldsymbol{x}_{i.t} = \boldsymbol{u}_{i.t} - \boldsymbol{u}_{i.t-1}$ and $q_{i.t} = p_{i.t+1} - p_{i.t}$, the resulting equation from Eq. (3) is

$$\boldsymbol{x}_{i.t+1} = \boldsymbol{A}_i\boldsymbol{x}_{i.t} + \boldsymbol{b}_i q_{i.t} \qquad (4)$$

Equation (4) becomes the state equation with respect to the incremental displacement vector $\boldsymbol{x}_{i.t}$.

3.2 *Geometrically constrained condition – the observation equation*

In order to improve the accuracy of GPS displacement measurements, an observation equation is introduced.

GPS surveying (relative positioning) is conducted to obtain the relative coordinates of measuring point a from two bench marks as shown in Figure 2. If the coordinate of the bench mark is represented by $X_i^{0(k)}$ ($k = 1, 2$: index of bench mark) and the relative coordinate of measuring point a from bench mark k by $\Delta X_{i.t}^{(k)}$, the following equation can be derived:

$$X_i^{0(k)} + \Delta X_{i.t}^{(k)} = X_{i.t} + w_i \quad (k = 1, 2) \qquad (5)$$

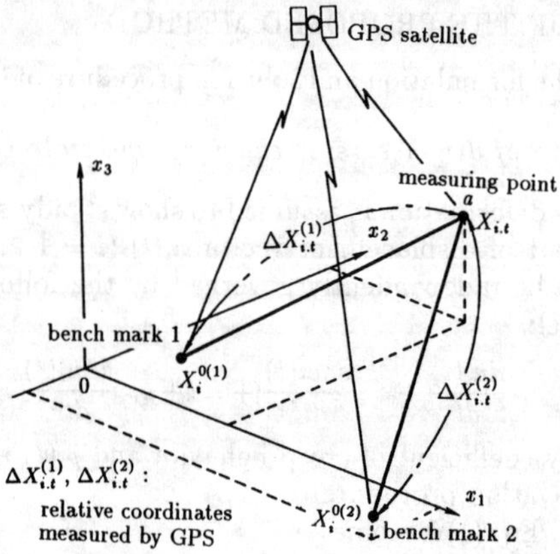

Figure 2 Measuring point and bench marks for GPS surveying

w_i is denoted by the measurement error. Equation (5) represents a geometrically constrained condition for measuring point a and two bench marks.

Obtaining the similar equation to Eq. (5) for $t = t-1$, subtracting it from Eq. (5) leads to

$$\Delta X_{i.t}^{(k)} - \Delta X_{i.t-1}^{(k)} = (u_{i.t} - u_{i.t-1}) + v_i \tag{6}$$

where, $u_{i.t} = X_{i.t} - X_{i.0}$ and measurement error is rewritten as v_i.

Equation (6) and the measurement vector $\boldsymbol{y}_{i.t}$ defined by

$$\boldsymbol{y}_{i.t} = \left\{ \begin{array}{ccc} \Delta X_{i.t}^{(1)} & - & \Delta X_{i.t-1}^{(1)} \\ \Delta X_{i.t}^{(2)} & - & \Delta X_{i.t-1}^{(2)} \\ \Delta X_{i.t-1}^{(1)} & - & \Delta X_{i.t-2}^{(1)} \\ \Delta X_{i.t-1}^{(2)} & - & \Delta X_{i.t-2}^{(2)} \\ & \cdots & \\ \Delta X_{i.t-n+1}^{(1)} & - & \Delta X_{i.t-n}^{(1)} \\ \Delta X_{i.t-n+1}^{(2)} & - & \Delta X_{i.t-n}^{(2)} \end{array} \right\}$$

give

$$\boldsymbol{y}_{i.t} = \boldsymbol{C}_i \boldsymbol{x}_{i.t} + \boldsymbol{v}_i \tag{7}$$

in which

$$\boldsymbol{C}_i = \begin{pmatrix} 1 & 0 & 0 & 0 & 0 \\ 1 & 0 & 0 & 0 & 0 \\ 0 & 1 & 0 & 0 & 0 \\ 0 & 1 & 0 & 0 & 0 \\ \cdots & \cdots & \cdots & \cdots & \cdots \\ 0 & 0 & 0 & 0 & 1 \\ 0 & 0 & 0 & 0 & 1 \end{pmatrix}, \quad \boldsymbol{v}_i = \left\{ \begin{array}{c} 1 \\ 1 \\ 1 \\ 1 \\ \cdots \\ 1 \\ 1 \end{array} \right\} v_i.$$

Equation (7) becomes the observation equation representing the relationship between measurement vector $\boldsymbol{y}_{i.t}$ and incremental displacement vector $\boldsymbol{x}_{i.t}$.

3.3 *Application of adaptive filtering*

Equations (4) and (7) give a set of equation for a linear system with respect to $\boldsymbol{x}_{i.t}$.

$$\begin{aligned} \boldsymbol{x}_{i.t+1} &= \boldsymbol{A}_i \boldsymbol{x}_{i.t} + \boldsymbol{b}_i q_{i.t} \\ \boldsymbol{y}_{i.t} &= \boldsymbol{C}_i \boldsymbol{x}_{i.t} + \boldsymbol{v}_i \end{aligned} \tag{8}$$

It should be noted that coefficient α_{ij} in matrix $\boldsymbol{A}_i$ and vector $\boldsymbol{b}_i$, which characterizes creep displacement behavior, is also an unknown parameter as well as $\boldsymbol{x}_{i.t}$ in our problem. Updating both $\boldsymbol{x}_{i.t}$ and α_{ij} from measurement results of $\boldsymbol{y}_{i.t}$ is called *adaptive filtering.*

A new unknown parameter $\boldsymbol{z}_{i.t}$ is defined by

$$\boldsymbol{z}_{i.t} = \{\boldsymbol{x}_{i.t}^T \ \boldsymbol{\theta}_{i.t}^T\}^T$$

in which

$$\boldsymbol{\theta}_{i.t} = \{-\frac{\alpha_{i1}}{\alpha_{i0}}, \ -\frac{\alpha_{i2}}{\alpha_{i0}}, \ \dots \ -\frac{\alpha_{in}}{\alpha_{i0}}, \ \frac{1}{\alpha_{i0}}\}^T.$$

Supposing that $\boldsymbol{\theta}_{i.t}$ is constant with respect to time t (i.e. $\boldsymbol{\theta}_{i.t+1} = \boldsymbol{\theta}_{i.t}$), Eq. (8) leads to following equations:

$$\begin{aligned} \boldsymbol{z}_{i.t+1} &= \boldsymbol{A}_i'\boldsymbol{z}_{i.t} + \boldsymbol{b}_i' q_{i.t} \\ \boldsymbol{y}_{i.t} &= \boldsymbol{C}_i'\boldsymbol{z}_{i.t} + \boldsymbol{v}_i \end{aligned} \tag{9}$$

$$\boldsymbol{A}_i' = \begin{pmatrix} \boldsymbol{A}_i & \mathbf{o} \\ \mathbf{o} & \boldsymbol{I} \end{pmatrix}, \quad \boldsymbol{b}_i' = \begin{Bmatrix} \boldsymbol{b}_i \\ \mathbf{o} \end{Bmatrix}, \quad \boldsymbol{C}_i' = (\boldsymbol{C}_i \ \mathbf{o})$$

where $\boldsymbol{I}$ represents the unit matrix.

In order to solve Eq. (9) the extended Kalman filtering algorithm is adopted. Denoting an estimate of the unknown parameter $\boldsymbol{z}_{i.t}$ as $\hat{\boldsymbol{z}}_{i.t/t}$, $\hat{\boldsymbol{z}}_{i.t/t}$ can be obtained from the filtering equation as follows:

$$\hat{\boldsymbol{z}}_{i.t/t} = \hat{\boldsymbol{z}}_{i.t/t-1} + \boldsymbol{K}_{i.t}[\boldsymbol{y}_{i.t} - \boldsymbol{C}_i'\hat{\boldsymbol{z}}_{i.t/t-1}], \quad \hat{\boldsymbol{z}}_{i.1/0} = \bar{\boldsymbol{z}}_{i.0} \ \text{(initial condition)} \tag{10}$$

where $\boldsymbol{K}_{i.t}$ is called Kalman gain as follows:

$$\boldsymbol{K}_{i.t} = \boldsymbol{P}_{i.t/t-1}\boldsymbol{H}_{i.t}^T[\boldsymbol{H}_{i.t}\boldsymbol{P}_{i.t/t-1}\boldsymbol{H}_{i.t}^T + \boldsymbol{R}_{i.t}]^{-1}, \quad \boldsymbol{R}_{i.t} = \boldsymbol{v}_i^T\boldsymbol{v}_i \ \text{(measurement error)}$$

$\boldsymbol{P}_{i.t/t-1}$ is denoted by covariance matrix of $\hat{\boldsymbol{z}}_{i.t/t-1}$ followed by

$$\boldsymbol{P}_{i.t+1/t} = \boldsymbol{F}_{i.t}\boldsymbol{P}_{i.t/t}\boldsymbol{F}_{i.t}^T$$

$$\boldsymbol{P}_{i.t/t} = \boldsymbol{P}_{i.t/t-1} - \boldsymbol{K}_{i.t}\boldsymbol{H}_{i.t}\boldsymbol{P}_{i.t/t-1}, \quad \boldsymbol{P}_{i.1/0} = \boldsymbol{\Sigma}_{i.0} \ \text{(initial condition)}$$

in which

$$\boldsymbol{F}_{i.t} = \left(\begin{array}{c|cc} \boldsymbol{A}_i & \begin{matrix} \boldsymbol{x}_{i.t}^T & q_{i.t} \\ \mathbf{o} & 0 \end{matrix} \\ \hline \mathbf{o} & \boldsymbol{I} \end{array}\right), \quad \boldsymbol{H}_{i.t} = (\boldsymbol{C}_i \ \mathbf{o}).$$

Once a new measurement value $\Delta X_{i.t}^{(k)}$ is obtained by GPS, the unknown parameter $\boldsymbol{\theta}_{i.t}$ as well as the incremental displacement $\boldsymbol{x}_{i.t}$ is updated by Eq. (9) and $\boldsymbol{y}_{i.t}$ computed from $\Delta X_{i.t}^{(k)}$.

Furthermore, using Eq. (4) and $\hat{\boldsymbol{z}}_{i.t/t}$ updated by Eq. (9), the displacement at $t + m\Delta t$ can be predicted by

$$\hat{\boldsymbol{u}}_{i.t+m/t} = \boldsymbol{u}_{i.0} + \sum_{j=1}^{t} \hat{\boldsymbol{x}}_{i.j/j} + \sum_{k=1}^{m} \hat{\boldsymbol{x}}_{i.t+k/t} \tag{11}$$

where $\hat{\boldsymbol{x}}_{i.t+1/t} = \boldsymbol{A}_i\hat{\boldsymbol{x}}_{i.t/t} + \boldsymbol{b}_i q_{i.t}$.

4 NUMERICAL SIMULATION

A numerical example is presented to demonstrate the applicability of the proposed method. The most simple model, i.e. $n = 1$ in Eq. (1) is adopted here. Supposing $p_{i.t}$ to be constant with respect to time leads to $q_{i.t} = 0$, then Eq. (9) gives a system equation of the model as follows:

$$\begin{aligned} \boldsymbol{z}_{i.t+1} &= \begin{pmatrix} \theta_{i.t} & 0 \\ 0 & 1 \end{pmatrix} \boldsymbol{z}_{i.t} \\ \boldsymbol{y}_{i.t} &= \begin{pmatrix} 1 & 0 \\ 1 & 0 \end{pmatrix} \boldsymbol{z}_{i.t} + \boldsymbol{v}_i \end{aligned} \tag{12}$$

where $\boldsymbol{z}_{i.t} = \{x_{i.t}\ \theta_{i.t}\}^T$, $x_{i.t} = u_{i.t} - u_{i.t-1}$ and $\boldsymbol{y}_{i.t} = \left\{ \begin{array}{l} \Delta X_{i.t}^{(1)} - \Delta X_{i.t-1}^{(1)} \\ \Delta X_{i.t}^{(2)} - \Delta X_{i.t-1}^{(2)} \end{array} \right\}$.

The final displacement predicted at time t (i.e. $\hat{u}_{i.\infty/t}$), can be computed by the following equation derived from Eq. (11).

$$\hat{u}_{i.\infty/t} = u_{i.0} + \sum_{j=1}^{t} \hat{x}_{i.j/j} + \frac{\theta_{i.t}}{1-\theta_{i.t}} \hat{x}_{i.t/t} \quad (0 < \theta_{i.t} < 1) \tag{13}$$

Figure 3 shows a plan view of a measuring point and two bench marks adopted in this example. Although the numerical result is described here only for the component of displacement in the direction of x_1 axis, other components are obtained in the same manner. The procedure of simulation is described below.

1. Relative coordinates $\Delta X_{1.t}^{(1)}$ and $\Delta X_{1.t}^{(2)}$ computed by generating random numbers based on normal distribution is used as measurement data in this example. The standard deviation of normal distribution is assumed to be 0.5 cm on the basis of practical measurement results (Sakurai & Shimizu 1992).

2. *True* displacement of the measuring point is assumed to be $u_{1.t} = 30(1 - exp(-t/30))$ cm (t: days).

3. GPS surveying is supposed to be conducted every two days. Let $X_1^{0(1)} + \Delta X_{1.t}^{(1)}$ and $X_1^{0(2)} + \Delta X_{1.t}^{(2)}$ be coordinate of the measuring point from bench marks 1 and 2, respectively. Those are given from the manner above mentioned, as shown in Figure 4.

4. Unknown parameter $\hat{\boldsymbol{z}}_{1.t/t}$ can be obtained by Eqs. (10) and (12).

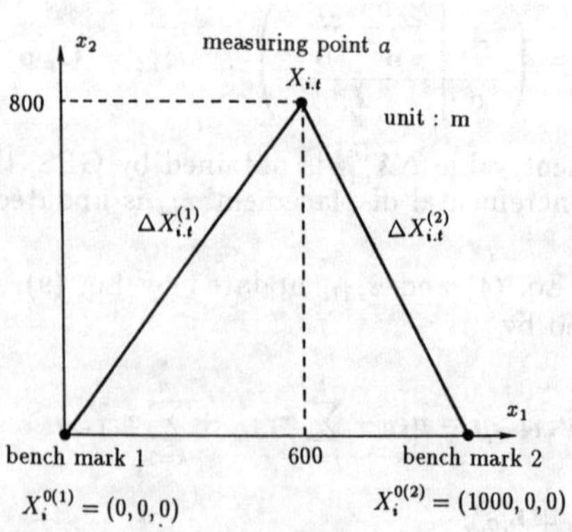

Figure 3 Plan view of a measuring point and bench marks in a numerical example

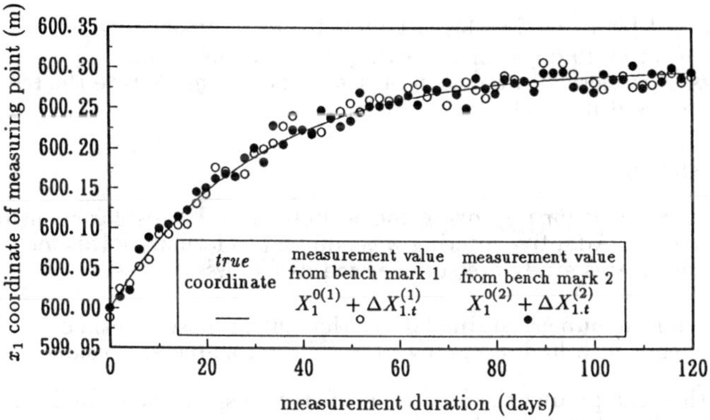

Figure 4 Input data of measurement coordinates in the direction of x_1 axis

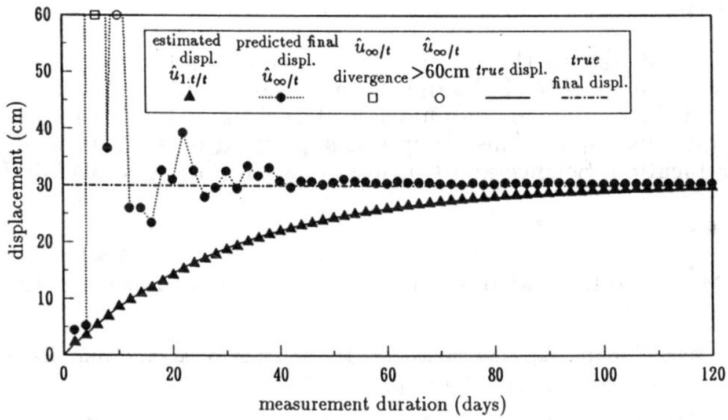

Figure 5 Numerical results during measurement duration

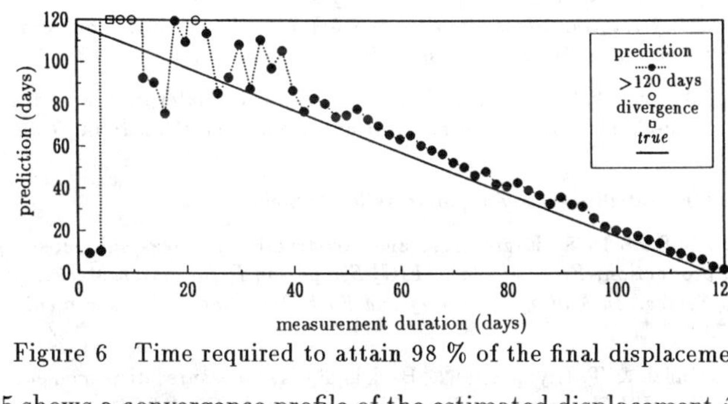

Figure 6 Time required to attain 98 % of the final displacement

Figure 5 shows a convergence profile of the estimated displacement $\hat{u}_{1.t/t}$ and the predicted value of the final displacement $\hat{u}_{1.\infty/t}$. $\hat{u}_{1.t/t}$ is denoted as the displacement estimated by using data measured until time t (▲ in Figure 5). It is recognized that $\hat{u}_{1.t/t}$ shows good agreement with exact value (solid line in Figure 5) in spite of some scattering observed for input data (see ∘, • in Figure 4). Therefore, it is concluded that the geometrically constrained condition of the second equations of Eq. (9) or (12) works efficiently to get accurate estimates of the displacement. In Figure 5, the error of the final displacement predicted at time t falls within 10 % after around 20 days since starting the GPS surveying. It then reduces to a few percent after 40 days.

Figure 6 shows a profile of time required to attain 98 % of the final displacement, which is predicted by using Eq. (13). Predicted time approaches to exact one (solid line in Figure 6) after 40 days since starting the GPS surveying.

The results of numerical simulation presented here demonstrate the fair applicability of the proposed method.

5 CONCLUSIONS

In this paper, a method for improving the accuracy of GPS displacement measurements was proposed. Adaptive filtering was adopted to formulate this method. The main conclusions of this study are summarized as follows :

1. It is shown from numerical simulations that the proposed method can estimate displacements with high accuracy from GPS measurement results.
2. The method can predict the final value of creep displacements and convergence time from the estimated value of the system parameters.

ACKNOWLEDGMENTS

The authors would like to thank Dr. A. Murakami of Kyoto University for his suggested discussion and Mr. T. Egami, former student of Yamaguchi University for his assistance in the numerical computation. Thanks are also due to Ms. S. Havard for proofreading this paper. This study was supported by a Grant-in-Aid from the Ministry of Education, Science and Culture of Japan (Grant No.04750477).

REFERENCES

Asaoka, A. & M. Matsuo 1980. An inverse problem approach to settlement prediction, Soils and Foundations, 20(4), pp.87–101.

Blodgett, J. C. 1990. Monitoring land subsidence in Sacramento valley, California, using GPS, *J. Surveying Engineering*, ASCE, 116(2), pp.112–130.

Chrzanowski, A., Y. Chen, R. Leeman, & J. Leal 1988. Integration of the Global Positioning System with geodetic levelling surveys in ground subsidence studies, *Proc. 5th Int. (FIG) Sympo. on Deformation Measurements and 5th Canadian Sympo. on Mining Surveying and Rock Deformation Measurements*, Fredericton, pp.142–155.

Gioda, G. & S. Sakurai. 1987 Back analysis procedures for the interpretation of field measurements in geomechanics, *Int. J. Numerical and Analytical Methods in Geomechanics*, 11, pp.555–583.

Leick, A. 1990. GPS satellite surveying, John Wiley & Sons.

Murria, J. & J. Abi Saab 1988. Engineering and construction in areas subjected to subsidence due to oil production, *Proc. 5th Int. (FIG) Sympo. on Deformation Measurements and 5th Canadian Sympo. on Mining Surveying and Rock Deformation Measurements*, Fredericton, pp.367–373.

Sakurai, S., N. Shimizu & T. Iriyama 1992. Back analysis of measured displacements in cut slopes, *Eurock'92, Int. ISRM Sympo. on Rock Characterization*, City of Chester, pp.378–383.

Sakurai, S. & N. Shimizu 1992. Monitoring of cut slopes by using the Global Positioning System (GPS). *Journal of Mining Research*, 1(3), pp.19–29.

Shimizu, N. & S. Sakurai 1992. Application of the Global Positioning System for monitoring a high cut slope. *6th Int. FIG-Sympo. on Deformation Measurements-Measurement, Monitoring and Prediction*, Hanover, Session 8.

Strange, W. E. 1989. GPS determination of groundwater withdrawal subsidence, *J. Surveying Engineering*, ASCE, 115(2), pp.198–217.

Wells, D. 1986. Guide to GPS positioning, Canadian GPS Associates.

Rock Mechanics, Nelson & Laubach (eds) © 1994 Balkema, Rotterdam, ISBN 90 5410 380 8

Stability analysis of a penstock slope

T.-C. Ke, B. Thapa & R. E. Goodman
Department of Civil Engineering, University of California, Berkeley, Calif., USA

ABSTRACT: A penstock rock slope was analyzed by two methods of stability analysis. They are a limiting equilibrium analysis initially developed by Goodman and Bray, and Discontinuous Deformation Analysis (DDA). Because the latter needs no assumption of a failure mode, and takes into account the deformability and real-time response of a slope, it proved to be more versatile and realistic.

1. INTRODUCTION

This paper compares two methods of stability used for a rock slope, on which a siphon constructed in 1968 requires remediation due to observed accumulated downslope movement. Field observations identified a toppling mode of failure, and seismic refraction surveys established the base of flexural cracking.

The first method of stability analysis used is a modification of the limit equilibrium method for toppling analysis, initially developed by Goodman and Bray (1976), in which the support force for the toe block is computed for a given friction angle. The second method is a new approach based on Discontinuous Deformation Analysis (DDA), which is able to compute the deformed geometry of a block system subjected to a general loading. In this method, the computed displacement of a selected point on the slope can be used to determine the limiting friction angle or the support force required for a given factor of safety.

2. SITE CHARACTERIZATION

The penstock slope is located at Belden, California, near the north fork of the Feather River approximately two miles north of the junction of Highway 70 with Caribou Road. Geophysical data suggest overlying colluvium (Layer I) is underlain by loosened or weathered phyllite (Layer II), which in turn is underlain by solid or unweathered rock (Layer III). Figure 1 depicts the siphon cross section of the slope which is under study in this paper. Line AB near the top hill is a tension crack. The foliation of the phyllite in Layer II is dipping into the slope at approximately 68^0 . The densities of Layers I and II are 2400 (kg/m^3) and 2480 (kg/m^3), respectively. The Young's modulus and Poisson's ratio of Layer II are estimated to be 100 (MPa) and 0.2. Field measurements suggest the friction angle of Layer II varies from 37.5^0 to 39^0

Extensometer measurements show greatest displacements at the slope crest, decreasing semicontinuously downward, indicating a toppling mode of failure. The formation of isolated columns within Layer II provides a reasonable configuration for toppling. Figure 2 shows an idealized profile of the slope section for analysis, in which Layer II has been replaced by 33 columnar blocks whose side walls are parallel to the foliation plane, and the portion on the right side of Line AB (see Figure 1) is assumed stable and neglected.

3 MODIFIED LIMIT EQUILIBRIUM ANALYSIS OF BLOCK TOPPLING

Goodman and Bray's (1977) limit equilibrium analysis (LEM) of block toppling was modified to allow a better modeling of the site geometry and loading conditions. The modified analysis accommodates blocks with bases not necessarily perpendicular to the sides. Loading from soil overburden and water pressure on the block sides are also included in the modified analysis.

Figure 3 shows the block geometry and loading conditions modeled by the analysis. The block sides are inclined at a constant angle of β from the horizontal while the block base is inclined at a variable angle of δ_n. Each block has a loading from a soil overburden of vertical thickness z_n, water pressure along the sides to depths h_n and h_{n-1}, and weight W_n.

As in Goodman and Bray's (1977) analysis, the force P_{n-1} is the force required to prevent either toppling or sliding, depending on which mode requires the higher value of P_{n-1}. In the case of toppling, $P_{n-1,t}$ is calculated by taking moments about point O in Fig. 1 (expressions for the loads on block n and other variables can be found in the appendix):

$$P_{n-1,t}=\frac{1}{y_{n-1}-\Delta_n}\left\{\begin{aligned}&P_n(y_n+g_n-\mu_n t)+\left[\frac{\gamma_w}{6}\sin\beta\Delta h[3h_n^2-3h_n\Delta h_n+\Delta h_n^2]+\frac{1}{2}\gamma_w\sin\beta h_n^2 g_n\right]\\&+\frac{\gamma_s z_n t_n\cos\alpha\sin\alpha}{\cos(90-\alpha-\beta)}\left([(y_n+g_n)\cos(\alpha-\delta+\rho)]-\left[\frac{t_n}{2}\sin(\alpha-\delta+\rho)\right]\right)\\&+\gamma_r y_n t_n\left[\sin(\delta-\rho)\left(\frac{y_n}{2}+g_n\right)-\cos(\delta-\rho)\right]+\gamma_r|g_n|\frac{t_n}{2}[\sin(\delta-\rho)a_1-\cos(\delta-\rho)a_2]\end{aligned}\right\}$$

---------------------------(1)

Forces parallel and perpendicular to the base of block n are summed to find the value of P_{n-1} needed to prevent sliding:

$$P_{n-1,s}=\frac{P_n(2\mu\sin\rho-\mu^2\cos\rho)+[(\mu\sin\rho+\cos\rho)U_n-U_{n-1}]-[D_s(\mu\cos\rho-\sin\rho)]-[W_n(\mu\cos\delta-\sin\delta)]}{(\cos\rho-\mu^2\cos\rho)+2\mu\sin\rho} \quad (2)$$

The analysis is implemented in a computer program that sequentially considers each block beginning from the uppermost block and ending with the toe block. For each block, the value of P_{n-1} is calculated for the given geometric configuration and friction angle as the greater of $P_{n-1,t}$ and $P_{n-1,s}$. P_o is the value of P_{n-1} for the toe block; P_o is the support force needed to stabilize the block system if it is greater than zero. Alternately, the minimum friction angle needed for stability can be found by repeating the analysis with modified μ until P_{n-1} is zero.

4 DISCONTINUOUS DEFORMATION ANALYSIS (DDA)

Shi and Goodman (1985) introduced the backward model of DDA to back-calculate the best fit deformed geometry of a block system from measured displacements and strains. Shi (1988), and Shi and Goodman (1989) further generalized the two-dimensional DDA to include the capability of analyzing the mechanical response of block systems in a forward model. Since then, DDA has become another numerical procedure in the family of discrete element methods to deal with complex interactions between disconnected blocks (Cundall & Hart 1989). (In the following, DDA only refers to the forward model of DDA.)

A system of simultaneous equations in DDA are formulated by minimizing the total potential energy of a block system. Therefore, the dynamic equilibrium of a block system is satisfied at any time. The inclusion of inertia terms also makes the global stiffness matrix positive definite and diagonally dominant, without numerical hazard. If a 'dynamic' mode is specified, DDA yields a real-time response of a block system. DDA incorporates a complete block kinematics which fulfills no interpenetration and no tension between blocks. This kinematical theory provides efficient rules for solving simultaneous equations with opening-closing iterations within each time step under the constraints of no penetration and no tension.

Shi's original DDA code can handle a system of linearly elastic, constant strain blocks subjected to large displacements, rotations and deformations, and general loading conditions. Figure 4 shows the deformed geometry of the block system in the slope section (Figure 2) with Layer I omitted and a dry condition at t=1sec and $\phi=0^0$ [1]. Since all block sides in this case are frictionless, the exclusive failure mode for each block is sliding as revealed in this figure. With adequate selection of a *slope performance indicator*, DDA is a useful slope stability analysis tool. As for example in Figure 4, the horizontal displacement of the toe can be treated as an indicator, i.e., the larger the horizontal displacement of the toe, the more unstable the slope. A limiting value of a slope parameter (e.g., friction angle) corresponds to the value of ϕ at which the indicator approaches zero. The idea of regarding DDA as a stability analyzer was initially derived from the work of Yeung (1991).

5 DETERMINATION OF LIMITING FRICTION ANGLE

Before applying both the modified Goodman and Bray's approach and DDA to the block system of concern, a simple example is first investigated as follows.

5.1 *Example of a deformable block*

A rectangular block, 1 m high and 2 m wide, sits on a slope at an inclination, $\beta=45^0$. The block has a density of 2480 (kg/m^3), and gravity is the only external force.

An assumption made in the limiting equilibrium analysis is that each block is rigid.

Here we examine the effect of block rigidity. Figure 5.a shows the P_o versus ϕ curve

---

[1] assume all block sides have the same friction angle

obtained from the modified limit equilibrium approach, which gives a ϕ_L of 45^0 at P_o=0 (which is obvious for this simple case). To presume the rigidity of the block, a very high value of Young's modulus of 5000 (MPa) was input in the DDA run. The solid line in Figure 5.b shows ϕ versus the computed horizontal displacement of the toe at t=1sec; this also yields the result $\phi_L = 45^0$ (as it must).

For a deformable block, one simply inputs a realistic value for Young's modulus; the value of ϕ_L now becomes larger than β. With reference of the dotted lines of Figure 5.b, ϕ_L is much higher than 45^0 when the block is very soft, with Young's modulus equal to 0.1 (MPa). This analysis shows that the effect of non-rigidity is to mix elastic and plastic displacements in the output, complicating the interpretation of the point of limiting equilibrium. One possible interpretative procedure is depicted in Figure 5.b.

5.2 *Analysis of the slope in figure 2*

Three cases are considered: Case 1 - with Layer I omitted and a dry condition; Case 2 - with Layer I included and a dry condition; Case 3 - with Layer I included and full saturation[2] with the phreatic surface conforming to the boundary between Layers I and II. In the DDA runs, the weight of a soil slice above each block was regarded as a concentrated load at the middle of the top of the block; similarly the water load acting on the block side walls was treated as a concentrated load. The initial stresses in the block system were assumed to be zero.

The analysis results of the modified limit equilibrium approach and DDA are summarized in Table 1. Determinations of ϕ_L for Case 2 by these two methods are presented in Figures 6.a and 6.b, respectively. Figure 6.b indicates that: (1) large displacements accrue rapidly when ϕ is below a limiting value; (2) the toe tends to hesitate at a small t, but responds sensitively at a larger t. The time of 0.6sec was assumed to be sufficient to define the limit between stable and unstable slope conditions.

Table 1: Results of LEM and DDA for Belden Slope

Case	Limiting ϕ for Limit Equilibrium	Limiting ϕ for DDA
1	32	37.5
2	34.5	38.5
3	36	40

Differences between the results of the two methods are attributed to the effects of: (1) the deformability of blocks; (2) the difference between dynamic and static equilibrium of blocks as discussed in Aydan et. al. (1989) and Yeung (1991); (3) the differences between a block system with a combined mode of toppling and sliding versus toppling only. DDA results seem more reliable, and they agree better with field measurements. Therefore DDA is preferred for the support design studies.

[2] water completely filling inter-block joints

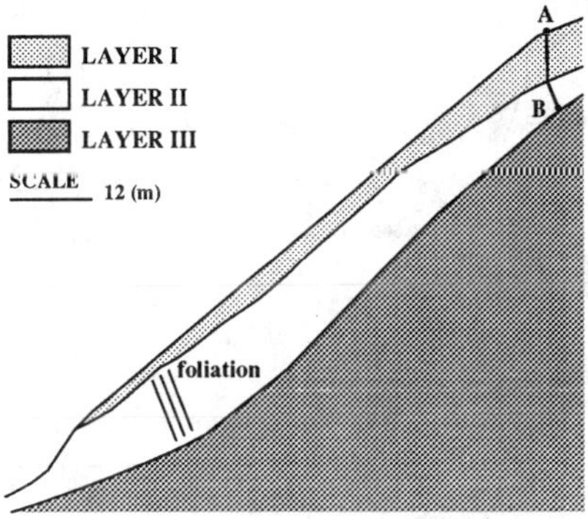

Figure 1 Cross section of the Belden siphon slope

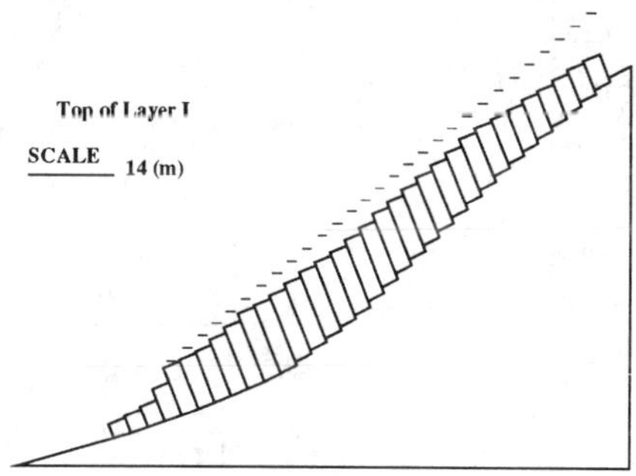

Figure 2 Idealized block geometry of the siphon section

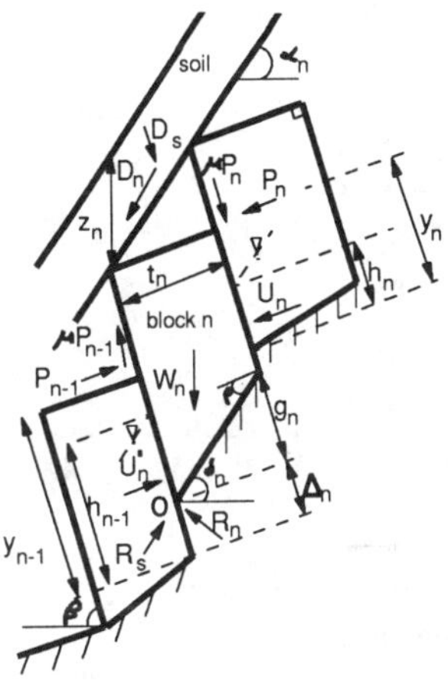

Figure 3 Parameters used in the modified Goodman & Bray's limiting equilibrium analysis

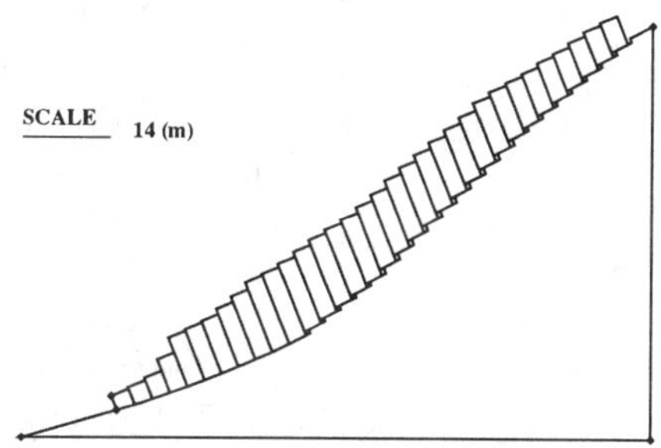

Figure 4 Deformed block geometry of the slope at t=0.6sec and ϕ=0°

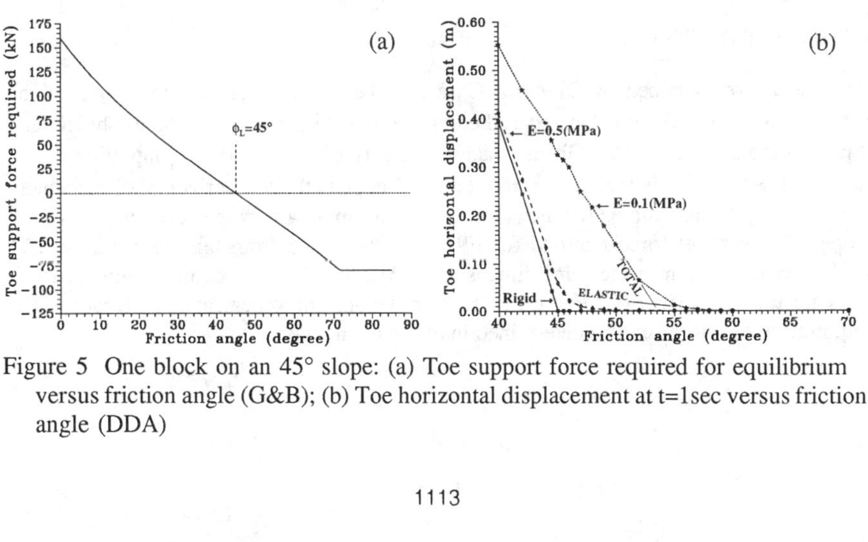

Figure 5 One block on an 45° slope: (a) Toe support force required for equilibrium versus friction angle (G&B); (b) Toe horizontal displacement at t=1sec versus friction angle (DDA)

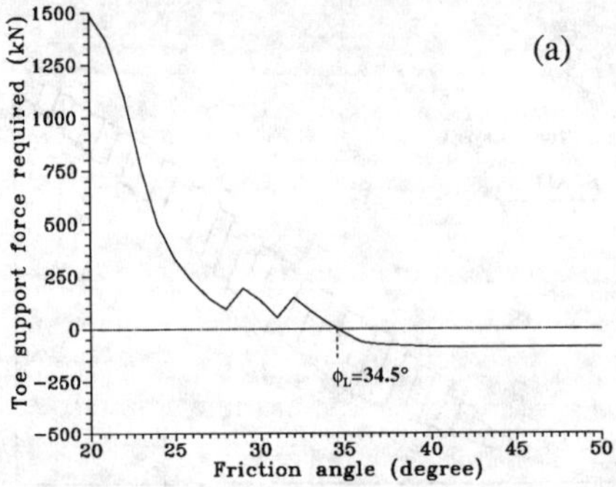

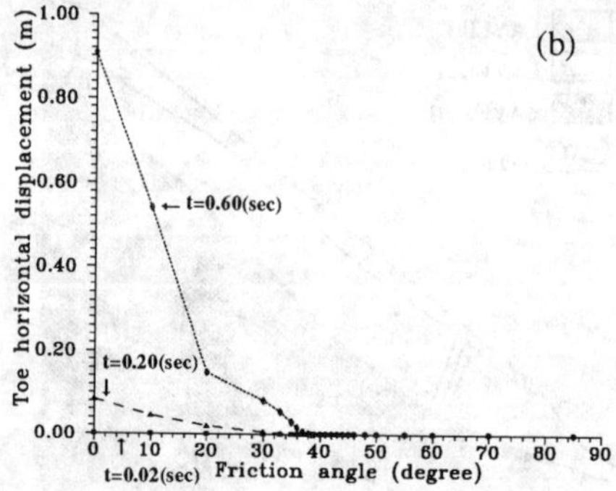

Figure 6 Case 2 of the slope: (a) Toe support force required for equilibrium versus friction angle (modified G&B); (b) Toe horizontal displacement versus friction angle (DDA)

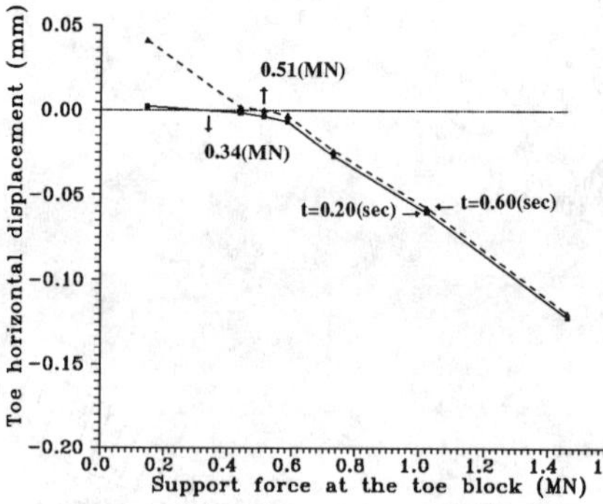

Figure 7 Toe horizontal displacement versus support force of Case 2 at FS=1.1 (DDA)

6 SUPPORT DESIGN

Assume the ϕ_L obtained by DDA for Case 2 to be the available friction angle of the slope. The support force is located at the upper-outer (left) corner of the toe block, and dips into the slope at 30^0. Given a factor of safety (FS) of 1.10 (i.e., inputting a $\phi = \tan^{-1}(\tan(38.5^0)/1.1)$ in the DDA runs), Figure 7 depicts the toe horizontal displacement versus support force curves for the slope. Again a time-lag is noted for response of the slope. The support force required for FS=1.1 is estimated from this figure to be 0.51 MN (per unit width in the third dimension). At FS=1.25, the required support force rises up to 0.89 (MN/m). For other support layouts or other factors of safety, the required support force can be determined in the same manner.

7 CONCLUSIONS

This paper presented a modified limiting equilibrium analysis for block toppling to accommodate more general geometry and loading conditions, and introduced DDA to perform slope stability analysis. Although the modified limit equilibrium approach is easier to use than DDA, its assumptions can give a significantly lower value of $\phi_{required}$ than DDA. DDA automatically handles the interactions between blocks, and considers the deformability of the slope. Therefore, DDA is expected to become a more promising alternative to analyze a variety of stability problems previously addressed only by limiting equilibrium analyses.

8 REFERENCES

Aydan, O, Shimizu, Y. & Ichikawa, Y. 1989. The effective failure modes and stability of slopes in rock mass with two discontinuity sets. *Rock Mechanics and Rock Engineering* (22): 163-188.

Cundall, P. A. & Hart, R. D. 1989. Numerical modelling of discontinua. *In* Mustoe et. al. (eds), *Proceedings of the 1st U. S. Conference on Discrent Element Methods.*

Goodman, R. E. & Bray, J. W. 1976. Toppling of rock slopes. *Proceedings of the Specialty Conference on Rock Engineering for Foundations and Slopes.* ASCE.

Shi, G. 1988. *Discontinuous Deformation Analysis - A New Numerical Model for the Statics and Dynamics of Block Systems*, Ph.D Dissertation, Department of Civil Engineering, University of California, Berkeley

Shi, G. & Goodman, R. E. 1985. Two-dimensional discontinuous deformation analysis. *International J. Numerical and Analytical Methods in Geomechanics* 9(6): 541-556.

Shi, G. & Goodman, R. E. 1989. Generalization of two-dimensional discontinuous deformation analysis for forward modeling. *International J. Numerical and Analytical Methods in Geomechanics* 13(4): 359-380.

Yeung, M. 1991. *Application of Shi's Discontinuous Deformation Analysis to Rock Slope.* Ph.D Dissertation. Department of Civil Engineering. University of California, Berkeley.

9 APPENDIX

The equation for $P_{n-1,t}$ is valid only when $\alpha > \delta$ and $\phi < \delta$. The constants a_1 and a_2 used in the expression for $P_{n-1,t}$ are:

$$a_1 = \begin{Bmatrix} \frac{2g_n}{3} & \text{if } g_n \geq 0 \\ \frac{g_n}{3} & \text{if } g_n < 0 \end{Bmatrix} \quad a_2 = \begin{Bmatrix} \frac{t_n}{3} & \text{if } g_n \geq 0 \\ \frac{2t_n}{3} & \text{if } g_n < 0 \end{Bmatrix}$$

The expression for the loads used to calculate $P_{n-1,t}$ and $P_{n-1,s}$ are:
Water forces on block n:

$U_n = \frac{1}{2}\gamma_w \sin\beta h_n^2$ acting at $\frac{h_n}{3} + g_n$ from the lower corner

$U_n^{'} = \frac{1}{2}\gamma_w \sin\beta\left(h_{n-1} - \Delta_n\right)^2$ acting at $\frac{\left(h_{n-1} - \Delta_n\right)}{3}$ from the lower corner

The forces exerted by the soil overburden, assuming no side forces on the soil slices, on block n are:

$$D_n = \gamma_s z_n t_n \cos\alpha \sin\alpha / \cos(90 - \alpha - \beta)$$

$$D_s = \gamma_s z_n t_n \cos\alpha \cos(\delta - \rho)$$

The weight of block n is:

$$W_n = \gamma_r \left(y_n t_n - \frac{t_n g_n}{2} \right)$$

Rock Mechanics, Nelson & Laubach (eds) © 1994 Balkema, Rotterdam, ISBN 90 5410 380 8

Estabilidad de taludes en los cortes del tramo Río Balsas-Chilpancingo de la nueva Autopista México-Acapulco

Roberto Ayala Muñiz & Alfredo Aguirre Ponce
Ingenieros Civiles Asociados, S.A. de C.V.

R E S U M E N La variación de las características generales en los macizos rocosos y en la disposición de sus sistemas de discontinuidades a lo largo del tramo Río Balsas-Chilpancingo de la nueva Autopista México-Acapulco, requirió la realización de un levantamiento geotécnico general, de todos y cada uno de los cortes, con objeto de establecer las recomendaciones generales de estabilización necesarias.

En el presente artículo se describe el trabajo geotécnico realizado a lo largo del tramo Río Balsas-Chilpancingo y se presentan los criterios de análisis de estabilidad y las recomendaciones generales de estabilización para los cortes de mayor interés técnico.

1. INTRODUCCIÓN

La nueva Autopista México-Acapulco ha sido construida respetando exigentes especificaciones de geometría y trazo, con una alta calidad de las obras y una construcción en tiempos restringidos; dichas condiciones hicieron necesaria la ejecución de grandes excavaciones, en general mediante procedimientos sistemáticos de voladura y con taludes comprendidos entre 0.5:1 y 0.75:1.

La diferencia en las condiciones geológicas a lo largo del trazo de la nueva autopista, así como la variación en el grado de alteración y en la disposición de los sistemas de discontinuidades en los macizos rocosos, generan una gran diversidad de condiciones de inestabilidad en una porción importante de los cortes.

La cantidad de taludes en corte comprendidos en el tramo en estudio y la restricción de tiempo para el análisis de las condiciones de estabilidad y la determinación de los tratamientos de estabilización, hicieron necesaria la estructuración de un criterio de evaluación de la inestabilidad de los taludes, tal que permitiera la selección de los cortes con inestabilidades de mayor riesgo, en los cuales se efectuaron análisis de estabilidad basados en la topografía final del talud y en ensayes de laboratorio del material muestreado en el propio corte.

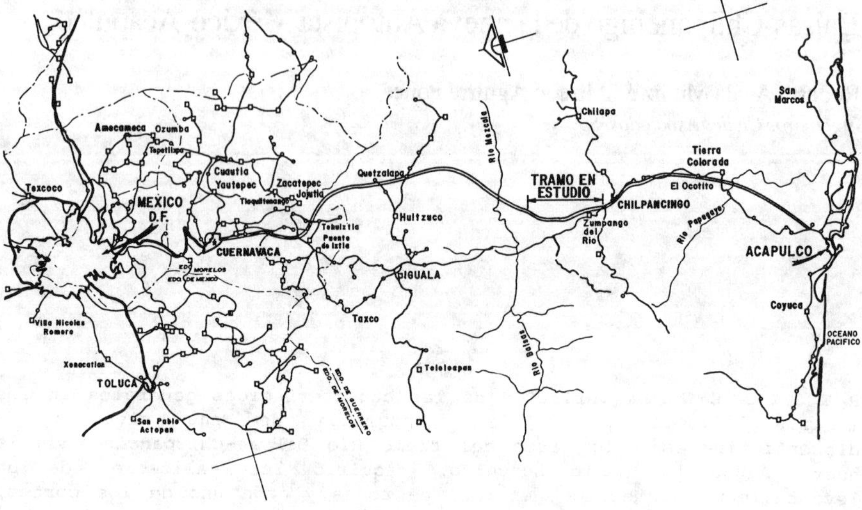

Fig. 1 - TRAZO GENERAL DE LA AUTOPISTA MEXICO - ACAPULCO

Los cortes estudiados y contenidos en el presente artículo corresponden al tramo Río Balsas - Chilpancingo, entre los cadenamientos 108+480 y 837+000 del proyecto ejecutivo, cuyo trazo se ha señalado en el plano de la Figura 1.

En el proyecto ejecutivo de la nueva autopista, el cadenamiento 108+480 corresponde al sitio donde la autopista intersecta el arroyo "Paso de Laguna"; para la presentación de este artículo, se han respetado los cadenamientos indicados en los planos del proyecto ejecutivo.

2. GEOLOGÍA

El panorama fisiográfico del área estudiada de acuerdo con Raisz (Ref. 1), se localiza en la porción central de la subprovincia denominada cuenca del Balsas - Mezcala, incluida dentro de la provincia Sierra Madre del Sur (Fig. 2).

Los rasgos geomorfológicos varían según el tipo de roca aflorante. Los carros y cercanías formados por calizas mesozoicas suelen ser los más prominentes de la región y muestran por lo general un aspecto redondeado. Donde afloran rocas mesozoicas más suaves, como lutitas, limolitas y areniscas interestratificadas, los cerros son más bajos, con una pared de drenaje dendrítico más denso con flancos menos inclinados, mostrando una etapa de madurez.

En contraste con las rocas sedimentarias mesozoicas, las rocas terciarias (brechas y tobas volcánicas) tienen un drenaje menos intenso, de tipo dendrítico y angular que representa, a su vez, una

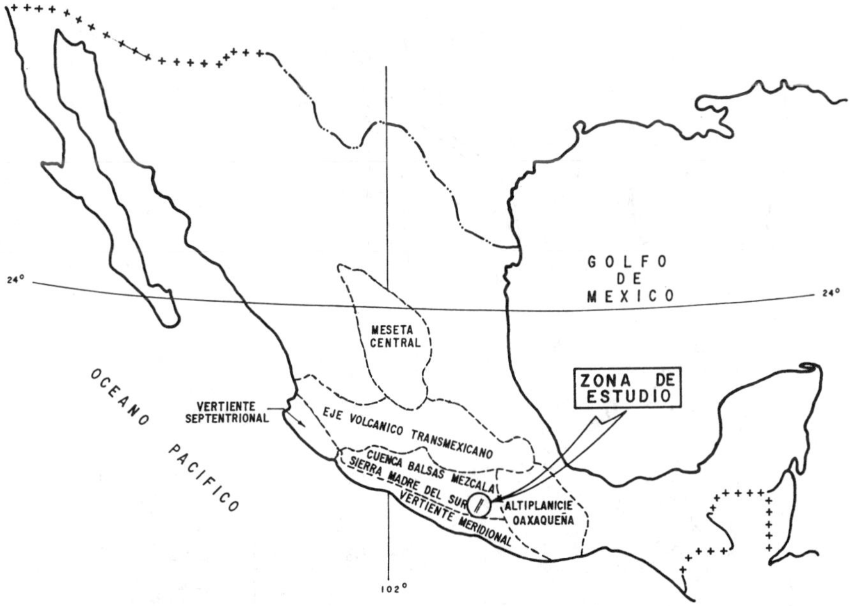

Fig. 2 – LOCALIZACION DE LAS PROVINCIAS FISIOGRAFICAS

etapa de juventud del ciclo geomorfológico; las sierras corresponden a pliegues anticlinales y los valles a sinclinales o fosas tectónicas orientadas Norte-Sur y Noroeste-Suroeste.

La cobertura sedimentaria y volcánica registra edades que fluctúan entre el Mesozoico al reciente, observándose deformada, metamorfoseada, plegada y fallada, diferenciándose claramentelas tendencias estructurales principales mostradas en la Figura 3: anticlinario Plan de las Cumbres, falla Huitziltepec, falla Mezcala y alineamiento NW-SE (Ref. 2).

3. LEVANTAMIENTO GEOTÉCNICO

Los cortes comprendidos en el tramo de interés fueron identificados físicamente en el sitio, conforme con los cadenamientos indicados en los planos del proyecto ejecutivo.

El levantamiento geotécnico se realizó mediante la observación directa de las condiciones de la roca en los cortes y la medición de las características geométricas de sus sistemas de discontinuidades, valorando posteriormente la calidad de los macizos rocosos, conforme con los criterios de clasificación propuestos por Romana (Ref. 3) y Barton (Ref. 4), cuyos resultados se incluyen en las tablas mencionadas.

A partir de la clasificación geotécnica de los taludes y tomando en

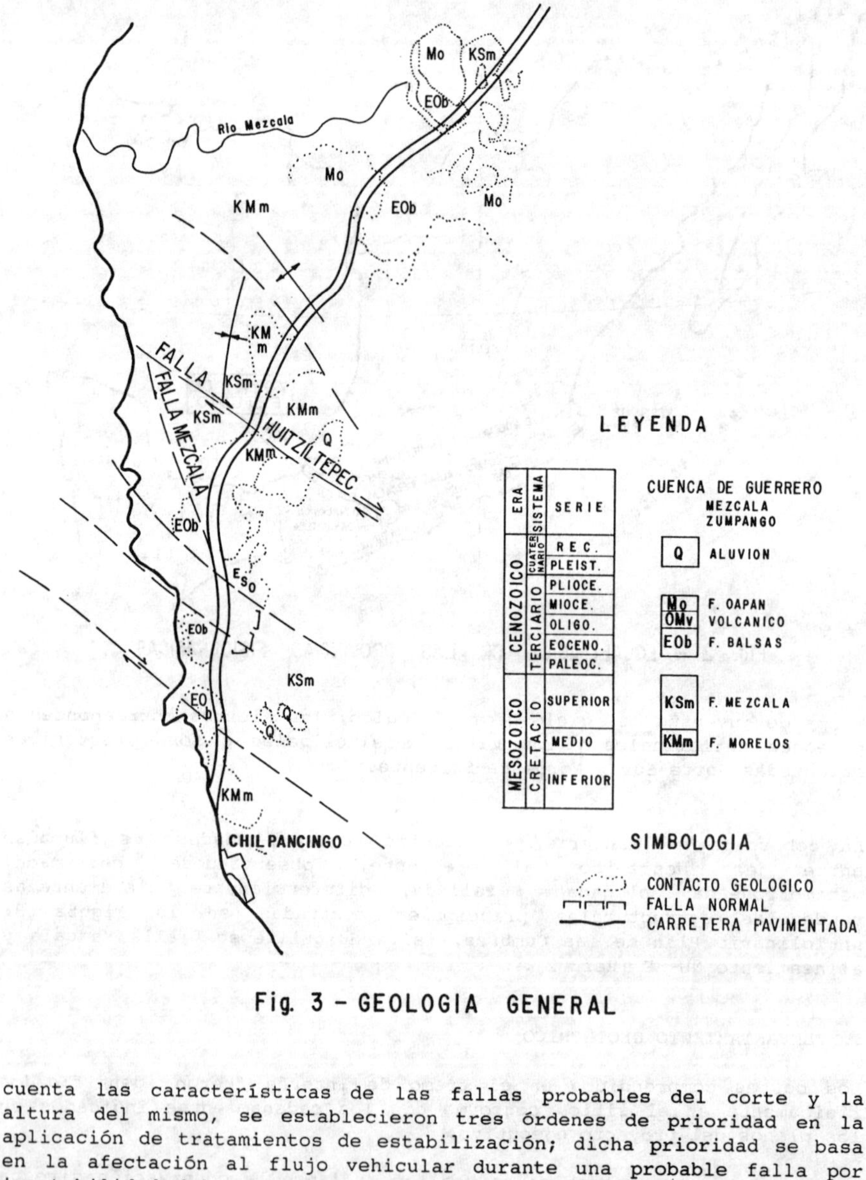

Fig. 3 - GEOLOGIA GENERAL

cuenta las características de las fallas probables del corte y la altura del mismo, se establecieron tres órdenes de prioridad en la aplicación de tratamientos de estabilización; dicha prioridad se basa en la afectación al flujo vehicular durante una probable falla por inestabilidad del corte. El primer orden corresponde a la prioridad inmediata, donde se prevé una suspensión del flujo vehicular; el segundo orden corresponde a la prioridad mediata, donde se prevé que puede alterarse el flujo vehicular, pero sin suspenderse; el tercer orden corresponde a desprendimientos que pueden retirarse sin alterar el flujo vehicular.

4. ANÁLISIS DE ESTABILIDAD

El levantamiento geotécnico descrito anteriormente, permitió la selección de los cortes de mayor riesgo, para los cuales se realizaron muestreos del material, pruebas de laboratorio y análisis de estabilidad. A continuación se describen de manera general las características geomecánicas y análisis de estabilidad de cinco taludes de mayor interés técnico, los cuales se han identificado con los cadenamientos señalados en los planos ejecutivos.

4.1 El talud en corte en el tramo 110+160 - 110+600.

El corte se excavó con taludes 0.5:1 y dos bermas de 4.0 m de ancho (Fig. 4). La formación corresponde a lutitas y limolitas de

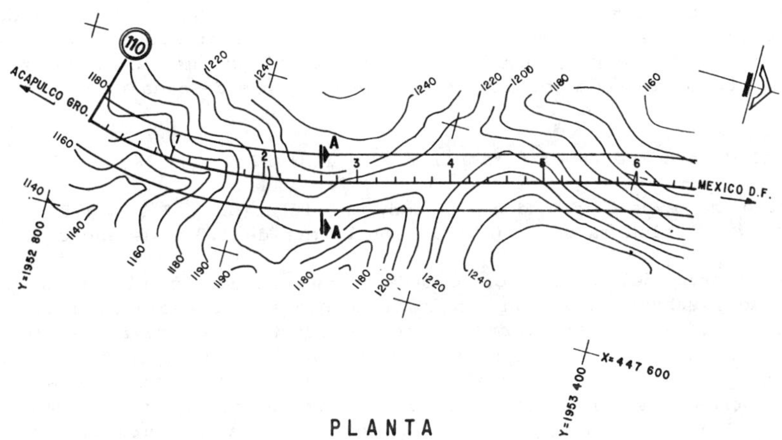

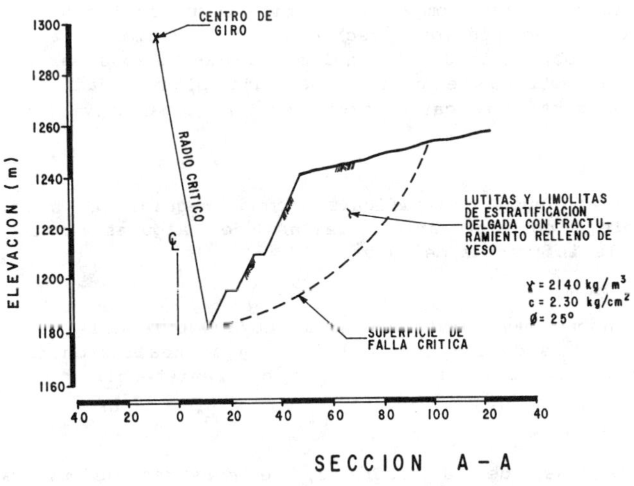

CONDICION	FACTOR DE SEGURIDAD
ESTATICA	1.6
SISMICA	1.5

Fig. 4 - 110+160 - 110+600

estratificación delgada con fracturas rellenas de mineral de yeso; la parte superior está constituida por un afloramiento de yeso muy fracturado con un espesor aproximado de 5.0 m.

El talud puede clasificarse como regular, con probables fallas de de algunas cuñas, principalmente bajo la influencia del agua.

Se realizaron ensayes a compresión simple y tenso-compresión, así como ensayes triaxiales en muestras obtenidas del corte, y un análisis para establecer los parámetros geomecánicos del material, adoptando un ángulo de fricción interna de 25° y una cohesión aparente de 2.3 kg/cm2, obteniéndose un factor de seguridad de 1.65 para condiciones estáticas y de 1.48 para condiciones sísmicas, las cuales garantizan una estabilidad general.

Debido a la susceptibilidad del material a los efectos del intemperismo, se recomendó únicamente, como tratamiento para el mejoramiento del corte, una protección sistemática mediante concreto lanzado, complementado con líneas de drenes al pie de cada talud.

4.2 Talud en corte en el tramo 113+710 - 113+990.

El corte se excavó con dos taludes, un inferior 0.5:1 y 18 m de altura, y un talud superior 0.75:1, con bermas de 4.0 m de ancho (Fig. 5).
La zona norte del corte corresponde a una formación riolítica muy fracturada, subyaciendo a un conglomerado que aflora hacia la zona sur del corte; en la primera formación se han identificado cuatro sistemas de discontinuidades principales; dos de las cuales presentan juntas abiertas y rellenas de material arcilloso-arenoso de consistencia dura y los otros dos sistemas muestran juntas cerradas, o ligeramente abiertas pero limpias y poco rugosas. La zona sur del corte corresponde a un conglomerado formado por arena media y fina, con grava, poco arcilloso y de compacto a muy compacto, con una cementación heterogénea y nódulos brechados. Subyaciendo a la formación riolítica y sobreyaciendo al conglomerado de la zona sur del corte se localiza un estrato de arcilla de baja plasticidad y de consistencia firme a dura, muy carbonatada, con un espesor variable entre 6 y 8 m.

En general, el talud puede clasificarse como regular o malo, parcialmente estable, con probables fallas de algunas cuñas, principalmente bajo la influencia del agua.

Con el objeto de conocer las características geomecánicas del talud, se extrajeron dos muestras de la pared del talud y se realizaron tres sondeos mixtos, donde se realizaron ensayes de resistencia en el laboratorio.

En la sección transversal de la Figura 5, se muestran de manera esquemática los diferentes materiales que integran el talud así como los parámetros geomecánicos adoptados para el análisis de estabilidad.

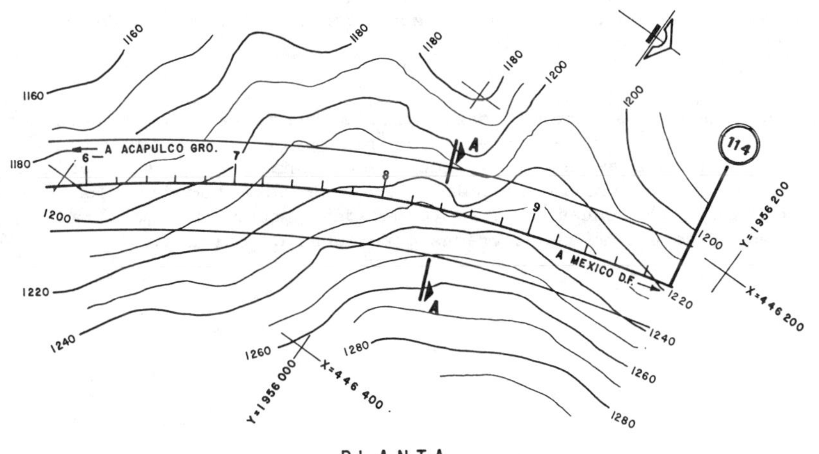

PLANTA

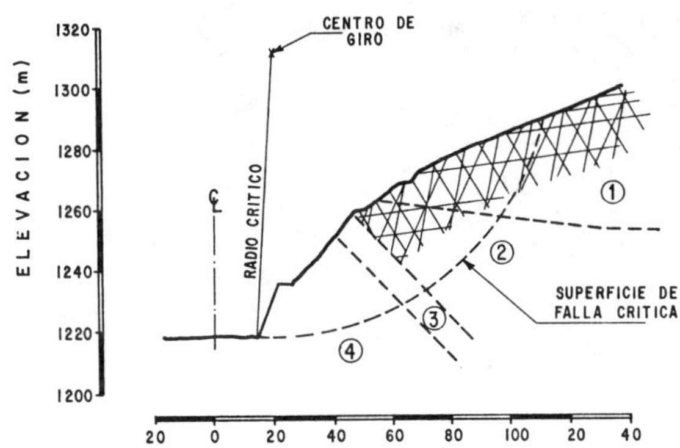

No.	MATERIAL	γ kg/m³	c' kg/cm²	ø °
1	TOBA RIOLITICA FRACTURADA	2.00	2.80	25
2	TOBA RIOLITICA INTEMPERIZADA	2.06	2.80	25
3	ARCILLA FIRME CARBONATADA	2.10	0.65	28
4	CONGLOMERADO CEMENTADO	2.15	3.20	29

CONDICION	FACTOR DE SEGURIDAD
ESTATICA	2.3
SISMICA	2.0

SECCION A - A

Fig. 5 - 113+710 - 113+990

En la misma Figura se ha marcado la línea que representa la superficie crítica de falla a partir de lo cual se obtuvieron factores de seguridad críticos de 2.3 y 2.0 para condiciones estáticas y sísmicas, los cuales resultan ampliamente aceptables. El tratamiento recomendado en este caso, incluye concreto lanzado sistemático, dos líneas de anclas bajo cada berma y líneas de drenes al pie de cada talud.

4.3 Talud en corte en el tramo 118+330 - 118+700.

El corte se excavó con talud 0.5:1 y una berma de 4.0 m de ancho (Fig. 6).
La formación corresponde a una caliza dolomitizada de fracturada a muy fracturada y se encuentra dividida en dos zonas, delimitadas por una

falla geológica localizada en la zona central del corte, siendo la zona sur la de mayor alteración.

De manera general se identificaron tres sistemas de discontinuidades; uno de los cuales corresponde a los planos de sedimentación, cuyas juntas son planas y ásperas, ligeramente intemperizadas y abiertas pero rellenas de arcilla arenosa y eventualmente húmedas; los otros dos sistemas presentan juntas onduladas, poco intemperizadas y generalmente se encuentran cerradas o rellenas de calcita.

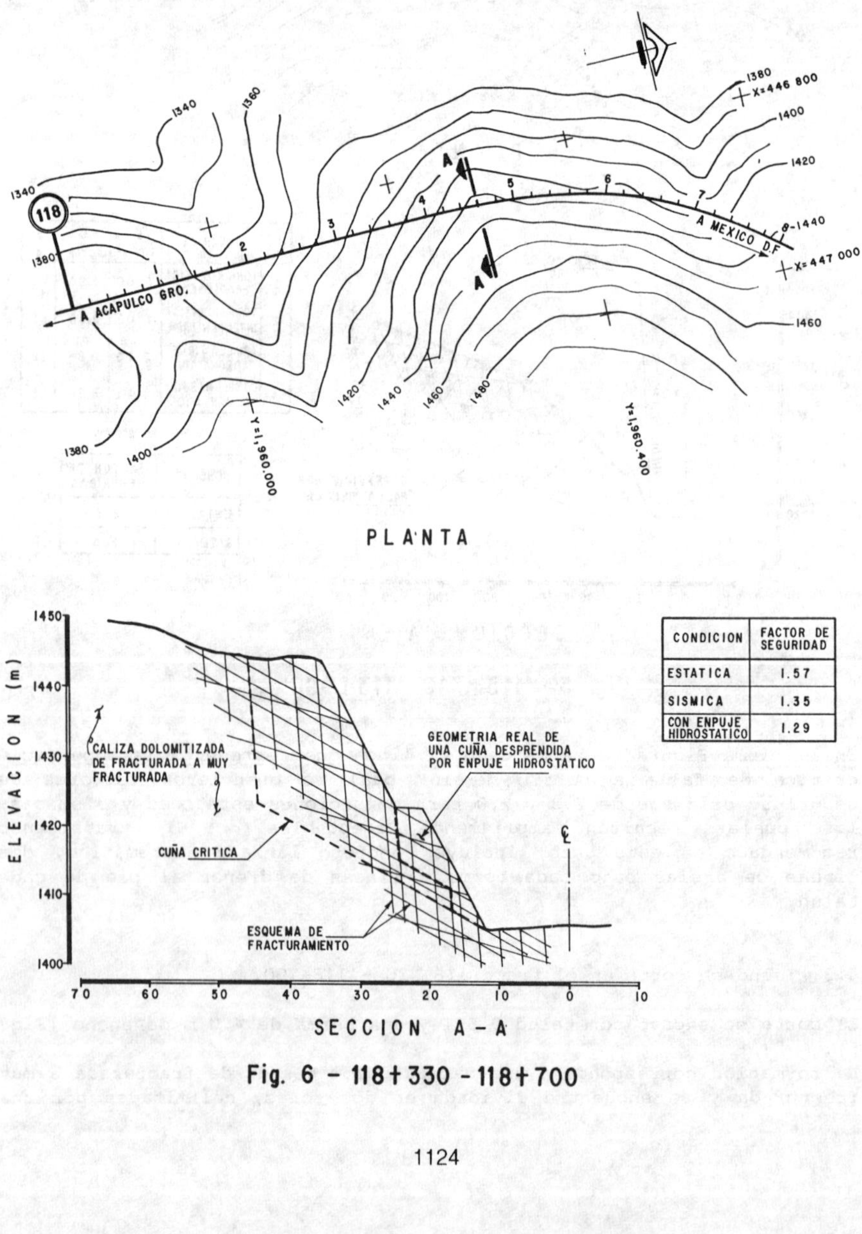

CONDICION	FACTOR DE SEGURIDAD
ESTATICA	1.57
SISMICA	1.35
CON ENPUJE HIDROSTATICO	1.29

Fig. 6 - 118+330 - 118+700

Con base en resultados de pruebas de laboratorio en muestras obtenidas del corte y con la geometría del desprendimiento marcado en el esquema de la Figura 6, se realizó un análisis para establecer los parámetros geomecánicos del material del material en las condiciones de una falla plana con empuje hidrostático a partir de la berma, adoptando un ángulo de fricción interna de 30º y considerando un factor de seguridad de 1.0 se obtuvo una cohesión aparente de 0.86 kg/cm2. Con los parámetros anteriores se realizaron análisis de estabilidad general del corte en condiciones drenadas, obteniéndose factores de seguridad de 1.62 y 1.38 para condiciones estáticas y sísmicas, se propuso como alternativa más recomendable la colocación sistemática de una retícula de drenes, complementando el tratamiento con la aplicación de concreto lanzado en las zonas de mayor fracturamiento y la colocación de una línea de anclaje en cada berma.

4.4 Talud en corte en el tramo 113+800 - 134+135.

El corte se excavó en ambos lados con un talud inferior 0.50:1 y dos taludes superiores 0.75:1, con bermas de 4 m de ancho (Fig. 7).

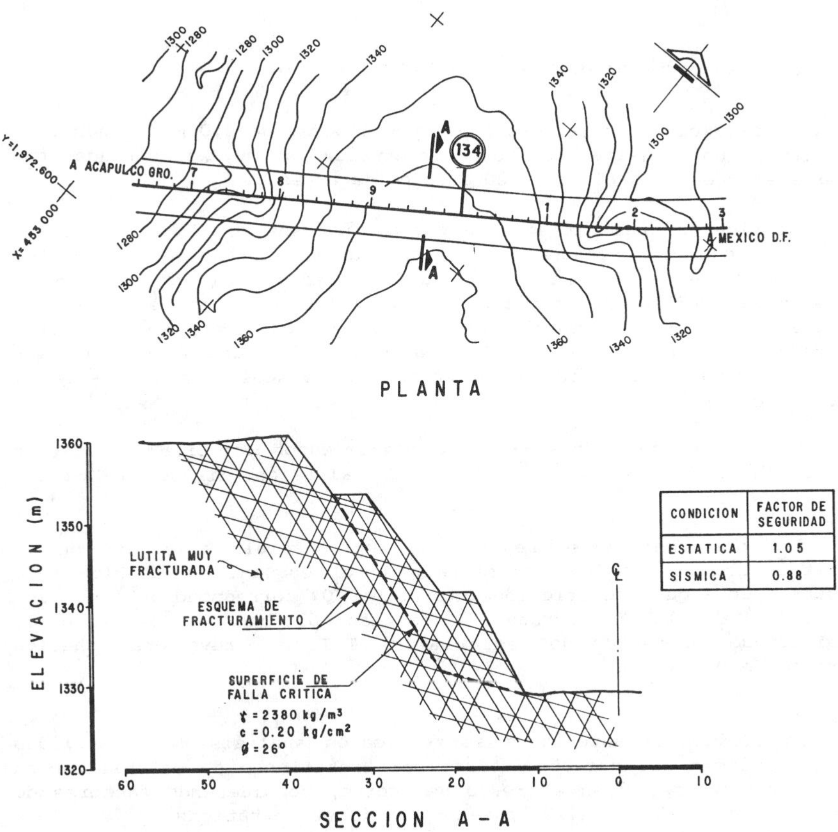

CONDICION	FACTOR DE SEGURIDAD
ESTATICA	1.05
SISMICA	0.88

Fig. 7 - 133+800 - 134+135

La formación corresponde a una lutita muy fracturada, en la cual se identificaron tres sistemas de discontinuidades principales; en forma general las juntas presentan separaciones entre 20 y 50 cm y son planas, lisas, cerradas y eventualmente rellenas de arcilla dura.

El corte se clasifica como regular y parcialmente estable, con desprendimientos probables de algunos bloques, principalmente durante la ocurrencia de lluvias.

El talud poniente resulta de mayor riesgo, y corresponde al esquema mostrado en la Figura 7.

Con base en los resultados de pruebas de laboratorio, se adoptaron como parámetros de resistencia característicos del macizo rocoso un ángulo de fricción interna de 26º y una cohesión aparente de 0.2 kg/cm2. Obteniéndose factores de seguridad de 1.1 y 0.9 para condiciones estáticas y sísmicas, respectivamente, los que resultan inaceptables para garantizar la estabilidad general del talud.

Se realizaron análisis de estabilidad para diferentes retículas de anclajes, hasta obtener factores de seguridad mínimos de 1.50 y 1.25 en condiciones estáticas y sísmicas respectivamente.

4.5 Talud en corte en el tramo 134+940 - 135+570.

El corte se excavó con taludes 0.5:1 y una berma de 4.0 m de ancho, lo constituye una caliza de poco fracturada a muy fracturada con intercalaciones de arcilla de 20 a 60 cm de espesor.

Se identificaron tres sistemas de discontinuidades en la masa rocosa; uno de los cuales corresponde los planos de sedimentación, con juntas planas y lisas, poco intemperizadas, cerradas pero ocasionalmente rellenas de arcilla-arenosa; las juntas de los otros dos sistemas son onduladas y abiertas, rellenas de calcita, eventualmente se encuentran alteradas y rellenas de arcilla arenosa; de manera general sus sistemas que se identificaron presentan una disposición favorable al corte.

La zona central del corte presenta plegamientos con una disposición desfavorable, asociados con un sistema de fallas con dirección NW-SE.

Para la zona central se realizó un retroanálisis a partir de la geometría del desprendimiento marcado en el esquema de la Figura 8, adoptando un ángulo de fricción interna de 30º correspondiente con los resultados de la interpretación de las pruebas de laboratorio; considerando un factor de seguridad de 1.0 se obtuvo una cohesión aparente de 0.4 kg/cm2.

Con los parámetros anteriores se realizaron análisis de estabilidad para establecer las características del tratamiento de estabilización en la zona central y en el resto del corte, considerando factores de seguridad de 1.5 y 1.25 para condiciones estáticas y sísmicas respectivamente.

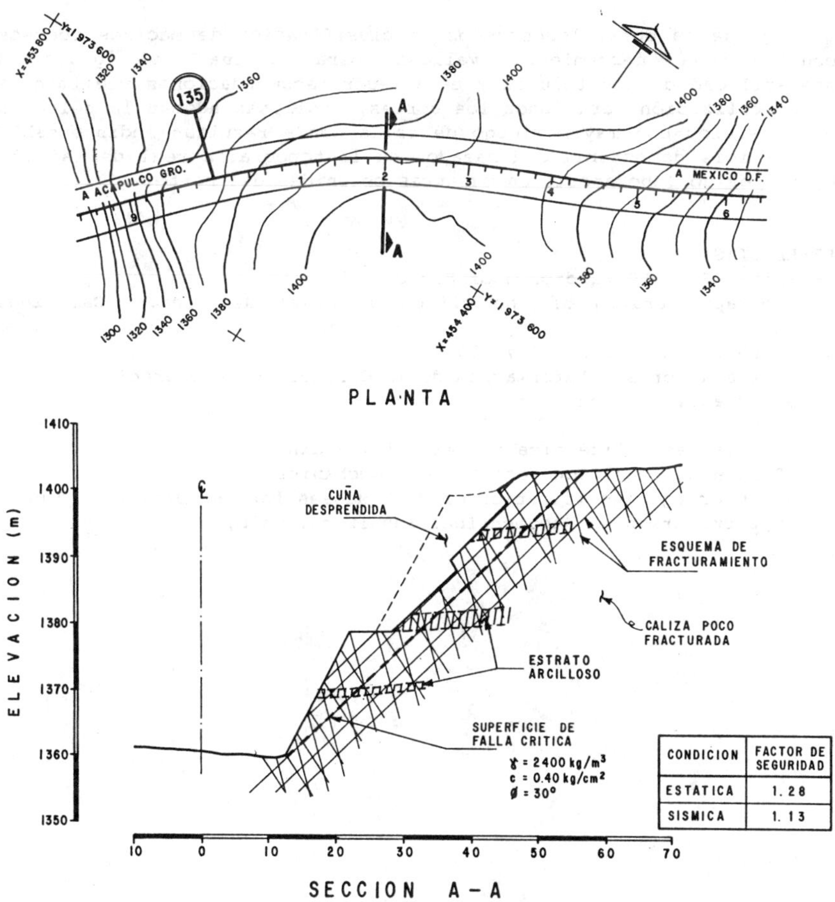

Fig. 8 - 134+940 - 135+570

La estabilización general del talud se recomendó mediante concreto lanzado en las zonas de mayor fracturamiento, con un anclaje local en zonas de desprendimiento, de grandes bloques y líneas de drenes a lo largo década talud.

En la zona central del corte se recomendó el abatimiento del talud, adoptando el perfil del plegamiento y la colocación sistemática de una retícula de anclaje, complementado el tratamiento con una protección de concreto lanzado y líneas de drenes al pie del talud.

5. CONCLUSIONES

En los proyectos geotécnicos de autopistas y obras de comunicación, donde puede esperarse una gran variedad en las condiciones geológicas y en la disposición y alteración de los sistema de discontinuidades a lo largo de la obra, los métodos de clasificación de macizos rocosos, resultan una herramienta valiosa para evaluar el grado de inestabilidad de los taludes y establecer recomendaciones preliminares de estabilización para todos los cortes, facilitan además la selección de los sitios con mayor riesgo de falla donde resultan indispensables los análisis de estabilidad basados en la topografía real del sitió y en ensayes de laboratorio en muestras obtenidas del corte.

REFERENCIAS

1. Raisz, E. 1959 Landforms of Mexico.
   Geography Branch of the office of Naval Research. Cambridge, Mass.
2. Petróleos Mexicanos, 1971-1976.
   Varios autores. Información de la Gerencia de Exploración.
3. SMR Classification.
   Romana, M.
   Universidad Politécnica de Valencia, Spain.
   7º International Congress on Rock Mechanics, 1991.
4. Engineering Classification of Rock Masses for the Design of Tunnel Support.Norwegian eotechnical Institute, Oslo,

Rock Mechanics, Nelson & Laubach (eds) © 1994 Balkema, Rotterdam, ISBN 90 5410 380 8

Estabilización de un talud excavado en riolitas

Enrique Padilla Corona
Secretaría de Comunicaciones y Transportes, México

RESUMEN Se presentan y analizan las soluciones aplicadas a mejorar las condiciones de estabilidad de un talud excavado, ubicado en el km 107+520 al 107+800, de la autopista Plan de Barrancas, situada en los límites de los estados de Jalisco y Nayarit en México. Esta excavación que se realizó hasta una profundidad de 65.0 m, en una formación riolítica muy heterogénea con presencia preponderante de agua subterránea, sufrió derrumbes desde el proceso constructivo por lo cual fue necesario revisar el proyecto para determinar las acciones necesarias para lograr su estabilización.

ABSTRACT Solutions are applied to improve the stability conditions of an excavated slope are shown and analyzed. This slope is situaded on km 107+250 to 107+800 on the Plan de Barrancas Highway in the limits of the states of Jalisco and Nayarit in Mexico. This excavation realized down to a maximum depth of 65.0 m in a formation basically composed of rhyolitic rock very heterogenous with the prevailing presence of groundwater, suffered landslides since the constructive process, for which was necessary to review the project to determine the necessary actions to obtain its stabilization.

1 INTRODUCCION

La autopista Plan de Barrancas, tiene una longitud de 18.0 km, cuatro carriles de circulación y constituye un libramiento de la gran Barranca de Mochitiltic, que limita a los estados de Jalisco y Nayarit y se localiza en el extremo occidental del Eje Neovolcánico Mexicano, zona de fuerte actividad sísmica.

Dentro de las excavaciones realizadas en esta vía, la más importante es la ubicada en el km 107+520 al 107+800, que se desarrolla en una curva horizontal; aquí, el proyecto señaló una excavación en cajón, con profundidad máxima de 54.0 m y taludes 1/4:1, con una berma a la mitad de la altura. La pared de menor profundidad desde al inicio quedó estable, no así el talud ladera arriba que desde el proceso constructivo sufrió derrumbes por lo cual fue necesario rediseñarlo. En este trabajo se presentan y se analizan las soluciones que se aplicaron para mejorar las condiciones de estabilidad del talud.

2 ESTUDIO Y PROYECTO INICIAL

En el proyecto, para esta excavación se realizó un estudio geoeléctrico con la modalidad de sondeo eléctrico vertical (SEV), con el uso del dispositivo de Schlumberger (Fig. 1). Además, se realizó un sondeo exploratorio con extracción de corazones con broca de diamante.

Los resultados obtenidos se usaron para definir la distribución estratigráfica , así como para hacer una estimación cualitativa de los materiales a excavar, a partir del estudio geológico, integrado con los datos de resistividad (Fig. 1), de las distintas rocas presentes. De acuerdo a esto, el macizo lo dividieron en cuatro unidades.

Unidad I, arena limosa, color grís claro, medianamente compacta y espesor de 1.5 m.

Unidad II, riolita muy fracturada y alterada, color grís y espesor de 3.5 m.

Unidad III. riolita fracturada y moderadamente alterada, color grís y espesor de 20.0 m.

Unidad IV, riolita moderadamente fracturada y poco alterada, color grís y espesor mayor a los 25.0 m.

Con base a estos resultados, se recomendó una pendiente de 1/4:1, en todo el talud (Fig.2 perfil A) y la formación de una berma de 6.0 m de ancho a la mitad de la altura. Este trabajo no indica algo en relación con la presencia de agua.

3 EXCAVACION Y COMPORTAMIENTO INICIAL

Desde al terminar la excavación inicial, empezaron los derrumbes en este corte del lado ladera arriba y especialmente en sus partes más profundas, debido a que su estructura geológica, tiene varios planos de fracturamiento con pendiente desfavorable

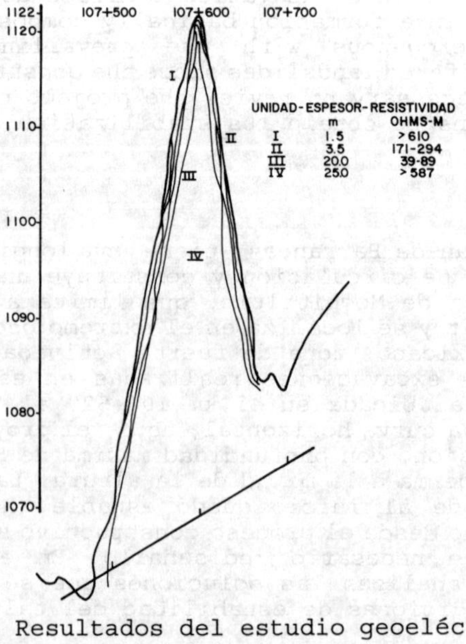

Fig. 1 Resultados del estudio geoeléctrico.

a la estabilidad, que aunados al alto grado de intemperización y heterogeneidad de la roca y a la presencia de agua subterránea, provocaron varias fallas por deslizamiento de cuñas o bloques con volumen cercano a los 100,000 m3 y dejaron aproximadamente el perfil marcado como B en la Fig. 2, por lo cual, aunque se pensó en abatir los taludes a 2/3:1, conservando la berma, se optó por suspender temporalmente los trabajos, mientras se realizaba un segundo análisis.

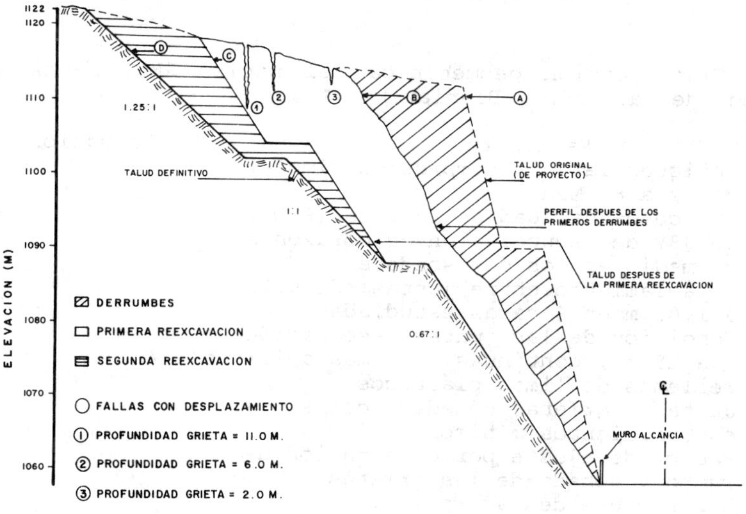

Fig. 2 Evolución del perfil del talud.

4 REVISION DEL PROYECTO

4.1 Estudios de Campo y laboratorio

Para facilitar la revisión, se hicieron dos sondeos adicionales hasta cerca de los 70.0 m de profundidad, con extracción de corazones de roca, de algunos de los cuales se elaboraron especímenes cilíndricos para ensayes de comprensión sin confinar. Además de campo, se llevaron muestras para labrar cubos que se probaron en el mismo ensaye.

También se hizo un levantamiento de campo, muy detallado que incluyó la determinación de rumbo y echado de los planos de fracturamiento, espaciamiento de juntas, condición de las mismas, rellenos entre ellas, detección de zonas húmedas, etc.

Con esta información y la inspección visual, se llegó a los resultados siguientes :

4.2 Estratigrafía

La formación geológica excavada; consiste en un estrato superficial de 3.0 m de espesor, formado por arena limosa, medianamente compacta, de color grís claro, le subyace hasta el final de los sondeos (70.0 m), roca riolítica, desde diques poco alterados y fracturados, hasta zonas de materiales muy intemperizados que llegan a la microfracturación; la coloración es diversa, desde grís claro hasta rojizo incluyendo roca clara con pórfidos oscuros. En la pared ladera arriba, se observaban huellas muy notorias de flujo de agua, hasta unos 8.0 m, a partir del fondo de la excavación.

Puede notarse que esta descripción difiere de la inicial especialmente en cuanto al estado de la roca y la presencia del agua.

4.3 Clasificación geomecánica del macizo RMR, (Rock Mass Rating) de acuerdo a Bieniawski (1989)

	Parámetro	Valuación
1.	Resistencia a la compresión simple, de 42 a 83 MPa	7
2.	Indice de calidad de la roca (RQD), 30-38% de recuperación de corazones promedio en los dos sondeos	8
3.	Espaciamiento entre juntas, de 0.20 a 1.05 m en 6 caras estudiadas	15
4.	Condición de las juntas, separación de 1 a 15 mm, continuas, las más abiertas rellenas de limos plásticos	10
5.	Juntas o grietas húmedas, con escurrimientos mínimos	4
6.	Factor de ajuste por orientación de rumbo y echado de las grietas ligeramente desfavorables.	-25
		19

El resultado obtenido de esta clasificación, indica que el macizo rocoso es de clase V, muy mala por lo que su cohesión debe ser inferior a 0.1 MPa y el ángulo de fricción interna menor a 30°. De acuerdo a esto el tratamiento adecuado para el talud era la reexcavación (Romana, 1992).

4.4 Análisis de Estabilidad del talud

Dado que lo visto mostraba que las fallas más importantes, se debían a traslaciones de bloques o cuñas sobre uno o varios planos de falla a partir de grietas de tensión y/o el efecto nocivo del agua, se optó por analizar el deslizamiento de bloques para lo cual se complementó la información obtenida, con lo disponible en la literatura existente (Comisión Federal de Electricidad 1972; Jumikis, 1983).

Se hicieron dos tipos de análisis, uno con el método bidimensional (Hoek and Bray, 1981), considerando que el bloque se desliza sobre una superficie plana que parte de una grieta de tensión; dos, con el método de la cuña, que se aplicó en dos casos aislados y considera que el bloque sólo puede moverse en una sola dirección, pero el deslizamiento ocurre en dos superficies de falla. En ambos casos, se analizaron por computadora diferentes planos de deslizamiento, incluyendo los medidos en el campo y varias situaciones diferentes tomando como la más crítica la que incluye presiones de agua y efecto sísmico.

Los resultados de estos análisis, para el caso de superficie de falla plana, se indican en la tabla, para la condición crítica.

Tabla Factores de seguridad para las condiciones críticas

Solución	Taludes			Número de bermas	Factor de Seguridad	Profundidad Máxima m
	Inferior	Medio	Superior			
I	1/4:1	-	1/4:1	1	0.76	56.0
II	2/3:1	-	2/3:1	1	1.50	58.0
III	2/3:1	2/3:1	2/3:1	2	1.69	62.0
IV	2/3:1	1:1	11/4:1	2	1.74	65.0

Ancho de las bermas : 6.0 m

De acuerdo a la tabla, este análisis hecho con una mayor cantidad de información, muestra que el proyecto inicial era inestable; la solución II era suficiente, pero dada la situación de no correr más riesgos se optó por la III.

El método de la cuña, se aplicó para dos bloques bien detectados y medidos; al analizarlos con los taludes del análisis bidimensional, dieron factores de seguridad mayores al 1.40 en la condición más crítica, sin embargo, uno de los bloques (señalado con un círculo en la Fig. 3), se deslizó en condiciones secas y sin sismo.

4.5 Recomendaciones

De acuerdo a los estudios realizados, para mejorar las condiciones de estabilidad del talud, se siguió una estrategia de acuerdo a los puntos siguientes :

a) Abatir los taludes de acuerdo a la solución III, asímismo

Fig. 3 Aspecto del talud después de la primera reexcavación.

ampliar la corona para construir una banqueta y un muro alcancía de 1.50 m de altura para retener caídos.

b) Aumentar las fuerzas resistentes, aminorando los efectos negativos del agua para lo cual se ordenaron drenes transversales de penetración en la parte baja de la pared, drenes longitudinales de zanja, cunetas y contracunetas revestidas e impermeabilización de las bermas.

c) Para evitar la rápida intemperización de la roca expuesta por la excavación, se recomendó proteger las zonas más críticas por medio de gunita (mortero lanzado).

También se analizó el uso de anclas, pero el fracturamiento tan intenso de la roca y la presencia del agua en la formación, le quitaba efectividad a estos dispositivos, lo que resultaba una solución muy costosa por lo cual se deshechó.

5 REEXCAVACION Y TRABAJOS COMPLEMENTARIOS

5.1 Reexcavacion

Los trabajos de abatimiento de taludes se llevaron a cabo de acuerdo a la solución III, de la tabla, durante el proceso se tuvo la falla de bloque, ya referida, en la parte superior del talud; antes de terminar los trabajos, se produjo un agrietamiento con desplazamiento que abarcó desde la parte superior del talud, hasta la primera berma (Fig. 3) por lo que fue necesario en esa zona extraer todo ese bloque, abatiendo aun mas el talud, como se muestra en las Figs. 2 y 4, y corresponde a la solución IV de la tabla.

Además se construyeron los drenes, cunetas, muros alcancía y otras obras complementarias.

5.2 Instrumentación

Al terminar los trabajos anteriores se colocaron dos inclinómetros de 30.0 m de longitud, con arreglo telescopiado los cuales, también (Fig. 5), se utilizaron como pozos de observación para medir niveles piezométricos. Además se instalaron referencias topográficas tanto en la zona del talud en dirección de la pendiente, como formando una red de precisión cubriendo toda el área en estudio.

6 COMPORTAMIENTO

6.1 Aspecto general

Después de cerca de tres años de operación de la vía, el talud sólo ha mostrado deslizamientos superficiales, que han sido colectados por el muro alcancía, cuyo material se ha utilizado para formar un sistema mecánico al pie del talud que aumenta su estabilidad.

El aspecto general se muestra en la Fig. 4 y se aprecia que el intemperismo esta afectando rápidamente la roca; por otro lado, de los drenes transversales colocados en la parte baja del talud, debido a los deslizamientos superficiales, solo trabajan aproximadamente la mitad de ellos.

Fig. 4 Aspecto actual del talud.

6.2 Instrumentación

a) De las 40 referencias topográficas instaladas, solo quedan 26 ya que las demás han desaparecido por los corrimientos superficiales. En algunas se han podido medir deslizamientos de más de 1.0 m, antes de perderlas.

b) Los inclinómetros colocados uno en cada berma (Fig. 5), han detectado movimientos muy pequeños, en el 1 de 19 mm a 3.6 m de profundidad y en el 2 de 12 mm a 7.8 m, los dos en casi tres años de observación.

c) En la Fig. 5, se presenta el perfil piezométrico máximo que se ha obtenido durante el período de observación, siendo dicho perfil, anterior a la colocación de los drenes transversales; la variación estacional es muy fuerte, ya que el acuífero se recarga en temporada de lluvias y se descarga en los meses

Fig. 5 Perfil piezométrico de la sección instrumentada.

siguientes, por lo cual, el efecto del agua es peligroso una gran parte del año.

7 TRABAJOS FUTUROS

Se tiene planeado, rehabilitar totalmente todos los drenes transversales de penetración, inhabilitados o destruidos ya que son básicos para la estabilidad del talud.

Por otra parte se tienen localizadas las áreas que se van a proteger con gunita, tarea que debió realizarse desde el principio.

Así mismo, se van a rehabilitar las bermas y todos los elementos que exhiban algún deterioro.

8 CONCLUSIONES

La excavación de este talud como en los demás, en esta zona montañosa de Plan de Barrancas, fue muy difícil debido a lo heterogéneo y al alto grado de intemperización de las estructuras geológicas atacadas, los cuales en muchos sitios tenían una estabilidad marginal, que fue rota por la construcción que, por una parte, reactivó muchas superficies de deslizamiento antiguas, que estaban sepultadas u ocultas y por otra parte, provocó cambios al sistema natural de drenaje superficial y subterráneo.

En este caso, los volúmenes adicionales de corte y extracción de derrumbes fueron del orden del 48% de los volúmenes del proyecto.

Esta experiencia nos muestra la importancia que tiene el realizar proyectos tomando como base la mayor información posible acerca de las condiciones geológicas de las zonas por excavar, ya que cualquier inversión que se realice en estudios, se justifica plenamente en los costos totales de realización del proyecto.

9 REFERENCIAS

Bieniawski, Z.T. (1989) "Engineering Rock Mass Classifications", Ed. J.Wiley and Sons, New York.

Comisión Federal de Electricidad (1972), "Clasificación de Suelos y Rocas", Manual de Diseño de Obras Civiles, Fascículo B.1.2.,México, D.F.

Hoek, E. and Bray, J.W. (1981), "Rock Slope Engineering" The Institution of Mining and Metallurgy, London, England.

Jumikis, A.R. (1983) "Rock Mechanics" Second Edition, Gulf Publishing Company, Houston, London, Tokyo.

Romana, M. (1992) "Métodos de corrección de taludes según la clasificación geomecánica SMR", III Simposio Nacional sobre Taludes y Laderas Inestables, La Coruña, España.

Rock Mechanics, Nelson & Laubach (eds) © 1994 Balkema, Rotterdam, ISBN 90 5410 380 8

Discrete element modeling for the Belden Siphon rock slope remediation project – Verification and reliability

Jon Y. Kaneshiro
Engineering-Science, Inc., San Diego, Calif., USA
Mihalis N. Madianos & Dario Rosidi
Geomatrix Consultants, Inc., San Francisco, Calif., USA
Robert K. White & Robert A. McManus
Pacific Gas and Electric, San Francisco, Calif., USA

ABSTRACT: The Belden Siphon slope was analyzed using a combination of two-dimensional numerical and limit-equilibrium methods. The analyses identified shallow and deep-seated sliding and toppling as the primary failure mechanisms to be considered in the tieback anchor design.

1 INTRODUCTION

The Belden Siphon, a 3.7-meter (m) penstock, spans the North Fork of the Feather River Canyon in Plumas County, California. Built in the late 1960s, it is part of the water supply system for the Belden hydroelectric powerhouse. It lies on a steep 122-meter-high (1 vertical: 1 to 1.5 horizontal) slope founded in phyllites of various degrees of weathering, fracturing, and competency.

Tension cracks as wide as 50 millimeters (mm) were observed in 1988 in the upper portions of the slope. A flexible coupling was installed to allow for greater displacements. Because of the critical nature of the facility, the slope and siphon were instrumented with a series of piezometers, inclinometers, survey points, and extensometers. Evidence of downhill creep and toppling was observed. A small portion of an adjacent slope failed, signs of movement were observed and the possibility of a deep-seated failure surface was indicated. A series of subhorizontal drains and relief wells to lower the groundwater level were installed and a series of tieback anchors to increase the stability during static and seismic loading were proposed.

The slope was analyzed using a combination of two-dimensional numerical and limit-equilibrium methods. Both the Universal Distinct Element Code (UDEC; Itasca, 1992) and the Discontinuous Deformation Analysis (DDA; Shi, 1989), which model toppling and treat the discontinuous nature of the rock, as well as the UTEXAS3 (Wright, 1992) slope stability program, were used.

Verification and reliability of models are established by everyday use in engineering practice through calibration of either simplified or detailed models with case histories. The intention is to compare discrete element analyses with limit equilibrium analyses and the field observations. This somewhat random approach, using various and complimentary analytical methods to verify analyses, combined with good engineering judgment is the best method for evaluating design reliability (Akky et al., 1994).

2 SITE CONDITIONS AND MATERIAL PROPERTIES

The Belden Siphon slope subsurface may be divided into three seismic velocity layers (Figure 1):(1) deeply weathered rock with an average seismic velocity of 427 meters per second (m/s); (2) weathered rock with an average seismic velocity of 1067 m/s; and (3) little-weathered rock with an average seismic velocity of 2900 m/s. The depth of weathering is about 26 meters. The rock varies from intensely fractured at the surface to moderately to little-fractured at depth. From borings drilled in 1988 and 1991, the average fracture spacing was estimated to be 0.15, 0.46, and 1.2 meters for each layer, respectively. The degree of fracturing is useful for determining the rock mass material properties, although modeling the slope with the actual joint configuration and spacing is impractical. A primary joint set, which is subparallel to the slope, is the exfoliation joint. Another joint set is the foliation, which strikes about perpendicular to the profile (Figure 1) and dips 42 degrees with a variation of about ±15 degrees into the hill. A loss of circulation during drilling occurred from 3 to 7 meters deep and from 16 to 17 meters deep in two of the exploration holes, indicating open fractures. The highest maximum historical piezometric surface, shown on Figure 1, was used in all the analyses. Dynamic slope stability was assessed using a pseudo-static horizontal earthquake acceleration of 0.1g.

The hillside shows evidence of creep and toppling. Contorted foliation planes, which were once much steeper, are now observed as shallow as 24 degrees. The inclinometer data also suggest creep or toppling (Figure 2). Reports specific to the Belden Siphon were reviewed, and back analyses in each of the analytical models were performed to obtain the joint strength and deformation properties of the phyllites. Little information is found in the literature on the strength of phyllites; most of the data are on intact rock properties.

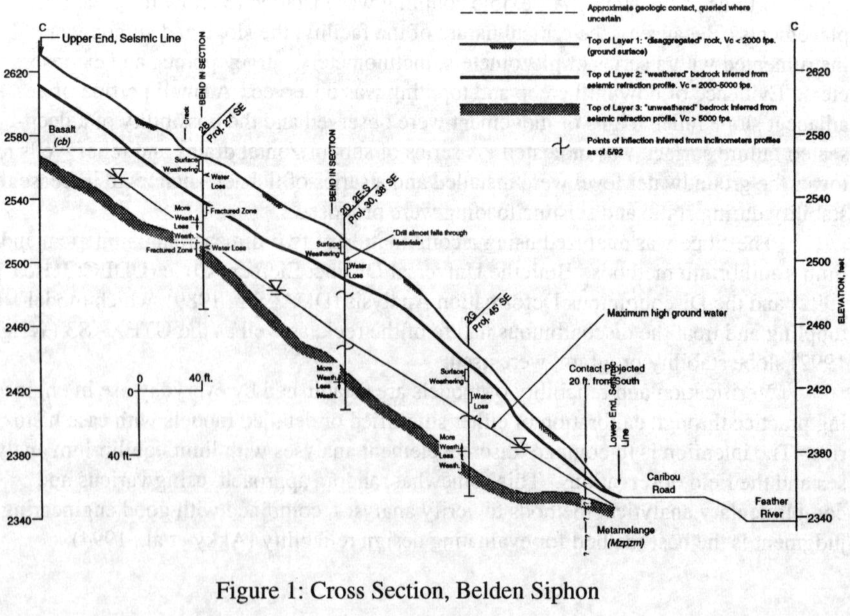

Figure 1: Cross Section, Belden Siphon

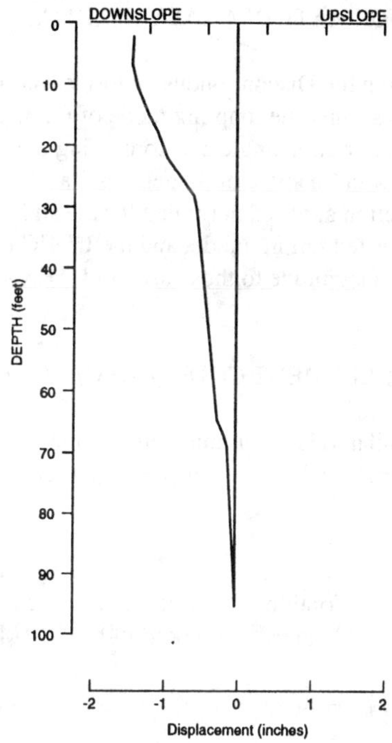

Figure 2: Typical Inclinometer Displacement Profile

3 LIMIT EQUILIBRIUM ANALYSES

The analyses using the limit equilibrium slope stability program UTEXAS3, did not model creep or toppling behavior, but modeled deep-seated failure (Figure 1). The unit weight, friction, and cohesion values used in the UTEXAS3 slope stability analyses were based on back analyses assuming existing conditions (Factor of Safety=1) with no anchors and reasonable values, as follows.

Layer	Unit weight (kN/m^3)	Friction (degrees)	Cohesion (kN/m^2)
1	23.6	29 to 29.5	0
2	24.4	36	0
3	25.1	42	105

The anchor forces incoporated into the UTEXAS3 analyses were 0.73 MN for the upper bench and 0.88 MN for the lower bench. With these anchors under static loading conditions, a factor of safety of 1.2 was computed. Under pseudostatic loading conditions, a minimum factor of safety of 1.0 was calculated.

4 DISCONTINUOUS DEFORMATION ANALYSIS (DDA)

A simplified joint model using the Discontinuous Deformation Analysis method (Shi, 1989) was performed to assess only the toppling mode of failure. A sensitivity analysis was performed by varying the friction angle and computing the corresponding displacements. A total critical joint friction angle of 42 degrees was determined. This is in contrast to the total joint friction angles determined for the sliding mode of failure in both the limit-equilibrium deep-seated failure model and the UDEC model. It should be noted that anchor forces of similar magnitude to those used in UTEXAS3 provided stability.

5 UNIVERSAL DISTINCT ELEMENT CODE (UDEC) ANALYSES

For the UDEC analyses, similar values for unit weight, joint friction, cohesion, shear stiffness, and normal stiffness based on back analyses and reasonable values consistent with the literature were used, as follows.

Layer	Unit Weight (kN/m^3)	Friction (degrees)	Cohesion (kN/m^2)	Shear Stiffness (MPa/m)	Normal Stiffness (MPa/m)
1	23.6	34 to 35	0	2,823	1,086
2	24.4	35	0	2,845	1,185
3	25.9	42	0	6,435	28,834

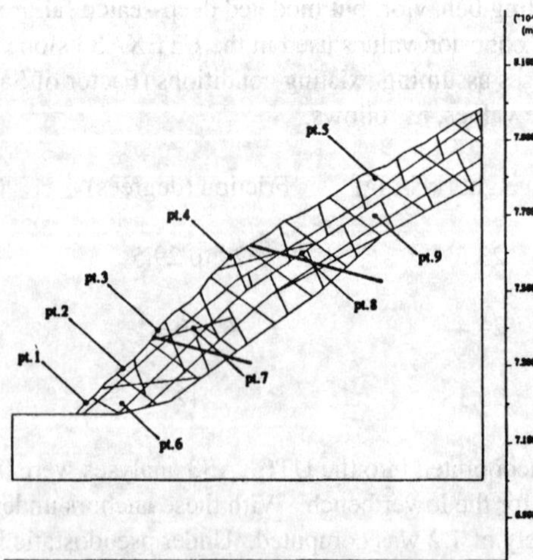

Figure 3: Block Model Geometry (UDEC)

Because the slope and the associated failure mechanism are at a relatively shallow depth, the rock mass was modeled using the rigid block assumption. The joint normal and shear stiffness were derived from the seismic velocity measurements and the calculated Young's and shear modulus of the rock mass. From the bulk and shear modulus and joint spacing, the joint normal and joint shear stiffness were calculated based on the equations in the UDEC manual. The joint strength and deformation properties used in the UDEC analysis, summarized above, are consistent with the literature.

The stability of each cross section was analyzed using a simplified jointed rock-slope model. The rock joints were generated using a combination of the random generating scheme in UDEC and prescribed fixed joint configurations to allow both shallow and deep failure mechanisms and rotation to occur. The model was set up to match as closely as possible the existing field data and observations, and allow for sliding and/or toppling behavior. About half of the foliation joints were incorporated as dipping steeply or vertically oriented to allow for toppling.

The model's joints (shown on Figure 3) are distributed as follows:

Joint Set	Number	Average Spacing (meters)	Description
1	11	10	Foliation, 35 to 55 degrees dip
2	11	10	Foliation, 70 to 90 degrees dip
3	9	5 to 15	Exfoliation joints

The critical "apparent" friction angle of the blocks was then estimated by gradually lowering its value from 38 to 34 degrees. Figure 4 shows displacement histories as a function of friction angle. As can been seen, total displacements are negligible at this point (at about 0.5 mm) up to the point where the friction angle is about 35 to 34 degrees.

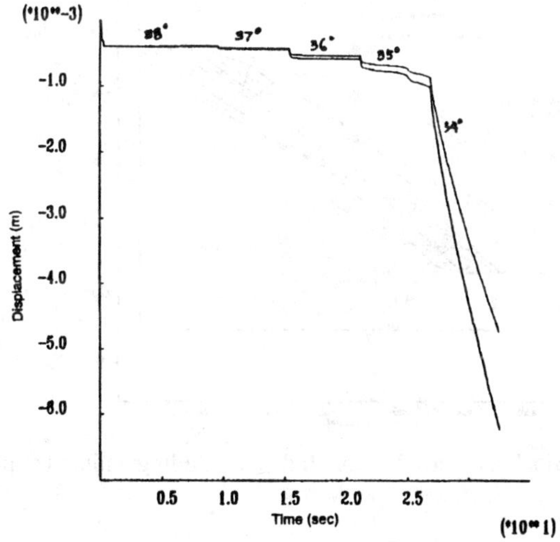

Figure 4: Displacement Histories as a Function of Joined Friction Angle

At 34 degrees, the displacement increases rapidly, indicating an unstable configuration. It should be noted that the slope has a geometric angle of about 36 degrees. The critical friction angle, therefore, is about 1 or 2 degrees less than the slope. In subsequent analyses, an apparent friction angle of 35 degrees was adopted.

After the critical friction angle was established, rock bolts were introduced into the analysis at the same elevations and capacities as used in the UTEXAS3 analyses. Anchor capacities also were varied, depending on the results, to check the adequacy of the anchors specified in UTEXAS3.

The stability under horizontal pseudo-static earthquake load was assessed by applying 0.1 g horizontal body force on the slope. The lower base block, composed of layer 3 and unjointed in the model, was fixed such that the applied horizontal forces are allowed to separate from the upper blocks. Displacement histories were monitored at nine points on the slope (Figure 3) to assess the stability of the model for each case.

The maximum joint shear displacements, which are present only between the first and second layer near the top of the slope, amount to only about 4 mm. Displacements at the randomly and regularly chosen points/blocks level off and are on the order of 2 to 3 mm in the top layer (points 1 to 5) and 1.5 mm in the second and third layers (points 6 to 9). The amount of displacement decreases with depth, following a trend similar to the inclinometer signature. As can be seen from Figure 5, the velocity vectors are random, and there is a sense of creep/sliding in the uppermost layer. In addition, near the top of the slope, there is a counter-clockwise sense of rotation or toppling.

We attempted to run the model with subtle joint geometry variations in the toe of the slope. With these slight variations, we were able to inhibit toppling and allow for sliding and vice versa.

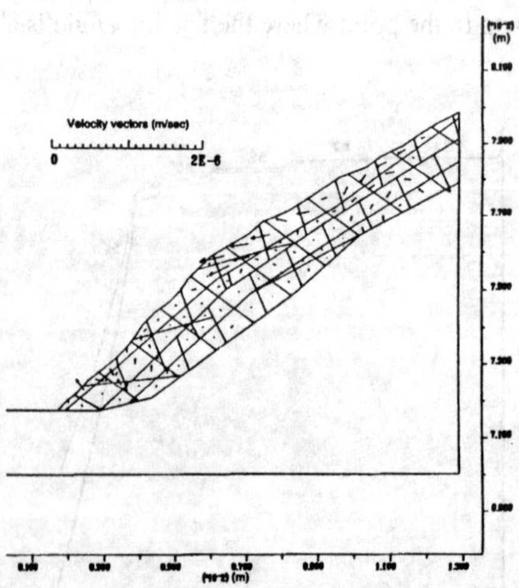

Figure 5: Block Velocity Vectors Indicated Sliding and/or Toppling

6 CONCLUSIONS

For design of a practical tieback anchor slope remediation scheme, simplified models combined with field evidence are appropriate for observing potential failure mechanisms as well as potential limitations of a given remediation scheme. These models, themselves may not realistically model the actual discontinuities and their distribution; however, they are believed to be representative of critical joint strengths and orientations.

The anchor forces and locations that were assumed in the limit equilibrium analyses and incorporated in the discrete element analyses resulted in stabilizing the slope. The limit equilibrium analyses focused on analyzing a deep-failure surface catastrophic to the entire slope. The tieback anchor solution produces stability on these deep-failure surfaces by providing increased confinement and added normal stress to the deep-failure plane. However, unlike limit equilibrium analyses, the UDEC analyses revealed other possible failure mechanisms and provided insight into other possible explanations for the inclinometer signatures besides toppling, such as shallow sliding or sliding creep. The UDEC analyses show that sliding of the uppermost layer and subtle joint geometries are also critical and could affect the penstock stability. These failure mechanisms revealed important considerations for the final design and construction of slope remediation measures for the Belden Siphon:

1) Tieback anchors, incorporating relatively large forces at two levels, do provide an increase in confinement on a relatively large percentage of a deep-failure plane. Although the stress transfer to a deep-failure plane can be very complex depending on the joint geometries, an increase in normal stress of about 20 KPa over about 50 to 75% of the deep-seated failure surface was estimated using simplified methods. However, widely spaced anchors provide less confinement for the shallower layers, which are moving and need to be stabilized. The solution included: (a) providing additional anchors at a third bench level and (b) removing of surficial layers and colluvium during benching for the tiebacks.

2) With the possibility of shallow failures there is a concern for the stability of the tieback anchor foundations themselves. Other slope stabilization measures and erosion control were considered to prevent undermining the foundations. Slope protection measures above, below, and around the tieback foundation further increase the stability to the uppermost layers against shallow failure.

3) Often subtle features in the rock joint geometry can affect the stability of the slope. Stabilization measures called for shotcreting of the rock at the toe to provide additional confinement to the toe area.

REFERENCES

Akky, M.R., D. Rosidi, M.N. Madianos, and J.Y. Kaneshiro, 1994, "Dynamic Analyses of Large Underground Caverns Using Discrete-Element Codes." Proceedings, International Congress on Tunneling and Ground Conditions, International Tunnelling Association, 20th Annual General Assembly, April, Cairo, Egypt.

Shi, G-H., 1989, "Discontinuous Deformation Analysis - A New Numerical Model for the Static and Dynamics of Block Systems, Ph.D. Thesis, University of California, August, Berkeley.

Technical and Ecological Services (TES), Civil Engineering Section for PG&E, 1992, Belden Siphon, Analysis of Geotechnical Instrumentation, 1988 to 1992, TES report number 431-92.4 prepared by Structural and Geotechnical Unit, December.
Technical and Ecological Services (TES), Civil Engineering Section for PG&E], 1991, Belden Siphon Slope Remediation, report number 431-91.3 prepared for Hydro Generation, June.
Universal Distinct Element Code (UDEC), 1992, Itasca Consulting Group, Inc., User's Manual
Wright S.G., 1992, "University of Texas Analysis of Slopes III (UTEXAS3), Geotechnical Engineering Software GS92-1, University of Texas, Austin.

Rock Mechanics, Nelson & Laubach (eds) © 1994 Balkema, Rotterdam, ISBN 90 5410 380 8

Estudio de estabilidad de taludes y propuesta para aumentar el grado de inclinacion de los mismos en la mina a cielo abierto de minera real de Angeles

Alberto S. Menache Varela
Proyectos Especializados de Ingenería, s.c., México

RESUMEN

El propósito de este trabajo es mostrar la utilidad y las repercusiones económicas que tiene el haber ejecutado un estudio de estabilidad de taludes en un tajo a cielo abierto para la extracción de minerales.

1. ANTECEDENTES

Minera Real de Angeles lleva 13 años desarrollando un corte a cielo abierto para la obtención de plata, zinc y plomo; en el estado de Zacatecas, México.

Al momento de iniciar los estudios los taludes tenían una inclinación general de 52º y una particular (entre bermas) de 70º, comportándose bien hasta el momento, en la mayor parte de la superficie expuesta, a excepción de ciertas áreas bien definidas por fallas.

Debido a la baja en el precio internacional de la plata, se requirió pensar en la forma de abaratar el proceso de excavación. Una de las alternativas era la de incrementar la inclinación de los taludes.

En base a un primer estereograma general que se dibujó con la información existente en el departamento de geología de la mina, se vislumbró la factibilidad de incrementar la inclinación de los taludes más o menos en 10º ó 15º.

Este panorama favorable sirvió para pensar en un estudio más detallado que seleccionara cuales taludes se podían incrementar en su inclinación y cuales no, ya que hay que aclarar que la mina tiene forma de cono circular invertido, y mientras el sistema de fracturamiento es favorable para unos taludes, es totalmente inestable para otros.

2. PANORAMA GEOLOGICO

Se mencionará brevemente el marco gelógico regional para ubicar el ambiente de estudio.

En el sitio la columna estratigráfica se encuentra constituida principalmente por areniscas de grano fino a muy fino de color gris pardo y por algunas limolitas de tonalidad gris pardo.

El área de interés se encuentra emplazada en el flanco occidental de un anticlinal cuyo eje es sensiblemente paralelo a la carretera que conduce a la mina desde Zacatecas o Aguascalientes.

Todas las unidades tienen sobrepuestos cuando menos tres sistemas de fracturas. El primer sistema presenta en promedio una alineación S45W con inclinación de 65 grados al NW. El segundo sistema exhibe un rumbo N70W, buzando 70 grados al NE y el tercero es sensiblemente EW con echados semiverticales. Un cuarto sistema adicional aleatorio es el NS con buzamientos de 65 a 70 grados tanto al E como al W. Existen sistemas de fracturamiento secundario.

La carta de regionalización sísmica de la República Mexicana, indica que la zona de estudio pertenece a la zona "A", por lo cual deberá considerarse un coeficiente sísmico de :

$$C = 0.08$$

3. CLASIFICACION MECANICA DEL MACIZO ROCOSO

Empleando el sistema de clasificación de macizos rocosos propuesto por Bieniawski, se encontraron las condiciones siguientes:

TIPO DE FORMACION	R.M.R.
Estratos gruesos con fisuramiento desfavorable	21
Estratos gruesos con fisuramiento medianamente desfavorable	46
Estratos gruesos con fisuramiento favorable	66
Estratos delgados con fisuramiento desfavorable	0
Estratificación delgada con fisuramiento medianamente desfavorable	16
Estratificación delgada con fisuramiento favorable	36

Como puede observarse, existen distintas calidades del macizo rocoso por zonas, lo cual es también función de la orientación del corte.

Se observa que si se parte del Noreste (NE), las condiciones de estabilidad disminuyen según se avanza en el sentido de las manecillas del reloj.

4. *FLUJO DE AGUA EN EL MACIZO ROCOSO*

De lo que pudo observarse en las visitas de campo, el nivel freático aparentemente está más abajo del pie de los taludes. Sin embargo, existen filtraciones pequeñas que implican la presencia de un bombeo continuo de achique. En la mina no se ha implementado un control de instrumentación piezométrica o medidas del nivel freático en los barrenos exploratorios. Se está haciendo la hipótesis de que los escurrimientos existentes pueden estar asociados a mantos colgados que al ser descubiertos por el corte del macizo rocoso drenan hacia el interior de la excavación.

Se está trabajando en establecer un programa de medidas periódicas de niveles del agua subterránea. En estos análisis se consideró que las fisuras se encuentran drenadas y no se acumula agua en su superficie.

Si por alguna causa se dejara de bombear y el nivel de aguas subiera, hay que tomar precauciones especiales sobre todo en el proceso de abatimiento. Mientras el agua sube, las presiones hidrostáticas se compensan tanto dentro como afuera del macizo rocoso, pero al bombear, el agua dentro de la excavación desciende a un ritmo mucho más rápido que el agua atrapada en las fisuras, ocasionando un efecto de vaciado rápido y la consecuente generación de presiones interiores que pueden causar la falla de los taludes. Por si este hecho llegara a presentarse, se recomendó recuperar los niveles con mucho cuidado y siempre apoyados en mediciones piezométricas en que el nivel de agua atrás de las cuñas de roca nunca sea mayor a 1 metro con respecto al nivel de agua en el interior de la excavación (esta especificación es función de la distancia del barreno a la cara del talud y deberá adecuarse en cada caso particular). Mientras más cerca esté el barreno del talud, la diferencia de presiones deberá ser menor.

5.*MECANISMOS DE FALLA*

Los sistemas de fracturas indicados en los capítulos anteriores se presentan en todas las estructuras rocosas, ya que se trata de un fenómeno regional y por tanto, todos los taludes existentes con la misma dirección tendrán problemas de estabilidad similares.

TALUDES CON RUMBO SENSIBLEMENTE N-S CON BUZAMIENTO AL ESTE

En el Talud Poniente se presenta la posibilidad de salida de cuñas tridimensionales de roca con dirección variable entre 70º y 90º en promedio al NE, cuando los taludes sobrepasan una inclinación de 66 grados con respecto a la horizontal. Lo mismo ocurre con las cuñas cuya salida es al SE, con direcciones comprendidas entre 92º y 135º aproximadamente. Es recomendable analizar la posibilidad de una cuña con deslizamiento en el plano, ya que el sistema con rumbo cercano al NS con buzamiento de

70 grados al E tiene su polo muy cercano al de la cara del talud. Las fracturas semiverticales con dirección sensible NS, con buzamiento al Poniente forman lajas de roca esbeltas que tienden al volteo de bloques.

TALUDES CON DIRECCION PREFERENCIAL E-W CON INCLINACION AL SUR

En el Talud Norte existe la posibilidad de formación de columnas o lajas de roca esbeltas que tiendan al volteo. La posibilidad de una falla en el plano es alta después de los 70º de inclinación y existen cuñas de roca tridimensionales con salida entre 97º y 135 º al SE después de los 65º de inclinación. Asímismo, también se presenta la posibilidad de salida de cuñas al SW con direcciones entre 230º y 250º después de 80º de inclinación de la cara del talud.

TALUDES CON DIRECCION PREFERENCIAL E-W CON INCLINACION AL NORTE

En el llamado Talud Sur, existe la posibilidad de formación de cuñas de roca tridimensionales, con salida hacia el NE con dirección variable entre 71º y 82º al sobrepasarse los 65º de inclinación. Existen cuñas de roca con salida hacia el NW con direcciones variables entre 291º y 355º, al alcanzarse los 49º de inclinación. El volteo de bloques parece probable, más la posibilidad de una falla en el plano después de los 70º de inclinación, la cual es recomendable analizar con más detalle.

TALUDES CON DIRECCION PREFERENCIAL NORTE-SUR, CON BUZAMIENTO AL PONIENTE

En los taludes ubicados al Este puede desarrollarse una superficie de falla plana, al acercarse el talud a los 70º de inclinación. El volteo de bloques es poco probable pero existe la posibilidad de presentarse en zonas aisladas. En cuanto a la formación de cuñas de roca tridimensionales, las hay con salida al SW al acercarse el talud a los 50º de inclinación. Y hacia el NW las salidas exhiben salidas entre los 291º y 355º al sobrepasarse los 50º de inclinación.

Para direcciones intermedias se pueden interpolar los comentarios aquí descritos.

6.ANALISIS DE ESTABILIDAD

La información proporcionada por el Departamento de Geología de la mina señala que el cuerpo mineral tiene unas dimensiones aproximadas de 600 por 300 m. La traza del cuerpo presenta una dirección promedio de N61ºW, con buzamiento de 45º hacia el NE, por lo cual la dirección de ataque es de 29º al SW. De esta forma la cara frontal del corte es un talud SW con inclinación al NE. Una de sus caras laterales es un talud NW con inclinación al SE y la otra es un talud SE con inclinación al NW.

El desarrollo del tajo se está llevando a cabo en forma escalonada, formando bermas de 15 m de altura por 11 m de ancho. En la actualidad se ha banqueado de la cota 2272 a la 1984, lo que equivale a una profundidad de 288 m y en los próximos tres años se pretende llegar a la cota 1940, con una profundidad total de 332 m. La inclinación general de los taludes se ha efectuado respetando un ángulo de 52º, con inclinacioness particulares de 70º en el banqueo, a excepción de la zona Sur, donde se han efectuado cortes en los bancos a 80º.

El factor más importante en la resistencia al corte de una roca fracturada es la magnitud del esfuerzo normal que actúa a través de las discontinuidades. El máximo esfuerzo normal que actúa sobre una discontinuidad localizada verticalmente debajo de la cresta, puede ser calculado a partir de la siguiente expresión:

En = &H(cotB - cotA) senBcosB

siendo "A", el buzamiento de la cara del talud, y "B", el buzamiento de la fractura. & Es el peso volumétrico.

En el proceso de falla por corte, la resistencia al corte en una discontinuidad, para un valor dado de la carga normal, depende de la magnitud de los desplazamientos relativos previos que se hayan producido entre las superficies rocosas.

Patton de las curvas de resistencia pico y residual, obtuvo la siguiente ecuación para determinar la resistencia al esfuerzo cortante, para esfuerzos normales bajos, donde hay movimientos verticales de dilatancia:

*Tau = N * Tan (Fir + i)*

donde *"i"*, es el ángulo que forman los dientes con la superficie de deslizamiento y se conoce con el nombre de *rugosidad efectiva*. Para esfuerzos normales elevados, donde ya no hay movimientos verticales de dilatancia y la mayoría de las irregularidades serían cortadas:

Tau = C + N Tan Fir

Para efecto de este trabajo, se considerará que la resistencia residual de las fracturas puede tener un ángulo de fricción interna variable entre 25 y 35 grados para rocas de mala calidad y entre 35 y 40 grados para rocas de mediana calidad. Para esfuerzos normales bajos, en diaclasas, se estima, que el ángulo de rugosidad efectiva "i", puede variar de acuerdo con la expresión presentada por Barton:

i = 20 log10 (Rc/N)

Este proceso es útil porque proporciona un medio para evaluar el probable límite superior de la influencia de la rugosidad de las superficies en las discontinuidades rocosas. En taludes muy altos y empinados, hay que proceder con mucho cuidado en la consideración de la suma de *Fir+ i*.

Se analizó la estabilidad de los taludes en condiciones drenadas (talud sin presiones hidrostáticas).

El método de análisis empleado fue el de vectores tridimensionales, desarrollado por Wittke para cuñas de roca bidimensionales y el de Londe para tetraedros tridimensionales. La parte dinámica se debe al desarrollo de Hendron. Los casos incluyen el macizo rocoso afectado por una, dos o tres discontinuidades. La resistencia dinámica se utiliza para predecir desplazamientos dinámicos asociados a movimientos telúricos, en el método de análisis dinámico presentado por Newmark (1965). En este trabajo se utilizó el método de Hoeck para definir los factores de seguridad en la totalidad del tajo y la zona de inestabilidad.

Las gráficas mostradas a continuación indican que para los taludes existentes en la Mina, la resistencia al esfuerzo cortante en las fracturas fluctúa entre:

$$2.5 < Tau < 20.2 \;;\; kg/cm2$$

Dependiendo de la profundidad del corte y de la inclinación tanto del talud como del fisuramiento.

La rugosidad efectiva quedó comprendida entre:

$$13 < i < 43$$

En base a las mismas variables que Tau; que para un Fir residual de 35 grados, implica:

$$48 < Fir + i < 78$$

La obtención de los factores de seguridad para todas las posibles combinaciones de fracturas permitieron definir la zona de inestabilidad. Se presentan a continuación las inclinaciones máximas que pueden tener los taludes, sin necesidad de soporte adicional (anclajes o concreto lanzado), en función de la dirección en que su cara libre se encuentran buzando.

INCLINACIONES MAXIMAS DE LOS TALUDES

TALUD CON SU CARA INCLINADA HACIA	ANGULO MAXIMO SIN NECESIDAD DE SOPORTE
NORTE	68º
NORESTE	62º
ESTE	56º
SURESTE	63º
SUR	70º
SUROESTE	65º
OESTE	42º
NOROESTE	45º

De esta tabla se puede ver que el ataque del cuerpo mineral con dirección al SW, implica que la cara frontal del corte, con buzamiento hacia el NE, puede excavarse dándole una inclinación de 62º. El talud lateral con buzamiento hacia el SE, se dejará a 63º y la otra cara lateral con inclinación al NW quedará a 50º. Se recomendó se continúe empleando el procedimiento del precorte para evitar dañar las paredes del talud.

7. *CONCLUSIONES*

La implementación de los resultados presentados permitieron que la compañía minera ahorrara, en un plazo de 5 meses, 10 millones de dólares, y le diera una vida adicional a la planta de beneficio de 4 meses. Lo anterior sin tomar en cuenta que el mineral pudo venderse antes de lo planeado, en un mercado que día a día se encontraba disminuyendo el precio de compra.

CONSIDERACIONES ADICIONALES

1. El estudio de estabilidad debe realizarse desde las primeras etapas de planeación de la mina, lo cual repercutirá en un ahorro considerable de tiempo y dinero; ya que adicionalmente al ahorro en la extracción de material esteril, debe tomarse en cuenta el beneficio de poner el mineral en el mercado cuando el precio es más favorable.
2. El estudio debe hacerse con el mayor cuidado posible, disponiendo de tiempo y recursos suficientes.
3. Los trabajos de gabinete deben complementarse con un programa de instrumentación para medir desplazamientos de la masa rocosa, y para el control de explosivos.
4. Debe contrase con un estudio paralelo de control de voladuras.

Author index